P9-DMU-098

ANNOTATED INSTRUCTOR'S EDITION

ELEMENTARY
STATISTICS

NINTH EDITION

MARIO F. TRIOLA

PEARSON

Addison
Wesley

Boston   San Francisco   New York
London   Toronto   Sydney   Tokyo   Singapore   Madrid
Mexico City   Munich   Paris   Cape Town   Hong Kong   Montreal

Publisher: Greg Tobin
Senior Acquisitions Editor: Deirdre Lynch
Executive Marketing Manager: Yolanda Cossio
Executive Project Manager: Christine O'Brien
Editorial Assistant: Keren Blankfeld
Senior Marketing Manager: Pamela Laskey
Managing Editor: Karen Guardino
Senior Production Supervisor: Peggy McMahon
Text and Cover Designer: Barbara T. Atkinson
Cover Photos: Pencils and vineyard © Getty;
　　　　　　　People in lab © Stone;
　　　　　　　Group at table and men at riverbank © Creatas;
　　　　　　　Women with tape measure © Corbis.
Associate Media Producer: Jennifer Kerber
Senior Manufacturing Buyer: Evelyn Beaton
Software Development: Kathleen Bowler and Jan Wann
Composition and Production Services: Nesbitt Graphics, Inc.
Printer: VonHoffman Press

Many of the designations used by manufacturers and sellers to distinguish their products are claimed as trademarks. Where those designations appear in this book, and Addison-Wesley was aware of a trademark claim, the designations have been printed in initial caps or all caps.

For permission to use copyrighted material, grateful acknowledgment is made to the copyright holders on pages 827–828 in the back of the book, which is hereby made part of this copyright page.

The Library of Congress has already cataloged the Student Edition as follows:

Library of Congress Cataloging-in-Publication Data

Triola, Mario F.
　　Elementary statistics / Mario F. Triola.—9th ed.
　　　　p.　cm.
　　Includes bibliographical references and index.
　　ISBN 0-201-77570-0
　　　　1. Statistics.　I. Title.

　QA276.12.T76 2003
　519.5—dc21　　　　　　　　　　　　　　　　　　　　　　　2002026028

ISBN 0-321-14963-7

2 3 4 5 6 7 8 9 10—VHP—060504

To Marc and Scott

About the author

Mario F. Triola is a Professor Emeritus of Mathematics at Dutchess Community College, where he has taught statistics for over 30 years. Marty is the author of *Essentials of Statistics*, *Elementary Statistics Using Excel*, *Mathematics in the Modern World*, and *Survey of Mathematics*. He is a co-author of *Statistical Reasoning for Everyday Life, Business Statistics,* and *Introduction to Technical Mathematics*. He designed the original STATDISK statistical software package, and he has written several manuals and workbooks for technology supporting statistics education. Outside of the classroom, Marty's consulting work includes the mathematical design of casino slot machines and fishing rods, and he has worked with attorneys in determining probabilities in paternity lawsuits, identifying salary inequities based on gender, and analyzing disputed election results. Marty has testified as an expert witness in New York State Supreme Court for an election dispute involving a former student. Marty was a recent writing team member of the Project Coalition with NASA and the American Mathematics Association of Two-Year Colleges.

When he's not working, Marty enjoys travel, golf, tennis, running, hiking, and anything that flies. He has a commercial pilot's license with an instrument rating, and has flown airplanes, helicopters, sail planes, hang gliders, and hot air balloons. His passion for flying has included parachute jumps, flying in a Goodyear blimp, and parasailing.

The Text and Academic Authors Association has awarded Mario F. Triola a "Texty" for Excellence for his work on *Elementary Statistics*.

About This Updated Edition

Some major improvements in technology have been implemented since the first printing of the Ninth Edition of *Elementary Statistics*. Although this Updated Edition includes the same examples, exercises, and statistical content as the original Ninth Edition, it also includes changes to reflect the following recent changes in technology.

Texas Instruments Calculators

The original printing of the Ninth Edition of *Elementary Statistics* includes many references to the TI-83 Plus calculator. The instructions, screen displays, programs, and the APP (application) for the TI-83 Plus calculator also apply to each of these calculators, which may also be used with this book:

- TI-83 Plus Silver Edition
- TI-84 Plus
- TI-84 Plus Silver Edition

Throughout this book, references to the *TI-83 Plus* calculator also apply to any of the above listed calculators.

The TI-89, TI-89 Titanium, TI-92 Plus, and Voyager 2000 have the ability to download the "Statistics with List Editor" application, so these calculators can have at least as much statistics functionality as the TI-83 Plus and TI-84 Plus family of calculators. However, procedures for using the "Statistics with List Editor" application are generally different from the calculator procedures described in this book. Consequently, the TI-89, TI-89 Titanium, TI-92 Plus, and Voyager 2000 calculators could be used for many procedures in this book, but instructions, screen displays, the data application, and programs are not compatible with this book.

Minitab

The original printing of the Ninth Edition of *Elementary Statistics* was based on Minitab Release 13. This Updated Edition includes updates for the newer Minitab Release 14. Among other improvements, Minitab Release 14 now includes more opportunities for using summary statistics, such as the sample size, sample mean, and sample standard deviation.

STATDISK

The original printing of the Ninth Edition of *Elementary Statistics* was based on STATDISK version 9.0, but dramatic improvements are now included in STATDISK version 9.5.2. (Check the Web site http://www.aw.com/triola for the latest version of STATDISK.) This Updated Edition includes updated instructions for using the latest version of STATDISK, which will be version 9.5.2 or later.

For Users of the Eighth Edition

Quick Summary of Major Changes in the Ninth Edition

STATISTICAL METHODS

1. **Normal vs. *t* Distributions:** In Chapters 6, 7, and 8, the criteria for choosing between the normal and *t* distributions have been updated.

   Ninth Edition (see Figure 6-6):

   Use normal if σ is known. (If $n \leq 30$, the population must be normal.)

   Use *t* if σ is unknown (and either the population is normal or $n > 30$).

   Eighth Edition

   Use normal if σ is known or $n > 30$. (If $n \leq 30$, population must be normal.)

   Use *t* if σ is unknown and $n \leq 30$ (and the population is normally distributed).

   Reasons for this change:

   i. The criteria used in this Ninth Edition are the same criteria used in the real world. Hundreds of articles in professional journals were surveyed, and they all use the same criteria included in this Ninth Edition. With minimal or no increase in difficulty, students will learn to apply those same criteria.

   ii. With σ not known, the distribution of $\dfrac{\bar{x} - \mu}{\frac{s}{\sqrt{n}}}$ is a *t* distribution, not a normal distribution. For very large sample sizes, the differences between the normal and *t* distributions are negligible, but the use of the *t* distribution in this edition generally yields better results.

   iii. Working with the *t* distribution is not much more difficult than working with the normal distribution. Confidence intervals are just as easy, test statistic calculations are the same, it's easy to find critical values, and only *P*-values are slightly more difficult to find. If technology is used, work with the *t* distribution is just as easy as work with the normal distribution. Also, the use of the *t* distribution table strengthens a skill in using tables that is important for other activities, such as determining tax amounts from income tax tables.

   iv. After taking an introductory statistics course, some students move on to take more advanced statistics courses that typically use the *t* distribution when σ is unknown. For those students, it would be better if they learned one procedure that can be used again in a later course, rather than learning a procedure that must be changed later.

2. Section 8-3: For inferences about the means of two independent populations, the method of *not pooling* the sample variances is now the recommended method, but the pooled method continues to be included. This updated procedure is easier for students, and it is the procedure used most often in real applications.

TABLES

Table A-2 is now on two pages (with both negative and positive values of z) in a format with all table entries representing *cumulative areas from the left*. This is the same format used by different technologies, and students generally find that it is easier to work with this format. Also, Table A-3 for the t distribution has been expanded to include more cases.

NOTATION

The Ninth Edition now uses only the equal symbol $=$ in statements of the null hypothesis (whereas the Eighth Edition used all of the symbols $=$ and $\leq$ and $\geq$).

ORGANIZATIONAL CHANGES

1. In Chapter 5, Sections 5-3 and 5-4 from the Eighth Edition (nonstandard normal distributions) are now combined in Section 5-3. This change is motivated by the new format of Table A-2, which makes it easier for students to work with normal distributions.

2. In Chapter 5, there is a new section discussing "Sampling Distributions and Estimators," so students can better understand some of the reasons behind the methods introduced in later chapters.

3. In Chapters 6, 7, and 8, proportions are now covered first, before means and standard deviations. Reasons for this change: Students see proportions frequently in the media, and they tend to be more interested in data that are expressed as proportions. Also, proportions are generally easier to work with than means, so students can better focus on the important principles of estimating parameters and testing hypotheses.

4. Section 6-4 from the Eighth Edition (Sample Size Required to Estimate μ) is now included in Section 6-3 (Estimating a Population Mean: σ Known) along with confidence intervals for estimating μ.

5. Because the topic of statistical process control is included less frequently than the topic of nonparametric statistics, those two topics have been switched so that Chapter 12 now covers methods of nonparametric statistics and Chapter 13 covers statistical process control.

Other Changes See the Preface for other important changes.

Preface

About This Book

Although much of this Ninth Edition of *Elementary Statistics* has been updated, the major objective remains the same: Provide the best possible introductory statistics book for both students and professors. This goal is realized through factors that include a friendly writing style, content that reflects the important features of a modern introductory statistics course, the use of the latest technology, interesting and real data sets, an abundance of pedagogical components, and an unmatched battery of supplements. This text reflects recommendations and guidelines from the American Statistical Association, the Mathematical Association of America, the American Mathematical Association of Two-Year Colleges, and the National Council of Teachers of Mathematics.

Audience/Prerequisites

Elementary Statistics is written for students majoring in any field. Although the use of algebra is minimal, students should have completed at least a high school or college elementary algebra course. In many cases, underlying theory is included, but this book does not stress the mathematical rigor more suitable for mathematics majors. Because the many examples and exercises cover a wide variety of different and interesting statistical applications, *Elementary Statistics* is appropriate for students pursuing majors in a wide variety of disciplines ranging from the social sciences of psychology and sociology to areas such as education, the allied health fields, business, economics, engineering, the humanities, the physical sciences, journalism, communications, and liberal arts.

Technology

Elementary Statistics, Ninth Edition, can be used easily without reference to any specific technology. Many instructors continue to use editions of this book with students using nothing more than a variety of different scientific calculators. However, for those who choose to supplement the course with specific technology, both in-text and supplemental materials are available.

Organization Changes

- In Chapter 5, Sections 5-3 and 5-4 from the Eighth Edition (nonstandard normal distributions) are now combined in Section 5-3. This change is motivated by the new format of Table A-2, which makes it easier for students to work with normal distributions.

- In Chapter 5 there is a new Section 5-4 describing "Sampling Distributions and Estimators."

- In Chapters 6, 7, and 8, confidence intervals and hypothesis testing procedures now begin with *proportions,* which students generally find more interesting than means. Also, the procedures for working with proportions are simpler, allowing students to better focus on the new methods of inferential statistics.

- Section 6-4 from the Eighth Edition (Sample Size Required to Estimate μ) is now included in Section 6-3 (Estimating a Population Mean: σ Known) along with confidence intervals for estimating a population mean μ.

- Because instructors include the topic of statistical process control less frequently than the topic of nonparametric statistics, those two topics have been switched so that Chapter 12 now covers methods of nonparametric statistics and Chapter 13 covers statistical process control.

Content Changes

- **Procedures** In Chapters 6, 7, and 8 there is a change from "$n > 30$" to "σ known" as a key criterion for choosing between the normal and t distributions. This change reflects the common practice used by working professionals, it provides more accurate results, and it is better for students moving on to other statistics courses, yet it is not much more difficult than using the "$n > 30$" criterion.

- **Tables** There is a new format for the important normal distribution in Table A-2: Cumulative areas from the left are now listed on two pages. Students generally find this format easier to use. Table A-3 has been expanded to include more of the larger sample sizes for the Student t distribution.

- **Notation** In hypothesis testing, the symbols $\geq$ and $\leq$ are no longer used in expressions of the null hypothesis. For claims about a specific value of a parameter, only the equal symbol $=$ is used. This change reflects the practice used by the overwhelming majority of professionals who use methods of statistics and report findings in professional journals.

- **Data Sets** Appendix B now includes 30 data sets (instead of 20), including 14 that are new.

- **Icons** Technology icons **T** are now used to identify exercises based on larger data sets from Appendix B, which are best completed by using software or a TI-83 Plus or TI-84 Plus calculator.

- **Interpreting Results** Throughout the book there is even greater emphasis on *interpreting* results. Instead of simply obtaining answers, the implications and consequences of answers are considered. For example, when discussing probability in Chapter 3, instead of simply finding probability values, we interpret them by differentiating between events that are usual and those that are unusual. With hypothesis testing, we don't simply end with a conclusion of rejecting or failing to reject a null hypothesis, we proceed to state a practical conclusion that addresses the real issue. Students are encouraged to *think* about the implications of results instead of cranking out cookbook results that make no real sense.

Flexible Syllabus

The organization of this book reflects the preferences of most statistics instructors, but there are two common variations that can be easily used with this Ninth Edition:

- **Early coverage of correlation/regression:** Some instructors prefer to cover the basics of correlation and regression early in the course, such as immediately following the topics of Chapter 2. *Sections 9-2 (Correlation) and 9-3 (Regression) can be covered early in the course.* Simply omit the subsection in Section 9-2 clearly identified as "Formal Hypothesis Test (Requires Coverage of Chapter 7)."

- **Minimum probability:** Some instructors feel strongly that coverage of probability should be extensive, while others feel just as strongly that coverage should be kept to a bare minimum. Instructors preferring minimum coverage can include Section 3-2 while skipping the remaining sections of Chapter 3, as they are not essential for the chapters that follow. Many instructors prefer to cover only the fundamentals of probability along with the basics of the addition rule and multiplication rule, and the coverage of the multiplication rule (Sections 3-4 and 3-5) now offers that flexibility.

Exercises

There are over 1500 exercises—more than *58 percent of them new!* In response to requests by users of the previous edition, there are now more of the simpler exercises that are based on small data sets. Many more of the exercises require interpretation of results. Because exercises are of such critical importance to any statistics book, great care has been taken to ensure their usefulness, relevance, and accuracy. Three statisticians have read carefully through the final stages of the book to verify accuracy of the text material and exercise answers. Exercises are arranged in order of increasing difficulty by dividing them into two groups: (1) Basic Skills and Concepts and (2) Beyond the Basics. The Beyond the Basics exercises address more difficult concepts or require a somewhat stronger mathematical background. In a few cases, these exercises also introduce a new concept.

   Real data: *64% of the exercises use real data.* Because the use of real data is such an important consideration for students, hundreds of hours have been devoted to finding real, meaningful, and interesting data. In addition to the real data included throughout the book, many exercises refer to the 30 data sets listed in Appendix B.

Hallmark Features

Beyond an interesting and accessible (and sometimes humorous) writing style, great care has been taken to ensure that each chapter of *Elementary Statistics* will help students understand the concepts presented. The following features are designed to help meet that objective:

- **Chapter-opening features:** A list of chapter sections previews the chapter for the student; a chapter-opening problem, using real data, then motivates the chapter material; and the first section is a chapter overview that provides a statement of the chapter's objectives.

- **End-of-chapter features:**

  A **Chapter Review** summarizes the key concepts and topics of the chapter;

  Review Exercises offer practice on the chapter concepts and procedures;

  Cumulative Review Exercises reinforce earlier material;

  From Data to Decision: Critical Thinking is a capstone problem that requires critical thinking and a writing component;

  Cooperative Group Activities encourage active learning in groups;

  Technology Projects are for use with STATDISK, Minitab, Excel, or a TI-83 Plus or TI-84 Plus calculator;

  Internet Projects involve students with Internet data sets and, in some cases, applets.

- **Margin Essays:** The text includes 120 margin essays, which illustrate uses and abuses of statistics in real, practical, and interesting applications. Topics include "Do Boys or Girls Run in the Family?," "Accuracy of Vote Counts," "Test of Touch Therapy," and "Picking Lottery Numbers."

- **Flowcharts:** These appear throughout the text to simplify and clarify more complex concepts and procedures.

- **Statistical Software:** STATDISK, Minitab, Excel, and TI-83 Plus (or TI-84 Plus) instructions and output appear throughout the text.

- **Real Data Sets:** These are used extensively throughout the entire book. Appendix B lists 30 data sets, 14 of which are new. These data sets are provided in printed form in Appendix B, and in electronic form on the Web site and the CD bound in the back of the book. The data sets include such varied topics as ages of Queen Mary stowaways, alcohol and tobacco use in animated children's movies, eruptions of the Old Faithful geyser, diamond prices and characteristics, and movie financial and rating data.

- **Interviews:** Every chapter of the text includes author-conducted interviews with professional men and women in a variety of fields who use statistics in their day-to-day work.

- **Quick-Reference Endpapers:** Table A-2 (the normal distribution) is reproduced on the front inside cover pages, and Table A-3 (*t* distribution) is reproduced on the back inside cover page. A symbol table is included at the back of the book for quick and easy reference to key symbols.

- **Detachable Formula/Table Card:** This insert, organized by chapter, gives students a quick reference for studying, or for use when taking tests (if allowed by the instructor).

- **CD-ROM:** The CD-ROM was prepared by Mario F. Triola and is packaged with every new copy of the text; it includes the data sets (except for Data Set 4) from Appendix B in the textbook. These data sets are stored as text files, Minitab worksheets, SPSS files, SAS files, Excel workbooks, and a TI-83 Plus application. The CD also includes programs for the TI-83 Plus® graphing calculator, STATDISK Statistical Software (Version 9.5.2 or later), and the Excel Add-In, which is designed to enhance the capabilities of Excel's statistics programs.

Supplements

The student and instructor supplements packages are intended to be the most complete and helpful learning system available for the introductory statistics course. Instructors should contact their local Addison-Wesley sales representative, or e-mail the company directly at exam@aw.com for examination copies.

FOR THE INSTRUCTOR

- ***Annotated Instructor's Edition,*** by Mario F. Triola, contains answers to all exercises in the margin, plus recommended assignments, and teaching suggestions. ISBN: 0-321-14963-7 (Student for-sale edition ISBN 0-201-77570-0).

- ***Instructor's Solutions Manual,*** by Mario F. Triola and Milton Loyer, contains solutions to all the exercises and sample course syllabi. ISBN: 0-321-12212-7.

- *New* **MyMathLab.com** is a complete online course that integrates interactive multimedia instruction with the textbook content. MyMathLab can be customized easily to suit the needs of students and instructors, and it provides a comprehensive and efficient online course management system that allows for diagnosis, assessment, and tracking of students' progress. MyMathLab has numerous useful features:

  Fully interactive textbooks are built in CourseCompass, a version of Blackboard™ designed specifically for Addison-Wesley.

  Chapter and section folders from the textbook contain a wide range of instructional content: software tools (such as STATDISK and the Excel Add-In), video clips, flash animations, and the data sets and formula table card from the text.

  Hyperlinks take you directly to online testing, diagnosis, tutorials, and gradebooks in MathXL for Statistics, Addison-Wesley's tutorial and testing system.

  Instructors can create, copy, edit, assign, and track all tests for their course, as well as track student tutorial and testing performance.

  With push-button ease, instructors can remove, hide, or annotate Addison-Wesley preloaded content, add their own course documents, or change the order in which material is presented.

  Using the communications tools found in MyMathLab, instructors can hold online office hours, host a discussion board, create communication groups within their class, send e-mail, and maintain a course calendar.

For more information, visit www.mymathlab.com or contact your Addison-Wesley sales representative for a demonstration.

- **Testing System:** Great care has been taken to ensure the strongest possible testing system for the new edition of *Elementary Statistics*. Not only is there a printed test bank, there is also a computerized test generator, **TestGen4.0 and Quizmaster3.0.,** that lets you view and edit test-bank questions, transfer them to tests, and print in a variety of formats. The program also offers many options for organizing and displaying test banks and tests. A built-in random number and test generator makes **TestGen-EQ** ideal for creating multiple versions of tests and provides more possible test items than printed test-bank questions. Powerful search and sort functions let the instructor easily locate questions and arrange them in the preferred order. Users can export tests as text files so they can be viewed with a Web browser. Additionally, tests created with TestGen can be used with **QuizMaster,** which enables students to take exams on a computer network. QuizMaster automatically grades the exams, stores results on disk, and allows the instructor to view or print a variety of reports for individual students, classes, or courses. **Printed Testbank** ISBN: 0-321-12214-3; TestGen-EQ for Mac and Windows ISBN: 0-321-12213-5.

- **PowerPoint® Lecture Presentation CD:** Free to qualified adopters, this classroom lecture presentation software is geared specifically to the sequence and philosophy of *Elementary Statistics*. Key graphics from the book are included. These slides are also available on the Triola Web site at www.aw.com/info/Triola. Mac and Windows ISBN: 0-321-12215-1.

FOR THE STUDENT

- **MathXL for Statistics** is a Web site that provides students with online homework, testing, and tutorial help. Students can take chapter tests correlated to the textbook, receive individualized study plans based on those test results, work practice problems and receive feedback on which text examples or exercises to review further, and take tests again to gauge their progress. Instructors can assign homework or customize tests using the online test bank or upload any tests they created using Addison-Wesley's TestGen testing software. All student test results, homework, study plans, and practice work are tracked and viewable in an online gradebook. The site is free to qualified adopters when an access code is bundled with a new text. MathXL for Statistics is also available as part of MyMathLab.com.

- **Videos** are designed to supplement many sections in the book, with some topics presented by the author. The videos feature all technologies in the book. This is an excellent resource for students who have missed class or wish to review a topic. It is also an excellent resource for instructors involved with distance learning, individual study, or self-paced learning programs. **Videotapes** ISBN: 0-321-12209-7; **Digital Video Tutor** (CD-ROM version) ISBN: 0-321-12231-3.

- **Triola *Elementary Statistics* Web Site:** This Web site may be accessed at http://www.aw.com/triola, and provides Internet projects keyed to every chapter of the text, and the book's data sets as they appear on the CD.

- ***Student's Solutions Manual,*** by Milton Loyer (Penn State University), provides detailed, worked-out solutions to all odd-numbered text exercises. ISBN: 0-321-12217-8.

The following technology manuals include instructions on and examples of the technology's use. Each one has been written to correspond with the text.

- *Excel® Student Laboratory Manual and Workbook,* written by Johanna Halsey and Ellena Reda (Dutchess Community College). ISBN: 0-321-12206-2.

- *Minitab® Student Laboratory Manual and Workbook,* written by Mario F. Triola. ISBN: 0-321-12205-4.

- *SAS Student Laboratory Manual and Workbook,* written by Joseph Morgan (DePaul University). ISBN: 0-321-12727-7.

- *SPSS® Student Laboratory Manual and Workbook,* written by Roger Peck (California State University, Bakersfield). ISBN: 0-321-12207-0.

- *STATDISK Student Laboratory Manual and Workbook,* written by Mario F. Triola. ISBN: 0-321-12216-X.

- *TI-83 Plus® Companion to Elementary Statistics,* by Marla Bell (Kennesaw State University). ISBN: 0-321-12208-9.

- **Triola Version of** *ActivStats®,* developed by Paul Velleman and Data Description, Inc., provides complete coverage of introductory statistics topics on CD-ROM, using a full range of multimedia. *ActivStats* integrates video, simulation, animation, narration, text, and interactive experiments, World Wide Web access, and Data Desk®, a statistical software package. Homework problems and data sets from the text are included on the CD-ROM. *ActivStats* for Windows and Macintosh ISBN: 0-201-77139-X. Also available in versions for Excel, JMP, Minitab and SPSS. See your Addison-Wesley sales representative for details or check the Web site at www.aw.com/activstats.

- *Addison-Wesley Tutor Center:* Free tutoring is available to students who purchase a new copy of the Ninth Edition of *Elementary Statistics* when bundled with an access code. The Addison-Wesley Math Tutor Center is staffed by qualified statistics and mathematics instructors who provide students with tutoring on text examples and any exercise with an answer in the back of the book. Tutoring assistance is provided by toll-free telephone, fax, e-mail, and whiteboard technology—which allows tutors and students to actually see the problems worked while they "talk" in real time over the Internet. This service is available five days a week, seven hours a day. For more information, please contact your Addison-Wesley sales representative.

- *The Student Edition of Minitab* is a condensed version of the professional release of Minitab Statistical Software. It offers students the full range of Minitab's statistical methods and graphical capabilities, along with worksheets that can include up to 5000 data points. It comes with a user's manual that includes case studies and hands-on tutorials, and is perfect for use in any introductory statistics course, including those in the life and social sciences. The currently available Student Edition is *Student Edition of Minitab, Release 12 for Windows 95/98 NT.* ISBN: 0-201-39715-3.

Any of these products can be purchased separately, or bundled with Addison-Wesley texts. Instructors can contact local sales representatives for details on purchasing and bundling supplements with the textbook or contact the company at exam@aw.com for examination copies of any of these items.

Acknowledgments

The success of **Elementary Statistics** is due to the efforts of many dedicated professionals. I wish to express my most sincere thanks to Deirdre Lynch, Christine O'Brien, Greg Tobin, Peggy McMahon, Barbara Atkinson, Brenda Bravener, Yolando Cossio, Karen Guardino, Jen Kerber, Sara Anderson, Keren Blankfeld, Joe Vetere, Joanne Ha, Weslie Lewis, Janet Nuciforo, and the entire Addison-Wesley staff.

I thank Paul Lorczak for his work on the Internet Projects. I thank Dale Peterson of the United States Air Force Academy for providing excellent suggestions for this new edition. I thank Don Puretz, Heather Carielli, and Ashley McConnaughay for providing help with data sets. I am grateful for the many suggestions and comments provided by instructors and students, and I am grateful to the hundreds of researchers who did studies, conducted surveys, and compiled interesting and meaningful data used in examples and exercises throughout this book.

I also take great pride in thanking my wife, Ginny, and my sons Marc and Scott for their help and encouragement.

I would also like to thank the following for their help with the Ninth Edition:

Text Accuracy Reviewers
David R. Lund, University of Wisconsin at Eau Claire
Tim Mogill
Kimberly Polly, Parkland College
John Ritschdorff, Marist College
Twin Prime Editorial
Tom Wegleitner

Reviewers of the Ninth Edition:

Dan Abbey, Broward Community College
Justine Baker, Peirce College
Donald Barrs, Pellissippi State Technical Community College
Marla Bell, Kennesaw State University
Wayne Ehler, Anne Arundel Community College
Joe Gallegos, Salt Lake Community College
Jim Graziose, Palm Beach Community College
Kristin Hartford, Long Beach City College

John Kozarski, Community College of Baltimore County-Catonsville
Tzong-Yow Lee, University of Maryland
Debra Loeffler, Community College of Baltimore County-Catonsville
Caren McClure, Santa Ana College
Carla Monticelli, Camden County Community College
Rick Moscatello, Southeastern Louisiana University
Nasser Ordoukhani, Barry University
Michael Oriolo, Herkimer County Community College
Kimberly Polly, Parkland College
Diann Reischman, Grand Valley State University
Don Robinson, Illinois State University
Ira Rosenthal, Palm Beach Community College—Eissey Campus
Adele Shapiro, Palm Beach Community College
Lewis Shoemaker, Millersville University
Consuelo Stewart, Howard Community College
David Stewart, Community College of Baltimore County-Dundalk
Sharon Testone, Onondaga Community College
Carol Yin, LeGrange College

I extend my sincere thanks for the suggestions made by the following reviewers and users of previous editions of the book:

Mary Abkemeier, Fontbonne College
William A. Ahroon, SUNY at Plattsburgh
Scott Albert, College of Du Page
Jules Albertini, Ulster County Community College
Tim Allen, Delta College
Stu Anderson, College of Du Page
Jeff Andrews, TSG Associates, Inc.
Mary Anne Anthony, Rancho Santiago Community College
William Applebaugh, University of Wisconsin at Eau Claire
James Baker, Jefferson Community College
Anna Bampton, Christopher Newport University
James Beatty, Burlington County College
Philip M. Beckman, Black Hawk College
Marian Bedee, BGSU, Firelands College
Don Benbow, Marshalltown Community College
Michelle Benedict, Augusta College
Kathryn Benjamin, Suffolk County Community College
Ronald Bensema, Joliet Junior College
David Bernklau, Long Island University
Maria Betkowski, Middlesex Community College
Shirley Blatchley, Brookdale Community College

David Blaueuer, University of Findlay
Randy Boan, Aims Community College
John Bray, Broward Community College-Central
Denise Brown, Collin County Community College
Patricia Buchanan, Pennsylvania State University
John Buchl, Rochester Community and Technical College
Michael Butler, Mt. San Antonio College
Jerome J. Cardell, Brevard Community College
Don Chambless, Auburn University
Rodney Chase, Oakland Community College
Bob Chow, Grossmont College
Philip S. Clarke, Los Angeles Valley College
Darrell Clevidence, Carl Sandburg College
Paul Cox, Ricks College
Susan Cribelli, Aims Community College
Imad Dakka, Oakland Community College
Arthur Daniel, Macomb Community College
Gregory Davis, University of Wisconsin, Green Bay
Tom E. Davis, III, Daytona Beach Community College
Charles Deeter, Texas Christian University
Joseph DeMaio, Kennesaw State University
Joe Dennin, Fairfield University
Nirmal Devi, Embry Riddle Aeronautical University

Richard Dilling, Grace College

Rose Dios, New Jersey Institute of Technology

Dennis Doverspike, University of Akron

Paul Duchow, Pasadena City College

Bill Dunn, Las Positas College

Marie Dupuis, Milwaukee Area Technical College

Evelyn Dwyer, Walters State Community College

Jane Early, Manatee Community College

Sharon Emerson-Stonnell, Longwood College

P. Teresa Farnum, Franklin Pierce College

Ruth Feigenbaum, Bergen Community College

Vince Ferlini, Keene State College

Maggie Flint, Northeast State Technical Community College

Bob France, Edmonds Community College

Christine Franklin, University of Georgia

Richard Fritz, Moraine Valley Community College

Maureen Gallagher, Hartwick College

Mahmood Ghamsary, Long Beach City College

Tena Golding, Southeastern Louisiana University

Elizabeth Gray, Southeastern Louisiana University

David Gurney, Southeastern Louisiana University

Francis Hannick, Mankato State University

Sr. Joan Harnett, Molloy College

Leonard Heath, Pikes Peak Community College

Peter Herron, Suffolk County Community College

Mary Hill, College of Du Page

Larry Howe, Rowan College of New Jersey

Lloyd Jaisingh, Morehead State University

Lauren Johnson, Inver Hills Community College

Martin Johnson, Gavilan College

Roger Johnson, Carleton College

Herb Jolliff, Oregon Institute of Technology

Francis Jones, Huntington College

Toni Kasper, Borough of Manhattan Community College

Alvin Kaumeyer, Pueblo Community College

William Keane, Boston College

Robert Keever, SUNY at Plattsburgh

Alice J. Kelly, Santa Clara University

Dave Kender, Wright State University

Michael Kern, Bismarck State College

John Klages, County College of Morris

Marlene Kovaly, Florida Community College at Jacksonville

Tomas Kozubowski, University of Tennessee

Shantra Krishnamachari, Borough of Manhattan Community College

Richard Kulp, David Lipscomb University

Linda Kurz, SUNY College of Technology

Christopher Jay Lacke, Rowan University

Tommy Leavelle, Mississippi College

R. E. Lentz, Mankato State University

Timothy Lesnick, Grand Valley State University

Dawn Lindquist, College of St. Francis

George Litman, National-Louis University

Benny Lo, Ohlone College

Sergio Loch, Grand View College

Vincent Long, Gaston College

Barbara Loughead, National-Louis University

David Lund, University of Wisconsin at Eau Claire

Rhonda Magel, North Dakota State University—Fargo

Gene Majors, Fullerton College

Hossein Mansouri, Texas State Technical College

Virgil Marco, Eastern New Mexico University

Joseph Mazonec, Delta College

Caren McClure, Rancho Santiago Community College

Phillip McGill, Illinois Central College

Marjorie McLean, University of Tennessee

Austen Meek, Canada College

Robert Mignone, College of Charleston

Glen Miller, Borough of Manhattan Community College

Kermit Miller, Florida Community College at Jacksonville

Kathleen Mittag, University of Texas—San Antonio

Mitra Moassessi, Santa Monica College

Charlene Moeckel, Polk Community College

Theodore Moore, Mohawk Valley Community College

Gerald Mueller, Columbus State Community College

Sandra Murrell, Shelby State Community College

Faye Muse, Asheville-Buncombe Technical Community College

Gale Nash, Western State College

Felix D. Nieves, Antillean Adventist University

Lyn Noble, Florida Community College at Jacksonville-South

DeWayne Nymann, University of Tennessee

Patricia Oakley, Seattle Pacific University

Keith Oberlander, Pasadena City College

Patricia Odell, Bryant College

James O'Donnell, Bergen Community College

Alan Olinksy, Bryant College

Ron Pacheco, Harding University

Lindsay Packer, College of Charleston

Kwadwo Paku, Los Medanos College

Deborah Paschal, Sacramento City College

S. A. Patil, Tennessee Technological University

Robin Pepper, Tri-County Technical College

David C. Perkins, Texas A&M University—Corpus Christi

Anthony Piccolino, Montclair State University

Richard J. Pulskamp, Xavier University

Vance Revennaugh, Northwestern College

C. Richard, Southeastern Michigan College

Sylvester Roebuck, Jr., Olive Harvey College

Kenneth Ross, Broward Community College

Charles M. Roy, Camden County College

Kara Ryan, College of Notre Dame

Fabio Santos, LaGuardia Community College

Richard Schoenecker, University of Wisconsin, Stevens Point

Nancy Schoeps, University of North Carolina, Charlotte

Jean Schrader, Jamestown Community College

A. L. Schroeder, Long Beach City College

Phyllis Schumacher, Bryant College

Sankar Sethuraman, Augusta College

Rosa Seyfried, Harrisburg Area Community College

Calvin Shad, Barstow College

Carole Shapero, Oakton Community College

Lewis Shoemaker, Millersville University

Joan Sholars, Mt. San Antonio College

Galen Shorack, University of Washington

Teresa Siak, Davidson County Community College

Cheryl Slayden, Pellissippi State Technical Community College

Arthur Smith, Rhode Island College

Marty Smith, East Texas Baptist University

Laura Snook, Blackhawk Community College

Aileen Solomon, Trident Technical College

Sandra Spain, Thomas Nelson Community College

Maria Spinacia, Pasco-Hernandez Community College

Paulette St. Ours, University of New England

W. A. Stanback, Norfolk State University

Carol Stanton, Contra Costra College

Richard Stephens, University of Alaska Southeast

W. E. Stephens, McNeese State University

Terry Stephenson, Spartanburg Methodist College

Consuelo Stewart, Howard Community College

Ellen Stutes, Louisiana State University at Eunice

Sr. Loretta Sullivan, University of Detroit Mercy

Tom Sutton, Mohawk College

Andrew Thomas, Triton College

Evan Thweatt, American River College

Judith A. Tully, Bunker Hill Community College

Gary Van Velsir, Anne Arundel Community College

Paul Velleman, Cornell University

Randy Villa, Napa Valley College

Hugh Walker, Chattanooga State Technical Community College

Charles Wall, Trident Technical College

Glen Weber, Christopher Newport College

David Weiner, Beaver College

Sue Welsch, Sierra Nevada College

Roger Willig, Montgomery County Community College

Gail Wiltse, St. Johns River Community College

Odell Witherspoon, Western Piedmont Community College

Jean Woody, Tulsa Junior College

Thomas Zachariah, Loyola Marymount University

Elyse Zois, Kean College of New Jersey

M.F.T.
LaGrange, New York
September, 2002

Contents

 Introduction to Statistics 2

1-1   Overview 4
1-2   Types of Data 5
1-3   Critical Thinking 11
1-4   Design of Experiments 20

Describing, Exploring, and Comparing Data 36

2-1   Overview 38
2-2   Frequency Distributions 39
2-3   Visualizing Data 46
2-4   Measures of Center 59
2-5   Measures of Variation 73
2-6   Measures of Relative Standing 92
2-7   Exploratory Data Analysis (EDA) 102

Probability 118

3-1   Overview 120
3-2   Fundamentals 120
3-3   Addition Rule 132
3-4   Multiplication Rule: Basics 139
3-5   Multiplication Rule: Complements and
        Conditional Probability 150
3-6   Probabilities Through Simulations 156
3-7   Counting 162

Probability Distributions 180

4-1   Overview 182
4-2   Random Variables 183
4-3   Binomial Probability Distributions 196
4-4   Mean, Variance, and Standard Deviation for the
        Binomial Distribution 207
4-5   The Poisson Distribution 212

Normal Probability Distributions 224

5-1 Overview 226
5-2 The Standard Normal Distribution 227
5-3 Applications of Normal Distributions 240
5-4 Sampling Distributions and Estimators 249
5-5 The Central Limit Theorem 259
5-6 Normal as Approximation to Binomial 271
5-7 Determining Normality 282

Estimates and Sample Sizes 296

6-1 Overview 298
6-2 Estimating a Population Proportion 298
6-3 Estimating a Population Mean: σ Known 318
6-4 Estimating a Population Mean: σ Not Known 330
6-5 Estimating a Population Variance 347

Hypothesis Testing 366

7-1 Overview 368
7-2 Basics of Hypothesis Testing 369
7-3 Testing a Claim About a Proportion 388
7-4 Testing a Claim About a Mean: σ Known 400
7-5 Testing a Claim About a Mean: σ Not Known 407
7-6 Testing a Claim About a Standard Deviation or Variance 419

Inferences from Two Samples 436

8-1 Overview 438
8-2 Inferences About Two Proportions 438
8-3 Inferences About Two Means: Independent Samples 452
8-4 Inferences from Matched Pairs 466
8-5 Comparing Variation in Two Samples 476

Correlation and Regression 494

9-1 Overview 496
9-2 Correlation 496
9-3 Regression 517
9-4 Variation and Prediction Intervals 531
9-5 Multiple Regression 541
9-6 Modeling 551

Multinomial Experiments and Contingency Tables 564

10-1 Overview 566
10-2 Multinomial Experiments: Goodness-of-Fit 567
10-3 Contingency Tables: Independence and Homogeneity 582

Analysis of Variance 602

11-1   Overview 604
11-2   One-Way ANOVA 606
11-3   Two-Way ANOVA 619

Nonparametric Statistics 636

12-1   Overview 638
12-2   Sign Test 640
12-3   Wilcoxon Signed-Ranks Test for Matched Pairs 650
12-4   Wilcoxon Rank-Sum Test for Two Independent Samples 656
12-5   Kruskal-Wallis Test 663
12-6   Rank Correlation 670
12-7   Runs Test for Randomness 679

Statistical Process Control 694

13-1   Overview 696
13-2   Control Charts for Variation and Mean 696
13-3   Control Charts for Attributes 710

Projects, Procedures, Perspectives 722

14-1   Projects 722
14-2   Procedure 726
14-3   Perspective 728

Appendices 730

Appendix A: Tables 731
Appendix B: Data Sets 747
Appendix C: TI-83 Plus Reference 783
Appendix D: Glossary 785
Appendix E: Bibliography 793
Appendix F: Answers to Odd-Numbered Exercises (and All Review Exercises and All Cumulative Review Exercises) 795

Credits 827

Index 829

Suggested Course Syllabus

This book contains much more material than can be covered in the typical one-semester introductory statistics course. Sample syllabi are provided on the following pages as an aid to instructors in planning their individual courses. The "class hour" referred to in these outlines consists of approximately one hour. In addition to the sample syllabi that follow, there are many other configurations that successfully satisfy individual needs and preferences.

When selecting the sections to be included, consider assigning some sections as out-of-class independent work. For example, after students understand the basic concepts of confidence intervals and hypothesis testing, consider assigning exercises from Chapter 8 (Inferences from Two Samples) to be done using a computer software package or a TI-83 Plus calculator. This is an excellent way to include important and useful concepts while encouraging students to extend their basic knowledge to different circumstances.

In planning a new course syllabus for the Ninth Edition of *Elementary Statistics,* users of the Eighth Edition should consider these major changes:

1. In Chapters 6, 7, and, 8, proportions are now covered first, before means and standard deviations. Reasons for this change: Students see proportions frequently in the media, and they tend to be more interested in data that are expressed as proportions. Also, proportions are generally easier to work with than means, so students can better focus on the important principles of estimating parameters and testing hypotheses.

2. In Chapter 5, there is a new section discussing "Sampling Distributions and Estimators," so students can better understand some of the reasons behind the methods introduced in later chapters.

3. In Chapter 5, Sections 5-3 and 5-4 from the Eighth Edition (nonstandard normal distributions) are now combined in Section 5-3. This change is motivated by the new format of Table A-2, which makes it easier for students to work with normal distributions.

4. Section 6-4 from the Eighth Edition (Sample Size Required to Estimate μ) is now included in Section 6-3 (Estimating a Population Mean: σ Known) along with confidence intervals for estimating μ.

5. Because the topic of statistical process control is included less frequently than the topic of nonparametric statistics, those two topics have been switched so that Chapter 12 now covers methods of nonparametric statistics and Chapter 13 covers statistical process control.

Syllabus Version A

Class Hour	Text Section	Topic
1	1-1, 1-2, 1-3	Overview; Nature of Data; Critical Thinking
2	1-4	Design of Experiments
3	2-1, 2-2, 2-3	Frequency Distributions and Visualizing Data
4	2-4	Measures of Center
5	2-5	Measures of Variation
6	2-5	Measures of Variation
7	2-6	Measures of Relative Standing
8	2-7	Exploratory Data Analysis
9	Ch. 1, 2	Review
10	Ch. 1, 2	Test 1
11	3-1, 3-2, 3-3	Fundamentals of Probability and Addition Rule
12	3-4	Multiplication Rule: Basics
13	4-1, 4-2	Random Variables
14	4-3	Binomial Probability Distributions
15	4-4	Mean, Variance, and Standard Deviation for the Binomial Distribution
16	Ch. 3, 4	Review
17	Ch. 3, 4	Test 2
18	5-1, 5-2	The Standard Normal Distribution
19	5-3	Applications of Normal Distributions
20	5-4	Sampling Distributions and Estimators
21	5-5	The Central Limit Theorem
22	6-1, 6-2	Estimating a Population Proportion
23	6-3	Estimating a Population Mean: σ Known
24	6-4	Estimating a Population Mean: σ Not Known
25	6-5	Estimating a Population Variance
26	Ch. 5, 6	Review
27	Ch. 5, 6	Test 3
28	7-1, 7-2	Basics of Hypothesis Testing
29	7-2, 7-3	Testing a Claim About a Proportion
30	7-3	Testing a Claim About a Proportion
31	7-4	Testing a Claim About a Mean: σ Known
32	7-5	Testing a Claim About a Mean: σ Not Known
33	7-6	Testing a Claim About Standard Deviation or Variance
34	Ch. 7	Review
35	Ch. 7	Test 4
36	9-1, 9-2	Correlation
37	9-2, 9-3	Correlation and Regression
38	9-3	Regression
39	10-3	Contingency Tables
40		Group Project Presentations
41		Group Project Presentations
42	Ch. 9, 10	Review
43	Ch. 9, 10	Test 5
44		Review for Comprehensive Final Exam
45		Review for Comprehensive Final Exam

Syllabus Version B: Relaxed Pace

Class Hour	Text Section	Topic
1	1-1, 1-2	Overview; Nature of Data
2	1-3	Critical Thinking
3	2-1, 2-2, 2-3	Frequency Distributions and Visualizing Data
4	2-4	Measures of Center
5	2-5	Measures of Variation
6	2-5	Measures of Variation
7	2-6	Measures of Relative Standing
8	2-7	Exploratory Data Analysis
9	Ch. 1, 2	Review
10	Ch. 1, 2	Test 1
11	3-1, 3-2	Fundamentals of Probability
12	3-2	Fundamentals of Probability
13	4-1, 4-2	Random Variables
14	4-3	Binomial Probability Distributions
15	4-4	Mean, Variance, and Standard Deviation for the Binomial Distribution
16	Ch. 3, 4	Review
17	Ch. 3, 4	Test 2
18	5-1, 5-2	The Standard Normal Distribution
19	5-3	Applications of Normal Distributions
20	5-3	Applications of Normal Distributions
21	5-5	The Central Limit Theorem
22	6-1, 6-2	Estimating a Population Proportion
23	6-2	Estimating a Population Proportion
24	6-3	Estimating a Population Mean: σ Known
25	6-4	Estimating a Population Mean: σ Not Known
26	Ch. 5, 6	Review
27	Ch. 5, 6	Test 3
28	7-1, 7-2	Basics of Hypothesis Testing
29	7-2, 7-3	Testing a Claim About a Proportion
30	7-3	Testing a Claim About a Proportion
31	7-4	Testing a Claim About a Mean: σ Known
32	7-5	Testing a Claim About a Mean: σ Not Known
33	7-5	Testing a Claim About a Mean: σ Not Known
34	Ch. 7	Review
35	Ch. 7	Test 4
36	9-1, 9-2	Correlation
37	9-2, 9-3	Correlation and Regression
38	9-3	Regression
39		Group Project Presentations
40		Group Project Presentations
41	Ch. 9	Review
42	Ch. 9	Test 5
43		Review for Comprehensive Final Exam
44		Review for Comprehensive Final Exam

Syllabus Version C: EARLY COVERAGE OF CORRELATION AND REGRESSION

Class Hour	Text Section	Topic
1	1-1, 1-2, 1-3	Overview; Nature of Data; Critical Thinking
2	1-4	Design of Experiments
3	2-1, 2-2, 2-3	Frequency Distributions and Visualizing Data
4	2-4	Measures of Center
5	2-5	Measures of Variation
6	2-5	Measures of Variation
7	2-6	Measures of Relative Standing
8	2-7	Exploratory Data Analysis
9	**9-1, 9-2**	**Correlation**
10	**9-3**	**Regression**
11	**9-3**	**Regression**
12	Ch. 1, 2, 9	Review
13	Ch. 1, 2, 9	Test 1
14	3-1, 3-2, 3-3	Fundamentals of Probability and Addition Rule
15	3-4	Multiplication Rule: Basics
16	4-1, 4-2	Random Variables
17	4-3	Binomial Probability Distributions
18	4-4	Mean, Variance, and Standard Deviation for the Binomial Distribution
19	5-1, 5-2	The Standard Normal Distribution
20	5-3	Applications of Normal Distributions
21	5-4	Sampling Distributions and Estimators
22	5-5	The Central Limit Theorem
23	5-6	Normal as Approximation to Binomial
24	Ch. 3, 4, 5	Review
25	Ch. 3, 4, 5	Test 2
26	6-1, 6-2	Estimating a Population Proportion
27	6-3	Estimating a Population Mean: σ Known
28	6-4	Estimating a Population Mean: σ Not Known
29	6-5	Estimating a Population Variance
30	7-1, 7-2	Basics of Hypothesis Testing
31	7-2, 7-3	Testing a Claim About a Proportion
32	7-3	Testing a Claim About a Proportion
33	7-4	Testing a Claim About a Mean: σ Known
34	7-5	Testing a Claim About a Mean: σ Not Known
35	7-6	Testing a Claim About Standard Deviation or Variance
36	10-3	Contingency Tables
37	Ch. 6, 7, 10	Review
38	Ch. 6, 7, 10	Test 3
39		Group Project Presentations
40		Group Project Presentations
41		Review for Comprehensive Final Exam
42		Review for Comprehensive Final Exam

Syllabus Version D: MINIMUM COVERAGE OF PROBABILITY

Class Hour	Text Section	Topic
1	1-1, 1-2, 1-3	Overview; Nature of Data; Critical Thinking
2	1-4	Design of Experiments
3	2-1, 2-2, 2-3	Frequency Distributions and Visualizing Data
4	2-4	Measures of Center
5	2-5	Measures of Variation
6	2-5	Measures of Variation
7	2-6	Measures of Relative Standing
8	2-7	Exploratory Data Analysis
9	Ch. 1, 2	Review
10	Ch. 1, 2	Test 1
11	**3-1, 3-2**	**Fundamentals of Probability**
12	4-1, 4-2	Random Variables
13	4-3	Binomial Probability Distributions
14	4-4	Mean, Variance, and Standard Deviation for the Binomial Distribution
15	5-1, 5-2	The Standard Normal Distribution
16	5-3	Applications of Normal Distributions
17	5-4	Sampling Distributions and Estimators
18	5-5	The Central Limit Theorem
19	Ch. 3, 4, 5	Review
20	Ch. 3, 4, 5	Test 2
21	6-1, 6-2	Estimating a Population Proportion
22	6-3	Estimating a Population Mean: σ Known
23	6-4	Estimating a Population Mean: σ Not Known
24	6-5	Estimating a Population Variance
25	7-1, 7-2	Basics of Hypothesis Testing
26	7-2, 7-3	Testing a Claim About a Proportion
27	7-3	Testing a Claim About a Proportion
28	7-4	Testing a Claim About a Mean: σ Known
29	7-5	Testing a Claim About a Mean: σ Not Known
30	7-6	Testing a Claim About Standard Deviation or Variance
31	Ch. 6, 7	Review
32	Ch. 6, 7	Test 3
33	9-1, 9-2	Correlation
34	9-2, 9-3	Correlation and Regression
35	9-3	Regression
36	10-1, 10-2	Multinomial Experiments
37	10-3	Contingency Tables
38	Ch. 9, 10	Review
39	Ch. 9, 10	Test 4
40		Group Project Presentations
41		Group Project Presentations
42		Review for Comprehensive Final Exam
43		Review for Comprehensive Final Exam

Syllabus Version E: INTEGRATION OF SOME NONPARAMETRIC STATISTICS

Class Hour	Text Section	Topic
1	1-1, 1-2, 1-3	Overview; Nature of Data; Critical Thinking
2	1-4	Design of Experiments
3	**12-7**	**Runs Test for Randomness**
4	2-1, 2-2, 2-3	Frequency Distributions and Visualizing Data
5	2-4	Measures of Center
6	2-5	Measures of Variation
7	2-5	Measures of Variation
8	2-6	Measures of Position
9	2-7	Exploratory Data Analysis
10	Ch. 1, 2, 12	Review
11	Ch. 1, 2, 12	Test 1
12	3-1, 3-2, 3-3	Fundamentals of Probability and Addition Rule
13	3-4	Multiplication Rule: Basics
14	4-1, 4-2	Random Variables
15	4-3	Binomial Probability Distributions
16	4-4	Mean, Variance, and Standard Deviation for the Binomial Distribution
17	5-1, 5-2	The Standard Normal Distribution
18	5-3	Applications of Normal Distributions
19	5-4	Sampling Distributions and Estimators
20	5-5	The Central Limit Theorem
21	5-6	Normal as Approximation to Binomial
22	Ch. 3, 4, 5	Review
23	Ch. 3, 4, 5	Test 2
24	6-1, 6-2	Estimating a Population Proportion
25	6-3	Estimating a Population Mean: σ Known
26	6-4	Estimating a Population Mean: σ Not Known
27	6-5	Estimating a Population Variance
28	7-1, 7-2	Basics of Hypothesis Testing
29	7-2, 7-3	Testing a Claim About a Proportion
30	7-3	Testing a Claim About a Proportion
31	7-4	Testing a Claim About a Mean: σ Known
32	7-5	Testing a Claim About a Mean: σ Not Known
33	7-6	Testing a Claim About Standard Deviation or Variance
34	Ch. 6, 7	Review
35	Ch. 6, 7	Test 3
36	9-1, 9-2	Correlation and Regression
37	9-3	Regression
38	**12-6**	**Rank Correlation**
39	**12-5**	**Kruskal-Wallis Test**
40	10-3	Contingency Tables
41	Ch. 9, 10, 12	Review
42	Ch. 9, 10, 12	Test 4
43		Group Project Presentations
44		Group Project Presentations
45		Review for Comprehensive Final Exam

Syllabus Version F: TWO SEMESTER COURSE

A two-semester sequence allows for more time to better incorporate computer usage, cooperative group activities, simulation techniques in probability, and bootstrap resampling techniques in estimating parameters and testing hypotheses.

First Semester:

Chapter 1 (Sections 1-1 through 1-4):	4 classes
Chapter 2 (Sections 2-1 through 2-7):	7 classes
Chapter 3 (Sections 3-1 through 3-7):	6 classes
Chapter 4 (Sections 4-1 through 4-5):	5 classes
Chapter 5 (Sections 5-1 through 5-7):	7 classes
Chapter 6 (Sections 6-1 through 6-5):	7 classes
Chapter 7 (Sections 7-1 through 7-6):	8 classes

Second Semester:

Chapter 8 (Sections 8-1 through 8-5):	8 classes
Chapter 9 (Sections 9-1 through 9-6):	9 classes
Chapter 10 (Sections 10-1 through 10-3):	4 classes
Chapter 11 (Sections 11-1 through 11-3):	6 classes
Chapter 12 (Sections 12-1 through 12-7):	4 classes
Chapter 13 (Sections 13-1 through 13-3):	9 classes
Chapter 14 (Section 14-1):	4 classes

ELEMENTARY
STATISTICS

NINTH EDITION

1

Introduction to Statistics

1-1 Overview

1-2 Types of Data

1-3 Critical Thinking

1-4 Design of Experiments

What can we learn from these surveys?

Here are brief descriptions of five different surveys:

1. In mid-December of a recent year, the Internet service provider America Online (AOL) ran a survey of its users. This question was asked about Christmas trees:

    Which do you prefer?
    - a real tree
    - a fake tree

    Among the 7073 responses received by the Internet users, 4650 preferred a real tree, and 2423 preferred a fake tree.

2. *Newsweek* magazine recently ran a survey about the controversial Napster Web site, which had been providing free access to copying music CDs. Readers were asked this question:

    Will you still use Napster if you have to pay a fee?

    Readers could register their responses on the Web site newsweek.msnbc.com. Among the 1873 responses received, 19% said yes, it is still cheaper than buying CDs. Another 5% said yes, they felt more comfortable using it with a charge.

3. *Good Housekeeping* magazine invited women visiting its Web site to complete a survey, and 1500 responses were recorded. When asked whether they would rather have more money or more sleep, 88% chose more money and 11% chose more sleep.

4. *USA Today* ran a ³⁄₄ page "Health Care Survey." Readers were asked to "please take a moment to fill out this survey and return it to us." Most questions asked about health conditions, tobacco use, and prescription drugs. Question 17 of the survey was this: "May we contact you again to participate in other surveys for *USA Today?*"

5. *USA Today* published a bar graph of results from a survey in which respondents were asked "Do you have plans for spring break?" Among the 4264 Internet users who chose to respond, 48% said that they had no plans yet, and 14% said that they planned to go to a beach.

What crucially important feature do these five surveys have in common? How does it affect what we can conclude about the general population based on the results obtained from such surveys? Can we conclude that most Americans prefer a real Christmas tree to a fake one? Can we conclude that the vast majority of American women prefer more money to more sleep? Can we conclude that the vast majority of women readers of *Good Housekeeping* prefer more money to more sleep? The answers to these questions are critically important for the evaluation of such survey results. The issue being considered here is the single most important issue of this entire chapter, and it may well be the single most important issue of the entire book.

In this chapter, we address important issues concerning the validity of surveys, such as the five surveys described. We will see that we can often form important conclusions by applying simple common sense. After completing this chapter, we should be able to identify the key issue affecting the validity of the five surveys, and we should have an important understanding of data collection methods in general.

1-1 Overview

The Chapter Problem on the previous page involves surveys. A survey is one of many tools that can be used for collecting data. A common goal of a survey is to collect data from a small part of a larger group so that we can learn something about the larger group. This is a common and important goal of the subject of statistics: Learn about a large group by examining data from some of its members. In this context, the terms *sample* and *population* become important. Formal definitions for these and other basic terms are given here.

Definitions

Data are observations (such as measurements, genders, survey responses) that have been collected.

Statistics is a collection of methods for planning experiments, obtaining data, and then organizing, summarizing, presenting, analyzing, interpreting, and drawing conclusions based on the data.

A **population** is the complete collection of all elements (scores, people, measurements, and so on) to be studied. The collection is complete in the sense that it includes all subjects to be studied.

A **census** is the collection of data from *every* member of the population.

A **sample** is a *subcollection* of members selected from a population.

For example, a Gallup Poll asked this of 1087 adults: "Do you have occasion to use alcoholic beverages such as liquor, wine, or beer, or are you a total abstainer?" The 1087 survey subjects constitute a *sample*, whereas the *population* consists of the entire collection of all 202,682,345 adult Americans. Every 10 years, the United States government attempts to obtain a *census* of every citizen, but fails because it is impossible to reach everyone. An ongoing controversy involves the attempt to use sound statistical methods to improve the accuracy of the Census, but political considerations are a key factor causing members of Congress to resist this improvement. Perhaps some readers of this text will one day be members of Congress with the wisdom to bring the Census into the twenty-first century.

An important activity of this book is to demonstrate how we can use sample data to form conclusions about populations. We will see that it is *extremely* critical to obtain sample data that are representative of the population from which the data are drawn. For example, if you survey the alumni who graduated from your college by asking them to write their annual income and mail it back to you, the responses are not likely to be representative of the population of all alumni. Those with low incomes will be less inclined to respond, and those who do respond may be inclined to exaggerate. As we proceed through this chapter, we should focus on these key concepts:

- **Sample data must be collected in an appropriate way, such as through a process of *random* selection.**
- **If sample data are not collected in an appropriate way, the data may be so completely useless that no amount of statistical torturing can salvage them.**

Above all else, we ask that you begin your study of statistics with an open mind. Don't assume that the study of statistics is comparable to a root canal procedure. It has been the author's experience that students are often surprised by the interesting nature of statistics, and they are also surprised by the fact that they can actually master the basic principles without much difficulty, even if they have not excelled in other mathematics courses. We are convinced that by the time you complete this introductory course, you will be firm in your belief that statistics is an interesting and rich subject with applications that are extensive, real, and meaningful. We are also convinced that with regular class attendance and diligence, you will succeed in mastering the basic concepts of statistics presented in this course.

1-2 Types of Data

In Section 1-1 we defined the terms *population* and *sample*. The following two terms are used to distinguish between cases in which we have data for an entire population, and cases in which we have data for a sample only.

Definitions

A **parameter** is a numerical measurement describing some characteristic of a *population*.

A **statistic** is a numerical measurement describing some characteristic of a *sample*.

EXAMPLES

1. **Parameter:** When Lincoln was first elected to the presidency, he received 39.82% of the 1,865,908 votes cast. If we consider the collection of all of those votes to be the population being considered, then the 39.82% is a *parameter,* not a statistic.

2. **Statistic:** Based on a sample of 877 surveyed executives, it was found that 45% of them would not hire someone with a typographical error on their job application. That figure of 45% is a *statistic* because it is based on a sample, not the entire population of all executives.

The State of Statistics

The word *statistics* is derived from the Latin word *status* (meaning "state"). Early uses of statistics involved compilations of data and graphs describing various aspects of a state or country. In 1662, John Graunt published statistical information about births and deaths. Graunt's work was followed by studies of mortality and disease rates, population sizes, incomes, and unemployment rates. Households, governments, and businesses rely heavily on statistical data for guidance. For example, unemployment rates, inflation rates, consumer indexes, and birth and death rates are carefully compiled on a regular basis, and the resulting data are used by business leaders to make decisions affecting future hiring, production levels, and expansion into new markets.

Some data sets consist of numbers (such as heights of 66 in. and 72 in.), while others are nonnumerical (such as eye colors of green and brown). The terms *quantitative data* and *qualitative data* are often used to distinguish between these types.

Definitions

Quantitative data consist of numbers representing counts or measurements.

Qualitative (or **categorical** or **attribute**) **data** can be separated into different categories that are distinguished by some nonnumerical characteristic.

EXAMPLES

1. **Quantitative Data:** The weights of supermodels.
2. **Qualitative Data:** The genders (male/female) of professional athletes.

When working with quantitative data, it is important to use the appropriate units of measurement, such as dollars, hours, feet, meters, and so on. We should be especially careful to observe such references as "all amounts are in *thousands of dollars*" or "all times are in *hundredths of a second*" or "units are in *kilograms*." To ignore such units of measurement could lead to very wrong conclusions. NASA lost its $125 million Mars Climate Orbiter when it crashed because the controlling software had acceleration data in *English* units, but they were incorrectly assumed to be in *metric* units.

Quantitative data can be further described by distinguishing between *discrete* and *continuous* types.

Definitions

Discrete data result when the number of possible values is either a finite number or a "countable" number. (That is, the number of possible values is 0 or 1 or 2 and so on.)

Continuous (numerical) data result from infinitely many possible values that correspond to some continuous scale that covers a range of values without gaps, interruptions, or jumps.

EXAMPLES

1. **Discrete Data:** The numbers of eggs that hens lay are *discrete* data because they represent counts.

Note to Instructor

The concept of "countable" is discussed very briefly here, but it could be very difficult for many students to understand well. Consider commenting that "countable" basically means that the item can be counted somehow. See the text example wherein eggs from a hen are counted, but you cannot count the milk from a cow.

Note to Instructor

The four levels of measurement will require some explanation, but students have most difficulty distinguishing between the interval and ratio levels of measurement. Suggest a "ratio" test: If one number is *twice* the other, is the quantity being measured also twice the other quantity? If yes, the data are at the ratio level. For example, $2 is twice as much money as $1, but 2°F is not twice as hot as 1°F. The money amounts are at the ratio level but the temperatures are at the interval level.

2. **Continuous Data:** The amounts of milk from cows are *continuous* data because they are measurements that can assume any value over a continuous span. During a given time interval, a cow might yield an amount of milk that can be any value between 0 gallons and 5 gallons. It would be possible to get 2.343115 gallons because the cow is not restricted to the discrete amounts of 0, 1, 2, 3, 4, or 5 gallons.

When describing relatively smaller amounts, correct grammar dictates that we use "fewer" for discrete amounts, and "less" for continuous amounts. For example, it is correct to say that we drank *fewer* cans of cola and, in the process, we drank *less* cola. The numbers of cans of cola are discrete data, whereas the actual volume amounts of cola are continuous data.

Another common way of classifying data is to use four levels of measurement: nominal, ordinal, interval, and ratio. In applying statistics to real problems, the level of measurement of the data is an important factor in determining which procedure to use. (See Figure 14-1 on page 727.) There will be some references to these levels of measurement in this book, but the important point here is based on common sense: Don't do computations and don't use statistical methods with data that are not appropriate. For example, it would not make sense to compute an average of social security numbers, because those numbers are data that are used for identification, and they don't represent measurements or counts of anything. For the same reason, it would make no sense to compute an average of the numbers sewn on the shirts of basketball players.

Definition

The **nominal level of measurement** is characterized by data that consist of names, labels, or categories only. The data cannot be arranged in an ordering scheme (such as low to high).

EXAMPLES The following examples illustrate sample data at the nominal level of measurement.

1. **Yes/no/undecided:** Survey responses of yes, no, and undecided
2. **Colors:** The colors of cars driven by college students (red, black, blue, white, and so on)

Because nominal data lack any ordering or numerical significance, they should not be used for calculations. Numbers are sometimes assigned to the different categories (especially when data are coded for computers), but these numbers have no real computational significance and any average calculated with them is meaningless.

Measuring Disobedience

How are data collected about something that doesn't seem to be measurable, such as people's level of disobedience? Psychologist Stanley Milgram devised the following experiment: A researcher instructed a volunteer subject to operate a control board that gave increasingly painful "electrical shocks" to a third person. Actually, no real shocks were given, and the third person was an actor. The volunteer began with 15 volts and was instructed to increase the shocks by increments of 15 volts. The disobedience level was the point at which the subject refused to increase the voltage. Surprisingly, two-thirds of the subjects obeyed orders even though the actor screamed and faked a heart attack.

Gambling for Science

Data are sometimes collected in very clever ways from some very unlikely sources. One example involves researchers investigating changes in climate. They found that each spring since 1917, the small town of Nenana, Alaska, had a lottery with people betting on the exact time that their Tanana River would break up. (The last prize was about $300,000.) A tripod was placed on the frozen river and it was connected to a clock. The clock stopped when the tripod was moved by the ice breaking up. The researchers were able to obtain the breakup times for each year since 1917, and the data were helpful in studying trends in climate.

Definition

Data are at the **ordinal level of measurement** if they can be arranged in some order, but differences between data values either cannot be determined or are meaningless.

EXAMPLES The following are examples of sample data at the ordinal level of measurement.

1. **Course Grades:** A college professor assigns grades of A, B, C, D, or F. These grades can be arranged in order, but we can't determine differences between such grades. For example, we know that A is higher than B (so there is an ordering), but we cannot subtract B from A (so the difference cannot be found).

2. **Ranks:** Based on several criteria, a magazine ranks cities according to their "livability." Those ranks (first, second, third, and so on) determine an ordering. However, the differences between ranks are meaningless. For example, a difference of "second minus first" might suggest $2 - 1 = 1$, but this difference of 1 is meaningless because it is not an exact quantity that can be compared to other such differences. The difference between the first city and the second city is not the same as the difference between the second city and the third city. Using the magazine rankings, the *difference* between New York City and Boston cannot be quantitatively compared to the *difference* between St. Louis and Philadelphia.

Ordinal data provide information about relative comparisons, but not the magnitudes of the differences. Usually, ordinal data should not be used for calculations such as an average, but this guideline is sometimes violated (such as when we use letter grades to calculate a grade point average).

Definition

The **interval level of measurement** is like the ordinal level, with the additional property that the difference between any two data values is meaningful. However, data at this level do not have a *natural* zero starting point (where *none* of the quantity is present).

EXAMPLES The following examples illustrate the interval level of measurement.

1. **Temperatures:** Body temperatures of 98.2°F and 98.6°F are examples of data at this interval level of measurement. Those values are ordered, and we can determine their difference of 0.4°F. However, there is no natural starting point. The value of 0°F might seem like a starting point, but it is arbitrary and does not represent the total absence of heat. Because 0°F is not

a natural zero starting point, it is wrong to say that 50°F is *twice* as hot as 25°F.

2. **Years:** The years 1000, 2000, 1776, and 1492. (Time did not begin in the year 0, so the year 0 is arbitrary instead of being a natural zero starting point representing "no time.")

Definition

The **ratio level of measurement** is the interval level with the additional property that there is also a natural zero starting point (where zero indicates that *none* of the quantity is present). For values at this level, differences and ratios are both meaningful.

EXAMPLES The following are examples of data at the ratio level of measurement. Note the presence of the natural zero value, and note the use of meaningful ratios of "twice" and "three times."

1. **Weights:** Weights (in carats) of diamond engagement rings (0 does represent no weight, and 4 carats is twice as heavy as 2 carats.)

2. **Prices:** Prices of college textbooks ($0 does represent no cost, and a $90 book is three times as costly as a $30 book).

This level of measurement is called the ratio level because the zero starting point makes ratios meaningful. Among the four levels of measurement, most difficulty arises with the distinction between the interval and ratio levels.

Hint: To simplify that distinction, use a simple "ratio test:" Consider two quantities where one number is twice the other, and ask whether "twice" can be used to correctly describe the quantities. Because a 200-lb weight is *twice* as heavy as a 100-lb weight, but 50°F is *not twice* as hot as 25°F, weights are at the ratio level while Fahrenheit temperatures are at the interval level. For a concise comparison and review, study Table 1-1 on the next page for the differences among the four levels of measurement.

1-2 Basic Skills and Concepts

In Exercises 1–4, determine whether the given value is a statistic or a parameter.

1. The current Senate of the United States consists of 87 men and 13 women.

2. A sample of students is selected and the average (mean) number of textbooks purchased this semester is 4.2.

3. A sample of students is selected and the average (mean) amount of time waiting in line to buy textbooks this semester is 0.65 hour.

4. In a study of all 2223 passengers aboard the *Titanic,* it was found that 706 survived when it sank.

1. Parameter
2. Statistic
3. Statistic
4. Parameter

Recommended Assignment

Exercises 1–20. Note that Exercises 17–20 ask the student to "determine whether the sample is likely to be representative of the population," so that they will be encouraged to think about the quality of the data. In Exercise 14, Social Security numbers are really different names for people. In Exercise 16, Zip codes are only names for locations; they do not follow a consistent ordering scheme. Also, consider pointing out to students that answers to odd-numbered exercises are found in Appendix F, and there is a *Student Solutions Manual* available for purchase. It includes worked solutions to many of the odd-numbered exercises.

Table 1-1	Levels of Measurement of Data	
Level	Summary	Example
Nominal	Categories only. Data cannot be arranged in an ordering scheme.	Student states: 5 Californians, 20 Texans, 40 New Yorkers } Categories or names only.
Ordinal	Categories are ordered, but differences can't be found or are meaningless.	Student cars: 5 compact, 20 mid-size, 40 full size } An order is determined by "compact, mid-size, full-size."
Interval	Differences are meaningful, but there is no natural starting point and ratios are meaningless.	Campus temperatures: 5°F, 20°F, 40°F } 0°F doesn't mean "no heat." 40°F is not twice as hot as 20°F.
Ratio	There is a natural zero starting point and ratios are meaningful.	Student commuting distances: 5 mi, 20 mi, 40 mi } 40 mi is *twice* as far as 20 miles.

In Exercises 5–8, determine whether the given values are from a discrete or continuous data set.

5. Discrete
6. Continuous
7. Discrete
8. Discrete
9. Ratio
10. Ordinal
11. Interval
12. Nominal
13. Ordinal
14. Nominal
15. Ratio

5. George Washington's presidential salary was $25,000 per year, and the current annual presidential salary is $400,000.

6. A statistics student obtains sample data and finds that the mean weight of cars in the sample is 3126 lb.

7. In a survey of 1059 adults, it is found that 39% of them have guns in their homes (based on a Gallup poll).

8. When 19,218 gas masks from branches of the U.S. military were tested, it was found that 10,322 of them were defective (based on data from *Time* magazine).

In Exercises 9–16, determine which of the four levels of measurement (nominal, ordinal, interval, ratio) is most appropriate.

9. Heights of women basketball players in the WNBA

10. Ratings of fantastic, good, average, poor, or unacceptable for blind dates

11. Current temperatures of the classrooms at your college

12. Numbers on the jerseys of women basketball players in the WNBA

13. *Consumer Reports* magazine ratings of "best buy, recommended, not recommended"

14. Social security numbers

15. The number of "yes" responses received when 1250 drivers are asked if they have ever used a cell phone while driving

16. Zip codes

In Exercises 17–20, identify the (a) sample and (b) population. Also, determine whether the sample is likely to be representative of the population.

17. A reporter for *Newsweek* stands on a street corner and asks 10 adults if they feel that the current president is doing a good job.

18. Nielsen Media Research surveys 5000 randomly selected households and finds that among the TV sets in use, 19% are tuned to *60 Minutes* (based on data from *USA Today*).

19. In a Gallup poll of 1059 randomly selected adults, 39% answered "yes" when asked "Do you have a gun in your home?"

20. A graduate student at the University of Newport conducts a research project about how adult Americans communicate. She begins with a survey mailed to 500 of the adults that she knows. She asks them to mail back a response to this question: "Do you prefer to use e-mail or snail mail (the U.S. Postal Service)?" She gets back 65 responses, with 42 of them indicating a preference for snail mail.

1-2 Beyond the Basics

21. Interpreting Temperature Increase In the "Born Loser" cartoon strip by Art Sansom, Brutus expresses joy over an increase in temperature from 1° to 2°. When asked what is so good about 2°, he answers that "It's twice as warm as this morning." Explain why Brutus is wrong yet again.

22. Interpreting Political Polling A pollster surveys 200 people and asks them their preference of political party. He codes the responses as 0 (for Democrat), 1 (for Republican), 2 (for Independent), or 3 (for any other responses). He then calculates the average (mean) of the numbers and gets 0.95. How can that value be interpreted?

23. Scale for Rating Food A group of students develops a scale for rating the quality of the cafeteria food, with 0 representing "neutral: not good and not bad." Bad meals are given negative numbers and good meals are given positive numbers, with the magnitude of the number corresponding to the severity of badness or goodness. The first three meals are rated as 2, 4, and −5. What is the level of measurement for such ratings? Explain your choice.

1-3 Critical Thinking

Success in the introductory statistics course typically requires more *common sense* than mathematical expertise (despite Voltaire's warning that "common sense is not so common"). Because we now have access to calculators and com-

16. Nominal
17. Sample: the 10 selected adults; population: all adults; not representative
18. Sample: the 5000 selected households; population: all households; representative
19. Sample: the 1059 selected adults; population: all adults; representative
20. Sample: the 65 respondents; population: all adult Americans; not representative
21. With no natural starting point, the temperatures are at the interval level of measurement; ratios such as "twice" are meaningless.
22. Such calculations with nominal data are meaningless.
23. Either ordinal or interval are reasonable answers, but ordinal makes more sense because differences between values are not likely to be meaningful. For example, the difference between a food rated 1 and a food rated 2 is not likely to be the same as the difference between a food rated 9 and a food rated 10.

Note to Instructor
The content of this section is easy enough for students to read on their own, but really stress the importance of good sampling techniques. Emphasize two of the very common misuses: (1) voluntary response samples, such as TV and radio surveys that ask audience members to call in, and (2) graphs such as the one shown in Figure 1-1 (b). Also emphasize that the method of sampling is critically important, and if the sampling is not done correctly, even very large samples may be totally worthless.

Consider assigning a small project of collecting current articles or ads that illustrate the misuses discussed in this section.

Should You Believe a Statistical Study?

In *Statistical Reasoning for Everyday Life*, 2nd edition, authors Jeff Bennett, William Briggs, and Mario Triola list the following eight guidelines for critically evaluating a statistical study. (1) Identify the goal of the study, the population considered, and the type of study. (2) Consider the source, particularly with regard to a possibility of bias. (3) Analyze the sampling method. (4) Look for problems in defining or measuring the variables of interest. (5) Watch out for confounding variables that could invalidate conclusions. (6) Consider the setting and wording of any survey. (7) Check that graphs represent data fairly, and conclusions are justified. (8) Consider whether the conclusions achieve the goals of the study, whether they make sense, and whether they have practical significance.

puters, modern applications of statistics no longer require us to master complex algorithms of mathematical manipulations. Instead, we can focus on *interpretation* of data and results. This section is designed to illustrate how common sense is used when we think critically about data and statistics.

About a century ago, statesman Benjamin Disraeli famously said, "There are three kinds of lies: lies, damned lies, and statistics." It has also been said that "figures don't lie; liars figure." Historian Andrew Lang said that some people use statistics "as a drunken man uses lampposts—for support rather than illumination." Political cartoonist Don Wright encourages us to "bring back the mystery of life: lie to a pollster." Author Franklin P. Jones wrote that "statistics can be used to support anything—especially statisticians." In *Esar's Comic Dictionary* we find the definition of a statistician to be "a specialist who assembles figures and then leads them astray." These statements refer to instances in which methods of statistics were misused in ways that were ultimately deceptive. There are two main sources of such deception: (1) evil intent on the part of dishonest persons and (2) unintentional errors on the part of people who don't know any better. Regardless of the source, as responsible citizens and as more valuable professional employees, we should have a basic ability to distinguish between statistical conclusions that are likely to be valid and those that are seriously flawed.

To keep this section in proper perspective, know that this is not a book about the misuses of statistics. The remainder of this book will be full of very meaningful uses of valid statistical methods. We will learn general methods for using sample data to make important inferences about populations. We will learn about polls and sample sizes. We will learn about important measures of key characteristics of data. Along with the discussions of these general concepts, we will see many specific real applications, such as the effects of secondhand smoke, the prevalence of alcohol and tobacco in cartoon movies for children, and the quality of consumer products including M&M candies, cereals, Coke, and Pepsi. But even in those meaningful and real applications, we must be careful to correctly interpret the results of valid statistical methods.

We begin our development of critical thinking by considering bad samples. These samples are bad in the sense that the sampling method dooms the sample so that it is likely to be biased (not representative of the population from which it has been obtained). The following section discusses more detail about methods of sampling, and the importance of *randomness* is described. The first example describes a sampling procedure that seriously lacks the randomness that is so important. The following definition refers to one of the most common and most serious misuses of statistics.

Definition

A **voluntary response sample** (or **self-selected sample**) is one in which the respondents themselves decide whether to be included.

For examples, refer to the Chapter Problem. When America Online or anyone else runs a poll on the Internet, individuals decide themselves whether to participate, so they constitute a voluntary response sample. But people with strong opinions are more likely to participate, so the responses are not representative of the whole population. Here are common examples of voluntary response samples which, by their very nature, are seriously flawed in the sense that we should not make conclusions about a population based on such a biased sample:

- Polls conducted through the Internet, where subjects can decide whether to respond
- Mail-in polls, where subjects can decide whether to reply
- Telephone call-in polls, where newspaper, radio, or television announcements ask that you voluntarily pick up a phone and call a special number to register your opinion

With such voluntary response samples, valid conclusions can be made only about the specific group of people who chose to participate, but a common practice is to incorrectly state or imply conclusions about a larger population. From a statistical viewpoint, such a sample is fundamentally flawed and should not be used for making general statements about a larger population.

Small Samples   Conclusions should not be based on samples that are far too small. As one example, the Children's Defense Fund published *Children Out of School in America* in which it was reported that among secondary school students suspended in one region, 67% were suspended at least three times. But that figure is based on a sample of only *three* students! Media reports failed to mention that this sample size was so small. (We will see in Chapters 6 and 7 that we can *sometimes* make some valuable inferences from small samples, but we should be careful to verify that the necessary requirements are satisfied.)

Sometimes a sample might seem relatively large (as in a survey of "2000 randomly selected adult Americans"), but if conclusions are made about subgroups, such as the 21-year-old male Republicans from Pocatello, such conclusions might be based on samples that are too small. Although it is important to have a sample that is sufficiently large, it is just as important to have sample data that have been collected in an appropriate way, such as random selection. Even large samples can be bad samples.

Graphs   Graphs—such as bar graphs and pie charts—can be used to exaggerate or understate the true nature of data. (In Chapter 2 we discuss a variety of different graphs.) The two graphs in Figure 1-1 on the next page depict the *same data* from the Bureau of Labor Statistics, but part (b) is designed to exaggerate the difference between the weekly salaries of men and women. By not starting the horizontal axis at zero, the graph in part (b) tends to produce a misleading subjective impression, causing readers to incorrectly believe that the difference is much worse than it really is. Figure 1-1 carries this important lesson: To correctly interpret a graph, we should analyze the *numerical* information given in the graph, so that we won't be misled by its general shape. (The term *median* used in Figure 1-1 will be clearly described in Section 2-4.)

Pictographs   Drawings of objects, called *pictographs*, may also be misleading. Some objects commonly used to depict data include three-dimensional objects, such as moneybags, stacks of coins, army tanks (for military expenditures), barrels (for oil production), and houses (for home construction). When drawing such

The Literary Digest Poll

In the 1936 presidential race, *Literary Digest* magazine ran a poll and predicted an Alf Landon victory, but Franklin D. Roosevelt won by a landslide. Maurice Bryson notes, "Ten million sample ballots were mailed to prospective voters, but only 2.3 million were returned. As everyone ought to know, such samples are practically always biased." He also states, "Voluntary response to mailed questionnaires is perhaps the most common method of social science data collection encountered by statisticians, and perhaps also the worst." (See Bryson's "The *Literary Digest* Poll: Making of a Statistical Myth," *The American Statistician*, Vol. 30, No. 4.)

FIGURE 1-1 **Weekly Salaries of Men and Women Aged 16–24**

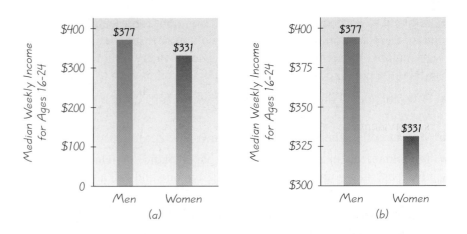

(a)

(b)

objects, artists can create false impressions that distort differences. If you double each side of a square, the area doesn't merely double; it increases by a factor of four. If you double each side of a cube, the volume doesn't merely double; it increases by a factor of eight, as shown in Figure 1-2. If taxes double over a decade, an artist may depict tax amounts with one moneybag for the first year and a second moneybag that is twice as deep, twice as tall, and twice as wide. Instead of appearing to double, taxes will appear to increase by a factor of eight, so the truth will be distorted by the drawing.

FIGURE 1-2 **Pictograph**
Double the length, width, and height of a cube and the volume increases by a factor of eight, as shown. If the smaller cube represents taxes in one year, and the larger cube represents a *doubling* of taxes at a later time, the later taxes appear to be eight times the original amount instead of twice the original amount.

Percentages   Misleading or unclear percentages are sometimes used. If you take 100% of some quantity, you are taking it *all*. (It shouldn't require a 110% effort to make sense of the preceding statement.) In referring to lost baggage, Continental Airlines ran ads claiming that this was "an area where we've already improved 100% in the last six months." In an editorial criticizing this statistic, the *New York Times* correctly interpreted the 100% improvement figure to mean that no baggage is now being lost—an accomplishment not enjoyed by Continental Airlines.

The following are a few of the key principles to be applied when dealing with percentages. These principles all use the basic notion that % or "percent" really means "divided by 100." The first principle will be used often in this book.

- **Percentage of:** To find some *percentage of* an amount, drop the % symbol and divide the percentage value by 100, then multiply. This example shows that 6% of 1200 is 72:

$$6\% \text{ of } 1200 \text{ responses} = \frac{6}{100} \times 1200 = 72$$

- **Fraction → Percentage:** *To convert from a fraction to a percentage,* divide the denominator into the numerator to get an equivalent decimal number, then multiply by 100 and affix the % symbol. This example shows that the fraction $^3/_4$ is equivalent to 75%:

$$\frac{3}{4} = 0.75 \rightarrow 0.75 \times 100\% = 75\%$$

- **Decimal → Percentage:** *To convert from a decimal to a percentage,* multiply by 100%. This example shows that 0.234 is equivalent to 23.4%:

$$0.234 \rightarrow 0.234 \times 100\% = 23.4\%$$

- **Percentage → Decimal:** *To convert from a percentage to a decimal number,* delete the % symbol and divide by 100. This example shows that 85% is equivalent to 0.85:

$$85\% = \frac{85}{100} = 0.85$$

Loaded Questions   There are many issues affecting survey questions. Survey questions can be "loaded" or intentionally worded to elicit a desired response. See the actual "yes" response rates for the different wordings of a question:

- 97% yes: "Should the President have the line item veto to eliminate waste?"
- 57% yes: "Should the President have the line item veto, or not?"

In *The Superpollsters*, David W. Moore describes an experiment in which different subjects were asked if they agree with the following statements:

- Too little money is being spent on welfare.
- Too little money is being spent on assistance to the poor.

Even though it is the poor who receive welfare, only 19% agreed when the word "welfare" was used, but 63% agreed with "assistance to the poor."

Order of Questions   Sometimes survey questions are unintentionally loaded by such factors as the order of the items being considered. See these questions from a poll conducted in Germany:

- Would you say that traffic contributes more or less to air pollution than industry?
- Would you say that industry contributes more or less to air pollution than traffic?

When traffic was presented first, 45% blamed traffic and 27% blamed industry; when industry was presented first, 24% blamed traffic and 57% blamed industry.

Refusals   When people are asked survey questions, some firmly refuse to answer. The refusal rate has been growing in recent years, partly because many persistent telemarketers try to sell goods or services by beginning with a sales pitch that sounds like it is part of an opinion poll. (This "selling under the guise" of a poll is often called *sugging.*) In *Lies, Damn Lies, and Statistics*, author Michael Wheeler correctly observes that "people who refuse to talk to pollsters are likely to be different from those who do not. Some may be fearful of strangers and others jealous of their privacy, but their refusal to talk demonstrates that their view of the world around them is markedly different from that of those people who will let poll-takers into their homes."

Statistics and Land Mines

The International Campaign to Ban Land Mines and the executive director of the Vietnam Veterans of America Foundation (VVAF) were recently awarded the Nobel Peace Prize. When VVAF asked for help in collecting data about land mines, a team of notable statisticians was assembled. Instead of working with intangible data, such as the value of a human life, they worked with tangible raw data, such as the area that a minefield makes unusable, and the cost of crops that cannot be grown. The data were included in *After the Guns Fall Silent: The Enduring Legacy of Landmines*, which became a key resource book in discussions of the land mine issue. The *AMSTAT News* quoted one of the book's editors: "This data-gathering and analysis effort is what made it possible to put the issue before policymakers. This work really made a difference."

Detecting Phony Data

A class is given the homework assignment of recording the results when a coin is tossed 500 times. One dishonest student decides to save time by just making up the results instead of actually flipping a coin. Because people generally cannot make up results that are really random, we can often identify such phony data. With 500 tosses of an actual coin, it is extremely likely that you will get a run of six heads or six tails, but people almost never include such a run when they make up results.

Another way to detect fabricated data is to establish that the results violate Benford's law: For many collections of data, the leading digits are not uniformly distributed. Instead, the leading digits of 1, 2, . . . , 9 occur with rates of 30%, 18%, 12%, 10%, 8%, 7%, 6%, 5%, and 5%, respectively. (See "The Difficulty of Faking Data" by Theodore Hill, *Chance*, Vol. 12, No. 3.)

Correlation and Causality   In Chapter 9 of this book we will discuss the statistical association between two variables, such as wealth and IQ. We will use the term *correlation* to indicate that the two variables are related. However, in Chapter 9 we make this important point: *Correlation does not imply causality*. This means that when we find a statistical association between two variables, we cannot conclude that one of the variables is the cause of (or directly affects) the other variable. If we find a correlation between wealth and IQ, we cannot conclude that a person's IQ directly affects his or her wealth, and we cannot conclude that a person's wealth directly affects his or her IQ score. It is quite common for the media to report about a newfound correlation with wording that directly indicates or implies that one of the variables is the cause of the other.

Self-Interest Study   Studies are sometimes sponsored by parties with interests to promote. For example, Kiwi Brands, a maker of shoe polish, commissioned a study that resulted in this statement printed in some newspapers: "According to a nationwide survey of 250 hiring professionals, scuffed shoes was the most common reason for a male job seeker's failure to make a good first impression." We should be very wary of such a survey in which the sponsor can enjoy monetary gains from the results. Of growing concern in recent years is the practice of pharmaceutical companies to pay doctors who conduct clinical experiments and report their results in prestigious journals, such as *Journal of the American Medical Association*.

Precise Numbers   "There are now 103,215,027 households in the United States." Because that figure is very precise, many people incorrectly assume that it is also *accurate*. In this case, that number is an estimate and it would be better to state that the number of households is about 103 million.

Partial Pictures   "Ninety percent of all our cars sold in this country in the last 10 years are still on the road." Millions of consumers heard that commercial message and didn't realize that 90% of the cars the advertiser sold in this country were sold within the last three years, so most of those cars on the road were quite new. The claim was technically correct, but it was very misleading in not presenting the complete results.

Deliberate Distortions   In the book *Tainted Truth*, Cynthia Crossen cites an example of the magazine *Corporate Travel* that published results showing that among car rental companies, Avis was the winner in a survey of people who rent cars. When Hertz requested detailed information about the survey, the actual survey responses disappeared and the magazine's survey coordinator resigned. Hertz sued Avis (for false advertising based on the survey) and the magazine; a settlement was reached.

In addition to the cases cited above, there are many other misuses of statistics. Some of those other cases can be found in books such as Darrell Huff's classic *How to Lie with Statistics,* Robert Reichard's *The Figure Finaglers,* and Cynthia

Crossen's *Tainted Truth.* Understanding these practices will be extremely helpful in evaluating the statistical data found in everyday situations.

1-3 Basic Skills and Concepts

In Exercises 1–4, use critical thinking to develop an alternative conclusion. For example, consider a media report that BMW drivers are healthier than adults who don't drive. The conclusion that BMW cars cause better health is probably wrong. Here is a better conclusion: BMW drivers tend to be wealthier than adults who don't drive, and greater wealth is associated with better health.

1. Weight and Trucks A study showed that truck drivers weigh more than adults who do not drive trucks. Conclusion: Trucks cause people to gain weight.

2. Homes and Longevity A study showed that homeowners tend to live longer than those who do not live in their own homes. Conclusion: Owning a home creates inner peace and harmony that causes people to be in better health and live longer.

3. Traffic Enforcement A study showed that in Orange County, more speeding tickets were issued to minorities than to whites. Conclusion: In Orange County, minorities speed more than whites.

4. Cold Remedy In a study of cold symptoms, every one of the study subjects with a cold was found to be improved two weeks after taking ginger pills. Conclusion: Ginger pills cure colds.

In Exercises 5–16, use critical thinking to address the key issue.

5. Chocolate Health Food The *New York Times* published an article that included these statements: "At long last, chocolate moves toward its rightful place in the food pyramid, somewhere in the high-tone neighborhood of red wine, fruits and vegetables, and green tea. Several studies, reported in the *Journal of Nutrition*, showed that after eating chocolate, test subjects had increased levels of antioxidants in their blood. Chocolate contains flavonoids, antioxidants that have been associated with decreased risk of heart disease and stroke. Mars Inc., the candy company, and the Chocolate Manufacturers Association financed much of the research." What is wrong with this study?

6. Census Data After the last national Census was conducted, the *Poughkeepsie Journal* ran this front-page headline: "281,421,906 in America." What is wrong with this headline?

7. Mail Survey When author Shere Hite wrote *Woman and Love: A Cultural Revolution in Progress*, she based conclusions on 4500 replies that she received after mailing 100,000 questionnaires to various women's groups. Are her conclusions likely to be valid in the sense that they can be applied to the general population of all women? Why or why not?

8. "900" Numbers In an ABC *Nightline* poll, 186,000 viewers each paid 50 cents to call a "900" telephone number with their opinion about keeping the United Nations in the United States. The results showed that 67% of those who called were in favor of moving the United Nations out of the United States. Interpret the results by identifying what we can conclude about the way the general population feels about keeping the United Nations in the United States.

Recommended Exercises

Exercises 1–16, but also assign Exercises 17–22 if you feel that your students need some review in working with percentages.

1. Truck drivers are often forced to dine in fast-food establishments, which provide diets that are higher in fat content. It is probably the fast-food diet that causes higher weights, not the trucks themselves. Avoid causality altogether by saying that driving trucks is associated with higher weights.

2. Homeowners tend to be wealthier and they can afford better health care, which leads to longer lives. Avoid causality altogether by saying that being a homeowner is associated with living longer.

3. One possible alternative: Racial profiling is used so that Orange County police tend to stop and ticket more minorities than whites.

4. Colds generally last less than two weeks, so anything would seem like a cure. There is no evidence to suggest that ginger pills cure colds.

5. Because the study was funded by a candy company and the Chocolate Manufacturers Association, there is a real possibility that researchers were somehow encouraged to obtain results favorable to the consumption of chocolate.

6. The headline suggests that the census count was determined with great precision, but the figure is likely to be in error by millions of people.

7. No, she used a voluntary response sample.

8. The voluntary response sample cannot be used to conclude anything about the general population. In fact, Ted Koppel reported that a "scientific" poll of 500 people showed that 72% of us want the United Nations to stay in the United States. In this poll of 500 people, respondents were randomly selected by the pollster so that the results are much more likely to reflect the true opinion of the general population.

9. People with unlisted numbers and without telephones are excluded.

10. Bus stops tend to be found more in cities, and crime rates are higher in cities than rural areas. Higher crime rates are associated with (not caused by) urban areas.

11. Motorcyclists who were killed

12. Because the responses are given to the Financial Consultant, the respondent might not be willing to give unfavorable responses that might jeopardize a working relationship. Better results are likely to be obtained with a commitment to confidentiality.

13. No. Each of the 29 cigarettes is given an equal weight, but some cigarettes are consumed in much greater numbers than others. Also, there are other cigarettes not included in Data Set 5.

14. It requires a calculation, which will result in some errors. Also, by asking for heights instead of measuring them, we tend to get desired values instead of actual values.

15. The results would not be good because you would be sampling only those people who died at a relatively young age. Many people born after 1945 are still alive.

9. **Telephone Surveys** The Hartford Insurance Company has hired you to poll a sample of adults about their car purchases. What is wrong with using people with telephone numbers listed in directories as the population from which the sample is drawn?

10. **Crime and Buses** The *Newport Chronicle* claims that bus stops cause crime, because a study showed that crime rates are higher in cities with bus stops than in rural areas that have no bus stops. What is wrong with that claim?

11. **Motorcycle Helmets** The Hawaii State Senate held hearings when it was considering a law requiring that motorcyclists wear helmets. Some motorcyclists testified that they had been in crashes in which helmets would not have been helpful. Which important group was not able to testify? (See "A Selection of Selection Anomalies" by Wainer, Palmer, and Bradlow in *Chance*, Volume 11, No. 2.)

12. **Merrill Lynch Client Survey** The author received a survey from the investment firm of Merrill Lynch. It was designed to gauge his satisfaction as a client, and it had specific questions for rating the author's personal Financial Consultant. The cover letter included this statement: "Your responses are extremely valuable to your Financial Consultant, Russell R. Smith, and to Merrill Lynch. . . . We will share your name and response with your Financial Consultant." What is wrong with this survey?

13. **Cigarette Nicotine** Refer to Data Set 5 in Appendix B and consider the nicotine content of the 29 different cigarette brands. The average (mean) of those amounts is 0.94 mg. Is this result likely to be a good estimate of the average (mean) of all cigarettes smoked in the United States? Why or why not?

14. **Bad Question** A survey includes this item: "Enter your height in inches." It is expected that actual heights of respondents can be obtained and analyzed, but there are two different major problems with this item. Identify them.

15. **Longevity** You need to conduct a study of longevity for people who were born after the end of World War II in 1945. If you were to visit graveyards and use the birth/death dates listed on tombstones, would you get good results? Why or why not?

16. **SIDS** In a letter to the editor in the *New York Times*, Moorestown, New Jersey, resident Jean Mercer criticized the statement that "putting infants in supine position has decreased deaths from SIDS." SIDS refers to sudden infant death syndrome, and the *supine* position is lying on the back with the face upward. She suggested that this statement is better: "Pediatricians advised the supine position during a time when the SIDS rate fell." What is wrong with saying that the supine position *decreased* deaths from SIDS?

In Exercises 17–22, answer the given questions that relate to percentages.

17. Percentages
    a. Convert the fraction 17/25 to an equivalent percentage.
    b. Convert 35.2% to an equivalent decimal.
    c. What is 57% of 1500?
    d. Convert 0.486 to an equivalent percentage.

18. Percentages
    a. What is 26% of 950?
    b. Convert 5% to an equivalent decimal.
    c. Convert 0.01 to an equivalent percentage.
    d. Convert the fraction 527/1200 to an equivalent percentage. Express the answer to the nearest tenth of a percent.

19. Percentages in a Gallup Poll
   a. In a Gallup poll, 52% of 1038 surveyed adults said that secondhand smoke is "very harmful." What is the actual number of adults who said that secondhand smoke is "very harmful"?
   b. Among the 1038 surveyed adults, 52 said that secondhand smoke is "not at all harmful." What is the percentage of people who chose "not at all harmful."

20. Percentages in a Study of Lipitor
   a. In a study of the cholesterol drug Lipitor, 270 patients were given a placebo, and 19 of those 270 patients reported headaches. What percentage of this placebo group reported headaches?
   b. Among the 270 patients in the placebo group, 3.0% reported back pains. What is the actual number of patients who reported back pains?

21. Percentages in Campus Crime In a study on college campus crimes committed by students high on alcohol or drugs, a mail survey of 1875 students was conducted. A *USA Today* article noted, "Eight percent of the students responding anonymously say they've committed a campus crime. And 62% of that group say they did so under the influence of alcohol or drugs." Assuming that the number of students responding anonymously is 1875, how many actually committed a campus crime while under the influence of alcohol or drugs?

22. Percentages in the Media
   a. A *New York Times* editorial criticized a chart caption that described a dental rinse as one that "reduces plaque on teeth by over 300%." What is wrong with that statement?
   b. In the *New York Times Magazine*, a report about the decline of Western investment in Kenya included this: "After years of daily flights, Lufthansa and Air France had halted passenger service. Foreign investment fell 500 percent during the 1990's." What is wrong with this statement?

1-3 Beyond the Basics

23. Phony Data A researcher at the Sloan-Kettering Cancer Research Center was once criticized for falsifying data. Among his data were figures obtained from 6 groups of mice, with 20 individual mice in each group. These values were given for the percentage of successes in each group: 53%, 58%, 63%, 46%, 48%, 67%. What is the major flaw?

24. What's Wrong with This Picture? Try to identify each of the four major flaws in the following. A daily newspaper ran a survey by asking readers to call in their response to this question: "Do you support the development of atomic weapons that could kill millions of innocent people?" It was reported that 20 readers responded and 87% said "no" while 13% said "yes."

25. Biased Wording Write a survey question that addresses a topic of your interest. First word the question objectively, then word it to encourage responses in one direction, then reword it a third time to sway responses in the opposite direction.

26. Graphs Currently, women earn about 74 cents for each dollar earned by men doing the same job. Draw a graph that depicts this information objectively, then draw a graph that exaggerates the difference. (*Hint:* See Figure 1-1.)

Design of Experiments

Note to Instructor
In addition to stressing the importance of a good design, again, stress the importance of good sampling techniques. If time is short and thorough coverage of this section is not possible, discuss the points made in the first paragraph and discuss the concept of a *simple random sample*.

Although this section contains much information, there are two major points, which are quite simple. We should understand that the method used to collect data is absolutely and critically important, and we should know that *randomness* is particularly important.

- **If sample data are not collected in an appropriate way, the data may be so completely useless that no amount of statistical torturing can salvage them.**
- *Randomness* **typically plays a crucial role in determining which data to collect.**

Statistical methods are driven by data. We typically obtain data from two distinct sources: *observational studies* and *experiments*.

> **Definitions**
>
> In an **observational study,** we observe and measure specific characteristics, but we don't attempt to *modify* the subjects being studied.
>
> In an **experiment,** we apply some *treatment* and then proceed to observe its effects on the subjects.

A Gallup poll is a good example of an observational study, whereas a clinical trial of the drug Lipitor is a good example of an experiment. The Gallup poll is observational in the sense that we merely observe people (often through interviews) without modifying them in any way. But the clinical trial of Lipitor involves treating some people with the drug, so the treated people are modified. There are different types of observational studies, as illustrated in Figure 1-3. These terms, commonly used in many different professional journals, are defined here.

> **Definitions**
>
> In a **cross-sectional study,** data are observed, measured, and collected at one point in time.
>
> In a **retrospective** (or **case-control**) **study,** data are collected from the past by going back in time (through examination of records, interviews, and so on).
>
> In a **prospective** (or **longitudinal** or **cohort**) **study,** data are collected in the future from groups (called *cohorts*) sharing common factors.

There is an important distinction between the sampling done in retrospective and prospective studies. In retrospective studies we go back in time to collect data about the resulting characteristic that is of concern, such as a group of drivers who died in car crashes and another group of drivers who did not die in car crashes. In

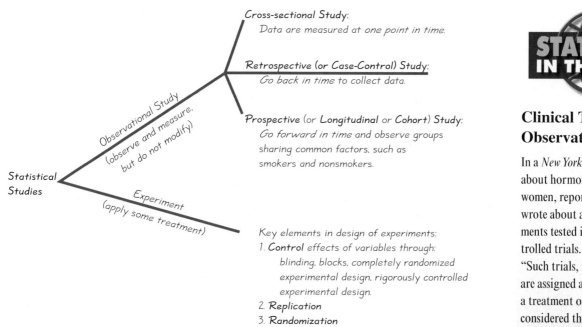

Cross-sectional Study:
Data are measured at *one point in time.*

Retrospective (or Case-Control) Study:
Go back in time to collect data.

Prospective (or Longitudinal or Cohort) Study:
*Go forward in time and observe groups
sharing common factors, such as
smokers and nonsmokers.*

*Observational Study
(observe and measure,
but do not modify)*

*Statistical
Studies*

*Experiment
(apply some treatment)*

Key elements in design of experiments:
1. ***Control*** *effects of variables through:
blinding, blocks, completely randomized
experimental design, rigorously controlled
experimental design.*
2. ***Replication***
3. ***Randomization***

FIGURE 1-3 **Elements of Statistical Studies**

prospective studies we go forward in time by following groups with a potentially causative factor and those without it, such as a group of drivers who use cell phones and a group of drivers who do not use cell phones.

These three definitions apply to observational studies, but we now shift our focus to experiments. Results of experiments are sometimes ruined because of *confounding*.

Definition

Confounding occurs in an experiment when the experimenter is not able to distinguish between the effects of different factors.

Try to plan the experiment so that confounding does not occur.

For example, suppose a Vermont professor experiments with a new attendance policy ("your course average drops one point for each class you cut"), but an exceptionally mild winter lacks the snow and cold temperatures that have hindered attendance in the past. If attendance does get better, we can't determine whether the improvement is attributable to the new attendance policy or to the mild winter. The effects of the attendance policy and the weather have been confounded.

Controlling Effects of Variables

Figure 1-3 shows that one of the key elements in the design of experiments is controlling effects of variables. We can gain that control by using such devices as blinding, blocks, a completely randomized experimental design, or a rigorously controlled experimental design, described as follows.

STATISTICS IN THE NEWS

Clinical Trials vs. Observational Studies

In a *New York Times* article about hormone therapy for women, reporter Denise Grady wrote about a report of treatments tested in randomized controlled trials. She stated that "Such trials, in which patients are assigned at random to either a treatment or a placebo, are considered the gold standard in medical research. By contrast, the observational studies, in which patients themselves decide whether to take a drug, are considered less reliable. . . . Researchers say the observational studies may have painted a falsely rosy picture of hormone replacement because women who opt for the treatments are healthier and have better habits to begin with than women who do not."

Hawthorne and Experimenter Effects

The well-known placebo effect occurs when an untreated subject incorrectly believes that he or she is receiving a real treatment and reports an improvement in symptoms. The Hawthorne effect occurs when treated subjects somehow respond differently, simply because they are part of an experiment. (This phenomenon was called the "Hawthorne effect" because it was first observed in a study of factory workers at Western Electric's Hawthorne plant.) An experimenter effect (sometimes called a Rosenthall effect) occurs when the researcher or experimenter unintentionally influences subjects through such factors as facial expression, tone of voice, or attitude.

Blinding   In 1954, a massive experiment was designed to test the effectiveness of the Salk vaccine in preventing the polio that killed or paralyzed thousands of children. In that experiment, a treatment group was given the actual Salk vaccine, while a second group was given a placebo that contained no drug at all. In experiments involving placebos, there is often a **placebo effect** that occurs when an untreated subject reports an improvement in symptoms. (The reported improvement in the placebo group may be real or imagined.) This placebo effect can be minimized or accounted for through the use of **blinding,** a technique in which the subject doesn't know whether he or she is receiving a treatment or a placebo. Blinding allows us to determine whether the treatment effect is significantly different from the placebo effect. The polio experiment was **double-blind,** meaning that blinding occurred at two levels: (1) the children being injected didn't know whether they were getting the Salk vaccine or a placebo, and (2) the doctors who gave the injections and evaluated the results did not know either.

Blocks   When designing an experiment to test the effectiveness of one or more treatments, it is important to put the subjects (often called *experimental units*) in different groups (or *blocks*) in such a way that those groups are very similar. A **block** is a group of subjects that are similar in the ways that might affect the outcome of the experiment.

> **When conducting an experiment of testing one or more different treatments, form blocks (or groups) of subjects with similar characteristics.**

Completely Randomized Experimental Design   When deciding how to assign the subjects to different blocks, you can use random selection or you can try to carefully control the assignment so that subjects within each block are similar. One approach is to use a **completely randomized experimental design,** whereby subjects are put into different blocks through a process of *random selection*. An example of a completely randomized experimental design is this feature of the polio experiment: Children were assigned to the treatment group or placebo group through a process of random selection (equivalent to flipping a coin).

Rigorously Controlled Design   Another approach for assigning subjects to blocks is to use a **rigorously controlled design,** in which subjects are very *carefully chosen* so that those in each block are similar in the ways that are important to the experiment. In an experiment testing the effectiveness of a drug designed to lower blood pressure, if the placebo group includes a 30-year-old overweight male smoker who drinks heavily and consumes an abundance of salt and fat, the treatment group should also include a person with similar characteristics (which, in this case, would be easy to find).

Replication and Sample Size

In addition to controlling effects of variables, another key element of experimental design is the size of the samples. Samples should be large enough so that the erratic behavior that is characteristic of very small samples will not disguise the true effects of different treatments. Repetition of an experiment is called **replication,** and replication is used effectively when we have enough subjects to recognize differences from different treatments. (In another context, *replication* refers to the repetition or duplication of an experiment so that results can be confirmed or verified.) With replication, large sample sizes increase the chance of recognizing different treatment effects. However, a large sample is not necessarily a good sample. Although it is important to have a sample that is sufficiently large, it is more important to have a sample in which data have been chosen in some appropriate way, such as random selection (described later).

> **Use a sample size that is large enough so that we can see the true nature of any effects, and obtain the sample using an appropriate method, such as one based on *randomness*.**

In the experiment designed to test the Salk vaccine, 200,000 children were given the actual Salk vaccine and 200,000 other children were given a placebo. Because the actual experiment used sufficiently large sample sizes, the effectiveness of the vaccine could be seen. Nevertheless, even though the treatment and placebo groups were very large, the experiment would have been a failure if subjects had not been assigned to the two groups in a way that made both groups similar in the ways that were important to the experiment.

Randomization and Other Sampling Strategies

In statistics as in life, one of the worst mistakes is to collect data in a way that is inappropriate. We cannot overstress this very important point:

> **If sample data are not collected in an appropriate way, the data may be so completely useless that no amount of statistical torturing can salvage them.**

In Section 1-3 we saw that a voluntary response sample is one in which the subjects themselves decide whether to respond. Such samples are very common, but their results are generally useless for making valid inferences about larger populations.

We now define some of the more common methods of sampling.

Definitions

In a **random sample** members from the population are selected in such a way that each *individual member* has an equal chance of being selected.

A **simple random sample** of size n subjects is selected in such a way that every possible *sample of the same size n* has the same chance of being chosen.

EXAMPLE **Random Sample and Simple Random Sample** Picture a classroom with 60 students arranged in six rows of 10 students each. Assume that the professor selects a sample of 10 students by rolling a die and selecting the row corresponding to the outcome. Is the result a random sample? Simple random sample?

SOLUTION   The sample is a random sample because each individual student has the same chance (one chance in six) of being selected. However, the sample is not a simple random sample because not all samples of size 10 have the same chance of being chosen. For example, this sampling design of using a die to select a row makes it impossible to select 10 students who are in different rows (but there is one chance in six of selecting the sample consisting of the 10 students in the first row).

Important: Throughout this book, we will use a variety of different statistical procedures, and we often have a requirement that we have collected a *simple random sample*, as defined above.

With random sampling we expect all components of the population to be (approximately) proportionately represented. Random samples are selected by many different methods, including the use of computers to generate random numbers. (Before computers, tables of random numbers were often used instead. For truly exciting reading, see this book consisting of one million digits that were randomly generated: *A Million Random Digits* published by Free Press. The *Cliff Notes* summary of the plot is not yet available.) Unlike careless or haphazard sampling, random sampling usually requires very careful planning and execution.

In addition to random sampling, there are other sampling techniques in use, and we describe the common ones here. See Figure 1-4 for an illustration depicting the different sampling approaches. Keep in mind that only random sampling and simple random sampling will be used throughout the remainder of the book.

Definitions

In **systematic sampling,** we select some starting point and then select every kth (such as every 50th) element in the population.

With **convenience sampling,** we simply use results that are very easy to get.

With **stratified sampling,** we subdivide the population into at least two different subgroups (or strata) that share the same characteristics (such as gender or age bracket), then we draw a sample from each subgroup (or stratum).

In **cluster sampling,** we first divide the population area into sections (or clusters), then randomly select some of those clusters, and then choose *all* the members from those selected clusters.

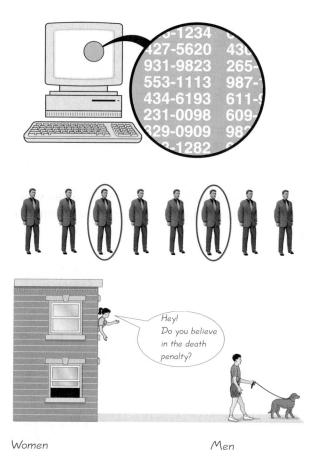

Random Sampling:
Each member of the population has an equal chance of being selected. Computers are often used to generate random telephone numbers.

Simple Random Sampling:
A sample of size n subjects is selected in such a way that every possible sample of the same size n has the same chance of being chosen.

Systematic Sampling:
Select some starting point, then select every kth (such as every 50th) element in the population.

Convenience Sampling:
Use results that are easy to get.

Stratified Sampling:
Subdivide the population into at least two different subgroups (or strata) that share the same characteristics (such as gender or age bracket), then draw a sample from each subgroup.

Cluster Sampling:
Divide the population area into sections (or clusters), then randomly select some of those clusters, and then choose all members from those selected clusters.

FIGURE 1-4 **Common Sampling Methods**

It is easy to confuse stratified sampling and cluster sampling, because they both involve the formation of subgroups. But cluster sampling uses *all* members from a *sample* of clusters, whereas stratified sampling uses a *sample* of members from *all* strata. An example of cluster sampling can be found in a pre-election poll, whereby we randomly select 30 election precincts from a large number of precincts, then survey all the people from each of those precincts. This is much faster and much less expensive than selecting one person from each of the many precincts in the population area. The results of stratified or cluster sampling can be adjusted or weighted to correct for any disproportionate representations of groups.

For a fixed sample size, if you randomly select subjects from different strata, you are likely to get more consistent (and less variable) results than by simply selecting a random sample from the general population. For that reason, stratified sampling is often used to reduce the variation in the results. Many of the methods discussed later in this book have a requirement that sample data be a *simple random sample*, and neither stratified sampling nor cluster sampling satisfies that requirement.

Figure 1-4 illustrates common methods of sampling. Professionals often collect data by using some combination of those methods. Here is one typical example of what is called a *multistage sample design*: First randomly select a sample of counties from all 50 states, then randomly select cities and towns in those counties, then randomly select residential blocks in each city or town, then randomly select households in each block, then randomly select someone from each household. We will not use such a sample design in this book. We should again stress that the methods of this book typically require that we have a *simple random sample*.

Sampling Errors

No matter how well you plan and execute the sample collection process, there is likely to be some error in the results. For example, randomly select 1000 adults, ask them if they graduated from high school, and record the sample percentage of "yes" responses. If you randomly select another sample of 1000 adults, it is likely that you will obtain a *different* sample percentage.

Definitions

A **sampling error** is the difference between a sample result and the true population result; such an error results from chance sample fluctuations.

A **nonsampling error** occurs when the sample data are incorrectly collected, recorded, or analyzed (such as by selecting a biased sample, using a defective measuring instrument, or copying the data incorrectly).

If we carefully collect a sample so that it is representative of the population, we can use methods in this book to analyze the sampling error, but we must exercise extreme care so that nonsampling error is minimized.

After reading through this section, it is easy to be somewhat overwhelmed by the variety of different definitions. But remember this main point: The method used to collect data is absolutely and critically important, and we should know that *randomness* is particularly important. If sample data are not collected in an appropriate way, the data may be so completely useless that no amount of statistical torturing can salvage them.

1-4 Basic Skills and Concepts

In Exercises 1–4, determine whether the given description corresponds to an observational study *or an* experiment.

Recommended Assignment
Exercises 1–26

1. Drug Testing Patients are given Lipitor to determine whether this drug has the effect of lowering high levels of cholesterol.

   1. Experiment

2. Treating Syphilis Much controversy arose over a study of patients with syphilis who were *not* given a treatment that could have cured them. Their health was followed for years after they were found to have syphilis.

   2. Observational study

3. Consumer Fraud The Dutchess County Bureau of Weights and Measures randomly selects gas stations and obtains 1 gallon of gas from each pump. The amount pumped is measured for accuracy.

   3. Observational study

4. Magnetic Bracelets Cruise ship passengers are given magnetic bracelets, which they agree to wear in an attempt to eliminate or diminish the effects of motion sickness.

   4. Experiment

In Exercises 5–8, identify the type of observational study (cross-sectional, retrospective, or prospective).

5. Medical Research A researcher from the New York University School of Medicine obtains data about head injuries by examining the hospital records from the past five years.

   5. Retrospective

6. Psychology of Trauma A researcher from Mt. Sinai Hospital in New York City plans to obtain data by following (to the year 2010) siblings of victims who perished in the World Trade Center terrorist attack of September 11, 2001.

   6. Prospective

7. Unemployment Statistics The U.S. Labor Department obtains current unemployment data by polling 50,000 people this month.

   7. Cross-sectional

8. Lottery Winners An economist collects data by interviewing people who won the lottery between the years of 1995 and 2000.

   8. Retrospective

In Exercises 9–20, identify which of these types of sampling is used: random, systematic, convenience, stratified, *or* cluster.

9. Convenience

9. Television News An NBC television news reporter gets a reaction to a breaking story by polling people as they pass the front of his studio.

10. Systematic

10. Jury Selection The Dutchess County Commissioner of Jurors obtains a list of 42,763 car owners and constructs a pool of jurors by selecting every 100th name on that list.

11. Random

11. Telephone Polls In a Gallup poll of 1059 adults, the interview subjects were selected by using a computer to randomly generate telephone numbers that were then called.

12. Stratified

12. Car Ownership A General Motors researcher has partitioned all registered cars into categories of subcompact, compact, mid-size, intermediate, and full-size. She is surveying 200 car owners from each category.

13. Cluster

13. Student Drinking Motivated by a student who died from binge drinking, the College of Newport conducts a study of student drinking by randomly selecting 10 different classes and interviewing all of the students in each of those classes.

14. Random

14. Marketing A marketing executive for General Motors finds that its public relations department has just printed envelopes with the names and addresses of all Corvette owners. She wants to do a pilot test of a new marketing strategy, so she thoroughly mixes all of the envelopes in a bin, then obtains a sample group by pulling 50 of those envelopes.

15. Systematic

15. Sobriety Checkpoint The author was an observer at a police sobriety checkpoint at which every 5th driver was stopped and interviewed. (He witnessed the arrest of a former student.)

16. Cluster

16. Exit Poll CNN is planning an exit poll in which 100 polling stations will be randomly selected and all voters will be interviewed as they leave the premises.

17. Stratified

17. Education and Salary An economist is studying the effect of education on salary and conducts a survey of 150 randomly selected workers from each of these categories: less than a high-school degree; high-school degree; more than a high-school degree.

18. Convenience

18. Anthropometrics A statistics student obtains height/weight data by interviewing family members.

19. Cluster

19. Medical Research A Johns Hopkins University researcher surveys all cardiac patients in each of 30 randomly selected hospitals.

20. Stratified

20. MTV Survey A marketing expert for MTV is planning a survey in which 500 people will be randomly selected from each age group of 10–19, 20–29, and so on.

Exercises 21–26 relate to random samples and simple random samples.

21. Yes; yes

21. Sampling Aspirin Tablets A pharmacist thoroughly mixes a container of 1000 Bufferin tablets, then scoops a sample of 50 tablets that are to be tested for the exact aspirin content. Does this sampling plan result in a random sample? Simple random sample? Explain.

22. Sampling Students A classroom consists of 30 students seated in five different rows, with six students in each row. The instructor rolls a die and the outcome is used to select a sample of the students in a particular row. Does this sampling plan result in a random sample? Simple random sample? Explain.

23. Convenience Sample A news reporter stands on a street corner and obtains a sample of city residents by selecting five passing adults about their smoking habits. Does this sampling plan result in a random sample? Simple random sample? Explain.

24. Systematic Sample A quality control engineer selects every 100th computer power supply unit that passes on a conveyor belt. Does this sampling plan result in a random sample? Simple random sample? Explain.

25. Stratified Sample General Foods plans to conduct a marketing survey of 100 men and 100 women in Orange County, which consists of an equal number of men and women. Does this sampling plan result in a random sample? Simple random sample? Explain.

26. Cluster Sample A market researcher randomly selects 10 blocks in the Village of Newport, then asks all adult residents of the selected blocks whether they own a DVD player. Does this sampling plan result in a random sample? Simple random sample? Explain.

1-4 Beyond the Basics

27. Sampling Design The Addison-Wesley Publishing Company has commissioned you to survey 100 students who use this book. Describe procedures for obtaining a sample of each type: random, systematic, convenience, stratified, cluster.

28. Confounding Give an example (different from the one in the text) illustrating how confounding occurs.

29. Random Selection Among the 50 states, one state is randomly selected. Then, a statewide voter registration list is obtained and one name is randomly selected. Does this procedure result in a randomly selected voter?

30. Sample Design In "Cardiovascular Effects of Intravenous Triiodothyronine in Patients Undergoing Coronary Artery Bypass Graft Surgery" (*Journal of the American Medical Association,* Vol. 275, No. 9), the authors explain that patients were assigned to one of three groups: (1) a group treated with triidothyronine, (2) a group treated with normal saline bolus and dopamine, and (3) a placebo group given normal saline. The authors summarize the sample design as a "prospective, randomized, double-blind, placebo-controlled trial." Describe the meaning of each of those terms in the context of this study.

31. Drivers with Cell Phones What are two major problems likely to be encountered in a prospective study in which some drivers have no cell phones while others are asked to use their cell phones while driving?

22. Yes; no
23. No; no
24. No; no
25. Yes; no
26. Yes; no
27. Answers vary.
28. Answers vary.
29. No, not every voter has the same chance of being selected. Voters in less populated states have a better chance of being selected.
30. Prospective: The experiment was begun and results were followed forward in time. Randomized: Subjects were assigned to the different groups through a process of random selection, whereby they had the same chance of belonging to each group. Double-blind: The subjects did not know which of the three groups they were in, and the people who evaluated results did not know either. Placebo-controlled: There was a group of subjects who were given a placebo; by comparing the placebo group to the two treatment groups, the effects of the treatments might be better understood.
31. Asking drivers to use cell phones might possibly place them in danger. The population of drivers not having cell phones might be fundamentally different from the population of drivers that have cell phones. The magnitude of cell phone use might vary considerably, so the effects of using a cell phone might not be clear. The cell phone users know that they are in the treatment group and might behave differently, and they might tend to blame driving problems or crashes on the presence of the cell phone.

Review

This chapter presented some important basics. There were fundamental definitions, such as *sample* and *population*, along with some very basic principles. Section 1-2 discussed different types of data. Section 1-3 dealt with the use of critical thinking in analyzing and evaluating statistical results. Section 1-4 introduced important elements in the design of experiments. On completing this chapter, you should be able to do the following:

- Distinguish between a population and a sample and distinguish between a parameter and a statistic.
- Identify the level of measurement (nominal, ordinal, interval, ratio) of a set of data.
- Understand the importance of good experimental design, including the control of variable effects, replication, and randomization.
- Recognize the importance of good sampling methods in general, and recognize the importance of a *simple random sample* in particular. Understand that if sample data are not collected in an appropriate way, the data may be so completely useless that no amount of statistical torturing can salvage them.

Note to Instructor
The answers to *all* Review Exercises are available to students in Appendix F. (In chapter sections, only answers to *odd*-numbered exercises are provided.)

1. No, because it is a voluntary response sample, it might not be representative of the population.
2. Answer varies.
3. a. Ratio
   b. Ordinal
   c. Nominal
   d. Interval
4. a. Discrete
   b. Ratio
   c. Stratified
   d. Statistic
   e. The largest values because they represent stockholders that could potentially gain control of the company.
   f. The voluntary response sample is likely to be biased.

Review Exercises

1. **Sampling** Shortly after the World Trade Center towers were destroyed by terrorists, America Online ran a poll of its Internet subscribers and asked this question: "Should the World Trade Center towers be rebuilt?" Among the 1,304,240 responses, 768,731 answered "yes," 286,756 answered "no," and 248,753 said that it was "too soon to decide." Given that this sample is extremely large, can the responses be considered to be representative of the population of the United States? Explain.

2. **Sampling Design** You have been hired by Visa to conduct a survey of credit card usage among the full-time students who attend your college. Describe a procedure for obtaining a sample of each type: random, systematic, convenience, stratified, cluster.

3. Identify the level of measurement (nominal, ordinal, interval, ratio) used in each of the following.
   a. The weights of people in a sample of elevator passengers
   b. A movie critic's ratings of "must see; recommended; not recommended; don't even think about going."
   c. A movie critic's classification of "drama; comedy; adventure."
   d. Bob, who is different in many ways, measures time in days, with 0 corresponding to his birth date. The day before his birth is -1, the day after his birth is $+1$, and so on. Bob has converted the dates of major historical events to his numbering system. What is the level of measurement of these numbers?

4. Coke The Coca Cola Company has 366,000 stockholders and a poll is conducted by randomly selecting 30 stockholders from each of the 50 states. The number of shares held by each sampled stockholder is recorded.

   a. Are the values obtained discrete or continuous?
   b. Identify the level of measurement (nominal, ordinal, interval, ratio) for the sample data.
   c. Which type of sampling (random, systematic, convenience, stratified, cluster) is being used?
   d. If the average (mean) number of shares is calculated, is the result a statistic or a parameter?
   e. If you are the chief executive officer of the Coca Cola Company, what characteristic of the data set would you consider to be extremely important?
   f. What is wrong with gauging stockholder views by mailing a questionnaire that stockholders could complete and mail back?

5. More Coke Identify the type of sampling (random, systematic, convenience, stratified, cluster) used when a sample of the 366,000 Coca Cola shareholders is obtained as described. Then determine whether the sampling scheme is likely to result in a sample that is representative of the population of all 366,000 shareholders.

   a. A complete list of all stockholders is compiled and every 500th name is selected.
   b. At the annual stockholders meeting, a survey is conducted of all who attend.
   c. Fifty different stockbrokers are randomly selected, and a survey is made of all their clients who own shares of Coca Cola.
   d. A computer file of all stockholders is compiled so that they are all numbered consecutively, then random numbers generated by computer are used to select the sample of stockholders.
   e. All of the stockholder zip codes are collected, and 5 stockholders are randomly selected from each zip code.

6. Design of Experiment You plan to conduct an experiment to test the effectiveness of Sleepeze, a new drug that is supposed to reduce the effect of insomnia. You will use a sample of subjects that are treated with the drug and another sample of subjects that are given a placebo.

   a. What is "blinding" and how might it be used in this experiment?
   b. Why is it important to use blinding in this experiment?
   c. What is a completely randomized block design?
   d. What is a rigorously controlled block design?
   e. What is replication, and why is it important?

5. a. Systematic; representative
   b. Convenience; not representative
   c. Cluster; not representative
   d. Random; representative
   e. Stratified; not representative

6. a. Design the experiment so that the subjects don't know whether they are using Sleepeze or a placebo, and also design it so that those who observe and evaluate the subjects do not know which subjects are using Sleepeze and which are using a placebo.
   b. Blinding will help to distinguish between the effectiveness of Sleepeze and the placebo effect, whereby subjects and evaluators tend to believe that improvements are occurring just because some treatment is given.
   c. Subjects are put into different groups through a process of *random selection*.
   d. Subjects are very *carefully chosen* for the different groups so that those groups are made to be similar in the ways that are important.
   e. Replication is used when the experiment is repeated. It is important to have a sample of subjects that is large enough so that we can see the true nature of any effects. It is important so that we are not misled by erratic behavior of samples that are too small.

Cumulative Review Exercises

The Cumulative Review Exercises in this book are designed to include topics from preceding chapters. For Chapters 2–13, the cumulative review exercises include topics from preceding chapters. For this chapter, we present calculator warm-up exercises with expressions similar to those found throughout this book. Use your calculator to find the indicated values.

1. Refer to Data Set 1 in Appendix B and consider only the weights of the first 10 males. What value is obtained when those 10 weights are added, and the total is then divided by 10? (This result, called the *mean*, is discussed in Chapter 2.)

2. $\dfrac{98.20 - 98.60}{0.62}$

3. $\dfrac{\frac{98.20 - 98.60}{0.62}}{\sqrt{106}}$

4. $\left[\dfrac{1.96 \cdot 15}{2}\right]^2$

5. $\sqrt{\dfrac{(5-7)^2 + (12-7)^2 + (4-7)^2}{3-1}}$

6. $\dfrac{(183 - 137.09)^2}{137.09} + \dfrac{(30 - 41.68)^2}{41.68}$

7. $\sqrt{\dfrac{10(513.27) - 71.5^2}{10(10-1)}}$

8. $\dfrac{8(151,879) - (516.5)(2176)}{\sqrt{8(34,525.75) - 516.5^2}\ \sqrt{8(728,520) - 2176^2}}$

In Exercises 9–12, the given expressions are designed to yield results expressed in a form of scientific notation. For example, the calculator-displayed result of 1.23E5 (or 1.23⁵ on some calculators) can be expressed as 123,000, and the result of 4.65E − 4 (or 4.65⁻⁴ on some calculators) can be expressed as 0.000456. Perform the indicated operation and express the result as an ordinary number that is not in scientific notation.

9. 0.95^{500} 10. 8^{14} 11. 9^{12} 12. 0.25^{17}

Cooperative Group Activities

1. In-class activity From the cafeteria, obtain 18 straws. Cut 6 of them in half, cut 6 of them into quarters, and leave the other 6 as they are. There should now be 42 straws of different lengths. Put them in a bag, mix them up, then select 1 straw, find its length, then replace it. Repeat this until 20 straws have been selected. *Important:* Select the straws without looking into the bag and select the first straw that is touched. Find the average (mean) of the sample of 20 straws. Now remove all of the straws and find the mean of the population. Did the sample provide an average that was close to the true population average? Why or why not?

2. In-class activity In mid-December of a recent year, the Internet service provider America Online (AOL) ran a survey of its users. This question was asked about Christmas trees: "Which do you prefer?" The response could be "a real tree" or "a fake tree." Among the 7073 responses received by the Internet users, 4650 indicated a real tree, and 2423 indicated a fake tree. We have already noted that because the sample is a voluntary response sample, no conclusions can be made about a population larger than the 7073 people who responded. Identify other problems with this survey question.

3. In-class activity Identify the problems with the following.

   - A recent televised report on *CNN Headline News* included a comment that crime in the United States fell in the 1980s because of the growth of abortions in the 1970s, which resulted in fewer unwanted children.

   - *Consumer Reports* magazine mailed an Annual Questionnaire about cars and other consumer products. Also included were a request for a voluntary contribution of money and a ballot for the Board of Directors. Responses were to be mailed back in envelopes that required postage stamps.

Technology Project

The objective of this project is to introduce the technology resources that you will be using in your statistics course. Refer to Data Set 14 in Appendix B and use the Flesch Reading Ease scores for J. K. Rowling's *Harry Potter and the Sorcerer's Stone*. Using your statistics software package or a TI-83 Plus calculator, enter those 12 values, then obtain a printout of them.

STATDISK	Click on **Data** at the top of the screen, then select **Sample Editor** and proceed to enter the data. To obtain a printout, click on the **Print data** button.
Minitab	Enter the data in the column C1, then click on **File,** and select **Print Worksheet.**
Excel	Enter the data in column A, then click on **File,** and select **Print.**
TI-83 Plus	Printing a TI-83 Plus screen display is possible only if you are using a Graphlink connection to a computer. The TI-84 Plus calculator uses a USB cable connection.

Note to Instructor

About the use of technology: You can require a particular technology, such as the TI-83 Plus calculator, or you can allow students to use a variety of different inexpensive scientific calculators. This book is flexible in that it does not require a particular technology. After making your decision, it is important to inform students as soon as possible, preferably during the first class. Be sure to notify the class about any calculators that are prohibited on tests. (Many instructors prohibit any calculator with a "QWERTY" keyboard, such as the TI-92.) Comment that calculators capable of handling two-variable statistics will simplify some pretty messy calculations that will be required later in the course.

STATDISK is free to colleges that adopt this textbook. Students using this book may make their own personal copy of STATDISK. This edition of the book is accompanied with a new version of STATDISK (version 9.0). Comment to the class that even if no statistical software package is used, there is value in interpreting the computer displays that appear throughout this book.

from DATA to DECISION

Critical Thinking

The Swiss physician H. C. Lombard once compiled longevity data for different professions. He used death certificates that included name, age at death, and profession. He then proceeded to compute the average (mean) length of life for the different professions, and he found that students were lowest with a mean of only 20.7 years! (See "A Selection of Selection Anomalies" by Wainer, Palmer, and Bradlow in *Chance*, Volume 11, No. 2.) Similar results would be obtained if the same data were collected today in the United States.

Analyzing the Results

Is being a student really more dangerous than being a police officer, a taxicab driver, or a postal employee? Explain.

INTERNET PROJECT   Web Site for *Elementary Statistics*

In this section of each chapter, you will be instructed to visit the home page on the Web for this textbook. From there you can reach the pages for all the Internet Projects accompanying *Elementary Statistics, Ninth Edition.* Go to this Web site now and familiarize yourself with all of the available features for the book.

Each Internet Project includes activities, such as exploring data sets, performing simulations, and researching true-to-life examples found at various Web sites. These activities will help you explore and understand the rich nature of statistics and its importance in our world. Visit the book site now and enjoy the explorations!

http://www.aw.com/triola

Statistics @ Work

"We use statistics to determine the degree of isolation between putative stocks."

Sarah Mesnick

Behavioral and Molecular Ecologist

Sarah Mesnick is a National Research Council postdoctoral fellow. In her work as a marine mammal biologist, she conducts research at sea as well as in the Laboratory of Molecular Ecology. Her research focuses on the social organization and population structure of sperm whales. She received her doctorate in evolutionary biology at the University of Arizona.

What do you do?

My research focuses on the relationship between sociality and population structure in sperm whales. We use this information to build better management models for the conservation of this, and other, endangered marine mammal species.

What concepts of statistics do you use?

Currently, I use chi-square and *F*-statistics to examine population structure and regression measures to estimate the degree of relatedness among individuals within whale pods. We use the chi-square and *F*-statistics to determine how many discrete populations of whales are in the Pacific. Discrete populations are managed as independent stocks. The regression analysis of relatedness is used to determine kinship within groups.

Could you cite a specific example illustrating the use of statistics?

I'm currently working with tissue samples obtained from three mass strandings of sperm whales. We use genetic markers to determine the degree of relatedness among individuals within the strandings. This is a striking behavior—entire pods swam up onto the beach following a young female calf, stranded, and subsequently all died. We thought that to do something as dramatic as this, the individuals involved must be very closely related. We're finding, however, that they are not.

The statistics enable us to determine the probability that two individuals are related given the number alleles that they share. Also, sperm whale—and many other marine mammal, bird, and turtle species—are injured or killed incidentally in fishing operations. We need to know the size of the population from which these animals are taken. If the population is small, and the incidental kill large, the marine mammal population may be threatened. We use statistics to determine the degree of isolation between putative stocks. If stocks are found to be isolated, we would use this information to prepare management plans specifically designed to conserve the marine mammals of the region. Human activities may need to protect the health of the marine environment and its inhabitants.

How do you approach your research?

We try not to have preconceived notions about how the animals are dispersed in their environment. In marine mammals in particular, because they are so difficult to study, there are generally accepted notions about what the animals are doing, yet these have not been critically investigated. In the case of relatedness among individuals within sperm whale groups, they were once thought to be matrilineal and accompanied by a "harem master." With the advent of genetic techniques, and dedicated field work, more open minds and more critical analyses—the statistics come in here—we're able to reassess these notions.

Describing, Exploring, and Comparing Data

2-1 Overview

2-2 Frequency Distributions

2-3 Visualizing Data

2-4 Measures of Center

2-5 Measures of Variation

2-6 Measures of Relative Standing

2-7 Exploratory Data Analysis (EDA)

Are nonsmoking people really affected by others who are smoking cigarettes, or is the effect of *secondhand smoke* really a myth?

Data Set 6 in Appendix B includes some of the latest available data from the National Institutes of Health. The data, reproduced below in Table 2-1, were obtained as part of the National Health and Nutrition Examination Survey. The data values consist of the measured levels of serum cotinine (in ng/ml) in people selected as study subjects. (The data were rounded to the nearest whole number, so a value of zero does not necessarily indicate the total absence of cotinine. In fact, all of the original values were greater than zero.) Cotinine is a metabolite of nicotine, meaning that when nicotine is absorbed by the body, cotinine is produced. Because it is known that nicotine is absorbed through cigarette smoking, we have a way of measuring the effective presence of cigarette smoke indirectly through cotinine.

Here are important issues: Should nonsmokers be concerned about their health because they are in the presence of smokers? In recent years, many regulations have been enacted to restrict smoking in public places. Are those regulations justified based on reasons of health, or are they simply causing unnecessary hardship for smokers?

Critical Thinking: A visual comparison of the numbers in the three groups in Table 2-1 might provide some insight, but in this chapter we present methods for gaining much greater understanding. We will be able to make intelligent and productive comparisons. We will learn techniques for describing, exploring, and comparing data sets such as the three groups of data in Table 2-1.

Table 2-1	Measured Cotinine Levels in Three Groups

Smoker: The subjects report tobacco use.

ETS: (Environmental Tobacco Smoke) Subjects are nonsmokers who are exposed to environmental tobacco smoke ("secondhand smoke") at home or work.

NOETS: (No Environmental Tobacco Smoke) Subjects are nonsmokers who are not exposed to environmental tobacco smoke at home or work. That is, the subjects do not smoke and are not exposed to secondhand smoke.

Smoker:	1	0	131	173	265	210	44	277	32	3
	35	112	477	289	227	103	222	149	313	491
	130	234	164	198	17	253	87	121	266	290
	123	167	250	245	48	86	284	1	208	173
ETS:	384	0	69	19	1	0	178	2	13	1
	4	0	543	17	1	0	51	0	197	3
	0	3	1	45	13	3	1	1	1	0
	0	551	2	1	1	1	0	74	1	241
NOETS:	0	0	0	0	0	0	0	0	0	0
	0	9	0	0	0	0	0	0	244	0
	1	0	0	0	90	1	0	309	0	0
	0	0	0	0	0	0	0	0	0	0

Overview

This chapter is extremely important because it presents the basic tools for measuring and describing different characteristics of a set of data. When describing, exploring, and comparing data sets, the following characteristics are usually extremely important.

Important Characteristics of Data

1. **Center:** A representative or average value that indicates where the middle of the data set is located.
2. **Variation:** A measure of the amount that the data values vary among themselves.
3. **Distribution:** The nature or shape of the distribution of the data (such as bell-shaped, uniform, or skewed).
4. **Outliers:** Sample values that lie very far away from the vast majority of the other sample values.
5. **Time:** Changing characteristics of the data over time.

Study Hint: Blind memorization is often ineffective for learning or remembering important information. However, the above five characteristics are so important, that they might be better remembered by using a mnemonic for their first letters CVDOT, such as "**C**omputer **V**iruses **D**estroy **O**r **T**erminate." (You might remember the names of the Great Lakes with the mnemonic *homes*, for Huron, Ontario, Michigan, Erie, and Superior.) Such memory devices have been found to be very effective in recalling important keywords that trigger key concepts.

Critical Thinking and Interpretation: Going Beyond Formulas

Statistics professors generally believe that it is not so important to memorize formulas or manually perform complex arithmetic calculations. Instead, they tend to focus on obtaining results by using some form of technology (calculator or software), then making practical sense of the results through critical thinking. Keep this in mind as you proceed through this chapter. For example, when studying the extremely important *standard deviation* in Section 2-5, try to see how the key formula serves as a measure of variation, then learn how to find values of standard deviations, but really work on *understanding* and *interpreting* values of standard deviations.

Although this chapter includes detailed steps for important procedures, it is not necessary to master those steps in all cases. However, we recommend that in each case you perform a few manual calculations before using your calculator or computer. Your understanding will be enhanced, and you will acquire a better appreciation for the results obtained from the technology.

The methods of this chapter are often called methods of **descriptive statistics,** because the objective is to summarize or *describe* the important characteristics of a set of data. Later in this book we will use methods of **inferential statistics** when we use sample data to make inferences (or generalizations) about a population. With inferential statistics, we are making an inference that goes beyond the known data. Descriptive statistics and inferential statistics are the two general divisions of the subject of statistics, and this chapter deals with basic concepts of descriptive statistics.

2-2   Frequency Distributions

When working with large data sets, it is often helpful to organize and summarize the data by constructing a table that lists the different possible data values (either individually or by groups) along with the corresponding frequencies, which represent the number of times those values occur.

Definition

A **frequency distribution** lists data values (either individually or by groups of intervals), along with their corresponding frequencies (or counts).

Table 2-2 is a frequency distribution summarizing the measured cotinine levels of the 40 smokers listed in Table 2-1. The **frequency** for a particular class is the number of original values that fall into that class. For example, the first class in Table 2–2 has a frequency of 11, indicating that 11 of the original data values are between 0 and 99 inclusive.

We first present some standard terms used in discussing frequency distributions, and then we describe how to construct and interpret them.

Table 2-2	
Frequency Distribution of Cotinine Levels of Smokers	
Cotinine	Frequency
0–99	11
100–199	12
200–299	14
300–399	1
400–499	2

Definitions

Lower class limits are the smallest numbers that can belong to the different classes. (Table 2-2 has lower class limits of 0, 100, 200, 300, and 400.)

Upper class limits are the largest numbers that can belong to the different classes. (Table 2-2 has upper class limits of 99, 199, 299, 399, and 499.)

Class boundaries are the numbers used to separate classes, but without the gaps created by class limits. They are obtained as follows: Find the size of the gap between the upper class limit of one class and the lower class limit of the next class. Add half of that amount to each upper class limit to find the upper class boundaries; subtract half of that amount from each lower class limit to find the lower class boundaries. (Table 2-2 has gaps of exactly 1 unit, so 0.5 is added to the upper class limits and subtracted from the lower class limits. The first class

continued

Note to Instructor

Students typically have difficulty with three items: (1) definition of class width (believing incorrectly that it's the difference between the upper and lower limits of a class), (2) determination of class boundaries, and (3) getting a lower class boundary that is negative when negative numbers don't make sense in the context of the data.

Growth Charts Updated

Pediatricians typically use standardized growth charts to compare their patient's weight and height to a sample of other children. Children are considered to be in the normal range if their weight and height fall between the 5th and 95th percentiles. If they fall outside of that range, they are often given tests to ensure that there are no serious medical problems. Pediatricians became increasingly aware of a major problem with the charts: Because they were based on children living between 1929 and 1975, the growth charts were found to be inaccurate. To rectify this problem, the charts were updated in 2000 to reflect the current measurements of millions of children. The weights and heights of children are good examples of populations that change over time. This is the reason for including changing characteristics of data over time as an important consideration for a population.

has boundaries of −0.5 and 99.5, the second class has boundaries of 99.5 and 199.5, and so on. The complete list of boundaries used for all classes is −0.5, 99.5, 199.5, 299.5, 399.5, and 499.5.)

Class midpoints are the midpoints of the classes. (Table 2-2 has class midpoints of 49.5, 149.5, 249.5, 349.5, and 449.5.) Each class midpoint can be found by adding the lower class limit to the upper class limit and dividing the sum by 2.

Class width is the difference between two consecutive lower class limits or two consecutive lower class boundaries. (Table 2-2 uses a class width of 100.)

The definitions of class width and class boundaries are tricky. Be careful to avoid the easy mistake of making the class width the difference between the lower class limit and the upper class limit. See Table 2-2 and note that the class width is 100, not 99. You can simplify the process of finding class boundaries by understanding that they basically fill the gaps between classes by splitting the difference between the end of one class and the beginning of the next class.

Procedure for Constructing a Frequency Distribution

Frequency distributions are constructed for these reasons: (1) Large data sets can be summarized, (2) we can gain some insight into the nature of data, and (3) we have a basis for constructing important graphs (such as *histograms,* introduced in the next section). Many uses of technology allow us to automatically obtain frequency distributions without manually constructing them, but here is the basic procedure:

1. Decide on the number of classes you want. The number of classes should be between 5 and 20, and the number you select might be affected by the convenience of using round numbers.

2. Calculate

$$\text{Class width} \approx \frac{(\text{highest value}) - (\text{lowest value})}{\text{number of classes}}$$

Round this result to get a convenient number. (Usually round *up*.) You might need to change the number of classes, but the priority should be to use values that are easy to understand.

3. Starting point: Begin by choosing a number for the lower limit of the first class. Choose either the lowest data value or a convenient value that is a little smaller.

4. Using the lower limit of the first class and the class width, proceed to list the other lower class limits. (Add the class width to the starting point to get the second lower class limit. Add the class width to the second lower class limit to get the third, and so on.)

5. List the lower class limits in a vertical column and proceed to enter the upper class limits, which can be easily identified.

6. Go through the data set putting a tally in the appropriate class for each data value. Use the tally marks to find the total frequency for each class.

When constructing a frequency distribution, be sure that classes do not overlap so that each of the original values must belong to exactly one class. Include all classes, even those with a frequency of zero. Try to use the same width for all classes, although it is sometimes impossible to avoid open-ended intervals, such as "65 years or older."

Note to Instructor
You need to determine how much of this detail you want your students to know and remember. Recommendation: Instead of introducing this procedure in class, assign exercises (such as Exercises 15–20) requiring the construction of frequency tables and let the students construct them by following these instructions.

EXAMPLE Cotinine Levels of Smokers Using the 40 cotinine levels for the *smokers* in Table 2-1, follow the above procedure to construct the frequency distribution shown in Table 2-2. Assume that you want 5 classes.

SOLUTION

Step 1: Begin by selecting 5 as the number of desired classes.

Step 2: Calculate the class width. In the following calculation, 98.2 is rounded up to 100, which is a more convenient number.

$$\text{class width} \approx \frac{(\text{highest value}) - (\text{lowest value})}{\text{number of classes}} = \frac{491 - 0}{5} = 98.2 \approx 100$$

Step 3: We choose a starting point of 0, which is the lowest value in the list and is also a convenient number.

Step 4: Add the class width of 100 to the starting point of 0 to determine that the second lower class limit is 100. Continue to add the class width of 100 to get the remaining lower class limits of 200, 300, and 400.

Step 5: List the lower class limits vertically, as shown in the margin. From this list, we can easily identify the corresponding upper class limits as 99, 199, 299, 399, and 499.

Step 6: After identifying the lower and upper limits of each class, proceed to work through the data set by entering a tally mark for each value. When the tally marks are completed, add them to find the frequencies shown in Table 2-2.

0 –
100 –
200 –
300 –
400 –

Relative Frequency Distribution

An important variation of the basic frequency distribution uses **relative frequencies,** which are easily found by dividing each class frequency by the total of all frequencies. A **relative frequency distribution** includes the same class limits as a frequency distribution, but relative frequencies are used instead of actual frequencies. The relative frequencies are sometimes expressed as percents.

$$\text{relative frequency} = \frac{\text{class frequency}}{\text{sum of all frequencies}}$$

In Table 2-3 the actual frequencies from Table 2-2 are replaced by the corresponding relative frequencies expressed as percents. The first class has a relative frequency of $11/40 = 0.275$, or 27.5%, which is often rounded to 28%. The

Table 2-3

Relative Frequency Distribution of Cotinine Levels in Smokers

Cotinine	Relative Frequency
0–99	28%
100–199	30%
200–299	35%
300–399	3%
400–499	5%

second class has a relative frequency of 12/40 = 0.3, or 30.0%, and so on. If constructed correctly, the sum of the relative frequencies should total 1 (or 100%), with some small discrepancies allowed for rounding errors. Because 27.5% was rounded to 28% and 2.5% was rounded to 3%, the sum of the relative frequencies in Table 2-3 is 101% instead of 100%.

Because they use simple proportions or percentages, relative frequency distributions make it easier for us to understand the distribution of the data and to compare different sets of data.

Cumulative Frequency Distribution

Another variation of the standard frequency distribution is used when cumulative totals are desired. The **cumulative frequency** for a class is the sum of the frequencies for that class and all previous classes. Table 2-4 is the cumulative frequency distribution based on the frequency distribution of Table 2-2. Using the original frequencies of 11, 12, 14, 1, and 2, we add 11 + 12 to get the second cumulative frequency of 23, then we add 11 + 12 + 14 = 37 to get the third, and so on. See Table 2-4 and note that in addition to using cumulative frequencies, the class limits are replaced by "less than" expressions that describe the new range of values.

Critical Thinking: Interpreting Frequency Distributions

The transformation of raw data to a frequency distribution is typically a means to some greater end. The following examples illustrate how frequency distributions can be used to describe, explore, and compare data sets. (The following section shows how the construction of a frequency distribution is often the first step in the creation of a graph that visually depicts the nature of the distribution.)

EXAMPLE **Describing Data** Refer to Data Set 1 in Appendix B for the pulse rates of 40 randomly selected adult males. Table 2-5 summarizes the *last digits* of those pulse rates. If the pulse rates are measured by counting the number of heartbeats in 1 minute, we expect that those last digits should occur with frequencies that are roughly the same. But note that the frequency distribution shows that the last digits are all *even* numbers; there are *no* odd numbers present. This suggests that the pulse rates were not counted for 1 minute. Perhaps they were counted for 30 seconds and the values were then doubled. (Upon further examination of the *original* pulse rates, we can see that every original value is a multiple of four, suggesting that the number of heartbeats was counted for 15 seconds, then that count was multiplied by four.) It's fascinating to learn something about the method of data collection by simply describing some characteristics of the data.

Table 2-4	
Cumulative Frequency Distribution of Cotinine Levels in Smokers	
Cotinine	Cumulative Frequency
Less than 100	11
Less than 200	23
Less than 300	37
Less than 400	38
Less than 500	40

Table 2-5	
Last Digits of Male Pulse Rates	
Last Digit	Frequency
0	7
1	0
2	6
3	0
4	11
5	0
6	9
7	0
8	7
9	0

EXAMPLE **Exploring Data** In studying the behavior of the Old Faithful geyser in Yellowstone National Park, geologists collect data for the times (in minutes) between eruptions. Table 2-6 summarizes actual data that were obtained. Examination of the frequency distribution reveals unexpected behavior: The distribution of times has two different peaks. This distribution led geologists to consider possible explanations.

Table 2-6	
Times (in minutes) Between Old Faithful Eruptions	
Time	Frequency
40–49	8
50–59	44
60–69	23
70–79	6
80–89	107
90–99	11
100–109	1

 EXAMPLE **Comparing Data Sets** The Chapter Problem given at the beginning of this chapter includes data sets consisting of measured cotinine levels from smokers, nonsmokers exposed to tobacco smoke, and nonsmokers not exposed to tobacco smoke. Table 2-7 shows

Table 2-7	Cotinine Levels for Three Groups		
Cotinine	Smokers	Nonsmokers Exposed to Smoke	Nonsmokers Not Exposed to Smoke
0–99	28%	85%	95%
100–199	30%	5%	0%
200–299	35%	3%	3%
300–399	3%	3%	3%
400–499	5%	0%	0%
500–599	0%	5%	0%

continued

Recommended Assignment
Exercises 1–18.

Due to limited space, only answers to even-numbered exercises are shown in this section. See Appendix F for answers to odd-numbered exercises.

2. Class width: 20. Class midpoints: 89.5, 109.5, 129.5, 149.5, 169.5, 189.5. Class boundaries: 79.5, 99.5, 119.5, 139.5, 159.5, 179.5, 199.5.

4. Class width: 6.0. Class midpoints: 17.95, 23.95, 29.95, 35.95, 41.95. Class boundaries: 14.95, 20.95, 26.95, 32.95, 38.95, 44.95.

6.

Systolic Blood Pressure of Women	Relative Frequency
80–99	22.5%
100–119	60.0%
120–139	12.5%
140–159	2.5%
160–179	0.0%
180–199	2.5%

Table for Exercise 13

Outcome	Frequency
1	27
2	31
3	42
4	40
5	28
6	32

Table for Exercise 14

Digit	Frequency
0	18
1	12
2	14
3	9
4	17
5	20
6	21
7	26
8	7
9	16

the relative frequencies for the three groups. By comparing those relative frequencies, it should be obvious that the frequency distribution for smokers is very different from the frequency distributions for the other two groups. Because the two groups of nonsmokers (exposed and not exposed) have such high frequency amounts for the first class, it might be helpful to further compare those data sets with a closer examination of those values.

2-2 Basic Skills and Concepts

In Exercises 1–4, identify the class width, class midpoints, and class boundaries for the given frequency distribution based on Data Set 1 in Appendix B.

1.

Systolic Blood Pressure of Men	Frequency
90–99	1
100–109	4
110–119	17
120–129	12
130–139	5
140–149	0
150–159	1

2.

Systolic Blood Pressure of Women	Frequency
80–99	9
100–119	24
120–139	5
140–159	1
160–179	0
180–199	1

3.

Cholesterol of Men	Frequency
0–199	13
200–399	11
400–599	5
600–799	8
800–999	2
1000–1199	0
1200–1399	1

4.

Body Mass Index of Women	Frequency
15.0–20.9	10
21.0–26.9	15
27.0–32.9	11
33.0–38.9	2
39.0–44.9	2

In Exercises 5–8, construct the relative frequency distribution that corresponds to the frequency distribution in the exercise indicated.

5. Exercise 1   **6.** Exercise 2   **7.** Exercise 3   **8.** Exercise 4

In Exercises 9–12, construct the cumulative frequency distribution that corresponds to the frequency distribution in the exercise indicated.

9. Exercise 1   **10.** Exercise 2   **11.** Exercise 3   **12.** Exercise 4

13. Loaded Die The author drilled a hole in a die and filled it with a lead weight, then proceeded to roll it 200 times. (Yes, the author has too much free time.) The results are given in the frequency distribution in the margin. Construct the corresponding relative frequency distribution and determine whether the die is significantly different from a fair die that has not been "loaded."

14. Lottery The frequency distribution in the margin is based on the Win Four numbers from the New York State Lottery, as listed in Data Set 26 in Appendix B. Construct the corresponding relative frequency distribution and determine whether the results appear to be selected in such a way that all of the digits are equally likely.

14. Change the heading of "Frequency" to "Relative Frequency," and enter these relative frequencies: 11.25%, 7.50%, 8.75%, 5.63%, 10.63%, 12.50%, 13.13%, 16.25%, 4.38%, 10.00%. The relative frequencies appear to vary by fairly large amounts. (Using methods described later in the book, the differences are significant.)

15. **Bears** Refer to Data Set 9 in Appendix B and construct a frequency distribution of the weights of bears. Use 11 classes beginning with a lower class limit of 0 and use a class width of 50 lb.

16. **Body Temperatures** Refer to Data Set 4 in Appendix B and construct a frequency distribution of the body temperatures for midnight on the second day. Use 8 classes beginning with a lower class limit of 96.5 and use a class width of 0.4°F. Describe two different notable features of the result.

17. **Head Circumferences** Refer to Data Set 3 in Appendix B. Construct a frequency distribution for the head circumferences of baby boys and construct a separate frequency distribution for the head circumferences of baby girls. In both cases, use the classes of 34.0–35.9, 36.0–37.9, and so on. Then compare the results and determine whether there appears to be a significant difference between the two genders.

18. **Animated Movies for Children** Refer to Data Set 7 in Appendix B. Construct a frequency distribution for the lengths of time that animated movies for children contain tobacco use and construct a separate frequency distribution for the lengths of time for alcohol use. In both cases, use the classes of 0–99, 100–199, and so on. Compare the results and determine whether there appears to be a significant difference.

19. **Marathon Runners** Refer to Data Set 8 in Appendix B. Construct a relative frequency distribution for the ages of the sample of males who finished the New York City marathon, then construct a separate relative frequency distribution for the ages of the females. In both cases, start the first class with a lower class limit of 19 and use a class width of 10. Compare the results and determine whether there appears to be any notable difference between the two groups.

20. **Regular Coke/Diet Coke** Refer to Data Set 17 in Appendix B. Construct a relative frequency distribution for the weights of regular Coke by starting the first class at 0.7900 lb and use a class width of 0.0050 lb. Then construct another relative frequency distribution for the weights of diet Coke by starting the first class at 0.7750 lb and use a class width of 0.0050 lb. Then compare the results and determine whether there appears to be a significant difference. If so, provide a possible explanation for the difference.

2-2 Beyond the Basics

21. **Interpreting Effects of Outliers** Refer to Data Set 20 in Appendix B for the axial loads of aluminum cans that are 0.0111 in. thick. The load of 504 lb is called an *outlier* because it is very far away from all of the other values. Construct a frequency distribution that includes the value of 504 lb, then construct another frequency distribution with the value of 504 lb excluded. In both cases, start the first class at 200 lb and use a class width of 20 lb. Interpret the results by stating a generalization about how much of an effect an outlier might have on a frequency distribution.

22. **Number of Classes** In constructing a frequency distribution, Sturges' guideline suggests that the ideal number of classes can be approximated by $1 + (\log n) / (\log 2)$, where n is the number of data values. Use this guideline to complete the table for determining the ideal number of classes.

22. 46–90, 91–181, 182–362, 363–724, 725–1448, 1449–2896

8.

Body Mass Index of Women	Relative Frequency
15.0–20.9	25.0%
21.0–26.9	37.5%
27.0–32.9	27.5%
33.0–38.9	5.0%
39.0–44.9	5.0%

10.

Systolic Blood Pressure of Women	Cumulative Frequency
Less than 100	9
Less than 120	33
Less than 140	38
Less than 160	39
Less than 180	39
Less than 200	40

12.

Body Mass Index of Women	Cumulative Frequency
Less than 21.0	10
Less than 27.0	25
Less than 33.0	36
Less than 39.0	38
Less than 45.0	40

16.

Temp. (°F)	Frequency
96.5–96.8	1
96.9–97.2	8
97.3–97.6	14
97.7–98.0	22
98.1–98.4	19
98.5–98.8	32
98.9–99.2	6
99.3–99.6	4

Table for Exercise 22

Number of Values	Ideal Number of Classes
16–22	5
23–45	6
	7
	8
	9
	10
	11
	12

See page 46 for the answers to Exercises 18 and 20.

18. There does not appear to be a significant difference.

Time (sec)	Tobacco	Alcohol
0–99	39	46
100–199	6	3
200–299	4	0
300–399	0	0
400–499	0	1
500–599	1	

20. The weights of regular Coke appear to be significantly higher, probably due to the sugar content.

Weight (lb)	Regular	Diet
0.7750–0.7799		11.1%
0.7800–0.7849		36.1%
0.7850–0.7899		41.7%
0.7900–0.7949	2.8%	11.1%
0.7950–0.7999	0.0%	
0.8000–0.8049	2.8%	
0.8050–0.8099	8.3%	
0.8100–0.8149	11.1%	
0.8150–0.8199	47.2%	
0.8200–0.8249	16.7%	
0.8250–0.8299	11.1%	

Note to Instructor

Minimum coverage for this section should include histograms. Suggestion: Discuss histograms and stem-and-leaf plots, and assign the others for reading.

Consider assigning a project requiring computer generation of a histogram. (See Cooperative Group Activity 2 near the end of this chapter.)

If you require or allow the use of calculators or computers for generating graphs, announce that labeling of axes need not follow the exact formats described in this section (such as class boundaries used in the horizontal axis for histograms).

2-3 Visualizing Data

Recall that the main objective of this chapter is to learn important techniques for investigating the "CVDOT" important characteristics of data sets: center, variation, distribution, outliers, and changes over time. In Section 2-2 we introduced frequency distributions as a tool for describing, exploring, or comparing *distributions* of data sets. In this section we continue the study of distributions by introducing graphs that are pictures of distributions. As you read through this section, keep in mind that the objective is not simply to construct graphs, but rather to learn something about data sets—that is, to understand the nature of their distributions.

Histograms

Among the different types of graphs presented in this section, the histogram is particularly important.

> **Definition**
>
> A **histogram** is a bar graph in which the horizontal scale represents classes of data values and the vertical scale represents frequencies. The heights of the bars correspond to the frequency values, and the bars are drawn adjacent to each other (without gaps).

We can construct a histogram after we have first completed a frequency distribution table for a data set. The cotinine levels of smokers are depicted in the histogram of Figure 2-1, which corresponds directly to the frequency distribution in Table 2-2 given in the preceding section. Each bar of the histogram is marked with

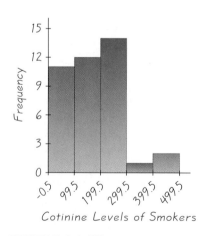

FIGURE 2-1 **Histogram**

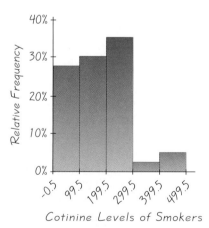

FIGURE 2-2 **Relative Frequency Histogram**

its lower class boundary at the left and its upper class boundary at the right. Instead of using class boundaries along the horizontal scale, it is often more practical to use class midpoint values centered below their corresponding bars. The use of class midpoint values is very common in software packages that automatically generate histograms.

Before constructing a histogram from a completed frequency distribution, we must give some thought to the scales used on the vertical and horizontal axes. The maximum frequency (or the next highest convenient number) should suggest a value for the top of the vertical scale; 0 should be at the bottom. In Figure 2-1 we designed the vertical scale to run from 0 to 15. The horizontal scale should be subdivided in a way that allows all the classes to fit well. Ideally, we should try to follow the rule of thumb that the vertical height of the histogram should be about three-fourths of the total width. Both axes should be clearly labeled.

Interpreting a Histogram   Remember that the objective is not simply to construct a histogram, but rather to learn something about the data. Analyze the histogram to see what can be learned about "CVDOT": the center of the data, the variation (which will be discussed at length in Section 2-5), the shape of the distribution, and whether there are any outliers (values far away from the other values). The histogram is not suitable for determining whether there are changes over time. Examining Figure 2-1, we see that the histogram is centered around 175, the values vary from around 0 to 500, and the shape of the distribution is heavier on the left.

Relative Frequency Histogram

A **relative frequency histogram** has the same shape and horizontal scale as a histogram, but the vertical scale is marked with relative frequencies instead of actual frequencies, as in Figure 2-2.

Frequency Polygon

A **frequency polygon** uses line segments connected to points located directly above class midpoint values. See Figure 2-3 on the next page for the frequency polygon corresponding to Table 2-2. The heights of the points correspond to the class frequencies, and the line segments are extended to the right and left so that the graph begins and ends on the horizontal axis.

Ogive

An **ogive** (pronounced "oh-jive") is a line graph that depicts *cumulative* frequencies, just as the cumulative frequency distribution (see Table 2-4 in the preceding section) lists cumulative frequencies. Figure 2-4 is an ogive corresponding to Table 2-4. Note that the ogive uses class boundaries along the horizontal scale, and the graph begins with the lower boundary of the first class and ends with the upper boundary of the last class. Ogives are useful for determining the number of

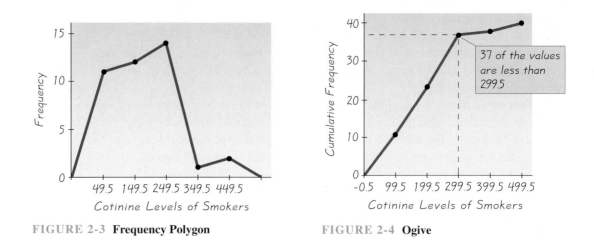

FIGURE 2-3 **Frequency Polygon**   FIGURE 2-4 **Ogive**

values below some particular value. For example, Figure 2-4 shows that 37 of the cotinine level values are less than 299.5.

Dotplots

A **dotplot** consists of a graph in which each data value is plotted as a point (or dot) along a scale of values. Dots representing equal values are stacked. See Figure 2-5, which represents the lengths of animated movies for children, as listed in Data Set 7 in Appendix B. The two leftmost dots, for example, represent the value of 64 min, which occurs twice in Data Set 7. We can see from this dotplot that the length of 120 min is very different from the other lengths.

Stem-and-Leaf Plots

A **stem-and-leaf plot** represents data by separating each value into two parts: the stem (such as the leftmost digit) and the leaf (such as the rightmost digit). The illustration on the next page shows a stem-and-leaf plot for the same movie lengths listed in Data Set 7 of Appendix B. Those lengths (in minutes) sorted according to increasing order are 64, 64, 69, 70, 71, 71, 71, 72, 73, . . . , 120. It is easy to see how the first value of 64 is separated into its stem of 6 and leaf of 4. Each of the remaining values is broken up in a similar way. Note that the leaves are arranged in increasing order, not the order in which they occur in the original list.

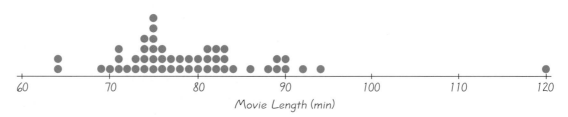

FIGURE 2-5 **Dotplot of Movie Lengths**

Stem-and-Leaf Plot

Stem (tens)	Leaves (units)	
6	449	← Values are 64, 64, 69.
7	011123344445555555666778899	
8	0011122233346899	
9	0024	
10		
11		
12	0	← Value is 120.

By turning the page on its side, we can see a distribution of these data. A great advantage of the stem-and-leaf plot is that we can see the distribution of data and yet retain all the information in the original list. If necessary, we could reconstruct the original list of values. Another advantage is that construction of a stem-and-leaf plot is a quick and easy way to *sort* data (arrange them in order), and sorting is required for some statistical procedures (such as finding a median, or finding percentiles).

The rows of digits in a stem-and-leaf plot are similar in nature to the bars in a histogram. One of the guidelines for constructing histograms is that the number of classes should be between 5 and 20, and the same guideline applies to stem-and-leaf plots for the same reasons. Better stem-and-leaf plots are often obtained by first rounding the original data values. Also, stem-and-leaf plots can be *expanded* to include more rows and can be *condensed* to include fewer rows. The stem-and-leaf plot in our example can be expanded by subdividing rows into those with leaves having digits of 0 through 4 and those with digits 5 through 9, as shown in the following diagram.

Expanded Stem-and-Leaf Plot

Stem	Leaves	
6	44	← For leaves of 0 through 4
6	9	← For leaves of 5 through 9
7	0111233444	
7	555555666778899	
8	001112223334	
8	6899	
9	0024	
9		
10		
10		
11		
11		
12	0	

The Growth of Statistics

Reporter Richard Rothstein wrote in the *New York Times* that the study of algebra, trigonometry, and geometry in high school "leaves too little room for the study of statistics and probability. Yet students need grounding in data analysis." He observed that calculus plays a prominent role in college studies, even though "only a few jobs, mostly in technical fields actually use it." Rothstein cited a study conducted by University of Massachusetts professor Clifford Konold, who counted data displays in the *New York Times*. In 1972 issues of the *Times*, Dr. Konold found four graphs or tables in each of 10 weekday editions (not including the sports and business sections), but in 1982 there were 8, and in 1992 there were 44, and "next year, he (Dr. Konold) could find more than 100." The growth of statistics as a discipline is fostered in part by the increasing use of such data displays in the media.

When it becomes necessary to *reduce* the number of rows, we can condense a stem-and-leaf plot by combining adjacent rows, as in the following illustration. Note that we insert an asterisk to separate digits in the leaves associated with the numbers in each stem. Every row in the condensed plot must include exactly one asterisk so that the shape of the plot is not distorted.

Condensed Stem-and-Leaf Plot

Stem	Leaves	
6-7	449*011123344445555555666778899	← 64, 64, 69, 70,
8-9	0011122233346899*0024	..., 79
10-11	*	
12-13	0*	← Value is 120.

Pareto Charts

The Federal Communications Commission monitors the quality of phone service in the United States. Complaints against phone carriers include slamming, which is changing a customer's carrier without the customer's knowledge, and cramming, which is the insertion of unauthorized charges. Recently, FCC data showed that complaints against U.S. phone carriers consisted of 4473 for rates and services, 1007 for marketing, 766 for international calling, 614 for access charges, 534 for operator services, 12,478 for slamming, and 1214 for cramming. If you were a print media reporter, how would you present that information? Simply writing the sentence with the numerical data would not really lead to understanding. A better approach is to use an effective graph, and a Pareto chart would be suitable here.

A **Pareto chart** is a bar graph for qualitative data, with the bars arranged in order according to frequencies. As in histograms, vertical scales in Pareto charts can represent frequencies or relative frequencies. The tallest bar is at the left, and the smaller bars are farther to the right. By arranging the bars in order of frequency, the Pareto chart focuses attention on the more important categories. Figure 2-6 is a Pareto chart clearly showing that slamming is by far the most serious issue in customer complaints about phone carriers.

Pie Charts

Pie charts are also used to visually depict qualitative data. Figure 2-7 is an example of a **pie chart,** which is a graph depicting qualitative data as slices of a pie. Figure 2-7 represents the same data as Figure 2-6. Construction of a pie chart involves slicing up the pie into the proper proportions. The category of slamming complaints represents 59% of the total, so the wedge representing slamming should be 59% of the total (with a central angle of $0.59 \times 360° = 212°$).

The Pareto chart (Figure 2-6) and the pie chart (Figure 2-7) depict the same data in different ways, but a comparison will probably show that the Pareto chart does a better job of showing the relative sizes of the different components. That helps explain why many companies, such as Boeing Aircraft, make extensive use of Pareto charts.

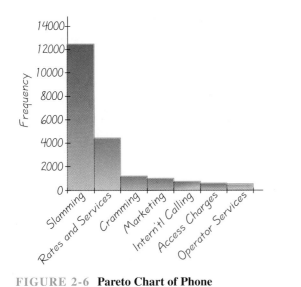

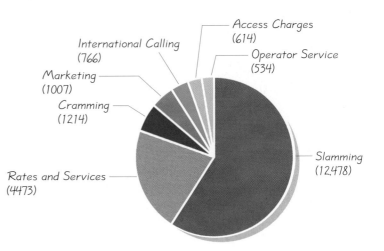

FIGURE 2-6 **Pareto Chart of Phone Company Complaints**

FIGURE 2-7 **Pie Chart of Phone Company Complaints**

Scatter Diagrams

A **scatter diagram** is a plot of paired (*x, y*) data with a horizontal *x*-axis and a vertical *y*-axis. The data are paired in a way that matches each value from one data set with a corresponding value from a second data set. To manually construct a scatter diagram, construct a horizontal axis for the values of the first variable, construct a vertical axis for the values of the second variable, then plot the points. The pattern of the plotted points is often helpful in determining whether there is some relationship between the two variables. (This issue is discussed at length when the topic of correlation is considered in Section 9-2.) Using the weight (in pounds) and waist circumference (in cm) for the males in Data Set 1 of Appendix B, we used Minitab to generate the scatter diagram shown here. On the basis of that graph, there does appear to be a relationship between weight and waist circumference, as shown by the pattern of the points.

Note to Instructor

Comment that scatter diagrams are discussed again in Chapter 9. Also, discuss the issue of *causation* by asking these questions: "Suppose grades and hours studied are recorded for a sample of students. What would the scatter diagram look like? If there is a clear pattern showing a relationship, can we conclude from the scatter diagram that more studying *causes* higher grades?" Section 9-2 makes the point that correlation does *not* imply causation.

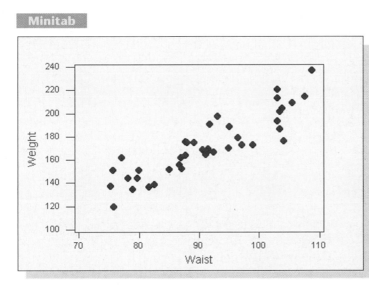

Florence Nightingale

Florence Nightingale (1820–1910) is known to many as the founder of the nursing profession, but she also saved thousands of lives by using statistics. When she encountered an unsanitary and undersupplied hospital, she improved those conditions and then used statistics to convince others of the need for more widespread medical reform. She developed original graphs to illustrate that, during the Crimean War, more soldiers died as a result of unsanitary conditions than were killed in combat. Florence Nightingale pioneered the use of social statistics as well as graphics techniques.

Time-Series Graph

Time-series data are data that have been collected at different points in time. For example, Figure 2-8 shows the numbers of screens at drive-in movie theaters for a recent period of 14 years (based on data from the National Association of Theater Owners). We can see that for this time period, there is a clear trend of decreasing values. A once significant part of Americana, especially to the author, is undergoing a decline. Fortunately, the rate of decline appears to be less than it was in the late 1980s. It is often critically important to know when population values change over time. Companies have gone bankrupt because they failed to monitor the quality of their goods or services and incorrectly believed that they were dealing with stable data. They did not realize that their products were becoming seriously defective as important population characteristics were changing. Chapter 13 introduces *control charts* as an effective tool for monitoring time-series data.

Other Graphs

Numerous pictorial displays other than the ones just described can be used to represent data dramatically and effectively. In Section 2-7 we present boxplots, which are very useful for revealing the spread of data. Pictographs depict data by using pictures of objects, such as soldiers, tanks, airplanes, stacks of coins, or moneybags.

The figure on page 53 has been described as possibly "the best statistical graphic ever drawn." This figure includes six different variables relevant to the march of Napoleon's army to Moscow and back in 1812–1813. The thick band at the left depicts the size of the army when it began its invasion of Russia from Poland. The lower band shows its size during the retreat, along with corresponding temperatures and dates. Although first developed in 1861 by Charles Joseph Minard, this graph is ingenious even by today's standards.

Another notable graph of historical importance is one developed by the world's most famous nurse, Florence Nightingale. This graph, shown in Figure 2-9 on page 54, is particularly interesting because it actually saved lives when Nightingale used it to convince British officials that military hospitals needed to improve sanitary conditions, treatment, and supplies. It is drawn somewhat like a pie chart, except that the central angles are all the same and different radii are used to show changes in the numbers of deaths each month. The outermost regions of Figure 2–9 represent

FIGURE 2-8 **Time-Series Data: Number of Screens at Drive-in Movie Theaters**

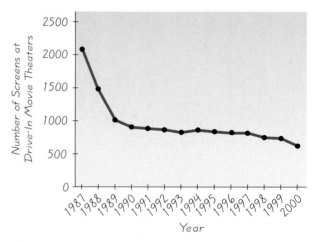

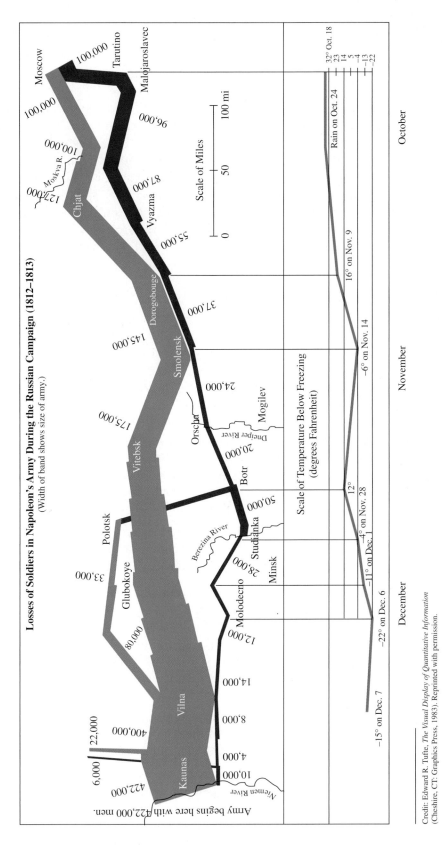

Losses of Soldiers in Napoleon's Army During the Russian Campaign (1812–1813)
(Width of band shows size of army.)

Credit: Edward R. Tufte, *The Visual Display of Quantitative Information* (Cheshire, CT: Graphics Press, 1983). Reprinted with permission.

FIGURE 2-9 **Deaths in British Military Hospitals During the Crimean War**

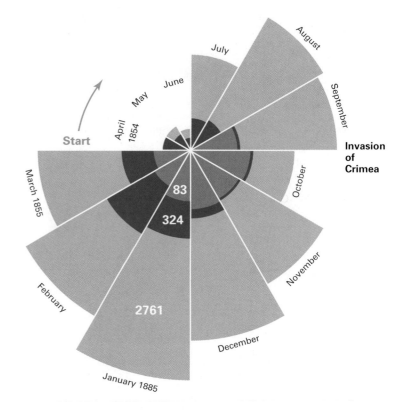

deaths due to preventable diseases, the innermost regions represent deaths from wounds, and the middle regions represent deaths from other causes.

Conclusion

The effectiveness of Florence Nightingale's graph illustrates well this important point: A graph is not in itself an end result; it is a tool for describing, exploring, and comparing data, which we consider as follows:

Describing data: In a histogram, for example, consider center, variation, distribution, and outliers (CVDOT without the last element of time). What is the approximate value of the center of the distribution, and what is the approximate range of values? Consider the overall shape of the distribution. Are the values evenly distributed? Is the distribution skewed (lopsided) to the right or left? Does the distribution peak in the middle? Identify any extreme values and any other notable characteristics.

Exploring data: We look for features of the graph that reveal some useful and/or interesting characteristics of the data set. In Figure 2-9, for example, we see that more soldiers were dying from inadequate hospital care than were dying from battle wounds.

Comparing data: Construct similar graphs that make it easy to compare data sets. For example, if you graph a frequency polygon for weights of men and another frequency polygon for weights of women on the same set of axes, the polygon for men should be farther to the right than the polygon for women, showing that men have higher weights.

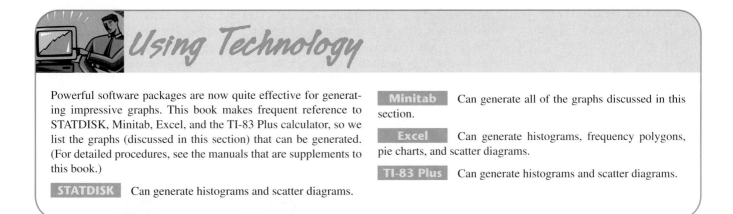

Powerful software packages are now quite effective for generating impressive graphs. This book makes frequent reference to STATDISK, Minitab, Excel, and the TI-83 Plus calculator, so we list the graphs (discussed in this section) that can be generated. (For detailed procedures, see the manuals that are supplements to this book.)

STATDISK   Can generate histograms and scatter diagrams.

Minitab   Can generate all of the graphs discussed in this section.

Excel   Can generate histograms, frequency polygons, pie charts, and scatter diagrams.

TI-83 Plus   Can generate histograms and scatter diagrams.

2-3 Basic Skills and Concepts

In Exercises 1–4, answer the questions by referring to the accompanying STATDISK-generated histogram, which represents the ages of all stowaways on the Queen Mary.

1. Center What is the approximate value of the center? That is, what age appears to be near the center of all ages?

2. Variation What are the lowest and highest possible ages? 10 years, 70 years

3. Percentage What percentage of the 131 stowaways were younger than 30 years of age?

4. Class Width What is the class width? 5 years

Recommended Assignment
Exercises 1–12, 17–22, 25, 26, 29–32

Due to limited space, only answers to even-numbered exercises are shown in this section. See Appendix F for answers to odd-numbered exercises.

STATDISK

In Exercises 5 and 6, refer to the accompanying pie chart of blood groups for a large sample of people (based on data from the Greater New York Blood Program).

5. Interpreting Pie Chart What is the approximate percentage of people with Group A blood? Assuming that the pie chart is based on a sample of 500 people, approximately how many of those 500 people have Group A blood?

6. Interpreting Pie Chart What is the approximate percentage of people with Group B blood? Assuming that the pie chart is based on a sample of 500 people, approximately how many of those 500 people have Group B blood? 10%; 50

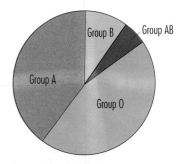

Table for Exercise 7

Age	Students	Faculty/Staff
0–2	23	30
3–5	33	47
6–8	63	36
9–11	68	30
12–14	19	8
15–17	10	0
18–20	1	0
21–23	0	1

Table for Exercise 8

Speed	Frequency
42–45	25
46–49	14
50–53	7
54–57	3
58–61	1

8. Although the posted limit is 30 mi/h, it appears that the police issue tickets only to those traveling at least 42 mi/h.

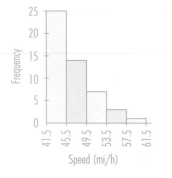

10. The mean appears to be lower than 98.6°F. Distribution is approximately bell-shaped.

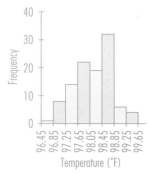

7. Student/Faculty Cars Samples of student cars and faculty/staff cars were obtained at the author's college, and their ages (in years) are summarized in the accompanying frequency distribution. Construct a relative frequency histogram for student cars and another relative frequency histogram for faculty cars. Compare the two relative frequency histograms. What are the noticeable differences?

8. Speeding Tickets The given frequency distribution describes the speeds of drivers ticketed by the Town of Poughkeepsie police. These drivers were traveling through a 30 mi/h speed zone on Creek Road, which passes the author's college. Construct a histogram corresponding to the given frequency distribution. What does the distribution suggest about the enforced speed limit compared to the posted speed limit?

9. Bears Exercise 15 in Section 2-2 referred to Data Set 9 in Appendix B. Using the frequency distribution of the weights of bears (with 11 classes beginning with a lower class limit of 0 and a class width of 50 lb), construct the corresponding histogram. What is the approximate weight that is at the center?

10. Body Temperatures Exercise 16 in Section 2-2 referred to Data Set 4 in Appendix B. Using the frequency distribution of the body temperatures for midnight on the second day (with 8 classes beginning with a lower class limit of 96.5 and a class width of 0.4°F), construct the corresponding histogram. What does the distribution suggest about the common belief that the average body temperature is 98.6°F? If the subjects are randomly selected, the temperatures should have a distribution that is approximately bell-shaped. Is it?

In Exercises 11–14, make the comparisons by constructing the indicated graphs.

11. Head Circumferences Exercise 17 in Section 2-2 referred to Data Set 3 in Appendix B. Using the frequency distribution for head circumferences of boys and the frequency distribution for head circumferences of girls (with the classes of 34.0–35.9, 36.0–37.9, and so on), use the same set of axes to construct the two corresponding frequency polygons. Compare the results and determine whether there appears to be a significant difference between the two genders.

12. Animated Movies for Children Exercise 18 in Section 2-2 referred to Data Set 7 in Appendix B. Using the frequency distribution for the lengths of time that animated movies for children contain tobacco use and the frequency distribution for the lengths of time for alcohol use (with classes of 0–99, 100–199, and so on), use the same set of axes to construct the two corresponding frequency polygons. Compare the results and determine whether there appears to be a significant difference.

13. Marathon Runners Exercise 19 in Section 2-2 referred to Data Set 8 in Appendix B. Using the relative frequency distribution for the ages of males and the relative frequency distribution for the ages of females (with a lower class limit of 19 and a class width of 10), construct the corresponding relative frequency histograms. Compare the results and determine whether there appear to be any notable differences between the two groups.

14. Regular Coke and Diet Coke Refer to Data Set 17 in Appendix B and use the weights of regular Coke and diet Coke. Using classes of 0.7750–0.7799, 0.7800–0.7849, . . . ,

0.8250–0.8299, construct the two frequency polygons on the same axes. Then compare the results and determine whether there appears to be a significant difference. What is a possible explanation for the difference?

In Exercises 15 and 16, list the original data represented by the given stem-and-leaf plots.

15.

Stem (tens)	Leaves (units)
20	0005
21	69999
22	2233333
23	
24	1177

16.

Stem (hundreds)	Leaves (tens and units)
50	12 12 12 55
51	
52	00 00 00 00
53	27 27 35
54	72

16. 5012, 5012, 5012, 5055, 5200, 5200, 5200, 5200, 5327, 5327, 5335, 5472

In Exercises 17 and 18, construct the dotplot for the data represented by the stem-and-leaf plot in the given exercise.

17. Exercise 15

18. Exercise 16

In Exercises 19 and 20, construct the stem-and-leaf plots for the given data sets found in Appendix B.

19. Bears The lengths (in inches) of the bears in Data Set 9. (*Hint:* First round the lengths to the nearest inch.)

20. Plastic Weights (in pounds) of plastic discarded by 62 households: Refer to Data Set 23, and start by rounding the listed weights to the nearest tenth of a pound (or one decimal place). (Use an expanded stem-and-leaf plot with about 11 rows.)

21. Jobs A study was conducted to determine how people get jobs. The table lists data from 400 randomly selected subjects. The data are based on results from the National Center for Career Strategies. Construct a Pareto chart that corresponds to the given data. If someone would like to get a job, what seems to be the most effective approach?

Job Sources of Survey Respondents	Frequency
Help-wanted ads	56
Executive search firms	44
Networking	280
Mass mailing	20

22. Jobs Refer to the data given in Exercise 21, and construct a pie chart. Compare the pie chart to the Pareto chart. Can you determine which graph is more effective in showing the relative importance of job sources?

12. There does not appear to be a significant difference.

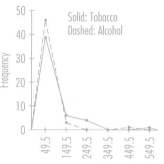

14. The weights of regular Coke appear to be significantly higher, probably due to the presence of sugar.

18.

20.

0.	234
0.	66777899999
1.	123444444
1.	55555566778
2.	001112223344
2.	7788999
3.	00144
3.	5
4.	4
4.	7
5.	3

22. The Pareto chart is more effective in showing the relative importance of job sources.

24. The Pareto chart appears to be more effective.

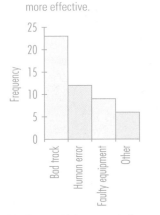

26. Bears with larger neck sizes also tend to weigh more.

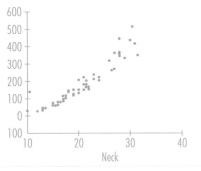

28. There appears to be a downward trend possibly associated with such factors as safer cars, seat belt laws, and drunk driving enforcement.

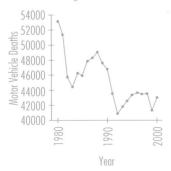

30. 22,000 of the 50,000 men died; 44%

32. 80,000 (out of 100,000) men died.

23. Train Derailments An analysis of train derailment incidents showed that 23 derailments were caused by bad track, 9 were due to faulty equipment, 12 were attributable to human error, and 6 had other causes (based on data from the Federal Railroad Administration). Construct a pie chart representing the given data.

24. Train Derailments Refer to the data given in Exercise 23, and construct a Pareto chart. Compare the Pareto chart to the pie chart. Can you determine which graph is more effective in showing the relative importance of the causes of train derailments.

In Exercises 25 and 26, use the given paired data from Appendix B to construct a scatter diagram.

25. Cigarette Tar/CO In Data Set 5, use cigarette tar for the horizontal scale and use carbon monoxide for the vertical scale. Determine whether there appears to be a relationship between cigarette tar and carbon monoxide. If so, describe the relationship.

26. Bear Neck/Weight In Data Set 9, use the distances around bear necks for the horizontal scale and use the bear weights for the vertical scale. Based on the result, what is the relationship between a bear's neck size and its weight?

In Exercises 27 and 28, use the given data from Appendix B to construct a time-series graph.

27. Investing in Stocks In Data Set 25, use the high values of the Dow Jones Industrial Average (DJIA) to construct a time-series graph, then determine whether there appears to be a trend. How might an investor profit from this trend?

28. Motor Vehicle Fatalities In Data Set 25, use the U.S. motor vehicle deaths to construct a time-series graph, then determine whether there appears to be a trend. If so, provide a possible explanation.

In Exercises 29–32, refer to the figure on page 53, which describes Napoleon's 1812 campaign to Moscow and back. The thick band at the left depicts the size of the army when it began its invasion of Russia from Poland, and the lower band describes Napoleon's retreat.

29. Find the percentage of men who survived the entire campaign.

30. Find the number of men and the percentage of men who died crossing the Berezina River.

31. How many men died on the return from Moscow during the time when the temperature dropped from 16°F to −6°F?

32. Of the men who made it to Moscow, how many died on the return trip between Moscow and Botr? (Note that 33,000 men did not go to Moscow, but they joined the returning men who did.)

2-3 Beyond the Basics

33. **a.** Refer to Data Set 20 in Appendix B and construct a histogram for the axial loads of cans that are 0.0111 in. thick. That data set includes an outlier of 504 lb. (An outlier is a value that is far away from the other values.)
    b. Repeat part (a) after excluding the outlier of 504 lb.
    c. How much of an effect does an outlier have on the shape of the histogram?

34. Oscars In "Ages of Oscar-winning Best Actors and Actresses" (*Mathematics Teacher* magazine) by Richard Brown and Gretchen Davis, stem-and-leaf plots are used to compare the ages of actors and actresses at the time they won Oscars. Here are the results for recent winners from each category.

Actors: 32   37   36   32   51   53   33   61   35   45   55   39

             76   37   42   40   32   60   38   56   48   48   40   43

             62   43   42   44   41   56   39   46   31   47   45   60

             46   40   36

Actresses: 50   44   35   80   26   28   41   21   61   38   49   33

             74   30   33   41   31   35   41   42   37   26   34   34

             35   26   61   60   34   24   30   37   31   27   39   34

             26   25   33

Actors' Ages (units)	Stem (tens)	Actresses' Ages (units)
	2	
7 2	3	
	4	4
	5	0
	6	
	7	
	8	

a. Construct a back-to-back stem-and-leaf plot for the given data. The first two ages from each group have been entered in the margin.

b. Using the results from part (a), compare the two different sets of data, and explain any differences.

34. a.

Actors' Ages	Stem	Actresses' Ages
	2	145666678
9987766532221	3	001133344445557789
8876655433221000	4	111249
66531	5	0
2100	6	011
6	7	4
	8	0

b. The ages of actresses when they win Oscars are significantly lower than the ages of actors.

2-4 Measures of Center

Remember that the main objective of this chapter is to master basic tools for measuring and describing different characteristics of a set of data. In Section 2-1 we noted that when describing, exploring, and comparing data sets, these characteristics are usually extremely important: center, variation, distribution, outliers, changes over time. The mnemonic of CVDOT ("Computer Viruses Destroy or Terminate") is helpful for remembering those characteristics. In Sections 2-2 and 2-3 we saw that frequency distributions and graphs, such as histograms, are helpful in investigating distribution. The focus of this section is the characteristic of *center*.

Note to Instructor

This is obviously a very important section. Students must recognize that there are different ways to define an "average" or "center." They should thoroughly understand the mean and median. This is a good time to remind students of the important characteristics of data (center, variation, distribution, outliers, time pattern). Sections 2-2 and 2-3 addressed the nature of the distribution; this section addresses "center."

Definition

A **measure of center** is a value at the center or middle of a data set.

There are several different ways to determine the center, so we have different definitions of measures of center, including the mean, median, mode, and midrange. We begin with the mean.

YOU

Six Degrees of Separation

Social psychologists, historians, political scientists, and communications specialists are interested in "the Small World Problem": Given any two people in the world, how many intermediate links are necessary to connect the two original people? Social psychologist Stanley Milgram conducted an experiment in which subjects tried to contact other target people by mailing an information folder to an acquaintance who they thought would be closer to the target. Among 160 such chains that were initiated, only 44 were completed. The number of intermediate acquaintances varied from 2 to 10, with a median of 5 (or "six degrees of separation"). The experiment has been criticized for including very social subjects and for no adjustments for many lost connections from people with lower incomes. Another mathematical study showed that if the missing chains were completed, the median would be slightly greater than 5.

Mean

The (arithmetic) mean is generally the most important of all numerical measurements used to describe data, and it is what most people call an *average*.

Definition

The **arithmetic mean** of a set of values is the measure of center found by adding the values and dividing the total by the number of values. This measure of center will be used often throughout the remainder of this text, and it will be referred to simply as the **mean.**

This definition can be expressed as Formula 2-1, which uses the Greek letter Σ (uppercase Greek sigma) to indicate that the data values should be added. That is, Σx represents the sum of all data values. The symbol n denotes the **sample size,** which is the number of values in the data set.

Formula 2-1
$$\text{mean} = \frac{\Sigma x}{n}$$

The mean is denoted by $\bar{x}$ (pronounced "x-bar") if the data set is a *sample* from a larger population; if all values of the population are used, then we denote the mean by μ (lowercase Greek mu). (Sample statistics are usually represented by English letters, such as $\bar{x}$, and population parameters are usually represented by Greek letters, such as μ.)

Notation

Σ	denotes the *addition* of a set of values.
x	is the *variable* usually used to represent the individual data values.
n	represents the *number of values* in a *sample*.
N	represents the *number of values* in a *population*.
$\bar{x} = \dfrac{\Sigma x}{n}$	is the mean of a set of *sample* values.
$\mu = \dfrac{\Sigma x}{N}$	is the mean of all values in a *population*.

EXAMPLE Monitoring Lead in Air Listed below are measured amounts of lead (in micrograms per cubic meter, or $\mu g/m^3$) in the air. The Environmental Protection Agency has established an air quality standard for lead: 1.5 $\mu g/m^3$. The measurements shown below were recorded at Building 5 of the World Trade Center site on different days immediately following the destruction caused by the terrorist attacks of September 11, 2001. After the collapse of the two World Trade Center buildings, there was considerable concern about the quality of the air. Find the mean for this sample of measured levels of lead in the air.

$$5.40 \quad 1.10 \quad 0.42 \quad 0.73 \quad 0.48 \quad 1.10$$

SOLUTION   The mean is computed by using Formula 2-1. First add the values, then divide by the number of values:

$$\bar{x} = \frac{\Sigma x}{n} = \frac{5.40 + 1.10 + 0.42 + 0.73 + 0.48 + 1.10}{6} = \frac{9.23}{6} = 1.538$$

The mean lead level is 1.538 μg/m^3. Apart from the value of the mean, it is also notable that the data set includes one value (5.40) that is very far away from the others. It would be wise to investigate such an "outlier." In this case, the lead level of 5.40 μg/m^3 was measured the day after the collapse of the two World Trade Center towers, and the excessive levels of dust and smoke provide a reasonable explanation for such an extreme value.

One disadvantage of the mean is that it is sensitive to every value, so one exceptional value can affect the mean dramatically. The median largely overcomes that disadvantage.

Median

Definition

The **median** of a data set is the measure of center that is the middle value when the original data values are arranged in order of increasing (or decreasing) magnitude. The median is often denoted by $\tilde{x}$ (pronounced "x-tilde").

To find the median, first sort the values (arrange them in order), then follow one of these two procedures:

1. If the number of values is odd, the median is the number located in the exact middle of the list.
2. If the number of values is even, the median is found by computing the mean of the two middle numbers.

Figure 2-10 demonstrates this procedure for finding the median.

EXAMPLE Monitoring Lead in Air Listed below are measured amounts of lead (in μg/m^3) in the air. Find the median for this sample.

$$5.40 \quad 1.10 \quad 0.42 \quad 0.73 \quad 0.48 \quad 1.10$$

SOLUTION   First sort the values by arranging them in order:

$$0.42 \quad 0.48 \quad 0.73 \quad 1.10 \quad 1.10 \quad 5.40$$

continued

Class Size Paradox

There are at least two ways to obtain the mean class size, and they can have very different results. At one college, if we take the numbers of students in 737 classes, we get a mean of 40 students. But if we were to compile a list of the class sizes for each student and use this list, we would get a mean class size of 147. This large discrepancy is due to the fact that there are many students in large classes, while there are few students in small classes. Without changing the number of classes or faculty, we could reduce the mean class size experienced by students by making all classes about the same size. This would also improve attendance, which is better in smaller classes.

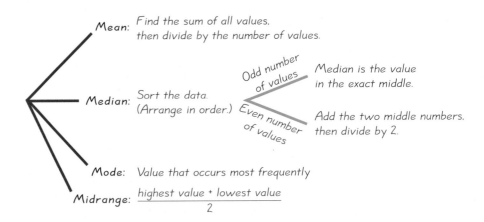

Mean: Find the sum of all values, then divide by the number of values.

Median: Sort the data. (Arrange in order.)

Odd number of values — Median is the value in the exact middle.

Even number of values — Add the two middle numbers, then divide by 2.

Mode: Value that occurs most frequently

Midrange: $\dfrac{\text{highest value + lowest value}}{2}$

FIGURE 2-10 **Procedures for Finding Measures of Center**

Because the number of values is an even number (6), the median is found by computing the mean of the two middle values of 0.73 and 1.10.

$$\text{Median} = \frac{0.73 + 1.10}{2} = \frac{1.83}{2} = 0.915$$

Because the number of values is an even number (6), the median is the number in the exact middle of the sorted list, so the median is 0.915 μg/m^3. Note that the median is very different from the mean of 1.538 μg/m^3 that was found from the same set of sample data in the preceding example. The reason for this large discrepancy is the effect that 5.40 had on the mean. If this extreme value were reduced to 1.20, the mean would drop from 1.538 μg/m^3 to 0.838 μg/m^3, but the median would not change.

EXAMPLE **Monitoring Lead in Air** Repeat the preceding example after including the measurement of 0.66 μg/m^3 recorded on another day. That is, find the median of these lead measurements:

5.40   1.10   0.42   0.73   0.48   1.10   0.66

SOLUTION   First arrange the values in order:

0.42   0.48   0.66   0.73   1.10   1.10   5.40

Because the number of values is an odd number (7), the median is the value in the exact middle of the sorted list: 0.73 μg/m^3.

After studying the preceding two examples, the procedure for finding the median should be clear. Also, it should be clear that the mean is dramatically affected by extreme values, whereas the median is not dramatically affected. Because the median is not so sensitive to extreme values, it is often used for data sets with a relatively small number of extreme values. For example, the U.S. Census Bureau recently reported that the *median* household income is $36,078 annually. The median was used because there is a small number of households with really high incomes.

Mode

Definition

The **mode** of a data set, often denoted by *M,* is the value that occurs most frequently.

- When two values occur with the same greatest frequency, each one is a mode and the data set is **bimodal.**
- When more than two values occur with the same greatest frequency, each is a mode and the data set is said to be **multimodal.**
- When no value is repeated, we say that there is no mode.

EXAMPLE Find the modes of the following data sets.

a. 5.40   1.10   0.42   0.73   0.48   1.10

b. 27   27   27   55   55   55   88   88   99

c. 1   2   3   6   7   8   9   10

SOLUTION

a. The number 1.10 is the mode because it is the value that occurs most often.

b. The numbers 27 and 55 are both modes because they occur with the same greatest frequency. This data set is bimodal because it has two modes.

c. There is no mode because no value is repeated.

In reality, the mode isn't used much with numerical data. But among the different measures of center we are considering, the mode is the only one that can be used with data at the nominal level of measurement. (Recall that the nominal level of measurement applies to data that consist of names, labels, or categories only.) For example, a survey of college students showed that 84% have TVs, 76% have VCRs, 60% have portable CD players, 39% have video game systems, and 35% have DVD players (based on data from the National Center for Education Statistics). Because TVs are most frequent, we can say that the mode is TV. We cannot find a mean or median for such data at the nominal level.

Midrange

Definition

The **midrange** is the measure of center that is the value midway between the highest and lowest values in the original data set. It is found by adding the highest data value to the lowest data value and then dividing the sum by 2, as in the following formula.

$$\text{midrange} = \frac{\text{highest value} + \text{lowest value}}{2}$$

Just an Average Guy

Men's Health magazine published statistics describing the "average guy," who is 34.4 years old, weighs 175 pounds, is about 5 ft 10 in. tall, and is named Mike Smith. The age, weight, and height are all mean values, but the name of Mike Smith is the mode that corresponds to the most common first and last names. Other notable statistics: The average guy sleeps about 6.9 hours each night, drinks about 3.3 cups of coffee each day, and consumes 1.2 alcoholic drinks per day. He earns about $36,100 per year, and has debts of $2563 from two credit cards. He has banked savings of $3100.

Note to Instructor
The midrange can be easily omitted. It is included mainly to emphasize the point that there are different ways of determining the "center" of a set of data.

The midrange is rarely used. Because it uses only the highest and lowest values, it is too sensitive to those extremes. However, the midrange does have three redeeming features: (1) It is easy to compute; (2) it helps to reinforce the important point that there are several different ways to define the center of a data set; (3) it is sometimes incorrectly used for the median, so confusion can be reduced by clearly defining the midrange along with the median.

EXAMPLE Monitoring Lead in Air Listed below are measured amounts of lead (in $\mu g/m^3$) in the air from the site of the World Trade Center on different days after September 11, 2001. Find the midrange for this sample.

$$5.40 \quad 1.10 \quad 0.42 \quad 0.73 \quad 0.48 \quad 1.10$$

SOLUTION The midrange is found as follows:

$$\frac{\text{highest value} + \text{lowest value}}{2} = \frac{5.40 + 0.42}{2} = 2.910$$

The midrange is 2.910 $\mu g/m^3$.

Unfortunately, the term *average* is sometimes used for any measure of center and is sometimes used for the mean. Because of this ambiguity, we should not use the term *average* when referring to a particular measure of center. Instead, we should use the specific term, such as mean, median, mode, or midrange. When encountering a reported value as being an *average*, we should know that the value could be the result of any of several different definitions.

In the spirit of describing, exploring, and comparing data, we provide Table 2-8 which summarizes the different measures of center for the cotinine levels listed in Table 2-1 in the Chapter Problem. Recall that cotinine is a metabolite of nicotine, so that when nicotine is absorbed by the body, cotinine is produced. A comparison of the measures of center suggests that cotinine levels are highest in smokers. Also, the cotinine levels in nonsmokers exposed to smoke are higher than nonsmokers not exposed. This suggests that "secondhand smoke" does have

Table 2-8	Comparison of Cotinine Levels in Smokers, Nonsmokers Exposed to Environmental Tobacco Smoke (ETS), and Nonsmokers Not Exposed to Environmental Tobacco Smoke (NOETS)		
	Smokers	ETS	NOETS
Mean	172.5	60.6	16.4
Median	170.0	1.5	0.0
Mode	1 and 173	1	0
Midrange	245.5	275.5	154.5

an effect. There are methods for determining whether such apparent differences are statistically significant, and we will consider some of those methods later in this text.

Note to Instructor
Try to encourage this good practice: Don't round off values in the middle of a calculation; instead, carry as many places as the calculator will handle. Only round off at the end. Large errors sometimes occur when intermediate results are rounded too much.

Round-Off Rule

A simple rule for rounding answers is this:

Carry one more decimal place than is present in the original set of values.

When applying this rule, round only the final answer, *not intermediate values that occur during calculations.* Thus the mean of 2, 3, 5, is 3.333333 . . . , which is rounded to 3.3. Because the original values are whole numbers, we rounded to the nearest tenth. As another example, the mean of 80.4 and 80.6 is 80.50 (one more decimal place than was used for the original values).

Mean from a Frequency Distribution

When data are summarized in a frequency distribution, we might not know the exact values falling in a particular class. To make calculations possible, we pretend that in each class, all sample values are equal to the class midpoint. Because each class midpoint is repeated a number of times equal to the class frequency, the sum of all sample values is $\Sigma(f \cdot x)$, where f denotes frequency and x represents the class midpoint. The total number of sample values is the sum of frequencies Σf. Formula 2-2 is used to compute the mean when the sample data are summarized in a frequency distribution. Formula 2-2 is not really a new concept; it is simply a variation of Formula 2-1.

Note to Instructor
If pressed for time or if you find that Chapter 2 is taking much longer than you had planned, consider omitting weighted mean and mean from a frequency table, or simply assign a few homework exercises for students to do on their own (such as Exercises 17–20).

First multiply each frequency and class midpoint, then add the products.
↓

Formula 2-2
$$\bar{x} = \frac{\Sigma(f \cdot x)}{\Sigma f}$$
(mean from frequency distribution)

↑
sum of frequencies

For example, see Table 2-9 on the following page. The first two columns duplicate the frequency distribution (Table 2-2) for the cotinine levels of smokers. Table 2-9 illustrates the procedure for using Formula 2-2 when calculating a mean from data summarized in a frequency distribution. In reality, software or calculators are generally used in place of manual calculations. Table 2-9 results in $\bar{x} = 177.0$, but we get $\bar{x} = 172.5$ if we use the original list of 40 values. Remember, the frequency distribution yields an approximation of $\bar{x}$, because it is not based on the exact original list of sample values.

Not At Home

Pollsters cannot simply ignore those who were not at home when they were called the first time. One solution is to make repeated call-back attempts until the person can be reached. Alfred Politz and Willard Simmons describe a way to compensate for those missing results without making repeated callbacks. They suggest weighting results based on how often people are not at home. For example, a person at home only two days out of six will have a 2/6 or 1/3 probability of being at home when called the first time. When such a person is reached the first time, his or her results are weighted to count three times as much as someone who is always home. This weighting is a compensation for the other similar people who are home two days out of six and were not at home when called the first time. This clever solution was first presented in 1949.

Table 2-9	Finding the Mean from a Frequency Distribution		
Cotinine Level	Frequency f	Class Midpoint x	$f \cdot x$
0–99	11	49.5	544.5
100–199	12	149.5	1794.0
200–299	14	249.5	3493.0
300–399	1	349.5	349.5
400–499	2	449.5	899.0
Totals:	$\Sigma f = 40$		$\Sigma(f \cdot x) = 7080.0$

$$\bar{x} = \frac{\Sigma(f \cdot x)}{\Sigma f} = \frac{7080}{40} = 177.0$$

Weighted Mean

In some cases, the values vary in their degree of importance, so we may want to weight them accordingly. We can then proceed to compute a **weighted mean,** which is a mean computed with the different values assigned different weights, as shown in Formula 2-3.

Formula 2-3   weighted mean: $\bar{x} = \dfrac{\Sigma(w \cdot x)}{\Sigma w}$

For example, suppose we need a mean of three test scores (85, 90, 75), but the first test counts for 20%, the second test counts for 30%, and the third test counts for 50% of the final grade. We can assign weights of 20, 30, and 50 to the test scores, then proceed to calculate the mean by using Formula 2-3 as follows:

$$\bar{x} = \frac{\Sigma(w \cdot x)}{\Sigma w}$$
$$= \frac{(20 \times 85) + (30 \times 90) + (50 \times 75)}{20 + 30 + 50} = \frac{8150}{100} = 81.5$$

As another example, college grade-point averages can be computed by assigning each letter grade the appropriate number of points (A = 4, B = 3, etc.), then assigning to each number a weight equal to the number of credit hours. Again, Formula 2-3 can be used to compute the grade-point average.

The Best Measure of Center

So far, we have considered the mean, median, mode, and midrange as measures of center. Which one of these is best? Unfortunately, there is no single best answer to that question because there are no objective criteria for determining the most representative measure for all data sets. The different measures of center have different advantages and disadvantages, some of which are summarized in Table 2-10.

Table 2-10	Comparison of Mean, Median, Mode, and Midrange					
Measure of Center	Definition	How Common?	Existence	Takes Every Value into Account?	Affected by Extreme Values?	Advantages and Disadvantages
Mean	$\bar{x} = \dfrac{\Sigma x}{n}$	most familiar "average"	always exists	yes	yes	used throughout this book; works well with many statistical methods
Median	middle value	commonly used	always exists	no	no	often a good choice if there are some extreme values
Mode	most frequent data value	sometimes used	might not exist; may be more than one mode	no	no	appropriate for data at the nominal level
Midrange	$\dfrac{\text{high} + \text{low}}{2}$	rarely used	always exists	no	yes	very sensitive to extreme values

General comments:
- For a data collection that is approximately symmetric with one mode, the mean, median, mode, and midrange tend to be about the same.
- For a data collection that is obviously asymmetric, it would be good to report both the mean and median.
- The mean is relatively *reliable*. That is, when samples are drawn from the same population, the sample means tend to be more consistent than the other measures of center (consistent in the sense that the means of samples drawn from the same population don't vary as much as the other measures of center).

An important advantage of the mean is that it takes every value into account, but an important disadvantage is that it is sometimes dramatically affected by a few extreme values. This disadvantage can be overcome by using a trimmed mean, as described in Exercise 21.

Skewness

A comparison of the mean, median, and mode can reveal information about the characteristic of skewness, defined below and illustrated in Figure 2-11.

Definition

A distribution of data is **skewed** if it is not symmetric and extends more to one side than the other. (A distribution of data is **symmetric** if the left half of its histogram is roughly a mirror image of its right half.)

FIGURE 2-11 **Skewness**

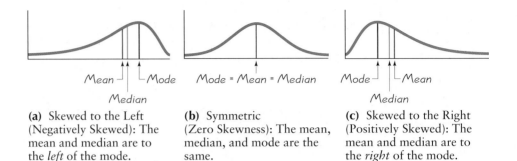

(a) Skewed to the Left (Negatively Skewed): The mean and median are to the *left* of the mode.

(b) Symmetric (Zero Skewness): The mean, median, and mode are the same.

(c) Skewed to the Right (Positively Skewed): The mean and median are to the *right* of the mode.

Data **skewed to the left** (also called *negatively skewed*) have a longer left tail, and the mean and median are to the left of the mode. Although not always predictable, data skewed to the left generally have a mean less than the median, as in Figure 2-11(a). Data **skewed to the right** (also called *positively skewed*) have a longer right tail, and the mean and median are to the right of the mode. Again, although not always predictable, data skewed to the right generally have the mean to the right of the median, as in Figure 2-11(c).

If we examine the histogram in Figure 2-1 for the cotinine levels of smokers, we see a graph that appears to be skewed to the right. In practice, many distributions of data are symmetric and without skewness. Distributions skewed to the right are more common than those skewed to the left because it's often easier to get exceptionally large values than values that are exceptionally small. With annual incomes, for example, it's impossible to get values below the lower limit of zero, but there are a few people who earn millions of dollars in a year. Annual incomes therefore tend to be skewed to the right, as in Figure 2-11(c).

Using Technology

The calculations of this section are fairly simple, but some of the calculations in the following sections require more effort. Many computer software programs allow you to enter a data set and use one operation to get several different sample statistics, referred to as *descriptive statistics*. (See Section 2-6 for sample displays resulting from STATDISK, Minitab, Excel, and the TI-83 Plus calculator.) Here are some of the procedures for obtaining such displays.

STATDISK Enter the data in the Data Window. Click on **Data** and select **Descriptive Statistics.** Click on **Evaluate** to get the various descriptive statistics, including the mean, median, midrange, and other statistics to be discussed in the following sections.

Minitab Enter the data in the column with the heading C1. Click on **Stat,** select **Basic Statistics,** then select **Display Descriptive Statistics.** The results will include the mean and median as well as other statistics.

Excel Enter the sample data in column A. Select **Tools,** then **Data Analysis,** then select **Descriptive Statistics** and click **OK.** In the dialog box, enter the input range (such as A1:A40 for 40 values in column A), click on **Summary Statistics,** then click **OK.** (If Data Analysis does not appear in the Tools menu, it must be installed by clicking on **Tools** and selecting **Add-Ins.**)

TI-83 Plus First enter the data in list L1 by pressing **STAT,** then selecting **Edit** and pressing the **ENTER** key. After the data values have been entered, press **STAT** and select **CALC,** then select **1-Var Stats** and press the **ENTER** key twice. The display will include the mean $\bar{x}$, the median, the minimum value, and the maximum value. Use the down-arrow key to view the results that don't fit on the initial display.

2-4 Basic Skills and Concepts

In Exercises 1–8, find the (a) mean, (b) median, (c) mode, and (d) midrange for the given sample data.

1. **Tobacco Use in Children's Movies** In "Tobacco and Alcohol Use in G-Rated Children's Animated Films," by Goldstein, Sobel, and Newman (*Journal of the American Medical Association,* Vol. 281, No. 12), the lengths (in seconds) of scenes showing tobacco use were recorded for animated movies from Universal Studios. The first six values included in Data Set 7 from Appendix B are listed below. Is there any problem with including scenes of tobacco use in animated children's films?

0   223   0   176   0   548

2. **Harry Potter** In an attempt to measure the reading level of a book, the Flesch Reading Ease ratings are obtained for 12 randomly selected pages from *Harry Potter and the Sorcerer's Stone* by J. K. Rowling. Those values, included in Data Set 14 from Appendix B, are listed below. Given that these ratings are based on 12 randomly selected pages, is the mean of this sample likely to be a reasonable estimate of the mean reading level of the whole book?

85.3   84.3   79.5   82.5   80.2   84.6
79.2   70.9   78.6   86.2   74.0   83.7

3. **Cereal** A dietitian obtains the amounts of sugar (in grams) from 1 gram in each of 16 different cereals, including Cheerios, Corn Flakes, Fruit Loops, Trix, and 12 others. Those values, included in Data Set 16 from Appendix B, are listed below. Is the mean of those values likely to be a good estimate of the mean amount of sugar in each gram of cereal consumed by the population of all Americans who eat cereal? Why or why not?

0.03   0.24   0.30   0.47   0.43   0.07   0.47   0.13
0.44   0.39   0.48   0.17   0.13   0.09   0.45   0.43

4. **Body Mass Index** As part of the National Health Examination, the body mass index is measured for a random sample of women. Some of the values included in Data Set 1 from Appendix B are listed below. Is the mean of this sample reasonably close to the mean of 25.74, which is the mean for all 40 women included in Data Set 1?

19.6   23.8   19.6   29.1   25.2   21.4   22.0   27.5
33.5   20.6   29.9   17.7   24.0   28.9   37.7

5. **Drunk Driving** The blood alcohol concentrations of a sample of drivers involved in fatal crashes and then convicted with jail sentences are given below (based on data from the U.S. Department of Justice). Given that current state laws prohibit driving with levels above 0.08 or 0.10, does it appear that these levels are significantly above the maximum that is allowed?

0.27   0.17   0.17   0.16   0.13   0.24   0.29   0.24
0.14   0.16   0.12   0.16   0.21   0.17   0.18

6. **Motorcycle Fatalities** Listed below are ages of motorcyclists when they were fatally injured in traffic crashes (based on data from the U.S. Department of Transportation). Do the results support the common belief that such fatalities are incurred by a greater proportion of younger drivers?

17   38   27   14   18   34   16   42   28
24   40   20   23   31   37   21   30   25

Recommended Assignment
Exercises 1–4, 9–12, 17, 18

1. $\bar{x} = 157.8$ sec; median = 88.0 sec; mode = 0 sec; midrange = 274.0 sec.
   Yes, young children should not be influenced by exposure to tobacco use.
2. $\bar{x} = 80.75$; median = 81.35; mode = none; midrange = 78.55.
   Yes, because each of the 12 pages contains many words and sentences.
3. $\bar{x} = 0.295$ g; median = 0.345 g; mode: 0.13 g, 0.43 g, 0.47 g; midrange = 0.255 g.
   Not necessarily. There are other cereals not listed, and Americans might consume much more of some brands than others.
4. $\bar{x} = 25.37$; median = 24.00; mode = 19.6; midrange = 27.70.
   Yes.
5. $\bar{x} = 0.187$ g; median = 0.170; mode: 0.16, 0.17; midrange = 0.205.
   Yes.
6. $\bar{x} = 26.9$; median = 26.0; mode = none; midrange = 28.0.
   Yes.
7. $\bar{x} = 18.3$; median = 18.0; mode = 17; midrange = 18.0.
   The results are very consistent, so the mean should be a good estimate.
8. $\bar{x} = 667.63$ mg; median = 668.60 mg; mode = 672.2 mg; midrange = 666.70 mg.
   Some tablets would be harmful because they contain too much aspirin, and others would be ineffective because they don't contain enough aspirin.
9. Jefferson Valley: $\bar{x} = 7.15$ min; median = 7.20 min; mode = 7.7 min; midrange = 7.10 min.
   Providence: same results as Jefferson Valley.
   Although the measures of center are the same, the Providence times are much more varied than the Jefferson Valley times.

10. Regular: $\bar{x}$ = 0.81907 lb; median = 0.81865 lb; mode: none;
midrange = 0.81985 lb.
Diet: $\bar{x}$ = 0.78333 lb;
median = 0.78525 lb; mode:
none; midrange = 0.78270 lb.
Diet appears to weigh less because it does not have the
sugar in regular Coke.

11. McDonald's: $\bar{x}$ = 186.3 sec;
median = 184.0 sec; mode =
none; midrange = 189.5 sec.
Jack in the Box: $\bar{x}$ = 262.5 sec;
median = 262.5 sec; mode =
109 sec; midrange = 277.5 sec.
McDonald's appears to be
significantly faster.

12. 4000 B.C.: $\bar{x}$ = 128.7;
median = 128.5; mode = 131;
midrange = 128.5.
150 A.D.: $\bar{x}$ = 133.3;
median = 133.5; mode = 126;
midrange = 133.5.
The head sizes appear to have
become larger, but it is difficult
to determine whether the difference is significant.

13. Males: $\bar{x}$ = 41.10 cm;
median = 41.10 cm
Females: $\bar{x}$ = 40.05 cm;
median = 40.20 cm
There does appear to be a small
difference.

14. Clancy: $\bar{x}$ = 6.50;
median = 5.75
Rowling: $\bar{x}$ = 5.08;
median = 5.05
Tolstoy: $\bar{x}$ = 8.43;
median = 8.30
Yes, Tolstoy appears to have the
highest rating and Rowling appears to have the lowest. No
big surprise there.

15. Thursday: $\bar{x}$ = 0.069 in.;
median = 0.000 in.
Sunday: $\bar{x}$ = 0.068 in.;
median = 0.000 in.
There does not appear to be a
substantial difference.

7. **Reaction Times** The author visited the Reuben H. Fleet Science Museum in San Diego and repeated an experiment of reaction times. The following times (in hundredths of a second) were obtained. How consistent are these results, and how does the consistency affect the usefulness of the sample mean as an estimate of the population mean?

19	20	17	21	21	21	19	18	19	19
17	17	15	17	18	17	18	18	18	17

8. **Bufferin Tablets** Listed below are the measured weights (in milligrams) of a sample of Bufferin aspirin tablets. What is a serious consequence of having weights that vary too much?

672.2	679.2	669.8	672.6	672.2	662.2
662.7	661.3	654.2	667.4	667.0	670.7

In Exercises 9–12, find the mean, median, mode, and midrange for each of the two samples, then compare the two sets of results.

9. **Customer Waiting Times** Waiting times (in minutes) of customers at the Jefferson Valley Bank (where all customers enter a single waiting line) and the Bank of Providence (where customers wait in individual lines at three different teller windows):

Jefferson Valley:	6.5	6.6	6.7	6.8	7.1	7.3	7.4	7.7	7.7	7.7
Providence:	4.2	5.4	5.8	6.2	6.7	7.7	7.7	8.5	9.3	10.0

Interpret the results by determining whether there is a difference between the two data sets that is not apparent from a comparison of the measures of center. If so, what is it?

10. **Regular Coke/Diet Coke** Weights (pounds) of samples of the contents in cans of regular Coke and diet Coke:

Regular:	0.8192	0.8150	0.8163	0.8211	0.8181	0.8247
Diet:	0.7773	0.7758	0.7896	0.7868	0.7844	0.7861

Does there appear to be a significant difference between the two data sets? How might such a difference be explained?

11. **Mickey D vs. Jack** When investigating times required for drive-through service, the following results (in seconds) are obtained (based on data from QSR Drive-Thru Time Study).

McDonald's:	287	128	92	267	176	240	192	118	153	254	193	136
Jack in the Box:	190	229	74	377	300	481	428	255	328	270	109	109

Which of the two fast-food giants appears to be faster? Does the difference appear to be significant?

12. **Skull Breadths** Maximum breadth of samples of male Egyptian skulls from 4000 B.C. and 150 A.D. (based on data from *Ancient Races of the Thebaid* by Thomson and Randall-Maciver):

4000 B.C.:	131	119	138	125	129	126	131	132	126	128	128	131
150 A.D.:	136	130	126	126	139	141	137	138	133	131	134	129

Changes in head sizes over time suggest interbreeding with people from other regions. Do the head sizes appear to have changed from 4000 B.C. to 150 A.D.?

In Exercises 13–16, refer to the data set in Appendix B. Use computer software or a calculator to find the means and medians, then compare the results as indicated.

T **13.** Head Circumferences In order to correctly diagnose the disorder of hydrocephalus, a pediatrician investigates head circumferences of 2-year-old males and females. Use the sample results listed in Data Set 3. Does there appear to be a difference between the two genders?

T **14.** Clancy, Rowling, Tolstoy A child psychologist investigates differences in reading difficulty and obtains data from *The Bear and the Dragon* by Tom Clancy, *Harry Potter and the Sorcerer's Stone* by J. K. Rowling, and *War and Peace* by Leo Tolstoy. Refer to Data Set 14 in Appendix B and use the Flesch-Kincaid Grade Level ratings for 12 pages randomly selected from each of the three books. Do the ratings appear to be different?

T **15.** Weekend Rainfall Using Data Set 11 in Appendix B, find the mean and median of the rainfall amounts in Boston on Thursday and find the mean and median of the rainfall amounts in Boston on Sunday. Media reports claimed that it rains more on weekends than during the week. Do these results appear to support that claim?

T **16.** Tobacco/Alcohol Use in Children's Movies In "Tobacco and Alcohol Use in G-Rated Children's Animated Films," by Goldstein, Sobel, and Newman (*Journal of the American Medical Association,* Vol. 281, No. 12), the lengths (in seconds) of scenes showing tobacco use and alcohol use were recorded for animated children's movies. Refer to Data Set 7 in Appendix B and find the mean and median for the tobacco times, then find the mean and median for the alcohol times. Does there appear to be a difference between those times? Which appears to be the larger problem: scenes showing tobacco use or scenes showing alcohol use?

In Exercises 17–20, find the mean of the data summarized in the given frequency distribution.

17. 74.4 min

17. Old Faithful Visitors to Yellowstone National Park consider an eruption of the Old Faithful geyser to be a major attraction that should not be missed. The given frequency distribution summarizes a sample of times (in minutes) between eruptions.

18. Loaded Die The author drilled a hole in a die and filled it with a lead weight, then proceeded to roll it 200 times. The results are given in the frequency distribution in the margin. Does the result appear to be very different from the result that would be expected with an unmodified die? 3.5; no

19. Speeding Tickets The given frequency distribution describes the speeds of drivers ticketed by the Town of Poughkeepsie police. These drivers were traveling through a 30 mi/h speed zone on Creek Road, which passes the authors' college. How does the mean compare to the posted speed limit of 30 mi/h?
19. 46.8 mi/h; the mean is significantly higher than the posted limit of 30 mi/h.

20. Body Temperatures The accompanying frequency distribution summarizes a sample of human body temperatures. (See the temperatures for midnight on the second day, as listed in Data Set 4 in Appendix B.) How does the mean compare to the value of 98.6°F, which is the value assumed to be the mean by most people?
20. 98.17°F; the mean appears to be substantially lower than 98.6°F.

2-4 Beyond the Basics

T **21.** Trimmed Mean Because the mean is very sensitive to extreme values, we say that it is not a *resistant* measure of center. The **trimmed mean** is more resistant. To find the

16. Tobacco: $\bar{x} = 57.4$ sec; median = 5.5 sec
Alcohol: $\bar{x} = 32.5$ sec; median = 1.5 sec
The tobacco times appear to have a significantly larger mean, so the depiction of tobacco use appears to be the larger problem.

Table for Exercise 17

Time	Frequency
40–49	8
50–59	44
60–69	23
70–79	6
80–89	107
90–99	11
100–109	1

Table for Exercise 18

Outcome	Frequency
1	27
2	31
3	42
4	40
5	28
6	32

Table for Exercise 19

Speed	Frequency
42–45	25
46–49	14
50–53	7
54–57	3
58–61	1

Table for Exercise 20

Temperature	Frequency
96.5–96.8	1
96.9–97.2	8
97.3–97.6	14
97.7–98.0	22
98.1–98.4	19
98.5–98.8	32
98.9–99.2	6
99.3–99.6	4

10% trimmed mean for a data set, first arrange the data in order, then delete the bottom 10% of the values and the top 10% of the values, then calculate the mean of the remaining values. For the weights of the bears in Data Set 9 from Appendix B, find (a) the mean; (b) the 10% trimmed mean; (c) the 20% trimmed mean. How do the results compare?

22. Mean of Means Using an almanac, a researcher finds the mean teacher's salary for each state. He adds those 50 values, then divides by 50 to obtain their mean. Is the result equal to the national mean teacher's salary? Why or why not?

23. Degrees of Freedom Ten values have a mean of 75.0. Nine of the values are 62, 78, 90, 87, 56, 92, 70, 70, and 93.
   a. Find the tenth value.
   b. We need to create a list of n values that have a specific known mean. We are free to select any values we desire for some of the n values. How many of the n values can be freely assigned before the remaining values are determined?

24. Censored Data An experiment is conducted to test the lives of car batteries. The experiment is run for a fixed time of five years. (The test is said to be *censored* at five years.) The sample results (in years) are 2.5, 3.4, 1.2, 5+, 5+ (where 5+ indicates that the battery was still working at the end of the experiment). What can you conclude about the mean battery life?

25. Weighted Mean Kelly Bell gets quiz grades of 65, 83, 80, and 90. She gets a 92 on her final exam. Find the weighted mean if the quizzes each count for 15% and the final counts for 40% of the final grade.

26. Transformed Data In each of the following, describe how the mean, median, mode, and midrange of a data set are affected.
   a. The same constant k is added to each value of the data set.
   b. Each value of the data set is multiplied by the same constant k.

27. The **harmonic mean** is often used as a measure of central tendency for data sets consisting of rates of change, such as speeds. It is found by dividing the number of values n by the sum of the *reciprocals* of all values, expressed as

$$\frac{n}{\sum \frac{1}{x}}$$

(No value can be zero.) Four students drive from New York to Florida (1200 miles) at a speed of 40 mi/h (yeah, right!). Because they need to make it to statistics class on time, they return at a speed of 60 mi/h. What is their average speed for the round trip? (The harmonic mean is used in averaging speeds.)

28. The **geometric mean** is often used in business and economics for finding average rates of change, average rates of growth, or average ratios. Given n values (all of which are positive), the geometric mean is the nth root of their product. The *average growth factor* for money compounded at annual interest rates of 10%, 8%, 9%, 12%, and 7% can be found by computing the geometric mean of 1.10, 1.08, 1.09, 1.12, and 1.07. Find that average growth factor.

29. The **quadratic mean** (or **root mean square**, or **RMS**) is usually used in physical applications. In power distribution systems, for example, voltages and currents are usually referred to in terms of their RMS values. The quadratic mean of a set of values is obtained by squaring each value, adding the results, dividing by the number of values n, and then taking the square root of that result, expressed as

$$\text{quadratic mean} = \sqrt{\frac{\Sigma x^2}{n}}$$

Find the RMS of these power supplies (in volts): 110, 0, −60, 12.

30. *Median* When data are summarized in a frequency distribution, the median can be found by first identifying the *median class* (the class that contains the median). We then assume that the values in that class are evenly distributed and we can interpolate. This process can be described by

$$(\text{lower limit of median class}) + (\text{class width})\left(\frac{\left(\dfrac{n+1}{2}\right) - (m+1)}{\text{frequency of median class}}\right)$$

where n is the sum of all class frequencies and m is the sum of the class frequencies that *precede* the median class. Use this procedure to find the median of the data set summarized in Table 2-2. How does the result compare to the median of the original list of data, which is 170? Which value of the median is better: the value computed for the frequency table or the value of 170?

2-5 Measures of Variation

Study Hint: Because this section introduces the concept of variation, which is so important in statistics, this is one of the most important sections in the entire book. First read through this section quickly and gain a general understanding of the characteristic of variation. Next, learn how to obtain measures of variation, especially the standard deviation. Finally, try to understand the reasoning behind the formula for standard deviation, but do not spend much time memorizing formulas or doing arithmetic calculations. Instead, place a high priority on learning how to *interpret* values of standard deviation.

For a visual illustration of variation, see Figure 2-12 on page 74, which includes samples of bolts from two different companies. Because these bolts are to be used for attaching wings to airliners, their quality is quite important. If we consider only the mean, we would not recognize any difference between the two samples, because they both have a mean of $\bar{x} = 2.000$ in. However, it should be obvious that the samples are very different in the amounts that the bolts *vary* in length. The bolts manufactured by the Precision Bolt Company appear to be very similar in length, whereas the lengths from the Ruff Bolt Company vary by large amounts. This is exactly the same issue that is so important in many different manufacturing processes. Better quality is achieved through lower variation. In this section, we want to develop the ability to measure and understand variation.

Another ideal situation illustrating the importance of variation can be seen in the waiting lines at banks. In times past, many banks required that customers wait in separate lines at each teller's window, but most have now changed to one single main waiting line. Why did they make that change? The mean waiting time didn't change, because the waiting-line configuration doesn't affect the efficiency of the

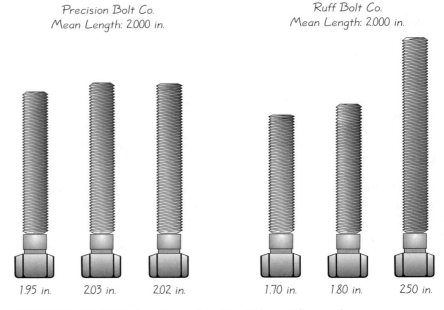

Precision Bolt Co.
Mean Length: 2.000 in.

Ruff Bolt Co.
Mean Length: 2.000 in.

1.95 in.   2.03 in.   2.02 in.   1.70 in.   1.80 in.   2.50 in.

FIGURE 2-12 **Bolts Manufactured by Two Different Companies**

tellers. They changed to the single line because customers prefer waiting times that are more *consistent* with less variation. Thousands of banks made a change that resulted in lower variation (and happier customers), even though the mean was not affected. Let's consider some specific waiting times (in minutes) of bank customers.

Salem Bank (single waiting line)	4	7	7
Mulberry Bank (multiple waiting lines)	1	3	14

It is easy to find that $\bar{x} = 6.0$ for both data sets. It is also easy to see through visual inspection that the waiting times of 4, 7, 7 vary much less than the waiting times of 1, 3, 14. Let's now proceed to develop some specific ways of actually measuring variation so that we can use specific numbers instead of subjective judgments. We begin with the range.

Range

Definition

The **range** of a set of data is the difference between the highest value and the lowest value.

$$\text{range} = \text{(highest value)} - \text{(lowest value)}$$

To compute the range, simply subtract the lowest value from the highest value. For the Salem Bank customers, the range is $7 - 4 = 3$ min. The Mulberry Bank has waiting times with a range of 13 min, and this larger value suggests greater variation.

The range is very easy to compute, but because it depends on only the highest and the lowest values, it isn't as useful as the other measures of variation that use every value. (See Exercise 35 for an example in which the range is misleading.)

Standard Deviation of a Sample

The standard deviation is the measure of variation that is generally the most important and useful. We define the standard deviation now, but in order to understand it fully you will need to study the subsection Interpreting and Understanding Standard Deviation found later in this section (see page 81).

Definition

The **standard deviation** of a set of sample values is a measure of variation of values about the mean. It is a type of average deviation of values from the mean that is calculated by using Formulas 2-4 or 2-5.

Formula 2-4
$$s = \sqrt{\frac{\Sigma(x - \bar{x})^2}{n - 1}}$$
sample standard deviation

Formula 2-5
$$s = \sqrt{\frac{n\Sigma(x^2) - (\Sigma x)^2}{n(n - 1)}}$$
shortcut formula for sample standard deviation

Later in this section we discuss the rationale for these formulas, but for now we recommend that you use Formula 2-4 for a few examples, then learn how to find standard deviation values using your calculator and by using a software program. (Most scientific calculators are designed so that you can enter a list of values and automatically get the standard deviation.) For now, we cite important properties that are consequences of the way in which the standard deviation is defined.

- The standard deviation is a measure of variation of all values from the *mean*.
- The value of the standard deviation s is usually positive. It is zero only when all of the data values are the same number. Also, larger values of s indicate greater amounts of variation.
- The value of the standard deviation s can increase dramatically with the inclusion of one or more outliers (data values that are very far away from all of the others).
- The units of the standard deviation s (such as minutes, feet, pounds, and so on) are the same as the units of the original data values.

More Stocks, Less Risk

In their book *Investments,* authors Zvi Bodie, Alex Kane, and Alan Marcus state that "the average standard deviation for returns of portfolios composed of only one stock was 0.554. The average portfolio risk fell rapidly as the number of stocks included in the portfolio increased." They note that with 32 stocks, the standard deviation is 0.325, indicating much less variation and risk. They make the point that with only a few stocks, a portfolio has a high degree of "firm-specific" risk, meaning that the risk is attributable to the few stocks involved. With more than 30 stocks, there is very little firm-specific risk; instead, almost all of the risk is "market risk," attributable to the stock market as a whole. They note that these principles are "just an application of the well-known law of averages."

Note to Instructor
Fewer instructors now require that their students show all steps for manually calculating a standard deviation. Most others simply require that students be able to compute values of standard deviations on their calculators. Suggestion: Fully develop the concept of standard deviation in class and show how it is calculated, but then move on and allow students to obtain values from their calculators. Strongly emphasize the *interpretation* of those values; see the subsection of *Interpreting and Understanding Standard Deviation*, found later in the section.

Procedure for Finding the Standard Deviation with Formula 2-4

Step 1:   Compute the mean $\bar{x}$.

Step 2:   Subtract the mean from each individual value to get a list of deviations of the form $(x - \bar{x})$.

Step 3:   Square each of the differences obtained from Step 2. This produces numbers of the form $(x - \bar{x})^2$.

Step 4:   Add all of the squares obtained from Step 3. This is the value of $\Sigma(x - \bar{x})^2$.

Step 5:   Divide the total from Step 4 by the number $(n - 1)$; that is, 1 less than the total number of values present.

Step 6:   Find the square root of the result of Step 5.

> **EXAMPLE Using Formula 2-4** Use Formula 2-4 to find the standard deviation of the Mulberry Bank customer waiting times. Those times (in minutes) are 1, 3, 14.
>
> **SOLUTION**   We will use the six steps in the procedure just given. Refer to those steps and refer to Table 2-11, which shows the detailed calculations.
>
> Step 1:   Obtain the mean of 6.0 by adding the values and then dividing by the number of values:
>
> $$\bar{x} = \frac{\Sigma x}{n} = \frac{18}{3} = 6.0 \text{ min}$$
>
> Step 2:   Subtract the mean of 6.0 from each value to get these values of $(x - \bar{x})$: $-5, -3, 8$.
>
> Step 3:   Square each value obtained in Step 2 to get these values of $(x - \bar{x})^2$: 25, 9, 64.
>
> Step 4:   Sum all of the preceding values to get the value of
>
> $$\Sigma(x - \bar{x})^2 = 98$$
>
> Step 5:   With $n = 3$ values, divide by 1 less than 3:
>
> $$\frac{98}{2} = 49.0$$
>
> Step 6:   Find the square root of 49.0. The standard deviation is
>
> $$\sqrt{49.0} = 7.0 \text{ min}$$

Ideally, we would now interpret the meaning of the result, but such interpretations will be discussed later in this section.

Table 2-11	Calculating Standard Deviation for Mulberry Bank Customer Times	
x	$x - \bar{x}$	$(x - \bar{x})^2$
1	-5	25
3	-3	9
14	8	64
Totals: 18		98

$$\bar{x} = \frac{18}{3} = 6.0 \text{ min} \qquad s = \sqrt{\frac{98}{3-1}} = \sqrt{49} = 7.0 \text{ min}$$

EXAMPLE Using Formula 2-5 The preceding example used Formula 2-4 for finding the standard deviation of the Mulberry Bank customer waiting times. Using the same data set, find the standard deviation by using Formula 2-5.

SOLUTION   Formula 2-5 requires that we first find values for n, Σx, and Σx^2.

$n = 3$ $\qquad$ (because there are 3 values in the sample)

$\Sigma x = 18$ $\qquad$ (found by adding the 3 sample values)

$\Sigma x^2 = 206$ $\qquad$ (found by adding the squares of the sample values, as in $1^2 + 3^2 + 14^2$)

Using Formula 2-5, we get

$$s = \sqrt{\frac{n(\Sigma x^2) - (\Sigma x)^2}{n(n-1)}} = \sqrt{\frac{3(206) - (18)^2}{3(3-1)}} = \sqrt{\frac{294}{6}} = 7.0 \text{ min}$$

A good activity is to stop here and calculate the standard deviation of the waiting times for Salem Bank. Follow the same procedures used in the preceding two examples and verify that, for Salem Bank, $s = 1.7$ min. (It will also become important to develop an ability to obtain values of standard deviations by using a calculator and software.) Although the interpretations of these standard deviations will be discussed later, we can now compare them to see that the standard deviation of the times for Salem Bank (1.7 min) is much lower than the standard deviation for Mulberry Bank (7.0 min). This supports our subjective conclusion that the waiting times at Salem Bank have much less variation than the times from Mulberry Bank.

Where Are the 0.400 Hitters?

The last baseball player to hit above 0.400 was Ted Williams, who hit 0.406 in 1941. There were averages above 0.400 in 1876, 1879, 1887, 1894, 1895, 1896, 1897, 1899, 1901, 1911, 1920, 1922, 1924, 1925, and 1930, but none since 1941. Are there no longer great hitters? The late Stephen Jay Gould of Harvard University noted that the mean batting average has been steady at 0.260 for about 100 years, but the standard deviation has been decreasing from 0.049 in the 1870s to 0.031, where it is now. He argued that today's stars are as good as those from the past, but consistently better pitchers now keep averages below 0.400.

Standard Deviation of a Population

In our definition of standard deviation, we referred to the standard deviation of *sample* data. A slightly different formula is used to calculate the standard deviation σ (lowercase Greek sigma) of a *population:* Instead of dividing by $n - 1$, divide by the population size N, as in the following expression:

$$\sigma = \sqrt{\frac{\Sigma(x - \mu)^2}{N}} \qquad \text{population standard deviation}$$

Because we generally deal with sample data, we will usually use Formula 2-4, in which we divide by $n - 1$. Many calculators give both the sample standard deviation and the population standard deviation, but they use a variety of different notations. Be sure to identify the notation used by your calculator, so that you get the correct result. (The TI-83 Plus uses Sx for the sample standard deviation and σx for the population standard deviation.)

Variance of a Sample and Population

We are using the term *variation* as a general description of the amount that values vary among themselves. (The terms *dispersion* and *spread* are sometimes used instead of *variation*.) The term *variance* refers to a specific definition.

> ### Definition
>
> The **variance** of a set of values is a measure of variation equal to the square of the standard deviation.
>
> Sample variance: Square of the standard deviation s.
> Population variance: Square of the population standard deviation σ.

The sample variance s^2 is said to be an **unbiased estimator** of the population variance σ^2, which means that values of s^2 tend to target the value of σ^2 instead of systematically tending to overestimate or underestimate σ^2. (See Exercise 41.)

EXAMPLE **Finding Variance** In the preceding example, we used the Mulberry Bank customer waiting times to find that the standard deviation is given by $s = 7.0$ min. Find the variance of that same sample.

SOLUTION   Because the variance is the square of the standard deviation, we get the result shown below. Note that the original data values are in units of minutes and the standard deviation is 7.0 minutes; the variance is given in units of min^2.

$$\text{sample variance} = s^2 = 7.0^2 = 49.0 \text{ min}^2$$

The variance is an important statistic used in some important statistical methods, such as analysis of variance discussed in Chapter 11. For our present purposes, the variance has this serious disadvantage: *The units of variance are dif-*

ferent than the units of the original data set. For example, if the original customer waiting times are in minutes, the units of the variance are in square minutes (min^2). What is a square minute? (Have some fun constructing a creative answer to that question.) Because the variance uses different units, it is extremely difficult to understand variance by relating it to the original data set. Because of this property, we will focus on the standard deviation as we try to develop an understanding of variation.

We now present the notation and round-off rule we are using.

Notation

s = *sample* standard deviation
s^2 = *sample* variance

σ = *population* standard deviation
σ^2 = *population* variance

Note: Articles in professional journals and reports often use SD for standard deviation and VAR for variance.

Round-Off Rule

We use the same round-off rule given in Section 2-4:

Carry one more decimal place than is present in the original set of data.

Round only the final answer, not values in the middle of a calculation. (If it becomes absolutely necessary to round in the middle, carry at least twice as many decimal places as will be used in the final answer.)

Note to Instructor
This is the same round-off rule used for measures of center. You might ask: "What is wrong with collecting heights of people (such as 65 in., 72 in., 60 in., and so on) and reporting that the mean is 66.234572 in. while the standard deviation is 2.5642217 in.?" That degree of precision is not justified by the original data set, and it is misleading.

Comparing Variation in Different Populations

We stated earlier that because the units of the standard deviation are the same as the units of the original data, it is easier to understand the standard deviation than the variance. However, that same property makes it difficult to compare variation for values taken from different populations. The *coefficient of variation* overcomes this disadvantage.

Definition

The **coefficient of variation** (or **CV**) for a set of sample or population data, expressed as a percent, describes the standard deviation relative to the mean, and is given by the following:

Sample	Population
$CV = \dfrac{s}{\bar{x}} \cdot 100\%$	$CV = \dfrac{\sigma}{\mu} \cdot 100\%$

EXAMPLE **Heights and Weights of Men** Using the sample height and weight data for the 40 males included in Data Set 1 in Appendix B, we find the statistics given in the table below. Find the coefficient of variation for heights, then find the coefficient of variation for weights, then compare the two results.

	Mean ($\bar{x}$)	Standard Deviation (s)
Height	68.34 in.	3.02 in.
Weight	172.55 lb	26.33 lb

SOLUTION   Because we have sample statistics, we find the two coefficients of variation as follows:

Heights: $CV = \dfrac{s}{\bar{x}} \cdot 100\% = \dfrac{3.02 \text{ in.}}{68.34 \text{ in.}} \cdot 100\% = 4.42\%$

Weights: $CV = \dfrac{s}{\bar{x}} \cdot 100\% = \dfrac{26.33 \text{ lb}}{172.55 \text{ lb}} \cdot 100\% = 15.26\%$

Although the difference in the units makes it impossible to compare the standard deviation of 3.02 in. to the standard deviation of 26.33 lb, we can compare the coefficients of variation, which have no units. We can see that heights (with $CV = 4.42\%$) have considerably less variation than weights (with $CV = 15.26\%$). This makes intuitive sense, because we routinely see that weights among men vary much more than heights. For example, it is very rare to see two adult men with one of them being twice as tall as the other, but it is much more common to see two men with one of them weighing twice as much as the other.

Finding Standard Deviation from a Frequency Distribution

We sometimes need to compute the standard deviation of a data set that is summarized in the form of a frequency distribution, such as Table 2-2 from Section 2-2. If the original list of sample values is available, use them with Formula 2-4 or 2-5 so that the result is more exact. If the original data are not available, use one of these two methods:

1. If the total number of values is not too large, use your calculator or software program by entering each class midpoint a number of times equal to the class frequency.
2. Calculate the standard deviation using Formula 2-6.

Formula 2-6 $s = \sqrt{\dfrac{n\left[\Sigma(f \cdot x^2)\right] - \left[\Sigma(f \cdot x)\right]^2}{n(n-1)}}$ standard deviation for frequency distribution

Table 2-12	Calculating Standard Deviation from a Frequency Distribution			
Cotinine	Frequency f	Class Midpoint, x	f · x	f · x²
0–99	11	49.5	544.5	26952.75
100–199	12	149.5	1794.0	268203.00
200–299	14	249.5	3493.0	871503.50
300–399	1	349.5	349.5	122150.25
400–499	2	449.5	899.0	404100.50
Totals:	$\Sigma f = 40$		$\Sigma(f \cdot x) = 7080$	$\Sigma(f \cdot x^2) = 1692910$

 EXAMPLE Cotinine Levels of Smokers Find the standard deviation of the 40 values summarized in the frequency distribution of Table 2-2, assuming that the original data set is not available.

SOLUTION

Method 1:   Table 2-12 has class midpoints of 49.5, 149.5, 249.5, 349.5, and 449.5. Using a calculator or software program, enter the value of 49.5 eleven times (because the frequency of the first class is 11), enter 149.5 12 times, and so on. Get the standard deviation of this set of 40 class midpoints. The result should be 106.2.

Method 2:   Use Formula 2-6. Application of Formula 2-6 requires that we first find the values of n, $\Sigma(f \cdot x)$, and $\Sigma(f \cdot x^2)$. After finding those values from Table 2-12, we apply Formula 2-6 as follows:

$$s = \sqrt{\frac{n\left[\Sigma(f \cdot x^2)\right] - \left[\Sigma(f \cdot x)\right]^2}{n(n-1)}} = \sqrt{\frac{40[1692910] - [7080]^2}{40(40-1)}}$$

$$= \sqrt{\frac{17{,}590{,}000}{1560}} = \sqrt{11275.64103} = 106.2$$

TI-83 Plus Calculator   Unlike most calculators, the TI-83 Plus can compute the standard deviation of values summarized in a frequency distribution. First enter the class midpoints in list L1, then enter the frequencies in list L2. Now press **STAT**, select **CALC**, select **1-VarStats**, and enter L1, L2 to obtain results that include the mean and standard deviation. Again, the sample standard deviation is identified as Sx, and the population standard deviation is identified as σx.

Interpreting and Understanding Standard Deviation

This subsection is extremely important, because we will now try to make some intuitive sense of the standard deviation. First, we should clearly understand that the standard deviation measures the variation among values. Values close together will yield a small standard deviation, whereas values spread farther apart will yield a larger standard deviation.

Because variation is such an important concept and because the standard deviation is such an important tool in measuring variation, we will consider three different ways of developing a sense for values of standard deviations. The first is the **range rule of thumb,** which is based on the principle that for many data sets, the vast majority (such as 95%) of sample values lie within two standard deviations of the mean. (We could improve the accuracy of this rule by taking into account such factors as the size of the sample and the nature of the distribution, but we prefer to sacrifice accuracy for the sake of simplicity. Also, we could use three or even four standard deviations instead of two standard deviations, which is a somewhat arbitrary choice. But we want a simple rule that will help us interpret values of standard deviations; later methods will produce more accurate results.)

Range Rule of Thumb

For Estimating a Value of the Standard Deviation s: To roughly estimate the standard deviation, use

$$s \approx \frac{\text{range}}{4}$$

where range = (highest value) − (lowest value).

For Interpreting a Known Value of the Standard Deviation s: If the standard deviation s is known, use it to find rough estimates of the minimum and maximum "usual" sample values by using

$$\text{minimum ``usual'' value} \approx (\text{mean}) - 2 \times (\text{standard deviation})$$

$$\text{maximum ``usual'' value} \approx (\text{mean}) + 2 \times (\text{standard deviation})$$

When calculating a standard deviation using Formula 2-4 or 2-5, you can use the range rule of thumb as a check on your result, but realize that although the approximation will get you in the general vicinity of the answer, it can be off by a fairly large amount.

 EXAMPLE Cotinine Levels of Smokers Use the range rule of thumb to find a rough estimate of the standard deviation of the sample of 40 cotinine levels of smokers, as listed in Table 2-1.

SOLUTION   In using the range rule of thumb to estimate the standard deviation of sample data, we find the range and divide by 4. By scanning the list of cotinine levels, we can see that the lowest is 0 and the highest is 491, so the range is 491. The standard deviation s is estimated as follows:

$$s \approx \frac{\text{range}}{4} = \frac{491}{4} = 122.75 \approx 123$$

INTERPRETATION This result is quite close to the correct value of 119.5 that is found by calculating the exact value of the standard deviation with Formula 2-4 or 2-5. Don't expect the range rule of thumb to work this well in other cases.

The next example is particularly important as an illustration of one way to *interpret* the value of a standard deviation.

EXAMPLE **Head Circumferences of Girls** Past results from the National Health Survey suggest that the head circumferences of 2-month-old girls have a mean of 40.05 cm and a standard deviation of 1.64 cm. Use the range rule of thumb to find the minimum and maximum "usual" head circumferences. (These results could be used by a physician who can identify "unusual" circumferences that might be the result of a disorder such as hydrocephalus.) Then determine whether a circumference of 42.6 cm would be considered "unusual."

SOLUTION With a mean of 40.05 cm and a standard deviation of 1.64 cm, we use the range rule of thumb to find the minimum and maximum usual heights as follows:

$$\text{minimum} \approx (\text{mean}) - 2 \times (\text{standard deviation})$$
$$= 40.05 - 2(1.64) = 36.77 \text{ cm}$$

$$\text{maximum} \approx (\text{mean}) + 2 \times (\text{standard deviation})$$
$$= 40.05 + 2(1.64) = 43.33 \text{ cm}$$

INTERPRETATION Based on these results, we expect that typical 2-month-old girls have head circumferences between 36.77 cm and 43.33 cm. Because 42.6 cm falls within those limits, it would be considered usual or typical, not unusual.

Empirical (or 68–95–99.7) Rule for Data with a Bell-Shaped Distribution

Another rule that is helpful in interpreting values for a standard deviation is the **empirical rule.** This rule states that *for data sets having a distribution that is approximately bell-shaped,* the following properties apply. (See Figure 2-13.)

- About 68% of all values fall within 1 standard deviation of the mean.
- About 95% of all values fall within 2 standard deviations of the mean.
- About 99.7% of all values fall within 3 standard deviations of the mean.

FIGURE 2-13 The
Empirical Rule

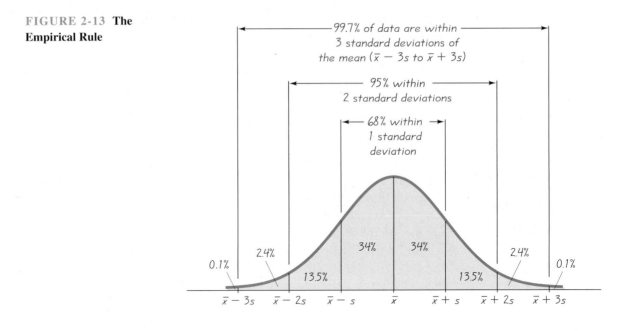

EXAMPLE **IQ Scores** IQ scores of normal adults on the Weschler test have a bell-shaped distribution with a mean of 100 and a standard deviation of 15. What percentage of adults have IQ scores between 55 and 145?

SOLUTION   The key to solving this problem is to recognize that 55 and 145 are each exactly 3 standard deviations away from the mean of 100, as shown below.

$$3 \text{ standard deviations} = 3s = 3(15) = 45$$

Therefore, 3 standard deviations from the mean is

$$100 - 45 = 55$$

or

$$100 + 45 = 145$$

The empirical rule tells us that about 99.7% of all values are within 3 standard deviations of the mean, so about 99.7% of all IQ scores are between 55 and 145.

Hint: Difficulty with applying the empirical rule usually stems from confusion with interpreting phrases such as "within 3 standard deviations of the mean." Stop here and review the preceding example until the meaning of that phrase becomes clear. Also, see the following general interpretations of such phrases.

Phrase	Meaning
Within 1 standard deviation of the mean	Between $(\bar{x} - s)$ and $(\bar{x} + s)$
Within 2 standard deviations of the mean	Between $(\bar{x} - 2s)$ and $(\bar{x} + 2s)$
Within 3 standard deviations of the mean	Between $(\bar{x} - 3s)$ and $(\bar{x} + 3s)$

A third concept that is helpful in understanding or interpreting a value of a standard deviation is **Chebyshev's theorem.** The preceding empirical rule applies only to data sets with a bell-shaped distribution. Instead of being limited to data sets with bell-shaped distributions, Chebyshev's theorem applies to *any* data set, but its results are very approximate.

Chebyshev's Theorem

The proportion (or fraction) of any set of data lying within K standard deviations of the mean is always *at least* $1 - 1/K^2$, where K is any positive number greater than 1. For $K = 2$ and $K = 3$, we get the following statements:

- At least 3/4 (or 75%) of all values lie within 2 standard deviations of the mean.
- At least 8/9 (or 89%) of all values lie within 3 standard deviations of the mean.

EXAMPLE **IQ Scores** IQ scores of normal adults taking the Weschler test have a mean of 100 and a standard deviation of 15. What can we conclude from Chebyshev's theorem?

SOLUTION   Applying Chebyshev's theorem with a mean of 100 and a standard deviation of 15, we can reach the following conclusions.

- At least 3/4 (or 75%) of all adults have IQ scores within 2 standard deviations of the mean (between 70 and 130).
- At least 8/9 (or 89%) of all adults have IQ scores within 3 standard deviations of the mean (between 55 and 145).

When trying to make sense of a value of a standard deviation, we should use one or more of the preceding three concepts. To gain additional insight into the nature of the standard deviation, we now consider the underlying rationale leading to Formula 2-4, which is the basis for its definition. (Formula 2-5 is simply another version of Formula 2-4, derived so that arithmetic calculations can be simplified.)

Rationale for Formula 2-4

The standard deviation of a set of sample data is defined by Formulas 2-4 and 2-5, which are equivalent in the sense that they will always yield the same result. Formula 2-4 has the advantage of reinforcing the concept that the standard deviation is a type of average deviation. Formula 2-5 has the advantage of being easier to use when you must calculate standard deviations on your own. Formula 2-5 also eliminates the intermediate rounding errors introduced in Formula 2-4 when the exact value of the mean is not used. Formula 2-5 is used in calculators and programs because it requires only three memory locations (for n, $\sum x$, and $\sum x^2$) instead of a memory location for every value in the data set.

Why define a measure of variation in the way described by Formula 2-4? In measuring variation in a set of sample data, it makes sense to begin with the individual amounts by which values deviate from the mean. For a particular value x, the amount of **deviation** is $x - \bar{x}$, which is the difference between the individual x value and the mean. For the Mulberry Bank waiting times of 1, 3, 14, the mean is 6.0 so the deviations away from the mean are -5, -3, and 8. It would be good to somehow combine those deviations into a single collective value. Simply adding the deviations doesn't work, because the sum will always be zero. To get a statistic that measures variation (instead of always being zero), we need to avoid the canceling out of negative and positive numbers. One approach is to add absolute values, as in $\Sigma|x - \bar{x}|$. If we find the mean of that sum, we get the **mean absolute deviation** (or **MAD**), which is the mean distance of the data from the mean.

$$\text{mean absolute deviation} = \frac{\Sigma|x - \bar{x}|}{n}$$

Because the Mulberry Bank waiting times of 1, 3, 14 have deviations of -5, -3, and 8, the mean absolute deviation is $(5 + 3 + 8)/3 = 16/3 = 5.3$.

Why Not Use the Mean Absolute Deviation?   Because the mean absolute deviation requires that we use absolute values, it uses an operation that is not algebraic. (The algebraic operations include addition, multiplication, extracting roots, and raising to powers that are integers or fractions, but absolute value is not included.) The use of absolute values creates algebraic difficulties in inferential methods of statistics. For example, Section 8-3 presents a method for making inferences about the means of two populations, and that method is built around an additive property of variances, but the mean absolute deviation has no such additive property. (Here is a simplified version of the additive property of variances: If you have two independent populations and you randomly selected one value from each population and add them, such sums will have a variance equal to the sum of the variances of the two populations.) That same additive property underlies the rationale for regression discussed in Chapter 9 and analysis of variance discussed in Chapter 11. Also, Exercise 42 shows that the mean absolute value is biased, meaning that when you find mean absolute values of samples, you do not tend to target the mean absolute value of the population. In contrast, the standard deviation uses only algebraic operations. Because it is based on the square root of a sum of squares, the standard deviation closely parallels distance formulas found in algebra. There are many instances where a statistical procedure is based on a similar sum of squares. Therefore, instead of using absolute values, we get a better measure of variation by making all deviations $(x - \bar{x})$ nonnegative by squaring them, and this approach leads to the standard deviation. For these reasons, scientific calculators typically include a standard deviation function, but they almost never include the mean absolute deviation.

Why Divide by $n - 1$?   After finding all of the individual values of $(x - \bar{x})^2$, we combine them by finding their sum, then we get an average by dividing by $n - 1$. We divide by $n - 1$ because there are only $n - 1$ independent values. That is, with a given mean, only $n - 1$ values can be freely assigned any

number before the last value is determined. See Exercise 41, which provides concrete numbers illustrating that division by $n - 1$ is better than division by n. That
exercise shows that if s^2 were defined with division by n, it would systematically
underestimate the value of σ^2, so we compensate by increasing its overall value
by making its denominator smaller (by using $n - 1$ instead of n). Exercise 41
shows how division by $n - 1$ causes the sample variance s^2 to target the value of
the population variance σ^2, whereas division by n causes the sample variance s^2 to
underestimate the value of the population variance σ^2.

Step 6 in Formula 2-4 for finding a standard deviation is to find a square root.
We take the square root to compensate for the squaring that took place in Step 3.
An important consequence of taking the square root is that the standard deviation
has the same units of measurement as the original values. For example, if customer waiting times are in minutes, the standard deviation of those times will also
be in minutes. If we were to stop at Step 5, the result would be in units of "square
minutes," which is an abstract concept having no direct link to reality.

After studying this section, you should understand that the standard deviation
is a measure of variation among values. Given sample data, you should be able to
compute the value of the standard deviation. You should be able to interpret the
values of standard deviations that you compute. You should know that for typical
data sets, it is unusual for a value to differ from the mean by more than 2 or 3 standard deviations.

2-5 Basic Skills and Concepts

*In Exercises 1–8, find the range, variance, and standard deviation for the given sample
data. (The same data were used in Section 2-4 where we found measures of center. Here
we find measures of variation.)*

Recommended Assignment
Exercises 1–4, 9–14, 17, 18, 21–24

1. **Tobacco Use in Children's Movies** In "Tobacco and Alcohol Use in G-Rated Children's Animated Films," by Goldstein, Sobel, and Newman (*Journal of the American
   Medical Association*, Vol. 281, No. 12), the lengths (in seconds) of scenes showing tobacco use were recorded for animated movies from Universal Studios. The first six
   values included in Data Set 7 from Appendix B are listed below. Do these times appear to be consistent, or do they vary widely?

   0   223   0   176   0   548

   1. range = 548.0 sec;
      $s^2 = 46308.2$ sec^2;
      $s = 215.2$ sec; they vary widely.

2. **Harry Potter** In an attempt to measure the reading level of a book, the Flesch Reading
   Ease ratings are obtained for 12 randomly selected pages from *Harry Potter and the
   Sorcerer's Stone* by J. K. Rowling. Those values, included in Data Set 14 from Appendix B, are listed below. Given that these ratings are based on 12 randomly selected
   pages, is the standard deviation of this sample likely to be a reasonable estimate of the
   standard deviation of the reading levels for all pages in the whole book?

85.3	84.3	79.5	82.5	80.2	84.6
79.2	70.9	78.6	86.2	74.0	83.7

   2. range = 15.3; $s^2 = 21.92$;
      $s = 4.68$
      The ratings don't vary much, so
      the estimate should be fairly
      good.

3. **Cereal** A dietitian obtains the amounts of sugar (in grams) from 1 gram in each of 16
   different cereals, including Cheerios, Corn Flakes, Fruit Loops, Trix, and 12 others.
   Those values, included in Data Set 16 from Appendix B, are listed below. Is the standard deviation of these values likely to be a good estimate of the standard deviation of

   3. range = 0.450 g; $s^2 = 0.028$ g^2;
      $s = 0.168$ g

the amounts of sugar in each gram of cereal consumed by the population of all Americans who eat cereal? Why or why not?

| 0.03 | 0.24 | 0.30 | 0.47 | 0.43 | 0.07 | 0.47 | 0.13 |
| 0.44 | 0.39 | 0.48 | 0.17 | 0.13 | 0.09 | 0.45 | 0.43 |

4. Body Mass Index As part of the National Health Examination, the body mass index is measured for a random sample of women. Some of the values included in Data Set 1 from Appendix B are listed below. Is the standard deviation of this sample reasonably close to the standard deviation of 6.17, which is the standard deviation for all 40 women included in Data Set 1?

| 19.6 | 23.8 | 19.6 | 29.1 | 25.2 | 21.4 | 22.0 | 27.5 |
| 33.5 | 20.6 | 29.9 | 17.7 | 24.0 | 28.9 | 37.7 |

5. Drunk Driving The blood alcohol concentrations of a sample of drivers involved in fatal crashes and then convicted with jail sentences are given below (based on data from the U.S. Department of Justice). When a state wages a campaign to "reduce drunk driving," is the campaign intended to lower the standard deviation?

| 0.27 | 0.17 | 0.17 | 0.16 | 0.13 | 0.24 | 0.29 | 0.24 |
| 0.14 | 0.16 | 0.12 | 0.16 | 0.21 | 0.17 | 0.18 |

6. Motorcycle Fatalities Listed below are ages of motorcyclists when they were fatally injured in traffic crashes (based on data from the U.S. Department of Transportation). How does the variation of these ages compare to the variation of ages of licensed drivers in the general population?

| 17 | 38 | 27 | 14 | 18 | 34 | 16 | 42 | 28 |
| 24 | 40 | 20 | 23 | 31 | 37 | 21 | 30 | 25 |

7. Reaction Times The author visited the Reuben H. Fleet Science Museum in San Diego and repeated an experiment of reaction times. The following times (in hundredths of a second) were obtained. How do the measures of variation reflect the fact that these times appear to be very consistent?

| 19 | 20 | 17 | 21 | 21 | 21 | 19 | 18 | 19 | 19 |
| 17 | 17 | 15 | 17 | 18 | 17 | 18 | 18 | 18 | 17 |

8. Bufferin Tablets Listed below are the measured weights (in milligrams) of a sample of Bufferin aspirin tablets. Given that this medication should be manufactured in a consistent way so that dosage amounts can be controlled, do the measures of variation seem to indicate that the variation is at an acceptable level?

| 672.2 | 679.2 | 669.8 | 672.6 | 672.2 | 662.2 |
| 662.7 | 661.3 | 654.2 | 667.4 | 667.0 | 670.7 |

In Exercises 9–12, find the range, variance, and standard deviation for each of the two samples, then compare the two sets of results. (The same data were used in Section 2-4.)

9. Customer Waiting Times Waiting times of customers at the Jefferson Valley Bank (where all customers enter a single waiting line) and the Bank of Providence (where customers wait in individual lines at three different teller windows):

| Jefferson Valley: | 6.5 | 6.6 | 6.7 | 6.8 | 7.1 | 7.3 | 7.4 | 7.7 | 7.7 | 7.7 |
| Providence: | 4.2 | 5.4 | 5.8 | 6.2 | 6.7 | 7.7 | 7.7 | 8.5 | 9.3 | 10.0 |

Answers (margin):

4. range = 20.00; $s^2 = 32.09$; $s = 5.66$. It is fairly close.

5. range = 0.170 sec; $s^2 = 0.003$; $s = 0.051$. No, the intent is to lower all of the individual values, which would result in a lower mean.

6. range = 28.0 years; $s^2 = 75.0$ years²; $s = 8.7$ years. The variation is likely to be less than the variation for the general population.

7. range = 6.0; $s^2 = 2.5$; $s = 1.6$. The measures of variation are low values.

8. range = 25.00 mg; $s^2 = 44.47$ mg²; $s = 6.67$ mg. The range of 25.00 appears to be a little larger than it should be.

9. Jefferson Valley: range = 1.20 min; $s^2 = 0.23$ min²; $s = 0.48$ min. Provident: range = 5.80 min; $s^2 = 3.32$ min²; $s = 1.82$ min

10. **Regular/Diet Coke** Weights (pounds) of samples of the contents in cans of regular Coke and diet Coke:

Regular: 0.8192   0.8150   0.8163   0.8211   0.8181   0.8247
Diet: 0.7773   0.7758   0.7896   0.7868   0.7844   0.7861

11. **Mickey D vs. Jack** When investigating times required for drive-through service, the following results (in seconds) are obtained (based on data from QSR Drive-Thru Time Study).

McDonald's: 287 128 92 267 176 240 192 118 153 254 193 136
Jack in the Box: 190 229 74 377 300 481 428 255 328 270 109 109

12. **Skull Breadths** Maximum breadth of samples of male Egyptian skulls from 4000 B.C. and 150 A.D. (based on data from *Ancient Races of the Thebaid* by Thomson and Randall-Maciver):

4000 B.C.: 131 119 138 125 129 126 131 132 126 128 128 131
150 A.D.:   136 130 126 126 139 141 137 138 133 131 134 129

In Exercises 13–16, refer to the data sets in Appendix B. Use computer software or a calculator to find the standard deviations, then compare the results.

T 13. **Head Circumferences** In order to correctly diagnose the disorder of hydrocephalus, a pediatrician investigates head circumferences of 2-year-old boys and girls. Use the sample results listed in Data Set 3. Does there appear to be a difference between the two genders?

T 14. **Clancy, Rowling, Tolstoy** A child psychologist investigates differences in reading difficulty and obtains data from *The Bear and the Dragon* by Tom Clancy, *Harry Potter and the Sorcerer's Stone* by J. K. Rowling, and *War and Peace* by Leo Tolstoy. Refer to Data Set 14 in Appendix B and use the Flesch-Kincaid Grade Level ratings for 12 pages randomly selected from each of the three books.
14. Clancy: 2.45; Rowling: 1.17; Tolstoy: 2.01.

T 15. **Weekend Rainfall** In Data Set 11 in Appendix B, use the rainfall amounts in Boston on Thursday and the rainfall amounts in Boston on Sunday.
15. Thursday: 0.167 in.; Sunday: 0.200 in.

T 16. **Tobacco/Alcohol Use in Children's Movies** In "Tobacco and Alcohol Use in G-Rated Children's Animated Films," by Goldstein, Sobel, and Newman (*Journal of the American Medical Association,* Vol. 281, No. 12), the lengths (in seconds) of scenes showing tobacco use and alcohol use were recorded for animated children's movies. In Data Set 7 in Appendix B, use the tobacco times, then the alcohol times.

In Exercises 17–20, find the standard deviation of the data summarized in the given frequency distribution. (The same frequency distributions were used in Section 2-4.)

17. **Old Faithful** Visitors to Yellowstone National Park consider an eruption of the Old Faithful geyser to be a major attraction that should not be missed. The given frequency distribution summarizes a sample of times (in minutes) between eruptions.
17. 14.7 min

18. **Loaded Die** The author drilled a hole in a die and filled it with a lead weight, then proceeded to roll it 200 times. The results are given in the frequency distribution in the margin. 1.6

19. **Speeding Tickets** The given frequency distribution describes the speeds of drivers ticketed by the Town of Poughkeepsie police. These drivers were traveling through a 30 mi/h speed zone on Creek Road, which passes the author's college. 4.1 mi/h

Answers (margin)

10. Regular: range = 0.00970 lb; $s^2 = 0.00001$ lb^2; $s = 0.00349$ lb. Diet: range = 0.01380 lb; $s^2 = 0.00003$ lb^2; $s = 0.00554$ lb
11. McDonald's: range = 195.0 sec; $s^2 = 4081.7$ sec^2; $s = 63.9$ sec. Jack in the Box: range = 407.0 sec; $s^2 = 16644.3$ sec^2; $s = 129.0$ sec
12. 4000 B.C.: range = 19.0; $s^2 = 21.5$; $s = 4.6$. 150 A.D.: range = 15.0; $s^2 = 25.2$; $s = 5.0$
13. Males: 1.50 cm; females: 1.64 cm; difference does not appear to be very substantial.
16. Tobacco: 104.0 sec; alcohol: 66.3 sec. The times for tobacco use vary much more.

Table for Exercise 17

Time	Frequency
40–49	8
50–59	44
60–69	23
70–79	6
80–89	107
90–99	11
100–109	1

Table for Exercise 18

Outcome	Frequency
1	27
2	31
3	42
4	40
5	28
6	32

Table for Exercise 19

Speed	Frequency
42–45	25
46–49	14
50–53	7
54–57	3
58–61	1

Table for Exercise 20

Temperature	Frequency
96.5–96.8	1
96.9–97.2	8
97.3–97.6	14
97.7–98.0	22
98.1–98.4	19
98.5–98.8	32
98.9–99.2	6
99.3–99.6	4

21. Approximately 12 years (based on minimum of 23 years and maximum of 70 years)

22. Approximately 12.5 (based on minimum of 50 and maximum of 100)

23. Minimum: 31.30 cm; maximum: 46.42 cm; yes

24. 58.6 in.; 68.6 in.; yes

25. a. 68%
   b. 99.7%

26. a. 68%
   b. 95%

27. Percentage is at least 75%.

28. Percentage is at least 89%.

29. Calories: 5.9%; sugar 56.9%. The sugar content has much greater variation when compared to calories.

30. Coke: 0.9%; Pepsi: 0.7%; no

31. All of the values are the same.

20. Body Temperatures The accompanying frequency distribution summarizes a sample of human body temperatures. (See the temperatures for midnight on the second day, as listed in Data Set 4 in Appendix B.) 0.62°F

21. Teacher Ages Use the range rule of thumb to estimate the standard deviation of ages of all teachers at your college.

22. Test Scores Use the range rule of thumb to estimate the standard deviation of the scores on the first statistics test in your class.

23. Leg Lengths For the sample data in Data Set 1 from Appendix B, the sample of 40 women have upper leg lengths with a mean of 38.86 cm and a standard deviation of 3.78 cm. Use the range rule of thumb to estimate the minimum and maximum "usual" upper leg lengths for women. Is a length of 47.0 cm considered unusual in this context?

24. Heights of Women Heights of women have a mean of 63.6 in. and a standard deviation of 2.5 in. (based on data from the National Health Survey). Use the range rule of thumb to estimate the minimum and maximum "usual" heights of women. In this context, is it unusual for a woman to be 6 ft tall?

25. Heights of Women Heights of women have a bell-shaped distribution with a mean of 63.6 in. and a standard deviation of 2.5 in. Using the empirical rule, what is the approximate percentage of women between
   a. 61.1 in. and 66.1 in.?
   b. 56.1 in. and 71.1 in.?

26. Weights of Regular Coke Using the weights of regular Coke listed in Data Set 17 from Appendix B, we find that the mean is 0.81682 lb, the standard deviation is 0.00751 lb, and the distribution is approximately bell-shaped. Using the empirical rule, what is the approximate percentage of cans of regular Coke with weights between
   a. 0.80931 lb and 0.82433 lb?
   b. 0.80180 lb and 0.83184 lb?

27. Heights of Women If heights of women have a mean of 63.6 in. and a standard deviation of 2.5 in., what can you conclude from Chebyshev's theorem about the percentage of women between 58.6 in. and 68.6 in.?

28. Weights of Regular Coke Using the weights of regular Coke listed in Data Set 17 from Appendix B, we find that the mean is 0.81682 lb and the standard deviation is 0.00751 lb. What can you conclude from Chebyshev's theorem about the percentage of cans of regular Coke with weights between 0.79429 lb and 0.83935 lb?

T 29. Coefficient of Variation for Cereal Refer to Data Set 16 in Appendix B. Find the coefficient of variation for the calories and find the coefficient of variation for the grams of sugar per gram of cereal. Compare the results.

T 30. Coefficient of Variation for Coke and Pepsi Refer to Data Set 17 in Appendix B. Find the coefficient of variation for the weights of regular Coke, then find the coefficient of variation for the weights of regular Pepsi. Compare the results. Does either company appear to have weights that are significantly more consistent?

31. Equality for All What do you know about the values in a data set having a standard deviation of $s = 0$?

32. *Understanding Units of Measurement* If a data set consists of the fines for speeding (in dollars), what are the units used for standard deviation? What are the units used for variance?

32. dollars; dollars²

33. *Comparing Car Batteries* The Everlast and Endurance brands of car battery are both labeled as lasting 48 months. In reality, they both have a mean life of 50 months, but the Everlast batteries have a standard deviation of 2 months, while the Endurance batteries have a standard deviation of 6 months. Which brand is the better choice? Why?

33. Everlast batteries are better in the sense that they are more consistent and predictable.

34. *Interpreting Outliers* A data set consists of 20 values that are fairly close together. Another value is included, but this new value is an outlier (very far away from the other values). How is the standard deviation affected by the outlier? No effect? A small effect? A large effect?

34. Large effect

2-5 Beyond the Basics

35. *Comparing Data Sets* Two different sections of a statistics class take the same quiz and the scores are recorded below. Find the range and standard deviation for each section. What do the range values lead you to conclude about the variation in the two sections? Why is the range misleading in this case? What do the standard deviation values lead you to conclude about the variation in the two sections?

| Section 1: | 1 | 20 | 20 | 20 | 20 | 20 | 20 | 20 | 20 | 20 | 20 |
| Section 2: | 2 | 3 | 4 | 5 | 6 | 14 | 15 | 16 | 17 | 18 | 19 |

35. Section 1: range = 19.0; $s = 5.7$
Section 2: range = 17.0; $s = 6.7$
The ranges suggest that Section 2 has less variation, but the standard deviations suggest that Section 1 has less variation.

36. *Transforming Data* In each of the following, describe how the range and standard deviation of a data set are affected.
  a. The same constant k is added to each value of the data set.
  b. Each value of the data set is multiplied by the same constant k.
  c. For the body temperature data listed in Data Set 4 of Appendix B (12 A.M. on day 2), $\bar{x} = 98.20°F$ and $s = 0.62°F$. Find the values of $\bar{x}$ and s after each temperature has been converted to the Celsius scale. [*Hint:* $C = 5(F - 32)/9$.]

36. a. They remain the same.
b. They are both multiplied by k.
c. $\bar{x} = 36.78$; $s = 0.34$

37. Genichi Taguchi developed a method of improving quality and reducing manufacturing costs through a combination of engineering and statistics. A key tool in the Taguchi method is the **signal-to-noise ratio.** The simplest way to calculate this ratio is to divide the mean by the standard deviation. Find the signal-to-noise ratio for the cotinine levels of smokers listed in Table 2-1.

37. 1.44

38. *Skewness* In Section 2-4 we introduced the general concept of skewness. Skewness can be measured by **Pearson's index of skewness:**

$$I = \frac{3(\bar{x} - \text{median})}{s}$$

If $I \geq 1.00$ or $I \leq -1.00$, the data can be considered to be *significantly skewed*. Find Pearson's index of skewness for the cotinine levels of smokers listed in Table 2-1, and then determine whether there is significant skewness.

38. 0.06; no significant skewness

39. *Understanding Standard Deviation* A sample consists of 10 test scores that fall between 70 and 100 inclusive. What is the largest possible standard deviation?

39. 15.8

40. No, every score must be within 20 points of the mean, so the highest possible score is 95.

41. a. 6
    b. 6
    c. 3.0
    d. $n - 1$
    e. No The mean of the sample standard variances (6) equals the population variance (6), but the mean of the sample standard deviations (1.9) does not equal the mean of the population standard deviation (2.4).

42. a. 2
    b. 1.3
    c. No; no

40. Phony Data? For any data set of n values with standard deviation s, every value must be within $s\sqrt{n-1}$ of the mean. A statistics teacher reports that the test scores in her class of 17 students had a mean of 75.0 and a standard deviation of 5.0. Kelly, the class's self-proclaimed best student, claims that she received a grade of 97. Could Kelly be telling the truth?

41. Why Divide by $n-1$? Let a population consist of the values 3, 6, 9. Assume that samples of two values are randomly selected *with replacement*.
   a. Find the variance σ^2 of the population $\{3, 6, 9\}$.
   b. List the nine different possible samples of two values selected with replacement, then find the sample variance s^2 (which includes division by $n - 1$) for each of them. If you repeatedly select two sample values, what is the mean value of the sample variance s^2?
   c. For each of the nine samples, find the variance by treating each sample as if it is a population. (Be sure to use the formula for population variance, which includes division by n.) If you repeatedly select 2 sample values, what is the mean value of the population variances?
   d. Which approach results in values that are better estimates of σ^2: Part (b) or part (c)? Why? When computing variances of samples, should you use division by n or $n - 1$?
   e. The preceding parts show that s^2 is an unbiased estimator of σ^2. Is s an unbiased estimator of σ?

42. Why Not Go MAD? Exercise 41 shows that the sample variance s^2 is an unbiased estimator of σ^2. Do the following with the same population of $\{3, 6, 9\}$ to show that the mean absolute deviation of a sample is a biased estimator of the mean absolute deviation of a population.
   a. Find the mean absolute deviation of the population $\{3, 6, 9\}$.
   b. List the nine different possible samples of two values selected with replacement, then find the mean absolute deviation for each of them. If you repeatedly select two sample values, what is the mean value of the mean absolute deviations?
   c. Based on the results of parts (a) and (b), does the mean absolute deviation of a sample tend to target the mean absolute deviation of the population? Does division by $n - 1$ instead of division by n make the mean absolute deviation an unbiased estimate of the mean absolute deviation for the population?

 Measures of Relative Standing

This section introduces measures that can be used to compare values from different data sets, or to compare values within the same data set. We introduce z scores (for comparing values from different data sets) and quartiles and percentiles (for comparing values within the same data set).

Note to Instructor
Students can generally read and understand z scores on their own; discuss percentiles and quartiles.

z Scores

A z score (or standard score) is found by converting a value to a standardized scale, as given in the following definition. We will use z scores extensively in Chapter 5 and later chapters, so they are extremely important.

Definition

A **standard score,** or **z score,** is the number of standard deviations that a given value x is above or below the mean. It is found using the following expressions:

$$\text{Sample} \qquad\qquad \text{Population}$$
$$z = \frac{x - \bar{x}}{s} \qquad \text{or} \qquad z = \frac{x - \mu}{\sigma}$$

(Round z to two decimal places.)

The following example illustrates how z scores can be used to compare values, even though they might come from different populations.

> **EXAMPLE   Comparing Heights** NBA superstar Michael Jordan is 78 in. tall and WNBA basketball player Rebecca Lobo is 76 in. tall. Jordan is obviously taller by 2 in., but which player is *relatively* taller? Does Jordan's height among men exceed Lobo's height among women? Men have heights with a mean of 69.0 in. and a standard deviation of 2.8 in.; women have heights with a mean of 63.6 in. and a standard deviation of 2.5 in. (based on data from the National Health Survey).
>
> **SOLUTION**   To compare the heights of Michael Jordan and Rebecca Lobo relative to the populations of men and women, we need to standardize those heights by converting them to z scores.
>
> $$\text{Jordan:} \qquad z = \frac{x - \mu}{\sigma} = \frac{78 - 69.0}{2.8} = 3.21$$
>
> $$\text{Lobo:} \qquad z = \frac{x - \mu}{\sigma} = \frac{76 - 63.6}{2.5} = 4.96$$
>
> **INTERPRETATION**   Michael Jordan's height is 3.21 standard deviations above the mean, but Rebecca Lobo's height is a whopping 4.96 standard deviations above the mean. Rebecca Lobo's height among women is relatively greater than Michael Jordan's height among men.

z Scores and Unusual Values

In Section 2-5 we used the range rule of thumb to conclude that a value is "unusual" if it is more than 2 standard deviations away from the mean. It follows that unusual values have z scores less than -2 or greater than $+2$. (See Figure 2-14 on page 94.) Using this criterion, both Michael Jordan and Rebecca Lobo are unusually tall because they both have heights with z scores greater than 2.

While considering professional basketball players with exceptional heights, another player is Mugsy Bogues, who was successful even though he is only 5 ft 3 in. tall. (We again use the fact that men have heights with a mean of 69.0 in. and a

Note to Instructor

The concept of "unusual values" is a good preparation for some of the concepts involved with hypothesis testing, introduced in Chapter 7.

FIGURE 2-14 Interpreting z scores

Unusual values are those with z scores less than 2.00 or greater than 2.00.

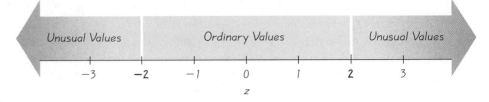

standard deviation of 2.8 in.) After converting 5 ft 3 in. to 63 in., we convert his height to a z score as follows:

$$\text{Bogues: } z = \frac{x - \mu}{\sigma} = \frac{63 - 69.0}{2.8} = -2.14$$

Let's be grateful to Mugsy Bogues for his many years of inspired play and for illustrating this principle:

Whenever a value is less than the mean, its corresponding z score is negative.

Ordinary values: $-2 \leq z \text{ score} \leq 2$

Unusual values: $z \text{ score} < -2 \quad or \quad z \text{ score} > 2$

z scores are measures of position in the sense that they describe the location of a value (in terms of standard deviations) relative to the mean. A z score of 2 indicates that a value is two standard deviations *above* the mean, and a z score of -3 indicates that a value is three standard deviations *below* the mean. Quartiles and percentiles are also measures of position, but they are defined differently than z scores and they are useful for comparing values within the same data set or between different sets of data.

Quartiles and Percentiles

Note to Instructor

Suggestion: Begin by discussing percentiles, then show how quartiles can be easily found by thinking of them as percentiles. Finding the value of Q_1 is really finding the value of P_{25}, and finding Q_3 is really finding P_{75}. Finding Q_2 is the same as finding the median.

Recall from Section 2-4 that the median of a data set is the middle value, so that 50% of the values are equal to or less than the median and 50% of the values are greater than or equal to the median. Just as the median divides the data into two equal parts, the three **quartiles,** denoted by Q_1, Q_2, and Q_3, divide the sorted values into four equal parts. (Values are *sorted* when they are arranged in order.) Here are descriptions of the three quartiles:

Q_1 **(First quartile):**   Separates the bottom 25% of the sorted values from the top 75% (To be more precise, at least 25% of the sorted values are less than or equal to Q_1, and at least 75% of the values are greater than or equal to Q_1.)

Q_2 **(Second quartile):**   Same as the median; separates the bottom 50% of the sorted values from the top 50%.

Q_3 **(Third quartile):**   Separates the bottom 75% of the sorted values from the top 25%. (To be more precise, at least 75% of the sorted values are less than or equal to Q_3, and at least 25% of the values are greater than or equal to Q_3.)

We will describe a procedure for finding quartiles after we discuss percentiles. There is not universal agreement on a single procedure for calculating quartiles, and different computer programs often yield different results. For example, if you use the data set of 1, 3, 6, 10, 15, 21, 28, and 36, you will get these results:

	Q_1	Q_2	Q_3
STATDISK	4.5	12.5	24.5
Minitab	3.75	12.5	26.25
Excel	5.25	12.5	22.75
TI-83 Plus	4.5	12.5	24.5

For this data set, STATDISK and the TI-83 Plus calculator agree, but they do not always agree. If you use a calculator or computer software for exercises involving quartiles, you may get results that differ slightly from the answers given in the back of the book.

Just as there are three quartiles separating a data set into four parts, there are also 99 **percentiles,** denoted $P_1, P_2, \ldots, P_{99}$, which partition the data into 100 groups with about 1% of the values in each group. (Quartiles and percentiles are examples of *quantiles*—or *fractiles*—which partition data into groups with roughly the same number of values.)

The process of finding the percentile that corresponds to a particular value x is fairly simple, as indicated in the following expression:

$$\text{percentile of value } x = \frac{\text{number of values less than } x}{\text{total number of values}} \cdot 100$$

EXAMPLE Cotinine Levels of Smokers Table 2-13 lists the 40 sorted cotinine levels of smokers included in Table 2-1. Find the percentile corresponding to the cotinine level of 112.

SOLUTION   From Table 2-13 we see that there are 12 values less than 112, so

$$\text{percentile of } 112 = \frac{12}{40} \cdot 100 = 30$$

INTERPRETATION The cotinine level of 112 is the 30th percentile.

The preceding example shows how to convert from a given sample value to the corresponding percentile. There are several different methods for the reverse procedure of converting a given percentile to the corresponding value in the data set. The procedure we will use is summarized in Figure 2-15, which uses the notation that follows the figure.

Cost of Laughing Index

There really is a Cost of Laughing Index (CLI), which tracks costs of such items as rubber chickens, Groucho Marx glasses, admission to comedy clubs, and 13 other leading humor indicators. This is the same basic approach used in developing the Consumer Price Index (CPI), which is based on a weighted average of goods and services purchased by typical consumers. While standard scores and percentiles allow us to compare different values, they ignore any element of time. Index numbers, such as the CLI and CPI, allow us to compare the value of some variable to its value at some base time period. The value of an index number is the current value, divided by the base value, multiplied by 100.

Table 2-13	*Sorted* Cotinine Levels of 40 Smokers								
0	1	1	3	17	32	35	44	48	86
87	103	112	121	123	130	131	149	164	167
173	173	198	208	210	222	227	234	245	250
253	265	266	277	284	289	290	313	477	491

FIGURE 2-15 Converting from the *k*th Percentile to the Corresponding Data Value

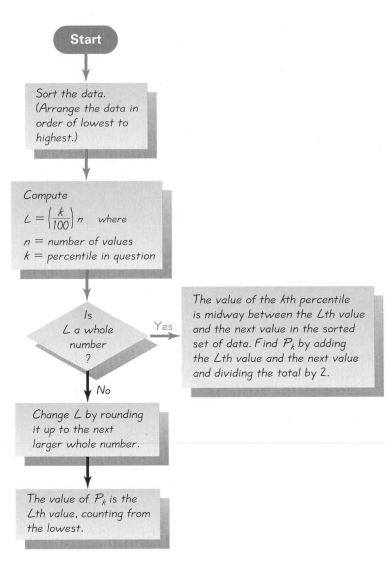

Notation

n = total number of values in the data set
k = percentile being used (Example: For the 25th percentile, $k = 25$.)
L = locator that gives the *position* of a value (Example: For the 12th value in the sorted list, $L = 12$.)
P_k = kth percentile (Example: P_{25} is the 25th percentile.)

EXAMPLE **Cotinine Levels of Smokers** Refer to the sorted cotinine levels of smokers in Table 2-13 and use Figure 2-15 to find the value of the 68th percentile, P_{68}.

SOLUTION   Referring to Figure 2-15, we see that the sample data are already sorted, so we can proceed to find the value of the locator L. In this computation we use $k = 68$ because we are trying to find the value of the 68th percentile. We use $n = 40$ because there are 40 data values.

$$L = \frac{k}{100} \cdot n = \frac{68}{100} \cdot 40 = 27.2$$

Next, we are asked if L is a whole number and we answer no, so we proceed to the next lower box where we change L by rounding it up from 27.2 to 28. (In this book we typically round off the usual way, but this is one of two cases where we round *up* instead of rounding *off*.) Finally, the bottom box shows that the value of P_{68} is the 28th value, counting up from the lowest. In Table 2-13, the 28th value is 234. That is, $P_{68} = 234$.

EXAMPLE Cotinine Levels of Smokers Refer to the sample of cotinine levels of smokers given in Table 2-13. Use Figure 2-15 to find the value of Q_1, which is the first quartile.

SOLUTION   First we note that Q_1 is the same as P_{25}, so we can proceed with the objective of finding the value of the 25th percentile. Referring to Figure 2-15, we see that the sample data are already sorted, so we can proceed to compute the value of the locator L. In this computation, we use $k = 25$ because we are attempting to find the value of the 25th percentile, and we use $n = 40$ because there are 40 data values.

$$L = \frac{k}{100} \cdot n = \frac{25}{100} \cdot 40 = 10$$

Next, we are asked if L is a whole number and we answer yes, so we proceed to the box located at the right. We now see that the value of the kth (25th) percentile is midway between the Lth (10th) value and the next value in the original set of data. That is, the value of the 25th percentile is midway between the 10th value and the 11th value. The 10th value is 86 and the 11th value is 87, so the value midway between them is 86.5. We conclude that the 25th percentile is $P_{25} = 86.5$. The value of the first quartile Q_1 is also 86.5.

The preceding example showed that when finding a quartile value (such as Q_1), we can use the equivalent percentile value (such as P_{25}) instead. See the margin for relationships relating quartiles to equivalent percentiles.

In earlier sections of this chapter we described several statistics, including the mean, median, mode, range, and standard deviation. Some other statistics are defined using quartiles and percentiles, as in the following:

$$\text{interquartile range (or IQR)} = Q_3 - Q_1$$

$$\text{semi-interquartile range} = \frac{Q_3 - Q_1}{2}$$

$$\text{midquartile} = \frac{Q_3 + Q_1}{2}$$

$$\text{10–90 percentile range} = P_{90} - P_{10}$$

$Q_1 = P_{25}$
$Q_2 = P_{50}$
$Q_3 = P_{75}$

After completing this section, you should be able to convert a value into its corresponding z score (or standard score) so that you can compare it to other values, which may be from different data sets. You should be able to convert a value into its corresponding percentile value so that you can compare it to other values in some data set. You should be able to convert a percentile to the corresponding data value. And finally, you should understand the meanings of quartiles and be able to relate them to their corresponding percentile values (as in $Q_3 = P_{75}$).

Using Technology

A variety of different computer programs and calculators can be used to find many of the statistics introduced so far in this chapter. In Section 2-4 we provided specific instructions for using STATDISK, Minitab, Excel, and the TI-83 Plus calculator. We noted that we can sometimes enter a data set and use one operation to get several different sample statistics, often referred to as

descriptive statistics. Examples of such results are shown in the following screen displays. These screen displays result from using the cotinine levels of the *smokers* given in Table 2-1 with the Chapter Problem. The TI-83 Plus results are shown on two screens because they do not all fit on one screen.

STATDISK

SD Descriptive Statistics

Which column of data would you like to evaluate?

1

Evaluate

Print

Help ?

Column 1 Descriptive Statistics

Sample Size, n:	40
Mean:	172.475
Median:	170
Midrange:	245.5
RMS:	208.9748
Variance, s^2:	14279.85
St Dev, s:	119.4983
Mean Abs Dev:	94.775
Range:	491
Minimum:	0
1st Quartile:	86.5
2nd Quartile:	170
3rd Quartile:	251.5
Maximum:	491
Sum:	6899
Sum Sq:	1.746819e+6

Minitab

Variable	Total Count	Mean	SE Mean	StDev	Minimum	Q1	Median	Q3
SMOKER	40	172.5	18.9	119.5	0.000000000	86.3	170.0	252.3

Variable	Maximum
SMOKER	491.0

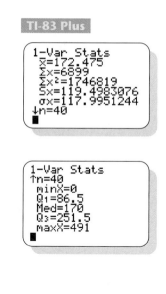

Column1	
Mean	172.475
Standard Error	18.89434
Median	170
Mode	1
Standard Deviation	119.4983
Sample Variance	14279.85
Kurtosis	0.519621
Skewness	0.587929
Range	491
Minimum	0
Maximum	491
Sum	6899
Count	40

TI-83 Plus

```
1-Var Stats
 x̄=172.475
 Σx=6899
 Σx²=1746819
 Sx=119.4983076
 σx=117.9951244
↓n=40
■
```

```
1-Var Stats
↑n=40
 minX=0
 Q₁=86.5
 Med=170
 Q₃=251.5
 maxX=491
■
```

2-6 Basic Skills and Concepts

In Exercises 1–4, express all z scores with two decimal places.

1. IQ Scores Stanford Binet IQ scores have a mean of 100 and a standard deviation of 16. Albert Einstein reportedly had an IQ of 160.
   a. What is the difference between Einstein's IQ and the mean?
   b. How many standard deviations is that [the difference found in part (a)]?
   c. Convert Einstein's IQ score to a z score.
   d. If we consider "usual" IQ scores to be those that convert to z scores between -2 and 2, is Einstein's IQ usual or unusual?

2. Pulse Rates of Adults Assume that adults have pulse rates (beats per minute) with a mean of 72.9 and a standard deviation of 12.3 (based on data from the National Health Examination). When this exercise question was written, the author's pulse rate was 48.
   a. What is the difference between the author's pulse rate and the mean?
   b. How many standard deviations is that [the difference found in part (a)]?
   c. Convert a pulse rate of 48 to a z score.
   d. If we consider "usual" pulse rates to be those that convert to z scores between -2 and 2, is a pulse rate of 48 usual or unusual? Can you explain why a pulse rate might be unusually low? (The reason for this low pulse rate is *not* that statistics textbook authors are usually in a state that could loosely be described as comatose.)

3. Heights of Men Adult males have heights with a mean of 69.0 in. and a standard deviation of 2.8 in. Find the z scores corresponding to the following:
   a. Actor Danny DeVito, who is 5 ft tall
   b. NBA basketball player Shaquille O'Neal, who is 7 ft 1 in. tall
   c. The author, who is a 69.72-in.-tall golf and tennis "player"

Recommended Assignment
(it's not really as long as it appears):
Exercises 1–10, 13–24.

1. a. 60
   b. 3.75
   c. 3.75
   d. Unusual

2. a. 24.9
   b. 2.02
   c. −2.02
   d. Unusual. Exercise, such as running, tends to lower the pulse rate.

3. a. −3.21
   b. 5.71
   c. 0.26

4. a. 2.90
   b. −2.00
   c. 0

4. Body Temperatures Human body temperatures have a mean of 98.20° and a standard deviation of 0.62°. Convert the given temperatures to z scores.
a. 100°   **b.** 96.96°   **c.** 98.20°

In Exercises 5–8, express all z scores with two decimal places. Consider a score to be unusual if its z score is less than −2.00 or greater than 2.00.

5. 2.56; unusual

5. Heights of Women The Beanstalk Club is limited to women and men who are very tall. The minimum height requirement for women is 70 in. Women's heights have a mean of 63.6 in. and a standard deviation of 2.5 in. Find the z score corresponding to a woman with a height of 70 in. and determine whether that height is unusual.

6. 2.67; unusual

6. Length of Pregnancy A woman wrote to *Dear Abby* and claimed that she gave birth 308 days after a visit from her husband, who was in the Navy. Lengths of pregnancies have a mean of 268 days and a standard deviation of 15 days. Find the z score for 308 days. Is such a length unusual? What do you conclude?

7. 4.52; yes; patient is ill.

7. Body Temperature Human body temperatures have a mean of 98.20° and a standard deviation of 0.62°. An emergency room patient is found to have a temperature of 101°. Convert 101° to a z score. Is that temperature unusually high? What does it suggest?

8. 1.99; no

8. Cholesterol Levels For men aged between 18 and 24 years, serum cholesterol levels (in mg/100 ml) have a mean of 178.1 and a standard deviation of 40.7 (based on data from the National Health Survey). Find the z score corresponding to a male, aged 18–24 years, who has a serum cholesterol level of 259.0 mg/100 ml. Is this level unusually high?

9. Psychology test, because
   $z = -0.50$ is greater than
   $z = -2.00$

9. Comparing Test Scores Which is relatively better: A score of 85 on a psychology test or a score of 45 on an economics test? Scores on the psychology test have a mean of 90 and a standard deviation of 10. Scores on the economics test have a mean of 55 and a standard deviation of 5.

10. The z scores are 0.47, 0.22, 0.60, so the score of 18 is the highest relative score.

10. Comparing Scores Three students take equivalent tests of a sense of humor and, after the laughter dies down, their scores are calculated. Which is the highest relative score?
a. A score of 144 on a test with a mean of 128 and a standard deviation of 34.
b. A score of 90 on a test with a mean of 86 and a standard deviation of 18.
c. A score of 18 on a test with a mean of 15 and a standard deviation of 5.

11. −3.56; yes

T 11. Weights of Coke Refer to Data Set 17 in Appendix B for the sample of 36 weights of regular Coke. Convert the weight of 0.7901 to a z score. Is 0.7901 an unusual weight for regular Coke?

12. 1.95; no

T 12. Green M&Ms Refer to Data Set 19 in Appendix B for the sample of weights of green M&M candies. Convert the weight of the heaviest green M&M candy to a z score. Is the weight of that heaviest green M&M an unusual weight for green M&Ms?

In Exercises 13–16, use the 40 sorted cotinine levels of smokers listed in Table 2-13. Find the percentile corresponding to the given cotinine level.

13. 149 43   **14.** 210 60   **15.** 35 15   **16.** 250 73

In Exercises 17–24, use the 40 sorted cotinine levels of smokers listed in Table 2-13. Find the indicated percentile or quartile.

17. P_{20} 46   **18.** Q_3 251.5   **19.** P_{75} 251.5   **20.** Q_2 170

21. P_{33} 121   **22.** P_{21} 48   **23.** P_1 0   **24.** P_{85} 280.5

(T) *In Exercises 25–28, use the cholesterol levels of females listed in Data Set 1 of Appendix B. Find the percentile corresponding to the given cholesterol level.*

25. 123 25   **26.** 309 75   **27.** 271 65   **28.** 126 30

(T) *In Exercises 29–36, use the cholesterol levels of females listed in Data Set 1 of Appendix B. Find the indicated percentile or quartile.*

29. P_{85} 415.5   **30.** P_{35} 138   **31.** Q_1 117.5   **32.** Q_3 305

33. P_{18} 98   **34.** P_{36} 146   **35.** P_{58} 254   **36.** P_{96} 600

2-6 Beyond the Basics

37. Units of Measurement When finding a z score for the height of a basketball player in the NBA, how is the result affected if, instead of using inches, all heights are expressed in centimeters? In general, how are z scores affected by the particular unit of measurement that is used?

37. The z score remains the same.

38. Converting a z Score Heights of women have a mean of 63.6 in. and a standard deviation of 2.5 in.
   a. Julia Roberts, who is one of the most successful actresses in recent years, has a height that converts to a z score of 2.16. How tall (in inches) is Julia Roberts?
   b. Female rapper Lil' Kim has a height that converts to a z score of −1.84. How tall (in inches) is Lil' Kim?

38. a. 69 in.
    b. 59 in.

39. Distribution of z Scores
   a. A data set has a distribution that is uniform. If all of the values are converted to z scores, what is the shape of the distribution of the z scores?
   b. A data set has a distribution that is bell-shaped. If all of the values are converted to z scores, what is the shape of the distribution of the z scores?
   c. In general, how is the shape of a distribution affected if all values are converted to z scores?

39. a. Uniform
    b. Bell-shaped
    c. The shape of the distribution remains the same.

40. Fibonacci Sequence Here are the first several terms of the famous Fibonacci sequence: 1, 1, 2, 3, 5, 8, 13.
   a. Find the mean $\bar{x}$ and standard deviation s, then convert each value to a z score. Don't round the z scores; carry as many places as your calculator can handle.
   b. Find the mean and standard deviation of the z scores found in part (a).
   c. If you use any other data set, will you get the same results obtained in part (b)?

40. a. −0.83958379, −0.83958379, −0.61354199, −0.38750021, 0.06458337, 0.74270874, 1.8729177
    b. 0, 1
    c. Yes

41. Cotinine Levels of Smokers Use the sorted cotinine levels of smokers listed in Table 2-13.
   a. Find the interquartile range.
   b. Find the midquartile.
   c. Find the 10–90 percentile range.
   d. Does $P_{50} = Q_2$? If so, does P_{50} *always* equal Q_2?
   e. Does $Q_2 = (Q_1 + Q_3)/2$? If so, does Q_2 *always* equal $(Q_1 + Q_3)/2$?

41. a. 165
    b. 169
    c. 279.5
    d. Yes; yes
    e. No; no

(T) **42.** Interpolation When finding percentiles using Figure 2-15, if the locator L is not a whole number, we round it up to the next larger whole number. An alternative to this procedure is to interpolate so that a locator of 23.75 leads to a value that is 0.75 (or 3/4) of the way between the 23rd and 24th values. Use this method of interpolation to find P_{35} and Q_1 for the weights of bears listed in Data Set 9 of Appendix B.

42. 115.8; 83

43. a. P_{10}, P_{50}, P_{80}
   b. 10, 46, 107.5, 130.5, 170, 209, 239.5 265.5, 289.5
   c. 46, 130.5, 209, 265.5

43. Deciles and Quintiles For a given data set, there are nine **deciles,** denoted by D_1, $D_2, \ldots, D_9$ which separate the sorted data into 10 groups, with about 10% of the values in each group. There are also four **quintiles,** which divide the sorted data into five groups, with about 20% of the values in each group. (Note the difference between quintiles and quantiles, which were described earlier in this section.)
   a. Which percentile is equivalent to D_1? D_5? D_8?
   b. Using the sorted cotinine levels of smokers in Table 2-13, find the nine deciles.
   c. Using the sorted cotinine levels of smokers in Table 2-13, find the four quintiles.

2-7 Exploratory Data Analysis (EDA)

This chapter presents the basic tools for describing, exploring, and comparing data, and the focus of this section is the exploration of data. We begin this section by first defining exploratory data analysis, then we introduce outliers, 5-number summaries, and boxplots.

> ### Definition
>
> **Exploratory data analysis** is the process of using statistical tools (such as graphs, measures of center, and measures of variation) to investigate data sets in order to understand their important characteristics.

Recall that in Section 2-1 we listed five important characteristics of data, and we began with (1) *center*, (2) *variation*, and (3) the nature of the *distribution*. These characteristics can be investigated by calculating the values of the mean and standard deviation, and by constructing a histogram. It is generally important to further investigate the data set to identify any notable features, especially those that could strongly affect results and conclusions. One such feature is the presence of outliers.

Outliers

An **outlier** is a value that is located very far away from almost all of the other values. Relative to the other data, an outlier is an *extreme* value. When exploring a data set, outliers should be considered because they may reveal important information, and they may strongly affect the value of the mean and standard deviation, as well as seriously distorting a histogram. The following example uses an incorrect entry as an example of an outlier, but not all outliers are errors; some outliers are correct values.

 EXAMPLE Cotinine Levels of Smokers When using computer software or a calculator, it is often easy to make keying errors. Refer to the cotinine levels of smokers listed in Table 2-1 with the

Chapter Problem and assume that the first entry of 1 is incorrectly entered as 11111 because you were distracted by a meteorite landing on your porch. The incorrect entry of 11111 is an outlier because it is located very far away from the other values. How does that outlier affect the mean, standard deviation, and histogram?

SOLUTION   When the entry of 1 is replaced by the outlier value of 11111, the mean changes from 172.5 to 450.2, so the effect of the outlier is very substantial. The incorrect entry of 11111 causes the standard deviation to change from 119.5 to 1732.7, so the effect of the outlier here is also substantial. Figure 2-1 in Section 2-3 depicts the histogram for the correct values of cotinine levels of smokers in Table 2-1, but the STATDISK display presented here shows the histogram that results from using the same data with the value of 1 replaced by the incorrect value of 11111. Compare this STATDISK histogram to Figure 2-1 and you can easily see that the presence of the outlier dramatically affects the shape of the distribution.

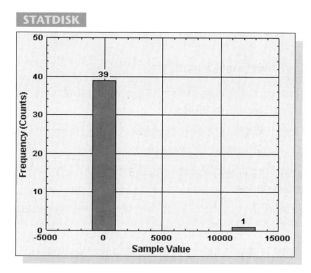

The preceding example illustrates these important principles:

1. **An outlier can have a dramatic effect on the mean.**

2. **An outlier can have a dramatic effect on the standard deviation.**

3. **An outlier can have a dramatic effect on the scale of the histogram so that the true nature of the distribution is totally obscured.**

An easy procedure for finding outliers is to examine a *sorted* list of the data. In particular, look at the minimum and maximum sample values and determine whether they are very far away from the other typical values. Some outliers are correct values and some are errors, as in the preceding example. If we are sure that an outlier is an error, we should correct it or delete it. If we include an outlier because we know that it is correct, we might study its effects by constructing graphs and calculating statistics with and without the outliers included.

An Outlier Tip

Outliers are important to consider because, in many cases, one extreme value can have a dramatic effect on statistics and conclusions derived from them. In some cases an outlier is a mistake that should be corrected or deleted. In other cases, an outlier is a valid data value that should be investigated for any important information. Students of the author collected data consisting of restaurant bills and tips, and no notable outliers were found among their sample data. However, one such outlier is the tip of $16,000 that was left for a restaurant bill of $8,899.78. The tip was left by an unidentified London executive to waiter Lenny Lorando at Nello's restaurant in New York City. Lorando said that he had waited on the customer before and "He's always generous, but never anything like that before. I have to tell my sister about him."

Good Advice for Journalists

Columnist Max Frankel wrote in the *New York Times* that "most schools of journalism give statistics short shrift and some let students graduate without any numbers training at all. How can such reporters write sensibly about trade and welfare and crime, or air fares, health care and nutrition? The media's sloppy use of numbers about the incidence of accidents or disease frightens people and leaves them vulnerable to journalistic hype, political demagoguery, and commercial fraud." He cites several cases, including an example of a full-page article about New York City's deficit with a promise by the mayor of New York City to close a budget gap of $2.7 billion; the entire article never once mentioned the *total* size of the budget, so the $2.7 billion figure had no context.

Boxplots

In addition to the graphs presented in Section 2-3, a boxplot is another graph that is used often. Boxplots are useful for revealing the center of the data, the spread of the data, the distribution of the data, and the presence of outliers. The construction of a boxplot requires that we first obtain the minimum value, the maximum value, and quartiles, as defined in the 5-number summary.

Definitions

For a set of data, the **5-number summary** consists of the minimum value; the first quartile, Q_1; the median (or second quartile, Q_2); the third quartile, Q_3; and the maximum value.

A **boxplot** (or **box-and-whisker diagram**) is a graph of a data set that consists of a line extending from the minimum value to the maximum value, and a box with lines drawn at the first quartile, Q_1; the median; and the third quartile, Q_3. (See Figure 2-16.)

Procedure for Constructing a Boxplot

1. Find the 5-number summary consisting of the minimum value, Q_1, the median, Q_3, and the maximum value.
2. Construct a scale with values that include the minimum and maximum data values.
3. Construct a box (rectangle) extending from Q_1 to Q_3, and draw a line in the box at the median value.
4. Draw lines extending outward from the box to the minimum and maximum data values.

Boxplots don't show as much detailed information as histograms or stem-and-leaf plots, so they might not be the best choice when dealing with a single data set. They are often great for comparing two or more data sets. When using two or more boxplots for comparing different data sets, it is important to use the same scale so that correct comparisons can be made.

EXAMPLE Cotinine Levels of Smokers Refer to the 40 cotinine levels of smokers in Table 2-1 (without the error of 11111 used in place of 1, as in the preceding example).

a. Find the values constituting the 5-number summary.

b. Construct a boxplot.

SOLUTION

a. The 5-number summary consists of the minimum, Q_1, median, Q_3, and maximum. To find those values, first sort the data (by arranging them in order from lowest to highest). The minimum of 0 and the maximum of 491 are easy to identify from the sorted list. Now proceed to find the quartiles. Using the flowchart of Figure 2-15, we get $Q_1 = P_{25} = 86.5$, which is located by calculating the locator $L = (25/100)40 = 10$ and finding the value midway between the 10th value and the 11th value in the sorted list. The median is 170, which is the value midway between the 20th and 21st values. We also find that $Q_3 = 251.5$ by using Figure 2-15 for the 75th percentile. The 5-number summary is therefore 0, 86.5, 170, 251.5, and 491.

b. In Figure 2-16 we graph the boxplot for the data. We use the minimum (0) and the maximum (491) to determine a scale of values, then we plot the values from the 5-number summary as shown.

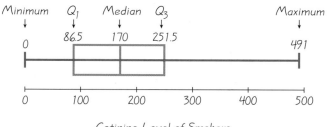

Cotinine Level of Smokers

FIGURE 2-16 **Boxplot**

In Figure 2-17 we show some generic boxplots along with common distribution shapes. It appears that the cotinine levels of smokers have a skewed distribution.

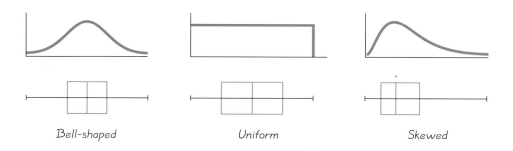

Bell-shaped   Uniform   Skewed

FIGURE 2-17 **Boxplots Corresponding to Bell-Shaped, Uniform, and Skewed Distributions**

To illustrate the use of boxplots to compare data sets, see the accompanying Minitab display of cholesterol levels for a sample of males and a sample of females, based on the National Health Examination data included in Data Set 1 of Appendix B. Based on the sample data, it appears that males have cholesterol levels that are generally higher than females, and the cholesterol levels of males appear to vary more than those of females.

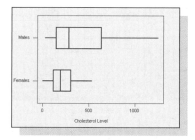

"Best" Colleges

Each year, *U.S. News and World Report* publishes an issue with a list of "America's Best Colleges and Universities." Sales typically jump 40% for that issue. The list has critics who argue against the criteria and method of collecting data. Common complaints: Too much emphasis is placed on the criteria of a college's wealth, reputation, College Board scores, alumni donations, and the opinions of college presidents; too little emphasis is placed on the satisfaction of students and effective educational practices. The *New York Times* interviewed Kenneth Auchincloss, who is editor of *How to Get Into College* (by Kaplan / Newsweek), and he said that "We have never been comfortable trying to quantify in numeric terms the various criteria that go into making a college good or less good, and we don't want to devote the resources to doing an elaborate statistical analysis that frankly we don't think is valid."

EXAMPLE Does It Rain More on Weekends? Refer to Data Set 11 in Appendix B, which lists rainfall amounts (in inches) in Boston for every day of a recent year. The collection of this data set was inspired by media reports that it rains more on weekends (Saturday and Sunday) than on weekdays. Later in this book we will describe important statistical methods that can be used to formally test that claim, but for now, let's explore the data set to see what can be learned. (Even if we already know how to apply those formal statistical methods, we should first explore the data before proceeding with the formal analysis.)

SOLUTION   Let's begin with an investigation into the key elements of center, variation, distribution, outliers, and characteristics over time (the same "CVDOT" list introduced in Section 2-1). Listed below are measures of center (mean), measures of variation (standard deviation), and the 5-number summary for the rainfall amounts for each day of the week. The accompanying STATDISK display shows boxplots for each of the seven days of the week, starting with Monday at the top. Because the histograms for all seven days are pretty much the same, we show only the histogram for the Monday rainfall amounts.

	Mean	Standard Deviation	Minimum	Q_1	Median	Q_3	Maximum
Monday	0.100	0.263	0.000	0.000	0.000	0.010	1.410
Tuesday	0.058	0.157	0.000	0.000	0.000	0.015	0.740
Wednesday	0.051	0.135	0.000	0.000	0.000	0.010	0.640
Thursday	0.069	0.167	0.000	0.000	0.000	0.040	0.850
Friday	0.095	0.228	0.000	0.000	0.000	0.040	0.960
Saturday	0.143	0.290	0.000	0.000	0.000	0.100	1.480
Sunday	0.068	0.200	0.000	0.000	0.000	0.010	1.280

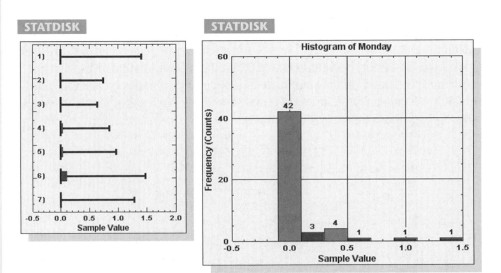

INTERPRETATION Examining and comparing the statistics and graphs, we make the following important observations.

- *Means:* The means vary from a low of 0.051 in. to a high of 0.143 in. The seven means vary by considerable amounts, and in later chapters of this book we will present methods for determining whether these differences are *significant*. (Later methods will show that the means do not differ by significant amounts.) If we list the means in order from low to high, we get this sequence of days: Wednesday, Tuesday, Sunday, Thursday, Friday, Monday, Saturday. There does not appear to be a pattern of higher rainfall on weekends (although the highest mean corresponds to Saturday). Also, see the Excel graph of the seven means, with the mean for Monday plotted first. The Excel graph does not support the claim of more rainfall on weekends (although it might be argued that there is more rainfall on Saturdays).

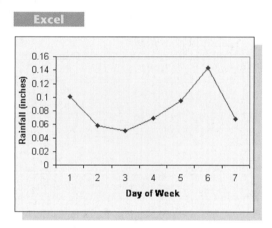

- *Variation*: The seven standard deviations vary from 0.135 in. to 0.290 in., but those values are not dramatically different. There does not appear to be anything highly unusual about the amounts of variation.

- The *minimums, first quartiles,* and *medians* are all 0.00 for each of the seven days. This is explained by the fact that for each day of the week, there are many days with no rain. The abundance of zeros is also seen in the boxplots and histograms, which show that the data have distributions that are heavy toward the low end (skewed right).

- *Outliers*: There are no outliers or unusual values. At the low end, there are many rainfall amounts of zero. At the high end, the sorted list of all 365 rainfall amounts ends with the high values of 0.92, 0.96, 1.28, 1.41, and 1.48.

- *Distributions:* The distributions of the rainfall amounts are skewed to the right. They are not bell-shaped, as we might have expected. If the use of a particular method of statistics requires normally distributed (bell-shaped) populations, that requirement is not satisfied for the rainfall amounts.

We now have considerable insight into the nature of the Boston rainfall amounts for different days of the week. Based on our exploration, we can conclude that Boston does not experience more rain on weekends than on the other days of the week (although we might argue that there is more rainfall on Saturdays).

Critical Thinking

Armed with a list of tools for investigating center, variation, distribution, outliers, and characteristics of data over time, we might be tempted to develop a rote and mindless procedure, but critical thinking is critically important. In addition to using the tools presented in this chapter, we should consider any other relevant factors that might be crucial to the conclusions we form. We might pose questions such as these: Is the sample likely to be representative of the population, or is the sample somehow biased? What is the source of the data, and might the source be someone with an interest that could affect the quality of the data? Suppose, for example, that we want to estimate the mean income of college students. Also suppose that we mail questionnaires to 500 students and receive 20 responses. We could calculate the mean, standard deviation, construct graphs, identify outliers, and so on, but the results will be what statisticians refer to as hogwash. The sample is a voluntary response sample, and it is not likely to be representative of the population of all college students. In addition to the specific statistical tools presented in this chapter, we should also *think!*

Using Technology

This section introduced outliers, 5-number summaries, and boxplots. To find outliers, sort the data in order from lowest to highest, then examine the highest and lowest values to determine whether they are far away from the other sample values. STAT-DISK, Minitab, Excel, and the TI-83 Plus calculator can provide values of quartiles, so the 5-number summary is easy to find. STATDISK, Minitab, Excel, and the TI-83 Plus calculator can be used to create boxplots, and we now describe the different procedures. (*Caution:* Remember that quartile values calculated by Minitab and the TI-83 Plus calculator may differ slightly from those calculated by applying Figure 2-15, so the boxplots may differ slightly as well.)

STATDISK   Enter the data in the Data Window, then click on **Data,** then **Boxplot.** Click on the columns you want to include, then click on **Plot.**

Minitab   Enter the data in column C1, then select **Graph,** then **Boxplot.** With Release 14 select a format from the first column. Enter C1 in the left upper entry cell, then click **OK.**

Excel   Although Excel is not designed to generate boxplots, they can be generated using the Data Desk XL add-in that is a supplement to this book. First enter the data in column A. Click on **DDXL** and select **Charts and Plots.** Under Function Type, select the option of **Boxplot.** In the dialog box, click on the pencil icon and enter the range of data, such as A1:A40 if you have 40 values listed in column A. Click on **OK.** The result is a modified boxplot as described in Exercise 13. The values of the 5-number summary are also displayed.

TI-83 Plus   Enter the sample data in list L1. Now select **STAT PLOT** by pressing the 2nd key followed by the key labeled Y =. Press the **ENTER** key, then select the option of **ON,** and select the boxplot type that is positioned in the middle of the second row. The Xlist should indicate L1 and the Freq value should be 1. Now press the **ZOOM** key and select option 9 for **ZoomStat.** Press the **ENTER** key and the boxplot should be displayed. You can use the arrow keys to move right or left so that values can be read from the horizontal scale.

Recommended Assignment
Exercises 1–4 and 9–12. Note that if students generate boxplots using calculators or computers, the quartile values may differ from those given as answers in this text.

2-7 Basic Skills and Concepts

1. *Lottery* Refer to Data Set 26 and use only the 40 digits in the first column of the Win 4 results from the New York State Lottery (9, 7, 0, and so on). Find the 5-number summary and construct a boxplot. What characteristic of the boxplot suggests that the digits are selected with a random and fair procedure?

2. Movie Budgets Refer to Data Set 21 in Appendix B for the budget amounts of the 15 movies that are R-rated. Find the 5-number summary and construct a boxplot. Determine whether the sample values are likely to be representative of movies made this year.

3. Cereal Calories Refer to Data Set 16 in Appendix B for the 16 values consisting of the calories per gram of cereal. Find the 5-number summary and construct a boxplot. Determine whether the sample values are likely to be representative of the cereals consumed by the general population.

4. Nicotine in Cigarettes Refer to Data Set 5 for the 29 amounts of nicotine (in mg per cigarette). Find the 5-number summary and construct a boxplot. Are the sample values likely to be representative of cigarettes smoked by an individual consumer?

5. Red M&Ms Refer to Data Set 19 for the 21 weights (in grams) of the red M&M candies. Find the 5-number summary and construct a boxplot. Are the red sample values likely to be representative of M&M candies of all colors?

T 6. Bear Lengths Refer to Data Set 9 for the lengths (in inches) of the 54 bears that were anesthetized and measured. Find the 5-number summary and construct a boxplot. Does the distribution of the lengths appear to be symmetric or does it appear to be skewed?

T 7. Alcohol in Children's Movies Refer to Data Set 7 for the 50 times (in seconds) of scenes showing alcohol use in animated children's movies. Find the 5-number summary and construct a boxplot. Based on the boxplot, does the distribution appear to be symmetric or is it skewed?

T 8. Body Temperatures Refer to Data Set 4 in Appendix B for the 106 body temperatures for 12 A.M. on day 2. Find the 5-number summary and construct a boxplot, then determine whether the sample values support the common belief that the mean body temperature is 98.6°F.

In Exercises 9–12, find 5-number summaries, construct boxplots, and compare the data sets.

9. Academy Awards In "Ages of Oscar-Winning Best Actors and Actresses" (*Mathematics Teacher* magazine) by Richard Brown and Gretchen Davis, the authors compare the ages of actors and actresses at the time they won Oscars. The results for winners from both categories are listed in the following table. Use boxplots to compare the two data sets.

Actors:	32	37	36	32	51	53	33	61	35	45	55
	39	76	37	42	40	32	60	38	56	48	48
	40	43	62	43	42	44	41	56	39	46	31
	47	45	60	46	40	36					
Actresses:	50	44	35	80	26	28	41	21	61	38	49
	33	74	30	33	41	31	35	41	42	37	26
	34	34	35	26	61	60	34	24	30	37	31
	27	39	34	26	25	33					

T 10. Regular/Diet Coke Refer to Data Set 17 in Appendix B and use the weights of regular Coke and the weights of diet Coke. Does there appear to be a significant difference? If so, can you provide an explanation?

1. 0, 2, 5, 7, 9. The separations in the boxplot are approximately the same, indicating that the values are equally likely.

2. 0.25, 8, 18.5, 70, 100. No, movie budgets change over time.

3. 3.3, 3.6, 3.75, 3.95, 4.1. No, cereal consumption is not uniformly distributed among the brands, so weighted values should be used.

4. 0.1, 0.8, 1.0, 1.2, 1.4. An individual smoker is likely to stick with one brand, so the sample values are not representative for an individual.

5. 0.870, 0.891, 0.908, 0.924, 0.983; yes

6. 36.0, 50.0, 60.75, 66.5, 76.5. Approximately symmetric.

7. 0, 0, 1.5, 39, 414. Skewed.

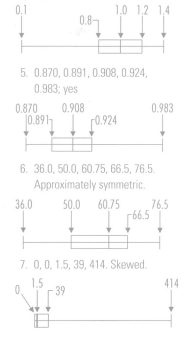

8. 96.5, 97.8, 98.4, 98.6, 99.6. The center does not appear to be 98.6.

9. Actors: 31, 37, 43, 51, 76.
Actresses: 21, 30, 34, 41, 80.
Actresses appear to be younger.

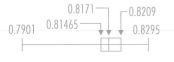

10. Regular: 0.7901, 0.81465, 0.8171, 0.8209, 0.8295.
Diet: 0.7758, 0.7822, 0.7852, 0.7879, 0.7923.
Diet Coke appears to weigh less.

Regular:

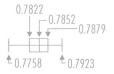

Diet:

11. ETS: 0, 1, 1.5, 32, 551. NOETS: 0, 0, 0, 0, 309. Differences are significant and show that cotinine increases with exposure or use of tobacco.

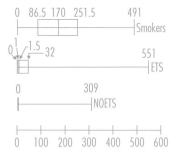

11. Cotinine Levels Refer to Table 2-1 located in the Chapter Problem. We have already found that the 5-number summary for the cotinine levels of smokers is 0, 86.5, 170, 251.5, and 491. Find the 5-number summaries for the other two groups, then construct the three boxplots using the same scale. Are there any apparent differences?

12. Clancy, Rowling, Tolstoy Refer to Data Set 14 in Appendix B and use the Flesch reading ease scores for the sample pages from Tom Clancy's *The Bear and the Dragon,* J. K. Rowling's *Harry Potter and the Sorcerer's Stone,* and Leo Tolstoy's *War and Peace.* (Higher scores indicate easier reading.) Does there appear to be a difference in ease of reading? Are the results consistent with your expectations?

2-7 Beyond the Basics

13. The boxplots discussed in this section are often called *skeletal* (or *regular*) boxplots. **Modified boxplots** are constructed as follows:

a. Find the *IQR*, which denotes the interquartile range defined by $IQR = Q_3 - Q_1$.

b. Draw the box with the median and quartiles as usual, but when drawing the lines to the right and left of the box, draw the lines only as far as the points corresponding to the largest and smallest values that are within 1.5 *IQR* of the box.

c. Mild outliers, plotted as *solid* dots, are values below Q_1 or above Q_3 by an amount that is greater than 1.5 *IQR* but not greater than 3 *IQR*. That is, mild outliers are values *x* such that

$$Q_1 - 3\ IQR \le x < Q_1 - 1.5\ IQR$$

or

$$Q_3 + 1.5\ IQR < x \le Q_3 + 3\ IQR$$

d. Extreme outliers, plotted as small *hollow* circles, are values that are either below Q_1 by more than 3 *IQR* or above Q_3 by more than 3 *IQR*. That is, extreme outliers are values *x* such that

$$x < Q_1 - 3\ IQR$$

or

$$x > Q_3 + 3\ IQR$$

The accompanying figure is an example of a modified boxplot. Refer to the cotinine levels of smokers in Table 2-1 included with the Chapter Problem. We have found that this data set has a 5-number summary of 0, 86.5, 170, 251.5, and 491. Identify the value of *IQR*, identify the ranges of values used to identify mild and extreme outliers, then identify any actual mild outliers or extreme outliers.

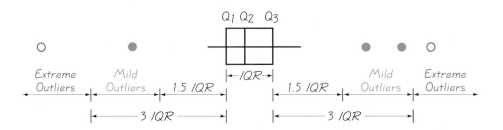

14. Refer to the accompanying STATDISK display of three boxplots that represent the measure longevity (in months) of samples of three different car batteries. If you are the manager of a fleet of cars and you must select one of the three brands, which boxplot represents the brand you should choose? Why?

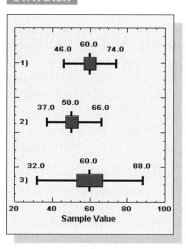

Review

In this chapter we considered methods for describing, exploring, and comparing data sets. When investigating a data set, these characteristics are generally very important:

1. *Center:* A representative or average value.

2. *Variation:* A measure of the amount that the values vary.

3. *Distribution:* The nature or shape of the distribution of the data (such as bell-shaped, uniform, or skewed).

4. *Outliers:* Sample values that lie very far away from the vast majority of the other sample values.

5. *Time:* Changing characteristics of the data over time.

After completing this chapter you should be able to do the following:

- Summarize data by constructing a frequency distribution or relative frequency distribution (Section 2-2)

- Visually display the nature of the distribution by constructing a histogram, dotplot, stem-and-leaf plot, pie chart, or Pareto chart (Section 2-3)

- Calculate measures of center by finding the mean, median, mode, and midrange (Section 2-4)

- Calculate measures of variation by finding the standard deviation, variance, and range (Section 2-5)

- Compare individual values by using *z* scores, quartiles, or percentiles (Section 2-6)

- Investigate and explore the spread of data, the center of the data, and the range of values by constructing a boxplot (Section 2-7)

12. Clancy: 43.9, 66.9, 73, 76.35, 89.2.
    Rowling: 70.9, 78.9, 81.35, 84.45, 86.2.
    Tolstoy: 51.9, 61.3, 68.95, 71.9, 74.4.

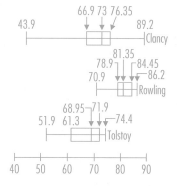

13. IQR = 165.
    Mild outliers: values x such as that $-408.5 \leq x < -161$ or $499 < x \leq 746.5$.
    Extreme outliers: values x such that $x < -408.5$ or $x > 746.5$. There are no mild outliers or extreme outliers.

14. Select the batteries represented by the top boxplot. They have the best combination of highest mean and lowest variation.

1. a. 54.8 years
   b. 55.0 years
   c. 51 years, 54 years
   d. 55.5 years
   e. 27.0 years
   f. 6.2 years
   g. 38.7 years^2
   h. 51 years
   i. 58 years
   j. 47 years
2. a. -1.90
   b. No, because the z score is within two standard deviations of the mean.
   c. 42, 68, 69
   d. Yes; yes

3.

Age	Frequency
40–44	2
45–49	6
50–54	13
55–59	12
60–64	7
65–69	3

4. Bell-shaped

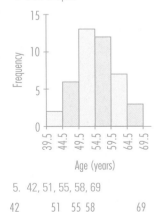

Age (years)

5. 42, 51, 55, 58, 69

In addition to creating these tables, graphs, and measures, you should be able to understand and interpret the results. For example, you should clearly understand that the standard deviation is a measure of how much data vary, and you should be able to use the standard deviation to distinguish between values that are usual and those that are unusual.

Review Exercises

1. Ages of Presidents Senator Hayes is considering a run for the U.S. presidency, but she is only 35 years of age, which is the minimum required age. While investigating this issue, she finds the ages of past presidents when they were inaugurated, and those ages are listed below. Using the ages listed, find the (a) mean; (b) median; (c) mode; (d) midrange; (e) range; (f) standard deviation; (g) variance; (h) Q_1; (i) Q_3; (j) P_{10}.

57	61	57	57	58	57	61	54	68	51	49	64	50	48
65	52	56	46	54	49	51	47	55	55	54	42	51	56
55	51	54	51	60	62	43	55	56	61	52	69	64	46
54													

2. a. John F. Kennedy was 43 years of age when he was inaugurated. Using the results from Exercise 1, convert his age to a z score.
   b. Is Kennedy's age of 43 years "unusual"? Why or why not?
   c. Using the range rule of thumb, identify any other listed ages that are unusual.
   d. Although the list of ages does not include an age of 35 years, would that age be unusual? Is it likely that a presidential candidate of age 35 would find that his or her age would be a major campaign issue?

3. Frequency Distribution Using the same ages listed in Exercise 1, construct a frequency distribution. Use six classes with 40 as the lower limit of the first class, and use a class width of 5.

4. Histogram Using the frequency distribution from Exercise 3, construct a histogram and identify the general nature of the distribution (such as uniform, bell-shaped, skewed).

5. Boxplot Using the same ages listed in Problem 1, construct a boxplot and identify the values constituting the 5-number summary.

6. Empirical Rule Assume the ages of past, present, and future presidents have a bell-shaped distribution with a mean of 54.8 years and a standard deviation of 6.2 years.
   a. What does the empirical rule say about the percentage of ages between 48.6 years and 61.0 years (or within 1 standard deviation of the mean)?
   b. What does the empirical rule say about the percentage of ages between 42.4 years and 67.2 years?

7. Comparing Scores An industrial psychologist for the Citation Corporation develops two different tests to measure job satisfaction. Which score is better: A score of 72 on the management test, which has a mean of 80 and a standard deviation of 12, or a score of 19 on the test for production employees, which has a mean of 20 and a standard deviation of 5. Explain.

8. a. Estimate the mean age of cars driven by students at your college.
   b. Use the range rule of thumb to make a rough estimate of the standard deviation of the ages of cars driven by students at your college.

9. Transforming Data A statistics professor has found that the times used by students on her final exam have a mean of 135 minutes and a standard deviation of 15 minutes. She plans to add a new question that will require 5 additional minutes from each student.

a. What is the mean after the new question is included?

b. What is the standard deviation after the new question is included?

c. What is the variance after the new question is included?

10. Air Passenger Complaints In a recent year, there were 23,000 complaints filed by air passengers. The complaint categories and frequencies provided by the Department of Transportation are as follows: customer care (4370); problems with the flight (9200); reservations, ticketing, boarding (1610); baggage (3450); obtaining refunds (1150); and other reasons (3220). Construct a Pareto chart summarizing the given data.

Cumulative Review Exercises

1. Wristwatch Errors As part of a project for a statistics class, a student collects data on wristwatch accuracy. The following error times (in seconds) are obtained. (Positive values represent watches that are ahead of the correct time, and negative values represent watches that are behind the correct time.)

140 −125 105 −241 −85 41 186 −151 325 80 27 20 20 30 −65

a. Find the mean, median, mode, and midrange.

b. Find the standard deviation, variance, and range.

c. Are the given times from a population that is discrete or continuous?

d. What is the level of measurement of these values? (nominal, ordinal, interval, ratio)

2. a. A set of data is at the nominal level of measurement and you want to obtain a representative data value. Which of the following is most appropriate: mean, median, mode, or midrange? Why?

b. A sample is obtained by telephoning the first 250 people listed in the local telephone directory. What type of sampling is being used (random, stratified, systematic, cluster, convenience)?

c. An exit poll is conducted by surveying everyone who leaves the polling booth at 50 randomly selected election precincts. What type of sampling is being used (random, stratified, systematic, cluster, convenience)?

d. A manufacturer makes refill cartridges for computer printers. A manager finds that the amounts of ink placed in the receptacles is not very consistent, so that some cartridges last longer than expected, but others run out too soon. She wants to improve quality by making the cartridges more consistent. When analyzing the amounts of ink, which of the following statistics is most relevant: mean, median, mode, midrange, standard deviation, first quartile, third quartile? Should the value of that statistic be raised, lowered, or left unchanged?

3. Energy Consumption Each year, the U.S. Energy Department publishes an *Annual Energy Review* that includes the per capita energy consumption (in millions of Btu) for each of the 50 states. If you calculate the mean of these 50 values, is the result the mean per capita energy consumption for the population in all 50 states combined? If it is not, explain how you would calculate the mean per capita energy consumption for all 50 states combined.

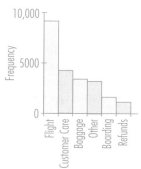

Cooperative Group Activities

1. **Out-of-class activity** Are estimates influenced by anchoring numbers? In the article "Weighing Anchors" in *Omni* magazine, author John Rubin observed that when people estimate a value, their estimate is often "anchored" to (or influenced by) a preceding number, even if that preceding number is totally unrelated to the quantity being estimated. To demonstrate this, he asked people to give a quick estimate of the value of $8 \times 7 \times 6 \times 5 \times 4 \times 3 \times 2 \times 1$. The average answer given was 2250, but when the order of the numbers was reversed, the average became 512. Rubin explained that when we begin calculations with larger numbers (as in $8 \times 7 \times 6$), our estimates tend to be larger. He noted that both 2250 and 512 are far below the correct product, 40,320. The article suggests that irrelevant numbers can play a role in influencing real estate appraisals, estimates of car values, and estimates of the likelihood of nuclear war.

   Conduct an experiment to test this theory. Select some subjects and ask them to quickly estimate the value of

   $$8 \times 7 \times 6 \times 5 \times 4 \times 3 \times 2 \times 1$$

   Then select other subjects and ask them to quickly estimate the value of

   $$1 \times 2 \times 3 \times 4 \times 5 \times 6 \times 7 \times 8$$

   Record the estimates along with the particular order used. Carefully design the experiment so that conditions are uniform and the two sample groups are selected in a way that minimizes any bias. Don't describe the theory to subjects until after they have provided their estimates. Compare the two sets of sample results by using the methods of this chapter. Provide a printed report that includes the data collected, the detailed methods used, the method of analysis, any relevant graphs and/or statistics, and a statement of conclusions. Include a critique of reasons why the results might not be correct and describe ways in which the experiment could be improved.

   A variation of the preceding experiment is to survey people about their knowledge of the population of Kenya. First ask half of the subjects whether they think the population is above 5 million or below 5 million, then ask them to estimate the population with an actual number. Ask the other half of the subjects whether they think the population is above 80 million or below 80 million, then ask them to estimate the population. (Kenya's population is 28 million.) Compare the two sets of results and identify the "anchoring" effect of the initial number that the survey subjects are given.

2. **Out-of-class activity** In each group of three or four students, collect an original data set of values at the interval or ratio levels of measurement. Provide the following: (1) a list of sample values, (2) printed computer results of descriptive statistics and graphs, and (3) a written description of the nature of the data, the method of collection, and important characteristics.

3. **In-class activity** Given below are the ages of motorcyclists at the time they were fatally injured in traffic accidents (based on data from the U.S. Department of Transportation). If your objective is to dramatize the dangers of motorcycles for young people, which would be most effective: histogram, Pareto chart, pie chart, dotplot, mean, median . . . ? Construct the graph and find the statistic that best meets that objective. Is it okay to deliberately distort data if the objective is one such as saving lives of motorcyclists?

17	38	27	14	18	34	16
42	28	24	40	20	23	31
37	21	30	25	17	28	33
25	23	19	51	18	29	

4. **Out-of-class activity** In each group of three or four students, select one of the following items and construct a graph that is effective in addressing the question:
   a. Is there a difference between the body mass index (BMI) values for men and for women? (See Data Set 1 in Appendix B.)
   b. Is there a relationship between the heights of sons (or daughters) and the heights of their fathers (or mothers)? (See Data Set 2 in Appendix B.)
   c. Do the digits generated by the New York State Win 4 lottery game appear to be randomly selected, or do they somehow appear to be biased? (See Data Set 26 in Appendix B.)

Technology Project

When dealing with larger data sets, manual entry of data can become quite tedious and time-consuming. There are better things to do with your time, such as mastering the aerodynamic principles of a Frisbee. Refer to Data Set 30 in Appendix B, which includes home-run distances of three exceptional baseball players: Barry Bonds (2001 season), Mark McGwire (1998 season), and Sammy Sosa (1998 season). Instead of manually entering the 209 distances in the three sets of data, use a TI-83 Plus calculator or STATDISK, Minitab, or Excel to load the data sets, which are available on the CD included with this book. Proceed to generate histograms and find appropriate statistics that allow you to compare the three sets of data. Are there any significant differences? Are there any outliers? Does it appear that more home runs are hit by players who hit farther? Why or why not? Analyze the last digits of the distances and determine whether the values appear to be estimates or measurements. Write a brief report including your conclusions and supporting graphs.

Data Set 30: Homerun Distances

Homerun distances are in feet for Mark McGwire (1998), Sammy Sosa (1998), and Barry Bonds (2001).

STATDISK: Data set name is Homeruns.
Minitab: Worksheet name is HOMERUNS.MTW.
Excel: Workbook name is HOMERUNS.XLS.
**TI-83 Plus: App name is HOMERUNS, and the file names are the same as for
    text files.**
Text file names: MCGWR, SOSA, BONDS.

McGwire

360	370	370	430	420	340	460	410	440	410
380	360	350	527	380	550	478	420	390	420
425	370	480	390	430	388	423	410	360	410
450	350	450	430	461	430	470	440	400	390
510	430	450	452	420	380	470	398	409	385
369	460	390	510	500	450	470	430	458	380
430	341	385	410	420	380	400	440	377	370

Sosa

371	350	430	420	430	434	370	420	440	410
420	460	400	430	410	370	370	410	380	340
350	420	410	415	430	380	380	366	500	380
390	400	364	430	450	440	365	420	350	420
400	380	380	400	370	420	360	368	430	433
388	440	414	482	364	370	400	405	433	390
480	480	434	344	410	420				

Bonds

420	417	440	410	390	417	420	410	380	430
370	420	400	360	410	420	391	416	440	410
415	436	430	410	400	390	420	410	420	410
410	450	320	430	380	375	375	347	380	429
320	360	375	370	440	400	405	430	350	396
410	380	430	415	380	375	400	435	420	420
488	361	394	410	411	365	360	440	435	454
442	404	385							

from DATA to DECISION

Critical Thinking

Car crash fatalities are devastating to the families involved, and they often involve lawsuits and large insurance payments. Listed below are the ages of 100 randomly selected drivers who were killed in car crashes. Also given is a frequency distribution of licensed drivers by age.

Ages (in years) of Drivers Killed in Car Crashes

37	76	18	81	28	29	18	18	27	20
18	17	70	87	45	32	88	20	18	28
17	51	24	37	24	21	18	18	17	40
25	16	45	31	74	38	16	30	17	34
34	27	87	24	45	24	44	73	18	44
16	16	73	17	16	51	24	16	31	38
86	19	52	35	18	18	69	17	28	38
69	65	57	45	23	18	56	16	20	22
77	18	73	26	58	24	21	21	29	51
17	30	16	17	36	42	18	76	53	27

Age	Licensed Drivers (millions)
16–19	9.2
20–29	33.6
30–39	40.8
40–49	37.0
50–59	24.2
60–69	17.5
70–79	12.7
80–89	4.3

Analysis

Convert the given frequency distribution to a relative frequency distribution, then create a relative frequency distribution for the ages of drivers killed in car crashes. Compare the two relative frequency distributions. Which age categories appear to have substantially greater proportions of fatalities than the proportions of licensed drivers? If you were responsible for establishing the rates for auto insurance, which age categories would you select for higher rates? Construct a graph that is effective in identifying age categories that are more prone to fatal car crashes.

INTERNET PROJECT   Data On The Internet

The Internet is host to a wealth of information and much of that information comes in the form of raw data, which can be studied and summarized using the statistics introduced in this chapter. For example, we found the following information with just a few clicks:

- The value of stock in Walt Disney Corporation tends to rise during the winter months but varies the rest of the year. In 2001, the maximum and minimum stock prices differed by over 17 points.
- Ichiro Suzuki, outfielder for the Seattle Mariners baseball team, had a slugging percentage of .457 in the 2001 season.

- The populations of California, New York, and Texas account for over 25% of the total U.S. population.

The Internet Project for this chapter, found at the *Elementary Statistics* Web site, will point you to data sets in the areas of sports, finance, and the weather. Once you have assembled a data set, you will apply the methods of this chapter to summarize and classify the data.

The Web site for this chapter can be found at

http://www.aw.com/triola

Statistics @ Work

"The journalist must be able to cast a critical eye on research, divine the true context and significance of the work."

Mark Fenton

Editor at Large Walking Magazine

Mark Fenton is also a walking advocate and is a champion walker. He was on the U.S. national racewalking team five times and has represented the United States in numerous international competitions. He has studied biomechanics and exercise physiology at the Olympic Training Center's Sports Science Laboratory in Colorado Springs, Colorado.

What concepts of statistics do you use?

I have to be familiar with all the common tools used in exercise science and public health research statistical analyses; from means and standard deviations, to statistical significance of differences, confidence intervals, and analyses of variance, as examples.

How do you use statistics in your job?

I regularly read medical research journals (*Medicine and Science in Sports and Exercise* and the *Journal of the American Medical Association* are two most notable) for exercise physiology, public health, and epidemiological research. Thus, I must be conversant in how studies are controlled and analyzed, and understand the relative power or value of a study's outcome. This information is used in my composition of articles and speeches on the findings of such work.

Please describe one specific example illustrating how the use of statistics was successful in improving a product or service.

I regularly read research papers, and have to be cautious of holes in the evidence—dangerously small sample sizes; statistical significance but small absolute differences between groups; extremely large standard deviations, which can obscure the relevance of a finding for the average person.

A simple example is the problem of saying that the AVERAGE maximum heart rate for a 36-year-old woman is 190 beats per minute (226 minus age). That's helpful in calculating target heart rate, until you find that the deviation is quite high, and so the resulting figure can be off by as much as 10 beats per minute for roughly 1/3 of the population—an error large enough to be problematic when exercising. The study that pointed this out has altered how I offer exercise intensity recommendations.

Is your use of probability and statistics increasing, decreasing, or remaining stable?

With my growing work in public health, and the increasing complexity of the statistical tools being used in large population analyses, my conversance in the field MUST continue growing if I am to keep up.

Do you recommend statistics for today's college students?

I absolutely recommend at least one course in statistics, as it is a useful tool in assessing the information with which we are bombarded—much of it without context—every day. At very least a comprehensive introductory course in statistics is a must for anyone interested in science journalism. The journalist must be able to cast a critical eye on research, divine the true context and significance of the work.

3

Probability

3-1 Overview

3-2 Fundamentals

3-3 Addition Rule

3-4 Multiplication Rule: Basics

3-5 Multiplication Rule: Complements and Conditional Probability

3-6 Probabilities Through Simulations

3-7 Counting

False positives and false negatives

All of us undergo a variety of health tests at different times in our lives. Some health tests are as simple as using a thermometer to determine whether body temperature is too high or low, or using a sphygmomanometer to determine whether blood pressure is too high or low. Some other tests involve the analysis of blood samples for identifying the presence of a disease. In this Chapter Problem we consider results from a clinical study of a test for pregnancy. It is important for a woman to know if she is pregnant so that she can discontinue any activities, medications, exposure to toxicants at work, smoking, or alcohol consumption that could be potentially harmful to the baby. Pregnancy tests, like almost all health tests, do not yield results that are 100% accurate. In clinical trials of a blood test for pregnancy, the results shown in Table 3-1 were obtained for the Abbot blood test (based on data from "Specificity and Detection Limit of Ten Pregnancy Tests," by Tiitinen and Stenman, *Scandinavian Journal of Clinical Laboratory Investigation,* Vol. 53, Supplement 216). There are factors that affect the accuracy of these tests, such as timing. Pregnancy tests are typically most reliable when taken at least two weeks after conception. Other tests are more reliable than the test with results given in Table 3-1. For example, the Abbot Testpack Plus is a urine test with a 0.2% false positive rate and a 0.6%

false negative rate. The terms *false positive* and *false negative* are included among these terms commonly used with health tests or screening procedures:

- **False positive:** Test incorrectly indicates pregnancy when the woman is not pregnant.
- **False negative:** Test incorrectly indicates that the woman is not pregnant when she really is.
- **True positive:** Test correctly indicates that the woman is pregnant when she really is.
- **True negative:** Test correctly indicates that the woman is not pregnant when she is not pregnant.
- **Test sensitivity:** The probability of a true positive.
- **Test specificity:** The probability of a true negative.

Based on the results in Table 3-1, what is the probability of a woman being pregnant if the test indicates a negative result? What is the probability of a false positive? We will address these questions in this chapter.

Table 3-1	Pregnancy Test Results	
	Positive Test Result (Pregnancy is indicated)	Negative Test Result (Pregnancy is not indicated)
Subject is pregnant	80	5
Subject is not pregnant	3	11

Overview

Probability is the underlying foundation on which the important methods of inferential statistics are built. As a simple example, suppose that you were to win the top prize in a state lottery five consecutive times. There would be investigations and accusations that you were somehow cheating. People know that even though there is a chance of someone winning five consecutive times with honest luck, that chance is so incredibly low that they would reject luck as a reasonable explanation. This is exactly how statisticians think: They reject explanations based on very low probabilities. Statisticians use the *rare event rule*.

Rare Event Rule for Inferential Statistics

If, under a given assumption (such as a lottery being fair), the probability of a particular observed event (such as five consecutive lottery wins) is extremely small, we conclude that the assumption is probably not correct.

The main objective in this chapter is to develop a sound understanding of probability values, which will be used in following chapters. A secondary objective is to develop the basic skills necessary to determine probability values in a variety of important circumstances.

3-2 Fundamentals

In considering probability, we deal with procedures (such as rolling a die, answering a multiple-choice test question, or undergoing a test for pregnancy) that produce outcomes.

Definitions

An **event** is any collection of results or outcomes of a procedure.

A **simple event** is an outcome or an event that cannot be further broken down into simpler components.

The **sample space** for a procedure consists of all possible *simple* events. That is, the sample space consists of all outcomes that cannot be broken down any further.

EXAMPLES

Procedure	Example of Event	Sample Space
Roll one die	5 (simple event)	{1, 2, 3, 4, 5, 6}
Roll two dice	7 (not a simple event)	{1-1, 1-2, . . . , 6-6}

Note to Instructor

Students typically find that this is the most difficult chapter in the book. Some statistics instructors prefer to cover this chapter thoroughly, and others prefer to include only minimum coverage. It is strongly recommended that at least Section 3-2 be included. Normally, Sections 3-1 through 3-4 could be covered, with Sections 3-5 through 3-7 optional. No sections are marked as "optional" because some instructors experienced student resistance to sections labeled that way.

An important feature of this chapter is the determination of whether some events are "unusual," which is a good introduction to methods of hypothesis testing introduced later in the text.

Note to Instructor

This section covers basic definitions and simple concepts of probability. Emphasize the importance of *understanding* the available data, because some students tend to develop this rule for finding probabilities: Find two numbers, then divide the smaller number by the larger one. For class examples, include at least one example where that rule doesn't work. For example, if a quality control test shows that there are 5 defective printers and there are 15 that are good, the probability of randomly selecting one that is defective is 5/20, not 5/15.

When rolling one die, 5 is a *simple event* because it cannot be broken down any further. When rolling two dice, 7 is *not a simple event* because it can be broken down into simpler events, such as 3-4 or 6-1. When rolling two dice, the *sample space* consists of 36 simple events: 1-1, 1-2, . . . , 6-6. When rolling two dice, the outcome of 3-4 is considered a simple event, because it is an outcome that cannot be broken down any further. We might incorrectly think that 3-4 can be further broken down into the individual results of 3 and 4, but 3 and 4 are not individual outcomes when two dice are rolled. When two dice are rolled, there are exactly 36 outcomes that are simple events: 1-1, 1-2, . . . , 6-6.

There are different ways to define the probability of an event, and we will present three approaches. First, however, we list some basic notation.

Notation for Probabilities

P denotes a probability.
A, B, and C denote specific events.
$P(A)$ denotes the probability of event A occurring.

Rule 1: Relative Frequency Approximation of Probability

Conduct (or observe) a procedure a large number of times, and count the number of times that event A actually occurs. Based on these actual results, $P(A)$ is *estimated* as follows:

$$P(A) = \frac{\text{number of times } A \text{ occurred}}{\text{number of times trial was repeated}}$$

Rule 2: Classical Approach to Probability (Requires Equally Likely Outcomes)

Assume that a given procedure has n different simple events and that each of those simple events has an *equal chance* of occurring. If event A can occur in s of these n ways, then

$$P(A) = \frac{\text{number of ways } A \text{ can occur}}{\text{number of different simple events}} = \frac{s}{n}$$

Rule 3: Subjective Probabilities

$P(A)$, the probability of event A, is found by simply guessing or estimating its value based on knowledge of the relevant circumstances.

STATISTICS IN THE NEWS

Killer Asteroids

The probability of our civilization being destroyed by an asteroid striking our planet has obvious importance to nearly all of us. In June 2002 a *New York Times* article reported that an asteroid "big enough to raze a major city came within 75,000 miles of earth . . . but was not spotted until three days later." When attempting to determine that probability, the relative frequency approach doesn't apply because we can't conduct trials and there is no usable history of such past destruction. The classical approach does not apply because the possible outcomes are not equally likely. Only the approach of subjective probabilities applies.

Based on the most current observations, astronomers estimate that there are 700,000 asteroids large enough and close enough to destroy us. Using that number with knowledge of asteroid orbits, astronomers have developed a subjective probability of our civilization being destroyed sometime during the next 100 years by an asteroid collision: It is approximately 1/5000.

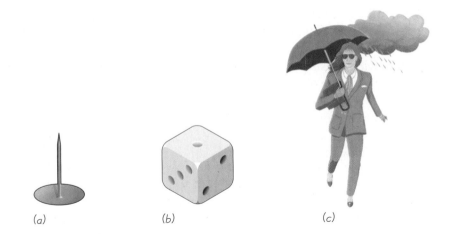

(a)   (b)   (c)

FIGURE 3-1   **Three Approaches to Finding Probability**

(a) Relative Frequency Approach (Rule 1): When trying to determine: P (tack lands point up), we must repeat the procedure of tossing the tack many times and then find the ratio of the number of times the tack lands with the point up to the number of tosses.

(b) Classical Approach (Rule 2): When trying to determine $P(2)$ with a balanced and fair die, each of the six faces has an equal chance of occurring.

$$P(2) = \frac{\text{number of ways 2 can occur}}{\text{total number of simple events}}$$

$$= \frac{1}{6}$$

(c) Subjective Probability (Rule 3): When trying to estimate the probability of rain tomorrow, meteorologists use their expert knowledge of weather conditions to develop an estimate of the probability.

It is very important to note that *the classical approach (Rule 2) requires equally likely outcomes.* If the outcomes are not equally likely, either we must use the relative frequency estimate or we must rely on our knowledge of the circumstances to make an *educated guess.* Figure 3-1 illustrates the three approaches.

When finding probabilities with the relative frequency approach (Rule 1), we obtain an *estimate* instead of an exact value. As the total number of observations increases, the corresponding estimates tend to get closer to the actual probability. This property is stated as a theorem commonly referred to as the *law of large numbers.*

Law of Large Numbers

As a procedure is repeated again and again, the relative frequency probability (from Rule 1) of an event tends to approach the actual probability.

The law of large numbers tells us that the relative frequency estimates from Rule 1 tend to get better with more observations. This law reflects a simple notion supported by common sense: A probability estimate based on only a few trials can be off by substantial amounts, but with a very large number of trials, the estimate tends to be much more accurate. For example, an opinion poll of only a dozen randomly selected people could easily be in error by large amounts, but a poll of thousands of randomly selected people will be fairly close to the true population values.

EXAMPLE   **Flying High**   Find the probability that a randomly selected adult has flown on a commercial airliner.

SOLUTION   The sample space consists of two simple events: The selected person has flown on a commercial airliner or has not. Because the sample space consists of events that are not equally likely, we can't use the classical approach (Rule 2). We can use the relative frequency approach (Rule 1) with these results from a Gallup poll: Among 855 randomly selected adults, 710 indicated that they have flown on commercial airliners. We get the following result.

$$P(\text{flew on commercial airliner}) = \frac{710}{855} \approx 0.830$$

EXAMPLE   **Roulette**   You plan to bet on the number 13 on the next spin of a roulette wheel. What is the probability that you will lose?

SOLUTION   A roulette wheel has 38 different slots, only one of which is the number 13. The roulette wheel is designed so that the 38 slots are equally likely. Among those 38 slots, there are 37 that result in a loss. Because the sample space includes equally likely outcomes, we use the classical approach (Rule 2) to get

$$P(\text{loss}) = \frac{37}{38}$$

EXAMPLE   **Crashing Meteorites**   What is the probability that your car will be hit by a meteorite this year?

SOLUTION   In the absence of historical data on meteorites hitting cars, we cannot use the relative frequency approach of Rule 1. There are two possible outcomes (hit or no hit), but they are not equally likely, so we cannot use the classical approach of Rule 2. That leaves us with Rule 3, whereby we make a subjective estimate. In this case, we all know that the probability in question is very, very small. Let's estimate it to be, say, 0.000000000001 (equivalent to 1 in a trillion). That subjective estimate, based on our general knowledge, is likely to be in the general ballpark of the true probability.

In basic probability problems of the type we are now considering, it is very important to examine the available information carefully and to identify the total number of possible outcomes correctly. In some cases, the total number of possible outcomes is given, but in other cases it must be calculated, as in the following example, which requires us to find the total number of possible outcomes.

Subjective Probabilities at the Racetrack

Researchers studied the ability of racetrack bettors to develop realistic subjective probabilities. (See "Racetrack Betting: Do Bettors Understand the Odds?", by Brown, D'Amato, and Gertner, *Chance* magazine, Vol. 7, No. 3.) After analyzing results for 4400 races, they concluded that although bettors slightly overestimate the winning probabilities of "long shots" and slightly underestimate the winning probabilities of "favorites," their general performance is quite good. The subjective probabilities were calculated from the payoffs, which are based on the amounts bet, and the actual probabilities were calculated from the actual race results.

Note to Instructor

Comment that drawing a model for visualizing the given information can be really helpful in understanding and solving probability problems (and many other problems as well). For the death penalty example, draw a box and write this on the inside: 319 for the death penalty; 133 against; 39 no opinion. Also write "491 total" below the box.

EXAMPLE **Death Penalty** Adults are randomly selected for a Gallup poll, and they are asked if they are in favor of the death penalty for a person convicted of murder. The responses include 319 who are for the death penalty, 133 who are against it, and 39 who have no opinion. Based on these results, estimate the probability that a randomly selected person is for the death penalty.

SOLUTION   *Hint:* Instead of trying to formulate an answer directly from the written statement, summarize the given information in a format that allows you to better understand it. For example:

$$
\begin{array}{rl}
319 & \text{for the death penalty} \\
133 & \text{against the death penalty} \\
\underline{39} & \underline{\text{no opinion}} \\
491 & \text{total}
\end{array}
$$

We can now use the relative frequency approach (Rule 1) as follows:

P(person favors the death penalty)

$$
= \frac{\text{number who favor the death penalty}}{\text{total}} = \frac{319}{491} = 0.650
$$

We estimate that there is a 0.650 probability that when an adult is randomly selected, he or she favors the death penalty for someone convicted of murder. As with all surveys, the accuracy of this result depends on the quality of the sampling method and the survey procedure. Because the poll was conducted by the Gallup organization, the results are likely to be reasonably accurate. Chapter 6 will include more advanced procedures for analyzing such survey results.

Note to Instructor

Students often need help in constructing a sample space such as the one listing the genders of babies. Suggest a systematic approach to constructing such a table, possibly referring to these patterns that are present: Going from right to left, the columns alternate every letter, then every 2 letters, then every 4 letters, and so on.

1st 2nd 3rd

boy-boy-boy

boy-boy-girl

exactly   boy-girl-boy

2 boys   boy-girl-girl

girl-boy-boy

girl-boy-girl

girl-girl-boy

girl-girl-girl

EXAMPLE **Gender of Children** Find the probability that when a couple has 3 children, they will have exactly 2 boys. Assume that boys and girls are equally likely and that the gender of any child is not influenced by the gender of any other child.

SOLUTION   The biggest obstacle here is correctly identifying the sample space. It involves more than working only with the numbers 2 and 3 that were given in the statement of the problem. The sample space consists of 8 different ways that 3 children can occur, and we list them in the margin. Those 8 outcomes are equally likely, so we use Rule 2. Of those 8 different possible outcomes, 3 correspond to exactly 2 boys, so

$$
P(\text{2 boys in 3 births}) = \frac{3}{8} = 0.375
$$

INTERPRETATION   There is a 0.375 probability that if a couple has 3 children, exactly 2 will be boys.

The statements of the three rules for finding probabilities and the preceding examples might seem to suggest that we should always use Rule 2 when a procedure has equally likely outcomes. In reality, many procedures are so complicated that the classical approach (Rule 2) is impractical to use. In the game of solitaire, for example, the outcomes (hands dealt) are all equally likely, but it is extremely frustrating to try to use Rule 2 to find the probability of winning. In such cases we can more easily get good estimates by using the relative frequency approach (Rule 1). Simulations are often helpful when using this approach. (A **simulation** of a procedure is a process that behaves in the same ways as the procedure itself, so that similar results are produced.) For example, it's much easier to use Rule 1 for estimating the probability of winning at solitaire—that is, to play the game many times (or to run a computer simulation)—than to perform the extremely complex calculations required with Rule 2.

> **EXAMPLE** **Thanksgiving Day** If a year is selected at random, find the probability that Thanksgiving Day will be on a (a) Wednesday, (b) Thursday.
>
> **SOLUTION**
>
> **a.** Thanksgiving Day always falls on the fourth Thursday in November. It is therefore impossible for Thanksgiving to be on a Wednesday. When an event is impossible, we say that its probability is 0.
>
> **b.** It is certain that Thanksgiving will be on a Thursday. When an event is certain to occur, we say that its probability is 1.

Because any event imaginable is impossible, certain, or somewhere in between, it follows that the mathematical probability of any event is 0, 1, or a number between 0 and 1 (see Figure 3-2).

- **The probability of an impossible event is 0.**
- **The probability of an event that is certain to occur is 1.**
- **$0 \le P(A) \le 1$ for any event A.**

In Figure 3-2, the scale of 0 through 1 is shown on the left, and the more familiar and common expressions of likelihood are shown on the right.

Complementary Events

Sometimes we need to find the probability that an event A does *not* occur.

Definition

The **complement** of event A, denoted by $\overline{A}$, consists of all outcomes in which event A does *not* occur.

> **EXAMPLE** **Birth Genders** In reality, more boys are born than girls. In one typical group, there are 205 newborn babies, 105 of whom are boys. If one baby is randomly selected from the group, what is the probability that the baby is *not* a boy?
>
> *continued*

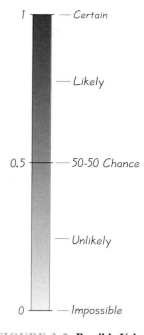

FIGURE 3-2 **Possible Values for Probabilities**

How Probable?

How do we interpret such terms as *probable, improbable,* or *extremely probable?* The FAA interprets these terms as follows. *Probable:* A probability on the order of 0.00001 or greater for each hour of flight. Such events are expected to occur several times during the operational life of each airplane. *Improbable:* A probability on the order of 0.00001 or less. Such events are not expected to occur during the total operational life of a single airplane of a particular type, but may occur during the total operational life of all airplanes of a particular type. *Extremely improbable:* A probabilty on the order of 0.000000001 or less. Such events are so unlikely that they need not be considered to ever occur.

Note to Instructor

In this edition, the distinction between actual odds (based on the likelihood of the event) and the payoff odds (based on the greed of the casino) is made more explicit through the definitions.

SOLUTION   Because 105 of the 205 babies are boys, it follows that 100 of them are girls, so

$$P(\text{not selecting a boy}) = P(\overline{\text{boy}}) = P(\text{girl}) = \frac{100}{205} = 0.488$$

Although it is difficult to develop a universal rule for rounding off probabilities, the following guide will apply to most problems in this text.

Rounding Off Probabilities

When expressing the value of a probability, either give the *exact* fraction or decimal or round off final decimal results to three significant digits. (*Suggestion:* When a probability is not a simple fraction such as 2/3 or 5/9, express it as a decimal so that the number can be better understood.)

All digits in a number are significant except for the zeros that are included for proper placement of the decimal point.

EXAMPLES

- The probability of 0.021491 has five significant digits (21491), and it can be rounded to three significant digits as 0.0215.
- The probability of 1/3 can be left as a fraction, or rounded to 0.333. Do *not* round to 0.3.
- The probability of heads in a coin toss can be expressed as 1/2 or 0.5; because 0.5 is exact, there's no need to express it as 0.500.
- The fraction 432/7842 is exact, but its value isn't obvious, so express it as the decimal 0.0551.

An important concept in this section is the mathematical expression of probability as a number between 0 and 1. This type of expression is fundamental and common in statistical procedures, and we will use it throughout the remainder of this text. A typical computer output, for example, may include a "*P*-value" expression such as "significance less than 0.001." We will discuss the meaning of *P*-values later, but they are essentially probabilities of the type discussed in this section. For now, you should recognize that a probability of 0.001 (equivalent to 1/1000) corresponds to an event so rare that it occurs an average of only once in a thousand trials.

Odds

Expressions of likelihood are often given as *odds,* such as 50:1 (or "50 to 1"). A serious disadvantage of odds is that they make many calculations extremely difficult. As a result, statisticians, mathematicians, and scientists prefer to use probabilities. The advantage of odds is that they make it easier to deal with money transfers associated with gambling, so they tend to be used in casinos, lotteries, and racetracks. Note that in the three definitions that follow, the actual odds against and the actual odds in favor describe the actual likelihood of some event, but the payoff odds describe the relationship between the bet and the amount of the payoff. Racetracks and casinos are in business to make a profit, so the payoff odds will not be the same as the actual odds.

Definition

The **actual odds against** event A occurring are the ratio $P(\overline{A})/P(A)$, usually expressed in the form of $a{:}b$ (or "a to b"), where a and b are integers having no common factors.

The **actual odds in favor** of event A are the reciprocal of the actual odds against that event. If the odds against A are $a{:}b$, then the odds in favor of A are $b{:}a$.

The **payoff odds** against event A represent the ratio of net profit (if you win) to the amount bet.

$$\text{payoff odds against event } A = (\text{net profit}) : (\text{amount bet})$$

EXAMPLE If you bet $5 on the number 13 in roulette, your probability of winning is 1/38 and the payoff odds are given by the casino as 35:1.

a. Find the actual odds against the outcome of 13.

b. How much net profit would you make if you win by betting on 13?

c. If the casino were operating just for the fun of it, and the payoff odds were changed to match the actual odds against 13, how much would you win if the outcome were 13?

SOLUTION

a. With $P(13) = 1/38$ and $P(\text{not } 13) = 37/38$, we get

$$\text{actual odds against } 13 = \frac{P(\text{not } 13)}{P(13)} = \frac{37/38}{1/38} = \frac{37}{1} \text{ or } 37{:}1$$

b. Because the payoff odds against 13 are $35{:}1$, we have

$$35{:}1 = (\text{net profit}){:}(\text{amount bet})$$

so that there is a $35 profit for each $1 bet. For a $5 bet, the net profit is $175. The winning bettor would collect $175 plus the original $5 bet.

c. If the casino were operating for fun and not for profit, the payoff odds would be equal to the actual odds against the outcome of 13. If the payoff odds were changed from 35:1 to 37:1, you would obtain a net profit of $37 for each $1 bet. If you bet $5, your net profit would be $185. (The casino makes its profit by paying only $175 instead of the $185 that would be paid with a roulette game that is fair instead of favoring the casino.)

3-2 Basic Skills and Concepts

In Exercises 1 and 2, express the indicated degree of likelihood as a probability value.

1. Identifying Probability Values
   a. "You have a 50-50 chance of choosing the correct road."
   b. "There is a 20% chance of rain tomorrow."
   c. "You have a snowball's chance in hell of marrying my daughter."

2. Identifying Probability Values
   a. "There is a 90% chance of snow tomorrow."
   b. "It will definitely become dark tonight."
   c. "You have one chance in ten of being correct."

3. Identifying Probability Values Which of the following values *cannot* be probabilities?

$$0, \quad 1, \quad -1, \quad 2, \quad 0.0123, \quad 3/5, \quad 5/3, \quad \sqrt{2}$$

4. Identifying Probability Values
   a. What is the probability of an event that is certain to occur?
   b. What is the probability of an impossible event?
   c. A sample space consists of 10 separate events that are equally likely. What is the probability of each?
   d. On a true/false test, what is the probability of answering a question correctly if you make a random guess?
   e. On a multiple-choice test with five possible answers for each question, what is the probability of answering a question correctly if you make a random guess?

5. Gender of Children In this section, we gave an example that included a list of the eight outcomes that are possible when a couple has three children. Refer to that list, and find the probability of each event.
   a. Among three children, there is exactly one girl.
   b. Among three children, there are exactly two girls.
   c. Among three children, all are girls.

6. Cell Phones and Brain Cancer In a study of 420,000 cell phone users in Denmark, it was found that 135 developed cancer of the brain or nervous system. Estimate the probability that a randomly selected cell phone user will develop such a cancer. Is the result very different from the probability of 0.000340 that was found for the general population? What does the result suggest about cell phones as a cause of such cancers, as has been claimed?

7. Probability of a Home Run Baseball player Barry Bonds broke a major record when he hit 73 home runs in the 2001 season. During that season, he was at bat 476 times. If one of those "at bats" is randomly selected, find the probability that it is one of the times he hit a home run. Is the result very different from the probability of 0.0715 that results from his 567 career home runs in 7932 "at bats"?

8. Being Struck by Lightning In a recent year, 389 of the 281,421,906 people in the United States were struck by lightning. Estimate the probability that a randomly selected person in the United States will be struck by lightning this year.

Using Probability to Identify Unusual Events. In Exercises 9–16, consider an event to be "unusual" if its probability is less than or equal to 0.05. (This is equivalent to the same

Probabilities That Challenge Intuition

In certain cases, our subjective estimates of probability values are dramatically different from the actual probabilities. Here is a classic example: If you take a deep breath, there is better than a 99% chance that you will inhale a molecule that was exhaled in dying Caesar's last breath. In that same morbid and unintuitive spirit, if Socrates' fatal cup of hemlock was mostly water, then the next glass of water you drink will likely contain one of those same molecules. Here's another less morbid example that can be verified: In classes of 25 students, there is better than a 50% chance that at least two students will share the same birthday.

Recommended Assignment:
Exercises 1–18, 25, 26. (Exercises 25 and 26 involve odds.)

1. a. 0.5
   b. 0.20
   c. 0

criterion commonly used in inferential statistics, but the value of 0.05 is not absolutely rigid, and other values such as 0.01 are sometimes used instead.)

9. Probability of a Wrong Result Table 3-1 shows that among 85 women who were pregnant, the test for pregnancy yielded the wrong conclusion 5 times.
   a. Based on the available results, find the probability of a wrong test conclusion for a woman who is pregnant.
   b. Is it "unusual" for the test conclusion to be wrong for women who are pregnant?

10. Probability of a Wrong Result Table 3-1 shows that among 14 women who were *not* pregnant, the test for pregnancy yielded the wrong conclusion 3 times.
   a. Based on the available results, find the probability of a wrong test conclusion for a woman who is not pregnant.
   b. Is it "unusual" for the test conclusion to be wrong for women who are not pregnant?

11. Smoking Survey In one Gallup poll, 1038 adults were asked about the effects of secondhand smoke, and 52 of them indicated that the effects are "not at all harmful."
   a. If you randomly select one of the surveyed adults, what is the probability of getting someone who feels that secondhand smoke is not at all harmful?
   b. Is it "unusual" for someone to believe that secondhand smoke is not at all harmful?

12. Cholesterol Reducing Drug In a clinical trial of Lipitor, a common drug used to lower cholesterol, one group of patients was given a treatment of 10 mg Atorvastatin tablets. That group consists of 19 patients who experienced flu symptoms and 844 patients who did not (based on data from Pfizer, Inc.).
   a. Estimate the probability that a patient taking the drug will experience flu symptoms.
   b. Is it "unusual" for a patient taking the drug to experience flu symptoms?

13. Bumping Airline Passengers In a recent year, 2624 American Airlines passengers were involuntarily denied boarding their flights, and 168,262 other passengers volunteered for bumping in exchange for cash or vouchers.
   a. Estimate the probability that a randomly selected bumped American Airlines passenger is one who was involuntarily bumped.
   b. Are involuntary bumpings "unusual"?

14. On-Time Flight Arrivals A study of 150 randomly selected American Airlines flights showed that 108 arrived on time (based on data from the Department of Transportation).
   a. What is the estimated probability of an American Airlines flight arriving *late*?
   b. Is it "unusual" for an American Airlines flight to arrive late?

15. Guessing Birthdays On their first date, Kelly asks Mike to guess the date of her birth, not including the year.
   a. What is the probability that Mike will guess correctly? (Ignore leap years.)
   b. Would it be "unusual" for him to guess correctly on his first try?
   c. If you were Kelly, and Mike did guess correctly on his first try, would you believe his claim that he made a lucky guess, or would you be convinced that he already knew when you were born?
   d. If Kelly asks Mike to guess her age, and Mike's guess is too high by 15 years, what is the probability that Mike and Kelly will have a second date?

16. Lottery In the old New York State Lottery, you had to select six numbers between 1 and 54 inclusive. There were 25,827,165 different possible six-number combinations,

2. a. 0.9
   b. 1
   c. 1/10 or 0.1
3. $-1, 2, 5/3, \sqrt{2}$
4. a. 1
   b. 0
   c. 1/10
   d. 1/2
   e. 1/5
5. a. 3/8
   b. 3/8
   c. 1/8
6. 0.000321; no. The result suggests that cell phones are not a cause of cancer of the brain or nervous system.
7. 0.153; yes
8. 0.00000138
9. a. 1/17 or 0.0588
   b. No
10. a. 3/14 or 0.214
    b. No
11. a. 0.0501
    b. No
12. a. 0.0220 (not 0.0225)
    b. Yes
13. a. 0.0154 (not 0.0156)
    b. Yes
14. a. 0.280
    b. No
15. a. 1/365
    b. Yes
    c. He already knew.
    d. 0
16. a. 0.0000000774
    b. Yes

and you had to select the correct combination of all six numbers to win the grand prize. For a $1 bet, you selected two different six-number combinations. (You could not select a single six-number combination; you had to select two.)

  a. If you placed a $1 bet and selected two different six-number combinations, what was the probability of winning the grand prize?

  b. Was it unusual to win the grand prize?

17. Probability of a Birthday

  a. If a person is randomly selected, find the probability that his or her birthday is October 18, which is National Statistics Day in Japan. Ignore leap years.

  b. If a person is randomly selected, find the probability that his or her birthday is in October. Ignore leap years.

  c. If a person is randomly selected, find the probability that he or she was born on a day of the week that ends with the letter *y*.

18. Probability of Brand Recognition

  a. In a study of brand recognition, 831 consumers knew of Campbell's Soup, and 18 did not (based on data from Total Research Corporation). Use these results to estimate the probability that a randomly selected consumer will recognize Campbell's Soup.

  b. *Estimate* the probability that a randomly selected adult American consumer will recognize the brand name of McDonald's, most notable as a fast-food restaurant chain.

  c. *Estimate* the probability that a randomly selected adult American consumer will recognize the brand name of Veeco Instruments, a manufacturer of microelectronic products.

19. Fruitcake Survey In a Bruskin-Goldring Research poll, respondents were asked how a fruitcake should be used. One hundred thirty-two respondents indicated that it should be used for a doorstop, and 880 other respondents cited other uses, including birdfeed, landfill, and a gift. If one of these respondents is randomly selected, what is the probability of getting someone who would use the fruitcake as a doorstop?

20. Probability of a Car Crash Among 400 randomly selected drivers in the 20–24 age bracket, 136 were in a car accident during the last year (based on data from the National Safety Council). If a driver in that age bracket is randomly selected, what is the approximate probability that he or she will be in a car accident during the next year? Is the resulting value high enough to be of concern to those in the 20–24 age bracket?

21. Probability of Winning Solitaire Refer to Data Set 27 in Appendix B and assume that the same Microsoft solitaire game is played.

  a. Estimate the probability of winning when a game is played.

  b. Estimate the probability of running the whole deck by winning $208.

22. Probability of an Adverse Drug Reaction When the drug Viagra was clinically tested, 117 patients reported headaches and 617 did not (based on data from Pfizer, Inc.). Use this sample to estimate the probability that a Viagra user will experience a headache. Is the probability high enough to be of concern to Viagra users?

23. Gender of Children: Constructing Sample Space Section 3-2 included a table summarizing the gender outcomes for a couple planning to have three children.

  a. Construct a similar table for a couple planning to have *two* children.

  b. Assuming that the outcomes listed in part (a) are equally likely, find the probability of getting two girls.

  c. Find the probability of getting exactly one child of each gender.

---

17. a. 1/365
    b. 31/365
    c. 1

18. a. 831/849 or 0.979
    b. Answer varies, but 0.99 is reasonable.
    c. Answer varies, but 0.001 is reasonable.

19. 0.130

20. 0.34; yes

21. a. 77/500 or 0.154
    b. 13/500 or 0.026

22. 117/734 or 0.159; yes

23. a. bb, bg, gb, gg
    b. 1/4
    c. 1/2

24. Genetics: Constructing Sample Space   Both parents have the brown/blue pair of eye-color genes, and each parent contributes one gene to a child. Assume that if the child has at least one brown gene, that color will dominate and the eyes will be brown. (The actual determination of eye color is somewhat more complicated.)

  a. List the different possible outcomes. Assume that these outcomes are equally likely.

  b. What is the probability that a child of these parents will have the blue/blue pair of genes?

  c. What is the probability that the child will have brown eyes?

25. Kentucky Derby Odds   When the horse Monarchos won the 127th Kentucky Derby, a $2 bet that Monarchos would win resulted in a return of $23.

  a. How much net profit was made from a $2 win bet on Monarchos?

  b. What were the payoff odds against a Monarchos win?

  c. Based on preliminary wagering before the race, bettors collectively believed that Monarchos had a 1/15 probability of winning. Assuming that 1/15 was the true probability of a Monarchos victory, what were the actual odds against his winning?

  d. If the payoff odds were the actual odds found in part (c), how much would a $2 ticket be worth after the Monarchos win?

26. Finding Odds in Roulette   A roulette wheel has 38 slots. One slot is 0, another is 00, and the others are numbered 1 through 36, respectively. You are placing a bet that the outcome is an odd number.

  a. What is your probability of winning?

  b. What are the actual odds against winning?

  c. When you bet that the outcome is an odd number, the payoff odds are 1:1. How much profit do you make if you bet $18 and win?

  d. How much profit would you make on the $18 bet if you could somehow convince the casino to change its payoff odds so that they are the same as the actual odds against winning? (*Recommendation:* Don't actually try to convince any casino of this; their sense of humor is remarkably absent when it comes to things of this sort.)

3-2 Beyond the Basics

27. Interpreting Effectiveness   A double-blind experiment is designed to test the effectiveness of the drug Statisticzene as a treatment for number blindness. When treated with Statisticzene, subjects seem to show improvement. Researchers calculate that there is a 0.04 probability that the treatment group would show improvement if the drug has no effect. What should you conclude about the effectiveness of Statisticzene?

28. Determining Whether a Jury Is Random   An attorney is defending a client accused of not meeting his alimony obligations. The pool of 20 potential jurors consists of all women, and the attorney calculates that there is a probability of 1/1,048,576 that 20 randomly selected people will be all women. Is there justification for arguing that the jury pool is unfair to his client?

29. Finding Probability from Odds   If the actual odds against event A are $a:b$, then $P(A) = b/(a + b)$. Find the probability of Millennium winning his next race, given that the actual odds against his winning are $3:5$.

24. a. brown/brown, brown/blue, blue/brown, blue/blue
   b. 1/4
   c. 3/4

25. a. $21
   b. 21:2
   c. 14:1
   d. $30

26. a. 18/38 = 0.474
   b. 10:9
   c. $18
   d. $20

27. Because the probability of showing improvement with an ineffective drug is so small (0.04), it appears that the drug is effective.

28. Yes. Because the probability of an all-women jury is so small, it appears that the jury was not randomly selected.

29. 5/8

30. 3.99; 4.55

30. Relative Risk and Odds Ratio In a clinical trial of 734 subjects treated with Viagra, 117 reported headaches. In a control group of 725 subjects not treated with Viagra, 29 reported headaches. Denoting the proportion of headaches in the treatment group by p_t and denoting the proportion of headaches in the control group by p_c, the **relative risk** is p_t/p_c. The relative risk is a measure of the strength of the effect of the Viagra treatment. Another such measure is the **odds ratio,** which is the ratio of the odds in favor of a headache for the treatment group to the odds in favor of a headache for the control group, found by evaluating the following:

$$\frac{p_t/(1 - p_t)}{p_c/(1 - p_c)}$$

The relative risk and odds ratios are commonly used in medicine and epidemiological studies. Find the relative risk and odds ratio for the headache data.

31. a. 4/1461
    b. 400/146,097

31. Leap Years and Guessing Birthdays In part (a) of Exercise 17, leap years were ignored in finding the probability that a randomly selected person will have a birthday on October 18.
  a. Recalculate this probability, assuming that a leap year occurs every four years. (Express your answer as an exact fraction.)
  b. Leap years occur in years evenly divisible by 4, except they are skipped in three of every four centesimal years (years ending in 00). The years 1700, 1800, and 1900 were not leap years, but 2000 was a leap year. Find the exact probability for this case, and express it as an exact fraction.

32. 1

32. Flies on an Orange If two flies land on an orange, find the probability that they are on points that are within the same hemisphere.

33. 1/4

33. Points on a Stick Two points along a straight stick are randomly selected. The stick is then broken at those two points. Find the probability that the three resulting pieces can be arranged to form a triangle. (This is possibly the most difficult exercise in this book.)

3-3 Addition Rule

Note to Instructor
Key points of this section: $P(A \text{ or } B)$ suggests the addition rule; associate "or" with adding. Avoid double counting of events that are not disjoint.

The main objective of this section is to introduce the *addition rule* as a device for finding probabilities that can be expressed as $P(A \text{ or } B)$, the probability that either event A occurs or event B occurs (or they both occur) as the single outcome of a procedure. The key word to remember is *or.* Throughout this text we use the *inclusive or,* which means either one or the other or both. (Except for Exercise 27, we will not consider the *exclusive or,* which means either one or the other but not both.)

In the previous section we presented the fundamentals of probability and considered events categorized as *simple.* In this and the following section we consider *compound events.*

> **Definition**
> A **compound event** is any event combining two or more simple events.

Notation for Addition Rule

$P(A \text{ or } B) = P(\text{event } A \text{ occurs or event } B \text{ occurs or they both occur})$

Probabilities play a very prominent role in genetics. See Figure 3-3, which depicts a sample of peas, like those Mendel used in his famous hybridization experiments. The peas shown have green or yellow pods and they have purple or white flowers. In that sample of 14 peas, how many of them have "green pods or purple flowers"? (Remember, "green pod or purple flower" really means "green pod, or purple flower, or both.") Examination of Figure 3-3 should show that a total of 12 peas have green pods or purple flowers. (*Important note*: It is *wrong* to add the 8 peas with green pods to the 9 peas with purple flowers, because this total of 17 would have counted 5 of the peas twice, but they are individuals that should be counted once each.) Because 12 of the 14 peas have "green pods or purple flowers," the probability of randomly selecting a pea that has a green pod or a purple flower can be expressed as $P(\text{green pod or purple flower}) = 12/14 = 6/7$.

This example suggests a general rule whereby we add the number of outcomes corresponding to each of the events in question:

When finding the probability that event *A* occurs or event *B* occurs, find the total of the number of ways *A* can occur and the number of ways *B* can occur, but *find that total in such a way that no outcome is counted more than once.*

One approach is to combine the number of ways event *A* can occur with the number of ways event *B* can occur and, if there is any overlap, compensate by subtracting the number of outcomes that are counted twice, as in the following rule.

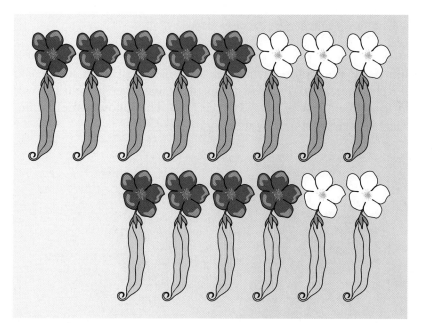

FIGURE 3-3   **Peas Used in a Genetics Study**

Formal Addition Rule

$$P(A \text{ or } B) = P(A) + P(B) - P(A \text{ and } B)$$

where $P(A$ and $B)$ denotes the probability that A and B both occur at the same time as an outcome in a trial of a procedure.

Although the formal addition rule is presented as a formula, it is generally better to understand the spirit of the rule and apply it intuitively, as follows.

Intuitive Addition Rule

To find $P(A$ or $B)$, find the sum of the number of ways event A can occur and the number of ways event B can occur, *adding in such a way that every outcome is counted only once*. $P(A$ or $B)$ is equal to that sum, divided by the total number of outcomes in the sample space.

Figure 3-4 shows a Venn diagram that provides a visual illustration of the formal addition rule. In this figure we can see that the probability of A or B equals the probability of A (left circle) plus the probability of B (right circle) minus the probability of A and B (football-shaped middle region). This figure shows that the addition of the areas of the two circles will cause double-counting of the football-shaped middle region. This is the basic concept that underlies the addition rule. Because of the relationship between the addition rule and the Venn diagram shown in Figure 3-4, the notation $P(A \cup B)$ is sometimes used in place of $P(A$ or $B)$. Similarly, the notation $P(A \cap B)$ is sometimes used in place of $P(A$ and $B)$ so the formal addition rule can be expressed as

$$P(A \cup B) = P(A) + P(B) - P(A \cap B)$$

The addition rule is simplified whenever A and B cannot occur simultaneously, so $P(A$ and $B)$ becomes zero. Figure 3-5 illustrates that when there is no overlapping of A and B, we have $P(A$ or $B) = P(A) + P(B)$. The following definition formalizes the lack of overlapping shown in Figure 3-5.

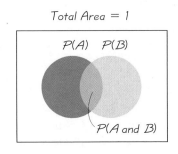

Total Area $= 1$

$P(A)$ $P(B)$

$P(A$ and $B)$

FIGURE 3-4 Venn Diagram Showing Overlapping Events

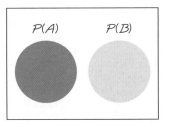

Total Area $= 1$

$P(A)$ $P(B)$

FIGURE 3-5 Venn Diagram Showing Nonoverlapping Events

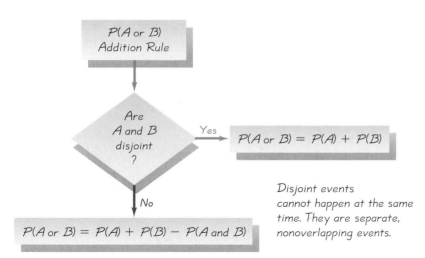

FIGURE 3-6 **Applying the Addition Rule**

Disjoint events cannot happen at the same time. They are separate, nonoverlapping events.

Definition

Events *A* and *B* are **disjoint** (or **mutually exclusive**) if they cannot both occur together.

Note to Instructor

Emphasize that "disjoint" or "mutually exclusive" essentially means "separate" or "not overlapping." Every time you use the term *disjoint*, say "or separate" until the meaning is well known.

The flowchart of Figure 3-6 shows how disjoint events affect the addition rule.

EXAMPLE Clinical Trials of Pregnancy Test Refer to Table 3-1, reproduced here for your convenience. Assuming that 1 person is randomly selected from the 99 people included in the study, apply the addition rule to find the probability of selecting a subject who is pregnant or had a positive test result.

Note to Instructor

Point out that a table such as table 3-1 is called a *two-way* table or *contingency* table, and such tables are very important in statistics because they arise frequently from analysis of survey results. Such tables are the focus of Section 10-3.

Table 3-1	Pregnancy Test Results	
	Positive Test Result (Pregnancy is indicated)	Negative Test Result (Pregnancy is not indicated)
Subject is pregnant	80	5
Subject is not pregnant	3	11

SOLUTION From the table we can easily see that there are 88 subjects who were pregnant or tested positive. We obtain that total of 88 by adding the pregnant subjects to the subjects who tested positive, being careful to count the 80 pregnant subjects who tested positive only once. It would be very wrong to add the 85 pregnant subjects to the 83 subjects who tested positive, because the total of 168 would count some subjects twice, even though they are individuals that should be counted only once. Dividing the correct total of 88 by the overall total of 99, we get this result: P(pregnant or positive) $= 88/99 = 8/9$ or 0.889.

Shakespeare's Vocabulary

According to Bradley Efron and Ronald Thisted, Shakespeare's writings included 31,534 different words. They used probability theory to conclude that Shakespeare probably knew at least another 35,000 words that he didn't use in his writings. The problem of estimating the size of a population is an important problem often encountered in ecology studies, but the result given here is another interesting application. (See "Estimating the Number of Unseen Species: How Many Words Did Shakespeare Know?", in *Biometrika*, Vol. 63, No. 3.)

In the preceding example, there are several strategies you could use for counting the subjects who were pregnant or tested positive. Any of the following would work.

- Color the cells representing subjects who are pregnant or tested positive, then add the numbers in those colored cells, being careful to add each number only once. This approach yields

$$3 + 80 + 5 = 88$$

- Add the 85 pregnant subjects to the 83 subjects who tested positive, but compensate for the double-counting by subtracting the 80 pregnant subjects who tested positive. This approach yields a result of

$$85 + 83 - 80 = 88$$

- Start with the total of 85 pregnant subjects, then add those subjects who tested positive and were not yet included in that total, to get a result of

$$85 + 3 = 88$$

Carefully study the preceding example, because it makes clear this essential feature of the addition rule: "Or" suggests addition, and the addition must be done without double-counting.

We can summarize the key points of this section as follows:

1. To find $P(A \text{ or } B)$, begin by associating "or" with addition.
2. Consider whether events A and B are disjoint; that is, can they happen at the same time? If they are not disjoint (that is, they can happen at the same time), be sure to avoid (or at least compensate for) double-counting when adding the relevant probabilities. If you understand the importance of not double-counting when you find $P(A \text{ or } B)$, you don't necessarily have to calculate the value of $P(A) + P(B) - P(A \text{ and } B)$.

Errors made when applying the addition rule often involve double-counting; that is, events that are not disjoint are treated as if they were. One indication of such an error is a total probability that exceeds 1; however, errors involving the addition rule do not always cause the total probability to exceed 1.

Complementary Events

In Section 3-2 we defined the complement of event A and denoted it by $\overline{A}$. We said that $\overline{A}$ consists of all the outcomes in which event A does *not* occur. Events A and $\overline{A}$ must be disjoint, because it is impossible for an event and its complement to occur at the same time. Also, we can be absolutely certain that A either does or does not occur, which implies that either A or $\overline{A}$ must occur. These observations led us to apply the addition rule for disjoint events as follows:

$$P(A \text{ or } \overline{A}) = P(A) + P(\overline{A}) = 1$$

We justify $P(A \text{ or } \overline{A}) = P(A) + P(\overline{A})$ by noting that A and $\overline{A}$ are disjoint; we justify the total of 1 by our certainty that A either does or does not occur. This result of the addition rule leads to the following three equivalent expressions.

Rule of Complementary Events

$$P(A) + P(\overline{A}) = 1$$
$$P(\overline{A}) = 1 - P(A)$$
$$P(A) = 1 - P(\overline{A})$$

Figure 3-7 visually displays the relationship between $P(A)$ and $P(\overline{A})$.

EXAMPLE   In reality, when a baby is born, $P(\text{boy}) = 0.5121$. Find $P(\overline{\text{boy}})$.

SOLUTION   Using the rule of complementary events, we get

$$P(\overline{\text{boy}}) = 1 - P(\text{boy}) = 1 - 0.5121 = 0.4879.$$

That is, the probability of not getting a boy, which is the probability of a girl, is 0.4879.

A major advantage of the *rule of complementary events* is that its use can greatly simplify certain problems. We will illustrate this advantage in Section 3-5.

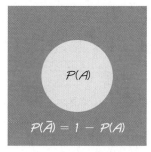

Total Area = 1

P(A)

$P(\overline{A}) = 1 - P(A)$

FIGURE 3-7   Venn Diagram for the Complement of Event *A*

3-3 Basic Skills and Concepts

Determining Whether Events Are Disjoint. For each part of Exercises 1 and 2, are the two events disjoint for a single trial? (Hint: Consider "disjoint" to be equivalent to "separate" or "not overlapping.")

1. a. Randomly selecting a cardiac surgeon
   Randomly selecting a female physician
   b. Randomly selecting a female college student
   Randomly selecting a college student who drives a motorcycle
   c. Randomly selecting someone treated with the cholesterol-reducing drug Lipitor
   Randomly selecting someone in a control group given no medication

2. a. Randomly selecting a head of household watching NBC on television at 8:15 tonight
   Randomly selecting a head of household watching CBS on television at 8:15 tonight
   b. Receiving a phone call from a volunteer survey subject who opposes all government taxation
   Receiving a phone call from a volunteer survey subject who approves of government taxation
   c. Randomly selecting a United States Senator currently holding office
   Randomly selecting a female elected official

3. Finding Complements
   a. If $P(A) = 0.05$, find $P(\overline{A})$.
   b. Based on data from the Census Bureau, when a woman over the age of 25 is randomly selected, there is a 0.218 probability that she has a bachelor's degree. If a woman over the age of 25 is randomly selected, find the probability that she does not have a bachelor's degree.

Recommended Assignment:
Exercises 1–24.

1. a. No
   b. No
   c. Yes

2. a. Yes
   b. Yes
   c. No

3. a. 0.95
   b. 0.782

4. a. 0.9825
   b. 0.39

5. 5/7 or 0.714

6. 11/14 or 0.786

7. 364/365 or 0.997

8. 334/365 or 0.915

9. 0.239

10. 0.929

11. 0.341

12. 0.825

13. 0.600

14. 0.140

15. 0.490

16. 0.500

17. 0.140

18. 0.880

19. 0.870

20. 0.970

4. Finding Complements
   a. Find $P(\overline{A})$ given that $P(A) = 0.0175$.
   b. A Reuters/Zogby poll showed that 61% of Americans say they believe that life exists elsewhere in the galaxy. What is the probability of randomly selecting someone not having that belief?

5. Using Addition Rule Refer to Figure 3-3. Find the probability of randomly selecting one of the peas and getting one with a green pod or a white flower.

6. Using Addition Rule Refer to Figure 3-3. Find the probability of randomly selecting one of the peas and getting one with a yellow pod or a purple flower.

7. National Statistics Day If someone is randomly selected, find the probability that his or her birthday is not October 18, which is National Statistics Day in Japan. Ignore leap years.

8. Birthday and Complement If someone is randomly selected, find the probability that his or her birthday is not in October. Ignore leap years.

In Exercises 9–12, use the data in the following table, which summarizes results from the sinking of the Titanic.

	Men	Women	Boys	Girls
Survived	332	318	29	27
Died	1360	104	35	18

9. *Titanic* Passengers If one of the *Titanic* passengers is randomly selected, find the probability of getting someone who is a woman or child.

10. *Titanic* Passengers If one of the *Titanic* passengers is randomly selected, find the probability of getting a man or someone who survived the sinking.

11. *Titanic* Passengers If one of the *Titanic* passengers is randomly selected, find the probability of getting a child or someone who survived the sinking.

12. *Titanic* Passengers If one of the *Titanic* passengers is randomly selected, find the probability of getting a woman or someone who did not survive the sinking.

Using the Addition Rule with Blood Categories. *In Exercises 13–20, refer to the accompanying figure, which describes the blood groups and Rh types of 100 people (based on data from the Greater New York Blood Program). In each case, assume that 1 of the 100 subjects is randomly selected, and find the indicated probability.*

13. P(not group A)

14. P(type Rh^-).

15. P(group A or type Rh^-)

16. P(group A or group B)

17. P(not type Rh^+)

18. P(group B or type Rh^+)

19. P(group AB or type Rh^+)

20. P(group A or O or type Rh^+)

Group B
8Rh^+
2Rh^-

Group AB
4Rh^+
1Rh^-

Group A
35Rh^+
5Rh^-

39Rh^+
6Rh^-

Group O

21. Car Data Refer to Data Set 22 in Appendix B. If one of the 20 cars is randomly selected, find the probability that it has a manual transmission or is a 6-cylinder car.

22. Smoking and Gender Refer to Data Set 4 in Appendix B. If one of the 107 study subjects is randomly selected, find the probability of getting a male or someone who smokes.

23. Poll Resistance Pollsters are concerned about declining levels of cooperation among persons contacted in surveys. A pollster contacts 84 people in the 18–21 age bracket and finds that 73 of them respond and 11 refuse to respond. When 275 people in the 22–29 age bracket are contacted, 255 respond and 20 refuse to respond (based on data from "I Hear You Knocking but You Can't Come In," by Fitzgerald and Fuller, *Sociological Methods and Research,* Vol. 11, No. 1). Assume that 1 of the 359 people is randomly selected. Find the probability of getting someone in the 18–21 age bracket or someone who refused to respond.

24. Poll Resistance Refer to the same data set as in Exercise 23. Assume that 1 of the 359 people is randomly selected, and find the probability of getting someone who is in the 18–21 age bracket or someone who responded.

21. 0.5

22. 103/107 or 0.963

23. 0.290

24. 0.944

3-3 Beyond the Basics

25. Determining Whether Events Are Disjoint
    a. If $P(A) = 3/11$, $P(B) = 4/11$ and $P(A \text{ or } B) = 7/11$, what do you know about events *A* and *B*?
    b. If $P(A) = 5/18$, $P(B) = 11/18$, and $P(A \text{ or } B) = 13/18$, what do you know about events *A* and *B*?

26. Disjoint Events If events *A* and *B* are disjoint and events *B* and *C* are disjoint, must events *A* and *C* be disjoint? Give an example supporting your answer.

27. Exclusive *Or* How is the addition rule changed if the *exclusive or* is used instead of the *inclusive or*? In this section it was noted that the *exclusive or* means either one or the other, but not both.

28. Extending the Addition Rule The formal addition rule included in this section expressed the probability of *A* or *B* as follows: $P(A \text{ or } B) = P(A) + P(B) - P(A \text{ and } B)$. Extend that formal rule to develop an expression for $P(A \text{ or } B \text{ or } C)$. (*Hint:* Draw a Venn diagram.)

25. a. They are disjoint.
    b. They are not disjoint.

26. No

27. $P(A \text{ or } B) = P(A) + P(B) - 2P(A \text{ and } B)$

28. $P(A \text{ or } B \text{ or } C) = P(A) + P(B) + P(C) - P(A \text{ and } B) - P(A \text{ and } C) - P(B \text{ and } C) + P(A \text{ and } B \text{ and } C)$

Note to Instructor
Key points of this section: $P(A \text{ and } B)$ suggests multiplication; adjust probabilities for *DEPENDENT* events. Notation: In this section, $P(A \text{ and } B)$ denotes that event *A* occurs in one trial and event *B* occurs in another trial; in Section 3-3 we used $P(A \text{ and } B)$ to denote that events *A* and *B* both occur in the same trial.

3-4 Multiplication Rule: Basics

In Section 3-3 we presented the addition rule for finding $P(A \text{ or } B)$, the probability that a trial has an outcome of *A* or *B* or both. The objective of this section is to develop a rule for finding $P(A \text{ and } B)$, the probability that event *A* occurs in a first trial and event *B* occurs in a second trial.

Convicted by Probability

A witness described a Los Angeles robber as a Caucasian woman with blond hair in a ponytail who escaped in a yellow car driven by an African-American male with a mustache and beard. Janet and Malcolm Collins fit this description, and they were convicted based on testimony that there is only about 1 chance in 12 million that any couple would have these characteristics. It was estimated that the probability of a yellow car is 1/10, and the other probabilities were estimated to be 1/4, 1/10, 1/3, 1/10, and 1/1000. The convictions were later overturned when it was noted that no evidence was presented to support the estimated probabilities or the independence of the events. However, because the couple was not randomly selected, a serious error was made in not considering the probability of *other* couples being in the same region with the same characteristics.

Notation

$P(A \text{ and } B) = P(\text{event } A \text{ occurs in a first trial and event } B \text{ occurs in a second trial})$

In Section 3-3 we associated *or* with addition; in this section we will associate *and* with multiplication. We will see that $P(A \text{ and } B)$ involves multiplication of probabilities and that we must sometimes adjust the probability of event B to reflect the outcome of event A.

Probability theory is used extensively in the analysis and design of standardized tests, such as the SAT, ACT, LSAT (for law), and MCAT (for medicine). For ease of grading, such tests typically use true/false or multiple-choice questions. Let's assume that the first question on a test is a true/false type, while the second question is a multiple-choice type with five possible answers (a, b, c, d, e). We will use the following two questions. Try them!

1. True or false: A pound of feathers is heavier than a pound of gold.
2. Among the following, which had the most influence on modern society?
   a. The remote control
   b. This book
   c. Computers
   d. Sneakers with heels that light up
   e. Hostess Twinkies

The answers to the two questions are T (for "true") and c. (The first question is true. Weights of feathers are expressed in avoirdupois pounds, but weights of gold are expressed in troy pounds.) Let's find the probability that if someone makes random guesses for both answers, the first answer will be correct *and* the second answer will be correct. One way to find that probability is to list the sample space as follows:

T,a	T,b	T,c	T,d	T,e
F,a	F,b	F,c	F,d	F,e

If the answers are random guesses, then the 10 possible outcomes are equally likely, so

$$P(\text{both correct}) = P(T \text{ and } c) = \frac{1}{10} = 0.1$$

Now note that $P(T \text{ and } c) = 1/10$, $P(T) = 1/2$, and $P(c) = 1/5$, from which we see that

$$\frac{1}{10} = \frac{1}{2} \cdot \frac{1}{5}$$

so that

$$P(T \text{ and } c) = P(T) \times P(c)$$

This suggests that, in general, $P(A \text{ and } B) = P(A) \cdot P(B)$, but let's consider another example before making that generalization.

For now, we note that tree diagrams are sometimes helpful in determining the number of possible outcomes in a sample space. A **tree diagram** is a picture of the possible outcomes of a procedure, shown as line segments emanating from one starting point. These diagrams are helpful in counting the number of possible outcomes if the number of possibilities is not too large. The tree diagram shown in Figure 3-8 summarizes the outcomes of the true/false and multiple-choice questions. From Figure 3-8 we see that if both answers are random guesses, all 10 branches are equally likely and the probability of getting the correct pair (T, c) is 1/10. For each response to the first question, there are 5 responses to the second. The total number of outcomes is 5 taken 2 times, or 10. The tree diagram in Figure 3-8 illustrates the reason for the use of multiplication.

Perfect SAT Score

If an SAT subject is randomly selected, what is the probability of getting someone with a perfect score? What is the probability of getting a perfect SAT score by guessing? These are two very different questions.

In a recent year, approximately 1.3 million people took the SAT, and only 587 of them received perfect scores of 1600, so there is a probability of 587 ÷ 1.3 million, or about 0.000452 of randomly selecting one of the test subjects and getting someone with a perfect score. Just one portion of the SAT consists of 35 multiple-choice questions, and the probability of answering all of them correct by guessing is $(1/5)^{35}$, which is so small that when written as a decimal, 24 zeros follow the decimal point before a nonzero digit appears.

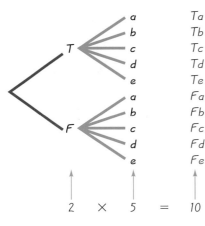

FIGURE 3-8 Tree Diagram of Test Answers

Our first example of the true/false and multiple-choice questions suggested that $P(A \text{ and } B) = P(A) \cdot P(B)$, but the next example will introduce another important element.

EXAMPLE Genetics Experiment Mendel's famous hybridization experiments involved peas, like those shown in Figure 3-3, introduced in Section 3-3 and reproduced on the next page. If two of the peas shown in Figure 3-3 are randomly selected *without replacement,* find the probability that the first selection has a green pod and the second selection has a yellow pod. (We can ignore the colors of the flowers on top.)

SOLUTION

First selection: $P(\text{green pod}) = 8/14$ (because there are 14 peas, 8 of which have green pods)

Second selection: $P(\text{yellow pod}) = 6/13$ (there are 13 peas remaining, 6 of which have yellow pods)

continued

Lottery Advice

New York Daily News columnist Stephen Allensworth recently provided tips for selecting numbers in New York State's Daily Numbers game. In describing a winning system, he wrote that "it involves double numbers matched with cold digits. (A cold digit is one that hits once or not at all in a seven-day period.)" Allensworth proceeded to identify some specific numbers that "have an excellent chance of being drawn this week."

Allensworth assumes that some numbers are "overdue," but the selection of lottery numbers is independent of past results. The system he describes has no basis in reality and will not work. Readers who follow such poor advice are being misled and they might lose more money because they incorrectly believe that their chances of winning are better.

With *P*(first pea with green pod) = 8/14 and *P*(second pea with yellow pod) = 6/13, we have

$$P(\text{1st pea with green pod and 2nd pea with yellow pod}) = \frac{8}{14} \cdot \frac{6}{13} \approx 0.264$$

The key point is that we must adjust the probability of the second event to reflect the outcome of the first event. Because the second pea is selected without replacement of the first pea, the second probability must take into account the result of a pea with a green pod for the first selection. After a pea with a green pod has been selected on the first trial, only 13 peas remain and 6 of them have yellow pods, so the second selection yields this: *P*(pea with yellow pod) = 6/13.

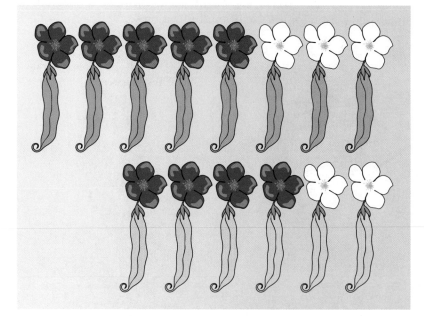

FIGURE 3-3 **Peas Used in a Genetics Study**

This example illustrates the important principle that *the probability for the second event* B *should take into account the fact that the first event* A *has already occurred.* This principle is often expressed using the following notation.

Notation for Conditional Probability

P(*B*|*A*) represents the probability of event *B* occurring after it is assumed that event *A* has already occurred. (We can read *B*|*A* as "*B* given *A*.")

Definitions

Two events *A* and *B* are **independent** if the occurrence of one does not affect the probability of the occurrence of the other. (Several events are similarly independent if the occurrence of any does not affect the probabilities of the occurrence of the others.) If *A* and *B* are not independent, they are said to be **dependent.**

For example, playing the California lottery and then playing the New York lottery are *independent* events because the result of the Californian lottery has absolutely no effect on the probabilities of the outcomes of the New York lottery. In contrast, the event of having your car start and the event of getting to class on time are *dependent* events, because the outcome of trying to start your car does affect the probability of getting to class on time.

Using the preceding notation and definitions, along with the principles illustrated in the preceding examples, we can summarize the key concept of this section as the following *formal multiplication rule*, but it is recommended that you work with the *intuitive multiplication rule*, which is more likely to reflect understanding instead of blind use of a formula.

Formal Multiplication Rule

$$P(A \text{ and } B) = P(A) \cdot P(B \mid A)$$

If A and B are independent events, $P(B \mid A)$ is really the same as $P(B)$. (For further discussion about determining whether events are independent or dependent, see the subsection "Testing for Independence" in Section 3-5. For now, try to understand the basic concept of independence and how it affects the computed probabilities.) See the following *intuitive multiplication rule*. (Also see Figure 3-9.)

Intuitive Multiplication Rule

When finding the probability that event A occurs in one trial and event B occurs in the next trial, multiply the probability of event A by the probability of event B, but be sure that the probability of event B takes into account the previous occurrence of event A.

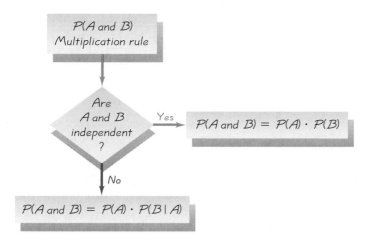

FIGURE 3-9 Applying the Multiplication Rule

Independent Jet Engines

Soon after departing from Miami, Eastern Airlines Flight 855 had one engine shut down because of a low oil pressure warning light. As the L-1011 jet turned to Miami for landing, the low pressure warning lights for the other two engines also flashed. Then an engine failed, followed by the failure of the last working engine. The jet descended without power from 13,000 ft to 4000 ft when the crew was able to restart one engine, and the 172 people on board landed safely. With independent jet engines, the probability of all three failing is only 0.0001^3, or about one chance in a trillion. The FAA found that the same mechanic who replaced the oil in all three engines failed to replace the oil plug sealing rings. The use of a single mechanic caused the operation of the engines to become dependent, a situation corrected by requiring that the engines be serviced by different mechanics.

EXAMPLE **Damaged Goods** Telektronics manufactures computers, televisions, CD players, and other electronics products. When shipped items are damaged, the causes of the damage are categorized as water (W), crushing (C), puncture (P), or carton marking (M). Listed below are the coded causes of five damaged items. A quality control analyst wants to randomly select two items for further investigation. Find the probability that the first selected item was damaged from crushing (C) and the second item was also damaged from crushing (C). Assume that the selections are made (a) with replacement; (b) without replacement.

W   C   C   P   M

SOLUTION

a. If the two items are selected with replacement, the two selections are independent because the second event is not affected by the first outcome. In each of the two selections there are two crushed (C) items among the five items, so we get

$$P(\text{first item is C and second item is C}) = \frac{2}{5} \cdot \frac{2}{5} = \frac{4}{25} \text{ or } 0.16$$

b. If the two items are selected without replacement, the two selections are dependent because the second event is affected by the first outcome. In the first selection, two of the five items were crushed (C). After selecting a crushed item on the first selection, we are left with four items including one that was crushed. We therefore get

$$P(\text{first item is C and second item is C}) = \frac{2}{5} \cdot \frac{1}{4} = \frac{2}{20} = \frac{1}{10} \text{ or } 0.1$$

Note that in this case, we adjust the second probability to take into account the selection of a crushed item (C) in the first outcome. After selecting C the first time, there would be one C among the four items that remain.

So far we have discussed two events, but the multiplication rule can be easily extended to several events. In general, the probability of any sequence of independent events is simply the product of their corresponding probabilities. For example, the probability of tossing a coin three times and getting all heads is $0.5 \cdot 0.5 \cdot 0.5 = 0.125$. We can also extend the multiplication rule so that it applies to several dependent events; simply adjust the probabilities as you go along. For example, the probability of drawing four different cards (without replacement) from a shuffled deck and getting all aces is

$$\frac{4}{52} \cdot \frac{3}{51} \cdot \frac{2}{50} \cdot \frac{1}{49} = 0.00000369$$

Part (b) of the last example involved selecting items without replacement, and we therefore treated the events as being dependent. However, it is a common practice to treat events as independent when *small samples* are drawn from *large populations*. In such cases, it is rare to select the same item twice. Here is a common guideline:

If a sample size is no more than 5% of the size of the population, treat the selections as being *independent* (even if the selections are made without replacement, so they are technically dependent).

Pollsters use this guideline when they survey roughly 1000 adults from a population of millions. They assume independence, even though they sample without replacement. The following example is another illustration of the above guideline. The following example also illustrates how probability can be used to test a claim made about a population. It gives us some insight into the important procedure of *hypothesis testing* that is introduced in Chapter 7.

Redundancy

Reliability of systems can be greatly improved with redundancy of critical components. Race cars in the NASCAR Winston Cup series have two ignition systems so that if one fails, the other can be used. Airplanes have two independent electrical systems, and aircraft used for instrument flight typically have two separate radios. The following is from a *Popular Science* article about stealth aircraft: "One plane built largely of carbon fiber was the Lear Fan 2100 which had to carry two radar transponders. That's because if a single transponder failed, the plane was nearly invisible to radar." Such redundancy is an application of the multiplication rule in probability theory. If one component has a 0.001 probability of failure, the probability of two independent components both failing is only 0.000001.

Note to Instructor
This example uses probability to test a stated claim, and it is actually a hypothesis test (see Chapter 7) without the formal notation.

EXAMPLE Quality Control A production manager for Telektronics claims that her new process for manufacturing DVDs is better because the rate of defects is lower than 3%, which had been the rate of defects in the past. To support her claim, she manufactures a batch of 5000 DVDs, then randomly selects 200 of them for testing, with the result that there are no defects among the 200 selected DVDs. Assuming that the new method has the same 3% defect rate as in the past, find the probability of getting no defects among the 200 DVDs. Based on the result, is there strong evidence to support the manager's claim that her new process is better?

SOLUTION   The probability of no defects is the same as the probability that all 200 DVDs are good. We therefore want to find P(all 200 DVDs are good). We want to assume a defect rate of 3% to see if the result of no defects could easily occur by chance with the old manufacturing process. If the defect rate is 3%, we have P(good DVD) = 0.97. The selected DVDs were chosen without replacement, but the sample of 200 DVDs is less than 5% of the population of 5000, so we will treat the events as if they are independent. We get this result:

P(1st is good and 2nd is good and 3rd is good . . . and 200th is good)

$= P$(good DVD) $\cdot$ P(good DVD) $\cdot$ . . . $\cdot$ P(good DVD)

$= 0.97 \cdot 0.97 \cdot \ldots \cdot 0.97$

$= 0.97^{200} = 0.00226$

The low probability of 0.00226 indicates that instead of getting a very rare outcome with a defect rate of 3%, a more reasonable explanation is that no defects occurred because the defect rate is actually less than 3%. Because there is such a small chance (0.00226) of getting all good DVDs with a sample size of 200 and a defect rate of 3%, we do have sufficient evidence to conclude that the new method is better.

We can summarize the fundamentals of the addition and multiplication rules as follows:

- In the addition rule, the word "or" in $P(A$ or $B)$ suggests addition. Add $P(A)$ and $P(B)$, being careful to add in such a way that every outcome is counted only once.

• In the multiplication rule, the word "and" in $P(A \text{ and } B)$ suggests multiplication. Multiply $P(A)$ and $P(B)$, but be sure that the probability of event B takes into account the previous occurrence of event A.

3-4 Basic Skills and Concepts

Recommended Assignment:
Exercises 1–24.

Identifying Events as Independent or Dependent. In Exercises 1 and 2, for each given pair of events, classify the two events as independent or dependent.

1. a. independent
   b. independent
   c. dependent

1. a. Rolling a die and getting a 5
      Flipping a coin and getting heads
   b. Randomly selecting a TV viewer who watches *Monday Night Football*
      Randomly selecting a second TV viewer who watches *Monday Night Football*
   c. Wearing plaid shorts with black socks and sandals
      Asking someone on a date and getting a positive response

2. a. independent
   b. dependent
   c. dependent

2. a. Finding that your calculator is not working
      Finding that your refrigerator is not working
   b. Finding that your kitchen light is not working
      Finding that your refrigerator is not working
   c. Drinking until your driving ability is impaired
      Being involved in a car crash

3. 1/12

3. Coin and Die Find the probability of getting the outcome of tails and 3 when a coin is tossed and a single die is rolled.

4. 1/260; no because the number of possibilities is too small.

4. Letter and Digit A new computer owner creates a password consisting of two characters. She randomly selects a letter of the alphabet for the first character and a digit (0, 1, 2, 3, 4, 5, 6, 7, 8, 9) for the second character. What is the probability that her password is "K9"? Would this password be effective as a deterrent against someone trying to gain access to her computer?

5. a. 9/49
   b. 1/7

5. Applying the Multiplication Rule If two of the items shown below are randomly selected, find the probability that both items are colored green. These items are used in tests of perception.

      red   yellow   green   red   blue   yellow

   a. Assume that the first item is replaced before the second item is selected.
   b. Assume that the first item is not replaced before the second item is selected.

6. a. 6/343 or 0.0175
   b. 1/35
7. a. 0.288
   b. 0.288
   c. Although the results are slightly different, they are the same when rounded to three decimal places.
   d. Sample without replacement so that duplication is avoided.

6. Applying the Multiplication Rule Using the same seven items from Exercise 5, find the probability of randomly selecting three items and getting one that is colored red on the first selection, one that is colored green on the second selection, and an item that is colored blue on the third selection.
   a. Assume that each item is replaced before the next one is selected.
   b. Assume that none of the selected items is replaced before the others are selected.

7. Defective Gas Masks *Time* magazine reported that when 19,218 gas masks from branches of the U.S. military were tested, it was found that 10,322 were defective (based on data from the World Health Organization). If further investigation begins with the random selection of two gas masks from this population, find the probability that they are both defective.

a. Assume that the first gas mask is replaced before the next one is selected.

b. Assume that the first gas mask is not replaced before the second item is selected.

c. Compare the results from parts (a) and (b).

d. Given a choice between selecting with replacement and selecting without replacement, which choice makes more sense in this situation? Why?

8. Wearing Hunter Orange A study of hunting injuries and the wearing of "hunter" orange clothing showed that among 123 hunters injured when mistaken for game, 6 were wearing orange (based on data from the Centers for Disease Control). If a follow-up study begins with the random selection of hunters from this sample of 123, find the probability that the first two selected hunters were both wearing orange.

a. Assume that the first hunter is replaced before the next one is selected.

b. Assume that the first hunter is not replaced before the second hunter is selected.

c. Given a choice between selecting with replacement and selecting without replacement, which choice makes more sense in this situation? Why?

9. Probability and Guessing A psychology professor gives a surprise quiz consisting of 10 true/false questions, and she states that passing requires at least 7 correct responses. Assume that an unprepared student adopts the questionable strategy of guessing for each answer.

a. Find the probability that the first 7 responses are correct and the last 3 are wrong.

b. Is the probability from part (a) equal to the probability of passing? Why or why not?

10. Selecting U.S. Senators In the 107th Congress, the Senate consists of 13 women and 87 men. If a lobbyist for the tobacco industry randomly selects three different senators, what is the probability that they are all women? Would a lobbyist be likely to use random selection in this situation?

11. Coincidental Birthdays

a. The author was born on November 27. What is the probability that two other randomly selected people are both born on November 27? (Ignore leap years.)

b. What is the probability that two randomly selected people have the same birthday? (Ignore leap years.)

12. Coincidental Birthdays

a. One couple attracted media attention when their three children, born in different years, were all born on July 4. Ignoring leap years, find the probability that three randomly selected people were all born on July 4. Is the probability low enough so that such an event is not likely to occur somewhere in the United States over the course of several years?

b. Ignoring leap years, find the probability that three randomly selected people all have the same birthday.

13. Acceptance Sampling With one method of a procedure called *acceptance sampling,* a sample of items is randomly selected without replacement and the entire batch is accepted if every item in the sample is okay. The Niko Electronics Company has just manufactured 5000 CDs, and 3% are defective. If 12 of these CDs are randomly selected for testing, what is the probability that the entire batch will be accepted?

8. a. 0.00238
   b. 0.00200
   c. If cases are to be selected for a follow-up study, it doesn't make much sense to select the same item twice, so select without replacement.

9. a. 1/1024
   b. No, because there are other ways to pass.

10. 0.00177; no

11. a. 1/133225 or 0.00000751
    b. 1/365

12. a. 0.0000000206; no
    b. 0.00000751

13. 0.694

14. 0.774; no

14. Poll Confidence Level It is common for public opinion polls to have a "confidence level" of 95%, meaning that there is a 0.95 probability that the poll results are accurate within the claimed margins of error. If five different organizations conduct independent polls, what is the probability that all five of them are accurate within the claimed margins of error? Does the result suggest that with a confidence level of 95%, we can expect that almost all polls will be within the claimed margin of error?

15. 1/1024; yes, because the probability of getting 10 girls by chance is so small.

15. Testing Effectiveness of Gender-Selection Method Recent developments appear to make it possible for couples to dramatically increase the likelihood that they will conceive a child with the gender of their choice. In a test of a gender-selection method, 10 couples try to have baby girls. If this gender-selection method has no effect, what is the probability that the 10 babies will be all girls? If there are actually 10 girls among 10 children, does this gender-selection method appear to be effective? Why?

16. 0.9936; yes

16. Nuclear Reactor Reliability Remote sensors are used to control each of two separate and independent valves, denoted by p and q, that open to provide water for emergency cooling of a nuclear reactor. Each valve has a 0.9968 probability of opening when triggered. For the given configuration, find the probability that when both sensors are triggered, water will get through the system so that cooling can occur. Is the result high enough to be considered safe?

Water   *Reactor*

17. 1/64

17. Flat Tire Excuse A classic excuse for a missed test is offered by four students who claim that their car had a flat tire. On the makeup test, the instructor asks the students to identify the particular tire that went flat. If they really didn't have a flat tire and randomly select one that supposedly went flat, what is the probability that they will all select the same tire?

18. 0.00000256; no

18. Voice Identification of Criminal In a Riverhead, New York, case, nine different crime victims listened to voice recordings of five different men. All nine victims identified the same voice as that of the criminal. If the voice identifications were made by random guesses, find the probability that all nine victims would select the same person. Does this constitute reasonable doubt?

19. 0.739 (or 0.738 assuming dependence); no

19. Quality Control A production manager for Telektronics claims that her new process for manufacturing CDs is better because the rate of defects is lower than 2%, which had been the rate of defects in the past. To support her claim, she manufactures a batch of 5000 CDs, then randomly selects 15 of them for testing, with the result that there are no defects among the 15 selected CDs. Assuming that the new method has the same 2% defect rate as in the past, find the probability of getting no defects among the 15 CDs. Based on the result, is there strong evidence to support the manager's claim that her new process is better?

20. a. 0.025
    b. 0.000625
    c. 0.999

20. Redundancy The principle of redundancy is used when system reliability is improved through redundant or backup components. Assume that your alarm clock has a 0.975 probability of working on any given morning. *continued*

a. What is the probability that your alarm clock will not work on the morning of an important final exam?

b. If you had two such alarm clocks, what is the probability that they both fail on the morning of an important final exam?

c. With one alarm clock, we have a 0.975 probability of being awakened. What is the probability of being awakened if we are using two alarm clocks?

(3) *Pregnancy Test Results In Exercises 21–24, use the data in Table 3-1, reproduced here.*

Table 3-1 Pregnancy Test Results		
	Positive Test Result (Pregnancy is indicated)	Negative Test Result (Pregnancy is not indicated)
Subject is pregnant	80	5
Subject is not pregnant	3	11

21. Positive Test Result If two different subjects are randomly selected, find the probability that they both test positive for pregnancy.

21. 0.702

22. Pregnant If one of the subjects is randomly selected, find the probability of getting someone who tests negative or someone who is not pregnant.

22. 19/99 or 0.192

23. Pregnant If two different subjects are randomly selected, find the probability that they are both pregnant.

23. 0.736

24. Negative Test Result If three different people are randomly selected, find the probability that they all test negative.

24. 0.00357

3-4 Beyond the Basics

25. Same Birthdays Find the probability that no two people have the same birthday when the number of randomly selected people is

a. 3   **b.** 5   **c.** 25

25. a. 0.992
    b. 0.973
    c. 0.431

26. Gender of Children If a couple plans to have eight children, find the probability that they are all of the same gender.

26. 0.00781

27. Drawing Cards Two cards are to be randomly selected without replacement from a shuffled deck. Find the probability of getting an ace on the first card and a spade on the second card.

27. 0.0192

28. Complements and the Addition Rule
   a. Develop a formula for the probability of not getting either A or B on a single trial. That is, find an expression for $P(\overline{A \text{ or } B})$.
   b. Develop a formula for the probability of not getting A or not getting B on a single trial. That is, find an expression for $P(\overline{A} \text{ or } \overline{B})$.
   c. Compare the results from parts (a) and (b). Does $P(\overline{A \text{ or } B}) = P(\overline{A} \text{ or } \overline{B})$?

28. a. $1 - P(A) - P(B) + P(A$ and $B)$
    b. $1 - P(A$ and $B)$
    c. Different

Note to Instructor

This section continues work with the multiplication rule. In the first subsection, students often have some difficulty with the key concept. Begin by clearly explaining the meaning of "at least 1." Then discuss the complement of "at least one." Then emphasize this key point: To find P(at least 1 something), it's usually better to first find the probability of the complement, then subtract from 1.

3-5 Multiplication Rule: Complements and Conditional Probability

Section 3-4 introduced the basic concept of the multiplication rule, but in this section we extend our use of that rule to two special applications. We begin with situations in which we want to find the probability that among several trials, *at least one* will result in some specified outcome.

Complements: The Probability of "At Least One"

The multiplication rule and the rule of complements can be used together to greatly simplify the solution to this type of problem: Find the probability that among several trials, *at least one* will result in some specified outcome. In such cases, it is critical that the meaning of the language be clearly understood:

- "At least one" is equivalent to "one or more."
- The complement of getting at least one item of a particular type is that you get no items of that type.

Suppose a couple plans to have three children and they want to know the probability of getting at least one girl. See the following interpretations:

At least 1 girl among 3 children = 1 or more girls.

The complement of "at least 1 girl" = no girls = all 3 children are boys.

We could easily find the probability from a list of the entire sample space of eight outcomes, but we want to illustrate the use of complements, which can be used in many other problems that cannot be solved so easily.

> **EXAMPLE** **Gender of Children** Find the probability of a couple having at least 1 girl among 3 children. Assume that boys and girls are equally likely and that the gender of a child is independent of the gender of any brothers or sisters.

SOLUTION

Step 1:   Use a symbol to represent the event desired. In this case, let A = at least 1 of the 3 children is a girl.

Step 2:   Identify the event that is the complement of A.

$\overline{A}$ = *not* getting at least 1 girl among 3 children

= all 3 children are boys

= boy and boy and boy

Step 3:   Find the probability of the complement.

$P(\overline{A}) = P(\text{boy and boy and boy})$

$$= \frac{1}{2} \cdot \frac{1}{2} \cdot \frac{1}{2} = \frac{1}{8}$$

Step 4:   Find $P(A)$ by evaluating $1 - P(\overline{A})$.

$$P(A) = 1 - P(\overline{A}) = 1 - \frac{1}{8} = \frac{7}{8}$$

INTERPRETATION   There is a 7/8 probability that if a couple has 3 children, at least 1 of them is a girl.

The principle used in this example can be summarized as follows:

> **To find the probability of *at least one* of something, calculate the probability of *none*, then subtract that result from 1. That is,**
>
> $$P(\text{at least one}) = 1 - P(\text{none}).$$

Conditional Probability

Next we consider the second major point of this section, which is based on the principle that the probability of an event is often affected by knowledge of circumstances. For example, if you randomly select someone from the general population, the probability of getting a male is 0.5, but if you know that the selected person frequently changes TV channels with a remote control, the probability is 0.999 (okay, that might be a slight exaggeration). A *conditional probability* of an event occurs when the probability is affected by the knowledge of other circumstances. The conditional probability of event B occurring, given that event A has already occurred, can be found by using the multiplication rule [$P(A \text{ and } B) = P(A) \cdot P(B \mid A)$] and solving for $P(B \mid A)$ by dividing both sides of the equation by $P(A)$.

> ### Definition
>
> A **conditional probability** of an event is a probability obtained with the additional information that some other event has already occurred. $P(B \mid A)$ denotes the conditional probability of event B occurring, given that event A has already occurred, and it can be found by dividing the probability of events A and B both occurring by the probability of event A:
>
> $$P(B \mid A) = \frac{P(A \text{ and } B)}{P(A)}$$

This formula is a formal expression of conditional probability, but we recommend the following intuitive approach.

> ### Intuitive Approach to Conditional Probability
>
> The conditional probability of B given A can be found by assuming that event A has occurred and, working under that assumption, calculating the probability that event B will occur.

Coincidences?

John Adams and Thomas Jefferson (the second and third presidents) both died on July 4, 1826. President Lincoln was assassinated in Ford's Theater; President Kennedy was assassinated in a Lincoln car made by the Ford Motor Company. Lincoln and Kennedy were both succeeded by vice presidents named Johnson. Fourteen years *before* the sinking of the *Titanic,* a novel described the sinking of the *Titan,* a ship that hit an iceberg; see Martin Gardner's *The Wreck of the Titanic Foretold?* Gardner states, "In most cases of startling coincidences, it is impossible to make even a rough estimate of their probability."

Note to Instructor
Students tend to find conditional probability extremely difficult when it is presented in its abstraction. Instead, emphasize careful reading and understanding of the available information and the probability being sought. Emphasize the intuitive approach described here.

Apply for Early Decision?

EXAMPLE Clinical Trials of Pregnancy Test Refer to Table 3-1, reproduced here for your convenience. Find the following:

a. If 1 of the 99 subjects is randomly selected, find the probability that the person tested positive given that she was pregnant.

b. If 1 of the 99 subjects is randomly selected, find the probability that she was pregnant, given that she tested positive.

Table 3-1	Pregnancy Test Results	
	Positive Test Result (Pregnancy is indicated)	Negative Test Result (Pregnancy is not indicated)
Subject is pregnant	80	5
Subject is not pregnant	3	11

SOLUTION   a. We want $P(\text{positive} \mid \text{pregnant})$, the probability of getting someone who tested positive, *given that the selected person was pregnant*. Here is the key point: If we assume that the selected person was pregnant, we are dealing only with the 85 subjects in the first row of Table 3-1. Among those 85 subjects, 80 tested positive, so

$$P(\text{positive} \mid \text{pregnant}) = \frac{80}{85} = 0.941$$

The same result can be found by using the formula given with the definition of conditional probability. In the following calculation, we use the fact that 80 of the 99 subjects were both pregnant and tested positive. Also, 85 of the 99 subjects were pregnant. We get

$$P(\text{positive} \mid \text{pregnant}) = \frac{P(\text{pregnant and positive})}{P(\text{pregnant})}$$
$$= \frac{80/99}{85/99} = 0.941$$

b. Here we want $P(\text{pregnant} \mid \text{positive})$. If we assume that the person selected tested positive, we are dealing with the 83 subjects in the first column of Table 3-1. Among those 83 subjects, 80 were pregnant, so

$$P(\text{pregnant} \mid \text{positive}) = \frac{80}{83} = 0.964$$

Again, the same result can be found by applying the formula for conditional probability:

$$P(\text{pregnant} \mid \text{positive}) = \frac{P(\text{positive and pregnant})}{P(\text{positive})}$$
$$= \frac{80/99}{83/99} = 0.964$$

By comparing the results from parts (a) and (b), we see that $P(\text{positive} \mid \text{pregnant})$ is not the same as $P(\text{pregnant} \mid \text{positive})$.

INTERPRETATION   The first result of $P(\text{positive} \mid \text{pregnant}) = 0.941$ indicates that a pregnant woman has a 0.941 probability of testing positive. This suggests that if a woman does not test positive, she cannot be confident that she is not pregnant, so she should pursue additional testing. The second result of $P(\text{pregnant} \mid \text{positive}) = 0.964$ indicates that for a woman who tests positive, there is a 0.964 probability that she is actually pregnant. A woman who tests positive would be wise to pursue additional testing.

Note that in the preceding example, $P(\text{positive} \mid \text{pregnant}) \neq P(\text{pregnant} \mid \text{positive})$. Although the two values of 0.941 and 0.964 are not too far apart in this example, such results can be very far apart in other cases. To incorrectly believe that $P(B \mid A) = P(A \mid B)$ is often called *confusion of the inverse*. Studies have shown that physicians sometimes give very misleading information when they suffer from confusion of the inverse.

Testing for Independence

In Section 3-4 we stated that events A and B are independent if the occurrence of one does not affect the probability of the occurrence of the other. In the multiplication rule for dependent events, if $P(B \mid A) = P(B)$, then the occurrence of event A has no effect on the probability of event B and the two events A and B are independent. This suggests a test for independence: If $P(B \mid A) = P(B)$, then A and B are independent events; however, if $P(B \mid A) \neq P(B)$, then A and B are dependent events. Another test for independence involves checking for the equality of $P(A \text{ and } B)$ and $P(A) \cdot P(B)$. If they are equal, events A and B are independent. If $P(A \text{ and } B) \neq P(A) \cdot P(B)$, then A and B are dependent events. These results are summarized as follows.

Two events A and B are *independent* if	Two events A and B are *dependent* if
$P(B \mid A) = P(B)$	$P(B \mid A) \neq P(B)$
or	or
$P(A \text{ and } B) = P(A) \cdot P(B)$	$P(A \text{ and } B) \neq P(A) \cdot P(B)$

Bayes' Theorem

Thomas Bayes (1702–1761) said that probabilities should be revised when we learn more about an event. Here's one form of Bayes' theorem:

$$P(A \mid B) = \frac{P(A) \cdot P(B \mid A)}{P(A) \cdot P(B \mid A) + P(\overline{A}) \cdot P(B \mid \overline{A})}$$

Suppose 60% of a company's computer chips are made in one factory (denoted by A) and 40% are made in its other factory (denoted by $\overline{A}$). For a randomly selected chip, the probability it came from factory A is 0.60. Suppose we learn that the chip is defective and the defect rates for the two factories are 35% (for A) and 25% (for $\overline{A}$). We can use the above formula to find that there is a 0.677 probability the defective chip came from factory A.

3-5 Basic Skills and Concepts

Describing Complements. In Exercises 1–4, provide a written description of the complement of the given event.

Recommended Assignment:
Exercises 1–22.

1. Blood Testing   When 10 students are tested for blood group, at least one of them has Group A blood.

2. Quality Control   When 50 HDTV units are shipped, all of them are free of defects.

1. None of the students has Group A blood.
2. At least one unit is defective.
3. At least one of the returns is found to be correct.
4. None of the women accepts. Poor Mike.
5. 0.97; no

6. 0.95; it is likely.

7. 31/32; yes

8. 4095/4096; either a very rare event has occurred or some genetic factor is causing the likelihood of a boy to be much greater than 0.5.
9. 0.410

10. 0.590

11. 0.5; no

12. 5/8

13. 11/14; get another test.

14. 11/16; no

15. 0.999999; yes because the likelihood of being awakened increases from 0.99 to 0.999999.

3. IRS Audits When an IRS agent selects 12 income tax returns and audits them, none of the returns are found to be correct.

4. A Hit with the Misses When Mike asks five different women for a date, at least one of them accepts.

5. Subjective Probability Use subjective probability to estimate the probability of randomly selecting an adult and getting a woman, given that the selected person has hair longer than 10 inches. Is the probability high enough to presume that someone with long hair is almost surely a woman?

6. Subjective Probability Use subjective probability to estimate the probability of randomly selecting an adult and getting a male, given that the selected person owns a motorcycle. If a criminal investigator finds that a motorcycle is registered to Pat Ryan, is it reasonable to believe that Pat is a male?

7. Probability of At Least One Girl If a couple plans to have five children, what is the probability that they will have at least one girl? Is that probability high enough for the couple to be very confident that they will get at least one girl in five children?

8. Probability of At Least One Girl If a couple plans to have 12 children (it could happen), what is the probability that there will be at least one girl? If the couple eventually has 12 children and they are all boys, what can the couple conclude?

9. At Least One Traffic Violation If you run a red traffic light at an intersection equipped with a camera monitor, there is a 0.1 probability that you will be given a traffic violation. If you run a red traffic light at this intersection five different times, what is the probability of getting at least one traffic violation?

10. At Least One Correct Answer If you make random guesses for four multiple-choice test questions (each with five possible answers), what is the probability of getting at least one correct? If a very nondemanding instructor says that passing the test occurs if there is at least one correct answer, can you reasonably expect to pass by guessing?

11. Probability of a Girl Find the probability of a couple having a baby girl when their third child is born, given that the first two children were both boys. Is the result the same as the probability of getting three girls among three children?

12. Mendelian Genetics Refer to Figure 3-3 in Section 3-3 for the peas used in a genetics experiment. If one of the peas is randomly selected and is found to have a green pod, what is the probability that it has a purple flower?

13. Clinical Trials of Pregnancy Test Refer to Table 3-1 and assume that one of the subjects is randomly selected. Find the probability of a negative test result given that the selected subject is not pregnant. What should a woman do if she uses this pregnancy test and obtains a negative result?

14. Clinical Trials of Pregnancy Test Refer to Table 3-1 and assume that one of the subjects is randomly selected. Find the probability that the selected subject is not pregnant, given that the test was negative. Is the result the same as the probability of a negative test result given that the selected subject is not pregnant?

15. Redundancy in Alarm Clocks A student misses many classes because of malfunctioning alarm clocks. Instead of using one alarm clock, he decides to use three. What is the probability that at least one of his alarm clocks works correctly if each individual alarm clock has a 99% chance of working correctly? Does the student really gain much by using three alarm clocks instead of only one?

16. Acceptance Sampling With one method of the procedure called *acceptance sampling*, a sample of items is randomly selected without replacement, and the entire batch is rejected if there is at least one defect. The Niko Electronics Company has just manufactured 5000 CDs, and 3% are defective. If 10 of the CDs are selected and tested, what is the probability that the entire batch will be rejected?

16. 0.263

17. Using Composite Blood Samples When doing blood testing for HIV infections, the procedure can be made more efficient and less expensive by combining samples of blood specimens. If samples from three people are combined and the mixture tests negative, we know that all three individual samples are negative. Find the probability of a positive result for three samples combined into one mixture, assuming the probability of an individual blood sample testing positive is 0.1 (the probability for the "at-risk" population, based on data from the New York State Health Department).

17. 0.271

18. Using Composite Water Samples The Orange County Department of Public Health tests water for contamination because of the presence of *E. coli* bacteria. To reduce laboratory costs, water samples from six public swimming areas are combined for one test, and further testing is done only if the combined sample fails. Based on past results, there is a 2% chance of finding *E. coli* bacteria in a public swimming area. Find the probability that a combined sample from six public swimming areas will reveal the presence of *E. coli* bacteria.

18. 0.114

Conditional Probabilities. In Exercises 19–22, use the Titanic *mortality data in the accompanying table.*

	Men	Women	Boys	Girls
Survived	332	318	29	27
Died	1360	104	35	18

19. If we randomly select someone who was aboard the *Titanic,* what is the probability of getting a man, given that the selected person died?

19. 0.897

20. If we randomly select someone who died, what is the probability of getting a man?

20. 0.897

21. What is the probability of getting a boy or girl, given that the randomly selected person is someone who survived?

21. 0.0793

22. What is the probability of getting a man or woman, given that the randomly selected person is someone who died?

22. 0.965

3-5 Beyond the Basics

23. Roller Coaster The Rock 'n' Roller Coaster at Disney-MGM Studios in Orlando has two seats in each of 12 rows. Riders are assigned to seats in the order that they arrive. If you ride this roller coaster once, what is the probability of getting the coveted first row? How many times must you ride in order to have at least a 95% chance of getting a first row seat at least once?

23. 1/12; 35

24. Whodunnit? The Atlanta plant of the Medassist Pharmaceutical Company manufactures 400 heart pacemakers, of which 3 are defective. The Baltimore plant of the same company manufactures 800 pacemakers, of which 2 are defective. If one of the 1200 pacemakers is randomly selected and is found to be defective, what is the probability that it was manufactured in Atlanta?

24. 3/5 or 0.6

25. a.

	Positive	Negative
HIV Infected	285	15
Not HIV Infected	4985	94,715

b. 0.0541

26. a. 0.431
b. 0.569

27. 1/3

25. Using a Two-Way Table The New York State Health Department reports a 0.3% HIV rate for the general population, and under certain conditions, preliminary screening tests for the HIV virus are correct 95% of the time (for both true positives and true negatives). Assume that the general population consists of 100,000 people.
a. Construct a table with a format similar to Table 3-1.
b. Using the table from part (a), find $P(\text{HIV} \mid \text{positive})$ for someone randomly selected from the general population. That is, find the probability of randomly selecting someone with HIV, given that this person tested positive.

26. Shared Birthdays Find the probability that of 25 randomly selected people,
a. no two share the same birthday.
b. at least two share the same birthday.

27. Unseen Coins A statistics professor tosses two coins that cannot be seen by any students. One student asks if one of the coins turned up heads. Given that the professor's response is "yes," find the probability that both coins turned up heads.

3-6 Probabilities Through Simulations

Students taking an introductory statistics course typically find that the topic of probability is the most difficult topic in the course. Some probability problems might sound simple but their solutions are incredibly complex. So far in this chapter we have identified several basic and important rules commonly used for finding probabilities, but in this section we introduce a very different approach that can overcome much of the difficulty encountered with the application of formal rules. This alternative approach consists of developing a simulation.

Note to Instructor
Simulation methods are now very important, and they provide an interesting alternative to traditional methods for finding probabilities. Carefully discuss simulation methods, such as those described in this section. Consider incorporating class activities. An alternative is to assign this section as reading to be done by the student.

Definition

A **simulation** of a procedure is a process that behaves the same way as the procedure, so that similar results are produced.

Consider the following examples to better understand how simulations can be used.

EXAMPLE Gender Selection When testing techniques of gender selection, medical researchers need to know probability values of different outcomes, such as the probability of getting at least 60 girls among 100 children. Assuming that male and female births are equally likely, describe a simulation that results in the genders of 100 newborn babies.

SOLUTION   One approach is simply to flip a coin 100 times, with heads representing females and tails representing males. Another approach is to use a calculator or computer to randomly generate 0s and 1s, with 0 representing a male and 1 representing a female. The numbers must be generated in such a way that they are equally likely.

lower right corner of this first cell, and pull it down the column until 25 cells are highlighted. When you release the mouse button, all 25 random numbers should be present. This display shows that the 1st and 3rd numbers are the same.

- **TI-83 Plus calculator:** Press the **MATH** key, select **PRB,** then choose **randInt** (and proceed to enter the minimum of 1, the maximum of 365, and 25 for the number of values. That is, enter randInt(1,365,25). See the TI-83 Plus screen display, which shows that we used **randInt** to generate the numbers, which were then stored in list L1, where they were sorted and displayed. This display shows that there are no matching numbers among the first few that can be seen. You can press **STAT** and select **Edit** to see the whole list of generated numbers.

```
TI-83 Plus

randInt(1,365,25
→L₁
{79 206 340 133…
SortA(L₁)
                Done
L₁
{17 34 46 70 79…
```

It is extremely important to construct a simulation so that it behaves just like the real procedure. In the next example we demonstrate the right way and a wrong way to construct a simulation.

EXAMPLE **Simulating Dice** Describe a procedure for simulating the rolling of a pair of dice.

SOLUTION   In the procedure of rolling a pair of dice, each of the two dice yields a number between 1 and 6 (inclusive), and those two numbers are then added. Any simulation should do the same thing. There is a right way and a wrong way to simulate rolling two dice.

The right way: Randomly generate one number between 1 and 6, randomly generate another number between 1 and 6, and then add the two results.

The wrong way: Randomly generate numbers between 2 and 12. This procedure is similar to rolling dice in the sense that the results are always between 2 and 12, but these outcomes between 2 and 12 are equally likely. With real dice, the values between 2 and 12 are *not* equally likely. This simulation would produce very misleading results.

Some probability problems can be solved only by estimating the probability from actual observations or constructing a simulation. The widespread availability of calculators and computers has made it very easy to use simulation methods, so that simulations are now used often for determining probability values.

The Random Secretary

One classic problem of probability goes like this: A secretary addresses 50 different letters and envelopes to 50 different people, but the letters are randomly mixed before being put into envelopes. What is the probability that at least one letter gets into the correct envelope? Although the probability might seem like it should be small, it's actually 0.632. Even with a million letters and a million envelopes, the probability is 0.632. The solution is beyond the scope of this text—way beyond.

3-6 Basic Skills and Concepts

46196
99438
72113
44044
86763
00151
64703
78907
19155
67640
98746
29910
82855
25259
14752
85446
75260
92532
87333
55848

In Exercises 1–8, use the list of randomly generated digits in the margin. (A similar list could be obtained by using calculators, computers, lottery results, or special tables of random numbers.)

1. Simulating Guesses Assume that you want to use the digits in the accompanying list to simulate guesses on a true/false test. If an odd digit represents "true" and an even digit represents "false," list the five answers corresponding to the first row of digits.

2. Simulating Dice Assume that you want to use the digits in the accompanying list to simulate the rolling of a single die. If the digits 1, 2, 3, 4, 5, 6 are used while all other digits are ignored, list the outcomes obtained from the first two rows.

3. Simulating Manufacturing The Telektronic Company is experimenting with a new process for manufacturing fuses, and the defect rate is 20%. We can simulate fuses by using 0 and 1 for defects, while 2, 3, 4, 5, 6, 7, 8, 9 represent good fuses (so that 20% are defects). Identify the defective and acceptable fuses corresponding to the first row of digits.

4. Simulating Birthdays In an example of this section, it was noted that birthdays can be simulated by generating integers between 1 and 365. If we use entries in a list of random digits, we can represent January 1 as 001, January 2 as 002, . . . , and December 31 as 365. All other triplets of digits should be ignored. Using this approach, the first row yields the valid birthday of 196. List the next five birthdays that can be obtained in this manner.

5. Simulating Families of Five Children Use the random digits in the margin for developing a simulation for finding the probability of getting at least two girls in a family of five children. Describe the simulation, then estimate the probability based on its results. How does the result compare to the correct result of 0.813? (*Hint:* Let the odd digits represent girls.)

6. Simulating Three Dice Use the random digits in the margin to develop a simulation for rolling three dice. Describe the simulation, then use it to estimate the probability of getting a total of 10 when three dice are rolled. How does the result compare to the correct result of 0.125? (*Hint:* Use only the digits 1, 2, 3, 4, 5, 6 and ignore all other digits.)

7. Simulating Left-Handedness Ten percent of us are left-handed. In a study of dexterity, people are randomly selected in groups of five. Use the random digits in the margin to develop a simulation for finding the probability of getting at least one left-handed person in a group of five. How does the probability compare to the correct result of 0.410, which can be found by using the probability rules in this chapter? (*Hint:* Because 10% of us are left-handed, let the digit 0 represent someone who is left-handed, and let the other digits represent someone who is not left-handed.)

8. Simulating Hybridization When Mendel conducted his famous hybridization experiments, he used peas with green pods and yellow pods. One experiment involved crossing peas in such a way that 25% of the offspring peas were expected to have yellow pods. Use the random digits in the margin to develop a simulation for finding the probability that when two offspring peas are produced, at least one of them has yellow pods. How does the result compare to the correct probability of 7/16, which can be found by using the probability rules of this chapter? (*Hint:* Because 25% of the offspring are expected to have yellow pods and the other 75% are expected to have

green pods, let the digit 1 represent yellow pods and let the digits 2, 3, 4 represent green pods, and ignore any other digits.)

Ⓣ *In Exercises 9–12, develop a simulation using a TI-83 Plus calculator, STATDISK, Minitab, Excel, or any other suitable calculator or program.*

9. Simulating Families of Five Children In Exercise 5 we used the digits in the margin to estimate the probability of getting at least two girls in a family of five children. Instead of using those same digits, develop your own simulation for finding the probability of getting at least two girls in a family of five children. Simulate 100 families. Describe the simulation, then estimate the probability based on its results.

10. Simulating Three Dice In Exercise 6 we used the digits in the margin to simulate the rolling of dice. Instead of using those digits, develop your own simulation for rolling three dice. Simulate the rolling of the three dice 100 times. Describe the simulation, then use it to estimate the probability of getting a total of 10 when three dice are rolled.

11. Simulating Left-Handedness In Exercise 7 we used the digits in the margin to simulate people that are either left-handed or right-handed. (Ten percent of us are left-handed.) Develop a simulation for finding the probability of getting at least one left-handed person in a group of five. Simulate 100 groups of five.

12. Simulating Hybridization In Exercise 8 we used the digits in the margin as a basis for simulating the hybridization of peas. Again assume that 25% of offspring peas are expected to have yellow pods, but develop your own simulation and generate 100 pairs of offspring. Based on the results, estimate the probability of getting at least one pea with yellow pods when two offspring peas are obtained.

3-6 Beyond the Basics

13. Simulating the Monty Hall Problem A problem that has attracted much attention is the *Monty Hall problem,* based on the old television game show "Let's Make a Deal," hosted by Monty Hall. Suppose you are a contestant who has selected one of three doors after being told that two of them conceal nothing, but that a new red Corvette is behind one of the three. Next, the host opens one of the doors you didn't select and shows that there is nothing behind it. He then offers you the choice of sticking with your first selection or switching to the other unopened door. Should you stick with your first choice or should you switch? Develop a simulation of this game and determine whether you should stick or switch. (According to *Chance* magazine, business schools at such institutions as Harvard and Stanford use this problem to help students deal with decision making.)

14. Simulating Birthdays
  a. Develop a simulation for finding the probability that when 50 people are randomly selected, at least two of them have the same birth date. Describe the simulation and estimate the probability.
  b. Develop a simulation for finding the probability that when 50 people are randomly selected, at least three of them have the same birth date. Describe the simulation and estimate the probability.

15. Genetics: Simulating Population Control A classic probability problem involves a king who wanted to increase the proportion of women by decreeing that after a

7. Among the 20 rows, there is at least one 0 in 7 rows, so the estimated probability is 7/20 or 0.35, which is reasonably close to the correct result of 0.410.

8. Thirteen rows have two simulated peas (with two values of 1, 2, 3, or 4). For each such row, use the first two simulated peas to get 6/13 or 0.462, which is reasonably close to the correct value of 7/16 or 0.438.

9. Approximately 0.813

10. Approximately 0.125

11. Approximately 0.410

12. Approximately 7/16 or 0.438

13. Switch: $P(\text{win}) = 2/3$; stick: $P(\text{win}) = 1/3$.

14. a. 0.970
    b. 0.127

15. No; no

mother gives birth to a son, she is prohibited from having any more children. The king reasons that some families will have just one boy, whereas other families will have a few girls and one boy, so the proportion of girls will be increased. Is his reasoning correct? Will the proportion of girls increase?

3-7 Counting

What is the probability that you will win the lottery? In Maine's lottery, which is typical, you must choose six numbers between 1 and 42 inclusive. If you get the same six-number combination that is randomly drawn by the lottery officials, you win the jackpot, which may be millions of dollars. There are some lesser prizes, but they are relatively insignificant. Using the classical approach to probability (because the outcomes are equally likely), the probability of winning the lottery is found by using $P(\text{win}) = s/n$, where s is the number of ways you can win and n is the total number of possible outcomes. With Maine's lottery $s = 1$, because there is only one way to win the grand prize: Choose the same six-number combination that is drawn in the lottery. Knowing that there is only one way to win, we now need to find n, the total number of outcomes; that is, how many six-number combinations are possible when you select numbers from 1 to 42? Writing a list of the possibilities would take about a year of nonstop work, so that approach would leave you with no time to study statistics. We need a more practical way of finding the total number of possibilities. This section introduces efficient methods for finding such numbers without directly listing and counting the possibilities. We will return to this lottery problem after we present some basic principles. We begin with the *fundamental counting rule*.

Fundamental Counting Rule

For a sequence of two events in which the first event can occur m ways and the second event can occur n ways, the events together can occur a total of $m \cdot n$ ways.

The fundamental counting rule easily extends to situations involving more than two events, as illustrated in the following examples.

EXAMPLE **Burglary Basics** The typical home alarm system has a code that consists of four digits. The digits (0 through 9) can be repeated, and they must be entered in the correct order. Assume that you plan to gain access by trying codes until you find the correct one. How many different codes are possible?

SOLUTION   There are 10 possible values for each of the four digits, so the number of different possible codes is $10 \cdot 10 \cdot 10 \cdot 10 = 10,000$. Although all of the 10,000 different possible codes can be tried in about 11 hours, alarm systems are usually designed so that the system rejects further attempts after only a few incorrect entries. Apart from the moral and legal problems of being a professional burglar, it appears that there is a mathematical issue that also suggests an alternative career path.

EXAMPLE **Cotinine in Smokers** Data Set 6 in Appendix B lists measured cotinine levels for a sample of people from each of three groups: smokers (denoted here by S), nonsmokers who were exposed to tobacco smoke (denoted by E), and nonsmokers not exposed to tobacco smoke (denoted by N). When nicotine is absorbed by the body, cotinine is produced. If we calculate the mean cotinine level for each of the three groups, then arrange those means in order from low to high, we get the sequence NES. An antismoking lobbyist claims that this is evidence that tobacco smoke is unhealthy, because the presence of cotinine increases as exposure to and use of tobacco increase. How many ways can the three groups denoted by N, E, and S be arranged? If an arrangement is selected at random, what is the probability of getting the sequence of NES? Is the probability low enough to conclude that the sequence of NES indicates that the presence of cotinine increases as exposure to and use of tobacco increase?

SOLUTION   In arranging sequences of the groups N, E, and S, there are 3 possible choices for the first group, 2 remaining choices for the second group, and only 1 choice for the third group. The total number of possible arrangements is therefore

$$3 \cdot 2 \cdot 1 = 6$$

There are six different ways to arrange the N, E, and S groups. (They can be listed as NES, NSE, ESN, ENS, SNE, and SEN.) If we randomly select one of the six possible sequences, there is a probability of 1/6 that the sequence NES is obtained. Because that probability of 1/6 is relatively high, we know that the sequence of NES could easily occur by chance. The probability is not low enough to conclude that the sequence of NES indicates that the presence of cotinine increases as exposure to and use of tobacco increase. We would need a smaller probability, such as 0.01.

In the preceding example, we found that three groups can be arranged $3 \cdot 2 \cdot 1 = 6$ different ways. This particular solution can be generalized by using the following notation for the symbol ! and the following *factorial rule*.

How Many Shuffles?

After conducting extensive research, Harvard mathematician Persi Diaconis found that it takes seven shuffles of a deck of cards to get a complete mixture. The mixture is complete in the sense that all possible arrangements are equally likely. More than seven shuffles will not have a significant effect, and fewer than seven are not enough. Casino dealers rarely shuffle as often as seven times, so the decks are not completely mixed. Some expert card players have been able to take advantage of the incomplete mixtures that result from fewer than seven shuffles.

Notation

The **factorial symbol !** denotes the product of decreasing positive whole numbers. For example, $4! = 4 \cdot 3 \cdot 2 \cdot 1 = 24$. By special definition, $0! = 1$. (Many calculators have a factorial key. On the TI-83 Plus calculator, first enter the number, then press **MATH** and select **PRB,** then select menu item 4 and press the **ENTER** key.)

Choosing Personal Security Codes

All of us use personal security codes for ATM machines, computer Internet accounts, and home security systems. The safety of such codes depends on the large number of different possibilities, but hackers now have sophisticated tools that can largely overcome that obstacle. Researchers found that by using variations of the user's first and last names along with 1800 other first names, they could identify 10% to 20% of the passwords on typical computer systems. When choosing a password, *do not* use a variation of any name, a word found in a dictionary, a password shorter than seven characters, telephone numbers, or social security numbers. Do include nonalphabetic characters, such as digits or punctuation marks.

Factorial Rule

A collection of *n* different items can be arranged in order *n*! different ways. (This **factorial rule** reflects the fact that the first item may be selected *n* different ways, the second item may be selected *n* − 1 ways, and so on.)

Routing problems often involve application of the factorial rule. AT&T wants to route telephone calls through the shortest networks. Federal Express wants to find the shortest routes for its deliveries. American Airlines wants to find the shortest route for returning crew members to their homes. See the following example.

EXAMPLE **Routes to All 50 Capitals** Because of your success in a statistics course, you have been hired by the Gallup Organization, and your first assignment is to conduct a survey in each of the 50 state capitals. As you plan your route of travel, you want to determine the number of different possible routes. How many different routes are possible?

SOLUTION By applying the factorial rule, we know that 50 items can be arranged in order 50! different ways. That is, the 50 state capitals can be arranged 50! ways, so the number of different routes is 50!, or

$$30,414,093,201,713,378,043,612,608,166,064,768,$$
$$844,377,641,568,960,512,000,000,000,000$$

Now there's a large number.

The preceding example is a variation of a classical problem called the *traveling salesman problem*. It is especially interesting because the large number of possibilities means that we can't use a computer to calculate the distance of each route. The time it would take even the fastest computer to calculate the shortest possible route is about

$$1,000,000,000,000,000,000,000,000,000,000,000,000,000,000 \text{ centuries}$$

Considerable effort is currently being devoted to finding efficient ways of solving such problems.

According to the factorial rule, *n* different items can be arranged *n*! different ways. Sometimes we have *n* different items, but we need to select *some* of them instead of all of them. If we must conduct surveys in state capitals, as in the preceding example, but we have time to visit only four capitals, the number of different possible routes is $50 \cdot 49 \cdot 48 \cdot 47 = 5,527,200$. Another way to obtain this same result is to evaluate

$$\frac{50!}{46!} = 50 \cdot 49 \cdot 48 \cdot 47 = 5,527,200$$

In this calculation, note that the factors in the numerator divide out with the factors in the denominator, except for the factors of 50, 49, 48, and 47 that remain.

We can generalize this result by noting that if we have n different items available and we want to select r of them, the number of different arrangements possible is $n!/(n - r)!$ as in $50!/46!$. This generalization is commonly called the *permutations rule*.

Permutations Rule (When Items Are All Different)

The number of **permutations** (or sequences) of r items selected from n available items (without replacement) is

$$_nP_r = \frac{n!}{(n - r)!}$$

Many calculators can evaluate expressions of $_nP_r$.

It is very important to recognize that the permutations rule requires the following conditions:

- We must have a total of n *different* items available. (This rule does not apply if some of the items are identical to others.)
- We must select r of the n items (without replacement).
- We must consider rearrangements of the same items to be different sequences.

When we use the term *permutations, arrangements,* or *sequences,* we imply that *order is taken into account* in the sense that different orderings of the same items are counted separately. The letters *ABC* can be arranged six different ways: *ABC, ACB, BAC, BCA, CAB, CBA*. (Later, we will refer to *combinations*, which do not count such arrangements separately.) In the following example, we are asked to find the total number of different sequences that are possible. That suggests use of the permutations rule.

EXAMPLE **Television Programming** You have just been hired to determine the programming for the Fox television network. When selecting the shows to be shown on Monday night, you find that you have 27 shows available and you must select 4 of them. Because of lead-in effects, the order of the shows is important. How many different sequences of 4 shows are possible when there are 27 shows available?

SOLUTION   We need to select $r = 4$ shows from $n = 27$ that are available. The number of different arrangements is found as shown:

$$_nP_r = \frac{n!}{(n - r)!} = \frac{27!}{(27 - 4)!} = 421,200$$

There are 421,200 different possible arrangements of 4 shows selected from the 27 that are available.

Making Cents of the Lottery

Many people spend large sums of money buying lottery tickets, even though they don't have a realistic sense for their chances of winning. Brother Donald Kelly of Marist College suggests this analogy: Winning the lottery is equivalent to correctly picking the "winning" dime from a stack of dimes that is 21 miles tall! Commercial aircraft typically fly at altitudes of 6 miles, so try to image a stack of dimes more than three times higher than those high-flying jets, then try to imagine selecting the one dime in that stack that represents a winning lottery ticket. Using the methods of this section, find the probability of winning your state's lottery, then determine the height of the corresponding stack of dimes.

The Phone Number Crunch

Telephone companies often split regions with one area code into regions with two or more area codes because new fax and Internet lines have nearly exhausted the possible numbers that can be listed under a single code. Because a seven-digit telephone number cannot begin with a 0 or 1, there are $8 \cdot 10 \cdot 10 \cdot 10 \cdot 10 \cdot 10 \cdot 10 = 8{,}000{,}000$ different possible telephone numbers.

Before cell phones, fax machines, and the Internet, all toll-free numbers had a prefix of 800. Those 800 numbers lasted for 29 years before they were all assigned. The 888 prefix was introduced to help meet the demand for toll-free numbers, but it was estimated that it would take only 2.5 years for the 888 numbers to be exhausted. Next up: toll-free numbers with a prefix of 877. The counting techniques of this section are used to determine the number of different possible toll-free numbers with a given prefix, so that future needs can be met.

We sometimes need to find the number of permutations when some of the items are identical to others. The following variation of the permutations rule applies to such cases.

Permutations Rule (When Some Items Are Identical to Others)

If there are n items with n_1 alike, n_2 alike, . . . , n_k alike, the number of permutations of all n items is

$$\frac{n!}{n_1! n_2! \cdots n_k!}$$

EXAMPLE **Investing in Stocks** The classic examples of the permutations rule are those showing that the letters of the word *Mississippi* can be arranged 34,650 different ways and that the letters of the word *statistics* can be arranged 50,400 ways. We will instead consider the letters *BBBBBAAAA*, which represent a sequence of recent years in which the Dow Jones Industrial Average was below (*B*) the mean or above (*A*) the mean. How many ways can we arrange the letters *BBBBBAAAA*? Does it appear that the sequence is random? Is there a pattern suggesting that it would be wise to invest in stocks?

SOLUTION In the sequence *BBBBBAAAA* we have $n = 9$ items, with $n_1 = 5$ alike and $n_2 = 4$ others that are alike. The number of permutations is computed as follows:

$$\frac{n!}{n_1! n_2!} = \frac{9!}{5! 4!} = \frac{362{,}880}{2880} = 126$$

There are 126 different ways that the letters *BBBBBAAAA* can be arranged. Because there are 126 different possible arrangements and only two of them (*BBBBBAAAA* and *AAAABBBBB*) result in the letters all grouped together, it appears that the sequence is not random. Because all of the *below* values occur at the beginning and all of the *above* values occur at the end, it appears that there is a pattern of increasing stock values. This suggests that it would be wise to invest in stocks. (See Section 13-7 for the *runs test for randomness*, which is a formal procedure often used to identify economic trends.)

The preceding example involved n items, each belonging to one of two categories. When there are only two categories, we can stipulate that x of the items are alike and the other $n - x$ items are alike, so the permutations formula simplifies to

$$\frac{n!}{(n - x)! x!}$$

This particular result will be used for the discussion of binomial probabilities, which are introduced in Section 4-3.

When we intend to select r items from n different items but *do not take order into account,* we are really concerned with possible combinations rather than permutations. That is, **when different orderings of the same items are counted separately, we have a permutation problem, but when different orderings of the same items are not counted separately, we have a combination problem** and may apply the following rule.

Combinations Rule

The number of **combinations** of r items selected from n different items is

$$_nC_r = \frac{n!}{(n-r)!r!}$$

Many calculators are designed to evaluate $_nC_r$.

It is very important to recognize that in applying the combinations rule, the following conditions apply:

- We must have a total of n different items available.
- We must select r of the n items (without replacement).
- We must consider rearrangements of the same items to be the same. (The combination ABC is the same as CBA.)

Because choosing between the permutations rule and the combinations rule can be confusing, we provide the following example, which is intended to emphasize the difference between them.

EXAMPLE Elected Offices The Board of Trustees at the author's college has 9 members. Each year, they elect a 3-person committee to oversee buildings and grounds. Each year, they also elect a chairperson, vice chairperson, and secretary.

a. When the board elects the buildings and grounds committee, how many different 3-person committees are possible?

b. When the board elects the 3 officers (chairperson, vice chairperson, and secretary), how many different slates of candidates are possible?

SOLUTION Note that order is irrelevant when electing the buildings and grounds committee. When electing officers, however, different orders are counted separately.

a. Because order does not count for the committees, we want the number of combinations of $r = 3$ people selected from the $n = 9$ available people. We get

$$_nC_r = \frac{n!}{(n-r)!r!} = \frac{9!}{(9-3)!3!} = \frac{362{,}880}{4320} = 84$$

continued

b. Because order does count with the slates of candidates, we want the number of sequences (or permutations) of $r = 3$ people selected from the $n = 9$ available people. We get

$$_nP_r = \frac{n!}{(n-r)!} = \frac{9!}{(9-3)!} = \frac{362{,}880}{720} = 504$$

There are 84 different possible committees of 3 board members, but there are 504 different possible slates of candidates.

The counting techniques presented in this section are sometimes used in probability problems. The following examples illustrate such applications.

EXAMPLE **Maine Lottery** In the Maine lottery, a player wins or shares in the jackpot by selecting the correct 6-number combination when 6 different numbers from 1 through 42 are drawn. If a player selects one particular 6-number combination, find the probability of winning the jackpot. (The player need not select the 6 numbers in the same order as they are drawn, so order is irrelevant.)

SOLUTION   Because 6 different numbers are selected from 42 different possibilities, the total number of combinations is

$$_{42}C_6 = \frac{42!}{(42-6)!6!} = \frac{42!}{36!6!} = 5{,}245{,}786$$

With only one combination selected, the player's probability of winning is $1/5{,}245{,}786$.

EXAMPLE **Powerball Lottery** The Powerball lottery is run in 21 states. You must select 5 numbers between 1 and 49, and you must also select another special Powerball number between 1 and 42. (Five balls are drawn from a drum with 49 white balls and 1 red ball is drawn from a drum with 42 red balls.) The special Powerball number may be the same as one of the 5 other numbers. In order to win or share in the jackpot, you must select the correct combination of 5 numbers *and* you must also select the correct Powerball number. Find the probability of winning or sharing in the jackpot.

SOLUTION   Let's break this problem down into three parts: (1) Get the correct 5-number combination; (2) get the correct Powerball number; and (3) combine the results to find the probability of winning or sharing the jackpot. First, we begin with the number of combinations that are possible when you select 5 numbers between 1 and 49, which is

$$_{49}C_5 = \frac{49!}{(49-5)!5!} = \frac{49!}{44!5!} = 1{,}906{,}884$$

The probability of getting the winning 5-number combination is therefore $1/1{,}906{,}884$.

Second, we must also select the correct Powerball number between 1 and 42. The probability of selecting the winning Powerball number is 1/42.

Third, because we must get the correct 5-number combination *and* the correct Powerball number, the probability of both events occurring is 1/1,906,884 × 1/42 = 1/80,089,128. This last result is an application of the multiplication rule introduced in Section 3-4. For a person buying one ticket, the probability of winning the Powerball lottery is 1/80,089,128.

3-7 Basic Skills and Concepts

Calculating Factorials, Combinations, Permutations. In Exercises 1–8, evaluate the given expressions and express all results using the usual format for writing numbers (instead of scientific notation).

1. 6!   **2.** 15!   **3.** $_{25}P_2$ **4.** $_{100}P_3$

5. $_{25}C_2$ **6.** $_{100}C_3$ **7.** $_{52}C_5$ **8.** $_{52}P_5$

Probability of Winning the Lottery. This section included an example showing that the probability of winning the Maine lottery is 1/5,245,786. In Exercises 9–12, find the probability of winning the indicated lottery.

9. Massachusetts Mass Millions: Select the winning six numbers from 1, 2, . . . , 49.

10. Pennsylvania Super 6 Lotto: Select the winning six numbers from 1, 2, . . . , 69.

11. New York Lotto: Select the winning six numbers from 1, 2, . . . , 59.

12. New York Take Five: Select the winning five numbers from 1, 2, . . . , 39.

13. Age Discrimination The Pitt Software Company reduced its sales staff from 32 employees to 28. The company claimed that four employees were randomly selected for job termination. However, the four employees chosen are the four oldest employees among the original sales force of 32. Find the probability that when four employees are randomly selected from a group of 32, the four oldest are selected. Is that probability low enough to charge that instead of using random selection, the Pitt Software Company actually fired the oldest employees?

14. Computer Design In designing a computer, if a *byte* is defined to be a sequence of 8 bits and each bit must be a 0 or 1, how many different bytes are possible? (A byte is often used to represent an individual character, such as a letter, digit, or punctuation symbol. For example, one coding system represents the letter *A* as 01000001.) Are there enough different bytes for the characters that we typically use, including lowercase letters, capital letters, digits, punctuation symbols, dollar sign, and so on?

15. Maine Lottery The probability of winning the Maine lottery is 1/5,245,786. What is the probability of winning if the rules are changed so that in addition to selecting the correct six numbers from 1 to 42, you must now select them in the same order as they are drawn?

16. Testing a Claim Mike claims that he has developed the ability to roll a 6 almost every time that he rolls a die. You test his claim by having Mike roll a die five times, and he gets a 6 each time. If Mike has no ability to affect the outcomes, find the probability that he will roll five consecutive 6s when a die is rolled five times. Is that probability low enough to support Mike's claim?

Recommended Assignment:
Exercises 1–16, 27, 28.

1. 720
2. 1,307,674,368,000
3. 600
4. 970,200
5. 300
6. 161,700
7. 2,598,960
8. 311,875,200
9. 1/13,983,816
10. 1/119,877,472
11. 1/45,057,474
12. 1/575,757
13. 1/35,960; it appears that the oldest employees were selected.
14. 256; yes, because the typical keyboard has about 94 characters.
15. 1/3,776,965,920
16. 1/7776; yes

17. Selection of Treatment Group Walton Pharmaceuticals wants to test the effectiveness of a new drug designed to relieve allergy symptoms. The initial test will be conducted by treating six people chosen from a pool of 15 volunteers. If the treatment group is randomly selected, what is the probability that it consists of the six youngest people in the pool? If the six youngest are selected, is there sufficient evidence to conclude that instead of being random, the selection was based on age?

18. He Did It His Way Singing legend Frank Sinatra recorded 381 songs. From a list of his top-10 songs, you must select three that will be sung in a medley as a tribute at the next MTV Music Awards ceremony. The order of the songs is important so that they fit together well. If you select three of Sinatra's top-10 songs, how many different sequences are possible?

19. Air Routes You have just started your own airline company called Air America (motto: "Where your probability of a safe flight is greater than zero"). You have one plane for a route connecting Austin, Boise, and Chicago. One route is Austin-Boise-Chicago and a second route is Chicago-Boise-Austin. How many other routes are possible? How many different routes are possible if service is expanded to include a total of eight cities?

20. Social Security Numbers Each social security number is a sequence of nine digits. What is the probability of randomly generating nine digits and getting *your* social security number?

21. Electrifying When testing for electrical current in a cable with five color-coded wires, the author used a meter to test two wires at a time. How many tests are required for every possible pairing of two wires?

22. Elected Board of Directors There are 12 members on the board of directors for the Newport General Hospital.
a. If they must elect a chairperson, first vice chairperson, second vice chairperson, and secretary, how many different slates of candidates are possible?
b. If they must form an ethics subcommittee of four members, how many different subcommittees are possible?

23. Jumble Puzzle Many newspapers carry "Jumble," a puzzle in which the reader must unscramble letters to form words. For example, the letters TAISER were included in newspapers on the day this exercise was written. How many ways can the letters of TAISER be arranged? Identify the correct unscrambling, then determine the probability of getting that result by randomly selecting an arrangement of the given letters.

24. Finding the Number of Possible Melodies In Denys Parsons' *Directory of Tunes and Musical Themes,* melodies for more than 14,000 songs are listed according to the following scheme: The first note of every song is represented by an asterisk *, and successive notes are represented by *R* (for repeat the previous note), *U* (for a note that goes up), or *D* (for a note that goes down). Beethoven's Fifth Symphony begins as *RRD. Classical melodies are represented through the first 16 notes. With this scheme, how many different classical melodies are possible?

25. Combination Locks A typical "combination" lock is opened with the correct sequence of three numbers between 0 and 49 inclusive. (A number can be used more than once.) What is the probability of guessing those three numbers and opening the lock with the first try?

26. Five Card Flush A standard deck of cards contains 13 clubs, 13 diamonds, 13 hearts, and 13 spades. If five cards are randomly selected, find the probability of getting a flush. (A flush is obtained when all five cards are of the same suit. That is, they are all clubs, or all diamonds, or all hearts, or all spades.)

27. Probabilities of Gender Sequences
   a. If a couple plans to have eight children, how many different gender sequences are possible?
   b. If a couple has four boys and four girls, how many different gender sequences are possible?
   c. Based on the results from parts (a) and (b), what is the probability that when a couple has eight children, the result will consist of four boys and four girls?

28. Is the Researcher Cheating? You become suspicious when a genetics researcher randomly selects groups of 20 newborn babies and seems to consistently get 10 girls and 10 boys. The researcher explains that it is common to get 10 boys and 10 girls in such cases.
   a. If 20 newborn babies are randomly selected, how many different gender sequences are possible?
   b. How many different ways can 10 boys and 10 girls be arranged in sequence?
   c. What is the probability of getting 10 boys and 10 girls when 20 babies are born?
   d. Based on the preceding results, do you agree with the researcher's explanation that it is common to get 10 boys and 10 girls when 20 babies are randomly selected?

29. Finding the Number of Area Codes *USA Today* reporter Paul Wiseman described the old rules for telephone area codes by writing about "possible area codes with 1 or 0 in the second digit. (Excluded: codes ending in 00 or 11, for toll-free calls, emergency services, and other special uses.)" Codes beginning with 0 or 1 should also be excluded. How many different area codes were possible under these old rules?

30. Cracked Eggs A carton contains 12 eggs, 3 of which are cracked. If we randomly select 5 of the eggs for hard boiling, what is the probability of the following events?
   a. All of the cracked eggs are selected.
   b. None of the cracked eggs are selected.
   c. Two of the cracked eggs are selected.

31. California Lottery In California's Super Lotto Plus lottery game, winning the jackpot requires that you select the correct 5 numbers between 1 and 47 and, in a separate drawing, you must also select the correct single number between 1 and 27. Find the probability of winning the jackpot.

32. N.C.A.A. Basketball Tournament Each year, 64 college basketball teams compete in the N.C.A.A. tournament. Sandbox.com recently offered a prize of $10 million to anyone who could correctly pick the winner in each of the tournament games. The president of that company also promised that, in addition to the cash prize, he would eat a bucket of worms. Yuck.
   a. How many games are required to get one championship team from the field of 64 teams?
   b. If someone makes random guesses for each game of the tournament, find the probability of picking the winner in each game.
   c. In an article about the $10 million prize, the *New York Times* wrote that "Even a college basketball expert who can pick games at a 70 percent clip has a 1 in _____ chance of getting all the games right." Fill in the blank.

26. 0.00198

27. a. 256
    b. 70
    c. $70/256 = 0.273$

28. a. 1,048,576
    b. 184,756
    c. 0.176
    d. No; although the result is common, it should not happen consistently.

29. 144

30. a. 0.0455
    b. 0.159
    c. 0.318

31. 1/41,416,353

32. a. 63
    b. $1/0.5^{63} \approx 1.08 \times 10^{-19}$
    c. 5,738,831,575 or about 5.7 billion

3-7 Beyond the Basics

33. *Finding the Number of Computer Variable Names* A common computer programming rule is that names of variables must be between 1 and 8 characters long. The first character can be any of the 26 letters, while successive characters can be any of the 26 letters or any of the 10 digits. For example, allowable variable names are A, BBB, and M3477K. How many different variable names are possible?

34. Handshakes and Round Tables
   a. Five managers gather for a meeting. If each manager shakes hands with each other manager exactly once, what is the total number of handshakes?
   b. If n managers shake hands with each other exactly once, what is the total number of handshakes?
   c. How many different ways can five managers be seated at a round table? (Assume that if everyone moves to the right, the seating arrangement is the same.)
   d. How many different ways can n managers be seated at a round table?

35. *Evaluating Large Factorials* Many calculators or computers cannot directly calculate 70! or higher. When n is large, $n!$ can be approximated by $n = 10^K$, where $K = (n + 0.5) \log n + 0.39908993 - 0.43429448n$.
   a. Evaluate 50! using the factorial key on a calculator and also by using the approximation given here.
   b. The Bureau of Fisheries once asked Bell Laboratories for help finding the shortest route for getting samples from 300 locations in the Gulf of Mexico. If you compute the number of different possible routes, how many digits are used to write that number?

36. *Computer Intelligence* Can computers "think"? According to the *Turing test,* a computer can be considered to think if, when a person communicates with it, the person believes he or she is communicating with another person instead of a computer. In an experiment at Boston's Computer Museum, each of 10 judges communicated with four computers and four other people and was asked to distinguish between them.
   a. Assume that the first judge cannot distinguish between the four computers and the four people. If this judge makes random guesses, what is the probability of correctly identifying the four computers and the four people?
   b. Assume that all 10 judges cannot distinguish between computers and people, so they make random guesses. Based on the result from part (a), what is the probability that all 10 judges make all correct guesses? (That event would lead us to conclude that computers cannot "think" when, according to the Turing test, they can.)

Review

We began this chapter with the basic concept of probability, which is so important for methods of inferential statistics introduced later in this book. We should know that a probability value, which is expressed as a number between 0 and 1, reflects the likelihood of some event. We should know that a value such as 0.01 represents an event that is very unlikely to occur. In Section 3-1 we introduced the rare event rule for inferential statistics: If, under a given assumption, the probability of a particular event is extremely small, we conclude that the assumption is probably not correct. As an example of the basic approach used, consider a test of someone's claim that a quarter used in a coin toss is fair. If we flip the quarter 10 times and get 10 consecutive heads, we can make one of two inferences from these sample results:

Sidebar answers (left margin):

33. 2,095,681,645,538 (about 2 trillion)

34. a. 10
    b. $n(n - 1)/2$
    c. 24
    d. $(n - 1)!$

35. a. Calculator: 3.0414093×10^{64}; approximation: 3.0363452×10^{64}
    b. 615

36. a. 1/70
    b. 3.54×10^{-19}

1. The coin is actually fair, and the string of 10 consecutive heads is a fluke.

2. The coin is not fair.

Statisticians use the rare event rule when deciding which inference is correct: In this case, the probability of getting 10 consecutive heads is so small (1/1024) that the inference of unfairness is the better choice. Here we can see the important role played by probability in the standard methods of statistical inference.

In Section 3-2 we presented the basic definitions and notation, including the representation of events by letters such as A. We defined probabilities of simple events as

$$P(A) = \frac{\text{number of times that } A \text{ occurred}}{\text{number of times trial was repeated}} \quad \text{(relative frequency)}$$

$$P(A) = \frac{\text{number of ways } A \text{ can occur}}{\text{number of different simple events}} = \frac{s}{n} \quad \text{(for equally likely outcomes)}$$

We noted that the probability of any impossible event is 0, the probability of any certain event is 1, and for any event A, $0 \leq P(A) \leq 1$. Also, $\overline{A}$ denotes the complement of event A. That is, $\overline{A}$ indicates that event A does *not* occur.

In Sections 3-3, 3-4, and 3-5 we considered compound events, which are events combining two or more simple events. We associate "or" with addition and associate "and" with multiplication. Always keep in mind the following key considerations:

- When conducting one trial, do we want the probability of event *A or B?* If so, use the addition rule, but be careful to avoid counting any outcomes more than once.

- When finding the probability that event *A* occurs on one trial *and* event *B* occurs on a second trial, use the multiplication rule. Multiply the probability of event *A* by the probability of event *B*. *Caution:* When calculating the probability of event *B*, be sure to take into account the fact that event *A* has already occurred.

In some probability problems, the biggest obstacle is finding the total number of possible outcomes. Section 3-7 was devoted to the following counting techniques:

- Fundamental counting rule
- Factorial rule
- Permutations rule (when items are all different)
- Permutations rule (when some items are identical to others)
- Combinations rule

Review Exercises

Lie Detectors. In Exercises 1–8, use the data in the accompanying table (based on data from the Office of Technology Assessment). The data reflect responses to a key question asked of 100 different subjects.

Note to Instructor

Reminder: The student version of this book includes *all* answers to Review Exercises and Cumulative Review Exercises. For the regular section Exercises, answers are given for odd-numbered exercises only.

	Polygraph Indicated Truth	Polygraph Indicated Lie
Subject actually told the truth	65	15
Subject actually told a lie	3	17

1. 0.2

1. If 1 of the 100 subjects is randomly selected, find the probability of getting someone who told a lie.

2. 0.32

2. If 1 of the 100 subjects is randomly selected, find the probability of getting someone for whom the polygraph test indicated that a lie was being told.

3. 0.35

3. If 1 of the 100 subjects is randomly selected, find the probability of getting someone who told a lie or had the polygraph test indicate that a lie was being told.

4. 0.83

4. If 1 of the 100 subjects is randomly selected, find the probability of getting someone who told the truth or had the polygraph test indicate that the truth was being told.

5. 0.638

5. If two different subjects are randomly selected, find the probability that they both told the truth.

6. 0.100

6. If two different subjects are randomly selected, find the probability that they both had the polygraph test indicate that a lie was being told.

7. 15/32 or 0.469

7. If one subject is randomly selected, find the probability that he or she told the truth, given that the polygraph test indicated that a lie was being told.

8. 15/80 = 3/16 or 0.188

8. If one subject is randomly selected, find the probability that he or she has a polygraph test indication that a lie was being told, given that the subject actually told the truth.

9. a. 0.248
   b. 0.0615
   c. 0.575

9. Probability of Computer Failures A *PC World* survey of 4000 personal computer owners showed that 992 of them broke down during the first two years. (The computers broke down, not the owners.) In choosing among several computer suppliers, a purchasing agent wants to know the probability of a personal computer breaking down during the first two years. Use the survey results to estimate that probability.
   a. If a personal computer is randomly selected, what is the probability that it will break down during the first two years?
   b. If two personal computers are randomly selected, what is the probability that they will both break down during the first two years?
   c. If three personal computers are randomly selected, what is the probability that at least one of them breaks down during the first two years?

10. 0.0777

10. Acceptance Sampling With one method of acceptance sampling, a sample of items is randomly selected without replacement and the entire batch is rejected if there is at least one defect. The Niko Electronics Company has just manufactured 2500 CDs, and 2% are defective. If 4 of the CDs are selected and tested, what is the probability that the entire batch will be rejected?

11. 1/4096; yes

11. Testing a Claim The Biogene Research Company claims that it has developed a technique for ensuring that a baby will be a girl. In a test of that technique, 12 couples all have baby girls. Find the probability of getting 12 baby girls by chance, assuming that boys and girls are equally likely and that the gender of any child is independent of the others. Does that result appear to support the company's claim?

12. a. 1/120
    b. 720

12. Selecting Members The board of directors for the Hartford Investment Fund has 10 members.
   a. If 3 members are randomly selected to oversee the auditors, find the probability that the three wealthiest members are selected.
   b. If members are elected to the positions of chairperson, vice chairperson, and treasurer, how many different slates are possible?

13. Roulette When betting on *even* in roulette, there are 38 equally likely outcomes, but only 2, 4, 6, . . . , 36 are winning outcomes.
  a. Find the probability of winning when betting on even.
  b. Find the actual odds against winning with a bet on even.
  c. Casinos pay winning bets according to odds described as 1:1. What is your net profit if you bet $5 on even and you win?

14. Is the Pollster Lying? A pollster claims that 12 voters were randomly selected from a population of 200,000 voters (30% of whom are Republicans), and all 12 were Republicans. The pollster claims that this could easily happen by chance. Find the probability of getting 12 Republicans when 12 voters are randomly selected from this population. Based on the result, does it seem that the pollster's claim is correct?

14. 0.000000531; no

15. Life Insurance The New England Life Insurance Company issues one-year policies to 12 men who are all 27 years of age. Based on data from the Department of Health and Human Services, each of these men has a 99.82% chance of living through the year. What is the probability that they all survive the year?

15. 0.979

16. Illinois Lotteries Illinois runs different lottery games. Find the probability of winning the jackpot for each game.
  a. Lotto: Select the winning six numbers from 1, 2, . . . , 52.
  b. Little Lotto: Select the winning five numbers from 1, 2, . . . , 30.
  c. The Big Game: Select the winning five numbers from 1, 2, . . . , 50 *and,* in a separate drawing, also select the winning single number from 1, 2, . . . , 36.

16. a. 1/20,358,520
     b. 1/142,506
     c. 1/76,275,360

Cumulative Review Exercises

1. Treating Chronic Fatigue Syndrome A sample of patients suffering from chronic fatigue syndrome were treated with medication, then their change in fatigue was measured on a scale from -7 to $+7$, with positive values representing improvement and 0 representing no change. The results are listed below (based on data from "The Relationship Between Neurally Mediated Hypotension and the Chronic Fatigue Syndrome," by Bou-Holaigah, Rowe, Kan, and Calkins, *Journal of the American Medical Association*, Volume 274, Number 12).

6   5   0   5   6   7   3   3   2   4   4   0   7   3   4   3   6   0   5   5   6

  a. Find the mean.
  b. Find the median.
  c. Find the standard deviation.
  d. Find the variance.
  e. Based on the results, does it appear that the treatment was effective?
  f. If one value is randomly selected from this sample, find the probability that it is positive.
  g. If two different values are randomly selected from this sample, find the probability that they are both positive.
  h. Ignore the three values of 0 and assume that only positive or negative values are possible. Assuming that the treatment is ineffective and that positive and negative values are equally likely, find the probability that 18 subjects all have positive values (as in this sample group). Is that probability low enough to justify rejection of

1. a. 4.0
    b. 4.0
    c. 2.2
    d. 4.7
    e. Yes
    f. 6/7
    g. 0.729
    h. 1/262,144; yes
2. a. 63.6 in.
    b. 1/4
    c. 3/4
    d. 1/16
    e. 5/16

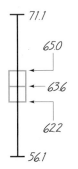

71.1

65.0

63.6

62.2

56.1

the assumption that the treatment is ineffective? Does the treatment appear to be effective?

2. Women's Heights The accompanying boxplot depicts heights (in inches) of a large collection of randomly selected adult women.
   a. What is the mean height of adult women?
   b. If one of these women is randomly selected, find the probability that her height is between 56.1 in. and 62.2 in.
   c. If one of these women is randomly selected, find the probability that her height is below 62.2 in. or above 63.6 in.
   d. If two women are randomly selected, find the probability that they both have heights between 62.2 in. and 63.6 in.
   e. If five women are randomly selected, find the probability that three of them are taller than the mean and the other two are shorter than the mean.

Cooperative Group Activities

1. In-class activity See Exercise 15 in Section 3-6. Divide into groups of three or four and use coin tossing to develop a simulation that emulates the kingdom that abides by this decree: After a mother gives birth to a son, she will not have any other children. If this decree is followed, does the proportion of girls increase?

2. In-class activity Divide into groups of three or four and use actual thumbtacks to estimate the probability that when dropped, a thumbtack will land with the point up. How many trials are necessary to get a result that appears to be reasonably accurate when rounded to the first decimal place?

3. Out-of-class activity Marine biologists often use the *capture-recapture* method as a way to estimate the size of a population, such as the number of fish in a lake. This method involves capturing a sample from the population, tagging each member in the sample, then returning them to the population. A second sample is later captured and the tagged members are counted along with the total size of this second sample. The results can be used to estimate the size of the population.

   Instead of capturing real fish, simulate the procedure using some uniform collection of items such as BB's, colored beads, M&Ms, Fruit Loop cereal pieces, or index cards. Start with a large collection of such items. Collect a sample of 50 and use a magic marker to "tag" each one. Replace the tagged items, mix the whole population, then select a second sample and proceed to estimate the population size. Compare the result to the actual population size obtained by counting all of the items.

4. In-class activity Divide into groups of two. Refer to Exercise 13 in Section 3-6 for a description of the "Monty Hall problem." Simulate the contest and record the results for sticking and switching, then determine which of those two strategies is better.

5. Out-of-class activity Divide into groups of two for the purpose of doing an experiment designed to show one approach to dealing with sensitive survey questions, such as those related to drug use, sexual activity (or inactivity), stealing, or cheating. Instead of actually using a controversial question that would reap wrath upon the author, we will use this innocuous question: "Were you born in a month that has the letter *r* in it?" About 2/3 of all responses should be "yes," but let's pretend that the question is very sensitive and that survey subjects are reluctant to answer honestly. Survey people by asking them to flip a coin and respond as follows:

   ● Answer "yes" if the coin turns up tails *or* you were born in a month containing the letter *r*.

   ● Answer "no" if the coin turns up heads *and* you were born in a month not containing the letter *r*.

Supposedly, respondents tend to be more honest because the coin flip protects their privacy. Survey people and analyze the results to determine the proportion of people born in a month containing the letter *r*. The accuracy of the results could be checked against their actual birth dates, which can be obtained from a second question. The experiment could be repeated with a question that is more sensitive, but such a question is not given here because the author already receives enough mail.

Technology Project

This project illustrates the law of large numbers described in Section 3-2. Use a computer or TI-83 Plus calculator to simulate 100 births. Accomplish this by randomly generating 100 numbers, each of which is 0 or 1 (where 0 = boy and 1 = girl). Use the results to fill in the following table. What happens to the proportion of girls as the sample size increases? How does this illustrate the law of large numbers?

Number of births	10	20	30	40	50	60	70	80	90	100
Proportion of girls										

Here are some details for the different applications of technology.

STATDISK Select **Data** from the main menu bar. Choose **Uniform Generator.** Make the entries for a sample of size 100, a minimum of 0, a maximum of 1, and 0 decimal places (because we want whole numbers). Click on the **Generate** button.

Minitab Select **Calc** from the main menu bar at the top. Select **Random Data,** then **Integer.** Proceed to enter 100 for the number of rows of data, C1 for the column in which to store the results, 0 for the minimum value, and 1 for the maximum value; then click **OK.**

Excel Position the cursor in cell A1. Click on the menu item of f_x, then select **Math & Trig,** then **RANDBETWEEN,** then click **OK.** In the dialog box, enter 0 for the bottom value, enter 1 for the top value, then click **OK.** Cell A1 should now contain either a 0 or 1. Click on the lower right corner of that cell and, while holding the mouse button down, drag the cursor downward to cell A100, then release it.

TI-83 Plus Press the **MATH** key. Select **PRB.** Select the 5th menu item: **randInt(.** Enter 0, 1, 100 and press the **ENTER** key. Press **STO** and **L1** to store the data in list L1. To view the generated births, press **STAT,** then select **Edit.**

from DATA to DECISION

Critical Thinking: When you apply for a job, should you be concerned about drug testing?

According to the American Management Association, about 70% of U.S. companies now test at least some employees and job applicants for drug use. The U.S. National Institute on Drug Abuse claims that about 15% of people in the 18–25 age bracket use illegal drugs. Allyn Clark, a 21-year-old college graduate, applied for a job at the Acton Paper Company, took a drug test, and was not offered a job. He suspected that he might have failed the drug test, even though he does not use drugs. In checking with the company's personnel department, he found that only 1% of drug users incorrectly test negative, and only 2% of nonusers are incorrectly identified as drug users. Allyn felt relieved by these figures because he believed that they reflected a very reliable test that usually provides good results—but is this really true?

Analyzing the Results

The accompanying table shows data for Allyn and 1999 other job applicants. Based on those results, find P(false positive); that is, find the probability of randomly selecting one of the subjects who tested positive and getting someone who does not use drugs. Also find P(false negative); that is, find the probability of randomly selecting someone who tested negative and getting someone who does use drugs. Are the probabilities of these wrong results low enough so that job applicants and the Acton Paper Company need not be concerned?

	Drug Users	Nonusers
Positive test result	297	34
Negative test result	3	1666

INTERNET PROJECT   Computing Probabilities

Finding probabilities when rolling dice is easy. With one die, there are six possible outcomes, so each outcome, such as a roll of 2, has probability 1/6. For a card game the calculations are more involved, but they are still manageable. But what about a more complicated game, such as the board game Monopoly? What is the probability of landing on a particular space on the board? The probability depends on the space your piece currently occupies, the roll of the dice, the drawing of cards, as well as other factors. Now consider a more true-to-life example, such as the probability of having an auto accident. The number of factors involved is too large

to even consider, yet such probabilities are nonetheless quoted, for example, by insurance companies.

The Internet Project for this chapter considers methods for computing probabilities in complicated situations. Go to the Internet Project for this chapter which can be found at this site:

http://www.aw.com/triola

You will be guided in the research of probabilities for a board game. Then you compute such probabilities yourself. Finally, you will estimate a health-related probability using empirical data.

Statistics @ Work

"Based on these statistics on preferred park use, we are now evaluating how we can better service the more diverse population."

Judy Shafer

Deputy Superintendent of Virgin Islands National Park

Judy has worked for the National Park Service for 17 years and in park management for the past 4 years.

As Deputy Superintendent of Virgin Islands National Park, do you use statistics in your work?

We use probability and methods of statistics in applications such as analyzing the sustainability of a particular coral species under the environmental stresses of visitor use, pollution, sedimentation, etc. We also use methods of statistics to determine which ethnic and racial populations are using the park, and to determine the manner in which the park is achieving its user-satisfaction goals.

There are so many uses of statistics that it is impossible to describe them all! Our Law Enforcement Rangers collect and use statistics for visitor and resource protection activities; our marine research scientists use statistics on a daily basis to determine the health and viability of various marine species and to determine the level of threat by visitor use and environmental agents, such as pollution and sedimentation. For example, some of the larger fish species, such as the Nassau Grouper, may be so over-fished that this species could be nearing extinction. The statistical data from research studies must guide the park's decisions in how to protect such species as they are inextricably linked to the larger coral reef ecosystem and ultimately to global warming.

Could you give a simple and specific example of how statistics is used?

In the realm of visitor use, the National Park Service has learned that different ethnic and racial groups use national parks in different ways. For example, we have learned that some groups like the traditional backcountry hiking while others prefer the sociability of group picnics. We are considering the incorporation of more group picnic areas to better accommodate visitors of different ethnic and racial backgrounds.

Should job applicants in your area have a statistics course in their background?

Our employees should have a minimum of at least a college-level course in statistics. Without that, you will be a dinosaur in a statistically based 21st century. Without statistics, you will not be as competitive as someone who does have that background.

What other skills are important for employees?

Vision. Leadership. Innovation. Creativity. And the willingness to take some risks to achieve your goals and protect the things that you feel passionate about.

4

Probability Distributions

4-1 Overview

4-2 Random Variables

4-3 Binomial Probability Distributions

4-4 Mean, Variance, and Standard Deviation for the Binomial Distribution

4-5 The Poisson Distribution

Determining whether a gender selection method is effective

Recent advances in medicine, genetics, and technology have been astounding. It now seems possible that a human being could be cloned. Once considered risky and dangerous, heart bypass operations are now routine. Laser surgery can correct vision so that many people can throw out their eyeglasses or contact lenses. Instead of relying solely on random chance, couples planning to have babies can use techniques for determining the gender of their children. Such advances are sometimes accompanied by considerable controversy. Some people believe that techniques of cloning or gender selection have serious moral ramifications and should be strictly avoided, regardless of the reason. Lisa Belkin wrote this in "Getting the Girl" (an article in the *New York Times Magazine*): "If we allow parents to choose the sex of their child today, how long will it be before they order up eye color, hair color, personality traits, and IQ?" There are some convincing arguments in favor of at least limited use of gender selection. One such argument asserts that some couples carry X-linked recessive genes with this result: Any male children have a 50% chance of inheriting a serious disorder, but none of the female children will inherit the disorder. Such couples may want to use gender selection as a way to ensure that they have baby girls, thereby ensuring that the serious disorder will not be inherited by any of their children.

The Genetics and IVF Institute in Fairfax, Virginia developed a technique called MicroSort, which supposedly increases the chances of a couple having a baby girl. In a preliminary test, 14 couples who wanted baby girls were found. After using the MicroSort technique, 13 of them had girls and one couple had a boy. These results provide us with an interesting question: Given that 13 out of 14 couples had girls, can we conclude that the MicroSort technique is effective, or might we explain the outcome as just a chance sample result? In answering that question, we will use principles of probability to determine whether the observed births differ significantly from results that we would expect from random chance. This example raises an issue that is at the very core of inferential statistics: How do we determine whether results are attributable to random chance or to some other factor, such as the use of the MicroSort technique of gender selection?

Note to Instructor

Suggested class activity: List two samples of data, each with about 20 values. Create one sample to have 20 single digits with approximately equal frequencies of 0, 1, 2, ..., 9 and select the other sample to be 20 digits that are mostly 0s and 5s with a few other digits thrown in. Inform the class that both samples are the last digits of recorded weights of people, but one of the samples came from *measured* weights whereas the other sample resulted from *asking* people what they weighed. Ask the class to distinguish between the reported weights and the measured weights. Emphasize that they can make a conclusion about the nature of the data by simply examining the distribution. Also ask them to construct an "ideal" distribution that would result from millions of people that were actually weighed; ask them to estimate the mean and standard deviation for this distribution. (The mean should be 4.5 and the standard deviation could be estimated using the range rule of thumb; the true mean is 4.5 and the true standard deviation is 2.9.)

 Overview

In this chapter we combine the methods of *descriptive statistics* presented in Chapter 2 and those of *probability* presented in Chapter 3. Figure 4-1 presents a visual summary of what we will accomplish in this chapter. As the figure shows, using the methods of Chapter 2, we would repeatedly roll the die to collect sample data, which then can be described with graphs (such as a histogram or boxplot), measures of center (such as the mean), and measures of variation (such as the standard deviation). Using the methods of Chapter 3, we could find the probability of each possible outcome. In this chapter we will combine those concepts as we develop probability distributions that describe what will *probably* happen instead of what actually *did* happen. In Chapter 2 we constructed frequency tables and histograms using *observed* sample values that were actually collected, but in this chapter we will construct probability distributions by presenting possible outcomes along with the relative frequencies we *expect*.

A casino "pit boss" knows how a die should behave. The table at the extreme right in Figure 4-1 represents a probability distribution that serves as a model of a theoretically perfect population frequency distribution. In essence, we can describe the relative frequency table for a die rolled an infinite number of times. With this knowledge of the population of outcomes, we are able to find its important characteristics, such as the mean and standard deviation. The remainder of this book and the very core of inferential statistics are based on some knowledge of probability distributions. We begin by examining the concept of a random variable, and then we consider important distributions that have many real applications.

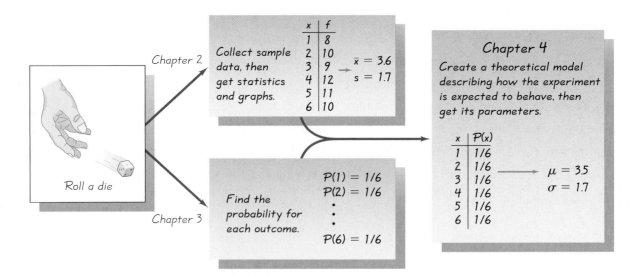

FIGURE 4-1   Combining Descriptive Methods and Probabilities to Form a Theoretical Model of Behavior

4-2 Random Variables

In this section we discuss random variables, probability distributions, procedures for finding the mean and standard deviation for a probability distribution, and methods for distinguishing between outcomes that are likely to occur by chance and outcomes that are "unusual." We begin with the related concepts of *random variable* and *probability distribution*.

Definitions

A **random variable** is a variable (typically represented by x) that has a single numerical value, determined by chance, for each outcome of a procedure.

A **probability distribution** is a graph, table, or formula that gives the probability for each value of the random variable.

 EXAMPLE **Gender of Children** A study consists of randomly selecting 14 newborn babies and counting the number of girls (as in the Chapter Problem). If we assume that boys and girls are equally likely and if we let

$$x = \text{number of girls among 14 babies}$$

then x is a random variable because its value depends on chance. The possible values of x are 0, 1, 2, . . . , 14. Table 4-1 lists the values of x along with the corresponding probabilities. (In Section 4-3 we will see how to find the probability values, such as those listed in Table 4-1.) Because Table 4-1 gives the probability for each value of the random variable x, that table describes a probability distribution.

In Section 1-2 we made a distinction between discrete and continuous data. Random variables may also be discrete or continuous, and the following two definitions are consistent with those given in Section 1-2.

Definitions

A **discrete random variable** has either a finite number of values or a countable number of values, where "countable" refers to the fact that there might be infinitely many values, but they can be associated with a counting process.

A **continuous random variable** has infinitely many values, and those values can be associated with measurements on a continuous scale in such a way that there are no gaps or interruptions.

Table 4-1 Probabilities of Girls	
x (girls)	$P(x)$
0	0.000
1	0.001
2	0.006
3	0.022
4	0.061
5	0.122
6	0.183
7	0.209
8	0.183
9	0.122
10	0.061
11	0.022
12	0.006
13	0.001
14	0.000

This chapter deals exclusively with discrete random variables, but the following chapters will deal with continuous random variables.

(a) Discrete Random
Variable: Count of the
number of movie patrons.

(b) Continuous Random
Variable: The measured
voltage of a smoke detector
battery.

**FIGURE 4-2 Devices Used
to Count and Measure Discrete
and Continuous Random
Variables**

Note to Instructor

Suggested class activity: Ask in
class: "If a satellite crashes at a random point on earth, what is the
probability it will crash on land?"
There are 54,225,000 square miles
of land and 142,715,000 square
miles of water. Answer: 0.275. This
shows that some probability problems can be solved using areas.
That's why it is important to see the
relationship between area and probability in Figure 4-3. That relationship will be used extensively in
Chapter 5 and later chapters.

EXAMPLES The following are examples of discrete and continuous random variables:

1. Let x = the number of eggs that a hen lays in a day. This is a *discrete* random variable because its only possible values are 0, or 1, or 2, and so on.
   No hen can lay 2.343115 eggs, which would have been possible if the data
   had come from a continuous scale.

2. The count of the number of patrons attending an 'N Sync concert is a
   whole number and is therefore a discrete random variable. The counting
   device shown in Figure 4-2(a) is capable of indicating only a finite number
   of values, so it is used to obtain values for a *discrete* random variable.

3. Let x = the amount of milk a cow produces in one day. This is a *continuous*
   random variable because it can have any value over a continuous span.
   During a single day, a cow might yield an amount of milk that can be any
   value between 0 gallons and 5 gallons. It would be possible to get 4.123456
   gallons, because the cow is not restricted to the discrete amounts of 0, 1, 2,
   3, 4, or 5 gallons.

4. The measure of voltage for a smoke-detector battery can be any value between 0 volts and 9 volts. It is therefore a continuous random variable. The
   voltmeter shown in Figure 4-2(b) is capable of indicating values on a continuous scale, so it can be used to obtain values for a *continuous* random
   variable.

Graphs

There are various ways to graph a probability distribution, but we will consider
only the **probability histogram.** Figure 4-3 is a probability histogram that is very
similar to the relative frequency histogram discussed in Chapter 2, but the vertical

**FIGURE 4-3 Probability
Histogram for Number of Girls
Among 14 Newborn Babies**

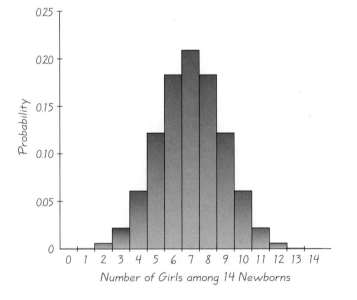

scale shows *probabilities* instead of relative frequencies based on actual sample results.

In Figure 4-3, note that along the horizontal axis, the values of 0, 1, 2, . . . , 14 are located at the centers of the rectangles. This implies that the rectangles are each 1 unit wide, so the areas of the rectangles are 0.000, 0.001, 0.006, and so on. The areas of these rectangles are the same as the *probabilities* in Table 4-1. We will see in Chapter 5 and future chapters that this correspondence between area and probability is very useful in statistics.

Every probability distribution must satisfy each of the following two requirements.

Requirements for a Probability Distribution

1. $\Sigma P(x) = 1$ where x assumes all possible values
2. $0 \le P(x) \le 1$ for every individual value of x

The first requirement states that the sum of the probabilities for all the possible values of the random variable must equal 1. This makes sense when we realize that the values of the random variable x represent all possible events in the entire sample space, so we are certain (with probability 1) that one of the events will occur. In Table 4-1, the sum of the probabilities is 0.999; it would be 1 if we eliminated the tiny rounding error by carrying more decimal places. Also the probability rule stating $0 \le P(x) \le 1$ for any event A (given in Section 3-2) implies that $P(x)$ must be between 0 and 1 for any value of x. Again, refer to Table 4-1 and note that each individual value of $P(x)$ does fall between 0 and 1 for any value of x. Because Table 4-1 does satisfy both of the requirements, it is an example of a probability distribution. A probability distribution may be described by a table, such as Table 4-1, or a graph, such as Figure 4-3, or a formula.

EXAMPLE Does Table 4-2 describe a probability distribution?

SOLUTION   To be a probability distribution, $P(x)$ must satisfy the preceding two requirements. But

$$\Sigma P(x) = P(0) + P(1) + P(2) + P(3)$$
$$= 0.2 + 0.5 + 0.4 + 0.3$$
$$= 1.4 \,[\text{showing that } \Sigma P(x) \ne 1]$$

Because the first requirement is not satisfied, we conclude that Table 4-2 does not describe a probability distribution.

Table 4-2

Probabilities for a Random Variable

x	$P(x)$
0	0.2
1	0.5
2	0.4
3	0.3

EXAMPLE Does $P(x) = x/3$ (where x can be 0, 1, or 2) determine a probability distribution?

SOLUTION For the given function we find that $P(0) = 0/3$, $P(1) = 1/3$, and $P(2) = 2/3$, so that

1. $\Sigma P(x) = \dfrac{0}{3} + \dfrac{1}{3} + \dfrac{2}{3} = \dfrac{3}{3} = 1$

2. Each of the $P(x)$ values is between 0 and 1.

Because both requirements are satisfied, the $P(x)$ function given in this example is a probability distribution.

Mean, Variance, and Standard Deviation

Recall that in Chapter 2 we described the following important characteristics of data (which can be remembered with the mnemonic of CVDOT for "**C**omputer **V**iruses **D**estroy **O**r **T**erminate"):

1. *Center:* A representative or average value that indicates where the middle of the data set is located.

2. *Variation:* A measure of the amount that the values vary among themselves.

3. *Distribution:* The nature or shape of the distribution of the data (such as bell-shaped, uniform, or skewed).

4. *Outliers:* Sample values that lie very far away from the vast majority of the other sample values.

5. *Time:* Changing characteristics of the data over time.

The probability histogram can give us insight into the nature or shape of the distribution. Also, we can often find the mean, variance, and standard deviation of data, which provide insight into the other characteristics. The mean, variance, and standard deviation for a probability distribution can be found by applying Formulas 4-1, 4-2, 4-3, and 4-4.

Formula 4-1	$\mu = \Sigma[x \cdot P(x)]$	mean for a probability distribution
Formula 4-2	$\sigma^2 = \Sigma[(x - \mu)^2 \cdot P(x)]$	variance for a probability distribution
Formula 4-3	$\sigma^2 = \Sigma[x^2 \cdot P(x)] - \mu^2$	variance for a probability distribution
Formula 4-4	$\sigma = \sqrt{\Sigma[x^2 \cdot P(x)] - \mu^2}$	standard deviation for a probability distribution

TI-83 Plus

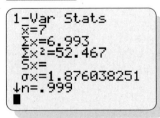

```
1-Var Stats
 x̄=7
 Σx=6.993
 Σx²=52.467
 Sx=
 σx=1.876038251
↓n=.999
```

Caution: Evaluate $\Sigma[x^2 \cdot P(x)]$ by first squaring each value of x, then multiplying each square by the corresponding $P(x)$, then adding.

The TI-83 Plus calculator can be used to find the mean and standard deviation. Shown here is the TI-83 Plus screen display for the probability distribution described by Table 4-1. In this TI-83 Plus display, the value shown as $\bar{x}$ is actually the value of the mean μ, and the value shown as σx is the value of the standard deviation σ. That is, $\mu = 7$ and $\sigma = 1.876038251$.

Rationale for Formulas 4-1 Through 4-4

Why do Formulas 4-1 through 4-4 work? Formula 4-1 accomplishes the same task as the formula for the mean of a frequency table. (Recall that f represents class frequency and N represents population size.) Rewriting the formula for the mean of a frequency table so that it applies to a population and then changing its form, we get

$$\mu = \frac{\Sigma(f \cdot x)}{N} = \Sigma \frac{f \cdot x}{N} = \Sigma x \cdot \frac{f}{N} = \Sigma x \cdot P(x)$$

In the fraction f/N, the value of f is the frequency with which the value x occurs and N is the population size, so f/N is the probability for the value of x.

Similar reasoning enables us to take the variance formula from Chapter 2 and apply it to a random variable for a probability distribution; the result is Formula 4-2. Formula 4-3 is a shortcut version that will always produce the same result as Formula 4-2. Although Formula 4-3 is usually easier to work with, Formula 4-2 is easier to understand directly. Based on Formula 4-2, we can express the standard deviation as

$$\sigma = \sqrt{\Sigma(x - \mu)^2 \cdot P(x)}$$

or as the equivalent form given in Formula 4-4.

When using Formulas 4-1 through 4-4, use this rule for rounding results.

Round-off Rule for μ, σ, and σ^2

Round results by carrying one more decimal place than the number of decimal places used for the random variable x. If the values of x are integers, round μ, σ, and σ^2 to one decimal place.

Note to Instructor
Remind your students that they should round only at the end of a computation, not at middle stages. Rounding in the middle can create errors that grow very large when used in subsequent steps.

It is sometimes necessary to use a different rounding rule because of special circumstances, such as results that require more decimal places to be meaningful. For example, with four-engine jets the mean number of jet engines working successfully throughout a flight is 3.999714286, which becomes 4.0 when rounded to one more decimal place than the original data. Here, 4.0 would be misleading because it suggests that all jet engines always work successfully. We need more precision to correctly reflect the true mean, such as the precision in the number 3.999714.

Identifying Unusual Results with the Range Rule of Thumb

The range rule of thumb (discussed in Section 2-5) may also be helpful in interpreting the value of a standard deviation. According to the range rule of thumb, most values should lie within 2 standard deviations of the mean; it is unusual for a

Note to Instructor
Stress that obtaining numerical answers is not the ultimate goal. We want to *interpret* results so that they can be applied in a meaningful way.

value to differ from the mean by more than 2 standard deviations. (The use of 2 standard deviations is not generally an absolutely rigid value, and other values such as 3 could be used instead.) We can therefore identify "unusual" values by determining that they lie outside of these limits:

$$\text{maximum usual value} = \mu + 2\sigma$$
$$\text{minimum usual value} = \mu - 2\sigma$$

EXAMPLE Table 4-1 describes the probability distribution for the number of girls among 14 randomly selected newborn babies. Assuming that we repeat the study of randomly selecting 14 newborn babies and counting the number of girls each time, find the mean number of girls (among 14), the variance, and the standard deviation. Use those results and the range rule of thumb to find the maximum and minimum usual values.

SOLUTION   In Table 4-3, the two columns at the left describe the probability distribution given earlier in Table 4-1, and we create the three columns at the right for the purposes of the calculations required.

Table 4-3	Calculating μ, σ, and σ^2 for a Probability Distribution			
x	$P(x)$	$x \cdot P(x)$	x^2	$x^2 \cdot P(x)$
0	0.000	0.000	0	0.000
1	0.001	0.001	1	0.001
2	0.006	0.012	4	0.024
3	0.022	0.066	9	0.198
4	0.061	0.244	16	0.976
5	0.122	0.610	25	3.050
6	0.183	1.098	36	6.588
7	0.209	1.463	49	10.241
8	0.183	1.464	64	11.712
9	0.122	1.098	81	9.882
10	0.061	0.610	100	6.100
11	0.022	0.242	121	2.662
12	0.006	0.072	144	0.864
13	0.001	0.013	169	0.169
14	0.000	0.000	196	0.000
Total		6.993		52.467
		↑		↑
		$\Sigma[x \cdot P(x)]$		$\Sigma[x^2 \cdot P(x)]$

Is Parachuting Safe?

About 30 people die each year as more than 100,000 people make about 2.25 million parachute jumps. In comparison, a typical year includes about 200 scuba diving fatalities, 7000 drownings, 900 bicycle deaths, 800 lightning deaths, and 1150 deaths from bee stings. Of course, these figures don't necessarily mean that parachuting is safer than bike riding or swimming. A fair comparison should involve fatality rates, not just the total number of deaths.

The author, with much trepidation, made two parachute jumps but quit after missing the spacious drop zone both times. He has also flown in a hang glider, hot air balloon, and Goodyear blimp.

Using Formulas 4-1 and 4-3 and the table results, we get

$$\mu = \Sigma[x \cdot P(x)] = 6.993 = 7.0 \qquad \text{(rounded)}$$
$$\sigma^2 = \Sigma[x^2 \cdot P(x)] - \mu^2$$
$$= 52.467 - 6.993^2 = 3.564951 = 3.6 \qquad \text{(rounded)}$$

The standard deviation is the square root of the variance, so

$$\sigma = \sqrt{3.564951} = 1.9 \qquad \text{(rounded)}$$

We now know that among groups of 14 newborn babies, the mean number of girls is 7.0, the variance is 3.6 "girls squared," and the standard deviation is 1.9 girls.

Using the range rule of thumb, we can now find the maximum and minimum usual values as follows:

$$\text{maximum usual value: } \mu + 2\sigma = 7.0 + 2(1.9) = 10.8$$
$$\text{minimum usual value: } \mu - 2\sigma = 7.0 - 2(1.9) = 3.2$$

INTERPRETATION Based on these results, we conclude that for groups of 14 randomly selected babies, the number of girls should usually fall between 3.2 and 10.8.

Identifying Unusual Results with Probabilities

Strong recommendation: The following discussion includes some difficult concepts, but it also includes an extremely important approach used often in statistics. You should make every effort to understand this discussion, even if it requires several readings. Keep in mind that this discussion is based on the *rare event rule* introduced in Section 3-2:

> **If, under a given assumption (such as the assumption that boys and girls are equally likely), the probability of a particular observed event (such as 13 girls in 14 births) is extremely small, we conclude that the assumption is probably not correct.**

In the Chapter Problem, we noted that with the MicroSort technique, there were 13 girls among 14 babies. Is this result unusual? Does this result really suggest that the technique is effective, or could we get 13 girls among 14 babies just by chance? To address this issue, we could use the range rule of thumb to find the minimum and maximum likely outcomes, but here we will consider another approach. We will find the probability of getting *13 or more* girls (not the probability of getting *exactly* 13 girls). It's difficult to see why this probability of 13 *or more girls* is the relevant probability, so let's try to clarify it with a more obvious example.

Suppose you were flipping a coin to determine whether it favors heads, and suppose 1000 tosses resulted in 501 heads. This is not evidence that the coin favors heads, because it is very easy to get a result like 501 heads in 1000 tosses just by chance. Yet, the probability of getting *exactly* 501 heads in 1000 tosses is actually quite small: 0.0252. This low probability reflects the fact that with 1000

Note to Instructor
In this text, we use the equal symbol (=) even though a result may be rounded, as in 1.88811 = 1.9 (rounded). We use the symbol ≈ for estimated values, such as $s \approx 1$ for a value of s estimated from the range rule of thumb. Also, many students tend to round expected values, such as "3.2 girls." They are thinking that you cannot have a fraction of a girl, but emphasize that the expected value is not the value expected when an experiment is conducted once; instead, it is a kind of average over a long run. When we say that the expected number of girls is 3.2, we don't mean that we expect any one trial to result in 3.2 girls; it means that among many such trials, the mean number of girls is 3.2.

Note to Instructor
Strongly consider elaborating on this discussion. It is very important and provides a great foundation for important concepts introduced when hypothesis testing is presented in Chapter 7.

tosses, any *specific* number of heads will have a very low probability. However, we do not consider 501 heads among 1000 tosses to be *unusual*, because the probability of getting *at least* 501 heads is high: 0.487. This principle can be generalized as follows:

Using Probabilities to Determine When Results Are Unusual

- **Unusually high:** x successes among n trials is an *unusually high* number of successes if $P(x$ or more$)$ is very small (such as 0.05 or less)

- **Unusually low:** x successes among n trials is an *unusually low* number of successes if $P(x$ or fewer$)$ is very small (such as 0.05 or less)

EXAMPLE Gender Selection Using the preceding two criteria based on probabilities, is it unusual to get 13 girls among 14 births? Does the MicroSort technique of gender selection appear to be effective?

SOLUTION Thirteen girls among 14 births is unusually high if $P(13$ or more girls$)$ is very small. If we refer to Table A-1, we get this result:

$$P(13 \text{ or more girls}) = P(13) + P(14)$$
$$= 0.001 + 0.000$$
$$= 0.001$$

INTERPRETATION Because the probability 0.001 is so low, we conclude that it is unusual to get 13 girls among 14 births. This suggests that the MicroSort technique of gender selection appears to be effective, because it is highly unlikely that the result of 13 girls among 14 births happened by chance.

Expected Value

The mean of a discrete random variable is the theoretical mean outcome for infinitely many trials. We can think of that mean as the *expected value* in the sense that it is the average value that we would expect to get if the trials could continue indefinitely. The uses of expected value (also called *expectation,* or *mathematical expectation*) are extensive and varied, and they play a very important role in an area of application called *decision theory.*

Definition

The **expected value** of a discrete random variable is denoted by E, and it represents the average value of the outcomes. It is obtained by finding the value of $\Sigma[x \cdot P(x)]$.

$$E = \Sigma[x \cdot P(x)]$$

From Formula 4-1 we see that $E = \mu$. That is, the mean of a discrete random variable is the same as its expected value. Repeat the procedure of flipping a coin five times, and the *mean* number of heads is 2.5; when flipping a coin five times, the *expected value* of the number of heads is also 2.5.

EXAMPLE **New Jersey Pick 3 Game** Years ago, members of organized crime groups ran numbers games. Similar games are now run by many members of organized governments, as well as some governments not so well organized. New Jersey's Pick 3 game closely resembles the older illegal games. A "straight" bet works like this: Bet 50¢ and select a three-digit number between 000 and 999. If your three digits match those that are randomly drawn, you collect $275 for a net gain of $274.50 (because your 50¢ bet is not returned). Suppose that you bet 50¢ on the number 007. What is your expected value of gain or loss?

SOLUTION   For this bet there are two simple outcomes: You win or you lose. Because you have selected the number 007 and there are 1000 possibilities (from 000 to 999), your probability of winning is 1/1000 (or 0.001) and your probability of losing is 999/1000 (or 0.999). Table 4-4 summarizes this situation.

From Table 4-4 we can see that when we bet 50¢ in New Jersey's Pick 3 game, our expected value is

$$E = \Sigma[x \cdot P(x)] = -22.5 \text{ cents}$$

Here is how the result can be interpreted: In the long run, for each 50¢ bet, we can expect to lose an average of 22.5¢. In any individual game, you will either lose 50¢ or you will win $274.50. Although you cannot lose 22.5¢ in one individual game, the expected value of -22.5¢ shows that in the long run of many games, the average loss per game is 22.5¢.

Table 4-4	New Jersey Pick 3 Game		
Event	x	$P(x)$	$x \cdot P(x)$
Win	$274.50	0.001	$0.2745
Lose	$-$0.50	0.999	$-$0.4995
Total			$-$0.225
			(or -22.5¢)

In this section we learned that a random variable has a numerical value associated with each outcome of some random procedure, and a probability distribution has a probability associated with each value of a random variable. We examined methods for finding the mean, variance, and standard deviation for a probability distribution. We saw that the expected value of a random variable is really the same as the mean. We also learned that we should not expect to get rich by playing New Jersey's Pick 3 lottery game.

Recommended Assignment
Exercises 1–12, 15–18.

4-2 Basic Skills and Concepts

Identifying Discrete and Continuous Random Variables. In Exercises 1 and 2, identify the given random variable as being discrete or continuous.

1. a. continuous
   b. discrete
   c. continuous
   d. discrete
   e. discrete

1. **a.** The height of a randomly selected basketball player in the NBA
   b. The number of points scored in a season by a randomly selected basketball player in the NBA
   c. The exact playing time of a randomly selected basketball player in the NBA
   d. The number of athletes who played in any NBA game in a season
   e. The salary of a randomly selected basketball player in the NBA

2. a. discrete
   b. discrete
   c. continuous
   d. discrete
   e. continuous

2. **a.** The cost of making a randomly selected movie
   b. The number of movies currently being shown in U.S. theaters
   c. The exact running time of a randomly selected movie
   d. The number of actors appearing in a randomly selected movie
   e. The weight of the lead actor in a randomly selected movie

Identifying Probability Distributions. In Exercises 3–10, determine whether a probability distribution is given. In those cases where a probability distribution is not described, identify the requirements that are not satisfied. In those cases where a probability distribution is described, find its mean and standard deviation.

3. $\mu = 1.5$, $\sigma = 0.9$

3. Gender Selection In a study of the MicroSort gender-selection method, couples in a control group are not given a treatment, and they each have three children. The probability distribution for the number of girls is given in the accompanying table.

x	$P(x)$
0	0.125
1	0.375
2	0.375
3	0.125

4. Not a probability distribution because $\Sigma P(x) = 0.977 \neq 1$.

4. Quality Control for DVDs When manufacturing DVDs for Sony, batches of DVDs are randomly selected and the number of defects x is found for each batch.

x	$P(x)$
0	0.502
1	0.365
2	0.098
3	0.011
4	0.001

5. Not a probability distribution because $\Sigma P(x) = 0.94 \neq 1$.

5. Videotape Rentals The accompanying table is constructed from data obtained in a study of the number of videotapes rented from Blockbuster.

x	$P(x)$
0	0.04
1	0.26
2	0.36
3	0.20
4	0.08

6. $\mu = 4.0$, $\sigma = 0.2$

6. Life Insurance The Telektronic Company provides life insurance policies for its top four executives, and the random variable x is the number of those employees who live through the next year.

x	$P(x)$
0	0.0000
1	0.0001
2	0.0006
3	0.0387
4	0.9606

7. Prior Sentences When randomly selecting a jail inmate convicted of DWI (driving while intoxicated), the probability distribution for the number x of prior DWI sentences is as described in the accompanying table (based on data from the U.S. Department of Justice).

x	$P(x)$
0	0.512
1	0.301
2	0.132
3	0.055

7. $\mu = 0.7$, $\sigma = 0.9$

8. Overbooked Flights Air America has a policy of routinely overbooking flights, because past experience shows that some passengers fail to show. The random variable x represents the number of passengers who cannot be boarded because there are more passengers than seats.

x	$P(x)$
0	0.805
1	0.113
2	0.057
3	0.009
4	0.002

8. Not a probability distribution because $\Sigma P(x) = 0.986 \neq 1$.

9. Number of Games in a Baseball World Series Based on past results found in the *Information Please Almanac*, there is a 0.1809 probability that a baseball World Series contest will last four games, a 0.2234 probability that it will last five games, a 0.2234 probability that it will last six games, and a 0.3723 probability that it will last seven games. Is it unusual for a team to "sweep" by winning in four games?

9. $\mu = 5.8$, $\sigma = 1.1$; no

10. Brand Recognition In a study of brand recognition of Sony, groups of four consumers are interviewed. If x is the number of people in the group who recognize the Sony brand name, then x can be 0, 1, 2, 3, or 4, and the corresponding probabilities are 0.0016, 0.0250, 0.1432, 0.3892, and 0.4096. Is it unusual to randomly select four consumers and find that none of them recognize the brand name of Sony?

10. Not a probability distribution because $\Sigma P(x) = 0.9686 \neq 1$.

11. Finding Expected Value in Craps When you give a casino $5 for a bet on the "pass line" in the game of craps, there is a 244/495 probability that you will win $5 and a 251/495 probability that you will lose $5. What is your expected value? In the long run, how much do you lose for each dollar bet?

11. -7.07¢; 1.4¢

12. Finding Expected Value in Roulette When you give a casino $5 for a bet on the number 7 in roulette, you have a 1/38 probability of winning $175 and a 37/38 probability of losing $5. If you bet $5 that the outcome is an odd number, the probability of winning $5 is 18/38, and the probability of losing $5 is 20/38.
a. If you bet $5 on the number 7, what is your expected value?
b. If you bet $5 that the outcome is an odd number, what is your expected value?
c. Which of these options is best: bet on 7, bet on odd, or don't bet? Why?

12. a. -26¢
    b. -26¢
    c. Don't bet, because the expected value with no bet is 0, which is better than -26¢.

13. Finding Expected Value for a Life Insurance Policy The CNA Insurance Company charges Mike $250 for a one-year $100,000 life insurance policy. Because Mike is a 21-year-old male, there is a 0.9985 probability that he will live for a year (based on data from U.S. National Center for Health Statistics).
a. From Mike's perspective, what are the values of the two different outcomes?
b. If Mike purchases the policy, what is his expected value?
c. What would be the cost of the insurance policy if the company just breaks even (in the long run with many such policies), instead of making a profit?
d. Given that Mike's expected value is negative (so the insurance company can make a profit), why should Mike or anyone else purchase life insurance?

13. a. Lives: $-$250 (a loss); dies: $99,750 (a gain)
    b. $-$100
    c. $150
    d. The negative expected value is a relatively small price to pay for insuring for the financial security of his heirs.

14. Finding Expected Value for a Magazine Sweepstakes *Reader's Digest* recently ran a sweepstakes in which prizes were listed along with the chances of winning:

14. a. 1.2¢
    b. 1.2¢ minus cost of stamp; no

$1,000,000 (1 chance in 90,000,000), $100,000 (1 chance in 110,000,000), $25,000 (1 chance in 110,000,000), $5000 (1 chance in 36,667,000), and $2500 (1 chance in 27,500,000).

a. Find the expected value of the amount won for one entry.

b. Find the expected value if the cost of entering this sweepstakes is the cost of a postage stamp. Is it worth entering this contest?

15. Finding Expected Value for New Jersey's Pick 4 Game In New Jersey's Pick 4 lottery game, you pay 50¢ to select a sequence of four digits, such as 7273. If you win by selecting the same sequence of four digits that are drawn, you collect $2788.

a. How many different selections are possible?

b. What is the probability of winning?

c. If you win, what is your net profit?

d. Find the expected value.

e. An example in this section showed that if you bet 50¢ in the New Jersey Pick 3 game, your expected value is −22.5¢. Which is the better game: New Jersey's Pick 3 game or Pick 4 game? Why?

16. Finding Expected Value for the Illinois Pick 3 Game In Illinois' Pick 3 game, you pay 50¢ to select a sequence of three digits, such as 911. If you win by selecting the same sequence of three digits that are drawn, you collect $250.

a. How many different selections are possible?

b. What is the probability of winning?

c. If you win, what is your net profit?

d. Find the expected value.

e. An example in this section showed that if you bet 50¢ in the New Jersey Pick 3 game, your expected value is −22.5¢. Which is the better game: The Illinois Pick 3 game or the New Jersey Pick 3 game? Why?

 17. Determining Whether Gender-Selection Technique Is Effective Assume that in a test of a gender-selection technique, a clinical trial results in 9 girls in 14 births. Refer to Table 4-1 and find the indicated probabilities.

a. Find the probability of exactly 9 girls in 14 births.

b. Find the probability of 9 or more girls in 14 births.

c. Which probability is relevant for determining whether 9 girls in 14 births is unusually high: the result from part (a) or part (b)?

d. Does 9 girls in 14 births suggest that the gender-selection technique is effective? Why or why not?

18. Determining Whether Gender-Selection Technique Is Effective Assume that in a test of a gender-selection technique, a clinical trial results in 12 girls in 14 births. Refer to Table 4-1 and find the indicated probabilities.

a. Find the probability of exactly 12 girls in 14 births.

b. Find the probability of 12 or more girls in 14 births.

c. Which probability is relevant for determining whether 12 girls in 14 births is unusually high: the result from part (a) or part (b)?

d. Does 12 girls in 14 births suggest that the gender-selection technique is effective? Why or why not?

 19. Determining Whether Gender-Selection Technique Is Effective Assume that in a test of a gender-selection technique, a clinical trial results in 11 girls in 14 births. Refer to Table 4-1 and find the indicated probability.

continued

15. a. 10,000
    b. 0.0001
    c. $2787.50
    d. −22.12¢
    e. Pick 4, because −22.12¢ is greater than −22.5¢.
16. a. 1000
    b. 0.001
    c. $249.50
    d. −25¢
    e. New Jersey, because −22.5¢ is greater than −25¢.
17. a. 0.122
    b. 0.212
    c. Part (b). The occurrence of 9 girls among 14 would be unusually high if the probability of 9 or more girls is very small (such as less than 0.05).
    d. No, because the probability of 9 or more girls is not very small (0.212). The result of 9 or more girls could easily occur by chance.
18. a. 0.006
    b. 0.007
    c. Part (b). The occurrence of 12 girls among 14 would be unusually high if the probability of 12 or more girls is very small (such as less than 0.05).
    d. Yes, because the probability of 12 or more girls is very small (0.007). The result of 12 or more girls could not easily occur by chance.
19. a. 0.029
    b. Yes, because the probability of 11 or more girls is very small (0.029). The result of 11 or more girls could not easily occur by chance.

a. What is the probability value that should be used for determining whether the result of 11 girls in 14 births is unusually high?

b. Does 11 girls in 14 births suggest that the gender-selection technique is effective? Why or why not?

④ 20. Determining Whether Gender-Selection Technique Is Effective   Assume that in a test of a gender-selection technique, a clinical trial results in 10 girls in 14 births. Refer to Table 4-1 and find the indicated probability.

a. What is the probability value that should be used for determining whether the result of 10 girls in 14 births is unusually high?

b. Does 10 girls in 14 births suggest that the gender-selection technique is effective? Why or why not?

21. Psychic Powers   Bob is a self-proclaimed "mentalist" who claims that he can pass true/false tests by guessing. In a test of Bob's claim, he is given 14 true/false questions, and Bob manages to correctly guess the answers to 8 of them. He claims that getting 8 correct answers among 14 is proof of his special powers, because he got more answers correct than the 7 correct answers that are expected with guessing. Is Bob's claim valid? Why or why not? (*Hint:* When finding the required probability, refer to Table 4-1 and change "girl" to "correct." The probabilities of girls will be the same as the probabilities of correct guesses.)

22. Psychic Powers   Bob is a self-proclaimed "mentalist" who claims that he can pass true/false tests by guessing. In a test of Bob's claim, he is given 14 true/false questions, and Bob manages to correctly guess the answers to 2 of them. Is the number of correct answers *unusual?* Why or why not? Does Bob's claim appear to be valid? (*Hint:* When finding the required probability, refer to Table 4-1 and change "girl" to "correct." The probabilities of girls will be the same as the probabilities of correct guesses.)

4-2 Beyond the Basics

23. Junk Bonds   Kim Hunter has $1000 to invest, and her financial analyst recommends two types of junk bonds. The A bonds have a 6% annual yield with a default rate of 1%. The B bonds have an 8% annual yield with a default rate of 5%. (If the bond defaults, the $1000 is lost.) Which of the two bonds is better? Why? Should she select either bond? Why or why not?

24. Finding Mean and Standard Deviation   Let the random variable x represent the number of girls in a family of four children. Construct a table describing the probability distribution, then find the mean and standard deviation. (*Hint:* List the different possible outcomes.)

25. Defective Parts: Finding Mean and Standard Deviation   The Sky Ranch is a supplier of aircraft parts. Included in stock are eight altimeters that are correctly calibrated and two that are not. Three altimeters are randomly selected without replacement. Let the random variable x represent the number that are not correctly calibrated. Find the mean and standard deviation for the random variable x.

26. Computer-Generated Numbers Transformed to z Scores   Computers are often used to randomly generate the last digits of telephone numbers of potential survey subjects. The digits are selected so that they are all equally likely. The random variable x is the selected digit.

continued

20. a. 0.090

    b. No, because the probability of 10 or more girls is not very small (0.090). The result of 10 or more girls could easily occur by chance.

21. Because the probability of correctly guessing 8 or more answers is 0.395, that outcome could easily occur, so there is no evidence that Bob has special powers.

22. Because the probability of 2 or fewer correct guesses is only 0.007, the outcome is unusual. Bob's claim is not valid because he got an unusually low number of correct answers instead of an unusually high number.

23. The A bonds are better because the expected value is $49.40, which is higher than the expected value of $26 for the B bonds. She should select the A bond because the expected value is positive, which indicates a likely gain.

24. The probabilities for 0, 1, 2, 3, 4 girls are 1/16, 4/16, 6/16, 4/16, 1/16. $\mu = 2.0$, $\sigma = 1.0$

25. $\mu = 0.6$, $\sigma = 0.6$

26. a. $\mu = 4.5$, $\sigma = 2.9$
    b. $\mu = 0$, $\sigma = 1$
    c. Yes

a. Find the mean and standard deviation of the random variable x.

b. Find the z score for each value of x, then find the mean and standard deviation of these z scores.

c. Will the same mean and standard deviation as you found in part (b) result from every probability distribution?

27. Equally Likely Integers: Mean and Standard Deviation Assume that a probability distribution is described by the discrete random variable x that can assume the values $1, 2, \ldots, n$, and those values are equally likely. This probability distribution has mean and standard deviation described as follows.

$$\mu = \frac{n + 1}{2} \quad \text{and} \quad \sigma = \sqrt{\frac{n^2 - 1}{12}}$$

a. Show that $\mu = (n + 1)/2$ for the case of $n = 5$.

b. Show that $\sigma = \sqrt{(n^2 - 1)/12}$ for the case of $n = 5$.

c. In testing someone who claims to have ESP, you randomly select whole numbers between 1 and 20, and the random variable x is the number selected. Find the mean and standard deviation for x.

28. Labeling Dice to Get a Uniform Distribution Assume that you have two blank dice, so that you can label the 12 faces with any numbers. Describe how the dice can be labeled so that, when the two dice are rolled, the totals of the two dice are uniformly distributed so that the outcomes of 1, 2, 3, . . . , 12 each have probability 1/12. (See "Can One Load a Set of Dice So That the Sum Is Uniformly Distributed?" by Chen, Rao, and Shreve, *Mathematics Magazine*, Vol. 70, No. 3.)

4-3 Binomial Probability Distributions

Note to Instructor
Although the binomial distribution is the main focus of this section, Exercises 37–39 deal with the geometric, hypergeometric, and multinomial distributions. Section 4-5 deals with the Poisson distribution.

In Section 4-2 we discussed several different discrete probability distributions, but in this section we will focus on one specific type: the binomial probability distribution. Binomial probability distributions are important because they allow us to deal with circumstances in which the outcomes belong to *two* relevant categories, such as acceptable/defective products or yes/no responses in a survey question. The Chapter Problem involves counting the number of girls in 14 births. The problem involves the two categories of boy/girl, so it has the required key element of "twoness." Other requirements are given in the following definition.

Definition

A **binomial probability distribution** results from a procedure that meets all the following requirements:

1. The procedure has a *fixed number of trials*.
2. The trials must be *independent*. (The outcome of any individual trial doesn't affect the probabilities in the other trials.)
3. Each trial must have all outcomes classified into *two categories*.
4. The probabilities must remain *constant* for each trial.

If a procedure satisfies these four requirements, the distribution of the random variable x is called a *binomial probability distribution* (or *binomial distribution*). The following notation is commonly used.

Notation for Binomial Probability Distributions

S and F (success and failure) denote the two possible categories of all outcomes; p and q will denote the probabilities of S and F, respectively, so

$$P(S) = p \qquad (p = \text{probability of a success})$$

$$P(F) = 1 - p = q \qquad (q = \text{probability of a failure})$$

n denotes the fixed number of trials.

x denotes a specific number of successes in n trials, so x can be any whole number between 0 and n, inclusive.

p denotes the probability of *success* in *one* of the n trials.

q denotes the probability of *failure* in *one* of the n trials.

$P(x)$ denotes the probability of getting exactly x successes among the n trials.

The word *success* as used here is arbitrary and does not necessarily represent something good. Either of the two possible categories may be called the success S as long as its probability is identified as p. Once a category has been designated as the success S, be sure that p is the probability of a success and x is the number of successes. That is, be sure that the values of p and x refer to the same category designated as a success. (The value of q can always be found by subtracting p from 1; if $p = 0.95$, then $q = 1 - 0.95 = 0.05$.) Here is an important hint for working with binomial probability problems:

Be sure that x and p both refer to the *same* category being called a success.

When selecting a sample for some statistical analysis, we usually sample without replacement. For example, when testing manufactured items or conducting surveys, we usually design the sampling process so that selected items cannot be selected a second time. Strictly speaking, sampling without replacement involves dependent events, which violates the second requirement in the above definition. However, the following rule of thumb is based on the fact that if the sample is very small relative to the population size, we can treat the trials as being independent (even though they are actually dependent) because the difference in results will be negligible.

When sampling without replacement, the events can be treated as if they were independent if the sample size is no more than 5% of the population size. (That is, $n \le 0.05N$.)

Prophets for Profits

Many books and computer programs claim to be helpful in predicting winning lottery numbers. Some use the theory that particular numbers are "due" (and should be selected) because they haven't been coming up often; others use the theory that some numbers are "cold" (and should be avoided) because they haven't been coming up often; and still others use astrology, numerology, or dreams. Because selections of winning lottery number combinations are independent events, such theories are worthless. A valid approach is to choose numbers that are "rare" in the sense that they are not selected by other people, so that if you win, you will not need to share your jackpot with many others. For this reason, the combination of 1, 2, 3, 4, 5, and 6 is a bad choice because many people use it, whereas 12, 17, 18, 33, 40, 46 is a much better choice, at least until it was published in this book.

EXAMPLE **Analysis of Multiple-Choice Answers** Because they are so easy to correct, multiple-choice questions are commonly used for class tests, SAT tests, MCAT tests for medical schools, LSAT tests for law schools, and many other circumstances. A professor teaching a course in Abnormal Psychology plans to give a surprise quiz consisting of 4 multiple-choice questions, each with 5 possible answers (a, b, c, d, e), one of which is correct. Let's assume that an unprepared student makes random guesses and we want to find the probability of exactly 3 correct responses to the 4 questions.

a. Does this procedure result in a binomial distribution?

b. If this procedure does result in a binomial distribution, identify the values of n, x, p, and q.

SOLUTION

a. This procedure does satisfy the requirements for a binomial distribution, as shown below.

1. The number of trials (4) is fixed.
2. The 4 trials are independent because a correct or wrong response for any individual question does not affect the probability of being correct or wrong on another question.
3. Each of the 4 trials has two categories of outcomes: The response is either correct or wrong.
4. For each response, there are 5 possible answers (a, b, c, d, e), one of which is correct, so the probability of a correct response is 1/5 or 0.2. That probability remains constant for each of the 4 trials.

b. Having concluded that the given procedure does result in a binomial distribution, we now proceed to identify the values of n, x, p, and q.

1. With 4 test questions, we have $n = 4$.
2. We want the probability of exactly 3 correct responses, so $x = 3$.
3. The probability of success (correct response) for one question is 0.2, so $p = 0.2$.
4. The probability of failure (wrong response) is 0.8, so $q = 0.8$.

Again, it is very important to be sure that x and p both refer to the same concept of "success." In this example, we use x to count the *correct* responses, so p must be the probability of a *correct* response. Therefore, x and p do use the same concept of success (correct response) here.

We will now present three methods for finding the probabilities corresponding to the random variable x in a binomial distribution. The first method involves calculations using the *binomial probability formula* and is the basis for the other two methods. The second method involves the use of Table A-1, and the third method involves the use of statistical software or a calculator. If you are using software or a calculator that automatically produces binomial probabilities, we recommend that you solve one or two exercises using Method 1 to ensure that you understand the basis for the calculations. Understanding is always infinitely better than blind application of formulas.

Method 1: Using the Binomial Probability Formula In a binomial distribution, probabilities can be calculated by using the binomial probability formula.

Formula 4-5
$$P(x) = \frac{n!}{(n - x)!x!} \cdot p^x \cdot q^{n-x} \qquad \text{for } x = 0, 1, 2, \ldots, n$$

where n = number of trials

x = number of successes among n trials

p = probability of success in any one trial

q = probability of failure in any one trial ($q = 1 - p$)

The factorial symbol !, introduced in Section 3-7, denotes the product of decreasing factors. Two examples of factorials are $3! = 3 \cdot 2 \cdot 1 = 6$ and $0! = 1$ (by definition). Many calculators have a factorial key, as well as a key labeled $_nC_r$ that can simplify the computations. For calculators with the $_nC_r$ key, use this version of the binomial probability formula (where n, x, p, and q are the same as in Formula 4-5):

$$P(x) = {}_nC_r \cdot p^x \cdot q^{n-x}$$

The TI-83 Plus calculator is designed to automatically calculate binomial probabilities using this formula. Details for using the TI-83 Plus calculator will be discussed later in this section.

Note to Instructor
Here's a good strategy for using Formula 4-5: Get a single number for $n!/(n-x)!x!$, get a single number for p^x and a single number for q^{n-x}; then simply multiply the three factors together. Also, point out the common calculator error of evaluating a/bc by entering $a \div b \times c$; the correct entry is $a \div b \div c$. Finally, point out that students often get wrong results when they round intermediate results too much.

EXAMPLE Analysis of Multiple-Choice Answers Use the binomial probability formula to find the probability of getting exactly 3 correct answers when random guesses are made for 4 multiple-choice questions. That is, find $P(3)$ given that $n = 4$, $x = 3$, $p = 0.2$, and $q = 0.8$.

SOLUTION Using the given values of n, x, p, and q in the binomial probability formula (Formula 4-5), we get

$$P(x) = \frac{4!}{(4 - 3)!3!} \cdot 0.2^3 \cdot 0.8^{4-3}$$

$$= \frac{4!}{1!3!} \cdot 0.008 \cdot 0.8$$

$$= (4)(0.008)(0.8) = 0.0256$$

The probability of getting exactly 3 correct answers out of 4 is 0.0256.

Calculation hint: When computing a probability with the binomial probability formula, it's helpful to get a single number for $n!/(n - x)!x!$, a single number for

From Table A-1:

n	x	p 0.20
4	0	0.410
	1	0.410
	2	0.154
	3	0.026
	4	0.002

↓

Binomial probability
distribution for $n = 4$
and $p = 0.2$

x	P(x)
0	0.410
1	0.410
2	0.154
3	0.026
4	0.002

p^x and a single number for q^{n-x}, then simply multiply the three factors together as shown at the end of the calculation for the preceding example. Don't round too much when you find those three factors; round only at the end.

Method 2: Using Table A-1 in Appendix A   In some cases, we can easily find binomial probabilities by simply referring to Table A-1 in Appendix A. First locate n and the corresponding value of x that is desired. At this stage, one row of numbers should be isolated. Now align that row with the proper probability of p by using the column across the top. The isolated number represents the desired probability. A very small probability, such as 0.000000345, is indicated by 0+.

Part of Table A-1 is shown in the margin. With $n = 4$ and $p = 0.2$ in a binomial distribution, the probabilities of 0, 1, 2, 3, and 4 successes are 0.410, 0.410, 0.154, 0.026, and 0.002, respectively.

EXAMPLE Use the portion of Table A-1 (for $n = 4$ and $p = 0.2$) shown in the margin to find the following:

a. The probability of exactly 3 successes
b. The probability of *at least* 3 successes

SOLUTION

a. The display from Table A-1 shows that when $n = 4$ and $p = 0.2$, the probability of $x = 3$ is given by $P(3) = 0.026$, which is the same value (except for rounding) computed with the binomial probability formula in the preceding example.

b. "At least" 3 successes means that the number of successes is 3 or 4.

$$P(\text{at least } 3) = P(3 \text{ or } 4)$$
$$= P(3) + P(4)$$
$$= 0.026 + 0.002$$
$$= 0.028$$

In part (b) of the preceding solution, if we wanted to find $P(\text{at least } 3)$ by using the binomial probability formula, we would need to apply that formula twice to compute two different probabilities, which would then be added. Given this choice between the formula and the table, it makes sense to use the table. Unfortunately, Table A-1 includes only limited values of n as well as limited values of p, so the table doesn't always work, and we must then find the probabilities by using the binomial probability formula, software, or a calculator, as in the following method.

Method 3: Using Technology   STATDISK, Minitab, Excel, and the TI-83 Plus calculator can all be used to find binomial probabilities. We will present details for using each of those technologies at the end of this section. For now, see the typical screen displays that list binomial probabilities for $n = 4$ and $p = 0.2$.

STATDISK

Binomial Probabilities

x	P(x)	P(x≤)	P(x≥)
0	0.40960	0.40960	1.00000
1	0.40960	0.81920	0.59040
2	0.15360	0.97280	0.18080
3	0.02560	0.99840	0.02720
4	0.00160	1.00000	0.00160

Num Trials, n: 4
Success Prob, p: 0.2

Evaluate   Help   Plot

Mean   0.800000
St Dev   0.800000
Variance 0.640000

Minitab

Binomial with n = 4 and p = 0.200000

x	P(X = x)
0.00	0.4096
1.00	0.4096
2.00	0.1536
3.00	0.0256
4.00	0.0016

Excel

	A	B
1	0	0.4096
2	1	0.4096
3	2	0.1536
4	3	0.0256
5	4	0.0016

TI-83 Plus

L1	L2	L3
0	.4096	------
1	.4096	
2	.1536	
3	.0256	
4	.0016	

L2(6) =

Voltaire Beats Lottery

In 1729, the philosopher Voltaire became rich by devising a scheme to beat the Paris lottery. The government ran a lottery to repay municipal bonds that had lost some value. The city added large amounts of money with the net effect that the prize values totaled more than the cost of all tickets. Voltaire formed a group that bought all the tickets in the monthly lottery and won for more than a year. A bettor in the New York State Lottery tried to win a share of an exceptionally large prize that grew from a lack of previous winners. He wanted to write a $6,135,756 check that would cover all combinations, but the state declined and said that the nature of the lottery would have been changed.

Given that we now have three different methods for finding binomial probabilities, here is an effective and efficient strategy:

1. Use computer software or a TI-83 Plus calculator, if available.
2. If neither software nor the TI-83 Plus calculator is available, use Table A-1, if possible.
3. If neither software nor the TI-83 Plus calculator is available and the probabilities can't be found using Table A-1, use the binomial probability formula.

Rationale for the Binomial Probability Formula

The binomial probability formula is the basis for all three methods presented in this section. Instead of accepting and using that formula blindly, let's see why it works.

Earlier in this section, we used the binomial probability formula for finding the probability of getting exactly three correct answers when random guesses are

Sensitive Surveys

Survey respondents are sometimes reluctant to honestly answer questions on a sensitive topic, such as employee theft or sex. Stanley Warner (York University, Ontario) devised a scheme that leads to more accurate results in such cases. As an example, ask employees if they stole within the past year and also ask them to flip a coin. The employees are instructed to answer no if they didn't steal and the coin turns up heads. Otherwise, they should answer yes. The employees are more likely to be honest because the coin flip helps protect their privacy. Probability theory can then be used to analyze responses so that more accurate results can be obtained.

made for four multiple-choice questions. For each question there are five possible answers so that the probability of a correct response is 1/5 or 0.2. If we use the multiplication rule from Section 3-4, we get the following result:

$$P(3 \text{ correct answers followed by 1 wrong answer})$$
$$= 0.2 \cdot 0.2 \cdot 0.2 \cdot 0.8$$
$$= 0.2^3 \cdot 0.8$$
$$= 0.0064$$

This result isn't correct because it assumes that the *first* three responses are correct and the *last* response is wrong, but there are other arrangements possible for three correct responses and one wrong response.

In Section 3-7 we saw that with three items identical to each other (such as *correct* responses) and one other item (such as a *wrong* response), the total number of arrangements (permutations) is $4!/(4 - 3)!3!$ or 4. Each of those 4 different arrangements has a probability of $0.2^3 \cdot 0.8$, so the total probability is as follows:

$$P(3 \text{ correct among } 4) = \frac{4!}{(4 - 3)!3!} \cdot 0.2^3 \cdot 0.8^1$$

Generalize this result as follows: Replace 4 with n, replace x with 3, replace 0.2 with p, replace 0.8 with q, and express the exponent of 1 as $4 - 3$, which can be replaced with $n - x$. The result is the binomial probability formula. That is, the binomial probability formula is a combination of the multiplication rule of probability and the counting rule for the number of arrangements of n items when x of them are identical to each other and the other $n - x$ are identical to each other. (See Exercises 9 and 10.)

The number of outcomes with exactly x successes among n trials   The probability of x successes among n trials for any one particular order

$$P(x) = \frac{n!}{(n - x)!x!} \cdot p^x \cdot q^{n-x}$$

Using Technology

Method 3 in this section involved the use of STATDISK, Minitab, Excel, or a TI-83 Plus calculator. Screen displays shown with Method 3 illustrated typical results obtained by applying the following procedures for finding binomial probabilities.

STATDISK   Select **Analysis** from the main menu, then select the **Binomial Probabilities** option. Enter the requested values for n and p, and the entire probability distribution will be displayed. Other columns represent cumulative probabilities that are obtained by adding the values of $P(x)$ as you go down or up the column.

Minitab   First enter a column C1 of the x values for which you want probabilities (such as 0, 1, 2, 3, 4), then select **Calc** from the main menu, and proceed to select the submenu items of **Probability Distributions** and **Binomial.** Select **Probabilities,** enter the number of trials, the probability of success, and C1 for the input column, then click on **OK.**

Excel   List the values of x in column A (such as 0, 1, 2, 3, 4). Click on cell B1, then click on f_x from the toolbar, and se-

lect the function category **Statistical** and then the function name **BINOMDIST.** In the dialog box, enter A1 for the number of successes, enter the number of trials, enter the probability, and enter 0 for the binomial distribution (instead of 1 for the cumulative binomial distribution). A value should appear in cell B1. Click and drag the lower right corner of cell B1 down the column to match the entries in column A, then release the mouse button.

TI-83 Plus   Press **2nd VARS** (to get **DISTR,** which denotes "distributions"), then select the option identified as **binompdf(.** Complete the entry of **binompdf(n, p, x)** with specific values for n, p, and x, then press **ENTER,** and the result will be the probability of getting x successes among n trials.

You could also enter **binompdf(n, p)** to get a list of *all* of the probabilities corresponding to $x = 0, 1, 2, \ldots, n$. You could store this list in L2 by pressing **STO→L2.** You could then enter the values of 0, 1, 2, . . . , n in list L1, which would allow you to calculate statistics (by entering **STAT, CALC,** then **L1, L2**) or view the distribution in a table format (by pressing **STAT,** then **EDIT**).

4-3 Basic Skills and Concepts

Identifying Binomial Distributions. In Exercises 1–8, determine whether the given procedure results in a binomial distribution. For those that are not binomial, identify at least one requirement that is not satisfied.

1. Surveying people by asking them what they think of the current president

2. Surveying 1012 people and recording whether there is a "should not" response to this question: "Do you think the cloning of humans should or should not be allowed?"

3. Rolling a fair die 50 times

4. Rolling a loaded die 50 times and finding the number of times that 5 occurs

5. Recording the genders of 250 newborn babies

6. Determining whether each of 3000 heart pacemakers is acceptable or defective

7. Spinning a roulette wheel 12 times

8. Spinning a roulette wheel 12 times and finding the number of times that the outcome is an odd number

9. Finding Probabilities When Guessing Answers Multiple-choice questions each have five possible answers, one of which is correct. Assume that you guess the answers to three such questions.

continued

a. Use the multiplication rule to find the probability that the first two guesses are wrong and the third is correct. That is, find P(WWC), where C denotes a correct answer and W denotes a wrong answer.

b. Beginning with WWC, make a complete list of the different possible arrangements of two wrong answers and one correct answer, then find the probability for each entry in the list.

c. Based on the preceding results, what is the probability of getting exactly one correct answer when three guesses are made?

10. Finding Probabilities When Guessing Answers A test consists of multiple-choice questions, each having four possible answers, one of which is correct. Assume that you guess the answers to six such questions.

a. Use the multiplication rule to find the probability that the first two guesses are wrong and the last four guesses are correct. That is, find P(WWCCCC), where C denotes a correct answer and W denotes a wrong answer.

b. Beginning with WWCCCC, make a complete list of the different possible arrangements of two wrong answers and four correct answers, then find the probability for each entry in the list.

c. Based on the preceding results, what is the probability of getting exactly four correct answers when six guesses are made?

Using Table A-1. In Exercises 11–16, assume that a procedure yields a binomial distribution with a trial repeated n times. Use Table A-1 to find the probability of x successes given the probability p of success on a given trial.

11. $n = 2, x = 0, p = 0.01$ **12.** $n = 7, x = 2, p = 0.01$

13. $n = 4, x = 3, p = 0.95$ **14.** $n = 6, x = 5, p = 0.99$

15. $n = 10, x = 4, p = 0.95$ **16.** $n = 11, x = 7, p = 0.05$

Using the Binomial Probability Formula. In Exercises 17–20, assume that a procedure yields a binomial distribution with a trial repeated n times. Use the binomial probability formula to find the probability of x successes given the probability p of success on a single trial.

17. $n = 6, x = 4, p = 0.55$ **18.** $n = 6, x = 2, p = 0.45$

19. $n = 8, x = 3, p = 1/4$ **20.** $n = 10, x = 8, p = 1/3$

Using Computer Results. In Exercises 21–24, refer to the Minitab display in the margin. The probabilities were obtained by entering the values of n = 6 and p = 0.723. There is a 0.723 probability that a randomly selected American Airlines flight will arrive on time (based on data from the Department of Transportation). In each case, assume that six American Airlines flights are randomly selected and find the indicated probability.

21. Find the probability that at least five American Airlines flights arrive on time. Is it unusual to have at least five of six American Airlines flights arriving on time?

22. Find the probability that at most two American Airlines flights arrive on time. Is it unusual to have at most two of six American Airlines flights arriving on time?

23. Find the probability that more than one American Airlines flight arrives on time. Is it unusual to *not* get more than one of six American Airlines flights arriving on time?

---

10. a. 0.00220
    b. Each of the 15 arrangements has probability 0.00220.
    c. 0.0330

11. 0.980
12. 0.002
13. 0.171
14. 0.057
15. 0+
16. 0+
17. 0.278
18. 0.278
19. 0.208
20. 0.003
21. 0.4711; no
22. 0.0538; no
23. 0.9925 (or 0.9924); yes

Binomial with $n = 6$ and $p = 0.723000$

x	$P(X = x)$
0.00	0.0005
1.00	0.0071
2.00	0.0462
3.00	0.1607
4.00	0.3145
5.00	0.3283
6.00	0.1428

24. Find the probability that at least one American Airlines flight arrives on time. Is it unusual to *not* get at least one of six American Airlines flights arriving on time?

25. Color Blindness Nine percent of men and 0.25% of women cannot distinguish between the colors red and green. This is the type of color blindness that causes problems with traffic signals. If six men are randomly selected for a study of traffic signal perceptions, find the probability that exactly two of them cannot distinguish between red and green.

26. Acceptance Sampling The Telektronic Company purchases large shipments of fluorescent bulbs and uses this acceptance sampling plan: Randomly select and test 24 bulbs, then accept the whole batch if there is only one or none that doesn't work. If a particular shipment of thousands of bulbs actually has a 4% rate of defects, what is the probability that this whole shipment will be accepted?

27. IRS Audits The Hemingway Financial Company prepares tax returns for individuals. (Motto: "We also write great fiction.") According to the Internal Revenue Service, individuals making $25,000–$50,000 are audited at a rate of 1%. The Hemingway Company prepares five tax returns for individuals in that tax bracket, and three of them are audited.
  a. Find the probability that when five people making $25,000–$50,000 are randomly selected, exactly three of them are audited.
  b. Find the probability that at least three are audited.
  c. Based on the preceding results, what can you conclude about the Hemingway customers? Are they just unlucky, or are they being targeted for audits?

28. Directory Assistance An article in *USA Today* stated that "Internal surveys paid for by directory assistance providers show that even the most accurate companies give out wrong numbers 15% of the time." Assume that you are testing such a provider by making 10 requests and also assume that the provider gives the wrong telephone number 15% of the time.
  a. Find the probability of getting one wrong number.
  b. Find the probability of getting at most one wrong number.
  c. If you do get at most one wrong number, does it appear that the rate of wrong numbers is not 15%, as claimed?

29. Overbooking Flights Air America has a policy of booking as many as 15 persons on an airplane that can seat only 14. (Past studies have revealed that only 85% of the booked passengers actually arrive for the flight.) Find the probability that if Air America books 15 persons, not enough seats will be available. Is this probability low enough so that overbooking is not a real concern for passengers?

30. Drug Reaction In a clinical test of the drug Viagra, it was found that 4% of those in a placebo group experienced headaches.
  a. Assuming that the same 4% rate applies to those taking Viagra, find the probability that among eight Viagra users, three experience headaches.
  b. Assuming that the same 4% rate applies to those taking Viagra, find the probability that among eight randomly selected users of Viagra, all eight experienced a headache.
  c. If all eight Viagra users were to experience a headache, would it appear that the headache rate for Viagra users is different than the 4% rate for those in the placebo group? Explain.

24. 0.9995 (or 0.9996); yes

25. 0.0833

26. 0.751

27. a. 0+ (or 0.00000980)
    b. 0+ (or 0.00000985)
    c. They are probably being targeted.

28. a. 0.347
    b. 0.544
    c. No, there is a good chance (0.544) of getting at most one wrong number if the rate is actually 15%.

29. 0.0874; no

30. a. 0.00292
    b. 0+ (or 0.00000000000655)
    c. Yes, because it is very unlikely (0+) that all 8 would get headaches by chance.

31. TV Viewer Surveys The CBS television show *60 Minutes* has been successful for many years. That show recently had a share of 20, meaning that among the TV sets in use, 20% were tuned to *60 Minutes* (based on data from Nielsen Media Research). Assume that an advertiser wants to verify that 20% share value by conducting its own survey, and a pilot survey begins with 10 households having TV sets in use at the time of a *60 Minutes* broadcast.
  a. Find the probability that none of the households are tuned to *60 Minutes*.
  b. Find the probability that at least one household is tuned to *60 Minutes*.
  c. Find the probability that at most one household is tuned to *60 Minutes*.
  d. If at most one household is tuned to *60 Minutes*, does it appear that the 20% share value is wrong? Why or why not?

32. Affirmative Action Programs A study was conducted to determine whether there were significant differences between medical students admitted through special programs (such as affirmative action) and medical students admitted through the regular admissions criteria. It was found that the graduation rate was 94% for the medical students admitted through special programs (based on data from the *Journal of the American Medical Association*).
  a. If 10 of the students from the special programs are randomly selected, find the probability that at least nine of them graduated.
  b. Would it be unusual to randomly select 10 students from the special programs and get only seven that graduate? Why or why not?

33. Identifying Gender Discrimination After being rejected for employment, Kim Kelly learns that the Bellevue Advertising Company has hired only two women among the last 20 new employees. She also learns that the pool of applicants is very large, with an approximately equal number of qualified men and women. Help her address the charge of gender discrimination by finding the probability of getting two or fewer women when 20 people are hired, assuming that there is no discrimination based on gender. Does the resulting probability really support such a charge?

34. Author's Slot Machine The author purchased a slot machine that is configured so that there is a 1/2000 probability of winning the jackpot on any individual trial. Although no one would seriously consider tricking the author, suppose that a guest claims that she played the slot machine five times and hit the jackpot twice.
  a. Find the probability of exactly two jackpots in five trials.
  b. Find the probability of at least two jackpots in five trials.
  c. Does the guest's claim of two jackpots in five trials seem valid? Explain.

35. Testing Effectiveness of Gender-Selection Technique The Chapter Problem describes the probability distribution for the number of girls *x* when 14 newborn babies are randomly selected. Assume that another clinical experiment involves 12 newborn babies. Using the same format as Table 4-1, construct a table for the probability distribution that results from 12 births, then determine whether a gender-selection technique appears to be effective if there are 9 girls and 3 boys.

36. Taking Courses after Graduation The Market Research Institute found that among employed college graduates aged 30–55 and out of college for at least 10 years, 57% have taken college courses after graduation (as reported in *USA Today*). If you randomly select five college graduates aged 30–55 and out of college for at least 10 years, and you find that only one of them has taken college courses after graduation, should you believe that the 57% rate is wrong? Explain.

4-3 Beyond the Basics

37. If a procedure meets all the conditions of a binomial distribution except that the number of trials is not fixed, then the **geometric distribution** can be used. The probability of getting the first success on the xth trial is given by $P(x) = p(1 - p)^{x-1}$ where p is the probability of success on any one trial. Assume that the probability of a defective computer component is 0.2. Find the probability that the first defect is found in the seventh component tested.

37. 0.0524

38. If we sample from a small finite population without replacement, the binomial distribution should not be used because the events are not independent. If sampling is done without replacement and the outcomes belong to one of two types, we can use the **hypergeometric distribution.** If a population has A objects of one type, while the remaining B objects are of the other type, and if n objects are sampled without replacement, then the probability of getting x objects of type A and $n - x$ objects of type B is

$$P(x) = \frac{A!}{(A - x)!x!} \cdot \frac{B!}{(B - n + x)!(n - x)!} \div \frac{(A + B)!}{(A + B - n)!n!}$$

In Lotto 54, a bettor selects six numbers from 1 to 54 (without repetition), and a winning six-number combination is later randomly selected. Find the probability of getting
a. All six winning numbers
b. Exactly five of the winning numbers
c. Exactly three of the winning numbers
d. No winning numbers

38. a. 0.0000000387
b. 0.0000112
c. 0.0134
d. 0.475

39. The binomial distribution applies only to cases involving two types of outcomes, whereas the **multinomial distribution** involves more than two categories. Suppose we have three types of mutually exclusive outcomes denoted by A, B, and C. Let $P(A) = p_1$, $P(B) = p_2$, and $P(C) = p_3$. In n independent trials, the probability of x_1 outcomes of type A, x_2 outcomes of type B, and x_3 outcomes of type C is given by

$$\frac{n!}{(x_1!)(x_2!)(x_3!)} \cdot p_1^{x_1} \cdot p_2^{x_2} \cdot p_3^{x_3}$$

A genetics experiment involves six mutually exclusive genotypes identified as A, B, C, D, E, and F, and they are all equally likely. If 20 offspring are tested, find the probability of getting exactly five A's, four B's, three C's, two D's, three E's, and three F's by expanding the above expression so that it applies to six types of outcomes instead of only three.

39. 0.000535

4-4 Mean, Variance, and Standard Deviation for the Binomial Distribution

We saw in Chapter 2 that when investigating collections of real data, these characteristics are usually very important: (1) the measure of center, (2) the measure of variation, (3) the nature of the distribution, (4) the presence of outliers, and (5) a pattern over time. (Recall that we used the key mnemonic "CVDOT" as a device for remembering those characteristics.) A key point of this chapter is that proba-

Do Boys or Girls Run in the Family?

The author of this book, his siblings, and his siblings' children consist of 11 males and only one female. Is this an example of a phenomenon whereby one particular gender runs in a family? This issue was studied by examining a random sample of 8770 households in the United States. The results were reported in the *Chance* magazine article "Does Having Boys or Girls Run in the Family?" by Joseph Rodgers and Debby Doughty. Part of their analysis involves use of the binomial probability distribution discussed in this section. Their conclusion is that "We found no compelling evidence that sex bias runs in the family."

Note to Instructor

It is recommended that *interpretation* of standard deviations be stressed and reviewed, as in part (b) of this example. The ultimate goal is not to simply obtain a numerical result, but to interpret that result in a practical and meaningful way.

bility distributions describe what will *probably* happen instead of what actually did happen. In Section 4-2, we learned methods for analyzing probability distributions by finding the mean, the standard deviation, and a probability histogram. Because a binomial distribution is a special type of probability distribution, we could use Formulas 4-1, 4-3, and 4-4 (from Section 4-2) for finding the mean, variance, and standard deviation. Fortunately, those formulas can be greatly simplified for binomial distributions, as shown below.

For Any Discrete Probability Distribution

Formula 4-1 $\mu = \Sigma[x \cdot P(x)]$

Formula 4-3 $\sigma^2 = \Sigma[x^2 \cdot P(x)] - \mu^2$

Formula 4-4 $\sigma = \sqrt{\Sigma[x^2 \cdot P(x)] - \mu^2}$

For Binomial Distributions

Formula 4-6 $\mu = np$

Formula 4-7 $\sigma^2 = npq$

Formula 4-8 $\sigma = \sqrt{npq}$

EXAMPLE Gender of Children In Section 4-2 we included an example illustrating calculations for μ and σ. We used the example of the random variable x representing the number of girls in 14 births. (See Table 4-2 on page 185 for the calculations that illustrate Formulas 4-1 and 4-4.) Use Formulas 4-6 and 4-8 to find the mean and standard deviation for the numbers of girls in groups of 14 births.

SOLUTION   Using the values $n = 14$, $p = 0.5$, and $q = 0.5$, Formulas 4-6 and 4-8 can be applied as follows:

$$\mu = np = (14)(0.5) = 7.0$$
$$\sigma = \sqrt{npq}$$
$$= \sqrt{(14)(0.5)(0.5)} = 1.9 \quad \text{(rounded)}$$

If you compare these calculations to those required in Table 4-3, it should be obvious that Formulas 4-6 and 4-8 are substantially easier to use.

Formula 4-6 for the mean makes sense intuitively. If we were to ask any statistics student how many girls are expected in 100 births, the usual response would be 50, which can be easily generalized as $\mu = np$. The variance and standard deviation are not so easily justified, and we will omit the complicated algebraic manipulations that lead to Formulas 4-7 and 4-8. Instead, refer again to the preceding example and Table 4-3 to verify that for a binomial distribution, Formulas 4-6, 4-7, and 4-8 will produce the same results as Formulas 4-1, 4-3, and 4-4.

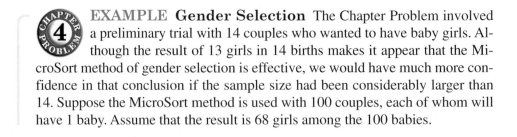

EXAMPLE Gender Selection The Chapter Problem involved a preliminary trial with 14 couples who wanted to have baby girls. Although the result of 13 girls in 14 births makes it appear that the MicroSort method of gender selection is effective, we would have much more confidence in that conclusion if the sample size had been considerably larger than 14. Suppose the MicroSort method is used with 100 couples, each of whom will have 1 baby. Assume that the result is 68 girls among the 100 babies.

a. Assuming that the MicroSort gender-selection method has no effect, find the mean and standard deviation for the numbers of girls in groups of 100 randomly selected babies.

b. *Interpret* the values from part (a) to determine whether this result (68 girls among 100 babies) supports a claim that the MicroSort method of gender selection is effective.

SOLUTION

a. Assuming that the MicroSort method has no effect and that girls and boys are equally likely, we have $n = 100$, $p = 0.5$, and $q = 0.5$. We can find the mean and standard deviation by using Formulas 4-6 and 4-8 as follows:

$$\mu = np = (100)(0.5) = 50$$
$$\sigma = \sqrt{npq} = \sqrt{(100)(0.5)(0.5)} = 5$$

For groups of 100 couples who each have a baby, the mean number of girls is 50 and the standard deviation is 5.

b. We must now interpret the results to determine whether 68 girls among 100 babies is a result that could easily occur by chance, or whether that result is so unlikely that the MicroSort method of gender selection seems to be effective. We will use the range rule of thumb as follows:

$$\text{maximum usual value: } \mu + 2\sigma = 50 + 2(5) = 60$$
$$\text{minimum usual value: } \mu - 2\sigma = 50 - 2(5) = 40$$

INTERPRETATION According to our range rule of thumb, values are considered to be usual if they are between 40 and 60, so 68 girls is an unusual result because it is not between 40 and 60. It is very unlikely that we will get 68 girls in 100 births just by chance. If we did get 68 girls in 100 births, we should look for an explanation that is an alternative to chance. If the 100 couples used the MicroSort method of gender selection, it would appear to be effective in increasing the likelihood that a child will be a girl.

You should develop the skills to calculate means and standard deviations using Formulas 4-6 and 4-8, but it is especially important to learn to *interpret* results by using those values. The range rule of thumb, as illustrated in part (b) of the preceding example, suggests that values are unusual if they lie outside of these limits:

$$\textbf{maximum usual value} = \boldsymbol{\mu + 2\sigma}$$
$$\textbf{minimum usual value} = \boldsymbol{\mu - 2\sigma}$$

4-4 Basic Skills and Concepts

Finding μ, σ, and Unusual Values. In Exercises 1–4, assume that a procedure yields a binomial distribution with n trials and the probability of success for one trial is p. Use the given values of n and p to find the mean μ and standard deviation σ. Also, use the

States Rig Lottery Selections

Many states run a lottery in which players select four digits, such as 1127 (the author's birthday). If a player pays $1 and selects the winning sequence in the correct order, a prize of $5000 is won. States monitor the number selections and, if one particular sequence is selected too often, players are prohibited from placing any more bets on it. The lottery machines are rigged so that once a popular sequence reaches a certain level of sales, the machine will no longer accept that particular sequence. This prevents states from paying out more than they take in. Critics say that this practice is unfair. According to William Thompson, a gambling expert at the University of Nevada in Las Vegas, "they're saying that they (the states) want to be in the gambling business, but they don't want to be gamblers. It just makes a sham out of the whole numbers game."

Recommended Assignment
Exercises 1–16.

1. $\mu = 80.0$, $\sigma = 8.0$,
   minimum $= 64.0$,
   maximum $= 96.0$
2. $\mu = 112.5$, $\sigma = 7.9$,
   minimum $= 96.7$,
   maximum $= 128.3$
3. $\mu = 1488.0$, $\sigma = 19.3$,
   minimum $= 1449.4$,
   maximum $= 1526.6$
4. $\mu = 127.8$, $\sigma = 10.3$,
   minimum $= 107.2$,
   maximum $= 148.4$
5. a. $\mu = 5.0$, $\sigma = 1.6$
   b. No, because 7 is within two
      standard deviations of the
      mean.
6. a. $\mu = 2.0$, $\sigma = 1.3$
   b. Yes, because 7 correct
      answers is more than two
      standard deviations above
      the mean.
7. a. $\mu = 2.6$, $\sigma = 1.6$
   b. No, because 0 wins is within
      two standard deviations of
      the mean.
8. a. $\mu = 2.5$, $\sigma = 1.5$
   b. No, because 5 is within two
      standard deviations of the
      mean.
9. a. The probabilities for 0, 1, 2,
      3, . . . , 15 are 0+, 0+,
      0.003, 0.014, . . . , 0+ (from
      Table A-1).
   b. $\mu = 7.5$, $\sigma = 1.9$
   c. No, because 10 is within
      two standard deviations of
      the mean. Also, P(10 or
      more girls) $= 0.151$, show-
      ing that it is easy to get 10
      or more girls by chance.
10. a. $\mu = 200.2$, $\sigma = 13.6$
    b. No, it is within two standard
       deviations of the mean.
11. a. $\mu = 27.2$, $\sigma = 5.1$
    b. Yes, it appears that the
       training program had an
       effect.

range rule of thumb to find the minimum usual value $\mu - 2\sigma$ and the maximum usual value $\mu + 2\sigma$.

1. $n = 400$, $p = 0.2$

2. $n = 250$, $p = 0.45$

3. $n = 1984$, $p = 3/4$

4. $n = 767$, $p = 1/6$

5. Guessing Answers Several students are unprepared for a surprise true/false test with 10 questions, and all of their answers are guesses.
   a. Find the mean and standard deviation for the number of correct answers for such students.
   b. Would it be unusual for a student to pass by guessing and getting at least 7 correct answers? Why or why not?

6. Guessing Answers Several students are unprepared for a multiple-choice quiz with 10 questions, and all of their answers are guesses. Each question has five possible answers, and only one of them is correct.
   a. Find the mean and standard deviation for the number of correct answers for such students.
   b. Would it be unusual for a student to pass by guessing and getting at least 7 correct answers? Why or why not?

7. Playing Roulette If you bet on any single number in roulette, your probability of winning is $1/38$. Assume that you bet on a single number in each of 100 consecutive spins.
   a. Find the mean and standard deviation for the number of wins.
   b. Would it be unusual to not win once in the 100 trials? Why or why not?

8. Left-Handed People Ten percent of American adults are left-handed. A statistics class has 25 students in attendance.
   a. Find the mean and standard deviation for the number of left-handed students in such classes of 25 students.
   b. Would it be unusual to survey a class of 25 students and find that 5 of them are left-handed? Why or why not?

9. Analyzing Results of Experiment in Gender Selection An experiment involving a gender-selection method includes a control group of 15 couples who are not given any treatment intended to influence the genders of their babies. Each of the 15 couples has one child.
   a. Construct a table listing the possible values of the random variable x (which represents the number of girls among the 15 births) and the corresponding probabilities.
   b. Find the mean and standard deviation for the numbers of girls in such groups of 15.
   c. If the couples have 10 girls and 5 boys, is that result unusual? Why or why not?

10. Deciphering Messages The Central Intelligence Agency has specialists who analyze the frequencies of letters of the alphabet in an attempt to decipher intercepted messages. In standard English text, the letter r is used at a rate of 7.7%.
   a. Find the mean and standard deviation for the number of times the letter r will be found on a typical page of 2600 characters.
   b. In an intercepted message sent to Iraq, a page of 2600 characters is found to have the letter r occurring 175 times. Is this unusual?

11. Determining Whether Complaints Are Lower After a Training Program The Newtower Department Store has experienced a 3.2% rate of customer complaints and attempts to lower this rate with an employee training program. After the program, 850 customers are tracked and it is found that only 7 of them filed complaints.

a. Assuming that the training program has no effect, find the mean and standard deviation for the number of complaints in such groups of 850 customers.

b. Based on the results from part (a), is the result of seven complaints unusual? Does it seem that the training program was effective in lowering the rate of complaints?

12. Are 10% of M&M Candies Blue? Mars, Inc., claims that 10% of its M&M plain candies are blue, and a sample of 100 such candies is randomly selected.

a. Find the mean and standard deviation for the number of blue candies in such groups of 100.

b. Data Set 19 in Appendix B consists of a random sample of 100 M&Ms in which only 5 are blue. Is this result unusual? Does it seem that the claimed rate of 10% is wrong?

13. Cell Phones and Brain Cancer In a study of 420,000 cell phone users in Denmark, it was found that 135 developed cancer of the brain or nervous system. If we assume that such cancer is not affected by cell phones, the probability of a person having such a cancer is 0.000340.

a. Assuming that cell phones have no effect on cancer, find the mean and standard deviation for the numbers of people in groups of 420,000 that can be expected to have cancer of the brain or nervous system.

b. Based on the results from part (a), is it unusual to find that among 420,000 people, there are 135 cases of cancer of the brain or nervous system? Why or why not?

c. What do these results suggest about the publicized concern that cell phones are a health danger because they increase the risk of cancer of the brain or nervous system?

14. Cholesterol Reducing Drug In a clinical trial of Lipitor, a common drug used to lower cholesterol, 863 patients were given a treatment of 10 mg Atorvastatin tablets. That group contains 19 patients who experienced flu symptoms (based on data from Pfizer, Inc.). The probability of flu symptoms for a person not receiving any treatment is 0.019.

a. Assuming that Lipitor has no effect on flu symptoms, find the mean and standard deviation for the numbers of people in groups of 863 that can be expected to have flu symptoms.

b. Based on the result from part (a), is it unusual to find that among 863 people, there are 19 who experience flu symptoms? Why or why not?

c. Based on the preceding results, do flu symptoms appear to be an adverse reaction that should be of concern to those who use Lipitor?

15. Opinions About Cloning A recent Gallup poll consisted of 1012 randomly selected adults who were asked whether "cloning of humans should or should not be allowed." Results showed that 89% of those surveyed indicated that cloning should not be allowed.

a. Among the 1012 adults surveyed, how many said that cloning should not be allowed?

b. If we assume that people are indifferent so that 50% believe that cloning of humans should not be allowed, find the mean and standard deviation for the numbers of people in groups of 1012 that can be expected to believe that such cloning should not be allowed.

c. Based on the preceding results, does the 89% result for the Gallup poll appear to be unusually higher than the assumed rate of 50%? Does it appear that an overwhelming majority of adults believe that cloning of humans should not be allowed?

12. a. $\mu = 10.0, \sigma = 3.0$
    b. No, it is within two standard deviations of the mean.

13. a. $\mu = 142.8, \sigma = 11.9$
    b. No, 135 is not unusual because it is within two standard deviations of the mean.
    c. Based on the given results, cell phones do not pose a health hazard that increases the likelihood of cancer of the brain or nervous system.

14. a. $\mu = 16.4, \sigma = 4.0$
    b. No, because 19 is within two standard deviations of the mean.
    c. No

15. a. 901
    b. $\mu = 506.0, \sigma = 15.9$
    c. Yes, because 901 is more than two standard deviations above the mean.

16. Car Crashes For drivers in the 20–24 age bracket, there is a 34% rate of car accidents in one year (based on data from the National Safety Council). An insurance investigator finds that in a group of 500 randomly selected drivers aged 20–24 living in New York City, 42% had accidents in the last year.

    a. How many drivers in the New York City group of 500 had accidents in the last year?

    b. Assuming that the same 34% rate applies to New York City, find the mean and standard deviation for the numbers of people in groups of 500 that can be expected to have accidents.

    c. Based on the preceding results, does the 42% result for the New York City drivers appear to be unusually high when compared to the 34% rate for the general population? Does it appear that higher insurance rates for New York City drivers are justified?

4-4 Beyond the Basics

17. Using the Empirical Rule and Chebyshev's Theorem An experiment is designed to test the effectiveness of the MicroSort method of gender selection, and 100 couples try to have baby girls using the MicroSort method. In an example included in this section, the range rule of thumb was used to conclude that among 100 births, the number of girls should usually fall between 40 and 60.

    a. The empirical rule (see Section 2-5) applies to distributions that are bell-shaped. Is the binomial probability distribution for this experiment (approximately) bell-shaped? How do you know?

    b. Assuming that the distribution is bell-shaped, how likely is it that the number of girls will fall between 40 and 60 (according to the empirical rule)?

    c. Assuming that the distribution is bell-shaped, how likely is it that the number of girls will fall between 35 and 65 (according to the empirical rule)?

    d. Using Chebyshev's theorem, what do we conclude about the likelihood that the number of girls will fall between 40 and 60?

18. Acceptable/Defective Products Mario's Pizza Parlor has just opened. Due to a lack of employee training, there is only an 0.8 probability that a pizza will be edible. An order for five pizzas has just been placed. What is the minimum number of pizzas that must be made in order to be at least 99% sure that there will be five that are edible?

4-5 The Poisson Distribution

The preceding two sections included the binomial distribution, which is one of many probability distributions that can be used for different situations. This section introduces the *Poisson distribution*. It is particularly important because it is often used as a mathematical model describing behavior such as radioactive decay, arrivals of people in a line, planes arriving at an airport, cars pulling into a gas station, diners arriving at a restaurant, students arriving at a bookstore line, and Internet users logging onto a Web site. For example, suppose your professor has an office hour scheduled every Monday at 11:00 and she finds that during that office hour, the mean number of students who come is 2.3. We can find the probability that for a randomly selected office hour on Monday at 11:00, exactly four students come. We use the Poisson distribution, defined as follows.

Definition

The **Poisson distribution** is a discrete probability distribution that applies to occurrences of some event *over a specified interval*. The random variable x is the number of occurrences of the event in an interval. The interval can be time, distance, area, volume, or some similar unit. The probability of the event occurring x times over an interval is given by Formula 4-9.

Formula 4-9 $$P(x) = \frac{\mu^x \cdot e^{-\mu}}{x!} \qquad \text{where } e \approx 2.71828$$

The Poisson distribution has the following requirements:

- The random variable x is the number of occurrences of an event *over some interval*.
- The occurrences must be *random*.
- The occurrences must be *independent* of each other.
- The occurrences must be *uniformly distributed* over the interval being used.

The Poisson distribution has these parameters:

- The mean is μ.
- The standard deviation is $\sigma = \sqrt{\mu}$.

A Poisson distribution differs from a binomial distribution in these fundamental ways:

1. The binomial distribution is affected by the sample size n and the probability p, whereas the Poisson distribution is affected only by the mean μ.
2. In a binomial distribution, the possible values of the random variable x are 0, 1, . . . , n, but a Poisson distribution has possible x values of 0, 1, 2, . . . , with no upper limit.

EXAMPLE World War II Bombs

In analyzing hits by V-1 buzz bombs in World War II, South London was subdivided into 576 regions, each with an area of 0.25 km². A total of 535 bombs hit the combined area of 576 regions.

a. If a region is randomly selected, find the probability that it was hit exactly twice.

b. Based on the probability found in part (a), how many of the 576 regions are expected to be hit exactly twice?

SOLUTION

a. The Poisson distribution applies because we are dealing with the occurrences of an event (bomb hits) over some interval (a region with area of 0.25 km²). The mean number of hits per region is

$$\mu = \frac{\text{number of bomb hits}}{\text{number of regions}} = \frac{535}{576} = 0.929$$

continued

Probability of an Event That Has Never Occurred

Some events are possible, but are so unlikely that they have never occurred. Here is one such problem of great interest to political scientists: Estimate the probability that your single vote will determine the winner in a U.S. presidential election. Andrew Gelman, Gary King, and John Boscardin write in the *Journal of the American Statistical Association* (Vol. 93, No. 441) that "the exact value of this probability is of only minor interest, but the number has important implications for understanding the optimal allocation of campaign resources, whether states and voter groups receive their fair share of attention from prospective presidents, and how formal 'rational choice' models of voter behavior might be able to explain why people vote at all." The authors show how the probability value of 1 in 10 million is obtained for close elections.

Because we want the probability of exactly two hits in a region, we let $x = 2$ and use Formula 4-9 as follows:

$$P(x) = \frac{\mu^x \cdot e^{-\mu}}{x!} = \frac{0.929^2 \cdot 2.71828^{-0.929}}{2!} = \frac{0.863 \cdot 0.395}{2} = 0.170$$

The probability of a particular region being hit exactly twice is $P(2) = 0.170$.

   b. Because there is a probability of 0.170 that a region is hit exactly twice, we expect that among the 576 regions, the number that are hit exactly twice is $576 \cdot 0.170 = 97.9$.

In the preceding example, we can also calculate the probabilities and expected values for 0, 1, 3, 4, and 5 hits. (We stop at $x = 5$ because no region was hit more than five times, and the probabilities for $x > 5$ are 0.000 when rounded to three decimal places.) Those probabilities and expected values are listed in Table 4-5. The fourth column of Table 4-5 describes the results that actually occurred during World War II. There were 229 regions that had no hits, 211 regions that were hit once, and so on. We can now compare the frequencies *predicted* with the Poisson distribution (third column) to the *actual* frequencies (fourth column) to conclude that there is very good agreement. In this case, the Poisson distribution does a good job of predicting the results that actually occurred. (Section 10-2 describes a statistical procedure for determining whether such expected frequencies constitute a good "fit" to the actual frequencies. That procedure does suggest that there is a good fit in this case.)

Table 4-5	V-1 Buzz Bomb Hits for 576 Regions in South London		
Number of Bomb Hits	Probability	Expected Number of Regions	Actual Number of Regions
0	0.395	227.5	229
1	0.367	211.4	211
2	0.170	97.9	93
3	0.053	30.5	35
4	0.012	6.9	7
5	0.002	1.2	1

Poisson as Approximation to Binomial

The Poisson distribution is sometimes used to approximate the binomial distribution when n is large and p is small. One rule of thumb is to use such an approximation when the following two conditions are both satisfied:

  1. $n \geq 100$

  2. $np \leq 10$

If these conditions are satisfied and we want to use the Poisson distribution as an approximation to the binomial distribution, we need a value for μ, and that value can be calculated by using Formula 4-6 (first presented in Section 4-4):

Formula 4-6 $\mu = np$

EXAMPLE **Illinois Pick 3 Game** In Illinois' Pick 3 game, you pay 50¢ to select a sequence of three digits, such as 911. If you play this game once every day, find the probability of winning exactly once in 365 days.

SOLUTION   Because the time interval is 365 days, $n = 365$. Because there is one winning set of numbers among the 1000 that are possible (from 000 to 999), $p = 1/1000$. The conditions $n \geq 100$ and $np \leq 10$ are both satisfied, so we can use the Poisson distribution as an approximation to the binomial distribution. We first need the value of μ, which is found as follows:

$$\mu = np = 365 \cdot \frac{1}{1000} = 0.365$$

Having found the value of μ, we can now find $P(1)$:

$$P(1) = \frac{\mu^{x} \cdot e^{-\mu}}{x!} = \frac{0.365^{1} \cdot 2.71828^{-0.365}}{1!} = 0.253$$

Using the Poisson distribution as an approximation to the binomial distribution, we find that there is a 0.253 probability of winning exactly once in 365 days. If we use the binomial distribution, we get the more accurate probability of 0.254, so we can see that the Poisson approximation is quite good here.

Using Technology

STATDISK   Select **Analysis** from the main menu bar, then select **Probability Distributions**, then **Poisson.** Enter the value of the mean, then click on the **Evaluate** button.

Minitab   First enter the desired value of x in column C1. Now select **Calc** from the main menu bar, then select **Probability Distributions**, then **Poisson.** Click on the **Probability** button. Enter the value of the mean μ and enter C1 for the input column.

Excel   Click on f_x on the main menu bar, then select the function category of **Statistical,** then select **POISSON,** then click **OK.** In the dialog box, enter the values for x and the mean, and enter 0 for "Cumulative." (Entering 1 for "Cumulative" results in the probability for values up to and including the entered value of x.)

TI-83 Plus   Press **2nd VARS** (to get **DISTR**), then select option B: **poissonpdf(.** Now press **ENTER,** then proceed to enter μ, x (including the comma). For μ, enter the value of the mean; for x, enter the desired number of occurrences.

4-5 Basic Skills and Concepts

Using a Poisson Distribution to Find Probability. In Exercises 1–4, assume that the Poisson distribution applies and proceed to use the given mean to find the indicated probability.

1. If $\mu = 2$, find $P(3)$.

2. If $\mu = 0.5$, find $P(2)$.

3. If $\mu = 100$, find $P(99)$.

4. If $\mu = 500$, find $P(512)$.

In Exercises 5–12, use the Poisson distribution to find the indicated probabilities.

5. Radioactive Decay Radioactive atoms are unstable because they have too much energy. When they release their extra energy, they are said to decay. When studying cesium 137, it is found that during the course of decay over 365 days, 1,000,000 radioactive atoms are reduced to 977,287 radioactive atoms.
   a. Find the mean number of radioactive atoms lost through decay in a day.
   b. Find the probability that on a given day, 50 radioactive atoms decayed.

6. Births Currently, 11 babies are born in the village of Westport (population 760) each year (based on data from the U.S. National Center for Health Statistics).
   a. Find the mean number of births per day.
   b. Find the probability that on a given day, there are no births.
   c. Find the probability that on a given day, there is at least one birth.
   d. Based on the preceding results, should medical birthing personnel be on permanent standby, or should they be called in as needed? Does this mean that Westport mothers might not get the immediate medical attention that they would be likely to get in a more populated area?

7. Deaths from Horse Kicks A classic example of the Poisson distribution involves the number of deaths caused by horse kicks of men in the Prussian Army between 1875 and 1894. Data for 14 corps were combined for the 20-year period, and the 280 corps-years included a total of 196 deaths. After finding the mean number of deaths per corps-year, find the probability that a randomly selected corps-year has the following numbers of deaths:
   a. 0   **b.** 1   **c.** 2   **d.** 3   **e.** 4

   The actual results consisted of these frequencies: 0 deaths (in 144 corps-years); 1 death (in 91 corps-years); 2 deaths (in 32 corps-years); 3 deaths (in 11 corps-years); 4 deaths (in 2 corps-years). Compare the actual results to those expected from the Poisson probabilities. Does the Poisson distribution serve as a good device for predicting the actual results?

8. Homicide Deaths In one year, there were 116 homicide deaths in Richmond, Virginia (based on "A Classroom Note on the Poisson Distribution: A Model for Homicidal Deaths in Richmond, VA for 1991," in *Mathematics and Computer Education*, by Winston A. Richards). For a randomly selected day, find the probability that the number of homicide deaths is
   a. 0   **b.** 1   **c.** 2   **d.** 3   **e.** 4

   Compare the calculated probabilities to these actual results: 268 days (no homicides); 79 days (1 homicide); 17 days (2 homicides); 1 day (3 homicides); no days with more than 3 homicides.

9. *Roulette* Scott bets on the number 7 for each of 200 spins of a roulette wheel. Because $P(7) = 1/38$, he expects to win about 5 times.
   a. Find the probability of no wins in the 200 spins.
   b. Find the probability of at least one win in the 200 spins.
   c. Scott will lose money if the number of wins is 0, 1, 2, 3, 4, or 5. Find the probability that Scott loses money after 200 spins.
   d. What is the probability that Scott will have a profit after 200 spins?

10. *Earthquakes* For a recent period of 100 years, there were 93 major earthquakes (at least 6.0 on the Richter scale) in the world (based on data from the *World Almanac and Book of Facts*). Assuming that the Poisson distribution is a suitable model, find the mean number of major earthquakes per year, then find the probability that the number of earthquakes in a randomly selected year is
    a. 0   b. 1   c. 2   d. 3   e. 4   f. 5   g. 6   h. 7

    Here are the actual results: 47 years (0 major earthquakes); 31 years (1 major earthquake); 13 years (2 major earthquakes); 5 years (3 major earthquakes); 2 years (4 major earthquakes); 0 years (5 major earthquakes); 1 year (6 major earthquakes); 1 year (7 major earthquakes). After comparing the calculated probabilities to the actual results, is the Poisson distribution a good model?

4-5 Beyond the Basics

11. *Poisson Approximation to Binomial* The Poisson distribution can be used to approximate a binomial distribution if $n \geq 100$ and $np \leq 10$. Assume that we have a binomial distribution with $n = 100$ and $p = 0.1$. It is impossible to get 101 successes in such a binomial distribution, but we *can* compute the probability that $x = 101$ with the Poisson approximation. Find that value. How does the result agree with the impossibility of having $x = 101$ with a binomial distribution?

12. *Poisson Approximation to Binomial* For a binomial distribution with $n = 10$ and $p = 0.5$, we should not use the Poisson approximation because both of the conditions $n \geq 100$ and $np \leq 10$ are not satisfied. Suppose we go way out on a limb and use the Poisson approximation anyway. Are the resulting probabilities unacceptable approximations? Why or why not?

Review

The concept of a probability distribution is a key element of statistics. A probability distribution describes the probability for each value of a random variable. This chapter included only discrete probability distributions, but the following chapters will include continuous probability distributions. The following key points were discussed:

- A *random variable* has values that are determined by chance.
- A *probability distribution* consists of all values of a random variable, along with their corresponding probabilities. A probability distribution must satisfy two requirements:

$$\Sigma P(x) = 1 \text{ and, for each value of } x, 0 \leq P(x) \leq 1.$$

9. a. 0.00518 (using binomial: 0.00483)
   b. 0.995
   c. 0.570
   d. 0.430

10. a. 0.395
    b. 0.367
    c. 0.171
    d. 0.0529
    e. 0.0123
    f. 0.00229
    g. 0.000355
    h. 0.0000471
    Using the computed probabilities, the expected frequencies are 39.5, 36.7, 17.1, 5.3, 1.2, 0.2, 0.0, and 0.0, and they agree reasonably well with the actual frequencies.

11. 4.82×10^{-64} is so small that, for all practical purposes, we can consider it to be zero.

12. The Poisson probabilities are not close enough to the binomial probabilities. For example, $P(6) = 0.205$ with the binomial distribution, but $P(6) = 0.146$ with the Poisson distribution.

- Important characteristics of a *probability distribution* can be explored by constructing a probability histogram and by computing its mean and standard deviation using these formulas:

$$\mu = \Sigma[x \cdot P(x)]$$

$$\sigma = \sqrt{\Sigma[x^2 \cdot P(x)] - \mu^2}$$

- In a *binomial distribution,* there are two categories of outcomes and a fixed number of independent trials with a constant probability. The probability of x successes among n trials can be found by using the binomial probability formula, or Table A-1, or software (such as STATDISK, Minitab, or Excel), or a TI-83 Plus calculator.

- In a binomial distribution, the mean and standard deviation can be easily found by calculating the values of $\mu = np$ and $\sigma = \sqrt{npq}$.

- A *Poisson probability distribution* applies to occurrences of some event over a specific interval, and probabilities can be computed with Formula 4-9.

- *Unusual outcomes:* This chapter stressed the importance of interpreting results by distinguishing between outcomes that are usual and those that are unusual. We used two different criteria. With the range rule of thumb we have

maximum usual value = $\mu + 2\sigma$

minimum usual value = $\mu - 2\sigma$

We can also determine whether outcomes are unusual by using probability values.

Unusually high: x successes among n trials is an *unusually high* number of successes if $P(x$ or more$)$ is very small (such as 0.05 or less).

Unusually low: x successes among n trials is an *unusually low* number of successes if $P(x$ or fewer$)$ is very small (such as 0.05 or less).

Review Exercises

1. **a.** What is a random variable?
   b. What is a probability distribution?
   c. A graph in *USA Today* listed the percentages for the number of days in a week that American adults cook at home in an average week. For example, 13% of American adults cook at home 3 days in an average week. The accompanying table is based on the graph. Does this table describe a probability distribution? Why or why not?
   d. Assuming that the table does describe a probability distribution, find its mean.
   e. Assuming that the table does describe a probability distribution, find its standard deviation.
   f. Is it unusual to randomly select an adult American and find that he or she does not cook at home in an average week? Why or why not?

2. TV Ratings The television show *West Wing* has a 15 share, meaning that while it is being broadcast, 15% of the TV sets in use are tuned to *West Wing* (based on data from Nielsen Media Research). A special focus group consists of 20 randomly selected households (each with one TV set in use during the time of a *West Wing* broadcast).

continued

1. a. A random variable is a variable that has a single numerical value (determined by chance) for each outcome of some procedure.
   b. A probability distribution gives the probability for each value of the random variable.
   c. Yes, because each probability value is between 0 and 1 and the sum of the probabilities is 1.
   d. 4.2 days
   e. 2.1 days
   f. No, because the probability of 0.08 shows that it easy to get 0 days by chance.

x	$P(x)$
0	0.08
1	0.05
2	0.10
3	0.13
4	0.15
5	0.21
6	0.09
7	0.19

a. What is the expected number of sets tuned to *West Wing?*
b. In such groups of 20, what is the mean number of sets tuned to *West Wing?*
c. In such groups of 20, what is the standard deviation for the number of sets tuned to *West Wing?*
d. For such a group of 20, find the probability that exactly 5 TV sets are tuned to *West Wing.*
e. For such a group of 20, would it be unusual to find that no sets are tuned to *West Wing?* Why or why not?

3. Employee Drug Testing Among companies doing highway or bridge construction, 80% test employees for substance abuse (based on data from the Construction Financial Management Association). A study involves the random selection of 10 such companies.
a. Find the probability that 5 of the 10 companies test for substance abuse.
b. Find the probability that at least half of the companies test for substance abuse.
c. For such groups of 10 companies, find the mean and standard deviation for the number (among 10) that test for substance abuse.
d. Would it be unusual to find that 6 of 10 companies test for substance abuse? Why or why not?

4. Reasons for Being Fired Inability to get along with others is the reason cited in 17% of worker firings (based on data from Robert Half International, Inc.). Concerned about her company's working conditions, the personnel manager at the Boston Finance Company plans to investigate the five employee firings that occurred over the past year.
a. Assuming that the 17% rate applies, find the probability that among those five employees, the number fired because of an inability to get along with others is at least four.
b. If the personnel manager actually does find that at least four of the firings are due to an inability to get along with others, does this company appear to be very different from other typical companies? Why or why not?

5. Deaths Currently, an average of 7 residents of the village of Westport (population 760) die each year (based on data from the U.S. National Center for Health Statistics).
a. Find the mean number of deaths per day.
b. Find the probability that on a given day, there are no deaths.
c. Find the probability that on a given day, there is one death.
d. Find the probability that on a given day, there is more than one death.
e. Based on the preceding results, should Westport have a contingency plan to handle more than one death per day? Why or why not?

Cumulative Review Exercises

1. Home Run Distances: Analysis of Last Digits The accompanying table lists the last digits of the 73 published distances (in feet) of the 73 home runs hit by Barry Bonds in 2001 when he set the record for the most home runs in a season (based on data from *USA Today*). The last digits of a data set can sometimes be used to determine whether the data have been measured or simply reported. The presence of disproportionately more 0s and 5s is often a sure indicator that the data have been reported instead of measured.

continued

2. a. 3.0
   b. 3.0
   c. 1.6
   d. 0.103
   e. Yes, because the probability of 0 sets is 0.0388, which shows that it is very unlikely that no sets would be tuned to *West Wing.*
3. a. 0.026
   b. 0.992 (or 0.994)
   c. $\mu = 8.0, \sigma = 1.3$
   d. No, because 6 is within two standard deviations of the mean.
4. a. 0.00361
   b. This company appears to be very different because the event of at least four firings is so unlikely with a probability of only 0.00361.
5. a. 7/365
   b. 0.981
   c. 0.0188
   d. 0.0002
   e. No, because the event is so rare.

x	f
0	47
1	3
2	1
3	0
4	3
5	11
6	3
7	3
8	1
9	1

a. Find the mean and standard deviation of those last digits.

b. Construct the relative frequency table that corresponds to the given frequency table.

c. Construct a table for the probability distribution of randomly selected digits that are all equally likely. List the values of the random variable x (0, 1, 2, ... , 9) along with their corresponding probabilities (0.1, 0.1, 0.1, ... , 0.1), then find the mean and standard deviation of this probability distribution.

d. Recognizing that sample data naturally deviate from the results we theoretically expect, does it seem that the given last digits roughly agree with the distribution we expect with random selection? Or does it seem that there is something about the sample data (such as disproportionately more 0s and 5s) suggesting that the given last digits are not random? (In Chapter 10, we will present a method for answering such questions much more objectively.)

2. Testing Cars for Pollution The Environmental Protection Agency used a tailpipe test to determine which of 116,667 cars generated too much pollution. It is estimated that 1% of cars fail such a test.

a. If we randomly select 20 cars from the group of 116,667, how many are expected to fail the tailpipe test?

b. Find the mean and standard deviation for the numbers of cars in groups of 20 that fail the tailpipe test.

c. Find the probability that in a randomly selected group of 20 cars, there is at least one that fails the tailpipe test.

d. Is it unusual to find that in a group of 20 randomly selected cars, there are 3 that fail the tailpipe test? Why or why not?

e. If two different cars are randomly selected, find the probability that they both fail the tailpipe test.

Cooperative Group Activities

1. In-class activity In Chapter 1 we gave several examples of misleading data sets. Suppose we want to identify the probability distribution for the number of children born to randomly selected couples. For each student in the class, find the number of brothers and sisters and record the total number of children (including the student) in each family. Construct the relative frequency table for the result obtained. (The values of the random variable x will be 1, 2, 3, ...) What is wrong with using this relative frequency table as an estimate of the probability distribution for the number of children born to randomly selected couples?

2. In-class activity Divide into groups of three. Select one person who will be tested for extrasensory perception (ESP) by trying to correctly identify a digit (0, 1, 2, ... , 9) randomly selected by another member of the group. Another group member should record the randomly selected digit, the digit guessed by the subject, and whether the guess was correct or wrong. Construct the table for the probability distribution of randomly generated digits, construct the relative frequency table for the random digits that were actually obtained, and construct a relative frequency table for the guesses. After comparing the three tables, what do you conclude? What proportion of guesses were correct? Does it seem that the subject has the ability to select the correct digit significantly more often than would be expected by chance?

Technology Project

American Air Flight 2705 from New York to San Francisco has seats for 340 passengers. An average of 5% of people with reservations don't show up, so American Air overbooks by accepting 350 reservations for the 340 seats. We can analyze this system by using a binomial distribution with $n = 350$ and $p = 0.95$ (the probability that someone with a reservation does show up).

Find the probability that when 350 reservations are accepted for a particular flight, there are more passengers than seats. That is, find the probability of at least 341 people showing up with reservations, assuming that 350 reservations were accepted. Because of the value of n, Table A-1 cannot be used, and calculations with the binomial probability formula would be extremely time-consuming and painfully tedious. The best approach is to use statistics software or a TI-83 Plus calculator. See Section 4-3 for instructions describing the use of STATDISK, Minitab, Excel, or a TI-83 Plus calculator. Is the probability of overbooking small enough so that it does not happen very often, or does it seem too high so that changes must be made to make it lower?

from DATA *to* DECISION

Critical Thinking: Is it better to play a slot machine or roulette?

The author purchased a Las Vegas Mills Golden Nugget slot machine for the purpose of determining how it works. Although this slot machine is based on a 1940 design, it works according to the same principles used by slot machines now present in Las Vegas casinos. This slot machine has three wheels that spin independently and, for each reel, one of 20 different positions is randomly selected each time the arm is pulled. The accompanying tables summarize the possible outcomes and the winning results. Each pull costs 25¢.

a. Using the information from the two tables, enter the probabilities in the second table. For example, there are 4 outcomes that result in a jackpot, and there are 8000 different possible outcomes, so $P(\text{jackpot}) = 4/8000$.
b. Does the second completed table describe a probability distribution? Why or why not?

c. Letting the random variable x represent the net gain or loss for a single pull of the slot machine, find the mean of this random variable. Based on this result, what is the average amount won or lost when a gambler inserts 25¢ for a pull? What is the average return for every $1 that is bet?
d. When betting $1 on the number 7 in roulette, there is a probability of 1/38 of winning, and a win results in a net profit of $35. What is the average return for every $1 that is bet?
e. Compare the results from parts (c) and (d) to determine whether it is better to play the slot machine or bet on 7 in roulette. Explain.

Frequency of Images on the Three Reels

	Reel 1	Reel 2	Reel 3
Golden Nugget	2	2	1
Lemon	0	0	4
Bell	1	7	7
Orange	7	2	5
Plum	7	2	3
Cherry	3	7	0

Possible Results for the Three Reels

	Net Profit	Probability
Jackpot (3 Golden Nuggets)	36.50	
Bell-Bell-Bell	4.25	
Bell-Bell-Golden Nugget	4.25	
Plum-Plum-Plum	3.25	
Plum-Plum-Golden Nugget	3.25	
Orange-Orange-Orange	2.25	
Orange-Orange-Golden Nugget	2.25	
Cherry-Cherry-anything	1.00	
Cherry-No cherry-anything	0.25	
Lose: any outcome not listed in the above 9 rows	−0.25	

INTERNET PROJECT   Probability Distributions and Simulation

Probability distributions are used to predict the outcome of the events they model. For example, if we toss a fair coin, the distribution for the outcome is a probability of 0.5 for heads and 0.5 for tails. If we toss the coin ten consecutive times, we expect five heads and five tails. We might not get this exact result, but in the long run, over hundreds or thousands of tosses, we expect the split between heads and tails to be very close to "50-50". Go to the Web site for this textbook:

http://www.aw.com/triola

Proceed to the Internet Project for Chapter 4 where you will find two explorations. In the first exploration you are asked to develop a probability distribution for a simple experiment, and use that distribution to predict the outcome of repeated trial runs of the experiment. In the second exploration, we will analyze a more complicated situation: the paths of rolling marbles as they move in pinball-like fashion through a set of obstacles. In each case, a dynamic visual simulation will allow you to compare the predicted results with a set of experimental outcomes.

"Our program is really an education program, but it has wide recognition because the results are released publicly"

Barbara Carvalho
Director of the Marist College Poll

Lee Miringoff
Director of the Marist College Institute for Public Opinion

Barbara Carvalho and Lee Miringoff report on their poll results in many interviews for print and electronic media, including news programs for NBC, CBS, ABC, FOX, and public television. Lee Miringoff appears regularly on NBC's *Today* show.

What do you do?

We do public polling. We survey public issues, approval ratings of public officials in New York City, New York State, and nationwide. We don't do partisan polling for political parties, political candidates, or lobby groups. We are independently funded by Marist College and we have no outside funding that in any way might suggest that we are doing research for any particular group on any one issue.

How do you select survey respondents?

For a statewide survey we select respondents in proportion to county voter registrations. Different counties have different refusal rates and if we were to select people at random throughout the state, we would get an uneven model of what the state looks like. We stratify by county and use random digit dialing so that we get listed and unlisted numbers.

You mentioned refusal rates. Are they a real problem?

One of the issues that we deal with extensively is the issue of people who don't respond to surveys. That has been increasing over time and there has been much attention from the survey research community. As a research center we do quite well

when compared to others. But when you do face-to-face interviews and have refusal rates of 25% to 50%, there's a real concern to find out who is refusing and why they are not responding, and the impact that has on the representativeness of the studies that we're doing.

Would you recommend a statistics course for students?

Absolutely. All numbers are not created equally. Regardless of your field of study or career interests, an ability to critically evaluate research information that is presented to you, to use data to improve services, or to interpret results to develop strategies is a very valuable asset. Surveys, in particular, are everywhere. It is vital that as workers, managers, and citizens we are able to evaluate their accuracy and worth. Statistics cuts across disciplines. Students will inevitably find it in their careers at some point.

Do you have any other recommendations for students?

It is important for students to take every opportunity to develop their communication and presentation skills. Sharpen not only your ability to speak and write, but also raise your comfort level with new technologies.

Normal Probability
Distributions

5-1 Overview

5-2 The Standard Normal Distribution

5-3 Applications of Normal Distributions

5-4 Sampling Distributions and Estimators

5-5 The Central Limit Theorem

5-6 Normal as Approximation to Binomial

5-7 Determining Normality

How do we fit in?

A relatively new discipline is *ergonomics,* the study of people fitting into their environments. Good ergonomic design results in an environment that is safe, functional, efficient, and comfortable. Ergonomics is used in applications including the design of car dashboards, cycle helmets, screw bottle tops, door handles, manhole covers, keyboards, air traffic control centers, and computer assembly lines. For example, at Vail, Colorado, a gondola carries skiers to the top of a mountain. It bears a plaque stating that the maximum capacity is 12 people or 2004 pounds. Reading that plaque causes many passengers to look around wondering whether they are in danger because there are too many fellow passengers or because there are 12 or fewer passengers who, by chance random variation, are all exceptionally heavy. How likely is it that 12 randomly selected people will have a total weight greater than 2004 pounds?

The ability to comfortably endure a long transcontinental flight is affected by the width of the seat we must occupy. Most U.S. commercial aircraft have seat widths of 17 in. to 18 in., which just barely accommodates the 16 in. required by the average passenger. Business and first class seats usually have widths between 19 in. and 21 in., so the extra room allows a greater degree of comfort. If American Airlines wants to gain business by increasing passenger comfort, what width should be used in redesigned seats?

When visiting preserved living quarters that are hundreds of years old, many people are struck by the fact that the doorways have openings that are too low for most adults living today. When walking through a modern doorway, most of us fit comfortably under the top threshold, which is typically 80 in. high. However, some people are exceptionally tall and must bend to avoid a bump on the head. What percent of people are too tall for the current doorway design standards?

In recent years, the United States Air Force recognized that women make perfectly good pilots of fighter jets. Cockpits of fighter jets were originally designed for men only, so various changes were required to better accommodate the new women pilots. One such change involved a redesign of the ACES-II ejection seats. Because they were originally designed for men who weighed between 140 and 211 pounds, the ejection seats provided a greater chance of injury for any women pilots weighing less than 140 pounds or more than 211 pounds. What weights should be used for the new cockpit design?

In this chapter we address questions such as those posed above.

 Overview

In Chapter 2 we considered important measures of data sets, including measures of center and variation, as well as the distribution of data. In Chapter 3 we discussed basic principles of probability, and in Chapter 4 we presented the following concepts:

- A *random variable* is a variable having a single numerical value, determined by chance, for each outcome of some procedure.
- A *probability distribution* describes the probability for each value of the random variable.
- A *discrete* random variable has either a finite number of values or a countable number of values. That is, the *number* of possible values that x can assume is 0, or 1, or 2, and so on.
- A *continuous* random variable has infinitely many values, and those values are often associated with measurements on a continuous scale with no gaps or interruptions.

In Chapter 4 we considered only *discrete* probability distributions, but in this chapter we present *continuous* probability distributions. Although we begin with a uniform distribution, most of the chapter focuses on *normal distributions*. Normal distributions are extremely important because they occur so often in real applications and they play such an important role in methods of inferential statistics. Normal distributions will be used often throughout the remainder of this text.

Definition

If a continuous random variable has a distribution with a graph that is symmetric and bell-shaped, as in Figure 5-1, and it can be described by the equation given as Formula 5-1, we say that it has a **normal distribution.**

Formula 5-1
$$y = \frac{e^{-\frac{1}{2}\left(\frac{x-\mu}{\sigma}\right)^2}}{\sigma\sqrt{2\pi}}$$

The complexity of Formula 5-1 causes many people to raise eyebrows while uttering the phrase "uh oh," or worse. But here is really good news: It isn't really

FIGURE 5-1 The Normal Distribution

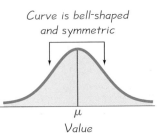

necessary for us to actually use Formula 5-1. That formula does show, however, that any particular normal distribution is determined by two parameters: the mean, μ, and standard deviation, σ. Once specific values are selected for μ and σ, we can graph Formula 5-1 as we would graph any equation relating x and y; the result is a continuous probability distribution with a bell shape.

5-2   The Standard Normal Distribution

The focus of this chapter is the concept of a normal probability distribution, but we begin with a *uniform distribution*. The uniform distribution makes it easier for us to see some very important properties, which will be used also with normal distributions.

Uniform Distributions

Definition

A continuous random variable has a **uniform distribution** if its values spread evenly over the range of possibilities. The graph of a uniform distribution results in a rectangular shape.

EXAMPLE Class Length A statistics professor plans classes so carefully that the lengths of her classes are uniformly distributed between 50.0 min and 52.0 min. (Because statistics classes are so interesting, they usually seem much shorter.) That is, any time between 50.0 min and 52.0 min is possible, and all of the possible values are equally likely. If we randomly select one of her classes and let x be the random variable representing the length of that class, then x has a distribution that can be graphed as in Figure 5-2.

When we discussed *discrete* probability distributions in Section 4-2, we identified two requirements: (1) $\Sigma P(x) = 1$ and (2) $0 \leq P(x) \leq 1$ for all values of x. Also in Section 4-2, we stated that the graph of a discrete probability distribution

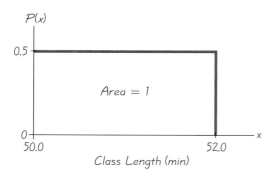

FIGURE 5-2 Uniform Distribution of Class Times

Changing Populations

Included among the five important data set characteristics listed in Chapter 2 is the changing pattern of data over time. Some populations change, and their important statistics change as well. Car seat belt standards haven't changed in 40 years, even though the weights of Americans have increased considerably since then. In 1960, 12.8% of adult Americans were considered obese, compared to 22.6% in 1994.

According to the National Highway Traffic Safety Administration, seat belts must fit a standard crash dummy (designed according to 1960 data) placed in the most forward position, with 4 in. to spare. In theory, 95% of men and 99% of women should fit into seat belts, but those percentages are now lower because of the increases in weight over the last half-century. Some car companies provide seat belt extenders, but some do not.

Sampling Rejected for the Census

It has been estimated that in the 2000 Census, 7 million people were not counted, while 4 million others were counted twice. These errors can be largely corrected by applying known methods of statistics, but there is a political issue. Population counts affect the number of seats in the House of Representatives, so Republicans oppose sampling because under-counted regions tend to be largely Democratic, while overcounted regions tend to be largely Republican. Democrats favor the use of sampling methods. Some people argue that the Constitution specifies that the Census be an "actual enumeration" (a head count), which does not allow for sampling methods, and the Supreme Court upheld that position. Statistical methods could be used to substantially improve Census results so that citizens could enjoy more equitable distributions of federal aid along with more equitable representation in Congress.

is called a *probability histogram.* The graph of a continuous probability distribution, such as that of Figure 5-2, is called a *density curve,* and it must satisfy two properties similar to the requirements for discrete probability distributions, as listed in the following definition.

Definition

A **density curve** (or **probability density function**) is a graph of a continuous probability distribution. It must satisfy the following properties:

1. The total area under the curve must equal 1.
2. Every point on the curve must have a vertical height that is 0 or greater. (That is, the curve cannot fall below the *x*-axis.)

By setting the height of the rectangle in Figure 5-2 to be 0.5, we force the enclosed area to be $2 \times 0.5 = 1$, as required. (In general, the area of the rectangle becomes 1 when we make its height equal to the value of 1/range.) This property (area = 1) makes it very easy to solve probability problems, so the following statement is important:

Because the total area under the density curve is equal to 1, there is a correspondence between *area* and *probability*.

EXAMPLE Class Length Kim, who has developed a habit of living on the edge, has scheduled a job interview immediately following her statistics class. If the class runs longer than 51.5 minutes, she will be late for the job interview. Given the uniform distribution illustrated in Figure 5-2, find the probability that a randomly selected class will last longer than 51.5 minutes.

SOLUTION   See Figure 5-3, where we shade the region representing times that are longer than 51.5 minutes. Because the total area under the density

FIGURE 5-3 Using Area to Find Probability

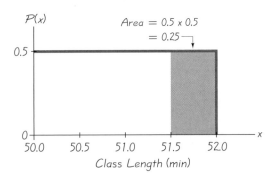

curve is equal to 1, there is a correspondence between area and probability. We can therefore find the desired probability by using areas as follows:

$$P(\text{class longer than 51.5 minutes}) = \text{area of shaded region in Figure 5-3}$$
$$= 0.5 \times 0.5$$
$$= 0.25$$

INTERPRETATION The probability of randomly selecting a class that runs longer than 51.5 minutes is 0.25. Because that probability is so high, Kim should consider making contingency plans that will allow her to get to her job interview on time. Nobody should ever be late for a job interview.

Standard Normal Distribution

The density curve of a uniform distribution is a horizontal line, so it's easy to find the area of any rectangular region: multiply width and height. The density curve of a normal distribution has the more complicated bell shape shown in Figure 5-1, so it's more difficult to find areas, but the basic principle is the same: *There is a correspondence between area and probability.*

Just as there are many different uniform distributions (with different ranges of values), there are also many different normal distributions, with each one depending on two parameters: the population mean, μ, and the population standard deviation, σ. (Recall from Chapter 1 that a *parameter* is a numerical measurement describing some characteristic of a *population.*) Figure 5-4 shows density curves for heights of adult women and men. Because men have a larger mean height, the peak of the density curve for men is farther to the right. Because men's heights have a slightly larger standard deviation, the density curve for men is slightly wider. Figure 5-4 shows two different possible normal distributions. There are infinitely many other possibilities, but one is of special interest.

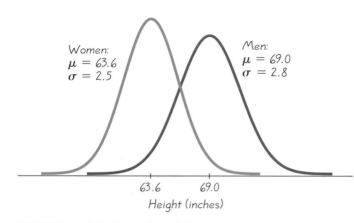

Women:
$\mu = 63.6$
$\sigma = 2.5$

Men:
$\mu = 69.0$
$\sigma = 2.8$

63.6   69.0
Height (inches)

FIGURE 5-4 **Heights of Adult Women and Men**

Reliability and Validity

The reliability of data refers to the consistency with which results occur, whereas the validity of data refers to how well the data measure what they are supposed to measure. The reliability of an IQ test can be judged by comparing scores for the test given on one date to scores for the same test given at another time. To test the validity of an IQ test, we might compare the test scores to another indicator of intelligence, such as academic performance. Many critics charge that IQ tests are reliable, but not valid; they provide consistent results, but don't really measure intelligence.

Note to Instructor
Before defining the standard normal distribution, ask the students which values of μ and σ they would select if they had to work with Formula 5-1. Students are usually quite good at seeing the advantages of choosing $\mu = 0$ and $\sigma = 1$.

Definition

The **standard normal distribution** is a normal probability distribution that has a mean of 0 and a standard deviation of 1, and the total area under its density curve is equal to 1. (See Figure 5-5.)

Suppose that we were hired to perform calculations using Formula 5-1. We would quickly see that the easiest values for μ and σ are $\mu = 0$ and $\sigma = 1$. By letting $\mu = 0$ and $\sigma = 1$, mathematicians have calculated many different areas under the curve. As shown in Figure 5-5, the area under the curve is 1, and this allows us to make the correspondence between area and probability, as we did in the preceding example with the uniform distribution.

Finding Probabilities When Given z Scores

Note to Instructor
Be sure to inform your students of the materials that will be available to them on tests. Note that the detachable Formula / Table card includes a copy of Table A-2. Also, constantly stress the difference between *areas* under the curve and *z* scores that are *distances* representing the number of standard deviations that a value is away from the mean.

Strongly encourage the drawing of a graph for each problem solved. Point out that the graph provides a visual understanding that can be of great help in solving the problems of this important chapter.

Using Table A-2 (in Appendix A and the *Formulas and Tables* insert card), we can find areas (or probabilities) for many different regions. Such areas can be found by using Table A-2, a TI-83 Plus calculator, or software such as STATDISK, Minitab, or Excel. The key features of the different methods are summarized in Table 5-1. It is not necessary to know all five methods; you only need to know the method you will be using for class and tests.

Because the following examples and exercises are based on Table A-2, it is essential to understand these points.

1. Table A-2 is designed only for the *standard* normal distribution, which has a mean of 0 and a standard deviation of 1.

2. Table A-2 is on two pages, with one page for *negative z* scores and the other page for *positive z* scores.

3. Each value in the body of the table is a *cumulative area from the left* up to a vertical boundary above a specific *z* score.

4. When working with a graph, avoid confusion between *z* scores and areas.

   z score: *Distance* **along the horizontal scale of the standard normal distribution; refer to the leftmost column and top row of Table A-2.**

   Area: *Region* **under the curve; refer to the values in the body of Table A-2.**

5. The part of the *z* score denoting hundredths is found across the top row of Table A-2.

The following example requires that we find the probability associated with a value less than 1.58. Begin with the *z* score of 1.58 by locating 1.5 in the left col-

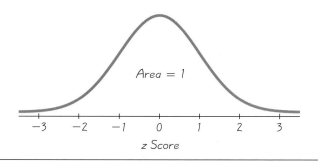

FIGURE 5-5 **Standard Normal Distribution:** $\mu = 0$ and $\sigma = 1$

 | Methods for Finding Normal Distribution Areas

Table A-2

Gives the cumulative area from the left up to a vertical line above a specific value of z.

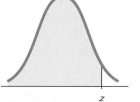

The procedure for using Table A-2 is described in the text.

STATDISK

Gives a few areas, including the cumulative area from the left and the cumulative area from the right.

Select **Analysis, Probability Distributions, Normal Distribution.** Enter the z score to find area, or enter the cumulative area from the left to find the z score.

Minitab

Gives the cumulative area from the left up to a vertical line above a specific value.

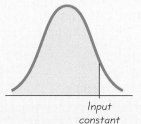

Select **Calc, Probability Distributions, Normal.** In the dialog box, select **Cumulative Probability, Input Constant.**

Excel

Gives the cumulative area from the left up to a vertical line above a specific value.

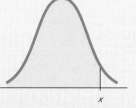

Select **fx, Statistical, NORMDIST.** In the dialog box, enter the value and mean, the standard deviation, and "true."

TI-83 Plus Calculator

Gives area bounded on the left and bounded on the right by vertical lines above any specific values.

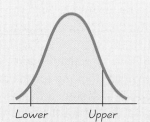

Press 2nd VARS [2: **normal cdf(**], then enter the two z scores separated by a comma, as in (left z score, right z score).

umn; next find the value in the adjoining row of probabilities that is directly below 0.08, as shown in this excerpt from Table A-2.

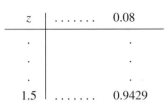

The area (or probability) value of 0.9429 indicates that there is a probability of 0.9429 of randomly selecting a z score less than 1.58. (The following sections will consider cases in which the mean is not 0 or the standard deviation is not 1.)

EXAMPLE Scientific Thermometers The Precision Scientific Instrument Company manufactures thermometers that are supposed to give readings of 0°C at the freezing point of water. Tests on a large sample of these instruments reveal that at the freezing point of water, some thermometers give readings below 0° (denoted by negative numbers) and some give readings above 0° (denoted by positive numbers). Assume that the mean reading is 0°C and the standard deviation of the readings is 1.00°C. Also assume that the readings are normally distributed. If one thermometer is randomly selected, find the probability that, at the freezing point of water, the reading is less than 1.58°.

SOLUTION   The probability distribution of readings is a standard normal distribution, because the readings are normally distributed with $\mu = 0$ and $\sigma = 1$. We need to find the area in Figure 5-6 below $z = 1.58$. The *area* below $z = 1.58$ is equal to the *probability* of randomly selecting a thermometer with a reading less than 1.58°. From Table A-2 we find that this area is 0.9429.

INTERPRETATION   The *probability* of randomly selecting a thermometer with a reading less than 1.58° (at the freezing point of water) is equal to the *area* of 0.9429 shown as the shaded region in Figure 5-6. Another way to interpret this result is to conclude that 94.29% of the thermometers will have readings below 1.58°.

FIGURE 5-6 Finding the Area Below $z = 1.58$

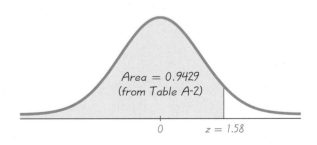

EXAMPLE Scientific Thermometers Using the thermometers from the preceding example, find the probability of randomly selecting one thermometer that reads (at the freezing point of water) above −1.23°.

SOLUTION   We again find the desired *probability* by finding a corresponding *area*. We are looking for the area of the region that is shaded in Figure 5-7, but Table A-2 is designed to apply only to cumulative areas from the *left*. Referring to Table A-2 for the page with *negative z* scores, we find that the cumulative area from the left up to $z = -1.23$ is 0.1093 as shown. Knowing that the total area under the curve is 1, we can find the shaded area by subtracting

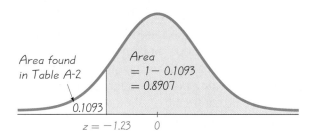

FIGURE 5-7 **Finding the Area Above** $z = -1.23$

0.1093 from 1. The result is 0.8907. Even though Table A-2 is designed only for cumulative areas from the left, we can use it to find cumulative areas from the right, as shown in Figure 5-7.

INTERPRETATION Because of the correspondence between probability and area, we conclude that the *probability* of randomly selecting a thermometer with a reading above $-1.23°$ at the freezing point of water is 0.8907 (which is the *area* above $z = -1.23$). In other words, 89.07% of the thermometers have readings above $-1.23°$.

The preceding example illustrates a way that Table A-2 can be used indirectly to find a cumulative area from the right. The following example illustrates another way that we can find an area by using Table A-2.

EXAMPLE **Scientific Thermometers** Once again, make a random selection from the same sample of thermometers. Find the probability that the chosen thermometer reads (at the freezing point of water) between $-2.00°$ and $1.50°$.

SOLUTION   We are again dealing with normally distributed values having a mean of $0°$ and a standard deviation of $1°$. The probability of selecting a thermometer that reads between $-2.00°$ and $1.50°$ corresponds to the shaded area in Figure 5-8. Table A-2 cannot be used to find that area directly, but we can use the table to find that $z = -2.00$ corresponds to the area of 0.0228, and $z = 1.50$ corresponds to the area of 0.9332, as shown in the figure. Refer to Figure 5-8 to see that the shaded area is the difference between 0.9332 and 0.0228. The shaded area is therefore $0.9332 - 0.0228 = 0.9104$.

continued

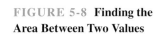

FIGURE 5-8 **Finding the Area Between Two Values**

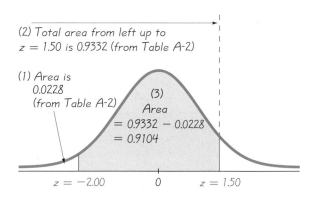

INTERPRETATION Using the correspondence between probability and area, we conclude that there is a probability of 0.9104 of randomly selecting one of the thermometers with a reading between $-2.00°$ and $1.50°$ at the freezing point of water. Another way to interpret this result is to state that if many thermometers are selected and tested at the freezing point of water, then 0.9104 (or 91.04%) of them will read between $-2.00°$ and $1.50°$.

The preceding example can be generalized as a rule stating that the area corresponding to the region between two specific z scores can be found by finding the difference between the two areas found in Table A-2. See Figure 5-9, which shows that the shaded region B can be found by finding the *difference* between two areas found from Table A-2: area A and B combined (found in Table A-2 as the area corresponding to z_{Right}) and area A (found in Table A-2 as the area corresponding to z_{Left}). Recommendation: Don't try to memorize a rule or formula for this case, because it is infinitely better to *understand* the procedure. Understand how Table A-2 works, then draw a graph, shade the desired area, and think of a way to find that area given the condition that Table A-2 provides only cumulative areas from the left.

The preceding example concluded with the statement that the probability of a reading between $-2.00°$ and $1.50°$ is 0.9104. Such probabilities can also be expressed with the following notation.

Note to Instructor
Many of us assume that terms such as "at most 2" or "at least 2" are clearly understood by all students, but that is often a wrong assumption. It could be very helpful to emphasize the importance of understanding and interpreting such terms. Ask them which job offer they would prefer: One that pays at most $90,000, one that pays no more than $90,000, or one that pays at least $90,000.

Notation

$P(a < z < b)$	denotes the probability that the z score is between a and b.
$P(z > a)$	denotes the probability that the z score is greater than a.
$P(z < a)$	denotes the probability that the z score is less than a.

Using this notation, we can express the result of the last example as: $P(-2.00 < z < 1.50) = 0.9104$, which states in symbols that the probability of a z score falling between -2.00 and 1.50 is 0.9104. With a continuous probability

FIGURE 5-9 **Finding the Area Between Two z Scores**

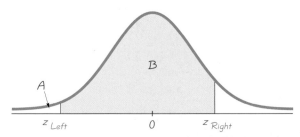

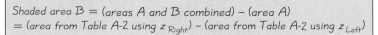

Shaded area B = (areas A and B combined) − (area A)
= (area from Table A-2 using z_{Right}) − (area from Table A-2 using z_{Left})

distribution such as the normal distribution, the probability of getting any single *exact* value is 0. That is, $P(z = a) = 0$. For example, there is a 0 probability of randomly selecting someone and getting a person whose height is exactly 68.12345678 in. In the normal distribution, any single point on the horizontal scale is represented not by a region under the curve, but by a vertical line above the point. For $P(z = 1.50)$ we have a vertical line above $z = 1.50$, but that vertical line by itself contains no area, so $P(z = 1.50) = 0$. With any continuous random variable, the probability of any one exact value is 0, and it follows that $P(a \leq z \leq b) = P(a < z < b)$. It also follows that the probability of getting a z score of *at most b* is equal to the probability of getting a z score *less than b*. It is important to correctly interpret key phrases such as *at most, at least, more than, no more than,* and so on.

Finding *z* Scores from Known Areas

So far, the examples of this section involving the standard normal distribution have all followed the same format: Given z scores, we found areas under the curve. These areas correspond to probabilities. In many other cases, we want a reverse process because we already know the area (or probability), but we need to find the corresponding z score. In such cases, it is very important to avoid confusion between z scores and areas. Remember, z scores are *distances* along the horizontal scale, and they are represented by the numbers in Table A-2 that are in the extreme left column and across the top row. Areas (or probabilities) are regions under the curve, and they are represented by the values in the body of Table A-2. Also, z scores positioned in the left half of the curve are always negative. If we already know a probability and want to determine the corresponding z score, we find it as follows.

Procedure for Finding a *z* Score from a Known Area

1. Draw a bell-shaped curve and identify the region under the curve that corresponds to the given probability. If that region is not a cumulative region from the left, work instead with a known region that is a cumulative region from the left.

2. Using the cumulative area from the left, locate the closest probability in the *body* of Table A-2 and identify the corresponding z score.

EXAMPLE **Scientific Thermometers** Use the same thermometers as earlier, with temperature readings at the freezing point of water that are normally distributed with a mean of 0°C and a standard deviation of 1°C. Find the temperature corresponding to P_{95}, the 95th percentile. That is, find the temperature separating the bottom 95% from the top 5%. See Figure 5-10.

SOLUTION   Figure 5-10 shows the z score that is the 95th percentile, with 95% of the area (or 0.95) below it. Important: Remember that when referring to Table A-2, the body of the table includes *cumulative areas from the left*. Referring to Table A-2, we search for the area of 0.95 *in the body* of the table and

continued

Note to Instructor

There are three important points that should be emphasized in this subsection: (1) The numbers in the body of Table A-2 are areas and the numbers in the extreme left column (and across the top row) are the *z* scores that are actually distances; (2) if a *z* score is to the *left* of the centerline, it must be negative; (3) it's always a good idea to check answers to be sure that they are reasonable, but that check becomes particularly important with the problems of this subsection.

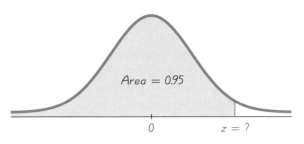

FIGURE 5-10 **Finding the 95th Percentile**

then find the corresponding z score. In Table A-2 we find the areas of 0.9495 and 0.9505, but there's an asterisk with a special note indicating that 0.9500 corresponds to a z score of 1.645. We can now conclude that the z score in Figure 5-10 is 1.645, so the 95th percentile is the temperature reading of 1.645°C.

INTERPRETATION When tested at freezing, 95% of the readings will be less than or equal to 1.645°C, and 5% of them will be greater than or equal to 1.645°C.

z Score	Cumulative Area from the Left
1.645	0.9500
−1.645	0.0500
2.575	0.9950
−2.575	0.0050

Note that in the preceding solution, Table A-2 led to a z score of 1.645, which is midway between 1.64 and 1.65. When using Table A-2, we can usually avoid interpolation by simply selecting the closest value. There are special cases listed in the accompanying table that are important because they are used so often in a wide variety of applications. (The value of $z = 2.576$ gives an area slightly closer to the area of 0.9950, but $z = 2.575$ has the advantage of being the value midway between $z = 2.57$ and $z = 2.58$.) Except in these special cases, we can select the closest value in the table. (If a desired value is midway between two table values, select the larger value.) Also, for z scores above 3.49, we can use 0.9999 as an approximation of the cumulative area from the left; for z scores below −3.49, we can use 0.0001 as an approximation of the cumulative area from the left.

EXAMPLE **Scientific Thermometers** Using the same thermometers, find the temperatures separating the bottom 2.5% and the top 2.5%.

SOLUTION Refer to Figure 5-11 where the required z scores are shown. To find the z score located to the left, refer to Table A-2 and search the *body of the table* for an area of 0.025. The result is $z = -1.96$. To find the z score located to the right, refer to Table A-2 and search the *body of the table* for an area of 0.975. (Remember that Table A-2 always gives cumulative areas from the *left*.) The result is $z = 1.96$. The values of $z = -1.96$ and $z = 1.96$ separate the bottom 2.5% and the top 2.5% as shown in Figure 5-11.

INTERPRETATION When tested at freezing, 2.5% of the thermometer readings will be equal to or less than −1.96°, and 2.5% of the readings will be equal to or greater than 1.96°. Another interpretation is that at the freezing level of water, 95% of all thermometer readings will fall between −1.96° and 1.96°.

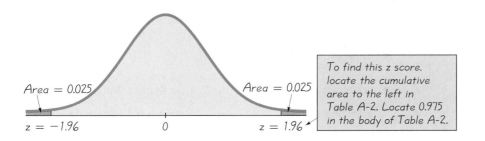

FIGURE 5-11   **Finding** z **Scores**

Area = 0.025   Area = 0.025

$z = -1.96$ 0 $z = 1.96$

> To find this z score, locate the cumulative area to the left in Table A-2. Locate 0.975 in the body of Table A-2.

The examples in this section were contrived so that the mean of 0 and the standard deviation of 1 coincided exactly with the parameters of the standard normal distribution. In reality, it is unusual to find such convenient parameters, because typical normal distributions involve means different from 0 and standard deviations different from 1. In the next section we introduce methods for working with such normal distributions, which are much more realistic.

Using Technology

STATDISK   Select **Analysis, Probability Distributions, Normal Distribution.** Enter the z score to find area, or enter the cumulative area from the left to find the z score.

Minitab

- To find the cumulative area to the left of a z score (as in Table A-2), select **Calc, Probability Distributions, Normal, Cumulative probabilities,** enter the mean of 0 and standard deviation of 1, then click on the **Input Constant** button and enter the z score.

- To find a z score corresponding to a known probability, select **Calc, Probability Distributions, Normal,** then select **Inverse cumulative probabilities** and the option **Input constant.** For the input constant, enter the total area to the left of the given value.

Excel

- To find the cumulative area to the left of a z score (as in Table A-2), click on *fx,* then select **Statistical, NORMSDIST,** and enter the z score.

- To find a z score corresponding to a known probability, select *fx,* **Statistical, NORMSINV,** and enter the total area to the left of the given value.

TI-83 Plus

- To find the area between two z scores, press **2nd, VARS, 2** (for normalcdf), then proceed to enter the two z scores separated by a comma, as in (left z score, right z score).

- To find a z score corresponding to a known probability, press **2nd, VARS, 3** (for invNorm), and proceed to enter the total area to the left of the value, the mean, and the standard deviation in the format of (total area to the left, mean, standard deviation) with the commas included.

5-2 Basic Skills and Concepts

Using a Continuous Uniform Distribution.   In Exercises 1–4, refer to the continuous uniform distribution depicted in Figure 5-2, assume that a class length between 50.0 min and 52.0 min is randomly selected, and find the probability that the given time is selected.

1. Less than 50.3 min

2. Greater than 51.0 min

3. Between 50.5 min and 50.8 min

4. Between 50.5 min and 51.8 min

Using a Continuous Uniform Distribution. In Exercises 5–8, assume that voltages in a circuit vary between 6 volts and 12 volts, and voltages are spread evenly over the range of possibilities, so that there is a uniform distribution. Find the probability of the given range of voltage levels.

5. 1/3

6. 5/6

7. 1/2

8. 1/4

5. Greater than 10 volts

6. Less than 11 volts

7. Between 7 volts and 10 volts

8. Between 6.5 volts and 8 volts

9. 0.4013

10. 0.0030

11. 0.5987

12. 0.9970

13. 0.0099

14. 0.0250

15. 0.9901

16. 0.9750

17. 0.2417

18. 0.0606

19. 0.1359

20. 0.0132

21. 0.8959

22. 0.8652

23. 0.6984

24. 0.8576

25. 0.0001

26. 0.0001

27. 0.5

28. 0.5

Using the Standard Normal Distribution. In Exercises 9–28, assume that the readings on the thermometers are normally distributed with a mean of $0°$ and a standard deviation of $1.00°C$. A thermometer is randomly selected and tested. In each case, draw a sketch, and find the probability of each reading in degrees.

9. Less than -0.25

10. Less than -2.75

11. Less than 0.25

12. Less than 2.75

13. Greater than 2.33

14. Greater than 1.96

15. Greater than -2.33

16. Greater than -1.96

17. Between 0.50 and 1.50

18. Between 1.50 and 2.50

19. Between -2.00 and -1.00

20. Between 2.00 and 2.34

21. Between -2.67 and 1.28

22. Between -1.18 and 2.15

23. Between -0.52 and 3.75

24. Between -3.88 and 1.07

25. Greater than 3.57

26. Less than -3.61

27. Greater than 0

28. Less than 0

Basis for Empirical Rule. In Exercises 29–32, find the indicated area under the curve of the standard normal distribution, then convert it to a percentage and fill in the blank. The results form the basis for the empirical rule introduced in Section 2-5.

29. 68.26%

30. 95.44%

31. 99.74%

32. 99.98%

29. About _____% of the area is between $z = -1$ and $z = 1$ (or within 1 standard deviation of the mean).

30. About _____% of the area is between $z = -2$ and $z = 2$ (or within 2 standard deviations of the mean).

31. About _____% of the area is between $z = -3$ and $z = 3$ (or within 3 standard deviations of the mean).

32. About _____% of the area is between $z = -3.5$ and $z = 3.5$ (or within 3.5 standard deviations of the mean).

Finding Probability. In Exercises 33–36, assume that the readings on the thermometers are normally distributed with a mean of $0°$ and a standard deviation of $1.00°$. Find the indicated probability, where z is the reading in degrees.

33. 0.9500

34. 0.9500

35. 0.9950

36. 0.0151

33. $P(-1.96 < z < 1.96)$

34. $P(z < 1.645)$

35. $P(z > -2.575)$

36. $P(1.96 < z < 2.33)$

Finding Temperature Values. *In Exercises 37–40, assume that the readings on the ther-mometers are normally distributed with a mean of 0° and a standard deviation of 1.00°C. A thermometer is randomly selected and tested. In each case, draw a sketch, and find the temperature reading corresponding to the given information.*

37. Find P_{90}, the 90th percentile. This is the temperature reading separating the bottom 90% from the top 10%.

37. 1.28

38. Find P_{20}, the 20th percentile.

38. −0.84

39. If 5% of the thermometers are rejected because they have readings that are too low, but all other thermometers are acceptable, find the reading that separates the rejected thermometers from the others.

39. −1.645

40. If 3.0% of the thermometers are rejected because they have readings that are too high and another 3.0% are rejected because they have readings that are too low, find the two readings that are cutoff values separating the rejected thermometers from the others.

40. −1.88, 1.88

5-2 Beyond the Basics

41. For a standard normal distribution, find the percentage of data that are
   a. within 1 standard deviation of the mean.
   b. within 1.96 standard deviations of the mean.
   c. between $\mu - 3\sigma$ and $\mu + 3\sigma$.
   d. between 1 standard deviation below the mean and 2 standard deviations above the mean.
   e. more than 2 standard deviations away from the mean.

41. a. 68.26%
   b. 95%
   c. 99.74%
   d. 81.85%
   e. 4.56%

42. If a continuous uniform distribution has parameters of $\mu = 0$ and $\sigma = 1$, then the minimum is $-\sqrt{3}$ and the maximum is $\sqrt{3}$.
   a. For this distribution, find $P(-1 < x < 1)$.
   b. Find $P(-1 < x < 1)$ if you incorrectly assume that the distribution is normal instead of uniform.
   c. Compare the results from parts (a) and (b). Does the distribution affect the results very much?

42. a. $1/\sqrt{3}$ or 0.5774
   b. 0.6826
   c. The results are very different.

43. Assume that z scores are normally distributed with a mean of 0 and a standard deviation of 1.
   a. If $P(0 < z < a) = 0.3907$, find a.
   b. If $P(-b < z < b) = 0.8664$, find b.
   c. If $P(z > c) = 0.0643$, find c.
   d. If $P(z > d) = 0.9922$, find d.
   e. If $P(z < e) = 0.4500$, find e.

43. a. 1.23
   b. 1.50
   c. 1.52
   d. −2.42
   e. −0.13

44. In a continuous uniform distribution,

$$\mu = \frac{\text{minimum} + \text{maximum}}{2} \qquad \text{and} \qquad \sigma = \frac{\text{range}}{\sqrt{12}}$$

Find the mean and standard deviation for the uniform distribution represented in Figure 5-2.

44. $\mu = 51.0$, $\sigma = 2/\sqrt{12}$ or 0.58

45. Sketch a graph representing a cumulative distribution for (a) a uniform distribution and (b) a normal distribution.

45. a.

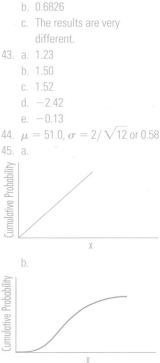

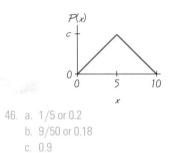

46. a. 1/5 or 0.2
     b. 9/50 or 0.18
     c. 0.9

46. Refer to the graph of the triangular probability distribution of the continuous random variable x. (See the margin graph.)
   a. Find the value of the constant c.
   b. Find the probability that x is between 0 and 3.
   c. Find the probability that x is between 2 and 9.

5-3 Applications of Normal Distributions

Because the examples and exercises in Section 5-2 all involved the *standard* normal distribution (with a mean of 0 and a standard deviation of 1), they were necessarily unrealistic. In this section we include nonstandard normal distributions so that we can work with applications that are real and practical. However, we can transform values from a nonstandard normal distribution to a standard normal distribution so that we can continue to use the same procedures from Section 5-2.

> **If we convert values to standard scores using Formula 5-2, then procedures for working with all normal distributions are the same as those for the standard normal distribution.**

Formula 5-2 $$z = \frac{x - \mu}{\sigma}$$ (round z scores to 2 decimal places)

Continued use of Table A-2 requires understanding and application of the above principle. (If you use certain calculators or software programs, the conversion to z scores is not necessary because probabilities can be found directly.) Regardless of the method used, you need to clearly understand the above basic principle, because it is an important foundation for concepts introduced in the following chapters.

Figure 5-12 illustrates the conversion from a nonstandard to a standard normal distribution. The area in any normal distribution bounded by some score x (as in Figure 5-12[a]) is the same as the area bounded by the equivalent z score in the standard normal distribution (as in Figure 5-12[b]). This means that when working with a nonstandard normal distribution, you can use Table A-2 the same way it was used in Section 5-2 as long as you first convert the values to z scores. When finding areas with a nonstandard normal distribution, use this procedure:

1. Sketch a normal curve, label the mean and the specific x values, then *shade* the region representing the desired probability.

2. For each relevant value x that is a boundary for the shaded region, use Formula 5-2 to convert that value to the equivalent z score.

FIGURE 5-12 Converting from Nonstandard to Standard Normal Distribution

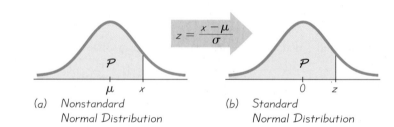

3. Refer to Table A-2 to find the area of the shaded region. This area is the desired probability.

The following example applies these three steps, and it illustrates the relationship between a typical nonnormal distribution and the standard normal distribution.

EXAMPLE **Designing Cars** The sitting height (from seat to top of head) of drivers must be considered in the design of a new car model. Men have sitting heights that are normally distributed with a mean of 36.0 in. and a standard deviation of 1.4 in. (based on anthropometric survey data from Gordon, Clauser, et al.). Engineers have provided plans that can accommodate men with sitting heights up to 38.8 in., but taller men cannot fit. If a man is randomly selected, find the probability that he has a sitting height less than 38.8 in. Based on that result, is the current engineering design feasible?

SOLUTION

Step 1:   See Figure 5-13, where we label the mean of 36.0, the maximum sitting height 38.8 in., and we shade the area representing the probability that we want. (We continue to use the same correspondence between *probability* and *area* that was first introduced in Section 5-2.)

Step 2:   To use Table A-2, we first must use Formula 5-2 to convert the distribution of heights to the standard normal distribution. The height of 38.8 in. is converted to a z score as follows:

$$z = \frac{x - \mu}{\sigma} = \frac{38.8 - 36.0}{1.4} = 2.00$$

This result shows that the sitting height of 38.8 in. is above the mean of 36.0 in. by 2.00 standard deviations.

Step 3:   Referring to Table A-2, we find that $z = 2.00$ corresponds to an area of 0.9772.

INTERPRETATION There is a probability of 0.9772 of randomly selecting a man with a sitting height less than 38.8 in. This can be expressed in symbols as

$$P(x < 38.8 \text{ in.}) = P(z < 2.00) = 0.9772$$

continued

Clinical Trial Cut Short

What do you do when you're testing a new treatment and, before your study ends, you find that it is clearly effective? You should cut the study short and inform all participants of the treatment's effectiveness. This happened when hydroxyurea was tested as a treatment for sickle cell anemia. The study was scheduled to last about 40 months, but the effectiveness of the treatment became obvious and the study was stopped after 36 months. (See "Trial Halted as Sickle Cell Treatment Proves Itself" by Charles Marwick, *Journal of the American Medical Association*, Vol. 273, No. 8.)

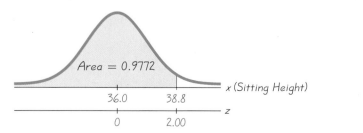

FIGURE 5-13 **Normal Distribution of Sitting Heights of Men**

Queues

Queuing theory is a branch of mathematics that uses probability and statistics. The study of queues, or waiting lines, is important to businesses such as supermarkets, banks, fast-food restaurants, airlines, and amusement parks. Grand Union supermarkets try to keep checkout lines no longer than three shoppers. Wendy's introduced the "Express Pak" to expedite servicing its numerous drive-through customers. Disney conducts extensive studies of lines at its amusement parks so that it can keep patrons happy and plan for expansion. Bell Laboratories uses queuing theory to optimize telephone network usage, and factories use it to design efficient production lines.

Another way to interpret this result is to conclude that 97.72% of men have sitting heights less than 38.8 in. An important consequence of that result is that 2.28% of men will not fit in the car. The manufacturer must now decide whether it can afford to lose 2.28% of all male car drivers.

EXAMPLE Jet Ejection Seats In the Chapter Problem, we noted that the Air Force had been using the ACES-II ejection seats designed for men weighing between 140 lb and 211 lb. Given that women's weights are normally distributed with a mean of 143 lb and a standard deviation of 29 lb (based on data from the National Health Survey), what percentage of women have weights that are within those limits?

SOLUTION   See Figure 5-14, which shows the shaded region for women's weights between 140 lb and 211 lb. We can't find that shaded area directly from Table A-2, but we can find it indirectly by using the basic procedures presented in Section 5-2. The way to proceed is first to find the cumulative area from the left up to 140 lb and the cumulative area from the left up to 211 lb, then find the difference between those two areas.

Finding the cumulative area up to 140 lb:

$$z = \frac{x - \mu}{\sigma} = \frac{140 - 143}{29} = -0.10$$

Using Table A-2, we find that $z = -0.10$ corresponds to an area of 0.4602, as shown in Figure 5-14.

Finding the cumulative area up to 211 lb:

$$z = \frac{x - \mu}{\sigma} = \frac{211 - 143}{29} = 2.34$$

Using Table A-2, we find that $z = 2.34$ corresponds to an area of 0.9904, as shown in Figure 5-14.

Finding the shaded area between 140 lb and 211 lb:

$$\text{Shaded area} = 0.9904 - 0.4602 = 0.5302$$

INTERPRETATION We found that 53.02% of women have weights between the ejection seat limits of 140 lb and 211 lb. This means that 46.98% of women do not have weights between the current limits, so far too many women pilots would risk serious injury if ejection became necessary.

FIGURE 5-14 Weights of Women and Ejection Seat Limits

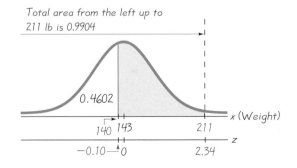

Finding Values from Known Areas

The preceding examples in this section are of the same type: We are given specific limit values and we must find an area (or probability or percentage). In many practical and real cases, the area (or probability or percentage) is known and we must find the relevant value(s). When finding values from known areas, be careful to keep these cautions in mind:

1. *Don't confuse z scores and areas.* Remember, z scores are *distances* along the horizontal scale, but areas are *regions* under the normal curve. Table A-2 lists z scores in the left columns and across the top row, but areas are found in the body of the table.

2. *Choose the correct (right/left) side of the graph.* A value separating the top 10% from the others will be located on the right side of the graph, but a value separating the bottom 10% will be located on the left side of the graph.

3. A z score must be *negative* whenever it is located in the *left* half of the normal distribution.

4. Areas (or probabilities) are positive or zero values, but they are never negative.

Graphs are extremely helpful in visualizing, understanding, and successfully working with normal probability distributions, so they should be used whenever possible.

Survey Medium Can Affect Results

In a survey of Catholics in Boston, the subjects were asked if contraceptives should be made available to unmarried women. In personal interviews, 44% of the respondents said yes. But among a similar group contacted by mail or telephone, 75% of the respondents answered yes to the same question.

Procedure for Finding Values Using Table A-2 and Formula 5-2

1. Sketch a normal distribution curve, enter the given probability or percentage in the appropriate region of the graph, and identify the x value(s) being sought.

2. Use Table A-2 to find the z score corresponding to the cumulative left area bounded by x. Refer to the *body* of Table A-2 to find the closest area, then identify the corresponding z score.

3. Using Formula 5-2, enter the values for μ, σ, and the z score found in Step 2, then solve for x. Based on the format of Formula 5-2, we can solve for x as follows:

$$x = \mu + (z \cdot \sigma) \quad \text{(another form of Formula 5-2)}$$
$$\uparrow$$

(If z is located to the left of the mean, be sure that it is a negative number.)

4. Refer to the sketch of the curve to verify that the solution makes sense in the context of the graph and in the context of the problem.

The following example uses the procedure just outlined.

EXAMPLE Hip Breadths and Airplane Seats In designing seats to be installed in commercial aircraft, engineers want to make the seats wide enough to fit 98% of all males. (Accommodating 100% of males would require very wide seats that would be much too expensive.) Men have hip breadths that are normally distributed with a mean of

continued

FIGURE 5-15 **Distribution of Hip Breadths of Men**

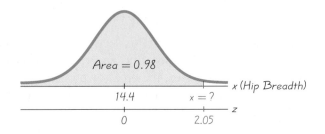

14.4 in. and a standard deviation of 1.0 in. (based on anthropometric survey data from Gordon, Clauser, et al.). Find P_{98}. That is, find the hip breadth of men that separates the bottom (no pun intended) 98% from the top 2%.

SOLUTION

Step 1: We begin with the graph shown in Figure 5-15. We have entered the mean of 14.4 in., shaded the area representing the bottom 98%, and identified the desired value as x.

Step 2: In Table A-2 we search for an area of 0.9800 in the *body* of the table. (The area of 0.98 shown in Figure 5-15 is a cumulative area from the left, and that is exactly the type of area listed in Table A-2.) The area closest to 0.98 is 0.9798, and it corresponds to a z score of 2.05.

Step 3: With $z = 2.05$, $\mu = 14.4$, and $\sigma = 1.0$, we solve for x by using Formula 5-2 directly or by using the following version of Formula 5-2:

$$x = \mu + (z \cdot \sigma) = 14.4 + (2.05 \cdot 1.0) = 16.45$$

Step 4: If we let $x = 16.45$ in Figure 5-15, we see that this solution is reasonable because the 98th percentile should be greater than the mean of 14.4.

INTERPRETATION The hip breadth of 16.5 in. (rounded to one decimal place, as in μ and σ) separates the lowest 98% from the highest 2%. That is, seats designed for a hip breadth up to 16.5 in. will fit 98% of men. This type of analysis was used to design the seats currently used in commercial aircraft.

EXAMPLE Designing Car Dashboards When designing the placement of a CD player in a new model car, engineers must consider the forward grip reach of the driver. If the CD player is placed beyond the forward grip reach, the driver must move his or her body in a way that could be distracting and dangerous. (We wouldn't want anyone injured while trying to hear the best of Barry Manilow.) Design engineers decide that the CD should be placed so that it is within the forward grip reach of 95% of women. Women have forward grip reaches that are normally distributed with a mean of 27.0 in. and a standard deviation of 1.3 in. (based on anthropometric survey data from Gordon, Churchill, et al.) Find the forward grip reach of women that separates the longest 95% from the others.

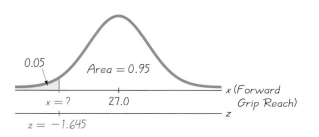

FIGURE 5-16 **Finding the
Value Separating the Top 95%**

SOLUTION

Step 1:   We begin with the graph shown in Figure 5-16. We have entered the mean of 27.0 in., and we have identified the area representing the longest 95% of forward grip reaches. Even though the problem refers to the top 95%, Table A-2 requires that we work with a cumulative *left* area, so we subtract 0.95 from 1 to get 0.05, which is shown as the shaded region.

Step 2:   In Table A-2 we search for an area of 0.05 in the *body* of the table. The areas closest to 0.05 are 0.0505 and 0.0495, but there is an asterisk indicating that an area of 0.05 corresponds to a z score of -1.645.

Step 3:   With $z = -1.645$, $\mu = 27.0$, and $\sigma = 1.3$, we solve for x by using Formula 5-2 directly or by using the following version of Formula 5-2:

$$x = \mu + (z \cdot \sigma) = 27.0 + (-1.645 \cdot 1.3) = 24.8615$$

Step 4:   If we let $x = 24.8615$ in Figure 5-16, we see that this solution is reasonable because the forward grip reach separating the top 95% from the bottom 5% should be less than the mean of 27.0 in.

INTERPRETATION The forward grip reach of 24.9 in. (rounded) separates the top 95% from the others, since 95% of women have forward grip reaches that are longer than 24.9 in. and 5% of women have forward grip reaches shorter than 24.9 in.

STATDISK Select **Analysis, Probability Distributions, Normal Distribution.** Enter the z score to find area, or enter the cumulative area from the left to find the z score.

Minitab
• To find the cumulative area to the left of a z score (as in Table A-2), select **Calc, Probability Distributions, Normal, Cumulative probabilities,** enter the mean and standard deviation, then click on the **Input Constant** button and enter the value.

• To find a value corresponding to a known area, select **Calc, Probability Distributions, Normal,** then select **Inverse cumulative probabilities** and enter the mean and standard deviation. Select the option **Input constant** and enter the total area to the left of the given value.

Excel
• To find the cumulative area to the left of a value (as in Table A-2), click on *fx*, then select **Statistical, NORMDIST.** In the dialog box, enter the value for x, enter the mean and standard deviation, and enter 1 in the "cumulative" space.

continued

- To find a value corresponding to a known area, select *fx,* **Statistical, NORMINV,** and proceed to make the entries in the dialog box. When entering the probability value, enter the total area to the left of the given value.

TI-83 Plus

- To find the area between two values, press **2nd, VARS, 2** (for normalcdf), then proceed to enter the two values, the mean, and the standard deviation, all separated by com-

mas, as in (left value, right value, mean, standard deviation).

- To find a value corresponding to a known area, press **2nd, VARS, 3** (for invNorm), and proceed to enter the total area to the left of the value, the mean, and the standard deviation in the format of (total area to the left, mean, standard deviation) with the commas included.

5-3 Basic Skills and Concepts

Recommended Assignment
Exercises 1–12, 17–20.

IQ Scores. In Exercises 1–8, assume that adults have IQ scores that are normally distributed with a mean of 100 and a standard deviation of 15 (as on the Weschler test). (Hint: *Draw a graph in each case.*)

1. 0.8413

1. Find the probability that a randomly selected adult has an IQ that is less than 115.

2. 0.0179

2. Find the probability that a randomly selected adult has an IQ greater than 131.5 (the requirement for membership in the Mensa organization).

3. 0.4972

3. Find the probability that a randomly selected adult has an IQ between 90 and 110 (referred to as the *normal* range).

4. 0.1596

4. Find the probability that a randomly selected adult has an IQ between 110 and 120 (referred to as *bright normal*).

5. 87.4

5. Find P_{20}, which is the IQ score separating the bottom 20% from the top 80%.

6. 112.6

6. Find P_{80}, which is the IQ score separating the bottom 80% from the top 20%.

7. 115.6

7. Find the IQ score separating the top 15% from the others.

8. 98.1

8. Find the IQ score separating the top 55% from the others.

9. a. 0.01%; yes
   b. 99.2°

9. Body Temperatures Based on the sample results in Data Set 4 of Appendix B, assume that human body temperatures are normally distributed with a mean of 98.20°F and a standard deviation of 0.62°F.

   a. Bellevue Hospital in New York City uses 100.6°F as the lowest temperature considered to be a fever. What percentage of normal and healthy persons would be considered to have a fever? Does this percentage suggest that a cutoff of 100.6°F is appropriate?

   b. Physicians want to select a minimum temperature for requiring further medical tests. What should that temperature be, if we want only 5.0% of healthy people to exceed it? (Such a result is a *false positive,* meaning that the test result is positive, but the subject is not really sick.)

10. a. 0.0038; either a very rare
    event has occurred or the
    husband is not the father.
    b. 242 days

10. Lengths of Pregnancies The lengths of pregnancies are normally distributed with a mean of 268 days and a standard deviation of 15 days.

   a. One classic use of the normal distribution is inspired by a letter to "Dear Abby" in which a wife claimed to have given birth 308 days after a brief visit from her husband, who was serving in the Navy. Given this information, find the probability of a pregnancy lasting 308 days or longer. What does the result suggest?

b. If we stipulate that a baby is *premature* if the length of pregnancy is in the lowest 4%, find the length that separates premature babies from those who are not premature. Premature babies often require special care, and this result could be helpful to hospital administrators in planning for that care.

11. SAT Test Requirement The combined math and verbal scores for females taking the SAT-I test are normally distributed with a mean of 998 and a standard deviation of 202 (based on data from the College Board). The College of Westport includes a minimum score of 1100 among its requirements.
   a. What percentage of females do *not* satisfy that requirement?
   b. If the requirement is changed to "a score that is in the top 40%," what is the minimum required score? What is a practical difficulty that would be created if the new requirement were announced as "the top 40%"?

12. Designing Helmets Engineers must consider the breadths of male heads when designing motorcycle helmets. Men have head breadths that are normally distributed with a mean of 6.0 in. and a standard deviation of 1.0 in. (based on anthropometric survey data from Gordon, Churchill, et al.). Due to financial constraints, the helmets will be designed to fit all men except those with head breadths that are in the smallest 2.5% or largest 2.5%. Find the minimum and maximum head breadths that will fit men.

13. TV Warranty Replacement times for TV sets are normally distributed with a mean of 8.2 years and a standard deviation of 1.1 years (based on data from "Getting Things Fixed," *Consumer Reports*).
   a. Find the probability that a randomly selected TV will have a replacement time less than 5.0 years.
   b. If you want to provide a warranty so that only 1% of the TV sets will be replaced before the warranty expires, what is the time length of the warranty?

14. CD Player Warranty Replacement times for CD players are normally distributed with a mean of 7.1 years and a standard deviation of 1.4 years (based on data from "Getting Things Fixed," *Consumer Reports*).
   a. Find the probability that a randomly selected CD player will have a replacement time less than 8.0 years.
   b. If you want to provide a warranty so that only 2% of the CD players will be replaced before the warranty expires, what is the time length of the warranty?

15. M&Ms Shown below is the Minitab display for the weights (in grams) of the 100 M&M candies listed in Data Set 19 of Appendix B. Although the mean and standard deviation are sample statistics, assume that they are population parameters for all M&Ms.
   a. Find the percentage of weights that are less than 0.88925 g. How does the result agree with the value of 0.88925 shown as Q_1, the first quartile?
   b. Find the value of Q_1. How does the result agree with the value of 0.88925 shown in the display?

Variable	N	Mean	Median	TrMean	StDev	SE Mean
M&M	100	0.91470	0.91050	0.91307	0.03691	0.00369

Variable	Minimum	Maximum	Q1	Q3
M&M	0.83800	1.03300	0.88925	0.93375

11. a. 69.15%
    b. 1049; if the top 40% of scores from the applicant pool are selected, nobody would know about acceptance or rejection until all applicant scores have been obtained.

12. 4.0 in.; 8.0 in.

13. a. 0.0018
    b. 5.6 years

14. a. 0.7389
    b. 4.2 years

15. a. 25% agrees
    b. 0.8895 g; close.

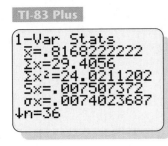

16. a. 0.1379
   b. 0.80211 g; 0.83154 g

17. 0.52%

18. 98.74%; 1.26% are being
   denied.

19. 0.1222; 12.22%; yes, they are
   all well above the mean.

20. 61.5 in.; 65.7 in.

21. a. The *z* scores are real numbers that have no units of
   measurement.
   b. $\mu = 0$; $\sigma = 1$; distribution
   is normal.
   c. $\mu = 64.9$ kg, $\sigma = 13.2$ kg,
   distribution is normal.

22. a. 0.3707
   b. 0.3557
   c. The values differ by 0.0150,
   which is small but not negligible.

16. **Weights of Regular Coke** Shown in the margin is the TI-83 Plus calculator display for the weights (in pounds) of regular Coke, as listed in Data Set 17 of Appendix B. Although the mean and standard deviation are sample statistics, assume that they are population parameters for all cans of regular Coke. (Use the value of Sx for the standard deviation.)
   a. If a can of regular Coke is randomly selected, find the probability that its contents weigh more than 0.8250 lb.
   b. For purposes of monitoring Coke production in the future, find the weights separating the bottom 2.5% and top 2.5%.

Heights of Women. In Exercises 17–20, assume that heights of women are normally distributed with a mean given by $\mu = 63.6$ in. and a standard deviation given by $\sigma = 2.5$ in. (based on data from the National Health Survey). In each case, draw a graph.

17. **Beanstalk Club Height Requirement** The Beanstalk Club, a social organization for tall people, has a requirement that women must be at least 70 in. (or 5 ft 10 in.) tall. What percentage of women meet that requirement?

18. **Height Requirement for Women Soldiers** The U.S. Army requires women's heights to be between 58 in. and 80 in. Find the percentage of women meeting that height requirement. Are many women being denied the opportunity to join the Army because they are too short or too tall?

19. **Height Requirement for Rockettes** In order to have a precision dance team with a uniform appearance, height restrictions are placed on the famous Rockette dancers at New York's Radio City Music Hall. Because women have grown taller, a more recent change now requires that a Rockette dancer must have a height between 66.5 in. and 71.5 in. If a woman is randomly selected, what is the probability that she meets this new height requirement? What percentage of women meet this new height requirement? Does it appear that Rockettes are generally taller than typical women?

20. **Height Requirement for Rockettes** Exercise 19 identified specific height requirements for Rockettes. Suppose that those requirements must be changed because too few women now meet them. What are the new minimum and maximum allowable heights if the shortest 20% and the tallest 20% are excluded?

5-3 Beyond the Basics

21. **Units of Measurement** Weights of women are normally distributed with a mean of 143 lb and a standard deviation of 29 lb.
   a. If weights of individual women are expressed in units of pounds, what are the units used for the *z* scores that correspond to individual weights?
   b. If weights of all women are converted to *z* scores, what are the mean, standard deviation, and distribution of these *z* scores?
   c. What are the distribution, mean, and standard deviation of women's weights after they have all been converted to kilograms (1 lb = 0.4536 kg)?

22. **Using Continuity Correction** There are many situations in which a normal distribution can be used as a good approximation to a random variable that has only *discrete* values. In such cases, we can use this *continuity correction:* Represent each whole number by the interval extending from 0.5 below the number to 0.5 above it. Assume

that IQ scores are all whole numbers having a distribution that is approximately normal with a mean of 100 and a standard deviation of 15.
   a. Without using any correction for continuity, find the probability of randomly selecting someone with an IQ score greater than 105.
   b. Using the correction for continuity, find the probability of randomly selecting someone with an IQ score greater than 105.
   c. Compare the results from parts (a) and (b).

23. Curving Test Scores   A teacher informs her psychology class that a test is very difficult, but the grades will be curved. Scores for the test are normally distributed with a mean of 25 and a standard deviation of 5.
   a. If she curves by adding 50 to each grade, what is the new mean? What is the new standard deviation?
   b. Is it fair to curve by adding 50 to each grade? Why or why not?
   c. If the grades are curved according to the following scheme (instead of adding 50), find the numerical limits for each letter grade.

   A: Top 10%

   B: Scores above the bottom 70% and below the top 10%

   C: Scores above the bottom 30% and below the top 30%

   D: Scores above the bottom 10% and below the top 70%

   F: Bottom 10%

   d. Which method of curving the grades is fairer: Adding 50 to each grade or using the scheme given in part (c)? Explain.

23. a. 75; 5
    b. No, the conversion should also account for variation.
    c. 31.4, 27.6, 22.4, 18.6
    d. Part (c), because variation is included in the conversion.

24. SAT Scores   According to data from the College Entrance Examination Board, scores on the SAT-I test have a mean of 1017, and Q_1 is 880. The scores have a distribution that is approximately normal. Find the standard deviation, then use that result to find P_{99}.

24. 204; 1492

25. SAT and ACT Tests   Scores by women on the SAT-I test are normally distributed with a mean of 998 and a standard deviation of 202. Scores by women on the ACT test are normally distributed with a mean of 20.9 and a standard deviation of 4.6. Assume that the two tests use different scales to measure the same aptitude.
   a. If a woman gets a SAT score that is the 67th percentile, find her actual SAT score and her equivalent ACT score.
   b. If a woman gets a SAT score of 1220, find her equivalent ACT score.

25. a. 1087; 22.9
    b. 26.0

5-4   Sampling Distributions and Estimators

We are beginning to embark on a journey that allows us to learn about populations by obtaining data from samples. Sections 5-5 and 5-6 provide important concepts revealing the behavior of sample means and sample proportions. Before considering those concepts, let's begin by focusing on the behavior of sample statistics in general. The main objective of this section is to learn what we mean by a *sampling distribution of a statistic*, and another important objective is to learn a basic principle about the sampling distribution of sample means and the sampling distribution of sample proportions.

Note to Instructor
This section is new to the 9th edition. Its objectives are to introduce the general concept of a sampling distribution of a statistic, and to demonstrate that some statistics (mean, variance, proportion) tend to target the population parameter while others do not.

Let's begin with sample means. Instead of getting too abstract, let's consider the *population* consisting of the values 1, 2, 5. Because of rude sales personnel, a poor business plan, ineffective advertising, and a poor location, the McGwire Electronics Center was in business for only three days. On the first day, 1 cell phone was sold, 2 were sold the second day, and only 5 were sold on the third day. Because 1, 2, 5 constitute the entire population, it is easy to find the values of the population parameters: $\mu = 2.7$ and $\sigma = 1.7$. When finding the value of the population standard deviation σ, we use

$$\sigma = \sqrt{\frac{\Sigma(x - \mu)^2}{N}} \text{ instead of } s = \sqrt{\frac{\Sigma(x - \bar{x})^2}{n - 1}}$$

It is rare that we know all values in an entire population. The more common case is that there is some really large unknown population that we want to investigate. Because it is not practical to survey every member of the population, we obtain a sample and, based on characteristics of the sample, we make estimates about characteristics of the population. For example, the Hartford Insurance Company might try to learn about the population of ages of all drivers by obtaining a sample of those ages.

Because the values of 1, 2, 5 constitute an entire population, let's consider samples of size 2. With only three population values, there are only 9 different possible samples of size 2, assuming that we sample with replacement. That is, each selected value is replaced before another selection is made.

Why sample *with* replacement? For small samples of the type that we are discussing, sampling *without replacement* has the very practical advantage of avoiding wasteful duplication whenever the same item is selected more than once. However, throughout this section, we are particularly interested in sampling *with replacement* for these reasons: (1) When selecting a relatively small sample from a large population, it makes no significant difference whether we sample with replacement or without replacement. (2) Sampling with replacement results in independent events that are unaffected by previous outcomes, and independent events are easier to analyze and they result in simpler formulas. We therefore focus on the behavior of samples that are randomly selected *with replacement*.

When we sample 2 values with replacement from the population of 1, 2, 5, each of the 9 possible samples is equally likely, and they each have probability 1/9. Table 5-2 lists the 9 possible samples of size 2 along with statistics for each sample. That table contains much information, but let's first consider the column of sample means. Because we have all possible values of $\bar{x}$ listed, and because the probability of each is known to be 1/9, we have a probability distribution. (Remember, a probability distribution describes the probability for each value of a random variable, and the random variable in this case is the value of the sample mean $\bar{x}$). Because several important methods of statistics begin with a sample mean that is subsequently used for making inferences about the population mean, it is important to understand the behavior of such sample means. Other important methods of statistics begin with a sample proportion that is subsequently used for making inferences about the population proportion, so it is also important to un-

Table 5-2	Sampling Distributions of Different Statistics (for Samples of Size 2 Drawn with Replacement from the Population 1, 2, 5)						

Sample	Mean $\bar{x}$	Median	Range	Variance s^2	Standard Deviation s	Proportion of Odd Numbers	Probability
1, 1	1.0	1.0	0	0.0	0.000	1	1/9
1, 2	1.5	1.5	1	0.5	0.707	0.5	1/9
1, 5	3.0	3.0	4	8.0	2.828	1	1/9
2, 1	1.5	1.5	1	0.5	0.707	0.5	1/9
2, 2	2.0	2.0	0	0.0	0.000	0	1/9
2, 5	3.5	3.5	3	4.5	2.121	0.5	1/9
5, 1	3.0	3.0	4	8.0	2.828	1	1/9
5, 2	3.5	3.5	3	4.5	2.121	0.5	1/9
5, 5	5.0	5.0	0	0.0	0.000	1	1/9
Mean of Statistic Values	2.7	2.7	1.8	2.9	1.3	0.667	
Population Parameter	2.7	2	4	2.9	1.7	0.667	
Does the sample statistic target the population parameter?	Yes	No	No	Yes	No	Yes	

derstand the behavior of such sample proportions. In general, it is often important to understand the behavior of sample statistics. The "behavior" of a statistic can be known by understanding its distribution.

Definition

The **sampling distribution of the mean** is the probability distribution of sample means, with all samples having the same sample size n. (In general, the sampling distribution of any statistic is the probability distribution of that statistic.)

EXAMPLE **Sampling Distribution of the Mean** A *population* consists of the values 1, 2, 5, and Table 5-2 lists all of the different possible samples of size $n = 2$. The probability of each sample is listed in Table 5-2 as 1/9. For samples of size $n = 2$ randomly selected with replacement from the population 1, 2, 5, identify the specific sampling distribution of the mean. Also, find the mean of this sampling distribution. Do the sample means target the value of the population mean?

continued

Table 5-3

Sampling Distribution of the Mean

Mean $\bar{x}$	Probability
1.0	1/9
1.5	1/9
3.0	1/9
1.5	1/9
2.0	1/9
3.5	1/9
3.0	1/9
3.5	1/9
5.0	1/9

This table lists the means from the samples in Table 5-2, but it could be condensed by listing 1.0, 1.5, 2.0, 3.0, 3.5, and 5.0 along with their corresponding probabilities of 1/9, 2/9, 1/9, 2/9, 2/9, and 1/9.

SOLUTION   The sampling distribution of the mean is the probability distribution describing the probability for each value of the mean, and all of those values are included in Table 5-2. The sampling distribution of the mean can therefore be described by Table 5-3. We could calculate the mean of the sampling distribution using two different approaches: (1) Use $\mu = \Sigma[x \cdot P(x)]$, which is Formula 4-2, or (2) because all of the 9 sample means are equally likely, we could simply find the mean of those 9 values. Because the population mean is also 2.7, it appears that the sample means "target" the value of the population mean, instead of systematically underestimating or overestimating the population mean.

From the preceding example we see that the mean of all of the different possible sample means is equal to the mean of the original population, which is $\mu = 2.7$. We can generalize this as a property of sample means: For a fixed sample size, the mean of all possible sample means is equal to the mean of the population. We will revisit this important property in the next section, but let's first make another obvious but important observation: *Sample means vary*. See Table 5-3 and observe how the sample means are different. The first sample mean is 1.0, the second sample mean is 1.5, and so on. This leads to the following definition.

Definition

The value of a statistic, such as the sample mean $\bar{x}$, depends on the particular values included in the sample, and it generally varies from sample to sample. This variability of a statistic is called **sampling variability.**

In Chapter 2 we introduced the important characteristics of a data set: center, variation, distribution, outliers, and pattern over time (summarized with the mnemonic of "CVDOT"). In examining the samples in Table 5-2, we have already identified a property describing the behavior of sample means: The mean of sample means is equal to the mean of the population. This property addresses the characteristic of center, and we will investigate other characteristics in the next section. We will see that as the sample size increases, the sampling distribution of sample means tends to become a *normal distribution*. (This isn't too surprising given that the title of this chapter is "Normal Probability Distributions.") Consequently, the normal distribution assumes an importance that goes far beyond the applications illustrated in Section 5-3. The normal distribution will be used for many cases in which we want to use a sample mean $\bar{x}$ for the purpose of making some inference about a population mean μ.

Sampling Distribution of Proportions

When making inferences about a population proportion, it is also important to understand the behavior of sample proportions. We define the distribution of sample proportions as follows.

Definition

The **sampling distribution of the proportion** is the probability distribution of sample proportions, with all samples having the same sample size n.

A typical use of inferential statistics is to find some sample proportion and use it to make an inference about the population proportion. Pollsters from the Gallup organization asked 491 randomly selected adults whether they are in favor of the death penalty for a person convicted of murder. Results showed that 319 (or 65%) of those surveyed were in favor. That sample result leads to the inference that "65% of all adults are in favor of the death penalty for a person convicted of murder." The sample proportion of 319/491 was used to estimate a population proportion p, but we can learn much more by understanding the sampling distribution of such sample proportions.

EXAMPLE **Sampling Distribution of Proportions** A *population* consists of the values 1, 2, 5, and Table 5-2 lists all of the different possible samples of size $n = 2$ selected with replacement. For each sample, consider the proportion of numbers that are *odd*. Identify the sampling distribution for the proportion of odd numbers, then find its mean. Do sample proportions target the value of the population proportion?

SOLUTION   See Table 5-2 where the nine sample proportions are listed as 1, 0.5, 1, 0.5, 0, 0.5, 1, 0.5, 1. Combining those sample proportions with their probabilities of 1/9 in each case, we get the sampling distribution of proportions summarized in Table 5-4. The mean of the *sample* proportions is 0.667. Because the population 1, 2, 5 contains two odd numbers, the *population* proportion of odd numbers is also 2/3 or 0.667. In general, sample proportions tend to target the value of the population proportion, instead of systematically tending to underestimate or overestimate that value.

Table 5-4	
Sampling Distribution of Proportions	
Proportion of Odd Numbers	Probability
1	1/9
0.5	1/9
1	1/9
0.5	1/9
0	1/9
0.5	1/9
1	1/9
0.5	1/9
1	1/9

This table lists the proportions from the samples in Table 5-2, but it could be condensed by listing the proportions of 0, 0.5, and 1 along with their corresponding probabilities of 1/9, 4/9, and 4/9.

The preceding example involves a fairly small population, so let's now consider the genders of the senators in the 107th Congress. Because there are only 100 members [13 females (F) and 87 males (M)], we can list the entire population:

```
M F M M F M M M M M M M F M M M M M M M
M M M M M M M M M M M M M F F M M M M M M
M M M F M M M M M F M M M M M M M M M M
F M M M M M M M M M M M M M M M M M F F F
M M M F M F M M M M M M M M M M M M M M
```

The population proportion of female senators is $p = 13/100 = 0.13$. Usually, we don't know all of the members of the population, so we must estimate it from a

Table 5-5 Results from 100 Samples	
Proportion of Female Senators	**Frequency**
0.0	26
0.1	41
0.2	24
0.3	7
0.4	1
0.5	1
Mean:	0.119
Standard Deviation:	0.100

sample. For the purpose of studying the behavior of sample proportions, we list a few samples of size $n = 10$:

Sample 1: M F M M F M M M M M → sample proportion is 0.2
Sample 2: M F M M M M M M M M → sample proportion is 0.1
Sample 3: M M M M M M F M M M → sample proportion is 0.1
Sample 4: M M M M M M M M M M → sample proportion is 0
Sample 5: M M M M M M M M F M → sample proportion is 0.1

Because there is a very large number of such samples, we cannot list all of them. The author randomly selected 95 additional samples before stopping to rotate his car tires. Combining these additional 95 samples with the five listed here, we get 100 samples summarized in Table 5-5.

We can see from Table 5-5 that the mean of the 100 sample proportions is 0.119, but if we were to include all other possible samples of size 10, the mean of the sample proportions would equal 0.13, which is the value of the population proportion. Figure 5-17 shows the distribution of the 100 sample proportions summarized in Table 5-5. The shape of that distribution is reasonably close to the shape that would have been obtained with all possible samples of size 10. We can see that the distribution depicted in Figure 5-17 is somewhat skewed to the right, but with a bit of a stretch, it might be approximated very roughly by a normal distribution. In Figure 5-18 we show the results obtained from 10,000 samples of size 50 randomly selected with replacement from the above list of 100 genders. Figure 5-18 very strongly suggests that the distribution is approaching the characteristic bell shape of a normal distribution. The results from Table 5-5 and Figure 5-18 therefore suggest the following.

Properties of the Distribution of Sample Proportions

- **Sample proportions tend to target the value of the population proportion.**
- **Under certain conditions, the distribution of sample proportions approximates a normal distribution.**

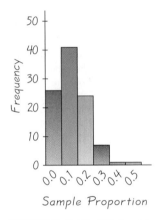

FIGURE 5-17 100 Sample Proportions with $n = 10$ in Each Sample

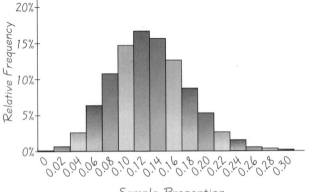

FIGURE 5-18 10,000 Sample Proportions with $n = 50$ in Each Sample

Which Statistics Make Good Estimators of Parameters?

Chapter 6 will introduce formal methods for using sample statistics to make estimates of the values of population parameters. Some statistics work much better than others, and we can judge their value by examining their sampling distributions, as in the following example.

EXAMPLE **Sampling Distributions** A *population* consists of the values 1, 2, 5. If we randomly select samples of size 2 with replacement, there are nine different possible samples, and they are listed in Table 5-2. Because the nine different samples are equally likely, each sample has probability 1/9.

a. For each sample, find the mean, median, range, variance, standard deviation, and the proportion of sample values that are odd. (For each statistic, this will generate nine values which, when associated with nine probabilities of 1/9 each, will combine to form a *sampling distribution* for the statistic.)

b. For each statistic, find the mean of the results from part (a).

c. Compare the means from part (b) to the corresponding population parameters, then determine whether each statistic targets the value of the population parameter. For example, the sample means tend to center about the value of the population mean, which is $8/3 = 2.7$, so the sample mean targets the value of the population mean.

SOLUTION

a. See Table 5-2. The individual statistics are listed for each sample.

b. The means of the sample statistics are shown near the bottom of Table 5-2. The mean of the sample means is 2.7, the mean of the sample medians is 2.7, and so on.

c. The bottom row of Table 5-2 is based on a comparison of the population parameter and results from the sample statistics. For example, the population mean of 1, 2, 5 is $\mu = 2.7$, and the sample means "target" that value of 2.7 in the sense that the mean of the sample means is also 2.7.

INTERPRETATION   Based on the results in Table 5-2, we can see that when using a sample statistic to estimate a population parameter, some statistics are good in the sense that they target the population parameter and are therefore likely to yield good results. Such statistics are called *unbiased estimators*. Other statistics are not so good (because they are *biased estimators*). Here is a summary.

- **Statistics that target population parameters:** Mean, Variance, Proportion

- **Statistics that do not target population parameters:** Median, Range, Standard Deviation

continued

Although the sample standard deviation does not target the population standard deviation, the bias is relatively small in large samples, so s is often used to estimate σ. Consequently, means, proportions, variances, and standard deviations will all be considered as major topics in following chapters, but the median and range will rarely be used.

The key point of this section is to introduce the concept of a sampling distribution of a statistic. Consider the goal of trying to find the mean body temperature of all adults. Because that population is so large, it is not practical to measure the temperature of every adult. Instead, we obtain a sample of body temperatures and use it to estimate the population mean. Data Set 4 in Appendix B includes a sample of 106 such body temperatures, and the mean for that sample is $\bar{x} = 98.20°F$. Conclusions that we make about the population mean temperature of all adults require that we understand the behavior of the sampling distribution of all such sample means. Even though it is not practical to obtain every possible sample and we are stuck with just one sample, we can form some very meaningful and important conclusions about the population of all body temperatures. A major goal of the following sections and chapters is to learn how we can effectively use a sample to form conclusions about a population. In Section 5-5 we consider more details about the sampling distribution of sample means, and in Section 5-6 we consider more details about the sampling distribution of sample proportions.

5-4 Basic Skills and Concepts

1. Survey of Voters Based on a random sample of $n = 400$ voters, the NBC news division predicts that the Democratic candidate for the presidency will get 49% of the vote, but she actually gets 51%. Should we conclude that the survey was done incorrectly? Why or why not?

2. Sampling Distribution of Harry Potter Data Set 14 in Appendix B includes a sample of measured reading levels for 12 pages randomly selected from J. K. Rowling's *Harry Potter and the Sorcerer's Stone*. The mean of the 12 Flesch-Kincaid grade level values is 5.08. The value of 5.08 is one value that is part of a sampling distribution. Describe that sampling distribution.

3. Sampling Distribution of Body Temperatures Data Set 4 in Appendix B includes a sample of 106 body temperatures of adults. If we were to construct a histogram to depict the shape of the distribution of that sample, would that histogram show the shape of a *sampling distribution* of sample means? Why or why not?

4. Sampling Distribution of Survey Results The Gallup organization conducted a poll of 1015 randomly selected students in grades K through 12 and found that 10% attended private or parochial schools.
   a. Is the 10% (or 0.10) result a statistic or a parameter? Explain.
   b. What is the sampling distribution suggested by the given data?
   c. Would you feel more confident in the results if the sample size were 2000 instead of 1015? Why or why not?

Recommended Assignment
Exercises 1–8.
1. No, because of sampling variability, sample proportions will naturally vary from the true population proportion, even if the sampling is done with a perfectly valid procedure.
2. It is the probability distribution for all such sample means. It consists of all values of sample means along with their corresponding probabilities.
3. No, the histogram represents the distribution shape of one sample, but a sampling distribution includes all possible samples of the same size, such as all of the means computed from all possible samples of 106 people.
4. a. Statistic, because it is calculated from a sample, not a population.
   b. It is the probability distribution of all sample proportions found for all possible samples of size 1015. It consists of all sample proportions along with their corresponding probabilities.
   c. Yes. As the sample size increases, the sample statistics tend to vary less and they tend to be closer to the population parameter.

5. Phone Center The Nome Ice Company was in business for only three days (guess why). Here are the numbers of phone calls received on each of those days: 10, 6, 5. Assume that samples of size 2 are randomly selected *with replacement* from this population of three values.

    a. List the 9 different possible samples and find the mean of each of them.

    b. Identify the probability of each sample and describe the sampling distribution of sample means (*Hint:* See Table 5-3.)

    c. Find the mean of the sampling distribution.

    d. Is the mean of the sampling distribution (from part [c]) equal to the mean of the population of the three listed values? Are those means *always* equal?

6. Telemarketing Here are the numbers of sales per day that were made by Kim Ryan, a courteous telemarketer who worked four days before being fired: 1, 11, 9, 3. Assume that samples of size 2 are randomly selected *with replacement* from this population of four values.

    a. List the 16 different possible samples and find the mean of each of them.

    b. Identify the probability of each sample, then describe the sampling distribution of sample means (*Hint:* See Table 5-3.)

    c. Find the mean of the sampling distribution.

    d. Is the mean of the sampling distribution (from part [c]) equal to the mean of the population of the four listed values? Are those means *always* equal?

7. Heights of L.A. Lakers Here are the heights (in inches) of the five starting basketball players for the L.A. Lakers: 85, 79, 82, 73, 78. Assume that samples of size 2 are randomly selected *with replacement* from the above population of five heights.

    a. After identifying the 25 different possible samples, find the mean of each of them.

    b. Describe the sampling distribution of means. (*Hint:* See Table 5-2.)

    c. Find the mean of the sampling distribution.

    d. Is the mean of the sampling distribution (from part [c]) equal to the mean of the population of the five heights listed above? Are those means *always* equal?

8. Military Presidents Here is the population of all five U.S. presidents who had professions in the military, along with their ages at inauguration: Eisenhower (62), Grant (46), Harrison (68), Taylor (64), and Washington (57). Assume that samples of size 2 are randomly selected *with replacement* from the population of five ages.

    a. After identifying the 25 different possible samples, find the mean of each of them.

    b. Describe the sampling distribution of means. (*Hint:* See Table 5-2.)

    c. Find the mean of the sampling distribution.

    d. Is the mean of the sampling distribution (from part [c]) equal to the mean of the population of the five heights listed above? Are those means *always* equal?

9. Genetics A genetics experiment involves a population of fruit flies consisting of 1 male named Mike and 3 females named Anna, Barbara, and Chris. Assume that two fruit flies are randomly selected *with replacement*.

    a. After identifying the 16 different possible samples, find the proportion of females in each of them.

    b. Describe the sampling distribution of proportions of females. (*Hint:* See Table 5-2.)

    c. Find the mean of the sampling distribution.

    d. Is the mean of the sampling distribution (from part [c]) equal to the population proportion of females? Does the mean of the sampling distribution of proportions *always* equal the population proportion?

---

5. a. 10-10; 10-6; 10-5; 6-10; 6-6; 6-5; 5-10; 5-6; 5-5; the means are listed in part (b).

    b. The means 10.0, 8.0, 7.5, 8.0, 6.0, 5.5, 7.5, 5.5, 5.0 each have a probability of 1/9.

    c. 7.0

    d. Yes; yes

6. a. 1-1; 1-11; 1-9; 1-3; 11-1; 11-11; 11-9; 11-3; 9-1; 9-11; 9-9; 9-3; 3-1; 3-11; 3-9; 3-3; the means are listed in part (b).

    b. The means 1, 6, 5, 2, 6, 11, 10, 7, 5, 10, 9, 6, 2, 7, 6, 3, each have a probability of 1/16.

    c. 6.0

    d. Yes; yes

7. a. Means: 85.0, 82.0, 83.5, 79.0, 81.5, 82.0, 79.0, 80.5, 76.0, 78.5, 83.5, 80.5, 82.0, 77.5, 80.0, 79.0, 76.0, 77.5, 73.0, 75.5, 81.5, 78.5, 80.0, 75.5, 78.0

    b. The probability of each mean is 1/25. The sampling distribution consists of the 25 sample means paired with the probability of 1/25.

    c. 79.4

    d. Yes; yes

8. a. Means: 62.0, 54.0, 65.0, 63.0, 59.5, 54.0, 46.0, 57.0, 55.0, 51.5, 65.0, 57.0, 68.0, 66.0, 62.5, 63.0, 55.0, 66.0, 64.0, 60.5, 59.5, 51.5, 62.5, 60.5, 57.0

    b. The sampling distribution consists of the 25 sample means paired with the probability of 1/25.

    c. 59.4

    d. Yes; yes

9. a. 0, 0.5, 0.5, 0.5, 0.5, 1, 1, 1, 0.5, 1, 1, 1, 0.5, 1, 1, 1

    b. The sampling distribution consists of the 16 proportions paired with the probability of 1/16.

    c. 0.75

    d. Yes; yes

10. a. 1, 1, 0.5, 0.5, 0.5, 1, 1, 0.5,
       0.5, 0.5, 0.5, 0.5, 0, 0, 0, 0.5,
       0.5, 0, 0, 0, 0.5, 0.5, 0, 0, 0
    b. The sampling distribution
       consists of the 25 propor-
       tions paired with the proba-
       bility of 1/25.
    c. 0.4
    d. Yes; yes

11. a. Answer varies.
    b. Answer varies, but it must
       be one of these: 0, 0.2, 0.4,
       0.6, 0.8, 1.
    c. Statistic
    d. No; no
    e. It must be 10/13 or 0.769.

12. The mean of the sampling dis-
    tribution is 2/3 and the popula-
    tion proportion is 2/3.

Sample	Sample Proportion	Probability
MM	1	1/3
MT	0.5	2/3

13. a. 59.4; 4.6
    b. 59.4; 3.1
    c. 59.4; 1.9
    d. Yes. Each sampling distribu-
       tion has a mean of 59.4,
       which is the mean of the
       population.
    e. As the sample size in-
       creases, the variation of the
       sampling distribution of
       sample means decreases.

10. Quality Control After constructing a new manufacturing machine, 5 prototype car headlights are produced and it is found that 2 are defective (D) and 3 are acceptable (A). Assume that two headlights are randomly selected *with replacement* from this population.

    a. After identifying the 25 different possible samples, find the proportion of defects in each of them.

    b. Describe the sampling distribution of proportions of defects. (*Hint:* See Table 5-2.)

    c. Find the mean of the sampling distribution.

    d. Is the mean of the sampling distribution (from part [c]) equal to the population proportion of defects? Does the mean of the sampling distribution of proportions *always* equal the population proportion?

11. Women Senators Let a population consist of the 10 Democrats and 3 Republicans who are women senators in the 107th Congress of the United States.

    a. Develop a procedure for randomly selecting (with replacement) a sample of size 5 from the population of 10 Democrats and 3 Republicans, then select such a sample and list the results.

    b. Find the proportion of Democrats in the sample from part (a).

    c. Is the proportion from part (b) a statistic or a parameter?

    d. Does the sample proportion from part (b) equal the population proportion of Democrats? Can any random sample of size 5 result in a sample proportion that is equal to the population proportion?

    e. Assume that all different possible samples of size 5 are listed, and the sample proportion is found for each of them. What can be concluded about the value of the mean of those sample proportions?

12. Women Senators Let a population consist of these home states of all three female Republican senators in the 107th Congress of the United States: Maine, Maine, Texas. Assuming that samples of size 2 are randomly selected from this population *without* replacement, list the different possible samples. Find the probability of each sample. Also, for each sample, find the proportion of senators from Maine. For example, the sample of "Maine and Texas" results in a sample proportion of 1/2 (because one of the two Senators is from Maine). Find the mean of the sampling distribution and verify that it is equal to the population proportion of senators from Maine.

5-4 Beyond the Basics

13. Here is the population of all five U.S. presidents who had professions in the military, along with their age at inauguration: Eisenhower (62), Grant (46), Harrison (68), Taylor (64), and Washington (57). Assume that all samples are selected *without replacement*.

    a. After listing all of the possible samples of size $n = 2$, find the mean and standard deviation of the sample means.

    b. After listing all of the possible samples of size $n = 3$, find the mean and standard deviation of the sample means.

    c. After listing all of the possible samples of size $n = 4$, find the mean and standard deviation of the sample means.

    d. When sampling without replacement, do sample means tend to target the value of the population mean?

    e. Based on the preceding results, how is the *variation* of the sampling distribution of sample means affected by increasing sample size?

14. Mean Absolute Deviation The population of 1, 2, 5 was used to develop Table 5-2. Identify the sampling distribution of the mean absolute deviation (defined in Section 2-5), then determine whether the mean absolute deviation of a sample is a good statistic for estimating the mean absolute deviation of the population.

15. Median as an Estimator In Table 5-2, the sampling distribution of the medians has a mean of 2.7. Because the population mean is also 2.7, it might appear that the median is a good statistic for estimating the value of the population mean. Using the population values 1, 2, 5, find the 27 samples of size $n = 3$ that can be selected with replacement, then find the median and mean for each of the 27 samples. After obtaining those results, find the mean of the sampling distribution of the median, and find the mean of the sampling distribution of the mean. Compare the results to the population mean of 2.7. What do you conclude?

5-5 The Central Limit Theorem

This section is *extremely* important because it presents the central limit theorem, which forms the foundation for estimating population parameters and hypothesis testing—topics discussed at length in the following chapters. As you study this section, try to avoid confusion caused by the fact that the central limit theorem involves two different distributions: the distribution of the original population and the distribution of the sample means.

These key terms and concepts were presented in earlier sections:

- A *random variable* is a variable that has a single numerical value, determined by chance, for each outcome of a procedure. (Section 4-2)

- A *probability distribution* is a graph, table, or formula that gives the probability for each value of a random variable. (Section 4-2)

- The *sampling distribution of the mean* is the probability distribution of sample means, with all samples having the same sample size n. (Section 5-4)

For specific examples of these abstract concepts, see the following example.

EXAMPLE Random Digits Consider the population of digits 0, 1, 2, 3, 4, 5, 6, 7, 8, 9, which are randomly selected with replacement.

a. *Random variable:* If we conduct trials that consist of randomly selecting a single digit, and if we represent the value of the selected digit by x, then x is a random variable (because its value depends on chance).

b. *Probability distribution:* Assuming that the digits are randomly selected, the probability of each digit is $1/10$, which can be expressed as the formula $P(x) = 1/10$. This is a probability distribution (because it describes the probability for each value of the random variable x).

c. *Sampling distribution*: Now suppose that we randomly select all of the different possible samples, each of size $n = 4$. (Remember, we are sampling with replacement, so any particular sample might have the same digit occurring more than once.) In each sample, we calculate the sample mean $\bar{x}$ (which is itself a random variable because its value depends on chance). The probability distribution of the sample means $\bar{x}$ is a sampling distribution.

14. The MAD values are 0, 0.5, 2, 0.5, 0, 1.5, 2, 1.5, 0. They each have probability 1/9. The mean of the sampling distribution is 0.89, but the MAD of the population is 1.6. Because the MAD values do not target the population MAD, a MAD statistic is not good for estimating the population parameter.

15. Medians: 2.5; means: 2.7. The sample means again target the population mean, but the medians do not. The median is not a good statistic for estimating the population mean.

Note to Instructor

The concepts of this section are extremely important for the chapters that follow. Here is a suggested class activity that could help motivate some of those concepts: Ask each student to announce the last four digits of his or her social security number. As they respond, construct a dotplot on the blackboard and show that it depicts a distribution that is approximately uniform. Then ask each student to compute the mean of the same four digits, and proceed to show that the sample means tend to have a bell-shaped distribution. Also, the mean of the sample means should be the same as the mean of the original list of digits. Finally, question them and try to draw out the fact that the amount of variation among the sample means is less than the amount of variation present in the original list of digits.

Part (c) of the example illustrates a specific sampling distribution of sample means. In Section 5-4 we saw that the mean of sample means is equal to the mean of the population, and as the sample size increases, the corresponding sample means tend to vary less. The *central limit theorem* tells us that if the sample size is large enough, the distribution of sample means can be approximated by a *normal distribution*, even if the original population is not normally distributed. Although we discuss a "theorem," we do not include rigorous proofs. Instead, we focus on the *concepts* and how to apply them. Here are the key points that form such an important foundation for the following chapters.

The Fuzzy Central Limit Theorem

In *The Cartoon Guide to Statistics* by Gonick and Smith, the authors describe the Fuzzy Central Limit Theorem as follows: "Data that are influenced by many small and unrelated random effects are approximately normally distributed. This explains why the normal is everywhere: stock market fluctuations, student weights, yearly temperature averages, SAT scores: All are the result of many different effects." People's heights, for example, are the results of hereditary factors, environmental factors, nutrition, health care, geographic region, and other influences which, when combined, produce normally distributed values.

The Central Limit Theorem and the Sampling Distribution of $\bar{x}$

Given:

1. The random variable x has a distribution (which may or may not be normal) with mean μ and standard deviation σ.
2. Simple random samples all of the same size n are selected from the population. (The samples are selected so that all possible samples of size n have the same chance of being selected.)

Conclusions:

1. The distribution of sample means $\bar{x}$ will, as the sample size increases, approach a *normal* distribution.
2. The mean of all sample means is the population mean μ. (That is, the normal distribution from Conclusion 1 has mean μ.)
3. The standard deviation of all sample means is $\sigma/\sqrt{n}$. (That is, the normal distribution from Conclusion 1 has standard deviation $\sigma/\sqrt{n}$.)

Practical Rules Commonly Used

1. If the original population is not itself normally distributed, here is a common guideline: For samples of size n greater than 30, the distribution of the sample means can be approximated reasonably well by a normal distribution. (There are exceptions, such as populations with very nonnormal distributions requiring sample sizes much larger than 30, but such exceptions are relatively rare.) The approximation gets better as the sample size n becomes larger.
2. If the original population is itself normally distributed, then the sample means will be normally distributed for *any* sample size n (not just the values of n larger than 30).

The central limit theorem involves two different distributions: the distribution of the original population and the distribution of the sample means. As in previous chapters, we use the symbols μ and σ to denote the mean and standard deviation of the original population, but we now need new notation for the mean and standard deviation of the distribution of sample means.

Notation for the Sampling Distribution of $\bar{x}$

If all possible random samples of size n are selected from a population with mean μ and standard deviation σ, the mean of the sample means is denoted by $\mu_{\bar{x}}$, so

$$\mu_{\bar{x}} = \mu$$

Also, the standard deviation of the sample means is denoted by $\sigma_{\bar{x}}$, so

$$\sigma_{\bar{x}} = \frac{\sigma}{\sqrt{n}}$$

$\sigma_{\bar{x}}$ is often called the **standard error of the mean.**

EXAMPLE **Random Digits** Again consider the population of digits 0, 1, 2, 3, 4, 5, 6, 7, 8, 9, which are randomly selected with replacement. Assume that we randomly select samples of size $n = 4$. In the original population of digits, all of the values are equally likely. Based on the "Practical Rules Commonly Used" (listed in the central limit theorem box), we *cannot* conclude that the sample means are normally distributed, because the original population does not have a normal distribution and the sample size of 4 is not larger than 30. However, we will explore the sampling distribution to see what can be learned.

Table 5-6 was constructed by recording the last four digits of social security numbers from each of 50 different students. The last four digits of social security numbers are random, unlike the beginning digits, which are used to code particular information. If we combine the four digits from each student into one big collection of 200 numbers, we get a mean of $\bar{x} = 4.5$, a standard deviation of $s = 2.8$, and a distribution with the graph shown in Figure 5-19. Now see what happens when we find the 50 sample means, as shown in Table 5-6. (For example, the first student has digits of 1, 8, 6, and 4, and the mean of these four digits is 4.75.) Even though the original collection of data does *not* have a normal distribution, the sample means have a distribution that is approximately *normal*. This can be a confusing concept, so you should stop right here and study this paragraph until its major point becomes clear: The original set of 200 individual numbers does *not* have a normal distribution (because the digits 0–9 occur with approximately equal frequencies), but the 50 sample means *do* have a distribution that is approximately normal. (One of the "Practical Rules Commonly Used" states that samples with $n > 30$ can be approximated with a normal distribution, but smaller samples, such as $n = 4$ in this example, can sometimes have a distribution that is approximately normal.) It's a truly fascinating and intriguing phenomenon in statistics that by sampling from any distribution, we can create a distribution of sample means that is normal or at least approximately normal.

Table 5-6

SSN digits				$\bar{x}$
1	8	6	4	4.75
5	3	3	6	4.25
9	8	8	8	8.25
5	1	2	5	3.25
9	3	3	5	5.00
4	2	6	2	3.50
7	7	1	6	5.25
9	1	5	4	4.75
5	3	3	9	5.00
7	8	4	1	5.00
0	5	6	1	3.00
9	8	2	2	5.25
6	1	5	7	4.75
8	1	3	0	3.00
5	9	6	9	7.25
6	2	3	4	3.75
7	4	0	7	4.50
5	7	5	6	5.75
4	1	5	7	4.25
1	2	0	6	2.25
4	0	2	8	3.50
3	1	2	5	2.75
0	3	4	0	1.75
1	5	1	0	1.75
9	7	4	0	5.00
7	3	1	1	3.00
9	1	1	3	3.50
8	6	5	9	7.00
5	6	4	1	4.00
9	3	9	5	6.50
6	0	7	3	4.00
8	2	9	6	6.25
0	2	8	6	4.00
2	0	9	7	4.50
5	8	9	0	5.50
6	5	4	9	6.00
4	8	7	6	6.25
7	1	2	0	2.50
2	9	5	0	4.00
8	3	2	2	3.75
2	7	1	6	4.00
6	7	7	1	5.25
2	3	3	9	4.25
2	4	7	5	4.50
5	4	3	7	4.75
0	4	3	8	3.75
2	5	8	6	5.25
7	1	3	4	3.75
8	3	7	0	4.50
5	6	6	7	6.00

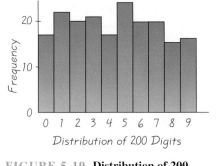

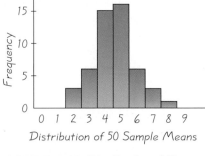

FIGURE 5-19 **Distribution of 200 Digits from Social Security Numbers**

FIGURE 5-20 **Distribution of 50 Sample Means**

Figure 5-20 shows that the distribution of the sample means from the preceding example is approximately normal, even though the original population does not have a normal distribution and the size of $n = 4$ for the individual samples does not exceed 30. If you closely examine Figure 5-20, you can see that it is not an exact normal distribution, but it would become closer to an exact normal distribution as the sample size increases far beyond 4.

As the sample size increases, the sampling distribution of sample means approaches a *normal* distribution.

Applying the Central Limit Theorem

Many important and practical problems can be solved with the central limit theorem. When working on such problems, remember that if the sample size is greater than 30, or if the original population is normally distributed, treat the distribution of sample means as if it were a normal distribution with mean μ and standard deviation $\sigma / \sqrt{n}$.

In the following example, part (a) involves an *individual* value, but part (b) involves the mean for a *sample* of 36 women, so we must use the central limit theorem in working with the random variable $\bar{x}$. Study this example carefully to understand the significant difference between the procedures used in parts (a) and (b). See how this example illustrates the following working procedure:

- **When working with an *individual* value from a normally distributed population, use the methods of Section 5-3. Use $z = \dfrac{x - \mu}{\sigma}$.**

- **When working with a mean for some *sample* (or group), be sure to use the value of $\sigma / \sqrt{n}$ for the standard deviation of the sample means. Use $z = \dfrac{\bar{x} - \mu}{\sigma / \sqrt{n}}$.**

EXAMPLE Ski Gondola Safety In the Chapter Problem we noted that a ski gondola in Vail, Colorado, carries skiers to the top of a mountain. It bears a plaque stating that the maximum capacity is 12 people or 2004 pounds. That capacity will be exceeded if 12 people have weights with a mean greater than $2004/12 = 167$ pounds. Because men tend to weigh more than women, a "worst case" scenario involves 12 passengers who are all men. Men have weights that are normally distributed with a mean of 172 lb and a standard deviation of 29 lb (based on data from the National Health Survey).

a. Find the probability that if an individual man is randomly selected, his weight will be greater than 167 pounds.

b. Find the probability that 12 randomly selected men will have a mean that is greater than 167 pounds (so that their total weight is greater than the gondola maximum capacity of 2004 lb).

SOLUTION

a. *Approach: Use the methods presented in Section 5-3 (because we are dealing with an individual value from a normally distributed population).* We seek the area of the green-shaded region in Figure 5-21(a). Before using Table A-2, we convert the weight of 167 to the corresponding z score:

$$z = \frac{x - \mu}{\sigma} = \frac{167 - 172}{29} = -0.17$$

We now refer to Table A-2 using $z = -0.17$ and find that the cumulative area to the left of 167 lb is 0.4325. The green-shaded region is therefore $1 - 0.4325 = 0.5675$. The probability of a randomly selected man weighing more than 167 lb is 0.5675.

b. *Approach: Use the central limit theorem (because we are dealing with the mean for a sample of 12 men, not an individual man).* Even though the sample size is not greater than 30, we use a normal distribution for this reason: The original population of men has a normal distribution, so samples of any size will yield means that are normally distributed. Because we are now dealing with a distribution of sample means, we must use the parameters $\mu_{\bar{x}}$ and $\sigma_{\bar{x}}$, which are evaluated as follows:

$$\mu_{\bar{x}} = \mu = 172$$

$$\sigma_{\bar{x}} = \frac{\sigma}{\sqrt{n}} = \frac{29}{\sqrt{12}} = 8.37158$$

continued

Note to Instructor

Instead of doing an example illustrating the central limit theorem alone, it is helpful for students to see how such an example is different from the examples of earlier sections. Consider doing a class example similar to the one given here. Stress that when dealing with the mean for a *group* of values, it is essential to modify the standard deviation by dividing by the square root of the sample size n.

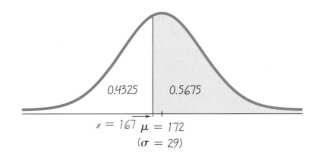

FIGURE 5-21(a) **Distribution of Individual Men's Weights**

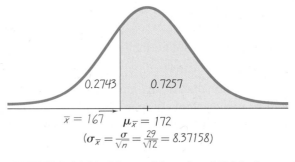

FIGURE 5-21(b) **Means of Samples of 12 Men's Weights**

Here is a really important point: We must use the computed standard deviation of 8.37158, not the original standard deviation of 29 (because we are working with the distribution of sample means for which the standard deviation is 8.37158, not the distribution of individual weights for which the standard deviation is 29). We want to find the green-shaded area shown in Figure 5-21(b). Using Table A-2, we find the relevant z score, which is calculated as follows:

$$z = \frac{\bar{x} - \mu_{\bar{x}}}{\sigma_{\bar{x}}} = \frac{167 - 172}{\dfrac{29}{\sqrt{12}}} = \frac{-5}{8.37158} = -0.60$$

Referring to Table A-2, we find that $z = -0.60$ corresponds to a cumulative left area of 0.2743, so the green-shaded region is $1 - 0.2743 = 0.7257$. The probability that the 12 men have a mean weight greater than 167 lb is 0.7257.

INTERPRETATION   There is a 0.5675 probability that an individual man will weigh more than 167 lb, and there is a 0.7257 probability that 12 men will have a mean weight of more than 167 lb. Given that the gondola maximum capacity is 2004 lb, it is likely (with probability 0.7257) to be overloaded if it is filled with 12 randomly selected men. However, passenger safety isn't quite so bad because of factors such as these: (1) Male skiers probably have a mean weight less than the mean of 172 lb for the general population of men; (2) women skiers are also likely to be passengers, and they tend to weigh less than men; (3) even though the maximum capacity is listed as 2004 lb, the gondola is designed to operate safely for weights well above that conservative load of 2004 lb. However, the gondola operators would be wise to avoid a load of 12 men, especially if they all appear to have high weights. The calculations used here are exactly the type of calculations used by engineers when they design ski lifts, elevators, escalators, airplanes, and other devices that carry people.

Interpreting Results

The next example illustrates another application of the central limit theorem, but carefully examine the conclusion that is reached. This example shows the type of thinking that is the basis for the important procedure of hypothesis testing (discussed in Chapter 7). This example illustrates the rare event rule for inferential statistics, first presented in Section 3-1.

Rare Event Rule

If, under a given assumption, the probability of a particular observed event is exceptionally small, we conclude that the assumption is probably not correct.

EXAMPLE Body Temperatures Assume that the population of human body temperatures has a mean of 98.6°F, as is commonly believed. Also assume that the population standard deviation is 0.62°F (based on data from University of Maryland researchers). If a sample of size $n = 106$ is randomly selected, find the probability of getting a mean of 98.2°F or lower. (The value of 98.2°F was actually obtained; see the midnight temperatures for Day 2 in Data Set 4 of Appendix B.)

SOLUTION   We weren't given the distribution of the population, but because the sample size $n = 106$ exceeds 30, we use the central limit theorem and conclude that the distribution of sample means is a normal distribution with these parameters:

$$\mu_{\bar{x}} = \mu = 98.6 \qquad \text{(by assumption)}$$

$$\sigma_{\bar{x}} = \frac{\sigma}{\sqrt{n}} = \frac{0.62}{\sqrt{106}} = 0.0602197$$

Figure 5-22 shows the shaded area (see the tiny left tail of the graph) corresponding to the probability we seek. Having already found the parameters that apply to the distribution shown in Figure 5-22, we can now find the shaded area by using the same procedures developed in Section 5-3. Using Table A-2, we first find the z score:

$$z = \frac{\bar{x} - \mu_{\bar{x}}}{\sigma_{\bar{x}}} = \frac{98.20 - 98.6}{0.0602197} = -6.64$$

continued

FIGURE 5-22 Distribution of Mean Body Temperatures for Samples of Size $n = 106$

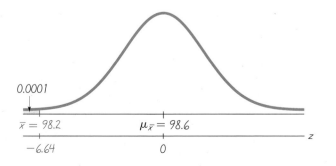

0.0001

$\bar{x} = 98.2$ $\mu_{\bar{x}} = 98.6$

-6.64 0 z

Referring to Table A-2, we find that $z = -6.64$ is off the chart, but for values of z below -3.49, we use an area of 0.0001 for the cumulative left area up to $z = -3.49$. We therefore conclude that the shaded region in Figure 5-22 is 0.0001. (More precise tables or TI-83 Plus calculator results indicate that the area of the shaded region is closer to 0.00000000002, but even those results are only approximations. We can safely report that the probability is quite small, such as less than 0.001.)

INTERPRETATION   The result shows that if the mean of our body temperatures is really 98.6°F, then there is an extremely small probability of getting a sample mean of 98.2°F or lower when 106 subjects are randomly selected. University of Maryland researchers did obtain such a sample mean, and there are two possible explanations: Either the population mean really is 98.6°F and their sample represents a chance event that is extremely rare, or the population mean is actually lower than 98.6°F and so their sample is typical. Because the probability is so low, it seems more reasonable to conclude that the population mean is lower than 98.6°F. This is the type of reasoning used in *hypothesis testing*, to be introduced in Chapter 7. For now, we should focus on the use of the central limit theorem for finding the probability of 0.0001, but we should also observe that this theorem will be used later in developing some very important concepts in statistics.

Correction for a Finite Population

In applying the central limit theorem, our use of $\sigma_{\bar{x}} = \sigma/\sqrt{n}$ assumes that the population has infinitely many members. When we sample with replacement (that is, put back each selected item before making the next selection), the population is effectively infinite. Yet many realistic applications involve sampling without replacement, so successive samples depend on previous outcomes. In manufacturing, quality-control inspectors typically sample items from a finite production run without replacing them. For such a finite population, we may need to adjust $\sigma_{\bar{x}}$. Here is a common rule of thumb:

When sampling without replacement and the sample size n is greater than 5% of the finite population size N (that is, $n > 0.05N$), adjust the standard deviation of sample means $\sigma_{\bar{x}}$ by multiplying it by the *finite population correction factor*:

$$\sqrt{\frac{N-n}{N-1}}$$

Except for Exercises 21 and 22, the examples and exercises in this section assume that the finite population correction factor does *not* apply, because we are sampling with replacement or the population is infinite or the sample size doesn't exceed 5% of the population size.

The central limit theorem is so important because it allows us to use the basic normal distribution methods in a wide variety of different circumstances. In Chapter 6, for example, we will apply the theorem when we use sample data to estimate means of populations. In Chapter 7 we will apply it when we use sample data to test claims made about population means. Such applications of estimating population parameters and testing claims are extremely important uses of statistics, and the central limit theorem makes them possible.

5-5 Basic Skills and Concepts

Using the Central Limit Theorem. *In Exercises 1–6, assume that men's weights are normally distributed with a mean given by* $\mu = 172$ *lb and a standard deviation given by* $\sigma = 29$ *lb (based on data from the National Health Survey).*

1. a. If 1 man is randomly selected, find the probability that his weight is less than 167 lb.
   b. If 36 men are randomly selected, find the probability that they have a mean weight less than 167 lb.

2. a. If 1 man is randomly selected, find the probability that his weight is greater than 180 lb.
   b. If 100 men are randomly selected, find the probability that they have a mean weight greater than 180 lb.

3. a. If 1 man is randomly selected, find the probability that his weight is between 170 lb and 175 lb.
   b. If 64 men are randomly selected, find the probability that they have a mean weight between 170 and 175 lb.

4. a. If 1 man is randomly selected, find the probability that his weight is between 100 lb and 165 lb.
   b. If 81 men are randomly selected, find the probability that they have a mean weight between 100 lb and 165 lb.

5. a. If 25 men are randomly selected, find the probability that they have a mean weight greater than 160 lb.
   b. Why can the central limit theorem be used in part (a), even though the sample size does not exceed 30?

6. a. If 4 men are randomly selected, find the probability that they have a mean weight between 160 lb and 180 lb.
   b. Why can the central limit theorem be used in part (a), even though the sample size does not exceed 30?

7. Redesign of Ejection Seats In the Chapter Problem, it was noted that engineers were redesigning fighter jet ejection seats to better accommodate women. Before women became fighter jet pilots, the ACES-II ejection seats were designed for men weighing between 140 lb and 211 lb. The population of women has normally distributed weights with a mean of 143 lb and a standard deviation of 29 lb (based on data from the National Health Survey).
   a. If 1 woman is randomly selected, find the probability that her weight is between 140 lb and 211 lb.
   b. If 36 different women are randomly selected, find the probability that their mean weight is between 140 lb and 211 lb.
   c. When redesigning the fighter jet ejection seats to better accommodate women, which probability is more relevant: the result from part (a) or the result from part (b)? Why?

Recommended Assignment
Exercises 1–16. (Exercises 21 and 22 require the finite population correction factor.)

1. a. 0.4325
   b. 0.1515
2. a. 0.3897
   b. 0.0029

3. a. 0.0677
   b. 0.5055

4. a. 0.3986
   b. 0.0149

5. a. 0.9808
   b. If the original population has a normal distribution, the central limit theorem provides good results for any sample size.
6. a. 0.5055
   b. If the original population has a normal distribution, the central limit theorem provides good results for any sample size.
7. a. 0.5302
   b. 0.7323
   c. Part (a), because the seats will be occupied by individual women, not groups of women.

8. a. 0.5793
   b. 0.9772
   c. Although the mean head
      breadth of 100 men is very
      likely to be less than 6.2 in.,
      there could be many individ-
      ual men that could not use
      the helmets because they
      have head breadths greater
      than 6.2 in. Based on the re-
      sult from part (a), these hel-
      mets would not fit about
      42% of men.
9. a. 0.0119
   b. No; yes

10. 0.0078; it would be unusual be-
    cause its probability is only
    0.0078.

11. a. 0.0001
    b. No, but consumers are not
       being cheated because the
       cans are being overfilled, not
       underfilled.

12. a. 0.0139
    b. 0.0001
    c. No, the mean can be 133
       while some individual scores
       are below 131.5.

13. a. 0.0051
    b. Yes

8. Designing Motorcycle Helmets Engineers must consider the breadths of male heads when designing motorcycle helmets. Men have head breadths that are normally distributed with a mean of 6.0 in. and a standard deviation of 1.0 in. (based on anthropometric survey data from Gordon, Churchill, et al.).
   a. If one male is randomly selected, find the probability that his head breadth is less than 6.2 in.
   b. The Safeguard Helmet company plans an initial production run of 100 helmets. Find the probability that 100 randomly selected men have a mean head breadth less than 6.2 in.
   c. The production manager sees the result from part (b) and reasons that all helmets should be made for men with head breadths less than 6.2 in., because they would fit all but a few men. What is wrong with that reasoning?

9. Designing a Roller Coaster The Rock 'n' Roller Coaster at Disney–MGM Studios in Orlando has two seats in each row. When designing that roller coaster, the total width of the two seats in each row had to be determined. In the "worst case" scenario, both seats are occupied by men. Men have hip breadths that are normally distributed with a mean of 14.4 in. and a standard deviation of 1.0 in. (based on anthropometric survey data from Gordon, Churchill, et al). Assume that two male riders are randomly selected.
   a. Find the probability that their mean hip width is greater than 16.0 in.
   b. If each row of two seats is designed to fit two men only if they have a mean hip breadth of 16.0 in. or less, would too many riders be unable to fit? Does this design appear to be acceptable?

10. Uniform Random-Number Generator The random-number generator on the TI-83 Plus calculator and many other calculators and computers yields numbers from a uniform distribution of values between 0 and 1, with a mean of 0.500 and a standard deviation of 0.289. If 100 random numbers are generated, find the probability that their mean is greater than 0.57. Would it be *unusual* to generate 100 such numbers and get a mean greater than 0.57? Why or why not?

11. Amounts of Coke Assume that cans of Coke are filled so that the actual amounts have a mean of 12.00 oz and a standard deviation of 0.11 oz.
   a. Find the probability that a sample of 36 cans will have a mean amount of at least 12.19 oz, as in Data Set 17 in Appendix B.
   b. Based on the result from part (a), is it reasonable to believe that the cans are actually filled with a mean of 12.00 oz? If the mean is not 12.00 oz, are consumers being cheated?

12. IQ Scores Membership in Mensa requires an IQ score above 131.5. Nine candidates take IQ tests, and their summary results indicated that their mean IQ score is 133. (IQ scores are normally distributed with a mean of 100 and a standard deviation of 15.)
   a. If 1 person is randomly selected, find the probability of getting someone with an IQ score of at least 133.
   b. If 9 people are randomly selected, find the probability that their mean IQ score is at least 133.
   c. Although the summary results are available, the individual IQ test scores have been lost. Can it be concluded that all 9 candidates have IQ scores above 133 so that they are all eligible for Mensa membership?

13. Mean Replacement Times The manager of the Portland Electronics store is concerned that his suppliers have been giving him TV sets with lower than average quality. His research shows that replacement times for TV sets have a mean of 8.2 years and a standard deviation of 1.1 years (based on data from "Getting Things Fixed," *Consumer Reports*). He then randomly selects 50 TV sets sold in the past and finds that the mean replacement time is 7.8 years.

   a. Assuming that TV replacement times have a mean of 8.2 years and a standard deviation of 1.1 years, find the probability that 50 randomly selected TV sets will have a mean replacement time of 7.8 years or less.

   b. Based on the result from part (a), does it appear that the Portland Electronics store has been given TV sets with lower than average quality?

14. Blood Pressure For women aged 18–24, systolic blood pressures (in mm Hg) are normally distributed with a mean of 114.8 and a standard deviation of 13.1 (based on data from the National Health Survey). Hypertension is commonly defined as a systolic blood pressure above 140.

   a. If a woman between the ages of 18 and 24 is randomly selected, find the probability that her systolic blood pressure is greater than 140.

   b. If 4 women in that age bracket are randomly selected, find the probability that their mean systolic blood pressure is greater than 140.

   c. Given that part (b) involves a sample size that is not larger than 30, why can the central limit theorem be used?

   d. If a physician is given a report stating that 4 women have a mean systolic blood pressure below 140, can she conclude that none of the women have hypertension (with a blood pressure greater than 140)?

15. Reduced Nicotine in Cigarettes The amounts of nicotine in Dytusoon cigarettes have a mean of 0.941 g and a standard deviation of 0.313 g (based on Data Set 5 in Appendix B). The Huntington Tobacco Company, which produces Dytusoon cigarettes, claims that it has now reduced the amount of nicotine. The supporting evidence consists of a sample of 40 cigarettes with a mean nicotine amount of 0.882 g.

   a. Assuming that the given mean and standard deviation have not changed, find the probability of randomly selecting 40 cigarettes with a mean of 0.882 g or less.

   b. Based on the result from part (a), is it valid to claim that the amount of nicotine is lower? Why or why not?

16. Coaching for the SAT Test Scores for men on the verbal portion of the SAT-I test are normally distributed with a mean of 509 and a standard deviation of 112 (based on data from the College Board). Randomly selected men are given the Columbia Review Course before taking the SAT test. Assume that the course has no effect.

   a. If 1 of the men is randomly selected, find the probability that his score is at least 590.

   b. If 16 of the men are randomly selected, find the probability that their mean score is at least 590.

   c. In finding the probability for part (b), why can the central limit theorem be used even though the sample size does not exceed 30?

   d. If the random sample of 16 men does result in a mean score of 590, is there strong evidence to support the claim that the course is actually effective? Why or why not?

17. Overloading of Waste Disposal Facility The town of Newport operates a rubbish waste disposal facility that is overloaded if its 4872 households discard waste with weights having a mean that exceeds 27.88 lb in a week. For many different weeks, it is found that the samples of 4872 households have weights that are normally distributed with a mean of 27.44 lb and a standard deviation of 12.46 lb (based on data

14. a. 0.0274
    b. 0.0001
    c. Because the original population is normally distributed, the sampling distribution of sample means will be normally distributed for any sample size.
    d. No, the mean can be less than 140 while individual values are above 140.

15. a. 0.1170
    b. No, because the probability of 0.1170 shows that it is easy to get a mean such as 0.882 g, assuming that the nicotine amounts have not been changed.

16. a. 0.2358
    b. 0.0019
    c. Because the original population is normally distributed, the sampling distribution of sample means will be normally distributed for any sample size.
    d. Yes, because the probability of 0.0019 shows that it is highly unlikely that by chance, a randomly selected group would get a mean as high as 590.

17. 0.0069; level is acceptable.

from the Garbage Project at the University of Arizona). What is the proportion of weeks in which the waste disposal facility is overloaded? Is this an acceptable level, or should action be taken to correct a problem of an overloaded system?

18. a. 0.5675
    b. 0.9999
    c. Yes

18. Labeling of M&M Packages M&M plain candies have a mean weight of 0.9147 g and a standard deviation of 0.0369 g (based on Data Set 19 in Appendix B). The M&M candies used in Data Set 19 came from a package containing 1498 candies, and the package label stated that the net weight is 1361 g. (If every package has 1498 candies, the mean weight of the candies must exceed $1361/1498 = 0.9085$ g for the net contents to weigh at least 1361 g.)
   a. If 1 M&M plain candy is randomly selected, find the probability that it weighs more than 0.9085 g.
   b. If 1498 M&M plain candies are randomly selected, find the probability their mean weight is at least 0.9085 g.
   c. Given these results, does it seem that the Mars Company is providing M&M consumers with the amount claimed on the label?

19. 2979 lb

19. Elevator Design Women's weights are normally distributed with a mean of 143 lb and a standard deviation of 29 lb, and men's weights are normally distributed with a mean of 172 lb and a standard deviation of 29 lb (based on data from the National Health Survey). You need to design an elevator for the Westport Shopping Center, and it must safely carry 16 people. Assuming a worst case scenario of 16 male passengers, find the maximum total allowable weight if we want a 0.975 probability that this maximum will not be exceeded when 16 males are randomly selected.

20. a. 267.5 in.
    b. Because college football
       players tend to be larger
       than randomly selected men,
       the given mean and standard
       deviation don't apply. The
       bench should be longer than
       the 267.5 in. length found in
       part (b).

20. Seating Design You need to build a bench that will seat 18 male college football players, and you must first determine the length of the bench. Men have hip breadths that are normally distributed with a mean of 14.4 in. and a standard deviation of 1.0 in.
   a. What is the minimum length of the bench if you want a 0.975 probability that it will fit the combined hip breadths of 18 randomly selected men?
   b. What would be wrong with actually using the result from part (a) as the bench length?

5-5 Beyond the Basics

21. a. 0.9750
    b. 1329 lb

21. Correcting for a Finite Population The Boston Women's club needs an elevator limited to 8 passengers. The club has 120 women members with weights that approximate a normal distribution with a mean of 143 lb and a standard deviation of 29 lb. (*Hint:* See the discussion of the finite population correction factor.)
   a. If 8 different members are randomly selected, find the probability that their total weight will not exceed the maximum capacity of 1300 lb.
   b. If we want a 0.99 probability that the elevator will not be overloaded whenever 8 members are randomly selected as passengers, what should be the maximum allowable weight?

22. a. $\mu = 8.0, \sigma = 5.4$
    b. 2,3 2,6 2,8 2,11 2,18 3,6
       3,8 3,11 3,18 6,8 6,11 6,18
       8,11 8,18 11,18
    c. 2.5, 4.0, 5.0, 6.5, 10.0,
       4.5, 5.5, 7.0, 10.5, 7.0, 8.5,
       12.0, 9.5, 13.0, 14.5
    d. $\mu_{\bar{x}} = 8.0, \sigma_{\bar{x}} = 3.4$

22. Population Parameters A *population* consists of these values: 2, 3, 6, 8, 11, 18.
   a. Find μ and σ.
   b. List all samples of size $n = 2$ that can be obtained without replacement.
   c. Find the population of all values of $\bar{x}$ by finding the mean of each sample from part (b).

d. Find the mean $\mu_{\bar{x}}$ and standard deviation $\sigma_{\bar{x}}$ for the population of sample means found in part (c).

e. Verify that

$$\mu_{\bar{x}} = \mu \quad \text{and} \quad \sigma_{\bar{x}} = \frac{\sigma}{\sqrt{n}} \sqrt{\frac{N-n}{N-1}}$$

23. Uniform Random-Number Generator In Exercise 10 it was noted that many calculators and computers have a random-number generator that yields numbers from a uniform distribution of values between 0 and 1, with a mean of 0.500 and a standard deviation of 0.289. If 100 random numbers are generated, find the probability that their mean is between 0.499 and 0.501. If we did generate 100 such numbers and found the mean to be between 0.499 and 0.501, can we conclude that the result is "unusual" so that the random number generator is somehow defective? Why or why not?

23. 0.0240. With $n = 100$, no sample mean between 0.499 and 0.501 meets the criterion of being "at least as extreme" as the value of the sample mean that was obtained, so we do not conclude that the random number generator is somehow defective.

5-6 Normal as Approximation to Binomial

Instead of "Normal as Approximation to Binomial," the proper title for this section should be "Using the Normal Distribution as an Approximation to the Binomial Distribution," but the shorter title has more pizzazz. The longer title does a better job of conveying the purpose of this section. Let's begin by reviewing the conditions required for a *binomial probability distribution*, which was introduced in Section 4-3:

1. The procedure must have a *fixed number of trials*.

2. The trials must be *independent*.

3. Each trial must have all outcomes classified into *two categories*.

4. The probabilities must remain *constant* for each trial.

In Section 4-3 we presented three methods for finding binomial probabilities: (1) using the binomial probability formula, (2) using Table A-1, and (3) using software (such as STATDISK, Minitab, or Excel) or a TI-83 Plus calculator. In many cases, however, none of those methods is practical, because the calculations require too much time and effort. We now present a new method, which uses a normal distribution as an approximation of a binomial distribution. The following box summarizes the key point of this section.

Note to Instructor

This section is another interesting application of the normal distribution, and it follows up on the earlier work done with the binomial distribution. Using appropriate software or a TI-83 Plus calculator, we can now solve many more binomial distribution problems directly without using a normal approximation, so the methods of this section are not as necessary as they once were.

Normal Distribution as Approximation to Binomial Distribution

If $np \geq 5$ and $nq \geq 5$, then the binomial random variable has a probability distribution that can be approximated by a normal distribution with the mean and standard deviation given as

$$\mu = np$$
$$\sigma = \sqrt{npq}$$

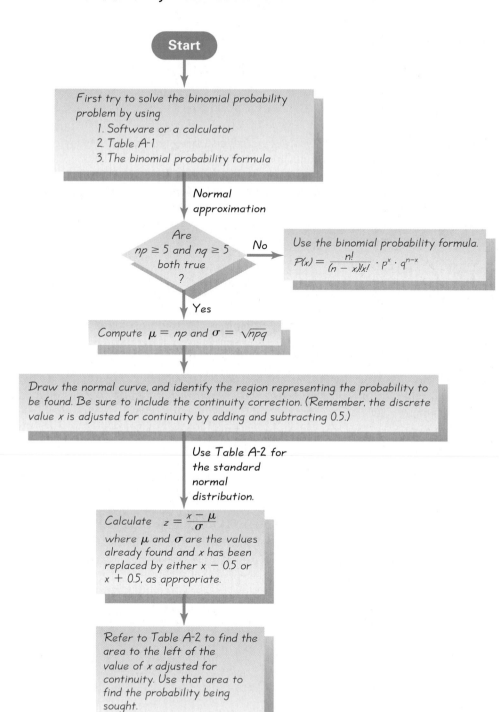

Start

First try to solve the binomial probability problem by using
    1. Software or a calculator
    2. Table A-1
    3. The binomial probability formula

Normal approximation

Are
$np \geq 5$ and $nq \geq 5$
both true
?

No → Use the binomial probability formula.
$$P(x) = \frac{n!}{(n - x)!x!} \cdot p^x \cdot q^{n-x}$$

Yes

Compute $\mu = np$ and $\sigma = \sqrt{npq}$

Draw the normal curve, and identify the region representing the probability to be found. Be sure to include the continuity correction. (Remember, the discrete value x is adjusted for continuity by adding and subtracting 0.5.)

Use Table A-2 for the standard normal distribution.

Calculate $z = \frac{x - \mu}{\sigma}$
where μ and σ are the values already found and x has been replaced by either $x - 0.5$ or $x + 0.5$, as appropriate.

Refer to Table A-2 to find the area to the left of the value of x adjusted for continuity. Use that area to find the probability being sought.

FIGURE 5-23 Using a Normal Approximation to a Binomial Distribution

To better understand how the normal distribution can be used to approximate a binomial distribution, refer to Figure 5-18 in Section 5-4. That figure is a relative frequency histogram for values of 10,000 sample *proportions*, where each of the 10,000 samples consists of 50 genders randomly selected with replacement from a population in which the proportion of females is 0.13. Those sample proportions can be considered to be binomial probabilities, so Figure 5-18 shows that under suitable conditions, binomial probabilities have a sampling distribution that is approximately normal. The formal justification that allows us to use the normal distribution as an approximation to the binomial distribution results from more advanced mathematics, but Figure 5-18 is a convincing visual argument supporting that approximation.

When solving binomial probability problems, first try to get more exact results by using computer software or a calculator, or Table A-1, or the binomial probability formula. If the binomial probability cannot be found using those more exact procedures, try the technique of using the normal distribution as an approximation to the binomial distribution. This approach involves the following procedure, which is also shown as a flowchart in Figure 5-23.

Procedure for Using a Normal Distribution to Approximate a Binomial Distribution

1. Establish that the normal distribution is a suitable approximation to the binomial distribution by verifying that $np \geq 5$ and $nq \geq 5$. (If these conditions are not both satisfied, then you must use software, or a calculator, or Table A-1, or calculations with the binomial probability formula.)

2. Find the values of the parameters μ and σ by calculating $\mu = np$ and $\sigma = \sqrt{npq}$.

3. Identify the discrete value x (the number of successes). Change the *discrete* value x by replacing it with the *interval* from $x - 0.5$ to $x + 0.5$. (For further clarification, see the discussion under the subheading "Continuity Corrections" found later in this section.) Draw a normal curve and enter the values of μ, σ, and either $x - 0.5$ or $x + 0.5$, as appropriate.

4. Change x by replacing it with $x - 0.5$ or $x + 0.5$, as appropriate.

5. Using $x - 0.5$ or $x + 0.5$ (as appropriate) in place of x, find the area corresponding to the desired probability by first finding the z score: $z = (x - \mu)/\sigma$. Now use that z score to find the area to the left of either $x - 0.5$ or $x + 0.5$, as appropriate. That area can now be used to identify the area corresponding to the desired probability.

We will illustrate this normal approximation procedure with the following example.

EXAMPLE **Loading Airliners** When an airliner is loaded with passengers, baggage, cargo, and fuel, the pilot must verify that the gross weight is below the maximum allowable limit, and the weight must be properly distributed so that the balance of the aircraft is within safe acceptable limits. Air America
continued

Boys and Girls Are Not Equally Likely

In many probability calculations, good results are obtained by assuming that boys and girls are equally likely to be born. In reality, a boy is more likely to be born (with probability 0.5117) than a girl (with probability 0.4883). These results are based on recent data from the National Center for Health Statistics, which showed that the 4,058,814 births in one year included 2,076,969 boys and 1,981,845 girls. Researchers monitor these probabilities for changes that might suggest such factors as changes in the environment and exposure to chemicals.

Multiple Lottery Winners

has established a procedure whereby extra cargo must be reduced whenever a plane filled with 200 passengers includes at least 120 men. Find the probability that among 200 randomly selected passengers, there are at least 120 men. Assume that the population of potential passengers consists of an equal number of men and women.

SOLUTION   Refer to Figure 5-23 for the procedure followed in this solution. The given problem does involve a binomial distribution with a fixed number of trials ($n = 200$), which are presumably independent, two categories (man, woman) of outcome for each trial, and a probability of a male ($p = 0.5$) that presumably remains constant from trial to trial.

We will assume that neither software nor a calculator is available. Table A-1 does not apply, because it stops at $n = 15$. The binomial probability formula is not practical, because we would have to use it 81 times (once for each value of x from 120 to 200 inclusive), and nobody in their right mind would want to do that.

Let's proceed with the five-step approach of using a normal distribution to approximate the binomial distribution.

Step 1: We must first verify that it is reasonable to approximate the binomial distribution by the normal distribution because $np \geq 5$ and $nq \geq 5$. With $n = 200$, $p = 0.5$, and $q = 1 - p = 0.5$, we verify the required conditions as follows:

$$np = 200 \cdot 0.5 = 100 \qquad \text{(Therefore } np \geq 5.)$$
$$nq = 200 \cdot 0.5 = 100 \qquad \text{(Therefore } nq \geq 5.)$$

Step 2: We now proceed to find the values for μ and σ that are needed for the normal distribution. We get the following:

$$\mu = np = 200 \cdot 0.5 = 100$$
$$\sigma = \sqrt{npq} = \sqrt{200 \cdot 0.5 \cdot 0.5} = 7.0710678$$

Step 3: The discrete value of 120 is represented by the vertical strip bounded by 119.5 and 120.5. (See the discussion of continuity corrections, which follows this example.)

Step 4: Because we want the probability of *at least* 120 men, we want the area representing the discrete number of 120 (the region bounded by 119.5 and 120.5), as well as the area to the right, as shown in Figure 5-24.

Step 5: We can now proceed to find the shaded area of Figure 5-24 by using the same methods used in Section 5-3. In order to use Table A-2 for the standard normal distribution, we must first convert 119.5 to a z score, then use the table to find the area to the left of 119.5, which is then subtracted from 1. The z score is found as follows:

$$z = \frac{x - \mu}{\sigma} = \frac{119.5 - 100}{7.0710678} = 2.76$$

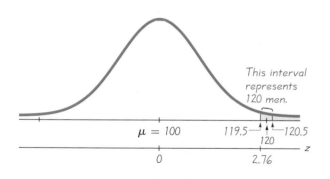

FIGURE 5-24 Finding the Probability of "At Least" 120 Men Among 200 Passengers

Using Table A-2, we find that $z = 2.76$ corresponds to an area of 0.9971, so the shaded region is $1 - 0.9971 = 0.0029$.

INTERPRETATION There is a 0.0029 probability of getting at least 120 men among 200 passengers. Because that probability is so small, we can conclude that a roster of 200 passengers will rarely include at least 120 men, so the reduction of extra cargo is not something to be very concerned about.

Continuity Corrections

The procedure for using a normal distribution to approximate a binomial distribution includes a step in which we change a discrete number to an interval that is 0.5 below and 0.5 above the discrete number. See the preceding solution, where we changed 120 to the interval between 119.5 and 120.5. This particular step, called a *continuity correction,* is usually difficult to understand, so we will now consider it in more detail.

Definition

When we use the normal distribution (which is a *continuous* probability distribution) as an approximation to the binomial distribution (which is *discrete*), a **continuity correction** is made to a discrete whole number x in the binomial distribution by representing the single value x by the *interval* from $x - 0.5$ to $x + 0.5$ (that is, adding and subtracting 0.5).

The following practical suggestions should help you use continuity corrections properly.

Procedure for Continuity Corrections

1. When using the normal distribution as an approximation to the binomial distribution, *always* use the continuity correction. (It is required because we are using the *continuous* normal distribution to approximate the *discrete* binomial distribution.)

2. In using the continuity correction, first identify the discrete whole number x that is relevant to the binomial probability problem. For example, if you're

trying to find the probability of getting at least 120 men in 200 randomly selected people, the discrete whole number of concern is $x = 120$. First focus on the x value itself, and temporarily ignore whether you want at least x, more than x, fewer than x, or whatever.

3. Draw a normal distribution centered about μ, then draw a *vertical strip area* centered over x. Mark the left side of the strip with the number equal to $x - 0.5$, and mark the right side with the number equal to $x + 0.5$. For $x = 120$, for example, draw a strip from 119.5 to 120.5. *Consider the entire area of the strip to represent the probability of the discrete number x itself.*

4. Now determine whether the value of x itself should be included in the probability you want. (For example, "at least x" does include x itself, but "more than x" does not include x itself.) Next, determine whether you want the probability of at least x, at most x, more than x, fewer than x, or exactly x. Shade the area to the right or left of the strip, as appropriate; also shade the interior of the strip itself *if and only if x itself* is to be included. This total shaded region corresponds to the probability being sought.

To see how this procedure results in continuity corrections, see the common cases illustrated in Figure 5-25. Those cases correspond to the statements in the following list.

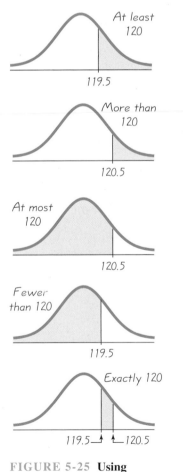

FIGURE 5-25 Using Continuity Corrections

Statement	Area
At least 120 (includes 120 and above)	To the *right* of 119.5
More than 120 (doesn't include 120)	To the *right* of 120.5
At most 120 (includes 120 and below)	To the *left* of 120.5
Fewer than 120 (doesn't include 120)	To the *left* of 119.5
Exactly 120	Between 119.5 and 120.5

EXAMPLE TV Ratings The CBS television show *60 Minutes* recently had a share of 20, meaning that among the TV sets in use, 20% were tuned to *60 Minutes* (based on data from Nielsen Media Research). An advertiser wants to verify that 20% share value by conducting its own survey of 200 households with TV sets in use at the time of a *60 Minutes* broadcast. The results show that among the 200 sets in use, 16% (or 32 sets) are tuned to *60 Minutes.* Assuming that the share value of 20% is correct, find the probability that in a survey of 200 households, *exactly* 32 sets are tuned to *60 Minutes.* Given that the sample result of 16% is less than the claimed share value of 20%, is there strong evidence to conclude that the share value of 20% is wrong?

SOLUTION   We have $n = 200$ independent trials, and $x = 32$ sets tuned to *60 Minutes*, and a population proportion of $p = 0.20$. For the purposes of this example, we assume that neither computer software nor a TI-83 Plus calculator is available. Table A-1 cannot be used, because $n = 200$ exceeds the largest table value of $n = 15$. If we were to use the binomial probability formula, we would need to evaluate an expression that includes 200!, but many calculators and software programs can't handle that. We therefore proceed by using a normal distribution to approximate the binomial distribution.

Step 1: We begin by checking to determine whether the approximation is suitable:

$$np = 200 \cdot 0.20 = 40 \qquad \text{(Therefore } np \geq 5.)$$
$$nq = 200 \cdot 0.80 = 160 \qquad \text{(Therefore } nq \geq 5.)$$

Step 2: We now proceed to find the values for μ and σ that are needed for the normal distribution. We get the following:

$$\mu = np = 200 \cdot 0.20 = 40$$
$$\sigma = \sqrt{npq} = \sqrt{200 \cdot 0.20 \cdot 0.80} = 5.6568542$$

Step 3: We draw the normal curve shown in Figure 5-26. The shaded region of the figure represents the probability we want. Use of the continuity correction results in the representation of 32 by the region between 31.5 and 32.5.

Step 4: Here is the approach used to find the shaded region in Figure 5-26: First find the total area to the left of 32.5, then find the total area to the left of 31.5, then find the *difference* between those two areas. Beginning with the total area to the left of 32.5, we must first find the z score corresponding to 32.5, then refer to Table A-2. We get

$$z = \frac{32.5 - 40}{5.6568542} = -1.33$$

We use Table A-2 to find that $z = -1.33$ corresponds to a probability of 0.0918, which is the total area to the left of 32.5. Now we proceed to find the area to the left of 31.5 by first finding the z score corresponding to 31.5:

$$z = \frac{31.5 - 40}{5.6568542} = -1.50$$

continued

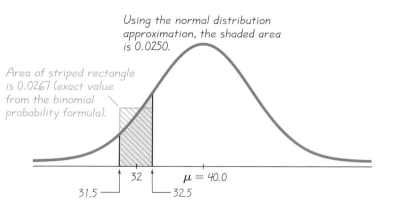

Using the normal distribution approximation, the shaded area is 0.0250.

Area of striped rectangle is 0.0267 (exact value from the binomial probability formula).

31.5   32   32.5 $\mu = 40.0$

FIGURE 5-26 **Using Continuity Correction for TV Market Share**

We use Table A-2 to find that $z = -1.50$ corresponds to a probability of 0.0668, which is the total area to the left of 31.5. The shaded area is $0.0918 - 0.0668 = 0.0250$.

INTERPRETATION The probability of exactly 32 sets tuned to *60 Minutes* (out of 200) is approximately 0.0250. The statement of the problem also asks us to determine whether the sample result of 16% constitutes strong evidence to conclude that the share value of 20% is wrong. However, instead of considering the probability of *exactly* 32 sets tuned to *60 Minutes*, we must consider the probability of *32 or fewer*. [In Section 4-2, we noted that x successes among n trials is an *unusually low* number of successes if $P(x$ or fewer$)$ is very small, such as 0.05 or less.] From the above solution, we see that the probability of 32 or fewer successes is P(less than 32.5), which is 0.0918. Because 0.0918 is not very small, we do *not* have strong evidence to conclude that the share value of 20% is wrong.

If we solve the preceding example using STATDISK, Minitab, or a calculator, we get a result of 0.0267, but the normal approximation method resulted in a value of 0.0250. The discrepancy of 0.0017 occurs because the use of the normal distribution results in an *approximate* value that is the area of the shaded region in Figure 5-26, whereas the exact correct area is a rectangle centered above 32. (Figure 5-26 illustrates this discrepancy.) The area of the rectangle is 0.0267, but the area of the approximating shaded region is 0.0250.

Interpreting Results

In reality, when we use a normal distribution as an approximation to a binomial distribution, our ultimate goal is not simply to find a probability number. We often need to make some *judgment* based on the probability value, as in the final conclusion of the preceding example. We should understand that low probabilities correspond to events that are very unlikely, whereas large probabilities correspond to likely events. The probability value of 0.05 is often used as a cutoff to distinguish between unlikely events and likely events. The following criterion (from Section 4-2) describes the use of probabilities for distinguishing between results that could easily occur by chance and those results that are highly unusual.

Using Probabilities to Determine When Results Are Unusual

- **Unusually high:** x successes among n trials is an *unusually high* number of successes if $P(x$ or more$)$ is very small (such as 0.05 or less)
- **Unusually low:** x successes among n trials is an *unusually low* number of successes if $P(x$ or fewer$)$ is very small (such as 0.05 or less)

5-6 Basic Skills and Concepts

Applying Continuity Correction. In Exercises 1–8, the given values are discrete. Use the continuity correction and describe the region of the normal distribution that corresponds to the indicated probability. For example, the probability of "more than 20 defective items" corresponds to the area of the normal curve described with this answer: "the area to the right of 20.5".

Recommended Assignment
Exercises 1–22.

1. Probability of more than 15 people in prison for removing caution labels from pillows

2. Probability of at least 24 students understanding continuity correction

3. Probability of fewer than 100 passengers on your next commercial flight

4. Probability that the number of working vending machines in the United States is exactly 27

5. Probability of no more than 4 absent students in a statistics class

6. Probability that the number of defective Wayne Newton CDs is between 15 and 20 inclusive

7. Probability that the number of absent U.S. senators is between 8 and 10 inclusive

8. Probability of exactly 3 "yes" responses to date requests

Using Normal Approximation. In Exercises 9–12, do the following. (a) Find the indicated binomial probability by using Table A-1 in Appendix A. (b) If np ≥ 5 and nq ≥ 5, also estimate the indicated probability by using the normal distribution as an approximation to the binomial distribution; if np < 5 or nq < 5, then state that the normal approximation is not suitable.

9. With $n = 14$ and $p = 0.5$, find $P(9)$.

10. With $n = 12$ and $p = 0.8$, find $P(7)$.

11. With $n = 15$ and $p = 0.9$, find P(at least 14).

12. With $n = 13$ and $p = 0.4$, find P(fewer than 3).

13. **Probability of More than 55 Girls** Estimate the probability of getting more than 55 girls in 100 births. Assume that boys and girls are equally likely. Is it unusual to get more than 55 girls in 100 births?

14. **Probability of at Least 65 Girls** Estimate the probability of getting at least 65 girls in 100 births. Assume that boys and girls are equally likely. Is it unusual to get at least 65 girls in 100 births?

15. **Probability of at Least Passing** Estimate the probability of passing a true/false test of 100 questions if 60% (or 60 correct answers) is the minimum passing grade and all responses are random guesses. Is that probability high enough to risk passing by using random guesses instead of studying?

16. **Multiple-Choice Test** A multiple-choice test consists of 25 questions with possible answers of a, b, c, d, and e. Estimate the probability that with random guessing, the number of correct answers is between 3 and 10 inclusive.

17. **Mendel's Hybridization Experiment** When Mendel conducted his famous hybridization experiments, he used peas with green pods and yellow pods. One experiment involved crossing peas in such a way that 25% (or 145) of the 580 offspring peas were expected to have yellow pods. Instead of getting 145 peas with yellow pods, he obtained 152. Assuming that Mendel's 25% rate is correct, estimate the probability of getting at least 152 peas with yellow pods among the 580 offspring peas. Is there strong evidence to suggest that Mendel's rate of 25% is wrong?

18. **Cholesterol Reducing Drug** The probability of flu symptoms for a person not receiving any treatment is 0.019. In a clinical trial of Lipitor, a common drug used to lower cholesterol, 863 patients were given a treatment of 10 mg Atorvastatin tablets, and 19 of those patients experienced flu symptoms (based on data from Pfizer, Inc.).

1. The area to the right of 15.5

2. The area to the right of 23.5

3. The area to the left of 99.5

4. The area between 26.5 and 27.5

5. The area to the left of 4.5

6. The area between 14.5 and 20.5

7. The area between 7.5 and 10.5

8. The area between 2.5 and 3.5

9. Table: 0.122; normal approximation: 0.1218

10. Table: 0.053; normal approximation is not suitable.

11. Table: 0.549; normal approximation is not suitable.

12. Table: 0.057; normal approximation: 0.0630

13. 0.1357; no

14. 0.0019; yes

15. 0.0287; no

16. 0.8914

17. 0.2676; no

18. 0.3015; Lipitor users do not appear to be more likely to have flu symptoms.

Assuming that these tablets have no effect on flu symptoms, estimate the probability that at least 19 of 863 people experience flu symptoms. What do these results suggest about flu symptoms as an adverse reaction to the drug?

19. Probability of at Least 50 Color-Blind Men Nine percent of men and 0.25% of women cannot distinguish between the colors red and green. This is the type of color blindness that causes problems with traffic signals. Researchers need at least 50 men with this type of color blindness, so they randomly select 600 men for a study of traffic-signal perceptions. Estimate the probability that at least 50 of the men cannot distinguish between red and green. Is the result high enough so that the researchers can be very confident of getting at least 50 men with red and green color blindness?

20. Cell Phones and Brain Cancer In a study of 420,000 cell phone users in Denmark, it was found that 135 developed cancer of the brain or nervous system. Assuming that cell phones have no effect, there is a 0.000340 probability of a person developing cancer of the brain or nervous system. We therefore expect about 143 cases of such cancer in a group of 420,000 randomly selected people. Estimate the probability of 135 or fewer cases of such cancer in a group of 420,000 people. What do these results suggest about media reports that cell phones cause cancer of the brain or nervous system?

21. Overbooking Flights Air America is considering a new policy of booking as many as 400 persons on an airplane that can seat only 350. (Past studies have revealed that only 85% of the booked passengers actually arrive for the flight.) Estimate the probability that if Air America books 400 persons, not enough seats will be available. Is that probability low enough to be workable, or should the policy be changed?

22. On-Time Flights Recently, American Airlines had 72.3% of its flights arriving on time (based on data from the U.S. Department of Transportation). In a check of 40 randomly selected American Airlines flights, 19 arrived on time. Estimate the probability of getting 19 or fewer on-time flights among 40, assuming that the 72.3% rate is correct. Is it unusual to get 19 or fewer on-time flights among 40 randomly selected American Airlines flights?

23. Identifying Gender Discrimination After being rejected for employment, Kim Kelly learns that the Bellevue Advertising Company has hired only 21 women among its last 62 new employees. She also learns that the pool of applicants is very large, with an equal number of qualified men and women. Help her in her charge of gender discrimination by estimating the probability of getting 21 or fewer women when 62 people are hired, assuming no discrimination based on gender. Does the resulting probability really support such a charge?

24. M&M Candies: Are 10% Blue? According to a consumer affairs representative from Mars (the candy company, not the planet), 10% of all M&M plain candies are blue. Data Set 19 in Appendix B shows that among 100 M&Ms chosen, 5 are blue. Estimate the probability of randomly selecting 100 M&Ms and getting 5 or fewer that are blue. Assume that the company's 10% blue rate is correct. Based on the result, is it very unusual to get 5 or fewer blue M&Ms when 100 are randomly selected?

25. Blood Group Forty-five percent of us have Group O blood, according to data provided by the Greater New York Blood Program. Providence Memorial Hospital is conducting a blood drive because its supply of Group O blood is low, and it needs 177 donors of Group O blood. If 400 volunteers donate blood, estimate the probability that the number with Group O blood is at least 177. Is the pool of 400 volunteers likely to be sufficient?

26. Acceptance Sampling We stated in Section 3-4 that some companies monitor quality by using a method of acceptance sampling, whereby an entire batch of items is re-

---

19. 0.7389; no, not very confident.

20. 0.2709; media reports appear to be wrong.

21. 0.0708; yes.

22. 0.0004; yes

23. 0.0080; yes

24. 0.0668; not very unusual

25. 0.6368; the pool is likely to be sufficient, but the probability should be much higher. It would be better to increase the pool of volunteers.

26. 0.9505; yes

jcctcd if a random sample of a particular size includes more than some specified number of defects. The Dayton Machine Company buys machine bolts in batches of 5000 and rejects a batch if, when 50 of them are sampled, at least 2 defects are found. Estimate the probability of rejecting a batch if the supplier is manufacturing the bolts with a defect rate of 10%. Is this monitoring plan likely to identify the unacceptable rate of defects?

27. Car Crashes For drivers in the 20–24 age bracket, there is a 34% rate of car accidents in one year (based on data from the National Safety Council). An insurance investigator finds that in a group of 500 randomly selected drivers aged 20–24 living in New York City, 40% had accidents in the last year. If the 34% rate is correct, estimate the probability that in a group of 500 randomly selected drivers, at least 40% had accidents in the last year. Based on that result, is there strong evidence supporting the claim that the accident rate in New York City is higher than 34%?

27. 0.0026; yes

28. Cloning Survey A recent Gallup poll consisted of 1012 randomly selected adults who were asked whether "cloning of humans should or should not be allowed." Results showed that 89% of those surveyed indicated that cloning should not be allowed. A news reporter wants to determine whether these survey results constitute strong evidence that the majority (more than 50%) of people are opposed to such cloning. Assuming that 50% of all people are opposed, estimate the probability of getting at least 89% opposed in a survey of 1012 randomly selected people. Based on that result, is there strong evidence supporting the claim that the majority is opposed to such cloning?

28. 0.0001; yes

5-6 Beyond the Basics

29. Winning at Roulette Marc Taylor plans to place 200 bets of $1 each on the number 7 at roulette. A win pays off with odds of 35:1 and, on any one spin, there is a probability of 1/38 that 7 will be the winning number. Among the 200 bets, what is the minimum number of wins needed for Marc to make a profit? Estimate the probability that Marc will make a profit.

29. 6; 0.4602

30. Replacement of TVs Replacement times for TV sets are normally distributed with a mean of 8.2 years and a standard deviation of 1.1 years (based on data from "Getting Things Fixed," *Consumer Reports*). Estimate the probability that for 250 randomly selected TV sets, at least 15 of them have replacement times greater than 10.0 years.

30. 0.2946

31. Joltin' Joe Assume that a baseball player hits .350, so his probability of a hit is 0.350. (Ignore the complications caused by walks.) Also assume that his hitting attempts are independent of each other.
    a. Find the probability of at least 1 hit in 4 tries in 1 game.
    b. Assuming that this batter gets up to bat 4 times each game, estimate the probability of getting a total of at least 56 hits in 56 games.
    c. Assuming that this batter gets up to bat 4 times each game, find the probability of at least 1 hit in each of 56 consecutive games (Joe DiMaggio's 1941 record).
    d. What minimum batting average would be required for the probability in part (c) to be greater than 0.1?

31. a. 0.821
    b. 0.9993
    c. 0.0000165
    d. 0.552

32. Overbooking Flights Vertigo Airlines works only with advance reservations and experiences a 7% rate of no-shows. How many reservations could be accepted for an airliner with a capacity of 250 if there is at least a 0.95 probability that all reservation holders who show will be accommodated?

32. 262

Determining Normality

The following chapters include some very important statistical methods requiring that the sample data be randomly selected from a population that has a *normal* distribution. It therefore becomes necessary to determine whether sample data appear to come from a population that is normally distributed. In this section we introduce the *normal quantile plot* as one tool that can help us determine whether the requirement of a normal distribution appears to be satisfied.

> ### Definition
>
> A **normal quantile plot** is a graph of points (x, y) where each x value is from the original set of sample data, and each y value is a z score corresponding to a quantile value of the standard normal distribution. (See Step 3 in the following procedure for details on finding these z scores.)

Procedure for Determining Whether Data Have a Normal Distribution

1. *Histogram:* Construct a histogram. Reject normality if the histogram departs dramatically from a bell shape.

2. *Outliers:* Identify outliers. Reject normality if there is more than one outlier present. (Just one outlier could be an error or the result of chance variation, but be careful, because even a single outlier can have a dramatic effect on results.)

3. *Normal quantile plot:* If the histogram is basically symmetric and there is at most one outlier, construct a *normal quantile plot*. The following steps describe the construction of a normal quantile plot, but the procedure is messy enough so that we usually use software or a calculator to generate the graph. The end of this section includes instructions for using STATDISK, Minitab, Excel, and a TI-83 Plus calculator to obtain normal quantile plots.

   a. First sort the data by arranging the values in order from lowest to highest.

   b. With a sample of size n, each value represents a proportion of $1/n$ of the sample. Using the known sample size n, identify the areas of $1/2n$, $3/2n$, $5/2n$, $7/2n$, and so on. These are the cumulative areas to the left of the corresponding sample values.

   c. Use the standard normal distribution (Table A-2) to find the z scores corresponding to the cumulative left areas found in Step (b).

   d. Match the original sorted data values with their corresponding z scores found in Step (c), then plot the points (x, y), where each x is an original sample value and y is the corresponding z score.

   e. Examine the normal quantile plot using these criteria: If the points do not lie close to a straight line, or if the points exhibit some systematic pattern that is not a straight-line pattern, then the data appear to come from a population that does *not* have a normal distribution. If the pattern of the points is reasonably close to a straight line, then the data appear to come from a population that has a normal distribution.

Steps 1 and 2 are straightforward, but we illustrate the construction of a normal quantile plot (Step 3) in the following example.

> **EXAMPLE Ages of Presidents** Exercises 8 and 13 in Section 5-4 include these five ages at inauguration of the U.S. presidents who had professions in the military: 62, 46, 68, 64, 57. Construct a normal quantile plot for those ages and determine whether they appear to come from a population that is normally distributed.
>
> **SOLUTION**   The following steps correspond to those listed in the above procedure for constructing a normal quantile plot.
>
> 1. First, sort the data by arranging them in order. We get 46, 57, 62, 64, 68.
> 2. With a sample of size $n = 5$, each value represents a proportion of $1/5$ of the sample, so we proceed to identify the cumulative areas to the left of the corresponding sample values. Those cumulative left areas, which are expressed in general as $1/2n, 3/2n, 5/2n, 7/2n$, and so on, become these specific areas for this example with $n = 5$: $1/10, 3/10, 5/10, 7/10$, and $9/10$. Those same cumulative left areas expressed in decimal form are 0.1, 0.3, 0.5, 0.7, and 0.9.
> 3. We now search the body of Table A-2 for the cumulative left areas of 0.1000, 0.3000, 0.5000, 0.7000, and 0.9000. We find these corresponding z scores: $-1.28, -0.52, 0, 0.52$, and 1.28.
> 4. We now pair the original sorted ages with their corresponding z scores, and we get these (x, y) coordinates, which are plotted in Figure 5-27: $(46, -1.28)$, $(57, -0.52)$, $(62, 0)$, $(64, 0.52)$, and $(68, 1.28)$.
>
> **INTERPRETATION** We examine the normal quantile plot in Figure 5-27. Because the points appear to lie reasonably close to a straight line, we conclude that the given age data appear to come from a normally distributed population.

Because the construction of a normal quantile plot requires that we sort the sample data, then use a somewhat complicated process to find corresponding z scores,

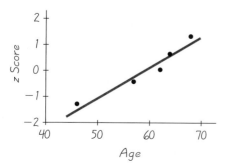

FIGURE 5-27 Normal Quantile Plot of Ages of Presidents

manual construction of the graph is difficult with large data sets. The next example illustrates the use of computer software.

EXAMPLE **Boston Rainfall** In Data Set 11 of Appendix B, use the 52 rainfall amounts listed for Sundays in Boston and test for normality.

SOLUTION

Step 1: Construct a histogram. The accompanying Minitab screen display includes the histogram of the 52 rainfall amounts, and that histogram is extremely skewed, suggesting that those amounts are not normally distributed.

Minitab   **Histogram of Boston Rainfall Amounts for Sunday**

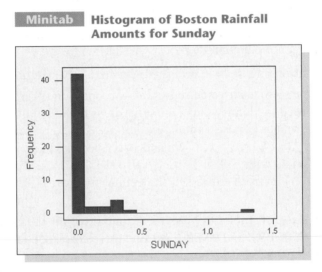

Step 2: Identify outliers. Examining the list of 52 rainfall amounts, we find that 1.28 in. appears to be the only outlier. Because there is only one outlier, we make no conclusion about the normality of the data based on outliers.

Step 3: Construct a normal quantile plot. The accompanying Minitab display includes a normal *probability* plot. (Because many data values are the

Minitab   **Normal Probability Plot of Boston Rainfall Amounts for Sunday**

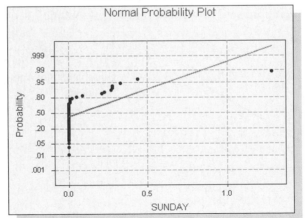

same, the original normal probability plot showed only 12 distinct points instead of 52, so the graph was modified to display all 52 points.) The normal probability plot is the same as a normal quantile plot, except for the scale shown on the vertical axis. A normal probability plot is interpreted with the same criteria as a normal quantile plot. Examination of the normal probability plot reveals a pattern that is very different from a straight-line pattern, suggesting that the data are not from a normally distributed population.

INTERPRETATION Because the histogram does not appear to be bell-shaped, and because the normal probability plot does not yield a pattern of points that reasonably approximates a straight line, we conclude that rainfall amounts in Boston on Sunday do not appear to be normally distributed. Some of the statistical procedures in later chapters require that sample data be normally distributed, but that requirement is not satisfied for the Boston rainfall amounts on Sunday, so those procedures cannot be used.

Here are a few final comments about procedures for determining whether data are from a normally distributed population:

- If the requirement of a normal distribution is not too strict, examination of a histogram and consideration of outliers may be all that you need to determine normality.

- Normal quantile plots can be difficult to construct on your own, but they can be generated with a TI-83 Plus calculator or suitable software, such as STATDISK, Minitab, and Excel.

- In addition to the procedures discussed in this section, there are other more advanced procedures, such as the chi-square goodness-of-fit test, the Kolmogorov-Smirnov test, and the Lilliefors test. (See "Beyond Basic Statistics with the Graphing Calculator, Part I: Assessing Goodness-of-Fit," by Calzada and Scariano, *Mathematics and Computer Education*.)

Using Technology

STATDISK   First enter the data in the STATDISK Data Window. Select **Data** from the main menu bar at the top, then select **Normal Quantile Plot.** Select the column containing the data, then click on the **Plot** button.

Minitab   Minitab can be used to generate a *normal probability plot,* which can be interpreted the same way as the normal quantile plot. That is, normally distributed data should lie close to a straight line. First enter the values in column C1, then select **Stat, Basic Statistics,** and **Normality Test.** Enter C1 for the variable, then click on **OK.**

Excel   The Data Desk XL add-in can be used to generate a *normal probability plot,* which can be interpreted the same

continued

way as a normal quantile plot. First enter the sample values in column A, then click on **DDXL.** (If DDXL does not appear on the Menu bar, install the Data Desk XL add-in.) Select **Charts and Plots,** then select the function type of **Normal Probability Plot.** Click on the pencil icon for "Quantitative Variable," then enter the range of values, such as A1:A36. Click **OK.**

TI-83 Plus   The TI-83 Plus calculator can be used to generate a normal quantile plot as follows: First enter the sample data in list L1, press **2nd** and the **Y =** key (for **STAT PLOT**), then press **ENTER.** Select **ON,** select the "type" item that is the last item in the second row of options, enter **L1** for the data list. After making all selections, press **ZOOM,** then **9.**

5-7 Basic Skills and Concepts

Recommended Assignment
Exercises 1–8. (Exercises 1–4 involve the interpretation of given graphs. Exercises 5–8 are based on raw data, but do not require the construction of normal quantile plots. Exercises 9–12 require the use of technology to generate normal quantile plots. Consider assigning Exercise 15, which involves a manual construction of a normal quantile plot.)

1. Not normal
2. Not normal
3. Not normal
4. Normal

Interpreting Normal Quantile Plots.   *In Exercises 1–4, examine the normal quantile plot and determine whether it depicts data that have a normal distribution.*

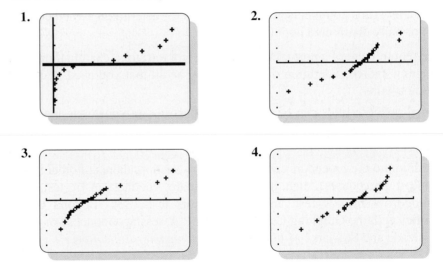

Determining Normality.   *In Exercises 5–8, refer to the indicated data set and determine whether the requirement of a normal distribution is satisfied. Assume that this requirement is loose in the sense that the population distribution need not be exactly normal, but it must be a distribution that is basically symmetric with only one mode.*

5. Not normal

6. Normal

7. Normal

8. Not normal

5. Boston Rainfall The Wednesday rainfall amounts for Boston, as listed in Data Set 11 in Appendix B.

6. Head Circumferences The head circumferences of males, as listed in Data Set 3 in Appendix B.

7. Weights of M&Ms The weights of the brown M&M candies, as listed in Data Set 19 in Appendix B.

8. Water Conductivity The Florida Everglades conductivity levels, as listed in Data Set 12 in Appendix B.

Generating Normal Quantile Plots. *In Exercises 9–12, use the data from the indicated exercise in this section. Use a TI-83 Plus calculator or software (such as STATDISK, Minitab, or Excel) capable of generating normal quantile plots or normal probability plots. Generate the graph, then determine whether the data come from a normally distributed population.*

9. Exercise 5

10. Exercise 6

11. Exercise 7

12. Exercise 8

13. Comparing Data Sets Using the heights of women and the cholesterol levels of women, as listed in Data Set 1 in Appendix B, analyze each of the two data sets and determine whether each appears to come from a normally distributed population. Compare the results and give a possible explanation for any notable differences between the two distributions.

14. Comparing Data Sets Using the systolic blood pressure levels and the elbow breadths of women, as listed in Data Set 1 in Appendix B, analyze each of the two data sets and determine whether each appears to come from a normally distributed population. Compare the results and give a possible explanation for any notable differences between the two distributions.

Constructing Normal Quantile Plots. *In Exercises 15 and 16, use the given data values and identify the corresponding z scores that are used for a normal quantile plot, then construct the normal quantile plot and determine whether the data appear to be from a population with a normal distribution.*

15. Heights of L.A. Lakers Use this sample of heights (in inches) of the players in the starting lineup for the L.A. Lakers professional basketball team: 85, 79, 82, 73, 78.

16. Monitoring Lead in Air On the days immediately following the destruction caused by the terrorist attacks on September 11, 2001, lead amounts (in micrograms per cubic meter) in the air were recorded at Building 5 of the World Trade Center site, and these values were obtained: 5.40, 1.10, 0.42, 0.73, 0.48, 1.10.

5-7 Beyond the Basics

17. Using Standard Scores When constructing a normal quantile plot, suppose that instead of finding z scores using the procedure described in this section, each value in a sample is converted to its corresponding standard score using $z = (x - \bar{x})/s$. If the (x, z) points are plotted in a graph, can this graph be used to determine whether the sample comes from a normally distributed population? Explain.

18. Lognormal Distribution The random variable x is said to have a *lognormal distribution* if the values of $\ln x$ are normally distributed. Test the following phone call times (in seconds) for normality, then test for normality of the natural logarithms of the times. What do you conclude?

31.5	75.9	31.8	87.4	54.1	72.2	138.1	47.9	210.6	127.7
160.8	51.9	57.4	130.3	21.3	403.4	75.9	93.7	454.9	55.1

9. Not normal
10. Normal
11. Normal
12. Not normal
13. Heights appear to be normal, but cholesterol levels do not appear to be normal. Cholesterol levels are strongly affected by diet, and diets might vary in dramatically different ways that do not yield normally distributed results.
14. Elbow breadths appear to be normal, but systolic blood pressure levels do not appear to be normal. Systolic blood pressure levels are affected by diet, and diets might vary in dramatically different ways that do not yield normally distributed results.
15. $-1.28, -0.52, 0, 0.52, 1.28$; normal

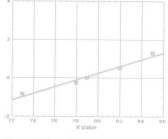

16. $-1.38, -0.67, -0.21, 0.21, 0.67, 1.38$; not normal

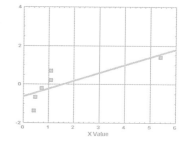

17. No, the transformation to z scores involves subtracting a constant and dividing by a constant, so the plot of the (x, z) points will always be a straight line, regardless of the nature of the distribution.
18. The original values are not from a normal distribution, but the logarithms of the values do appear to come from a normal distribution.

Review

We introduced the concept of probability distributions in Chapter 4, but included only *discrete* distributions. In this chapter we introduced *continuous* probability distributions and focused on the most important category: normal distributions. Normal distributions will be used extensively in the following chapters.

Normal distributions are approximately bell-shaped when graphed. The total area under the density curve of a normal distribution is 1, so there is a convenient correspondence between areas and probabilities. Specific areas can be found using Table A-2 or a TI-83 Plus calculator or software. (We do not use Formula 5-1, the equation that is used to define the normal distribution.)

In this chapter we presented important methods for working with normal distributions, including those that use the standard score $z = (x - \mu)/\sigma$ for solving problems such as these:

- Given that IQ scores are normally distributed with $\mu = 100$ and $\sigma = 15$, find the probability of randomly selecting someone with an IQ above 90.
- Given that IQ scores are normally distributed with $\mu = 100$ and $\sigma = 15$, find the IQ score separating the bottom 85% from the top 15%.

In Section 5-4 we introduced the concept of a sampling distribution. The sampling distribution of the mean is the probability distribution of sample means, with all samples having the same sample size n. The sampling distribution of the proportion is the probability distribution of sample proportions, with all samples having the same sample size n. In general, the sampling distribution of any statistic is the probability distribution of that statistic.

In Section 5-5 we presented the following important points associated with the central limit theorem:

1. The distribution of sample means will, as the sample size n increases, approach a normal distribution.
2. The mean of the sample means is the population mean μ.
3. The standard deviation of the sample means is $\sigma/\sqrt{n}$.

In Section 5-6 we noted that we can sometimes approximate a binomial probability distribution by a normal distribution. If both $np \geq 5$ and $nq \geq 5$, the binomial random variable x is approximately normally distributed with the mean and standard deviation given as $\mu = np$ and $\sigma = \sqrt{npq}$. Because the binomial probability distribution deals with discrete data and the normal distribution deals with continuous data, we apply the continuity correction, which should be used in normal approximations to binomial distributions.

Finally, in Section 5-7 we presented a procedure for determining whether sample data appear to come from a population that has a normal distribution. Some of the statistical methods covered later in this book have a loose requirement of a normally distributed population. In such cases, examination of a histogram and outliers might be all that is needed. In other cases, normal quantile plots might be necessary because of a very strict requirement that the population must have a normal distribution.

Review Exercises

1. High Cholesterol Levels The serum cholesterol levels in men aged 18–24 are normally distributed with a mean of 178.1 and a standard deviation of 40.7. Units are mg/100 mL, and the data are based on the National Health Survey.
   a. If one man aged 18–24 is randomly selected, find the probability that his serum cholesterol level is greater than 260, a value considered to be "moderately high."
   b. If one man aged 18–24 is randomly selected, find the probability that his serum cholesterol level is between 170 and 200.
   c. If 9 men aged 18–24 are randomly selected, find the probability that their mean serum cholesterol level is between 170 and 200.
   d. The Providence Health Maintenance Organization wants to establish a criterion for recommending dietary changes if cholesterol levels are in the top 3%. What is the cutoff for men aged 18–24?

2. Babies at Risk Weights of newborn babies in the United States are normally distributed with a mean of 3420 g and a standard deviation of 495 g (based on data from "Birth Weight and Prenatal Mortality," by Wilcox et al., *Journal of the American Medical Association,* Vol. 273, No. 9).
   a. A newborn weighing less than 2200 g is considered to be at risk, because the mortality rate for this group is at least 1%. What percentage of newborn babies are in the "at-risk" category? If the Chicago General Hospital has 900 births in a year, how many of those babies are in the "at-risk" category?
   b. If we redefine a baby to be at risk if his or her birth weight is in the lowest 2%, find the weight that becomes the cutoff separating at-risk babies from those who are not at risk.
   c. If 16 newborn babies are randomly selected, find the probability that their mean weight is greater than 3700 g.
   d. If 49 newborn babies are randomly selected, find the probability that their mean weight is between 3300 g and 3700 g.

3. Blue Genes Some couples have genetic characteristics configured so that one-quarter of all their offspring have blue eyes. A study is conducted of 100 couples believed to have those characteristics, with the result that 19 of their 100 offspring have blue eyes. Assuming that one-quarter of all offspring have blue eyes, estimate the probability that among 100 offspring, 19 or fewer have blue eyes. Based on that probability, does it seem that the one-quarter rate is wrong? Why or why not?

4. Marine Corps Height Requirements for Men The U.S. Marine Corps requires that men have heights between 64 in. and 78 in. (The National Health Survey shows that heights of men are normally distributed with a mean of 69.0 in. and a standard deviation of 2.8 in.)
   a. Find the percentage of men meeting those height requirements. Are too many men denied the opportunity to join the Marines because they are too short or too tall?
   b. If you are appointed to be the Secretary of Defense and you want to change the requirement so that only the shortest 2% and tallest 2% of all men are rejected, what are the new minimum and maximum height requirements?
   c. If 64 men are randomly selected, find the probability that their mean height is greater than 68.0 in.

1. a. 0.0222
   b. 0.2847
   c. 0.6720
   d. 254.6

2. a. 0.69% of 900 = 6.21 babies
   b. 2405 g
   c. 0.0119
   d. 0.9553

3. 0.1020; no, assuming that the correct rate is 25%, there is a high probability (0.1020) that 19 or fewer offspring will have blue eyes. Because the observed event could easily occur by chance, there is not strong evidence against the 25% rate.

4. a. 0.9626
   b. 63.3 in., 74.7 in.
   c. 0.9979

5. Uniform Distribution The San Francisco Supply Company has designed a machine that fills containers of coffee in such a way that the contents are *uniformly* distributed with a minimum of 11.8 oz and a maximum of 12.2 oz. If one container is randomly selected, find the probability that the amount of coffee is
a. less than 12.0 oz.
b. between 11.2 oz and 12.7 oz.
c. greater than 12.2 oz.
d. between 11.9 oz and 12.0 oz.

6. Sampling Distributions
a. Many different samples of size 100 are randomly selected from the weights of cars currently registered in the United States. What can be concluded about the shape of the distribution of the means of the different samples?
b. If the weights of all cars registered in the United States have a standard deviation of 512 lb, what is the standard deviation of the sample means that are found from many different samples of size 100?
c. Many different samples of size 1200 are randomly selected from the population of all adults in the United States. In each sample, the proportion of people who voted in the last election is recorded. What can be concluded about the shape of the distribution of those sample proportions?

7. Gender Discrimination When several women are not hired at the Telektronics Company, they do some research and find that among the many people who applied, 30% were women. However, the 20 people who were hired consist of only 2 women and 18 men. Find the probability of randomly selecting 20 people from a large pool of applicants (30% of whom are women) and getting 2 or fewer women. Based on the result, does it appear that the company is discriminating based on gender?

8. Testing for Normality Refer to the weights of sugar packets, as listed in Data Set 28 in Appendix B. Do those weights appear to come from a population that has a normal distribution? Explain.

Cumulative Review Exercises

1. Eye Measurement Statistics The listed sample distances (in millimeters) were obtained by using a pupilometer to measure the distances between the pupils of adults (based on data collected by a student of the author).

    67   66   59   62   63   66   66   55

a. Find the mean $\bar{x}$ of the distances in this sample.
b. Find the median of the distances in this sample.
c. Find the mode of the distances in this sample.
d. Find the standard deviation s of this sample.
e. Convert the distance of 59 mm to a z score.
f. Find the actual percentage of these sample values that exceed 59 mm.
g. Assuming a normal distribution, find the percentage of *population* distances that exceed 59 mm. Use the sample values of $\bar{x}$ and s as estimates of μ and σ.
h. What level of measurement (nominal, ordinal, interval, ratio) describes this data set?
i. The listed measurements appear to be rounded to the nearest millimeter, but are the exact unrounded distances discrete data or continuous data?

2. Left-Handedness According to data from the American Medical Association, 10% of us are left-handed.

   a. If three people are randomly selected, find the probability that they are all left-handed.

   b. If three people are randomly selected, find the probability that at least one of them is left-handed.

   c. Why can't we solve the problem in part (b) by using the normal approximation to the binomial distribution?

   d. If groups of 50 people are randomly selected, what is the mean number of left-handed people in such groups?

   e. If groups of 50 people are randomly selected, what is the standard deviation for the numbers of left-handed people in such groups?

   f. Would it be unusual to get 8 left-handed people in a randomly selected group of 50 people? Why or why not?

2. a. 0.001

   b. 0.271

   c. The requirement that $np \geq 5$ is not satisfied, indicating that the normal approximation would result in errors that are too large.

   d. 5.0

   e. 2.1

   f. No, 8 is within two standard deviations of the mean and is within the range of values that could easily occur by chance.

Cooperative Group Activities

1. Out-of-class activity Divide into groups of three or four students. In each group, develop an original procedure to illustrate the central limit theorem. The main objective is to show that when you randomly select samples from a population, the means of those samples tend to be *normally* distributed, regardless of the nature of the population distribution. In Section 5-5, for example, we used the last four digits of social security numbers as a source of samples from a population of digits that are equally likely; we proceeded to show that even though the original population did not have a normal distribution, the sample means tended to be normally distributed.

2. In-class activity Divide into groups of three or four students. Using a coin to simulate births, each individual group member should simulate 25 births and record the number of simulated girls. Combine all the results in the group and record n = total number of births and x = number of girls. Given batches of n births, compute the mean and standard deviation for the number of girls. Is the simulated result usual or unusual? Why?

3. In-class activity Divide into groups of three or four students. Locate the Lotto numbers in Data Set 26 in Appendix B. There are six randomly selected numbers for each of 40 different Lotto games. Combine the 240 numbers into one big data set and test for normality. Then find the 40 means corresponding to the 40 different Lotto games and test for normality. What do you conclude? What really important concept is illustrated in this project?

4. In-class activity Divide into groups of three or four students. Select a set of data from Appendix B (excluding Data Sets 1, 3, 11, 12, 19, and 28, which were used in examples or exercises in Section 5-7). Use the methods of Section 5-7 and construct a histogram and normal quantile plot, then determine whether the data set appears to come from a normally distributed population.

Technology Project

In Section 5-3 we included an example that involved the design of cars. The solution in that example showed that 97.72% of men have sitting heights less than 38.8 in. That solution involved theoretical calculations based on the assumption that men have sitting heights that are normally distributed with a mean of 36.0 in. and a standard deviation of 1.4 in. (based on anthropometric survey data from Gordon, Clauser, et al.). This project describes a different method of solution that is based on a *simulation* technique: We will use a computer or a TI-83 Plus calculator to randomly generate 500 men's sitting heights (from a normally distributed population with $\mu = 36.0$ and $\sigma = 1.4$), then we will find the percentage of those simulated weights that are less than 38.8 in. The STATDISK, Minitab, Excel, and TI-83 Plus procedures are described as follows.

STATDISK   Select **Data** from the main menu bar, then choose the option of **Normal Generator.** Proceed to generate 500 values with a mean of 36.0 and a standard deviation of 1.4. (Use one decimal place.) Next, sort the data by copying the data to the Data Window, clicking on **Data tools,** and using the **Sort** option. With this sorted list, it becomes quite easy to count the number of sitting heights less than 38.8 in. Divide that number by 500 to find the percentage of these simulated sitting heights that are less than 38.8 in. Compare the result to the theoretical value of 97.72% that was found in Section 5-3.

Minitab   Select the options of **Calc,** then **Random Data,** then **Normal.** Proceed to enter 500 for the number of rows, C1 for the column in which to store the data, 36.0 for the value of the mean, and 1.4 for the value of the standard deviation. Now select the option of **Data** or **Manip,** then **Sort,** and proceed to sort column C1, with the sorted column stored in column C1, and with the sorting to be done by column C1. Examine the values in column C1 and determine the number of simulated sitting heights that are less than 38.8, then divide that number by 500 to find the percentage that are less than 38.8 in. Compare the result to the theoretical value of 97.72% that was found in Section 5-3.

Excel   Select **Tools** from the main menu bar, then select **Data Analysis** and **Random Number Generation.** After clicking OK, use the dialog box to enter 1 for the number of variables and 500 for the number of random numbers, then select "normal" for the type of distribution. Enter 36.0 for the mean and 1.4 for the standard deviation. Examine the displayed values and determine the number of simulated sitting heights that are less than 38.8, then divide that number by 500 to find the percentage that are less than 38.8 in. Compare the result to the theoretical value of 97.72% that was found in Section 5-3.

TI-83 Plus   Press **MATH,** then select **PRB,** then enter **randNorm** (36.0, 1.4, 500) to generate 500 values from a normally distributed population with $\mu = 36.0$ and $\sigma = 1.4$. Enter **STO→L1** to store the data in list L1. Now press **STAT,** then enter **SortA(L1)** to arrange the data in order. Examine the entries in list L1 to determine the number of simulated sitting heights that are less than 38.8, then divide that number by 500 to find the percentage that are less than 38.8 in. Compare the result to the theoretical value of 97.72% that was found in Section 5-3.

from DATA *to* DECISION

Critical Thinking: Designing a car seat

If a car seat is too low or too high, it will be uncomfortable and possibly dangerous. Most seats are made to be adjustable, so that men and women with different sitting heights can select a comfortable setting. When designing car seats, the sitting knee heights of men and women become very relevant. Men have sitting knee heights that are normally distributed with a mean of 22.0 in. and a standard deviation of 1.1 in., and women have sitting knee heights that are normally distributed with a mean of 20.3 in. and a standard deviation of 1.0 in. (based on anthropometric survey data from Gordon, Churchill, et al.). Find minimum and maximum sitting knee heights that include at least 95% of all men and at least 95% of all women, but try to find cost-effective limits that are as close together as possible. Find the percentage of men with sitting knee heights between the limits you have determined, then find the percentage of women with sitting knee heights between those same limits. Do your limits favor one gender at the expense of the other? Why is it not practical to simply design car seats so that they accommodate everyone? If you were a design engineer for General Motors, what percentage of the population would you be willing to exclude in your design of car seats? Apart from sitting knee height, what is another important design component that should be considered when determining the adjustment range of car seats?

INTERNET PROJECT   Exploring the Central Limit Theorem

The Central Limit Theorem is one of the most important results in statistics. It also may be one of the most surprising. Informally, the Central Limit Theorem says that the normal distribution is everywhere. No matter what probability distribution underlies an experiment, there is a corresponding distribution of means that will be approximately normal in shape.

http://www.aw.com/triola

The best way to both understand and appreciate the Central Limit Theorem is to see it in action. The Internet Project for this chapter, found at the *Elementary Statistics* Web site, will allow you to do just that. You will be asked to view, interpret, and discuss a demonstration of the Central Limit Theorem as part of a dice rolling experiment. In addition, you will be guided in a search through the Internet for other such demonstrations.

Statistics @ Work

"It is possible to be a journalist and not be comfortable with statistics, but you're definitely limited in what you can do."

Joel B. Obermayer

Newspaper reporter for The News & Observer

Joel B. Obermayer writes about medical issues and health affairs for *The News & Observer,* a newspaper that covers the eastern half of North Carolina. He reports on managed care, public health, and research at academic medical centers including Duke University and the University of North Carolina at Chapel Hill.

What concepts of statistics do you use?

I use ideas like statistical significance, error rates, and probability. I don't need to do anything incredibly sophisticated, but I need to be very comfortable with the math and with asking questions about it.

I use statistics to look at medical research to decide whether different studies are significant and to decide how to write about them. Mostly, I need to be able to read statistics and understand them, rather than develop them myself. I use statistics to develop good questions and to bolster the arguments I make in print. I also use statistics to decide if someone is trying to give me a positive spin on something that might be questionable. For example, someone at a local university once sent me a press release about miracle creams that supposedly help slim you down by dissolving fat cells. Well, I doubt that those creams work. The study wasn't too hot either. They were trying to make claims based on a study of only 11 people. The researcher argued that 11 people were enough to make good empirical health claims. Not too impressive. People try to manipulate the media all the time. Good verifiable studies with good verifiable statistical bases make it easier to avoid being manipulated.

Is your use of probability and statistics increasing, decreasing, or remaining stable?

Increasing. People's interest in new therapies that may only be in the clinical trial stage is increasing, partly because of the emphasis on AIDS research and on getting new drugs approved and delivered to patients sooner. It's more important than ever for a medical writer to use statistics to make sure that the studies really prove what the public relations people say they prove.

Should prospective employees have studied some statistics?

It is possible to be a journalist and not be comfortable with statistics, but you're definitely limited in what you can do. If you write about government-sponsored education programs and whether they're effective, or if you write about the dangers of particular contaminants in the environment, you're going to need to use statistics.

In my field, editors often don't think about statistics in the interview process. They worry more about writing skills. A knowledge of statistics is more important for what you can do once you are on the job.

Military Time 1750 07-07-97 (DD-M

Interval
Photo Time 0.50 T/O R009 Light
red 0.9 se

Violation
Number 008 O C V#58 Spee
(MPH

6

Estimates and
Sample Sizes

6-1 Overview

6-2 Estimating a Population Proportion

6-3 Estimating a Population Mean: σ Known

6-4 Estimating a Population Mean: σ Not Known

6-5 Estimating a Population Variance

"Photo-cop" survey results: What do they tell us?

The *Star Tribune,* a Minneapolis-St. Paul newspaper, sponsored a poll designed to reveal opinions about the "photo-cop," which consists of cameras positioned to catch drivers who run red lights. The cameras photograph the license plates of cars passing through red lights, and those car owners are later mailed traffic violations. The newspaper sponsored the poll because of pending Minnesota legislation that would approve the use of cameras for issuing traffic violations. (Thanks to Beth Hentges who provided the newspaper information.)

Pollsters surveyed 829 adult Minnesotans and found that 51% were opposed to the photo-cop legislation. These survey results, like most survey results, raise some interesting questions, such as these:

- Given that only 829 adults were surveyed, can we conclude anything about the population of all adult Minnesotans? If so, what can we conclude?

- Given that only 829 adults were surveyed, how accurate are the results?

- Is the sample size of 829 large enough to yield meaningful results?

- How were the survey respondents selected? Were they selected in such a way that they are representative of the population?

Surveys have become an important component of American life. They directly affect the television shows we watch, the products we buy, the officials we elect, and the clothes that we wear. Although surveys are now an integral part of all of our lives, it is unfortunate that most of us are not able to correctly interpret survey results. This chapter presents the statistical concepts that we need for such interpretations. We will address questions such as those listed above. We will analyze the *Star Tribune* survey results and, in the process, we will learn much about surveys in general.

Note to Instructor

This chapter is the first introduction to one of the two major activities of inferential statistics. Every introductory statistics course should include at least Sections 6-2 and 6-3, and most include Sections 6-2 through 6-4. Section 6-5 can be omitted if there is not sufficient time to include it. Here are the major changes from the previous edition:

1. When estimating μ, the criteria for choosing between the normal and t distributions have been updated. Instead of using the normal distribution whenever $n > 30$, the use of the normal distribution is now based on whether σ is known. See Figure 6-6 or Table 6-1 in Section 6-4.

2. Proportions have been moved to Section 6-2. They are common and interesting statistics and they are easier to work with so that basic concepts can be presented with fewer complications.

3. Instead of a separate section on sample size required for estimating μ, that topic is now included with confidence intervals for estimating μ.

Note to Instructor

This section discusses estimates of p, as well as determining sample size required to estimate p.

Overview

In this chapter we begin working with the true core of inferential statistics as we use sample data to make inferences about populations. Specifically, we will use sample data to make estimates of population parameters. For example, the Chapter Problem includes survey results consisting of 829 adult Minnesotans, 51% of whom are opposed to using cameras for issuing traffic tickets. Based on the sample statistic of 51%, we will estimate the percentage of adults in the population of all Minnesotans who are opposed to the photo-cop legislation.

The two major applications of inferential statistics involve the use of sample data to (1) estimate the value of a population parameter, and (2) test some claim (or hypothesis) about a population. In this chapter we introduce methods for estimating values of these important population parameters: proportions, means, and variances. We also present methods for determining the sample sizes necessary to estimate those parameters. In Chapter 7 we will introduce the basic methods for testing claims (or hypotheses) that have been made about a population parameter.

This chapter and Chapter 7 include important inferential methods involving population proportions, population means, and population variances (or standard deviations). In both chapters we begin with proportions for the following reasons:

1. We all see proportions frequently in the media.
2. People generally tend to be more interested in data that are expressed as proportions.
3. Proportions are generally easier to work with than means or variances, so we can better focus on the important principles of estimating parameters and testing hypotheses when those principles are introduced.

6-2 Estimating a Population Proportion

A Study Strategy: This section contains much information and introduces many concepts. The time devoted to this section will be well spent because we introduce the concept of a confidence interval, and that same general concept will be applied to the following sections of this chapter. We suggest that you use this study strategy: First, read this section with the limited objective of simply trying to understand what confidence intervals are, what they accomplish, and why they are needed. Second, try to develop the ability to construct confidence interval estimates of population proportions. Third, learn how to interpret a confidence interval correctly. Fourth, read the section once again and try to understand the underlying theory. You will always enjoy much greater success if you understand what you are doing, instead of blindly applying mechanical steps in order to obtain an answer that may or may not make any sense.

Here is the main objective of this section: Given a sample proportion, estimate the value of the population proportion p. For example, the Chapter Problem includes poll results consisting of 829 surveyed adults, 51% of whom are opposed to the photo-cop system of using cameras to ticket drivers who run red lights. The sample statistic of 51% can be represented as the sample proportion of 0.51. By using the sample size of $n = 829$ and the sample proportion of 0.51, we will proceed to estimate the proportion p of *all* adult Minnesotans who oppose the photo-cop legislation.

This section will consider only cases in which the normal distribution can be used to approximate the sampling distribution of sample proportions. In Section 5-6 we noted that in a binomial procedure with n trials and probability p, if $np \geq 5$ and $nq \geq 5$, then the binomial random variable has a probability distribution that can be approximated by a normal distribution. (Remember that $q = 1 - p$.) Those conditions are included among the following assumptions that apply to the methods of this section.

Assumptions

1. The sample is a simple random sample.

2. The conditions for the binomial distribution are satisfied. That is, there is a fixed number of trials, the trials are independent, there are two categories of outcomes, and the probabilities remain constant for each trial. (See Section 4-3.)

3. The normal distribution can be used to approximate the distribution of sample proportions because $np \geq 5$ and $nq \geq 5$ are both satisfied. (Because p and q are unknown, we use the sample proportion to estimate their values. Also, there are procedures for dealing with situations in which the normal distribution is not a suitable approximation. See Exercise 48.)

Recall from Section 1-4 that a simple random sample of n values is obtained if every possible sample of size n has the same chance of being selected. This requirement of random selection means that the methods of this section cannot be used with some other types of sampling, such as stratified, cluster, and convenience sampling. We should be especially clear about this important point:

Data collected carelessly can be absolutely worthless, even if the sample is quite large.

We know that different samples naturally produce different results. The methods of this section assume that those sample differences are due to chance random fluctuations, not some unsound method of sampling. If you were to conduct a survey of opinions about drunk-driving laws by selecting a sample of bar patrons, you should not use the results for making some estimate of the proportion of all adult Americans. The sample of bar patrons is very likely to be a biased sample in the sense that it is not representative of all Americans.

Assuming that we have a simple random sample and the other assumptions listed above are satisfied, we can now proceed with our major objective: using the sample as a basis for estimating the value of the population proportion p. We introduce the new notation $\hat{p}$ (called "p hat") for the sample proportion.

Accuracy of Vote Counts

The 2000 presidential election became the closest presidential election in American history, and it also became the first such election to be decided by court decisions. The election was held on November 7, 2000, but George W. Bush was not determined to be the winner until December 12, 2000. The delay was largely due to contested votes in the swing state of Florida, where recounts raised serious issues of accuracy. On November 17, 2000, *New York Times* reporters Ford Fessenden and Christopher Drew wrote that "The people who sell the voting systems . . . say the machines can be, in ideal conditions, 99.99 percent accurate. . . . The maker of one type of card reader said the accuracy of his machine would be 99.9 percent . . ." Although the final tally of 50,996,582 votes for Bush and 50,456,062 votes for Al Gore seem very precise, the voting systems caused those totals to be *estimates*, not precise counts.

Notation for Proportions

$p = population$ proportion

$\hat{p} = \dfrac{x}{n}$ sample proportion of x *successes* in a sample of size n

$\hat{q} = 1 - \hat{p} =$ sample proportion of *failures* in a sample of size n

Small Sample

The Children's Defense Fund was organized to promote the welfare of children. The group published *Children Out of School in America,* which reported that in one area, 37.5% of the 16- and 17-year-old children were out of school. This statistic received much press coverage, but it was based on a sample of only 16 children. Another statistic was based on a sample size of only 3 students. (See "First-hand Report: How Flawed Statistics Can Make an Ugly Picture Look Even Worse," *American School Board Journal*, Vol. 162.)

Proportion, Probability, and Percent   Although this section focuses on the population proportion p, the procedures discussed here can also be applied to probabilities or percentages, but percentages must be converted to proportions by dropping the percent sign and dividing by 100. For example, 51% can be expressed in decimal form as 0.51. The symbol p may therefore represent a proportion, a probability, or the decimal equivalent of a percent. For example, if you survey 200 statistics students and find that 80 of them have purchased TI-83 Plus calculators, then the sample proportion is $\hat{p} = x/n = 80/200 = 0.400$ and $\hat{q} = 0.600$ (calculated from $1 - 0.400$). Instead of computing the value of x/n, the value of $\hat{p}$ is sometimes already known because the sample proportion or percentage is given directly. For example, if it is reported that 829 adult Minnesotans are surveyed and 51% of them are opposed to photo-cop legislation, then $\hat{p} = 0.51$ and $\hat{q} = 0.49$.

If we want to estimate a population proportion with a single value, the best estimate is $\hat{p}$. Because $\hat{p}$ consists of a single value, it is called a point estimate.

Definition

A **point estimate** is a single value (or point) used to approximate a population parameter.

The sample proportion $\hat{p}$ is the best point estimate of the population proportion p.

We use $\hat{p}$ as the point estimate of p because it is unbiased and is the most consistent of the estimators that could be used. It is unbiased in the sense that the distribution of sample proportions tends to center about the value of p; that is, sample proportions $\hat{p}$ do not systematically tend to underestimate p, nor do they systematically tend to overestimate p. (See Section 5-4.) The sample proportion $\hat{p}$ is the most consistent estimator in the sense that the standard deviation of sample proportions tends to be smaller than the standard deviation of any other unbiased estimators.

EXAMPLE Photo-Cop Survey Responses In the Chapter Problem we noted that 829 adult Minnesotans were surveyed, and 51% of them are opposed to the use of the photo-cop for issuing traffic tickets. Using these survey results, find the best point estimate of the proportion of all adult Minnesotans opposed to photo-cop use.

SOLUTION   Because the sample proportion is the best point estimate of the population proportion, we conclude that the best point estimate of p is 0.51. When using the survey results to estimate the percentage of all adult Minnesotans that are opposed to photo-cop use, our best estimate is 51%.

Why Do We Need Confidence Intervals?

In the preceding example we saw that 0.51 was our *best* point estimate of the population proportion p, but we have no indication of just how *good* our best estimate was. If we had a sample of only 20 adult Minnesotans and 12 are opposed to photo-cop use, our best point estimate would be the sample proportion of $12/20 = 0.6$, but we wouldn't expect this point estimate to be very good because it is based on such a small sample. Because the point estimate has the serious flaw of not revealing anything about how good it is, statisticians have cleverly developed another type of estimate. This estimate, called a *confidence interval* or *interval estimate,* consists of a range (or an interval) of values instead of just a single value.

Definition

A **confidence interval** (or **interval estimate**) is a range (or an interval) of values used to estimate the true value of a population parameter. A confidence interval is sometimes abbreviated as CI.

A confidence interval is associated with a confidence level, such as 0.95 (or 95%). The confidence level gives us the success rate of the procedure used to construct the confidence interval. The confidence level is often expressed as the probability or area $1 - \alpha$ (lowercase Greek alpha). The value of α is the complement of the *confidence level.* For a 0.95 (or 95%) confidence level, $\alpha = 0.05$. For a 0.99 (or 99%) confidence level, $\alpha = 0.01$.

Definition

The **confidence level** is the probability $1 - \alpha$ (often expressed as the equivalent percentage value) that is the proportion of times that the confidence interval actually does contain the population parameter, assuming that the estimation process is repeated a large number of times. (The confidence level is also called the **degree of confidence,** or the **confidence coefficient.**)

The most common choices for the confidence level are 90% (with $\alpha = 0.10$), 95% (with $\alpha = 0.05$), and 99% (with $\alpha = 0.01$). The choice of 95% is most common because it provides a good balance between precision (as reflected in the width of the confidence interval) and reliability (as expressed by the confidence level).

Here's an example of a confidence interval based on the sample data of 829 surveyed adult Minnesotans, 51% of whom are opposed to use of the photo-cop:

The 0.95 (or 95%) confidence interval estimate of the population proportion p is $0.476 < p < 0.544$.

Push Polling

"Push polling" is the practice of political campaigning under the guise of a poll. Its name is derived from its objective of pushing voters away from opposition candidates by asking loaded questions designed to discredit them. Here's an example of one such question that was used: "Please tell me if you would be more likely or less likely to vote for Roy Romer if you knew that Gov. Romer appoints a parole board which has granted early release to an average of four convicted felons per day every day since Romer took office." The National Council on Public Polls characterizes push polls as unethical, but some professional pollsters do not condemn the practice as long as the questions do not include outright lies.

⑥ Interpreting a Confidence Interval

We must be careful to interpret confidence intervals correctly. There is a correct interpretation and many different and creative wrong interpretations of the confidence interval $0.476 < p < 0.544$.

Correct: "We are 95% confident that the interval from 0.476 to 0.544 actually does contain the true value of p." This means that if we were to select many different samples of size 829 and construct the corresponding confidence intervals, 95% of them would actually contain the value of the population proportion p. (Note that in this correct interpretation, the level of 95% refers to the success rate of the *process* being used to estimate the proportion, and it does not refer to the population proportion itself.)

Wrong: "There is a 95% chance that the true value of p will fall between 0.476 and 0.544."

At any specific point in time, there is a fixed and constant value of p, the proportion of all adult Minnesotans opposed to use of the photo-cop. If we use sample data to find specific limits, such as 0.476 and 0.544, those limits either enclose the population proportion p or do not, and we cannot determine whether they do or do not without knowing the true value of p. But it's wrong to say that p has a 95% chance of falling within the specific limits of 0.476 and 0.544, because p is a fixed (but unknown) constant, not a random variable. Either p will fall within these limits or it won't; there's no probability involved. This is a confusing concept, so consider the easier example in which we want to find the probability of a baby being born a girl. If the baby has already been born, but the doctor hasn't yet announced the gender, we can't say that there is a 0.5 probability that the baby is a girl, because the baby is already a girl or is not. There is no chance involved, because the gender has been determined. Similarly, a population proportion p is already determined, and the confidence interval limits either contain p or do not, so it's wrong to say that there is a 95% chance that p will fall between 0.476 and 0.544.

A confidence level of 95% tells us that the *process* we are using will, in the long run, result in confidence interval limits that contain the true population proportion 95% of the time. Suppose that the true proportion of all adult Minnesotans opposed to the photo-cop is $p = 0.520$. Then the confidence interval obtained from the given sample data would contain the population proportion, because the true population proportion 0.520 is between 0.476 and 0.544. This is illustrated in Figure 6-1. Figure 6-1 shows the first confidence interval for the real survey data given in the Chapter Problem (with 51% of 829 surveyed people opposed to the photo-cop), but the other 19 confidence intervals represent hypothetical samples. With 95% confidence, we expect that 19 out of 20 samples should result in confidence intervals that do contain the true value of p, and Figure 6-1 illustrates this with 19 of the confidence intervals containing p, while one confidence interval does not contain p.

Caution: Confidence intervals can be used informally to compare different data sets, but the overlapping of confidence intervals should not be used for mak-

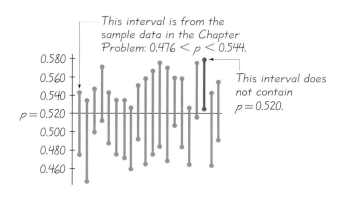

This interval is from the sample data in the Chapter Problem: $0.476 < p < 0.544$.

$p = 0.520$

This interval does not contain $p = 0.520$.

FIGURE 6-1 **Confidence Intervals from 20 Different Samples**

ing formal and final conclusions about equality of proportions. The analysis of overlap between two individual confidence intervals is associated with pitfalls described in later chapters. (See "On Judging the Significance of Differences by Examining the Overlap Between Confidence Intervals" by Schenker and Gentleman, *The American Statistician*, Vol. 55, No. 3.) In the following chapters we will describe procedures for determining whether populations have equal proportions, and those procedures will not have the pitfalls associated with conclusions based on overlapping of confidence intervals.

Do not use the *overlapping* of confidence intervals as the basis for making final conclusions about the equality of proportions.

Critical Values

The methods of this section and many of the other statistical methods found in the following chapters include the use of a standard z score that can be used to distinguish between sample statistics that are likely to occur and those that are unlikely. Such a z score is called a *critical value* (defined below). Critical values are based on the following observations.

1. We know from Section 5-6 that under certain conditions, the sampling distribution of sample proportions can be approximated by a normal distribution, as in Figure 6-2.

2. Sample proportions have a relatively small chance (with probability denoted by α) of falling in one of the red tails of Figure 6-2.

3. Denoting the area of each shaded tail by $\alpha/2$, we see that there is a total probability of α that a sample proportion will fall in either of the two red tails.

4. By the rule of complements (from Chapter 3), there is a probability of $1 - \alpha$ that a sample proportion will fall within the inner green-shaded region of Figure 6-2.

5. The z score separating the right-tail region is commonly denoted by $z_{\alpha/2}$, and is referred to as a *critical value* because it is on the borderline separating sample proportions that are likely to occur from those that are unlikely to occur.

These observations can be formalized with the following notation and definition.

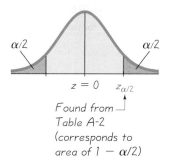

$\alpha/2$ $\alpha/2$

$z = 0$ $z_{\alpha/2}$

Found from Table A-2 (corresponds to area of $1 - \alpha/2$)

FIGURE 6-2 **Critical Value $z_{\alpha/2}$ in the Standard Normal Distribution**

Note to Instructor
This is a new notation for a concept already introduced in Chapter 5. Some books use the notation of z_c or z^* or $z(\alpha/2)$ instead of $z_{\alpha/2}$. The notation of $z_{\alpha/2}$ and the concept of a critical value are used in hypothesis testing, which begins in Chapter 7.

Notation for Critical Value

The critical value $z_{\alpha/2}$ is the positive z value that is at the vertical boundary separating an area of $\alpha/2$ in the right tail of the standard normal distribution. (The value of $-z_{\alpha/2}$ is at the vertical boundary for the area of $\alpha/2$ in the left tail.) The subscript $\alpha/2$ is simply a reminder that the z score separates an area of $\alpha/2$ in the right tail of the standard normal distribution.

Definition

A **critical value** is the number on the borderline separating sample statistics that are likely to occur from those that are unlikely to occur. The number $z_{\alpha/2}$ is a critical value that is a z score with the property that it separates an area of $\alpha/2$ in the right tail of the standard normal distribution. (See Figure 6-2.)

EXAMPLE Finding a Critical Value Find the critical value $z_{\alpha/2}$ corresponding to a 95% confidence level.

SOLUTION   *Caution:* To find the critical z value for a 95% confidence level, do *not* look up 0.95 in the body of Table A-2. A 95% confidence level corresponds to $\alpha = 0.05$. See Figure 6-3, where we show that the area in each of the red-shaded tails is $\alpha/2 = 0.025$. We find $z_{\alpha/2} = 1.96$ by noting that all of the area to its left must be $1 - 0.025$, or 0.975. We can refer to Table A-2 and find that the area of 0.9750 (found in the *body* of the table) corresponds exactly to a z score of 1.96. For a 95% confidence level, the critical value is therefore $z_{\alpha/2} = 1.96$. Bottom line: To find the critical z score for a 95% confidence level, look up 0.9750 in the body of Table A-2, not 0.95.

FIGURE 6-3 **Finding** $z_{\alpha/2}$
for a 95% Confidence Level

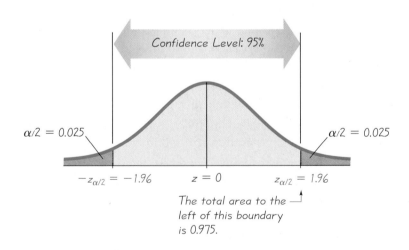

Confidence Level: 95%

$\alpha/2 = 0.025$ $\alpha/2 = 0.025$

$-z_{\alpha/2} = -1.96$ $z = 0$ $z_{\alpha/2} = 1.96$

The total area to the
left of this boundary
is 0.975.

The preceding example showed that a 95% confidence level results in a critical value of $z_{\alpha/2} = 1.96$. This is the most common critical value, and it is listed with two other common values in the table that follows.

Confidence Level	α	Critical Value, $z_{\alpha/2}$
90%	0.10	1.645
95%	0.05	1.96
99%	0.01	2.575

Margin of Error

When we collect a set of sample data, such as the survey data given in the Chapter Problem (with 51% of 829 respondents opposed to the photo-cop), we can calculate the sample proportion $\hat{p}$ and that sample proportion is typically different from the population proportion p. The difference between the sample proportion and the population proportion can be thought of as an error. We now define the *margin of error E* as follows.

Definition

When data from a simple random sample are used to estimate a population proportion p, the **margin of error,** denoted by **E,** is the maximum likely (with probability $1 - \alpha$) difference between the observed sample proportion $\hat{p}$ and the true value of the population proportion p. The margin of error E is also called the *maximum error of the estimate* and can be found by multiplying the critical value and the standard deviation of sample proportions, as shown in Formula 6-1.

Formula 6-1
$$E = z_{\alpha/2}\sqrt{\frac{\hat{p}\hat{q}}{n}}$$
margin of error for proportions

Given the way that the margin of error E is defined, there is a probability of $1 - \alpha$ that a sample proportion will be in error (different from the population proportion p) by no more than E, and there is a probability α that the sample proportion will be in error by more than E.

Confidence Interval (or Interval Estimate) for the Population Proportion p

$$\hat{p} - E < p < \hat{p} + E \qquad \text{where} \qquad E = z_{\alpha/2}\sqrt{\frac{\hat{p}\hat{q}}{n}}$$

The confidence interval is often expressed in the following equivalent formats:

$$\hat{p} \pm E$$

or

$$(\hat{p} - E, \hat{p} + E)$$

In Chapter 3, when probabilities were given in decimal form, we rounded to three significant digits. We use that same rounding rule here.

Round-Off Rule for Confidence Interval Estimates of *p*

Round the confidence interval limits for *p* to three significant digits.

Based on the preceding results, we can summarize the procedure for constructing a confidence interval estimate of a population proportion *p* as follows.

Procedure for Constructing a Confidence Interval for *p*

1. Verify that the required assumptions are satisfied. (The sample is a simple random sample, the conditions for the binomial distribution are satisfied, and the normal distribution can be used to approximate the distribution of sample proportions because $np \geq 5$ and $nq \geq 5$ are both satisfied.)
2. Refer to Table A-2 and find the critical value $z_{\alpha/2}$ that corresponds to the desired confidence level. (For example, if the confidence level is 95%, the critical value is $z_{\alpha/2} = 1.96$.)
3. Evaluate the margin of error $E = z_{\alpha/2} \sqrt{\hat{p}\hat{q}/n}$.
4. Using the value of the calculated margin of error E and the value of the sample proportion $\hat{p}$, find the values of $\hat{p} - E$ and $\hat{p} + E$. Substitute those values in the general format for the confidence interval:

$$\hat{p} - E < p < \hat{p} + E$$
or
$$\hat{p} \pm E$$
or
$$(\hat{p} - E, \hat{p} + E)$$

5. Round the resulting confidence interval limits to three significant digits.

EXAMPLE Photo-Cop Survey Responses In the Chapter Problem we noted that 829 adult Minnesotans were surveyed, and 51% of them are opposed to the use of the photo-cop for issuing traffic tickets. In a previous example we noted that the best point estimate of the population proportion is 0.51. Use these same survey results for the following.

a. Find the margin of error *E* that corresponds to a 95% confidence level.
b. Find the 95% confidence interval estimate of the population proportion *p*.
c. Based on the results, can we safely conclude that the majority of adult Minnesotans oppose use of the photo-cop?

SOLUTION   We should first verify that the required assumptions are satisfied. Assuming that the sample is a simple random sample, we see that the conditions for a binomial distribution are satisfied. With $n = 829$ and $\hat{p} = 0.51$, we have $n\hat{p} = 422.79 \geq 5$ and $n\hat{q} = 406.21 \geq 5$, so the normal distribution can be used to approximate the binomial distribution.

a. The margin of error is found by using Formula 6-1 with $z_{\alpha/2} = 1.96$ (as found in the preceding example), $\hat{p} = 0.51$, $\hat{q} = 1 - 0.51 = 0.49$, and $n = 829$. Extra digits are used so that rounding error will be minimized in the confidence interval limits found in part (b).

$$E = z_{\alpha/2} \sqrt{\frac{\hat{p}\hat{q}}{n}} = 1.96 \sqrt{\frac{(0.51)(0.49)}{829}} = 0.03403000$$

b. Constructing the confidence interval is quite easy now that we have the values of $\hat{p}$ and E. We simply substitute those values to obtain this result:

$$\hat{p} - E < p < \hat{p} + E$$

$$0.51 - 0.03403000 < p < 0.51 + 0.03403000$$

$$0.476 < p < 0.544 \text{ (rounded to three significant digits)}$$

This same result could be expressed in the format of 0.51 ± 0.034 or (0.476, 0.544). If we want the 95% confidence interval for the true population *percentage,* we could express the result as $47.6\% < p < 54.4\%$. This confidence interval is often reported with a statement such as this: "It is estimated that 51% of adult Minnesotans are opposed to use of the photo-cop, with a margin of error of plus or minus 3.4 percentage points." That statement is a verbal expression of this format for the confidence interval: $51\% \pm 3.4\%$. The level of confidence should also be reported, but it rarely is in the media. The media typically use a 95% confidence level but omit any reference to it. However, information provided by the *Star Tribune* about this poll included a statement that "the maximum margin of sampling error for percentages based on 829 is 3.4 percentage points, plus or minus, at a 95% confidence level, if one does not include the effect of sample design." Way to go, *Star Tribune!*

c. Based on the survey results, we are 95% confident that the limits of 47.6% and 54.4% contain the true percentage of adult Minnesotans opposed to the photo-cop. The percentage of opposed adult Minnesotans is likely to be any value between 47.6% and 54.4%. However, a majority requires a percentage greater than 50%, so we *cannot* safely conclude that the majority is opposed (because the *entire* confidence interval is not greater than 50%).

Rationale for the Margin of Error   Because the sampling distribution of proportions is approximately normal (because the conditions $np \geq 5$ and $nq \geq 5$ are both satisfied), we can use results from Section 5-6 to conclude that μ and σ are given by $\mu = np$ and $\sigma = \sqrt{npq}$. Both of these parameters pertain to n trials, but we convert them to a per-trial basis by dividing by n as follows:

$$\text{Mean of sample proportions: } \mu = \frac{np}{n} = p$$

$$\text{Standard deviation of sample proportions: } \sigma = \frac{\sqrt{npq}}{n} = \sqrt{\frac{npq}{n^2}} = \sqrt{\frac{pq}{n}}$$

The first result may seem trivial, because we already stipulated that the true population proportion is p. The second result is nontrivial and is useful in describing the margin of error E, but we replace the product pq by $\hat{p}\hat{q}$ because we don't yet know the value of p (it is the value we are trying to estimate). Formula 6-1 for the margin of error reflects the fact that $\hat{p}$ has a probability of $1 - \alpha$ of being within $z_{\alpha/2}\sqrt{pq/n}$ of p. The confidence interval for p, as given previously, reflects the fact that there is a probability of $1 - \alpha$ that $\hat{p}$ differs from p by less than the margin of error $E = z_{\alpha/2}\sqrt{pq/n}$.

Determining Sample Size

Suppose we want to collect sample data with the objective of estimating some population proportion. How do we know *how many* sample items must be obtained? If we take the expression for the margin of error E (Formula 6-1), then solve for n, we get Formula 6-2. Formula 6-2 requires $\hat{p}$ as an estimate of the population proportion p, but if no such estimate is known (as is usually the case), we replace $\hat{p}$ by 0.5 and replace $\hat{q}$ by 0.5, with the result given in Formula 6-3.

Sample Size for Estimating Proportion p

When an estimate $\hat{p}$ is known: **Formula 6-2** $n = \dfrac{[z_{\alpha/2}]^2 \, \hat{p}\hat{q}}{E^2}$

When no estimate $\hat{p}$ is known: **Formula 6-3** $n = \dfrac{[z_{\alpha/2}]^2 \cdot 0.25}{E^2}$

Round-Off Rule for Determining Sample Size

In order to ensure that the required sample size is at least as large as it should be, if the computed sample size is not a whole number, round it up to the next *higher* whole number.

Use Formula 6-2 when reasonable estimates of $\hat{p}$ can be made by using previous samples, a pilot study, or someone's expert knowledge. When no such estimate can be made, we assign the value of 0.5 to each of $\hat{p}$ and $\hat{q}$ so the resulting sample size will be at least as large as it should be. The underlying reason for the assignment of 0.5 is this: The product $\hat{p} \cdot \hat{q}$ has 0.25 as its largest possible value, which occurs when $\hat{p} = 0.5$ and $\hat{q} = 0.5$. (Try experimenting with different values of $\hat{p}$ to verify that $\hat{p} \cdot \hat{q}$ has 0.25 as the largest possible value.) Note that Formulas 6-2 and 6-3 do not include the population size N, so the size of the population is irrelevant. (*Exception:* When sampling is without replacement from a relatively small finite population. See Exercise 46.)

> **EXAMPLE Sample Size for E-Mail Survey** The ways that we communicate have been dramatically affected by the use of answering machines, fax machines, voice mail, and e-mail. Suppose a sociologist wants to determine the current percentage of U.S. households using e-mail. How many households must be surveyed in order to be 95% confident that the sample percentage is in error by no more than four percentage points?
>
> **a.** Use this result from an earlier study: In 1997, 16.9% of U.S. households used e-mail (based on data from *The World Almanac and Book of Facts*).
>
> **b.** Assume that we have no prior information suggesting a possible value of $\hat{p}$.
>
> **SOLUTION**
>
> **a.** The prior study suggests that $\hat{p} = 0.169$, so $\hat{q} = 0.831$ (found from $\hat{q} = 1 - 0.169$). With a 95% level of confidence, we have $\alpha = 0.05$, so $z_{\alpha/2} = 1.96$. Also, the margin of error is $E = 0.04$ (the decimal equivalent of "four percentage points"). Because we have an estimated value of $\hat{p}$, we use Formula 6-2 as follows:
>
> $$n = \frac{[z_{\alpha/2}]^2\,\hat{p}\hat{q}}{E^2} = \frac{[1.96]^2(0.169)(0.831)}{0.04^2}$$
>
> $$= 337.194 = 338 \qquad \text{(rounded up)}$$
>
> We must survey at least 338 randomly selected households.
>
> **b.** As in part (a), we again use $z_{\alpha/2} = 1.96$ and $E = 0.04$, but with no prior knowledge of $\hat{p}$ (or $\hat{q}$), we use Formula 6-3 as follows.
>
> $$n = \frac{[z_{\alpha/2}]^2 \cdot 0.25}{E^2} = \frac{[1.96]^2 \cdot 0.25}{0.04^2}$$
>
> $$= 600.25 = 601 \qquad \text{(rounded up)}$$
>
> **INTERPRETATION** To be 95% confident that our sample percentage is within four percentage points of the true percentage for all households, we should randomly select and survey 601 households. By comparing this result to the sample size of 338 found in part (a), we can see that if we have no knowledge of a prior study, a larger sample is required to achieve the same results as when the value of $\hat{p}$ can be estimated. But now let's use some common sense: We know that the use of e-mail is growing so rapidly that the 1997 estimate is too old to be of much use. Today, substantially more than 16.9% of households use e-mail. Realistically, we need a sample larger than 338 households. Assuming that we don't really know the current rate of e-mail usage, we should randomly select 601 households. With 601 households, we will be 95% confident that we are within four percentage points of the true percentage of households using e-mail.

Common Errors   When calculating sample size using Formula 6-2 or 6-3, be sure to substitute the critical z score for $z_{\alpha/2}$. For example, if you are working with 95% confidence, be sure to replace $z_{\alpha/2}$ with 1.96. (Here is the logical sequence: 95% $\Rightarrow \alpha = 0.05 \Rightarrow z_{\alpha/2} = 1.96$ found from Table A-2.) Don't make the mistake of replacing $z_{\alpha/2}$ with 0.95 or 0.05. Also, don't make the mistake of using $E = 4$ as the margin of error corresponding to "four percentage points." When using Formula 6-2 or 6-3, the value of E never exceeds 1. The error of using $E = 4$ instead of $E = 0.04$ causes the sample size to be 1/10,000th of what it should be, so that you might end up with a sample size of only 1 when the answer is rounded up. You really can't estimate a population proportion by surveying only one person (even though there are individuals who claim to know everything).

Population Size   Part (b) of the preceding example involved application of Formula 6-3, the same formula frequently used by Nielsen, Gallup, and other professional pollsters. Many people incorrectly believe that the sample size should be some percentage of the population, but Formula 6-3 shows that the population size is irrelevant. (In reality, the population size is sometimes used, but only in cases in which we sample without replacement from a relatively small population. See Exercise 46.) Most of the polls featured in newspapers, magazines, and broadcast media involve sample sizes in the range of 1000 to 2000. Even though such polls may involve a very small percentage of the total population, they can provide results that are quite good. When Nielsen surveys 4000 TV households from a population of 104 million households, only 0.004% of the households are surveyed; still, we can be 95% confident that the sample percentage will be within one percentage point of the true population percentage.

Finding the Point Estimate and E from a Confidence Interval   Sometimes we want to better understand a confidence interval that might have been obtained from a journal article, or it might have been generated using software or a calculator. If we already know the confidence interval limits, the sample proportion $\hat{p}$ and the margin of error E can be found as follows:

Point estimate of p:

$$\hat{p} = \frac{(\text{upper confidence limit}) + (\text{lower confidence limit})}{2}$$

Margin of error:

$$E = \frac{(\text{upper confidence limit}) - (\text{lower confidence limit})}{2}$$

EXAMPLE   The article "High-Dose Nicotine Patch Therapy" by Dale, Hurt, et al. (*Journal of the American Medical Association,* Vol. 274, No. 17) includes this statement: "Of the 71 subjects, 70% were abstinent from smoking at 8 weeks (95% confidence interval [CI], 58% to 81%)." Use that statement to find the point estimate $\hat{p}$ and the margin of error E.

SOLUTION   From the given statement, we see that the 95% confidence interval is $0.58 < p < 0.81$. The point estimate $\hat{p}$ is the value midway between the upper and lower confidence interval limits, so we get

$$\hat{p} = \frac{(\text{upper confidence limit}) + (\text{lower confidence limit})}{2}$$

$$= \frac{0.81 + 0.58}{2} = 0.695$$

The margin of error can be found as follows:

$$E = \frac{(\text{upper confidence limit}) - (\text{lower confidence limit})}{2}$$

$$= \frac{0.81 - 0.58}{2} = 0.115$$

Using Technology for Confidence Intervals

STATDISK   Select **Analysis,** then **Confidence Intervals,** then **Population Proportion,** and proceed to enter the requested items.

Minitab   Select **Stat, Basic Statistics,** then **1 Proportion.** In the dialog box, click on the button for **Summarized Data.** Also click on the **Options** button, enter the desired confidence level (the default is 95%), and click on the box with this statement: "Use test and interval based on normal distribution."

Excel   Use the Data Desk XL add-in that is a supplement to this book. First enter the number of successes in cell A1, then enter the total number of trials in cell B1. Click on **DDXL** and select **Confidence Intervals,** then select **Summ 1 Var Prop Interval** (which is an abbreviated form of "confidence interval for a proportion using summary data for one variable"). Click on the pencil icon for "Num successes" and enter A1. Click on the pencil icon for "Num trials" and enter B1. Click **OK.** In the dialog box, select the level of confidence, then click on **Compute Interval.**

TI-83 Plus   Press **STAT,** select **TESTS,** then select **1-PropZInt** and proceed to enter the required items.

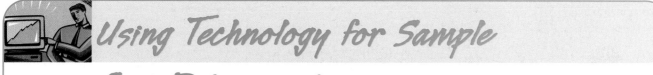

Using Technology for Sample Size Determination

STATDISK   Select **Analysis,** then **Sample Size Determination,** then **Estimate Proportion.** Proceed to enter the required items in the dialog box.

Sample size determination is not available as a built-in function with Minitab, Excel, or the TI-83 Plus calculator.

1. 2.575
2. 1.645
3. 2.33
4. 1.75
5. $p = 0.250 \pm 0.030$
6. $p = 0.476 \pm 0.020$
7. $p = 0.654 \pm 0.050$
8. $0.712 < p < 0.772$
9. $\hat{p} = 0.464$; $E = 0.020$
10. $\hat{p} = 0.308$; $E = 0.030$
11. $\hat{p} = 0.655$; $E = 0.023$
12. $\hat{p} = 0.907$; $E = 0.020$

13. 0.0300

14. 0.0350

15. 0.0405

16. 0.0351

17. $0.708 < p < 0.792$

18. $0.139 < p < 0.194$

19. $0.0887 < p < 0.124$

20. $0.876 < p < 0.899$

21. 461

22. 666

23. 232

24. 222

25. a. There is 95% confidence
        that the limits of 0.0489 and
        0.0531 contain the popula-
        tion proportion.
    b. Yes, about 5% of males
        aged 18–20 drive while
        impaired.
    c. 5.31%

6-2 Basic Skills and Concepts

Finding Critical Values. *In Exercises 1–4, find the critical value $z_{\alpha/2}$ that corresponds to the given confidence level.*

1. 99%

2. 90%

3. 98%

4. 92%

5. Express the confidence interval $0.220 < p < 0.280$ in the form of $\hat{p} \pm E$.

6. Express the confidence interval $0.456 < p < 0.496$ in the form of $\hat{p} \pm E$.

7. Express the confidence interval $(0.604, 0.704)$ in the form of $\hat{p} \pm E$.

8. Express the confidence interval 0.742 ± 0.030 in the form of $\hat{p} - E < p < \hat{p} + E$.

Interpreting Confidence Interval Limits. *In Exercises 9–12, use the given confidence interval limits to find the point estimate $\hat{p}$ and the margin of error E.*

9. $(0.444, 0.484)$

10. $0.278 < p < 0.338$

11. $0.632 < p < 0.678$

12. $0.887 < p < 0.927$

Finding Margin of Error. *In Exercises 13–16, assume that a sample is used to estimate a population proportion p. Find the margin of error E that corresponds to the given statistics and confidence level.*

13. $n = 800$, $x = 200$, 95% confidence

14. $n = 1200$, $x = 400$, 99% confidence

15. 99% confidence; the sample size is 1000, of which 45% are successes.

16. 95% confidence; the sample size is 500, of which 80% are successes.

Constructing Confidence Intervals. *In Exercises 17–20, use the sample data and confidence level to construct the confidence interval estimate of the population proportion p.*

17. $n = 400$, $x = 300$, 95% confidence

18. $n = 1200$, $x = 200$, 99% confidence

19. $n = 1655$, $x = 176$, 98% confidence

20. $n = 2001$, $x = 1776$, 90% confidence

Determining Sample Size. *In Exercises 21–24, use the given data to find the minimum sample size required to estimate a population proportion or percentage.*

21. Margin of error: 0.060; confidence level: 99%; $\hat{p}$ and $\hat{q}$ unknown

22. Margin of error: 0.038; confidence level: 95%; $\hat{p}$ and $\hat{q}$ unknown

23. Margin of error: five percentage points; confidence level: 95%; from a prior study, $\hat{p}$ is estimated by the decimal equivalent of 18.5%.

24. Margin of error: three percentage points; confidence level: 90%; from a prior study, $\hat{p}$ is estimated by the decimal equivalent of 8%.

25. Interpreting Calculator Display The Insurance Institute of America wants to estimate the percentage of drivers aged 18–20 who drive a car while impaired because of alcohol consumption. In a large study, 42,772 males aged 18–20 were surveyed, and 5.1% of them said that they drove in the last month while being impaired from alcohol (based on data from "Prevalence of Alcohol-Impaired Driving," by Liu, Siegel, et al, *Journal of the American Medical Association*, Vol. 277, No. 2). Using the sample data and a 95% confidence level, the TI-83 Plus calculator display is as shown.

a. Write a statement that interprets the confidence interval.

b. Based on the preceding result, does alcohol-impaired driving appear to be a problem for males aged 18–20? (All states now prohibit the sale of alcohol to persons under the age of 21.)

c. When setting insurance rates for male drivers aged 18–24, what percentage of alcohol-impaired driving would you use if you are working for the insurance company and you want to be conservative by using the likely worst case scenario?

26. Interpreting Calculator Display In 1920 only 35% of U.S. households had telephones, but that rate is now much higher. A recent survey of 4276 randomly selected households showed that 4019 of them had telephones (based on data from the U.S. Census Bureau). Using those survey results and a 99% confidence level, the TI-83 Plus calculator display is as shown.

a. Write a statement that interprets the confidence interval.

b. Based on the preceding result, should pollsters be concerned about results from surveys conducted by telephone?

27. Internet Shopping In a Gallup poll, 1025 randomly selected adults were surveyed and 29% of them said that they used the Internet for shopping at least a few times a year.

a. Find the point estimate of the *percentage* of adults who use the Internet for shopping.

b. Find a 99% confidence interval estimate of the *percentage* of adults who use the Internet for shopping.

c. If a traditional retail store wants to estimate the percentage of adult Internet shoppers in order to determine the maximum impact of Internet shoppers on its sales, what percentage of Internet shoppers should be used?

28. Death Penalty Survey In a Gallup poll, 491 randomly selected adults were asked whether they are in favor of the death penalty for a person convicted of murder, and 65% of them said that they were in favor.

a. Find the point estimate of the *percentage* of adults who are in favor of this death penalty.

b. Find a 95% confidence interval estimate of the *percentage* of adults who are in favor of this death penalty.

c. Can we safely conclude that the majority of adults are in favor of this death penalty? Explain.

29. Mendelian Genetics When Mendel conducted his famous genetics experiments with peas, one sample of offspring consisted of 428 green peas and 152 yellow peas.

a. Find a 95% confidence interval estimate of the percentage of yellow peas.

b. Based on his theory of genetics, Mendel expected that 25% of the offspring peas would be yellow. Given that the percentage of offspring yellow peas is not 25%, do the results contradict Mendel's theory? Why or why not?

30. Misleading Survey Responses In a survey of 1002 people, 701 said that they voted in a recent presidential election (based on data from ICR Research Group). Voting records show that 61% of eligible voters actually did vote.

a. Find a 99% confidence interval estimate of the proportion of people who say that they voted.

b. Are the survey results consistent with the actual voter turnout of 61%? Why or why not?

TI-83 Plus

```
1-PropZInt
 (.04891,.05308)
 p̂=.0509913027
 n=42772
```

TI-83 Plus

```
1-PropZInt
 (.93053,.94926)
 p̂=.9398971001
 n=4276
```

26. a. There is 95% confidence that the limits of 0.931 and 0.949 contain the population proportion.
    b. Yes. Based on the result from part (a), about 5% to 7% of the population does not have a telephone, so those people are missed.
27. a. 29%
    b. $25.4\% < p < 32.6\%$
    c. 32.6%
28. a. 65%
    b. $60.8\% < p < 69.2\%$
    c. Yes

29. a. $22.6\% < p < 29.8\%$
    b. No, the confidence interval limits include 25%.

30. a. $0.662 < p < 0.737$
    b. No, because the confidence interval limits do not include a rate of 61%.

31. a. $0.134\% < p < 6.20\%$ using $x = 7, n = 221$

b. Ziac does not appear to cause dizziness as an adverse reaction.

32. a. $15.1\% < p < 21.6\%$

b. Yes

33. 4145

34. a. 2653

b. 1353

c. Results from a voluntary response sample are not valid.

35. a. 473

b. 982

c. Because they are based on a voluntary response sample, the results would not be valid.

31. Drug Testing The drug Ziac is used to treat hypertension. In a clinical test, 3.2% of 221 Ziac users experienced dizziness (based on data from Lederle Laboratories).

   a. Construct a 99% confidence interval estimate of the percentage of all Ziac users who experience dizziness.

   b. In the same clinical test, people in the placebo group didn't take Ziac, but 1.8% of them reported dizziness. Based on the result from part (a), what can we conclude about dizziness as an adverse reaction to Ziac?

32. Smoking and College Education The tobacco industry closely monitors all surveys that involve smoking. One survey showed that among 785 randomly selected subjects who completed four years of college, 18.3% smoke (based on data from the American Medical Association).

   a. Construct the 98% confidence interval for the true percentage of smokers among all people who completed four years of college.

   b. Based on the result from part (a), does the smoking rate for those with four years of college appear to be substantially different than the 27% rate for the general population?

33. Sample Size for Internet Purchases Many states are carefully considering steps that would help them collect sales taxes on items purchased through the Internet. How many randomly selected sales transactions must be surveyed to determine the percentage that transpired over the Internet? Assume that we want to be 99% confident that the sample percentage is within two percentage points of the true population percentage for all sales transactions.

34. Sample Size for Left-Handed Golfers As a manufacturer of golf equipment, the Spalding Corporation wants to estimate the proportion of golfers who are left-handed. (The company can use this information in planning for the number of right-handed and left-handed sets of golf clubs to make.) How many golfers must be surveyed if we want 99% confidence that the sample proportion has a margin of error of 0.025?

   a. Assume that there is no available information that could be used as an estimate of $\hat{p}$.

   b. Assume that we have an estimate of $\hat{p}$ found from a previous study that suggests that 15% of golfers are left-handed (based on a *USA Today* report).

   c. Assume that instead of using randomly selected golfers, the sample data are obtained by asking TV viewers of the golfing channel to call an "800" phone number to report whether they are left-handed or right-handed. How are the results affected?

35. Sample Size for Motor Vehicle Ownership You have been hired by the Ford Motor Company to do market research, and you must estimate the percentage of households in which a vehicle is owned. How many households must you survey if you want to be 94% confident that your sample percentage has a margin of error of three percentage points?

   a. Assume that a previous study suggested that vehicles are owned in 86% of households.

   b. Assume that there is no available information that can be used to estimate the percentage of households in which a vehicle is owned.

   c. Assume that instead of using randomly selected households, the sample data are obtained by asking readers of the *Washington Post* newspaper to mail in a survey form. How are the results affected?

36. Sample Size for Weapons on Campus Concerned about campus safety, college officials want to estimate the percentage of students who carry a gun, knife, or other such weapon. How many randomly selected students must be surveyed in order to be 95% confident that the sample percentage has a margin of error of three percentage points?
  a. Assume that another study indicated that 7% of college students carry weapons (based on a study by Cornell University).
  b. Assume that there is no available information that can be used to estimate the percentage of college students carrying weapons.

37. Color Blindness In a study of perception, 80 men are tested and 7 are found to have red/green color blindness (based on data from *USA Today*).
  a. Construct a 90% confidence interval estimate of the proportion of all men with this type of color blindness.
  b. What sample size would be needed to estimate the proportion of male red/green color blindness if we wanted 96% confidence that the sample proportion is in error by no more than 0.03? Use the sample proportion as a known estimate.
  c. Women have a 0.25% rate of red/green color blindness. Based on the result from part (a), can we safely conclude that women have a lower rate of red/green color blindness than men?

38. TV Ratings The CBS television show *60 Minutes* has been successful for many years. That show recently had a share of 20, meaning that among the TV sets in use, 20% were tuned to *60 Minutes* (based on data from Nielsen Media Research). Assume that this is based on a sample size of 4000 (typical for Nielsen surveys).
  a. Construct a 97% confidence interval estimate of the proportion of all sets in use that were tuned to *60 Minutes* at the time of the broadcast.
  b. What sample size would be required to estimate the percentage of sets tuned to *60 Minutes* if we wanted 99% confidence that the sample percentage is in error by no more than one-half of one percentage point? (Assume that we have no estimate of the proportion.)
  c. At the time of this particular *60 Minutes* broadcast, ABC ran "Exposed: Pro Wrestling," and that show received a share of 11. Based on the result from part (a), can we conclude that *60 Minutes* had a greater proportion of viewers? Did professional wrestling really need to be exposed?
  d. How is the confidence interval in part (a) affected if, instead of randomly selected subjects, the survey data are based on 4000 television viewers volunteering to call an "800" number to register their responses?

39. Cell Phones and Cancer A study of 420,000 Danish cell phone users found that 135 of them developed cancer of the brain or nervous system. Prior to this study of cell phone use, the rate of such cancer was found to be 0.0340% for those not using cell phones. The data are from *Journal of the National Cancer Institute*.
  a. Use the sample data to construct a 95% confidence interval estimate of the percentage of cell phone users who develop cancer of the brain or nervous system.
  b. Do cell phone users appear to have a rate of cancer of the brain or nervous system that is different from the rate of such cancer among those not using cell phones? Why or why not?

40. Pilot Fatalities Researchers studied crashes of general aviation (noncommercial and nonmilitary) airplanes and found that pilots died in 5.2% of 8411 crash landings (based on data from "Risk Factors for Pilot Fatalities in General Aviation Airplane

36. a. 278
    b. 1068

37. a. $0.0355 < p < 0.139$
    b. 373
    c. Yes

38. a. $0.186 < p < 0.214$
    b. 66,307
    c. Yes; no
    d. The results would not be valid.

39. a. $0.0267\% < p < 0.0376\%$
    b. No, because 0.0340% is in the confidence interval.

40. a. $4.72\% < p < 5.67\%$
    b. Yes. Explosion or fire is a serious risk factor.
    c. 5.67%

41. a. $1.07\% < p < 8.68\%$
    b. $70.1\% < p < 75.3\%$
    c. Yes, if wearing orange had no effect, we would expect that the percentage of injured hunters wearing orange should be between 70.1% and 75.3%, but it is much lower.

42. a. $92.5\% < p < 95.5\%$
    b. $72.2\% < p < 77.8\%$
    c. Yes, because the confidence intervals do not overlap and the likely range of percentages is higher for sloppy dressers.

43. $13.0\% < p < 29.0\%$; yes

44. a. $42.2\% < p < 69.8\%$
    b. $36.1\% < p < 63.9\%$
    c. Neither appears significantly more than the other.
    d. The duration of the tobacco use or alcohol use in each movie.

Crash Landings," by Rostykus, Cummings, and Mueller, *Journal of the American Medical Association*, Vol. 280, No. 11).

   a. Construct a 95% confidence interval estimate of the percentage of pilot deaths in all general aviation crashes.

   b. Among crashes with an explosion or fire on the ground, the pilot fatality rate is estimated by the 95% confidence interval of (15.5%, 26.9%). Is this result substantially different from the result from part (a)? What can you conclude about an explosion or fire as a risk factor?

   c. In planning for the allocation of federal funds to help with medical examinations of deceased pilots, what single percentage should be used? (We want to be reasonably sure that we have enough resources for the worst case scenario.)

41. Wearing Hunter Orange A study of hunting injuries and the wearing of "hunter" orange clothing showed that among 123 hunters injured when mistaken for game, 6 were wearing orange. Among 1115 randomly selected hunters, 811 reported that they routinely wear orange. The data are from the Centers for Disease Control.

   a. Construct a 95% confidence interval estimate of the percentage of injured hunters who are wearing orange.

   b. Construct a 95% confidence interval estimate of the percentage of hunters who routinely wear orange.

   c. Do these results indicate that a hunter who wears orange is less likely to be injured because of being mistaken for game? Why or why not?

42. Appearance Counts A Sales and Marketing Management survey included 651 sales managers, and 94% of them said that being a sloppy dresser can make a sales representative's job more difficult. For that same group, 75% said that being an unstylish dresser can make a sales representative's job more difficult.

   a. Construct a 90% confidence interval estimate of the percentage of sales managers who say that being a sloppy dresser can make a sales representative's job more difficult.

   b. Construct a 90% confidence interval estimate of the percentage of sales managers who say that being an unstylish dresser can make a sales representative's job more difficult.

   c. Given that sample proportions naturally vary, can we conclude that when sales managers state reasons for a sales representative's job becoming more difficult, the percentage is higher for sloppy dressing than for unstylish dressing? Why or why not?

43. Red M&M Candies Refer to Data Set 19 in Appendix B and find the sample proportion of M&Ms that are red. Use that result to construct a 95% confidence interval estimate of the population percentage of M&Ms that are red. Is the result consistent with the 20% rate that is reported by the candy maker Mars?

44. Alcohol and Tobacco Use in Children's Movies Refer to Data Set 7 in Appendix B.

   a. Construct a 95% confidence interval estimate of the percentage of animated children's movies showing any tobacco use.

   b. Construct a 95% confidence interval estimate of the percentage of animated children's movies showing any alcohol use.

   c. Compare the preceding results. Does either tobacco or alcohol appear in a greater percentage of animated children's movies?

   d. In using the results from parts (a) and (b) as measures of the depiction of unhealthy habits, what important characteristic of the data is not included?

6-2 Beyond the Basics

45. *Probing for Precision* An example of this section used the photo-cop survey data with $n = 829$ and $\hat{p} = 0.51$ to construct the 95% confidence interval of $0.476 < p < 0.544$. However, $\hat{p}$ cannot be exactly 0.51 because 51% of 829 people is 422.79 people, which is not possible. The sample statistic of 51% has been rounded to the nearest whole number. Find the minimum and maximum values of x for which $x/829$ is rounded to 0.51, then construct the confidence intervals corresponding to those two values of x. Do the results differ substantially from the confidence interval of $0.476 < p < 0.544$ that was found using 0.51?

46. *Using Finite Population Correction Factor* This section presented Formulas 6-2 and 6-3, which are used for determining sample size. In both cases we assumed that the population is infinite or very large and that we are sampling with replacement. When we have a relatively small population with size N and sample without replacement, we modify E to include the *finite population correction factor* shown here, and we can solve for n to obtain the result given here. Use this result to repeat part (b) of Exercise 38, assuming that we limit our population to a town with 10,000 television sets in use.

$$E = z_{\alpha/2}\sqrt{\frac{\hat{p}\hat{q}}{n}}\sqrt{\frac{N - n}{N - 1}} \qquad n = \frac{N\hat{p}\hat{q}[z_{\alpha/2}]^2}{\hat{p}\hat{q}[z_{\alpha/2}]^2 + (N - 1)E^2}$$

47. *One-Sided Confidence Interval* A *one-sided confidence interval* for p can be expressed as $p < \hat{p} + E$ or $p > \hat{p} - E$, where the margin of error E is modified by replacing $z_{\alpha/2}$ with z_{α}. If Air America wants to report an on-time performance of at least x percent with 95% confidence, construct the appropriate one-sided confidence interval and then find the percent in question. Assume that a simple random sample of 750 flights results in 630 that are on time.

48. *Confidence Interval from Small Sample* Special tables are available for finding confidence intervals for proportions involving small numbers of cases, where the normal distribution approximation cannot be used. For example, given $x = 3$ successes among $n = 8$ trials, the 95% confidence interval found in *Standard Probability and Statistics Tables and Formulae* (CRC Press) is $0.085 < p < 0.755$. Find the confidence interval that would result if you were to use the normal distribution incorrectly as an approximation to the binomial distribution. Are the results reasonably close?

49. *Interpreting Confidence Interval Limits* Assume that a coin is modified so that it favors heads, and 100 tosses result in 95 heads. Find the 99% confidence interval estimate of the proportion of heads that will occur with this coin. What is unusual about the results obtained by the methods of this section? Does common sense suggest a modification of the resulting confidence interval?

50. *Rule of Three* Suppose n trials of a binomial experiment result in no successes. According to the *Rule of Three*, we have 95% confidence that the true population proportion has an upper bound of $3/n$. (See "A Look at the Rule of Three," by Jovanovic and Levy, *American Statistician*, Vol. 51, No. 2.)
   a. If n independent trials result in no successes, why can't we find confidence interval limits by using the methods described in this section?
   b. If 20 patients are treated with a drug and there are no adverse reactions, what is the 95% upper bound for p, the proportion of all patients who experience adverse reactions to this drug?

45. $x = 419$ results in the confidence interval (0.471, 0.539) and $x = 426$ results in the confidence interval (0.480, 0.548). They do not differ by substantial amounts.

46. 8690

47. $p > 0.818$; 81.8%

48. $0.0395 < p < 0.710$; no

49. $0.894 < p < 1.006$; the upper confidence interval limit exceeds 1; use an upper limit of 1.

50. a. The assumption of $np \geq 5$ is not satisfied, so the normal distribution cannot be used.
    b. 0.15

51. 602

52. 9604

51. Women's Heights Women's heights are normally distributed with a mean of 63.6 in. and a standard deviation of 2.5 in. How many women must be surveyed if we want to estimate the percentage who are taller than 5 ft? Assume that we want 98% confidence that the error is no more than 2.5 percentage points. (*Hint:* The answer is substantially smaller than 2172.)

52. Poll Accuracy A *New York Times* article about poll results states, "In theory, in 19 cases out of 20, the results from such a poll should differ by no more than one percentage point in either direction from what would have been obtained by interviewing all voters in the United States." Find the sample size suggested by this statement.

6-3   Estimating a Population Mean: σ Known

In Section 6-2 we introduced the point estimate and confidence interval as tools for using a sample proportion to estimate a population proportion. We also showed how to determine the minimum sample size required to estimate a population proportion. In this section we again discuss point estimate, confidence interval, and sample size determination, but we now consider the objective of estimating a population mean μ.

Assumptions

1. The sample is a simple random sample. (All samples of the same size have an equal chance of being selected.)

2. The value of the population standard deviation σ is known.

3. Either or both of these conditions is satisfied: The population is normally distributed or $n > 30$.

In the above assumptions, we see that we want to estimate an unknown population mean μ, but we must know the value of the population standard deviation σ. It would be an unusual set of circumstances that would allow us to know σ without knowing μ. After all, the only way to find the value of σ is to compute it from all of the known population values, so the computation of μ would also be possible and, if we can find the true value of μ, there is no need to estimate it. Although the confidence interval methods of this section are not very realistic, they do reveal the basic concepts of important statistical reasoning, and they form the foundation for sample size determination discussed later in this section.

Assumption of Normality   In this section we will use the assumptions that we have a simple random sample, the value of σ is known, and either the population is normally distributed or $n > 30$. Technically, the population need not have a distribution that is exactly normal, but it should be approximately normal, meaning that the distribution is somewhat symmetric with one mode and no outliers. Investigate normality by using the sample data to construct a histogram, then determine whether it is approximately bell-shaped. A normal quantile plot (Section 5-7) could be constructed, but the methods of this section are said to be *robust*, which means that these methods are not strongly affected by departures from normality, provided that those departures are not too extreme. We can usually consider the population to be normally distributed after using the sample data to confirm that there are no outliers and the histogram has a shape that is not too far from a normal distribution.

Assumption of Required Sample Size   This section uses the normal distribution as the distribution of sample means. If the original population is itself normally distributed, then the means from samples of any size will be normally distributed. If the original population is not itself normally distributed, then we say that means of samples with size $n > 30$ have a distribution that can be approximated by a normal distribution. The condition that the sample size is $n > 30$ is commonly used as a guideline, but it is not possible to identify a specific minimum sample size that is sufficient for all cases. The minimum sample size actually depends on how much the population distribution departs from a normal distribution. Sample sizes of 15 to 30 are adequate if the population appears to have a distribution that is not far from being normal, but some other populations have distributions that are extremely far from normal and sample sizes of 50 or even 100 or higher might be necessary. We will use the simplified criterion of $n > 30$ as justification for treating the distribution of sample means as a normal distribution.

In Section 6-2 we saw that the sample proportion $\hat{p}$ is the best point estimate of the population proportion p. For similar reasons, the sample mean $\bar{x}$ is the best point estimate of the population mean μ.

The sample mean $\bar{x}$ is the best point estimate of the population mean.

Although we could use another statistic such as the sample median, midrange, or mode as an estimate of the population mean μ, studies have shown that the sample mean $\bar{x}$ usually provides the best estimate, for the following two reasons:

1. For many populations, the distribution of sample means $\bar{x}$ tends to be more consistent (with *less variation*) than the distributions of other sample statistics. (That is, if you use sample means to estimate the population mean μ, those sample means will have a smaller standard deviation than would other sample statistics, such as the median or the mode. The differences between $\bar{x}$ and μ therefore tend to be smaller than the differences obtained with some other statistic, such as the median.)

2. For all populations, the sample mean $\bar{x}$ is an **unbiased estimator** of the population mean μ, meaning that the distribution of sample means tends to center about the value of the population mean μ. (That is, sample means do not systematically tend to overestimate the value of μ, nor do they systematically tend to underestimate μ. Instead, they tend to target the value of μ itself. See Section 5-4 where we illustrated the principle that sample means tend to target the value of the population mean.)

Captured Tank Serial Numbers Reveal Population Size

During World War II, Allied intelligence specialists wanted to determine the number of tanks Germany was producing. Traditional spy techniques provided unreliable results, but statisticians obtained accurate estimates by analyzing serial numbers on captured tanks. As one example, records show that Germany actually produced 271 tanks in June 1941. The estimate based on serial numbers was 244, but traditional intelligence methods resulted in the extreme estimate of 1550. (See "An Empirical Approach to Economic Intelligence in World War II," by Ruggles and Brodie, *Journal of the American Statistical Association*, Vol. 42.)

*Estimating
Wildlife
Population Sizes*

The National Forest Management Act protects endangered species, including the northern spotted owl, with the result that the forestry industry was not allowed to cut vast regions of trees in the Pacific Northwest. Biologists and statisticians were asked to analyze the problem, and they concluded that survival rates and population sizes were decreasing for the female owls, known to play an important role in species survival. Biologists and statisticians also studied salmon in the Snake and Columbia Rivers in Washington, and penguins in New Zealand. In the article "Sampling Wildlife Populations" (*Chance*, Vol. 9, No. 2), authors Bryan Manly and Lyman McDonald comment that in such studies, "biologists gain through the use of modeling skills that are the hallmark of good statistics. Statisticians gain by being introduced to the reality of problems by biologists who know what the crucial issues are."

EXAMPLE **Body Temperatures** Data Set 4 in Appendix B includes 106 body temperatures taken at 12 AM on day 2. Here are statistics for that sample: $n = 106$, $\bar{x} = 98.20°F$, and $s = 0.62°F$. Use this sample to find the best point estimate of the population mean μ of all body temperatures.

SOLUTION   For the sample data, $\bar{x} = 98.20°F$. Because the sample mean $\bar{x}$ is the best point estimate of the population mean μ, we conclude that the best point estimate of the population mean μ of all body temperatures is 98.20°F.

Confidence Intervals

We saw in Section 6-2 that although a point estimate is the *best* single value for estimating a population parameter, it does not give us any indication of just how *good* the best estimate is. Statisticians developed the confidence interval or interval estimate, which consists of a range (or an interval) of values instead of just a single value. The confidence interval is associated with a confidence level, such as 0.95 (or 95%). The confidence level gives us the success rate of the procedure used to construct the confidence interval. As described in Section 6-2, the confidence level is often expressed as the probability or area $1 - \alpha$ (lowercase Greek alpha), where α is the complement of the *confidence level*. For a 0.95 (or 95%) confidence level, $\alpha = 0.05$. For a 0.99 (or 99%) confidence level, $\alpha = 0.01$.

Margin of Error   When we collect a set of sample data, such as the set of 106 body temperatures listed for 12 AM of day 2 in Data Set 4 in Appendix B, we can calculate the sample mean $\bar{x}$ and that sample mean is typically different from the population mean μ. The difference between the sample mean and the population mean is an error. In Section 5-5 we saw that $\sigma/\sqrt{n}$ is the standard deviation of sample means. Using $\sigma/\sqrt{n}$ and the $z_{\alpha/2}$ notation introduced in Section 6-2, we now use the *margin of error E* expressed as follows.

Formula 6-4 $E = z_{\alpha/2} \cdot \dfrac{\sigma}{\sqrt{n}}$ margin of error for mean (based on known σ)

Formula 6-4 reflects the fact that the sampling distribution of sample means $\bar{x}$ is *exactly* a normal distribution with mean μ and standard deviation $\sigma/\sqrt{n}$ whenever the population has a normal distribution with mean μ and standard deviation σ. If the population is not normally distributed, large samples yield sample means with a distribution that is *approximately* normal.

Given the way that the margin of error E is defined, there is a probability of $1 - \alpha$ that a sample mean will be in error (different from the population mean μ) by no more than E, and there is a probability of α that the sample mean will be in error by more than E. The calculation of the margin of error E as given in Formula 6-4 requires that you know the population standard deviation σ, but Section 6-4 will present a method for calculating the margin of error E when σ is not known.

Using the margin of error E, we can now identify the confidence interval for the population mean μ (if the assumed conditions for this section are satisfied). The three commonly used formats for expressing the confidence interval are shown in the following box.

Confidence Interval Estimate of the Population Mean μ (With σ Known)

$$\bar{x} - E < \mu < \bar{x} + E \qquad \text{where } E = z_{\alpha/2} \cdot \frac{\sigma}{\sqrt{n}}$$

or

$$\bar{x} \pm E$$

or

$$(\bar{x} - E, \bar{x} + E)$$

Definition

The two values $\bar{x} - E$ and $\bar{x} + E$ are called **confidence interval limits.**

Procedure for Constructing a Confidence Interval for μ (with Known σ)

1. Verify that the required assumptions are satisfied. (We have a simple random sample, σ is known, and either the population appears to be normally distributed or $n > 30$.)

2. Refer to Table A-2 and find the critical value $z_{\alpha/2}$ that corresponds to the desired confidence level. (For example, if the confidence level is 95%, the critical value is $z_{\alpha/2} = 1.96$.)

3. Evaluate the margin of error $E = z_{\alpha/2}\sigma/\sqrt{n}$.

4. Using the value of the calculated margin of error E and the value of the sample mean $\bar{x}$, find the values of $\bar{x} - E$ and $\bar{x} + E$. Substitute those values in the general format for the confidence interval:

$$\bar{x} - E < \mu < \bar{x} + E$$

or

$$\bar{x} \pm E$$

or

$$(\bar{x} - E, \bar{x} + E)$$

5. Round the resulting values by using the following round-off rule.

Round-Off Rule for Confidence Intervals Used to Estimate μ

1. When using the *original set of data* to construct a confidence interval, round the confidence interval limits to one more decimal place than is used for the original set of data.

2. When the original set of data is unknown and only the *summary statistics* $(n, \bar{x}, s)$ are used, round the confidence interval limits to the same number of decimal places used for the sample mean.

Note to Instructor

This is basically the same round-off rule used in Chapter 2 for statistics such as the sample mean.

Interpreting a Confidence Interval   As in Section 6-2, we must be careful to interpret confidence intervals correctly. After obtaining a confidence interval estimate of the population mean μ, such as a 95% confidence interval of $98.08 < \mu < 98.32$, there is a correct interpretation and many wrong interpretations.

Correct: "We are 95% confident that the interval from 98.08 to 98.32 actually does contain the true value of μ." This means that if we were to select many different samples of the same size and construct the corresponding confidence intervals, in the long run 95% of them would actually contain the value of μ. (As in Section 6-2, this correct interpretation refers to the success rate of the *process* being used to estimate the population mean.)

Wrong: Because μ is a fixed constant, it would be wrong to say "there is a 95% chance that μ will fall between 98.08 and 98.32." The confidence interval does not describe the behavior of individual sample values, so it would also be wrong to say that "95% of all data values are between 98.08 and 98.32." Also, the confidence interval does not describe the behavior of individual sample means, so it would also be wrong to say that "95% of sample means fall between 98.08 and 98.32."

EXAMPLE   **Body Temperatures**   For the sample of body temperatures in Data Set 4 in Appendix B (for 12 AM on day 2), we have $n = 106$ and $\bar{x} = 98.20°F$. Assume that the sample is a simple random sample and that σ is somehow known to be 0.62°F. Using a 0.95 confidence level, find both of the following:

a. The margin of error E

b. The confidence interval for μ.

SOLUTION   First verify that the required assumptions are satisfied. The value of σ is assumed to be known (0.62°F) and the sample size $n = 106$ is greater than 30. Also, there are no outliers. (Because $n > 30$, there is no need to check that the sample comes from a normally distributed population, but a histogram of the 106 body temperatures would show that the sample data have a distribution that is approximately bell-shaped, suggesting that the population of body temperatures is normally distributed.) The required assumptions are therefore satisfied and we can proceed with the methods of this section.

a. The 0.95 confidence level implies that $\alpha = 0.05$, so $z_{\alpha/2} = 1.96$ (as was shown in an example in Section 6-2). The margin of error E is calculated by using Formula 6-4 as follows. Extra decimal places are used to minimize rounding errors in the confidence interval found in part (b).

$$E = z_{\alpha/2} \cdot \frac{\sigma}{\sqrt{n}} = 1.96 \cdot \frac{0.62}{\sqrt{106}} = 0.118031$$

b. With $\bar{x} = 98.20$ and $E = 0.118031$, we construct the confidence interval as follows:

$$\bar{x} - E < \mu < \bar{x} + E$$

$$98.20 - 0.118031 < \mu < 98.20 + 0.118031$$

$$98.08 < \mu < 98.32 \quad \text{(rounded to two decimal places as in } \bar{x}\text{)}$$

INTERPRETATION This result could also be expressed as 98.20 ± 0.12 or as $(98.08, 98.32)$. Based on the sample with $n = 106$, $\bar{x} = 98.20$, and σ assumed to be 0.62, the confidence interval for the population mean μ is $98.08°\text{F} < \mu < 98.32°\text{F}$ and this interval has a 0.95 confidence level. This means that if we were to select many different samples of size 106 and construct the confidence intervals as we did here, 95% of them would actually contain the value of the population mean μ. Note that the confidence interval limits of $98.08°\text{F}$ and $98.32°\text{F}$ do not contain $98.6°\text{F}$, the value generally believed to be the mean body temperature. Based on these results, it seems very unlikely that $98.6°\text{F}$ is the correct value of μ.

Rationale for the Confidence Interval The basic idea underlying the construction of confidence intervals relates to the central limit theorem, which indicates that if we have a simple random sample from a normally distributed population, or a simple random sample of size $n > 30$ from any population, the distribution of sample means is approximately normal with mean μ and standard deviation $\sigma/\sqrt{n}$. The confidence interval format is really a variation of the equation that was already used with the central limit theorem. In the expression $z = (\bar{x} - \mu_{\bar{x}})/\sigma_{\bar{x}}$, replace $\sigma_{\bar{x}}$ with $\sigma/\sqrt{n}$, replace $\mu_{\bar{x}}$ with μ, then solve for μ to get

$$\mu = \bar{x} - z\frac{\sigma}{\sqrt{n}}$$

Using the positive and negative values for z results in the confidence interval limits we are using.

Let's consider the specific case of a 95% confidence level, so $\alpha = 0.05$ and $z_{\alpha/2} = 1.96$. For this case, there is a probability of 0.05 that a sample mean will be more than 1.96 standard deviations (or $z_{\alpha/2}\sigma/\sqrt{n}$ which we denote by E) away from the population mean μ. Conversely, there is a 0.95 probability that a sample mean will be within 1.96 standard deviations (or $z_{\alpha/2}\sigma/\sqrt{n}$) of μ. (See Figure 6-4 on the next page.) If the sample mean $\bar{x}$ is within $z_{\alpha/2}\sigma/\sqrt{n}$ of the population mean μ, then μ must be between $\bar{x} - z_{\alpha/2}\sigma/\sqrt{n}$ and $\bar{x} + z_{\alpha/2}\sigma/\sqrt{n}$; this is expressed in the general format of our confidence interval (with $z_{\alpha/2}\sigma/\sqrt{n}$ denoted as E): $\bar{x} - E < \mu < \bar{x} + E$.

Alternative Method (not used in this book) When constructing a confidence interval estimate of the population mean μ, an alternative method *not used*

FIGURE 6-4 Distribution of Sample Means with Known σ

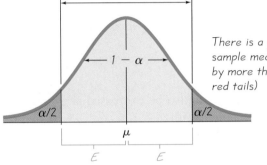

There is a $1 - \alpha$ probability that a sample mean will be in error by less than E or $z_{\alpha/2}\sigma/\sqrt{n}$

There is a probability of α that a sample mean will be in error by more than E (in one of the red tails)

in this book is to use the procedures described above even if σ is not known, but the sample is a simple random sample with $n > 30$. With this alternative method, we use s as an estimate of σ provided that $n > 30$. In Section 6-4 we list reasons why this alternative method is not used in this book, and those reasons include the fact that this alternative method is not commonly used in the real world. The real world uses the methods described in this book.

A key feature of the methods we are using in this section is that we want to estimate an unknown population mean μ, and the population standard deviation σ is known. In the next section we present a method for estimating an unknown population mean μ when the population standard deviation is not known. The conditions of the following section are much more likely to occur in real circumstances. Although the methods of this section are unrealistic because they are based on knowledge of the population standard deviation σ, they do enable us to see the basic method for constructing a confidence interval estimate of μ by using the same normal distribution that has been used often in Chapter 5 and Section 6-2. Also, the methods we have discussed so far in this section lead to a very practical method for determining sample size.

Determining Sample Size Required to Estimate μ

We now want to address this key question: When we plan to collect a simple random sample of data that will be used to estimate a population mean μ, *how many* sample values must be obtained? In other words, we will find the sample size n that is required to estimate the value of a population mean. For example, suppose we want to estimate the mean weight of airline passengers (an important value for reasons of safety). How many passengers must be randomly selected and weighed? Determining the size of a simple random sample is a very important issue, because samples that are needlessly large waste time and money, and samples that are too small may lead to poor results. In many cases we can find the minimum sample size needed to estimate some parameter, such as the population mean μ.

If we begin with the expression for the margin of error E (Formula 6-4) and solve for the sample size n, we get the following.

Sample Size for Estimating Mean μ

Formula 6-5
$$n = \left[\frac{z_{\alpha/2}\sigma}{E} \right]^2$$

where $z_{\alpha/2}$ = critical z score based on the desired confidence level
  E = desired margin of error
  σ = population standard deviation

Formula 6-5 is remarkable because it shows that the sample size does not depend on the size (N) of the population; the sample size depends on the desired confidence level, the desired margin of error, and the value of the standard deviation σ. (See Exercise 33 for dealing with cases in which a relatively large sample is selected without replacement from a finite population.)

The sample size must be a whole number, because it represents the number of sample values that must be found. However, when we use Formula 6-5 to calculate the sample size n, we usually get a result that is not a whole number. When this happens, we use the following round-off rule. (It is based on the principle that when rounding is necessary, the required sample size should be rounded *upward* so that it is at least adequately large as opposed to slightly too small.)

Round-Off Rule for Sample Size n

When finding the sample size n, if the use of Formula 6-5 does not result in a whole number, always *increase* the value of n to the next *larger* whole number.

Dealing with Unknown σ When Finding Sample Size   When applying Formula 6-5, there is one very practical dilemma: The formula requires that we substitute some value for the population standard deviation σ, but in reality, it is usually unknown. When determining a required sample size (not constructing a confidence interval), here are some ways that we can work around this problem:

1. Use the range rule of thumb (see Section 2-5) to estimate the standard deviation as follows: $\sigma \approx$ range/4. (With a sample of 87 or more values randomly selected from a normally distributed population, range/4 will yield a value that is greater than or equal to σ at least 95% of the time. See "Using the Sample Range as a Basis for Calculating Sample Size in Power Calculations" by Richard Browne, *The American Statistician,* Vol. 55, No. 4.)

2. Conduct a pilot study by starting the sampling process. Based on the first collection of at least 31 randomly selected sample values, calculate the sample standard deviation s and use it in place of σ. The estimated value of σ can then be improved as more sample data are obtained.

3. Estimate the value of σ by using the results of some other study that was done earlier.

In addition, we can sometimes be creative in our use of other known results. For example, IQ tests are typically designed so that the mean is 100 and the standard deviation is 15. Statistics professors have IQ scores with a mean greater than 100 and a standard deviation less than 15 (because they are a more homogeneous group than people randomly selected from the general population). We do not know the specific value of σ for statistics professors, but we can play it safe by using $\sigma = 15$. Using a value for σ that is larger than the true value will make the sample size larger than necessary, but using a value for σ that is too small would result in a sample size that is inadequate. *When calculating the sample size n, any errors should always be conservative in the sense that they make n too large instead of too small.*

> EXAMPLE **IQ Scores of Statistics Professors** Assume that we want to estimate the mean IQ score for the population of statistics professors. How many statistics professors must be randomly selected for IQ tests if we want 95% confidence that the sample mean is within 2 IQ points of the population mean?
>
> SOLUTION   The values required for Formula 6-5 are found as follows:
>
> $z_{\alpha/2} = 1.96$ (This is found by converting the 95% confidence level to $\alpha = 0.05$, then finding the critical z score as described in Section 6-2.)
>
> $E = 2$ (Because we want the sample mean to be within 2 IQ points of μ, the desired margin of error is 2.)
>
> $\sigma = 15$ (See the discussion in the paragraph that immediately precedes this example.)
>
> With $z_{\alpha/2} = 1.96$, $E = 2$, and $\sigma = 15$, we use Formula 6-5 as follows:
>
> $$ n = \left[\frac{z_{\alpha/2}\sigma}{E} \right]^2 = \left[\frac{1.96 \cdot 15}{2} \right]^2 = 216.09 = 217 \quad \text{(rounded } up\text{)} $$
>
> INTERPRETATION Among the thousands of statistics professors, we need to obtain a simple random sample of at least 217 of them, then we need to get their IQ scores. With a simple random sample of only 217 statistics professors, we will be 95% confident that the sample mean $\bar{x}$ is within 2 IQ points of the true population mean μ.

If we are willing to settle for less accurate results by using a larger margin of error, such as 4, the sample size drops to 54.0225, which is rounded *up* to 55.

Doubling the margin of error causes the required sample size to decrease to one-fourth its original value. Conversely, halving the margin of error quadruples the sample size. Consequently, if you want more accurate results, the sample size must be substantially increased. Because large samples generally require more time and money, there is often a need for a trade-off between the sample size and the margin of error E.

Using Technology

Confidence Intervals See the end of Section 6-4 for the confidence interval procedures that apply to the methods of this section as well as those of Section 6-4. STATDISK, Minitab, Excel, and the TI-83 Plus calculator can all be used to find confidence intervals when we want to estimate a population mean and the assumptions of this section (including a known value of σ) are all satisfied.

Sample Size Determination Sample size calculations are not included with the TI-83 Plus calculator, or Minitab, or Excel. The STATDISK procedure for determining the sample size required to estimate a population mean μ is described below.

STATDISK   Select **Analysis** from the main menu bar at the top, then select **Sample Size Determination,** followed by **Estimate Mean.** You must now enter the confidence level (such as 0.95), the error E, and the population standard deviation σ. There is also an option that allows you to enter the population size N, assuming that you are sampling without replacement from a finite population. (See Exercise 34.)

6-3 Basic Skills and Concepts

Recommended Assignment:
Exercises 1–28. (These exercises require the determination of sample size: 13–16, 25–32.)

Finding Critical Values. In Exercises 1–4, find the critical value $z_{\alpha/2}$ that corresponds to the given confidence level.

1. 98%

2. 95%

3. 96%

4. 99.5%

1. 2.33
2. 1.96
3. 2.05
4. 2.81

Verifying Assumptions. In Exercises 5–8, determine whether the given conditions justify using the margin of error $E = z_{\alpha/2}\sigma/\sqrt{n}$ when finding a confidence interval estimate of the population mean μ.

5. The sample size is $n = 200$ and $\sigma = 15$.

5. Yes

6. The sample size is $n = 5$ and σ is not known.

6. No

7. The sample size is $n = 5$, $\sigma = 12.4$, and the original population is normally distributed.

7. Yes

8. The sample size is $n = 9$, σ is not known, and the original population is normally distributed.

8. No

Finding Margin of Error and Confidence Interval. In Exercises 9–12, use the given confidence level and sample data to find (a) the margin of error E and (b) a confidence interval for estimating the population mean μ.

9. Salaries of statistics professors: 95% confidence; $n = 100$, $\bar{x} = \$95,000$ (we wish), and σ is known to be $\$12,345$

9. $2419.62; $92,580 < \mu <$ $97,420

10. 1.68 years; 78.8 years $< \mu <$ 82.2 years

11. 0.823 sec; 4.42 sec $< \mu <$ 6.06 sec

12. $3667; $42,011 $< \mu <$ $49,345

13. 62
14. 166
15. 250
16. 1380

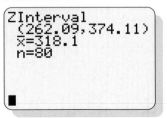

17. 318.1
18. 262.09 $< \mu <$ 374.11
19. $\mu = 318.10 \pm 56.01$
20. There is 95% confidence that the limits of 262.09 and 374.11 contain the true value of the population mean μ.
21. 30.0ºC $< \mu <$ 30.8ºC; it is unrealistic to know σ.
22. 140.2 lb $< \mu <$ 225.6 lb; it is unrealistic to know σ.
23. 141.4 $< \mu <$ 203.6; it is unrealistic to know σ.
24. 40.2 cm $< \mu <$ 41.0 cm; it is unrealistic to know σ.

10. Ages of drivers occupying the passing lane while driving 25 mi/h with the left signal flashing: 99% confidence; $n = 50$, $\bar{x} = 80.5$ years, and σ is known to be 4.6 years

11. Times between uses of a TV remote control by males during commercials: 90% confidence; $n = 25$, $\bar{x} = 5.24$ sec, the population is normally distributed, and σ is known to be 2.50 sec

12. Starting salaries of college graduates who have taken a statistics course: 95% confidence; $n = 28$, $\bar{x} = \$45,678$, the population is normally distributed, and σ is known to be $9900

Finding Sample Size. In Exercises 13–16, use the given margin of error, confidence level, and population standard deviation σ to find the minimum sample size required to estimate an unknown population mean μ.

13. Margin of error: $125, confidence level: 95%, $\sigma = \$500$

14. Margin of error: 3 lb, confidence level: 99%, $\sigma = 15$ lb

15. Margin of error: 5 min, confidence level: 90%, $\sigma = 48$ min

16. Margin of error: $500, confidence level: 94%, $\sigma = \$9877$

Interpreting Results. In Exercises 17–20, refer to the accompanying TI-83 Plus calculator display of a 95% confidence interval generated using the methods of this section. The sample display results from using a sample of 80 measured cholesterol levels of randomly selected adults.

17. Identify the value of the point estimate of the population mean μ.

18. Express the confidence interval in the format of $\bar{x} - E < \mu < \bar{x} + E$.

19. Express the confidence interval in the format of $\bar{x} \pm E$.

20. Write a statement that interprets the 95% confidence interval.

21. Everglades Temperatures In order to monitor the ecological health of the Florida Everglades, various measurements are recorded at different times. The bottom temperatures are recorded at the Garfield Bight station and the mean of 30.4°C is obtained for 61 temperatures recorded on 61 different days. Assuming that $\sigma = 1.7$°C, find a 95% confidence interval estimate of the population mean of all such temperatures. What aspect of this problem is not realistic?

22. Weights of Bears The health of the bear population in Yellowstone National Park is monitored by periodic measurements taken from anesthetized bears. A sample of 54 bears has a mean weight of 182.9 lb. Assuming that σ is known to be 121.8 lb, find a 99% confidence interval estimate of the mean of the population of all such bear weights. What aspect of this problem is not realistic?

23. Cotinine Levels of Smokers When people smoke, the nicotine they absorb is converted to cotinine, which can be measured. A sample of 40 smokers has a mean cotinine level of 172.5. Assuming that σ is known to be 119.5, find a 90% confidence interval estimate of the mean cotinine level of all smokers. What aspect of this problem is not realistic?

24. Head Circumferences In order to help identify baby growth patterns that are unusual, we need to construct a confidence interval estimate of the mean head circumference of all babies that are two months old. A random sample of 100 babies is obtained, and the mean head circumference is found to be 40.6 cm. Assuming that the population standard deviation is known to be 1.6 cm, find a 99% confidence interval estimate of the mean head circumference of all two-month-old babies. What aspect of this problem is not realistic?

25. *Sample Size for Mean IQ of Statistics Students* The Weschler IQ test is designed so that the mean is 100 and the standard deviation is 15 for the population of normal adults. Find the sample size necessary to estimate the mean IQ score of statistics students. We want to be 95% confident that our sample mean is within 2 IQ points of the true mean. The mean for this population is clearly greater than 100. The standard deviation for this population is probably less than 15 because it is a group with less variation than a group randomly selected from the general population; therefore, if we use $\sigma = 15$, we are being conservative by using a value that will make the sample size at least as large as necessary. Assume then that $\sigma = 15$ and determine the required sample size.

25. 217

26. *Sample Size for Weights of Quarters* The Tyco Video Game Corporation finds that it is losing income because of slugs used in its video games. The machines must be adjusted to accept coins only if they fall within set limits. In order to set those limits, the mean weight of quarters in circulation must be estimated. A sample of quarters will be weighed in order to determine the mean. How many quarters must we randomly select and weigh if we want to be 99% confident that the sample mean is within 0.025 g of the true population mean for all quarters? Based on results from the sample of quarters in Data Set 29 in Appendix B, we can estimate that the population standard deviation is 0.068 g.

26. 50

27. *Sample Size for Estimating Income* An economist wants to estimate the mean income for the first year of work for college graduates who have had the profound wisdom to take a statistics course. How many such incomes must be found if we want to be 95% confident that the sample mean is within $500 of the true population mean? Assume that a previous study has revealed that for such incomes, $\sigma = \$6250$.

27. 601

28. *Sample Size for Television Viewing* Nielsen Media Research wants to estimate the mean amount of time (in minutes) that full-time college students spend watching television each weekday. Find the sample size necessary to estimate that mean with a 15-minute margin of error. Assume that a 96% confidence level is desired. Also assume that a pilot study showed that the standard deviation is estimated to be 112.2 min.

28. 236

29. *Sample Size Using Range Rule of Thumb* You have just been hired by the marketing division of General Motors to estimate the mean amount of money now being spent on the purchase of new cars in the United States. First use the range rule of thumb to make a rough estimate of the standard deviation of the amounts spent. It is reasonable to assume that typical amounts range from $12,000 to about $70,000. Then use that estimated standard deviation to determine the sample size corresponding to 95% confidence and a $100 margin of error. Is the sample size practical? If not, what should be changed to get a practical sample size?

29. 80,770; no, increase the margin of error.

30. *Sample Size Using Range Rule of Thumb* Estimate the minimum and maximum ages for typical textbooks currently used in college courses, then use the range rule of thumb to estimate the standard deviation. Next, find the size of the sample required to estimate the mean age (in years) of textbooks currently used in college courses. Use a 90% confidence level and assume that the sample mean will be in error by no more than 0.25 year.

30. Answer varies, but using 0 and 8 for minimum and maximum yields 174.

31. The range is 40, so $\sigma \approx 40/4 = 10$ by the range rule of thumb, and the sample size is 97. $s = 11.3$, which results in a sample size of 123. The sample size of 123 is likely to be better because s is a better estimate of σ than range/4.

32. The range is 61, so $\sigma \approx 15.25$ by the range rule of thumb, and the sample size is 100. $s = 11.6$, which results in a sample size of 58. The sample size of 58 is likely to be better because s is a better estimate of σ than range/4.

(T) 31. Sample Size Using Sample Data You want to estimate the mean pulse rate of adult males. Refer to Data Set 1 in Appendix B and find the maximum and minimum pulse rates for males, then use those values with the range rule of thumb to estimate σ. How many adult males must you randomly select and test if you want to be 95% confident that the sample mean pulse rate is within 2 beats (per minute) of the true population mean μ? If, instead of using the range rule of thumb, the standard deviation of the male pulse rates in Data Set 1 is used as an estimate of σ, is the required sample size very different? Which sample size is likely to be closer to the correct sample size?

(T) 32. Sample Size Using Sample Data You want to estimate the mean diastolic blood pressure level of adult females. Refer to Data Set 1 in Appendix B and find the maximum and minimum diastolic blood pressure level for females, then use those values with the range rule of thumb to estimate σ. How many adult females must you randomly select and test if you want to be 95% confident that the sample mean diastolic blood pressure level is within 3 mm Hg of the true population mean μ? If, instead of using the range rule of thumb, the standard deviation of the female diastolic blood pressure levels in Data Set 1 is used as an estimate of σ, is the required sample size very different? Which sample size is likely to be closer to the correct sample size?

6-3 Beyond the Basics

33. $105 < \mu < 115$

33. Confidence Interval with Finite Population Correction Factor The standard error of the mean is $\sigma / \sqrt{n}$ provided that the population size is infinite. If the population size is finite and is denoted by N, then the correction factor $\sqrt{(N - n)/(N - 1)}$ should be used whenever $n > 0.05N$. This correction factor multiplies the margin of error E given in Formula 6-4, so that the margin of error is as shown below. Find the 95% confidence interval for the mean of 250 IQ scores if a sample of 35 of those scores produces a mean of 110. Assume that $\sigma = 15$.

$$E = z_{\alpha/2} \frac{\sigma}{\sqrt{n}} \sqrt{\frac{N - n}{N - 1}}$$

34. 105

34. Sample Size with Finite Population Correction Factor In Formula 6-4 for the margin of error E, we assume that the population is infinite, that we are sampling with replacement, or that the population is very large. If we have a relatively small population and sample without replacement, we should modify E to include a *finite population correction factor*, so that the margin of error is as shown in Exercise 33, where N is the population size. That expression for the margin of error can be solved for n to yield

$$n = \frac{N\sigma^2(z_{\alpha/2})^2}{(N - 1)E^2 + \sigma^2(z_{\alpha/2})^2}$$

Repeat Exercise 25, assuming that the statistics students are randomly selected without replacement from a population of $N = 200$ statistics students.

Note to Instructor

This section introduces the t distribution that is used when σ is not known. Comment that the methods of this section require that the parent population has a distribution that is essentially normal (even when the t distribution is used). If the original population is not normally distributed, the bootstrap method is an interesting alternative, especially for students with great enthusiasm for computers. Consider the Technology Project with an oral and written report as an extra-credit assignment.

6-4 Estimating a Population Mean: σ Not Known

In Section 6-3 we presented methods for constructing a confidence interval estimate of an unknown population mean μ, but we considered only cases in which the population standard deviation σ is known. We noted that the assumption of a known σ is not very realistic, because the calculation of σ requires that we know all of the population values, but if we know all of the population values we can easily find the value of the population mean μ, so there is no need to estimate μ. In this section we present a method for constructing confidence interval estimates of μ without the requirement that σ is known. The usual procedure is to collect sample data and find the value of the statistics n, $\bar{x}$, and s. Because the methods of this section are based on those statistics and σ is not required, the methods of this section are very realistic, practical, and they are used often. Note that the following assumptions for the methods of this section do *not* include a requirement that σ is known.

Assumptions

1. The sample is a simple random sample.
2. Either the sample is from a normally distributed population or $n > 30$.

As in Section 6-3, the requirement of a normally distributed population is not a strict requirement. We can usually consider the population to be normally distributed after using the sample data to confirm that there are no outliers and the histogram has a shape that is not very far from a normal distribution. Also, as in Section 6-3, the requirement that the sample size is $n > 30$ is commonly used as a guideline, but the minimum sample size actually depends on how much the population distribution departs from a normal distribution. We will use the simplified criterion of $n > 30$ as justification for treating the distribution of sample means as a normal distribution. The sampling distribution of sample means $\bar{x}$ is *exactly* a normal distribution with mean μ and standard deviation $\sigma / \sqrt{n}$ whenever the population has a normal distribution with mean μ and standard deviation σ. If the population is not normally distributed, large samples yield sample means with a distribution that is *approximately* normal with mean μ and standard deviation $\sigma / \sqrt{n}$.

As in Section 6-3, the sample mean $\bar{x}$ is the best point estimate (or single-valued estimate) of the population mean μ. As in Section 6-3, the distribution of sample means $\bar{x}$ tends to be more consistent (with *less variation*) than the distributions of other sample statistics, and the sample mean $\bar{x}$ is an *unbiased estimator* that targets the population mean μ.

The sample mean $\bar{x}$ is the best point estimate of the population mean μ.

In Sections 6-2 and 6-3 we noted that there is a serious limitation to the usefulness of a point estimate: The single value of a point estimate does not reveal how good that estimate is. Confidence intervals give us much more meaningful information by providing a range of values associated with a degree of likelihood that the range actually does contain the true value of μ.

Here is the key point of this section: If σ is not known, but the above conditions are satisfied, instead of using the normal distribution, we use the *Student t distribution* developed by William Gosset (1876–1937). Gosset was a Guinness Brewery employee who needed a distribution that could be used with small samples. The Irish brewery where he worked did not allow the publication of research results, so Gosset published under the pseudonym *Student*.

Estimating Sugar in Oranges

In Florida, members of the citrus industry make extensive use of statistical methods. One particular application involves the way in which growers are paid for oranges used to make orange juice. An arriving truckload of oranges is first weighed at the receiving plant, then a sample of about a dozen oranges is randomly selected. The sample is weighed and then squeezed, and the amount of sugar in the juice is measured. Based on the sample results, an estimate is made of the total amount of sugar in the entire truckload. Payment for the load of oranges is based on the estimate of the amount of sugar because sweeter oranges are more valuable than those less sweet, even though the amounts of juice may be the same.

Excerpts from a Department of Transportation Circular

The following excerpts from a Department of Transportation circular concern some of the accuracy requirements for navigation equipment used in aircraft. Note the use of the confidence interval. "The total of the error contributions of the airborne equipment, when combined with the appropriate flight technical errors listed, should not exceed the following with a 95% confidence (2-sigma) over a period of time equal to the update cycle." "The system of airways and routes in the United States has widths of route protection used on a VOR system with accuracy of ± 4.5 degrees on a 95% probability basis."

Because we do not know the value of σ, we estimate it with the value of the sample standard deviation s, but this introduces another source of unreliability, especially with small samples. In order to keep a confidence interval at some desired level, such as 95%, we compensate for this additional unreliability by making the confidence interval wider: We use critical values larger than the critical values of $z_{\alpha/2}$ that were used in Section 6-3 where σ was known. Instead of using critical values of $z_{\alpha/2}$, we use the larger critical values of $t_{\alpha/2}$ found from the Student t distribution.

Student t Distribution

If the distribution of a population is essentially normal (approximately bell-shaped), then the distribution of

$$t = \frac{\bar{x} - \mu}{\frac{s}{\sqrt{n}}}$$

is essentially a **Student t distribution** for all samples of size n. The Student t distribution, often referred to as the t **distribution**, is used to find critical values denoted by $t_{\alpha/2}$.

We will soon discuss some of the important properties of the t distribution, but we will first present the components needed for the construction of confidence intervals. Let's start with the critical value denoted by $t_{\alpha/2}$. A value of $t_{\alpha/2}$ can be found in Table A-3. To find a critical value $t_{\alpha/2}$ in Table A-3, locate the appropriate number of *degrees of freedom* in the left column and proceed across the corresponding row until reaching the number directly below the appropriate area at the top.

Definition

The number of **degrees of freedom** for a collection of sample data set is the number of sample values that can vary after certain restrictions have been imposed on all data values.

For example, if 10 students have quiz scores with a mean of 80, we can freely assign values to the first 9 scores, but the 10th score is then determined. The sum of the 10 scores must be 800, so the 10th score must equal 800 minus the sum of the first 9 scores. Because those first 9 scores can be freely selected to be any values, we say that there are 9 degrees of freedom available. For the applications of this section, the number of degrees of freedom is simply the sample size minus 1.

$$\text{degrees of freedom} = n - 1$$

EXAMPLE **Finding a Critical Value** A sample of size $n = 15$ is a simple random sample selected from a normally distributed population. Find the critical value $t_{\alpha/2}$ corresponding to a 95% confidence level.

SOLUTION   Because $n = 15$, the number of degrees of freedom is given by $n - 1 = 14$. Using Table A-3, we locate the 14th row by referring to the column at the extreme left. As in Section 6-2, a 95% confidence level corresponds to $\alpha = 0.05$, so we find the column listing values for an *area of 0.05 in two tails*. The value corresponding to the row for 14 degrees of freedom and the column for an area of 0.05 in two tails is 2.145, so $t_{\alpha/2} = 2.145$.

Now that we know how to find critical values denoted by $t_{\alpha/2}$ we can go on to describe the margin of error E and the confidence interval.

Margin of Error E for the Estimate of μ (With σ Not Known)

Formula 6-6 $$E = t_{\alpha/2}\frac{s}{\sqrt{n}}$$

where $t_{\alpha/2}$ has $n - 1$ degrees of freedom

Confidence Interval for the Estimate of μ (With σ Not Known)

$$\bar{x} - E < \mu < \bar{x} + E$$

where $$E = t_{\alpha/2}\frac{s}{\sqrt{n}}$$

The following procedure uses the above margin of error in the construction of confidence interval estimates of μ.

Procedure for Constructing a Confidence Interval for μ (With σ Unknown)

1. Verify that the required assumptions are satisfied. (We have a simple random sample, and either the population appears to be normally distributed or $n > 30$.)

2. Using $n - 1$ degrees of freedom, refer to Table A-3 and find the critical value $t_{\alpha/2}$ that corresponds to the desired confidence level.

3. Evaluate the margin of error $E = t_{\alpha/2}s/\sqrt{n}$.

4. Using the value of the calculated margin of error E and the value of the sample mean $\bar{x}$, find the values of $\bar{x} - E$ and $\bar{x} + E$. Substitute those values in the general format for the confidence interval:

$$\bar{x} - E < \mu < \bar{x} + E$$

or

$$\bar{x} \pm E$$

or

$$(\bar{x} - E, \bar{x} + E)$$

5. Round the resulting confidence interval limits. If using the original set of data, round to one more decimal place than is used for the original set of data. If using summary statistics $(n, \bar{x}, s)$, round the confidence interval limits to the same number of decimal places used for the sample mean.

EXAMPLE Constructing a Confidence Interval In Section 6-3 we included an example illustrating the construction of a confidence interval to estimate μ. We used the sample of body temperatures in Data Set 4 in Appendix B (for 12 AM on day 2), with $n = 106$ and $\bar{x} = 98.20°F$, and we also assumed that the sample is a simple random sample and that σ is "somehow known to be 0.62°F." In reality, σ is not known. Using the statistics $n = 106$, $\bar{x} = 98.20°F$, and $s = 0.62°F$ (with σ not known) obtained from a simple random sample, find both of the following by again using a 95% confidence level:

a. The margin of error E

b. The confidence interval for μ.

SOLUTION

1. We should first verify that the two assumptions for this section are satisfied. We do have a simple random sample and $n > 30$. (Because $n > 30$, there is no need to check that the sample appears to come from a normally distributed population.) We therefore proceed to construct a 95% confidence interval by using the t distribution.

2. Next we find the critical value of $t_{\alpha/2} = 1.984$. It is found in Table A-3 as the critical value corresponding to $n - 1 = 105$ degrees of freedom (left column of Table A-3) and an area in two tails of 0.05. (Remember, a 95% confidence level corresponds to $\alpha = 0.05$, which is divided equally between the two tails.) Table A-3 does not include 105 degrees of freedom, so we select the closest number of degrees of freedom, which is 100. The correct value of $t_{\alpha/2}$ for 105 degrees of freedom is 1.983, so using the closest Table A-3 value of 1.984 produces a negligible error here.

3. *Find the margin of error E:* The margin of error $E = 0.11947593$ is computed using Formula 6-2 as shown below, with extra decimal places used to minimize rounding error in the confidence interval found in Step 4.

$$E = t_{\alpha/2} \frac{s}{\sqrt{n}} = 1.984 \cdot \frac{0.62}{\sqrt{106}} = 0.11947593$$

4. *Find the confidence interval:* The confidence interval can now be found by using $\bar{x} = 98.20$ and $E = 0.11947593$ as shown below:

$$\bar{x} - E < \mu < \bar{x} + E$$

$$98.20 - 0.11947593 < \mu < 98.20 + 0.11947593$$

$$98.08052407 < \mu < 98.31947593$$

5. Round the confidence interval limits. Because the sample mean of 98.20 uses two decimal places, round the result to two decimal places to get this result: $98.08 < \mu < 98.32$.

INTERPRETATION This result could also be expressed in the format of 98.20 ± 0.12 or $(98.08, 98.32)$. On the basis of the given sample results, we are 95% confident that the limits of 98.08°F and 98.32°F actually do contain the value of the population mean μ. Note that the confidence interval limits do not contain 98.6°F, the value commonly believed to be the mean body temperature. Based on these results, it appears that the commonly believed value of 98.6°F is *wrong*.

The confidence interval found in the preceding example appears to be the same as the one found in Section 6-3, where we used the normal distribution and the assumption that σ is known to be 0.62°F. Actually, the two confidence intervals are the same only after rounding. Without rounding, the confidence interval in Section 6-3 is (98.08196934, 98.31803066) and the confidence interval found here is (98.08052407, 98.31947593). In some other cases, the differences might be much greater.

We now list the important properties of the t distribution that we are using in this section.

Important Properties of the Student t Distribution

1. The Student t distribution is different for different sample sizes. (See Figure 6-5 for the cases $n = 3$ and $n = 12$.)

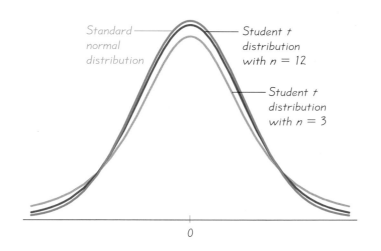

Standard normal distribution

Student t distribution with $n = 12$

Student t distribution with $n = 3$

0

FIGURE 6-5 **Student t Distributions for $n = 3$ and $n = 12$**

The Student t distribution has the same general shape and symmetry as the standard normal distribution, but it reflects the greater variability that is expected with small samples.

2. The Student t distribution has the same general symmetric bell shape as the standard normal distribution, but it reflects the greater variability (with wider distributions) that is expected with small samples.

3. The Student t distribution has a mean of $t = 0$ (just as the standard normal distribution has a mean of $z = 0$).

4. The standard deviation of the Student t distribution varies with the sample size, but it is greater than 1 (unlike the standard normal distribution, which has $\sigma = 1$).

5. As the sample size n gets larger, the Student t distribution gets closer to the standard normal distribution.

Choosing the Appropriate Distribution

It is sometimes difficult to decide whether to use the standard normal z distribution or the Student t distribution. The flowchart in Figure 6-6 and the accompanying Table 6-1 both summarize the key points to be considered when constructing confidence intervals for estimating μ, the population mean. In Figure 6-6 or Table 6-1, note that if we have a small ($n \leq 30$) sample drawn from a distribution that differs dramatically from a normal distribution, we can't use the methods described in this chapter. One alternative is to use nonparametric methods (see Chapter 12), and another alternative is to use the computer bootstrap method. In both of those approaches, no assumptions are made about the original population.

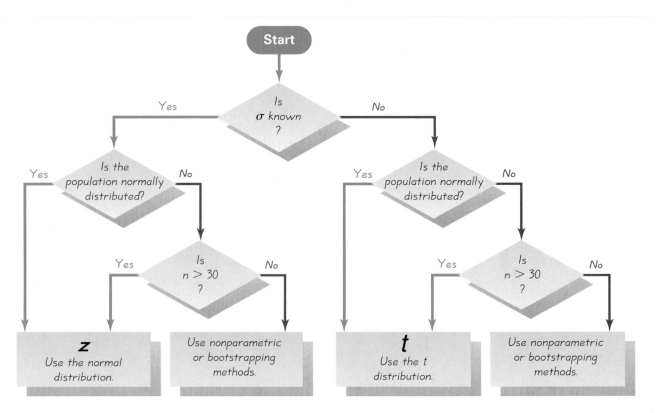

FIGURE 6-6 **Choosing Between z and t**

Table 6-1	Choosing between z and t

Method	Conditions
Use normal (z) distribution.	σ known and normally distributed population
	or
	σ known and $n > 30$
Use t distribution.	σ not known and normally distributed population
	or
	σ not known and $n > 30$
Use a nonparametric method or bootstrapping.	Population is not normally distributed and $n \leq 30$

Notes: 1. **Criteria for deciding whether the population is normally distributed:** Population need not be exactly normal, but it should appear to be somewhat symmetric with one mode and no outliers.

2. **Sample size $n > 30$:** This is a commonly used guideline, but sample sizes of 15 to 30 are adequate if the population appears to have a distribution that is not far from being normal and there are no outliers. For some population distributions that are extremely far from normal, the sample size might need to be larger than 50 or even 100.

The bootstrap method is described in the Technology Project at the end of this chapter.

The following example focuses on choosing the correct approach by using the methods of this section and Section 6-3.

EXAMPLE **Choosing Distributions** Assuming that you plan to construct a confidence interval for the population mean μ, use the given data to determine whether the margin of error E should be calculated using a critical value of $z_{\alpha/2}$ (from the normal distribution), a critical value of $t_{\alpha/2}$ (from a t distribution), or neither (so that the methods of Sections 6-3 and this section cannot be used).

a. $n = 150, \bar{x} = 100, s = 15$, and the population has a skewed distribution.

b. $n = 8, \bar{x} = 100, s = 15$, and the population has a normal distribution.

c. $n = 8, \bar{x} = 100, s = 15$, and the population has a very skewed distribution.

d. $n = 150, \bar{x} = 100, \sigma = 15$, and the distribution is skewed. (This situation almost never occurs.)

e. $n = 8, \bar{x} = 100, \sigma = 15$, and the distribution is extremely skewed. (This situation almost never occurs.)

continued

SOLUTION   Refer to Figure 6-6 or Table 6-1 to determine the following:

a. Because the population standard deviation σ is not known and the sample is large ($n > 30$), the margin of error is calculated using $t_{\alpha/2}$ in Formula 6-6.

b. Because the population standard deviation σ is not known and the population is normally distributed, the margin of error is calculated using $t_{\alpha/2}$ in Formula 6-6.

c. Because the sample is small and the population does not have a normal distribution, the margin of error E should not be calculated using a critical value of $z_{\alpha/2}$ or $t_{\alpha/2}$. The methods of Section 6-3 and this section do not apply.

d. Because the population standard deviation σ is known and the sample is large ($n > 30$), the margin of error is calculated using $z_{\alpha/2}$ in Formula 6-4.

e. Because the population is not normally distributed and the sample is small ($n \leq 30$), the margin of error E should not be calculated using a critical value of $z_{\alpha/2}$ or $t_{\alpha/2}$. The methods of Section 6-3 and this section do not apply.

EXAMPLE   **Confidence Interval for *Harry Potter*** Data Set 14 in Appendix B includes the Flesch ease of reading scores for 12 different pages randomly selected from J. K. Rowling's *Harry Potter and the Sorcerer's Stone*. Use the simple random sample of those 12 values to construct a 95% confidence interval estimate of μ, the mean Flesch ease of reading score for all pages in the book.

SOLUTION

1. We should first verify that the two assumptions for this section are satisfied. We do have a simple random sample. Because the sample size $n = 12$ does not exceed 30, we must verify that the population has a distribution that is approximately normal. The accompanying STATDISK screen dis-

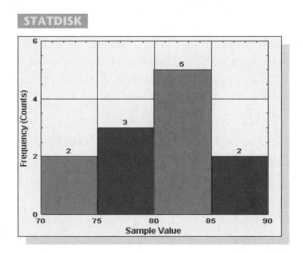

play shows that the 12 sample values result in a histogram that is approximately bell-shaped, so we can assume that the population has a distribution that is approximately normal. While running STATDISK, we also found that $\bar{x} = 80.75$ and $s = 4.68$ for the sample of 12 reading scores. With σ not known and a normally distributed population, we now proceed to construct a 95% confidence interval by using the t distribution.

2. Next we find the critical value of $t_{\alpha/2} = 2.201$. It is found in Table A-3 as the critical value corresponding to $n - 1 = 11$ degrees of freedom (left column of Table A-3) and an area in two tails of 0.05. (Remember, a 95% confidence level corresponds to $\alpha = 0.05$, which is divided equally between the two tails.)

3. *Find the margin of error E:* The margin of error $E = 2.97355$ is computed using Formula 6-6 as shown below, with extra decimal places used to minimize rounding error in the confidence interval found in Step 4.

$$E = t_{\alpha/2} \frac{s}{\sqrt{n}} = 2.201 \cdot \frac{4.68}{\sqrt{12}} = 2.97355$$

4. *Find the confidence interval:* The confidence interval can now be found by using $\bar{x} = 80.75$ and $E = 2.97355$ as shown below:

$$\bar{x} - E < \mu < \bar{x} + E$$

$$80.75 - 2.97355 < \mu < 80.75 + 2.97355$$

$$77.77645 < \mu < 83.72355$$

5. Round the confidence interval limits. Because the original sample data use one decimal place, the result is rounded to one *additional* place to yield this result with two decimal places: $77.78 < \mu < 83.72$.

INTERPRETATION On the basis of the sample data, we are 95% confident that the limits of 77.78 and 83.72 actually do contain the value of the mean Flesch ease of reading score for all pages in *Harry Potter and the Sorcerer's Stone.*

Finding Point Estimate and *E* from a Confidence Interval

Later in this section we will describe how software and calculators can be used to find a confidence interval. A typical usage requires that you enter a confidence level and sample statistics, and the display shows the confidence interval limits. The sample mean $\bar{x}$ is the value midway between those limits, and the margin of error E is one-half the difference between those limits (because the upper limit is $\bar{x} + E$ and the lower limit is $\bar{x} - E$, the distance separating them is $2E$).

Point estimate of μ:

$$\bar{x} = \frac{(\text{upper confidence limit}) + (\text{lower confidence limit})}{2}$$

Margin of error:

$$E = \frac{(\text{upper confidence limit}) - (\text{lower confidence limit})}{2}$$

EXAMPLE **Ages of Stowaways** In analyzing the ages of all Queen Mary stowaways listed in Data Set 15 in Appendix B, the Minitab display shown below is obtained. Use the given confidence interval to find the point estimate $\bar{x}$ and the margin of error E. Treat the values as sample data randomly selected from a large population.

```
                95.0% CI

            ( 24.065, 27.218)
```

SOLUTION   In the following calculations, results are rounded to one decimal place, which is one additional decimal place beyond the rounding used for the original list of ages.

$$\bar{x} = \frac{(\text{upper confidence limit}) + (\text{lower confidence limit})}{2}$$

$$= \frac{27.218 + 24.065}{2} = 25.6 \text{ years}$$

$$E = \frac{(\text{upper confidence limit}) - (\text{lower confidence limit})}{2}$$

$$= \frac{27.218 - 24.065}{2} = 1.6 \text{ years}$$

Using Confidence Intervals to Describe, Explore, or Compare Data

In some cases, we might use a confidence interval to achieve an ultimate goal of estimating the value of a population parameter. For the body temperature data used in this section, an important goal might be to estimate the mean body temperature of healthy adults, and our results strongly suggest that the commonly used value of 98.6°F is incorrect (because we have 95% confidence that the limits of 98.08°F and 98.32°F contain the true value of the population mean). In other cases, a confidence interval might be one of several different tools used to describe, explore, or compare data sets.

Caution: As in Sections 6-2 and 6-3, confidence intervals can be used informally to compare different data sets, but the overlapping of confidence intervals should not be used for making formal and final conclusions about equality of means. Later chapters will include procedures for deciding whether two populations have equal means, and those methods will not have the pitfalls associated with comparisons based on the overlap of confidence intervals.

Do not use the overlapping of confidence intervals as the basis for making formal conclusions about the equality of means.

Consider three different data sets consisting of the Flesch ease of reading scores for 12 randomly selected pages from each of three books: Tom Clancy's *Bear and the Dragon*, J. K. Rowling's *Harry Potter and the Sorcerer's Stone*, and Leo Tolstoy's *War and Peace*. The Flesch scores are on a scale of 1 to 100 with higher scores for works that are easier to read. (See Data Set 14 in Appendix B for the lists of sample scores.) Histograms and normal quantile plots suggest that the three distributions are not dramatically far from normal distributions. The accompanying descriptive statistics are used to find the 95% confidence intervals. (In each case, $n = 12$, so the critical value $t_{\alpha/2} = 2.201$ is found using 11 degrees of freedom, and the margin of error is $E = 2.201 s / \sqrt{12}$.) In Figure 6-7 we graph the three confidence intervals so that they can be better compared.

Author	Descriptive Statistics	95% Confidence Interval
Clancy	$n = 12, \bar{x} = 70.73, s = 11.33$	$63.53 < \mu < 77.93$
Rowling	$n = 12, \bar{x} = 80.75, s = 4.68$	$77.78 < \mu < 83.72$
Tolstoy	$n = 12, \bar{x} = 66.15, s = 7.86$	$61.16 < \mu < 71.14$

Comparing the descriptive statistics from the three samples, we see that the means all appear to be very different. However, Figure 6-7 shows that there is some overlapping among the confidence intervals. Because the confidence intervals for Tolstoy and Rowling do not overlap at all, it appears that those authors have very different writing levels, with Rowling being easier to read. Clancy and Rowling just barely overlap, but the overlap suggests that their population means are not significantly different, so we should not conclude that Rowling has a mean Flesch score higher than Clancy's mean. However, *all of these conclusions based on the overlap of confidence intervals should be considered as tentative indications, not definite conclusions.* Later chapters will introduce better and more reliable methods for determining whether population means are equal.

Alternative Method (not used in this book)   In this section we presented a method for constructing a confidence interval estimate of the population mean

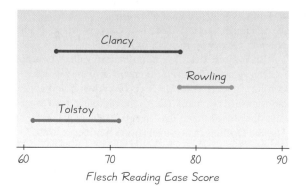

FIGURE 6-7 **Comparing Confidence Intervals**

μ, and this method assumes that the value of σ is not known. An alternative method *not used in this book* is to substitute s for σ whenever $n > 30$, then proceed as if σ is known (as in Section 6-3). In this book, the criteria for choosing between the normal and t distributions are based on the following important considerations:

1. The criteria for choosing between the normal and t distributions used in this book are the same criteria used in the real world. Hundreds of articles in professional journals were surveyed, and they all use the same criteria used in this book.

2. With σ not known, the distribution of $\dfrac{\bar{x} - \mu}{\dfrac{s}{\sqrt{n}}}$ is a t distribution, not a normal distribution. For very large sample sizes, the differences between the normal and t distributions are negligible, but the use of the t distribution generally yields better results.

3. After taking an introductory statistics course, some students go on to take more advanced statistics courses that typically use the t distribution when σ is unknown. For those students, it would be better if they learned one procedure that can be used again in a later course, rather than learning a procedure that must be changed later.

4. Working with the t distribution is not much more difficult than working with the normal distribution, especially if software or a TI-83 Plus calculator is available. Also, the use of Table A-3 helps strengthen skills in using tables that are important for other activities, such as determining tax amounts from income tax tables.

Using Technology

The following procedures apply to confidence intervals for estimating a mean μ, and they include the confidence intervals described in Section 6-3 as well as the confidence intervals presented in this section. Before using software or a calculator to generate a confidence interval, be sure to first check that the required assumptions are satisfied. See the assumptions listed near the beginning of this section and Section 6-3.

STATDISK   You must first find the sample size n, the sample mean $\bar{x}$ and the sample standard deviation s. (See the STATDISK procedure described in Section 2-4.) Select **Analysis** from the main menu bar, select **Confidence Intervals,** then select **Population Mean.** Proceed to enter the items in the dialog box, then click the **Evaluate** button. The confidence interval will be displayed.

Minitab   Minitab Release 14 now allows you to use either the summary statistics of n, $\bar{x}$ and s or a list of the original sample values. (Earlier versions do not allow the use of summary statistics. See Section 5-5 of the *Minitab Manual,* which supplements this textbook, for a way to work around this limitation.) Select **Stat** and **Basic Statistics.** If σ is not known, select **1-sample t** and enter the summary statistics or enter **C1** in the box located at the top right. (If σ is known, select **1–sample Z** and enter the summary statistics or enter **C1** in the box located at the top right. Also enter the value of σ in the "Standard deviation" or "Sigma" box.) Use the **Options** button to enter the confidence level.

Excel   Use the Data Desk XL add-in that is a supplement to this book. Click on **DDXL** and select **Confidence Intervals.** Under the Function Type options, select **1 Var t Interval** if

σ is not known. (If σ is known, select **1 Var z Interval.**) Click on the pencil icon and enter the range of data, such as A1:A12 if you have 12 values listed in column A. Click **OK.** In the dialog box, select the level of confidence. (If using 1 Var z Interval, also enter the value of σ.) Click on **Compute Interval** and the confidence interval will be displayed.

The use of Excel's tool for finding confidence intervals is not recommended. It assumes that σ is known, and you must first find the sample size n and the sample standard deviation s (which can be found using *fx*, **Statistical, STDEV**). Instead of generating the completed confidence interval with specific limits, this tool calculates only the margin of error E. You must then subtract this result from $\bar{x}$ and add it to $\bar{x}$ so that you can identify the actual confidence interval limits. To use this tool when σ is known,

click on *fx*, select the function category of **Statistical**, then select the item of **CONFIDENCE.** In the dialog box, enter the value of α (called the *significance level*), the standard deviation, and the sample size. The result will be the value of the margin of error E.

TI-83 Plus   The TI-83 Plus calculator can be used to generate confidence intervals for original sample values stored in a list, or you can use the summary statistics n, $\bar{x}$ and s. Either enter the data in list L1 or have the summary statistics available, then press the **STAT** key. Now select **TESTS** and choose **TInterval** if σ is not known. (Choose **ZInterval** if σ is known.) After making the required entries, the calculator display will include the confidence interval in the format of $(\bar{x} - E, \bar{x} + E)$.

6-4 Basic Skills and Concepts

Using Correct Distribution In Exercises 1–8, do one of the following, as appropriate: (a) Find the critical value $z_{\alpha/2}$, (b) find the critical value $t_{\alpha/2}$, (c) state that neither the normal nor the t distribution applies.

1. 95%; $n = 5$; σ is unknown; population appears to be normally distributed.

2. 95%; $n = 10$; σ is unknown; population appears to be normally distributed.

3. 99%; $n = 15$; σ is known; population appears to be very skewed.

4. 99%; $n = 45$; σ is known; population appears to be very skewed.

5. 90%; $n = 92$; σ is unknown; population appears to be normally distributed.

6. 90%; $n = 9$; $\sigma = 4.2$; population appears to be very skewed.

7. 98%; $n = 7$; $\sigma = 27$; population appears to be normally distributed.

8. 98%; $n = 37$; σ is unknown; population appears to be normally distributed.

Finding Confidence Intervals. In Exercises 9 and 10, use the given confidence level and sample data to find (a) the margin of error and (b) the confidence interval for the population mean μ. Assume that the population has a normal distribution.

9. Math SAT scores for women: 95% confidence; $n = 15$, $\bar{x} = 496$, $s = 108$

10. Elbow to fingertip length of men: 99% confidence; $n = 32$, $\bar{x} = 14.50$ in., $s = 0.70$ in.

Interpreting Calculator Display. In Exercises 11 and 12, use the given data and the corresponding TI-83 Plus calculator display to express the confidence interval in the format of $\bar{x} - E < \mu < \bar{x} + E$. Also write a statement that interprets the confidence interval.

11. IQ scores of statistics students: 95% confidence; $n = 32$, $\bar{x} = 117.2$, $s = 12.1$

12. Heights of NBA players: 99% confidence; $n = 16$, $\bar{x} = 77.875$ in., $s = 3.50$ in.

Constructing Confidence Intervals. In Exercises 13–24, construct the confidence interval.

13. Destroying Dodge Vipers With *destructive testing*, sample items are destroyed in the process of testing them. Crash testing of cars is one very expensive example of de-

Recommended Assignment:
Exercises 1–18, 23, 24.

1. $t_{\alpha/2} = 2.776$
2. $t_{\alpha/2} = 2.262$
3. Neither the normal nor the t distibution applies.
4. $z_{\alpha/2} = 2.575$
5. $t_{\alpha/2} = 1.662$
6. Neither the normal nor the t distibution applies.
7. $z_{\alpha/2} = 2.33$
8. $t_{\alpha/2} = 2.434$

TI-83 Plus Exercise 11

```
TInterval
(112.84,121.56)
x̄=117.2
Sx=12.1
n=32
```

TI-83 Plus Exercise 12

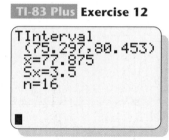

```
TInterval
(75.297,80.453)
x̄=77.875
Sx=3.5
n=16
```

9. 60; $436 < \mu < 556$

10. 0.34; $14.16 < \mu < 14.84$

11. $112.84 < \mu < 121.56$; there is 95% confidence that the interval from 112.84 to 121.56 contains the true value of the population mean μ.

12. $75.297 < \mu < 80.453$; there is 99% confidence that the interval from 75.297 to 80.453 contains the true value of the population mean μ.

13. $\$16,142 < \mu < \$36,312$; there is 95% confidence that the interval from $16,142 to $36,312 contains the true value of the population mean μ.

14. a. $\$5403 < \mu < \$12,605$
    b. $12,605

15. a. $-2.248° < \mu < 1.410°$
    b. The CI does include 0°. The claim does not appear to be valid because a mean of 0° represents no difference between the actual high temperatures and the three-day forecast high temperatures, and the CI does include 0°, indicating that 0° is a very possible value of the difference.

16. a. 1.7 in. $< \mu < 7.1$ in.
    b. The CI does not include 0 in. The claim is supported because the CI shows that the difference is very likely to be positive, indicating that the husbands tend to be taller than their wives.

17. $0.075 < \mu < 0.168$; no, it is possible that the requirement is being met, but it is also very possible that the mean is not less than 0.165 grams/mile.

18. $-0.471 < \mu < 3.547$; the confidence interval is likely to be a poor estimate because the value of 5.40 appears to be an outlier, suggesting that the assumption of a normally distributed population is not correct.

structive testing. Twelve Dodge Viper sports cars (list price: $59,300) are crash tested under a variety of conditions that simulate typical collisions. Analysis of the 12 damaged cars results in repair costs having a distribution that appears to be bell-shaped, with a mean of $\bar{x} = \$26,227$ and a standard deviation of $s = \$15,873$ (based on data from the Highway Loss Data Institute). Find the 95% interval estimate of μ, the mean repair cost for all Dodge Vipers involved in collisions, and interpret the result.

14. Crash Hospital Costs A study was conducted to estimate hospital costs for accident victims who wore seat belts. Twenty randomly selected cases have a distribution that appears to be bell-shaped with a mean of $9004 and a standard deviation of $5629 (based on data from the U.S. Department of Transportation).
    a. Construct the 99% confidence interval for the mean of all such costs.
    b. If you are a manager for an insurance company that provides lower rates for drivers who wear seat belts, and you want a conservative estimate for a worst case scenario, what amount should you use as the possible hospital cost for an accident victim who wears seat belts?

15. Forecast and Actual Temperatures Data Set 10 in Appendix B includes a list of actual high temperatures and the corresponding list of three-day forecast high temperatures. If the difference for each day is found by subtracting the three-day forecast high temperature from the actual high temperature, the result is a list of 31 values with a mean of $-0.419°$ and a standard deviation of $3.704°$.
    a. Construct a 99% confidence interval estimate of the mean difference between all actual high temperatures and three-day forecast high temperatures.
    b. Does the confidence interval include 0°? If a meteorologist claims that three-day forecast high temperatures tend to be too high because the mean difference of the sample is $-0.419°$, does that claim appear to be valid? Why or why not?

16. Heights of Parents Data Set 2 in Appendix B includes the heights of parents of 20 males. If the difference in height for each set of parents is found by subtracting the mother's height from the father's height, the result is a list of 20 values with a mean of 4.4 in. and a standard deviation of 4.2 in. A histogram and a normal quantile plot suggest that the population has a distribution that is not far from normal.
    a. Construct a 99% confidence interval estimate of the mean difference between the heights of the mothers and fathers.
    b. Does the confidence interval include 0 in.? If a sociologist claims that women tend to marry men who are taller than themselves, does the confidence interval support that claim? Why or why not?

17. Estimating Car Pollution In a sample of seven cars, each car was tested for nitrogen-oxide emissions (in grams per mile) and the following results were obtained: 0.06, 0.11, 0.16, 0.15, 0.14, 0.08, 0.15 (based on data from the Environmental Protection Agency). Assuming that this sample is representative of the cars in use, construct a 98% confidence interval estimate of the mean amount of nitrogen-oxide emissions for all cars. If the Environmental Protection Agency requires that nitrogen-oxide emissions be less than 0.165 grams/mile, can we safely conclude that this requirement is being met?

18. Monitoring Lead in Air Listed below are measured amounts of lead (in micrograms per cubic meter or $\mu g/m^3$) in the air. The Environmental Protection Agency has established an air quality standard for lead: $1.5 \ \mu g/m^3$. The measurements shown below were recorded at Building 5 of the World Trade Center site on different days immediately following the destruction caused by the terrorist attacks of September 11, 2001.

After the collapse of the two World Trade Center buildings, there was considerable concern about the quality of the air. Use the given values to construct a 95% confidence interval estimate of the mean amount of lead in the air. Is there anything about this data set suggesting that the confidence interval might not be very good? Explain.

5.40   1.10   0.42   0.73   0.48   1.10

19. Shoveling Heart Rates Because cardiac deaths appear to increase after heavy snowfalls, an experiment was designed to compare cardiac demands of snow shoveling to those of using an electric snow thrower. Ten subjects cleared tracts of snow using both methods, and their maximum heart rates (beats per minute) were recorded during both activities. The following results were obtained (based on data from "Cardiac Demands of Heavy Snow Shoveling," by Franklin et al., *Journal of the American Medical Association,* Vol. 273, No. 11):

> Manual Snow Shoveling Maximum Heart Rates: $n = 10, \bar{x} = 175, s = 15$
> Electric Snow Thrower Maximum Heart Rates: $n = 10, \bar{x} = 124, s = 18$

a. Find the 95% confidence interval estimate of the population mean for those people who shovel snow manually.
b. Find the 95% confidence interval estimate of the population mean for those people who use the electric snow thrower.
c. If you are a physician with concerns about cardiac deaths fostered by manual snow shoveling, what single value in the confidence interval from part (a) would be of greatest concern?
d. Compare the confidence intervals from parts (a) and (b) and interpret your findings.

20. Pulse Rates A physician wants to develop criteria for determining whether a patient's pulse rate is atypical, and she wants to determine whether there are significant differences between males and females. Using the sample pulse rates in Data Set 1 in Appendix B, the male pulse rates can be summarized with the statistics $n = 40, \bar{x} = 69.4$, $s = 11.3$. For females, the statistics are $n = 40, \bar{x} = 76.3, s = 12.5$.
a. Construct a 95% confidence interval estimate of the mean pulse rate for males.
b. Construct a 95% confidence interval estimate of the mean pulse rate for females.
c. Compare the preceding results. Can we conclude that the population means for males and females are different? Why or why not?

21. Skull Breadths Maximum breadth of samples of male Egyptian skulls from 4000 B.C. and 150 A.D. (based on data from *Ancient Races of the Thebaid* by Thomson and Randall-Maciver):

4000 B.C.:   131   119   138   125   129   126   131   132   126   128   128   131
150 A.D.:   136   130   126   126   139   141   137   138   133   131   134   129

Changes in head sizes over time suggest interbreeding with people from other regions. Use confidence intervals to determine whether the head sizes appear to have changed from 4000 B.C. to 150 A.D. Explain your result.

T 22. Head Circumferences In order to correctly diagnose the disorder of hydrocephalus, a pediatrician investigates head circumferences of two-month-old males and females. Use the sample data from Data Set 3 to construct confidence intervals, then determine whether there appears to be a difference between the two genders.

---

19. a. $164 < \mu < 186$
   b. $111 < \mu < 137$
   c. 186
   d. The CIs do not overlap at all, suggesting that the two population means are likely to be significantly different, and the mean for those who manually shovel snow appears to be higher than the mean for those who use the electric snow thrower.

20. a. $65.8 < \mu < 73.0$
   b. $72.3 < \mu < 80.3$
   c. Because the two CIs overlap, we cannot conclude that the two population means are different.

21. 95% CI for 4000 B.C.:
   $125.7 < \mu < 131.6$
   95% CI for 150 A.D.:
   $130.1 < \mu < 136.5$
   The two CIs overlap, so it is possible that the two population means are the same, and we cannot conclude that head sizes appear to have changed.

22. 95% CI for males:
   40.67 cm $< \mu < 41.52$ cm
   95% CI for females:
   39.58 cm $< \mu < 40.51$ cm
   The CIs do not overlap and there appears to be a difference.

23. a. $0.82217 \text{ lb} < \mu < 0.82603$ lb
    b. $0.78238 \text{ lb} < \mu < 0.78533$ lb
    c. The cans of diet Pepsi appear to have a mean weight that is significantly less than the mean weight of cans of regular Pepsi, probably due to the sugar content.
24. a. $24.53 < \mu < 27.47$
    b. $23.10 < \mu < 28.38$
    c. Men and women appear to have very similar mean body mass indexes.

25. $-12.244 < \mu < 29.613$; the CI is dramatically different with the outlier. The CI limits are very sensitive to outliers. Outliers should be carefully examined and discarded if they are found to be errors.

26. The CI limits will be closer together than they should be.

27. a. E is multiplied by 5/9.
    b. $\frac{5}{9}(a - 32), \frac{5}{9}(b - 32)$
    c. Yes

28. a. No, a single value has no variation.
    b. The critical t value cannot be obtained because df $= 0$ is not available in the table of critical t values.
    c. $-27.78 \text{ ft} < \mu < 34.18 \text{ ft}$; no

T 23. Comparing Regular and Diet Pepsi Refer to Data Set 17 in Appendix B and use the sample data.
   a. Construct a 95% confidence interval estimate of the mean weight of cola in cans of regular Pepsi.
   b. Construct a 95% confidence interval estimate of the mean weight of cola in cans of diet Pepsi.
   c. Compare the results from parts (a) and (b) and interpret them.

T 24. Body Mass Index Refer to Data Set 1 in Appendix B and use the sample data.
   a. Construct a 99% confidence interval estimate of the mean body mass index for men.
   b. Construct a 99% confidence interval estimate of the mean body mass index for women.
   c. Compare and interpret the results. We know that men have a mean weight that is greater than the mean for women, and the mean height of men is greater than the mean height of women, but do men also have a mean body mass index that is greater than the mean body mass index of women?

6-4 Beyond the Basics

25. Effect of an Outlier Test the effect of an outlier as follows: Use the sample data from Exercise 17 to find a 95% confidence interval estimate of the population mean, after changing the first value from 0.06 grams/mile to 60 grams/mile. This value is not realistic, but such an error can easily occur during a data entry process. Compare the two confidence intervals. Are the confidence interval limits sensitive to outliers? How should you handle outliers when they are found in sample data sets that will be used for the construction of confidence intervals?

26. Using the Wrong Distribution Assume that a small simple random sample is selected from a normally distributed population for which σ is unknown. Construction of a confidence interval should use the t distribution, but how are the confidence interval limits affected if the normal distribution is incorrectly used instead?

27. Effects of Units of Measurement A confidence interval is constructed for a small simple random sample of temperatures (in degrees Fahrenheit) selected from a normally distributed population for which σ is unknown.
   a. How is the margin of error E affected if each temperature is converted to the Celsius scale? $\left[C = \frac{5}{9}(F - 32) \right]$
   b. If the confidence interval limits are denoted by a and b, find expressions for the confidence interval limits after the original temperatures have been converted to the Celsius scale.
   c. Based on the results from part (b), can confidence interval limits for the Celsius temperatures be found by simply converting the confidence interval limits from the Fahrenheit scale to the Celsius scale?

28. Confidence Interval for Sample of Size $n = 1$ When a lone alien lands on earth, he is measured and is found to have a height of 3.2 ft. It is reasonable to expect that the height of all such aliens are normally distributed?

continued

a. The methods of this chapter require information about the variation of a variable. If only one sample value is available, can it give us any information about the variation of the variable?

b. When using the methods of this section, what happens when you try to use the single height in constructing a 95% confidence interval?

c. Based on the article "An Effective Confidence Interval for the Mean with Samples of Size One and Two" (by Wall, Boen, and Tweedie, *The American Statistician*, Vol. 55, No. 2), a 95% confidence interval for μ can be found (using methods not discussed in this book) for a sample of size $n = 1$ randomly selected from a normally distributed population, and it can be expressed as $x \pm 9.68|x|$. Use this result to construct a 95% confidence interval using the single sample value of 3.2 ft, and express it in the format of $\bar{x} - E < \mu < \bar{x} + E$. Based on the result, is it likely that some other randomly selected alien might be 50 ft tall?

6-5   Estimating a Population Variance

In this section we consider the same three concepts introduced earlier in this chapter: (1) point estimate, (2) confidence interval, and (3) determining the required sample size. Whereas the preceding sections applied these concepts to estimates of proportions and means, this section applies them to the population variance σ^2 or standard deviation σ. Here are the main objectives of this section:

1. Given sample values, estimate the population standard deviation σ or the population variance σ^2.

2. Determine the sample size required to estimate a population standard deviation or variance.

Many real situations, such as quality control in a manufacturing process, require that we estimate values of population variances or standard deviations. In addition to making products with measurements yielding a desired mean, the manufacturer must make products of *consistent* quality that do not run the gamut from extremely good to extremely poor. As this consistency can often be measured by the variance or standard deviation, these become vital statistics in maintaining the quality of products and services.

Assumptions

1. The sample is a simple random sample.

2. The population must have normally distributed values (even if the sample is large).

The assumption of a normally distributed population was made in earlier sections, but that requirement is more critical here. For the methods of this section, departures from normal distributions can lead to gross errors. Consequently, the requirement of having a normal distribution is much stricter, and we should check the distribution of data by constructing histograms and normal quantile plots, as described in Section 5-7.

When we considered estimates of proportions and means, we used the normal and Student t distributions. When developing estimates of variances or standard

Note to Instructor

Many instructors omit this section because of time limitations. The chi-square distribution is introduced here, but it can be introduced in Chapter 7. (Section 7-6 is written so that the chi-square distribution can be introduced there.) Many other instructors feel strongly that the importance of variation requires inclusion of this section.

deviations, we use another distribution, referred to as the chi-square distribution. We will examine important features of that distribution before proceeding with the development of confidence intervals.

Chi-Square Distribution

In a normally distributed population with variance σ^2, we randomly select independent samples of size n and compute the sample variance s^2 (see Formula 2-5) for each sample. The sample statistic $\chi^2 = (n - 1)s^2/\sigma^2$ has a distribution called the **chi-square distribution.**

Chi-Square Distribution

Formula 6-7

$$\chi^2 = \frac{(n - 1)s^2}{\sigma^2}$$

where

n = sample size
s^2 = sample variance
σ^2 = population variance

We denote chi-square by χ^2, pronounced "kigh square." (The specific mathematical equations used to define this distribution are not given here because they are beyond the scope of this text.) To find critical values of the chi-square distribution, refer to Table A-4. The chi-square distribution is determined by the number of degrees of freedom, and in this chapter we use $n - 1$ degrees of freedom.

$$\text{degrees of freedom} = n - 1$$

In later chapters we will encounter situations in which the degrees of freedom are not $n - 1$, so we should not make the incorrect generalization that the number of degrees of freedom is always $n - 1$.

Properties of the Distribution of the Chi-Square Statistic

1. The chi-square distribution is not symmetric, unlike the normal and Student t distributions (see Figure 6-8). (As the number of degrees of freedom increases, the distribution becomes more symmetric, as Figure 6-9 illustrates.)

FIGURE 6-8 **Chi-Square Distribution**

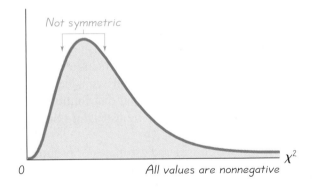

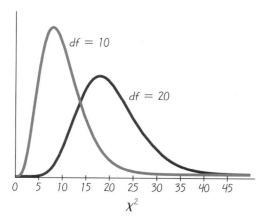

FIGURE 6-9 **Chi-Square Distribution for df = 10 and df = 20**

2. The values of chi-square can be zero or positive, but they cannot be negative (see Figure 6-8).

3. The chi-square distribution is different for each number of degrees of freedom (see Figure 6-9), and the number of degrees of freedom is given by df $= n - 1$ in this section. As the number of degrees of freedom increases, the chi-square distribution approaches a normal distribution.

Because the chi-square distribution is skewed instead of symmetric, the confidence interval does not fit a format of $s^2 \pm E$ and we must do separate calculations for the upper and lower confidence interval limits. There is a different procedure for finding critical values, illustrated in the following example. Note the following essential feature of Table A-4:

> **In Table A-4, each critical value of χ^2 corresponds to an area given in the top row of the table, and that area represents the *total region located to the right* of the critical value.**

Table A-2 for the standard normal distribution provides cumulative areas from the *left*, but Table A-4 for the chi-square distribution provides cumulative areas from the *right*.

Note to Instructor

Consider stopping here and illustrating the use of Table A-4 in class.

EXAMPLE Critical Values Find the critical values of χ^2 that determine critical regions containing an area of 0.025 in each tail. Assume that the relevant sample size is 10 so that the number of degrees of freedom is $10 - 1$, or 9.

SOLUTION   See Figure 6-10 and refer to Table A-4. The critical value to the right ($\chi^2 = 19.023$) is obtained in a straightforward manner by locating 9 in the degrees-of-freedom column at the left and 0.025 across the top. The critical value of $\chi^2 = 2.700$ to the left once again corresponds to 9 in the degrees-of-freedom column, but we must locate 0.975 (found by subtracting 0.025 from 1) across the top because the values in the top row are always *areas to the right* of the critical value. Refer to Figure 6-10 and see that the total area to the right of $\chi^2 = 2.700$ is 0.975. Figure 6-10 shows that, for a sample of 10 values taken from a normally distributed population, the chi-square statistic $(n - 1)s^2/\sigma^2$ has a 0.95 probability of falling between the chi-square critical values of 2.700 and 19.023.

FIGURE 6-10 **Critical Values of the Chi-Square Distribution**

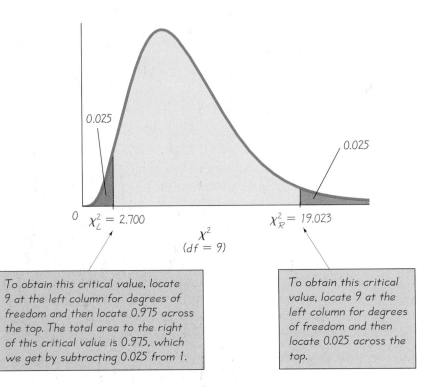

0.025

0.025

0 $\chi_L^2 = 2.700$ $\chi_R^2 = 19.023$

χ^2
$(df = 9)$

To obtain this critical value, locate 9 at the left column for degrees of freedom and then locate 0.975 across the top. The total area to the right of this critical value is 0.975, which we get by subtracting 0.025 from 1.

To obtain this critical value, locate 9 at the left column for degrees of freedom and then locate 0.025 across the top.

When obtaining critical values of χ^2 from Table A-4, note that the numbers of degrees of freedom are consecutive integers from 1 to 30, followed by 40, 50, 60, 70, 80, 90, and 100. When a number of degrees of freedom (such as 52) is not found on the table, you can usually use the closest critical value. For example, if the number of degrees of freedom is 52, refer to Table A-4 and use 50 degrees of freedom. (If the number of degrees of freedom is exactly midway between table values, such as 55, simply find the mean of the two χ^2 values.) For numbers of degrees of freedom greater than 100, use the equation given in Exercise 22, or a more detailed table, or a statistical software package.

Estimators of σ^2

In Section 5-4 we showed that sample variances s^2 (found by using Formula 2-5) tend to target (or center on) the value of the population variance σ^2, so we say that s^2 is an *unbiased estimator* of σ^2. That is, sample variances s^2 do not systematically tend to overestimate the value of σ^2, nor do they systematically tend to underestimate σ^2. Instead, they tend to target the value of σ^2 itself. Also, the values of s^2 tend to produce smaller errors by being closer to σ^2 than do other unbiased measures of variation. For these reasons, the value of s^2 is generally used to estimate σ^2.

> **The sample variance s^2 is the best point estimate of the population variance σ^2.**

Because s^2 is an unbiased estimator of σ^2, we might expect that s would be an unbiased estimator of σ, but this is not the case. (See Section 5-4.) If the sample size is large, however, the bias is small so that we can use s as a reasonably good estimate of σ. Even though it is a biased estimate, s is often used as a point estimate of σ.

The sample standard deviation s is commonly used as a point estimate of σ (even though it is a biased estimate).

Although s^2 is the best point estimate of σ^2, there is no indication of how good it actually is. To compensate for that deficiency, we develop an interval estimate (or confidence interval) that is more informative.

Confidence Interval (or Interval Estimate) for the Population Variance σ^2

$$\frac{(n-1)s^2}{\chi_R^2} < \sigma^2 < \frac{(n-1)s^2}{\chi_L^2}$$

This expression is used to find a confidence interval for the variance σ^2, but the confidence interval (or interval estimate) for the standard deviation σ is found by taking the square root of each component, as shown below.

$$\sqrt{\frac{(n-1)s^2}{\chi_R^2}} < \sigma < \sqrt{\frac{(n-1)s^2}{\chi_L^2}}$$

The notations χ_R^2 and χ_L^2 in the preceding expressions are described as follows. (Note that some other texts use $\chi_{a/2}^2$ in place of χ_R^2 and they use $\chi_{1-a/2}^2$ in place of χ_L^2.)

Notation

With a total area of α divided equally between the two tails of a chi-square distribution, χ_L^2 denotes the left-tailed critical value and χ_R^2 denotes the right-tailed critical value (as illustrated in Figure 6-11.)

Based on the preceding results, we can summarize the procedure for constructing a confidence interval estimate of σ or σ^2 as follows.

Procedure for Constructing a Confidence Interval for σ or σ^2

1. Verify that the required assumptions are satisfied. (The sample is a simple random sample and a histogram or normal quantile plot suggests that the population has a distribution that is very close to a normal distribution.)

2. Using $n-1$ degrees of freedom, refer to Table A-4 and find the critical values χ_R^2 and χ_L^2 that correspond to the desired confidence level.

FIGURE 6-11 **Chi-Square Distribution with Critical Values** χ_L^2 **and** χ_R^2

The critical values χ_L^2 and χ_R^2 separate the extreme areas corresponding to sample variances that are unlikely (with probability α).

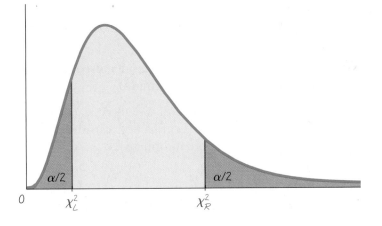

3. Evaluate the upper and lower confidence interval limits using this format of the confidence interval:

$$\frac{(n-1)s^2}{\chi_R^2} < \sigma^2 < \frac{(n-1)s^2}{\chi_L^2}$$

4. If a confidence interval estimate of σ is desired, take the square root of the upper and lower confidence interval limits and change σ^2 to σ.

5. Round the resulting confidence interval limits. If using the original set of data, round to one more decimal place than is used for the original set of data. If using the sample standard deviation or variance, round the confidence interval limits to the same number of decimal places.

Caution: Confidence intervals can be used informally to compare different data sets, but the overlapping of confidence intervals should not be used for making formal and final conclusions about equality of variances or standard deviations. Later chapters will include procedures for deciding whether two populations have equal variances or standard deviations, and those methods will not have the pitfalls associated with comparisons based on the overlap of confidence intervals.

Do not use the overlapping of confidence intervals as the basis for making definitive conclusions about the equality of variances or standard deviations.

EXAMPLE **Body Temperatures** Data Set 4 in Appendix B lists 106 body temperatures (at 12:00 AM on day 2) obtained by University of Maryland researchers. Use the following characteristics of the data set to construct a 95% confidence interval estimate of σ, the standard deviation of the body temperatures of the whole population:

a. As revealed by a histogram of the sample data, the population appears to have a normal distribution.

b. The sample mean is 98.20°F.

c. The sample standard deviation is $s = 0.62$°F.

d. The sample size is $n = 106$.

e. There are no outliers.

SOLUTION   We begin by finding the critical values of χ^2. With a sample of 106 values, we have 105 degrees of freedom. This isn't too far away from the 100 degrees of freedom found in Table A-4, so we will go with that. (See Exercise 22 for a method that will yield more accurate critical values.) For a 95% confidence level, we divide $\alpha = 0.05$ equally between the two tails of the chi-square distribution, and we refer to the values of 0.975 and 0.025 across the top row of Table A-4. The critical values of χ^2 are $\chi_L^2 = 74.222$ and $\chi_R^2 = 129.561$. Using these critical values, the sample standard deviation of $s = 0.62$, and the sample size of 106, we construct the 95% confidence interval by evaluating the following:

$$\frac{(106 - 1)(0.62)^2}{129.561} < \sigma^2 < \frac{(106 - 1)(0.62)^2}{74.222}$$

This becomes $0.31 < \sigma^2 < 0.54$. Finding the square root of each part (before rounding) yields $0.56°F < \sigma < 0.74°F$.

INTERPRETATION   Based on this result, we have 95% confidence that the limits of 0.56°F and 0.74°F contain the true value of σ. We are 95% confident that the standard deviation of body temperatures of all healthy people is between 0.56°F and 0.74°F.

The confidence interval $0.56 < \sigma < 0.74$ can also be expressed as (0.56, 0.74), but the format of $s \pm E$ *cannot* be used because the confidence interval does not have s at its center.

Instead of approximating the critical values by using 100 degrees of freedom, we could use software or the method described in Exercise 22, and the confidence interval becomes $0.55°F < \sigma < 0.72°F$, which is very close to the result obtained here.

Rationale   We now explain why the confidence intervals for σ and σ^2 have the forms just given. If we obtain samples of size n from a population with variance σ^2, the distribution of the $(n - 1)s^2/\sigma^2$ values will be as shown in Figure 6-11. For a simple random sample, there is a probability of $1 - \alpha$ that the statistic $(n - 1)s^2/\sigma^2$ will fall between the critical values of χ_L^2 and χ_R^2. In other words (and symbols), there is a $1 - \alpha$ probability that both of the following are true:

$$\frac{(n - 1)s^2}{\sigma^2} < \chi_R^2 \qquad \text{and} \qquad \frac{(n - 1)s^2}{\sigma^2} > \chi_L^2$$

If we multiply both of the preceding inequalities by σ^2 and divide each inequality by the appropriate critical value of χ^2, we see that the two inequalities can be expressed in the equivalent forms:

$$\frac{(n - 1)s^2}{\chi_R^2} < \sigma^2 \qquad \text{and} \qquad \frac{(n - 1)s^2}{\chi_L^2} > \sigma^2$$

Meta-Analysis

The term *meta-analysis* refers to a technique of doing a study that essentially combines results of other studies. It has the advantage that separate smaller samples can be combined into one big sample, making the collective results more meaningful. It also has the advantage of using work that has already been done. Meta-analysis has the disadvantage of being only as good as the studies that are used. If the previous studies are flawed, the "garbage in, garbage out" phenomenon can occur. The use of meta-analysis is currently popular in medical research and psychological research. As an example, a study of migraine headache treatments was based on data from 46 other studies. (See "Meta-Analysis of Migraine Headache Treatments: Combining Information from Heterogeneous Designs" by Dominici et al., *Journal of the American Statistical Association*, Vol. 94, No. 445.)

These last two inequalities can be combined into one inequality:

$$\frac{(n-1)s^2}{\chi_R^2} < \sigma^2 < \frac{(n-1)s^2}{\chi_L^2}$$

There is a probability of $1 - \alpha$ that these confidence interval limits contain the population variance σ^2. Remember that we should be very careful when interpreting such a confidence interval. It is wrong to say that there is a probability of $1 - \alpha$ that σ^2 will fall between the two confidence interval limits. Instead, we should say that we have $1 - \alpha$ confidence that the limits contain σ^2. Also remember that the required assumptions are very important. If the sample data were collected in an inappropriate way, the resulting confidence interval may be very wrong.

Determining Sample Size

The procedures for finding the sample size necessary to estimate σ^2 are much more complex than the procedures given earlier for means and proportions. Instead of using very complicated procedures, we will use Table 6-2. STATDISK also provides sample sizes. With STATDISK, select **Analysis, Sample Size Determination,** and then **Estimate St Dev.** Minitab, Excel, and the TI-83 Plus calculator do not provide such sample sizes.

Note to Instructor

Almost every other textbook ignores the topic of determining sample sizes required to estimate σ or σ^2, even after covering sample sizes required for estimating means and proportions. In many cases, the standard deviation is the most important parameter, so its estimation is critically important. That is why we include sample sizes in this section.

Table 6-2 Sample Size for σ^2		Sample Size for σ	
To be 95% confident that s^2 is within	of the value of σ^2, the sample size n should be at least	To be 95% confident that s is within	of the value of σ, the sample size n should be at least
1%	77,207	1%	19,204
5%	3,148	5%	767
10%	805	10%	191
20%	210	20%	47
30%	97	30%	20
40%	56	40%	11
50%	37	50%	7
To be 99% confident that s^2 is within	of the value of σ^2, the sample size n should be at least	To be 99% confident that s is within	of the value of σ, the sample size n should be at least
1%	133,448	1%	33,218
5%	5,457	5%	1,335
10%	1,401	10%	335
20%	368	20%	84
30%	171	30%	37
40%	100	40%	21
50%	67	50%	13

EXAMPLE We want to estimate σ, the standard deviation of all body temperatures. We want to be 95% confident that our estimate is within 10% of the true value of σ. How large should the sample be? Assume that the population is normally distributed.

SOLUTION   From Table 6-2, we can see that 95% confidence and an error of 10% for σ correspond to a sample of size 191. We should randomly select 191 values from the population of body temperatures.

Using Technology for Confidence Intervals

STATDISK   First obtain the descriptive statistics and verify that the distribution is normal by using a histogram or normal quantile plot. Next, select **Analysis** from the main menu, then select **Confidence Intervals,** and **Population StDev.** Proceed to enter the required data.

Minitab   With Minitab Release 14, enter the data in column C1, click on **Stat,** click on **Basic Statistics,** and select **Graphical Summary** (or **Display Descriptive Statistics** in ear-

lier versions). Enter C1 in the Variables box. (With earlier versions of Minitab, click on **Graphs,** click on **Graphical Summary.**) Enter the confidence level. Click **OK.**

Excel   Excel does not provide confidence intervals for σ or σ^2.

TI-83 Plus   The TI-83 Plus calculator does not provide confidence intervals for σ or σ^2.

6-5 Basic Skills and Concepts

Finding Critical Values.   In Exercises 1–4, find the critical values χ_L^2 and χ_R^2 that correspond to the given confidence level and sample size.

1. 95%; $n = 16$

2. 95%; $n = 51$

3. 99%; $n = 80$

4. 90%; $n = 40$

Finding Confidence Intervals.   In Exercises 5–8, use the given confidence level and sample data to find a confidence interval for the population standard deviation σ. In each case, assume that a simple random sample has been selected from a population that has a normal distribution.

5. Salaries of statistics professors: 95% confidence; $n = 20$, $\bar{x} = \$95,000$, $s = \$12,345$

6. Ages of drivers occupying the passing lane while driving 25 mi/h with the left signal flashing: 99% confidence; $n = 27$, $\bar{x} = 80.5$ years, $s = 4.6$ years

7. $2.06 \text{ sec} < \sigma < 3.20 \text{ sec}$

7. Times between uses of a TV remote control by males during commercials: 90% confidence; $n = 30, \bar{x} = 5.24 \text{ sec}, s = 2.50 \text{ sec}$

8. $\$8283 < \sigma < \$12,307$

8. Starting salaries of college graduates who have taken a statistics course: 95% confidence; $n = 51, \bar{x} = \$45,678, s = \$9900$

Determining Sample Size. *In Exercises 9–12, assume that each sample is a simple random sample obtained from a normally distributed population.*

9. 191

9. Find the minimum sample size needed to be 95% confident that the sample standard deviation s is within 10% of σ.

10. 20

10. Find the minimum sample size needed to be 95% confident that the sample standard deviation s is within 30% of σ.

11. 133,448; no

11. Find the minimum sample size needed to be 99% confident that the sample variance is within 1% of the population variance. Is such a sample size practical in most cases?

12. 210

12. Find the minimum sample size needed to be 95% confident that the sample variance is within 20% of the population variance.

Finding Confidence Intervals. *In Exercises 13–20, assume that each sample is a simple random sample obtained from a population with a normal distribution.*

13. $\$11,244 < \sigma < \$26,950$; there is 95% confidence that the limits of $\$11,244$ and $\$26,950$ contain the true value of σ.

13. Destroying Dodge Vipers   With *destructive testing,* sample items are destroyed in the process of testing them. Crash testing of cars is one very expensive example of destructive testing. Twelve Dodge Viper sports cars (list price: $59,300) are crash tested under a variety of conditions that simulate typical collisions. Analysis of the 12 damaged cars results in repair costs having a distribution that appears to be bell-shaped, with a mean of $\bar{x} = \$26,227$ and a standard deviation of $s = \$15,873$ (based on data from the Highway Loss Data Institute). Find a 95% interval estimate of σ, the standard deviation of repair costs for all Dodge Vipers involved in collisions, and interpret the result.

14. $38.2 \text{ mL} < \sigma < 95.7 \text{ mL}$; no, the fluctuation appears to be too high.

14. Car Antifreeze   A container of car antifreeze is supposed to hold 3785 mL of the liquid. Realizing that fluctuations are inevitable, the quality-control manager wants to be quite sure that the standard deviation is less than 30 mL. Otherwise, some containers would overflow while others would not have enough of the coolant. She selects a simple random sample, with the results given here. Use these sample results to construct the 99% confidence interval for the true value of σ. Does this confidence interval suggest that the fluctuations are at an acceptable level?

3761	3861	3769	3772	3675	3861	$n = 18$
3888	3819	3788	3800	3720	3748	$\bar{x} = 3787.0$
3753	3821	3811	3740	3740	3839	$s = 55.4$

15. $1.195 < \sigma < 4.695$; yes, the confidence interval is likely to be a poor estimate because the value of 5.40 appears to be an outlier, suggesting that the assumption of a normally distributed population is not correct.

15. Monitoring Lead in Air   Listed below are measured amounts of lead (in micrograms per cubic meter or $\mu g/m^3$) in the air. The Environmental Protection Agency has established an air quality standard for lead: $1.5 \ \mu g/m^3$. The measurements shown below were recorded at Building 5 of the World Trade Center site on different days immediately following the destruction caused by the terrorist attacks of September 11, 2001. After the collapse of the two World Trade Center buildings, there was considerable concern about the quality of the air. Use the given values to construct a 95% confidence interval estimate of the standard deviation of the amounts of lead in the air. Is there anything about this data set suggesting that the confidence interval might not be very good? Explain.

5.40   1.10   0.42   0.73   0.48   1.10

16. Quality Control of Doughnuts The Hudson Valley Bakery makes doughnuts that are packaged in boxes with labels stating that there are 12 doughnuts weighing a total of 42 oz. If the variation among the doughnuts is too large, some boxes will be underweight (cheating consumers) and others will be overweight (lowering profit). A consumer would not be happy with a doughnut so small that it can be seen only with an electron microscope, nor would a consumer be happy with a doughnut so large that it resembles a tractor tire. The quality-control supervisor has found that he can stay out of trouble if the doughnuts have a mean of 3.50 oz and a standard deviation of 0.06 oz or less. Twelve doughnuts are randomly selected from the production line and weighed, with the results given here (in ounces). Construct a 95% confidence interval for σ, then determine whether the quality-control supervisor is in trouble.

3.43   3.37   3.58   3.50   3.68   3.61   3.42   3.52   3.66   3.50   3.36   3.42

17. Shoveling Heart Rates Because cardiac deaths appear to increase after heavy snowfalls, an experiment was designed to compare cardiac demands of snow shoveling to those of using an electric snow thrower. Ten subjects cleared tracts of snow using both methods, and their maximum heart rates (beats per minute) were recorded during both activities. The following results were obtained (based on data from "Cardiac Demands of Heavy Snow Shoveling," by Franklin et al., *Journal of the American Medical Association,* Vol. 273, No. 11):

Manual Snow Shoveling Maximum Heart Rates: $n = 10, \bar{x} = 175, s = 15$

Electric Snow Thrower Maximum Heart Rates: $n = 10, \bar{x} = 124, s = 18$

a. Construct a 95% confidence interval estimate of the population standard deviation σ for those who did manual snow shoveling.

b. Construct a 95% confidence interval estimate of the population standard deviation σ for those who used the automated electric snow thrower.

c. Compare and interpret the results. Does the variation appear to be different for the two groups?

18. Pulse Rates A medical researcher wants to determine whether male pulse rates vary more or less than female pulse rates. Using the sample pulse rates in Data Set 1 in Appendix B, the male pulse rates can be summarized with the statistics $n = 40, \bar{x} = 69.4$, $s = 11.3$. For females, the statistics are $n = 40, \bar{x} = 76.3, s = 12.5$.

a. Construct a 95% confidence interval estimate of the population standard deviation σ of pulse rates for males.

b. Construct a 95% confidence interval estimate of the population standard deviation σ of pulse rates for females.

c. Compare the preceding results. Does it appear that the population standard deviations for males and females are different? Why or why not?

19. a. Comparing Waiting Lines The listed values are waiting times (in minutes) of customers at the Jefferson Valley Bank, where customers enter a single waiting line that feeds three teller windows. Construct a 95% confidence interval for the population standard deviation σ.

6.5   6.6   6.7   6.8   7.1   7.3   7.4   7.7   7.7   7.7

b. The listed values are waiting times (in minutes) of customers at the Bank of Providence, where customers may enter any one of three different lines that have formed at three teller windows. Construct a 95% confidence interval for the population standard deviation σ.

4.2   5.4   5.8   6.2   6.7   7.7   7.7   8.5   9.3   10.0

continued

16. 0.077 oz $< \sigma <$ 0.185 oz; supervisor is in trouble.

17. a. $10 < \sigma < 27$
    b. $12 < \sigma < 33$
    c. The variation does not appear to be significantly different.

18. a. $9.2 < \sigma < 14.3$
    b. $10.1 < \sigma < 15.8$
    c. The population standard deviations do not appear to be significantly different.

19. a. 0.33 min $< \sigma <$ 0.87 min
    b. 1.25 min $< \sigma <$ 3.33 min
    c. The variation appears to be significantly lower with a single line. The single line appears to be better.

c. Interpret the results found in parts (a) and (b). Do the confidence intervals suggest a difference in the variation among waiting times? Which arrangement seems better: the single-line system or the multiple-line system?

20. Body Mass Index Refer to Data Set 1 in Appendix B and use the sample data.
a. Construct a 99% confidence interval estimate of the standard deviation of body mass indexes for men.
b. Construct a 99% confidence interval estimate of the standard deviation of body mass indexes for women.
c. Compare and interpret the results.

6-5 Beyond the Basics

21. Finding Missing Data A journal article includes a graph showing that sample data are normally distributed.
a. The confidence level is inadvertently omitted when this confidence interval is given: $2.8 < \sigma < 6.0$. Find the confidence level for these given sample statistics: $n = 20$, $\bar{x} = 45.2$, and $s = 3.8$.

b. This 95% confidence interval is given: $19.1 < \sigma < 45.8$. Given $n = 12$, find the value of the standard deviation s, which was omitted from the article.

22. Finding Critical Values In constructing confidence intervals for σ or σ^2, we use Table A-4 to find the critical values χ_L^2 and χ_R^2, but that table applies only to cases in which $n \leq 101$, so the number of degrees of freedom is 100 or fewer. For larger numbers of degrees of freedom, we can approximate χ_L^2 and χ_R^2 by using

$$\chi^2 = \frac{1}{2}\left[\pm z_{\alpha/2} + \sqrt{2k - 1} \right]^2$$

where k is the number of degrees of freedom and $z_{\alpha/2}$ is the critical z score first described in Section 6-2. Construct the 95% confidence interval for σ by using the following sample data: The measured heights of 772 men between the ages of 18 and 24 have a standard deviation of 2.8 in. (based on data from the National Health Survey).

Review

The two main activities of inferential statistics are estimating population parameters and testing claims made about population parameters. In this chapter we introduced basic methods for finding *estimates* of population proportions, means, and variances and developed procedures for finding each of the following:

- point estimate
- confidence interval
- required sample size

We discussed point estimate (or single-valued estimate) and formed these conclusions:

- Proportion: The best point estimate of p is $\hat{p}$.
- Mean: The best point of estimate of μ is $\bar{x}$.

- Variation: The value of s is commonly used as a point estimate of σ, even though it is a biased estimate. Also, s^2 is the best point estimate of σ^2.

Because the above point estimates consist of single values, they have the serious disadvantage of not revealing how good they are, so confidence intervals (or interval estimates) are commonly used as more revealing and useful estimates. We also considered ways of determining the sample sizes necessary to estimate parameters to within given margins of error. This chapter also introduced the Student t and chi-square distributions. We must be careful to use the correct probability distribution for each set of circumstances. This chapter used the following criteria for selecting the appropriate distribution:

Confidence interval for proportion p: Use the *normal* distribution (assuming that the required assumptions are satisfied and $np \geq 5$ and $nq \geq 5$ so that the normal distribution can be used to approximate the binomial distribution).

Confidence interval for μ: See Figure 6-6 or Table 6-1 to choose between the *normal* or t distributions (or conclude that neither applies).

Confidence interval for σ or σ^2:   Use the *chi-square* distribution (assuming that the required assumptions are satisfied).

For the confidence interval and sample size procedures in this chapter, it is very important to verify that the required assumptions are satisfied. If they are not, then we cannot use the methods of this chapter and we may need to use other methods, such as the bootstrap method described in the Technology Project at the end of this chapter or nonparametric methods, such as those discussed in Chapter 12.

Review Exercises

1. Estimating Theme Park Attendance Each year, billions of dollars are spent at theme parks owned by Disney, Universal Studios, Sea World, Busch Gardens, and others. A survey of 1233 people who took trips revealed that 111 of them included a visit to a theme park (based on data from the Travel Industry Association of America).
   a. Find the point estimate of the *percentage* of people who visit a theme park when they take a trip.
   b. Find a 95% confidence interval estimate of the *percentage* of all people who visit a theme park when they take a trip.
   c. The survey was conducted among people who took trips, but no information was given about the percentage of people who take trips for pleasure. If you want to estimate the percentage of adults who take a pleasure trip in a year, how many people must you survey if you want to be 99% confident that your sample percentage is within 2.5 percentage points of the correct population percentage?

2. Estimating Length of Car Ownership A NAPA Auto Parts supplier wants information about how long car owners plan to keep their cars. A simple random sample of 25 car owners results in $\bar{x} = 7.01$ years and $s = 3.74$ years, respectively (based on data from a Roper poll). Assume that the sample is drawn from a normally distributed population.

continued

1. a. 9.00%
   b. 7.40% $< p <$ 10.6%
   c. 2653

2. a. 5.47 years $< \mu <$ 8.55
       years
   b. 2.92 years $< \sigma <$ 5.20
       years
   c. 1484
   d. No, the sample would not be
       representative of the popu-
       lation of all car owners.

a. Find a 95% confidence interval estimate of the population mean.
b. Find a 95% confidence interval estimate of the population standard deviation.
c. If several years pass and you want to conduct a new survey to estimate the mean length of time that car owners plan to keep their cars, how many randomly selected car owners must you survey? Assume that you want 99% confidence that the sample mean is within 0.25 year (or 3 months) of the population mean, and also assume that $\sigma = 3.74$ years (based on the latest result).
d. When conducting the survey described in part (c), you find that the survey process can be simplified with a substantially reduced cost if you use an available database consisting of people who purchased a General Motors car within the past 10 years. Would good results be obtained from this population?

3. a. 50.4%
   b. 45.7% $< p <$ 55.1%
   c. No, perhaps respondents are
       trying to impress the poll-
       sters, or perhaps their
       memories have a tendency
       to indicate that they voted
       for the winner.

3. Estimates from Voter Surveys In a recent presidential election, 611 voters were surveyed and 308 of them said that they voted for the candidate who won (based on data from the ICR Survey Research Group).
a. Find the point estimate of the *percentage* of voters who said that they voted for the candidate who won.
b. Find a 98% confidence interval estimate of the *percentage* of voters who said that they voted for the candidate who won.
c. Of those who voted, 43% actually voted for the candidate who won. Is this result consistent with the survey results? How might a discrepancy be explained?

4. a. 4.94 $< \mu <$ 8.06
   b. 4.33 $< \mu <$ 5.82
   c. 7.16 $< \mu <$ 9.71
   d. Tolstoy has a significantly
       higher mean, so his work is
       more difficult to read than
       Clancy or Rowling.

4. Estimates of Ease of Reading Refer to Data Set 14 for the Flesch-Kincaid Grade Level ratings for 12 randomly selected pages taken from books by Tom Clancy, J. K. Rowling, and Leo Tolstoy.
a. Construct a 95% confidence interval estimate of the mean Flesch-Kincaid Grade Level rating for the population of all pages from *The Bear and the Dragon* by Tom Clancy.
b. Construct a 95% confidence interval estimate of the mean Flesch-Kincaid Grade Level rating for the population of all pages from *Harry Potter and the Sorcerer's Stone* by J. K. Rowling.
c. Construct a 95% confidence interval estimate of the mean Flesch-Kincaid Grade Level rating for the population of all pages from *War and Peace* by Leo Tolstoy.
d. Compare the preceding confidence intervals. What do you conclude about the grade level readability ratings?

5. 65

5. Estimating Ease of Reading Data Set 14 in Appendix B includes Flesch-Kincaid Grade Level ratings for works by Tom Clancy, J. K. Rowling, and Leo Tolstoy. If you want to estimate the mean Flesch-Kincaid Grade Level rating for pages from J. R. R. Tolkien's *Lord of the Rings*, how many pages must you randomly select if you want to be 90% confident that the sample mean is within 0.5 of the population mean? Because the samples of pages from Clancy, Rowling, and Tolstoy in Data Set 14 in Appendix B have Flesch-Kincaid Grade Level ratings with standard deviations of 2.45, 1.17, and 2.01, assume that $\sigma = 2.45$ for *Lord of the Rings*.

6. 0.83 $< \sigma <$ 1.99

6. Estimating Variation Data Set 14 in Appendix B includes the Flesch-Kincaid Grade Level ratings for 12 pages randomly selected from J. K. Rowlings's *Harry Potter and the Sorcerer's Stone*, and those 12 ratings have a standard deviation of 1.17, and they appear to come from a normally distributed population. Construct a 95% confidence interval estimate of the standard deviation σ for the Flesch-Kincaid Grade Level ratings for all pages from *Harry Potter and the Sorcerer's Stone*.

7. 2944

7. Determining Sample Size You want to estimate the percentage of U.S. statistics students who get grades of B or higher. How many such students must you survey if you want 97% confidence that the sample percentage is off by no more than two percentage points?

8. **Alcohol Service Policy: Determining Sample Size** In a Gallup poll of 1004 adults, 93% indicated that restaurants and bars should refuse service to patrons who have had too much to drink. If you plan to conduct a new poll to confirm that the percentage continues to be correct, how many randomly selected adults must you survey if you want 98% confidence that the margin of error is four percentage points?

8. 221

Cumulative Review Exercises

1. **Analyzing Weights of Supermodels** Supermodels are sometimes criticized on the grounds that their low weights encourage unhealthy eating habits among young women. Listed below are the weights (in pounds) of nine randomly selected supermodels.

125 (Taylor)	119 (Auermann)	128 (Schiffer)	128 (MacPherson)
119 (Turlington)	127 (Hall)	105 (Moss)	123 (Mazza)
115 (Hume)			

Find each of the following:

a. mean
b. median
c. mode
d. midrange
e. range
f. variance
g. standard deviation
h. Q_1
i. Q_2
j. Q_3
k. What is the level of measurement of these data (nominal, ordinal, interval, ratio)?
l. Construct a boxplot for the data.
m. Construct a 99% confidence interval for the population mean.
n. Construct a 99% confidence interval for the standard deviation σ.
o. Find the sample size necessary to estimate the mean weight of all supermodels so that there is 99% confidence that the sample mean is in error by no more than 2 lb. Use the sample standard deviation s from part (g) as an estimate of the population standard deviation σ.
p. When women are randomly selected from the general population, their weights are normally distributed with a mean of 143 lb and a standard deviation of 29 lb (based on data from the National Health and Examination Survey). Based on the given sample values, do the weights of supermodels appear to be substantially less than the weights of randomly selected women? Explain.

2. **X-Linked Recessive Disorders** A genetics expert has determined that for certain couples, there is a 0.25 probability that any child will have an X-linked recessive disorder.
   a. Find the probability that among 200 such children, at least 65 have the X-linked recessive disorder.
   b. A subsequent study of 200 actual births reveals that 65 of the children have the X-linked recessive disorder. Based on these sample results, construct a 95% confidence interval for the proportion of all such children having the disorder.
   c. Based on parts (a) and (b), does it appear that the expert's determination of a 0.25 probability is correct? Explain.

1. a. 121.0 lb
   b. 123.0 lb
   c. 119 lb, 128 lb
   d. 116.5 lb
   e. 23.0 lb
   f. 56.8 lb^2
   g. 7.5 lb
   h. 119.0 lb
   i. 123.0 lb
   j. 127.0 lb
   k. ratio
   l.

   m. 112.6 lb $< \mu <$ 129.4 lb
   n. 4.5 lb $< \sigma <$ 18.4 lb
   o. 95
   p. The individual supermodel weights do not appear to be considerably different from weights of randomly selected women, because they are all within 1.31 standard deviations of the mean of 143 lb. However, when considered as a group, their mean is significantly less than the mean of 143 lb (see part [m]).
2. a. 0.0089
   b. 0.260 $< p <$ 0.390
   c. Because the confidence interval limits do not contain 0.25, it is unlikely that the expert is correct.

3. Analyzing Survey Results In a Gallup poll, adult survey subjects were asked "Do you have a gun in your home?" The respondents consist of 413 people who answered "yes," and 646 other respondents who either answered "no" or had no opinion.
  a. What percentage of respondents answered "yes"?
  b. Construct a 95% confidence interval estimate of the percentage of all adults who answer "yes" when asked if they have a gun in their home.
  c. Can we safely conclude that less than 50% of adults answer "yes" when asked if they have a gun in their home? Why or why not?
  d. What is a sensible response to the criticism that the Gallup poll cannot provide good results because the sample size is only 1059 adults selected from a huge population with more than 200 million adults?

Cooperative Group Activities

1. Out-of-class activity Collect sample data, and use the methods of this chapter to construct confidence interval estimates of population parameters. Here are some suggestions for parameters:

  - Proportion of students at your college who can raise one eyebrow without raising the other eyebrow [These sample results are easy to obtain because survey subjects tend to raise one eyebrow (if they can) when they are approached by someone asking questions.]
  - Mean age of cars driven by statistics students and/or the mean age of cars driven by faculty
  - Mean age of math books and mean age of science books in your college library (based on the copyright dates)
  - Mean length of words in *New York Times* editorials and mean length of words in editorials found in your local newspaper
  - Mean lengths of words in *Time* magazine, *Newsweek* magazine, and *People* magazine
  - Proportion of students at your college who can correctly identify the president, and vice president, and the secretary of state

  - Proportion of students at your college who are over the age of 18 and are registered to vote
  - Mean age of full-time students at your college
  - Proportion of motor vehicles in your region that are cars

2. In-class activity Divide into groups of three or four. Examine a current magazine such as *Time* or *Newsweek,* and find the proportion of pages that include advertising. Based on the results, construct a 95% confidence interval estimate of the percentage of all such pages that have advertising. Compare results with other groups.

3. In-class activity Divide into groups of two. First find the sample size required to estimate the proportion of times that a coin turns up heads when tossed, assuming that you want 80% confidence that the sample proportion is within 0.08 of the true population proportion. Then toss a coin the required number of times and record your results. What percentage of such confidence intervals should actually contain the true value of the population proportion, which we know is $p = 0.5$? Verify this last result by comparing your confidence interval with the confidence intervals found in other groups.

Technology Project

Bootstrap Resampling The *bootstrap method* can be used to construct confidence intervals for situations in which traditional methods cannot (or should not) be used. For example, the following sample of 10 values was randomly selected from a population with a distribution that is very far from normal, so any methods requiring a normal distribution cannot be used.

2.9   564.2   1.4   4.7   67.6   4.8   51.3   3.6   18.0   3.6

If we want to use the above sample data for the construction of a confidence interval estimate of the population mean μ, we note that the sample is small, and a normal quantile plot indicates that we should not assume that the sample is from a population with a normal distribution. Consequently, the methods of this chapter do not apply. The bootstrap method, which makes no assumptions about the original population, typically requires a computer to build a bootstrap population by replicating (duplicating) a sample many times. We can draw from the sample with replacement, thereby creating an approximation of the original population. In this way, we pull the sample up "by its own bootstraps" to simulate the original population. Using the sample data given above, construct a 95% confidence interval estimate of the population mean μ by using the following bootstrap method with STATDISK or Minitab.

a. Create 500 new samples, each of size 10, by selecting 10 values with replacement from the 10 sample values given above, then find the mean of each of those 500 samples.
- STATDISK: First enter the sample values in column 1 of the STATDISK Data Window. Click on **Analysis** and select **Bootstrap Resampling.** Enter 500 for the number of resamplings, then click **Resample.** The 500 sample means will be listed in column 2 of the Data Window.
- Minitab: Enter the sample values in column C1, then enter probabilities of 0.1, 0.1, . . . , 0.1 (ten

times) in column C2. Now select **Calc** from the main menu bar, select **Random Data,** followed by **Discrete.** Proceed to generate 500 rows of data, stored in columns C11–C20, with the values in C1 and probabilities in C2, and then click **OK.** Find the means of the 500 bootstrap samples by selecting **Calc, Row Statistics,** and **Mean,** enter input variables of C11–C20 with results to be stored in C21, and then click **OK.**

b. Sort the 500 means (arrange them in order).
- STATDISK: Click on the **Data tools** button in the Data Window. Select **Sort a column,** and proceed to sort the means listed in column 2.
- Minitab: Select **Data** or **Manip** from the main menu bar, choose the option of **Sort,** and proceed to sort column C21. Store the sorted column in C21, and sort by column C21. Click **OK.**

c. Find the percentiles $P_{2.5}$ and $P_{97.5}$ for the sorted means that result from the preceding step. ($P_{2.5}$ is the mean of the 12th and 13th scores in the sorted list of sample means; $P_{97.5}$ is the mean of the 487th and 488th scores in the sorted list of sample means.) Identify the resulting confidence interval by substituting the values for $P_{2.5}$ and $P_{97.5}$ in $P_{2.5} < \mu < P_{97.5}$. Does this confidence interval contain the true value of μ, which is 148?

Now use the bootstrap method to find a 95% confidence interval for the population standard deviation σ. (Use the same steps listed above, but use *standard deviations* instead of means.) Does the bootstrap procedure yield a confidence interval for σ that contains 232.1, verifying that the bootstrap method is effective?

An alternative to using STATDISK or Minitab is to use special software designed specifically for bootstrap resampling methods. The author recommends Resampling Stats, available from Resampling Stats, Inc., 612 N. Jackson St., Arlington, VA, 22201, telephone number: (703) 522-2713.

from DATA *to* DECISION

Critical Thinking: Redesigning the standard keyboard

Most keyboards currently used have the keys configured in a standard pattern called the Qwerty arrangement, named for the position of the letters QWERTY on the top row of letters. Developed in 1872, the QWERTY configuration was supposed to force typists to slow down so that their typewriters would jam less often. The Dvorak keyboard was developed in 1936 as a more efficient configuration with keys arranged according to their frequency of use. An article in *Discover* magazine suggests that you can measure the ease of typing by using this point rating system: Count each letter on the top row of letters as 1, each letter on the "home" or middle row as 0, and each letter on the bottom row as 2. (See "Typecasting" by Scott Kim, *Discover.*) Using this rating system with each of the 52 words in the Preamble to the Constitution, we get these statistics for each of the keyboard configurations:

QWERTY configuration: $n = 52, \bar{x} = 4.4, s = 2.8$
Dvorak configuration: $n = 52, \bar{x} = 1.7, s = 1.8$

a. Use the given sample data with methods of this chapter to show that the Dvorak configuration has significantly lower ratings, indicating that the Dvorak keyboard configuration is easier to use.

b. Is there anything about the rating system or the choice of words for the sample that might affect the conclusion about which keyboard configuration is easier to use?

c. Write a brief report of your findings and conclusions.

d. If the Dvorak keyboard configuration is actually easier to use, why is it not being adopted by almost everyone who now uses a keyboard? How might obstacles to adoption of the Dvorak configuration be overcome so that those of us who use keyboards can all become more efficient?

INTERNET PROJECT   Confidence Intervals

The confidence intervals in this chapter illustrate an important point in the science of statistical estimation. Namely, estimations based on sample data are made with certain degrees of confidence. In the Internet Project for this chapter, you will use confidence intervals to make a statement about the temperature where you live. Go to the Web site for this textbook:

http://www.aw.com/triola

Locate the project for this chapter. There you will find instructions on how to use the Internet to find temperature data collected by the weather station nearest your home. With this data in hand, you will construct confidence intervals for temperatures during different time periods and attempt to draw conclusions about temperature change in your area. In addition, you will learn more about the relationship between confidence and probability.

Statistics @ Work

"For research and teaching in the field of ecology, animal behavior and ecotoxicology, knowledge of statistics is essential to obtain a good job, and hold it."

Joanna Burger

Distinguished Professor of Biology at Rutgers University, and member of the Environmental and Occupational Health Sciences Institute.

Joanna Burger teaches, does research, and serves on many national and international environmental committees dealing with endangered species, contaminants in wildlife, the effects of chemicals on animal behavior, and the effects of people on ecosystems.

What concepts of statistics do you use in your work?

I use a variety of statistical approaches including both parametric and nonparametric methods. Without a firm understanding of statistics I would not be able to test whether environmental factors affect reproductive success. I use statistics to test hypotheses that I generate by observing animals within their natural environments. While observation leads to hypotheses, only with the use of well-designed experiments and statistical tests is it possible to answer the questions. For research and teaching in the field of ecology, animal behavior and ecotoxicology, knowledge of statistics is essential to obtain a good job, and hold it.

Could you give a specific example of how you have used statistics in the past?

Statistics is very helpful in identifying which factors are important in influencing animal behavior. Birds nest in particular habitats, but we questioned whether they nest randomly or select very specific nest sites. This is important because conservation requires knowing what animals need in order to create, protect and/or manage that habitat. I tested the hypothesis that common terns were selecting particular salt marsh islands. By comparing statistically a wide range of environmental factors (such as island height, island size, and vegetation type and density) on all islands, with the same set of factors on islands used for nesting by terns, we could show that terns actually selected a very specific set of characteristics. Although there were over 250

islands in the bay where this study was conducted, only 36 met the criteria used by the terns. The characteristics they were selecting actually resulted in their choosing islands that were high enough to avoid summer storm tides, but low enough so that predators could not survive over the winter. Islands that are high enough to avoid winter storm tides can have viable predator populations, such as fox and raccoons, which will eat the eggs and chicks of the terns.

Is a knowledge of statistics critical for your work?

A firm understanding of statistics is absolutely essential to conducting research with humans and animals. Using hypothesis testing, and multiple regression analysis, it is possible to begin to identify and evaluate the factors that affect behavior, such as fishing and consumption behavior of people, foraging behavior of shorebirds, and nesting behavior of seabirds.

In terms of statistics, what would you recommend for prospective employees?

Anyone who wants to be in the field of conservation biology, ecotoxicology, animal behavior or ecology needs a wide range of statistical skills. Two or three courses would be best, including general statistics, regression, and nonparametric. The nature of each problem and the characteristics of the data will determine the statistics that are required, and one should not be limited by a lack of statistical knowledge.

Hypothesis Testing

7-1 Overview

7-2 Basics of Hypothesis Testing

7-3 Testing a Claim About a Proportion

7-4 Testing a Claim About a Mean: σ Known

7-5 Testing a Claim About a Mean: σ Not Known

7-6 Testing a Claim About a Standard Deviation or Variance

Do most of us run red lights?

In Chapter 6 we used survey results to estimate the proportion of Minnesotans who are opposed to the "photo-cop" system of using cameras to issue traffic tickets to drivers who run red lights. The sample data consisted of 829 randomly selected adult Minnesotans, with 51% of them opposed to legislation that would implement the photo-cop system in their state. Although 51% of the 829 respondents were opposed to the photo-cop legislation, the *Star Tribune* newspaper ran the headline of "Respondents divided on 'photo-cop' proposal." The newspaper headline claimed that respondents were *divided*, but 51% of them were opposed, so why can't we say that the *majority* of Minnesotans are opposed?

In a separate nationwide survey of 880 randomly selected drivers, 56% admitted that they run red lights. In writing an article distributed by the Associated Press, reporter Sonja Barisic wrote this: "Nearly all American drivers agree that running red lights is dangerous, but more than half admit they've done it, mostly because they were simply in a hurry, a survey found." This statement includes the claim that the *majority* (more than 50%) of all Americans run red lights. Do the survey results really support that claim?

In this chapter we present standard methods for testing claims such as the following two claims based on the above information:

- Is there sufficient sample evidence to support a claim that the proportion of all adult Minnesotans opposed to photo-cop legislation is greater than 0.5? That is, does a sample of $n = 829$ randomly selected adult Minnesotans with 51% (or $\hat{p} = 0.51$) opposed to photo-cop legislation provide sufficient evidence to support the claim that $p > 0.5$?

- Is there sufficient sample evidence to support a claim that the proportion of all adult American drivers who admit running red lights is greater than 0.5? That is, does a sample of $n = 880$ randomly selected adult American drivers with 56% (or $\hat{p} = 0.56$) admitting that they run red lights constitute sufficient evidence to support the claim that $p > 0.5$?

There is a very standard procedure for testing such claims, and this chapter will describe that procedure.

 Overview

This chapter describes the statistical procedure for testing hypotheses, which is a very standard procedure that is commonly used by professionals in a wide variety of different disciplines. Professional publications such as the *Journal of the American Medical Association*, *American Journal of Psychiatry*, and *International Journal of Advertising* routinely include the same basic procedures presented in this chapter. Consequently, the work done in studying the methods of this chapter is really work that applies to all disciplines, not just statistics.

The two major activities of inferential statistics are the estimation of population parameters (introduced in Chapter 6) and hypothesis testing (introduced in this chapter). A hypothesis test is a standard procedure for testing some *claim*.

> **Definition**
>
> In statistics, a **hypothesis** is a claim or statement about a property of a population.
>
> A **hypothesis test** (or **test of significance**) is a standard procedure for testing a claim about a property of a population.

The following statements are typical of the hypotheses (claims) that can be tested by the procedures we develop in this chapter.

- A reporter claims that the majority of American drivers run red lights.
- Medical researchers claim that the mean body temperature of healthy adults is not equal to 98.6°F.
- When new equipment is used to manufacture aircraft altimeters, the new altimeters are better because the variation in the errors is reduced so that the readings are more consistent.

Before beginning to study this chapter, you should recall—and understand clearly—this basic rule, first introduced in Section 3-1.

Rare Event Rule for Inferential Statistics

If, under a given assumption, the probability of a particular observed event is exceptionally small, we conclude that the assumption is probably not correct.

Following this rule, we test a claim by analyzing sample data in an attempt to distinguish between results that can *easily occur by chance* and results that are *highly unlikely to occur by chance*. We can explain the occurrence of highly unlikely results by saying that either a rare event has indeed occurred or that the underlying assumption is not true. Let's apply this reasoning in the following example.

EXAMPLE Gender Selection ProCare Industries, Ltd., once provided a product called "Gender Choice," which, according to advertising claims, allowed couples to "increase your chances of having a boy up to 85%, a girl up to 80%." Gender Choice was available in blue packages for couples wanting a baby boy and (you guessed it) pink packages for couples wanting a baby girl. Suppose we conduct an experiment with 100 couples who want to have baby girls, and they all follow the Gender Choice "easy-to-use in-home system" described in the pink package. For the purpose of testing the claim of an increased likelihood for girls, we will assume that Gender Choice has no effect. Using common sense and no formal statistical methods, what should we conclude about the assumption of no effect from Gender Choice if 100 couples using Gender Choice have 100 babies consisting of

a. 52 girls?

b. 97 girls?

SOLUTION

a. We normally expect around 50 girls in 100 births. The result of 52 girls is close to 50, so we should not conclude that the Gender Choice product is effective. If the 100 couples used no special methods of gender selection, the result of 52 girls could easily occur by chance. The assumption of no effect from Gender Choice appears to be correct. There isn't sufficient evidence to say that Gender Choice is effective.

b. The result of 97 girls in 100 births is extremely unlikely to occur by chance. We could explain the occurrence of 97 girls in one of two ways: Either an *extremely* rare event has occurred by chance, or Gender Choice is effective. The extremely low probability of getting 97 girls is strong evidence against the assumption that Gender Choice has no effect. It does appear to be effective.

The key point of the preceding example is that we should conclude that the product is effective only if we get *significantly* more girls than we would normally expect. Although the outcomes of 52 girls and 97 girls are both "above average," the result of 52 girls is not significant, whereas 97 girls is a significant result.

This brief example illustrates the basic approach used in testing hypotheses. The formal method involves a variety of standard terms and conditions incorporated into an organized procedure. We suggest that you begin the study of this chapter by first reading Sections 7-2 and 7-3 casually to obtain a general idea of their concepts and then rereading Section 7-2 more carefully to become familiar with the terminology.

 ## Basics of Hypothesis Testing

In this section we describe the formal components used in hypothesis testing: null hypothesis, alternative hypothesis, test statistic, critical region, significance level, critical value, *P*-value, type I error, and type II error. *The focus in this section is on*

Note to Instructor

Instead of beginning with a sequence of mechanical steps, it is really helpful to begin hypothesis testing with an overview of the basic concept used. Focus on the issue of *significance*: Do the sample results differ from the claim by an amount that is statistically significant? Recommendation: Assign Section 7-1 for reading to be completed before this chapter is discussed in class.

Data Mining

The term *data mining* is commonly used to describe the now popular practice of analyzing an existing large set of data for the purpose of finding relationships, patterns, or any interesting results that were not found in the original studies of the data set. Some statisticians express concern about ad hoc inference—a practice in which a researcher goes on a fishing expedition through old data, finds something significant, and then identifies an important question that has already been answered. Robert Gentleman, a column editor for *Chance* magazine, writes that "there are some interesting and fundamental statistical issues that data mining can address. We simply hope that its current success and hype don't do our discipline (statistics) too much damage before its limitations are discussed."

the individual components of the hypothesis test, whereas the following sections will bring those components together in comprehensive procedures. Here are the objectives for this section.

Objectives for this Section

- Given a claim, identify the null hypothesis and the alternative hypothesis, and express them both in symbolic form.
- Given a claim and sample data, calculate the value of the test statistic.
- Given a significance level, identify the critical value(s).
- Given a value of the test statistic, identify the *P*-value.
- State the conclusion of a hypothesis test in simple, nontechnical terms.
- Identify the type I and type II errors that could be made when testing a given claim.

You should study the following example until you thoroughly understand it. Once you do, you will have captured a major concept of statistics.

EXAMPLE **Gender Selection and Probability** Let's again refer to the Gender Choice product that was once distributed by ProCare Industries. In Section 7-1 we noted that the pink packages of Gender Choice were intended to help couples increase the likelihood that when a couple has a baby, the baby will be a girl. ProCare Industries claimed that couples using the pink packages of Gender Choice would have girls at a rate that is greater than 50% or 0.5. Let's again consider an experiment whereby 100 couples use Gender Choice in an attempt to have a baby girl; let's assume that the 100 babies include exactly 52 girls, and let's formalize some of the analysis.

Under normal circumstances the proportion of girls is 0.5, so a claim that Gender Choice is effective can be expressed as $p > 0.5$. The outcome of 52 girls supports that claim if the probability of getting *at least 52 girls* is small, such as less than or equal to 0.05. [*Important note:* The probability of getting *exactly* 52 girls or any other specific number of girls is relatively small, but we need the probability of getting a result that is *at least as extreme* as the result of 52 girls. If this point is confusing, review the subsection of "Using Probabilities to Determine When Results Are Unusual" in Section 4-2, where we noted that "*x* successes among *n* trials is an *unusually high* number of successes if $P(x \text{ or more})$ is very small (such as 0.05 or less)." Using that criterion, the result of 52 girls in 100 births would be an unusually high number of girls if $P(52 \text{ or more girls}) \leq 0.05$.]

Using a normal distribution as an approximation to the binomial distribution (see Section 5-6), we find $P(52 \text{ or more girls in 100 births}) = 0.3821$. Because we need to determine whether a result of at least 52 girls has a small probability under normal circumstances, we assume that the probability of a girl is 0.5. Figure 7-1 shows that with a probability of 0.5, the outcome of 52 girls in 100 births is not unusual, so we do *not* reject random chance as a reasonable explanation. We conclude that the proportion of girls born to couples

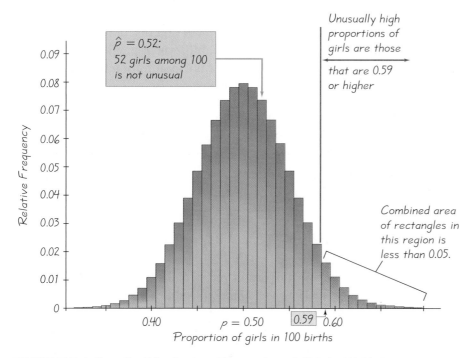

FIGURE 7-1   **Sampling Distribution of Proportions of Girls in 100 Births**

Lie Detectors

Why not require all criminal suspects to take lie detector tests and dispense with trials by jury? The Council of Scientific Affairs of the American Medical Association states, "It is established that classification of guilty can be made with 75% to 97% accuracy, but the rate of false positives is often sufficiently high to preclude use of this (polygraph) test as the sole arbiter of guilt or innocence." A "false positive" is an indication of guilt when the subject is actually innocent. Even with accuracy as high as 97%, the percentage of false positive results can be 50%, so half of the innocent subjects incorrectly appear to be guilty.

using Gender Choice is not significantly greater than the number that we would expect by random chance. Here are the key points:

- Claim: For couples using Gender Choice, the proportion of girls is $p > 0.5$.

- Working assumption: The proportion of girls is $p = 0.5$ (with no effect from Gender Choice).

- The sample resulted in 52 girls among 100 births, so the sample proportion is $\hat{p} = 52/100 = 0.52$.

- Assuming that $p = 0.5$, we use a normal distribution as an approximation to the binomial distribution to find that P(at least 52 girls in 100 births) = 0.3821. (Using the methods of Section 5-6 with the normal distribution as an approximation to the binomial distribution, we have $n = 100$, $p = 0.5$. The observed value of 52 girls is changed to 51.5 as a correction for continuity, and 51.5 converts to $z = 0.30$.)

- There are two possible explanations for the result of 52 girls in 100 births: Either a random chance event (with probability 0.3821) has occurred, or the proportion of girls born to couples using Gender Choice is greater than 0.5. Because the probability of getting at least 52 girls by chance is so high (0.3821), we go with random chance as a reasonable explanation. There isn't sufficient evidence to support a claim that Gender Choice is effective in producing more girls than expected by chance. (It was actually this type of analysis that led to the removal of Gender Choice from the market.)

The preceding example illustrates well the basic method of reasoning we will use throughout this chapter. Focus on the use of the rare event rule of inferential statistics: **If, under a given assumption, the probability of a particular observed event is exceptionally small, we conclude that the assumption is probably not correct.** But if the probability of a particular observed sample result is *not* very small, then we do *not* have sufficient evidence to reject the assumption.

In Section 7-3 we will describe the specific steps used in hypothesis testing, but let's first describe the components of a formal **hypothesis test**, or **test of significance.** These terms are often used in a wide variety of disciplines when statistical methods are required.

Components of a Formal Hypothesis Test

Null and Alternative Hypotheses

Note to Instructor

Here is a change in notation from the previous edition: The null hypothesis will be expressed in terms of equality only, so expressions such as $p \leq 0.5$ or $p \geq 0.5$ will not be used for the null hypothesis. (The reason for this change is largely based on the author's observation that the overwhelming majority of journal articles use only equality for null hypotheses.)

Comment that students will encounter several new terms in this section. These special terms are not unique to this book, or statistics books in general. Instead, these terms are commonly used by medical researchers, manufacturers, psychologists, educators, and many other people who use methods of statistics in their professions. When students learn these terms, they are becoming familiar with a language used in many different disciplines.

- The **null hypothesis** (denoted by H_0) is a statement that the value of a population parameter (such as proportion, mean, or standard deviation) is *equal to* some claimed value. Here are some typical null hypotheses of the type considered in this chapter:

$$H_0: p = 0.5 \qquad H_0: \mu = 98.6 \qquad H_0: \sigma = 15$$

We test the null hypothesis directly in the sense that we assume it is true and reach a conclusion to either reject H_0 or fail to reject H_0.

- The **alternative hypothesis** (denoted by H_1 or H_a) is the statement that the parameter has a value that somehow differs from the null hypothesis. For the methods of this chapter, the symbolic form of the alternative hypothesis must use one of these symbols: $<$ or $>$ or $\neq$. Here are nine different examples of alternative hypotheses involving proportions, means, and standard deviations:

Proportions:	$H_1: p > 0.5$	$H_1: p < 0.5$	$H_1: p \neq 0.5$
Means:	$H_1: \mu > 98.6$	$H_1: \mu < 98.6$	$H_1: \mu \neq 98.6$
Standard Deviations:	$H_1: \sigma > 15$	$H_1: \sigma < 15$	$H_1: \sigma \neq 15$

Note About Always Using the Equal Symbol in H_0: Some textbooks use the symbols $\leq$ and $\geq$ in the null hypothesis H_0, but most professional journals use only the equal symbol for equality. We conduct the hypothesis test by assuming that the proportion, mean, or standard deviation is *equal to* some specified value so that we can work with a single distribution having a specific value. (Where this textbook now uses an expression such as $p = 0.5$ for a null hypothesis, some other textbooks might use $p \leq 0.5$ or $p \geq 0.5$ instead.)

Note About Forming Your Own Claims (Hypotheses): If you are conducting a study and want to use a hypothesis test to *support* your claim, the claim must be worded so that it becomes the alternative hypothesis. This means that your claim must be expressed using only these symbols: $<$ or $>$ or $\neq$. You cannot use a hypothesis test to support a claim that some parameter is *equal to* some specified value.

For example, suppose you have developed a magic potion that raises IQ scores so that the mean becomes greater than 100. If you want to provide evidence of the potion's effectiveness, you must state the claim as $\mu > 100$. (In this context

FIGURE 7-2   Identifying H_0 and H_1

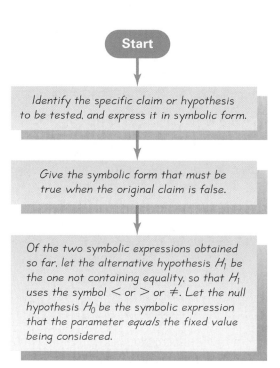

Start

Identify the specific claim or hypothesis to be tested, and express it in symbolic form.

Give the symbolic form that must be true when the original claim is false.

Of the two symbolic expressions obtained so far, let the alternative hypothesis H_1 be the one not containing equality, so that H_1 uses the symbol $<$ or $>$ or $\neq$. Let the null hypothesis H_0 be the symbolic expression that the parameter *equals* the fixed value being considered.

of trying to support the goal of the research, the alternative hypothesis is sometimes referred to as the *research hypothesis*. Also in this context, the null hypothesis of $\mu = 100$ is assumed to be true for the purpose of conducting the hypothesis test, but it is hoped that the conclusion will be rejection of the null hypothesis so that the claim of $\mu > 100$ is supported.)

Note About Identifying H_0 and H_1: Figure 7-2 summarizes the procedures for identifying the null and alternative hypotheses. Note that the original statement could become the null hypothesis, it could become the alternative hypothesis, or it might not correspond exactly to either the null hypothesis or the alternative hypothesis.

For example, we sometimes test the validity of someone else's claim, such as the claim of the Coca Cola Bottling Company that "the mean amount of Coke in cans is at least 12 oz." That claim can be expressed in symbols as $\mu \geq 12$. In Figure 7-2 we see that if that original claim is false, then $\mu < 12$. The alternative hypothesis becomes $\mu < 12$, but the null hypothesis is $\mu = 12$. We will be able to address the original claim after determining whether there is sufficient evidence to reject the null hypothesis of $\mu = 12$.

EXAMPLE   Identifying the Null and Alternative Hypotheses Refer to Figure 7-2 and use the given claims to express the corresponding null and alternative hypotheses in symbolic form.

a. The proportion of drivers who admit to running red lights is greater than 0.5.

b. The mean height of professional basketball players is at most 7 ft.

c. The standard deviation of IQ scores of actors is equal to 15.

continued

Note to Instructor
Some mathematicians acknowledge only the research role of hypothesis testing and insist that *every* claim must be stated as an alternative hypothesis. However, many textbooks and many real applications reflect the comments made here about testing the validity of someone else's claim. Specifically, some of the claims we consider will become null hypotheses, while some others become alternative hypotheses. Some textbooks completely dodge the issue by wording all claims so that they all become alternative hypotheses.

Large Sample Size Isn't Good Enough

Biased sample data should not be used for inferences, no matter how large the sample is. For example, in *Women and Love: A Cultural Revolution in Progress*, Shere Hite bases her conclusions on 4500 replies that she received after mailing 100,000 questionnaires to various women's groups. A *random* sample of 4500 subjects would usually provide good results, but Hite's sample is biased. It is criticized for overrepresenting women who join groups and women who feel strongly about the issues addressed. Because Hite's sample is biased, her inferences are not valid, even though the sample size of 4500 might seem to be sufficiently large.

SOLUTION   See Figure 7-2, which shows the three-step procedure.

a. In Step 1 of Figure 7-2, we express the given claim as $p > 0.5$. In Step 2 we see that if $p > 0.5$ is false, then $p \leq 0.5$ must be true. In Step 3, we see that the expression $p > 0.5$ does not contain equality, so we let the alternative hypothesis H_1 be $p > 0.5$, and we let H_0 be $p = 0.5$.

b. In Step 1 of Figure 7-2, we express "a mean of at most 7 ft" in symbols as $\mu \leq 7$. In Step 2 we see that if $\mu \leq 7$ is false, then $\mu > 7$ must be true. In Step 3, we see that the expression $\mu > 7$ does not contain equality, so we let the alternative hypothesis H_1 be $\mu > 7$, and we let H_0 be $\mu = 7$.

c. In Step 1 of Figure 7-2, we express the given claim as $\sigma = 15$. In Step 2 we see that if $\sigma = 15$ is false, then $\sigma \neq 15$ must be true. In Step 3, we let the alternative hypothesis H_1 be $\sigma \neq 15$, and we let H_0 be $\sigma = 15$.

Test Statistic

- The **test statistic** is a value computed from the sample data, and it is used in making the decision about the rejection of the null hypothesis. The test statistic is found by converting the sample statistic (such as the sample proportion $\hat{p}$, or the sample mean $\bar{x}$, or the sample standard deviation s) to a score (such as z, t, or χ^2) with the assumption that the null hypothesis is true. The test statistic can therefore be used for determining whether there is significant evidence against the null hypothesis. In this chapter, we consider hypothesis tests involving proportions, means, and standard deviations (or variances). Based on results from preceding chapters about the sampling distributions of proportions, means, and standard deviations, we use the following test statistics:

Test statistic for proportion
$$z = \frac{\hat{p} - p}{\sqrt{\dfrac{pq}{n}}}$$

Test statistic for mean
$$z = \frac{\bar{x} - \mu}{\dfrac{\sigma}{\sqrt{n}}} \quad \text{or} \quad t = \frac{\bar{x} - \mu}{\dfrac{s}{\sqrt{n}}}$$

Test statistic for standard deviation
$$\chi^2 = \frac{(n-1)s^2}{\sigma^2}$$

The above test statistic for a proportion is based on the results given in Section 5-6, but it does not include the continuity correction that we usually use when approximating a binomial distribution by a normal distribution. When working with proportions in this chapter, we will work with large samples, so the continuity correction can be ignored because its effect is small. Also, the test statistic for a mean can be based on the normal or Student t distribution, depending on the conditions that are satisfied. When choosing between the normal or Student t distributions, this chapter will use the same criteria described in Section 6-4. (See Figure 6-6 and Table 6-1.)

EXAMPLE Finding the Test Statistic A survey of $n = 880$ randomly selected adult drivers showed that 56% (or $\hat{p} = 0.56$) of those respondents admitted to running red lights. Find the value of the test statistic for the claim that the majority of all adult drivers admit to running red lights. (In Section 7-3 we will see that there are assumptions that must be verified. For this example, assume that the required assumptions are satisfied and focus on finding the indicated test statistic.)

SOLUTION   The preceding example showed that the given claim results in the following null and alternative hypotheses: H_0: $p = 0.5$ and H_1: $p > 0.5$. Because we work under the assumption that the null hypothesis is true with $p = 0.5$, we get the following test statistic:

$$z = \frac{\hat{p} - p}{\sqrt{\dfrac{pq}{n}}} = \frac{0.56 - 0.5}{\sqrt{\dfrac{(0.5)(0.5)}{880}}} = 3.56$$

INTERPRETATION We know from previous chapters that a z score of 3.56 is exceptionally large. It appears that in addition to being "more than half," the sample result of 56% is *significantly* more than 50%. See Figure 7-3 where we show that the sample proportion of 0.56 (from 56%) does fall within the range of values considered to be significant because they are so far above 0.5 that they are not likely to occur by chance (assuming that the population proportion is $p = 0.5$).

Critical Region, Significance Level, Critical Value, and *P*-Value

- The **critical region** (or **rejection region**) is the set of all values of the test statistic that cause us to reject the null hypothesis. For example, see the red-shaded region in Figure 7-3.

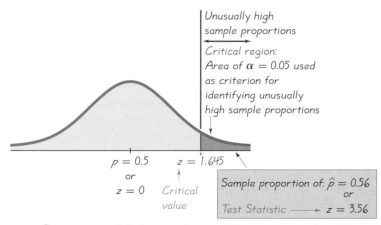

Proportion of adult drivers admitting that they run red lights

FIGURE 7-3 Critical Region, Critical Value, Test Statistic

- The **significance level** (denoted by α) is the probability that the test statistic will fall in the critical region when the null hypothesis is actually true. If the test statistic falls in the critical region, we will reject the null hypothesis, so α is the probability of making the mistake of rejecting the null hypothesis when it is true. This is the same α introduced in Section 6-2, where we defined the confidence level for a confidence interval to be the probability $1-\alpha$. Common choices for α are 0.05, 0.01, and 0.10, with 0.05 being most common.

- A **critical value** is any value that separates the critical region (where we reject the null hypothesis) from the values of the test statistic that do not lead to rejection of the null hypothesis. The critical values depend on the nature of the null hypothesis, the sampling distribution that applies, and the significance level α. See Figure 7-3 where the critical value of $z = 1.645$ corresponds to a significance level of $\alpha = 0.05$. (Critical values were also discussed in Chapter 6.)

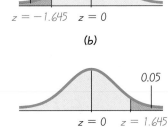

EXAMPLE **Finding Critical Values** Using a significance level of $\alpha = 0.05$, find the critical z values for each of the following alternative hypotheses (assuming that the normal distribution can be used to approximate the binomial distribution):

a. $p \neq 0.5$ (so the critical region is in *both* tails of the normal distribution)

b. $p < 0.5$ (so the critical region is in the *left* tail of the normal distribution)

c. $p > 0.5$ (so the critical region is in the *right* tail of the normal distribution)

FIGURE 7-4 **Finding Critical Values**

SOLUTION

a. See Figure 7-4(a). The shaded tails contain a total area of $\alpha = 0.05$, so each tail contains an area of 0.025. Using the methods of Section 5-2, the values of $z = 1.96$ and $z = -1.96$ separate the right and left tail regions. The critical values are therefore $z = 1.96$ and $z = -1.96$.

b. See Figure 7-4(b). With an alternative hypothesis of $p < 0.5$, the critical region is in the left tail. With a left-tail area of 0.05, the critical value is found to be $z = -1.645$ (by using the methods of Section 5-2).

c. See Figure 7-4(c). With an alternative hypothesis of $p > 0.5$, the critical region is in the right tail. With a right-tail area of 0.05, the critical value is found to be $z = 1.645$ (by using the methods of Section 5-2).

Note to Instructor

Students usually like the hint that the symbol $<$ suggests a left-tailed test and the symbol $>$ suggests a right-tailed test. Many will want to use only that criterion, but they should also *understand* the reasoning used to determine whether a test is left-tailed, right-tailed, or two-tailed. They should not rely blindly on a rote process that is devoid of logical reasoning.

Two-Tailed, Left-Tailed, Right-Tailed　　The *tails* in a distribution are the extreme regions bounded by critical values. Some hypothesis tests are two-tailed, some are right-tailed, and some are left-tailed.

- **Two-tailed test:** The critical region is in the two extreme regions (tails) under the curve.

- **Left-tailed test:** The critical region is in the extreme left region (tail) under the curve.

- **Right-tailed test:** The critical region is in the extreme right region (tail) under the curve.

In two-tailed tests, the significance level α is divided equally between the two tails that constitute the critical region. For example, in a two-tailed test with a significance level of $\alpha = 0.05$, there is an area of 0.025 in each of the two tails. In tests that are right- or left-tailed, the area of the critical region in one tail is α (See Figure 7-4.)

By examining the alternative hypothesis, we can determine whether a test is right-tailed, left-tailed, or two-tailed. The tail will correspond to the critical region containing the values that would conflict significantly with the null hypothesis. A useful check is summarized in the margin figures (see Figure 7-5), which show that the inequality sign in H_1 points in the direction of the critical region. The symbol $\neq$ is often expressed in programming languages as $<>$, and this reminds us that an alternative hypothesis such as $p \neq 0.5$ corresponds to a two-tailed test.

- The **P-value** (or **p-value** or **probability value**) is the probability of getting a value of the test statistic that is *at least as extreme* as the one representing the sample data, assuming that the null hypothesis is true. The null hypothesis is rejected if the P-value is very small, such as 0.05 or less. P-values can be found by using the procedure summarized in Figure 7-6 on the next page.

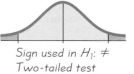
Sign used in H_1: $\neq$
Two-tailed test

Sign used in H_1: $<$
Left-tailed test

Sign used in H_1: $>$
Right-tailed test

FIGURE 7-5 Two-Tailed, Left-Tailed, Right-Tailed Tests

Decisions and Conclusions

We have seen that the original claim sometimes becomes the null hypothesis and sometimes becomes the alternative hypothesis. However, our standard procedure of hypothesis testing requires that we always test the null hypothesis, so our initial conclusion will always be one of the following:

1. Reject the null hypothesis.
2. Fail to reject the null hypothesis.

Decision Criterion: The decision to reject or fail to reject the null hypothesis is usually made using either the traditional method (or classical method) of testing hypotheses, the *P*-value method, or the decision is sometimes based on confidence intervals. In recent years, use of the traditional method has been declining.

Traditional method: *Reject H_0 if the test statistic falls within the critical region.*

Fail to reject H_0 if the test statistic does not fall within the critical region.

P-value method: *Reject H_0 if P-value $\leq \alpha$ (where α is the significance level, such as 0.05).*

Fail to reject H_0 if P-value $> \alpha$.

Another option: Instead of using a significance level such as $\alpha = 0.05$, simply identify the *P*-value and leave the decision to the reader.

Confidence intervals: Because a confidence interval estimate of a population parameter contains the likely values of that parameter, reject a claim that the population parameter has a value that is not included in the confidence interval.

Many statisticians consider it good practice to always select a significance level *before* doing a hypothesis test. This is a particularly good procedure when using the *P*-value method because we may be tempted to adjust the significance level based on the results. For example, with a 0.05 significance level and a *P*-value of 0.06, we should fail to reject the null hypothesis, but it is sometimes tempting to say that a probability of 0.06 is small enough to warrant rejection of the null hypothesis. Other statisticians argue that prior selection of a significance level reduces the usefulness of *P*-values. They contend that no significance level should be specified and that the conclusion should be left to the reader. We will use the decision criterion that involves a comparison of a significance level and the *P*-value.

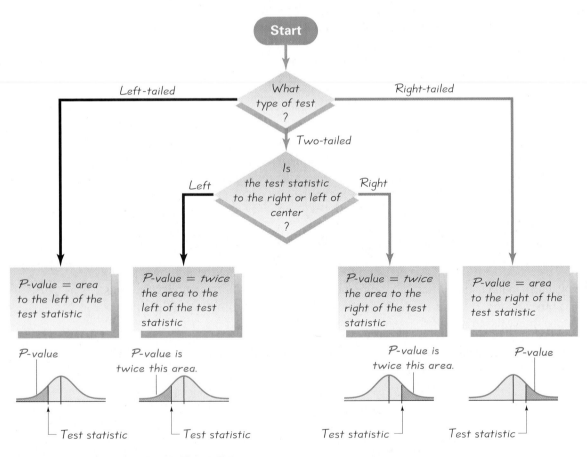

FIGURE 7-6   **Procedure for Finding *P*-Values**

EXAMPLE **Finding *P*-Values** First determine whether the given conditions result in a right-tailed test, a left-tailed test, or a two-tailed test, then use Figure 7-6 to find the *P*-value, then state a conclusion about the null hypothesis.

a. A significance level of $\alpha = 0.05$ is used in testing the claim that $p > 0.25$, and the sample data result in a test statistic of $z = 1.18$.

b. A significance level of $\alpha = 0.05$ is used in testing the claim that $p \neq 0.25$, and the sample data result in a test statistic of $z = 2.34$.

SOLUTION

a. With a claim of $p > 0.25$, the test is right-tailed (see Figure 7-5.) We can find the *P*-value by using Figure 7-6. Because the test is right-tailed, Figure 7-6 shows that the *P*-value is the area to the right of the test statistic $z = 1.18$. Using the methods of Section 5-2, we refer to Table A-2 and find that the area to the *right* of $z = 1.18$ is 0.1190. The *P*-value of 0.1190 is greater than the significance level $\alpha = 0.05$, so we fail to reject the null hypothesis. The *P*-value of 0.1190 is relatively large, indicating that the sample results could easily occur by chance.

b. With a claim of $p \neq 0.25$, the test is two-tailed (see Figure 7-5). We can find the *P*-value by using Figure 7-6. Because the test is two-tailed, and because the test statistic of $z = 2.34$ is to the right of the center, Figure 7-6 shows that the *P*-value is *twice* the area to the right of $z = 2.34$. Using the methods of Section 5-2, we refer to Table A-2 and find that the area to the right of $z = 2.34$ is 0.0096, so *P*-value $= 2 \times 0.0096 = 0.0192$. The *P*-value of 0.0192 is less than or equal to the significance level, so we reject the null hypothesis. The small *P*-value of 0.0192 shows that the sample results are not likely to occur by chance.

Wording the Final Conclusion: The conclusion of rejecting the null hypothesis or failing to reject it is fine for those of us with the wisdom to take a statistics course, but we should use simple, nontechnical terms in stating what the conclusion really means. Figure 7-7 on the next page summarizes a procedure for wording of the final conclusion. Note that only one case leads to wording indicating that the sample data actually *support* the conclusion. If you want to support some claim, state it in such a way that it becomes the alternative hypothesis, and then hope that the null hypothesis gets rejected. For example, to support the claim that the mean body temperature is different from 98.6°, make the claim that $\mu \neq 98.6°$. This claim will be an alternative hypothesis that will be supported if you reject the null hypothesis, H_0: $\mu = 98.6°$. If, on the other hand, you claim that $\mu = 98.6°$, you will either reject or fail to reject the claim; in either case, you will never *support* the claim that $\mu = 98.6°$.

Accept/Fail to Reject: Some texts say "accept the null hypothesis" instead of "fail to reject the null hypothesis." Whether we use the term *accept* or *fail to reject*, we should recognize that *we are not proving the null hypothesis;* we are merely saying that the sample evidence is not strong enough to warrant rejection

Note to Instructor
Students typically have some degree of difficulty with the correct statement of the final conclusions. Stress that the final conclusion should address the original claim that was made, and the precise wording of the final conclusion is very important. Differences between terms such as "support" and "fail to reject" are very important. Show how Figure 7-7 can be used to form the wording of the final conclusion.

Also, some students have trouble clearly understanding the meaning of "fail to reject the null hypothesis." You might also use "don't reject" instead of "fail to reject."

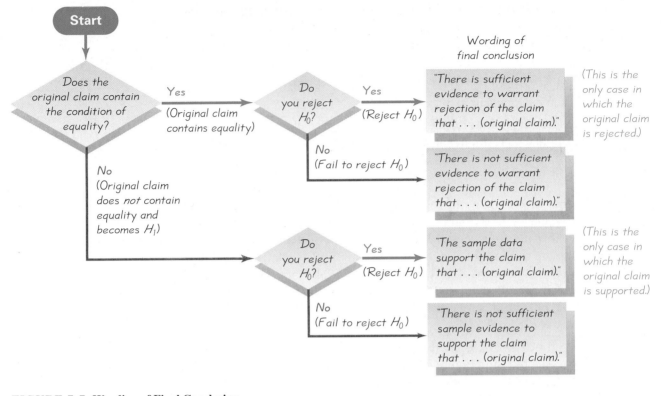

FIGURE 7-7 **Wording of Final Conclusion**

of the null hypothesis. It's like a jury's saying that there is not enough evidence to convict a suspect. The term *accept* is somewhat misleading, because it seems to imply incorrectly that the null hypothesis has been proved. (It is misleading to state that "there is sufficient evidence to accept the null hypothesis.") The phrase *fail to reject* says more correctly that the available evidence isn't strong enough to warrant rejection of the null hypothesis. In this text we will use the terminology *fail to reject the null hypothesis*, instead of *accept the null hypothesis*.

Multiple Negatives: When stating the final conclusion in nontechnical terms, it is possible to get correct statements with up to three negative terms. (Example: "There is *not* sufficient evidence to warrant *rejection* of the claim of *no* difference between 0.5 and the population proportion.") Such conclusions with so many negative terms can be confusing, so it would be good to restate them in a way that makes them understandable, but care must be taken to not change the meaning. For example, instead of saying that "there is not sufficient evidence to warrant rejection of the claim of no difference between 0.5 and the population proportion," better statements would be these:

- Fail to reject the claim that the population proportion is equal to 0.5.
- Until stronger evidence is obtained, assume that the population proportion is equal to 0.5.

EXAMPLE **Stating the Final Conclusion** Suppose a reporter claims that "more than half" of all adult American drivers admit to running red lights. This claim of $p > 0.5$ becomes the alternative hypothesis, while the null hypothesis becomes $p = 0.5$. Further suppose that the sample evidence causes us to reject the null hypothesis of $p = 0.5$. State the conclusion in simple, nontechnical terms.

SOLUTION   Refer to Figure 7-7. The original claim does not contain the condition of equality, and we do reject the null hypothesis. The wording of the final conclusion should therefore be as follows: "The sample data support the claim that more than half of all adult American drivers admit to running red lights."

Type I and Type II Errors   When testing a null hypothesis, we arrive at a conclusion of rejecting it or failing to reject it. Such conclusions are sometimes correct and sometimes wrong (even if we do everything correctly). Table 7-1 summarizes the two different types of errors that can be made, along with the two different types of correct decisions. We distinguish between the two types of errors by calling them type I and type II errors.

- **Type I error:** The mistake of rejecting the null hypothesis when it is actually true. The symbol α (alpha) is used to represent the probability of a type I error.

- **Type II error:** The mistake of failing to reject the null hypothesis when it is actually false. The symbol β (beta) is used to represent the probability of a type II error.

Table 7-1   Type I and Type II Errors

		True State of Nature	
		The null hypothesis is true.	The null hypothesis is false.
Decision	We decide to reject the null hypothesis.	Type I error (rejecting a true null hypothesis) α	Correct decision
	We fail to reject the null hypothesis.	Correct decision	Type II error (failing to reject a false null hypothesis) β

Notation

α (alpha) = probability of a type I error (the probability of rejecting the null hypothesis when it is true)

β (beta) = probability of a type II error (the probability of failing to reject a null hypothesis when it is false)

Because students usually find it difficult to remember which error is type I and which is type II, we recommend a mnemonic device, such as "ROUTINE FOR FUN." Using only the consonants from those words (**R**ou**T**i**N**e **F**o**R** **F**u**N**), we can easily remember that a type I error is RTN: reject true null (hypothesis), whereas a type II error is FRFN: failure to reject a false null (hypothesis).

EXAMPLE **Identifying Type I and Type II Errors** Assume that we are conducting a hypothesis test of the claim that $p > 0.5$. Here are the null and alternative hypotheses:

$$H_0: p = 0.5$$
$$H_1: p > 0.5$$

Give statements identifying

a. a type I error.

b. a type II error.

SOLUTION

a. A type I error is the mistake of rejecting a true null hypothesis, so this is a type I error: Conclude that there is sufficient evidence to support $p > 0.5$, when in reality $p = 0.5$.

b. A type II error is the mistake of failing to reject the null hypothesis when it is false, so this is a type II error: Fail to reject $p = 0.5$ (and therefore fail to support $p > 0.5$) when in reality $p > 0.5$.

Controlling Type I and Type II Errors: One step in our standard procedure for testing hypotheses involves the selection of the significance level α, which is the probability of a type I error. However, we don't select β [P(type II error)]. It would be great if we could always have $\alpha = 0$ and $\beta = 0$, but in reality that is not possible, so we must attempt to manage the α and β error probabilities. Mathematically, it can be shown that α, β, and the sample size n are all related, so when you choose or determine any two of them, the third is automatically determined. The usual practice in research and industry is to select the values of α and n, so the value of β is determined. Depending on the seriousness of a type I error, try to use the largest α that you can tolerate. For type I errors with more serious consequences, select smaller values of α. Then choose a sample size n as large as is reasonable, based on considerations of time, cost, and other relevant factors. (Sample size determinations were discussed in Chapter 6.) The following practical considerations may be relevant:

1. For any fixed α, an increase in the sample size n will cause a decrease in β. That is, a larger sample will lessen the chance that you make the error of not rejecting the null hypothesis when it's actually false.

2. For any fixed sample size n, a decrease in α will cause an increase in β. Conversely, an increase in α will cause a decrease in β.

3. To decrease both α and β, increase the sample size.

To make sense of these abstract ideas, let's consider M&Ms (produced by Mars, Inc.) and Bufferin brand aspirin tablets (produced by Bristol-Myers Products).

- The mean weight of the M&M candies is supposed to be at least 0.9085 g (in order to conform to the weight printed on the package label).

- The Bufferin tablets are supposed to have a mean weight of 325 mg of aspirin.

Because M&Ms are candies used for enjoyment, whereas Bufferin tablets are drugs used for treatment of health problems, we are dealing with two very different levels of seriousness. If the M&Ms don't have a mean weight of 0.9085 g, the consequences are not very serious, but if the Bufferin tablets don't contain a mean of 325 mg of aspirin, the consequences could be very serious, possibly including consumer lawsuits and actions on the part of the Federal Drug Administration. Consequently, in testing the claim that $\mu = 0.9085$ g for M&Ms, we might choose $\alpha = 0.05$ and a sample size of $n = 100$; in testing the claim that $\mu = 325$ mg for Bufferin tablets, we might choose $\alpha = 0.01$ and a larger sample size of $n = 500$. (The larger sample size allows us to decrease β while we are also decreasing α.) The smaller significance level α and larger sample size n are chosen because of the more serious consequences associated with testing a commercial drug.

Power of a Test: We use β to denote the probability of failing to reject a false null hypothesis (type II error). It follows that $1 - \beta$ is the probability of rejecting a false null hypothesis. Statisticians refer to this probability as the *power* of a test, and they often use it to gauge the test's effectiveness in recognizing that a null hypothesis is false.

Note to Instructor
The calculations required to determine the power of a test are usually quite difficult, so we include only Exercise 46 that deals with power.

Definition

The **power** of a hypothesis test is the probability $(1 - \beta)$ of rejecting a false null hypothesis, which is computed by using a particular significance level α and a particular value of the population parameter that is an alternative to the value assumed true in the null hypothesis. That is, the power of a hypothesis test is the probability of supporting an alternative hypothesis that is true.

Suppose we are using a 0.05 significance level to test the null hypothesis that the mean height of men is 6 ft (or 72 in.). Given sample data and given the alter-

FIGURE 7-8 **Traditional Method**   FIGURE 7-9 *P-Value Method*

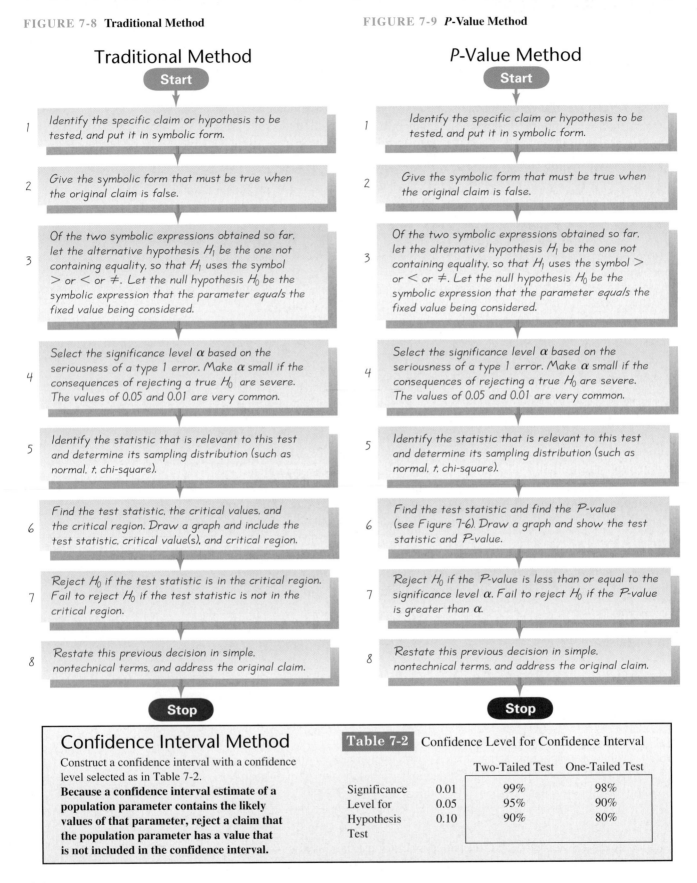

Traditional Method

Start

1   Identify the specific claim or hypothesis to be tested, and put it in symbolic form.

2   Give the symbolic form that must be true when the original claim is false.

3   Of the two symbolic expressions obtained so far, let the alternative hypothesis H_1 be the one not containing equality, so that H_1 uses the symbol $>$ or $<$ or $\neq$. Let the null hypothesis H_0 be the symbolic expression that the parameter *equals* the fixed value being considered.

4   Select the significance level α based on the seriousness of a type 1 error. Make α small if the consequences of rejecting a true H_0 are severe. The values of 0.05 and 0.01 are very common.

5   Identify the statistic that is relevant to this test and determine its sampling distribution (such as normal, t, chi-square).

6   Find the test statistic, the critical values, and the critical region. Draw a graph and include the test statistic, critical value(s), and critical region.

7   Reject H_0 if the test statistic is in the critical region. Fail to reject H_0 if the test statistic is not in the critical region.

8   Restate this previous decision in simple, nontechnical terms, and address the original claim.

Stop

P-Value Method

Start

1   Identify the specific claim or hypothesis to be tested, and put it in symbolic form.

2   Give the symbolic form that must be true when the original claim is false.

3   Of the two symbolic expressions obtained so far, let the alternative hypothesis H_1 be the one not containing equality, so that H_1 uses the symbol $>$ or $<$ or $\neq$. Let the null hypothesis H_0 be the symbolic expression that the parameter *equals* the fixed value being considered.

4   Select the significance level α based on the seriousness of a type 1 error. Make α small if the consequences of rejecting a true H_0 are severe. The values of 0.05 and 0.01 are very common.

5   Identify the statistic that is relevant to this test and determine its sampling distribution (such as normal, t, chi-square).

6   Find the test statistic and find the *P*-value (see Figure 7-6). Draw a graph and show the test statistic and *P*-value.

7   Reject H_0 if the *P*-value is less than or equal to the significance level α. Fail to reject H_0 if the *P*-value is greater than α.

8   Restate this previous decision in simple, nontechnical terms, and address the original claim.

Stop

Confidence Interval Method

Construct a confidence interval with a confidence level selected as in Table 7-2.
Because a confidence interval estimate of a population parameter contains the likely values of that parameter, reject a claim that the population parameter has a value that is not included in the confidence interval.

Table 7-2   Confidence Level for Confidence Interval

		Two-Tailed Test	One-Tailed Test
Significance	0.01	99%	98%
Level for	0.05	95%	90%
Hypothesis	0.10	90%	80%
Test			

native height of 69 in., we can compute the power of the test to reject $\mu = 72$. If our sample consists of only a few observations, the power will be low, but if it consists of hundreds of observations, the power will be much higher. (In addition to increasing the sample size, there are other ways to increase the power, such as increasing the significance level, using a more extreme value for the population mean, or decreasing the standard deviation.) Just as 0.05 is a common choice for a significance level, a power of at least 0.80 is a common requirement for determining that a hypothesis test is effective. (Some statisticians argue that the power should be higher, such as 0.85 or 0.90.) Because the calculations of power are really tough, only Exercise 46 deals with power.

Comprehensive Hypothesis Test   In this section we described the individual components used in a hypothesis test, but the following sections will combine those components in comprehensive procedures. We can test claims about population parameters by using the traditional method summarized in Figure 7-8, the P-value method summarized in Figure 7-9, or we can use a confidence interval (described in Chapter 6). For two-tailed hypothesis tests construct a confidence interval with a confidence level of $1 - \alpha$; but for a one-tailed hypothesis test with significance level α, construct a confidence interval with a confidence level of $1 - 2\alpha$. (See Table 7-2 for common cases.) After constructing the confidence interval, use this criterion:

> **A confidence interval estimate of a population parameter contains the likely values of that parameter. We should therefore reject a claim that the population parameter has a value that is not included in the confidence interval.** *Caution:* **In some cases, a conclusion based on a confidence interval may be different from a conclusion based on a hypothesis test. See the comments in the individual sections.**

The exercises for this section involve isolated components of hypothesis tests, but the following sections will involve complete and comprehensive hypothesis tests.

7-2 Basic Skills and Concepts

Stating Conclusions about Claims. In Exercises 1–4, what do you conclude? (Don't use formal procedures and exact calculations. Use only the rare event rule described in Section 7-1, and make subjective estimates to determine whether events are likely.)

1. Claim: A gender selection method is effective in helping couples have baby girls and, among 50 babies, 26 are girls.

2. Claim: A gender selection method is effective in helping couples have baby girls and, among 50 babies, 49 are girls.

3. Claim: The majority of adult Americans like pizza, and a survey of 500 randomly selected adult Americans shows that 475 of them like pizza.

4. Claim: People born on February 29 have IQ scores that vary less than the general population for which $\sigma = 15$, and a random sample of 50 people born on February 29 results in IQ scores with $s = 14.99$.

Recommended Assignment:
Even-numbered exercises 2–40.

1. There is not sufficient evidence to support the claim that the gender selection method is effective.
2. The gender selection method does appear to be effective.
3. There does appear to be sufficient evidence to support the claim that the majority of adult Americans like pizza.
4. There is not sufficient evidence to support the claim that the IQ scores vary less for people born on February 29.

Identifying H_0 and H_1. *In Exercises 5–12, examine the given statement, then express the null hypothesis H_0 and alternative hypothesis H_1 in symbolic form. Be sure to use the correct symbol (μ, p, σ) for the indicated parameter.*

5. H_0: $\mu = \$50,000$. H_1: $\mu >$ $\$50,000$.

6. H_0: $\mu = 110$. H_1: $\mu < 110$.

7. H_0: $p = 0.5$. H_1: $p > 0.5$.

8. H_0: $p = 0.7$. H_1: $p \neq 0.7$.

9. H_0: $\sigma = 2.8$. H_1: $\sigma < 2.8$.

10. H_0: $p = 0.24$. H_1: $p \neq 0.24$.

11. H_0: $\mu = 12$. H_1: $\mu < 12$.

12. H_0: $\sigma = 3000$. H_1: $\sigma > 3000$.

5. The mean annual income of workers who have studied statistics is greater than $\$50,000$.

6. The mean IQ of statistics students is at least 110.

7. More than one-half of all Internet users make on-line purchases.

8. The percentage of men who watch golf on TV is not 70%, as is claimed by the Madison Advertising Company.

9. Women's heights have a standard deviation less than 2.8 in., which is the standard deviation for men's heights.

10. The percentage of viewers tuned to *60 Minutes* is equal to 24%.

11. The mean amount of Coke in cans is at least 12 oz.

12. Salaries among women business analysts have a standard deviation greater than $\$3000$.

Finding Critical Values. *In Exercises 13–20, find the critical z values. In each case, assume that the normal distribution applies.*

13. $z = \pm 1.96$

14. $z = \pm 2.575$

15. $z = 2.33$

16. $z = -1.645$

17. $z = \pm 1.645$

18. $z = 1.28$

19. $z = -2.05$

20. $z = \pm 2.81$

13. Two-tailed test; $\alpha = 0.05$.

14. Two-tailed test; $\alpha = 0.01$.

15. Right-tailed test; $\alpha = 0.01$.

16. Left-tailed test; $\alpha = 0.05$.

17. $\alpha = 0.10$; H_1 is $p \neq 0.17$.

18. $\alpha = 0.10$; H_1 is $p > 0.18$.

19. $\alpha = 0.02$; H_1 is $p < 0.19$.

20. $\alpha = 0.005$; H_1 is $p \neq 0.20$.

Finding Test Statistics. *In Exercises 21–24, find the value of the test statistic z using*

$$z = \frac{\hat{p} - p}{\sqrt{\dfrac{pq}{n}}}$$

21. -13.45

22. 0.67

23. 1.85

21. Gallup Poll The claim is that the proportion of adults who shop using the Internet is less than 0.5 (or 50%), and the sample statistics include $n = 1025$ subjects with 29% saying that they use the Internet for shopping.

22. Genetics Experiment The claim is that the proportion of peas with yellow pods is equal to 0.25 (or 25%), and the sample statistics include $n = 580$ peas with 26.2% of them having yellow pods.

23. Safety Study The claim is that the proportion of choking deaths of children attributable to balloons is more than 0.25, and the sample statistics include $n = 400$ choking deaths of children with 29.0% of them attributable to balloons.

24. *Police Practices* The claim is that the proportion of drivers stopped by police in a year is different from the 10.3% rate reported by the Department of Justice. Sample statistics include $n = 800$ randomly selected drivers with 12% of them stopped in the past year.

Finding P-values. *In Exercises 25–32, use the given information to find the P-value. (Hint: See Figure 7-6.)*

25. The test statistic in a right-tailed test is $z = 0.55$.

26. The test statistic in a left-tailed test is $z = -1.72$.

27. The test statistic in a two-tailed test is $z = 1.95$.

28. The test statistic in a two-tailed test is $z = -1.63$.

29. With H_1: $p > 0.29$, the test statistic is $z = 1.97$.

30. With H_1: $p \neq 0.30$, the test statistic is $z = 2.44$.

31. With H_1: $p \neq 0.31$, the test statistic is $z = 0.77$.

32. With H_1: $p < 0.32$, the test statistic is $z = -1.90$.

Stating Conclusions. *In Exercises 33–36, state the final conclusion in simple nontechnical terms. Be sure to address the original claim. (Hint: See Figure 7-7.)*

33. Original claim: The proportion of married women is greater than 0.5.
Initial conclusion: Reject the null hypothesis.

34. Original claim: The proportion of college graduates who smoke is less than 0.27.
Initial conclusion: Reject the null hypothesis.

35. Original claim: The proportion of fatal commercial aviation crashes is different from 0.038.
Initial conclusion: Fail to reject the null hypothesis.

36. Original claim: The proportion of M&Ms that are blue is equal to 0.10.
Initial conclusion: Reject the null hypothesis.

Identifying Type I and Type II Errors. *In Exercises 37–40, identify the type I error and the type II error that correspond to the given hypothesis.*

37. The proportion of married women is greater than 0.5.

38. The proportion of college graduates who smoke is less than 0.27.

39. The proportion of fatal commercial aviation crashes is different from 0.038.

40. The proportion of M&Ms that are blue is equal to 0.10.

7-2 Beyond the Basics

41. *Unnecessary Test* When testing a claim that the majority of adult Americans are against the death penalty for a person convicted of murder, a random sample of 491 adults is obtained, and 27% of them are against the death penalty (based on data from a Gallup poll). Find the *P*-value. Why is it not necessary to go through the steps of conducting a formal hypothesis test?

24. 1.58
25. 0.2912
26. 0.0427
27. 0.0512
28. 0.1032
29. 0.0244
30. 0.0146
31. 0.4412
32. 0.0287
33. There is sufficient evidence to support the claim that the proportion of married women is greater than 0.5.
34. There is sufficient evidence to support the claim that the proportion of college graduates who smoke is less than 0.27.
35. There is not sufficient evidence to support the claim that the proportion of fatal commercial aviation crashes is different from 0.038.
36. There is sufficient evidence to warrant rejection of the claim that the proportion of M&Ms that are blue is equal to 0.10.
37. Type I: Support $p > 0.5$ when in reality $p = 0.5$. Type II: Fail to reject $p = 0.5$ (and fail to support $p > 0.5$) when in reality $p > 0.5$.
38. Type I: Support $p < 0.27$ when in reality $p = 0.27$. Type II: Fail to reject $p = 0.27$ (and fail to support $p < 0.27$) when in reality $p < 0.27$.
39. Type I: Support $p \neq 0.038$ when in reality $p = 0.038$. Type II: Fail to reject $p = 0.038$ (and therefore fail to support $p \neq 0.038$) when in reality $p \neq 0.038$.
40. Type I: Reject $p = 0.10$ when in reality $p = 0.10$. Type II: Fail to reject $p = 0.10$ when in reality $p \neq 0.10$.
41. *P*-value = 0.9999. With an alternative hypothesis of $p > 0.5$, it is impossible for a sample statistic of 0.27 to fall in the critical region. No sample proportion less than 0.5 can ever support a claim that $p > 0.5$.

42. Significance Level If a null hypothesis is rejected with a significance level of 0.05, is it also rejected with a significance level of 0.01? Why or why not?

43. P-Value Assume that you have just developed a new manufacturing process that you believe reduces the rate of defects when microchips are produced. You plan to justify your claim of a lower rate of defects by using a hypothesis test. Which P-value would you prefer: 0.10, 0.05, 0.01? Why?

44. Proving Claims You are the quality control manager at Mars, Inc., and you want to prove the company's claim that 10% of all M&M candies are blue. Is it possible to prove that claim using methods of hypothesis testing? Why or why not?

45. Why Not Let $\alpha = 0$? Someone suggests that in testing hypotheses, you can eliminate a type I error by making $\alpha = 0$. In a two-tailed test, what critical values correspond to $\alpha = 0$? If $\alpha = 0$, will the null hypothesis ever be rejected?

46. Power of a Test Assume that you are using a significance level of $\alpha = 0.05$ to test the claim that $p < 0.5$ and that your sample is a simple random sample of size $n = 1998$ with $\hat{p} = 0.48$.
   a. Find β, the probability of making a type II error, given that the population proportion is actually 0.45. (*Hint:* First find the values of the sample proportions that do not lead to rejection of H_0. Then, assuming that $p = 0.45$, find the probability of getting a sample proportion with one of those values.)
   b. Find $1 - \beta$, which is the *power* of the test. If β is the probability of *failing* to reject a false null hypothesis, describe the probability of $1 - \beta$.

7-3 Testing a Claim About a Proportion

In Section 7-2 we presented the isolated components of a hypothesis test, but in this section we combine those components in comprehensive hypothesis tests of claims made about population proportions. The proportions can also represent probabilities or the decimal equivalents of percents. The following are examples of the types of claims we will be able to test.

- Fewer than 1/4 of all college graduates smoke.
- Subjects taking the cholesterol-reducing drug Lipitor experience headaches at a rate that is greater than the 7% rate for people who do not take Lipitor.
- The percentage of late-night television viewers who watch *The Late Show with David Letterman* is equal to 18%.
- Based on early exit polls, the Republican candidate for the presidency will win a majority (more than 50%) of the votes.

The required assumptions, notation, and test statistic are all given below. Basically, claims about a population proportion are usually tested by using a normal distribution as an approximation to the binomial distribution, as we did in Section 5-6. Instead of using the same exact methods of Section 5-6, we use a different but equivalent form of the test statistic shown below, and we don't include the correction for continuity (because its effect tends to be very small with large samples). If the given assumptions are not all satisfied, we may be able to use other methods not described in this section. In this section, all examples and exercises involve cases in which the assumptions are satisfied, so the sampling distribution of sample proportions can be approximated by the normal distribution.

Testing Claims About a Population Proportion p

Assumptions

1. The sample observations are a simple random sample. (Never forget the critical importance of sound sampling methods.)

2. The conditions for a *binomial distribution* are satisfied. (There are a fixed number of independent trials having constant probabilities, and each trial has two outcome categories of "success" and "failure.")

3. The conditions $np \geq 5$ and $nq \geq 5$ are both satisfied, so **the binomial distribution of sample proportions can be approximated by a normal distribution with $\mu = np$ and $\sigma = \sqrt{npq}$** (as described in Section 5-6).

Notation

n = sample size or number of trials

$\hat{p} = \dfrac{x}{n}$ (*sample* proportion)

p = population proportion (used in the null hypothesis)

$q = 1 - p$

Test Statistic for Testing a Claim About a Proportion

$$z = \frac{\hat{p} - p}{\sqrt{\dfrac{pq}{n}}}$$

P-values: Use the standard normal distribution (Table A-2) and refer to Figure 7-6.
Critical values: Use the standard normal distribution (Table A-2).

EXAMPLE Survey of Drivers In the Chapter Problem we noted that an article distributed by the Associated Press included these results from a nationwide survey: Of 880 randomly selected drivers, 56% admitted that they run red lights. Reporter Sonja Barisic wrote this: "Nearly all American drivers agree that running red lights is dangerous, but more than half admit they've done it, . . . , a survey found." This statement includes the claim that the *majority* (more than half) of all Americans run red lights. Here is a summary of the claim and the sample data:

Claim: More than half (of all Americans) admit to running red lights. That is, $p > 0.5$.

Sample data: $n = 880$ and $\hat{p} = 0.56$

continued

Ethics in Reporting

The American Association for Public Opinion Research developed a voluntary code of ethics to be used in news reports of survey results. This code requires that the following be included: (1) identification of the survey sponsor, (2) date the survey was conducted, (3) size of the sample, (4) nature of the population sampled, (5) type of survey used, (6) exact wording of survey questions. Surveys funded by the U.S. government are subject to a prescreening that assesses the risk to those surveyed, the scientific merit of the survey, and the guarantee of the subject's consent to participate.

We will illustrate the hypothesis test using the traditional method, the popular *P*-value method, and confidence intervals. Before proceeding, however, we should verify that the required assumptions are satisfied. The sample is a simple random sample, there is a fixed number (880) of independent trials with two categories (respondent admits to running red lights or does not), and $np \geq 5$ and $nq \geq 5$ are both satisfied with $n = 880$, $p = 0.5$, and $q = 0.5$. (Technically, the trials are not independent, but they can be treated as independent by using this guideline presented in Section 4-3: "When sampling without replacement, the events can be treated as if they were independent if the sample size is no more than 5% of the population size. That is, $n \leq 0.05N$.") With the required assumptions all satisfied, we can now proceed to conduct formal hypothesis tests. The traditional method, the *P*-value method, and the use of confidence intervals are illustrated in the following discussion.

The Traditional Method

Note to Instructor
Refer to Figure 7-8 and suggest that students use it regularly as a basic guide for including all of the steps of a complete hypothesis test. Describe exactly what you expect from them. Recommendation: Require the statements of H_0 and H_1; a graph showing the normal distribution with the test statistic and either the *P*-value or the critical value(s) and critical region; a statement of either "reject H_0" or "fail to reject H_0"; and a summary statement of the conclusion in nontechnical terms.

The traditional method requires a decision based on a comparison of the test statistical and critical value, but the TI-83 Plus calculator and some software packages (such as Minitab) might not include critical values with their results; they typically do provide *P*-values, so the *P*-value approach may be more suitable.

The traditional method of testing hypotheses is summarized in Figure 7-8. When testing the claim $p > 0.5$ given in the preceding example, the following steps correspond to the procedure in Figure 7-8:

Step 1:   The original claim in symbolic form is $p > 0.5$.

Step 2:   The opposite of the original claim is $p \leq 0.5$.

Step 3:   Of the preceding two symbolic expressions, the expression $p > 0.5$ does not contain equality, so it becomes the alternative hypothesis. The null hypothesis is the statement that p equals the fixed value of 0.5. We can therefore express H_0 and H_1 as follows:

$$H_0: p = 0.5$$
$$H_1: p > 0.5$$

Step 4:   In the absence of any special circumstances, we will select $\alpha = 0.05$ for the significance level.

Step 5:   Because we are testing a claim about a population proportion p, the sample statistic $\hat{p}$ is relevant to this test, and the sampling distribution of sample proportions $\hat{p}$ is approximated by a *normal distribution*.

Step 6:   The test statistic is evaluated using $n = 880$ and $\hat{p} = 0.56$. In the null hypothesis we are assuming that $p = 0.5$, so $q = 1 - 0.5 = 0.5$. The test statistic is

$$z = \frac{\hat{p} - p}{\sqrt{\dfrac{pq}{n}}} = \frac{0.56 - 0.5}{\sqrt{\dfrac{(0.5)(0.5)}{880}}} = 3.56$$

This is a right-tailed test, so the critical region is an area of $\alpha = 0.05$ in the right tail. Referring to Table A-2 and applying the methods of Section 5-2, we find that the critical value of $z = 1.645$ is at the boundary of the critical region. See Figure 7-3 on page 375, which shows the critical region, critical value, and test statistic.

Step 7: Because the test statistic falls within the critical region, we reject the null hypothesis.

Step 8: We conclude that there is sufficient sample evidence to support the claim that the majority of Americans admit to running red lights. (See Figure 7-7 for help with wording this final conclusion.)

The *P*-Value Method

The *P*-value method of testing hypotheses is summarized in Figure 7-9, and it requires the *P*-value that is found using the procedure summarized in Figure 7-6. A comparison of Figures 7-8 and 7-9 shows that the first five steps of the traditional method are the same as the first five steps of the *P*-value method. For the hypothesis test described in the preceding example, the first five steps of the *P*-value method are the same as those shown in the above traditional method, so we now continue with Step 6.

Step 6: The test statistic is $z = 3.56$ as shown in the preceding traditional method. We now find the *P*-value (instead of the critical value) by using the following procedure, which is shown in Figure 7-6:

Right-tailed test:	*P*-value =	area to right of test statistic z
Left-tailed test:	*P*-value =	area to left of test statistic z
Two-tailed test:	*P*-value =	*twice* the area of the extreme region bounded by the test statistic z

Because the hypothesis test we are considering is right-tailed with a test statistic of $z = 3.56$, the *P*-value is the area to the right of $z = 3.56$. Referring to Table A-2, we see that for values of $z = 3.50$ and higher, we use 0.9999 for the cumulative area to the *left* of the test statistic. The area to the right of $z = 3.56$ is therefore $1 - 0.9999 = 0.0001$. We now know that the *P*-value is 0.0001. Figure 7-10 shows the test statistic and *P*-value for this example.

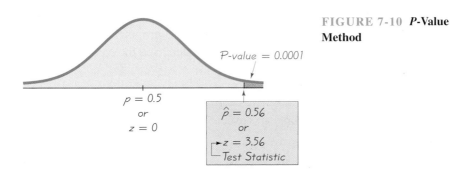

FIGURE 7-10 *P*-Value Method

Note to Instructor

The *P*-value topic is important because TI-83 Plus calculators and so many computer programs generate *P*-values, but not critical values. Also, studying hypothesis testing from another perspective may be helpful in understanding the general approach.

When defining *P*-value, reinforce the importance of getting a value *at least as extreme* as the one found. Refer to Figure 7-6 for the procedure used to find *P*-values. A copy of Figure 7-6 is included with the detachable Formula/Table card.

Step 7:   Because the *P*-value of 0.0001 is less than or equal to the significance level of $\alpha = 0.05$, we reject the null hypothesis.

Step 8:   As with the traditional method, we conclude that there is sufficient sample evidence to support the claim that the majority of Americans admit to running red lights. (See Figure 7-7 for help with wording this final conclusion.)

Confidence Interval Method

Note to Instructor
This subsection requires Chapter 6 as a prerequisite, so skip it if Chapter 6 has not been covered. This subsection discusses the correspondence between confidence intervals and hypothesis testing, but this approach is not emphasized in the following chapters.

For two-tailed hypothesis tests construct a confidence interval with a confidence level of $1 - \alpha$; but for a one-tailed hypothesis test with significance level α, construct a confidence interval with a confidence level of $1 - 2\alpha$. (See Table 7-2 for common cases.) For example, the claim of $p > 0.5$ can be tested with a 0.05 significance level by constructing a 90% confidence interval.

Let's now use the confidence interval method to test the claim of $p > 0.5$, with sample data consisting of $n = 880$ and $\hat{p} = 0.56$ (from the example near the beginning of this section). If we want a significance level of $\alpha = 0.05$ in a right-tailed test, we use a 90% confidence level with the methods of Section 6-2 to get this result: $0.533 < p < 0.588$. Because we are 90% confident that the true value of p is contained within the limits of 0.533 and 0.588, we have sufficient evidence to support the claim that $p > 0.5$.

Caution: When testing claims about a population proportion, the traditional method and the *P*-value method are equivalent in the sense that they always yield the same results, but the confidence interval method is somewhat different. Both the traditional method and *P*-value method use the same standard deviation based on the *claimed proportion p*, but the confidence interval uses an estimated standard deviation based on the *sample proportion $\hat{p}$*. Consequently, it is possible that in some cases, the traditional and *P*-value methods of testing a claim about a proportion might yield a different conclusion than the confidence interval method. (See Exercise 21.) If different conclusions are obtained, realize that the traditional and *P*-value methods use an *exact* standard deviation based on the assumption that the population proportion has the value given in the null hypothesis. However, the confidence interval is constructed using a standard deviation based on an *estimated* value of the population proportion. If you want to estimate a population proportion, do so by constructing a confidence interval, but if you want to test a hypothesis, use the *P*-value method or the traditional method.

When testing a claim about a population proportion p, be careful to identify correctly the sample proportion $\hat{p}$. The sample proportion $\hat{p}$ is sometimes given directly, but in other cases it must be calculated. See the examples below.

Given Statement	Finding $\hat{p}$
10% of the observed sports cars are red.	$\hat{p}$ is given directly: $\hat{p} = 0.10$
96 surveyed households have cable TV and 54 do not.	$\hat{p}$ must be calculated using $\hat{p} = x/n$. $$\hat{p} = \frac{x}{n} = \frac{96}{96 + 54} = 0.64$$

Caution: When a calculator or computer display of $\hat{p}$ results in many decimal places, use all of those decimal places when evaluating the z test statistic. Large errors can result from rounding $\hat{p}$ too much.

EXAMPLE Mendel's Genetics Experiments When Gregor Mendel conducted his famous hybridization experiments with peas, one such experiment resulted in offspring consisting of 428 peas with green pods and 152 peas with yellow pods. According to Mendel's theory, 1/4 of the offspring peas should have yellow pods. Use a 0.05 significance level with the *P*-value method to test the claim that the proportion of peas with yellow pods is equal to 1/4.

SOLUTION   After verifying that the assumptions are all satisfied, we begin with the *P*-value method summarized in Figure 7-9 found in Section 7-2. Note that $n = 428 + 152 = 580$, $\hat{p} = 152/580 = 0.262$, and, for the purposes of the test, we assume that $p = 0.25$.

Step 1:   The original claim is that the proportion of peas with yellow pods is equal to 1/4. We express this in symbolic form as $p = 0.25$.

Step 2:   The opposite of the original claim is $p \neq 0.25$.

Step 3:   Because $p \neq 0.25$ does not contain equality, it becomes H_1. We get

H_0: $p = 0.25$ (null hypothesis and original claim)
H_1: $p \neq 0.25$ (alternative hypothesis)

Step 4:   The significance level is $\alpha = 0.05$.

Step 5:   Because the claim involves the proportion p, the statistic relevant to this test is the sample proportion $\hat{p}$, and the sampling distribution of sample proportions is approximated by the normal distribution (provided that the required assumptions are satisfied). (The requirements $np \geq 5$ and $nq \geq 5$ are both satisfied with $n = 580$, $p = 0.25$, and $q = 0.75$.)

Step 6:   The test statistic of $z = 0.67$ is found as follows:

$$z = \frac{\hat{p} - p}{\sqrt{\dfrac{pq}{n}}} = \frac{0.262 - 0.25}{\sqrt{\dfrac{(0.25)(0.75)}{580}}} = 0.67$$

Refer to Figure 7-6 for the procedure for finding the *P*-value. Figure 7-6 shows that for this two-tailed test with the test statistic located to the right of the center (because $z = 0.67$ is positive), the *P*-value is *twice* the area to the right of the test statistic. Using Table A-2, $z = 0.67$ has an area of 0.7486 to its left, so the area to the right of $z = 0.67$ is $1 - 0.7486 = 0.2514$, which we double to get 0.5028.

Step 7:   Because the *P*-value of 0.5028 is greater than the significance level of 0.05, we fail to reject the null hypothesis.

continued

Polls and Psychologists

Poll results can be dramatically affected by the wording of questions. A phrase such as "over the last few years" is interpreted differently by different people. Over the last few years (actually, since 1980), survey researchers and psychologists have been working together to improve surveys by decreasing bias and increasing accuracy. In one case, psychologists studied the finding that 10 to 15 percent of those surveyed say they voted in the last election when they did not. They experimented with theories of faulty memory, a desire to be viewed as responsible, and a tendency of those who usually vote to say that they voted in the most recent election, even if they did not. Only the last theory was actually found to be part of the problem.

Test of Touch Therapy

At the age of nine, Emily Rosa entered a school science fair with a project designed to test touch therapy. Instead of actually touching subjects, touch therapists move their hands a few inches away from the subject's body so that they can improve the human energy field. Emily Rosa tested 21 touch therapists by sitting on one side of a cardboard shield while the therapists placed their hands through the shield. Emily placed her hand above one of the therapist's hands (selected with a coin toss), then the therapist tried to identify the selected hand without seeing Emily's hand. A 50% success rate would be expected with random guesses, but the touch therapists were correct only 44% of the time. Emily Rosa became the youngest author in the *Journal of the American Medical Association* when this article was published: "A Close Look at Therapeutic Touch" by L. Rosa, E. Rosa, L. Sarner, and S. Barrett, Vol. 279, No. 1005.

INTERPRETATION The methods of hypothesis testing never allow us to support a claim of equality, so we cannot conclude that the proportion of peas with yellow pods is equal to 1/4. Here is the correct conclusion: There is not sufficient evidence to warrant rejection of the claim that 1/4 of the offspring peas have yellow pods.

Traditional Method: If we were to repeat the preceding example using the traditional method of testing hypotheses, we would see that in Step 6, the critical values are found to be $z = -1.96$ and $z = 1.96$. In Step 7, we would fail to reject the null hypothesis because the test statistic of $z = 0.67$ would not fall within the critical region. See the accompanying STATDISK display. We would reach the same conclusion from the *P*-value method: There is not sufficient evidence to warrant rejection of the claim that 1/4 of the offspring peas have yellow pods.

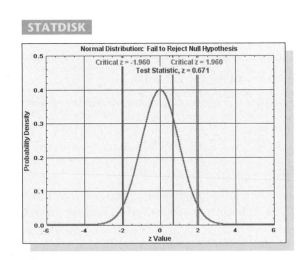

Confidence Interval Method: If we were to repeat the preceding example using the confidence interval method, we would obtain this 95% confidence interval: $0.226 < p < 0.298$. Because the confidence interval limits do contain the claimed value of 0.25, we conclude that there is not sufficient evidence to warrant rejection of the claim that 1/4 of the offspring peas have yellow pods. In this case, the *P*-value method, traditional method, and confidence interval method all lead to the same conclusion. In some other relatively rare cases, the *P*-value method and the traditional method might lead to a conclusion that is different from the conclusion reached through the confidence interval method.

Rationale for the Test Statistic: The test statistic used in this section is justified by noting that when using the normal distribution to approximate a binomial distribution, we use $\mu = np$ and $\sigma = \sqrt{npq}$ to get

$$z = \frac{x - \mu}{\sigma} = \frac{x - np}{\sqrt{npq}}$$

We used the above expression in Section 5-6 along with a correction for continuity, but when testing claims about a population proportion, we make two modifications. First, we don't use the correction for continuity because its effect is usually very small for the large samples we are considering. Also, instead of using the above expression to find the test statistic, we use an equivalent expression obtained by dividing the numerator and denominator by n, and we replace x/n by the symbol $\hat{p}$ to get the test statistic we are using. The end result is that the test statistic is simply the same standard score (from Section 2-5) of $z = (x - \mu)/\sigma$, but modified for the binomial notation.

Using Technology

STATDISK   Select **Analysis, Hypothesis Testing, Proportion-One Sample,** then proceed to enter the data in the dialog box.

Minitab   Select **Stat, Basic Statistics, 1 Proportion,** then click on the button for "Summarized data." Enter the sample size and number of successes, then click on **Options** and proceed to enter the data in the dialog box.

Excel   First enter the number of successes in cell A1, and enter the total number of trials in cell B1. Use the Data Desk XL add-in by clicking on **DDXL**, then select **Hypothesis Tests.**

Under the function type options, select **Summ 1 Var Prop Test** (for testing a claimed proportion using summary data for one variable). Click on the pencil icon for "Num successes" and enter A1. Click on the pencil icon for "Num trials" and enter B1. Click **OK.** Follow the four steps listed in the dialog box. After clicking on **Compute** in Step 4, you will get the P-value, test statistic, and conclusion.

TI-83 Plus   Press **STAT,** select **TESTS,** and then select **1-PropZTest.** Enter the claimed value of the population proportion for p0, then enter the values for x and n, and then select the type of test. Highlight **Calculate,** then press the **ENTER** key.

7-3 Basic Skills and Concepts

1. Mendel's Hybridization Experiments In one of Mendel's famous hybridization experiments, 8023 offspring peas were obtained, and 24.94% of them had green flowers. The others had white flowers. Consider a hypothesis test that uses a 0.05 significance level to test the claim that green-flowered peas occur at a rate of 25%.
   a. What is the test statistic?
   b. What are the critical values?
   c. What is the P-value?
   d. What is the conclusion?
   e. Can a hypothesis test be used to "prove" that the rate of green-flowered peas is 25%, as claimed?

2. Survey of Drinking In a Gallup survey, 1087 randomly selected adults were asked "Do you have occasion to use alcoholic beverages such as liquor, wine, or beer, or are you a total abstainer?" Sixty-two percent of the subjects said that they used alcoholic beverages. Consider a hypothesis test that uses a 0.05 significance level to test the claim that the majority (more than 50%) of adults use alcoholic beverages.
   a. What is the test statistic?
   b. What is the critical value?

continued

Recommended Assignment:
Exercises 1–8, 17, 18. Note that the directions for Exercises 3–20 stipulate that the P-value method should be used "unless your instructor specifies otherwise." The P-value method is recommended, but instructors can stipulate that the traditional method be used for some or all of the assigned exercises. Most Appendix F answers include both P-values and critical values.

1. a. $z = -0.12$
   b. $z = \pm 1.96$
   c. 0.9044
   d. There is not sufficient evidence to warrant rejection of the claim that green-flowered peas occur at a rate of 25%.
   e. No, a hypothesis test cannot be used to prove that a proportion is equal to some claimed value.

2. a. $z = 7.91$
   b. $z = 1.645$
   c. 0.0001
   d. There is sufficient evidence to support the claim that the majority of adults use alcoholic beverages.
   e. No, 62% might not be significantly greater than 50% for some samples, such as one with $n = 50$.

3. TS: $z = -2.06$. CV: $z = -2.33$. P-value: 0.0197. Fail to reject H_0.

4. TS: $z = -6.51$. CV: $z = \pm 2.575$. P-value: 0.0002. Reject H_0.

5. TS: $z = 1.60$. CV: $z = 1.645$. P-value: 0.0548. Fail to reject H_0.

6. TS: $z = 80.87$. CV: $z = 2.33$. P-value: 0.0001. Reject H_0.

7. TS: $z = 0.58$. CV: $z = 1.28$. P-value: 0.2810. Fail to reject H_0.

8. TS: $z = -1.06$. CV: $z = -1.645$. P-value: 0.1446. Fail to reject H_0.

9. TS: $z = 2.19$. CV: $z = \pm 1.96$. P-value: 0.0286. Reject H_0.

c. What is the P-value?

d. What is the conclusion?

e. Based on the preceding results, can we conclude that 62% is significantly greater than 50% for all such hypothesis tests? Why or why not?

Testing Claims About Proportions. In Exercises 3–20, test the given claim. Identify the null hypothesis, alternative hypothesis, test statistic, P-value or critical value(s), conclusion about the null hypothesis, and final conclusion that addresses the original claim. Use the P-value method unless your instructor specifies otherwise.

3. *Glamour* Magazine Survey *Glamour* magazine sponsored a survey of 2500 prospective brides and found that 60% of them spent less than $750 on their wedding gown. Use a 0.01 significance level to test the claim that less than 62% of brides spend less than $750 on their wedding gown. How are the results affected if it is learned that the responses were obtained from magazine readers who decided to respond to the survey through an Internet Web site?

4. Federal Drug Offenses In a recent year, of the 109,857 arrests for Federal offenses, 29.1% were for drug offenses (based on data from the U.S. Department of Justice). Use a 0.01 significance level to test the claim that the drug offense rate is equal to 30%. How can the result be explained, given that 29.1% appears to be so close to 30%?

5. Percentage of E-Mail Users Technology is dramatically changing the way we communicate. In 1997, a survey of 880 U.S. households showed that 149 of them use e-mail (based on data from *The World Almanac and Book of Facts*). Use those sample results to test the claim that more than 15% of U.S. households use e-mail. Use a 0.05 significance level. Is the conclusion valid today? Why or why not?

6. Percentage of Telephone Users A recent survey of 4276 randomly selected households showed that 4019 of them had telephones (based on data from the U.S. Census Bureau). Use those survey results to test the claim that the percentage of households is now greater than the 35% rate that was found in 1920. Use a 0.01 significance level. The current rate of 4019/4276 (or 94%) appears to be significantly greater than the 1920 rate of 35%, but is there sufficient evidence to support that claim?

7. Photo-Cop Legislation The Chapter Problem included this question: "Is there sufficient sample evidence to support a claim that the proportion of all adult Minnesotans opposed to photo-cop legislation is greater than 0.5?" Use a 0.10 significance level to test the claim that the proportion is greater than 0.5. Sample evidence consists of $n = 829$ randomly selected adult Minnesotans with 51% opposed to photo-cop legislation. Given that the sample includes only Minnesotans, does the conclusion apply to all adult Americans?

8. Cloning Survey In a Gallup poll of 1012 randomly selected adults, 9% said that cloning of humans should be allowed. Use a 0.05 significance level to test the claim that less than 10% of all adults say that cloning of humans should be allowed. Can a newspaper run a headline that "less than 10% of all adults are opposed to cloning of humans"?

9. Store Checkout-Scanner Accuracy In a study of store checkout-scanners, 1234 items were checked and 20 of them were found to be overcharges (based on data from "UPC Scanner Pricing Systems: Are They Accurate?" by Goodstein, *Journal of Marketing,* Vol. 58). Use a 0.05 significance level to test the claim that with scanners, 1%

of sales are overcharges. (Before scanners were used, the overcharge rate was estimated to be about 1%.) Based on these results, do scanners appear to help consumers avoid overcharges?

10. Drug Testing of Job Applicants   In 1990, 5.8% of job applicants who were tested for drugs failed the test. At the 0.01 significance level, test the claim that the failure rate is now lower if a simple random sample of 1520 current job applicants results in 58 failures (based on data from the American Management Association). Does the result suggest that fewer job applicants now use drugs?

10. TS: $z = -3.31$. CV: $z = -2.33$.
P-value: 0.0001. Reject H_0.

11. Umpire Strike Rate   In a recent year, some professional baseball players complained that umpires were calling more strikes than the average rate of 61.0% called the previous year. At one point in the season, umpire Dan Morrison called strikes in 2231 of 3581 pitches (based on data from *USA Today*). Use a 0.05 significance level to test the claim that his strike rate is greater than 61.0%.

11. TS: $z = 1.60$. CV: $z = 1.645$.
P-value: 0.0548. Fail to reject H_0.

12. Testing Lipitor for Cholesterol Reduction   In clinical tests of the drug Lipitor (generic name, atorvastatin), 863 patients were treated with 10 mg doses of atorvastatin, and 19 of those patients experienced flu symptoms (based on data from Parke-Davis). Use a 0.01 significance level to test the claim that the percentage of treated patients with flu symptoms is greater than the 1.9% rate for patients not given the treatments. Does it appear that flu symptoms are an adverse reaction of the treatment?

12. TS: $z = 0.65$. CV: $z = 2.33$.
P-value: 0.2578. Fail to reject H_0.

13. Cell Phones and Cancer   In a study of 420,095 Danish cell phone users, 135 subjects developed cancer of the brain or nervous system (based on data from the *Journal of the National Cancer Institute* as reported in *USA Today*). Test the claim of a once popular belief that such cancers are affected by cell phone use. That is, test the claim that cell phone users develop cancer of the brain or nervous system at a rate that is different from the rate of 0.0340% for people who do not use cell phones. Because this issue has such great importance, use a 0.005 significance level. Should cell phone users be concerned about cancer of the brain or nervous system?

13. TS: $z = -0.66$. CV: $z = \pm 2.81$.
P-value: 0.5092. Fail to reject H_0.

14. Testing Effectiveness of Nicotine Patches   In one study of smokers who tried to quit smoking with nicotine patch therapy, 39 were smoking one year after the treatment, and 32 were not smoking one year after the treatment (based on data from "High-Dose Nicotine Patch Therapy," by Dale et al., *Journal of the American Medical Association,* Vol. 274, No. 17). Use a 0.10 significance level to test the claim that among smokers who try to quit with nicotine patch therapy, the majority are smoking a year after the treatment. Do these results suggest that the nicotine patch therapy is ineffective?

14. TS: $z = 0.83$. CV: $z = 1.28$.
P-value: 0.2033. Fail to reject H_0.

15. Smoking and College Education   One survey showed that among 785 randomly selected subjects who completed four years of college, 144 smoke and 641 do not smoke (based on data from the American Medical Association). Use a 0.01 significance level to test the claim that the rate of smoking among those with four years of college is less than the 27% rate for the general population. Why would college graduates smoke at a lower rate than others?

15. TS: $z = -5.46$. CV: $z = -2.33$.
P-value: 0.0001. Reject H_0.

16. TV Ratings   A simple random sample of households with TV sets in use shows that 1024 of them were tuned to *60 Minutes* while 3836 were tuned to some other show. Use a 0.025 significance level to test the claim of a CBS executive that "*60 Minutes* gets more than a 20 share," which means that more than 20% of the sets in use are tuned to *60 Minutes.* If you are a commercial advertiser and you are trying to negotiate lower costs, what would you argue?

16. TS: $z = 1.86$. CV: $z = 1.96$.
P-value: 0.0314. Fail to reject H_0.

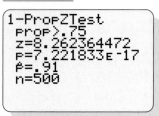

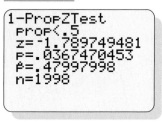

17. Interpreting Calculator Display   The Federal Aviation Administration will fund research on spatial disorientation of pilots if there is sufficient sample evidence (at the 0.01 significance level) to conclude that among aircraft accidents involving such disorientation, more than three-fourths result in fatalities. A study of 500 aircraft accidents involving spatial disorientation of the pilot found that 91% of those accidents resulted in fatalities (based on data from the U.S. Department of Transportation). The accompanying TI-83 Plus calculator display is obtained. Interpret that display. Based on these sample results, will the funding be approved?

18. Interpreting Calculator Display   A television executive claims that "fewer than half of all adults are annoyed by the violence shown on television." Sample data from a Roper poll showed that 48% of 1,998 surveyed adults indicated their annoyance with television violence. The accompanying TI-83 Plus calculator display is obtained. Use a 0.05 significance level and interpret that display. Is the executive's claim supported by the sample data?

19. Using M&M Data   Refer to Data Set 19 in Appendix B and find the sample proportion of M&Ms that are blue. Use that result to test the claim of Mars, Inc., that 10% of its plain M&M candies are blue.

20. Alcohol and Tobacco Use in Animated Children's Movies   Using results listed in Data Set 7 in Appendix B, test the claim that the majority of animated children's movies show the use of alcohol or tobacco (or both). Use a 0.05 significance level.

17. TS: $z = 8.26$. CV: $z = 2.33$.
    P-value: 0.0000 when rounded.
    Reject H_0.

18. TS: $z = -1.79$. CV: $z = -1.645$.
    P-value: 0.0367. Reject H_0.

19. TS: $z = -1.67$. CV assuming
    $\alpha = 0.05$: $z = \pm 1.96$. *P*-value:
    0.0950. Fail to reject H_0.

20. TS: $z = 2.55$. CV: $z = 1.645$.
    P-value: 0.0054. Reject H_0.

21. a. TS: $z = 2.00$. CV: $z = \pm 1.96$.
        Reject H_0.

    b. TS: $z = 2.00$. *P*-value:
        0.0456. Reject H_0.

    c. $0.0989 < p < 0.139$;
        because 0.1 is contained
        within the confidence interval, fail to reject H_0.

    d. The traditional and *P*-value
        methods both lead to rejection of the claim, but the CI
        method does not lead to
        rejection.

22. TS: $z = 2.40$. *P*-value: 0.0082.
    Reject H_0. *P*-value is slightly
    larger because the use of the
    continuity correction adds a
    small region (between 33.5 and
    34) not included in Exercise 20.

7-3 Beyond the Basics

21. Using Confidence Intervals to Test Hypotheses   When analyzing the last digits of telephone numbers in Port Jefferson, it is found that among 1000 randomly selected digits, 119 are zeros. If the digits are randomly selected, the proportion of zeros should be 0.1.
   a. Use the traditional method with a 0.05 significance level to test the claim that the proportion of zeros equals 0.1.
   b. Use the *P*-value method with a 0.05 significance level to test the claim that the proportion of zeros equals 0.1.
   c. Use the sample data to construct a 95% confidence interval estimate of the proportion of zeros. What does the confidence interval suggest about the claim that the proportion of zeros equals 0.1?
   d. Compare the results from the traditional method, the *P*-value method, and the confidence interval method. Do they all lead to the same conclusion?

22. Using the Continuity Correction   Repeat Exercise 20, but include the correction for continuity that was described in Section 5-6. How are the results affected by including the continuity correction?

23. Proving Claims   In the *USA Today* article "Power Lines Not a Cancer Risk for Kids," the first sentence states that "Children who live near high voltage power lines appear to be no more likely to get leukemia than other kids, doctors report today in the most extensive study of one of the most controversial issues ever done." Representing the rate of leukemia for children not living near high voltage power lines by the constant *c*, write the claim in symbolic form, then identify the null and alternative hypotheses suggested by this statement. Given that we either reject or fail to reject the null hypothesis, what possible conclusions can be made about the original claim? Can the sample data support the claim that children who live near high voltage power lines are no more likely to get leukemia than other children?

24. *Alternative Method of Testing a Claim About p* In a study of perception, 80 men are tested and 7 are found to have red/green color blindness (based on data from *USA Today*). We want to use a 0.01 significance level to test the claim that men have a red/green color-blindness rate that is greater than the 0.25% rate for women.
    a. Why can't we use the methods of this section?
    b. Assuming that the red/green color-blindness rate for men is equal to the 0.25% rate for women, find the probability that among 80 randomly selected men, at least 7 will have that type of color blindness. Describe the method used to find that probability.
    c. Based on the result from part (b), what do you conclude?

25. *Coping with No Successes* In a simple random sample of 50 plain M&M candies, it is found that none of them are blue. We want to use a 0.01 significance level to test the claim of Mars, Inc., that the proportion of M&M candies that are blue is equal to 0.10. Can the methods of this section be used? If so, test the claim. If not, explain why not.

26. *Misleading with Statistics* Chemco, a supplier of chemical-waste containers, finds that 3% of a sample of 500 units are defective. Being fundamentally dishonest, the Chemco production manager wants to make a claim that the rate of defective units is no more than some specified percentage, and he doesn't want that claim rejected at the 0.05 significance level if the sample data are used. What is the *lowest* defective rate he can claim under these conditions?

27. *False Claim* A researcher claimed that when 20 mice were treated, the success rate was equal to 47%. What is the basis for rejecting that claim?

28. *Probability of Type II Error* For a hypothesis test with a specified significance level α, the probability of a type I error is α, whereas the probability β of a type II error depends on the particular value of p that is used as an alternative to the null hypothesis. Refer to Exercise 20. Assuming that the true value of p is 0.45, find β, the probability of a type II error. Use the following procedure. [*Hint:* In Step 3, use the values $p = 0.45$ and $pq/n = (0.45)(0.55)/50$.]

    Step 1:   Find the value(s) of the sample statistic $\hat{p}$ that correspond to the critical value(s). In

    $$z = \frac{\hat{p} - p}{\sqrt{\dfrac{pq}{n}}}$$

    substitute the critical value(s) for *z*, enter the values for *p* (from the null hypothesis) and *q*, then solve for $\hat{p}$.

    Step 2:   Given a particular value of *p* that is an alternative to the value given in the null hypothesis, draw the normal curve with this new alternative value of *p* at the center. Also plot the value(s) of $\hat{p}$ found in Step 1.

    Step 3:   Refer to the graph in Step 2, and find the area of the new critical region bounded by the value(s) of $\hat{p}$ found in Step 1. (Be sure to use the standard deviation based on the new value of p.) This is the probability of rejecting the null hypothesis, given that the new value of p is correct.

Step 4:   The value of β is 1 minus the area from Step 3. This is the probability of failing to reject the null hypothesis, given that the new value of p is correct.

The preceding steps allow you to find the probability of failing to reject H_0 when it is false. You are determining the area under the curve that excludes the critical region in which you reject H_0; this area corresponds to a failure to reject a false H_0, and we know that H_0 is false because we are using an alternative value that is assumed to be the correct population proportion.

 7-4 Testing a Claim About a Mean: σ Known

In this section we consider methods of testing claims made about a population mean μ, and we assume that the population standard deviation σ is known. It would be an unusual set of circumstances that would allow us to know σ without knowing μ, but Section 7-5 deals with cases in which σ is not known. Although this section involves cases that are less realistic than those in Section 7-5, this section is important in describing the same general method used in the following section. Also, there are cases in which the specific value of σ is unknown, but some information about σ could be used. The example presented in this section includes the unrealistic assumption that σ is known to be 0.62°F. The test statistic in that example is found to be $z = -6.64$, which leads to rejection of the common belief that the mean body temperature is equal to 98.6°F. If we analyze the variation of body temperatures, it becomes obvious that σ can't possibly be as high as 2°F, yet using $\sigma = 2$°F would result in a test statistic of $z = -2.05$, which again leads to rejection of the claim that $\mu = 98.6$°F. Because σ must be less than 2°F for body temperatures, the test statistic must be at least as extreme as $z = -2.05$. (See Exercise 17.) This shows that even though we might not know a specific value of σ, there are cases in which the use of very conservative values of σ will allow us to form some meaningful conclusions.

The assumptions, test statistic, critical values, and *P*-value are summarized as follows.

Testing Claims About a Population Mean (with σ Known)
Assumptions
1. The sample is a simple random sample. (Remember this very important point made in Chapter 1: *Data carelessly collected may be so completely useless that no amount of statistical torturing can salvage them.*)
2. The value of the population standard deviation σ is known.
3. Either or both of these conditions is satisfied: The population is normally distributed or $n > 30$.

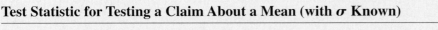

Test Statistic for Testing a Claim About a Mean (with σ Known)

$$z = \frac{\bar{x} - \mu_{\bar{x}}}{\frac{\sigma}{\sqrt{n}}}$$

P-values: Use the standard normal distribution (Table A-2) and refer to Figure 7-6.
Critical values: Use the standard normal distribution (Table A-2).

Before starting the hypothesis testing procedure, we should first *explore* the data set. Using the methods introduced in Chapter 2, investigate center, variation, and distribution by drawing a graph; finding the mean, standard deviation, and 5-number summary; and identifying any outliers. We should verify that the required assumptions are satisfied. For the sample of 106 body temperatures used in the following example, a histogram shows that the sample data appear to come from a normally distributed population. Also, there are no outliers. The issue of normality is not too important in this example because the sample is so large, but it is important to know that there are no outliers that would dramatically affect results.

EXAMPLE *P-Value Method* Data Set 4 in Appendix B lists a sample of 106 body temperatures having a mean of 98.20°F. Assume that the sample is a simple random sample and that the population standard deviation σ is known to be 0.62°F. Use a 0.05 significance level to test the common belief that the mean body temperature of healthy adults is equal to 98.6°F. Use the P-value method by following the procedure outlined in Figure 7-9.

SOLUTION   Refer to Figure 7-9 and follow these steps:

Step 1:   The claim that the mean is equal to 98.6 is expressed in symbolic form as $\mu = 98.6$.

Step 2:   The alternative (in symbolic form) to the original claim is $\mu \neq 98.6$.

Step 3:   Because the statement $\mu \neq 98.6$ does not contain the condition of equality, it becomes the alternative hypothesis. The null hypothesis is the statement that $\mu = 98.6$.

$$H_0: \mu = 98.6 \quad \text{(original claim)}$$
$$H_1: \mu \neq 98.6$$

Step 4:   As specified in the statement of the problem, the significance level is $\alpha = 0.05$.

Step 5:   Because the claim is made about the *population mean* μ, the sample statistic most relevant to this test is the *sample mean* $\bar{x} = 98.20$. Because σ is assumed to be known (0.62) and $n > 30$, the central limit

continued

theorem indicates that the distribution of sample means can be approximated by a *normal* distribution.

Step 6:   The test statistic is calculated as follows:

$$z = \frac{\bar{x} - \mu_{\bar{x}}}{\frac{\sigma}{\sqrt{n}}} = \frac{98.20 - 98.6}{\frac{0.62}{\sqrt{106}}} = -6.64$$

Using the test statistic of $z = -6.64$, we now proceed to find the *P*-value. See Figure 7-6 for the flowchart summarizing the procedure for finding *P*-values. This is a two-tailed test and the test statistic is to the left of the center (because $z = -6.64$ is less than $z = 0$), so the *P*-value is *twice* the area to the left of $z = -6.64$. We now refer to Table A-2 to find that the area to the left of $z = -6.64$ is 0.0001, so the *P*-value is 2(0.0001) = 0.0002. (More precise results show that the *P*-value is actually much less than 0.0002.) See Figure 7-11.

Step 7:   Because the *P*-value of 0.0002 is less than the significance level of $\alpha = 0.05$, we reject the null hypothesis.

INTERPRETATION   The *P*-value of 0.0002 is the probability of getting a sample mean as extreme as 98.20°F (with a sample size of $n = 106$) by chance, assuming that $\mu = 98.6°F$ and $\sigma = 0.62°F$. Because that probability is so small, we reject random chance as a likely explanation, and we conclude that the assumption of $\mu = 98.6°F$ must be wrong. We refer to Figure 7-7 in Section 7-2 for help in correctly stating the final conclusion. We are rejecting the null hypothesis, which is the original claim, so we conclude that there is sufficient evidence to warrant rejection of the claim that the mean body temperature of healthy adults is 98.6°F. There is sufficient evidence to conclude that the mean body temperature of all healthy adults differs from 98.6°F.

Traditional Method   If the traditional method of testing hypotheses is used for the preceding example, the first five steps would be the same. In Step 6 we would find the critical values of $z = -1.96$ and $z = 1.96$ instead of finding the

FIGURE 7-11   *P*-Value
Method of Testing H_0: $\mu = 98.6$

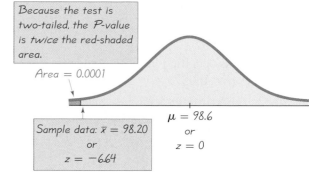

P-value. We would again reject the null hypothesis because the test statistic of $z = -6.64$ would fall in the critical region. The final conclusion will be the same.

Confidence Interval Method   We can use a confidence interval for testing a claim about μ when σ is known. For a two-tailed hypothesis test with a 0.05 significance level, we construct a 95% confidence interval. If we use the sample data in the preceding example ($n = 106$ and $\bar{x} = 98.20$) and assume that $\sigma = 0.62$ is known, we can test the claim that $\mu = 98.6$ by using the methods of Section 6-3 to construct this 95% confidence interval: $98.08 < \mu < 98.32$. Because the claimed value of $\mu = 98.6$ is not contained within the confidence interval, we reject that claim. We are 95% confident that the limits of 98.08 and 98.32 contain the true value of μ, so it appears that 98.6 cannot be the true value of μ.

In Section 7-3 we saw that when testing a claim about a population proportion, the traditional method and *P*-value method are equivalent, but the confidence interval method is somewhat different. When testing a claim about a population mean, there is no such difference, and all three methods are equivalent.

Caution: When testing a claim about μ using a confidence interval, be sure to use the confidence level that is appropriate for a specified significance level. With two-tailed tests, it is easy to see that a 0.05 significance level corresponds to a 95% confidence level, but it gets tricky with one-tailed tests. To test the claim that $\mu < 98.6$ with a 0.05 significance level, construct a 90% confidence interval. To test the claim that $\mu > 98.6$ with a 0.01 significance level, construct a 98% confidence interval.

In the remainder of the text, we will apply methods of hypothesis testing to other circumstances. It is easy to become entangled in a complex web of steps without ever understanding the underlying rationale of hypothesis testing. The key to that understanding lies in the rare event rule for inferential statistics: **If, under a given assumption, there is an exceptionally small probability of getting sample results at least as extreme as the results that were obtained, we conclude that the assumption is probably not correct.** When testing a claim, we make an assumption (null hypothesis) of equality. We then compare the assumption and the sample results to form one of the following conclusions:

- If the sample results (or more extreme results) can easily occur when the assumption (null hypothesis) is true, we attribute the relatively small discrepancy between the assumption and the sample results to chance.

- If the sample results (or more extreme results) cannot easily occur when the assumption (null hypothesis) is true, we explain the relatively large discrepancy between the assumption and the sample results by concluding that the assumption is not true, so we reject the assumption.

Alternative Method (not used in this book)   An alternative method not used in this book is to use *s* as an estimate of the unknown σ, provided that the sample is large ($n > 30$). That is, if the sample size *n* is greater than 30, replace the unknown σ with the sample standard deviation *s*, then use the methods of this section by proceeding as if σ is known. Section 7-5 lists reasons why this alternative method is not used in this book.

STATDISK If working with a list of the original sample values, first find the sample size, sample mean, and sample standard deviation by using the STATDISK procedure described in Section 2-4. After finding the values of n, $\bar{x}$ and s, proceed to select the main menu bar item **Analysis**, then select **Hypothesis Testing**, followed by **Mean-One Sample.**

Minitab Minitab Release 14 allows you to use either the summary statistics or a list of the original sample values. (Earlier versions do not allow the use of summary statistics. See Section 5-5 of the *Minitab Manual* for a way to work around this limitation.) Select **Stat, Basic Statistics,** then **1-sample Z.** Enter the summary statistics or enter **C1** in the box located at the top right. Also enter the value of σ in the "Standard deviation" or "Sigma" box. Use the **Options** button to change the alternative hypothesis.

Excel Excel's built-in ZTEST function is extremely tricky to use, because the generated P-value is not always the same standard P-value used by the rest of the world. Instead, use the Data Desk XL add-in that is a supplement to this book. First enter the sample data in column A. Select **DDXL**, then **Hypothesis Tests.** Under the function type options, select **1 Var z Test.** Click on the pencil icon and enter the range of data values, such as A1:A106 if you have 106 values listed in column A. Click on **OK.** Follow the four steps listed in the dialog box. After clicking on **Compute** in Step 4, you will get the P-value, test statistic, and conclusion.

TI-83 Plus If using a TI-83 Plus calculator, press **STAT**, then select **TESTS** and choose the first option, **Z-Test.** You can use the original data or the summary statistics **(Stats)** by providing the entries indicated in the window display. The first three items of the TI-83 Plus results will include the alternative hypothesis, the test statistic, and the P-value.

7-4 Basic Skills and Concepts

Recommended Assignment:
Even-numbered Exercises 2–16.

1. Yes
2. No
3. No
4. Yes
5. $z = 1.18$, *P*-value: 0.1190; CV: $z = 1.645$. There is not sufficient evidence to support the claim that the mean is greater than 118.
6. $z = -6.64$, *P*-value: 0.0001; CV: $z = -2.33$. There is sufficient evidence to support the claim that the mean is less than 98.6°F.
7. $z = 0.89$, *P*-value: 0.3734; CV: $z = \pm 2.575$. There is not sufficient evidence to warrant rejection of the claim that the mean equals 5.00 sec.

Verifying Assumptions. In Exercises 1–4, determine whether the given conditions justify using the methods of this section when testing a claim about the population mean μ.

1. The sample size is $n = 25$, $\sigma = 6.44$, and the original population is normally distributed.

2. The sample size is $n = 7$, σ is not known, and the original population is normally distributed.

3. The sample size is $n = 11$, σ is not known, and the original population is normally distributed.

4. The sample size is $n = 47$, $\sigma = 12.6$, and the original population is not normally distributed.

Finding Test Components. In Exercises 5–8, find the test statistic, P-value, critical value(s), and state the final conclusion.

5. Claim: The mean IQ score of statistics professors is greater than 118.
   Sample data: $n = 50$, $\bar{x} = 120$. Assume that $\sigma = 12$ and the significance level is $\alpha = 0.05$.

6. Claim: The mean body temperature of healthy adults is less than 98.6°F.
   Sample data: $n = 106$, $\bar{x} = 98.20$°F. Assume that $\sigma = 0.62$ and the significance level is $\alpha = 0.01$.

7. Claim: The mean time between uses of a TV remote control by males during commercials equals 5.00 sec.

continued

Sample data: $n = 80$, $\bar{x} = 5.25$ sec. Assume that $\sigma = 2.50$ sec and the significance level is $\alpha = 0.01$.

8. **Claim:** The mean starting salary for college graduates who have taken a statistics course is equal to $46,000.
   Sample data: $n = 65$, $\bar{x} = \$45,678$. Assume that $\sigma = \$9900$ and the significance level is $\alpha = 0.05$.

Testing Hypotheses. In Exercises 9–12, test the given claim. Identify the null hypothesis, alternative hypothesis, test statistic, P-value or critical value(s), conclusion about the null hypothesis, and final conclusion that addresses the original claim. Use the P-value method unless your instructor specifies otherwise.

9. **Everglades Temperatures** In order to monitor the ecological health of the Florida Everglades, various measurements are recorded at different times. The bottom temperatures are recorded at the Garfield Bight station and the mean of 30.4°C is obtained for 61 temperatures recorded on 61 different days. Assuming that $\sigma = 1.7$°C, test the claim that the population mean is greater than 30.0°C. Use a 0.05 significance level.

10. **Weights of Bears** The health of the bear population in Yellowstone National Park is monitored by periodic measurements taken from anesthetized bears. A sample of 54 bears has a mean weight of 182.9 lb. Assuming that σ is known to be 121.8 lb, use a 0.10 significance level to test the claim that the population mean of all such bear weights is less than 200 lb.

11. **Cotinine Levels of Smokers** When people smoke, the nicotine they absorb is converted to cotinine, which can be measured. A sample of 40 smokers has a mean cotinine level of 172.5. Assuming that σ is known to be 119.5, use a 0.01 significance level to test the claim that the mean cotinine level of all smokers is equal to 200.0.

12. **Head Circumferences** A random sample of 100 babies is obtained, and the mean head circumference is found to be 40.6 cm. Assuming that the population standard deviation is known to be 1.6 cm, use a 0.05 significance level to test the claim that the mean head circumference of all two-month-old babies is equal to 40.0 cm.

Interpreting Computer and Calculator Displays. In Exercises 13–16, use the computer or calculator display to form a conclusion.

13. **Mean Weight of M&M Candies** A package of M&M plain candies is labeled as containing 1361 g, and there are 1498 candies, so the mean weight of the individual candies should be 1361/1498, or 0.9085 g. The Mars Company wants to produce the M&Ms with weights that don't cheat consumers, nor do they want to waste production money by having a mean that is significantly greater than is necessary. In a test of the claim that $\mu \neq 0.9085$ g, a sample of 100 M&Ms is randomly selected. (See Data Set 19 in Appendix B.) When the 100 weights are used with Minitab, the display is as shown here (assuming that σ is known to be 0.03691 g). Interpret those results. Are consumers being cheated? Is money being wasted by making the M&Ms heavier than necessary?

```
Test of mu = 0.9085 vs mu not = 0.9085
The assumed sigma = 0.03691

Variable      N       Mean      StDev     SE Mean
M&M          100     0.91470   0.03691   0.00369

Variable          95.0% CI           Z         P
M&M         ( 0.90747, 0.92193)     1.68      0.093
```

8. $z = -0.26$, P-value: 0.7948; CV: $z = \pm 1.96$. There is not sufficient evidence to warrant rejection of the claim that the mean equals $46,000.

9. TS: $z = 1.84$. P-value: 0.0329. (CV: $z = 1.645$.) Reject H_0.

10. TS: $z = -1.03$. P-value: 0.1515. (CV: $z = -1.28$.) Fail to reject H_0.

11. TS: $z = -1.46$. P-value: 0.1442. (CV: $z = \pm 2.575$.) Fail to reject H_0.

12. TS: $z = 3.75$. P-value: 0.0002. (CV: $z = \pm 1.96$.) Reject H_0.

13. TS: $z = 1.68$. P-value: 0.093. Fail to reject H_0.

14. TS: $z = -8.18$. *P*-value: 0.000.
    Reject H_0.
15. TS: $z = -0.63$. *P*-value: 0.5288.
    Fail to reject H_0.
16. TS: $z = -8.80$. *P*-value: 0.
    Reject H_0.

14. Analysis of Last Digits Analysis of the last digits of sample data values sometimes reveals whether the data have been accurately measured and reported. When single digits 0 through 9 are randomly selected with replacement, the mean should be 4.50 and the standard deviation should be 2.87. Reported data (such as weights or heights) are often rounded so that the last digits include disproportionately more 0s and 5s. The last digits in the reported lengths (in feet) of the 73 home runs hit by Barry Bonds in 2001 are used to test the claim that they come from a population with a mean of 4.50 (based on data from *USA Today*). When Minitab is used to test that claim, the display is as shown here. Using a 0.05 significance level, interpret the Minitab results. Does it appear that the distances were accurately measured?

```
Test of mu = 4.5 vs mu not = 4.5
The assumed sigma = 2.87
```

Variable	N	Mean	StDev	SE Mean
BONDS	73	1.753	2.650	0.336

Variable	95.0% CI	Z	P
BONDS	(1.095, 2.412)	-8.18	0.000

15. Differences Between Forecast and Actual High Temperatures Data Set 10 in Appendix B lists the actual high temperatures and the three-day forecast high temperatures. One way to investigate the accuracy of those forecast temperatures is to find the differences between the actual and the forecast temperatures. The 31 differences (actual high − three-day forecast high) have a mean of $-0.419°$. Assuming that $\sigma = 3.704°$, we get the results shown in the accompanying TI-83 Plus calculator display. Interpret those results. Does the mean difference appear to be close to $0°$, or does there appear to be a significant difference? What do these results suggest about the accuracy of the three-day forecast high temperatures?

TI-83 Plus

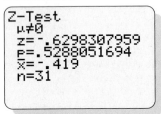

16. Are Thinner Aluminum Cans Weaker? Data Set 20 in Appendix B includes the measured axial loads (in pounds) of 175 cola cans that use aluminum 0.0109 in. thick. Before obtaining these sample results, the standard cans had a thickness of 0.0111 in. and the mean axial load was 281.81 lb. When using the axial loads of the thinner cans in a test of the claim that the mean axial load is less than 281.81 lb, the TI-83 Plus calculator provides the accompanying display. (The display is based on the assumption that σ is known to be 22.11 lb.) Assume that we are using a 0.01 significance level. Interpret the results. Do the thinner cans appear to have a mean axial load less than 281.81 lb?

TI-83 Plus

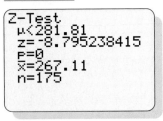

7-4 Beyond the Basics

17. a. It is not likely that σ is
       known.
    b. 2.10
    c. No, so the assumption that
       $\sigma = 0.62$ is a safe assumption in the sense that if σ is actually some value other than 0.62, it is very unlikely that the result of the hypothesis test would be affected.

17. Testing the Assumed σ In the example included in this section, we rejected H_0: $\mu = 98.6$ and supported H_1: $\mu \neq 98.6$ given the assumption that $\sigma = 0.62$ and the sample data consist of $n = 106$ values with $\bar{x} = 98.20$.
    a. What aspect of this example is not realistic?
    b. Find the largest value of σ that results in the same conclusion that was reached assuming that $\sigma = 0.62$.
    c. Given that the 106 body temperatures have a standard deviation of 0.62, is there any reasonable chance that the true value of σ is greater than the value found in part (b)? What does this imply about the assumption that $\sigma = 0.62$?

18. *Finding Standard Deviation* A journal article reported that a null hypothesis of $\mu = 100$ was rejected because the *P*-value was less than 0.01. The sample size was given as 62, and the sample mean was given as 103.6. Find the largest possible standard deviation.

19. *Finding Probability of Type II Error* For a hypothesis test with a given significance level α, the probability of a type I error is the fixed value α, but the probability β of a type II error depends on the particular value of μ that is used as an alternative to the null hypothesis. For hypothesis tests of the type found in this section, we can find β as follows:

Step 1: Find the value(s) of $\bar{x}$ that correspond to the critical value(s). In

$$z = \frac{\bar{x} - \mu_{\bar{x}}}{\dfrac{\sigma}{\sqrt{n}}}$$

substitute the critical value(s) for z, enter the values for $\mu_{\bar{x}}$, σ, and n, then solve for $\bar{x}$.

Step 2: Given a particular value of μ that is an alternative to the value given in the null hypothesis, draw the normal curve with this new value of μ at the center. Also plot the value(s) of $\bar{x}$ found in Step 1.

Step 3: Refer to the graph in Step 2, and find the area of the new critical region bounded by the value(s) of $\bar{x}$ found in Step 1. This is the probability of rejecting the null hypothesis, given that the new value of μ is correct while the value of μ given in the null hypothesis is false.

Step 4: The value of β is 1 minus the area from Step 3. This is the probability of failing to reject the null hypothesis, given that the new value of μ is correct.

The preceding steps allow you to find the probability of failing to reject the null hypothesis when it is false. You are determining the area under the curve that excludes the critical region in which you reject H_0; this area corresponds to a failure to reject a false H_0 because we use a particular value of μ that goes against H_0. Refer to the body-temperature example discussed in this section and find β (the probability of a type II error) corresponding to the following:
a. $\mu = 98.7$
b. $\mu = 98.4$

20. *Power of Test* The *power* of a test, expressed as $1 - \beta$, is the probability of rejecting a false null hypothesis. Assume that in testing the claim that $\mu < 98.6$, the sample data are $n = 106$ and $\bar{x} = 98.20$. Assume that $\sigma = 0.62$ and a 0.05 significance level is used. If the test of the claim $\mu < 98.6$ has a power of 0.8, find the mean μ that is being used as an alternative to the value given in H_0 (see Exercise 19).

7-5 Testing a Claim About a Mean: σ Not Known

One great advantage of learning the methods of hypothesis testing described in the earlier sections of this chapter is that those same methods can be easily modified for use in many other circumstances, such as those discussed in this section. The main objective of this section is to develop the ability to test claims made about population means when the population standard deviation σ is not known.

Better Results with Smaller Class Size

An experiment at the State University of New York at Stony Brook found that students did significantly better in classes limited to 35 students than in large classes with 150 to 200 students. For a calculus course, failure rates were 19% for the small classes compared to 50% for the large classes. The percentages of As were 24% for the small classes and 3% for the large classes. These results suggest that students benefit from smaller classes, which allow for more direct interaction between students and teachers.

Section 7-4 presented methods for testing claims about μ when σ is known, but it is rare that we do not know the value of μ while we do know the value of σ. The methods of this section are much more practical and realistic because they assume that σ is not known, as is usually the case. The assumptions, test statistic, P-value, and critical values are summarized as follows.

Testing Claims About a Population Mean (with σ Not Known)

Assumptions

1. The sample is a simple random sample.

2. The value of the population standard deviation σ is *not* known.

3. Either or both of these conditions is satisfied: The population is normally distributed or $n > 30$.

Test Statistic for Testing a Claim About a Mean (with σ Not Known)

$$t = \frac{\bar{x} - \mu_{\bar{x}}}{\frac{s}{\sqrt{n}}}$$

P-values and critical values: Use Table A-3 and use df $= n - 1$ for the number of degrees of freedom. (See Figure 7-6 for P-value procedures.)

The requirement of a normally distributed population is not a strict requirement, and we can usually consider the population to be normally distributed after using the sample data to confirm that there are no outliers and the histogram has a shape that is not very far from a normal distribution. Also, we use the simplified criterion of $n > 30$ as justification for treating the distribution of sample means as a normal distribution, but the minimum sample size actually depends on how much the population distribution departs from a normal distribution. Because we do not know the value of σ, we estimate it with the value of the sample standard deviation s, but this introduces another source of unreliability, especially with small samples. We compensate for this added unreliability by finding P-values and critical values using the t distribution instead of the normal distribution that was used in Section 7-4 where σ was known. Here are the important properties of the Student t distribution:

Important Properties of the Student t Distribution

1. The Student t distribution is different for different sample sizes (see Figure 6-5 in Section 6-4).

2. The Student t distribution has the same general bell shape as the standard normal distribution; its wider shape reflects greater variability that is expected when s is used to estimate σ.

3. The Student *t* distribution has a mean of $t = 0$ (just as the standard normal distribution has a mean of $z = 0$).

4. The standard deviation of the Student *t* distribution varies with the sample size and is greater than 1 (unlike the standard normal distribution, which has $\sigma = 1$).

5. As the sample size *n* gets larger, the Student *t* distribution gets closer to the standard normal distribution.

Choosing the Appropriate Distribution

When testing claims made about population means, sometimes the normal distribution applies, sometimes the Student *t* distribution applies, and sometimes neither applies, so we must use nonparametric methods or bootstrap resampling techniques. (Nonparametric methods, which do not require a particular distribution, are discussed in Chapter 12; the bootstrap resampling technique is described in the Technology Project at the end of Chapter 6.) See pages 336–337 where Figure 6-6 and Table 6-1 both summarize the decisions to be made in choosing between the normal and Student *t* distributions. They show that when testing claims about population means, the Student *t* distribution is used under these conditions:

Use the Student *t* distribution when σ is not known and either or both of these conditions is satisfied:

The population is normally distributed or $n > 30$.

EXAMPLE Body Temperatures A premed student in a statistics class is required to do a class project. Intrigued by the body temperatures in Data Set 4 of Appendix B, she plans to collect her own sample data to test the claim that the mean body temperature is less than 98.6°F, as is commonly believed. Because of time constraints imposed by other courses and the desire to maintain a social life that goes beyond talking in her sleep, she finds that she has time to collect data from only 12 people. After carefully planning a procedure for obtaining a simple random sample of 12 healthy adults, she measures their body temperatures and obtains the results listed below. Use a 0.05 significance level to test the claim that these body temperatures come from a population with a mean that is less than 98.6°F.

98.0   97.5   98.6   98.8   98.0   98.5   98.6   99.4   98.4   98.7   98.6   97.6

SOLUTION   Before jumping to the hypothesis test, let's first explore the sample data. There are no outliers and, based on a histogram and normal quantile plot, we can assume that the data are from a population with a normal distribution. We use the sample data to find these statistics: $n = 12$, $\bar{x} = 98.39$, $s = 0.535$. The sample mean of $\bar{x} = 98.39$ is less than 98.6, but we need to determine whether it is *significantly* less than 98.6. Let's proceed with a formal hypothesis test. We will use the traditional method of hypothesis testing summarized in Figure 7-8.

continued

Death Penalty as Deterrent

A common argument supporting the death penalty is that it discourages others from committing murder. Jeffrey Grogger of the University of California analyzed daily homicide data in California for a four-year period during which executions were frequent. Among his conclusions published in the *Journal of the American Statistical Association* (Vol. 85, No. 410): "The analyses conducted consistently indicate that these data provide no support for the hypothesis that executions deter murder in the short term." This is a major social policy issue, and the efforts of people such as Professor Grogger help to dispel misconceptions so that we have accurate information with which to address such issues.

Step 1:   The original claim that "the mean body temperature is less than 98.6°F" can be expressed symbolically as $\mu < 98.6$.

Step 2:   The opposite of the original claim is $\mu \geq 98.6$.

Step 3:   Of the two symbolic expressions obtained so far, the expression $\mu < 98.6$ does not contain equality, so it becomes the alternative hypothesis H_1. The null hypothesis is the assumption that $\mu = 98.6$.

$$H_0: \mu = 98.6$$
$$H_1: \mu < 98.6 \qquad \text{(original claim)}$$

Step 4:   The significance level is $\alpha = 0.05$.

Step 5:   In this test of a claim about the population mean, the most relevant statistic is the sample mean. In selecting the correct distribution, we refer to Figure 6-6 or Table 6-1. We select the t distribution because of these conditions: we have a simple random sample, the value of σ is not known, and the sample data appear to come from a population that is normally distributed.

Step 6:   The test statistic is

$$t = \frac{\bar{x} - \mu_{\bar{x}}}{\dfrac{s}{\sqrt{n}}} = \frac{98.39 - 98.6}{\dfrac{0.535}{\sqrt{12}}} = -1.360$$

The critical value of $t = -1.796$ is found by referring to Table A-3. First locate $n - 1 = 11$ degrees of freedom in the column at the left. Because this test is left-tailed with $\alpha = 0.05$, refer to the column indicating an area of 0.05 in one tail. The test statistic and critical value are shown in the accompanying STATDISK display.

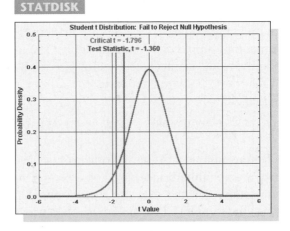

Step 7:   Because the test statistic of $t = -1.360$ does not fall in the critical region, we fail to reject H_0.

INTERPRETATION (Refer to Figure 7-7 for help in wording the final conclusion.) There is not sufficient evidence to support the claim that the sample comes from a population with a mean less than 98.6°F. This does not "prove" that the mean is 98.6°F. In fact, μ may well be less than 98.6°F, but the 12 sample values do not provide evidence strong enough to support that claim. If we use the 106 body temperatures included in Data Set 4 in Appendix B, we would find that there is sufficient evidence to support the claim that the mean body temperature is less than 98.6°F, but the 12 sample values included in this example do not support that claim.

The critical value in the preceding example was $t = -1.796$, but if the normal distribution was being used, the critical value would have been $z = -1.645$. The Student t critical value is farther to the left, showing that with the Student t distribution, the sample evidence must be *more extreme* before we consider it to be significant.

Finding *P*-Values with the Student *t* Distribution

The preceding example followed the traditional approach to hypothesis testing, but STATDISK, Minitab, the TI-83 Plus calculator, and many articles in professional journals will display *P*-values. For the preceding example, STATDISK displays a *P*-value of 0.1023, Minitab and Excel display a *P*-value of 0.102, and the TI-83 Plus calculator displays a *P*-value of 0.1022565104. With a significance level of 0.05 and a *P*-value greater than 0.05, we fail to reject the null hypothesis, as we did using the traditional method in the preceding example. If software or a TI-83 Plus calculator is not available, we can use Table A-3 to identify a *range of values* containing the *P*-value. We recommend this strategy for finding *P*-values using the *t* distribution:

Note to Instructor
Point out that the *P*-value approach is more difficult in this section than it was in Section 7-3. However, many computer programs and the TI-83 Plus calculator overcome that disadvantage by automatically providing *P*-values. Also, we can use Table A-3 to identify limits that contain the *P*-value.

1. Use software or a TI-83 Plus calculator.

2. If the technology is not available, use Table A-3 to identify a range of *P*-values. (See the following example.)

EXAMPLE **Finding *P*-Values** Assuming that neither software nor a TI-83 Plus calculator is available, use Table A-3 to find a range of values for the *P*-value corresponding to the given results.

a. In a left-tailed hypothesis test, the sample size is $n = 12$ and the test statistic is $t = -2.007$.

b. In a right-tailed hypothesis test, the sample size is $n = 12$ and the test statistic is $t = 1.222$.

c. In a two-tailed hypothesis test, the sample size is $n = 12$ and the test statistic is $t = -3.456$.

SOLUTION   Recall from Figure 7-6 that the *P*-value is the area determined as follows:

continued

Left-tailed test: The *P*-value is the area to the *left* of the test statistic.

Right-tailed test: The *P*-value is the area to the *right* of the test statistic.

Two-tailed test: The *P*-value is *twice* the area in the tail bounded by the test statistic.

In each of parts (a), (b), and (c), the sample size is $n = 12$, so the number of degrees of freedom is df $= n - 1 = 11$. See the accompanying portion of Table A-3 for 11 degrees of freedom along with the boxes describing the procedures for finding *P*-values.

a. The test is a left-tailed test with test statistic $t = -2.007$, so the *P*-value is the area to the left of -2.007. Because of the symmetry of the *t* distribution, that is the same as the area to the right of $+2.007$. See the accompanying illustration showing that any test statistic between 2.201 and 1.796 has

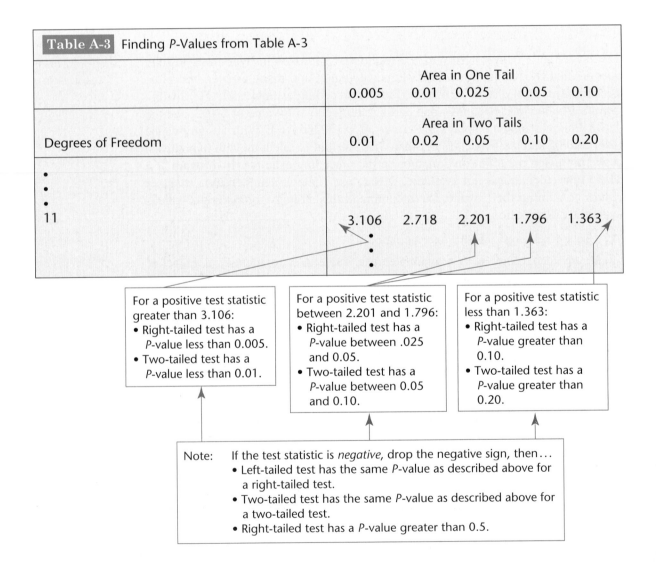

a right-tailed *P*-value that is between 0.025 and 0.05. We conclude that 0.025 < *P*-value < 0.05. (The exact *P*-value found using software is 0.0350.)

b. The test is a right-tailed test with test statistic $t = 1.222$, so the *P*-value is the area to the right of 1.222. See the accompanying illustration showing that any test statistic less than 1.363 has a right-tailed *P*-value that is greater than 0.10. We conclude that the *P*-value is > 0.10. (The exact *P*-value found using software is 0.124.)

c. The test is a two-tailed test with test statistic $t = -3.456$. The *P*-value is *twice* the area to the left of -3.456, but with the symmetry of the *t* distribution, that is the same as twice the area to the right of $+3.456$. See the accompanying illustration showing that any test statistic greater than 3.106 has a two-tailed *P*-value that is less than 0.01. We conclude that the *P*-value is < 0.01. (The exact *P*-value found using software is 0.00537.)

Once the format of Table A-3 is understood, it is not difficult to find a range of numbers for *P*-values. Check your results to be sure that they follow the same patterns shown in Table A-3. From left to right, the areas *increase* while the *t* values *decrease*. For example, in part (b), the test statistic of $t = 1.222$ is less than 1.363, so the right-tailed area is *greater than* 0.10.

Remember, *P*-values can be easily found by using software or a TI-83 Plus calculator. Also, the traditional method of testing hypotheses can be used instead of the *P*-value method.

Confidence Interval Method   We can use a confidence interval for testing a claim about μ when σ is not known. For a two-tailed hypothesis test with a 0.05 significance level, we construct a 95% confidence interval, but for a one-tailed hypothesis test with a 0.05 significance level we construct a 90% confidence interval. (See Table 7-2.) Using the sample data from the first example in this section ($n = 12$ and $\bar{x} = 98.39$, $s = 0.535$) with σ not known, and using a 0.05 significance level, we can test the claim that $\mu < 98.6$ by using the confidence interval method. Construct this 90% confidence interval: $98.11 < \mu < 98.67$ (see Section 6-4). Because the assumed value of $\mu = 98.6$ is contained within the confidence interval, we cannot reject that assumption. Based on the 12 sample values given in the example, we do not have sufficient evidence to support the claim that the mean body temperature is less than 98.6°F. Based on the confidence interval, the true value of μ is likely to be any value between 98.11 and 98.67, including 98.6.

In Section 7-3 we saw that when testing a claim about a population proportion, the traditional method and *P*-value method are equivalent, but the confidence interval method is somewhat different. When testing a claim about a population mean, there is no such difference, and all three methods are equivalent.

Alternative Method (not used in this book)   When testing a claim about the population mean μ using a simple random sample from a normally distributed population with unknown σ, an alternative method (not used in this book) is to use the methods of this section if the sample is small ($n \leq 30$), but if the sample is large ($n > 30$) substitute s for σ and proceed as if σ is known (as in Section 7-4). This alternative method is not used in this book for the following reasons (also cited in Section 6-4): (1) The criteria for choosing between the normal and t distributions used in this book are the same criteria used in the real world. (2) With σ not known, the distribution of $(\overline{x} - \mu) \div (s/\sqrt{n})$ is a t distribution, not a normal distribution; for very large sample sizes, the differences between the normal and t distributions are negligible, but the use of the t distribution generally yields better results. (3) For those students taking more statistics courses, it would be better if they learned one procedure that can be used later, rather than learning a procedure that must be changed later. (4) Working with the t distribution is not much more difficult than working with the normal distribution, especially if software or a TI-83 Plus calculator is available.

Using Technology

STATDISK   If working with a list of the original sample values, first find the sample size, sample mean, and sample standard deviation by using the STATDISK procedure described in Section 2-4. After finding the values of n, $\overline{x}$, and s, proceed to select the main menu bar item **Analysis,** then select **Hypothesis Testing,** followed by **Mean-One Sample.**

Minitab   Minitab Release 14 allows you to use either the summary statistics or a list of the original sample values. (Earlier versions do not allow the use of summary statistics. See Section 5-5 of the *Minitab Manual,* which is a supplement to this book, for a way to work around this limitation.) Select **Stat, Basic Statistics,** then **1-sample t.** Enter the summary statistics or enter **C1** in the box located at the top right. Use the **Options** button to change the format of the alternative hypothesis.

Excel   Excel does not have a built-in function for a t test, so use the Data Desk XL add-in that is a supplement to this book. First enter the sample data in column A. Select **DDXL,** then **Hypothesis Tests.** Under the function type options, select **1 Var t Test.** Click on the pencil icon and enter the range of data values, such as A1:A12 if you have 12 values listed in column A. Click on **OK.** Follow the four steps listed in the dialog box. After clicking on **Compute** in Step 4, you will get the P-value, test statistic, and conclusion.

TI-83 Plus   If using a TI-83 Plus calculator, press **STAT,** then select **TESTS** and choose the second option, **T-Test.** You can use the original data or the summary statistics (**Stats**) by providing the entries indicated in the window display. The first three items of the TI-83 Plus results will include the alternative hypothesis, the test statistic, and the P-value.

Recommended Assignment:
Exercises 1–12, 17–22. The directions for Exercises 9–28 state that unless stipulated by the instructor, either the traditional method or the P-value method can be used, so indicate a preference if you have one.

7-5 Basic Skills and Concepts

Using Correct Distribution. In Exercises 1–4, determine whether the hypothesis test involves a sampling distribution of means that is a normal distribution, Student t distribution, or neither. (Hint: See Figure 6-6 and Table 6-1.)

1. Claim: $\mu = 100$. Sample data: $n = 15$, $\bar{x} = 102$, $s = 15.3$. The sample data appear to come from a normally distributed population with unknown μ and σ.

2. Claim: $\mu = 75$. Sample data: $n = 25$, $\bar{x} = 102$, $s = 15.3$. The sample data appear to come from a population with a distribution that is very far from normal, and σ is unknown.

3. Claim: $\mu = 980$. Sample data: $n = 25$, $\bar{x} = 950$, $s = 27$. The sample data appear to come from a normally distributed population with $\sigma = 30$.

4. Claim: $\mu = 2.80$. Sample data: $n = 150$, $\bar{x} = 2.88$, $s = 0.24$. The sample data appear to come from a population with a distribution that is not normal, and σ is unknown.

Finding P-values. *In Exercises 5–8, use the given information to find a range of numbers for the P-value. (Hint: See the example and its accompanying display in the subsection of "Finding P-Values with the Student t Distribution.")*

5. Right-tailed test with $n = 12$ and test statistic $t = 2.998$

6. Left-tailed test with $n = 12$ and test statistic $t = -0.855$

7. Two-tailed test with $n = 16$ and test statistic $t = 4.629$

8. Two-tailed test with $n = 9$ and test statistic $t = -1.577$

Finding Test Components. *In Exercises 9–12 assume that a simple random sample has been selected from a normally distributed population. Find the test statistic, P-value, critical value(s), and state the final conclusion.*

9. Claim: The mean IQ score of statistics professors is greater than 118.
   Sample data: $n = 20$, $\bar{x} = 120$, $s = 12$. The significance level is $\alpha = 0.05$.

10. Claim: The mean body temperature of healthy adults is less than 98.6°F.
    Sample data: $n = 35$, $\bar{x} = 98.20$°F, $s = 0.62$. The significance level is $\alpha = 0.01$.

11. Claim: The mean time between uses of a TV remote control by males during commercials equals 5.00 sec.
    Sample data: $n = 81$, $\bar{x} = 5.25$ sec, $s = 2.50$ sec. The significance level is $\alpha = 0.01$.

12. Claim: The mean starting salary for college graduates who have taken a statistics course is equal to $46,000.
    Sample data: $n = 27$, $\bar{x} = \$45,678$, $s = \$9900$. The significance level is $\alpha = 0.05$.

Testing Hypotheses. *In Exercises 13–32, assume that a simple random sample has been selected from a normally distributed population and test the given claim. Unless specified by your instructor, use either the traditional method or P-value method for testing hypotheses.*

13. Harry Potter and Reading Level Data Set 14 in Appendix B lists measured reading levels for 12 pages randomly selected from J. K. Rowling's *Harry Potter and the Sorcerer's Stone*. The Flesch-Kincaid Grade Level measurements are summarized with these statistics: $n = 12$, $\bar{x} = 5.075$, $s = 1.168$. Teachers at the West Park School District will not use the book unless the reading level for a typical page can be shown to be above grade 4. Use a 0.05 significance level to test the claim that the mean is greater than 4. Will the teachers use the book?

14. Sugar in Cereal Data Set 16 in Appendix B lists the sugar content (grams of sugar per gram of cereal) for a sample of different cereals. Those amounts are summarized with

---

1. Student t

2. Neither

3. Normal

4. Student t

5. Between 0.005 and 0.01

6. Greater than 0.10

7. Less than 0.01

8. Between 0.10 and 0.20

9. $t = 0.745$. *P*-value is greater than 0.10. CV: $t = 1.729$.

10. $t = -3.817$; *P*-value is less than 0.005. CV: $t = -2.441$.

11. $t = 0.900$; *P*-value is greater than 0.20. CV: $t = \pm2.639$.

12. $t = -0.169$; *P*-value is greater than 0.20. CV: $t = \pm2.056$.

13. TS: $t = 3.188$. *P*-value is less than 0.005. CV: $t = 1.796$.

14. TS: $t = -0.119$. *P*-value is greater than 0.10. CV: $t = -1.752$.

these statistics: $n = 16$, $\bar{x} = 0.295$ g, $s = 0.168$ g. Use a 0.05 significance level to test the claim of a cereal lobbyist that the mean for all cereals is less than 0.3 g.

15. TS: $t = -0.63$. *P*-value is greater than 0.20. CV: $t = \pm 2.042$.

15. Actual and Forecast Temperatures Data Set 10 in Appendix B includes a list of actual high temperatures and the corresponding list of three-day forecast high temperatures. If the difference for each day is found by subtracting the three-day forecast high temperature from the actual high temperature, the result is a list of 31 values with a mean of $-0.419°$ and a standard deviation of $3.704°$. Use a 0.05 significance level to test the claim that the mean difference is different from $0°$. Based on the result, does it appear that the three-day forecast high temperatures are reasonably accurate?

16. TS: $t = 4.685$. *P*-value is less than 0.005. CV: $t = \pm 2.539$.

16. Heights of Parents Data Set 2 in Appendix B includes the heights of parents of 20 males. If the difference in height for each set of parents is found by subtracting the mother's height from the father's height, the result is a list of 20 values with a mean of 4.4 in. and a standard deviation of 4.2 in. Use a 0.01 significance level to test the claim that the mean difference is greater than 0. Do the results support a sociologist's claim that women tend to marry men who are taller than themselves?

17. TS: $t = 4.010$. *P*-value is less than 0.01. CV: $t = \pm 2.704$.

17. Testing Wristwatch Accuracy Students of the author randomly selected 40 people and measured the accuracy of their wristwatches, with positive errors representing watches that are ahead of the correct time and negative errors representing watches that are behind the correct time. The 40 values have a mean of 117.3 sec and a standard deviation of 185.0 sec. Use a 0.01 significance level to test the claim that the population of all watches has a mean equal to 0 sec. What can be concluded about the accuracy of people's wristwatches?

18. TS: $t = -0.932$. *P*-value is greater than 0.10. CV: $t = -1.753$, assuming $\alpha = 0.05$.

18. Textbook Prices Heather Carielli is a former student of the author who earned a Master's degree in statistics at the University of Massachusetts. When she randomly selected 16 new textbooks in the college bookstore, she found that they had prices with a mean of $70.41 and a standard deviation of $19.70. Is there sufficient evidence to warrant rejection of a claim in the college catalog that the mean price of a textbook at this college is less than $75?

19. TS: $t = 2.652$. *P*-value is between 0.005 and 0.01. CV: $t = 1.691$.

19. Conductor Life Span A *New York Times* article noted that the mean life span for 35 male symphony conductors was 73.4 years, in contrast to the mean of 69.5 years for males in the general population. Assuming that the 35 males have life spans with a standard deviation of 8.7 years, use a 0.05 significance level to test the claim that male symphony conductors have a mean life span that is greater than 69.5 years. Does it appear that male symphony conductors live longer than males from the general population? Why doesn't the experience of being a male symphony conductor cause men to live longer? (*Hint*: Are male symphony conductors born, or do they become conductors at a much later age?)

20. TS: $t = -0.601$. *P*-value is greater than 0.20. CV: $t = \pm 2.021$ (approx.).

20. Baseballs In previous tests, baseballs were dropped 24 ft onto a concrete surface, and they bounced an average of 92.84 in. In a test of a sample of 40 new balls, the bounce heights had a mean of 92.67 in. and a standard deviation of 1.79 in. (based on data from Brookhaven National Laboratory and *USA Today*). Use a 0.05 significance level to determine whether there is sufficient evidence to support the claim that the new balls have bounce heights with a mean different from 92.84 in. Does it appear that the new baseballs are different?

21. TS: $t = -1.83$. *P*-value: 0.071.

21. BMW Crash Tests Because of the expense involved, car crash tests often use small samples. When five BMW cars are crashed under standard conditions, the repair costs (in dollars) are used to test the claim that the mean repair cost for all BMW cars is less than $1000. The Minitab results of this hypothesis test are shown below. Based on the accompanying Minitab results of this hypothesis test, would BMW be justified in advertising that under the standard conditions, the repair costs average less than $1000?

```
Test of mu = 1000 vs mu < 1000

Variable        N       Mean      StDev     SE Mean
Cost            5       767       285         127

Variable    95.0% Upper Bound       T         P
Cost              1039           -1.83      0.071
```

22. **Reliability of Aircraft Radios** The mean time between failures (in hours) for a Telektronic Company radio used in light aircraft is 420 h. After 15 new radios were modified in an attempt to improve reliability, tests were conducted to measure the times between failures. When Minitab is used to test the claim that the modified radios have a mean greater than 420 h, the results are as shown here. Does it appear that the modifications improved reliability?

```
Test of mu = 420 vs mu > 420

Variable        N       Mean      StDev     SE Mean
Time           15      442.2      44.0        11.4

Variable    95.0% Lower Bound       T         P
Time              422.2            1.95      0.035
```

23. **Effect of Vitamin Supplement on Birth Weight** The birth weights (in kilograms) are recorded for a sample of male babies born to mothers taking a special vitamin supplement (based on data from the New York State Department of Health). When testing the claim that the mean birth weight for all male babies of mothers given vitamins is equal to 3.39 kg, which is the mean for the population of all males, the TI-83 Plus calculator yields the results shown. Based on those results, does the vitamin supplement appear to have an effect on birth weight?

24. **Pulse Rates** The author, at the peak of an exercise program, claimed that his pulse rate was lower than the mean pulse rate of statistics students. The author's pulse rate was measured to be 60 beats per minute, and the 20 students in his class measured their pulse rates. When testing the claim that statistics students have a mean pulse rate greater than 60 beats per minute, the accompanying TI-83 Plus calculator display was obtained. Based on those results, is there sufficient evidence to support the claim that the mean pulse rate of statistics students is greater than 60 beats per minute?

25. **Monitoring Lead in Air** Listed below are measured amounts of lead (in micrograms per cubic meter or $\mu g/m^3$) in the air. The Environmental Protection Agency has established an air quality standard for lead: 1.5 $\mu g/m^3$. The measurements shown below were recorded at Building 5 of the World Trade Center site on different days immediately following the destruction caused by the terrorist attacks of September 11, 2001. After the collapse of the two World Trade Center buildings, there was considerable concern about the quality of the air. Use a 0.05 significance level to test the claim that the sample is from a population with a mean greater than the EPA standard of 1.5 $\mu g/m^3$. Is there anything about this data set suggesting that the assumption of a normally distributed population might not be valid?

5.40   1.10   0.42   0.73   0.48   1.10

22. TS: $t = 1.95$. P-value: 0.035. The claim of improved reliability is supported at a 0.05 significance level, but it is not supported at a 0.01 significance level.
23. TS: $t = 1.734$. P-value: 0.1034.
24. TS: $t = 6.393$. P-value: 0.000002.

TI-83 Plus

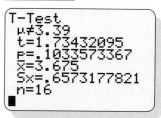

TI-83 Plus

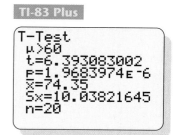

25. TS: $t = 0.049$. P-value is greater than 0.10. CV: $t = 2.015$.

26. Treating Chronic Fatigue Syndrome Patients with chronic fatigue syndrome were tested, then retested after being treated with fludrocortisone. Listed below are the changes in fatigue after the treatment (based on data from "The Relationship Between Neurally Mediated Hypotension and the Chronic Fatigue Syndrome" by Bou-Holaigah, Rowe, Kan, and Calkins, *Journal of the American Medical Association,* Vol. 274, No. 12). A standard scale from -7 to $+7$ was used, with positive values representing improvements. Use a 0.01 significance level to test the claim that the mean change is positive. Does the treatment appear to be effective?

6   5   0   5   6   7   3   3   2   6   5   5   0   6   3   4   3   7   0   4   4

27. Olympic Winners Listed below are the winning times (in seconds) of men in the 100-meter dash for consecutive summer Olympic games, listed in order by row. Assuming that these results are sample data randomly selected from the population of all past and future Olympic games, test the claim that the mean time is less than 10.5 sec. What do you observe about the precision of the numbers? What extremely important characteristic of the data set is not considered in this hypothesis test? Do the results from the hypothesis test suggest that future winning times should be around 10.5 sec, and is such a conclusion valid?

12.0   11.0   11.0   11.2   10.8   10.8   10.8   10.6   10.8   10.3   10.3   10.3
10.4   10.5   10.2   10.0   9.95   10.14   10.06   10.25   9.99   9.92   9.96

28. Nicotine in Cigarettes The Carolina Tobacco Company advertised that its best-selling nonfiltered cigarettes contain at most 40 mg of nicotine, but *Consumer Advocate* magazine ran tests of 10 randomly selected cigarettes and found the amounts (in mg) shown in the accompanying list. It's a serious matter to charge that the company advertising is wrong, so the magazine editor chooses a significance level of $\alpha = 0.01$ in testing her belief that the mean nicotine content is greater than 40 mg. Using a 0.01 significance level, test the editor's belief that the mean is greater than 40 mg.

47.3   39.3   40.3   38.3   46.3   43.3   42.3   49.3   40.3   46.3

T 29. Tom Clancy Reading Level Refer to Data Set 14 in Appendix B and use the Flesch-Kincaid Grade Level measurements for *The Bear and the Dragon* by Tom Clancy. A high school teacher wants to assign the book for a reading assignment, but he requires a book with a reading level above grade 6. Is there sufficient evidence to support the claim that Clancy's book meets that requirement?

T 30. Tobacco Use in Children's Movies Refer to Data Set 7 in Appendix B and use only those movies that show some use of tobacco. Test the claim of a movie critic that "among those movies that show the use of tobacco, the mean exposure time is 2 minutes." Given the sample data, is that claim deceptive?

T 31. Coke Volumes Data Set 17 in Appendix B includes the volumes (in ounces) of the regular Coke in a sample of 36 different cans that are all labeled 12 oz. A line manager claims that the mean amount of regular Coke is greater than 12 oz, causing lower company profits. Using a 0.01 significance level, test the manager's claim that the mean is greater than 12 oz. Should the production process be adjusted?

T 32. Sodium in Cereal Refer to Data Set 16 in Appendix B and test a nutritionist's claim that "the average box of cereal contains more than 6 mg of sodium per gram of cereal." If 6 mg of sodium per gram of cereal is considered to be excessive, can we say that cereal is unhealthy because of the high sodium content?

7-5 Beyond the Basics

33. *Using Computer Results* Refer to the Minitab display in Exercise 22. If the claim is changed from "greater than 420 h" to "not equal to 420 h," how are the test statistic, *P*-value, and conclusion affected?

34. *Using the Wrong Distribution* When testing a claim about a population mean with a simple random sample selected from a normally distributed population with unknown σ, the Student t distribution should be used for finding critical values and/or a *P*-value. If the standard normal distribution is incorrectly used instead, does that mistake make you more or less likely to reject the null hypothesis, or does it not make a difference? Explain.

35. *Effect of an Outlier* Repeat Exercise 25 after changing the first value from 5.40 to 540. Based on the results, describe the effect of an outlier on a t test.

36. *Finding Critical t Values* When finding critical values, we sometimes need significance levels other than those available in Table A-3. Some computer programs approximate critical t values by calculating

$$t = \sqrt{df \cdot (e^{A^2/df} - 1)}$$

where $df = n - 1$, $e = 2.718$, $A = z(8 \cdot df + 3)/(8 \cdot df + 1)$, and z is the critical z score. Use this approximation to find the critical t score corresponding to $n = 10$ and a significance level of 0.05 in a right-tailed case. Compare the results to the critical t value found in Table A-3.

37. *Probability of Type II Error* Refer to Exercise 28 and assume that you're testing the claim that $\mu > 40$ mg. Find β, the probability of a type II error, given that the actual value of the population mean is $\mu = 45.0518$ mg. (See Exercise 19 in Section 7-4.)

33. The *P*-value becomes 0.070. The test statistic does not change. The claim that the mean is different from 420 h is not rejected at the 0.05 significance level.

34. More likely to reject the null hypothesis because the critical *t* values are more extreme than the corresponding critical *z* values.

35. The test statistic changes to $t = 0.992$ and the *P*-value changes to 0.182. An outlier can change the test statistic and *P*-value substantially. Although the conclusion does not change here, it could change in other cases.

36. With $z = 1.645$, the table and approximation both result in $t = 1.833$.

37. 0.10

7-6 Testing a Claim About a Standard Deviation or Variance

The industrial world shares this common goal: Improve quality by reducing variation. Quality-control engineers want to ensure that a product has an acceptable mean, but they also want to produce items of *consistent* quality so that there will be few defects. For example, the consistency of aircraft altimeters is governed by Federal Aviation Regulation 91.36, which requires that aircraft altimeters be tested and calibrated to give a reading "within 125 feet (on a 95-percent probability basis)." Even if the mean altitude reading is exactly correct, an excessively large standard deviation will result in individual readings that are dangerously low or high. Consistency is improved by reducing the standard deviation. In the preceding sections of this chapter we described methods for testing claims made about population means and proportions. This section focuses on variation, which is critically important in many applications, including quality control. The main objective of this section is to present methods to test claims made about a population standard deviation σ or variance σ^2. The assumptions, test statistic, *P*-value, and critical values are summarized as follows.

Note to Instructor

Before covering this section, you might first tell the class that the general population has IQ scores with a mean of 100 and a standard deviation around 15, then ask about the value of the standard deviation for the population of all statistics professors. Because this group is more homogeneous, the standard deviation should be lower.

> ## Testing Claims About σ or σ^2
>
> ### Assumptions
>
> 1. The sample is a simple random sample.
> 2. The population has a normal distribution. (This is a much stricter requirement than the requirement of a normal distribution when testing claims about means, as in Sections 7-4 and 7-5.)
>
> ### Test Statistic for Testing a Claim about σ or σ^2
>
> $$\chi^2 = \frac{(n-1)s^2}{\sigma^2}$$
>
> **P-values and Critical values:** Use Table A-4 with df $= n - 1$ for the number of degrees of freedom. (Table A-4 is based on *cumulative areas from the right*.)

In Sections 7-4 and 7-5 we saw that the methods of testing claims about means require a normally distributed population, and those methods work reasonably well as long as the population distribution is not very far from being normal. However, tests of claims about standard deviations or variances are not as *robust,* meaning that the results can be very misleading if the population does not have a normal distribution. The condition of a normally distributed population is therefore a much stricter requirement in this section. If the population has a distribution that is far from normal and you use the methods of this section to reject a null hypothesis, you don't really know if the standard deviation is not as assumed or if the rejection is due to the lack of normality.

Don't be confused by reference to both the normal and the chi-square distributions. After verifying that the sample data appear to come from a normally distributed population, we should then shift gears and think in terms of the "chi-square" distribution. The chi-square distribution was introduced in Section 6-5, where we noted the following important properties.

Properties of the Chi-Square Distribution

1. All values of χ^2 are nonnegative, and the distribution is not symmetric (see Figure 7-12).
2. There is a different χ^2 distribution for each number of degrees of freedom (see Figure 7-13).
3. The critical values are found in Table A-4 using

$$\text{degrees of freedom} = n - 1$$

Table A-4 is based on cumulative areas from the *right* (unlike the entries in Table A-2, which are cumulative areas from the left). Critical values are found in Table A-4 by first locating the row corresponding to the appropriate number of

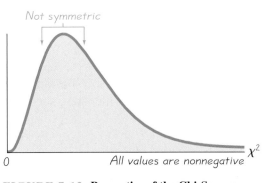

FIGURE 7-12 **Properties of the Chi-Square Distribution**

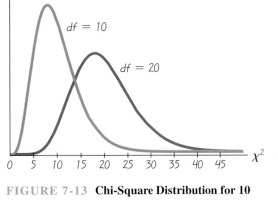

FIGURE 7-13 **Chi-Square Distribution for 10 and 20 Degrees of Freedom**

degrees of freedom (where df $= n - 1$). Next, the significance level α is used to determine the correct column. The following examples are based on a significance level of $\alpha = 0.05$, but any other significance level can be used in a similar manner. Note that in each case, the key area is the region to the *right* of the critical value(s).

Right-tailed test: Because the area to the *right* of the critical value is 0.05, locate 0.05 at the top of Table A-4.

Left-tailed test: With a left-tailed area of 0.05, the area to the *right* of the critical value is 0.95, so locate 0.95 at the top of Table A-4.

Two-tailed test: Divide the significance level of 0.05 between the left and right tails, so the areas to the *right* of the two critical values are 0.975 and 0.025, respectively. Locate 0.975 and 0.025 at the top of Table A-4. (See Figure 6-10 and the example on pages 349–350.)

EXAMPLE IQ Scores of Statistics Professors For a simple random sample of adults, IQ scores are normally distributed with a mean of 100 and a standard deviation of 15. A simple random sample of 13 statistics professors yields a standard deviation of $s = 7.2$. A psychologist is quite sure that statistics professors have IQ scores that have a mean greater than 100. He doesn't understand the concept of standard deviation very well and does not realize that the standard deviation should be lower than 15 (because statistics professors have less variation than the general population). Instead, he claims that statistics professors have IQ scores with a standard deviation equal to 15, the same standard deviation for the general population. Assume that IQ scores of statistics professors are normally distributed and use a 0.05 significance level to test the claim that $\sigma = 15$. Based on the result, what do you conclude about the standard deviation of IQ scores for statistics professors?

continued

Ethics in Experiments

Sample data can often be obtained by simply observing or surveying members selected from the population. Many other situations require that we somehow manipulate circumstances to obtain sample data. In both cases ethical questions may arise. Researchers in Tuskegee, Alabama, withheld the effective penicillin treatment to syphilis victims so that the disease could be studied. This experiment continued for a period of 27 years!

SOLUTION   We will use the traditional method of testing hypotheses as outlined in Figure 7-8.

Step 1:   The claim is expressed in symbolic form as $\sigma = 15$.

Step 2:   If the original claim is false, then $\sigma \neq 15$.

Step 3:   The expression $\sigma \neq 15$ does not contain equality, so it becomes the alternative hypothesis. The null hypothesis is the statement that $\sigma = 15$.

$$H_0: \sigma = 15 \qquad \text{(original claim)}$$
$$H_1: \sigma \neq 15$$

Step 4:   The significance level is $\alpha = 0.05$.

Step 5:   Because the claim is made about σ we use the chi-square distribution.

Step 6:   The test statistic is

$$\chi^2 = \frac{(n-1)s^2}{\sigma^2} = \frac{(13-1)(7.2)^2}{15^2} = 2.765$$

The critical values of 4.404 and 23.337 are found in Table A-4, in the 12th row (degrees of freedom $= n - 1 = 12$) in the columns corresponding to 0.975 and 0.025. See the test statistic and critical values shown in Figure 7-14.

Step 7:   Because the test statistic is in the critical region, we reject the null hypothesis.

INTERPRETATION There is sufficient evidence to warrant rejection of the claim that the standard deviation is equal to 15. It appears that statistics professors have IQ scores with a standard deviation that is significantly different than the standard deviation of 15 for the general population.

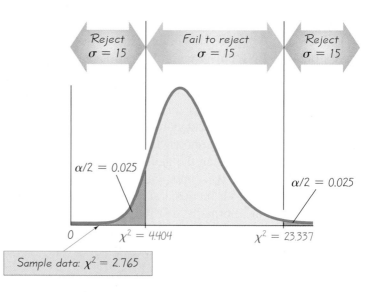

FIGURE 7-14 Testing the Claim That $\sigma = 15$

The *P*-Value Method

Instead of using the traditional approach to hypothesis testing for the preceding example, we can also use the *P*-value approach summarized in Figures 7-6 and 7-9. If STATDISK is used for the preceding example, the *P*-value of 0.0060 will be found. If we use Table A-4, we usually cannot find *exact P*-values because that chi-square distribution table includes only selected values of α. (Because of this limitation, testing claims about σ or σ^2 with Table A-4 is easier with the traditional method than the *P*-value method.) If using Table A-4, we can identify limits that contain the *P*-value. The test statistic from the last example is $\chi^2 = 2.765$ and we know that the test is two-tailed with 12 degrees of freedom. Refer to the 12th row of Table A-4 and see that the test statistic of 2.765 is less than every entry in that row, which means that the area to the left of the test statistic is less than 0.005. The *P*-value for a two-tailed test is *twice* the tail area bounded by the test statistic, so we double 0.005 to conclude that the *P*-value is less than 0.01. Because the *P*-value is less than the significance level of $\alpha = 0.05$, we reject the null hypothesis. Again, the traditional method and the *P*-value method are equivalent in the sense that they always lead to the same conclusion.

The Confidence Interval Method

The preceding example can also be solved with the confidence interval method of testing hypotheses. Using the methods described in Section 6-5, we can use the sample data ($n = 13$, $s = 7.2$) to construct this 95% confidence interval: $5.2 < \sigma < 11.9$. Because the claimed value of $\sigma = 15$ is *not* contained within the confidence interval, we reject the claim that $\sigma = 15$, and we reach the same conclusion from the traditional and *P*-value methods.

1. TS: $\chi^2 = 8.444$. CV: $\chi^2 = 8.907, 32.852$. *P*-value: Between 0.02 and 0.05. Reject H_0.
2. TS: $\chi^2 = 9.000$. CV: $\chi^2 = 13.277$. *P*-value: Between 0.05 and 0.10. Fail to reject H_0.

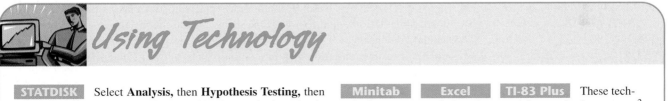

Using Technology

STATDISK   Select **Analysis,** then **Hypothesis Testing,** then **StDev-One Sample.** Proceed to provide the required entries in the dialog box, then click on **Evaluate.** STATDISK will display the test statistic, critical values, *P*-value, conclusion, and confidence interval.

Minitab   **Excel**   **TI-83 Plus**   These technologies are not yet designed to test claims made about σ or σ^2.

7-6 Basic Skills and Concepts

Finding Critical Values. In Exercises 1–4, find the test statistic, then use Table A-4 to find critical value(s) of χ^2 and limits containing the P-value, then determine whether there is sufficient evidence to support the given alternative hypothesis.

1. H_1: $\sigma \neq 15$, $\alpha = 0.05$, $n = 20$, $s = 10$.

2. H_1: $\sigma > 12$, $\alpha = 0.01$, $n = 5$, $s = 18$.

Recommended Assignment:
Even-numbered exercises 2–12. Because *P*-values are usually difficult to find, the directions stipulate that the traditional method of testing hypotheses should be used. If you prefer, stipulate that the *P*-value method could or should be used instead.

3. TS: $\chi^2 = 10.440$. CV: $\chi^2 = 14.257$. P-value < 0.005. Reject H_0.

4. TS: $\chi^2 = 110.450$. CV: $\chi^2 = 57.153, 106.629$. P-value: Between 0.02 and 0.05. Reject H_0.

5. TS: $\chi^2 = 2342.438$. CV: $\chi^2 = 63.691$ (approx.). (P-value < 0.005.) Reject H_0.

6. TS: $\chi^2 = 9.720$. CV: $\chi^2 = 4.575$. (P-value: Between 0.10 and 0.90.) Fail to reject H_0.

7. TS: $\chi^2 = 114.586$. CV: $\chi^2 = 57.153, 106.629$. (P-value: Between 0.01 and 0.02.) Reject H_0.

8. TS: $\chi^2 = 11.311$. CV: $\chi^2 = 12.198$. (P-value: Between 0.005 and 0.01.) Reject H_0.

9. TS: $\chi^2 = 9.016$. CV: $\chi^2 = 13.848$. (P-value: Between 0.005 and 0.01.) Reject H_0.

10. $\chi^2 = 9.066$. CV: $\chi^2 = 67.328$ (approx.). (P-value < 0.005.) Reject H_0.

3. H_1: $\sigma < 50$, $\alpha = 0.01$, $n = 30$, $s = 30$.

4. H_1: $\sigma \neq 4.0$, $\alpha = 0.05$, $n = 81$, $s = 4.7$

Testing Claims About Variation. In Exercises 5–16, test the given claim. Assume that a simple random sample is selected from a normally distributed population. Use the traditional method of testing hypotheses unless your instructor indicates otherwise.

5. Variation in Peanut M&Ms Use a 0.01 significance level to test the claim that peanut M&M candies have weights that vary more than the weights of plain M&M candies. The standard deviation for the weights of plain M&M candies is 0.04 g. A sample of 40 peanut M&Ms has weights with a standard deviation of 0.31 g. Why should peanut M&M candies have weights that vary more than the weights of plain M&M candies?

6. Variation in Pistons When designing a piston to be used for a pump for transferring liquid solutions, engineers specified a mean of 0.1 in. as the target for the piston radius. The maximum standard deviation is specified as 0.0005 in. (based on data from Taylor Industries). When 12 pistons are randomly selected from the production line and measured, their radii have a standard deviation of 0.00047 in. Is there sufficient evidence to support the claim that the pistons are being manufactured with radii that have a standard deviation less than the specified maximum of 0.0005 in.? Use a 0.05 significance level.

7. Manufacturing Aircraft Altimeters The Stewart Aviation Products Company uses a new production method to manufacture aircraft altimeters. A simple random sample of 81 altimeters is tested in a pressure chamber, and the errors in altitude are recorded as positive values (for readings that are too high) or negative values (for readings that are too low). The sample has a standard deviation of $s = 52.3$ ft. At the 0.05 significance level, test the claim that the new production line has errors with a standard deviation different from 43.7 ft, which was the standard deviation for the old production method. If it appears that the standard deviation has changed, does the new production method appear to be better or worse than the old method?

8. Statistics Test Scores Tests in the author's past statistics classes have scores with a standard deviation equal to 14.1. One of his recent classes has 27 test scores with a standard deviation of 9.3. Use a 0.01 significance level to test the claim that this current class has less variation than past classes. Does a lower standard deviation suggest that the current class is doing better?

9. Bank Customer Waiting Times With individual lines at its various windows, the Jefferson Valley Bank found that the standard deviation for normally distributed waiting times on Friday afternoons was 6.2 min. The bank experimented with a single main waiting line and found that for a simple random sample of 25 customers, the waiting times have a standard deviation of 3.8 min. Use a 0.05 significance level to test the claim that a single line causes lower variation among the waiting times. Why would customers prefer waiting times with less variation? Does the use of a single line result in a shorter wait?

10. Body Temperatures In Section 7-4, we tested the claim that the mean body temperature is equal to 98.6°F, and we used sample data given in Data Set 4 in Appendix B. The body temperatures taken at 12:00 AM on day 2 can be summarized with these statistics: $n = 106$, $\bar{x} = 98.20°F$, $s = 0.62°F$, and a histogram shows that the values have a distribution that is approximately normal. In Section 7-4 we assumed that $\sigma = 0.62°F$, which is an unrealistic assumption. However, the test statistic will cause rejection of $\mu = 98.6°F$ as long as the standard deviation is less than 2.11°F. Use the sample statistics and a 0.005 significance level to test the claim that $\sigma < 2.11°F$.

11. **Supermodel Weights** Use a 0.01 significance level to test the claim that weights of female supermodels vary less than the weights of women in general. The standard deviation of weights of the population of women is 29 lb. Listed below are the weights (in pounds) of nine randomly selected supermodels.

125 (Taylor)   119 (Auermann)   128 (Schiffer)   128 (MacPherson)
119 (Turlington)   127 (Hall)   105 (Moss)   123 (Mazza)
115 (Hume)

12. **Supermodel Heights** Use a 0.05 significance level to test the claim that heights of female supermodels vary less than the heights of women in general. The standard deviation of heights of the population of women is 2.5 in. Listed below are the heights (in inches) of randomly selected supermodels (Taylor, Harlow, Mulder, Goff, Evangelista, Avermann, Schiffer, MacPherson, Turlington, Hall, Crawford, Campbell, Herzigova, Seymour, Banks, Moss, Mazza, Hume).

71   71   70   69   69.5   70.5   71   72   70
70   69   69.5   69   70   70   66.5   70   71

T 13. **Volumes of Pepsi** A new production manager claims that the volumes of cans of regular Pepsi have a standard deviation less than 0.10 oz. Use a 0.05 significance level to test that claim with the sample results listed in Data Set 17 in Appendix B. What problems are caused by a mean that is not 12 oz? What problems are caused by a standard deviation that is too high?

T 14. **Systolic Blood Pressure for Women** Systolic blood pressure results from contraction of the heart. Based on past results from the National Health Survey, it is claimed that women have systolic blood pressures with a mean and standard deviation of 130.7 and 23.4, respectively. Use the systolic blood pressures of women listed in Data Set 1 in Appendix B and test the claim that the sample comes from a population with a standard deviation of 23.4.

T 15. **Weights of Men** Anthropometric survey data are used to publish values that can be used in designing products that are suitable for use by adults. According to Gordon, Churchill, et al., men have weights with a mean of 172.0 lb and a standard deviation of 28.7 lb. Using the sample of weights of men in Data Set 1 in Appendix B, test the claim that the standard deviation is 28.7 lb. Use a 0.05 significance level. When designing elevators, what would be a consequence of believing that weights of men vary less than they really vary?

T 16. **Heights of Women** Anthropometric survey data are used to publish values that can be used in designing products that are suitable for use by adults. According to Gordon, Churchill, et al., women have heights with a mean of 64.1 in. and a standard deviation of 2.52 in. Using the sample of heights of women in Data Set 1 in Appendix B, test the claim that the standard deviation is 2.52 in. Use a 0.05 significance level. When designing car seats for women, what would be a consequence of believing that heights of women vary less than they really vary?

7-6 Beyond the Basics

17. **Controlling Variation in Cans of Pepsi** Refer to Exercise 13 and, for a sample of size $n = 36$ and a significance level of 0.05, find the largest sample standard deviation that can be used to support the claim that $\sigma < 0.10$ oz.

11. TS: $\chi^2 = 0.540$. CV: $\chi^2 = 1.646$. (P-value < 0.005.) Reject H_0.

12. TS: $\chi^2 = 3.831$. CV: $\chi^2 = 8.672$. (P-value < 0.005.) Reject H_0.

13. TS: $\chi^2 = 28.750$. CV of χ^2 is between 18.493 and 26.509. Fail to reject H_0.

14. TS: $\chi^2 = 20.861$. CV: $\chi^2 = 24.433, 59.342$ (approx). (P-value: Between 0.01 and 0.02.) Reject H_0.

15. TS: $\chi^2 = 32.818$. CV: $\chi^2 = 24.433, 59.342$ (approx). (P-value > 0.20.) Fail to reject H_0. Reject H_0.

16. TS: $\chi^2 = 46.148$. CV: $\chi^2 = 24.433, 59.342$ (approx). (P-value > 0.20.) Fail to reject H_0.

17. Using interpolation, estimate the CV as $\chi^2 = 22.501$ to get $s = 0.08$ oz.

18. a. Estimated values: 73.772, 129.070; Table A-5 values: 74.222, 129.561.
   b. 116.643, 184.199

19. a. Estimated values: 74.216, 129.565; Table A-5 values: 74.222, 129.561.
   b. 117.093, 184.690

20. Yes, because an outlier dramatically affects the value of the sample standard deviation.

21. a. The standard deviation will be lower.
   b. The requirement of a normally distributed population is not satisfied.

22. Approximately 0.10

18. Finding Critical Values of χ^2 For large numbers of degrees of freedom, we can approximate critical values of χ^2 as follows:

$$\chi^2 = \frac{1}{2}\left(z + \sqrt{2k - 1}\right)^2$$

Here k is the number of degrees of freedom and z is the critical value, found in Table A-2. For example, if we want to approximate the two critical values of χ^2 in a two-tailed hypothesis test with $\alpha = 0.05$ and a sample size of 150, we let $k = 149$ with $z = -1.96$ followed by $k = 149$ and $z = 1.96$.

a. Use this approximation to estimate the critical values of χ^2 in a two-tailed hypothesis test with $n = 101$ and $\alpha = 0.05$. Compare the results to those found in Table A-4.

b. Use this approximation to estimate the critical values of χ^2 in a two-tailed hypothesis test with $n = 150$ and $\alpha = 0.05$.

19. Finding Critical Values of χ^2 Repeat Exercise 18 using this approximation (with k and z as described in Exercise 18):

$$\chi^2 = k\left(1 - \frac{2}{9k} + z\sqrt{\frac{2}{9k}}\right)^3$$

20. Effect of Outlier When using the hypothesis testing procedure of this section, will the result be dramatically affected by the presence of an outlier? Describe how you arrived at your response.

21. Last-Digit Analysis The last digits of sample data are sometimes used in an attempt to determine whether the data have been measured or simply reported by the subject. Reported data often have last digits with disproportionately more 0s and 5s. Measured data tend to have last digits with a mean of 4.5, a standard deviation of about 3, and the digits should occur with roughly the same frequency.

a. How is the standard deviation of the data affected if there are disproportionately more 0s and 5s?

b. Why can't we use the methods of this section to test that the last digits of the sample data have a standard deviation equal to 3?

22. Probability of a Type II Error Refer to Exercise 9. Assuming that σ is actually 4.0, find β, which denotes the probability of a type II error. See Exercise 19 in Section 7-4, and modify the procedure so that it applies to a hypothesis test involving σ instead of μ.

Review

This chapter presented basic methods for testing claims about a population proportion, population mean, or population standard deviation (or variance). The methods of this chapter are used by professionals in a wide variety of disciplines, as illustrated in their many professional journals.

In Section 7-2 we presented the fundamental concepts of a hypothesis test: null hypothesis, alternative hypothesis, test statistic, critical region, significance level, critical value, P-value, type I error, and type II error. We also discussed two-tailed tests, left-tailed tests, right-tailed tests, and the statement of conclusions. We used those components in identifying three different methods for testing hypotheses:

1. The traditional method (summarized in Figure 7-8)

2. The P-value method (summarized in Figure 7-9)

3. Confidence intervals (discussed in Chapter 6)

		Table 7-3	Hypothesis Tests

Parameter	Conditions	Distribution and Test Statistic	Critical and P-Values
Proportion	$np \geq 5$ and $nq \geq 5$	Normal: $z = \dfrac{\hat{p} - p}{\sqrt{\dfrac{pq}{n}}}$	Table A-2
Mean	σ known and normally distributed population *or* σ known and $n > 30$	Normal: $z = \dfrac{\bar{x} - \mu_{\bar{x}}}{\dfrac{\sigma}{\sqrt{n}}}$	Table A-2
	σ not known and normally distributed population *or* σ not known and $n > 30$	Student t: $t = \dfrac{\bar{x} - \mu_{\bar{x}}}{\dfrac{s}{\sqrt{n}}}$	Table A-3
	Population not normally distributed and $n \leq 30$	Use a nonparametric method or bootstrapping.	
Standard Deviation or Variance	Population normally distributed	Chi-Square: $\chi^2 = \dfrac{(n-1)s^2}{\sigma^2}$	Table A-4

In Sections 7-3 through 7-6 we discussed specific methods for dealing with different parameters. Because it is so important to be correct in selecting the distribution and test statistic, we provide Table 7-3, which summarizes the hypothesis testing procedures of this chapter.

Review Exercises

1. a. You have just collected a very large ($n = 2575$) sample of responses obtained from adult Americans who mailed responses to a questionnaire printed in *Fortune* magazine. A hypothesis test conducted at the 0.01 significance level leads to the conclusion that most (more than 50%) adults are opposed to estate taxes. Can we conclude that most adult Americans are opposed to estate taxes? Why or why not?

b. When testing a diet control drug, a hypothesis test based on 5000 randomly selected subjects shows that the mean weight loss of 0.2 lb is significant at the 0.01 level. Should this drug be used by subjects wanting to lose weight? Why or why not?

c. You have just developed a new cure for the common cold and you plan to conduct a formal test to justify its effectiveness. Which *P*-value would you most prefer: 0.99, 0.05, 0.5, 0.01, or 0.001?

d. In testing the claim that the mean amount of cola in cans is greater than 12 oz, you fail to reject the null hypothesis. State the final conclusion that addresses the original claim.

e. Complete the statement: "A type I error is the mistake of . . ."

1. a. No, the sample is a voluntary response sample, so the results do not necessarily apply to the population of adult Americans.
b. No, even though there does appear to be statistically significant weight loss, the average amount of lost weight is so small that the drug is not practical.
c. 0.001, because this *P*-value corresponds to results that provide the most support for the effectiveness of the cure.
d. There is not sufficient evidence to support the claim that the mean is greater than 12 oz.
e. rejecting a true null hypothesis.

2. a. H_1: $\mu <$ \$10,000; Student t distribution.
   b. H_1: $\sigma >$ 1.8 sec; chi-square distribution.
   c. H_1: $p >$ 0.5; normal distribution.
   d. H_1: $\mu \neq$ 100; normal distribution.

3. a. TS: $z = -0.75$. P-value: 0.4532. (CV: $z = \pm 1.645$.) Fail to reject H_0.
   b. TS: $t = -0.694$. P-value > 0.20. (CV: $t = \pm 1.676$ approx.) Fail to reject H_0.
   c. TS: $\chi^2 = 57.861$. CV: $\chi^2 = 34.764, 67.505$. (P-value > 0.20.) Fail to reject H_0.
   d. Yes

4. TS: $z = -1.47$. CV: $z = -1.645$. P-value: 0.0708. Fail to reject H_0.

5. TS: $t = -4.991$. P-value < 0.01. CV: $t = \pm 2.678$ approx. Reject H_0.

6. TS: $t = -0.277$. P-value is greater than 0.10. CV: $t = -2.132$. Fail to reject H_0.

7. TS: $z = -1.17$. CV: $z = -1.645$. P-value: 0.1210. Fail to reject H_0.

2. **Identifying Hypotheses and Distributions** Based on the given conditions, identify the alternative hypothesis and the sampling distribution (normal, t, chi-square) of the test statistic.
   a. Claim: The mean annual income of full-time college students is below \$10,000. Sample data: For 750 randomly selected college students, the mean is \$3662 and the standard deviation is \$2996.
   b. Claim: With manual assembly of telephone parts, the assembly times vary more than the times for automated assembly, which are known to have a mean of 27.6 sec and a standard deviation of 1.8 sec.
   c. Claim: The majority of college students are women. Sample data: Of 500 randomly selected college students, 58% are women.
   d. Claim: When a group of adult survey respondents is randomly selected, their mean IQ is equal to 100. Sample data: $n = 150$ and $\bar{x} = 98.8$. It is reasonable to assume that $\sigma = 15$.

3. **Random Generation of Data** The TI-83 Plus calculator can be used to generate random data from a normally distributed population. The command **rand-Norm(100,15,50)** generates 50 values from a normally distributed population with $\mu = 100$ and $\sigma = 15$. One such generated sample of 50 values has a mean of 98.4 and a standard deviation of 16.3.
   a. Use a 0.10 significance level to test the claim that the sample actually does come from a population with a mean equal to 100. Assume that $\sigma = 15$.
   b. Repeat part (a) assuming that σ is unknown.
   c. Use a 0.10 significance level to test the claim that this sample actually does come from a population with a standard deviation equal to 15. What does the result say about the variation among the generated sample values?
   d. Based on the preceding results, does it appear that the calculator's random number generator is working correctly?

4. **Interviewing Mistakes** An Accountemps survey of 150 executives showed that 44% of them say that "little or no knowledge of the company" is the most common mistake made by candidates during job interviews (based on data from *USA Today*). Use a 0.05 significance level to test the claim that less than half of all executives identify that error as being the most common job interviewing error.

5. **Weights of Quarters** If we refer to the weights (in grams) of quarters listed in Data Set 29 in Appendix B, we find 50 weights with a mean of 5.622 g and a standard deviation of 0.068 g. The U.S. Department of the Treasury claims that the procedure it uses to mint quarters yields a mean weight of 5.670 g. Use a 0.01 significance level to test the claim that the mean weight of quarters in circulation is 5.670 g. If the claim is rejected, what is a possible explanation for the discrepancy?

6. **Weights of Blue M&Ms** Using the weights of only the *blue* M&Ms listed in Data Set 19 in Appendix B, test the claim that the mean is at least 0.9085 g, the mean value necessary for the 1498 M&Ms to produce a total of 1361 g as the package indicates. Use a 0.05 significance level. For the blue M&Ms, $\bar{x} = 0.9014$ g and $s = 0.0573$ g. Based on the result, can we conclude that the package contents do not agree with the claimed weight printed on the label?

7. **Percentage of Theme Park Visits** Each year, billions of dollars are spent at theme parks owned by Disney, Universal Studios, Sea World, Busch Gardens, and others. A survey of 1233 people who took trips revealed that 111 of them included a visit to a theme park (based on data from the Travel Industry Association of America). Based

on those survey results, management consultant Laura Croft claims that less than 10% of trips include a theme park visit. Use a 0.05 significance level to test her claim. Would it be wise for her to use that claim in trying to convince theme park management to increase advertising spending?

8. Voting for the Winning Candidate In a recent presidential election, 611 voters were surveyed, and 308 of them said that they voted for the candidate who won (based on data from the ICR Survey Research Group). Use a 0.04 significance level to test the claim that among all voters, 43% say that they voted for the winning candidate. (Voting records showed that the actual percentage who voted for the winning candidate was equal to 43%.) What does the result suggest about voter perceptions?

9. Are Consumers Being Cheated? The Orange County Bureau of Weights and Measures received complaints that the Windsor Bottling Company was cheating consumers by putting less than 12 oz of root beer in its cans. When 24 cans are randomly selected and measured, the amounts are found to have a mean of 11.4 oz and a standard deviation of 0.62 oz. The company president, Harry Windsor, claims that the sample is too small to be meaningful. Use the sample data to test the claim that consumers are being cheated. Does Harry Windsor's argument have any validity?

10. Percentage Believing That Elvis Is Alive *USA Today* ran a report about a University of North Carolina poll of 1248 adults from the southern United States. It was reported that 8% of those surveyed believe that Elvis Presley still lives. The article began with the claim that "almost 1 out of 10" Southerners still thinks Elvis is alive. At the 0.01 significance level, test the claim that the true percentage is less than 10%. Based on the result, determine whether the 8% sample result justifies the phrase "almost 1 out of 10."

11. Is the New Machine Better? The Medassist Pharmaceutical Company uses a machine to pour cold medicine into bottles in such a way that the standard deviation of the weights is 0.15 oz. A new machine is tested on 71 bottles, and the standard deviation for this sample is 0.12 oz. The Dayton Machine Company, which manufactures the new machine, claims that it fills bottles with lower variation. At the 0.05 significance level, test the claim made by the Dayton Machine Company. If Dayton's machine is being used on a trial basis, should its purchase be considered?

T **12.** Weights of Sugar Packets Refer to Data Set 28 in Appendix B and test the claim that the weights of the sugar packets have a mean equal to 3.5 g, as indicated on the label. If the mean does not appear to equal 3.5 g, what do you conclude?

Cumulative Review Exercises

1. Monitoring Dioxin Listed below are measured amounts of dioxin in the air at the site of the World Trade Center on the days immediately following the terrorist attacks of September 11, 2001. Dioxin includes a group of chemicals produced from burning and some types of manufacturing. The listed amounts are in nanograms per cubic meter (ng/m^3) and they are in order with the earliest recorded values at the left. The data are provided by the U.S. Environmental Protection Agency.

0.161   0.175   0.176   0.032   0.0524   0.044   0.018   0.0281   0.0268

a. Find the mean of this sample.
b. Find the median.
c. Find the standard deviation.

continued

8. TS: $z = 3.70$. CV: $z = \pm 2.05$. P-value: 0.0002. Reject H_0.

9. TS: $t = -4.741$. P-value is less than 0.005. CV: $t = -1.714$ (for $\alpha = 0.05$). Reject H_0.

10. TS: $z = -2.36$. CV: $z = -2.33$. P-value: 0.0091. Reject H_0.

11. TS: $\chi^2 = 44.800$. CV: $\chi^2 = 51.739$. (P-value: Between 0.005 and 0.01.) Reject H_0.

12. TS: $t = 9.720$. P-value < 0.01. CV: $t = \pm 1.994$ (approx., $\alpha = 0.05$). Reject H_0.

1. a. 0.0793 ng/m^3
   b. 0.044 ng/m^3
   c. 0.0694 ng/m^3
   d. 0.0048
   e. 0.158 ng/m^3
   f. $0.0259 < \mu < 0.1326$
   g. TS: $t = -3.491$. P-value is less than 0.005. CV: $t = -1.860$. Reject H_0.
   h. Yes, the data are listed in order and there appears to be a trend of decreasing values. The population is changing over time.

d. Find the variance.

e. Find the range.

f. Construct a 95% confidence interval estimate of the population mean.

g. The EPA uses 0.16 ng/m^3 as its "screening level," which is "set to protect against significantly increased risks of cancer and other adverse health effects." Use a 0.05 significance level to test the claim that this sample comes from a population with a mean less than 0.16 ng/m^3.

h. Is there any important characteristic of the data not addressed by the preceding results? If so, what is it?

2. SAT Math Scores of Women   The math SAT scores for women are normally distributed with a mean of 496 and a standard deviation of 108.

a. If a woman who takes the math portion of the SAT is randomly selected, find the probability that her score is above 500.

b. If five math SAT scores are randomly selected from the population of women who take the test, find the probability that all five of the scores are above 500.

c. If five women who take the math portion of the SAT are randomly selected, find the probability that their mean is above 500.

d. Find P_{90}, the score separating the bottom 90% from the top 10%.

3. ESP   A student majoring in psychology designs an experiment to test for extrasensory perception (ESP). In this experiment, a card is randomly selected from a shuffled deck, and the blindfolded subject must guess the suit (clubs, diamonds, hearts, spades) of the card selected. The experiment is repeated 25 times, with the card replaced and the deck reshuffled each time.

a. For subjects who make random guesses with no ESP, find the mean number of correct responses.

b. For subjects who make random guesses with no ESP, find the standard deviation for the numbers of correct responses.

c. For subjects who make random guesses with no ESP, find the probability of getting more than 12 correct responses.

d. If a subject gets more than 12 correct responses, test the claim that they made random guesses. Use a 0.05 significance level.

e. You want to conduct a survey to estimate the percentage of adult Americans who believe that some people have ESP. How many people must you survey if you want 90% confidence that your sample percentage is in error by no more than four percentage points?

Cooperative Group Activities

1. In-class activity Each student should estimate the length of the classroom. The values should be based on visual estimates, with no actual measurements being taken. After the estimates have been collected, measure the length of the room, then test the claim that the sample mean is equal to the actual length of the classroom. Is there a "collective wisdom," whereby the class mean is approximately equal to the actual room length?

2. Out-of-class activity Using a wristwatch that is reasonably accurate, set the time to be exact. Use a radio station or telephone time report which states that "at the tone, the time is . . . ." If you cannot set the time to the nearest second, record the error for the watch you are using. Now compare the time on your watch to the time on others. Record the errors with positive signs for watches that are ahead of the actual time and negative signs for those watches that are behind the actual time. Use the data to test the claim that the mean error of all wristwatches is equal to 0. Do we collectively run on time, or are we early or late? Also test the claim that the standard deviation of errors is less than 1 min. What are the practical implications of a standard deviation that is excessively large?

3. In-class activity In a group of three or four people, conduct an ESP experiment by selecting one of the group members as the subject. Draw a circle on one small piece of paper and draw a square on another sheet of the same size. Repeat this experiment 20 times: Randomly select the circle or the square and place it in the subject's hand behind his or her back so that it cannot be seen, then ask the subject to identify the shape (without looking at it); record whether the response is correct. Test the claim that the subject has ESP because the proportion of correct responses is greater than 0.5.

4. In-class activity After dividing into groups with sizes between 10 and 20 people, each group member should record the number of heartbeats in a minute. After calculating $\bar{x}$ and s, each group should proceed to test the claim that the mean is greater than 60, which is the author's result. (When people exercise, they tend to have lower pulse rates, and the author runs five miles a few times each week. What a guy.)

5. Out-of-class activity As part of a Gallup poll, subjects were asked "Are you in favor of the death penalty for persons convicted of murder?" Sixty-five percent of the respondents said that they were in favor, while 27% were against and 8% had no opinion. Use the methods of Section 6-2 to determine the sample size necessary to estimate the proportion of students at your college who are in favor. The class should agree on a confidence level and margin of error. Then divide the sample size by the number of students in the class, and conduct the survey by having each class member ask the appropriate number of students at the college. Analyze the results to determine whether the students differ significantly from the results found in the Gallup poll.

Technology Project

STATDISK, Minitab, Excel, the TI-83 Plus calculator, and many other tools can be used to randomly generate data from a normally distributed population with a given mean and standard deviation.

a. Use such a tool to randomly generate five values from a normally distributed population with a mean of 100 and a standard deviation of 15 (the parameters for typical IQ tests).

b. Using the five sample values generated in part (a), test the claim that the sample is from a population with a mean equal to 100. Use a 0.10 significance level.

c. Repeat parts (a) and (b) nine more times, so that a total of 10 different samples have been generated, and 10 different hypothesis tests have been executed.

d. With a 0.10 significance level, there is 0.10 probability of making a type I error (rejecting the null hypothesis when it is true). Because of the way that the sample data are generated, we know that 100 is the true population mean, so we do make a type I error in this experiment anytime that we reject the null hypothesis of $\mu = 100$. For the 10 trials of this experiment, how often was the null hypothesis actually rejected? When we conduct 10 such trials, how many times do we expect to reject the null hypothesis? Are the actual results consistent with the theoretical results? Explain.

Here are specific instructions for steps (a) and (b) using STATDISK, Minitab, Excel, and the TI-83 Plus calculator.

STATDISK

a. Click on **Data,** then click on **Normal Generator.** In the dialog box, enter a sample size of 5, a mean of 100, a standard deviation of 15, and enter 0 for the number of decimal places. Click on **Generate.**

Copy the generated values to the **Statdisk Data Window,** then use **Data** and **Descriptive Statistics** to find the mean and standard deviation.

b. Click on the main menu item of **Analysis,** then click on **Hypothesis Testing.** Select **Mean-One Sample.** In the dialog box, enter a significance level of 0.10, a claimed mean of 100, a sample size of 5, and enter the values of the sample mean and standard deviation that were recorded in step (a). Click on **Evaluate** and record the result (reject the null hypothesis or fail to reject the null hypothesis).

Minitab

a. Click on the main menu item of **Calc,** then click on **Random Data,** then **Normal.** In the dialog box, enter 5 for the number of rows to be generated, enter C1 for the column in which data will be stored, enter a mean of 100, and enter a standard deviation of 15. Click **OK.**

b. Click on the main menu item of **Stat,** select **Basic Statistics,** then select **1-Sample t.** In the dialog box, enter C1 in the top right box, click on Test Mean, enter 100 in the adjacent box, then click **OK.** Interpret the displayed results and record the conclusion (reject the null hypothesis or fail to reject the null hypothesis).

Excel

a. Click on **Tools,** select **Data Analysis,** then select **Random Number Generation** and click **OK.** In the dialog box, enter 1 for the number of variables, enter 5 for the number of random numbers, select the distribution option of Normal, enter a mean of 100, enter a standard deviation of 15, then click OK.

b. Click **DDXL** and select **Hypothesis Tests.** Select the item of **1 Var t Test.** Click on the pencil icon and enter the range of cells containing the generated sample data. For example, enter A1:A5 for five values in rows 1 through 5 of column A. Click **OK.** In the dialog box, click on the bar in Step 1 and proceed to enter a claimed mean of 100. Proceed to click on the bars in the remaining steps. After clicking **Compute,** record the conclusion (reject the null hypothesis or fail to reject the null hypothesis).

TI-83 Plus

a. First clear list L1 by pressing **STAT,** then **4:ClrList,** then **L1.** Now press **MATH,** then select **PRB** and select the menu item of **6:randNorm(.** Press **ENTER,** then proceed to enter 100,15,5 and press **ENTER.** Store the generated sample data by pressing **STO L1** followed by the **ENTER** key.

b. Press **STAT,** select **TESTS,** then select **2:T-Test** and press **ENTER.** Select **Data** (because we have the generated data in list L1), enter 100 for the claimed mean, and proceed to obtain the results of the hypothesis test. Interpret the displayed results and record the conclusion (reject the null hypothesis or fail to reject the null hypothesis).

from DATA *to* DECISION

Critical Thinking: Questioning survey results

Because surveys are now so pervasive throughout our society, each of us should develop the ability to think critically about them. We should question the process of selecting survey subjects, the wording of the question, the significance of the results, the objectivity of the survey sponsor and the group conducting the survey, as well as other issues. Consider the following hypothetical newspaper article.

"Survey Proves That Most Americans Don't Cheat on Taxes"   In *Every Day is Saturday*, a magazine for retired people, it was reported that a survey of 250 randomly selected subjects included 55% who said that they don't cheat when it comes to paying their taxes. The magazine printed the survey in its March issue, and the 250 responses were randomly selected from the surveys that were mailed in by readers. The first question on the survey asked "Do you cheat on your taxes or are you honest?" The magazine stated that "It's encouraging to learn that most Americans are honest—at least when it comes to filling out their tax forms."

Analyzing the Results

a. Use the methods of this chapter to test the claim that "most Americans don't cheat on taxes." What do you conclude? Is it possible for any survey results to prove that most Americans don't cheat on their taxes?

b. Apart from the results of the hypothesis test, there are at least four other important issues that affect the validity of the survey results. Identify the other issues and describe how they affect the validity of the results.

INTERNET PROJECT   Hypothesis Testing

This chapter introduced the methodology behind hypothesis testing, an essential technique of inferential statistics. This Internet Project will have you conduct tests using a variety of data sets in different areas of study. For each subject, you will be asked to

- Collect data available on the Internet.

- Formulate a null and alternative hypothesis based on a given question.

- Conduct a hypothesis test at a specified level of significance.

- Summarize your conclusion.

Go to the *Elementary Statistics* Web site at

http://www.aw.com/triola

and locate the internet Project for this chapter. There you will find guided investigations in the fields of education, economics, and sports, and a classic example from the physical sciences.

Statistics @ Work

Michael Saccucci

Director of Statistics and Quality Management for Consumers Union, which tests products and services and provides ratings and recommendations to consumers in Consumer Reports magazine

"It is extremely important for everyone to have an understanding of statistics in order to effectively process the huge amounts of information that we are presented with each day in our professional and personal lives."

Author's Note: The author met with Mike Saccucci and the other statisticians at Consumers Union: Keith Newsom–Stewart, Martin Romm, and Eric Rosenberg. The author toured the product-testing facilities and observed several different experiments in progress. He was extremely impressed with the involvement of statisticians at the various stages of product testing, the extreme and detailed care taken in the design of experiments, and the careful and effective use of statistical analyses of test results.

What statistical concepts and procedures do you use at Consumers Union?

On any given day, the statisticians may have to use any number of statistical procedures, many of which are discussed in this text. For example, in a recent study to evaluate the quality and safety of chicken, we developed a complex sampling scheme so that the different manufacturers were fairly represented. In a recent study of sunscreens, we used the normal distribution to help determine the appropriate number of replicates necessary to fairly evaluate the products. Depending on the type of test, the statistician might have to construct a completely randomized design, a randomized block design, or some other type of experimental design to ensure that our results are accurate and unbiased. During the analysis phase, the statistician might use any number of techniques, such as analysis of variance, regression analysis, time series analysis, categorical analysis, and/or non-parametric analysis.

What do the statisticians do at Consumers Union?

Statisticians perform a variety of tasks. Early in a project, the statistician works with the project team to develop the test protocol and help select which products to test. Next, the statistician helps develop an appropriate experimental design for use during the testing. Once the test data have been obtained, the statistician analyzes the results and presents findings in a statistical

report. The statistician also gets involved with a variety of special projects, depending on the needs of the organization. Consumers rely on the information we provide, so it's important that we use proper statistical techniques to make sure that our ratings are correct.

What steps do you take to ensure objectivity in your testing procedures?

It is Consumers Union's policy that all tests must be performed in an objective, scientific manner, and with due regard for the safety of test personnel. We go to great lengths to adhere to this policy. For example, we don't accept any type of outside advertising in our publications. We employ anonymous shoppers located throughout the United States to purchase our test samples in ways normally available to consumers. We don't accept free samples from anyone, including vendors. And we don't test unsolicited samples sent from a manufacturer. In addition, technicians use randomized experimental designs to ensure that our testing is done with scientific integrity and objectivity. When practical, tested items are blind coded so that the testers do not know which brands they are evaluating.

Are the ratings and recommendations in *Consumer Reports* magazine based on statistical significance alone?

No. The information we provide must be useful to consumers. Our technicians perform a variety of tests to evaluate a product's performance. These tests are designed to simulate conditions of predictable consumer usage. If it turns out that there is a statistically significant, but not meaningful difference in test results, we would not rate one brand over another. In testing water sealants, for example, we might find that there is a statistically significant difference between the amounts of water that seep through two different brands of sealant. However, if the difference amounts to a few drops of water, we would rate the products similarly for that attribute.

Do you feel that job applicants are viewed more favorably if they have studied some statistics?

Given the level of innumeracy that exists today, I believe that a basic knowledge of statistics would be viewed favorably for just about any field of study. This would be especially true of quantitative fields, such as the sciences, engineering, and business. It is extremely important for everyone to have an understanding of statistics in order to effectively process the huge amounts of information that we are presented with each day in our professional and personal lives. A focus on statistical thinking would be especially useful.

How critical do you find your background for performing your responsibilities with excellence?

Consumers Union's mission is to advance the interests of consumers by providing in-formation and advice about products and services and about issues affecting their welfare, and by advocating a consumer point of view. To stay competitive, we've had to look for more efficient ways to provide more information to consumers in less time. My background in both statistics and quality management have been extremely valuable in helping Consumers Union achieve this mission.

While a college student, did you expect to be using statistics on the job?

I started out as a math major and didn't really get interested in statistics until my senior year. It wasn't until graduate school, while working under the direction of Professor Hoerl at the University of Delaware, that I realized how interesting a career in statistics would be. And despite the negative feelings a lot of students have about statistics, I think I have one of the most interesting jobs. I never know what to expect on a given day. One day I may be sitting in on the training session for wine tasting to learn about the testing procedures. On another day, I may be discussing various ways to test paints. On most days, though, I spend a significant amount of time using a computer to help design an upcoming study or to sift through large amounts of data that will ultimately be used as the basis for product ratings.

8

Inferences from
Two Samples

8-1 Overview

8-2 Inferences About Two Proportions

8-3 Inferences About Two Means: Independent Samples

8-4 Inferences from Matched Pairs

8-5 Comparing Variation in Two Samples

Using statistics to identify racial profiling

Racial profiling is the controversial practice of targeting someone for criminal behavior on the basis of the person's race, national origin, or ethnicity. Examples of racial profiling occur when disproportionately more blacks are stopped for traffic violations than whites, or disproportionately more people of Middle Eastern origins are stopped for extensive searches at airports.

Consider the Table 8-1 data for randomly selected drivers stopped by police in a recent year (based on data from the U.S. Department of Justice, Bureau of Justice Statistics). It could be argued that many more whites were stopped than blacks. However, the population includes many more whites than blacks, so it doesn't make much sense to compare the 147 stopped white drivers to the 24 stopped black drivers. Any comparison should involve the *rates* at which whites and blacks are stopped.

It could also be claimed that the 12.0% rate for blacks is not *significantly* greater than the 10.5% rate for whites. That claim will be tested in this chapter. We will test the claim that based on the sample proportions of 24/200 for blacks and 147/1400 for whites, it appears that the proportion of blacks who are stopped is *greater* than the proportion of whites who are stopped. We will use statistical procedures that are so important for issues such as those that involve racial profiling.

Table 8-1	Racial Profiling Data	
	Race and Ethnicity	
	Black and Non-Hispanic	White and Non-Hispanic
Drivers stopped by police	24	147
Total number of observed drivers	200	1400
Percent Stopped by Police	**12.0%**	**10.5%**

 Overview

Note to Instructor

Changes from the previous edition:

1. The section on proportions has been moved to Section 8-2.

2. The sections dealing with independent means for large samples and for small samples have been combined in Section 8-3. The criteria for choosing between the normal and t distributions have also been updated.

3. In Section 8-3, when using the t distribution for inferences about μ, the criteria for using a pooled estimate of the standard error have been updated. See Figure 8-3 in Section 8-3.

Some instructors cover the entire chapter while others omit it. Even if this chapter is not discussed in class, strongly consider assigning some exercises from it. Once students understand the basic concepts of hypothesis testing and confidence intervals, they can use software or a TI-83 Plus calculator to do the number crunching, then they can focus on the interpretation of the results. Sections 8-2, 8-3, and 8-4 are especially important, so consider covering those sections or assigning exercises from them.

Chapter 6 introduced an important activity of inferential statistics: Sample data were used to construct confidence interval *estimates* of population parameters. Chapter 7 introduced a second important activity of inferential statistics: Sample data were used to *test hypotheses* about population parameters. In Chapters 6 and 7, all examples and exercises involved the use of *one* sample to form an inference about *one* population. In reality, however, there are many important and meaningful situations in which it becomes necessary to compare *two* sets of sample data. The following are examples typical of those found in this chapter, which presents methods for using sample data from two populations so that inferences can be made about those populations.

- When testing a claim of racial profiling, determine whether the proportion of black drivers stopped by police is greater than the proportion of white drivers stopped by police.

- When testing the effectiveness of the Salk vaccine in preventing paralytic polio, determine whether the treatment group had a lower incidence of polio than the group given a placebo.

- When investigating the accuracy of heights reported by people, determine whether there is a significant difference between the heights they report and the actual measured heights.

Chapters 6 and 7 included methods that were applied to proportions, means, and measures of variation (standard deviation and variance), and this chapter will address those same parameters. This chapter extends the same methods introduced in Chapters 6 and 7 to situations involving comparisons of two samples instead of only one.

8-2   Inferences About Two Proportions

There are many real and important situations in which it is necessary to use sample data to compare two population proportions. In fact, a strong argument could be made that this section is one of the most important sections in the book because this is where we describe methods for dealing with two sample proportions. Although this section is based on proportions, we can deal with probabilities or we can deal with percentages by using the corresponding decimal equivalents. For example, we might want to determine whether there is a difference between the percentage of adverse reactions in a placebo group and the percentage of adverse reactions in a drug treatment group. We can convert the percentages to their corresponding decimal values and proceed to use the methods of this section.

When testing a hypothesis made about two population proportions or when constructing a confidence interval for the difference between two population proportions, we make the following assumptions and use the following notation.

Assumptions

1. We have proportions from two simple random samples that are *independent*, which means that the sample values selected from one population are not related to or somehow paired or matched with the sample values selected from the other population.

2. For both samples, the conditions $np \geq 5$ and $nq \geq 5$ are satisfied. That is, there are at least five successes and five failures in each of the two samples. (In many cases, we will test the claim that two populations have equal proportions so that $p_1 - p_2 = 0$. Because we assume that $p_1 - p_2 = 0$, it is not necessary to specify the particular value that p_1 and p_2 have in common. In such cases, the conditions $np \geq 5$ and $nq \geq 5$ can be checked by replacing p with the estimated pooled proportion $\bar{p}$, which will be described later.)

Notation for Two Proportions

For population 1 we let

$p_1 = population$ proportion

$n_1 = $ size of the sample

$x_1 = $ number of successes in the sample

$\hat{p} = \dfrac{x_1}{n_1}$ (the *sample* proportion)

$\hat{q}_1 = 1 - \hat{p}_1$

The corresponding meanings are attached to p_2, n_2, x_2, $\hat{p}_2$, and $\hat{q}_2$, which come from population 2.

Finding the Numbers of Successes x_1 and x_2: The calculations for hypothesis tests and confidence intervals require that we have specific values for x_1, n_1, x_2, and n_2. Sometimes the available sample data include those specific numbers, but sometimes it is necessary to calculate the values of x_1 and x_2.

For example, consider the statement that "when 734 men were treated with Viagra, 16% of them experienced headaches." From that statement we can see that $n_1 = 734$ and $\hat{p}_1 = 0.16$, but the actual number of successes x_1 is not given. However, from $\hat{p}_1 = x_1/n_1$, we know that

$$x_1 = n_1 \cdot \hat{p}_1$$

so that $x_1 = 734 \cdot 0.16 = 117.44$. But you cannot have 117.44 men who experienced headaches, because everyone either experiences a headache or does not, and the number of successes x_1 must therefore be a whole number. We can round 117.44 to 117. We can now use $x_1 = 117$ in the calculations that require its value. It's really quite simple: 16% of 734 means 0.16×734, which results in 117.44, which we round to 117.

Note to Instructor

Many students experience difficulty finding the actual number of successes from a statement such as this: "Among the 734 subjects, 16% experienced headaches." A brief review of the procedure described here can help remove that obstacle. Also, assign Exercises 1–4, which deal with this issue.

Hypothesis Tests

In Section 7-2 we discussed tests of hypotheses made about a single population proportion. We will now consider tests of hypotheses made about two population proportions, but *we will be testing only claims that $p_1 = p_2$*, and we will use the following pooled (or combined) estimate of the value that p_1 and p_2 have in common. (For claims that the difference between p_1 and p_2 is equal to a nonzero constant, see Exercise 34 in this section.) You can see from the form of the pooled estimate $\overline{p}$ that it basically combines the two different samples into one big sample.

Note to Instructor

Point out that the concept of a pooled estimate of proportions applies only to cases in which we assume that $p_1 = p_2$. If we do not assume that $p_1 = p_2$, as in the construction of a confidence interval for their difference, we do not use a pooled estimate of proportions.

Pooled Estimate of p_1 and p_2

The **pooled estimate of p_1 and p_2** is denoted by $\overline{p}$ and is given by

$$\overline{p} = \frac{x_1 + x_2}{n_1 + n_2}$$

We denote the complement of $\overline{p}$ by $\overline{q}$, so $\overline{q} = 1 - \overline{p}$.

Test Statistic for Two Proportions (with H_0: $p_1 = p_2$)

$$z = \frac{(\hat{p}_1 - \hat{p}_2) - (p_1 - p_2)}{\sqrt{\dfrac{\overline{p}\,\overline{q}}{n_1} + \dfrac{\overline{p}\,\overline{q}}{n_2}}}$$

where　　$p_1 - p_2 = 0$　(assumed in the null hypothesis)

$$\hat{p}_1 = \frac{x_1}{n_1} \quad \text{and} \quad \hat{p}_2 = \frac{x_2}{n_2}$$

$$\overline{p} = \frac{x_1 + x_2}{n_1 + n_2}$$

$$\overline{q} = 1 - \overline{p}$$

P-value:　Use Table A-2. (Use the computed value of the test statistic z and find the P-value by following the procedure summarized in Figure 7-6.)

Critical values:　Use Table A-2. (Based on the significance level α, find critical values by using the procedures introduced in Section 7-2.)

Once again, the test statistic fits the common format of

$$\frac{(\text{sample statistic}) - (\text{claimed value of parameter})}{(\text{standard deviation of sample statistics})}$$

The following example will help clarify the roles of x_1, n_1, $\hat{p}_1$, $\bar{p}$, and so on. In particular, you should recognize that under the assumption of equal proportions, the best estimate of the common proportion is obtained by pooling both samples into one big sample, so that $\bar{p}$ becomes a more obvious estimate of the common population proportion.

EXAMPLE **Racial Profiling** For the sample data listed in Table 8-1, use a 0.05 significance level to test the claim that the proportion of black drivers stopped by the police is greater than the proportion of white drivers who are stopped.

SOLUTION   For notation purposes, we stipulate that Sample 1 is the group of black drivers, and Sample 2 is the group of white drivers. We can summarize the sample data as follows.

Black Drivers	White Drivers
$n_1 = 200$	$n_2 = 1400$
$x_1 = 24$	$x_2 = 147$
$\hat{p}_1 = \dfrac{x_1}{n_1} = \dfrac{24}{200} = 0.120$	$\hat{p}_2 = \dfrac{x_2}{n_2} = \dfrac{147}{1400} = 0.105$

We will now use the *P*-value method of hypothesis testing, as summarized in Figure 7-9.

Step 1: The claim of a greater rate for black drivers can be represented by $p_1 > p_2$.

Step 2: If $p_1 > p_2$ is false, then $p_1 \leq p_2$.

Step 3: Because our claim of $p_1 > p_2$ does not contain equality, it becomes the alternative hypothesis. The null hypothesis is the statement of equality, so we have

$$H_0: p_1 = p_2 \qquad H_1: p_1 > p_2 \quad \text{(original claim)}$$

Step 4: The significance level is $\alpha = 0.05$.

Step 5: We will use the normal distribution (with the test statistic previously given) as an approximation to the binomial distribution. We have two independent samples, and the conditions $np \geq 5$ and $nq \geq 5$ are satisfied for each of the two samples. To check this, we note that in conducting this test, we assume that $p_1 = p_2$, where their common value is the pooled estimate $\bar{p}$ calculated as shown below, with extra decimal places used to minimize rounding errors in later calculations.

$$\bar{p} = \frac{x_1 + x_2}{n_1 + n_2} = \frac{24 + 147}{200 + 1400} = 0.106875$$

With $\bar{p} = 0.106875$, it follows that $\bar{q} = 1 - 0.106875 = 0.893125$.

continued

Polio Experiment

In 1954 an experiment was conducted to test the effectiveness of the Salk vaccine as protection against the devastating effects of polio. Approximately 200,000 children were injected with an ineffective salt solution, and 200,000 other children were injected with the vaccine. The experiment was "double blind" because the children being injected didn't know whether they were given the real vaccine or the placebo, and the doctors giving the injections and evaluating the results didn't know either. Only 33 of the 200,000 vaccinated children later developed paralytic polio, whereas 115 of the 200,000 injected with the salt solution later developed paralytic polio. Statistical analysis of these and other results led to the conclusion that the Salk vaccine was indeed effective against paralytic polio.

We verify that $np \geq 5$ and $nq \geq 5$ for both samples as shown below, with p estimated by $\bar{p}$ and with q estimated by $\bar{q}$.

Sample 1	Sample 2
$n_1 p = (200)(0.106875) = 21.375 \geq 5$	$n_2 p = (1400)(0.106875)$ $= 149.625 \geq 5$
$n_1 q = (200)(0.893125) = 178.625 \geq 5$	$n_2 q = (1400)(0.893125)$ $= 1250.375 \geq 5$

Step 6:   We can now find the value of the test statistic.

$$z = \frac{(\hat{p}_1 - \hat{p}_2) - (p_1 - p_2)}{\sqrt{\dfrac{\bar{p}\,\bar{q}}{n_1} + \dfrac{\bar{p}\,\bar{q}}{n_2}}}$$

$$= \frac{\left(\dfrac{24}{200} - \dfrac{147}{1400}\right) - 0}{\sqrt{\dfrac{(0.106875)(0.893125)}{200} + \dfrac{(0.106875)(0.893125)}{1400}}} = 0.64$$

The P-value of 0.2611 is found as follows: This is a right-tailed test, so the P-value is the area to the right of the test statistic $z = 0.64$. (See Figure 7-6.) Refer to Table A-2 and find that the area to the left of the test statistic $z = 0.64$ is 0.7389, so the P-value is $1 - 0.7389 = 0.2611$. (Software shows that a more exact P-value is 0.2603.) The test statistic and P-value are shown in Figure 8-1(a).

Step 7:   Because the P-value of 0.2611 is greater than the significance level of $\alpha = 0.05$, we fail to reject the null hypothesis of $p_1 = p_2$.

INTERPRETATION   We must address the original claim that black drivers get stopped at a greater rate than white drivers. Because we fail to reject the null

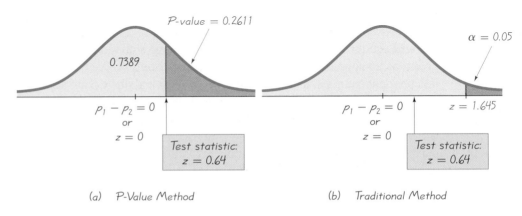

(a)   P-Value Method   (b)   Traditional Method

FIGURE 8-1
Testing Claim That $p_1 > p_2$

hypothesis, we conclude that there is not sufficient evidence to support the claim that the proportion of black drivers stopped by police is greater than that for white drivers. (See Figure 7-7 for help in wording the final conclusion.) This does *not* mean that racial profiling has been disproved. It means only that the evidence is not yet strong enough to conclude that the 12.0% rate for stopping black drivers is significantly greater than the 10.5% rate for stopping white drivers. The evidence might be strong enough with more data. (See Exercise 33.) In fact, data sets larger than those used in this example do suggest that racial profiling has been in effect.

Traditional Method of Testing Hypotheses

The preceding example illustrates the *P*-value approach to hypothesis testing, but it would be quite easy to use the traditional approach instead. In Step 6, instead of finding the *P*-value, we would find the critical value. With a significance level of $\alpha = 0.05$ in a right-tailed test based on the normal distribution, refer to Table A-2 to find that an area of $\alpha = 0.05$ in the right tail corresponds to the critical value of $z = 1.645$. See Figure 8-1(b) where we can see that the test statistic does not fall in the critical region bounded by the critical value of $z = 1.645$. We again fail to reject the null hypothesis. Again, we conclude that there is not sufficient evidence to support the claim that black drivers are stopped at a greater rate than white drivers.

Confidence Intervals

We can construct a confidence interval estimate of the difference between population proportions ($p_1 - p_2$) by using the format given on page 444. If a confidence interval estimate of $p_1 - p_2$ does not include 0, we have evidence suggesting that p_1 and p_2 have different values. However, we recommend against using a confidence interval estimate of $p_1 - p_2$ as the basis for testing the claim that $p_1 = p_2$, for the following reasons.

Don't use a confidence interval to test the claim that $p_1 = p_2$ (because the standard deviation used for confidence intervals is different from the standard deviation used for the hypothesis test that uses the *P*-value method or the traditional method). When testing claims about the difference between two population proportions, the traditional method and the *P*-value method are equivalent in the sense that they always yield the same results, but the confidence interval estimate of the difference might suggest a different conclusion. (See Exercise 32.) If different conclusions are obtained, realize that the traditional and *P*-value methods use an *exact* standard deviation based on the assumption that there is no difference between the population proportions (as stated in the null hypothesis). However, the confidence interval is constructed using a standard deviation based on *estimated* values of the two population proportions. Use this strategy: If you want to estimate the difference between two population proportions, do so by constructing a confidence interval, but if you want to test some claim about two population proportions, use the *P*-value method or the traditional method.

Does Aspirin Help Prevent Heart Attacks?

In a recent study of 22,000 male physicians, half were given regular doses of aspirin while the other half were given placebos. The study ran for six years at a cost of $4.4 million. Among those who took the aspirin, 104 suffered heart attacks. Among those who took the placebos, 189 suffered heart attacks. (The figures are based on data from *Time* and the *New England Journal of Medicine,* Vol. 318, No. 4.) This is a classic experiment involving a treatment group (those who took the aspirin) and a placebo group (those who took pills that looked and tasted like the aspirin pills, but no aspirin was present). We can use methods presented in this chapter to address the issue of whether the results show a statistically significant lower rate of heart attacks among the sample group who took aspirin.

Author as a Witness

The author was asked to testify in New York State Supreme Court by a former student who was contesting a lost reelection to the office of Dutchess County Clerk. The author testified by using statistics to show that the voting behavior in one contested district was significantly different from the behavior in all other districts. When the opposing attorney asked about results of a confidence interval, he asked if the 5% error (from a 95% confidence level) could be added to the three percentage point margin of error to get a total error of 8%, thereby indicating that he did not understand the basic concept of a confidence interval. The judge cited the author's testimony, upheld the claim of the former student, and ordered a new election in the contested district. That judgment was later overturned by the appellate court on the grounds that the ballot irregularities should have been contested before the election, not after.

Also, *don't test for equality of two population proportions by determining whether there is an overlap between two individual confidence interval estimates of the two individual population proportions.* When compared to the confidence interval estimate of $p_1 - p_2$, the analysis of overlap between two individual confidence intervals is more conservative (by rejecting equality less often), and it has less power (because it is less likely to reject $p_1 = p_2$ when in reality $p_1 \neq p_2$). (See "On Judging the Significance of Differences by Examining the Overlap Between Confidence Intervals" by Schenker and Gentleman, *The American Statistician*, Vol. 55, No. 3.) See Exercise 31.

Confidence Interval Estimate of $p_1 - p_2$

The confidence interval estimate of the difference $p_1 - p_2$ is:

$$(\hat{p}_1 - \hat{p}_2) - E < (p_1 - p_2) < (\hat{p}_1 - \hat{p}_2) + E$$

where the margin of error E is given by

$$E = z_{\alpha/2}\sqrt{\frac{\hat{p}_1\hat{q}_1}{n_1} + \frac{\hat{p}_2\hat{q}_2}{n_2}}$$

EXAMPLE   Racial Profiling Use the sample data given in Table 8-1 to construct a 90% confidence interval estimate of the difference between the two population proportions. (The confidence level of 90% is comparable to the significance level of $\alpha = 0.05$ used in the preceding right-tailed hypothesis test. See Table 7-2 in Section 7-2.)

SOLUTION   With a 90% confidence level, $z_{\alpha/2} = 1.645$ (from Table A-2). We first calculate the value of the margin of error E as shown.

$$E = z_{\alpha/2}\sqrt{\frac{\hat{p}_1\hat{q}_1}{n_1} + \frac{\hat{p}_2\hat{q}_2}{n_2}} = 1.645\sqrt{\frac{\left(\frac{24}{200}\right)\left(\frac{176}{200}\right)}{200} + \frac{\left(\frac{147}{1400}\right)\left(\frac{1253}{1400}\right)}{1400}} = 0.040$$

With $\hat{p}_1 = 24/200 = 0.120$, $\hat{p}_2 = 147/1400 = 0.105$, and $E = 0.040$, the confidence interval is evaluated as follows.

$$(\hat{p}_1 - \hat{p}_2) - E < (p_1 - p_2) < (\hat{p}_1 - \hat{p}_2) + E$$
$$(0.120 - 0.105) - 0.040 < (p_1 - p_2) < (0.120 - 0.105) + 0.040$$
$$-0.025 < (p_1 - p_2) < 0.055$$

INTERPRETATION The confidence interval limits do contain 0, suggesting that there is *not* a significant difference between the two proportions. However, if the goal is to test for equality of the two population proportions, we should use the *P*-value or traditional method of hypothesis testing; we should not base the decision on the confidence interval.

Rationale: Why Do the Procedures of This Section Work? The test statistic given for hypothesis tests is justified by the following:

1. With $n_1 p_1 \geq 5$ and $n_1 q_1 \geq 5$, the distribution of $\hat{p}_1$ can be approximated by a normal distribution with mean p_1 and standard deviation $\sqrt{p_1 q_1 / n_1}$ and variance $p_1 q_1 / n_1$. These conclusions are based on Sections 5-6 and 6-2, and they also apply to the second sample.

2. Because $\hat{p}_1$ and $\hat{p}_2$ are each approximated by a normal distribution, $\hat{p}_1 - \hat{p}_2$ will also be approximated by a normal distribution with mean $p_1 - p_2$ and variance

$$\sigma^2_{(\hat{p}_1 - \hat{p}_2)} = \sigma^2_{\hat{p}_1} + \sigma^2_{\hat{p}_2} = \frac{p_1 q_1}{n_1} + \frac{p_2 q_2}{n_2}$$

(The above result is based on this property: The variance of the *differences* between two independent random variables is the *sum* of their individual variances. See Exercise 38.)

3. Because the values of p_1, q_1, p_2, and q_2 are typically unknown and from the null hypothesis we assume that $p_1 = p_2$, we can pool (or combine) the sample data. The pooled estimate of the common value of p_1 and p_2 is $\bar{p} = (x_1 + x_2)/(n_1 + n_2)$. If we replace p_1 and p_2 by $\bar{p}$ and replace q_1 and q_2 by $\bar{q} = 1 - \bar{p}$, the variance from Step 2 leads to the following standard deviation.

$$\sigma_{(\hat{p}_1 - \hat{p}_2)} = \sqrt{\frac{\bar{p}\,\bar{q}}{n_1} + \frac{\bar{p}\,\bar{q}}{n_2}}$$

4. We now know that the distribution of $p_1 - p_2$ is approximately normal, with mean $p_1 - p_2$ and standard deviation as shown in Step 3, so that the z test statistic has the form given earlier.

The form of the confidence interval requires an expression for the variance different from the one given in Step 3. In Step 3 we are assuming that $p_1 = p_2$, but if we don't make that assumption (as in the construction of a confidence interval), we estimate the variance of $p_1 - p_2$ as

$$\sigma^2_{(\hat{p}_1 - \hat{p}_2)} = \sigma^2_{\hat{p}_1} + \sigma^2_{\hat{p}_2} = \frac{\hat{p}_1 \hat{q}_1}{n_1} + \frac{\hat{p}_2 \hat{q}_2}{n_2}$$

and the standard deviation becomes

$$\sqrt{\frac{\hat{p}_1 \hat{q}_1}{n_1} + \frac{\hat{p}_2 \hat{q}_2}{n_2}}$$

In the test statistic

$$z = \frac{(\hat{p}_1 - \hat{p}_2) - (p_1 - p_2)}{\sqrt{\dfrac{\hat{p}_1 \hat{q}_1}{n_1} + \dfrac{\hat{p}_2 \hat{q}_2}{n_2}}}$$

use the positive and negative values of z (for two tails) and solve for $p_1 - p_2$. The results are the limits of the confidence interval given earlier.

The Lead Margin of Error

Authors Stephen Ansolabehere and Thomas Belin wrote in their article "Poll Faulting" (*Chance* magazine) that "our greatest criticism of the reporting of poll results is with the margin of error of a single proportion (usually ±3%) when media attention is clearly drawn to the *lead* of one candidate." They point out that the lead is really the *difference* between two proportions $(p_1 - p_2)$ and go on to explain how they developed the following rule of thumb: The lead is approximately $\sqrt{3}$ times larger than the margin of error for any one proportion. For a typical preelection poll, a reported ±3% margin of error translates to about ±5% for the lead of one candidate over the other. They write that the margin of error for the lead should be reported.

Using Technology

STATDISK   Select **Analysis** from the main menu bar, then select **Hypothesis Testing,** then **Proportion-Two Samples.** Enter the required items in the dialog box. Confidence interval limits are included with the hypothesis test results.

Minitab   Minitab can now handle summary statistics for two samples. Select **Stat** from the main menu bar, then select **Basic Statistics,** then **2 Proportions.** Click on the button for **Summarize data.** Click on the **Options** bar. Enter the desired confidence level, enter the claimed value of $p_1 - p_2$, select the format for the alternative hypothesis, and click on the box to use the pooled estimate of p for the test. Click **OK** twice.

Excel   You must use the Data Desk XL add-in, which is a supplement to this book. First make these entries: In cell A1 enter the number of successes for Sample 1, in cell B1 enter the

number of trials for Sample 1, in cell C1 enter the number of successes for Sample 2, and in cell D1 enter the number of trials for Sample 2. Click on **DDXL.** Select **Hypothesis Tests** and **Summ 2 Var Prop Test** or select **Confidence Intervals** and **Summ 2 Var Prop Interval.** In the dialog box, click on the four pencil icons and enter A1, B1, C1, and D1 in the four input boxes. Click **OK.** Proceed to complete the new dialog box.

TI-83 Plus   The TI-83 Plus calculator can be used for hypothesis tests and confidence intervals. Press **STAT** and select **TESTS.** Then choose the option of **2-PropZTest** (for a hypothesis test) or **2-PropZInt** (for a confidence interval). When testing hypotheses, the TI-83 Plus calculator will display a P-value instead of critical values, so the P-value method of testing hypotheses is used.

8-2 Basic Skills and Concepts

Recommended Assignment:
Exercises 1–10, 15–18.

1. 30
2. 151
3. 85
4. 49
5. a. 0.417
   b. 2.17
   c. ±1.96
   d. 0.0300
6. a. 0.00986
   b. 3.38
   c. ±1.96
   d. 0.0008
7. $H_0: p_1 = p_2$. $H_1: p_1 > p_2$.
   TS: $z = 2.17$. P-value: 0.0150.
   CV: $z = 1.645$. Reject H_0. There
   is sufficient evidence to support
   the claim that for those saying
   that monitoring e-mail is seriously unethical, the proportion
   of employees is greater than
   the proportion of bosses.

Finding Number of Successes. In Exercises 1–4, find the number of successes x suggested by the given statement.

1. From the Arizona Department of Weights and Measures: Among 37 inspections at NAPA Auto Parts stores, 81% failed.

2. From the *New York Times*: Among 240 vinyl gloves subjected to stress tests, 63% leaked viruses.

3. From *Sociological Methods and Research*: When 294 central-city residents were surveyed, 28.9% refused to respond.

4. From a Time/CNN survey: 24% of 205 single women said that they "definitely want to get married."

Calculations for Testing Claims. In Exercises 5 and 6, assume that you plan to use a significance level of $\alpha = 0.05$ to test the claim that $p_1 = p_2$. Use the given sample sizes and numbers of successes to find (a) the pooled estimate $\bar{p}$, (b) the z test statistic, (c) the critical z values, and (d) the P-value.

5.

Workers	Bosses
$n_1 = 436$	$n_2 = 121$
$x_1 = 192$	$x_2 = 40$

6.

Low Activity	High Activity
$n_1 = 10{,}239$	$n_2 = 9877$
$x_1 = 101$	$x_2 = 56$

7. E-Mail and Privacy   A survey of 436 workers showed that 192 of them said that it was seriously unethical to monitor employee e-mail. When 121 senior-level bosses were surveyed, 40 said that it was seriously unethical to monitor employee e-mail (based on data from a Gallup poll). Use a 0.05 significance level to test the claim that for those saying that monitoring e-mail is seriously unethical, the proportion of employees is greater than the proportion of bosses.

8. **E-Mail and Privacy** Refer to the sample data given in Exercise 7 and construct a 90% confidence interval estimate of the difference between the two population proportions. Is there a substantial gap between the employees and bosses?

9. **Exercise and Coronary Heart Disease** In a study of women and coronary heart disease, the following sample results were obtained: Among 10,239 women with a low level of physical activity (less than 200 kcal/wk), there were 101 cases of coronary heart disease. Among 9877 women with physical activity measured between 200 and 600 kcal/wk, there were 56 cases of coronary heart disease (based on data from "Physical Activity and Coronary Heart Disease in Women" by Lee, Rexrode, et al., *Journal of the American Medical Association,* Vol. 285, No. 11). Construct a 90% confidence interval estimate for the difference between the two proportions. Does the difference appear to be substantial? Does it appear that physical activity corresponds to a lower rate of coronary heart disease?

10. **Exercise and Coronary Heart Disease** Refer to the sample data in Exercise 9 and use a 0.05 significance level to test the claim that the rate of coronary heart disease is higher for women with the lower levels of physical activity. What does the conclusion suggest?

11. **Instant Replay in Football** In the 2000 football season, 247 plays were reviewed by officials using instant video replays, and 83 of them resulted in reversal of the original call. In the 2001 football season, 258 plays were reviewed and 89 of them were reversed (based on data from "Referees Turn to Video Aid More Often" by Richard Sandomir, *New York Times*). Is there a significant difference in the two reversal rates? Does it appear that the reversal rate was the same in both years?

12. **Effectiveness of Smoking Bans** The Joint Commission on Accreditation of Healthcare Organizations mandated that hospitals ban smoking by 1994. In a study of the effects of this ban, subjects who smoke were randomly selected from two different populations. Among 843 smoking employees of hospitals with the smoking ban, 56 quit smoking one year after the ban. Among 703 smoking employees from workplaces without a smoking ban, 27 quit smoking a year after the ban (based on data from "Hospital Smoking Bans and Employee Smoking Behavior" by Longo, Brownson, et al., *Journal of the American Medical Association*, Vol. 275, No. 16). Is there a significant difference between the two proportions at a 0.05 significance level? Is there a significant difference between the two proportions at a 0.01 significance level? Does it appear that the ban had an effect on the smoking quit rate?

13. **Testing Effectiveness of Vaccine** In a *USA Today* article about an experimental nasal spray vaccine for children, the following statement was presented: "In a trial involving 1602 children only 14 (1%) of the 1070 who received the vaccine developed the flu, compared with 95 (18%) of the 532 who got a placebo." The article also referred to a study claiming that the experimental nasal spray "cuts children's chances of getting the flu." Is there sufficient sample evidence to support the stated claim?

14. **Color Blindness in Men and Women** In a study of red/green color blindness, 500 men and 2100 women are randomly selected and tested. Among the men, 45 have red/green color blindness. Among the women, 6 have red/green color blindness (based on data from *USA Today*).

continued

8. $0.0293 < p_1 - p_2 < 0.190$; the gap appears to be statistically significant, and it's likely value is between 2.93% and 19.0%, so it could be substantial.

9. $0.00216 < p_1 - p_2 < 0.00623$; it appears that physical activity corresponds to a lower rate of coronary heart disease.

10. H_0: $p_1 = p_2$. H_1: $p_1 > p_2$. TS: $z = 3.38$. P-value: 0.0004. CV: $z = 1.645$. Reject H_0. There is sufficient evidence to support the claim that the rate of coronary heart disease is higher for women with the lower levels of physical activity.

11. H_0: $p_1 = p_2$. H_1: $p_1 \neq p_2$. TS: $z = -0.21$. P-value: 0.8336. CV: $z = \pm 1.96$. Fail to reject H_0. There is not sufficient evidence to warrant rejection of the claim that the reversal rate is the same in both years.

12. H_0: $p_1 = p_2$. H_1: $p_1 \neq p_2$. TS: $z = 2.43$. P-value: 0.0150. CV for $\alpha = 0.05$: $z = \pm 1.96$. CV for $\alpha = 0.01$: $z = \pm 2.575$. Difference is significant at the 0.05 level, but not at the 0.01 level.

13. H_0: $p_1 = p_2$. H_1: $p_1 < p_2$. TS: $z = -12.39$. P-value: 0.0001. CV for $\alpha = 0.05$: $z = -1.645$. Reject H_0. There is sufficient sample evidence to support the stated claim.

14. a. H_0: $p_1 = p_2$. H_1: $p_1 \neq p_2$. TS: $z = 12.63$. P-value: 0.0001. CV: $z = 2.33$. Reject H_0. There is sufficient evidence to support the claim that men have a higher rate of red/green color blindness than women.

b. $0.0573 < p_1 - p_2 < 0.117$; because the difference in percentages is estimated to be between 5.73% and 11.7%, there does appear to be a substantial difference.

c. We need a large sample for women so that the prerequisite conditions of $np \geq 5$ and $nq \geq 5$ are both satisfied.

15. $H_0: p_1 = p_2$. $H_1: p_1 > p_2$. TS: $z = 1.07$. P-value: 0.1423. CV: $z = 1.645$. Fail to reject H_0. No. Based on the available evidence, delay taking any action.

16. $H_0: p_1 = p_2$. $H_1: p_1 > p_2$. TS: $z = 3.86$. P-value: 0.0001. Reject H_0. CV: $z = 2.33$. There is sufficient evidence to support the claim that the proportion of drinkers among convicted arsonists is greater than the proportion of drinkers convicted of fraud.

17. With a test statistic of $z = 4.94$ and a P-value of 0.000, reject H_0. There is sufficient evidence to support the claim that the conviction rate for wives is less than the conviction rate for husbands.

18. With a test statistic of $z = -6.74$ and a P-value of 0.0000 (rounded), reject H_0 and conclude that there is sufficient evidence to support the claim that the polio rate is lower for those given the Salk vaccine.

TI-83 Plus

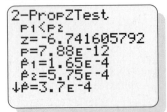

19. $H_0: p_1 = p_2$. $H_1: p_1 \neq p_2$. TS: $z = -2.01$. P-value: 0.0444. CV: $z = \pm 1.96$. Reject H_0. There appears to be a significant difference. Because the Autozone failure rate is lower, it appears to be the better choice.

20. a. $H_0: p_1 = p_2$. $H_1: p_1 \neq p_2$. TS: $z = 10.42$. P-value: 0.0002. CV: $z = \pm 1.96$ (assuming a

a. Is there sufficient evidence to support the claim that men have a higher rate of red/green color blindness than women? Use a 0.01 significance level.

b. Construct the 98% confidence interval for the difference between the color blindness rates of men and women. Does there appear to be a substantial difference?

c. Why would the sample size for women be so much larger than the sample size for men?

15. Seat Belts and Hospital Time A study was made of 413 children who were hospitalized as a result of motor vehicle crashes. Among 290 children who were not using seat belts, 50 were injured severely. Among 123 children using seat belts, 16 were injured severely (based on data from "Morbidity Among Pediatric Motor Vehicle Crash Victims: The Effectiveness of Seat Belts," by Osberg and Di Scala, *American Journal of Public Health,* Vol. 82, No. 3). Is there sufficient sample evidence to conclude, at the 0.05 significance level, that the rate of severe injuries is lower for children wearing seat belts? Based on these results, what action should be taken?

16. Drinking and Crime Karl Pearson, who developed many important concepts in statistics, collected crime data in 1909. Of those convicted of arson, 50 were drinkers and 43 abstained. Of those convicted of fraud, 63 were drinkers and 144 abstained. Use a 0.01 significance level to test the claim that the proportion of drinkers among convicted arsonists is greater than the proportion of drinkers among those convicted of fraud. Does it seem reasonable that drinking might have had an effect on the type of crime? Why?

17. Interpreting a Computer Display A U.S. Department of Justice report (NCJ-156831) included the claim that "in spouse murder cases, wife defendants were less likely to be convicted than husband defendants." Sample data consisted of 277 convictions among 318 husband defendants, and 155 convictions among 222 wife defendants. Test the stated claim and identify one possible explanation for the result. The Minitab results are shown here.

Sample	X	N	Sample p
1	277	318	0.871069
2	155	222	0.698198

Estimate for p(1) − p(2): 0.172871

95% lower bound for p(1) − p(2): 0.113511

Test for p(1) − p(2) = 0 (vs > 0): Z = 4.94 P-value = 0.000

18. Effectiveness of Salk Vaccine for Polio In initial tests of the Salk vaccine, 33 of 200,000 vaccinated children later developed polio. Of 200,000 children vaccinated with a placebo, 115 later developed polio. The TI-83 Plus calculator display is shown here. At the 0.01 significance level, test the claim that the Salk vaccine is effective in lowering the polio rate. Does it appear that the vaccine is effective?

19. Failed Inspections When conducting tests of auto parts stores, the Arizona Department of Weights and Measures conducted 100 inspections of Autozone stores and found that 63% of those inspections failed. Among 37 inspections at NAPA Auto Parts stores, 81% failed. Use a 0.05 significance level to determine whether there is a significant difference between those two rates of failures. Does it appear that either store is a better choice for consumers?

20. Airline Load Factor In a recent year, Southwest Airlines had 3,131,727 aircraft seats available on all of its flights, and 2,181,604 of them were occupied by passengers. America West had 2,091,859 seats available, and 1,448,255 of them were occupied. The percentage of seats occupied is called the *load factor*, so these results show that the load factor is 69.7% (rounded) for Southwest Airlines and 69.2% (rounded) for America West. (The data are from the U.S. Department of Transportation.) Answer the following by assuming that the results are from randomly selected samples.
  a. Test the claim that both airlines have the same load factor.
  b. Given that 69.7% and 69.2% appear to be so obviously close, how do you explain the results from part (a)?
  c. Generalize the key point of this example by completing the following statement: "If two sample sizes are extremely large, even seemingly small differences in sample proportions . . ."

21. Attitudes Toward Marriage In a Time/CNN survey, 24% of 205 single women said that they "definitely want to get married." In the same survey, 27% of 260 single men gave that same response. Construct a 99% confidence interval estimate of the difference between the proportions of single women and single men who definitely want to get married. Is there a gender gap on this issue?

22. Attitudes Toward Marriage Refer to the same sample data in Exercise 21 and use a 0.01 significance level to test the claim that there is a difference between the proportion of men and the proportion of women who definitely want to get married. Does there appear to be a difference?

23. Violent Crime and Age Group The newly appointed head of the state mental health agency claims that a smaller proportion of the crimes committed by persons younger than 21 years of age are violent crimes (when compared to the crimes committed by persons 21 years of age or older). Of 2750 randomly selected arrests of criminals younger than 21 years of age, 4.25% involve violent crimes. Of 2200 randomly selected arrests of criminals 21 years of age or older, 4.55% involve violent crimes (based on data from the *Uniform Crime Reports*). Construct a 95% confidence interval for the difference between the two proportions of violent crimes. Does the confidence interval indicate that there isn't a significant difference between the two rates of violent crimes?

24. Testing Laboratory Gloves The *New York Times* ran an article about a study in which Professor Denise Korniewicz and other Johns Hopkins researchers subjected laboratory gloves to stress. Among 240 vinyl gloves, 63% leaked viruses. Among 240 latex gloves, 7% leaked viruses. At the 0.005 significance level, test the claim that vinyl gloves have a larger virus leak rate than latex gloves.

25. Written Survey and Computer Survey In a study of 1700 teens aged 15–19, half were given written surveys and half were given surveys using an anonymous computer program. Among those given the written surveys, 7.9% say that they carried a gun within the last 30 days. Among those given the computer surveys, 12.4% say that they carried a gun within the last 30 days (based on data from the Urban Institute).
  a. The sample percentages of 7.9% and 12.4% are obviously not equal, but is the difference significant? Explain.
  b. Construct a 99% confidence interval estimate of the difference between the two population percentages, and interpret the result.

26. Adverse Drug Reactions The drug Viagra has become quite well known, and it has had a substantial economic impact on its producer, Pfizer Pharmaceuticals. In prelim-

---

0.05 significance level). Reject H_0. There is sufficient evidence to reject the claim that both airlines have the same load factor.
  b. The large sample sizes make the differences highly significant.
  c. . . . can be significantly different.
21. $-0.135 < p_1 - p_2 < 0.0742$ (using $x_1 = 49$ and $x_2 = 70$); there does not appear to be a gender gap.
22. H_0: $p_1 = p_2$. H_1: $p_1 \neq p_2$. TS: $z = -0.74$. P-value: 0.4592. CV: $z = \pm 2.575$. Fail to reject H_0. There does not appear to be a significant difference.
23. $-0.0144 < p_1 - p_2 < 0.0086$; yes
24. H_0: $p_1 = p_2$. H_1: $p_1 > p_2$. TS: $z = 12.82$. P-value: 0.0001. CV: $z = 2.575$. Reject H_0. There is sufficient evidence to support the claim that vinyl gloves have a higher virus leak rate than latex gloves.
25. a. H_0: $p_1 = p_2$. H_1: $p_1 \neq p_2$. TS: $z = -3.06$. CV: $z = \pm 2.575$ assuming a 0.01 significance level). P-value: 0.0022. Reject H_0: $p_1 = p_2$. There is sufficient evidence to support the claim that the two population percentages are different.
  b. $-0.0823 < p_1 - p_2 < -0.00713$; because the confidence interval limits do not contain 0, there appears to be a significant difference (although it would be better to use a hypothesis test of the null hypothesis $p_1 = p_2$).
26. a. H_0: $p_1 = p_2$. H_1: $p_1 > p_2$. TS: $z = 7.60$. P-value: 0.0001. CV: $z = 2.33$. Reject H_0. There is sufficient evidence to support the claim that the Viagra group has headaches at a rate greater than the rate for those who do not take Viagra.

inary tests for adverse reactions, it was found that when 734 men were treated with Viagra, 16% of them experienced headaches. (There's some real irony there.) Among 725 men in a placebo group, 4% experienced headaches (based on data from Pfizer Pharmaceuticals).

a. Using a 0.01 significance level, is there sufficient evidence to support the claim that among those men who take Viagra, headaches occur at a rate that is greater than the rate for those who do not take Viagra?

b. Construct a 99% confidence interval estimate of the difference between the rate of headaches among Viagra users and the headache rate for those who are given a placebo. What does the confidence interval suggest about the two rates?

27. Poll Refusal Rate Professional pollsters are becoming concerned about the growing rate of refusals among potential survey subjects. In analyzing the problem, there is a need to know if the refusal rate is universal or if there is a difference between the rates for central-city residents and those not living in central cities. Specifically, it was found that when 294 central-city residents were surveyed, 28.9% refused to respond. A survey of 1015 residents not living in a central city resulted in a 17.1% refusal rate (based on data from "I Hear You Knocking But You Can't Come In," by Fitzgerald and Fuller, *Sociological Methods and Research*, Vol. 11, No. 1). At the 0.01 significance level, test the claim that the central-city refusal rate is the same as the refusal rate in other areas.

28. Home Field Advantage When games were sampled from throughout a season, it was found that the home team won 127 of 198 professional *basketball* games, and the home team won 57 of 99 professional *football* games (based on data from "Predicting Professional Sports Game Outcomes from Intermediate Game Scores," by Cooper et al., *Chance*, Vol. 5, No. 3–4). Construct a 95% confidence interval for the difference between the proportions of home wins. Does there appear to be a significant difference between the proportions of home wins? What do you conclude about the home field advantage?

29. Alcohol and Tobacco in Children's Movies Test the claim that the proportion of 25 of 50 randomly selected children's movies showing some use of alcohol is significantly less than the sample proportion of 28 of 50 other such movies showing some use of tobacco. Do the results apply to Data Set 7?

30. Health Survey Refer to Data Set 1 in Appendix B and use the sample data to test the claim that the proportion of men over the age of 30 is equal to the proportion of women over the age of 30.

8-2 Beyond the Basics

31. Interpreting Overlap of Confidence Intervals In the article "On Judging the Significance of Differences by Examining the Overlap Between Confidence Intervals" by Schenker and Gentleman (*The American Statistician*, Vol. 55, No. 3), the authors consider sample data in this statement: "Independent simple random samples, each of size 200, have been drawn, and 112 people in the first sample have the attribute, whereas 88 people in the second sample have the attribute."

a. Use the methods of this section to construct a 95% confidence interval estimate of the difference $p_1 - p_2$. What does the result suggest about the equality of p_1 and p_2?

b. Use the methods of Section 6-2 to construct individual 95% confidence interval estimates for each of the two population proportions. After comparing the overlap between the two confidence intervals, what do you conclude about the equality of p_1 and p_2?

c. Use a 0.05 significance level to test the claim that the two population proportions are equal. What do you conclude?

d. Based on the preceding results, what should you conclude about equality of p_1 and p_2? Which of the three preceding methods is least effective in testing for equality of p_1 and p_2?

32. *Equivalence of Hypothesis Test and Confidence Interval* Two different simple random samples are drawn from two different populations. The first sample consists of 20 people with 10 having a common attribute. The second sample consists of 2000 people with 1404 of them having the same common attribute. Compare the results from a hypothesis test of $p_1 = p_2$ (with a 0.05 significance level) and a 95% confidence interval estimate of $p_1 - p_2$.

33. *Same Proportions with Larger Samples* This section used the sample data in Table 8-1 to test the claim that $p_1 = p_2$ and to construct a confidence interval estimate of $p_1 - p_2$. How are the results affected if the sample data in Table 8-1 are modified so that p_1 becomes 240/2000 instead of 24/200, and p_2 becomes 1470/14,000 instead of 147/1400? Note that both sample proportions remain the same, but the sample sizes are larger. Is there now sufficient evidence to support the claim that the proportion of black drivers stopped by the police is greater than the proportion of white drivers who are stopped?

34. *Testing for Constant Difference* To test the null hypothesis that the difference between two population proportions is equal to a nonzero constant c, use the test statistic

$$z = \frac{(\hat{p}_1 - \hat{p}_2) - c}{\sqrt{\dfrac{\hat{p}_1(1 - \hat{p}_1)}{n_1} + \dfrac{\hat{p}_2(1 - \hat{p}_2)}{n_2}}}$$

As long as n_1 and n_2 are both large, the sampling distribution of the test statistic z will be approximately the standard normal distribution. Refer to Exercise 26 and use a 0.05 significance level to test the claim that the headache rate of Viagra users is 10 percentage points more than the percentage for those who are given a placebo.

35. *Transitivity of Hypothesis Tests* Sample data are randomly drawn from three independent populations, each of size 100. The sample proportions are $\hat{p}_1 = 40/100$, $\hat{p}_2 = 30/100$, and $\hat{p}_3 = 20/100$.

a. At the 0.05 significance level, test $H_0: p_1 = p_2$.

b. At the 0.05 significance level, test $H_0: p_2 = p_3$.

c. At the 0.05 significance level, test $H_0: p_1 = p_3$.

d. In general, if hypothesis tests lead to the conclusions that $p_1 = p_2$ and $p_2 = p_3$ are reasonable, does it follow that $p_1 = p_3$ is also reasonable? Why or why not?

36. *Determining Sample Size* The sample size needed to estimate the difference between two population proportions to within a margin of error E with a confidence level of $1 - \alpha$ can be found as follows. In the expression

$$E = z_{\alpha/2}\sqrt{\frac{p_1q_1}{n_1} + \frac{p_2q_2}{n_2}}$$

replace n_1 and n_2 by n (assuming that both samples have the same size) and replace each of p_1, q_1, p_2, and q_2 by 0.5 (because their values are not known). Then solve for n.

continued

b. $0.491 < p_1 < 0.629$; $0.371 < p_2 < 0.509$; because the confidence intervals do overlap, it appears that $p_1 = p_2$ cannot be rejected.

c. $H_0: p_1 = p_2$. $H_1: p_1 \neq p_2$. TS: $z = 2.40$. P-value: 0.0164. CV: $z \pm 1.96$. Reject H_0. There is sufficient evidence to reject $p_1 = p_2$.

d. Reject $p_1 = p_2$. Least effective: Using the overlap between the individual confidence intervals.

32. Hypothesis test: With a test statistic of $z = -1.9615$, reject $p_1 = p_2$. Confidence interval: $-0.422 < p_1 - p_2 < 0.0180$, which suggests that we should not reject $p_1 = p_2$ (because 0 is included). The hypothesis test and confidence interval lead to different conclusions about the equality of $p_1 = p_2$.

33. The test statistic changes to $z = 2.03$ and the 90% confidence interval limits change to 0.00231 and 0.0277, so there is now sufficient evidence to support the given claim.

34. $H_0: p_1 - p_2 = 0.10$. $H_1: p_1 - p_2 \neq 0.10$. TS: $z = 1.26$. P-value: 0.2076. CV: $z = \pm 1.96$. Fail to reject H_0. There is not sufficient evidence to warrant rejection of the claim that the headache rate for Viagra users is 10 percentage points more than the percentage for those who are given a placebo.

35. a. TS: $z = 1.48$. P-value: 0.1388. CV: $z = \pm 1.96$. Fail to reject $H_0: p_1 = p_2$.
b. Test statistic: $z = 1.63$. P-value: 0.1032. Critical values: $z = \pm 1.96$. Fail to reject $H_0: p_2 = p_3$.
c. Test statistic: $z = 3.09$. P-value: 0.0020. Critical values: $z = \pm 1.96$. Reject $H_0: p_1 = p_3$.
d. No

36. 2135

37. a. No, because the conditions
np ≥ 5 and nq ≥ 5 are not
satisfied for both samples.
b. With 144 people in the
placebo group, 1.8% is not a
possible result.

38. a. For the proportions 0, 0.5,
0.5, 1, $\mu = 0.5$ and $\sigma^2 =$
1/8.
b. For the differences 0, −0.5,
−0.5, −1, 0.5, 0, 0, −0.5,
0.5, 0, 0, −0.5, 1, 0.5, 0.5, 0,
$\mu = 0$ and $\sigma^2 = 1/4$.
c. $\sigma^2_{\hat{p}_D} + \sigma^2_{\hat{p}_Q} = 1/8 + 1/8 =$
1/4, and $\sigma^2_{(\hat{p}_D - \hat{p}_Q)} = 1/4$.

Use this approach to find the size of each sample if you want to estimate the difference between the proportions of men and women who own cars. Assume that you want 95% confidence that your error is no more than 0.03.

37. *Interpreting Drug Test Results* Ziac is a Lederle Laboratories drug developed to treat hypertension. Lederle Laboratories reported that when 221 people were treated with Ziac, 3.2% of them experienced dizziness. It was also reported that among the 144 people in the placebo group, 1.8% experienced dizziness.
    a. Can you use the methods of this section to test the claim that there is a significant difference between the two rates of dizziness? Why or why not?
    b. Can the given information be correct? Why or why not?

38. *Verifying Property of Variances* When discussing the rationale for the methods of this section, it was stated that because $\hat{p}_1$ and $\hat{p}_2$ are each approximated by a normal distribution, $\hat{p}_1 - \hat{p}_2$ will also be approximated by a normal distribution with mean $p_1 - p_2$ and variance $\sigma^2_{(\hat{p}_1 - \hat{p}_2)} = \sigma^2_{\hat{p}_1} + \sigma^2_{\hat{p}_2}$. Do the following to verify that the variance of the *difference* between two independent random variables is the *sum* of their individual variances.
    a. Assuming that two dimes are tossed, list the sample space of four simple events, then find the proportion of heads in each of the four cases. Use the formula $\sigma^2 = \Sigma (x - \mu)^2/N$ to find the variance for the population of the four proportions.
    b. Assuming that two quarters are tossed, the sample space and variance will be the same as in part (a). List the 16 differences in proportions $(\hat{p}_D - \hat{p}_Q)$ that are possible when every outcome of the two dimes is matched with every possible outcome of the two quarters. Find the variance of σ^2 of the population of 16 differences in proportions.
    c. Use the preceding results to verify that the *difference* between two independent random variables is the *sum* of their individual variances.

8-3   Inferences About Two Means: Independent Samples

In this section we consider methods for using sample data from two independent samples to test hypotheses made about two population means or to construct confidence interval estimates of the difference between two population means. We begin by formally defining *independent* and *dependent* samples.

Definitions

Two samples are **independent** if the sample values selected from one population are not related to or somehow paired or matched with the sample values selected from the other population. If there is some relationship so that each value in one sample is paired with a corresponding value in the other sample, the samples are **dependent.** Dependent samples are often referred to as **matched pairs,** or **paired samples.** (We will use *matched pairs*, which best describes the nature of the data.)

EXAMPLE Drug Testing

Independent samples: One group of subjects is treated with the cholesterol-reducing drug Lipitor, while a second and separate group of subjects is given a placebo. These two sample groups are independent because the individuals in the treatment group are in no way paired or matched with corresponding members in the placebo group.

Matched pairs (or dependent samples): The effectiveness of a diet is tested using weights of subjects measured before and after the diet treatment. Each "before" value is matched with the "after" value because each before/after pair of measurements comes from the same person.

This section considers two independent samples, and the following section addresses matched pairs. When using two independent samples to test a claim about the difference $\mu_1 - \mu_2$, or to construct a confidence interval estimate of $\mu_1 - \mu_2$, use the following.

Assumptions

1. The two samples are *independent*.

2. Both samples are *simple random samples*.

3. Either or both of these conditions is satisfied: The two sample sizes are both *large* (with $n_1 > 30$ and $n_2 > 30$) or both samples come from populations having normal distributions. (For small samples, the normality requirement is loose in the sense that the procedures perform well as long as there are no outliers and there isn't strong skewness.)

Hypothesis Test Statistic for Two Means: Independent Samples

$$t = \frac{(\bar{x}_1 - \bar{x}_2) - (\mu_1 - \mu_2)}{\sqrt{\dfrac{s_1^2}{n_1} + \dfrac{s_2^2}{n_2}}}$$

Degrees of Freedom: When finding critical values or *P*-values, use the following for determining the number of degrees of freedom, denoted by df. (Although these two methods typically result in different numbers of degrees of freedom, the conclusion of a hypothesis test is rarely affected by the choice.)

1. In this book we use this simple and conservative estimate: df = smaller of $n_1 - 1$ and $n_2 - 1$.

continued

Note to Instructor

Observant students might wonder why the test statistic uses a standard deviation that includes a *sum* of two terms, whereas the numerator uses a *difference*. Explain that the variance of the differences between two independent random variables equals the variance of the first random variable *plus* the variance of the second random variable. This is discussed later in the section.

2. Statistical software packages typically use the more accurate but more difficult estimate given in Formula 8-1. (We will not use Formula 8-1 for the examples and exercises in this book.)

Formula 8-1
$$\text{df} = \frac{(A + B)^2}{\dfrac{A^2}{n_1 - 1} + \dfrac{B^2}{n_2 - 1}}$$

where
$$A = \frac{s_1^2}{n_1} \quad \text{and} \quad B = \frac{s_2^2}{n_2}$$

***P*-values:** Refer to Table A-3. Use the procedure summarized in Figure 7-6. (See also the subsection of "Finding *P*-Values with the Student *t* Distribution" in Section 7-5.)

Critical values: Refer to Table A-3.

Confidence Interval Estimate of $\mu_1 - \mu_2$: Independent Samples

The confidence interval estimate of the difference $\mu_1 - \mu_2$ is
$$(\bar{x}_1 - \bar{x}_2) - E < (\mu_1 - \mu_2) < (\bar{x}_1 - \bar{x}_2) + E$$

$$\text{where} \quad E = t_{\alpha/2}\sqrt{\frac{s_1^2}{n_1} + \frac{s_2^2}{n_2}}$$

and the number of degrees of freedom df is as described above for hypothesis tests. (In this book, we use df = smaller of $n_1 - 1$ and $n_2 - 1$.)

Because the hypothesis test and confidence interval use the same distribution and standard error, they are equivalent in the sense that they result in the same conclusions. Consequently, the null hypothesis of $\mu_1 = \mu_2$ (or $\mu_1 - \mu_2 = 0$) can be tested by determining whether the confidence interval includes 0. For two-tailed hypothesis tests construct a confidence interval with a confidence level of $1 - \alpha$; but for a one-tailed hypothesis test with significance level α, construct a confidence interval with a confidence level of $1 - 2\alpha$. (See Table 7-2 for common cases.) For example, the claim of $\mu_1 > \mu_2$ can be tested with a 0.05 significance level by constructing a 90% confidence interval.

We will discuss the rationale for the above expressions later in this section. For now, note that the listed assumptions do not include the conditions that the population standard deviations σ_1 and σ_2 must be known, nor do we assume that the two populations have the same standard deviation. Alternative methods based on these additional assumptions are discussed later in the section.

Exploring the Data Sets

We should verify the required assumptions when using two independent samples to make inferences about two population means. Instead of immediately conducting a hypothesis test or constructing a confidence interval, we should first *explore* the two samples using the methods described in Chapter 2. For each of the two samples, we should investigate center, variation, distribution, outliers, and whether the population appears to be changing over time (CVDOT). It could be very helpful to do the following:

- Find descriptive statistics for both data sets, including n, $\bar{x}$, and s.
- Create boxplots of both data sets, drawn on the same scale so that they can be compared.
- Create histograms of both data sets, so that their distributions can be compared.
- Identify any outliers.

EXAMPLE **Hypothesis Test of Bonds and McGwire Home Run Distances** Data Set 30 in Appendix B includes the distances of the home runs hit in record-setting seasons by Mark McGwire and Barry Bonds. Sample statistics, histograms, and boxplots are shown below. Assume that we have simple random samples from large populations and use a 0.05 significance level to test the claim that the distances come from populations with different means.

	McGwire	Bonds
n	70	73
$\bar{x}$	418.5	403.7
s	45.5	30.6

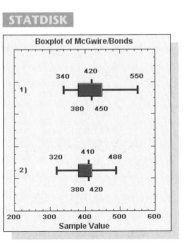

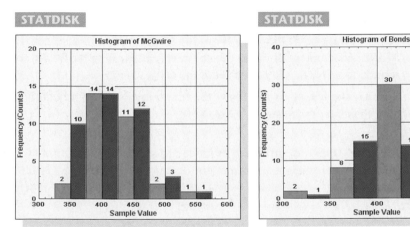

continued

The Placebo Effect

It has long been believed that placebos actually help some patients. In fact, some formal studies have shown that when given a placebo (a treatment with no medicinal value), many test subjects show some improvement. Estimates of improvement rates have typically ranged between one-third and two-thirds of the patients. However, a more recent study suggests that placebos have no real effect. An article in the *New England Journal of Medicine* (Vol. 334, No. 21) was based on research of 114 medical studies over 50 years. The authors of the article concluded that placebos appear to have some effect only for relieving pain, but not for other physical conditions. They concluded that apart from clinical trials, the use of placebos "cannot be recommended."

SOLUTION   In the spirit of exploring the two data sets, we see that the sample means are different, the histograms suggest that the populations have distributions that are approximately normal, and the boxplots seem to display a difference. There do not appear to be any outliers. Let's proceed with a formal hypothesis test to determine whether the difference between the two sample means is really significant. Because it's a bit tricky to find the *P*-value in this example, we will use the traditional method of hypothesis testing.

Step 1:   The claim of different means can be expressed symbolically as $\mu_1 \neq \mu_2$.

Step 2:   If the original claim is false, then $\mu_1 = \mu_2$.

Step 3:   The alternative hypothesis is the expression not containing equality, and the null hypothesis is an expression of equality, so we have

$$H_0: \mu_1 = \mu_2 \qquad H_1: \mu_1 \neq \mu_2 \quad \text{(original claim)}$$

We now proceed with the assumption that $\mu_1 = \mu_2$, or $\mu_1 - \mu_2 = 0$.

Step 4:   The significance level is $\alpha = 0.05$.

Step 5:   Because we have two independent samples and we are testing a claim about the two population means, we use a *t* distribution with the test statistic given earlier in this section.

Step 6:   The test statistic is calculated as follows:

$$t = \frac{(\bar{x}_1 - \bar{x}_2) - (\mu_1 - \mu_2)}{\sqrt{\dfrac{s_1^2}{n_1} + \dfrac{s_2^2}{n_2}}} = \frac{(418.5 - 403.7) - 0}{\sqrt{\dfrac{45.5^2}{70} + \dfrac{30.6^2}{73}}} = 2.273$$

Because we are using a *t* distribution, the critical values of $t = \pm 1.994$ are found from Table A-3. (With an area of 0.05 in two tails, we want the *t* value corresponding to 69 degrees of freedom, which is the smaller of $n_1 - 1$ and $n_2 - 1$ [or the smaller of 69 and 72]. Table A-3 doesn't include 69 degrees of freedom, so we use the closest value of 70 degrees of freedom to get critical *t* values of ± 1.994.) The test statistic, critical values, and critical region are shown in Figure 8-2.

Using STATDISK, Minitab, Excel, or a TI-83 Plus calculator, we can also find that the *P*-value is 0.0248 and the more accurate critical values are $t = \pm 1.995$ (based on df = 69). We could also use Table A-3 to find that with df = 69, the test statistic of $t = 2.273$ corresponds to a *P*-value between 0.02 and 0.05.

Step 7:   Because the test statistic falls within the critical region, reject the null hypothesis $\mu_1 = \mu_2$ (or $\mu_1 - \mu_2 = 0$).

INTERPRETATION   There is sufficient evidence to support the claim that there is a difference between the mean home run distances of Mark McGwire and Barry Bonds.

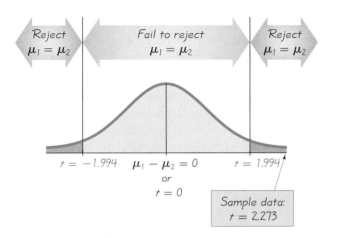

FIGURE 8-2
Distribution of $\bar{x}_1 - \bar{x}_2$ Values

Commercials

Television networks have their own clearance departments for screening commercials and verifying claims. The National Advertising Division, a branch of the Council of Better Business Bureaus, investigates advertising claims. The Federal Trade Commission and local district attorneys also become involved. In the past, Firestone had to drop a claim that its tires resulted in 25% faster stops, and Warner Lambert had to spend $10 million informing customers that Listerine doesn't prevent or cure colds. Many deceptive ads are voluntarily dropped, and many others escape scrutiny simply because the regulatory mechanisms can't keep up with the flood of commercials.

EXAMPLE Confidence Interval for Bonds and McGwire Home Run Distances Using the sample data given in the preceding example, construct a 95% confidence interval estimate of the difference between the mean home run distance of Mark McGwire and the mean home run distance of Barry Bonds.

SOLUTION   We first find the value of the margin of error E. We use $t_{\alpha/2} = 1.994$, which is found in Table A-3 as the t score corresponding to an area of 0.05 in two tails and df = 70. (As in the preceding example, we want the t score corresponding to 69 degrees of freedom, which is the smaller of $n_1 - 1$ and $n_2 - 1$ [or the smaller of 69 and 72]. Table A-3 doesn't include 69 degrees of freedom, so we use the closest value of 70 degrees of freedom.)

$$E = t_{\alpha/2}\sqrt{\frac{s_1^2}{n_1} + \frac{s_1^2}{n_2}} = 1.994\sqrt{\frac{45.5^2}{70} + \frac{30.6^2}{73}} = 13.0$$

We now find the desired confidence interval as follows:

$$(\bar{x}_1 - \bar{x}_2) - E < (\mu_1 - \mu_2) < (\bar{x}_1 - \bar{x}_2) + E$$
$$(418.5 - 403.7) - 13.0 < (\mu_1 - \mu_2) < (418.5 - 403.7) + 13.0$$
$$1.8 < (\mu_1 - \mu_2) < 27.8$$

If we use statistics software or the TI-83 Plus calculator to obtain more accurate results, we get the confidence interval of $1.9 < (\mu_1 - \mu_2) < 27.7$, so we can see that the above confidence interval is quite good.

INTERPRETATION We are 95% confident that the limits of 1.8 ft and 27.8 ft actually do contain the difference between the two population means. This result could be more clearly presented by stating that μ_1 exceeds μ_2 by an amount that is between 1.8 ft and 27.8 ft. Because those limits do not contain 0, this confidence interval suggests that it is very unlikely that the two population means are equal.

Using Statistics to Identify Thieves

Methods of statistics can be used to determine that an employee is stealing, and they can also be used to estimate the amount stolen. The following are some of the indicators that have been used. For comparable time periods, samples of sales have means that are significantly different. The mean sale amount decreases significantly. There is a significant increase in the proportion of "no sale" register openings. There is a significant decrease in the ratio of cash receipts to checks. Methods of hypothesis testing can be used to identify such indicators. (See "How To Catch a Thief" by Manly and Thomson, *Chance*, Vol. 11, No. 4.)

Rationale: Why Do the Test Statistic and Confidence Interval Have the Particular Forms We Have Presented? If the given assumptions are satisfied, the sampling distribution of $\bar{x}_1 - \bar{x}_2$ can be approximated by a t distribution with mean equal to $\mu_1 - \mu_2$ and standard deviation equal to $\sqrt{s_1^2/n_1 + s_2^2/n_2}$. This last expression for the standard deviation is based on the property that the variance of the *differences* between two independent random variables equals the variance of the first random variable *plus* the variance of the second random variable. That is, the variance of sample values $\bar{x}_1 - \bar{x}_2$ will tend to equal $s_1^2/n_1 + s_2^2/n_2$ provided that $\bar{x}_1$ and $\bar{x}_2$ are independent. (See Exercise 31.)

Alternative Method: σ_1 and σ_2 Are Known. In reality, the population standard deviations σ_1 and σ_2 are almost never known, but if they are known, the test statistic and confidence interval are based on the normal distribution instead of the t distribution. See the following.

Test statistic:
$$z = \frac{(\bar{x}_1 - \bar{x}_2) - (\mu_1 - \mu_2)}{\sqrt{\dfrac{\sigma_1^2}{n_1} + \dfrac{\sigma_2^2}{n_2}}}$$

Confidence interval:
$$(\bar{x}_1 - \bar{x}_2) - E < (\mu_1 - \mu_2) < (\bar{x}_1 - \bar{x}_2) + E$$

where
$$E = z_{\alpha/2}\sqrt{\frac{\sigma_1^2}{n_1} + \frac{\sigma_2^2}{n_2}}$$

One alternative method (not used in this book) is to use the above expressions if σ_1 and σ_2 are not known but both samples are large (with $n_1 > 30$ and $n_2 > 30$). This alternative method is used with σ_1 replaced by s_1 and σ_2 replaced by s_2. Because σ_1 and σ_2 are rarely known in reality, this book will not use this alternative method. See Figure 8-3.

Alternative Method: Assume That $\sigma_1 = \sigma_2$ and *Pool* the Sample Variances. Even when the specific values of σ_1 and σ_2 are not known, if it can be assumed that they have the *same* value, the sample variances s_1^2 and s_2^2 can be *pooled* to obtain an estimate of the common population variance σ^2. The **pooled estimate of σ^2** is denoted by s_p^2 and is a weighted average of s_1^2 and s_2^2, which is included in the following box.

Assumptions

1. The two populations have the same standard deviation. That is, $\sigma_1 = \sigma_2$.
2. The two samples are *independent*.
3. Both samples are *simple random samples*.
4. Either or both of these conditions is satisfied: The two sample sizes are both *large* (with $n_1 > 30$ and $n_2 > 30$) or both samples come from populations having normal distributions. (For small samples, the normality requirement is loose in the sense that the procedures perform well as long as there are no outliers and there isn't strong skewness.)

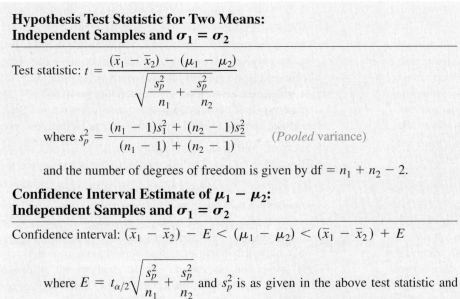

Hypothesis Test Statistic for Two Means:
Independent Samples and $\sigma_1 = \sigma_2$

Test statistic: $t = \dfrac{(\bar{x}_1 - \bar{x}_2) - (\mu_1 - \mu_2)}{\sqrt{\dfrac{s_p^2}{n_1} + \dfrac{s_p^2}{n_2}}}$

where $s_p^2 = \dfrac{(n_1 - 1)s_1^2 + (n_2 - 1)s_2^2}{(n_1 - 1) + (n_2 - 1)}$ *(Pooled* variance)

and the number of degrees of freedom is given by df $= n_1 + n_2 - 2$.

Confidence Interval Estimate of $\mu_1 - \mu_2$:
Independent Samples and $\sigma_1 = \sigma_2$

Confidence interval: $(\bar{x}_1 - \bar{x}_2) - E < (\mu_1 - \mu_2) < (\bar{x}_1 - \bar{x}_2) + E$

where $E = t_{\alpha/2}\sqrt{\dfrac{s_p^2}{n_1} + \dfrac{s_p^2}{n_2}}$ and s_p^2 is as given in the above test statistic and

the number of degrees of freedom is given by df $= n_1 + n_2 - 2$.

Inferences About Two Independent Means

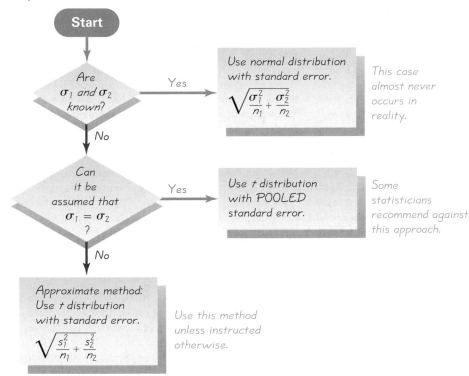

FIGURE 8-3 **Methods for Inferences About Two Independent Means**

If we want to use this method, how do we determine that $\sigma_1 = \sigma_2$? One approach is to use a hypothesis test of the null hypothesis $\sigma_1 = \sigma_2$, as given in Section 8-5, but that approach is not recommended and, in this book, we will not use the preliminary test of $\sigma_1 = \sigma_2$. In the article "Homogeneity of Variance in the Two-Sample Means Test" (by Moser and Stevens, *The American Statistician,* Vol. 46, No. 1), the authors note that we rarely know that $\sigma_1 = \sigma_2$. They analyze the performance of the difference tests by considering sample sizes and powers of the tests. They conclude that more effort should be spent learning the method given near the beginning of this section, and less emphasis should be placed on the method based on the assumption of $\sigma_1 = \sigma_2$. Unless instructed otherwise, we use the following strategy, which is consistent with the recommendations in the article by Moser and Stevens:

Assume that σ_1 and σ_2 are unknown, do *not* assume that $\sigma_1 = \sigma_2$, and use the test statistic and confidence interval given near the beginning of this section. (See Figure 8-3.)

Using Technology

STATDISK   Select the menu items **Analysis, Hypothesis Testing,** and **Mean-Two Independent Samples.** Enter the required values in the dialog box. You have the options of "Not Eq vars: NO POOL," "Eq vars: POOL," or "Prelim F Test." The option of **Not Eq vars: NO POOL** is recommended. (The *F* test is described in Section 8-5.)

Minitab   Minitab Release 14 allows the use of summary statistics or original lists of sample data. If the original sample values are known, enter them in columns C1 and C2. (Earlier versions don't allow the use of summary statistics. If you don't have the original sample values, see the *Minitab Manual* for a way to work around this constraint.) Select **Stat, Basic Statistics,** and **2-Sample t.** Make the required entries in the window that pops up. Use the **Options** button to select a confidence level, enter a claimed value of the difference, or select a format for the alternative hypothesis. The Minitab display includes confidence interval limits.

If the two population variances appear to be equal, Minitab does allow use of a pooled estimate of the common variance. There will be a box next to **Assume equal variances,** and click on that box only if you want to assume that the two populations have equal variances. This approach is not recommended.

Excel   Enter the data for the two samples in columns A and B.

To use the Data Desk XL add-in, click on **DDXL.** Select **Hypothesis Tests** and **2 Var t Test** or select **Confidence Intervals** and **2 Var t Interval.** In the dialog box, click on the pencil icon for the first quantitative column and enter the range of values for the first sample, such as A1:A14. Click on the pencil icon for the second quantitative column and enter the range of values for the second sample. Click on **OK.** Now complete the new dialog box by following the indicated steps. In Step 1, select **2-sample** for the assumption of unequal population variances. (You can also select **Pooled** for the assumption of equal population variances, but this method is not recommended.)

To use Excel's Data Analysis add-in, click on **Tools** and select **Data Analysis.** Select one of the following two items (we recommend the assumption of *unequal variances*):

t-test: Two-Sample Assuming Equal Variances
t-test: Two-Sample Assuming Unequal Variances

Proceed to enter the range for the values of the first sample (such as A1:A14) and then the range of values for the second sample. Enter a value for the claimed difference between the two population means, which will often be 0. Enter the significance level in the Alpha box and click on **OK.** (Excel does not provide a confidence interval.)

TI-83 Plus   The TI-83 Plus calculator does give you the option of using "pooled" variances (if you believe that $\sigma_1^2 = \sigma_2^2$) or not pooling the variances, but we recommend that the variances not be pooled. To conduct tests of the type found in this section, press **STAT**, then select **TESTS** and choose **2-SampTTest** (for a hypothesis test) or **2-SampTInt** (for a confidence interval).

8-3 Basic Skills and Concepts

Independent Samples and Matched Pairs. In Exercises 1–4, determine whether the samples are independent or consist of matched pairs.

Recommended Assignment:
Exercises 1–10, 17–20.

1. The effectiveness of Prilosec for treating heartburn is tested by measuring gastric acid secretion in a group of patients treated with Prilosec and another group of patients given a placebo.

   1. Independent samples

2. The effectiveness of Prilosec for treating heartburn is tested by measuring gastric acid secretion in patients before and after the drug treatment. The data consist of the before/after measurements for each patient.

   2. Matched pairs

3. The accuracy of verbal responses is tested in an experiment in which subjects report their weights and they are then weighed on a physician's scale. The data consist of the reported weight and measured weight for each subject.

   3. Matched pairs

4. The effect of sugar as an ingredient is tested with a sample of cans of regular Coke and another sample of cans of diet Coke.

   4. Independent samples

In Exercises 5–24, assume that the two samples are independent simple random samples selected from normally distributed populations. Do not assume that the population standard deviations are equal.

5. Hypothesis Test for Effect of Marijuana Use on College Students **Many studies have been conducted to test the effects of marijuana use on mental abilities. In one such study, groups of light and heavy users of marijuana in college were tested for memory recall, with the results given below (based on data from "The Residual Cognitive Effects of Heavy Marijuana Use in College Students" by Pope and Yurgelun-Todd, *Journal of the American Medical Association*, Vol. 275, No. 7). Use a 0.01 significance level to test the claim that the population of heavy marijuana users has a lower mean than the light users. Should marijuana use be of concern to college students?**

   Items sorted correctly by light marijuana users: $n = 64$, $\bar{x} = 53.3$, $s = 3.6$

   Items sorted correctly by heavy marijuana users: $n = 65$, $\bar{x} = 51.3$, $s = 4.5$

   5. TS: $t = 2.790$. CV: $t = 2.660$.
      P-value < 0.01.

6. Confidence Interval for Effects of Marijuana Use on College Students **Refer to the sample data used in Exercise 5 and construct a 98% confidence interval for the difference between the two population means. Does the confidence interval include zero? What does the confidence interval suggest about the equality of the two population means?**

   6. $0.3 < \mu_1 - \mu_2 < 3.7$

7. Confidence Interval for Bipolar Depression Treatment **In clinical experiments involving different groups of independent samples, it is important that the groups be similar in the important ways that affect the experiment. In an experiment designed to test the effectiveness of paroxetine for treating bipolar depression, subjects were measured using the Hamilton depression scale with the results given below (based on data from "Double-Blind, Placebo-Controlled Comparison of Imipramine and Paroxetine**

   7. $-0.65 < \mu_1 - \mu_2 < 3.03$

in the Treatment of Bipolar Depression" by Nemeroff et al., *American Journal of Psychiatry*, Vol. 158, No. 6). Construct a 95% confidence interval for the difference between the two population means. Based on the results, does it appear that the two populations have different means? Should paroxetine be recommended as a treatment for bipolar depression?

$$\text{Placebo group:} \qquad n = 43, \bar{x} = 21.57, s = 3.87$$
$$\text{Paroxetine treatment group:} \quad n = 33, \bar{x} = 20.38, s = 3.91$$

8. Hypothesis Test for Bipolar Depression Treatment Refer to the sample data in Exercise 7 and use a 0.05 significance level to test the claim that the treatment group and placebo group come from populations with the same mean. What does the result of the hypothesis test suggest about paroxetine as a treatment for bipolar depression?

9. Hypothesis Test for Magnet Treatment of Pain People spend huge sums of money (currently around $5 billion annually) for the purchase of magnets used to treat a wide variety of pains. Researchers conducted a study to determine whether magnets are effective in treating back pain. Pain was measured using the visual analog scale, and the results given below are among the results obtained in the study (based on data from "Bipolar Permanent Magnets for the Treatment of Chronic Lower Back Pain: A Pilot Study" by Collacott, Zimmerman, White, and Rindone, *Journal of the American Medical Association*, Vol. 283, No. 10). Use a 0.05 significance level to test the claim that those treated with magnets have a greater reduction in pain than those given a sham treatment (similar to a placebo). Does it appear that magnets are effective in treating back pain? Is it valid to argue that magnets might appear to be effective if the sample sizes are larger?

$$\text{Reduction in pain level after magnet treatment:} \quad n = 20, \bar{x} = 0.49, s = 0.96$$
$$\text{Reduction in pain level after sham treatment:} \quad n = 20, \bar{x} = 0.44, s = 1.4$$

10. Confidence Interval for Magnet Treatment of Pain Refer to the sample data from Exercise 9 and construct a 90% confidence interval estimate of the difference between the mean reduction in pain for those treated with magnets and the mean reduction in pain for those given a sham treatment. Based on the result, does it appear that the magnets are effective in reducing pain?

Regular Coke	Diet Coke
$n_1 = 36$	$n_2 = 36$
$\bar{x}_1 = 0.81682$	$\bar{x}_2 = 0.78479$
$s_1 = 0.007507$	$s_2 = 0.004391$

11. Inferences from Samples of Regular Coke and Diet Coke Using Data Set 17 in Appendix B, we find the sample statistics for the weights (in pounds) of regular Coke and diet Coke as listed in the margin.
   a. Use a 0.01 significance level to test the claim that cans of regular Coke and diet Coke have the same mean weight. If there appears to be a difference, try to provide an explanation.
   b. Construct a 99% confidence interval estimate of $\mu_1 - \mu_2$, the difference between the mean weight of regular Coke and the mean weight of diet Coke.

Nicotine (mg)	
Filtered Kings	Nonfiltered Kings
$n_1 = 21$	$n_2 = 8$
$\bar{x}_1 = 0.94$	$\bar{x}_2 = 1.65$
$s_1 = 0.31$	$s_2 = 0.16$

12. Cigarette Filters and Nicotine Refer to the sample results listed in the margin for the measured nicotine contents of randomly selected filtered and nonfiltered king-size cigarettes. All measurements are in milligrams, and the data are from the Federal Trade Commission.
   a. Use a 0.05 significance level to test the claim that king-size cigarettes with filters have a lower mean amount of nicotine than the mean amount of nicotine in nonfiltered king-size cigarettes.

b. Construct a 90% confidence interval estimate of the difference between the two population means.

c. Do cigarette filters appear to be effective in reducing nicotine?

13. Hypothesis Test for Identifying Psychiatric Disorders Are severe psychiatric disorders related to biological factors that can be physically observed? One study used x-ray computed tomography (CT) to collect data on brain volumes for a group of patients with obsessive-compulsive disorders and a control group of healthy persons. Sample results for volumes (in mL) follow for the right cordate (based on data from "Neuroanatomical Abnormalities in Obsessive-Compulsive Disorder Detected with Quantitative X-Ray Computed Tomography," by Luxenberg et al., *American Journal of Psychiatry*, Vol. 145, No. 9). Construct a 99% confidence interval estimate of the difference between the mean brain volume for the healthy control group and the mean brain volume for the obsessive-compulsive group. What does the confidence interval suggest about the difference between the two population means? Based on this result, does it seem that obsessive-compulsive disorders have a biological basis?

$$\text{Control group:} \qquad n = 10, \bar{x} = 0.45, s = 0.08$$
$$\text{Obsessive-compulsive patients:} \quad n = 10, \bar{x} = 0.34, s = 0.08$$

14. Confidence Interval for Identifying Psychiatric Disorders Refer to the sample data in Exercise 13 and use a 0.01 significance level to test the claim that there is a difference between the two population means. Based on the result, does it seem that obsessive-compulsive disorders have a biological basis?

15. Confidence Interval for Effects of Alcohol An experiment was conducted to test the effects of alcohol. The errors were recorded in a test of visual and motor skills for a treatment group of people who drank ethanol and another group given a placebo. The results are shown in the accompanying table (based on data from "Effects of Alcohol Intoxication on Risk Taking, Strategy, and Error Rate in Visuomotor Performance," by Streufert et al., *Journal of Applied Psychology*, Vol. 77, No. 4). Construct a 95% confidence interval estimate of the difference between the two population means. Do the results support the common belief that drinking is hazardous for drivers, pilots, ship captains, and so on? Why or why not?

16. Hypothesis Test for Effects of Alcohol Refer to the sample data in Exercise 15 and use a 0.05 significance level to test the claim that there is a difference between the treatment group and control group. If there is a significant difference, can we conclude that the treatment causes a decrease in visual and motor skills?

17. Hypothesis Test for Eyewitness Accuracy of Police Does stress affect the recall ability of police eyewitnesses? This issue was studied in an experiment that tested eyewitness memory a week after a nonstressful interrogation of a cooperative suspect and a stressful interrogation of an uncooperative and belligerent suspect. The numbers of details recalled a week after the incident are summarized in the margin (based on data from "Eyewitness Memory of Police Trainees for Realistic Role Plays," by Yuille et al., *Journal of Applied Psychology*, Vol. 79, No. 6). Use a 0.01 significance level to test the claim in the article that "stress decreased the amount recalled."

18. Confidence Interval for Eyewitness Accuracy of Police Using the sample data from Exercise 17, construct a 98% confidence interval estimate of the difference between the two population means. Does the result support the claim in the article that "stress decreased the amount recalled"? Why or why not?

13. $-0.01 < \mu_1 - \mu_2 < 0.23$; TI-83 calculator: df = 18 and $0.01 < \mu_1 - \mu_2 < 0.21$.

14. TS: $t = 3.075$. CV: $t = \pm 3.250$. *P*-value is between 0.01 and 0.02. TI-83: df = 18, *P*-value = 0.007.

Treatment Group	Placebo Group
$n_1 = 22$	$n_2 = 22$
$\bar{x}_1 = 4.20$	$\bar{x}_2 = 1.71$
$s_1 = 2.20$	$s_2 = 0.72$

15. $1.46 < \mu_1 - \mu_2 < 3.52$
16. TS: $t = 5.045$. CV: $t = \pm 2.080$. *P*-value = 0.000.
17. TS: $t = 2.879$. CV: $t = 2.429$. *P*-value = 0.006 approx.

Nonstress	Stress
$n_1 = 40$	$n_2 = 40$
$\bar{x}_1 = 53.3$	$\bar{x}_2 = 45.3$
$s_1 = 11.6$	$s_2 = 13.2$

18. $1.3 < \mu_1 - \mu_2 < 14.7$

19. TS: $t = 1.130$. CV: $t = \pm 1.983$.
    P-value $= 0.261$.
20. TS: $t = -2.831$. P-value $=$
    0.013.
21. Filtered: $n_1 = 21$, $\bar{x}_1 = 13.3$,
    $s_1 = 3.7$. Nonfiltered: $n_2 = 8$,
    $\bar{x}_2 = 24.0$, $s_2 = 1.7$. TS: $t =$
    -10.585. CV: $t = -1.895$.
    P-value $= 0.000$.

19. Queen Mary Stowaways Data Set 15 in Appendix B lists the ages of stowaway passengers on westbound and eastbound trips of the *Queen Mary*. When Excel is used with those two sets of ages, the results are as shown below. Is there a significant difference between the ages of stowaway passengers on westbound trips of the *Queen Mary* and the ages of stowaways on eastbound trips?

Excel

t-Test: Two-Sample Assuming Unequal Variances

	Variable 1	Variable 2
Mean	26.71428571	24.84
Variance	103.2987013	67.81189189
Observations	56	75
Hypothesized Mean Difference	0	
df	104	
t Stat	1.130487967	
P(T<=t) one-tail	0.130435525	
t Critical one-tail	1.659636837	
P(T<=t) two-tail	0.26087105	
t Critical two-tail	1.983034963	

TI-83 Plus

```
2-SampTTest
 µ1≠µ2
 t=-2.831291425
 p=.0128658011
 df=14.65189403
 x̄1=70.73333333
↓x̄2=80.75
■
```

20. Reading Levels When a TI-83 Plus calculator is used with the Flesch reading ease scores for Tom Clancy's *The Bear and the Dragon* and J. K. Rowling's *Harry Potter and the Sorcerer's Stone*, the accompanying results are obtained. (The sample data are listed in Data Set 14 in Appendix B.) Is there sufficient evidence to conclude that the mean Flesch reading ease score for Clancy is different than the mean for Rowling?

T 21. Tar and Cigarettes Refer to the sample data listed below and use a 0.05 significance level to test the claim that the mean amount of tar in filtered king-size cigarettes is *less than* the mean amount of tar in nonfiltered king-size cigarettes. All measurements are in milligrams, and the data are from the Federal Trade Commission.

Filtered	16 15 16 14 16 1 16 18 10 14 12
	11 14 13 13 13 16 16 8 16 11
Nonfiltered	23 23 24 26 25 26 21 24

22. $n_1 = 25$, $\bar{x}_1 = 27.1152$, $s_1 =$
    6.8571. $n_2 = 16$, $\bar{x}_2 = 31.7281$,
    $s_2 = 4.2602$. TS: $t = -2.657$.
    CV: $t = \pm 2.131$ (assuming a
    0.05 significance level).
    P-value $= 0.016$ approx.

T 22. Blanking Out on Tests Many students have had the unpleasant experience of panicking on a test because the first question was exceptionally difficult. The arrangement of test items was studied for its effect on anxiety. The following scores are measures of "debilitating test anxiety," which most of us call panic or blanking out (based on data from "Item Arrangement, Cognitive Entry Characteristics, Sex and Test Anxiety as Predictors of Achievement in Examination Performance," by Klimko, *Journal of Experimental Education,* Vol. 52, No. 4.) Is there sufficient evidence to support the claim that the two populations of scores have the same mean? Is there sufficient evidence to support the claim that the arrangement of the test items has an effect on the score?

Questions Arranged from Easy to Difficult					Questions Arranged from Difficult to Easy			
24.64	39.29	16.32	32.83	28.02	33.62	34.02	26.63	30.26
33.31	20.60	21.13	26.69	28.90	35.91	26.68	29.49	35.32
26.43	24.23	7.10	32.86	21.06	27.24	32.34	29.34	33.53
28.89	28.71	31.73	30.02	21.96	27.62	42.91	30.20	32.54
25.49	38.81	27.85	30.29	30.72				

T **23.** BMI of Men and Women Refer to Data Set 1 in Appendix B and test the claim that the mean body mass index (BMI) of men is equal to the mean body mass index of women.

T **24.** Marathon Runners Refer to Data Set 8 in Appendix B and test the claim that the mean age of a male runner in the New York City marathon is equal to the mean age of a female runner in that marathon.

In Exercises 25–28, assume that the two samples are independent simple random samples selected from normally distributed populations. Also assume that the population standard deviations are equal ($\sigma_1 = \sigma_2$) so that the standard error of the differences between means is obtained by pooling *the sample variances.*

25. Confidence Interval with Pooling Do Exercise 7 with the additional assumption that $\sigma_1 = \sigma_2$. How are the results affected by this additional assumption?

26. Hypothesis Test with Pooling Do Exercise 8 with the additional assumption that $\sigma_1 = \sigma_2$. How are the results affected by this additional assumption?

27. Hypothesis Test with Pooling Do Exercise 9 with the additional assumption that $\sigma_1 = \sigma_2$. How are the results affected by this additional assumption?

28. Confidence Interval with Pooling Do Exercise 10 with the additional assumption that $\sigma_1 = \sigma_2$. How are the results affected by this additional assumption?

8-3 Beyond the Basics

29. Effects of an Outlier
 a. Refer to Exercise 19 and include an outlier consisting of a 90-year-old stowaway on a westbound crossing of the *Queen Mary*. Is the hypothesis test dramatically affected by the presence of the outlier?
 b. Refer to Exercise 19 and include an outlier consisting of a 5000-year-old stowaway on a westbound crossing of the *Queen Mary*. Why does the *t* test statistic *decrease* instead of increasing?

30. Effects of Units of Measurement How are the results of Exercise 12 affected if the amounts of nicotine are all converted from milligrams to ounces? In general, does the choice of the scale affect the conclusions about equality of the two population means, and does the choice of scale affect the confidence interval?

31. Verifying a Property of Variances
 a. Find the variance for this *population* of *x* values: 5, 10, 15. (See Section 2-5 for the variance σ^2 of a population.)

23. Men: $n_1 = 40$, $\bar{x}_1 = 25.9975$, $s_1 = 3.4307$. Women: $n_2 = 40$, $\bar{x}_2 = 25.7400$, $s_2 = 6.1656$. TS: $t = 0.231$. CV: $t = \pm 2.024$ (assuming a 0.05 significance level). P-value = 0.842 approx.

24. Men: $n_1 = 111$, $\bar{x}_1 = 39.6126$, $s_1 = 9.4899$. Women: $n_2 = 39$, $\bar{x}_2 = 36.7436$, $s_2 = 11.6838$. TS: $t = 1.382$. CV: $t = \pm 2.024$ (assuming a 0.05 significance level). P-value = 0.170 approx.

25. $-0.62 < \mu_1 - \mu_2 < 3.00$

26. TS: $t = 1.323$. CV: $t = \pm 1.992$.

27. CV: $t = 1.686$. P-value: 0.460 approx.

28. $-0.59 < \mu_1 - \mu_2 < 0.69$

29. a. TS changes substantially from $t = 1.130$ to $t = 1.508$, but it is not enough of a change to cause a change in the conclusion.
 b. The numerator of the test statistic does increase substantially because the sample means do have a much greater difference, but the denominator also increases dramatically because of the increase in the variance of the first sample.

30. The hypothesis test results will not change.

31. a. 50/3
 b. 2/3
 c. 50/3 + 2/3 = 52/3
 d. The range of the $x - y$ values equals the range of the x values plus the range of the y values.

b. Find the variance for this *population* of y values: 1, 2, 3.

c. List the *population* of all possible differences $x - y$, and find the variance of this population.

d. Use the results from parts (a), (b), and (c) to verify that the variance of the *differences* between two independent random variables is the *sum* of their individual variances ($\sigma^2_{x-y} = \sigma^2_x + \sigma^2_y$). (This principle is used to derive the test statistic and confidence interval given in this section.)

e. How is the *range* of the differences $x - y$ related to the range of the x values and the range of the y values?

32. Effect of No Variation in Sample An experiment was conducted to test the effects of alcohol. The breath alcohol levels were measured for a treatment group of people who drank ethanol and another group given a placebo. The results are given in the accompanying table. Use a 0.05 significance level to test the claim that the two sample groups come from populations with the same mean. The given results are based on data from "Effects of Alcohol Intoxication on Risk Taking, Strategy, and Error Rate in Visuomotor Performance," by Streufert et al., *Journal of Applied Psychology*, Vol. 77, No. 4.

Treatment Group	Placebo Group
$n_1 = 22$	$n_2 = 22$
$\bar{x}_1 = 0.049$	$\bar{x}_2 = 0.000$
$s_1 = 0.015$	$s_2 = 0.000$

33. Calculating Degrees of Freedom How is the number of degrees of freedom for Exercises 13 and 14 affected if Formula 8-1 is used instead of selecting the smaller of $n_1 - 1$ and $n_2 - 1$? If Formula 8-1 is used for the number of degrees of freedom instead of the smaller of $n_1 - 1$ and $n_2 - 1$, how are the P-value and the width of the confidence interval affected? In what sense is "df = smaller of $n_1 - 1$ and $n_2 - 1$" a more conservative estimate of the number of degrees of freedom than the estimate obtained with Formula 8-1?

32. TS: $t = 15.322$. CV: $t = \pm 2.080$. P-value = 0.000.

33. df = 18 (instead of 9), CV becomes $t = \pm 2.878$ (instead of ± 3.250), and the confidence interval limits become 0.007 and 0.213, and the P-value is less than 0.01 (instead of between 0.01 and 0.02).

8-4 Inferences from Matched Pairs

In Section 8-3 we defined two samples to be *independent* if the sample values selected from one population are not related to or somehow paired or matched with the sample values selected from the other population. Section 8-3 dealt with inferences about the means of two independent populations, and this section focuses on dependent samples, which we refer to as *matched pairs*. With matched pairs, there is some relationship so that each value in one sample is paired with a corresponding value in the other sample. Here are some typical examples of matched pairs:

- When conducting an experiment to test the effectiveness of a low-fat diet, the weight of each subject is measured once before the diet and once after the diet.

- The effectiveness of an SAT coaching program is tested by giving each subject an SAT test before the program and another equivalent SAT test after the program.

- The accuracy of reported weights is analyzed with a sample of people when, for each person, the reported weight is recorded and the actual weight is measured.

When dealing with inferences about the means of matched pairs, summaries of the relevant assumptions, notation, hypothesis test statistic, and confidence interval are given below. Because the hypothesis test and confidence interval use the same distribution and standard error, they are equivalent in the sense that they result in the same conclusions. Consequently, the null hypothesis that the mean difference equals 0 can be tested by determining whether the confidence interval includes 0. (For two-tailed hypothesis tests construct a confidence interval with a confidence level of $1 - \alpha$; but for a one-tailed hypothesis test with significance level α, construct a confidence interval with a confidence level of $1 - 2\alpha$. [See Table 7-2 for common cases.] For example, the claim that the mean difference is greater than 0 can be tested with a 0.05 significance level by constructing a 90% confidence interval.)

Assumptions

1. The sample data consist of matched pairs.
2. The samples are simple random samples.
3. Either or both of these conditions is satisfied: The number of matched pairs of sample data is large ($n > 30$) or the pairs of values have differences that are from a population having a distribution that is approximately normal. (If there is a radical departure from a normal distribution, we should not use the methods given in this section, but we may be able to use nonparametric methods discussed in Chapter 12.)

Notation for Matched Pairs

d = individual difference between the two values in a single matched pair
μ_d = mean value of the differences d for the *population* of all matched pairs
$\overline{d}$ = mean value of the differences d for the paired *sample* data (equal to the mean of the $x - y$ values)
s_d = standard deviation of the differences d for the paired *sample* data
n = number of *pairs* of data

Hypothesis Test Statistic for Matched Pairs

$$t = \frac{\overline{d} - \mu_d}{\frac{s_d}{\sqrt{n}}}$$

where degrees of freedom = $n - 1$.

P-values and **Critical values:** Table A-3 (t distribution)

continued

Note to Instructor

Point out that the test statistic follows the same basic format of many test statistics: The difference between the sample statistic and the claimed population parameter is divided by the standard deviation of the sample statistics.

Confidence Intervals for Matched Pairs

$$\overline{d} - E < \mu_d < \overline{d} + E$$

where
$$E = t_{\alpha/2}\frac{s_d}{\sqrt{n}}$$

Critical values of $t_{\alpha/2}$: Use Table A-3 with $n - 1$ degrees of freedom.

Exploring the Data Sets

As always, we should avoid the mindless application of any statistical procedure. We should begin by exploring the data to see what might be learned. We should consider the center, variation, distribution, outliers, and any changes that take place over time (CVDOT). Because we want to illustrate the methods of this section with easy calculations, the following examples use sample data consisting of only five matched pairs. We can see that the actual low temperatures appear to be substantially different from the low temperatures that were predicted five days earlier. A normal quantile plot of these five sample differences suggests that they have a distribution that is approximately normal. (These five matched pairs are taken from Data Set 10 in Appendix B, and a histogram of the complete list of 31 differences indicates that the population of differences has a distribution that is approximately normal.) We can see that there aren't any outliers. It is particularly important to consider outliers because their presence can dramatically affect results.

EXAMPLE **Are Forecast Temperatures Accurate?** Table 8-2 consists of five actual low temperatures and the corresponding low temperatures that were predicted five days earlier. The data consist of matched pairs, because each pair of values represents the same day. The forecast temperatures appear to be very different from the actual temperatures, but is there sufficient evidence to conclude that the mean difference is not zero? Use a 0.05 significance level to test the claim that there is a difference between the actual low temperatures and the low temperatures that were forecast five days earlier.

SOLUTION We will follow the same basic method of hypothesis testing that was introduced in Chapter 7, but we will use the above test statistic for matched pairs.

Step 1: The claim that there is a difference between the actual low temperatures and the five day predicted low temperatures can be expressed as $\mu_d \neq 0$.

Step 2: If the original claim is not true, we have $\mu_d = 0$.

Step 3: The null hypothesis must express equality and the alternative hypothesis cannot include equality, so we have

$$H_0: \mu_d = 0 \qquad H_1: \mu_d \neq 0 \quad \text{(original claim)}$$

Table 8-2	Actual and Forecast Temperature				
Actual low	1	−5	−5	23	9
Low forecast five days earlier	16	16	20	22	15
Difference d = actual − predicted	−15	−21	−25	1	−6

Step 4: The significance level is $\alpha = 0.05$.

Step 5: We use the Student t distribution because the required assumptions are satisfied. (We are testing a claim about matched pairs of data, we have two simple random samples, and a normal quantile plot of the sample differences shows that they have a distribution that is approximately normal.)

Step 6: Before finding the value of the test statistic, we must first find the values of $\bar{d}$, and s_d. Refer to Table 8-2 and use the differences of −15, −21, −25, 1, −6 to find these sample statistics: $\bar{d} = -13.2$ and $s_d = 10.7$. Using these sample statistics and the assumption of the hypothesis test that $\mu_d = 0$, we can now find the value of the test statistic.

$$t = \frac{\bar{d} - \mu_d}{\frac{s_d}{\sqrt{n}}} = \frac{-13.2 - 0}{\frac{10.7}{\sqrt{5}}} = -2.759$$

The critical values of $t = \pm 2.776$ are found from Table A-3 as follows: Use the column for 0.05 (Area in Two Tails), and use the row with degrees of freedom of $n - 1 = 4$. Figure 8-4 shows the test statistic, critical values, and critical region.

Step 7: Because the test statistic does not fall in the critical region, we fail to reject the null hypothesis.

continued

Crest and Dependent Samples

In the late 1950s, Procter & Gamble introduced Crest toothpaste as the first such product with fluoride. To test the effectiveness of Crest in reducing cavities, researchers conducted experiments with several sets of twins. One of the twins in each set was given Crest with fluoride, while the other twin continued to use ordinary toothpaste without fluoride. It was believed that each pair of twins would have similar eating, brushing, and genetic characteristics. Results showed that the twins who used Crest had significantly fewer cavities than those who did not. This use of twins as dependent samples allowed the researchers to control many of the different variables affecting cavities.

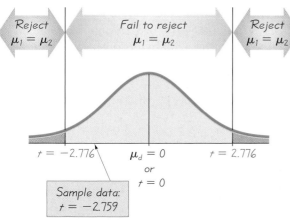

FIGURE 8-4 Distribution of Differences d Between Values in Matched Pairs

Do Air Bags Save Lives?

The National Highway Transportation Safety Administration reported that for a recent year, 3,448 lives were saved because of air bags. It was reported that for car drivers involved in frontal crashes, the fatality rate was reduced 31%; for passengers, there was a 27% reduction. It was noted that "calculating lives saved is done with a mathematical analysis of the real-world fatality experience of vehicles with air bags compared with vehicles without air bags. These are called double-pair comparison studies, and are widely accepted methods of statistical analysis."

Note to Instructor

It would be helpful to review the *interpretation* of confidence intervals in general.

INTERPRETATION The sample data in Table 8-2 do not provide sufficient evidence to support the claim that actual and five-day forecast low temperatures are different. This does *not* establish that the actual and forecast temperatures are equal. Perhaps additional sample data might provide the necessary evidence to conclude that the actual and forecast low temperatures are different. (See Exercise 19 where results for 31 days are used.)

P-Value Method. The preceding example used the traditional method, but the P-value approach could be used by modifying Steps 6 and 7. In Step 6, use the test statistic of $t = -2.759$ and refer to the 4th row of Table A-3 to find that the test statistic (without the negative sign) is between 2.776 and 2.132, indicating that the P-value is between 0.05 and 0.10. Using STATDISK, Excel, Minitab, and a TI-83 Plus calculator, the P-value is found to be 0.0507. We again fail to reject the null hypothesis, because the P-value is greater than the significance level of $\alpha = 0.05$.

EXAMPLE Are Forecast Temperatures Accurate? Using the same sample matched pairs in Table 8-2, construct a 95% confidence interval estimate of μ_d, which is the mean of the differences between actual low temperatures and five-day forecast low temperatures. Interpret the result.

SOLUTION   We use the values of $\bar{d} = -13.2$, $s_d = 10.7$, $n = 5$, and $t_{\alpha/2} = 2.776$ (found from Table A-3 with $n - 1 = 4$ degrees of freedom and an area of 0.05 in two tails). We first find the value of the margin of error E.

$$E = t_{\alpha/2} \frac{s_d}{\sqrt{n}} = 2.776 \cdot \frac{10.7}{\sqrt{5}} = 13.3$$

The confidence interval can now be found.

$$\bar{d} - E < \mu_d < \bar{d} + E$$
$$-13.2 - 13.3 < \mu_d < -13.2 + 13.3$$
$$-26.5 < \mu_d < 0.1$$

INTERPRETATION The result is sometimes expressed as -13.2 ± 13.3 or as $(-26.5, 0.1)$. In the long run, 95% of such samples will lead to confidence interval limits that actually do contain the true population mean of the differences. Note that the confidence interval limits do contain 0, indicating that the true value of μ_d is not significantly different from 0. We cannot conclude that there is a significant difference between the actual and forecast low temperatures.

Using Technology

STATDISK   First enter the data in columns of the Statdisk Data Window. Select **Analysis,** then **Hypothesis Testing,** then **Mean-Matched Pairs.** In the dialog box, choose the format of the claim, enter a significance level, select the columns of data, and then click on **Evaluate.** STATDISK automatically provides confidence interval limits.

Minitab   Enter the paired sample data in columns C1 and C2. Click on **Stat,** select **Basic Statistics,** then select **Paired t.** Enter C1 for the first sample, enter C2 for the second sample, then click on the **Options** box to change the confidence level or form of the alternative hypothesis.

Excel   Enter the paired sample data in columns A and B. To use the Data Desk XL add-in, click on **DDXL.** Select **Hypotheses Tests** and **Paired t Test** or select **Confidence Intervals** and **2 Var t Interval.** In the dialog box, click on the pencil icon for the first quantitative column and enter the range of values for the first sample, such as A1:A14. Click on the pencil icon for the second quantitative column and enter the range of values for the second sample. Click on **OK.** Now complete the new dialog box by following the indicated steps.

To use Excel's Data Analysis add-in, click on **Tools,** found on the main menu bar, then select **Data Analysis,** and proceed to select **t-test Paired Two Sample for Means.** In the dialog box, enter the range of values for each of the two samples, enter the desired population mean difference, and enter the significance level. The displayed results will include the test statistic, the *P*-values for a one-tailed test and a two-tailed test, and the critical values for a one-tailed test and a two-tailed test.

TI-83 Plus   *Caution:* Do not use the menu item **2-SampTTest** because it applies to *independent* samples. Instead, enter the data for the first variable in list L1, enter the data for the second variable in list L2, then clear the screen and enter **L1 − L2 → L3.** Next press **STAT,** then select **TESTS,** and choose the option of **T-Test.** Using the input option of **Data,** enter the indicated data, including list L3, and press **ENTER** when done. A confidence interval can also be found by pressing **STAT,** then selecting **TESTS,** then **TInterval.**

8-4 Basic Skills and Concepts

Calculations for Matched Pairs. In Exercises 1 and 2, assume that you want to use a 0.05 significance level to test the claim that the paired sample data come from a population for which the mean difference is $\mu_d = 0$. Find (a) $\bar{d}$, (b) s_d, (c) the t test statistic, and (d) the critical values.

1.

x	1	1	3	5	4
y	0	2	5	8	0

2.

x	5	3	7	9	2	5
y	5	1	2	6	6	4

3. Using the sample paired data in Exercise 1, construct a 95% confidence interval for the population mean of all differences $x - y$.

4. Using the sample paired data in Exercise 2, construct a 99% confidence interval for the population mean of all differences $x - y$.

5. Self-Reported and Measured Female Heights   As part of the National Health and Nutrition Examination Survey conducted by the Department of Health and Human Services, self-reported heights and measured heights were obtained for females aged 12–16. Listed below are sample results.

Recommended Assignment:
Exercises 1–8, 13–16. (Note: Exercises 5–12, 14, 17–20 each require both a hypothesis test and the construction of a confidence interval.)

1. a. −0.2
   b. 2.8
   c. $t = -0.161$
   d. ±2.776
2. a. 1.2
   b. 3.1
   c. $t = 0.934$
   d. ±2.571
3. $-3.6 < \mu_d < 3.2$

4. $-3.9 < \mu_d < 6.2$

5. a. TS: $t = -0.831$.
      CV: $t = \pm 2.201$.
      P-value: 0.440 approx.
   b. $-1.7 < \mu_d < 0.8$

a. Is there sufficient evidence to support the claim that there is a difference between self-reported heights and measured heights of females aged 12–16? Use a 0.05 significance level.

b. Construct a 95% confidence interval estimate of the mean difference between reported heights and measured heights. Interpret the resulting confidence interval, and comment on the implications of whether the confidence interval limits contain 0.

Reported height	53	64	61	66	64	65	68	63	64	64	64	67
Measured height	58.1	62.7	61.1	64.8	63.2	66.4	67.6	63.5	66.8	63.9	62.1	68.5

6. Self-Reported and Measured Male Heights As part of the National Health and Nutrition Examination Survey conducted by the Department of Health and Human Services, self-reported heights and measured heights were obtained for males aged 12–16. Listed below are sample results.

a. Is there sufficient evidence to support the claim that there is a difference between self-reported heights and measured heights of males aged 12–16? Use a 0.05 significance level.

b. Construct a 95% confidence interval estimate of the mean difference between reported heights and measured heights. Interpret the resulting confidence interval, and comment on the implications of whether the confidence interval limits contain 0.

Reported height	68	71	63	70	71	60	65	64	54	63	66	72
Measured height	67.9	69.9	64.9	68.3	70.3	60.6	64.5	67.0	55.6	74.2	65.0	70.8

7. Effectiveness of SAT Course Refer to the data in the table that lists SAT scores before and after the sample of 10 students took a preparatory course (based on data from the College Board and "An Analysis of the Impact of Commercial Test Preparation Courses on SAT Scores," by Sesnowitz, Bernhardt, and Knain, *American Educational Research Journal,* Vol. 19, No. 3.)

a. Is there sufficient evidence to conclude that the preparatory course is effective in raising scores? Use a 0.05 significance level.

b. Construct a 95% confidence interval estimate of the mean difference between the before and after scores. Write a statement that interprets the resulting confidence interval.

Student	A	B	C	D	E	F	G	H	I	J
SAT score before course (x)	700	840	830	860	840	690	830	1180	930	1070
SAT score after course (y)	720	840	820	900	870	700	800	1200	950	1080

8. Before/After Treatment Results Captopril is a drug designed to lower systolic blood pressure. When subjects were tested with this drug, their systolic blood pressure readings (in mm of mercury) were measured before and after the drug was taken, with the results given in the accompanying table (based on data from "Essential Hypertension: Effect of an Oral Inhibitor of Angiotensin-Converting Enzyme," by MacGregor et al., *British Medical Journal,* Vol. 2).

a. Use the sample data to construct a 99% confidence interval for the mean difference between the before and after readings.

b. Is there sufficient evidence to support the claim that captopril is effective in lowering systolic blood pressure?

Subject	A	B	C	D	E	F	G	H	I	J	K	L
Before	200	174	198	170	179	182	193	209	185	155	169	210
After	191	170	177	167	159	151	176	183	159	145	146	177

9. Effectiveness of Hypnotism in Reducing Pain A study was conducted to investigate the effectiveness of hypnotism in reducing pain. Results for randomly selected subjects are given in the accompanying table (based on "An Analysis of Factors That Contribute to the Efficacy of Hypnotic Analgesia," by Price and Barber, *Journal of Abnormal Psychology,* Vol. 96, No. 1). The values are before and after hypnosis; the measurements are in centimeters on a pain scale.

a. Construct a 95% confidence interval for the mean of the "before–after" differences.

b. Use a 0.05 significance level to test the claim that the sensory measurements are lower after hypnotism.

c. Does hypnotism appear to be effective in reducing pain?

Subject	A	B	C	D	E	F	G	H
Before	6.6	6.5	9.0	10.3	11.3	8.1	6.3	11.6
After	6.8	2.4	7.4	8.5	8.1	6.1	3.4	2.0

10. Measuring Intelligence in Children Mental measurements of young children are often made by giving them blocks and telling them to build a tower as tall as possible. One experiment of block building was repeated a month later, with the times (in seconds) listed in the accompanying table (Based on data from "Tower Building," by Johnson and Courtney, *Child Development,* Vol. 3).

a. Is there sufficient evidence to support the claim that there is a difference between the two times? Use a 0.01 significance level.

b. Construct a 99% confidence interval for the mean of the differences. Do the confidence interval limits contain 0, indicating that there is not a significant difference between the times of the first and second trials?

Child	A	B	C	D	E	F	G	H	I	J	K	L	M	N	O
First trial	30	19	19	23	29	178	42	20	12	39	14	81	17	31	52
Second trial	30	6	14	8	14	52	14	22	17	8	11	30	14	17	15

11. Testing Corn Seeds In 1908, William Gosset published the article "The Probable Error of a Mean" under the pseudonym of "Student" (*Biometrika,* Vol. 6, No. 1). He included the data listed below for two different types of corn seed (regular and kiln dried) that were used on adjacent plots of land. The listed values are the yields of head corn in pounds per acre.

a. Using a 0.05 significance level, test the claim that there is no difference between the yields from the two types of seed.

b. Construct a 95% confidence interval estimate of the mean difference between the yields from the two types of seed.

c. Does it appear that either type of seed is better?

Regular	1903	1935	1910	2496	2108	1961	2060	1444	1612	1316	1511
Kiln dried	2009	1915	2011	2463	2180	1925	2122	1482	1542	1443	1535

12. Parent's Heights Refer to Data Set 2 in Appendix B and use only the data corresponding to male children. Use the paired data consisting of the mother's height and the father's height.

a. Use a 0.01 significance level to test the claim that mothers of male children are shorter than the fathers.

b. Construct a 98% confidence interval estimate of the mean of the differences between the heights of mothers and the heights of fathers.

9. a. $0.69 < \mu_d < 5.56$
b. TS: $t = 3.036$.
   CV: $t = 1.895$.
   P-value: 0.007.
c. Yes

10. a. TS: $t = 2.631$. CV: $t = \pm 2.977$. *P*-value: between 0.01 and 0.02.
b. $-2.9 < \mu_d < 47.5$; yes

11. a. TS: $t = -1.690$. CV: $t = \pm 2.228$. *P*-value: 0.120.
b. $-78.2 < \mu_d < 10.7$
c. No

12. a. TS: $t = -4.716$. CV: $t = -2.539$. *P*-value: 0.000.
b. $-6.8 < \mu_d < -2.0$

13. a. TS: $t = -0.41$. P-value:
    0.691.
  b. 0.3455
14. a. Yes. The TS of $t = 6.431$
    and the P-value of 0.000334
    indicate that the weight
    losses are significant.
  b. 5.1 lb
  c. $3.2 \text{ lb} < \mu_d < 7.1 \text{ lb}$

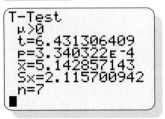

T-Test
μ>0
t=6.431306409
P=3.340322E-4
x̄=5.142857143
Sx=2.115700942
n=7

15. TS: $t = -0.501$. CV: $t = \pm 2.201$. P-value: 0.626.
16. TS: $t = 2.701$. CV: $t = 1.8125$.
    P-value: 0.0111.

13. Treating Motion Sickness The following Minitab display resulted from an experiment in which 10 subjects were tested for motion sickness before and after taking the drug astemizole. The Minitab data column C3 consists of differences in the number of head movements that the subjects could endure without becoming nauseous. (The differences were obtained by subtracting the "after" values from the "before" values.)

  a. Use a 0.05 significance level to test the claim that astemizole has an effect (for better or worse) on vulnerability to motion sickness. Based on the result, would you use astemizole if you were concerned about motion sickness while on a cruise ship?

  b. Instead of testing for some effect (for better or worse), suppose we want to test the claim that astemizole is effective in *preventing* motion sickness? What is the P-value, and what do you conclude?

```
95% CI for mean difference: (-48.8, 33.8)
T-Test of mean difference = 0 (vs not = 0):
T-Value = -0.41 P-Value = 0.691
```

14. Dieting: Interpreting Calculator Display Researchers obtained weight loss data from a sample of dieters using the New World Athletic Club facilities. The before and after weights are recorded, then the differences (before − after) are computed. The TI-83 Plus calculator results are shown for a test of the claim that the diet is effective.

  a. Is there sufficient evidence to support the claim that the diet is effective? Explain.

  b. What is the mean weight loss? Is it large enough to make the diet practical for someone wanting to lose weight?

  c. Use the displayed results to construct a 95% confidence interval for the mean weight loss.

15. Self-Reported and Measured Weights of Males Refer to the Excel display of the results obtained when testing the claim that there is no difference between self-reported weights and measured weights of males aged 12–16. Is there sufficient evidence to support the claim that there is a difference? The data are from the National Health and Nutrition Examination Survey conducted by the Department of Health and Human Services.

Excel

t-Test: Paired Two Sample for Means		
	Variable 1	Variable 2
Mean	133.75	134.7333333
Variance	291.4772727	280.6151515
Observations	12	12
Pearson Correlation	0.919502265	
Hypothesized Mean Difference	0	
df	11	
t Stat	-0.501440942	
P(T<=t) one-tail	0.312972232	
t Critical one-tail	1.795883691	
P(T<=t) two-tail	0.625944463	
t Critical two-tail	2.200986273	

Sample Size, n	11
Difference Mean, $\bar{x}_d$	0.6727
Difference St Dev, s_d	0.8259
Test Statistic, t	2.7014
Critical t	1.8125
P-Value	0.0111

16. Self-Reported and Measured Heights of Male Statistics Students Male statistics students were given a survey that included a question asking them to report their height in inches. They weren't told that their height would be measured, but heights were accurately measured after the survey was completed. Anonymity was maintained, with code numbers used instead of names, so that no personal information would be publicly announced and nobody would be embarrassed by the results. STATDISK results for the claim $\mu_d > 0$ are shown for a 0.05 significance level. Is there sufficient evidence to support a claim that male statistics students exaggerate their heights?

T **17.** Morning and Night Body Temperatures   Refer to Data Set 4 in Appendix B. Use the paired data consisting of body temperatures of women at 8:00 AM and at 12:00 AM on Day 2.
   a. Construct a 95% confidence interval for the mean difference of 8 AM temperatures minus 12 AM temperatures.
   b. Using a 0.05 significance level, test the claim that for those temperatures, the mean difference is 0. Based on the results, do morning and night body temperatures appear to be about the same?

T **18.** Tobacco and Alcohol in Children's Movies   Refer to Data Set 7 in Appendix B. Use the paired data consisting of times that the movies showed tobacco use and the times that they showed alcohol use.
   a. Is there sufficient evidence to conclude that the times are different?
   b. Construct a 99% confidence interval estimate of the mean of the differences between the times of tobacco use and alcohol use. Based on the result, is there a significant difference in the times that children are exposed to tobacco use and the times that they are exposed to alcohol use?

T **19.** Forecast and Actual Temperatures   The examples in this section used only five pairs of sample data so that the calculations would be easy. Refer to Data Set 10 in Appendix B and use all of the actual low temperatures and the low temperatures that were forecast five days earlier.
   a. Using a 0.05 significance level, test the claim that there is no difference between the actual low temperatures and the low temperatures that were forecast five days earlier.
   b. Construct a 95% confidence interval estimate of the mean difference between the actual low temperatures and the low temperatures that were forecast five days earlier.
   c. Compare the results to those obtained in the examples of this section. Does it appear that the forecast low temperatures are accurate?

T **20.** Forecast and Actual Temperatures   Refer to Data Set 10 in Appendix B and use all of the actual low temperatures and the low temperatures that were forecast one day earlier.
   a. Using a 0.05 significance level, test the claim that there is no difference between the actual low temperatures and the low temperatures that were forecast one day earlier.
   b. Construct a 95% confidence interval estimate of the mean difference between the actual low temperatures and the low temperatures that were forecast one day earlier.
   c. Compare the results to those obtained in Exercise 19. Do the one-day forecast values appear to be better than the five-day forecast values?

8-4 Beyond the Basics

21. Effects of an Outlier and Units of Measurement
   a. When using the methods of this section, can an outlier have a dramatic effect on the hypothesis test and confidence interval?

17. a. $-1.40 < \mu_d < -0.17$
    b. TS: $t = -2.840$. CV: $t = \pm 2.228$. P-value < 0.02.

18. a. TS: $t = 1.617$. CV: $t = \pm 2.009$ (assuming a 0.05 significance level). P-value is between 0.10 and 0.20.
    b. $-16.4 < \mu_d < 66.3$

19. a. TS: $t = -2.966$. CV: $t = \pm 2.042$. P-value: 0.006.
    b. $-10.1 < \mu_d < -1.9$
    c. With the larger data set, there is sufficient evidence to conclude that there is a significant difference.

20. a. TS: $t = -1.658$. CV: $t = \pm 2.042$. P-value: 0.100.
    b. $-7.8 < \mu_d < 0.8$
    c. Yes

21. a. Yes
    b. The test is not affected. The CI limits will change from the Fahrenheit scale to equivalent values on the Celsius scale.

b. The examples in this section used temperatures measured in degrees Fahrenheit. If we convert all sample temperatures from Fahrenheit degrees to Celsius degrees, is the hypothesis test affected by such a change in units? Is the confidence interval affected by such a change in units? How?

22. 0.025

22. Confidence Intervals and One-Sided Tests The 95% confidence interval for a collection of paired sample data is $0.0 < \mu_d < 1.2$. Based on this confidence interval, the traditional method of hypothesis testing leads to the conclusion that the claim of $\mu_d > 0$ is supported. What is the smallest possible value of the significance level of the hypothesis test?

23. a. TS: $t = 1.861$. CV: $t = 1.833$. P-value: 0.045
    b. TS: $t = 1.627$. CV: $t = 1.833$. P-value: 0.072.
    c. Yes

23. Using the Correct Procedure
   a. Consider the sample data given below to be matched pairs and use a 0.05 significance level to test the claim that $\mu_d > 0$.
   b. Consider the sample data given below to be two independent samples. Use a 0.05 significance level to test the claim that $\mu_1 > \mu_2$.
   c. Compare the results from parts (a) and (b). Is it critical that the correct method be used? Why or why not?

x	1	3	2	2	1	2	3	3	2	1
y	1	2	1	2	1	2	1	2	1	2

 8-5 Comparing Variation in Two Samples

Because the characteristic of variation among data is extremely important, this section presents a method for using two samples to compare the variances of the two populations from which the samples are drawn. In Section 2-5 we saw that variation in a sample can be measured by the standard deviation, variance, and other measures such as the range and mean absolute deviation. Because standard deviation is a very effective measure of variation, and because it is easier to understand than variance, the early chapters of this book have stressed the use of standard deviation instead of variance. Although the basic procedure of this section is designed for variances, we can also use it for standard deviations. Let's briefly review this relationship between standard deviation and variance: The variance is the square of the standard deviation.

> **Measures of Variation**
>
> s = standard deviation of *sample* s^2 = variance of *sample* (sample standard deviation squared)
> σ = standard deviation of *population* σ^2 = variance of *population* (population standard deviation squared)

The computations of this section will be greatly simplified if we designate the two samples so that s_1^2 represents the *larger* of the two sample variances. Mathematically, it doesn't really matter which sample is designated as Sample 1, so life will be better if we let s_1^2 represent the larger of the two sample variances, as in the test statistic included in the summary box.

Assumptions

1. The two populations are *independent* of each other. (Recall from Section 8-2 that two samples are independent if the sample selected from one population is not related to the sample selected from the other population. The samples are not matched or paired.)

2. The two populations are each *normally distributed*. (This assumption is important because the methods of this section are extremely sensitive to departures from normality.)

Notation for Hypothesis Tests with Two Variances or Standard Deviations

s_1^2 = *larger* of the two sample variances
n_1 = size of the sample with the *larger* variance
σ_1^2 = variance of the population from which the sample with the *larger* variance was drawn

The symbols s_2^2, n_2, and σ_2^2 are used for the other sample and population.

Test Statistic for Hypothesis Tests with Two Variances

$$F = \frac{s_1^2}{s_2^2} \qquad \text{(where } s_1^2 \text{ is the } larger \text{ of the two sample variances)}$$

Critical values: Use Table A-5 to find critical F values that are determined by the following:

1. The significance level α (Table A-5 has four pages of critical values for $\alpha = 0.025$ and 0.05.)
2. **Numerator degrees of freedom** = $n_1 - 1$
3. **Denominator degrees of freedom** = $n_2 - 1$

Note to Instructor

By making s_1^2 the *larger* variance, we avoid the tricky problem of finding critical F values for left-tailed cases. Note, however, that the distribution of s_1^2 / s_2^2 is the F distribution illustrated in Figure 8-5 only if we haven't yet imposed the condition that s_1^2 is the larger of the two sample variances. Once we impose that condition, the ratio of s_1^2 / s_2^2 must be 1 or greater.

Note to Instructor

There will be questions about finding critical values when the number of degrees of freedom is not one of those found in Table A-5. We can usually use the nearest values; it usually doesn't make a difference if we use the value above or below the desired missing value. It only makes a difference if the test statistic is between the table values, in which case you could interpolate, but it's best to run the problem using software or a TI-83 Plus calculator.

For two normally distributed populations with equal variances (that is, $\sigma_1^2 = \sigma_2^2$), the sampling distribution of the test statistic $F = s_1^2 / s_2^2$ is the **F distribution** shown in Figure 8-5 with critical values listed in Table A-5. If you continue to repeat an experiment of randomly selecting samples from two normally distributed populations with equal variances, the distribution of the ratio s_1^2 / s_2^2 of the sample variances is the F distribution.

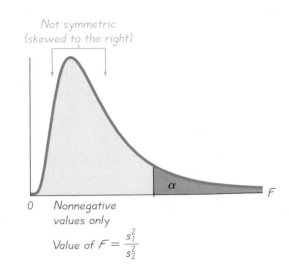

FIGURE 8-5 *F* Distribution

There is a different *F* distribution for each different pair of degrees of freedom for the numerator and the denominator.

In Figure 8-5, note these properties of the *F* distribution:

- The *F* distribution is not symmetric.
- Values of the *F* distribution cannot be negative.
- The exact shape of the *F* distribution depends on two different degrees of freedom.

Critical Values: To find a critical value, first refer to the part of Table A-5 corresponding to α (for a one-tailed test) or $\alpha/2$ (for a two-tailed test), then intersect the column representing the degrees of freedom for s_1^2 with the row representing the degrees of freedom for s_2^2. Because we are stipulating that the larger sample variance is s_1^2, all one-tailed tests will be right-tailed and all two-tailed tests will require that we find only the critical value located to the right. Good news: We have no need to find a critical value separating a left-tailed critical region. (Because the *F* distribution is not symmetric and has only nonnegative values, a left-tailed critical value cannot be found by using the negative of the right-tailed critical value; instead, a left-tailed critical value is found by using the reciprocal of the right-tailed value with the numbers of degrees of freedom reversed. See Exercise 19.)

We often have numbers of degrees of freedom that are not included in Table A-5. We could use linear interpolation to approximate the missing values, but in most cases that's not necessary because the *F* test statistic is either less than the lowest possible critical value or greater than the largest possible critical value. For example, Table A-5 shows that for $\alpha = 0.025$ in the right tail, 20 degrees of freedom for the numerator, and 34 degrees of freedom for the denominator, the critical *F* value is between 2.0677 and 2.1952. Any *F* test statistic below 2.0677 will result in failure to reject the null hypothesis, any *F* test statistic above 2.1952 will result in rejection of the null hypothesis, and interpolation is necessary only if the *F* test statistic happens to fall between 2.0677 and 2.1952. The use of a statistical software package such as STATDISK or Minitab eliminates this problem by providing critical values or *P*-values.

Interpreting the *F* Test Statistic: If the two populations really do have equal variances, then the ratio s_1^2/s_2^2 tends to be close to 1 because s_1^2 and s_2^2 tend to be close in value. But if the two populations have radically different variances, s_1^2 and s_2^2 tend to be very different numbers. Denoting the larger of the sample variances by s_1^2, we see that the ratio s_1^2/s_2^2 will be a large number whenever s_1^2 and s_2^2 are far apart in value. Consequently, a value of *F* near 1 will be evidence in favor of the conclusion that $\sigma_1^2 = \sigma_2^2$, but a large value of *F* will be evidence against the conclusion of equality of the population variances.

Large **values of *F* are evidence against $\sigma_1^2 = \sigma_2^2$.**

Claims about Standard Deviations: The *F* test statistic applies to a claim made about two variances, but we can also use it for claims about two population standard deviations. Any claim about two population standard deviations can be restated in terms of the corresponding variances.

Exploring the Data

Because the requirement of normal distributions is so important and so strict, we should begin by comparing the two sets of sample data by using tools such as histograms, boxplots, and normal quantile plots (see Section 5-7), and we should search for outliers (see Exercise 17). We should find the values of the sample statistics, especially the standard deviations. For example, consider the 36 weights of regular Coke in 36 different cans. (The weights are listed in Data Set 17 in Appendix B.) Shown here are a histogram from a TI-83 Plus calculator and a normal probability plot from Minitab. The histogram shows that the data have a distribution that is approximately normal and that there is one value that is a potential outlier. The normal probability plot, which can be interpreted as if it were a normal quantile plot, shows that the points are reasonably close to a straight line, but they don't fit the straight line perfectly. This data set clearly satisfies a requirement of a distribution that is *approximately* normal, but it isn't so clear that this data set satisfies the stricter requirements of normality that apply to the methods of this section.

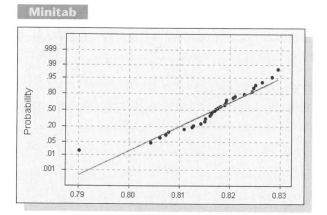

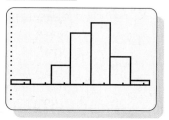

Lower Variation, Higher Quality

Ford and Mazda were producing similar transmissions that were supposed to be made with the same specifications. But the American-made transmissions required more warranty repairs than the Japanese-made transmissions. When investigators inspected samples of the Japanese transmission gearboxes, they first thought that their measuring instruments were defective because they weren't detecting any variability among the Mazda transmission gearboxes. They realized that although the American transmissions were within the specifications, the Mazda transmissions were not only within the specifications, but consistently close to the desired value. By reducing variability among transmission gearboxes, Mazda reduced the costs of inspection, scrap, rework, and warranty repair.

EXAMPLE **Coke Versus Pepsi** Data Set 17 in Appendix B includes the weights (in pounds) of samples of regular Coke and regular Pepsi. The sample statistics are summarized in the accompanying table. Use a 0.05 significance level to test the claim that the weights of regular Coke and the weights of regular Pepsi have the same standard deviation.

	Regular Coke	Regular Pepsi
n	36	36
$\bar{x}$	0.81682	0.82410
s	0.007507	0.005701

SOLUTION   Instead of using the sample standard deviations to test the claim of equal population standard deviations, we will use the sample variances to test the claim of equal population variances. Because we stipulate in this section that the larger variance is denoted by s_1^2, we let $s_1^2 = 0.007507^2$, $n_1 = 36$, $s_2^2 = 0.005701^2$, and $n_2 = 36$. We now proceed to use the traditional method of testing hypotheses as outlined in Figure 7-8.

Step 1:   The claim of equal standard deviations is equivalent to a claim of equal variances, which we express symbolically as $\sigma_1^2 = \sigma_2^2$.

Step 2:   If the original claim is false, then $\sigma_1^2 \neq \sigma_2^2$.

Step 3:   Because the null hypothesis is the statement of equality and because the alternative hypothesis cannot contain equality, we have

$$H_0: \sigma_1^2 = \sigma_2^2 \quad \text{(original claim)} \qquad H_1: \sigma_1^2 \neq \sigma_2^2$$

Step 4:   The significance level is $\alpha = 0.05$.

Step 5:   Because this test involves two population variances, we use the F distribution.

Step 6:   The test statistic is

$$F = \frac{s_1^2}{s_2^2} = \frac{0.007507^2}{0.005701^2} = 1.7339$$

For the critical values, first note that this is a two-tailed test with 0.025 in each tail. As long as we are stipulating that the larger variance is placed in the numerator of the F test statistic, we need to find only the right-tailed critical value. From Table A-5 we see that the critical value of F is between 1.8752 and 2.0739, which we find by referring to 0.025 in the right tail, with 35 degrees of freedom for the numerator and 35 degrees of freedom for the denominator. (STATDISK and Excel provide a critical value of 1.9611.)

Step 7:   Figure 8-6 shows that the test statistic $F = 1.7339$ does not fall within the critical region, so we fail to reject the null hypothesis of equal variances.

INTERPRETATION There is not sufficient evidence to warrant rejection of the claim that the two variances are equal. However, we should recognize that the

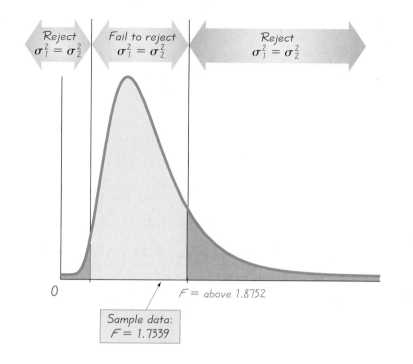

FIGURE 8-6 **Distribution of s_1^2/s_2^2 for Weights of Regular Coke and Regular Pepsi**

F test is extremely sensitive to distributions that are not normally distributed, so this conclusion might make it appear that there is no significant difference between the population variances when there really is a difference that was hidden by nonnormal distributions.

In the preceding example we used a two-tailed test for the claim of equal variances. A right-tailed test would yield the same test statistic of $F = 1.7339$, but a different critical value of *F*.

We have described the traditional method of testing hypotheses made about two population variances. Exercise 18 deals with the *P*-value approach, and Exercise 20 deals with the construction of confidence intervals.

Using Technology

STATDISK Select **Analysis** from the main menu, then select **Hypothesis Testing,** then **StDev-Two Samples.** Enter the required items in the dialog box.

Minitab Either obtain the summary statistics for each sample, or enter the individual sample values in two columns. (If using Minitab Release 13 or earlier, you must use the original lists of raw data. If you know only the summary statistics, generate two lists of sample data as described in the *Minitab Manual.*) Select **Stat,** then **Basic Statistics,** then **2 Variances.** A dialog box will appear. Either select the option of "Samples in different columns" and proceed to enter the column names, or select "Summarized data" and proceed to enter the summary statistics (if using Minitab Release 14). Click on the **Options** bar and enter the confidence level. (Enter 95 for a 0.05 significance level.) Click **OK,** then click **OK** in the main dialog box.

Excel First enter the data from the first sample in the first column A, then enter the values of the second sample in column B. Select **Tools, Data Analysis,** and then **F-Test Two-Sample for Variances.** In the dialog box, enter the range of values for the first sample (such as A1:A36) and the range of values for the second sample. Enter the value of the significance level in the "Alpha" box. Excel will provide the F test statistic, the P-value for the one-tailed case, and the critical F value for the one-tailed case. For a two-tailed test, double the P-value given by Excel.

TI-83 Plus Press the **STAT** key, then select **TESTS,** then **2-SampFTEST.** You can use the summary statistics or you can use the data that are entered as lists.

8-5 Basic Skills and Concepts

Recommended Assignment:
Exercises 1–4, 6, 7, 13, 14.

Hypothesis Test of Equal Variances. In Exercises 1 and 2, test the given claim. Use a significance level of $\alpha = 0.05$ and assume that all populations are normally distributed. Use the traditional method of testing hypotheses outlined in Figure 7-8.

1. TS: $F = 2.2500$. Upper CV: $F = 2.1540$.

1. Claim: The treatment population and the placebo population have different variances.

   Treatment group: $n = 25, \bar{x} = 98.6, s = 0.78$
   Placebo group: $n = 30, \bar{x} = 98.2, s = 0.52$

2. TS: $F = 1.9172$. CV: $F = 2.7186$.

2. Claim: Heights of male statistics students have a larger variance than female statistics students.

   Males: $n = 16, \bar{x} = 68.4, s = 0.54$
   Women: $n = 12, \bar{x} = 63.2, s = 0.39$

3. TS: $F = 2.1267$. CV is between 2.1555 and 2.2341.

3. Hypothesis Test for Magnet Treatment of Pain Researchers conducted a study to determine whether magnets are effective in treating back pain, with results given below (based on data from "Bipolar Permanent Magnets for the Treatment of Chronic Lower Back Pain: A Pilot Study" by Collacott, Zimmerman, White, and Rindone, *Journal of the American Medical Association*, Vol. 283, No. 10). The values represent measurements of pain using the visual analog scale. Use a 0.05 significance level to test the claim that those given a sham treatment (similar to a placebo) have pain reductions that vary more than the pain reductions for those treated with magnets.

   Reduction in pain level after sham treatment: $n = 20, \bar{x} = 0.44, s = 1.4$
   Reduction in pain level after magnet treatment: $n = 20, \bar{x} = 0.49, s = 0.96$

4. Hypothesis Test for Effect of Marijuana Use on College Students In a study of the effects of marijuana use, light and heavy users of marijuana in college were tested for memory recall, with the results given below (based on data from "The Residual Cognitive Effects of Heavy Marijuana Use in College Students" by Pope and Yurgelun-Todd, *Journal of the American Medical Association*, Vol. 275, No. 7). Use a 0.05 significance level to test the claim that the population of heavy marijuana users has a standard deviation different from that of light users.

Items sorted correctly by light marijuana users: $n = 64$, $\bar{x} = 53.3$, $s = 3.6$

Items sorted correctly by heavy marijuana users: $n = 65$, $\bar{x} = 51.3$, $s = 4.5$

5. Weights of Regular Coke and Diet Coke This section included an example about a hypothesis test of the claim that weights of regular Coke and regular Pepsi have the same standard deviation. Test the claim that regular Coke and diet Coke have weights with different standard deviations. Sample weights are found in Data Set 17 in Appendix B, but here are the summary statistics: The sample of 36 weights of regular Coke have a standard deviation of 0.007507 lb, and the sample of 36 weights of diet Coke have a standard deviation of 0.004391 lb. Use a 0.05 significance level. If the results were to show that the standard deviations are significantly different, what would be an important factor that might explain the difference?

6. Axial Loads of Aluminum Cans Data Set 20 in Appendix B includes axial loads (in pounds) of a sample of 175 aluminum cans that are 0.0109 in. thick and another sample of 175 aluminum cans that are 0.0111 in. thick. (An axial load is the maximum weight that the sides can support. It is measured by using a plate to apply increasing pressure to the top of the can until it collapses.) The sample of 0.0109-in. cans has axial loads with a mean of 267.1 lb and a standard deviation of 22.1 lb. The sample of 0.0111-in. cans has axial loads with a mean of 281.8 lb and a standard deviation of 27.8 lb. Use a 0.05 significance level to test the claim that the samples come from populations with the same standard deviation.

7. Cigarette Filters and Nicotine Refer to the sample results listed in the margin for the measured nicotine contents of randomly selected filtered and nonfiltered kingsize cigarettes. All measurements are in milligrams, and the data are from the Federal Trade Commission. Use a 0.05 significance level to test the claim that king-size cigarettes with filters have amounts of nicotine that vary more than the amounts of nicotine in nonfiltered king-size cigarettes.

8. Effects of Alcohol An experiment was conducted to test the effects of alcohol. The errors were recorded in a test of visual and motor skills for a treatment group of people who drank ethanol and another group given a placebo. The results are shown in the accompanying table (Based on data from "Effects of Alcohol Intoxication on Risk Taking, Strategy, and Error Rate in Visuomotor Performance," by Streufert et al., *Journal of Applied Psychology*, Vol. 77, No. 4). Use a 0.05 significance level to test the claim that the treatment group has scores that vary more than the scores of the placebo group.

9. Ages of Faculty and Student Cars Students at the author's college randomly selected 217 student cars and found that they had ages with a mean of 7.89 years and a standard deviation of 3.67 years. They also randomly selected 152 faculty cars and found that they had ages with a mean of 5.99 years and a standard deviation of 3.65 years. Is there sufficient evidence to support the claim that the ages of faculty cars vary less than the ages of student cars?

---

4. TS: $F = 1.5625$. Upper CV is between 1.4327 and 1.6668, but it should be much closer to 1.6668.

5. TS: $F = 2.9228$. Upper CV is between 1.8752 and 2.0739.

6. TS: $F = 1.5824$. Upper CV is approx. 1.4327.

7. TS: $F = 3.7539$. CV: $F = 3.4445$.

8. TS: $F = 9.3364$. CV is between 2.0540 and 2.0960.

Nicotine (mg)	
Filtered Kings	Nonfiltered Kings
$n_1 = 21$	$n_2 = 8$
$\bar{x}_1 = 0.94$	$\bar{x}_2 = 1.65$
$s_1 = 0.31$	$s_2 = 0.16$

Treatment Group	Placebo Group
$n_1 = 22$	$n_2 = 22$
$\bar{x}_1 = 4.20$	$\bar{x}_2 = 1.71$
$s_1 = 2.20$	$s_2 = 0.72$

9. TS: $F = 1.0110$. CV is less than 1.3519 (assuming a 0.05 significance level).

Zinc Supplement Group	Placebo Group
$n = 294$	$n = 286$
$\bar{x} = 3214$	$\bar{x} = 3088$
$s = 669$	$s = 728$

10. TS: $F = 1.1842$. CV is between 1.0000 and 1.3519, but it is much closer to 1.3519.

11. a. TS: $F = 2.1722$. Upper CV is between 1.6668 and 1.8752.
    b. Because they have so many zeros as the lowest values, neither the Wednesday rainfall amounts nor the Sunday rainfall amounts are normally distributed.
    c. Because the populations do not appear to be normally distributed, the conclusion given in part (a) is not necessarily valid.

12. a. TS: $F = 2.4606$. Upper CV is between 1.6668 and 1.8752.
    b. Because they have so many zeros, neither the tobacco times nor the alcohol times are normally distributed.
    c. The conclusion given in part (a) is not necessarily valid.

13. TS: $F = 1.2478$. P-value: 0.6852.

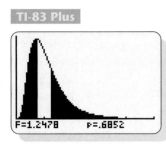

TI-83 Plus

F=1.2478   P=.6852

10. Testing Effects of Zinc A study of zinc-deficient mothers was conducted to determine effects of zinc supplementation during pregnancy. Sample data are listed in the margin (based on data from "The Effect of Zinc Supplementation on Pregnancy Outcome," by Goldenberg et al., *Journal of the American Medical Association*, Vol. 274, No. 6). The weights were measured in grams. Using a 0.05 significance level, is there sufficient evidence to support the claim that the variation of birth weights for the placebo population is greater than the variation for the population treated with zinc supplements?

11. Rainfall on Weekends *USA Today* and other newspapers reported on a study that supposedly showed that it rains more on weekends. The study referred to areas on the East Coast near the ocean. Data Set 11 in Appendix B lists the rainfall amounts in Boston for one year. The 52 rainfall amounts for Wednesday have a mean of 0.0517 in. and a standard deviation of 0.1357 in. The 52 rainfall amounts for Sunday have a mean of 0.0677 in. and a standard deviation of 0.2000 in.
   a. Assuming that we want to use the methods of this section to test the claim that Wednesday and Sunday rainfall amounts have the same standard deviation, identify the F test statistic, critical value, and conclusion. Use a 0.05 significance level.
   b. Consider the prerequisite of normally distributed populations. Instead of constructing histograms or normal quantile plots, simply examine the numbers of days with no rainfall. Are Wednesday rainfall amounts normally distributed? Are Sunday rainfall amounts normally distributed?
   c. What can be concluded from the results of parts (a) and (b)?

12. Tobacco and Alcohol Use in Animated Children's Movies Data Set 7 in Appendix B lists times (in seconds) that animated children's movies show tobacco use and alcohol use. The 50 times of tobacco use have a mean of 57.4 sec and a standard deviation of 104.0 sec. The 50 times of alcohol use have a mean of 32.46 sec and a standard deviation of 66.3 sec.
   a. Assuming that we want to use the methods of this section to test the claim that the times of tobacco use and the times of alcohol use have different standard deviations, identify the F test statistic, critical value, and conclusion. Use a 0.05 significance level.
   b. Consider the prerequisite of normally distributed populations. Instead of constructing histograms or normal quantile plots, simply examine the numbers of movies showing no tobacco or alcohol use. Are the times for tobacco use normally distributed? Are the times for alcohol use normally distributed?
   c. What can be concluded from the results of parts (a) and (b)?

13. Calcium and Blood Pressure Sample data were collected in a study of calcium supplements and their effects on blood pressure. A placebo group and a calcium group began the study with blood pressure measurements (based on data from "Blood Pressure and Metabolic Effects of Calcium Supplementation in Normotensive White and Black Men," by Lyle et al., *Journal of the American Medical Association*, Vol. 257, No. 13). Sample values are listed and a TI-83 Plus display is shown. At the 0.05 significance level, test the claim that the two sample groups come from populations with the same standard deviation. If the experiment requires groups with equal standard deviations, are these two groups acceptable?

Placebo:	124.6	104.8	96.5	116.3	106.1	128.8	107.2	123.1
	118.1	108.5	120.4	122.5	113.6			

Calcium:	129.1	123.4	102.7	118.1	114.7	120.9	104.4	116.3
	109.6	127.7	108.0	124.3	106.6	121.4	113.2	

14. Blanking Out on Tests Many students have had the unpleasant experience of pan-icking on a test because the first question was exceptionally difficult. The arrange-ment of test items was studied for its effect on anxiety. Sample values consisting of measures of "debilitating test anxiety" (which most of us call panic or blanking out) are obtained for a group of subjects with test questions arranged from easy to diffi-cult, and another group with test questions arranged from difficult to easy. (See the list of test scores in Exercise 22 in Section 8-3.) The Excel display is shown below (based on data from "Item Arrangement, Cognitive Entry Characteristics, Sex and Test Anxiety as Predictors of Achievement in Examination Performance," by Klimko, *Journal of Experimental Education,* Vol. 52, No. 4). Use a 0.05 signifi-cance level to test the claim that the two samples come from populations with the same variance.

14. TS: $F = 2.5908$. *P*-value: 0.0599.

	A	B	C
1	F-Test Two-Sample for Variances		
2		Variable 1	Variable 2
3	Mean	27.1152	31.72813
4	Variance	47.01983	18.1489
5	Observations	25	16
6	df	24	15
7	F	2.590782	
8	P(F<=f) one-tail	0.029928	
9	F Critical one-tail	2.287827	

T 15. Comparing Readability of J. K. Rowling and Leo Tolstoy Refer to Data Set 14 in Ap-pendix B, and use a 0.05 significance level to test the claim that for Flesch Reading Ease scores, pages from J. K. Rowling's *Harry Potter and the Sorcerer's Stone* have the same variation as pages from Leo Tolstoy's *War and Peace.*

15. TS: $F = 2.8176$. Upper CV is between 3.5257 and 3.4296.

T 16. Comparing Ages of Marathon Runners Refer to Data Set 8 in Appendix B, and use a 0.05 significance level to test the claim that for the runners in the New York City marathon, men and women have ages with different amounts of variation.

16. TS: $F = 1.5159$. Upper CV of F is approx. 1.6141.

8-5 Beyond the Basics

17. Effect of an Outlier The methods of this section have a fairly strict requirement that the two populations have normal distributions. The presence of an outlier is evidence against a normally distributed population. Repeat Exercise 6 after deleting the outlier of 504 lb in the sample of axial loads of cans that are 0.0111 in. thick. After deleting that outlier, the 174 values have a mean of 280.5 lb and a standard deviation of 22.1 lb. Does the outlier have much of an effect on the results?

17. TS changes from $F = 1.5824$ to $F = 1.0000$.

18. Determining *P*-Values To test a claim about two population variances by using the *P*-value approach, first find the F test statistic, then refer to Table A-5 to determine how it compares to the critical values listed for $\alpha = 0.025$ and $\alpha = 0.05$. Referring to Exercise 5, what can be concluded about the *P*-value?

18. *P*-value < 0.05

19. a. $F_L = 0.2484$, $F_R = 4.0260$
    b. $F_L = 0.2315$, $F_R = 5.5234$
    c. $F_L = 0.1810$, $F_R = 4.3197$

19. Finding Lower Critical F Values In this section, for hypothesis tests that were two-tailed, we found only the upper critical value. Let's denote that value by F_R, where the subscript indicates the critical value for the right tail. The lower critical value F_L (for the left tail) can be found as follows: First interchange the degrees of freedom, and then take the reciprocal of the resulting F value found in Table A-5. (F_R is sometimes denoted by $F_{\alpha/2}$ and F_L is sometimes denoted by $F_{1-\alpha/2}$.) Find the critical values F_L and F_R for two-tailed hypothesis tests based on the following values.
   a. $n_1 = 10$, $n_2 = 10$, $\alpha = 0.05$
   b. $n_1 = 10$, $n_2 = 7$, $\alpha = 0.05$
   c. $n_1 = 7$, $n_2 = 10$, $\alpha = 0.05$

20. $0.41 < \sigma_1^2/\sigma_2^2 < 3.96$ (approximately)

20. Constructing Confidence Intervals In addition to testing claims involving σ_1^2 and σ_2^2, we can also construct confidence interval estimates of the ratio σ_1^2/σ_2^2 using the following expression.

$$\left(\frac{s_1^2}{s_2^2} \cdot \frac{1}{F_R}\right) < \frac{\sigma_1^2}{\sigma_2^2} < \left(\frac{s_1^2}{s_2^2} \cdot \frac{1}{F_L}\right)$$

Here F_L and F_R are as described in Exercise 19. Refer to the data in Exercise 13 and construct a 95% confidence interval estimate for the ratio of the placebo group variance to the calcium-supplement group variance.

Review

In Chapters 6 and 7 we introduced two major concepts of inferential statistics: the estimation of population parameters and the methods of testing hypotheses made about population parameters. Chapters 6 and 7 considered only cases involving a single population, but this chapter considered two samples drawn from two populations.

- Section 8-2 considered inferences made about two population proportions.
- Section 8-3 considered inferences made about the means of two independent populations. Section 8-3 included three different methods, but one method is rarely used because it requires that the two population standard deviations be known. Another method involves pooling the two sample standard deviations to develop an estimate of the standard error, but this method is based on the assumption that the two population standard deviations are known to be equal, and that assumption is often risky. See Figure 8-3 for help in determining which method to apply.
- Section 8-4 considered inferences made about the mean difference for a population consisting of matched pairs.
- Section 8-5 presented methods for testing claims about the equality of two population standard deviations or variances.

Review Exercises

1. Warmer Surgical Patients Recover Better? An article published in *USA Today* stated that "in a study of 200 colorectal surgery patients, 104 were kept warm with blankets and intravenous fluids; 96 were kept cool. The results show: Only 6 of those warmed developed wound infections vs. 18 who were kept cool."
  a. Use a 0.05 significance level to test the claim of the article's headline: "Warmer surgical patients recover better." If these results are verified, should surgical patients be routinely warmed?
  b. If a confidence interval is to be used for testing the claim in part (a), what confidence level should be used?
  c. Using the confidence level from part (b), construct a confidence interval estimate of the difference between the two population proportions.
  d. In general, if a confidence interval estimate of the difference between two population proportions is used to test some claim about the proportions, will the conclusion based on the confidence interval always be the same as the conclusion from a standard hypothesis test?

2. Historical Data Set In 1908, "Student" (William Gosset) published the article "The Probable Error of a Mean" (*Biometrika*, Vol. 6, No. 1). He included the data listed below for two different types of straw seed (regular and kiln dried) that were used on adjacent plots of land. The listed values are the yields of straw in cwt per acre.
  a. Using a 0.05 significance level, test the claim that there is no difference between the yields from the two types of seed.
  b. Construct a 95% confidence interval estimate of the mean difference between the yields from the two types of seed.
  c. Does it appear that either type of seed is better?

Regular	19.25	22.75	23	23	22.5	19.75	24.5	15.5	18	14.25	17
Kiln Dried	25	24	24	28	22.5	19.5	22.25	16	17.25	15.75	17.25

3. Brain Volume and Psychiatric Disorders A study used x-ray computed tomography (CT) to collect data on brain volumes for a group of patients with obsessive-compulsive disorders and a control group of healthy persons. Sample results (in mL) are given below for total brain volumes (based on data from "Neuroanatomical Abnormalities in Obsessive-Compulsive Disorder Detected with Quantitative X-Ray Computed Tomography," by Luxenberg et al., *American Journal of Psychiatry*, Vol. 145, No. 9).
  a. Construct a 95% confidence interval for the difference between the mean brain volume of obsessive-compulsive patients and the mean brain volume of healthy persons. Assume that the two populations have *unequal* variances.
  b. Assuming that the population variances are *unequal*, use a 0.05 significance level to test the claim that there is no difference between the mean for obsessive-compulsive patients and the mean for healthy persons.
  c. Based on the results from parts (a) and (b), does it appear that the total brain volume can be used as an indicator of obsessive-compulsive disorders?

$$\text{Obsessive-compulsive patients:} \quad n = 10, \bar{x} = 1390.03, s = 156.84$$
$$\text{Control group:} \quad\quad\quad\quad\quad\quad n = 10, \bar{x} = 1268.41, s = 137.97$$

1. a. TS: $z = -2.82$. CV: $z = -1.645$. *P*-value: 0.0024.
   b. 90%
   c. $-0.205 < p_1 - p_2 < -0.0543$
   d. No, the conclusions may be different.

2. a. TS: $t = -1.532$. CV: $t = \pm 2.228$. *P*-value: 0.164 approx.
   b. $-2.7 < \mu_d < 0.5$
   c. No, there is not a significant difference.

3. a. $-27.80 < \mu_1 - \mu_2 < 271.04$; TI-83 uses df = 17.7 to get limits of -17.32 and 260.56.
   b. TS: $t = 1.841$. CV: $t = \pm 2.262$. *P*-value: 0.106 approx.
   c. No

4. TS: $F = 1.2922$. Upper CV: $F = 4.0260$.

5. TS: $t = -3.500$. CV: $t = \pm 2.365$. P-value: 0.010.

6. TS: $t = 2.169$. CV: $t = \pm 1.968$ approx. P-value is between 0.01 and 0.025.

Zinc Supplement Group	Placebo Group
$n = 294$	$n = 286$
$\bar{x} = 3214$	$\bar{x} = 3088$
$s = 669$	$s = 728$

7. TS: $z = 2.41$. CV: $z = 1.645$. P-value: 0.0080.

8. a. TS: $t = 2.301$. CV: $t = \pm 2.262$ (assuming a 0.05 significance level). P-value: 0.046.

   b. $0.0 < \mu_d < 4.0$

4. **Variation of Brain Volumes** Use the same sample data given in Exercise 3 with a 0.05 significance level to test the claim that the populations of total brain volumes for obsessive-compulsive patients and the control group have different amounts of variation.

5. **Carbon Monoxide and Cigarettes** Refer to the given data for the measured amounts of carbon monoxide (CO) from samples of filtered and nonfiltered king-size cigarettes. All measurements are in milligrams, and the data are from the Federal Trade Commission. Use a 0.05 significance level to test the claim that the mean amount of carbon monoxide in filtered king-size cigarettes is equal to the mean amount of carbon monoxide for nonfiltered king-size cigarettes. Based on this result, are cigarette filters effective in reducing carbon monoxide?

Filtered:	14 12 14 16 15 2 14 16 11 13 13 12 13 12 13 14 14 14 9 17 12
Nonfiltered:	14 15 17 17 16 16 14 16

6. **Zinc for Mothers** A study of zinc-deficient mothers was conducted to determine whether zinc supplementation during pregnancy results in babies with increased weights at birth. Sample data are listed in the margin (based on data from "The Effect of Zinc Supplementation on Pregnancy Outcome," by Goldenberg et al., *Journal of the American Medical Association*, Vol. 274, No. 6). The weights were measured in grams. Using a 0.05 significance level, is there sufficient evidence to support the claim that zinc supplementation does result in increased birth weights?

7. **People Helping People** In a study of people who stop to help drivers with disabled cars, researchers hypothesized that more people would stop to help someone if they first saw another driver with a disabled car getting help. In one experiment, 2000 drivers first saw a woman being helped with a flat tire and then saw a second woman who was alone, farther down the road, with a flat tire; 2.90% of those 2000 drivers stopped to help the second woman. Among 2000 other drivers who did not see the first woman being helped, only 1.75% stopped to help (based on data from "Help on the Highway," by McCarthy, *Psychology Today*). At the 0.05 significance level, test the claim that the percentage of people who stop after first seeing a driver with a disabled car being helped is greater than the percentage of people who stop without first seeing someone else being helped.

8. **Testing Effects of Physical Training** A study was conducted to investigate some effects of physical training. Sample data are listed below, with all weights given in kilograms. (See "Effect of Endurance Training on Possible Determinants of VO_2 During Heavy Exercise," by Casaburi et al., *Journal of Applied Physiology*, Vol. 62, No. 1.)
   a. Is there sufficient evidence to conclude that there is a difference between the pretraining and posttraining weights? What do you conclude about the effect of training on weight?
   b. Construct a 95% confidence interval for the mean of the differences between pretraining and posttraining weights.

Pretraining:	99	57	62	69	74	77	59	92	70	85
Posttraining:	94	57	62	69	66	76	58	88	70	84

Cumulative Review Exercises

1. **Speeding Tickets for Men and Women** The data in the accompanying table were obtained through a survey of randomly selected subjects (based on data from R. H. Bruskin Associates).
   a. If one of the survey subjects is randomly selected, find the probability of getting someone ticketed for speeding.
   b. If one of the survey subjects is randomly selected, find the probability of getting a man or someone ticketed for speeding.
   c. Find the probability of getting someone ticketed for speeding, given that the selected person is a man.
   d. Find the probability of getting someone ticketed for speeding, given that the selected person is a woman.
   e. Use a 0.05 significance level to test the claim that the percentage of women ticketed for speeding is less than the percentage of men. Can we conclude that men generally speed more than women?

	Ticketed for Speeding Within the Last Year?	
	Yes	No
Men	26	224
Women	27	473

2. **Cell Phones and Crashes: Analyzing Newspaper Report** In an article from the Associated Press, it was reported that researchers "randomly selected 100 New York motorists who had been in an accident and 100 who had not. Of those in accidents, 13.7 percent owned a cellular phone, while just 10.6 percent of the accident-free drivers had a phone in the car." Analyze these results.

3. **Clinical Tests of Viagra** In clinical tests of adverse reactions to the drug Viagra, 4.0% of the 734 subjects in the treatment group experienced nasal congestion, but 2.1% of the 725 subjects in the placebo group experienced nasal congestion (based on data from Pfizer Pharmaceuticals).
   a. Construct a 95% confidence interval estimate of the proportion of Viagra users who experience nasal congestion.
   b. Construct a 95% confidence interval estimate of the proportion of placebo users who experience nasal congestion.
   c. Construct a 95% confidence interval estimate of the difference between the two population proportions.
   d. When attempting to determine whether there is a significant difference between the two population proportions, which of the following methods is best?
      i. Determine whether the confidence intervals in parts (a) and (b) overlap.
      ii. Determine whether the confidence interval in part (c) contains the value of zero.
      iii. Conduct a hypothesis test of the null hypothesis $p_1 = p_2$, using a 0.05 significance level.
      iv. The methods in parts (i), (ii), and (iii) are all equally good.

T 4. **Marathon Finishers** Refer to the results for female marathon runners in Data Set 8 in Appendix B.
   a. Find the proportion of female runners who finished the New York City marathon, then test the claim that the proportion is less than 0.5.
   b. For the times of the female finishers, find the mean, median, standard deviation, describe the nature of the distribution, and identify any outliers.
   c. Using a 0.05 significance level, test the claim that female finishers have a mean running time less than 5 hours.

continued

1. a. 0.0707
   b. 0.369
   c. 0.104
   d. 0.0540
   e. TS: $z = -2.52$. CV: $z = -1.645$. *P*-value: 0.0059.
2. There must be an error, because the rates of 13.7% and 10.6% are not possible with sample sizes of 100.
3. a. $0.0254 < p < 0.0536$ (using $x = 29$)
   b. $0.0103 < p < 0.0311$ (using $x = 15$)
   c. $0.00133 < p_1 - p_2 < 0.0363$
   d. method (iii)
4. a. TS: $z = -5.88$. CV: $z = -1.645$ (assuming a 0.05 significance level). *P*-value: 0.0001.
   b. $\bar{x} = 17{,}198.3$ sec; median = 16,792 sec; $s = 3107.2$; distribution is approximately normal; no outliers
   c. TS: $t = -1.611$. CV: $t = -1.686$. *P*-value: 0.059.
   d. TS: $t = 3.101$. CV: $t = \pm 2.024$. *P*-value: 0.004.
   e. Using the sample proportions of 39/150 and 111/150 results in the pooled value of $\bar{p} = 150/300$, which assumes that the total sample size is 300 instead of 150.

d. Given that the male runners included in Data Set 8 have a mean time of 15,415.2 sec and a standard deviation of 3036.8 sec, use a 0.05 significance level to test the claim that the mean time of males is different from the mean time of females.

e. Identify the proportions of female runners and male runners. What is wrong with using those two sample proportions with the methods of Section 8-2 in a test of the claim that the population proportions of females and males are different?

Cooperative Group Activities

1. Out-of-class activity Are estimates influenced by anchoring numbers? Refer to the related Chapter 2 Cooperative Group Activity. In Chapter 2 we noted that, according to author John Rubin, when people must estimate a value, their estimate is often "anchored" to (or influenced by) a preceding number. In that Chapter 2 activity, some subjects were asked to quickly estimate the value of $8 \times 7 \times 6 \times 5 \times 4 \times 3 \times 2 \times 1$, and others were asked to quickly estimate the value of $1 \times 2 \times 3 \times 4 \times 5 \times 6 \times 7 \times 8$. In Chapter 2, we could compare the two sets of results by using statistics (such as the mean) and graphs (such as boxplots). The methods of Chapter 8 now allow us to compare the results with a formal hypothesis test. Specifically, collect your own sample data and test the claim that when we begin with larger numbers (as in $8 \times 7 \times 6$), our estimates tend to be larger.

2. In-class activity Divide into groups according to gender, with about 10 or 12 students in each group. Each group member should record his or her pulse rate by counting the number of heartbeats in 1 minute, and the group statistics (n, $\bar{x}$, s) should be calculated. The groups should test the null hypothesis of no difference between their mean pulse rate and the mean of the pulse rates for the population from which subjects of the same gender were selected for Data Set 1 in Appendix B.

3. Out-of-class activity Randomly select a sample of male students and a sample of female students and ask each selected person whether they support a death penalty for people convicted of murder. Use a formal hypothesis test to determine whether there is a gender gap on this issue. Also, keep a record of the responses according to the gender of the person asking the question. Does the response appear to be influenced by the gender of the interviewer?

4. Out-of-class activity Use a watch to record the waiting times of a sample of McDonald's customers and the waiting times of a sample of Burger King customers. Use a hypothesis test to determine whether there is a significant difference.

Technology Project

STATDISK, Minitab, Excel, and the TI-83 Plus calculator are all capable of generating normally distributed data drawn from a population with a specified mean and standard deviation. Generate two sets of sample data that represent simulated IQ scores, as shown below.

IQ Scores of Treatment Group: Generate 10 sample values from a normally distributed population with mean 100 and standard deviation 15.

IQ Scores of Placebo Group: Generate 12 sample values from a normally distributed population with mean 100 and standard deviation 15.

Statdisk	Select **Data,** then select **Normal Generator.**
Minitab	Select **Calc, Random Data, Normal.**
Excel	Select **Tools, Data Analysis, Random Number Generator,** and be sure to select **Normal** for the distribution.
TI-83 Plus	Press **MATH,** select **PRB,** then select **6:randNorm(** and proceed to enter the mean, the standard deviation, and the number of scores (such as 100, 15, 10).

You can see from the way the data are generated that both data sets really come from the same population, so there should be no difference between the two sample means.

a. After generating the two data sets, use a 0.10 significance level to test the claim that the two samples come from populations with the same mean.

b. If this experiment is repeated many times, what is the expected percentage of trials leading to the conclusion that the two population means are different? How does this relate to a type I error?

c. If your generated data should lead to the conclusion that the two population means are different, would this conclusion be correct or incorrect in reality? How do you know?

d. If part (a) is repeated 20 times, what is the probability that none of the hypothesis tests leads to rejection of the null hypothesis?

e. Repeat part (a) 20 times. How often was the null hypothesis of equal means rejected? Is this the result you expected?

from DATA to DECISION

Critical Thinking: The fear of flying

The lives of many people are affected by a fear that prevents them from flying. Sports announcer John Madden gained notoriety as he crossed the country by rail or motor home, traveling from one football stadium to another. The Marist Institute for Public Opinion conducted a poll of 1014 adults, 48% of whom were men. The results are depicted in the accompanying illustration from *USA Today*. The poll results show that 12% of the men and 33% of the women fear flying.

Analyzing the Results

1. Is there sufficient evidence to conclude that there is a significant difference between the percentage of men and the percentage of women who fear flying?

2. Construct a 95% confidence interval estimate of the difference between the percentage of men and the percentage of women who fear flying. Do the confidence interval limits con-

Just plane scared
While 47% of adults believe flying is the safest way to travel (vs. 39% who say cars and 14% for trains), about 41 million are afraid. Percent of Americans who fear flying:

All adults 22%
Men 12%
Women 33%

Source: Marist Institute for Public Opinion   By Scott Boeck and Genevieve Lynn, USA TODAY

USA Today Snapshot. "A look at statistics that shape the nation."

continued

tain 0, and what is the significance of whether they do or do not?

3. Construct a 95% confidence interval for the percentage of men who fear flying.

4. Based on the result from Exercise 3, complete the following statement, which is typical of the statement that would be reported in a newspaper or magazine: "Based on the Marist Institute for Public Opinion poll, the percentage of men who fear flying is 22% with a margin of error of _____."

5. Examine the completed statement in Exercise 4. What important piece of information should be included, but is not included?

6. In a separate Gallup poll, 1001 randomly selected adults were asked this question: "If you had to fly on an airplane tomorrow, how would you describe your feelings about fly-

ing? Would you be—very afraid, somewhat afraid, not very afraid, or not afraid at all?" Here are the responses: Very afraid (18%), somewhat afraid (26%), not very afraid (17%), not afraid at all (38%), and no opinion (1%). Are these Gallup poll results consistent with those obtained by the survey conducted by the Marist Institute for Public Opinion? Explain. Can discrepancies be explained by the fact that the Gallup survey was conducted after the terrorist attacks of September 11, 2001, whereas the other survey was conducted before that date?

7. Does the *USA Today* illustration do a good job of depicting the survey results? Construct a graph which more clearly illustrates the survey results.

INTERNET PROJECT   Comparing Populations

The previous chapter showed you methods for testing hypotheses about a single population. This chapter expanded on those ideas, allowing you to test hypotheses about the relationships between two populations. In a similar fashion, the Internet Project for this chapter differs from that of the previous chapter in that you will need data for two populations or groups to conduct investigations. Go to the Internet Project for this chapter at

http://www.aw.com/triola

There you will find several hypothesis-testing problems involving multiple populations. In these problems, you will analyze salary fairness, population demographics, and a traditional superstition. In each case you will formulate the problem as a hypothesis test, collect relevant data, then conduct and summarize the appropriate test.

Statistics @ Work

*"It would be impossible to conduct
archaeological research without at least a working
knowledge of basic statistics."*

Mark T. Lycett

Kathleen Morrison

Mark T. Lycett and Kathleen Morrison are both on the faculty of the Department of Anthropology at the University of Chicago.

Dr. Lycett's research deals with issues of economic, social, and political transformation associated with Spanish colonialism in the southwestern United States, and Dr. Morrison's research in southern India deals with problems of agricultural change, imperialism, and regional economic organization.

How important is the use of statistics in archaeology?

It would be impossible to conduct archaeological research without at least a working knowledge of basic statistics.

What concepts of statistics do you use?

Archaeologists make extensive use of both descriptive and inferential statistics on a daily basis. Exploratory data analysis using a variety of graphical and numerical summaries is increasingly common in modern archaeology. Archaeological problems routinely include studies of association for categorical variables, hypothesis testing for both 2-sample and *k*-sample data, correlation and regression problems, and a suite of nonparametric approaches.

Please give a specific example illustrating the use of statistics in your work.

We have explored the size distribution of ancient grass pollen grains to investigate changes in agriculture in both the New and Old Worlds during the first centuries of European colonial expansion. Although almost all important crops are grasses with morphologically similar pollen, New World staple crops (corn) have much larger pollen grains than wild grasses, and Old World crops (principally wheat, barley, and rice) are intermediate in size. By studying the size distribution of reference samples of these staple crops as well as fossil grass pollen from archaeological contexts, we

have been able to specify the range of crops introduced and grown at colonial period sites in New Mexico and India.

Our data have been used to make inferences about the number and kind of archaeological sites that existed in our study areas; to reconstruct ancient patterns of vegetation, agriculture, and economy; and to study the effects of colonialism and imperialism on local social, economic, and religious practices.

Is your use of probability and statistics increasing, decreasing, or remaining stable?

Both the number and variety of statistical applications in archaeology is increasing, particularly as more sophisticated spatial databases become available through the widespread use of Geographic Information Systems technology.

In terms of statistics, what would you recommend for prospective employees?

When we were college students, we understood that statistics would be a part of our professional lives, but we never imagined the degree to which we would use it on a daily basis. Undergraduates interested in archaeology should begin with an introductory course in probability and statistics. Those with professional or academic goals should consider more advanced undergraduate or graduate level course work in quantitative data analysis.

Correlation and Regression

9-1 Overview

9-2 Correlation

9-3 Regression

9-4 Variation and Prediction Intervals

9-5 Multiple Regression

9-6 Modeling

Should boating restrictions be imposed to reduce the killing of manatees?

Manatees, also called "sea cows," are large mammals that live underwater, often in or near waterways with considerable boat traffic. In Florida, manatee deaths from powerboat encounters have been at the center of considerable controversy between environmentalists and boat operators. Recently, Andrew Revkin wrote the *New York Times* article "How Endangered a Species?" He wrote that "The [manatee] deaths from boating have continued despite the creation of a network of refuges and slow-speed boating zones, and a result is one of the country's most intense debates over an endangered species." The article included two side-by-side graphs similar to those shown in Figure 9-1. (The *New York Times* graphs included data from 1976 through 2000.) The graphs in Figure 9-1 reflect the data in Table 9-1 (based on data from the Florida Department of Highway Safety and Motor Vehicles and the Florida Marine Research Institute).

Here are some key issues addressed in this chapter:

- In comparison to Figure 9-1, is there another graph that better illustrates the relationship between the number of registered boats and the number of manatee deaths from boats?
- How can methods of statistics be used to objectively determine whether there is a relationship between two variables, such as the numbers of registered boats and the numbers of manatees killed by boats?
- If there is a relationship between the numbers of registered boats and the numbers of manatees killed by boats, how can it be described? Is there an equation that can be used to predict manatee deaths from boats, given a particular number of registered boats?
- In addition to the numbers of registered boats, are there other important variables that affect the numbers of manatees killed by boats?

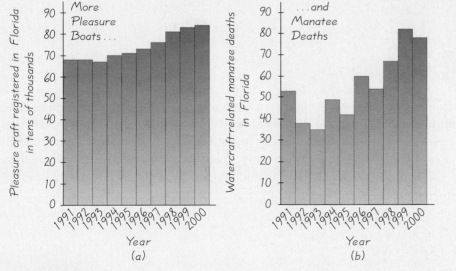

FIGURE 9-1 **Registered Florida Pleasure Craft (in tens of thousands) and Watercraft-Related Manatee Deaths**

Table 9-1	Registered Florida Pleasure Craft (in tens of thousands) and Watercraft-Related Manatee Deaths

Year	1991	1992	1993	1994	1995	1996	1997	1998	1999	2000
x: Boats	68	68	67	70	71	73	76	81	83	84
y: Manatee Deaths	53	38	35	49	42	60	54	67	82	78

 Overview

This chapter introduces important methods for making inferences based on sample data that come in *pairs*. Section 8-4 used matched pairs, but the inferences in Section 8-4 dealt with differences between two population means. This chapter has the objective of determining whether there is a relationship between the two variables and, if such a relationship exists, we want to describe it with an equation that can be used for predictions.

We begin in Section 9-2 by considering the concept of correlation, which is used to determine whether there is a statistically significant relationship between two variables. We investigate correlation using the scatterplot (a graph) and the linear correlation coefficient (a measure of the direction and strength of linear association between two variables). In Section 9-3 we investigate regression analysis; we describe the relationship between two variables with an equation that relates them and show how to use that equation to predict values of one variable when we know values of the other variable.

In Section 9-4 we analyze the differences between predicted values and actual observed values of a variable. Sections 9-2 through 9-4 involve relationships between *two* variables, but in Section 9-5 we use concepts of multiple regression to describe the relationship among three or more variables. Finally, in Section 9-6 we describe some basic methods for developing a mathematical model that can be used to describe the relationship between two variables. Although Section 9-3 is limited to linear relationships, Section 9-6 includes some common nonlinear relationships.

9-2 Correlation

The main objective of this section is to analyze a collection of paired sample data (sometimes called **bivariate data**) and determine whether there appears to be a relationship between the two variables. In statistics, we refer to such a relationship as a *correlation.* (We will consider only *linear* relationships, which means that when graphed, the points approximate a straight-line pattern. Also, we consider only quantitative data.)

> **Definition**
>
> A **correlation** exists between two variables when one of them is related to the other in some way.

Table 9-1, for example, consists of paired boat/manatee data for each year of the past decade. We will determine whether there is a correlation between the variable *x* (number of registered boats) and the variable *y* (number of manatees killed by boats).

Exploring the Data

Before working with the more formal computational methods of this section, we should first explore the data set to see what we can learn. We can often see a relationship between two variables by constructing a graph called a *scatterplot*, or *scatter diagram*.

> ### Definition
>
> A **scatterplot** (or **scatter diagram**) is a graph in which the paired (*x*, *y*) sample data are plotted with a horizontal *x*-axis and a vertical *y*-axis. Each individual (*x*, *y*) pair is plotted as a single point.

Note to Instructor

Because of their subjective nature, some students will feel that conclusions based on scatterplots are effectively worthless. Point out that although analyses of scatter diagrams are largely subjective, they are extremely helpful in detecting patterns that do not fit a straight line.

As an example, see the Excel display of the 10 pairs of data listed in Table 9-1. When we examine such a scatterplot, we should study the overall pattern of the plotted points. If there is a pattern, we should note its direction. That is, as one variable increases, does the other seem to increase or decrease? We should observe whether there are any outliers, which are points that lie very far away from all of the other points. The Excel-generated scatterplot does appear to reveal a pattern showing that more registered boats appear to be associated with more manatee deaths from boats. The scatterplot does a much better job of visually depicting the association between registered boats and manatee deaths than the side-by-side bar charts shown in Figure 9-1. Because they are arranged according to a time sequence, the side-by-side bar charts in Figure 9-1 do a good job of showing the long-term trends of the number of registered boats and the number of manatee deaths from boats, but the scatterplot does a better job of illustrating the relationship between those two variables.

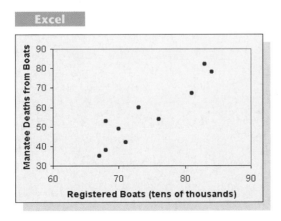

Student Ratings of Teachers

Many colleges equate high student ratings with good teaching—an equation often fostered by the fact that student evaluations are easy to administer and measure. However, one study that compared student evaluations of teachers with the amount of material learned found a strong *negative* correlation between the two factors. Teachers rated highly by students seemed to induce less learning. In a related study, an audience gave a high rating to a lecturer who conveyed very little information but was interesting and entertaining. (See "Rating the Teachers" by Miriam Rodin, *Center Magazine,* Vol. VIII, No. 5.)

Other examples of scatterplots are shown in Figure 9-2. The graphs in Figure 9-2(a), (b), and (c) depict a pattern of increasing values of *y* that correspond to increasing values of *x*. As you proceed from (a) to (c), the dot pattern becomes closer to a straight line, suggesting that the relationship between *x* and *y* becomes stronger. The scatterplots in (d), (e), and (f) depict patterns in which the *y*-values decrease as the *x*-values increase. Again, as you proceed from (d) to (f), the relationship becomes stronger. In contrast to the first six graphs, the scatterplot of (g) shows no pattern and suggests that there is no correlation (or relationship) be-

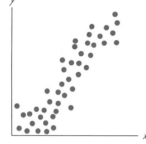 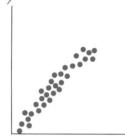 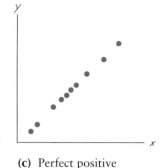

(a) Positive correlation between *x* and *y*   **(b)** Strong positive correlation between *x* and *y*   **(c)** Perfect positive correlation between *x* and *y*

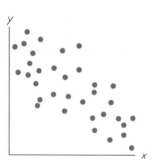 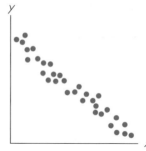

(d) Negative correlation between *x* and *y*   **(e)** Strong negative correlation between *x* and *y*   **(f)** Perfect negative correlation between *x* and *y*

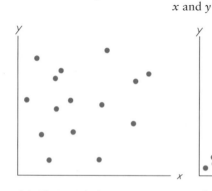 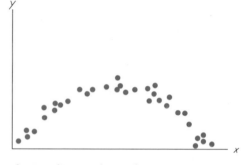

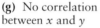

(g) No correlation between *x* and *y*   **(h)** Nonlinear relationship between *x* and *y*

FIGURE 9-2 **Scatterplots**

tween *x* and *y*. Finally, the scatterplot of (h) shows a pattern, but it is not a straight-line pattern.

Linear Correlation Coefficient

Because visual examinations of scatterplots are largely subjective, we need more precise and objective measures. We use the linear correlation coefficient *r*, which is useful for detecting straight-line patterns.

"Stocks Skid on Superstition of Patriot Win"

The above *New York Post* headline is a statement about the *Super Bowl omen,* which states that a Super Bowl victory by a team with NFL origins is followed by a year in which the New York Stock Exchange index rises; otherwise, it falls. (In 1970, the NFL and AFL merged into the current NFL.) This indicator has been correct in 29 of the past 35 years, largely due to the fact that NFL teams win more often, and the stock market tends to rise over time. Forecasting and predicting are important goals of statistics and investment advisors, but common sense suggests that no one should base investments on the outcome of one football game. Other indicators used to forecast stock market performance include rising skirt hemlines, aspirin sales, limousines on Wall Street, orders for cardboard boxes, sales of beer versus wine, and elevator traffic at the New York Stock Exchange.

Definition

The **linear correlation coefficient** *r* measures the strength of the linear relationship between the paired *x*- and *y*-quantitative values in a *sample*. Its value is computed by using Formula 9-1, included in the accompanying box. [The linear correlation coefficient is sometimes referred to as the **Pearson product moment correlation coefficient** in honor of Karl Pearson (1857–1936), who originally developed it.]

Because the linear correlation coefficient *r* is calculated using sample data, it is a sample statistic used to measure the strength of the linear correlation between *x* and *y*. If we had every pair of population values for *x* and *y*, the result of Formula 9-1 would be a population parameter, represented by ρ (Greek rho). The accompanying box includes the required assumptions, notation, and Formula 9-1.

Assumptions

1. The sample of paired (x, y) data is a *random* sample of quantitative data.
2. The pairs of (x, y) data have a **bivariate normal distribution.** (Normal distributions are discussed in Chapter 5, but this assumption basically requires that for any fixed value of *x*, the corresponding values of *y* have a distribution that is bell-shaped, and for any fixed value of *y*, the values of *x* have a distribution that is bell-shaped.) This assumption is usually difficult to check, but a partial check can be made by determining whether the values of both *x* and *y* have distributions that are basically bell-shaped.

Notation for the Linear Correlation Coefficient

n	represents the number of pairs of data present.
Σ	denotes the addition of the items indicated.
Σx	denotes the sum of all *x*-values.
Σx^2	indicates that each *x*-value should be squared and then those squares added.
$(\Sigma x)^2$	indicates that the *x*-values should be added and the total then squared. It is extremely important to avoid confusing Σx^2 and $(\Sigma x)^2$.

continued

Σxy indicates that each x-value should first be multiplied by its corresponding y-value. After obtaining all such products, find their sum.

r represents the linear correlation coefficient for a *sample*.

ρ represents the linear correlation coefficient for a *population*.

Formula 9-1 $r = \dfrac{n\Sigma xy - (\Sigma x)(\Sigma y)}{\sqrt{n(\Sigma x^2) - (\Sigma x)^2}\sqrt{n(\Sigma y^2) - (\Sigma y)^2}}$

Interpreting r Using Table A-6: If the absolute value of the computed value of r exceeds the value in Table A-6, conclude that there is a significant linear correlation. Otherwise, there is not sufficient evidence to support the conclusion of a significant linear correlation.

Note to Instructor
Formula 9-1 is the "shortcut" version of the formula for r given later in this section, along with an explanation of why it is constructed as it is. Consider doing an in-class example with six pairs of data. You might randomly select six students and get their pulse rates and heights.
    Some instructors require that students use Formula 9-1 with all work shown in manual calculations, and others allow their students to obtain values of r by using the programs in their calculators. Be sure to inform your class of your preference. (Suggestion: Demonstrate the manual calculation of r, but then move on and assume that values of r will be found by using calculator programs. This reflects the trend in statistics to have less number crunching and more thinking, analysis, and interpretation.)

Rounding the Linear Correlation Coefficient

Round the linear correlation coefficient r to three decimal places (so that its value can be directly compared to critical values in Table A-6). When calculating r and other statistics in this chapter, rounding in the middle of a calculation often creates substantial errors, so try using your calculator's memory to store intermediate results and round off only at the end. Many inexpensive calculators have Formula 9-1 built in so that you can automatically evaluate r after entering the sample data.

> **EXAMPLE Calculating r** Using the data given below, find the value of the linear correlation coefficient r.

x	1	1	3	5
y	2	8	6	4

SOLUTION   For the given sample of paired data, $n = 4$ because there are four pairs of data. The other components required in Formula 9-1 are found from the calculations in Table 9-2. Note how this vertical format makes the calculations easier.

Using the calculated values and Formula 9-1, we can now evaluate r as follows:

$$r = \dfrac{n\Sigma xy - (\Sigma x)(\Sigma y)}{\sqrt{n(\Sigma x^2) - (\Sigma x)^2}\sqrt{n(\Sigma y^2) - (\Sigma y)^2}}$$

$$= \dfrac{4(48) - (10)(20)}{\sqrt{4(36) - (10)^2}\sqrt{4(120) - (20)^2}}$$

$$= \dfrac{-8}{\sqrt{44}\sqrt{80}} = -0.135$$

Table 9-2	Finding Statistics Used to Calculate r				
	x	y	$x \cdot y$	x^2	y^2
	1	2	2	1	4
	1	8	8	1	64
	3	6	18	9	36
	5	4	20	25	16
Total	10	20	48	36	120
	$\uparrow$	$\uparrow$	$\uparrow$	$\uparrow$	$\uparrow$
	Σx	Σy	Σxy	Σx^2	Σy^2

These calculations get quite messy with large data sets, so it's fortunate that the linear correlation coefficient can be found automatically with many different calculators and computer programs. See "Using Technology" at the end of this section for comments about STATDISK, Minitab, Excel, and the TI-83 Plus calculator.

Interpreting the Linear Correlation Coefficient

We need to interpret a calculated value of r, such as the value of -0.135 found in the preceding example. Given the way that Formula 9-1 is constructed, the value of r must always fall between -1 and $+1$ inclusive. If r is close to 0, we conclude that there is no significant linear correlation between x and y, but if r is close to -1 or $+1$ we conclude that there is a significant linear correlation between x and y. Interpretations of "close to" 0 or 1 or -1 are vague, so we use the following very specific decision criterion:

> **If the absolute value of the computed value of r exceeds the value in Table A-6, conclude that there is a significant linear correlation. Otherwise, there is not sufficient evidence to support the conclusion of a significant linear correlation.**

When there really is no linear correlation between x and y, Table A-6 lists values that are "critical" in this sense: They separate *usual* values of r from those that are *unusual*. For example, Table A-6 shows us that with $n = 10$ pairs of sample data, the critical values are 0.632 (for $\alpha = 0.05$) and 0.765 (for $\alpha = 0.01$). Critical values and the role of α are carefully described in Chapters 6 and 7. Here's how we interpret those numbers: With 10 pairs of data and no linear correlation between x and y, there is a 5% chance that the absolute value of the computed linear correlation coefficient r will exceed 0.632. With $n = 10$ and no linear correlation, there is a 1% chance that $|r|$ will exceed 0.765.

EXAMPLE **Boats and Manatees** Given the sample data in Table 9-1, find the value of the linear correlation coefficient r; then refer to Table A-6 to determine whether there is a significant linear correlation between the numbers of registered boats and the numbers of manatees killed by boats. In Table A-6, use the critical value for $\alpha = 0.05$. (With $\alpha = 0.05$ we conclude that there is a significant linear correlation only if the sample is unlikely in this sense: If there is no linear correlation between the two variables, such a value of r occurs 5% of the time or less.)

SOLUTION   Using the same procedure illustrated in the preceding example, or using technology, we can find that the 10 pairs of boat/manatee data in Table 9-1 result in $r = 0.922$. Here is the Minitab display:

> **Minitab**
>
> ```
> Pearson correlation of Boats and Manatees = 0.922
> P-Value = 0.000
> ```

Referring to Table A-6, we locate the row for which $n = 10$ (because there are 10 pairs of data). That row contains the critical values of 0.632 (for $\alpha = 0.05$) and 0.765 (for $\alpha = 0.01$). Using the critical value for $\alpha = 0.05$, we see that there is less than a 5% chance that with no linear correlation, the absolute value of the computed r will exceed 0.632. Because $r = 0.922$, its absolute value does exceed 0.632, so we conclude that there is a significant linear correlation between the number of registered boats and the number of manatee deaths from boats.

We have already noted that Formula 9-1 requires that the calculated value of r always fall between -1 and $+1$ inclusive. We list that property along with other important properties.

> **Properties of the Linear Correlation Coefficient r**
>
> 1. The value of r is always between -1 and $+1$ inclusive. That is,
>
> $$-1 \leq r \leq +1$$
>
> 2. *The value of r does not change if all values of either variable are converted to a different scale.*
> 3. *The value of r is not affected by the choice of x or y.* Interchange all x- and y-values and the value of r will not change.
> 4. *r measures the strength of a linear relationship.* It is not designed to measure the strength of a relationship that is not linear.

Note to Instructor
The concept of explained variation is introduced briefly here, but it is considered later in Section 9-4.

Interpreting r: Explained Variation

If we conclude that there is a significant linear correlation between x and y, we can find a linear equation that expresses y in terms of x, and that equation can be used to predict values of *y* for given values of *x*. In Section 9-3 we will describe a procedure for finding such equations and show how to predict values of *y* when given values of *x*. But a predicted value of *y* will not necessarily be the exact result, because in addition to *x*, there are other factors affecting *y*, such as random variation and other characteristics not included in the study. In Section 9-4 we will present a rationale and more details about this important principle:

> **The value of r^2 is the proportion of the variation in *y* that is explained by the linear relationship between *x* and *y*.**

EXAMPLE Boats and Manatees Using the boat/manatee data in Table 9-1, we have found that the linear correlation coefficient is $r = 0.922$. What proportion of the variation in the manatee deaths can be explained by the variation in the number of boat registrations?

SOLUTION With $r = 0.922$, we get $r^2 = 0.850$. (Using the unrounded value of *r* yields $r^2 = 0.849$.)

INTERPRETATION We conclude that 0.850 (or about 85%) of the variation in manatee deaths from boats can be explained by the linear relationship between the number of boat registrations and the number of manatee deaths from boats. This implies that about 15% of the variation in such manatee deaths cannot be explained by the number of boat registrations. Another really important factor is the size of the manatee population. In fact, there is evidence indicating that the manatee population is steadily growing. Some people argue that the increase in manatee deaths from boats can be explained by the fact that the growing manatee population results in many more manatees in the water, and the growing death rate from boats is a symptom of a healthy and growing manatee population, so additional boating restrictions are unnecessary. Because the manatee population estimates are based on aerial observations, others argue that the estimates of the manatee population size are unreliable. Researcher Thomas Fraser suggested in a report that "a vigorous capture–tag and recapture program should be implemented by the State to gain much better information about population size and change." (See the third Cooperative Group Activity in Chapter 3.)

Common Errors Involving Correlation

We now identify three of the most common sources of errors made in interpreting results involving correlation:

1. *A common error is to conclude that correlation implies causality.* Using the sample data in Table 9-1, we can conclude that there is a correlation between the number of registered boats and the number of manatees killed by boats, but we cannot conclude that more registered boats *cause* more manatee deaths. The manatee deaths from boats may be affected by some other variable

Note to Instructor

The first common error should be strongly emphasized in class: "Correlation does not imply causality." Here's a classic example: There is a correlation between per capita beer consumption and teachers' salaries, but teachers don't use salary increases to buy more beer. Consider asking the class for a likely explanation of this correlation.

lurking in the background. (A **lurking variable** is one that affects the variables being studied, but is not included in the study.) For example, warmer temperatures may well affect the number of boats and the number of manatees that are killed by boats. The temperature would then be a lurking variable.

2. *Another error arises with data based on averages.* Averages suppress individual variation and may inflate the correlation coefficient. One study produced a 0.4 linear correlation coefficient for paired data relating income and education among individuals, but the linear correlation coefficient became 0.7 when regional averages were used.

3. *A third error involves the property of linearity.* A relationship may exist between *x* and *y* even when there is no significant linear correlation. The data depicted in Figure 9-3 result in a value of $r = 0$, which is an indication of no *linear* correlation between the two variables. However, we can easily see from looking at the figure that there is a pattern reflecting a very strong *nonlinear* relationship. (Figure 9-3 is a scatterplot that depicts the relationship between distance above ground and time elapsed for an object thrown upward.)

Formal Hypothesis Test (Requires Coverage of Chapter 7)

We present two methods (summarized in the accompanying box and in Figure 9-4) for using a formal hypothesis test to determine whether there is a significant linear correlation between two variables. Some instructors prefer Method 1 because it reinforces concepts introduced in earlier chapters. Others prefer Method 2 because it involves easier calculations. Method 1 uses the Student *t* distribution with a test statistic having the form $t = (r - \mu_r)/s_r$, where μ_r and s_r denote the claimed value of the mean and the sample standard deviation of *r* values. The test statistic given in the box (for Method 1) reflects the fact that the standard deviation of *r* values can be expressed as $\sqrt{(1 - r^2)/(n - 2)}$.

Figure 9-4 shows that the decision criterion is to reject the null hypothesis of $\rho = 0$ if the absolute value of the test statistic exceeds the critical values; rejection

Note to Instructor

Here is a good exercise: Ask students to identify the coordinates of the points in Figure 9-3, then use those values to calculate *r*. Based on the result, what do you conclude about the relationship between time and distance? What does the graph itself suggest about that relationship?

Note to Instructor

As the title of this subsection indicates, this material requires prior coverage of Chapter 7. If you are including Sections 9-2 and 9-3 early (such as following Chapter 2), skip this subsection. If you do include this subsection, be sure to inform the class that you prefer Method 1 or Method 2.

FIGURE 9-3 **Scatterplot Showing a Pattern That Is Nonlinear**

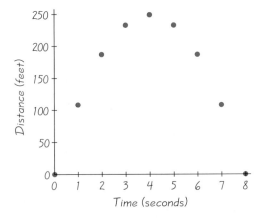

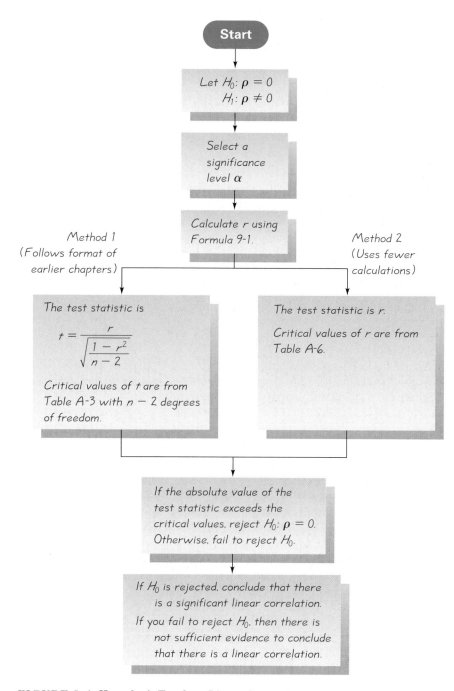

Start

Let $H_0: \rho = 0$
$H_1: \rho \neq 0$

Select a
significance
level α

Calculate r using
Formula 9-1.

Method 1
(Follows format of
earlier chapters)

Method 2
(Uses fewer
calculations)

The test statistic is

$$t = \frac{r}{\sqrt{\dfrac{1-r^2}{n-2}}}$$

Critical values of t are from
Table A-3 with $n-2$ degrees
of freedom.

The test statistic is r.

Critical values of r are from
Table A-6.

If the absolute value of the
test statistic exceeds the
critical values, reject $H_0: \rho = 0$.
Otherwise, fail to reject H_0.

If H_0 is rejected, conclude that there
is a significant linear correlation.
If you fail to reject H_0, then there is
not sufficient evidence to conclude
that there is a linear correlation.

FIGURE 9-4 Hypothesis Test for a Linear Correlation

of $\rho = 0$ means that there is sufficient evidence to support a claim of a linear correlation between the two variables. If the absolute value of the test statistic does not exceed the critical values, then we fail to reject $\rho = 0$; that is, there is not sufficient evidence to conclude that there is a linear correlation between the two variables.

Hypothesis Test for Correlation (See Figure 9-4.)

$H_0: \rho = 0$ (There is no linear correlation.)
$H_1: \rho \neq 0$ (There is a linear correlation.)

Method 1: Test Statistic Is t

Test statistic: $t = \dfrac{r - \mu_r}{\sqrt{\dfrac{1 - r^2}{n - 2}}}$

where μ_r denotes the claimed value of the mean of the r values. Let $\mu_r = 0$ when testing the null hypothesis of $\rho = 0$.

Critical values: Use Table A-3 with $n - 2$ degrees of freedom.

P-value: Use Table A-3 with $n - 2$ degrees of freedom.

Conclusion: If $|t| >$ critical value from Table A-3, reject H_0 and conclude that there is a linear correlation. If $|t| \leq$ critical value, fail to reject H_0; there is not sufficient evidence to conclude that there is a linear correlation.

Method 2: Test Statistic Is r

Test statistic: r

Critical values: Refer to Table A-6.

Conclusion: If $|r| >$ critical value from Table A-6, reject H_0 and conclude that there is a linear correlation. If $|r| \leq$ critical value, fail to reject H_0; there is not sufficient evidence to conclude that there is a linear correlation.

EXAMPLE Boats and Manatees Using the sample data in Table 9-1, test the claim that there is a linear correlation between the number of registered boats and the number of manatee deaths from boats. For the test statistic, use both (a) Method 1 and (b) Method 2.

SOLUTION   Refer to Figure 9-4. To claim that there is a significant linear correlation is to claim that the population linear correlation coefficient ρ is different from 0. We therefore have the following hypotheses:

$H_0: \rho = 0$ (There is no linear correlation.)
$H_1: \rho \neq 0$ (There is a linear correlation.)

No significance level was specified, so use $\alpha = 0.05$.

In a preceding example we already found that $r = 0.922$. With that value, we now find the test statistic and critical value, using each of the two methods just described.

a. *Method 1:* The test statistic is

$$t = \dfrac{r}{\sqrt{\dfrac{1 - r^2}{n - 2}}} = \dfrac{0.922}{\sqrt{\dfrac{1 - 0.922^2}{10 - 2}}} = 6.735$$

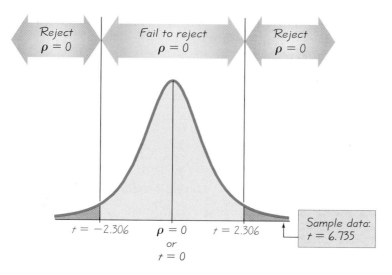

FIGURE 9-5 Testing H_0: $\rho = 0$ with Method 1

The critical values of $t = \pm 2.306$ are found in Table A-3, where 2.306 corresponds to an area of 0.05 divided between two tails and the number of degrees of freedom is $n - 2 = 8$. See Figure 9-5 for the graph that includes the test statistic and critical values.

b. *Method 2:* The test statistic is $r = 0.922$. The critical values of $r = \pm 0.632$ are found in Table A-6 with $n = 10$ and $\alpha = 0.05$. See Figure 9-6 for a graph that includes this test statistic and critical values.

Using either of the two methods, we find that the absolute value of the test statistic does exceed the critical value (Method 1: $6.735 > 2.306$. Method 2: $0.922 > 0.632$); that is, the test statistic falls in the critical region. We therefore reject H_0: $\rho = 0$. There is sufficient evidence to support the claim of a linear correlation between the number of registered boats and the number of manatee deaths from boats.

One-tailed Tests: The preceding example and Figures 9-5 and 9-6 illustrate a two-tailed hypothesis test. The examples and exercises in this section will gener-

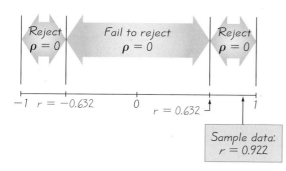

FIGURE 9-6 Testing H_0: $\rho = 0$ with Method 2

ally involve only two-tailed tests, but one-tailed tests can occur with a claim of a positive linear correlation or a claim of a negative linear correlation. In such cases, the hypotheses will be as shown here.

Claim of *Negative* Correlation (Left-tailed test)	Claim of *Positive* Correlation (Right-tailed test)
H_0: $\rho = 0$ H_1: $\rho < 0$	H_0: $\rho = 0$ H_1: $\rho > 0$

For these one-tailed tests, Method 1 can be handled as in earlier chapters. For Method 2, either calculate the critical value as described in Exercise 31 or modify Table A-6 by replacing the column headings of $\alpha = 0.05$ and $\alpha = 0.01$ by the one-sided critical values of $\alpha = 0.025$ and $\alpha = 0.005$, respectively.

Rationale: We have presented Formula 9-1 for calculating r and have illustrated its use; we will now give a justification for it. Formula 9-1 simplifies the calculations used in this equivalent formula:

$$r = \frac{\Sigma(x - \bar{x})(y - \bar{y})}{(n - 1)s_x s_y}$$

We will temporarily use this latter version of Formula 9-1 because its form relates more directly to the underlying theory. We now consider the following paired data, which are depicted in the scatterplot shown in Figure 9-7.

x	1	1	2	4	7
y	4	5	8	15	23

Figure 9-7 includes the point $(\bar{x}, \bar{y}) = (3, 11)$, which is called the *centroid* of the sample points.

Definition

Given a collection of paired (x, y) data, the point $(\bar{x}, \bar{y})$ is called the **centroid.**

The statistic r, sometimes called the *Pearson product moment,* was first developed by Karl Pearson. It is based on the sum of the products of the moments $(x - \bar{x})$ and $(y - \bar{y})$; that is, on the statistic $\Sigma(x - \bar{x})(y - \bar{y})$. In any scatterplot, vertical and horizontal lines through the centroid $(\bar{x}, \bar{y})$ divide the diagram into four quadrants, as in Figure 9-7. If the points of the scatterplot tend to approximate an uphill line (as in the figure), individual values of the product $(x - \bar{x})(y - \bar{y})$ tend to be positive because most of the points are found in the first and third quadrants, where the products of $(x - \bar{x})$ and $(y - \bar{y})$ are positive. If the points of the

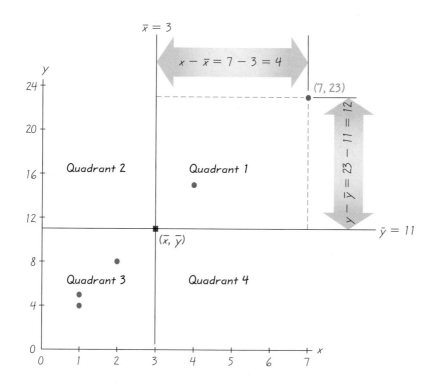

FIGURE 9-7 **Scatterplot Partitioned into Quadrants**

scatterplot approximate a downhill line, most of the points are in the second and fourth quadrants, where $(x - \bar{x})$ and $(y - \bar{y})$ are opposite in sign, so $\Sigma(x - \bar{x})(y - \bar{y})$ is negative. Points that follow no linear pattern tend to be scattered among the four quadrants, so the value of $\Sigma(x - \bar{x})(y - \bar{y})$ tends to be close to 0.

The sum $\Sigma(x - \bar{x})(y - \bar{y})$ depends on the magnitude of the numbers used. For example, if you change x from inches to feet, that sum will change. To make r independent of the particular scale used, we include the sample standard deviations as follows:

$$r = \frac{\Sigma(x - \bar{x})(y - \bar{y})}{(n - 1)s_x s_y}$$

This expression can be algebraically manipulated into the equivalent form of Formula 9-1.

In preceding chapters we discussed methods of inferential statistics by addressing methods of hypothesis testing, as well as methods for constructing confidence interval estimates. A similar procedure may be used to find confidence intervals for ρ. However, because the construction of such confidence intervals involves somewhat complicated transformations, that process is presented in Exercise 33 (Beyond the Basics).

We can use the linear correlation coefficient to determine whether there is a linear relationship between two variables. Using the data in Table 9-1, we have concluded that there is a linear correlation between the number of registered boats and the number of manatee deaths from boats. Having concluded that a relationship exists, we would like to determine what that relationship is so that we can predict the number of manatee deaths for a given number of registered boats. This next stage of analysis is addressed in the following section.

Using Technology

STATDISK   First enter the paired data in the columns of the Statdisk Data Window. Select **Analysis** from the main menu bar, then use the option **Correlation and Regression.** Enter a value for the significance level and select the columns of data. Click on the **Evaluate** button. The STATDISK display will include the value of the linear correlation coefficient along with the critical value of r, the conclusion, and other results to be discussed in later sections. A scatterplot can be obtained by clicking on the **Scatterplot** button.

Minitab   Enter the paired data in columns C1 and C2, then select **Stat** from the main menu bar, choose **Basic Statistics,** followed by **Correlation,** and proceed to enter C1 and C2 for the columns to be used. Minitab will provide the value of the linear correlation coefficient r as well as a P-value. To obtain a scatterplot, select **Graph,** followed by **Scatterplot** or **Plot,** then enter C1 and C2 for X and Y, and click **OK.**

Excel   Excel has a function that calculates the value of the linear correlation coefficient. First enter the paired sample data in columns A and B. Click on the *fx* function key located on the main menu bar. Select the function category **Statistical** and the function name **CORREL,** then click **OK.** In the dialog box, enter the cell range of values for x, such as A1:A10. Also enter the cell range of values for y, such as B1:B10. To obtain a scatterplot, click on the Chart Wizard on the main menu, then select the chart type identified as **XY(Scatter).** In the dialog box, enter the input range of the data, such as A1:B10. Click **Next** and proceed to use the dialog boxes to modify the graph as desired.

The Data Desk XL add-in can also be used. Click on **DDXL** and select **Regression,** then click on the Function Type box and select **Correlation.** In the dialog box, click on the pencil icon for the X-Axis Variable and enter the range of values for the variable x, such as A1:A10. Click on the pencil icon for the Y-Axis Variable and enter the range of values for y. Click **OK.** A scatter diagram and the correlation coefficient will be displayed.

TI-83 Plus   Enter the paired data in lists L1 and L2, then press **STAT** and select **TESTS.** Using the option of **LinRegTTest** will result in several displayed values, including the value of the linear correlation coefficient r.

To obtain a scatterplot, press **2nd,** then **Y=** (for STAT PLOT). Press **Enter** twice to turn Plot 1 on, then select the first graph type, which resembles a scatterplot. Set the X list and Y list labels to L1 and L2 and press the **ZOOM** key, then select **ZoomStat** and press the **Enter** key.

9-2 Basic Skills and Concepts

In Exercises 1–4, use a significance level of $\alpha = 0.05$.

1. Chest Sizes and Weights of Bears When eight bears were anesthetized, researchers measured the distances (in inches) around the bears' chests and weighed the bears (in pounds). Minitab was used to find that the value of the linear correlation coefficient is $r = 0.993$.
   a. Is there a significant linear correlation between chest size and weight? Explain.
   b. What proportion of the variation in weight can be explained by the linear relationship between weight and chest size?

2. Guns and Murder Rate   Using data collected from the FBI and the Bureau of Alcohol, Tobacco, and Firearms, the number of registered automatic weapons and the murder rate (in murders per 100,000 people) was obtained for each of eight randomly selected states. STATDISK was used to find that the value of the linear correlation coefficient is $r = 0.885$.

a. Is there a significant linear correlation between the number of registered automatic weapons and the murder rate? Explain.

b. What proportion of the variation in the murder rate can be explained by the linear relationship between the murder rate and the number of registered automatic weapons?

3. Stocks and Super Bowl   Data Set 25 in Appendix B includes pairs of data for the Dow-Jones Industrial Average (DJIA) high value and the total number of points scored in the Super Bowl for 21 different years. Excel was used to find that the value of the linear correlation coefficient is $r = -0.133$.

a. Is there a significant linear correlation between DJIA high value and Super Bowl points? Explain.

b. What proportion of the variation in Super Bowl points can be explained by the variation in the high value of the DJIA?

4. Car Sales and Sunspots   Data Set 25 in Appendix B includes pairs of data for the sunspot number and the number of U.S. car sales for 21 different years. A TI-83 Plus calculator was used to find that the value of the linear correlation coefficient is $r = -0.284$.

a. Is there a significant linear correlation between the sunspot number and the number of U.S. car sales? Explain.

b. What proportion of the variation in the number of U.S. car sales can be explained by the variation in the sunspot number?

Testing for a Linear Correlation. In Exercises 5 and 6, use a scatterplot and the linear correlation coefficient r to determine whether there is a correlation between the two variables.

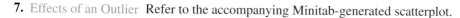

5.

x	0	1	2	3	4
y	4	1	0	1	4

6.

x	1	2	2	5	6
y	2	5	4	15	15

7. Effects of an Outlier   Refer to the accompanying Minitab-generated scatterplot.

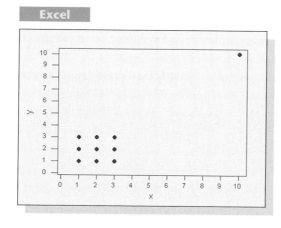

continued

2. a. Yes, because the absolute value of the test statistic exceeds the critical values $r = \pm 0.707$.
   b. 0.783

3. a. No, because the absolute value of the test statistic does not exceed the critical values $r = \pm 0.444$ (approx).
   b. 0.0177

4. a. No, because the absolute value of the test statistic does not exceed the critical values $r = \pm 0.444$ (approx).
   b. 0.0807

5. The scatterplot suggests that there is a correlation, but it is not linear. With $r = 0$ and critical values of $r = \pm 0.878$ (for a 0.05 significance level), there is not a significant linear correlation.

6. With $r = 0.985$ and critical values of $r = \pm 0.878$ (for a 0.05 significance level), there is a significant linear correlation.

7. a. There appears to be a linear correlation.
   b. $r = 0.906$. CV: $r = \pm 0.632$ (for a 0.05 significance level). Significant linear correlation.
   c. $r = 0$. CV: $r = \pm 0.666$ (for a 0.05 significance level). No significant linear correlation.
   d. The effect from a single pair of values can be very substantial, and it can change the conclusion.

a. Examine the pattern of all 10 points and subjectively determine whether there appears to be a correlation between *x* and *y*.

b. After identifying the 10 pairs of coordinates corresponding to the 10 points, find the value of the correlation coefficient *r* and determine whether there is a significant linear correlation.

c. Now remove the point with coordinates (10, 10) and repeat parts (a) and (b).

d. What do you conclude about the possible effect from a single pair of values?

Testing for a Linear Correlation. In Exercises 8–14, construct a scatterplot, find the value of the linear correlation coefficient r and use a significance level of $\alpha = 0.05$ to determine whether there is a significant linear correlation between the two variables. Save your work because the same data sets will be used in the next section.

8. $r = 0.517$. CV: $r = \pm0.602$. No significant linear correlation. No, in addition to no significant linear correlation, other data would be needed to address the stated argument.

8. Fires and Acres Burned Listed below are the numbers of fires (in thousands) and the acres that were burned (in millions) in 11 western states in each year of the last decade (based on data from *USA Today*). Is there a correlation? The data were listed under a headline of "Loggers seize on fires to argue for more cutting." Do the data support the argument that as loggers remove more trees, the risk of fire decreases because the forests are less dense?

Fires	73	69	58	48	84	62	57	45	70	63	48
Acres burned	6.2	4.2	1.9	2.7	5.0	1.6	3.0	1.6	1.5	2.0	3.7

9. $r = -0.118$. CV: $r = \pm0.707$. No significant linear correlation. Lowest: Susan Lucci. Highest: Kelsey Grammer.

9. Buying a TV Audience The *New York Post* published the annual salaries (in millions) and the number of viewers (in millions), with results given below for Oprah Winfrey, David Letterman, Jay Leno, Kelsey Grammer, Barbara Walters, Dan Rather, James Gandolfini, and Susan Lucci, repsectively. Is there a correlation between salary and number of viewers? Which of the listed stars has the lowest cost per viewer? Highest cost per viewer?

Salary	100	14	14	35.2	12	7	5	1
Viewers	7	4.4	5.9	1.6	10.4	9.6	8.9	4.2

10. $r = 0.796$. CV: $r = \pm0.666$. Significant linear correlation. No, the population of all women is different from the population of supermodels, so inferences about supermodels do not necessarily apply to all women.

10. Supermodel Heights and Weights Listed below are heights (in inches) and weights (in pounds) for supermodels Niki Taylor, Nadia Avermann, Claudia Schiffer, Elle MacPherson, Christy Turlington, Bridget Hall, Kate Moss, Valerie Mazza, and Kristy Hume. Is there a correlation between height and weight? If there is a correlation, does it mean that there is a correlation between height and weight of all adult women?

Height (in.)	71	70.5	71	72	70	70	66.5	70	71
Weight (lb)	125	119	128	128	119	127	105	123	115

11. $r = 0.658$. CV: $r = \pm0.532$. Significant linear correlation. Measurements appear to vary widely. A study might be conducted to determine whether the subject's blood pressure really does vary considerably.

11. Blood Pressure Measurements Fourteen different second-year medical students took blood pressure measurements of the same patient and the results are listed below (data provided by Marc Triola, MD). Is there a correlation between systolic and diastolic values? Apart from correlation, is there some other method that might be used to address an important issue suggested by the data?

Systolic	138	130	135	140	120	125	120	130	130	144	143	140	130	150
Diastolic	82	91	100	100	80	90	80	80	80	98	105	85	70	100

12. $r = 0.183$. CV: $r = \pm0.707$. No significant linear correlation. Winning times do not appear to be affected by temperature.

12. Temperatures and Marathons In "The Effects of Temperature on Marathon Runner's Performance" by David Martin and John Buoncristiani (*Chance*, Vol. 12, No. 4), high temperatures and times (in minutes) were given for women who won the New York City marathon in recent years. Results are listed below. Is there a correlation between temperature and winning time? Does it appear that winning times are affected by temperature?

x (temperature)	55	61	49	62	70	73	51	57
y (time)	145.283	148.717	148.300	148.100	147.617	146.400	144.667	147.533

13. Smoking and Nicotine When nicotine is absorbed by the body, cotinine is produced. A measurement of cotinine in the body is therefore a good indicator of how much a person smokes. Listed below are the reported numbers of cigarettes smoked per day and the measured amounts of nicotine (in ng/mL). (The values are from randomly selected subjects in the National Health Examination Survey.) Is there a significant linear correlation? Explain the result.

x (cigarettes per day)	60	10	4	15	10	1	20	8	7	10	10	20
y (cotinine)	179	283	75.6	174	209	9.51	350	1.85	43.4	25.1	408	344

14. Tree Circumference and Height Listed below are the circumferences (in feet) and the heights (in feet) of trees in Marshall, Minnesota (based on data from "Tree Measurements" by Stanley Rice, *American Biology Teacher*, Vol. 61, No. 9). Is there a correlation? Why should there be a correlation?

x (circ.)	1.8	1.9	1.8	2.4	5.1	3.1	5.5	5.1	8.3	13.7	5.3	4.9	3.7	3.8
y (ht)	21.0	33.5	24.6	40.7	73.2	24.9	40.4	45.3	53.5	93.8	64.0	62.7	47.2	44.3

Testing for a Linear Correlation. In Exercises 15–24, use the data from Appendix B to construct a scatterplot, find the value of the linear correlation coefficient r, and use a significance level of $\alpha = 0.05$ to determine whether there is a significant linear correlation between the two variables. Save your work because the same data sets will be used in the next section.

T 15. Cereal Killers Refer to Data Set 16 in Appendix B and use the amounts of fat and the measured calorie counts. Is there a correlation?

T 16. Tobacco and Alcohol in Children's Movies Refer to Data Set 7 in Appendix B and use the times that the animated children's movies showed tobacco use and alcohol use. Is there a correlation between the times for tobacco and the times for alcohol?

T 17. Cholesterol and Body Mass Index Refer to Data Set 1 in Appendix B and use the cholesterol levels and body mass index values of the 40 women. Is there a correlation between cholesterol level and body mass index?

T 18. Readability Levels Refer to Data Set 14 in Appendix B and use the Flesch Reading Ease scores and the Flesch-Kincaid Grade Level values for Tom Clancy's *The Bear and the Dragon*. Given that both scores are designed to measure readability, we would expect a correlation between them. Is there a correlation? How can the negative value of the correlation coefficient be explained?

T 19. Home Selling Prices, List Prices, and Taxes Refer to Data Set 24 in Appendix B.
  a. Use the paired data consisting of home list price and selling price. We might expect that these variables would be related, but is there sufficient evidence to support that expectation?
  b. Use the paired data consisting of home selling price and the amount of taxes. The tax bill is supposed to be based on the value of the house. Is it? Explain.

13. $r = 0.262$. CV: $r = \pm 0.576$. No significant linear correlation. Because the cigarette counts were reported, subjects may have given wrong values. Subjects may have been exposed to varying levels of secondhand smoke.

14. $r = 0.828$. CV: $r = \pm 0.532$. Significant linear correlation. Taller trees tend to have larger trunks.

15. $r = 0.359$. CV: $r = \pm 0.497$. No significant linear correlation.

16. $r = 0.238$. CV: $r = \pm 0.279$. No significant linear correlation.

17. $r = 0.482$. CV: $r = \pm 0.312$. Significant linear correlation.

18. $r = -0.913$. CV: $r = \pm 0.576$. Significant linear correlation.

19. a. $r = 0.997$. CV: $r = \pm 0.279$. Significant linear correlation.
   b. $r = 0.899$. CV: $r = \pm 0.279$. Significant linear correlation.

20. a. $r = 0.961$. CV: $r = \pm 0.361$ approx. Significant linear correlation. Yes.
    b. $r = 0.863$. CV: $r = \pm 0.361$ approx. Significant linear correlation. Yes.
    c. Tar, because it has a higher correlation with nicotine.
21. a. $r = 0.574$. CV: $r = \pm 0.361$ approx. Significant linear correlation. No, there can be a high correlation even though the forecast temperatures are very inaccurate.
    b. $r = 0.685$. CV: $r = \pm 0.361$ approx. Significant linear correlation. No, there can be a high correlation even though the forecast temperatures are very inaccurate.
    c. The one-day forecast temperatures are better because they have a higher correlation with the actual temperatures. However, a high correlation does not imply that the forecast temperatures are accurate.
22. a. $r = 0.121$. CV: $r = \pm 0.254$ approx. No significant linear correlation.
    b. $r = -0.119$. CV: $r = \pm 0.254$ approx. No significant linear correlation.
    c. In both cases, there is not a significant linear correlation.
23. a. $r = 0.870$. CV: $r = \pm 0.279$. Significant linear correlation.
    b. $r = -0.010$. CV: $r = \pm 0.279$. No significant linear correlation.
    c. Duration, because it has a significant linear correlation with interval.
24. a. $r = 0.767$. CV: $r = \pm 0.361$. Significant linear correlation.
    b. $r = -0.441$. CV: $r = \pm 0.361$. Significant linear correlation.
    c. Carat weight, because it has a higher correlation with price.

T 20. Tar and Nicotine Refer to Data Set 5 in Appendix B.
  a. Use the paired data consisting of tar and nicotine. Based on the result, does there appear to be a significant linear correlation between cigarette tar and nicotine? If so, can researchers reduce their laboratory expenses by measuring only one of these two variables?
  b. Use the paired data consisting of carbon monoxide and nicotine. Based on the result, does there appear to be a significant linear correlation between cigarette carbon monoxide and nicotine? If so, can researchers reduce their laboratory expenses by measuring only one of these two variables?
  c. Assume that researchers want to develop a method for predicting the amount of nicotine, and they want to measure only one other item. In choosing between tar and carbon monoxide, which is the better choice? Why?

T 21. Forecasting Weather Refer to Data Set 10 in Appendix B.
  a. Use the five-day forecast high temperatures and the actual high temperatures. Is there a correlation? Does a significant linear correlation imply that the five-day forecast temperatures are accurate?
  b. Use the one-day forecast high temperatures and the actual high temperatures. Is there a correlation? Does a significant linear correlation imply that the one-day forecast temperatures are accurate?
  c. Which would you expect to have a higher correlation with the actual high temperatures: the five-day forecast high temperatures or the one-day forecast high temperatures? Are the results from parts (a) and (b) what you would expect? If there is a very high correlation between forecast temperatures and actual temperatures, does it follow that the forecast temperatures are accurate?

T 22. Florida Everglades Refer to Data Set 12 in Appendix B.
  a. Use the bottom temperatures and the conductivity measurements. Is there a correlation?
  b. Use the rainfall amounts and the conductivity measurements. Is there a correlation?
  c. When conductivity values are paired with measures of salinity (salt content), the correlation coefficient is nearly 1. What can you conclude about the correlation between bottom temperature and salinity? What can you conclude about the correlation between rainfall amount and salinity?

T 23. Old Faithful Refer to Data Set 13 in Appendix B.
  a. Use the paired data for durations and intervals after eruptions of the geyser. Is there a significant linear correlation, suggesting that the interval after an eruption is related to the duration of the eruption?
  b. Use the paired data for heights of eruptions and intervals after eruptions of the Old Faithful geyser. Is there a significant linear correlation, suggesting that the interval after an eruption is related to the height of the eruption?
  c. Assume that you want to develop a method for predicting the time interval to the next eruption. Based on the results from parts (a) and (b), which factor would be more relevant: eruption duration or eruption height? Why?

T 24. Diamond Prices, Carats, and Color Refer to Data Set 18 in Appendix B.
  a. Use the paired data consisting of carat (weight) and price. Is there a significant linear correlation between the weight of a diamond in carats and its price?
  b. Use the paired color/price data. Is there a significant linear correlation between the color of a diamond and its price?

c. Assume that you are planning to buy a diamond engagement ring. In considering the value of a diamond, which characteristic should you consider to be more important: the carat weight or the color? Why?

Identifying Correlation Errors. In Exercises 25–28, describe the error in the stated conclusion. (See the list of common errors included in this section.)

25. *Given:* The paired sample data of the ages of subjects and their scores on a test of reasoning result in a linear correlation coefficient very close to 0.
Conclusion: Younger people tend to get higher scores.

26. *Given:* There is a significant linear correlation between personal income and years of education.
Conclusion: More education causes a person's income to rise.

27. *Given:* Subjects take a test of verbal skills and a test of manual dexterity, and those pairs of scores result in a linear correlation coefficient very close to 0.
Conclusion: Scores on the two tests are not related in any way.

28. *Given:* There is a significant linear correlation between state average tax burdens and state average incomes.
Conclusion: There is a significant linear correlation between individual tax burdens and individual incomes.

9-2 Beyond the Basics

29. Using Data from Scatterplot Sometimes, instead of having numerical data, we have only graphical data. The accompanying Excel scatterplot is similar to one that was included in "The Prevalence of Nosocomial Infection in Intensive Care Units in Europe," by Vincent et al., *Journal of the American Medical Association,* Vol. 274, No. 8. Each point represents a different European country. Estimate the value of the linear correlation coefficient, and determine whether there is a significant linear correlation between the mortality rate and the rate of infections acquired in intensive care units.

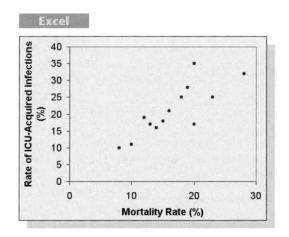

30. Correlations with Transformed Data In addition to testing for a linear correlation between *x* and *y*, we can often use *transformations* of data to explore for other relation-

25. With a linear correlation coefficient very close to 0, there does not appear to be a correlation, but the conclusion suggests that there is a correlation.
26. The presence of a significant linear correlation does not necessarily mean that one of the variables *causes* the other. Correlation does not imply causality.
27. Although there is no *linear* correlation, the variables may be related in some other *nonlinear* way.
28. Averages tend to suppress variation among individuals, so a correlation among *averages* does not necessarily mean that there is a correlation among *individuals.*
29. *r* = 0.819 (approx). CV: *r* = ±0.553. Significant linear correlation.
30. a. 0.972
    b. 0.905
    c. 0.999 (largest)
    d. 0.992
    e. −0.984

ships. For example, we might replace each x value by x^2 and use the methods of this section to determine whether there is a linear correlation between y and x^2. Given the paired data in the accompanying table, construct the scatterplot and then test for a linear correlation between y and each of the following. Which case results in the largest value of r?

a. x **b.** x^2 **c.** $\log x$ **d.** $\sqrt{x}$ **e.** $1/x$

x	1.3	2.4	2.6	2.8	2.4	3.0	4.1
y	0.11	0.38	0.41	0.45	0.39	0.48	0.61

31. Finding Critical *r*-Values The critical values of r in Table A-6 are found by solving

$$t = \frac{r}{\sqrt{\dfrac{1 - r^2}{n - 2}}}$$

for r to get

$$r = \frac{t}{\sqrt{t^2 + n - 2}}$$

where the t value is found from Table A-3 by assuming a two-tailed case with $n - 2$ degrees of freedom. Table A-6 lists the results for selected values of n and α. Use the formula for r given here and Table A-3 (with $n - 2$ degrees of freedom) to find the critical values of r for the given cases.

a. $H_1: \rho \neq 0, n = 50, \alpha = 0.05$
b. $H_1: \rho \neq 0, n = 75, \alpha = 0.10$
c. $H_1: \rho < 0, n = 20, \alpha = 0.05$
d. $H_1: \rho > 0, n = 10, \alpha = 0.05$
e. $H_1: \rho > 0, n = 12, \alpha = 0.01$

32. Including Categorical Data in a Scatterplot It sometimes becomes important to include categorical data in a scatterplot. Consider the sample data listed below, where weight is in pounds and the "remote" values consist of the number of times the subject used the television remote control during a period of 1 hour. Minitab was used to generate the scatterplot, with the characters F (for females) and M (for males) used to identify gender.

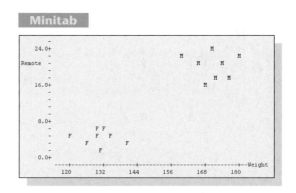

Gender	F	F	F	F	F	F	F	F	M	M	M	M	M	M	M	M
Weight	120	126	129	130	131	132	134	140	160	166	168	170	172	174	176	180
Remote	5	3	6	4	2	7	4	3	23	20	16	24	18	21	17	22

Answers (left margin):

31. a. ±0.279
    b. ±0.191
    c. −0.378
    d. 0.549
    e. 0.658

32. a. There appears to be a correlation.
    b. $r = 0.915$. CV: $r = \pm0.497$. Significant linear correlation.
    c. $r = -0.213$. CV: $r = \pm0.707$. No significant linear correlation.
    d. $r = -0.164$. CV: $r = \pm0.707$. No significant linear correlation.
    e. Combining different groups in the same data set could make it appear that there is a significant linear correlation between two variables, even though there is no correlation.

a. Before doing any calculations, examine the Minitab-generated scatterplot. What do you conclude about the correlation between weight and remote control use?

b. Using all 16 pairs of data, is there a correlation between weight and use of the remote?

c. Using only the eight females, is there a correlation between weight and use of the remote?

d. Using only the eight males, is there a correlation between weight and use of the remote?

e. Based on the preceding results, what do you conclude?

33. Constructing Confidence Intervals for ρ Given n pairs of data from which the linear correlation coefficient r can be found, use the following procedure to construct a confidence interval about the population parameter ρ.

Step a.   Use Table A-2 to find $z_{\alpha/2}$ that corresponds to the desired degree of confidence.

Step b.   Evaluate the interval limits w_L and w_R:

$$w_L = \frac{1}{2} \ln \left(\frac{1 + r}{1 - r} \right) - z_{\alpha/2} \cdot \frac{1}{\sqrt{n - 3}}$$

$$w_R = \frac{1}{2} \ln \left(\frac{1 + r}{1 - r} \right) + z_{\alpha/2} \cdot \frac{1}{\sqrt{n - 3}}$$

Step c.   Now evaluate the confidence interval limits in the expression below.

$$\frac{e^{2w_L} - 1}{e^{2w_L} + 1} < \rho < \frac{e^{2w_R} - 1}{e^{2w_R} + 1}$$

Use this procedure to construct a 95% confidence interval for ρ, given 50 pairs of data for which $r = 0.600$.

33. $0.386 < \rho < 0.753$

9-3   Regression

In Section 9-2 we analyzed paired data with the goal of determining whether there is a linear correlation between two variables. The main objective of this section is to describe the relationship between two variables by finding the graph and equation of the straight line that represents the relationship. This straight line is called the *regression line,* and its equation is called the *regression equation.* Sir Francis Galton (1822–1911) studied the phenomenon of heredity and showed that when tall or short couples have children, the heights of those children tend to *regress,* or revert to the more typical mean height for people of the same gender. We continue to use Galton's "regression" terminology, even though our data do not involve the same height phenomena studied by Galton.

The accompanying box includes the definition of regression equation and regression line, as well as the notation and formulas we are using. The regression equation expresses a relationship between x (called the **independent variable,** or **predictor variable,** or **explanatory variable**) and $\hat{y}$ (called the **dependent variable,** or **response variable**). The typical equation of a straight line $y = mx + b$ is expressed in the form $\hat{y} = b_0 + b_1 x$, where b_0 is the y-intercept and b_1 is the slope. The given notation shows that b_0 and b_1 are sample statistics used to estimate the

Note to Instructor

It will be helpful to briefly review the $y = mx + b$ format of the equation of a straight line. What does m represent? What does b represent? What if the equation is changed to a format of $y = b_0 + b_1 x$? What is the slope? What is the y-intercept?

population parameters β_0 and β_1. We will use paired sample data to estimate the regression equation. Using only sample data, we can't find the exact values of the population parameters β_0 and β_1, but we can use the sample data to estimate them with b_0 and b_1, which are found by using Formulas 9-2 and 9-3.

Assumptions

1. We are investigating only *linear* relationships.

2. For each *x*-value, *y* is a random variable having a normal (bell-shaped) distribution. All of these *y* distributions have the same variance. Also, for a given value of *x*, the distribution of *y*-values has a mean that lies on the regression line. (Results are not seriously affected if departures from normal distributions and equal variances are not too extreme.)

Definitions

Given a collection of paired sample data, the **regression equation**

$$\hat{y} = b_0 + b_1 x$$

algebraically describes the relationship between the two variables. The graph of the regression equation is called the **regression line** (or *line of best fit*, or *least-squares line*).

Notation for Regression Equation

	Population Parameter	Sample Statistic
y-intercept of regression equation	β_0	b_0
Slope of regression equation	β_1	b_1
Equation of the regression line	$y = \beta_0 + \beta_1 x$	$\hat{y} = b_0 + b_1 x$

Finding the slope b_1 and *y*-intercept b_0 in the regression equation $\hat{y} = b_0 + b_1 x$

Formula 9-2   **Slope:** $b_1 = \dfrac{n(\Sigma xy) - (\Sigma x)(\Sigma y)}{n(\Sigma x^2) - (\Sigma x)^2}$

Formula 9-3   ***y*-intercept:** $b_0 = \bar{y} - b_1 \bar{x}$

The *y*-intercept b_0 can also be found using the formula shown below, but it is much easier to use Formula 9-3 instead.

$$b_0 = \frac{(\Sigma y)(\Sigma x^2) - (\Sigma x)(\Sigma xy)}{n(\Sigma x^2) - (\Sigma x)^2}$$

Formulas 9-2 and 9-3 might look intimidating, but they are programmed into many calculators and computer programs, so the values of b_0 and b_1 can be easily found. (See "Using Technology" at the end of this section.) In those cases when we must use formulas instead of a calculator or computer, the required computations will be much easier if we keep in mind the following facts:

1. If the linear correlation coefficient r has been computed using Formula 9-1, the values of Σx, Σy, Σx^2, and Σxy have already been found, and they can be used again in Formula 9-2. (Also, the numerator for r in Formula 9-1 is the same numerator for b_1 in Formula 9-2; the denominator for r includes the denominator for b_1. If the calculation for r is set up carefully, the calculation for b_1 requires the simple division of one known number by another.)

2. If you use Formula 9-2 to find the slope b_1 first, it is easy to use Formula 9-3 to find the y-intercept b_0. [The regression line always passes through the centroid $(\bar{x}, \bar{y})$ so that $\bar{y} = b_0 + b_1\bar{x}$ must be true, and this equation can be expressed as Formula 9-3.]

Once we have evaluated b_1 and b_0, we can identify the estimated regression equation, which has the following special property: *The regression line fits the sample points best.* (The specific criterion used to determine which line fits "best" is the least-squares property, which will be described later.) We will now briefly discuss rounding and then illustrate the procedure for finding and applying the regression equation.

Rounding the Slope b_1 and the y-Intercept b_0

It's difficult to provide a simple universal rule for rounding values of b_1 and b_0, but we usually try to round each of these values to *three significant digits* or use the values provided by STATDISK, Minitab, Excel, or a TI-83 Plus calculator. Because these values are very sensitive to rounding at intermediate steps of calculations, try to carry at least six significant digits (or use exact values) in the intermediate steps. Depending on how you round, this book's answers to examples and exercises may be slightly different from your answers.

EXAMPLE **Finding the Regression Equation** In Section 9-2 we used the values listed below to find that the linear correlation coefficient of $r = -0.135$. (Using the methods of Section 9-2, there is not a significant linear correlation between x and y.) Use the given sample data to find the regression equation.

x	1	1	3	5
y	2	8	6	4

SOLUTION   We will find the regression equation by using Formulas 9-2 and 9-3 and these values already found in Table 9-2 in Section 9-2:

$$n = 4 \qquad \Sigma x = 10 \qquad \Sigma y = 20$$
$$\Sigma x^2 = 36 \qquad \Sigma y^2 = 120 \qquad \Sigma xy = 48$$

continued

1° Forecast Error = $1 Billion

Although the prediction of forecast temperatures might seem to be an inexact science, many companies are working feverishly to obtain more accurate estimates. *USA Today* reporter Del Jones wrote that "the annual cost of electricity could decrease by at least $1 billion if the accuracy of weather forecasts improved by 1 degree Fahrenheit." When referring to the Tennessee Valley Authority, he states that "forecasts over its 80,000 square miles have been wrong by an average of 2.35 degrees the last 2 years, fairly typical of forecasts nationwide. Improving that to within 1.35 degrees would save TVA as much as $100,000 a day, perhaps more." Forecast temperatures are used to determine the allocation of power from generators, nuclear plants, hydroelectric plants, coal, natural gas, and wind. Statistical forecasting techniques are being refined so that money and natural resources can be saved.

Cell Phones and Crashes

Because some countries have banned the use of cell phones in cars while other countries are considering such a ban, researchers studied the issue of whether the use of cell phones while driving increases the chance of a crash. A sample of 699 drivers was obtained. Members of the sample group used cell phones and were involved in crashes. Subjects completed questionnaires and their telephone records were checked. Telephone usage was compared to the time interval immediately preceding a crash to a comparable time period the day before. Conclusion: Use of a cell phone was associated with a crash risk that was about four times as high as the risk when a cell phone was not used. (See "Association between Cellular-Telephone Calls and Motor Vehicle Collisions," by Redelmeier and Tibshirani, *New England Journal of Medicine*, Vol. 336, No. 7.)

First find the slope b_1 by using Formula 9-2:

$$b_1 = \frac{n(\Sigma xy) - (\Sigma x)(\Sigma y)}{n(\Sigma x^2) - (\Sigma x)^2}$$

$$= \frac{4(48) - (10)(20)}{4(36) - (10)^2} = \frac{-8}{44} = -0.181818 = -0.182$$

Next, find the y-intercept b_0 by using Formula 9-3 (with $\bar{y} = 20/4 = 5$ and $\bar{x} = 10/4 = 2.5$):

$$b_0 = \bar{y} - b_1\bar{x}$$
$$= 5 - (-0.181818)(2.5) = 5.45$$

Knowing the slope b_1 and y-intercept b_0, we can now express the estimated equation of the regression line as

$$\hat{y} = 5.45 - 0.182x$$

We should realize that this equation is an *estimate* of the true regression equation $y = \beta_0 + \beta_1 x$. This estimate is based on one particular set of sample data, but another sample drawn from the same population would probably lead to a slightly different equation.

 EXAMPLE **Boats and Manatees** Using the boat/manatee data in Table 9-1, we have found that the linear correlation coefficient is $r = 0.922$. Using the same sample data, find the equation of the regression line.

SOLUTION   Using the same procedure illustrated in the preceding example, or using technology, we can find that the 10 pairs of boat/manatee data in Table 9-1 result in $b_0 = -113$ and $b_1 = 2.27$. The Minitab display is shown on the next page. Substituting the computed values for b_0 and b_1, we express the regression equation as $\hat{y} = -113 + 2.27x$. Also shown below is the Minitab-generated scatterplot with the regression line included. We can see that the regression line fits the data well.

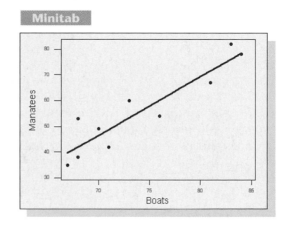

```
Minitab
The regression equation is

Manatees = -113 + 2.27 Boats

Predictor          Coef      SE Coef         T         P
Constant        -112.71        25.19     -4.47     0.002
Boats            2.2741        0.3388      6.71     0.000

S = 6.612       R-Sq = 84.9%    R-Sq(adj) = 83.0%
```

Using the Regression Equation for Predictions

Regression equations can be helpful when used for *predicting* the value of one variable, given some particular value of the other variable. If the regression line fits the data quite well, then it makes sense to use its equation for predictions, provided that we don't go beyond the scope of the available values. However, *we should use the equation of the regression line only if r indicates that there is a linear correlation. In the absence of a linear correlation, we should not use the regression equation for projecting or predicting; instead, our best estimate of the second variable is simply its sample mean.*

> **In predicting a value of y based on some given value of x . . .**
> 1. **If there is *not* a linear correlation, the best predicted y-value is ȳ.**
> 2. **If there is a linear correlation, the best predicted y-value is found by substituting the x-value into the regression equation.**

Figure 9-8 on the next page summarizes this process, which is easier to understand if we think of r as a measure of how well the regression line fits the sample data. If r is near -1 or $+1$, then the regression line fits the data well, but if r is near 0, then the regression line fits poorly (and should not be used for predictions).

Note to Instructor
Students typically experience some difficulty in determining the best predicted value of a variable. Discuss this subsection with reference to examples. Once again, the ultimate goal is not to obtain a regression equation. Instead, we want to use the regression equation in appropriate and meaningful ways.

EXAMPLE Predicting Manatee Deaths Using the sample data in Table 9-1, we found that there is a significant linear correlation between the number of registered boats and the number of manatees killed by boats, and we also found that the regression equation is $\hat{y} = -113 + 2.27x$. Assume that in 2001 there were 850,000 registered boats. Because Table 9-1 lists the numbers of registered boats in tens of thousands, this means that for 2001 we have $x = 85$. Given $x = 85$, find the best predicted value of y, the number of manatee deaths from boats.

SOLUTION   There's a strong temptation to jump in and substitute 85 for x in the regression equation, but we should first consider whether there is a linear correlation that justifies the use of that equation. In this example, we do have a

continued

Pizza Correlates with Crisis

When former President Clinton was threatened with impeachment by Congress, government employees worked late and ordered record numbers of pizzas. Frank Meeks, owner of 59 Domino's Pizza outlets in Washington, D.C., reported that on the Saturday during the height of the impeachment crisis, Capitol Hill pizza deliveries exceeded $10,000 while White House pizza deliveries totaled $3,000. Meeks noted that pizza sales also peaked during the Persian Gulf War, and they peak annually during budget debates.

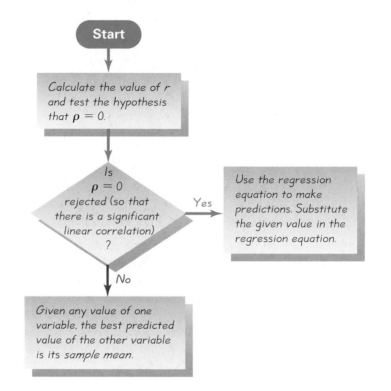

FIGURE 9-8 **Procedure for Predicting**

significant linear correlation (with $r = 0.922$), so our predicted value is found as follows:

$$\hat{y} = -113 + 2.27x$$
$$= -113 + 2.27(85) = 80.0$$

The predicted number of manatee deaths from 850,000 registered boats is 80.0. (If there had not been a significant linear correlation, our best predicted value would have been $\bar{y} = 558/10 = 55.8$.) The actual number of manatee deaths from boats in 2001 was 82, so the predicted value of 80.0 is quite close.

EXAMPLE **Hat Size and IQ** There is obviously no linear correlation between hat sizes and IQ scores of adults. Given that an individual has a hat size of 7, find the best predicted value of this person's IQ score.

SOLUTION   Because there is no linear correlation, we do not use a regression equation. There is no need to collect paired sample data consisting of hat size and IQ score for a sample of randomly selected adults. Instead, the best predicted IQ score is simply the mean IQ of all adults, which is 100.

Carefully compare the solutions to the preceding two examples and note that we used the regression equation when there was a linear correlation, but in the absence of such a correlation, the best predicted value of y is simply the value of the sample mean $\bar{y}$. A common error is to use the regression equation for making a prediction when there is no linear correlation. That error violates the first of the following guidelines.

Guidelines for Using the Regression Equation

1. *If there is no linear correlation, don't use the regression equation to make predictions.*

2. *When using the regression equation for predictions, stay within the scope of the available sample data.* If you find a regression equation that relates women's heights and shoe sizes, it's absurd to predict the shoe size of a woman who is 10 ft tall.

3. *A regression equation based on old data is not necessarily valid now.* The regression equation relating used-car prices and ages of cars is no longer usable if it's based on data from the 1970s.

4. *Don't make predictions about a population that is different from the population from which the sample data were drawn.* If we collect sample data from men and develop a regression equation relating age and TV remote-control usage, the results don't necessarily apply to women. If we use state *averages* to develop a regression equation relating SAT math scores and SAT verbal scores, the results don't necessarily apply to *individuals*.

Interpreting the Regression Equation: Marginal Change

We can use the regression equation to see the effect on one variable when the other variable changes by some specific amount.

> **Definition**
>
> In working with two variables related by a regression equation, the **marginal change** in a variable is the amount that it changes when the other variable changes by exactly one unit. The slope b_1 in the regression equation represents the marginal change in y that occurs when x changes by one unit.

For the boat/manatee data of Table 9-1, the regression line has a slope of 2.27, which shows that if we increase x (the number of registered boats in tens of thousands) by 1, the predicted number of deaths will increase by 2.27 manatees. That is, for every additional 10,000 registered boats, we expect about 2.27 additional manatee deaths from boats.

Outliers and Influential Points

A correlation/regression analysis of bivariate (paired) data should include an investigation of *outliers* and *influential points*, defined as follows.

Note to Instructor

If time does not allow discussion of outliers and influential points, this subsection can be easily omitted. Only Exercises 25 and 26 require this subsection.

> **Definitions**
>
> In a scatterplot, an **outlier** is a point lying far away from the other data points.
>
> Paired sample data may include one or more **influential points,** which are points that strongly affect the graph of the regression line.

An outlier is easy to identify: Examine the scatterplot and identify a point that is far away from the others. Here's how to determine whether a point is an influential point: Graph the regression line resulting from the data with the point included, then graph the regression line resulting from the data with the point excluded. If the graph changes by a considerable amount, the point is influential. Influential points are often found by identifying those outliers that are *horizontally* far away from the other points.

For example, refer to the preceding Minitab display. Suppose that we include this additional pair of data: $x = 200$, $y = 5$ (in a year with 2,000,000 registered boats, only 5 manatees were killed by boats). This additional point would be an influential point because the graph of the regression line would change considerably, as shown by the Minitab display here. Compare this regression line to the one shown in the preceding Minitab display, and you will see clearly that the addition of that one pair of values has a very dramatic effect on the regression line.

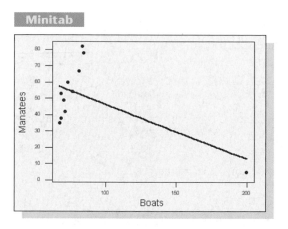

Residuals and the Least-Squares Property

We have stated that the regression equation represents the straight line that fits the data "best," and we will now describe the criterion used in determining the line that is better than all others. This criterion is based on the vertical distances between the original data points and the regression line. Such distances are called *residuals*.

Definition

For a sample of paired (x, y) data, a **residual** is the difference $(y - \hat{y})$ between an observed sample y-value and the value of $\hat{y}$, which is the value of y that is predicted by using the regression equation. That is,

$$\text{residual} = \text{observed } y - \text{predicted } y = y - \hat{y}$$

This definition might seem as clear as tax-form instructions, but you can easily understand residuals by referring to Figure 9-9, which corresponds to the paired sample data listed below. In Figure 9-9, the residuals are represented by the dashed lines. For a specific example, see the residual indicated as 7, which is directly above $x = 5$. If we substitute $x = 5$ into the regression equation $\hat{y} = 5 + 4x$, we get a predicted value of $\hat{y} = 25$. When $x = 5$, the *predicted* value of y is $\hat{y} = 25$, but the actual *observed* sample value is $y = 32$. The difference $y - \hat{y} = 32 - 25 = 7$ is a residual.

x	1	2	4	5
y	4	24	8	32

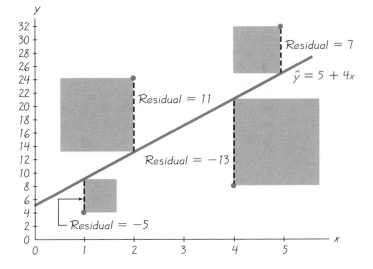

FIGURE 9-9 **Residuals and Squares of Residuals**

The regression equation represents the line that fits the points "best" according to the following *least-squares property.*

Definition

A straight line satisfies the **least-squares property** if the sum of the squares of the residuals is the smallest sum possible.

Note to Instructor

Point out that the standard deviation is one of many different statistics that are based on a sum of squares, so it is not too surprising to see that the criterion for the line of best fit is also based on a sum of squares.

From Figure 9-9, we see that the residuals are -5, 11, -13, and 7, so the sum of their squares is

$$(-5)^2 + 11^2 + (-13)^2 + 7^2 = 364$$

We can visualize the least-squares property by referring to Figure 9-9, where the squares of the residuals are represented by the red-square areas. The sum of the red-square areas is 364, which is the smallest sum possible. Use any other straight line, and the red squares will combine to produce an area larger than the combined red area of 364.

Fortunately, we need not deal directly with the least-squares property when we want to find the equation of the regression line. Calculus has been used to build the least-squares property into Formulas 9-2 and 9-3. Because the derivations of these formulas require calculus, we don't include them in this text.

Using Technology

Because of the messy calculations involved, the linear correlation coefficient r and the slope and y-intercept of the regression line are usually found by using a calculator or computer software.

STATDISK   First enter the paired data in columns of the Statdisk Data Window. Select **Analysis** from the main menu bar, then use the option **Correlation and Regression.** Enter a value for the significance level and select the columns of data. Click on the **Evaluate** button. The STATDISK display will include the value of the linear correlation coefficient along with the critical value of r, the conclusion about correlation, and the intercept and slope of the regression equation, as well as some other results. Click on **Scatterplot** to get a graph of the scatterplot with the regression line included.

Minitab   First enter the x values in column C1 and enter the y values in column C2. In Section 9-2 we saw that we could find the value of the linear correlation coefficient r by selecting **Stat/Basic Statistics/Correlation.** To get the equation of the regression line, select **Stat/Regression/Regression,** and enter C2 for "response" and C1 for "predictor." To get the graph of the scatterplot with the regression line, select **Stat/Regression/ Fitted Line Plot,** then enter C2 for the response variable and C1 for the predictor variable. Select the "linear" model.

Excel   Enter the paired data in columns A and B. Use Excel's Data Analysis add-in by selecting **Tools** from the main menu, then selecting **Data Analysis** and **Regression,** then clicking **OK.** Enter the range for the y values, such as B1:B10. Enter the range for the x values, such as A1:A10. Click on the box adjacent to Line Fit Plots, then click **OK.** Among all of the information provided by Excel, the slope and intercept of the regression equation can be found under the table heading "Coefficient." The displayed graph will include a scatterplot of the original sample points along with the points that would be predicted by the regression equation. You can easily get the regression line by connecting the "predicted y" points.

To use the Data Desk XL add-in, click on **DDXL** and select **Regression,** then click on the Function Type box and select **Simple Regression.** Click on the pencil icon for the response variable and enter the range of values for the y (or dependent) variable. Click on the pencil icon for the explanatory variable and enter the range of values for the x (or independent) variable. Click **OK.** The slope and intercept of the regression equation can be found under the table heading "Coefficient."

TI-83 Plus   Enter the paired data in lists L1 and L2, then press **STAT** and select **TESTS,** then choose the option **LinRegTTest.** The displayed results will include the y-intercept and slope of the regression equation. Instead of b_0 and b_1, the TI-83 display represents these values as a and b.

9-3 Basic Skills and Concepts

Recommended Assignment:
Exercises 1–12. (Exercises 19 and 20 deal with outliers and influential points.)

Making Predictions. In Exercises 1–4, use the given data to find the best predicted value of the dependent variable. Be sure to follow the prediction procedure described in this section.

1. In each of the following cases, find the best predicted value of y given that $x = 3.00$. The given statistics are summarized from paired sample data.
   a. $r = 0.987$, $\bar{y} = 5.00$, $n = 20$, and the equation of the regression line is $\hat{y} = 6.00 + 4.00x$.
   b. $r = 0.052$, $\bar{y} = 5.00$, $n = 20$, and the equation of the regression line is $\hat{y} = 6.00 + 4.00x$.

2. In each of the following cases, find the best predicted value of y given that $x = 2.00$. The given statistics are summarized from paired sample data.
   a. $r = -0.123$, $\bar{y} = 8.00$, $n = 30$, and the equation of the regression line is $\hat{y} = 7.00 - 2.00x$.
   b. $r = -0.567$, $\bar{y} = 8.00$, $n = 30$, and the equation of the regression line is $\hat{y} = 7.00 - 2.00x$.

3. **Chest Sizes and Weights of Bears** When eight bears were anesthetized, researchers measured the distances (in inches) around the bears' chests and weighed the bears (in pounds). Minitab was used to find that the value of the linear correlation coefficient is $r = 0.993$ and the equation of the regression line is $\hat{y} = -187 + 11.3x$, where x represents chest size. Also, the mean weight of the eight bears is 234.5 lb. What is the best predicted weight of a bear with a chest size of 52 in.?

4. **Stocks and Super Bowl** Data Set 25 in Appendix B includes pairs of data for the Dow-Jones Industrial Average (DJIA) high value and the total number of points scored in the Super Bowl for 21 different years. Excel was used to find that the value of the linear correlation coefficient is $r = -0.133$ and the regression equation is $\hat{y} = 53.3 - 0.000442x$, where x is the high value of the DJIA. Also, the mean number of Super Bowl points is 51.4. What is the best predicted value for the total number of Super Bowl points scored in a year with a DJIA high of 1200?

Finding the Equation of the Regression Line. In Exercises 5 and 6, use the given data to find the equation of the regression line.

5.
x	0	1	2	3	4
y	4	1	0	1	4

6.
x	1	2	2	5	6
y	2	5	4	15	15

7. **Effects of an Outlier** Refer to the Minitab-generated scatterplot given in Exercise 7 of Section 9-2.
   a. Using the pairs of values for all 10 points, find the equation of the regression line.
   b. After removing the point with coordinates (10, 10), use the pairs of values for the remaining nine points and find the equation of the regression line.
   c. Compare the results from parts (a) and (b).

1. a. 18.00
   b. 5.00

2. a. 8.00
   b. 3.00

3. 401 lb

4. 51.4

5. $\hat{y} = 2 + 0x$ (or $\hat{y} = 2$)

6. $\hat{y} = -0.957 + 2.86x$

7. a. $\hat{y} = 0.264 + 0.906x$
   b. $\hat{y} = 2 + 0x$ (or $\hat{y} = 2$)
   c. The results are very different, so one point can dramatically affect the regression equation.

Finding the Equation of the Regression Line and Making Predictions. Exercises 8–24 *use the same data sets as the exercises in Section 9-2. In each case, find the regression equation, letting the first variable be the independent (x) variable. Find the indicated predicted values. Caution: When finding predicted values, be sure to follow the prediction procedure described in this section.*

8. $\hat{y} = -1.13 + 0.0677x$; 3.04 acres

8. Fires and Acres Burned Find the best predicted value for the number of acres burned given that there were 80 fires.

Fires	73	69	58	48	84	62	57	45	70	63	48
Acres burned	6.2	4.2	1.9	2.7	5.0	1.6	3.0	1.6	1.5	2.0	3.7

9. $\hat{y} = 6.76 - 0.0111x$; 6.5 million. The predicted value of 6.5 million is very far from the actual value of 24 million.

9. Buying a TV Audience Find the best predicted value for the number of viewers (in millions), given that the salary (in millions of dollars) of television star Jennifer Anniston is $16 million. How does the predicted value compare to the actual number of viewers, which was 24 million?

Salary	100	14	14	35.2	12	7	5	1
Viewers	7	4.4	5.9	1.6	10.4	9.6	8.9	4.2

10. $\hat{y} = -152 + 3.88x$; 116 lb

10. Supermodel Heights and Weights Find the best predicted weight of a supermodel who is 69 in. tall.

Height (in.)	71	70.5	71	72	70	70	66.5	70	71
Weight (lb)	125	119	128	128	119	127	105	123	115

11. $\hat{y} = -14.4 + 0.769x$; 79

11. Blood Pressure Measurements Find the best predicted diastolic blood pressure for a person with a systolic reading of 122.

Systolic	138	130	135	140	120	125	120	130	130	144	143	140	130	150
Diastolic	82	91	100	100	80	90	80	80	80	98	105	85	70	100

12. $\hat{y} = 145 + 0.0316x$; 147.0771

12. Temperatures and Marathons Find the best predicted winning time for the 1990 marathon given that the temperature was 73 degrees. How does the predicted value compare to the actual winning time of 150.750 min?

x (temperature)	55	61	49	62	70	73	51	57
y (time)	145.283	148.717	148.300	148.100	147.617	146.400	144.667	147.533

13. $\hat{y} = 139 + 2.48x$; 175.2

13. Smoking and Nicotine Find the best predicted level of cotinine for a person who smokes 40 cigarettes per day.

x (cigarettes per day)	60	10	4	15	10	1	20	8	7	10	10	20
y (cotinine)	179	283	75.6	174	209	9.51	350	1.85	43.4	25.1	408	344

14. $\hat{y} = 22.5 + 5.34x$; 44 ft. Easier to measure circumference.

14. Tree Circumference and Height Find the best predicted height of a tree that has a circumference of 4.0 ft. What is an advantage of being able to determine the height of a tree from its circumference?

x (circ.)	1.8	1.9	1.8	2.4	5.1	3.1	5.5	5.1	8.3	13.7	5.3	4.9	3.7	3.8
y (ht)	21.0	33.5	24.6	40.7	73.2	24.9	40.4	45.3	53.5	93.8	64.0	62.7	47.2	44.3

T **15.** Cereal Killers Refer to Data Set 16 in Appendix B and use the amounts of fat (x) and the measured calorie counts (y). Find the best predicted calorie count for a cereal with 0.05 grams of fat per gram of cereal.

T **16.** Tobacco and Alcohol in Children's Movies Refer to Data Set 7 in Appendix B and use the times that the animated children's movies showed tobacco use (x) and alcohol use (y). Find the best predicted time for alcohol use, given that a movie does not show any tobacco use.

T **17.** Cholesterol and Body Mass Index Refer to Data Set 1 in Appendix B and use the cholesterol levels (x) and body mass index values (y) of the 40 women. What is the best predicted value for the body mass index of a woman having a cholesterol level of 500?

T **18.** Readability Levels Refer to Data Set 14 in Appendix B and use the Flesch Reading Ease scores (x) and the Flesch-Kincaid Grade Level values (y) for Tom Clancy's *The Bear and the Dragon*. Find the best predicted Flesch-Kincaid Grade Level value for a page with a Flesch Reading Ease score of 50.0.

T **19.** Home Selling Prices, List Prices, and Taxes Refer to Data Set 24 in Appendix B. *Caution:* The sample values of list prices and selling prices are in thousands of dollars, but the tax amounts are in dollars.
   a. Use the paired data consisting of home list price (x) and selling price (y). What is the best predicted selling price of a home with a list price of $200,000?
   b. Use the paired data consisting of home selling price (x) and the amount of taxes (y). What is the best predicted tax bill for a home that sold for $400,000?

T **20.** Tar and Nicotine Refer to Data Set 5 in Appendix B.
   a. Use the paired data consisting of tar (x) and nicotine (y). What is the best predicted nicotine level for a cigarette with 15 mg of tar?
   b. Use the paired data consisting of carbon monoxide (x) and nicotine (y). What is the best predicted nicotine level for a cigarette with 15 mg of carbon monoxide?

T **21.** Forecasting Weather Refer to Data Set 10 in Appendix B.
   a. Use the five-day forecast high temperatures (x) and the actual high temperatures (y). What is the best predicted actual high temperature if the five-day forecast high temperature is 28°?
   b. Use the one-day forecast high temperatures (x) and the actual high temperatures (y). What is the best predicted actual high temperature if the one-day forecast high temperature is 28°?
   c. Which predicted value is better: the result from part (a) or the result from part (b)? Why?

T **22.** Florida Everglades Refer to Data Set 12 in Appendix B.
   a. Use the bottom temperatures (x) and the conductivity measurements (y). What is the best predicted conductivity measurement for a time when the bottom temperature is 30.0°C?
   b. Use the rainfall amounts (x) and the conductivity measurements (y). What is the best predicted conductivity measurement for a time when the rainfall amount is 0.00 in.?
   c. After identifying the best predicted conductivity measurement from parts (a) and (b), is either of the predicted values likely to be accurate? Why or why not?

15. $\hat{y} = 3.68 + 3.78x$; 3.76

16. $\hat{y} = 23.7 + 0.152x$; 32.5

17. $\hat{y} = 21.9 + 0.0160x$; 29.9

18. $\hat{y} = 20.5 - 0.197x$; 10.7

19. a. $\hat{y} = 7.26 + 0.914x$; $190,060
    b. $\hat{y} = 380 + 19.5x$; $8180

20. a. $\hat{y} = 0.154 + 0.0651x$; 1.1
    b. $\hat{y} = 0.192 + 0.0606x$; 1.1

21. a. $\hat{y} = 13.8 + 0.611x$; 31°
    b. $\hat{y} = 13.8 + 0.634x$; 32°
    c. Part (b), because r is higher.

22. a. $\hat{y} = 28.6 + 0.611x$; 47.12
    b. $\hat{y} = 47.6 - 1.84x$; 47.12
    c. No, because neither case involves a significant linear correlation.

23. a. $\hat{y} = 41.9 + 0.179x$; 79 min
b. $\hat{y} = 81.9 - 0.009x$; 81 min
c. Part (a), because there is a significant linear correlation between durations and intervals, but not between heights and intervals.

24. a. $\hat{y} = -6158 + 12201x$; $12,144
b. $\hat{y} = 28,647 - 3044x$; $19,515
c. Part (a), because r is higher.

25. Yes; no. The point is far away from the others, but it doesn't have a dramatic effect on the regression line.

26. Yes; yes. The point is far away from the others, and it does have a dramatic effect on the regression line.

27. $\hat{y} = -182 + 0.000351x$; $\hat{y} = -182 + 0.351x$. Slope is multiplied by 1000 and y-intercept doesn't change. If each y entry is divided by 1000, the slope and the y-intercept are both divided by 1000.

28. Residuals are -7, 10, -12, and 9, and the sum of their squares is 374.

x	1	2	4	5
y	4	24	8	32

29. The equation $\hat{y} = -49.9 + 27.2x$ is better because it has $r = 0.997$, which is higher than $r = 0.963$ for $\hat{y} = -103.2 + 134.9 \ln x$.

T 23. Old Faithful Refer to Data Set 13 in Appendix B.
   a. Use the paired data for durations (x) and intervals (y) after eruptions of the geyser. What is the best predicted time before the next eruption if the previous eruption lasted for 210 sec?
   b. Use the paired data for heights of eruptions (x) and intervals (y) after eruptions of the Old Faithful geyser. What is the best predicted time before the next eruption if the previous eruption had a height of 275 ft?
   c. Which predicted time is better: the result from part (a) or the result from part (b)? Why?

T 24. Diamond Prices, Carats, and Color Refer to Data Set 18 in Appendix B.
   a. Use the paired data consisting of the carat weight (x) and the price (y). What is the best predicted price of a diamond with a weight of 1.5 carats?
   b. Use the paired color (x) and price (y) data. What is the best predicted price of a diamond with a color rating of 3?
   c. Which predicted price is better: the result from part (a) or the result from part (b)? Why?

25. Identifying Outliers and Influential Points Refer to the sample data listed in Table 9-1. If we include another pair of values consisting of $x = 120$ (for 1,200,000 boats) and $y = 160$ (manatee deaths from boats), is the new point an outlier? Is it an influential point?

26. Identifying Outliers and Influential Points Refer to the sample data listed in Table 9-1. If we include another pair of values consisting of $x = 120$ (for 1,200,000 boats) and $y = 10$ (manatee deaths from boats), is the new point an outlier? Is it an influential point?

9-3 Beyond the Basics

27. How Is a Regression Equation Affected by Change in Scale? Large numbers, such as those in the accompanying table, often cause computational problems. First use the given data to find the equation of the regression line, then find the equation of the regression line after each x-value has been divided by 1000. How are the results affected by the change in x? How would the results be affected if each y-value were divided by 1000?

x	924,736	832,985	825,664	793,427	857,366
y	142	111	109	95	119

28. Testing Least-Squares Property According to the least-squares property, the regression line minimizes the sum of the squares of the residuals. We noted that with the paired data in the margin, the regression equation is $\hat{y} = 5 + 4x$ and the sum of the squares of the residuals is 364. Show that the equation $\hat{y} = 8 + 3x$ results in a sum of squares of residuals that is greater than 364.

29. Using Logarithms to Transform Data If a scatterplot reveals a nonlinear (not a straight line) pattern that you recognize as another type of curve, you may be able to apply the methods of this section. For the data given in the margin, find the linear equation ($y = b_0 + b_1x$) that best fits the sample data, and find the logarithmic equation ($y = a + b \ln x$) that best fits the sample data. (*Hint:* Begin by replacing each x-value with $\ln x$.) Which of these two equations fits the data better? Why?

x	2.0	2.5	4.2	10.0
y	12.0	18.7	53.0	225.0

30. *Equivalent Hypothesis Tests* Explain why a test of the null hypothesis H_0: $\rho = 0$ is equivalent to a test of the null hypothesis H_0: $\beta_1 = 0$ where ρ is the linear correlation coefficient for a population of paired data, and β_1 is the slope of the regression line for that same population.

31. *Residual Plot*   A scatterplot is a plot of the paired (x, y) sample data. A **residual plot** is a graph of the points with the same x-coordinates, but the corresponding y-coordinates are the residual values. To construct a residual plot, use the same x-axis as the scatterplot, but use a vertical axis of residual values. Draw a horizontal reference line through the residual value of 0, then plot the paired values of (x, residual). Residual plots are helpful in identifying patterns suggesting that the relationship between the variables is not linear, or that the assumption of constant variances is not satisfied. Construct a residual plot for the data in Table 9-1. Are there any noticeable patterns?

30. With $\beta_1 = 0$, the regression line is horizontal so that different values of x result in the same y value, and there is no correlation between x and y.

31. No. See Appendix F for the residual plot.

9-4   Variation and Prediction Intervals

So far, we have used paired sample data to test for a linear correlation between x and y, and to identify the regression equation. In this section we continue to analyze paired (x, y) data as we proceed to consider different types of variation that can be used for two major applications:

Note to Instructor
The calculations in this section will seriously challenge students without software or calculators capable of dealing with two-variable statistics.

1. To determine the proportion of the variation in y that can be explained by the linear relationship between x and y.

2. To construct interval estimates of predicted y-values. Such intervals are called *prediction intervals,* which are formally defined later in the section.

Explained and Unexplained Variation

In Section 9-2 we introduced the concept of correlation and used the linear correlation coefficient r in determining whether there is a significant linear correlation between two variables, denoted by x and y. In addition to serving as a measure of the linear correlation between two variables, the value of r can also provide us with additional information about the variation of sample points about the regression line. We begin with a sample case, which leads to an important definition (coefficient of determination).

Suppose we have a large collection of paired data, with these results:

- There is a significant linear correlation.
- The equation of the regression line is $\hat{y} = 3 + 2x$.
- The mean of the y-values is given by $\bar{y} = 9$.
- One of the pairs of sample data is $x = 5$ and $y = 19$.
- The point (5, 13) is one of the points on the regression line, because substi-

Wage Gender Gap

Although a recent report by *Working Woman* magazine states that the earnings gap based on gender is narrowing, men still hold the higher paying jobs for the most part. The most recent data indicate that, on average, full-time female workers earn about 73¢ for each $1 earned by full-time male workers. Researchers at the Institute for Social Research at the University of Michigan analyzed the effects of various key factors and found that about one-third of the discrepancy between female and male earnings can be explained by differences in education, seniority, work interruptions, and job choices. The other two-thirds remains unexplained by such labor factors.

tuting $x = 5$ into the regression equation yields $\hat{y} = 13$.

$$\hat{y} = 3 + 2x = 3 + 2(5) = 13$$

Figure 9-10 shows that the point (5, 13) lies on the regression line, but the point (5, 19) is from the original data set and does not lie on the regression line because it does not satisfy the regression equation. Take time to examine Figure 9-10 carefully and note the differences defined as follows.

Unexplained, Explained, and Total Deviation

Definitions

Assume that we have a collection of paired data containing the sample point (x, y), that $\hat{y}$ is the predicted value of y (obtained by using the regression equation), and that the mean of the sample y-values is $\overline{y}$.

The **total deviation** (from the mean) of the particular point (x, y) is the vertical distance $y - \overline{y}$, which is the distance between the point (x, y) and the horizontal line passing through the sample mean $\overline{y}$.

The **explained deviation** is the vertical distance $\hat{y} - \overline{y}$, which is the distance between the predicted y-value and the horizontal line passing through the sample mean $\overline{y}$.

The **unexplained deviation** is the vertical distance $y - \hat{y}$, which is the vertical distance between the point (x, y) and the regression line. (The distance $y - \hat{y}$ is also called a *residual*, as defined in Section 9-3.)

FIGURE 9-10 Unexplained, Explained, and Total Deviation

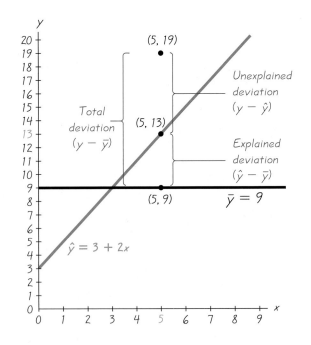

For the specific data under consideration, we get these results:

Total deviation of $(5, 19) = y - \bar{y} = 19 - 9 = 10$

Explained deviation of $(5, 19) = \hat{y} - \bar{y} = 13 - 9 = 4$

Unexplained deviation of $(5, 19) = y - \hat{y} = 19 - 13 = 6$

If we were totally ignorant of correlation and regression concepts and wanted to predict a value of y given a value of x and a collection of paired (x, y) data, our best guess would be $\bar{y}$. But we are not totally ignorant of correlation and regression concepts: We know that in this case (with a significant linear correlation) the way to predict the value of y when $x = 5$ is to use the regression equation, which yields $\hat{y} = 13$, as calculated earlier. We can explain the discrepancy between $\bar{y} = 9$ and $\hat{y} = 13$ by simply noting that there is a significant linear correlation best described by the regression line. Consequently, when $x = 5$, y *should be* 13, not the mean value of 9. But whereas y should be 13, *it is* 19. The discrepancy between 13 and 19 cannot be explained by the regression line, and it is called an *unexplained deviation,* or a *residual.* The specific case illustrated in Figure 9-10 can be generalized as follows:

$$
\begin{array}{ccccc}
\text{(total deviation)} & = & \text{(explained deviation)} & + & \text{(unexplained deviation)} \\
\text{or} \qquad (y - \bar{y}) & = & (\hat{y} - \bar{y}) & + & (y - \hat{y})
\end{array}
$$

This last expression applies to a particular point (x, y), and the same relationship applies to the sums of squares shown in Formula 9-4, even though this last expression is not algebraically equivalent to Formula 9-4. In Formula 9-4, the **total variation** is expressed as the sum of the squares of the total deviation values, the **explained variation** is the sum of the squares of the explained deviation values, and the **unexplained variation** is the sum of the squares of the unexplained deviation values.

Formula 9-4

$$
\begin{array}{ccccc}
\text{(total variation)} & = & \text{(explained variation)} & + & \text{(unexplained variation)} \\
\text{or} \qquad \Sigma(y - \bar{y})^2 & = & \Sigma(\hat{y} - \bar{y})^2 & + & \Sigma(y - \hat{y})^2
\end{array}
$$

Coefficient of Determination

The components of Formula 9-4 are used in the following important definition.

Definition

The **coefficient of determination** is the amount of the variation in y that is explained by the regression line. It is computed as

$$
r^2 = \frac{\text{explained variation}}{\text{total variation}}
$$

We can compute r^2 by using the definition just given with Formula 9-4, or we can simply square the linear correlation coefficient r, which is found by using the methods described in Section 9-2. For example, in Section 9-2 we noted that if $r = 0.922$, then $r^2 = 0.850$, which means that *85.0% of the total variation in y can be explained by the linear relationship between x and y (as described by the regression equation). It follows that 15.0% of the total variation in y remains unexplained.*

EXAMPLE **Diamonds** In Exercise 24(a) in Section 9-2, we find that for the paired data consisting of weights (in carats) of diamonds and the prices of the diamonds, the linear correlation coefficient is given by $r = 0.767$. Find the percentage of the variation in y (price) that can be explained by the linear relationship between the weight and the price.

SOLUTION   The coefficient of determination is $r^2 = 0.767^2 = 0.588$, indicating that the ratio of explained variation in y to total variation in y is 0.588. We can now state that 58.8% of the total variation in y can be explained by the regression equation. We interpret this to mean that 58.8% of the total variation in diamond prices can be explained by the variation in their weights; the other 41.2% is attributable to other factors, such as color, clarity, and random chance. But remember that these results are estimates based on the given sample data. Other sample data will likely result in different estimates.

Prediction Intervals

In Section 9-3 we used the Table 9-1 sample data to find the regression equation $\hat{y} = -113 + 2.27x$, where $\hat{y}$ represents the predicted number of manatee deaths and x represents the number of boats (in tens of thousands). We then used that equation to predict the y-value, given that $x = 85$ (for 850,000 boats). We found that the best predicted number of manatee deaths is 80.0. If we use the unrounded values for slope and intercept, we get the more accurate result of 80.6 manatee deaths. Because 80.6 is a single value, it is referred to as a *point estimate.* In Chapter 6 we saw that point estimates have the serious disadvantage of not giving us any information about how accurate they might be. Here, we know that 80.6 is the best predicted value, but we don't know how accurate that value is. In Chapter 6 we developed confidence interval estimates to overcome that disadvantage, and in this section we follow that precedent. We will use a **prediction interval,** which is an interval estimate of a predicted value of y.

The development of a prediction interval requires a measure of the spread of sample points about the regression line. Recall that the unexplained deviation (or residual) is the vertical distance between a sample point and the regression line, as illustrated in Figure 9-10. The *standard error of estimate* is a collective measure of the spread of the sample points about the regression line, and it is formally defined as follows.

Definition

The **standard error of estimate,** denoted by s_e, is a measure of the differences (or distances) between the observed sample y-values and the predicted values $\hat{y}$ that are obtained using the regression equation. It is given as

$$s_e = \sqrt{\frac{\Sigma(y - \hat{y})^2}{n - 2}} \qquad \text{(where } \hat{y} \text{ is the predicted } y\text{-value)}$$

or as the following equivalent formula:

Formula 9-5
$$s_e = \sqrt{\frac{\Sigma y^2 - b_0 \Sigma y - b_1 \Sigma xy}{n - 2}}$$

Note to Instructor

The calculation of the value of s_e is challenging, but this statistic is important enough to be included among the values provided by STATDISK, Minitab, Excel, and the TI-83 Plus calculator.

STATDISK, Minitab, Excel, and the TI-83 Plus calculator are all designed to automatically compute the value of s_e. See "Using Technology" at the end of this section.

The development of the standard error of estimate s_e closely parallels that of the ordinary standard deviation introduced in Chapter 2. Just as the standard deviation is a measure of how values deviate from their mean, the standard error of estimate s_e is a measure of how sample data points deviate from their regression line. The reasoning behind dividing by $n - 2$ is similar to the reasoning that led to division by $n - 1$ for the ordinary standard deviation. It is important to note that relatively smaller values of s_e reflect points that stay close to the regression line, and relatively larger values occur with points farther away from the regression line.

Formula 9-5 is algebraically equivalent to the other expression in the definition, but Formula 9-5 is generally easier to work with because it doesn't require that we compute each of the predicted values $\hat{y}$ by substitution in the regression equation. However, Formula 9-5 does require that we find the y-intercept b_0 and the slope b_1 of the estimated regression line.

EXAMPLE Use Formula 9-5 to find the standard error of estimate s_e for the boat/manatee sample data listed in Table 9-1.

SOLUTION   Using the sample data in Table 9-1, we find these values:

$$n = 10 \qquad \Sigma y^2 = 33{,}456 \qquad \Sigma y = 558 \qquad \Sigma xy = 42{,}214$$

In Section 9-3 we used the Table 9-1 sample data to find the y-intercept and the slope of the regression line. Those values are given here with extra decimal places for greater precision.

$$b_0 = -112.7098976 \qquad b_1 = 2.274087687$$

continued

We can now use these values in Formula 9-5 to find the standard error of estimate s_e.

$$s_e = \sqrt{\frac{\Sigma y^2 - b_0\Sigma y - b_1\Sigma xy}{n - 2}}$$

$$= \sqrt{\frac{33{,}456 - (-112.7098976)(558) - (2.274087687)(42{,}214)}{10 - 2}}$$

$$= 6.6123487 = 6.61 \quad \text{(rounded)}$$

We can measure the spread of the sample points about the regression line with the standard error of estimate $s_e = 6.61$.

We can use the standard error of estimate s_e to construct interval estimates that will help us see how dependable our point estimates of y really are. Assume that for each fixed value of x, the corresponding sample values of y are normally distributed about the regression line, and those normal distributions have the same variance. The following interval estimate applies to an *individual* y-value. (For a confidence interval used to predict the *mean* of all y-values for some given x-value, see Exercise 24.)

Note to Instructor

Minitab can be used to generate 95% prediction intervals, which are displayed and identified as "95% P. I." The default is 95%, but it can be changed as desired.

Prediction Interval for an Individual y

Given the fixed value x_0, the prediction interval for an individual y is

$$\hat{y} - E < y < \hat{y} + E$$

where the margin of error E is

$$E = t_{\alpha/2}s_e \sqrt{1 + \frac{1}{n} + \frac{n(x_0 - \bar{x})^2}{n(\Sigma x^2) - (\Sigma x)^2}}$$

and x_0 represents the given value of x, $t_{\alpha/2}$ has $n - 2$ degrees of freedom, and s_e is found from Formula 9-5.

EXAMPLE Boats and Manatees For the paired boat/manatee data in Table 9-1, we have found that when $x = 85$ (for 850,000 boats), the best predicted number of manatee deaths is 80.0, but we obtain a predicted value of 80.6 when we use more precise values of the y-intercept b_0 and slope b_1. Construct a 95% prediction interval for the number of manatee deaths from boats, given that the number of boats is 850,000 (so that $x = 85$). This will provide a sense of how accurate the predicted value of 80.6 really is.

SOLUTION   In previous sections we have shown that there is a significant linear correlation (at the 0.05 significance level), and the regression equation is

$\hat{y} = -113 + 2.27x$. In the preceding example we found that $s_e = 6.6123487$, and the following statistics are obtained from the Table 9-1 sample data:

$$n = 10 \qquad \bar{x} = 74.1 \qquad \Sigma x = 741 \qquad \Sigma x^2 = 55{,}289$$

From Table A-3 we find $t_{\alpha/2} = 2.306$. (We used $10 - 2 = 8$ degrees of freedom with $\alpha = 0.05$ in two tails.) We first calculate the margin of error E by letting $x_0 = 85$, because we want the prediction interval of the number of manatee deaths given that $x = 85$ (for 850,000 boats).

$$E = t_{\alpha/2}s_e \sqrt{1 + \frac{1}{n} + \frac{n(x_0 - \bar{x})^2}{n(\Sigma x^2) - (\Sigma x)^2}}$$

$$= (2.306)(6.6123487)\sqrt{1 + \frac{1}{10} + \frac{10(85 - 74.1)^2}{10(55{,}289) - (741)^2}}$$

$$= (2.306)(6.6123487)(1.1882420) = 18.1$$

With $\hat{y} = 80.6$ and $E = 18.1$, we get the prediction interval as follows:

$$\hat{y} - E < y < \hat{y} + E$$
$$80.6 - 18.1 < y < 80.6 + 18.1$$
$$62.5 < y < 98.7$$

That is, for $x = 85$ (for 850,000 boats), we are 95% certain that the number of manatee deaths is between 62.5 and 98.7. That's a relatively large range. (One factor contributing to the large range is that the sample size is very small because we are using only 10 pairs of sample data.)

Minitab can be used to find the prediction interval limits. If Minitab is used here, it will provide the result of (62.5, 98.7) below the heading "95.0% P.I." This corresponds to the same prediction interval found above.

In addition to knowing that for $x = 85$, the predicted number of manatee deaths is 80.6, we now have a sense of how reliable that estimate really is. The 95% prediction interval found in this example shows that the actual value of y can vary substantially from the predicted value of 80.6.

Using Technology

STATDISK   STATDISK can be used to find the linear correlation coefficient r, the equation of the regression line, the standard error of estimate s_e, the total variation, the explained variation, the unexplained variation, and the coefficient of determination. Enter the paired data in columns of the Statdisk Data Window, select **Analysis,** then select **Correlation and Regression.** Enter a value for the significance level and select the columns of data. Click on the **Evaluate** button. The STATDISK display will include the linear correlation coefficient, the coefficient of determination, the regression equation, and the value of the standard error of estimate s_e.

Minitab   Minitab can be used to find the regression equation, the standard error of estimate s_e (labeled S), the value of the coefficient of determination (labeled R-sq), and the limits of a

continued

prediction interval. Enter the *x*-data in column C1 and the *y*-data in column C2, then select the options **Stat, Regression,** and **Regression.** Enter C2 in the box labeled "Response" and enter C1 in the box labeled "Predictors." If you want a prediction interval for some given value of *x*, click on the **Options** box and enter the desired value of x_0 in the box labeled "Prediction intervals for new observations."

Excel   Excel can be used to find the regression equation, the standard error of estimate s_e, and the coefficient of determination (labeled as R square). Enter the paired data in columns A and B.

To use Excel's Data Analysis add-in, select **Tools** from the main menu, then select **Data Analysis,** followed by **Regression,** and then click **OK.** Enter the range for the *y* values, such as B1:B10. Enter the range for the *x* values, such as A1:A10. Click **OK.**

To use the Data Desk XL add-in, click **DDXL** and select **Regression,** then click on the Function Type box and select **Simple Regression.** Click on the pencil icon for the response variable and enter the range of values for the *y* (or dependent) variable. Click on the pencil icon for the explanatory variable and enter the range of values for the *x* (or independent) variable. Click on **OK.**

TI-83 Plus   The TI-83 Plus calculator can be used to find the linear correlation coefficient *r*, the equation of the regression line, the standard error of estimate s_e, and the coefficient of determination (labeled as r^2). Enter the paired data in lists L1 and L2, then press **STAT** and select **TESTS,** and then choose the option **LinRegTTest.**

9-4 Basic Skills and Concepts

Recommended Assignment:
Exercises 1–10, 17–20. (Exercises 1–4 are quick and easy. Exercises 5–8 simply require interpretation of a given computer display.)

1. 0.64; 64%
2. 0.36; 36%
3. 0.253; 25.3%
4. 0.404; 40.4%

Interpreting the Coefficient of Determination. In Exercises 1–4, use the value of the linear correlation coefficient r to find the coefficient of determination and the percentage of the total variation that can be explained by the linear relationship between the two variables.

1. $r = 0.8$

2. $r = -0.6$

3. $r = -0.503$

4. $r = 0.636$

Interpreting a Computer Display. In Exercises 5–8, refer to the Minitab display that was obtained by using the paired data consisting of tar and nicotine for a sample of 29 cigarettes, as listed in Data Set 5 in Appendix B. Along with the paired sample data, Minitab was also given a tar amount of 17 mg to be used for predicting the amount of nicotine.

Minitab

```
The regression equation is
Nicotine = 0.154 + 0.0651 Tar

Predictor         Coef       SE Coef         T         P
Constant       0.15403       0.04635       3.32     0.003
Tar           0.065052      0.003585      18.15     0.000

S = 0.08785    R-Sq = 92.4%    R-Sq(adj) = 92.1%

Predicted Values for New Observations

New Obs     Fit SE Fit       95.0% CI          95.0% PI
1        1.2599 0.0240 ( 1.2107, 1.3091)( 1.0731, 1.4468)
```

5. Testing for Correlation Using the information provided in the display determine the value of the linear correlation coefficient. Given that there are 29 pairs of data, is there a significant linear correlation between the amount of tar and the amount of nicotine in a cigarette?

5. 0.961; yes

6. Identifying Total Variation What percentage of the total variation in nicotine can be explained by the linear relationship between tar and nicotine?

6. 92.4%

7. Predicting Nicotine Amount If a cigarette has 17 mg of tar, what is the single value that is the best predicted amount of nicotine? (Assume that there is a significant linear correlation between tar and nicotine.)

7. 1.3

8. Finding Prediction Interval For a given tar amount of 17 mg, identify the 95% prediction interval estimate of the amount of nicotine, and write a statement interpreting that interval.

8. $1.1 < y < 1.4$; we have 95% confidence that the limits of 1.1 and 1.4 contain the true amount of nicotine.

Finding Measures of Variation. In Exercises 9–12, find the (a) explained variation, (b) unexplained variation, (c) total variation, (d) coefficient of determination, and (e) standard error of estimate s_e. In each case, there is a significant linear correlation so that it is reasonable to use the regression equation when making predictions.

9. Supermodel Heights and Weights Listed below are heights (in inches) and weights (in pounds) for supermodels Niki Taylor, Nadia Avermann, Claudia Schiffer, Elle MacPherson, Christy Turlington, Bridget Hall, Kate Moss, Valerie Mazza, and Kristy Hume.

Height (in.)	71	70.5	71	72	70	70	66.5	70	71
Weight (lb)	125	119	128	128	119	127	105	123	115

9. a. 287.37026
b. 166.62974
c. 454
d. 0.63297415
e. 4.8789597

10. Blood Pressure Measurements Fourteen different second-year medical students took blood pressure measurements of the same patient and the results are listed below (data provided by Marc Triola, MD).

Systolic	138	130	135	140	120	125	120	130	130	144	143	140	130	150
Diastolic	82	91	100	100	80	90	80	80	80	98	105	85	70	100

10. a. 628.96027
b. 824.25402
c. 1453.2143
d. 0.43280628
e. 8.287812413

11. Tree Circumference and Height Listed below are the circumferences (in feet) and the heights (in feet) of trees in Marshall, Minnesota (based on data from "Tree Measurements" by Stanley Rice, *American Biology Teacher*, Vol. 61, No. 9).

x (circ.)	1.8	1.9	1.8	2.4	5.1	3.1	5.5	5.1	8.3	13.7	5.3	4.9	3.7	3.8
y (ht)	21.0	33.5	24.6	40.7	73.2	24.9	40.4	45.3	53.5	93.8	64.0	62.7	47.2	44.3

11. a. 3696.9263
b. 1690.5830
c. 5387.5093
d. 0.68620324
e. 11.869369

12. Readability Levels Refer to Data Set 14 in Appendix B and use the characters per word (*x*) and the Flesch Reading Ease scores (*y*) for J. K. Rowling's *Harry Potter and the Sorcerer's Stone*.

12. a. 137.46857
b. 103.60143
c. 241.07
d. 0.57024338
e. 3.2187176

13. Effect of Variation on Prediction Interval Refer to the data given in Exercise 9 and assume that the necessary conditions of normality and variance are met.
a. Find the predicted weight of a supermodel who is 69 in. tall.
b. Find a 95% prediction interval estimate of the weight of a supermodel who is 69 in. tall.

13. a. 116 lb
b. $103.7 \text{ lb} < y < 128.8 \text{ lb}$

14. a. 78
    b. $57.7 < y < 98.2$

15. a. 44 ft
    b. $6.2 \text{ ft} < y < 81.4 \text{ ft}$

16. a. 87.3
    b. $75.25 < y < 99.37$

17. $54.2 < y < 107.0$
18. $66.0 < y < 95.2$
19. $71.7 < y < 112.2$
20. $62.5 < y < 121.4$

21. $-170.8 < \beta_0 < -54.6;$
    $1.5 < \beta_1 < 3.1$

14. Finding Predicted Value and Prediction Interval Refer to Exercise 10 and assume that the necessary conditions of normality and variance are met.
   a. Find the predicted diastolic reading given that the systolic reading is 120.
   b. Find a 95% prediction interval estimate of the diastolic reading given that the systolic reading is 120.

15. Finding Predicted Value and Prediction Interval Refer to the data given in Exercise 11 and assume that the necessary conditions of normality and variance are met.
   a. Find the predicted height of a tree that has a circumference of 4.0 ft.
   b. Find a 99% prediction interval estimate of the height of a tree that has a circumference of 4.0 ft.

16. Finding Predicted Value and Prediction Interval Refer to the data described in Exercise 12 and assume that the necessary conditions of normality and variance are met.
   a. Find the predicted Flesch Reading Ease score for a page that has an average of 4.0 characters per word.
   b. Find a 99% prediction interval estimate of the Flesch Reading Ease score for a page that has an average of 4.0 characters per word.
   c. How do the results from parts (a) and (b) compare to the observed pair of data consisting of 4.0 characters per word and a Flesch Reading Ease score of 86.2?

Finding a Prediction Interval. In Exercises 17–20, refer to the Table 9-1 sample data. Let x represent the number of registered boats (in tens of thousands) and let y represent the number of manatee deaths from boats. Use the given number of registered boats (in tens of thousands) and the given confidence level to construct a prediction interval estimate of the number of manatee deaths from boats. (See the example in this section.)

17. $x = 85$ (for 850,000 boats); 99% confidence

18. $x = 85$ (for 850,000 boats); 90% confidence

19. $x = 90$ (for 900,000 boats); 95% confidence

20. $x = 90$ (for 900,000 boats); 99% confidence

9-4 Beyond the Basics

21. Confidence Intervals for β_0 and β_1 Confidence intervals for the y-intercept β_0 and slope β_1 for a regression line ($y = \beta_0 + \beta_1 x$) can be found by evaluating the limits in the intervals below.

$$b_0 - E < \beta_0 < b_0 + E$$

where

$$E = t_{\alpha/2} s_e \sqrt{\frac{1}{n} + \frac{\bar{x}^2}{\Sigma x^2 - \frac{(\Sigma x)^2}{n}}}$$

$$b_1 - E < \beta_1 < b_1 + E$$

where

$$E = t_{\alpha/2} \cdot \frac{s_e}{\sqrt{\Sigma x^2 - \frac{(\Sigma x)^2}{n}}}$$

In these expressions, the y-intercept b_0 and the slope b_1 are found from the sample data and $t_{\alpha/2}$ is found from Table A-3 by using $n - 2$ degrees of freedom. Using the boat/manatee data in Table 9-1, find the 95% confidence interval estimates of β_0 and β_1.

22. Understanding Variation
   a. If a collection of paired data includes at least three pairs of values, what do you know about the linear correlation coefficient if $s_e = 0$?
   b. If a collection of paired data is such that the total explained variation is 0, what do you know about the slope of the regression line?

23. Understanding Variation
   a. Find an expression for the unexplained variation in terms of the sample size n and the standard error of estimate s_e.
   b. Find an expression for the explained variation in terms of the coefficient of determination r^2 and the unexplained variation.
   c. Suppose we have a collection of paired data for which $r^2 = 0.900$ and the regression equation is $\hat{y} = 3 - 2x$. Find the linear correlation coefficient.

24. Finding Confidence Interval for Mean Predicted Value From the expression given in this section for the margin of error corresponding to a prediction interval for y, we can get the expression

$$s_{\hat{y}} = s_e \sqrt{1 + \frac{1}{n} + \frac{n(x_0 - \bar{x})^2}{n(\Sigma x^2) - (\Sigma x)^2}}$$

which is the *standard error of the prediction* when predicting for a *single y*, given that $x = x_0$. When predicting for the *mean* of all values of y for which $x = x_0$, the point estimate $\hat{y}$ is the same, but $s_{\hat{y}}$ is as follows:

$$s_{\hat{y}} = s_e \sqrt{\frac{1}{n} + \frac{n(x_0 - \bar{x})^2}{n(\Sigma x^2) - (\Sigma x)^2}}$$

Use the data from Table 9-1 and extend the last example of this section to find a point estimate and a 95% confidence interval estimate of the mean number of manatee deaths from boats, given that the number of registered boats is 850,000 (so that $x = 85$).

9-5 Multiple Regression

So far, we have used methods of correlation and regression to investigate relationships between exactly *two* variables, but some circumstances require more than two variables. In predicting the price of a diamond, for example, we might consider variables such as weight (in carats), color, and clarity, so that a total of four variables are involved. This section presents a method for analyzing such relationships involving *more than two* variables. We will focus on three key elements: (1) the multiple regression equation, (2) the value of adjusted R^2, and (3) the P-value. As in the previous sections of this chapter, we will work with *linear* relationships only. We begin with the *multiple regression equation*.

22. a. $r = 1$ or $r = -1$
    b. The slope is 0.

23. a. $(n - 2)s_e^2$
    b. $\dfrac{r^2 \cdot (\text{unexplained variation})}{1 - r^2}$
    c. $r = -0.949$

24. $70.8 < \mu_y < 90.4$

Note to Instructor
Because the formulas required for manual calculations of multiple regression equations are monstrous, this section emphasizes computer usage and interpretation of computer results. Some instructors include this section by assigning exercises as extra credit or out-of-class work.
   Students may have wondered why we used the format of $\hat{y} = b_0 + b_1x_1$ for the equation of the regression line in Section 9-3 (instead of using $y = mx + b$). The format of $y = mx + b$ does not extend easily to equations with more than one independent variable, but the format of $\hat{y} = b_0 + b_1x_1$ can be extended in a very natural way.

NBA Salaries and Performance

Researcher Matthew Weeks investigated the correlation between NBA salaries and basketball game statistics. In addition to salary (S), he considered minutes played (M), assists (A), rebounds (R), and points scored (P), and he used data from 30 players. The multiple regression equation is $S = -0.716 - 0.0756M - 0.425A + 0.0536R + 0.742P$ with $R^2 = 0.458$. Because of a high correlation between minutes played (M) and points scored (P), and because points scored had a higher correlation with salary, the variable of minutes played was removed from the multiple regression equation. Also, the variables of assists (A) and rebounds (R) were not found to be significant, so they were removed as well. The single variable of points scored appeared to be the best choice for predicting NBA salaries, but the predictions were found to be not very accurate because of other variables not considered, such as popularity of the player.

Multiple Regression Equation

Definition

A **multiple regression equation** expresses a linear relationship between a dependent variable y and two or more independent variables $(x_1, x_2, \ldots, x_k)$. The general form of a multiple regression equation is $\hat{y} = b_0 + b_1x_1 + b_2x_2 + \ldots + b_kx_k$.

We will use the following notation, which follows naturally from the notation used in Section 9-3.

Notation

$\hat{y} = b_0 + b_1x_1 + b_2 x_2 + \ldots + b_k x_k$ (General form of the estimated multiple regression equation)

n = sample size

k = number of *independent* variables. (The independent variables are also called **predictor variables** or x variables.)

$\hat{y}$ = predicted value of the dependent variable y (computed by using the multiple regression equation)

$x_1, x_2, \ldots, x_k$ are the independent variables

β_0 = the y-intercept, or the value of y when all of the predictor variables are 0 (This value is a population parameter.)

b_0 = estimate of β_0 based on the sample data (b_0 is a sample statistic.)

$\beta_1, \beta_2, \ldots, \beta_k$ are the coefficients of the independent variables $x_1, x_2, \ldots, x_k$

$b_1, b_2, \ldots, b_k$ are the sample estimates of the coefficients $\beta_1, \beta_2, \ldots, \beta_k$

The computations required for multiple regression are so complicated that a statistical software package *must* be used, so we will focus on *interpreting* computer displays. Instructions for using STATDISK, Minitab, Excel, and a TI-83 Plus calculator are included at the end of this section.

EXAMPLE **Bears** For reasons of safety, a study of bears involved the collection of various measurements that were taken after the bears were anesthetized. When obtaining measurements from an anesthetized bear in the wild, it is relatively easy to use a tape measure for finding values such as the chest size, neck size, and overall length, but it is difficult to find the weight because the bear must be lifted. Instead of actually weighing a bear, can we predict its weight based on other measurements that are easier to find? Data Set 9 in Appendix B has measurements taken from 54 bears, but we will consider the data from only eight of those bears, as listed in Table 9-3. Using the data in Table

Table 9-3	Data from Anesthetized Male Bears									
Variable	Minitab Column	Name	Sample Data							
y	C1	WEIGHT	80	344	416	348	262	360	332	34
x_2	C2	AGE	19	55	81	115	56	51	68	8
x_3	C3	HEADLEN	11.0	16.5	15.5	17.0	15.0	13.5	16.0	9.0
x_4	C4	HEADWDTH	5.5	9.0	8.0	10.0	7.5	8.0	9.0	4.5
x_5	C5	NECK	16.0	28.0	31.0	31.5	26.5	27.0	29.0	13.0
x_6	C6	LENGTH	53.0	67.5	72.0	72.0	73.5	68.5	73.0	37.0
x_7	C7	CHEST	26	45	54	49	41	49	44	19

9-3, find the multiple regression equation in which the dependent (y) variable is weight and the independent variables are head length (HEADLEN) and total overall length (LENGTH).

SOLUTION   Using Minitab, we obtain the results shown in the display below. The multiple regression equation is shown as

WEIGHT = −374 + 18.8 HEADLEN + 5.87 LENGTH

Using our notation presented earlier in this section, we could write this equation as

$$\hat{y} = -374 + 18.8x_3 + 5.87x_6$$

Minitab

```
The regression equation is
WEIGHT = −374 + 18.8 HEADLEN + 5.87 LENGTH    ← ① Multiple regression equation

Predictor        Coef        Stdev       t-ratio         p
Constant        −374.3       134.1        −2.79       0.038
HEADLEN          18.82       23.15          0.81       0.453
LENGTH           5.875        5.065         1.16       0.299

s = 68.56   R-sq = 82.8%  R-sq(adj) = 75.9%
                          ↖                    ↖
Analysis of Variance   $R^2 = 0.828$        ② Adjusted $R^2 = 0.759$

SOURCE          DF          SS          MS          F          p
Regression       2      113142       56571      12.03       0.012
Error            5       23506        4701                    ↑
Total            7      136648                  ③ Overall significance of
                                                multiple regression equation
```

Making Music with Multiple Regression

Sony manufactures millions of compact discs in Terre Haute, Indiana. At one step in the manufacturing process, a laser exposes a photographic plate so that a musical signal is transferred into a digital signal coded with 0s and 1s. This process was statistically analyzed to identify the effects of different variables, such as the length of exposure and the thickness of the photographic emulsion. Methods of multiple regression showed that among all of the variables considered, four were most significant. The photographic process was adjusted for optimal results based on the four critical variables. As a result, the percentage of defective discs dropped and the tone quality was maintained. The use of multiple regression methods led to lower production costs and better control of the manufacturing process.

If a multiple regression equation fits the sample data well, it can be used for predictions. For example, if we determine that the equation is suitable for predictions, and if we have a bear with a 14.0-in. head length and a 71.0-in. overall length, we can predict its weight by substituting those values into the regression equation to get a predicted weight of 306 lb. Also, the coefficients $b_3 = 18.8$ and $b_6 = 5.87$ can be used to determine marginal change, as described in Section 9-3. For example, the coefficient $b_3 = 18.8$ shows that when the overall length of a bear remains constant, the predicted weight increases by 18.8 lb for each 1-in. increase in the length of the head.

Adjusted R^2

R^2 denotes the **multiple coefficient of determination,** which is a measure of how well the multiple regression equation fits the sample data. A perfect fit would result in $R^2 = 1$, and a very good fit results in a value near 1. A very poor fit results in a value of R^2 close to 0. The value of $R^2 = 0.828$ in the Minitab display indicates that 82.8% of the variation in bear weight can be explained by the head length x_3 and the overall length x_6. However, the multiple coefficient of determination R^2 has a serious flaw: As more variables are included, R^2 increases. (R^2 could remain the same, but it usually increases.) The largest R^2 is obtained by simply including all of the available variables, but the best multiple regression equation does not necessarily use all of the available variables. Because of that flaw, comparison of different multiple regression equations is better accomplished with the adjusted coefficient of determination, which is R^2 adjusted for the number of variables and the sample size.

Definition

The **adjusted coefficient of determination** is the multiple coefficient of determination R^2 modified to account for the number of variables and the sample size. It is calculated by using Formula 9-6.

Formula 9-6 $$\text{adjusted } R^2 = 1 - \frac{(n-1)}{[n-(k+1)]}(1 - R^2)$$

where
$$n = \text{sample size}$$
$$k = \text{number of independent } (x) \text{ variables}$$

The Minitab display for the data in Table 9-3 shows the adjusted coefficient of determination as `R-sq(adj) = 75.9%`. If we use Formula 9-6 with the R^2 value of 0.828, $n = 8$ and $k = 2$, we find that the adjusted R^2 value is 0.759, confirming Minitab's displayed value of 75.9%. For the weight, head length, and length data in Table 9-3, the R^2 value of 82.8% indicates that 82.8% of the variation in weight can be explained by the head length x_3 and overall length x_6, but when we compare this multiple regression equation to others, it is better to use the adjusted R^2 of 75.9% (or 0.759).

P-Value

The P-value is a measure of the overall significance of the multiple regression equation. The displayed Minitab P-value of 0.012 is small, indicating that the multiple regression equation has good overall significance and is usable for predictions. That is, it makes sense to predict weights of bears based on their head lengths and overall lengths. Like the adjusted R^2, this P-value is a good measure of how well the equation fits the sample data. The value of 0.012 results from a test of the null hypothesis that $\beta_3 = \beta_6 = 0$. Rejection of $\beta_3 = \beta_6 = 0$ implies that at least one of β_3 and β_6 is not 0, indicating that this regression equation is effective in determining weights of bears. A complete analysis of the Minitab results might include other important elements, such as the significance of the individual coefficients, but we will limit our discussion to the three key components—multiple regression equation, adjusted R^2, and P-value.

Finding the Best Multiple Regression Equation

Table 9-3 includes seven different variables of measurement for eight different bears. The Minitab display is based on the selection of weight as the dependent variable and the selection of head length and overall length as the independent variables. But if we want to predict the weight of a bear, is there some other combination of variables that might be better than head length and overall length? Table 9-4 lists a few of the combinations of variables, and we are now confronted with the important objective of finding the *best* multiple regression equation. Because determination of the best multiple regression requires a good dose of judgment, there is no exact and automatic procedure that can be used. *Determination of the best multiple regression equation is often quite difficult and beyond the scope of this book,* but the following guidelines should provide some help.

Note to Instructor
Be sure to point out that determination of the best multiple regression equation is beyond the scope of this book.

Guidelines for Finding the Best Multiple Regression Equation

1. *Use common sense and practical considerations to include or exclude variables.* For example, we might exclude the variable of age because inexperienced researchers might not know how to determine the age of a bear and, when questioned, bears are reluctant to reveal their ages. Unlike the other independent variables, the age of a bear cannot be easily obtained with a tape measure. It therefore makes sense to exclude age as an independent variable.

Table 9-4	Searching for the Best Multiple Regression Equation				
	LENGTH	CHEST	HEADLEN/ LENGTH	AGE/NECK/ LENGTH/CHEST	AGE/HEADLEN/HEADWDTH/ NECK/LENGTH/CHEST
R^2	0.805	0.983	0.828	0.999	0.999
Adjusted R^2	0.773	0.980	0.759	0.997	0.996
Overall significance	0.002	0.000	0.012	0.000	0.046

Predictors for Success

When a college accepts a new student, it would like to have some positive indication that the student will be successful in his or her studies. College admissions deans consider SAT scores, standard achievement tests, rank in class, difficulty of high school courses, high school grades, and extracurricular activities. In a study of characteristics that make good predictors of success in college, it was found that class rank and scores on standard achievement tests are better predictors than SAT scores. A multiple regression equation with college grade-point average predicted by class rank and achievement test score was not improved by including another variable for SAT score. This particular study suggests that SAT scores should not be included among the admissions criteria, but supporters argue that SAT scores are useful for comparing students from different geographic locations and high school backgrounds.

2. *Consider the P-value.* Select an equation having overall significance, as determined by the *P*-value found in the computer display. For example, see the values of overall significance in Table 9-4. The use of all six independent variables results in an overall significance of 0.046, which is just barely significant at the $\alpha = 0.05$ level; we're better off with the single variable CHEST, which has an overall significance of 0.000.

3. *Consider equations with high values of adjusted R^2, and try to include only a few variables.* Instead of including almost every available variable, try to include relatively few independent (x) variables. Use these guidelines:

   - Select an equation having a value of adjusted R^2 with this property: If an additional independent variable is included, the value of adjusted R^2 does not increase by a substantial amount. For example, Table 9-4 shows that if we use only the independent variable CHEST, the adjusted R^2 is 0.980, but when we include all six variables, the adjusted R^2 increases to 0.996. Using six variables instead of only one is too high a price to pay for such a small increase in the adjusted R^2. We're better off using the single independent variable CHEST than using all six independent variables.

   - For a given number of independent (x) variables, select the equation with the largest value of adjusted R^2.

   - In weeding out independent variables that don't have much of an effect on the dependent variable, it might be helpful to find the linear correlation coefficient r for each pair of variables being considered. For example, using the data in Table 9-3, we will find that there is a 0.955 linear correlation for the paired NECK/HEADLEN data. Because there is such a high correlation between neck size and head length, there is no need to include both of those variables. In choosing between NECK and HEADLEN, we should select NECK for this reason: NECK is a better predictor of WEIGHT because the NECK/WEIGHT paired data have a linear correlation coefficient of $r = 0.971$, which is higher than $r = 0.884$ for the paired HEADLEN/WEIGHT data.

Using these guidelines in an attempt to find the best equation for predicting weights of bears, we find that for the data of Table 9-3, the best regression equation uses the single independent variable of chest size (CHEST). The best regression equation appears to be

$$\text{WEIGHT} = -195 + 11.4 \,\text{CHEST}$$

or

$$\hat{y} = -195 + 11.4 x_7$$

Some statistical software packages include a program for performing **stepwise regression,** whereby computations are performed with different combinations of independent variables, but there are some serious problems associated with it, including these: Stepwise regression will not necessarily yield the best model if some predictor variables are highly correlated; it yields inflated values of R^2; it uses too much paper; and it allows us to not *think* about the problem. As always, we should be careful to use computer results as a tool that helps us make intelligent decisions; we should not let the computer become the decision maker.

Instead of relying solely on the result of a computer stepwise regression program, consider the preceding factors when trying to identify the best multiple regression equation.

If we eliminate the variable AGE (as in guideline 1) and then run Minitab's stepwise regression program, we will get a display suggesting that the best regression equation is the one in which CHEST is the only independent variable. (If we include all six independent variables, Minitab selects a regression equation with the independent variables AGE, NECK, LENGTH, and CHEST, with an adjusted R^2 value of 0.997 and overall significance of 0.000.) It appears that we can estimate the weight of a bear based on its chest size, and the regression equation leads to this rule: The weight of a bear (in pounds) is estimated to be 11.4 times the chest size (in inches) minus 195.

When we discussed regression in Section 9-3, we listed four common errors that should be avoided when using regression equations to make predictions. These same errors should be avoided when using multiple regression equations. Be especially careful about concluding that a cause-effect relationship exists.

Using Technology

STATDISK   First enter the sample data in columns of the Statdisk Data Window. Select **Analysis,** then **Multiple Regression.** Select the columns to be included, and also identify the column corresponding to the dependent variable. Click on **Evaluate** and you will get the multiple regression equation along with other items, including the multiple coefficient of determination R^2, the adjusted R^2, and the *P*-value.

Minitab   First enter the values in different columns. To avoid confusion among the different variables, enter a name for each variable in the box atop its column of data. Select the main menu item **Stat,** then select **Regression,** then **Regression** once again. In the dialog box, enter the variable to be used for the response (*y*) variable, and enter the variables you want included as *x*-variables. Click **OK.** The display will include the multiple regression equation, along with other items, including the multiple coefficient of determination R^2 and the adjusted R^2.

Excel   First enter the sample data in columns. Select **Tools** from the main menu, then select **Data Analysis** and **Regression.** In the dialog box, enter the range of values for the dependent *Y*-variable, then enter the range of values for the independent *X*-variables, which must be in adjacent columns. (Use Copy/Paste to move columns as desired.) The display will include the multiple coefficient of determination R^2, the adjusted

R^2, and a list of the intercept and coefficient values used for the multiple regression equation.

TI-83 Plus   The TI-83 Plus program A2MULREG can be downloaded from the CD-ROM included with this book. Select the *software* folder, then select *TI83PlusPRGMS*. The program must be downloaded to your calculator, then the sample data must first be entered as matrix D, with the first column containing the values of the dependent (*y*) variable. Press **2nd,** and the x^{-1} key, scroll to the right for **EDIT,** scroll down for **[D],** then press **ENTER** and proceed to enter the number of values listed for each variable followed by the total number of variables (including the dependent variable). Now press **PRGM,** select **A2MULREG** and press **ENTER** three times, then select **MULT REGRESSION** and press **ENTER.** When prompted, enter the number of independent variables, then enter the column numbers of the independent variables that you want to include. The screen will provide a display that includes the *P*-value and the value of the adjusted R^2. Press **ENTER** to see the values to be used in the multiple regression equation. Press **ENTER** again to get a menu that includes options for generating confidence intervals, prediction intervals, residuals, or quitting. If you want to generate confidence and prediction intervals, use the displayed number of degrees of freedom, go to Table A-3 and look up the corresponding critical *t* value, enter it, then proceed to enter the values to be used for the independent variables. Press **ENTER** to select the **QUIT** option.

1. $\hat{y} = -272 - 0.870x_1 + 0.554x_2 + 12.2\ x_3$
2. a. 0.000 (rounded)
   b. 0.928
   c. 0.924
3. Yes, because the *P*-value is 0.000 and the adjusted R^2 value is 0.924.
4. a. 364 lb
   b. The error of 44 lb is fairly large.
5. Highway fuel consumption, because it has the highest adjusted R^2.
6. Highway fuel consumption and weight, because that combination has the highest adjusted R^2.
7. The regression equation with highway fuel consumption and weight has the highest adjusted R^2 of 0.861, but a good argument can be made for using the single independent variable of highway fuel consumption, because its adjusted R^2 is 0.853, which is only slightly less.
8. Using highway fuel consumption and weight results in a predicted value of 26.8 mi/gal, with a 95% prediction interval of 23.2 mi/gal $< y <$ 30.4 mi/gal. Using the single independent variable of highway fuel consumption results in a predicted value of 27.0 mi/gal, with a 95% prediction interval of 23.3 mi/gal $< y <$ 30.6 mi/gal. Because the sample size is only 20 and because there is a fair amount of variation, the predicted value is not likely to be very accurate.

9-5 Basic Skills and Concepts

Interpreting a Computer Display. In Exercises 1–4, refer to the Minitab display given here and answer the given questions or identify the indicated items. The Minitab display is based on the sample of 54 bears listed in Data Set 9 in Appendix B.

1. Bear Measurements Identify the multiple regression equation that expresses weight in terms of head length, length, and chest size.

2. Bear Measurements Identify the following.
   a. The *P*-value corresponding to the overall significance of the multiple regression equation
   b. The value of the multiple coefficient of determination R^2
   c. The adjusted value of R^2

3. Bear Measurements Is the multiple regression equation usable for predicting a bear's weight based on its head length, length, and chest size? Why or why not?

4. Bear Measurements A bear is found to have a head length of 14.0 in., a length of 70.0 in., and a chest size of 50.0 in.
   a. Find the predicted weight of the bear.
   b. The bear in question actually weighed 320 lb. How accurate is the predicted weight from part (a)?

Minitab

```
The regression equation is
WEIGHT = -272 - 0.87 HEADLEN + 0.55 LENGTH + 12.2 CHEST
Predictor        Coef      SE Coef        T         P
Constant       -271.71       31.62     -8.59     0.000
HEADLEN          -0.870       5.676     -0.15     0.879
LENGTH            0.554       1.259      0.44     0.662
CHEST            12.153       1.116     10.89     0.000

S = 33.66        R-Sq = 92.8%       R-Sq(adj) = 92.4%

Analysis of Variance

Source           DF         SS        MS         F        P
Regression        3     729645    243215    214.71    0.000
Residual Error   50      56638      1133
Total            53     786283
```

Car Data: Finding the Best Multiple Regression Equation. In Exercises 5–8, refer to the accompanying table, which was obtained by using Data Set 22 in Appendix B. The dependent variable is the city fuel consumption (in mi/gal), and the independent variables are listed in the table. HWY denotes the highway fuel consumption, WT denotes the weight of a car, and DISP denotes the car's engine displacement.

Independent Variables	P-value	R^2	Adjusted R^2	Regression Equation
HWY, WT, DISP	0.000	0.882	0.860	$\hat{y} = 5.9 + 0.742x_1 - 0.00162x_2 - 0.441x_3$
HWY, WT	0.000	0.876	0.861	$\hat{y} = 4.6 + 0.794x_1 - 0.00209x_2$
HWY, DISP	0.000	0.873	0.859	$\hat{y} = -3.23 + 0.892x_1 - 0.626x_2$
WT, DISP	0.000	0.788	0.763	$\hat{y} = 41.5 - 0.00535x_1 - 0.950x_2$
HWY	0.000	0.860	0.853	$\hat{y} = -9.73 + 1.05x$
WT	0.000	0.759	0.746	$\hat{y} = 44.2 - 0.00708x$
DISP	0.000	0.620	0.599	$\hat{y} = 29.5 - 2.74x$

5. If only one independent variable is used to predict the city fuel consumption amount (in mi/gal), which single variable is best? Why?

6. If exactly two independent variables are to be used to predict the city fuel consumption amount, which two variables should be chosen? Why?

7. Which regression equation is best for predicting the city fuel consumption amount? Why?

8. If a car has a highway fuel consumption rate of 35 mi/gal, a weight of 2675 lb, and an engine displacement of 3.8 L, what is the best predicted value of city fuel consumption rate? Is that predicted value likely to be a good estimate? Is that predicted value likely to be very accurate?

T 9. Heights of Parents and Children Refer to Data Set 2 in Appendix B.
   a. Find the regression equation that expresses the dependent variable of a child's height in terms of the independent variable of the height of the mother.
   b. Find the regression equation that expresses the dependent variable of a child's height in terms of the independent variable of the height of the father.
   c. Find the regression equation that expresses the dependent variable of a child's height in terms of the independent variables of the height of the mother and the height of the father.
   d. For the regression equations found in parts (a), (b), and (c), which is the best equation for predicting the height of a child? Why?
   e. Is the best regression equation identified in part (d) a *good* equation for predicting a child's height? Why or why not?

T 10. Readability of Harry Potter Refer to Data Set 14 in Appendix B and use the values for J. K. Rowling's *Harry Potter and the Sorcerer's Stone*.
   a. Find the regression equation that expresses the dependent variable of Flesch Reading Ease score in terms of the independent variable of the words per sentence.
   b. Find the regression equation that expresses the dependent variable of Flesch Reading Ease score in terms of the independent variable of the characters per word.
   c. Find the regression equation that expresses the dependent variable of Flesch Reading Ease score in terms of the independent variables of the words per sentence and characters per word.
   d. For the regression equations found in parts (a), (b), and (c), which is the best equation for predicting the Flesch Reading Ease score? Why?
   e. Is the best regression equation identified in part (d) a *good* equation for predicting a Flesch Reading Ease score? Why or why not?

T 11. Cereal and Calories Refer to Data Set 16 in Appendix B.
   a. Find the regression equation that expresses the dependent variable of calories in terms of the independent variable of the amount of fat.

9. a. $\hat{y} = 21.6 + 0.690x$
   b. $\hat{y} = 45.7 + 0.293x$
   c. $\hat{y} = 9.80 + 0.658x_1 + 0.200x_2$ (where x_1 = mother's height)
   d. Part (c), because the adjusted R^2 is highest
   e. No, because the highest value of adjusted R^2 is only 0.366, which isn't very high.
10. a. $\hat{y} = 91.8 - 0.843x$
   b. $\hat{y} = 180 - 23.1x$
   c. $\hat{y} = 172 - 0.547x_1 - 19.6x_2$ (where x_1 = words per sentence)
   d. Part (c), because the adjusted R^2 is highest
   e. Yes, but the predicted values will not necessarily be very accurate.
11. a. $\hat{y} = 3.68 + 3.78x$
   b. $\hat{y} = 3.46 + 1.01x$
   c. $\hat{y} = 3.40 + 3.20x_1 + 0.982x_2$ (where x_1 = fat)
   d. Part (c), because the adjusted R^2 is highest
   e. Yes, but the predicted values will not necessarily be very accurate.

b. Find the regression equation that expresses the dependent variable of calories in terms of the independent variable of the amount of sugar.

c. Find the regression equation that expresses the dependent variable of calories in terms of the independent variables of the amount of fat and the amount of sugar.

d. For the regression equations found in parts (a), (b), and (c), which is the best equation for predicting the number of calories? Why?

e. Is the best regression equation identified in part (d) a *good* equation for predicting the number of calories? Why or why not?

12. a. $\hat{y} = 2.84 + 0.180x$
   b. $\hat{y} = 1.08 + 1.38x$
   c. $\hat{y} = 1.14 - 0.0298x_1 +$
     $1.42x_2$ (where x_1 = food)
   d. Part (b), because the adjusted R^2 is highest
   e. Yes, but the predicted values will not necessarily be very accurate.

13. $\hat{y} = 0.154 + 0.0651x$, where x is the amount of tar. With close values of R^2, it is better to select the equation with one independent variable instead of two.

14. All three independent variables: adjusted $R^2 = 0.797$. Carat and color: adjusted $R^2 =$ of 0.778. Because these are close, it is better to use only two independent variables. Use $\hat{y} = 8221$ $+ 12{,}297x_1 - 3116x_2$, where x_1 represents the carat weight.

15. $\hat{y} = 7.26 + 0.914x_1$ (where x_1 represents list price); by using more variables, adjusted R^2 can be raised from 0.995 to 0.996, but the small increase in adjusted R^2 does not justify the inclusion of additional variables.

16. $\hat{y} = 2.17 + 2.44x + 0.464x^2$; because $R^2 = 0.9997$, the parabola fits almost perfectly.

x	1	3	4	7	5
y	5	14	19	42	26

T 12. Using Garbage to Predict Population Size Refer to Data Set 23 in Appendix B.

a. Find the regression equation that expresses the dependent variable of household size in terms of the independent variable of the weight of discarded food.

b. Find the regression equation that expresses the dependent variable of household size in terms of the independent variable of the weight of discarded plastic.

c. Find the regression equation that expresses the dependent variable of household size in terms of the independent variables of the weight of discarded food and the weight of discarded plastic.

d. For the regression equations found in parts (a), (b), and (c), which is the best equation for predicting the household size? Why?

e. Is the best regression equation identified in part (d) a *good* equation for predicting the household size? Why or why not?

9-5 Beyond the Basics

T 13. Cigarette Nicotine: Finding Best Multiple Regression Equation Refer to Data Set 5 in Appendix B and find the best multiple regression equation with nicotine as the dependent variable. Is this "best" equation good for predicting the amount of nicotine in a cigarette based on the amount of tar and carbon monoxide?

T 14. Price of a Diamond: Finding Best Multiple Regression Equation Refer to Data Set 18 in Appendix B.

a. Using only the traditional "three C's" of carat, color, and clarity, find the best multiple regression equation that could be used to predict the price of a diamond.

b. The variables of depth and table describe the cut of a diamond, which supposedly affects its color. Is there a significant linear relationship between the dependent variable of color and the independent variables of depth and table? If there is no significant linear relationship, does it mean that the color is not affected by depth and table?

T 15. Home Selling Price: Finding Best Multiple Regression Equation Refer to Data Set 24 in Appendix B and find the best multiple regression equation with selling price as the dependent variable. Is this "best" equation good for predicting the selling price of a home?

T 16. Using Multiple Regression for Equation of Parabola In some cases, the best-fitting multiple regression equation is of the form $\hat{y} = b_0 + b_1x + b_2x^2$. The graph of such an equation is a parabola. Using the data set listed in the margin, let $x_1 = x$, let $x_2 = x^2$, and find the multiple regression equation for the parabola that best fits the given data. Based on the value of the multiple coefficient of determination, how well does this equation fit the data?

9-6 Modeling

No, not that kind of modeling. This section introduces some basic concepts of developing a **mathematical model,** which is a mathematical function that "fits" or describes real-world data. For example, we might want a mathematical model consisting of an equation relating a variable for population size to another variable representing time. This is much like the methods of regression discussion in Section 9-3, except that we are no longer restricted to a model that must be linear. Also, instead of using randomly selected sample data, we will consider data collected periodically over time or some other basic unit of measurement. There are some powerful statistical methods that we could discuss (such as *time series*), but the main objective of this section is to describe briefly how technology can be used to find a good mathematical model.

The following are some generic models as listed in a menu from the TI-83 Plus calculator (press **STAT,** then select **CALC**):

Linear: $y = a + bx$ Quadratic: $y = ax^2 + bx + c$

Logarithmic: $y = a + b \ln x$ Exponential: $y = ab^x$

Power: $y = ax^b$ Logistic: $y = \dfrac{c}{1 + ae^{-bx}}$

The particular model that you select depends on the nature of the sample data, and a scatterplot can be very helpful in making that determination. The illustrations that follow are graphs of some common models displayed on a TI-83 Plus calculator.

Note to Instructor
This section can be omitted if time is not available. It requires access to some form of technology. A TI-83 Plus calculator is ideal, but STATDISK, Minitab, or Excel can also be used. See the comments in the "Using Technology" box at the end of this section.

TI-83 Plus

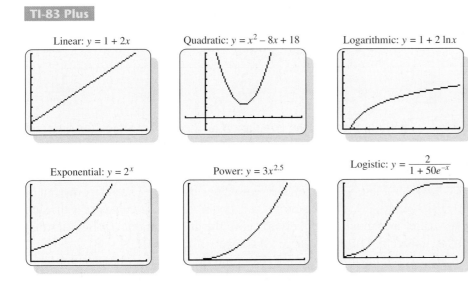

Linear: $y = 1 + 2x$ Quadratic: $y = x^2 - 8x + 18$ Logarithmic: $y = 1 + 2 \ln x$

Exponential: $y = 2^x$ Power: $y = 3x^{2.5}$ Logistic: $y = \dfrac{2}{1 + 50e^{-x}}$

Here are three basic rules for developing a good mathematical model:

1. *Look for a pattern in the graph.* Examine the graph of the plotted points and compare the basic pattern to the known generic graphs of a linear function, quadratic function, exponential function, power function, and so on. (Refer to the accompanying graphs shown in the examples of the TI-83 Plus calculator displays.) When trying to select a model, consider only those functions that visually appear to fit the observed points reasonably well.

2. *Find and compare values of R^2.* For each model being considered, use computer software or a TI-83 Plus calculator to find the value of the coefficient of determination R^2. Values of R^2 can be interpreted here the same way that they were interpreted in Section 9-5. When narrowing your possible models, select functions that result in larger values of R^2, because such larger values correspond to functions that better fit the observed points. However, don't place much importance on small differences, such as the difference between $R^2 = 0.984$ and $R^2 = 0.989$. (Another measurement used to assess the quality of a model is the sum of squares of the residuals. See Exercise 10.)

3. *Think.* Use common sense. Don't use a model that leads to predicted values known to be totally unrealistic. Use the model to calculate future values, past values, and values for missing years, then determine whether the results are realistic.

Table 9-5	Population (in millions) of the United States										
Year	1800	1820	1840	1860	1880	1900	1920	1940	1960	1980	2000
Coded year	1	2	3	4	5	6	7	8	9	10	11
Population	5	10	17	31	50	76	106	132	179	227	281

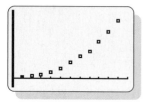

TI-83 Plus

EXAMPLE Table 9-5 lists the population of the United States for different years. Find a good mathematical model for the population size, then predict the size of the U.S. population in the year 2020.

SOLUTION First, we "code" the year values by using 1, 2, 3 . . . instead of 1800, 1820, 1840. . . . The reason for this coding is to use values of x that are much smaller and much less likely to cause the computational difficulties that are likely to occur with really large x-values.

Look for a pattern in the graph. Examine the pattern of the data values in the TI-83 Plus display (shown in the margin) and compare that pattern to the generic models shown earlier in this section. The pattern of those points is clearly not a straight line, so we rule out a linear model. We rule out a logistic model because the points don't show the "S" pattern of that graph, with a flattening of the graph occurring at the right. Good candidates for the model appear to be the quadratic, exponential, and power functions.

Find and compare values of R^2. The following displays show the TI-83 Plus results based on the quadratic, exponential, and power models. Comparing the values of the coefficient R^2, it appears that the quadratic model is best because it has the highest value of 0.9992, but the other displayed values are also quite high. If we select the quadratic function as the best model, we conclude that the equation $y = 2.77x^2 - 6.00x + 10.01$ best describes the relationship between the year x (coded with $x = 1$ representing 1800, $x = 2$ representing 1820, and so on) and the population y (in millions).

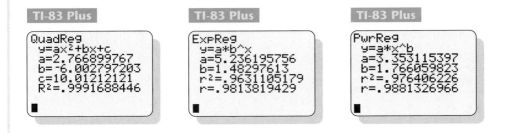

To predict the U.S. population for the year 2020, first note that the year 2020 is coded as $x = 12$ (see Table 9-5). Substituting $x = 12$ into the quadratic model of $y = 2.77x^2 - 6.00x + 10.01$ results in $y = 337$, which indicates that the U.S. population is estimated to be 337 million in the year 2020.

Think. The forecast result of 337 million in 2020 seems reasonable. (A U.S. Bureau of the Census projection suggests that the population in 2020 will be around 325 million.) However, there is considerable danger in making estimates for times that are beyond the scope of the available data. For example, the quadratic model suggests that in 1492, the U.S. population was 671 million—an absurd result. For future estimates, only the logistic model shows this behavior typical of growing populations: The population begins to stabilize when it approaches the *carrying capacity of the environment*—the maximum population that can be supported by the limited resources. The quadratic model appears to be good for the available data (1800–2000), but other models might be better if it is absolutely necessary to make population estimates far beyond that time frame.

In "Modeling the U.S. Population" (*AMATYC Review,* Vol. 20, No. 2), Sheldon Gordon uses more data than Table 9-5, and he uses much more advanced techniques to find better population models. In that article, he makes this important point:

"The best choice (of a model) depends on the set of data being analyzed and requires an exercise in judgment, not just computation."

Using Technology

Any system capable of handling multiple regression can be used to generate some of the models described in this section. For example, STATDISK is not designed to work directly with the quadratic model, but its multiple regression feature can be used with the data in Table 9-5 to generate the quadratic model as follows: First enter the population values in column 1 of the Statdisk Data Window. Enter 1, 2, 3, . . . , 11 in column 2 and enter 1, 4, 9, . . . , 121 in column 3. Click on **Analysis,** then select **Multiple Regression.** Use columns 1, 2, 3 with column 1 as the dependent variable. After clicking on **Evaluate,** STATDISK generates the equation $y = 10.012 - 6.0028x + 2.7669x^2$ along with $R^2 = 0.99917$, which are the same results obtained from the TI-83 Plus calculator.

Minitab   First enter the matched data in columns C1 and C2, then select **Stat, Regression,** and **Fitted Line Plot.** You can choose a linear model, quadratic model, or cubic model. Displayed results include the equation, the value of R^2, and the sum of squares of the residuals.

TI-83 Plus   First turn on the diagnostics feature as follows: Press **2nd CATALOG,** then scroll down to **DiagnosticON** and press the **ENTER** key twice. Enter the matched data in lists L1 and L2. Press **STAT,** select **CALC,** and then select the desired model from the available options. Press **ENTER,** then enter L1, L2 (with the comma), and press **ENTER** again. The display includes the format of the equation along with the coefficients used in the equation; also the value of R^2 is included for many of the models.

9-6 Basic Skills and Concepts

T *Finding the Best Model.* *In Exercises 1–8, construct a scatterplot and identify the mathematical model that best fits the given data. Assume that the model is to be used only for the scope of the given data, and consider only linear, quadratic, logarithmic, exponential, and power models.*

1.

x	1	2	3	4	5	6
y	8	2	0	2	8	18

2.

x	1	2	3	4	5	6
y	3	8	13	18	23	28

3.

x	1	2	3	4	5	6
y	3	9	27	80	245	725

4.

x	1	2	3	4	5	6
y	2.000	2.828	3.464	4.000	4.472	4.899

T 5. Manatee Deaths from Boats The accompanying table lists the number of Florida manatee deaths related to encounters with watercraft (based on data from *The New York Times*). What is the best predicted value for 2001? In 2001, there were 82 watercraft-related manatee deaths. How does the predicted value compare to the actual value?

Year	1980	1981	1982	1983	1984	1985	1986	1987	1988	1989	1990
Deaths	16	24	20	15	34	33	33	39	43	50	47

Year	1991	1992	1993	1994	1995	1996	1997	1998	1999	2000	
Deaths	53	38	35	49	42	60	54	67	82	78	

6. Stock Market Refer to the annual high values of the Dow-Jones Industrial Average that are listed in Data Set 25 in Appendix B. What is the best predicted value for the year 2001? Given that the actual high value in 2001 was 11,350, how good was the predicted value? What does the pattern suggest about the stock market for investment purposes? (Acts of terrorism and bad economic conditions caused substantial stock market losses in 2002.)

7. Target Stores The accompanying table lists the number of Target department stores in the United States (based on data from Target). What is the best predicted value for the number of Target stores in 2005?

Year	1990	1991	1992	1993	1994	1995	1996	1997	1998	1999	2000
Stores	420	463	506	554	611	670	736	796	851	914	984

8. Return on Investment Kendra Korbin, owner and operator of the Cyber Video Game Store, records her business costs and revenue for different years, with the results listed below.

Amount invested (thousands of dollars)	1	2	5	11	20	31	41	46	48
Revenue (dollars)	2001	2639	3807	5219	6629	7899	8834	9250	9409

9-6 Beyond the Basics

9. Moore's Law In 1965, Intel co-founder Gordon Moore initiated what has since become known as *Moore's law:* the number of transistors per square inch on integrated circuits will double approximately every 18 months. Here are data describing the number of transistors (in thousands) for different years: 1971: 2.3; 1978: 31; 1982: 110; 1985: 280; 1989: 1200; 1993: 3100; 1995: 5500; 1999: 14,000. Let 1971 be the base year represented by $x = 1$.
   a. Assuming that Moore's law is correct and transistors double every 18 months, which mathematical model best describes this law: linear, quadratic, logarithmic, exponential, power, logistic? What specific function describes Moore's law?
   b. Which mathematical model best fits the listed sample data?
   c. Compare the results from parts (a) and (b). Does Moore's law appear to be working reasonably well?

10. Using the Sum of Squares Criterion It was noted that in addition to the value of R^2, another measurement used to assess the quality of a model is the *sum of squares of the residuals.* A residual is the difference between an observed y value and the value of y predicted from the model, which is denoted as $\hat{y}$. Better models have smaller sums of squares. Refer to the example in this section.
   a. Find $\Sigma(y - \hat{y})^2$, the sum of squares of the residuals resulting from the linear model.
   b. Find the sum of squares of residuals resulting from the quadratic model.
   c. Verify that according to the sum of squares criterion, the quadratic model is better than the linear model.

6. Exponential: $y = (762.701)$ $(1.13456)^x$, where x is coded as 1 for 1980. Predicted value: 12,262, which misses by a fairly large amount.

7. Quadratic: $y = 1.21445x^2 +$ $42.4084x + 371.958$, where x is coded as 1 for 1990. Predicted value: 1361.

8. Power: $y = 2000.43x^{0.399937}$. $R^2 = 0.999999938$, so model fits almost perfectly.

9. a. Exponential: $y = 2^{\frac{2}{3}(x-1)}$ [or $y = (0.629961)(1.587401)^x$ for an initial value of 1 that doubles every 1.5 years]
    b. Exponential: $y = (2.32040)(1.36587)^x$
    c. Moore's law does appear to be working reasonably well.

10. a. 6641.8
    b. 73.2
    c. The quadratic sum of squares of residuals (73.2) is less than the sum of squares of residuals from the linear model (6641.8).

11. a. 189.1
    b. 0.9979
    c. The quadratic R^2 value of
       0.9992 is higher than $R^2 =$
       0.9979 for the logistic
       model, and the sum of
       squares of the residuals is
       lower for the quadratic
       model (73.2) than for the lo-
       gistic model (189.1).

11. Finding Sum of Squares and R^2 Using the data from Table 9-5, the logistic model is

$$y = \frac{465.9305}{1 + 72.5260e^{-0.425483x}}$$

a. Find $\Sigma(y - \hat{y})^2$, the sum of squares of the residuals.

b. Find

$$R^2 = 1 - \frac{\Sigma(y - \hat{y})^2}{\Sigma(y - \bar{y})^2}$$

c. After comparing the R^2 values and the sums of the squares of the residuals, determine whether the logistic model is better than the quadratic model.

Review

This chapter presents basic methods for investigating relationships or correlations between two or more variables.

- Section 9-2 used scatter diagrams and the linear correlation coefficient to decide
  whether there is a linear correlation between two variables.

- Section 9-3 presented methods for finding the equation of the regression line that
  (by the least-squares criterion) best fits the paired data. When there is a significant
  linear correlation, the regression equation can be used to predict the value of a variable, given some value of the other variable.

- Section 9-4 introduced the concept of total variation, with components of explained and unexplained variation. We defined the coefficient of determination r^2
  to be the quotient obtained by dividing explained variation by total variation. We
  also developed methods for constructing prediction intervals, which are helpful in
  judging the accuracy of predicted values.

- In Section 9-5 we considered multiple regression, which allows us to investigate
  relationships among several variables. We discussed procedures for obtaining a
  multiple regression equation, as well as the value of the multiple coefficient of determination R^2, the adjusted R^2, and a P-value for the overall significance of the
  equation.

- In Section 9-6 we explored basic concepts of developing a mathematical model,
  which is a function that can be used to describe a relationship between two variables. Unlike the preceding sections of this chapter, Section 9-6 included several
  nonlinear functions.

Review Exercises

1. $r = -0.069$. CV: $r = \pm 0.707$
   (assuming a 0.05 significance
   level). No significant linear correlation.

1. DWI and Jail A study was conducted to investigate the relationship between age (in
years) and BAC (blood alcohol concentration) measured when convicted DWI (driving while intoxicated) jail inmates were first arrested. Sample data are given below
for randomly selected subjects (based on data from the Dutchess County STOP-DWI
Program). Based on the result, does the BAC level seem to be related to the age of the
person tested?

Age	17.2	43.5	30.7	53.1	37.2	21.0	27.6	46.3
BAC	0.19	0.20	0.26	0.16	0.24	0.20	0.18	0.23

2. Tipping Many of us have heard that the tip should be 15% of the bill. The accompanying table lists some sample data collected from the author's students. Use the sample data for the following.
  a. Is there sufficient evidence to conclude that there is a relationship between the amount of the bill and the amount of the tip?
  b. If there is a relationship, how do we use it to determine how much of a tip should be left?

Bill ($)	33.46	50.68	87.92	98.84	63.60	107.34
Tip ($)	5.50	5.00	8.08	17.00	12.00	16.00

Ice Cream Data: Understanding Correlation and Regression. In Exercises 3–6, use the data in the accompanying table (based on data from Kadiyala, Econometrica, *Vol. 38). The data come from a study of ice cream consumption that spanned the springs and summers of three years. The ice cream consumption is in pints per capita per week, price of the ice cream is in dollars, family income of consumers is in dollars per week, and temperature is in degrees Fahrenheit.*

Consumption	0.386	0.374	0.393	0.425	0.406	0.344	0.327	0.288	0.269	0.256
Price	1.35	1.41	1.39	1.40	1.36	1.31	1.38	1.34	1.33	1.39
Income	351	356	365	360	342	351	369	356	342	356
Temperature	41	56	63	68	69	65	61	47	32	24

3. a. Use a 0.05 significance level to test for a linear correlation between consumption and price.
  b. What percentage of the variation in price can be explained by the linear relationship between price and consumption?
  c. Find the equation of the regression line that expresses consumption (y) in terms of price (x).
  d. What is the best predicted consumption amount if the price is $1.38?

4. a. Use a 0.05 significance level to test for a linear correlation between consumption and income.
  b. What percentage of the variation in consumption can be explained by the linear relationship between consumption and income?
  c. Find the equation of the regression line that expresses consumption (y) in terms of income (x).
  d. What is the best predicted consumption amount if the income is $365?

5. a. Use a 0.05 significance level to test for a linear correlation between consumption and temperature.
  b. What percentage of the variation in consumption can be explained by the linear relationship between consumption and temperature?
  c. Find the equation of the regression line that expresses consumption (y) in terms of temperature (x).
  d. What is the best predicted consumption amount if the temperature is 32°F?

6. Use software such as STATDISK or Minitab or Excel to find the multiple regression equation of the form $\hat{y} = b_0 + b_1x_1 + b_2x_2 + b_3x_3$, where the dependent variable y represents consumption, x_1 represents price, x_2 represents income, and x_3 represents temperature. Also identify the value of the multiple coefficient of determination R^2, the adjusted R^2, and the P-value representing the overall significance of the multiple regression equation. Can the regression equation be used to predict ice cream consumption? Are any of the equations from Exercises 3–5 better?

2. a. $r = 0.828$. CV: $r = \pm0.811$ (assuming a 0.05 significance level). Significant linear correlation.
  b. The regression equation of $\hat{y} = -0.347 + 0.149x$ (with x representing the bill) shows that the predicted amount of tip is 35 cents less than 15% of the bill, which is approximately 15% of the bill.

3. a. $r = 0.338$. CV: $r = \pm0.632$; no significant linear correlation.
  b. 11%
  c. $\hat{y} = -0.488 + 0.611x$
  d. 0.347 pints per capita per week

4. a. $r = 0.116$. CV: $r = \pm0.632$; no significant linear correlation.
  b. 1.3%
  c. $\hat{y} = 0.0657 + 0.000792x$
  d. 0.347 pints per capita per week

5. a. $r = 0.777$. CV: $r = \pm0.632$. Significant linear correlation.
  b. 60%
  c. $\hat{y} = 0.193 + 0.00293x$
  d. 0.286 pints per capita per week

6. $\hat{y} = -0.0526 + 0.747x_1 - 0.00220x_2 + 0.00303x_3$; $R^2 = 0.726$; adjusted $R^2 = 0.589$; P-value $= 0.040$. Equation can be used to predict ice cream consumption. Best regression equation: use temperature as the only independent variable.

Cumulative Review Exercises

1. a. $r = -0.884$. CV: $r =$
      ± 0.576 (assuming a 0.05
      significance level). Signifi-
      cant linear correlation.
   b. $\hat{y} = 95.3 - 3.46x$
   c. It is possible to conduct the
      test of two equal population
      means, but the test would
      make no sense because the
      two variables measure read-
      ability using different crite-
      ria and different scales.
   d. $61.16 < \mu < 71.14$

2. a. $\bar{x} = 99.1$, $s = 8.5$
   b. $\bar{x} = 102.8$, $s = 8.7$
   c. No, but a better comparison
      would involve treating the
      data as matched pairs.
   d. TS: $t = 0.546$. CV: $t =$
      ± 2.069. P-value > 0.20.
      There is not sufficient evi-
      dence to support the claim
      that mean IQ score of twins
      reared apart is different
      from the mean IQ of 100.
   e. Yes. $r = 0.702$. CV: $r =$
      ± 0.576 (assuming a 0.05
      significance level). Signifi-
      cant linear correlation.

1. Leo Tolstoy's *War and Peace* Refer to the sample data from 12 randomly selected pages from Leo Tolstoy's *War and Peace* as listed in Data Set 14 in Appendix B.
   a. The Flesch Reading Ease scores and the Flesch-Kincaid Grade Level scores are both designed to measure readability. Test for a correlation between those two variables.
   b. Find the regression equation in which the Flesch Reading Ease score is the dependent variable and the Flesch-Kincaid Grade Level score is the independent variable.
   c. Is it possible to test the claim that for the population of all pages in *War and Peace*, the mean Flesch Reading Ease score is equal to the mean Flesch-Kincaid Grade Level score? Would such a test make sense?
   d. Construct a 95% confidence interval estimate for the mean Flesch Reading Ease score for the population of all pages in *War and Peace*.

2. Effects of Heredity and Environment on IQ In studying the effects of heredity and environment on intelligence, it has been helpful to analyze the IQs of identical twins who were separated soon after birth. Identical twins share identical genes inherited from the same fertilized egg. By studying identical twins raised apart, we can eliminate the variable of heredity and better isolate the effects of the environment. The accompanying table shows the IQs of pairs of identical twins (older twins are x) raised apart (based on data from "IQs of Identical Twins Reared Apart," by Arthur Jensen, *Behavioral Genetics*). The sample data are typical of those obtained from other studies.
   a. Find the mean and standard deviation of the sample of older twins.
   b. Find the mean and standard deviation of the sample of younger twins.
   c. Based on the results from parts (a) and (b), does there appear to be a difference between the means of the two populations? In exploring the relationship between IQs of twins, is such a comparison of the two sample means the best approach? Why or why not?
   d. Combine all of the sample IQ scores, then use a 0.05 significance level to test the claim that the mean IQ score of twins reared apart is different from the mean IQ of 100.
   e. Is there a relationship between IQs of twins who were separated soon after birth? What method did you use? Write a summary statement about the effect of heredity and environment on intelligence, and note that your conclusions will be based on this relatively small sample of 12 pairs of identical twins.

x	107	96	103	90	96	113	86	99	109	105	96	89
y	111	97	116	107	99	111	85	108	102	105	100	93

Cooperative Group Activities

1. In-class activity Divide into groups of 8 to 12 people. For each group member, measure the person's height and also measure his or her navel height, which is the height from the floor to the navel. Is there a correlation between height and navel height? If so, find the regression equation with height expressed in terms of navel height. According to an old theory, the average person's ratio of height to navel height is the golden ratio: $(1 + \sqrt{5})/2 \approx 1.6$. Does this theory appear to be reasonably accurate?

2. In-class activity Divide into groups of 8 to 12 people. For each group member, *measure* height and arm span. For the arm span, the subject should stand with arms extended, like the wings on an airplane. It's easy to mark the height and arm span on a chalkboard, then measure the distances there. Using the paired sample data, is there a correlation between height and arm span? If so, find the regression equation with height expressed in terms of arm span. Can arm span be used as a reasonably good predictor of height?

3. In-class activity Divide into groups of 8 to 12 people. For each group member, use a string and ruler to measure head circumference and forearm length. Is there a relationship between these two variables? If so, what is it?

4. In-class activity Divide into groups of three or four people. Appendix B includes many data sets not yet addressed by the methods of this chapter. For example, using Data Set 25, we can investigate the correlation between high values of the Dow Jones Industrial Average and the numbers of U.S. cars sales. Search Appendix B for a pair of variables of interest, then investigate correlation and regression. State your conclusions and try to identify practical applications.

5. Out-of-class activity Divide into groups of three or four people. Investigate the relationship between two variables by collecting your own paired sample data and using the methods of this chapter to determine whether there is a significant linear correlation. Also identify the regression equation and describe a procedure for predicting values of one of the variables when given values of the other variable. Suggested topics:

   - Is there a relationship between taste and cost of different brands of chocolate chip cookies (or colas)? Taste can be measured on some number scale, such as 1 to 10.
   - Is there a relationship between salaries of professional baseball (or basketball, or football) players and their season achievements?
   - Rates versus weights: Is there a relationship between car fuel consumption rates and car weights? If so, what is it?
   - Is there a relationship between the lengths of men's (or women's) feet and their heights?
   - Is there a relationship between student grade-point averages and the amount of television watched? If so, what is it?

6. In-class activity Divide into groups of 3 or 4 people. The Chapter Problem and the Data to Decision problem deal with the issue of manatee deaths from boats. Identify two other ecological or environmental variables that might be related, such as the whale population and the number of fishing boats. Conduct research, collect and analyze data, and state conclusions. Based on the results, state recommendations for improving our world.

Technology Project

In Exercise 2 of the Cumulative Review Exercises, we noted that when studying the effects of heredity and environment on intelligence, it has been helpful to analyze the IQs of identical twins who were separated soon after birth. In this project, we will simulate 100 sets of twin births, but we will generate their IQ scores in a way that has no common genetic or environmental influences. Using the same procedure described in the Technology Project at the end of Chapter 5, generate a list of 100 simulated IQ scores randomly selected from a normally distributed population having a mean of 100 and a standard deviation of 15. Now use the same procedure to generate a second list of 100 simulated IQ scores that are also randomly selected from a normally distributed population with a mean of 100 and a standard deviation of 15. Even though the two lists were independently generated, treat them as paired data, so that the first score from each list represents the first set of twins, the

second score from each list represents the second set of twins, and so on. Before doing any calculations, first estimate a value of the linear correlation coefficient that you would expect. Now use the methods of Section 9-2 with a 0.05 significance level to test for a significant linear correlation and state your results.

Consider the preceding procedure to be one trial. Given the way that the sample data were generated, what proportion of such trials should lead to the incorrect conclusion that there is a significant linear correlation? By repeating the trials, we can verify that the proportion is approximately correct. Either repeat the trial or combine your results with others to verify that the proportion is approximately correct. Note that a type I error is the mistake of rejecting a true null hypothesis which, in this case, means that we conclude that there is a significant linear correlation when there really is no such linear correlation.

from DATA to DECISION

Critical Thinking: Should more regulations be imposed to save manatees?

Using the sample data in Table 9-1, we have found that there is a significant linear correlation between the number of Florida boat registrations and the number of manatee deaths from encounters with boats in Florida. Consequently, manatee refuges have been created, boat speed limits have been imposed in some zones, and construction of new marinas has been delayed. Given that more boats correspond to more manatee deaths, should more regulations be imposed?

The issue seems clear for environmentalists. After all, autopsies clearly prove that manatees are being killed by boats. However, others argue that in addition to considering the number of boat registrations and the number of manatee deaths from boats, it is also important to consider the changing size of the manatee population. Identifying manatee population values is difficult because they often reside in water with poor visibility, and they often rest on the bottom of deep water. Aerial surveys have been criticized as being very unreliable. We have noted in Section 9-2 that researcher Thomas Fraser suggested in a report that "a vigorous capture-tag and recapture program should be implemented by the State to gain much better information about population size and change." Cathy Beck, a biologist working with the U. S. Geological Survey has said that using statistical methods with the scar patterns on the backs of manatees could be helpful in understanding the manatee population.

Analyze the Results

The accompanying list includes boat registration numbers (in tens of thousands), manatee deaths from boats, and aerial survey counts of manatees for various years. Use the data to investigate and analyze the relevant issues. Determine whether more regulations should be imposed to save manatees.

Year	Boats	Manatee Deaths	Manatee Population
1976	41	10	738
1977	42	13	
1978	43	21	
1979	45	24	
1980	46	16	
1981	48	24	
1982	47	20	
1983	50	15	
1984	52	34	
1985	55	33	1200
1986	64	33	
1987	62	39	
1988	64	43	
1989	66	50	
1990	67	47	
1991	68	53	1267
1992	68	38	1856
1993	67	35	
1994	70	49	
1995	71	42	1443
1996	73	60	2639
1997	76	54	2229
1998	81	67	2022
1999	83	82	1873
2000	84	78	2223
2001		81	3276

INTERNET PROJECT   Linear Regression

The linear correlation coefficient is a tool that is used to measure the strength of the linear relationship between two sets of measurements. From a strictly computational point of view, the correlation coefficient may be found for any two data sets of paired values, regardless of what the data values represent. For this reason, certain questions should be asked whenever a correlation is being investigated. Is it reasonable to expect a linear correlation? Could a perceived correlation be caused by a third quantity related to each of the variables being studied? Go to the Web site for this textbook:

http://www.aw.com/triola

The Internet Project for this chapter will guide you to several sets of paired data in the fields of sports, medicine, and economics. You will then apply the methods of this chapter, computing correlation coefficients and determining regression lines, while considering the true relationships between the variables involved.

Statistics @ Work

"Prospective employees should have a fundamental grasp of statistics and its implications in the business world."

Angela Gillespie

Traffic Analyst, Lycos.com

As a Traffic Analyst for Lycos, Inc., Angela reports on major and minor traffic metrics. She monitors changes in trends and behavior patterns for use, enhancing the site to increase reach and stickiness (the amount of time people spend online at any particular Web site).

What is your job at Lycos?

I produce traffic reports of our site's activities each week. These are reviewed by our product group teams and senior management. They see what is increasing, what is decreasing, and make decisions about where our resources are spent.

My reports basically analyze trends on the sites and give projections for where we will be in a year or in any given time frame.

What concepts of statistics do you use?

Regression analysis and *R*-squared values.

How do you use statistics on the job?

To determine what is working and what is not working for our users. To determine the effectiveness of advertising campaigns, and to project future growth.

Please describe one specific example illustrating how the use of statistics was successful in improving a product or service.

At the end of our last fiscal year our CEO, Bob Davis, presented the company with an average daily pageview goal to be reached by the end of the next fiscal year. Using two years' worth of pageview data, I put together a projection showing where we would be at the end of the next fiscal

year if things remained static. Using an *R*-squared value gave these charts the oomph I needed to be effective. I updated the charts each week and presented them to the Product Management team. The data helped them understand what adjustments to make to their products and each week they got closer and closer to their goals. When Bob first presented the pageview goal, we all thought he had gone mad, but I am happy to say that at the end of the next fiscal year, we will have either reached our goal or be within 98% of it. Without the representation I supplied, Product Management would not have known where to focus their energy and resources. Because they were an efficient team, we have reached our unreachable goal.

Is your use of probability and statistics increasing, decreasing, or remaining stable?

As Lycos gets more sophisticated, they (management) expect more and more sophisticated reporting. It is increasing.

Do you feel job applicants are viewed more favorably if they have studied some statistics?

Absolutely, and not just within Lycos Reporting, but also in Product Marketing and Finance.

Multinomial Experiments and Contingency Tables

10-1 Overview

10-2 Multinomial Experiments: Goodness-of-Fit

10-3 Contingency Tables: Independence and Homogeneity

Using statistics to detect fraud

In the *New York Times* article "Following Benford's Law, or Looking Out for No. 1," Malcolm Browne writes that "the income tax agencies of several nations and several states, including California, are using detection software based on Benford's Law, as are a score of large companies and accounting businesses." According to Benford's law, a variety of different data sets include numbers with leading (first) digits that follow the distribution shown in the first two rows of Table 10-1. Data sets with values having leading digits that conform to Benford's law include stock market values, population sizes, numbers appearing on the front page of a newspaper, amounts on tax returns, lengths of rivers, and check amounts.

When working for the Brooklyn District Attorney, investigator Robert Burton used Benford's law to identify fraud by analyzing the leading digits on 784 checks. If the 784 checks follow Benford's law perfectly, 30.1% of the checks should have amounts with a leading digit of 1. The expected number of checks with amounts having a leading digit of 1 is 235.984 (because 30.1% of 784 is 235.984). The other expected frequencies are listed in the third row of Table 10-1. The bottom row of Table 10-1 lists the frequencies of the leading digits from amounts on 784 checks issued by seven different companies. A quick visual comparison shows that there appear to be major discrepancies between the frequencies expected by Benford's law and the frequencies observed in the check amounts, but how do we measure that disagreement? Are those discrepancies *significant*? Is there enough evidence to justify the conclusion that fraud has been committed? Is the evidence beyond a "reasonable doubt"? We will address these questions in this chapter.

Table 10-1	Benford's Law: Distribution of Leading Digits							
Leading Digit	1	2	3	4	5	6	7	8	9
Frequency according to Benford's law	30.1%	17.6%	12.5%	9.7%	7.9%	6.7%	5.8%	5.1%	4.6%
Leading digits of 784 checks following Benford's law	235.984	137.984	98.000	76.048	61.936	52.528	45.472	39.984	36.064
Leading digits of 784 checks analyzed for fraud	0	15	0	76	479	183	8	23	0

Note to Instructor
Coverage of at least Section 10-3 is strongly recommended. Analysis of survey results often involves the use of contingency tables, so the concepts of Section 10-3 are used frequently in real applications. Both Sections 10-2 and 10-3 require the χ^2 distribution, so if it has not yet been introduced, be sure to provide a thorough description of it. Basic concepts of the χ^2 distribution are included here as well as in Section 6-5 and again in Section 7-6.

 Overview

In this chapter we continue to apply inferential methods to different configurations of data. Recall from Chapter 1 that categorical (or qualitative, or attribute) data are those data that can be separated into different categories (often called **cells**) that are distinguished by some nonnumeric characteristic. For example, we might separate a sample of M&Ms into the color categories of red, orange, yellow, brown, blue, and green. After finding the frequency count for each category, we might proceed to test the claim that the frequencies fit (or agree with) the color distribution claimed by the manufacturer (Mars, Inc.). The main objective of this chapter is to test claims about categorical data consisting of frequency counts for different categories. In Section 10-2 we will consider multinomial experiments, which consist of observed frequency counts arranged in a single row or column (called a one-way frequency table), and we will test the claim that the observed frequency counts agree with some claimed distribution. In Section 10-3 we will consider contingency tables (or two-way frequency tables), which consist of frequency counts arranged in a table with at least two rows and two columns. We will use contingency tables for two types of very similar tests: (1) tests of independence, which test the claim that the row and column variables are independent; and (2) tests of homogeneity, which test the claim that different populations have the same proportion of some specified characteristic.

We will see that the methods of this chapter use the same χ^2 (chi-square) distribution that was first introduced in Section 6-5. Here are important properties of the chi-square distribution:

1. Unlike the normal and Student t distributions, the chi-square distribution is not symmetric. (See Figure 10-1.)

2. The values of the chi-square distribution can be 0 or positive, but they cannot be negative. (See Figure 10-1.)

3. The chi-square distribution is different for each number of degrees of freedom. (See Figure 10-2.)

Critical values of the chi-square distribution are found in Table A-4.

FIGURE 10-1 The **Chi-Square Distribution**

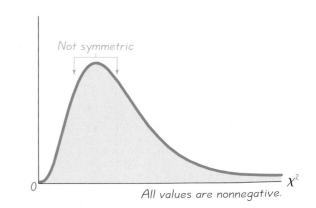

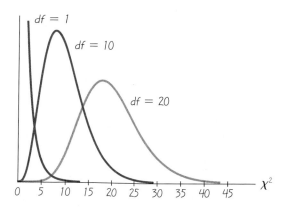

FIGURE 10-2 Chi-Square Distribution for 1, 10, and 20 Degrees of Freedom

10-2 Multinomial Experiments: Goodness-of-Fit

Each data set in this section consists of data that have been separated into different categories. The main objective is to determine whether the distribution agrees with or "fits" some claimed distribution. We define a *multinomial experiment* the same way we defined a binomial experiment (Section 4-3), except that a multinomial experiment has more than two categories (unlike a binomial experiment, which has exactly two categories).

Definition

A **multinomial experiment** is an experiment that meets the following conditions:

1. The number of trials is fixed.
2. The trials are independent.
3. All outcomes of each trial must be classified into exactly one of several different categories.
4. The probabilities for the different categories remain constant for each trial.

EXAMPLE Last-Digit Analysis of Home Run Distances In 2001, Barry Bonds hit 73 home runs and became the new record holder as the baseball player who hit the most home runs in one season. Data Set 30 in Appendix B lists the distances that were recorded for those home runs, and Table 10-2 summarizes the *last digits* of those distances. If such distances are actually measured, we usually expect that the last digits will occur with relative frequencies (or probabilities) that are roughly the same. In contrast, estimated values tend to have 0 or 5 occurring much more often as last digits. In Table 10-2, it appears that there are many more 0s than we would get with actual measurements. Later, we will analyze the data, but for now, simply verify that the four conditions of a multinomial experiment are satisfied.

continued

Table 10-2	
Last Digits of Barry Bonds's Home Run Distances	
Last Digit	Frequency
0	47
1	3
2	1
3	0
4	3
5	11
6	3
7	3
8	1
9	1

Note to Instructor

The analysis of the last digits of data is often used to detect data that have been reported instead of measured. In the National Health Examination, for example, government procedures require that doctors actually measure the heights of survey subjects, and the last digits of those heights are examined using methods of this section.

SOLUTION   Here is the verification that the four conditions of a multinomial experiment are all satisfied:

1. The number of trials (last digits) is the fixed number 73.
2. The trials are independent, because the last digit of the length of a home run does not affect the last digit of the length of any other home run.
3. Each outcome (last digit) is classified into exactly 1 of 10 different categories. The categories are identified as 0, 1, 2, . . . , 9.
4. Finally, if we assume that the home-run distances are measured, the last digits should be equally likely, so that each possible digit has a probability of 1/10.

In this section we are presenting a method for testing a claim that in a multinomial experiment, the frequencies observed in the different categories fit a particular distribution. Because we test for how well an observed frequency distribution fits some specified theoretical distribution, this method is often called a *goodness-of-fit test*.

Definition

A **goodness-of-fit test** is used to test the hypothesis that an observed frequency distribution fits (or conforms to) some claimed distribution.

For example, using the data in Table 10-2, we can test the hypothesis that the data fit a uniform distribution, with all of the digits being equally likely. Our goodness-of-fit tests will incorporate the following notation.

Notation

O represents the *observed frequency* of an outcome.

E represents the *expected frequency* of an outcome.

k represents the *number of different categories* or outcomes.

n represents the total *number of trials*.

Note to Instructor

It would be helpful to briefly review the concept of *expected value,* introduced in Section 4-2. Present a few simple and obvious examples such as this one: Find the expected number of girls born in groups of 100 babies. When students respond with the correct answer of 50, ask them to describe the exact thought process that led to the answer. They will respond that they found 1/2 of 100, which can be generalized as $p \cdot n$, which leads to $E = np$. Also point out that the expected number of girls in 3 babies is 1.5, so the expected value need not be an integer.

Also, emphasize that the methods of this section require that each *expected frequency* must be at least 5, but there is no requirement that *observed frequencies* must be at least 5.

Finding Expected Frequencies

In Table 10-2 we see that the observed frequencies O are denoted by 47, 3, 1, 0, 3, 11, 3, 3, 1, and 1. The sum of the observed frequencies is 73, so $n = 73$. If we assume that the 73 digits were obtained from a population in which all digits are equally likely, then we *expect* that each digit should occur in 1/10 of the 73 trials, so each of the 10 expected frequencies is given by $E = 7.3$. If we generalize this result, we get an easy procedure for finding expected frequencies whenever we are assuming that all of the expected frequencies are equal: Simply divide the total number of observations by the number of different categories ($E = n/k$). In other cases where the expected frequencies are not all equal, we can often find the expected frequency for each category by multiplying the sum of all observed frequencies and the probability p for the category, so $E = np$. We summarize these two procedures here.

- **If all expected frequencies are equal, then each expected frequency is the sum of all observed frequencies divided by the number of categories, so that $E = n/k$.**

- **If the expected frequencies are not all equal, then each expected frequency is found by multiplying the sum of all observed frequencies by the probability for the category, so $E = np$ for each category.**

As good as these two formulas for E might be, it would be better to use an informal approach based on an understanding of the circumstances. Just ask yourself, "How can the observed frequencies be split up among the different categories so that there is perfect agreement with the claimed distribution?" Also, recognize that the *observed* frequencies must all be whole numbers because they represent actual counts, but *expected* frequencies need not be whole numbers. For example, when rolling a single die 33 times, the expected frequency for each possible outcome is $33/6 = 5.5$. The expected frequency for the number of 3s occurring is 5.5, even though it is impossible to have the outcome of 3 occur exactly 5.5 times.

We know that sample frequencies typically deviate somewhat from the values we theoretically expect, so we now present the key question: Are the differences between the actual *observed* values O and the theoretically *expected* values E statistically significant? We need a measure of the discrepancy between the O and E values, so we use the test statistic that is given with the assumptions and critical values. (Later, we will explain how this test statistic was developed, but you can see that it has differences of $O - E$ as a key component.)

Assumptions

1. The data have been randomly selected.
2. The sample data consist of frequency counts for each of the different categories.
3. For each category, the *expected* frequency is at least 5. (The expected frequency for a category is the frequency that would occur if the data actually have the distribution that is being claimed. There is no requirement that the *observed* frequency for each category must be at least 5.)

Test Statistic for Goodness-of-Fit Tests in Multinomial Experiments

$$\chi^2 = \sum \frac{(O - E)^2}{E}$$

Critical Values

1. Critical values are found in Table A-4 by using $k - 1$ degrees of freedom, where k = number of categories.
2. Goodness-of-fit hypothesis tests are always *right-tailed*.

Note to Instructor

About the χ^2 notation: We usually use Greek letters for population parameters, but here we use χ^2 for a test statistic. Although not consistent, this is a very common notation in this context. Someone once said that "consistency is the hobgoblin of small minds."

Try to lead the class to a habit of developing their own reasoning process for why the tests of this section are all right-tailed. Ask the class these questions: If there are large discrepancies between the observed frequencies and those that are expected, what do we know about the $O - E$ values? The $(O - E)^2$ values? The value of the χ^2 test statistic? Where on the χ^2 distribution do large discrepancies fall?

The form of the χ^2 test statistic is such that *close agreement* between observed and expected values will lead to a *small* value of χ^2 and a *large P*-value. A large discrepancy between observed and expected values will lead to a *large* value of χ^2 and a *small P*-value. The hypothesis tests of this section are therefore always right-tailed, because the critical value and critical region are located at the extreme right of the distribution. These relationships are summarized and illustrated in Figure 10-3.

Once we know how to find the value of the test statistic and the critical value, we can test hypotheses by using the procedure introduced in Chapter 7 and summarized in Figure 7-8.

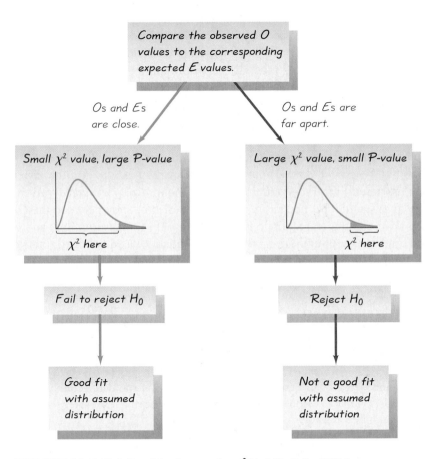

FIGURE 10-3 **Relationships Among the χ^2 Test Statistic, *P*-Value, and Goodness-of-Fit**

EXAMPLE **Last-Digit Analysis of Home Runs: Equal Expected Frequencies** Let's again refer to Table 10-2 for the last digits of Barry Bonds's home run distances. The value of 0 seems to occur considerably more often, but is that really significant? Test the claim that the digits do not occur with the same frequency.

SOLUTION   The claim that the digits do not occur with the same frequency is equivalent to the claim that the relative frequencies or probabilities of the 10 cells ($p_0, p_1, \ldots, p_9$) are not all equal. We will apply our standard procedure for testing hypotheses.

Step 1: The original claim is that the digits do not occur with the same frequency. That is, at least one of the probabilities $p_0, p_1, \ldots, p_9$ is different from the others.

Step 2: If the original claim is false, then all of the probabilities are the same. That is, $p_0 = p_1 = \ldots = p_9$.

Step 3: The null hypothesis must contain the condition of equality, so we have

$$H_0: \quad p_0 = p_1 = p_2 = p_3 = p_4 = p_5 = p_6 = p_7 = p_8 = p_9$$
$$H_1: \quad \text{At least one of the probabilities is different from the others.}$$

Step 4: No significance level was specified, so we select $\alpha = 0.05$, a very common choice.

Step 5: Because we are testing a claim about the distribution of the last digits being a uniform distribution, we use the goodness-of-fit test described in this section. The χ^2 distribution is used with the test statistic given earlier.

Step 6: The observed frequencies O are listed in Table 10-2, and each corresponding expected frequency E is equal to 7.3 (if the 73 digits were uniformly distributed through the 10 categories). Table 10-3 on the next page shows the computation of the χ^2 test statistic. The test statistic is $\chi^2 = 251.521$ (rounded). The critical value is $\chi^2 = 16.919$ (found in Table A-4 with $\alpha = 0.05$ in the right tail and degrees of freedom equal to $k - 1 = 9$). The test statistic and critical value are shown in Figure 10-4 on the next page.

Step 7: Because the test statistic falls within the critical region, there is sufficient evidence to reject the null hypothesis.

Step 8: There is sufficient evidence to support the claim that the last digits do not occur with the same relative frequency. We now have very strong evidence suggesting that the home run distances were not actually measured. It is reasonable to speculate that the distances are estimates instead of actual measurements.

The techniques in this section can be used to test whether an observed frequency distribution is a good fit with some theoretical frequency distribution. The preceding example tested for goodness-of-fit with a uniform distribution. Because many statistical analyses require a normally distributed population, we can use the chi-square test in this section to help determine whether given samples are drawn from normally distributed populations (see Exercise 25).

The preceding example dealt with the null hypothesis that the probabilities for the different categories are all equal. The methods of this section can also be used when the hypothesized probabilities (or frequencies) are different, as shown in the next example.

Safest Airplane Seats

Many of us believe that the rear seats are safest in an airplane crash. Safety experts do not agree that any particular part of an airplane is safer than others. Some planes crash nose first when they come down, but others crash tail first on takeoff. Matt McCormick, a survival expert for the National Transportation Safety Board, told *Travel* magazine that "there is no one safe place to sit." Goodness-of-fit tests can be used with a null hypothesis that all sections of an airplane are equally safe. Crashed airplanes could be divided into the front, middle, and rear sections. The observed frequencies of fatalities could then be compared to the frequencies that would be expected with a uniform distribution of fatalities. The χ^2 test statistic reflects the size of the discrepancies between observed and expected frequencies, and it would reveal whether some sections are safer than others.

Table 10-3 Calculating the χ^2 Test Statistic for the Last Digits of Home Run Distances

Last Digit	Observed Frequency O	Expected Frequency E	$O - E$	$(O - E)^2$	$\dfrac{(O - E)^2}{E}$
0	47	7.3	39.7	1576.09	215.9027
1	3	7.3	−4.3	18.49	2.5329
2	1	7.3	−6.3	39.69	5.4370
3	0	7.3	−7.3	53.29	7.3000
4	3	7.3	−4.3	18.49	2.5329
5	11	7.3	3.7	13.69	1.8753
6	3	7.3	−4.3	18.49	2.5329
7	3	7.3	−4.3	18.49	2.5329
8	1	7.3	−6.3	39.69	5.4370
9	1	7.3	−6.3	39.69	5.4370
	73	73		$\chi^2 = \Sigma \dfrac{(O - E)^2}{E} = 251.5206$	

(These two totals must agree.)

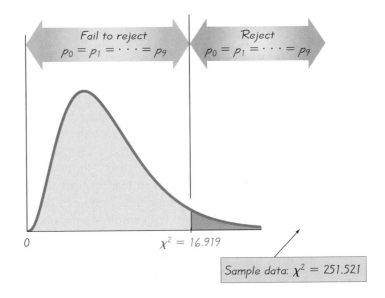

Sample data: $\chi^2 = 251.521$

$\chi^2 = 16.919$

Fail to reject
$p_0 = p_1 = \cdots = p_9$

Reject
$p_0 = p_1 = \cdots = p_9$

FIGURE 10-4 **Test of $p_0 = p_1 = p_2 = p_3 = p_4 = p_5 = p_6 = p_7 = p_8 = p_9$**

EXAMPLE **Detecting Fraud** In the Chapter Problem, it was noted that statistics is sometimes used to detect fraud. The second row of Table 10-1 lists percentages for leading digits as expected from Benford's law, and the third row lists the frequencies expected when the Benford's law percentages are applied to 784 leading digits. The bottom row of Table 10-1 lists the observed frequencies of the leading digits from amounts on 784 checks issued by seven different companies. Test the claim that there is a significant discrepancy between the leading digits expected from Benford's law and the leading digits observed on the 784 checks. Use a significance level of 0.01.

Note to Instructor
Students often have difficulty finding the expected frequencies when the proportions are not all equal, as in this example. It would be helpful to carefully explain how this is done.

SOLUTION   In testing the given claim, Steps 1, 2, and 3 result in the following hypotheses:

H_0: The distribution of leading digits is the distribution described by Benford's law. That is, $p_1 = 0.301$ and $p_2 = 0.176$ and $p_3 = 0.125$ and $p_4 = 0.097$ and $p_5 = 0.079$ and $p_6 = 0.067$ and $p_7 = 0.058$ and $p_8 = 0.051$ and $p_9 = 0.046$. (The proportions are the decimal equivalent values of the percentages listed for Benford's Law in Table 10-1.)

H_1: At least one of the above proportions is different from the claimed value.

Steps 4, 5, and 6 lead us to use the goodness-of-fit test with a 0.01 significance level and a test statistic calculated from Table 10-4.

continued

Table 10-4	Observed Frequencies and Frequencies Expected with Benford's Law				
Digit	Observed Frequency	Expected Frequency	$O - E$	$(O - E)^2$	$\dfrac{(O - E)^2}{E}$
1	0	235.984	−235.984	55688.4483	235.9840
2	15	137.984	−122.984	15125.0643	109.6146
3	0	98.000	−98.000	9604.0000	98.0000
4	76	76.048	−0.048	0.0023	0.0000
5	479	61.936	417.064	173942.3801	2808.4213
6	183	52.528	130.472	17022.9428	324.0737
7	8	45.472	−37.472	1404.1508	30.8795
8	23	39.984	−16.984	288.4563	7.2143
9	0	36.064	−36.064	1300.6121	36.0640

$$\text{Total: } \chi^2 = \sum \frac{(O - E)^2}{E} = 3650.2514$$

The test statistic is $\chi^2 = 3650.251$. The critical value of χ^2 is 20.090, and it is found in Table A-4 (using $\alpha = 0.01$ in the right tail with $k - 1 = 8$ degrees of freedom). The test statistic and critical value are shown in Figure 10-5. Because the test statistic falls within the critical region, there is sufficient evidence to warrant rejection of the null hypothesis. There is sufficient evidence to support the claim that there is a significant discrepancy between the leading digits expected from Benford's law and the leading digits observed on the 784 checks.

In Figure 10-6(a) we graph the claimed proportions of 0.301, 0.176, 0.125, 0.097, 0.079, 0.067, 0.058, 0.051, and 0.046 along with the observed proportions of 0.000, 0.019, 0.000, 0.097, 0.611, 0.233, 0.010, 0.029, and 0.000, so that we can visualize the discrepancy between the Benford's law distribution that was claimed and the frequencies that were observed. The points along the red line represent the claimed proportions, and the points along the green line represent the observed proportions. The corresponding pairs of points are far apart, showing that the expected frequencies are very different from the corresponding observed frequencies. The great disparity between the green line for observed frequencies and the red line for expected frequencies suggests that the check amounts are not the result of typical transactions. It appears that fraud may be involved. In fact, the Brooklyn District Attorney charged fraud by using this line of reasoning. For comparison, see Figure 10-6(b), which is based on the leading digits from the amounts on the last 200 checks written by the author. Note how the observed proportions from the author's checks agree quite well with the proportions expected with Benford's law. The author's checks appear to be typical instead of showing a pattern that might suggest fraud. In general, graphs such as Figure 10-6 are helpful in visually comparing expected frequencies and observed frequencies, as well as suggesting which categories result in the major discrepancies.

FIGURE 10-5 **Testing for Agreement Between Observed Frequencies and Frequencies Expected with Benford's Law**

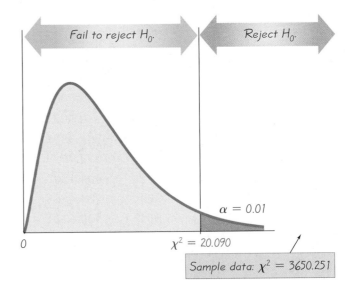

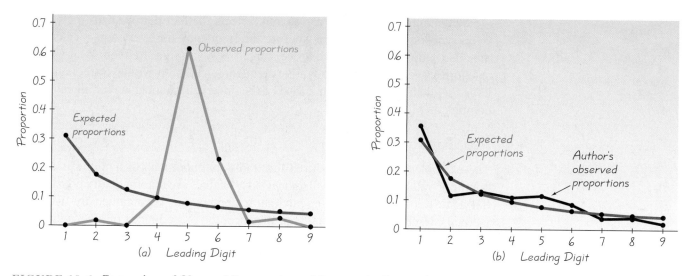

FIGURE 10-6 Comparison of Observed Frequencies and Frequencies Expected with Benford's Law

Rationale for the Test Statistic: The preceding examples should be helpful in developing a sense for the role of the χ^2 test statistic. It should be clear that we want to measure the amount of disagreement between observed and expected frequencies. Simply summing the differences between observed and expected values does not result in an effective measure because that sum is always 0, as shown below.

$$\Sigma(O - E) = \Sigma O - \Sigma E = n - n = 0$$

Squaring the $O - E$ values provides a better statistic, which reflects the differences between observed and expected frequencies. (The reasons for squaring the $O - E$ values are essentially the same as the reasons for squaring the $x - \bar{x}$ values in the formula for standard deviation.) The value of $\Sigma(O - E)^2$ measures only the magnitude of the differences, but we need to find the magnitude of the differences relative to what was expected. This relative magnitude is found through division by the expected frequencies, as in the test statistic.

The theoretical distribution of $\Sigma(O - E)^2/E$ is a discrete distribution because the number of possible values is limited to a finite number. The distribution can be approximated by a chi-square distribution, which is continuous. This approximation is generally considered acceptable, provided that all expected values E are at least 5. We included this requirement with the assumptions that apply to this section. In Section 5-6 we saw that the continuous normal probability distribution can reasonably approximate the discrete binomial probability distribution, provided that np and nq are both at least 5. We now see that the continuous chi-square distribution can reasonably approximate the discrete distribution of $\Sigma(O - E)^2/E$ provided that all values of E are at least 5. (There are ways of circumventing the problem of an expected frequency that is less than 5, such as combining categories so that all expected frequencies are at least 5.)

The number of degrees of freedom reflects the fact that we can freely assign frequencies to $k - 1$ categories before the frequency for every category is determined. (Although we say that we can "freely" assign frequencies to $k - 1$ categories, we cannot have negative frequencies nor can we have frequencies so large that their sum exceeds the total of the observed frequencies for all categories combined.)

P-Values

The examples in this section used the traditional approach to hypothesis testing, but the P-value approach can also be used. P-values are automatically provided by STATDISK or the TI-83 Plus calculator, or they can be obtained by using the methods described in Chapter 7. For instance, the preceding example resulted in a test statistic of $\chi^2 = 3650.251$. That example had $k = 9$ categories, so there were $k - 1 = 8$ degrees of freedom. Referring to Table A-4, we see that for the row with 8 degrees of freedom, the test statistic of 3650.251 is greater than the highest value in the row (21.955). Because the test statistic of $\chi^2 = 3650.251$ is farther to the right than 21.955, the P-value is less than 0.005. If the calculations for the preceding example are run on STATDISK, the display will include a P-value of 0.0000. The small P-value suggests that the null hypothesis should be rejected. (Remember, we reject the null hypothesis when the P-value is equal to or less than the significance level.) While the traditional method of testing hypotheses led us to reject the claim that the 784 check amounts have leading digits that conform to Benford's law, the P-value of 0.0000 indicates that the probability of getting leading digits like those that were obtained is extremely small. This appears to be evidence "beyond a reasonable doubt" that the check amounts are not the result of typical transactions.

Using Technology

STATDISK   First enter the observed frequencies in a column of the Data Window. If the expected frequencies are not all equal, also enter another column that includes either expected proportions or actual expected frequencies. Select **Analysis** from the main menu bar, then select the option **Multinomial Experiments.** Choose between "equal expected frequencies" and "unequal expected frequencies" and enter the data in the dialog box.

TI-83 Plus   The methods of this section are not available as a direct procedure on the TI-83 Plus calculator, but there is a simple trick that can be used. (Thanks to Rich Stephens of the University of Alaska.) First identify the observed and expected frequencies, then enter them as a *matrix*. Press **2nd x^{-1}** to get the **MATRIX** menu (or the **MATRIX** key on the TI-83). Select

EDIT, and press **ENTER.** Enter the dimensions of the matrix (2 rows by the number of columns) and proceed to enter the observed frequencies in the top row. For the bottom row, enter the expected frequencies multiplied by a really large number, such as 10^{30}. (For expected frequencies of 25, 15, 50, enter 25E30, 15E30, and 50E30.) When finished, press **STAT,** select **TESTS,** and then select the option χ^2**-Test.** Be sure that the observed matrix is the one you entered, such as matrix A. Scroll down to **Calculate** and press **ENTER** to get the test statistic, P-value, and number of degrees of freedom.

Minitab   **Excel**   The methods of this section are not available as built-in procedures.

10-2 Basic Skills and Concepts

Recommended Assignment:
Exercises 1–6, 8, 10, 16, 19, 20.

1. Testing for Equally Likely Categories Here are the observed frequencies from four categories: 5, 6, 8, 13. Assume that we want to use a 0.05 significance level to test the claim that the four categories are all equally likely.
  a. What is the null hypothesis?
  b. What is the expected frequency for each of the four categories?
  c. What is the value of the test statistic?
  d. What is the critical value?
  e. What do you conclude about the given claim?

1. a. $H_0: p_1 = p_2 = p_3 = p_4$
   b. 8, 8, 8, 8
   c. $\chi^2 = 4.750$
   d. $\chi^2 = 7.815$
   e. There is not sufficient evidence to reject the claim that the four categories are equally likely.

2. Testing for Categories with Different Proportions Here are the observed frequencies from five categories: 9, 8, 13, 14, 6. Assume that we want to use a 0.05 significance level to test the claim that the five categories have proportions of 0.2, 0.2, 0.2, 0.3, and 0.1, respectively.
  a. What is the null hypothesis?
  b. What are the expected frequencies for the five categories?
  c. What is the value of the test statistic?
  d. What is the critical value?
  e. What do you conclude about the given claim?

2. a. $H_0: p_1 = p_2 = p_3 = p_4 = p_5$
   b. 10, 10, 10, 15, 5
   c. $\chi^2 = 1.667$
   d. $\chi^2 = 9.488$
   e. There is not sufficient evidence to reject the claim that the five categories are equally likely.

3. Testing Fairness of Roulette Wheel The author observed 500 spins of a roulette wheel at the Mirage Resort and Casino. (To the IRS: Isn't that Las Vegas trip now a tax deduction?) For each spin, the ball can land in any one of 38 different slots that are supposed to be equally likely. When STATDISK was used to test the claim that the slots are in fact equally likely, the test statistic $\chi^2 = 38.232$ was obtained.
  a. Find the critical value assuming that the significance level is 0.10.
  b. STATDISK displayed a *P*-value of 0.41331, but what do you know about the *P*-value if you must use only Table A-4 along with the given test statistic of 38.232, which results from the 38 spins?
  c. Write a conclusion about the claim that the 38 results are equally likely.

3. a. df = 37, so $\chi^2 = 51.805$ (approximately).
   b. $0.10 < P\text{-value} < 0.90$
   c. There is not sufficient evidence to reject the claim that the roulette slots are equally likely.

4. Testing a Slot Machine The author purchased a slot machine (Bally Model 809), and tested it by playing it 1197 times. When testing the claim that the observed outcomes agree with the expected frequencies, a test statistic of $\chi^2 = 8.185$ was obtained. There are ten different categories of outcome, including no win, win jackpot, win with three bells, and so on.
  a. Find the critical value assuming that the significance level is 0.05.
  b. What can you conclude about the *P*-value from Table A-4 if you know that the test statistic is $\chi^2 = 8.185$ and there are 10 categories?
  c. State a conclusion about the claim that the observed outcomes agree with the expected frequencies. Does the author's slot machine appear to be working correctly?

4. a. $\chi^2 = 16.919$
   b. $0.10 < P\text{-value} < 0.90$
   c. There is not sufficient evidence to reject the claim that the observed outcomes agree with the expected outcomes.

5. Loaded Die The author drilled a hole in a die and filled it with a lead weight, then proceeded to roll it 200 times. Here are the observed frequencies for the outcomes of 1, 2, 3, 4, 5, and 6, respectively: 27, 31, 42, 40, 28, 32. Use a 0.05 significance level to test the claim that the outcomes are not equally likely. Does it appear that the loaded die behaves differently than a fair die?

5. $\chi^2 = 5.860$. CV: $\chi^2 = 11.071$. Loaded die does not appear to behave differently from a fair die.

6. Flat Tire and Missed Class A classic tale involves four car-pooling students who missed a test and gave as an excuse a flat tire. On the makeup test, the instructor asked the students to identify the particular tire that went flat. If they really didn't have a flat

6. $\chi^2 = 4.600$. CV: $\chi^2 = 7.815$. Students do not have the ability to select the same tire.

tire, would they be able to identify the same tire? The author asked 41 other students to identify the tire they would select. The results are listed in the following table (except for one student who selected the spare). Use a 0.05 significance level to test the author's claim that the results fit a uniform distribution. What does the result suggest about the ability of the four students to select the same tire when they really didn't have a flat?

Tire	Left front	Right front	Left rear	Right rear
Number selected	11	15	8	6

7. Do Car Crashes Occur on Different Days with the Same Frequency? It is a common belief that more fatal car crashes occur on certain days of the week, such as Friday or Saturday. A sample of motor vehicle deaths for a recent year in Montana is randomly selected. The numbers of fatalities for the different days of the week are listed in the accompanying table. At the 0.05 significance level, test the claim that accidents occur with equal frequency on the different days.

Day	Sun	Mon	Tues	Wed	Thurs	Fri	Sat
Number of fatalities	31	20	20	22	22	29	36

Based on data from the Insurance Institute for Highway Safety.

8. Are DWI Fatalities the Result of Weekend Drinking? Many people believe that fatal DWI crashes occur because of casual drinkers who tend to binge on Friday and Saturday nights, whereas others believe that fatal DWI crashes are caused by people who drink every day of the week. In a study of fatal car crashes, 216 cases are randomly selected from the pool in which the driver was found to have a blood alcohol content over 0.10. These cases are broken down according to the day of the week, with the results listed in the accompanying table. At the 0.05 significance level, test the claim that such fatal crashes occur on the different days of the week with equal frequency. Does the evidence support the theory that fatal DWI car crashes are due to casual drinkers or that they are caused by those who drink daily?

Day	Sun	Mon	Tues	Wed	Thurs	Fri	Sat
Number	40	24	25	28	29	32	38

Based on data from the Dutchess County STOP-DWI Program.

9. Testing for Uniformly Distributed Industrial Accidents A study was made of 147 industrial accidents that required medical attention. Among those accidents, 31 occurred on Monday, 42 on Tuesday, 18 on Wednesday, 25 on Thursday, and 31 on Friday (based on results from "Counted Data CUSUM's," by Lucas, *Technometrics,* Vol. 27, No. 2). Test the claim that accidents occur with equal proportions on the five workdays. If the proportions are not the same, what factors might explain the differences?

10. Grade and Seating Location Do "A" students tend to sit in a particular part of the classroom? The author recorded the locations of the students who received grades of A, with these results: 17 sat in the front, 9 sat in the middle, and 5 sat in the back of the classroom. Is there sufficient evidence to support the claim that the "A" students are not evenly distributed throughout the classroom? If so, does that mean you can increase your likelihood of getting an A by sitting in the front?

11. Post Position and Winning Horse Races Many people believe that when a horse races, it has a better chance of winning if its starting line-up position is closer to the rail on the inside of the track. The starting position of 1 is closest to the inside rail, followed by position 2, and so on. The accompanying table lists the numbers of wins for horses in the different starting positions. Test the claim that the probabilities of winning in the different post positions are not all the same.

Starting Position	1	2	3	4	5	6	7	8
Number of wins	29	19	18	25	17	10	15	11

Based on data from the *New York Post*.

12. Measuring Pulse Rates An example in this section was based on the principle that when certain quantities are measured, the last digits tend to be uniformly distributed, but if they are estimated or reported, the last digits tend to have disproportionately more 0s or 5s. Refer to Data Set 1 in Appendix B and use the last digits of the pulse rates of the 80 men and women. Those pulse rates were obtained as part of the National Health and Examination Survey. Test the claim that the last digits of 0, 1, 2, ..., 9 occur with the same frequency. Based on the observed digits, what can be inferred about the procedure used to obtain the pulse rates?

13. Are the Win 4 Numbers Random? Refer to Data Set 26 in Appendix B and use the 160 digits that were selected in New York State's Win 4 lottery game. Using a 0.05 significance level, test the claim that those digits are selected in such a way that the ten possible digits are all equally likely. Does the conclusion change if the significance level of 0.01 is used instead of 0.05? What would be an implication of the conclusion that the digits are not equally likely?

14. Are Violent Crimes Distributed Uniformly? Based on data from the Federal Bureau of Investigation, violent crimes in a recent year occurred with the distribution given in the accompanying table. Violent crimes include murder, nonnegligent manslaughter, rape, robbery, and aggravated assault. The listed percentages are based on a total of 1,424,287 cases of violent crime. Use a 0.01 significance level to test the claim that violent crimes are distributed equally among the 12 months. How do you explain the conclusion, given that the listed percentages do not appear to be dramatically different? Is there a reasonable explanation of why violent crimes might not be distributed equally among the 12 months?

Month	Jan.	Feb.	March	April	May	June	July	Aug.	Sept.	Oct.	Nov.	Dec.
Percent	7.7	7.4	8.4	8.3	9.2	8.6	9.0	8.9	8.6	8.7	7.6	7.7

15. M&M Candies Mars, Inc. claims that its M&M plain candies are distributed with the following color percentages: 30% brown, 20% yellow, 20% red, 10% orange, 10% green, and 10% blue. Refer to Data Set 19 in Appendix B and use the sample data to test the claim that the color distribution is as claimed by Mars, Inc. Use a 0.05 significance level.

16. Car Crashes and Age Brackets Among drivers who have had a car crash in the last year, 88 are randomly selected and categorized by age, with the results listed in the accompanying table. If all ages have the same crash rate, we would expect (because of the age distribution of licensed drivers) the given categories to have 16%, 44%, 27%, and 13% of the subjects, respectively. At the 0.05 significance level, test the

11. $\chi^2 = 16.333$. CV: $\chi^2 = 14.067$ (assuming a 0.05 significance level). Support the claim that the probabilities of winning in the different post positions are not all the same.

12. $\chi^2 = 80.750$. CV: $\chi^2 = 16.919$ (assuming a 0.05 significance level). Reject the claim that the digits occur with the same frequency. All digits are even numbers, suggesting that pulse rates were recorded for 30 seconds, then doubled.

13. $\chi^2 = 18.500$. CV: $\chi^2 = 16.919$. Reject the claim that the digits occur with the same frequency. The conclusion changes with a 0.01 significance level.

14. $\chi^2 = 6853.765$. CV: $\chi^2 = 24.725$. Summer months appear to have more crimes. The differences are highly significant because of the very large sample size.

15. $\chi^2 = 5.950$. CV: $\chi^2 = 11.071$. There is not sufficient evidence to reject the distribution claimed by Mars, Inc.

16. $\chi^2 = 53.051$. CV: $\chi^2 = 7.815$. Drivers under 21 appear to have disproportionately more crashes.

claim that the distribution of crashes conforms to the distribution of ages. Does any age group appear to have a disproportionate number of crashes?

Age	Under 25	25–44	45–64	Over 64
Drivers	36	21	12	19

Based on data from the Insurance Information Institute.

17. Distribution of Digits in the Irrational Number Pi The number π is an irrational number with the property that when we try to express it in decimal form, it requires an infinite number of decimal places and there is no pattern of repetition. In the decimal representation of π, the first 100 digits occur with the frequencies described in the accompanying table. At the 0.05 significance level, test the claim that the digits are uniformly distributed.

Digit	0	1	2	3	4	5	6	7	8	9
Frequency	8	8	12	11	10	8	9	8	12	14

18. Distribution of Digits in the Rational Number 22/7 The number 22/7 is similar to π in the sense that they both require an infinite number of decimal places. However, 22/7 is a rational number because it can be expressed as the ratio of two integers, whereas π cannot. When rational numbers such as 22/7 are expressed in decimal form, there is a pattern of repetition. In the decimal representation of 22/7, the first 100 digits occur with the frequencies described in the accompanying table. At the 0.05 significance level, test the claim that the digits are uniformly distributed. How does the result differ from that found in Exercise 17?

Digit	0	1	2	3	4	5	6	7	8	9
Frequency	0	17	17	1	17	16	0	16	16	0

19. Author's Check Amounts and Benford's Law Figure 10-6(b) illustrates the observed frequencies of the leading digits from the amounts of the last 200 checks that the author wrote. The observed frequencies of those leading digits are listed below. Using a 0.05 significance level, test the claim that they come from a population of leading digits that conform to Benford's law. (See the first two rows of Table 10-1 included in the Chapter Problem.)

Leading Digit	1	2	3	4	5	6	7	8	9
Frequency	72	23	26	20	21	18	8	8	4

20. Do World War II Bomb Hits Fit a Poisson Distribution? In analyzing hits by V-1 buzz bombs in World War II, South London was subdivided into regions, each with an area of 0.25 km^2. In Section 4-5 we presented an example and included a table of actual frequencies of hits and the frequencies expected with the Poisson distribution. Use the values listed here and test the claim that the actual frequencies fit a Poisson distribution. Use a 0.05 significance level.

Number of bomb hits	0	1	2	3	4 or more
Actual number of regions	229	211	93	35	8
Expected number of regions (from Poisson distribution)	227.5	211.4	97.9	30.5	8.7

10-2 Beyond the Basics

21. Testing Effects of Outliers In conducting a test for the goodness-of-fit as described in this section, does an outlier have much of an effect on the value of the χ^2 test statistic? Test for the effect of an outlier by repeating Exercise 6 after changing the frequency for the right rear tire from 6 to 60. Describe the general effect of an outlier.

22. Detecting Altered Experimental Data When Gregor Mendel conducted his famous hybridization experiments with peas, it appears that his gardening assistant knew the results that Mendel expected, and he altered the results to fit Mendel's expectations. Subsequent analysis of the results led to the conclusion that there is a probability of only 0.00004 that the expected results and reported results would agree so closely. How could the methods of this section be used to detect such results that are just too perfect to be realistic?

23. Equivalent Test In this exercise we will show that a hypothesis test involving a multinomial experiment with only two categories is equivalent to a hypothesis test for a proportion (Section 7-3). Assume that a particular multinomial experiment has only two possible outcomes, A and B, with observed frequencies of f_1 and f_2, respectively.
  a. Find an expression for the χ^2 test statistic, and find the critical value for a 0.05 significance level. Assume that we are testing the claim that both categories have the same frequency, $(f_1 + f_2)/2$.
  b. The test statistic $z = (\hat{p} - p)/\sqrt{pq/n}$ is used to test the claim that a population proportion is equal to some value p. With the claim that $p = 0.5$, $\alpha = 0.05$, and $\hat{p} = f_1/(f_1 + f_2)$, show that z^2 is equivalent to χ^2 [from part (a)]. Also show that the square of the critical z score is equal to the critical χ^2 value from part (a).

24. Testing Goodness-of-Fit with a Binomial Distribution An observed frequency distribution is as follows:

Number of successes	0	1	2	3
Frequency	89	133	52	26

  a. Assuming a binomial distribution with $n = 3$ and $p = 1/3$, use the binomial probability formula to find the probability corresponding to each category of the table.
  b. Using the probabilities found in part (a), find the expected frequency for each category.
  c. Use a 0.05 significance level to test the claim that the observed frequencies fit a binomial distribution for which $n = 3$ and $p = 1/3$.

25. Testing Goodness-of-Fit with a Normal Distribution An observed frequency distribution of sample IQ scores is as follows:

IQ score	Less than 80	80–95	96–110	111–120	More than 120
Frequency	20	20	80	40	40

  a. Assuming a normal distribution with $\mu = 100$ and $\sigma = 15$, use the methods given in Chapter 5 to find the probability of a randomly selected subject belonging to each class. (Use class boundaries of 79.5, 95.5, 110.5, and 120.5.)
  b. Using the probabilities found in part (a), find the expected frequency for each category.
  c. Use a 0.01 significance level to test the claim that the IQ scores were randomly selected from a normally distributed population with $\mu = 100$ and $\sigma = 15$.

21. The test statistic changes from 4.600 to 76.638, so the outlier has a dramatic effect.
22. Instead of a right-tailed test, use a left-tailed test. If the sample data fit the claimed distribution almost perfectly, the test statistic will be located near the extreme left position of zero.
23. a. CV is $\chi^2 = 3.841$ and test statistic is
$$\chi^2 = \frac{\left(f_1 - \frac{f_1 + f_2}{2}\right)^2}{\frac{f_1 + f_2}{2}} + \frac{\left(f_2 - \frac{f_1 + f_2}{2}\right)^2}{\frac{f_1 + f_2}{2}}$$
$$= \frac{(f_1 - f_2)^2}{f_1 + f_2}$$
  b. The χ^2 critical value is 3.841 and it is approximately equal to the square of $z = 1.96$.
24. a. 0.296, 0.444, 0.222, 0.037
  b. 88.9, 133.3, 66.7, 11.1
  c. $\chi^2 = 23.241$. CV: $\chi^2 = 7.815$. Reject the claim that the observed frequencies fit a binomial distribution with $n = 3$ and $p = 1/3$.
25. a. 0.0853, 0.2968, 0.3759, 0.1567, 0.0853
  b. 17.06, 59.36, 75.18, 31.34, 17.06
  c. $\chi^2 = 60.154$. CV: $\chi^2 = 13.277$. Reject the claim that the IQ scores were randomly selected from a normally distributed population with mean 100 and standard deviation 15.

10-3 Contingency Tables: Independence and Homogeneity

In Section 10-2 we considered categorical data summarized with frequency counts listed in a single row or column. Because the cells of the single row or column correspond to categories of a single variable (such as color), the tables in Section 10-2 are sometimes called *one-way frequency tables*. In this section we again consider categorical data summarized with frequency counts, but the cells correspond to two different variables. The tables we consider in this section are called *contingency tables,* or *two-way frequency tables*.

> **Definitions**
>
> A **contingency table** (or **two-way frequency table**) is a table in which frequencies correspond to two variables. (One variable is used to categorize rows, and a second variable is used to categorize columns.)

Table 10-5, which summarizes the fate of the passengers and crew when the *Titanic* sank on Monday, April 15, 1912, has two variables: a row variable, which indicates whether the person survived or died; and a column variable, which lists the demographic categories—men, women, boys, girls.

Contingency tables are especially important because they are often used to analyze survey results. For example, we might ask subjects one question in which they identify their gender (male/female), and we might ask another question in which they describe the frequency of their use of TV remote controls (often/ sometimes/never). The methods of this section can then be used to determine whether the use of TV remote controls is independent of gender. (We probably already know the answer to that one.) Applications of this type are very numerous, so the methods presented in this section are among those most often used.

This section presents two types of hypothesis testing based on contingency tables. We first consider tests of independence, used to determine whether a contingency table's row variable is independent of its column variable. We then consider tests of homogeneity, used to determine whether different populations have the same proportions of some characteristic. Good news: Both types of hypothesis testing use the *same* basic methods. We begin with tests of independence.

Table 10-5	*Titanic* Mortality				
	Men	Women	Boys	Girls	**Total**
Survived	332	318	29	27	**706**
Died	1360	104	35	18	**1517**
Total	**1692**	**422**	**64**	**45**	**2223**

Test of Independence

One of the two tests included in this section is a *test of independence* between the row variable and column variable.

> ## Definition
>
> A **test of independence** tests the null hypothesis that there is no association between the row variable and the column variable in a contingency table. (For the null hypothesis, we will use the statement that "the row and column variables are independent.")

It is very important to recognize that in this context, the word *contingency* refers to dependence, but this is only a statistical dependence, and it cannot be used to establish a direct cause-and-effect link between the two variables in question. For example, after analyzing the data in Table 10-5, we might conclude that whether a person survived the sinking of the Titanic is dependent on whether that person was a man, woman, boy, or girl, but that doesn't mean that the gender/age category has some direct causative effect on surviving.

When testing the null hypothesis of independence between the row and column variables in a contingency table, the assumptions, test statistic, and critical values are described in the following box.

Assumptions

1. The sample data are randomly selected.
2. The null hypothesis H_0 is the statement that the row and column variables are *independent;* the alternative hypothesis H_1 is the statement that the row and column variables are dependent.
3. For every cell in the contingency table, the *expected* frequency E is at least 5. (There is no requirement that every *observed* frequency must be at least 5. Also, there is no requirement that the population must have a normal distribution or any other specific distribution.)

Test Statistic for a Test of Independence

$$\chi^2 = \sum \frac{(O - E)^2}{E}$$

Critical values

1. The critical values are found in Table A-4 by using
$$\text{degrees of freedom} = (r - 1)(c - 1)$$
where r is the number of rows and c is the number of columns.
2. In a test of independence with a contingency table, the critical region is located in the *right tail only.*

Delaying Death

University of California sociologist David Phillips has studied the ability of people to postpone their death until after some important event. Analyzing death rates of Jewish men who died near Passover, he found that the death rate dropped dramatically in the week before Passover, but rose the week after. He found a similar phenomenon occurring among Chinese-American women; their death rate dropped the week before their important Harvest Moon Festival, then rose the week after.

Note to Instructor

It will be obvious that the test statistic given here is identical to the χ^2 test statistic given in Section 10-2. The hypothesis tests of this section are all right-tailed, as in Section 10-2. Differences between this section and Section 10-2 are found in the method used for finding expected values E and the calculation of the number of degrees of freedom.

The test statistic allows us to measure the degree of disagreement between the frequencies actually observed and those that we would theoretically expect when the two variables are independent. Small values of the χ^2 test statistic result from close agreement between frequencies observed and frequencies expected with independent row and column variables. Large values of the χ^2 test statistic are in the rightmost region of the chi-square distribution, and they reflect significant differences between observed and expected frequencies. In repeated large samplings, the distribution of the test statistic χ^2 can be approximated by the chi-square distribution, provided that all expected frequencies are at least 5. The number of degrees of freedom $(r-1)(c-1)$ reflects the fact that because we know the total of all frequencies in a contingency table, we can freely assign frequencies to only $r-1$ rows and $c-1$ columns before the frequency for every cell is determined. [However, we cannot have negative frequencies or frequencies so large that any row (or column) sum exceeds the total of the observed frequencies for that row (or column).]

In the preceding section we knew the corresponding probabilities and could easily determine the expected values, but the typical contingency table does not come with the relevant probabilities. For each cell in the frequency table, the expected frequency E can be calculated by applying the multiplication rule of probability for independent events. Assuming that the row and column variables are independent (which is assumed in the null hypothesis), the probability of a value being in a particular cell is the probability of being in the row containing the cell (namely, the row total divided by the sum of all frequencies) multiplied by the probability of being in the column containing the cell (namely, the column total divided by the sum of all frequencies) multiplied by the sum of all frequencies. Sound too complicated? The expected frequency for a cell can be simplified to the following equation.

Note to Instructor
Instead of simply presenting the formula for the calculation of expected value *E*, strongly consider discussing the rationale for that expression. Refer to Table 10-5 and ask the class these questions "in the spirit of reviewing a few of the important and basic principles of probabilities:" (1) If one of the *Titanic* passengers is randomly selected, find the probability of getting a man; (2) Find the probability of getting someone who survived; (3) Find the probability of getting a man and someone who survived, assuming that surviving or dying is *independent* of being a man, woman, boy, or girl.

Expected Frequency for a Contingency Table

$$\text{expected frequency} = \frac{(\text{row total})(\text{column total})}{(\text{grand total})}$$

Here *grand total* refers to the total of all observed frequencies in the table. For example, the expected frequency for the upper left cell of Table 10-6 (a duplicate of Table 10-5 with expected frequencies inserted in parentheses) is 537.360, which is found by noting that the total of all frequencies for the first row is 706, the total of the column frequencies is 1692, and the sum of all frequencies in the table is 2223, so we get an expected frequency of

$$E = \frac{(\text{row total})(\text{column total})}{(\text{grand total})} = \frac{(706)(1692)}{2223} = 537.360$$

Table 10-6	Observed Frequencies (and Expected Frequencies)				
	Gender/Age Category				Row totals
	Men	Women	Boys	Girls	
Survived	332	318	29	27	706
	(537.360)	(134.022)	(20.326)	(14.291)	
Died	1360	104	35	18	1517
	(1154.640)	(287.978)	(43.674)	(30.709)	
Column totals:	1692	422	64	45	Grand total: 2223

EXAMPLE **Finding Expected Frequency** The expected frequency for the upper left cell of Table 10-6 is 537.360. Find the expected frequency for the lower left cell, assuming independence between the row variable (whether the person survived) and the column variable (whether the person is a man, woman, boy, or girl).

SOLUTION   The lower left cell lies in the second row (with total 1517) and the first column (with total 1692). The expected frequency is

$$E = \frac{(\text{row total})(\text{column total})}{(\text{grand total})} = \frac{(1517)(1692)}{2223} = 1154.640$$

INTERPRETATION To interpret this result for the lower left cell, we can say that although 1360 men actually died, we would have expected 1154.640 men to die if survivability is independent of whether the person is a man, woman, boy, or girl. There is a discrepancy between $O = 1360$ and $E = 1154.640$, and such discrepancies are key components of the test statistic.

To better understand the rationale for finding expected frequencies with this procedure, let's pretend that we know only the row and column totals and that we must fill in the cell expected frequencies by assuming independence (or no relationship) between the two variables involved—that is, we pretend that we know only the row and column totals shown in Table 10-6. Let's begin with the cell in the upper left corner. Because 706 of the 2223 persons survived, we have $P(\text{survived}) = 706/2223$. Similarly, 1692 of the people were men, so $P(\text{man}) = 1692/2223$. Because we are assuming independence between survivability and the column gender/age category, we can use the multiplication rule of probability to get

$$P(\text{survived and man}) = P(\text{survived}) \cdot P(\text{man}) = \frac{706}{2223} \cdot \frac{1692}{2223}$$

Home Field Advantage

In the *Chance* magazine article "Predicting Professional Sports Game Outcomes from Intermediate Game Scores," authors Harris Cooper, Kristina DeNeve, and Frederick Mosteller used statistics to analyze two common beliefs: Teams have an advantage when they play at home, and only the last quarter of professional basketball games really counts. Using a random sample of hundreds of games, they found that for the four top sports, the home team wins about 58.6% of games. Also, basketball teams ahead after 3 quarters go on to win about 4 out of 5 times, but baseball teams ahead after 7 innings go on to win about 19 out of 20 times. The statistical methods of analysis included the chi-square distribution applied to a contingency table.

This equation is an application of the multiplication rule for independent events, which is expressed in general as follows: $P(A \text{ and } B) = P(A) \cdot P(B)$. Knowing the probability of being in the upper left cell, we can now find the *expected value* for that cell, which we get by multiplying the probability for that cell by the total number of people, as shown in the following equation:

$$E = n \cdot p = 2223\left[\frac{706}{2223} \cdot \frac{1692}{2223}\right] = 537.360$$

The form of this product suggests a general way to obtain the expected frequency of a cell:

$$\text{Expected frequency } E = (\text{grand total}) \cdot \frac{(\text{row total})}{(\text{grand total})} \cdot \frac{(\text{column total})}{(\text{grand total})}$$

This expression can be simplified to

$$E = \frac{(\text{row total}) \cdot (\text{column total})}{(\text{grand total})}$$

We can now proceed to use contingency table data for testing hypotheses, as in the following example.

EXAMPLE *Titanic* Sinking   Refer to the *Titanic* mortality data in Table 10-5. We will treat the 2223 people aboard the *Titanic* as a *sample*. We could take the position that the *Titanic* data constitute a *population* and therefore should not be treated as a sample, so that methods of inferential statistics do not apply. Let's stipulate that the data are sample data randomly selected from the population of all theoretical people who would find themselves in the same conditions. Realistically, no other people will actually find themselves in the same conditions, but we will make that assumption for the purposes of this discussion and analysis. We can then determine whether the observed differences have statistical significance. (See also Paul Velleman's *ActivStats* software for the example involving the *Titanic*.)

Using a 0.05 significance level, test the claim that when the *Titanic* sank, whether someone survived or died is independent of whether the person is a man, woman, boy, or girl.

SOLUTION   The null hypothesis and alternative hypothesis are as follows:

H_0:   Whether a person survived is independent of whether the person is a man, woman, boy, or girl.

H_1:   Surviving the *Titanic* sinking and being a man, woman, boy, or girl are dependent.

The significance level is $\alpha = 0.05$.

Because the data are in the form of a contingency table, we use the χ^2 distribution with this test statistic:

$$\chi^2 = \sum \frac{(O - E)^2}{E}$$

$$= \frac{(332 - 537.360)^2}{537.360} + \frac{(318 - 134.022)^2}{134.022} + \frac{(29 - 20.326)^2}{20.326}$$

$$+ \frac{(27 - 14.291)^2}{14.291} + \frac{(1360 - 1154.640)^2}{1154.640} + \frac{(104 - 287.978)^2}{287.978}$$

$$+ \frac{(35 - 43.674)^2}{43.674} + \frac{(18 - 30.709)^2}{30.709}$$

$$= 78.481 + 252.555 + 3.702 + 11.302$$

$$+ 36.525 + 117.536 + 1.723 + 5.260$$

$$= 507.084$$

(The more accurate test statistic of 507.080 is obtained by carrying more decimal places in the intermediate calculations. STATDISK, Minitab, and the TI-83 Plus calculator all agree that 507.080 is a better result.) The critical value is $\chi^2 = 7.815$ and it is found from Table A-4 by noting that $\alpha = 0.05$ in the right tail and the number of degrees of freedom is given by $(r - 1)(c - 1) = (2 - 1)(4 - 1) = 3$. The test statistic and critical value are shown in Figure 10-7. Because the test statistic falls within the critical region, we reject the null hypothesis that whether a person survived is independent of whether the person is a man, woman, boy, or girl. It appears that whether a person survived the *Titanic* and whether that person is a man, woman, boy, or girl are dependent variables.

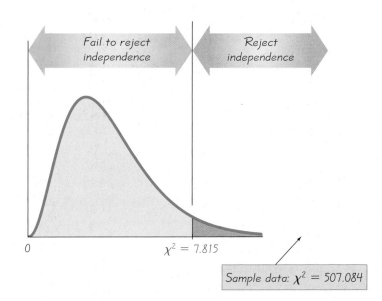

FIGURE 10-7 **Test of Independence for the *Titanic* Mortality Data**

P-Values

The preceding example used the traditional approach to hypothesis testing, but we can easily use the *P*-value approach. STATDISK, Minitab, Excel, and the TI-83 Plus calculator all provide *P*-values for tests of independence in contingency tables. If you don't have a suitable calculator or statistical software package, you can estimate *P*-values from Table A-4 in Appendix A. Locate the appropriate number of degrees of freedom to isolate a particular row in that table. Find where the test statistic falls in that row, and you can identify a range of possible *P*-values by referring to the areas given at the top of each column. In the preceding example, there are 3 degrees of freedom, so go to the third row of Table A-4. Now use the test statistic of $\chi^2 = 507.084$ to see that this test statistic is greater than (and farther to the right of) every critical value of χ^2 that is in the third row, so the *P*-value is less than 0.005. On the basis of this small *P*-value, we again reject the null hypothesis and conclude that there is sufficient sample evidence to warrant rejection of the null hypothesis of independence.

As in Section 10-2, if observed and expected frequencies are close, the χ^2 test statistic will be small and the *P*-value will be large. If observed and expected frequencies are far apart, the χ^2 test statistic will be large and the *P*-value will be small. These relationships are summarized and illustrated in Figure 10-8.

FIGURE 10-8 Relationships Among Key Components in Test of Independence

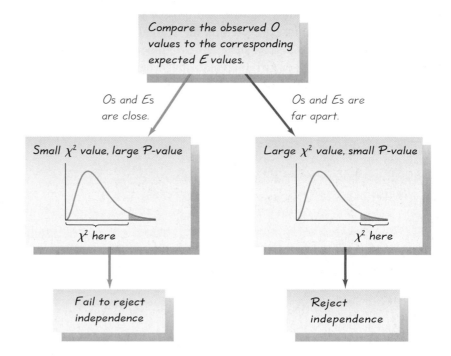

Note to Instructor

Emphasize that the test of homogeneity uses the same procedures already presented in this section. Because a test of homogeneity is a test of the claim that different populations have the same proportions of some characteristics, we use data sampled from different populations, and we therefore have predetermined totals for either the rows or the columns in the contingency table.

Test of Homogeneity

In the preceding example, we illustrated a test of independence by using a sample of 2223 people who were aboard the *Titanic*. We were treating those 2223 people as a random sample drawn from *one* hypothetical population of all people who would find themselves in similar circumstances. However, some other samples are drawn from *different* populations, and we want to determine whether those populations have the same proportions of the characteristics being considered. The *test of homogeneity* can be used in such cases. (The word *homogeneous* means "having the same quality," and in this context, we are testing to determine whether the proportions are the same.)

Definition

In a **test of homogeneity,** we test the claim that *different populations* have the same proportions of some characteristics.

In conducting a test of homogeneity, we can use the same procedures already presented in this section, as illustrated in the following example.

EXAMPLE   Influence of Gender   Does a pollster's gender have an effect on poll responses by men? A *U.S. News & World Report* article about polls stated: "On sensitive issues, people tend to give 'acceptable' rather than honest responses; their answers may depend on the gender or race of the interviewer." To support that claim, data were provided for an Eagleton Institute poll in which surveyed men were asked if they agreed with this statement: "Abortion is a private matter that should be left to the woman to decide without government intervention." We will analyze the effect of gender on male survey subjects only. Table 10-7 is based on the responses of surveyed men. Assume that the survey was designed so that male interviewers were instructed to obtain 800 responses from male subjects, and female interviewers were instructed to obtain 400 responses from male subjects. Using a 0.05 significance level, test the claim that the proportions of agree/disagree responses are the same for the subjects interviewed by men and the subjects interviewed by women.

SOLUTION   Because we have two separate populations (subjects interviewed by men and subjects interviewed by women), we test for homogeneity with these hypotheses:

H_0:   The proportions of agree/disagree responses are the same for the subjects interviewed by men and the subjects interviewed by women.

H_1:   The proportions are different.

Table 10-7	Gender and Survey Responses	
	Gender of Interviewer	
	Man	Woman
Men who agree	560	308
Men who disagree	240	92

continued

The significance level is $\alpha = 0.05$. We use the same χ^2 test statistic described earlier, and it is calculated by using the same procedure. Instead of listing the details of that calculation, we provide the Minitab display that results from the data in Table 10-7.

Minitab

```
Expected counts are printed below observed counts

                    C1            C2         Total
         1          560           308         868
                    578.67        289.33

         2          240           92          332
                    221.33        110.67

      Total         800           400         1200

   Chi-Sq = 0.602 + 1.204 + 1.574
            + 3.149 = 6.529

   DF = 1, P-Value = 0.011
```

The Minitab display shows the expected frequencies of 578.67, 289.33, 221.33, and 110.67. The display also includes the test statistic of $\chi^2 = 6.529$ and the P-value of 0.011. Using the P-value approach to hypothesis testing, we reject the null hypothesis of equal (homogeneous) proportions (because the P-value of 0.011 is less than 0.05). There is sufficient evidence to warrant rejection of the claim that the proportions are the same. It appears that response and the gender of the interviewer are dependent. Although this statistical analysis cannot be used to justify any statement about causality, it does appear that men are influenced by the gender of the interviewer.

Using Technology

STATDISK   First enter the observed frequencies in columns of the Data Window. Select **Analysis** from the main menu bar, then select **Contingency Tables,** and proceed to identify the columns that contain the frequencies. Click on **Evaluate.** The results include the test statistic, critical value, P-value, and conclusion.

Minitab   First enter the observed frequencies in columns, then select **Stat** from the main menu bar. Next select the option **Tables,** then select **Chi Square Test** and proceed to enter the names of the columns containing the observed frequencies, such as C1 C2 C3 C4. Minitab provides the test statistic and P-value.

TI-83 Plus   First enter the contingency table as a *matrix* by pressing **2nd x^{-1}** to get the **MATRIX** menu (or the **MATRIX** key on the TI-83). Select **EDIT,** and press **ENTER.** Enter the di-

mensions of the matrix (rows by columns) and proceed to enter the individual frequencies. When finished, press **STAT,** select **TESTS,** and then select the option χ^2**-Test.** Be sure that the observed matrix is the one you entered, such as matrix A. The expected frequencies will be automatically calculated and stored in the separate matrix identified as "Expected." Scroll down to **Calculate** and press **ENTER** to get the test statistic, P-value, and number of degrees of freedom.

Excel   You must enter the observed frequencies, and you must also determine and enter the expected frequencies. When finished, click on the *fx* icon in the menu bar, select the function category **Statistical,** and then select the function name **CHITEST.** You must enter the range of values for the observed frequencies and the range of values for the expected frequencies. Only the P-value is provided.

10-3 Basic Skills and Concepts

Recommended Assignment:
Exercises 1–4, 9, 10, 20.

1. Is there Racial Profiling? *Racial profiling* is the controversial practice of targeting someone for criminal behavior on the basis of the person's race, national origin, or ethnicity. The accompanying table summarizes results for randomly selected drivers stopped by police in a recent year (based on data from the U.S. Department of Justice, Bureau of Justice Statistics). Using the data in this table results in the Minitab display. Use a 0.05 significance level to test the claim that being stopped is independent of race and ethnicity. Based on the available evidence, can we conclude that racial profiling is being used?

1. $\chi^2 = 0.413$. *P*-value: 0.521. There is not sufficient evidence to support a claim of racial profiling.

Race and Ethnicity

	Black and Non-Hispanic	White and Non-Hispanic
Stopped by police	24	147
Not stopped by police	176	1253

Minitab

```
Chi-Sq = 0.322 + 0.046 +
              0.039 + 0.006 = 0.413
DF = 1, P-Value = 0.521
```

2. Testing Effectiveness of Bicycle Helmets A study was conducted of 531 persons injured in bicycle crashes, and randomly selected sample results are summarized in the accompanying table. The TI-83 Plus results also are shown. At the 0.05 significance level, test the claim that wearing a helmet has no effect on whether facial injuries are received. Based on these results, does a helmet seem to be effective in helping to prevent facial injuries in a crash?

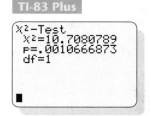

TI-83 Plus

```
χ²-Test
 χ²=10.7080789
 P=.0010666873
 df=1
```

	Helmet Worn	No Helmet
Facial injuries received	30	182
All injuries nonfacial	83	236

Based on data from "A Case-Control Study of the Effectiveness of Bicycle Safety Helmets in Preventing Facial Injury," by Thompson, Thompson, Rivara, and Wolf, *American Journal of Public Health,* Vol. 80, No. 12.

2. $\chi^2 = 10.708$. *P*-value: 0.001. The helmets appear to be effective.

3. E-Mail and Privacy Workers and senior-level bosses were asked if it was seriously unethical to monitor employee e-mail, and the results are summarized in the table (based on data from a Gallup poll). Use a 0.05 significance level to test the claim that the response is independent of whether the subject is a worker or a senior-level boss. Does the conclusion change if a significance level of 0.01 is used instead of 0.05? Do workers and bosses appear to agree on this issue?

3. $\chi^2 = 4.698$. CV: $\chi^2 = 3.841$. Conclusion changes if the significance level of 0.01 is used.

	Yes	No
Workers	192	244
Bosses	40	81

4. Accuracy of Polygraph Tests The data in the accompanying table summarize results from tests of the accuracy of polygraphs (based on data from the Office of Technology Assessment). Use a 0.05 significance level to test the claim that whether the sub-

4. $\chi^2 = 32.273$. CV: $\chi^2 = 3.841$. It appears that the polygraph is effective, but it is not 100% accurate.

ject lies is independent of the polygraph indication. What do the results suggest about the effectiveness of polygraphs?

	Polygraph Indicated Truth	Polygraph Indicated Lie
Subject actually told the truth	65	15
Subject actually told a lie	3	17

5. $\chi^2 = 51.458$. CV: $\chi^2 = 6.635$. Reject the claim that the proportions of agree/disagree responses are the same for the subjects interviewed by men and the subjects interviewed by women.

5. Testing Influence of Gender Table 10-7 summarizes data for male survey subjects, but the accompanying table summarizes data for a sample of women. Using a 0.01 significance level, and assuming that the sample sizes of 800 men and 400 women are predetermined, test the claim that the proportions of agree/disagree responses are the same for the subjects interviewed by men and the subjects interviewed by women.

	Gender of Interviewer	
	Man	Woman
Women who agree	512	336
Women who disagree	288	64

Based on data from the Eagleton Institute.

6. $\chi^2 = 12.321$. CV: $\chi^2 = 3.841$. The test does appear to discriminate.

6. Testing for Discrimination In the judicial case *United States* v. *City of Chicago*, fair employment practices were challenged. A minority group (group A) and a majority group (group B) took the Fire Captain Examination. Assume that the study began with predetermined sample sizes of 24 minority candidates (Group A) and 562 majority candidates (Group B), with the results as shown in the table. At the 0.05 significance level, test the claim that the proportion of minority candidates who pass is the same as the proportion of majority candidates who pass. Based on the results, does the test appear to discriminate?

	Pass	Fail
Group A	10	14
Group B	417	145

7. $\chi^2 = 63.908$. CV: $\chi^2 = 3.841$. Reject the claim that gender is independent of the fear of flying.

7. Fear of Flying Gender Gap The Marist Institute for Public Opinion conducted a poll of 1014 adults, 48% of whom were men. The poll results show that 12% of the men and 33% of the women fear flying. After constructing a contingency table that summarizes the data in the form of frequency counts, use a 0.05 significance level to test the claim that gender is independent of the fear of flying.

8. $\chi^2 = 2.900$. CV: $\chi^2 = 3.841$. The evidence does not suggest that either method is significantly better.

8. No Smoking The accompanying table summarizes successes and failures when subjects used different methods in trying to stop smoking. The determination of smoking or not smoking was made five months after the treatment was begun, and the data are based on results from the Centers for Disease Control and Prevention. Use a 0.05 significance level to test the claim that success is independent of the method used. If someone wants to stop smoking, does the choice of the method make a difference?

	Nicotine Gum	Nicotine Patch
Smoking	191	263
Not smoking	59	57

9. No Smoking Repeat Exercise 8 after including the additional data shown in the table.

	Nicotine Gum	Nicotine Patch	Nicotine Inhaler
Smoking	191	263	95
Not smoking	59	57	27

10. Smoking in China The table below summarizes results from a survey of males aged 15 or older living in the Minhang District of China (based on data from "Cigarette Smoking in China" by Gong, Koplan, Feng, et al., *Journal of the American Medical Association,* Vol. 274, No. 15). Using a 0.05 significance level, test the claim that smoking is independent of education level. What do you conclude about the relationship between smoking and education in China?

	Primary School	Middle School	College
Smoker	606	1234	100
Never smoked	205	505	137

11. Occupational Hazards Use the data in the table to test the claim that occupation is independent of whether the cause of death was homicide. The table is based on data from the U.S. Department of Labor, Bureau of Labor Statistics. Does any particular occupation appear to be most prone to homicides? If so, which one?

	Police	Cashiers	Taxi Drivers	Guards
Homicide	82	107	70	59
Cause of death other than homicide	92	9	29	42

12. Is Scanner Accuracy the Same for Specials? In a study of store checkout scanning systems, samples of purchases were used to compare the scanned prices to the posted prices. The accompanying table summarizes results for a sample of 819 items. When stores use scanners to check out items, are the error rates the same for regular-priced items as they are for advertised-special items? How might the behavior of consumers change if they believe that disproportionately more overcharges occur with advertised-special items?

	Regular-Priced Items	Advertised-Special Items
Undercharge	20	7
Overcharge	15	29
Correct price	384	364

Based on data from "UPC Scanner Pricing Systems: Are They Accurate?" by Ronald Goodstein, *Journal of Marketing,* Vol. 58.

13. Survey Refusals and Age Bracket A study of people who refused to answer survey questions provided the randomly selected sample data shown in the table. At the 0.01 significance level, test the claim that the cooperation of the subject (response or refusal) is independent of the age category. Does any particular age group appear to be particularly uncooperative?

	Age					
	18–21	22–29	30–39	40–49	50–59	60 and over
Responded	73	255	245	136	138	202
Refused	11	20	33	16	27	49

Based on data from "I Hear You Knocking But You Can't Come In," by Fitzgerald and Fuller, *Sociological Methods and Research,* Vol. 11, No. 1.

9. $\chi^2 = 3.062$. CV: $\chi^2 = 5.991$. The evidence does not suggest that any method is significantly better than the others.

10. $\chi^2 = 95.725$. CV: $\chi^2 = 5.991$. It appears that in China there is an association between education and smoking.

11. $\chi^2 = 65.524$. CV: $\chi^2 = 7.815$ (assuming a 0.05 significance level). Cashiers appear to be most vulnerable to homicide.

12. $\chi^2 = 10.814$. CV: $\chi^2 = 5.991$. Reject the claim that error rates are the same for regular-priced items and advertised specials.

13. $\chi^2 = 20.271$. CV: $\chi^2 = 15.086$. Reject the claim that cooperation of the subject is independent of the age category.

14. $\chi^2 = 19.645$. CV: $\chi^2 = 3.841$. Reject the claim that formal training is independent of how firearms are stored.

14. Firearm Training and Safety Does firearm training result in safer practices by gun owners? In one study, randomly selected subjects were surveyed with the results given in the accompanying table. Use a 0.05 significance level to test the claim that formal firearm training is independent of how firearms are stored. Does the formal training appear to have a positive effect?

	Guns Stored Loaded and Unlocked?	
	Yes	No
Had formal firearm training	122	329
Had no formal firearm training	49	299

Based on data from "Firearm Training and Storage," by Hemenway, Solnick, Azrael, *Journal of the American Medical Association*, Vol. 273, No. 1.

15. $\chi^2 = 119.330$. CV: $\chi^2 = 5.991$. Reject the claim that the type of crime is independent of whether the criminal is a stranger.

15. Crime and Strangers The accompanying table lists survey results obtained from a random sample of different crime victims. At the 0.05 significance level, test the claim that the type of crime is independent of whether the criminal is a stranger. How might the results affect the strategy police officers use when they investigate crimes?

	Homicide	Robbery	Assault
Criminal was a stranger	12	379	727
Criminal was an acquaintance or relative	39	106	642

Based on data from the U.S. Department of Justice.

16. Test statistic: $\chi^2 = 1.358$. CV: $\chi^2 = 7.815$ (assuming a 0.05 significance level). There is not sufficient evidence to warrant rejection of the claim that the amount of smoking is independent of seat belt use.

16. Is Seat Belt Use Independent of Cigarette Smoking? A study of seat belt users and nonusers yielded the randomly selected sample data summarized in the given table. Test the claim that the amount of smoking is independent of seat-belt use. A plausible theory is that people who smoke more are less concerned about their health and safety and are therefore less inclined to wear seat belts. Is this theory supported by the sample data?

	Number of Cigarettes Smoked per Day			
	0	1–14	15–34	35 and over
Wear seat belts	175	20	42	6
Don't wear seat belts	149	17	41	9

Based on data from "What Kinds of People Do Not Use Seat Belts?" by Helsing and Comstock, *American Journal of Public Health*, Vol. 67, No. 11.

17. Test statistic: $\chi^2 = 42.557$. CV: $\chi^2 = 3.841$. Reject the claim that the sentence is independent of the plea. The results encourage pleas for guilty defendants.

17. Is Sentence Independent of Plea? Many people believe that criminals who plead guilty tend to get lighter sentences than those who are convicted in trials. The accompanying table summarizes randomly selected sample data for San Francisco defendants in burglary cases. All of the subjects had prior prison sentences. At the 0.05 significance level, test the claim that the sentence (sent to prison or not sent to prison) is independent of the plea. If you were an attorney defending a guilty defendant, would these results suggest that you should encourage a guilty plea?

	Guilty Plea	Not Guilty Plea
Sent to prison	392	58
Not sent to prison	564	14

Based on data from "Does It Pay to Plead Guilty? Differential Sentencing and the Functioning of the Criminal Courts," by Brereton and Casper, *Law and Society Review*, Vol. 16, No. 1.

18. **Is the Home Field Advantage Independent of the Sport?** Winning team data were collected for teams in different sports, with the results given in the accompanying table. Use a 0.10 significance level to test the claim that home/visitor wins are independent of the sport. Given that among the four sports included here, baseball is the only sport in which the home team can modify field dimensions to favor its own players, does it appear that baseball teams are effective in using this advantage?

	Basketball	Baseball	Hockey	Football
Home team wins	127	53	50	57
Visiting team wins	71	47	43	42

Based on data from "Predicting Professional Sports Game Outcomes from Intermediate Game Scores," by Copper, DeNeve, and Mosteller, *Chance,* Vol. 5, No. 3–4.

19. **Clinical Test of Lipitor** The cholesterol-reducing drug Lipitor consists of atorvastatin calcium, and results summarizing headaches as an adverse reaction in clinical tests are given in the table (based on data from Parke-Davis). Using a 0.05 significance level, test the claim that getting a headache is independent of the amount of atorvastatin used as a treatment. (*Hint:* Because not all of the expected values are 5 or greater, combine the results for the treatments consisting of 20 mg of atorvastatin and 40 mg of atorvastatin.)

	Placebo	10 mg Atorvastatin	20 mg Atorvastatin	40 mg Atorvastatin	80 mg Atorvastatin
Headache	19	47	6	2	6
No headache	251	816	30	77	88

20. **Exercise and Smoking** A study of the effects of exercise by women included results summarized in the table (based on data from "Physical Activity and Coronary Heart Disease in Women" by Lee, Rexrode, Cook, Manson, and Buring, *Journal of the American Medical Association,* Vol. 285, No. 11). Exercise values are in kilocalories of physical activity per week. Use a 0.05 significance level to test the claim that the level of smoking is independent of the level of exercise.

	Below 200	200–599	600–1499	1500 or greater
Never smoked	4997	5205	5784	4155
Smoke less than 15 cigarettes per day	604	484	447	359
Smoke 15 or more cigarettes per day	1403	830	644	350

10-3 Beyond the Basics

21. **Using Yates' Correction for Continuity** The chi-square distribution is continuous, whereas the test statistic used in this section is discrete. Some statisticians use *Yates' correction for continuity* in cells with an expected frequency of less than 10 or in all cells of a contingency table with two rows and two columns. With Yates' correction, we replace

$$\sum \frac{(O - E)^2}{E} \quad \text{with} \quad \sum \frac{(|O - E| - 0.5)^2}{E}$$

Given the contingency table in Exercise 1, find the value of the χ^2 test statistic with and without Yates' correction. What effect does Yates' correction have?

22. a. Row totals are $a + b$ and $c + d$. Column totals are $a + c$ and $b + d$. Grand total is $a + b + c + d$. Use these values to find the expected frequencies and then calculate the test statistic using $\Sigma(O - E)^2/E$ to obtain the given expression.

22. Equivalent Tests Assume that a contingency table has two rows and two columns with frequencies of a and b in the first row and frequencies of c and d in the second row.

a. Verify that the test statistic can be expressed as

$$\chi^2 = \frac{(a + b + c + d)(ad - bc)^2}{(a + b)(c + d)(b + d)(a + c)}$$

b. Let $\hat{p}_1 = a/(a + c)$ and let $\hat{p}_2 = b/(b + d)$. Show that the test statistic

$$z = \frac{(\hat{p}_1 - \hat{p}_2) - 0}{\sqrt{\dfrac{\overline{p}\,\overline{q}}{n_1} + \dfrac{\overline{p}\,\overline{q}}{n_2}}}$$

where

$$\overline{p} = \frac{a + b}{a + b + c + d}$$

and

$$\overline{q} = 1 - \overline{p}$$

is such that $z^2 = \chi^2$ [the same result as in part (a)]. This result shows that the chi-square test involving a 2 × 2 table is equivalent to the test for the difference between two proportions, as described in Section 8-2.

Review

In this chapter we worked with data summarized as frequency counts for different categories. In Section 10-2 we described methods for testing goodness-of-fit in a multinomial experiment, which is similar to a binomial experiment except that there are more than two categories of outcomes. Multinomial experiments result in frequency counts arranged in a single row or column, and we tested to determine whether the observed sample frequencies agree with (or "fit") some claimed distribution.

In Section 10-3 we described methods for testing claims involving contingency tables (or two-way frequency tables), which have at least two rows and two columns. Contingency tables incorporate two variables: One variable is used for determining the row that describes a sample value, and the second variable is used for determining the column that describes a sample value. Section 10-3 included two types of hypothesis test: (1) a test of independence between the row and column variables; (2) a test of homogeneity to decide whether different populations have the same proportions of some characteristics. The following are some key components of the methods discussed in this chapter.

- *Section 10-2 (Test for goodness-of-fit):*

$$\text{Test statistic is } \chi^2 = \sum \frac{(O - E)^2}{E}$$

  Test is right-tailed with $k - 1$ degrees of freedom. All expected frequencies must be at least 5.

- *Section 10-3 (Contingency table test of independence or homogeneity):*

$$\text{Test statistic is } \chi^2 = \sum \frac{(O - E)^2}{E}$$

  Test is right-tailed with $(r - 1)(c - 1)$ degrees of freedom. All expected frequencies must be at least 5.

Review Exercises

1. Call Center Data The table lists the calls received by a call center during one week in a recent year. (The data are from a large U.S. electronics producer that wishes to remain anonymous.) Use a 0.05 significance level to test the claim that calls are uniformly distributed over the days of the business week. What does the result suggest about staffing requirements at this call center?

	Mon	Tues	Wed	Thurs	Fri
Calls	98	68	89	64	56

2. Do Gunfire Deaths Occur More Often on Weekends? When *Time* magazine tracked U.S. deaths by gunfire during a one-week period, the results shown in the accompanying table were obtained. At the 0.05 significance level, test the claim that gunfire death rates are the same for the different days of the week. Is there any support for the theory that more gunfire deaths occur on weekends when more people are at home?

Weekday	Mon	Tues	Wed	Thurs	Fri	Sat	Sun
Number of deaths by gunfire	74	60	66	71	51	66	76

3. Is Drinking Independent of Type of Crime? The accompanying table lists sample data that statistician Karl Pearson used in 1909. Does the type of crime appear to be related to whether the criminal drinks or abstains? Are there any crimes that appear to be associated with drinking?

	Arson	Rape	Violence	Stealing	Coining (Counterfeiting)	Fraud
Drinker	50	88	155	379	18	63
Abstainer	43	62	110	300	14	144

4. Testing for Independence Between Early Discharge and Rehospitalization of Newborn Is it safe to discharge newborns from the hospital early after their births? The accompanying table shows results from a study of this issue. Use a 0.05 significance level to test the claim that whether the newborn was discharged early or late is independent of whether the newborn was rehospitalized within a week of discharge. Does the conclusion change if the significance level is changed to 0.01?

	Rehospitalized within Week of Discharge?	
	Yes	No
Early discharge (less than 30 hours)	622	3997
Late discharge (30–78 hours)	631	4660

Based on data from "The Safety of Newborn Early Discharge," by Liu and others, *Journal of the American Medical Association*, Vol. 278, No. 4.

---

1. $\chi^2 = 16.747$. CV: $\chi^2 = 9.488$. Reject the claim that calls are uniformly distributed over the days of the business week.

2. $\chi^2 = 6.780$. CV: $\chi^2 = 12.592$. There is not sufficient evidence to support the theory that more gunfire deaths occur on weekends.

3. $\chi^2 = 49.731$. CV: $\chi^2 = 11.071$ (assuming a 0.05 significance level). Support the claim that the type of crime is related to whether the criminal drinks or abstains.

4. $\chi^2 = 5.297$. CV: $\chi^2 = 3.841$. Reject the claim that whether a newborn is discharged early or late is independent of whether the newborn was rehospitalized within a week of discharge. The conclusion changes if the significance level is changed to 0.01.

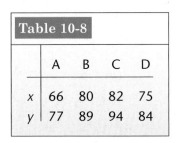

Table 10-8

	A	B	C	D
x	66	80	82	75
y	77	89	94	84

1. $\bar{x} = 80.9$; median: 81.0; range: 28.0; $s^2 = 75.4$; $s = 8.6$; 5-number summary: 66, 76.0, 81.0, 86.5, 94.

2. a. 0.272
   b. 0.468
   c. 0.614
   d. 0.282

3. Contingency table; see Section 10-3. TS: $\chi^2 = 0.055$. CV: $\chi^2 = 7.815$ (assuming a 0.05 significance level). There is not sufficient evidence to reject the claim that men and women choose the different answers in the same proportions.

4. Use correlation; see Section 9-2. TS: $r = 0.978$. CV: $r = \pm 0.950$ (assuming a 0.05 significance level). Support the claim that there is relationship between the memory and reasoning scores.

Cumulative Review Exercises

1. Finding Statistics Assume that in Table 10-8, the row and column titles have no meaning so that the table contains test scores for eight randomly selected prisoners who were convicted of removing labels from pillows. Find the mean, median, range, variance, standard deviation, and 5-number summary.

2. Finding Probability Assume that in Table 10-8, the letters A, B, C, and D represent the choices on the first question of a multiple-choice quiz. Also assume that x represents men and y represents women and that the table entries are frequency counts, so 66 men chose answer A, 77 women chose answer A, 80 men chose answer B, and so on.
   a. If one response is randomly selected, find the probability that it is response C.
   b. If one response is randomly selected, find the probability that it was made by a man.
   c. If one response is randomly selected, find the probability that it is C or was made by a man.
   d. If two different responses are randomly selected, find the probability that they were both made by a woman.

3. Testing for Equal Proportions Using the same assumptions as in Exercise 2, test the claim that men and women choose the different answers in the same proportions.

4. Testing for a Relationship Assume that Table 10-8 lists test scores for four people, where the x-score is from a test of memory and the y-score is from a test of reasoning. Test the claim that there is a relationship between the x- and y-scores.

5. Testing for Effectiveness of Training Assume that Table 10-8 lists test scores for four people, where the x-score is from a pretest taken before a training session on memory improvement and the y-score is from a posttest taken after the training. Test the claim that the training session is effective in raising scores.

6. Testing for Equality of Means Assume that in Table 10-8, the letters A, B, C, and D represent different versions of the same test of reasoning. The x-scores were obtained by four randomly selected men and the y-scores were obtained by four randomly selected women. Test the claim that men and women have the same mean score.

Cooperative Group Activities

1. Out-of-class activity Divide into groups of four or five students. See the first two rows of Table 10-1 in the Chapter Problem for the distribution of leading digits expected with Benford's law. Collect data and use the methods of Section 10-2 to verify that the data conform reasonably well to Benford's law. Here are some possibilities that might be considered:
   - The amounts on the checks you wrote
   - The prices of stocks
   - Populations of counties in the United States

2. Out-of-class activity Divide into groups of four or five students and collect past results from a state lottery.

Such results are often available on Web sites for individual state lotteries. Use the methods of Section 10-2 to test that the numbers are selected in such a way that all possible outcomes are equally likely.

3. Out-of-class activity Divide into groups of four or five students. Each group member should survey at least 15 male students and 15 female students at the same college by asking two questions: (1) Which political party does the subject favor most? (2) If the subject were to make up an absence excuse of a flat tire, which tire would he or she say went flat if the instructor asked? (See Exercise 6 in Section 10-2.) Ask the subject to

write the two responses on an index card, and also record the gender of the subject and whether the subject wrote with the right or left hand. Use the methods of this chapter to analyze the data collected. Include these tests:

- Political party choice is independent of the gender of the subject.
- The tire identified as being flat is independent of the gender of the subject.
- Political party choice is independent of whether the subject is right- or left-handed.
- The tire identified as being flat is independent of whether the subject is right- or left-handed.
- Gender is independent of whether the subject is right- or left-handed.
- Political party choice is independent of the tire identified as being flat.

4. Out-of-class activity Divide into groups of four or five students. Each group member should select about 15 other students and first ask them to "randomly" select four digits each. After the four digits have been recorded, ask each subject to write the last four digits of his or her social security number. Take the "random" sample results and mix them into one big sample, then mix the social security digits into a second big sample. Using the "random" sample set, test the claim that students select digits randomly. Then use the social se-

curity digits to test the claim that they come from a population of random digits. Compare the results. Does it appear that students can randomly select digits? Are they likely to select any digits more often than others? Are they likely to select any digits less often than others? Do the last digits of social security numbers appear to be randomly selected?

5. In-class activity Divide into groups of three or four students. Each group should be given a die along with the instruction that it should be tested for "fairness." Is the die fair or is it biased? Describe the analysis and results.

6. Out-of-class activity Divide into groups of two or three students. Some examples and exercises of this chapter were based on the analysis of last digits of values. (See the Barry Bonds examples in Section 10-2 and Exercise 12 in Section 10-2.) It was noted that the analysis of last digits can sometimes reveal whether values are the results of actual measurements or whether they are reported estimates. Refer to an almanac and find the lengths of rivers in the world, then analyze the last digits to determine whether those lengths appear to be actual measurements or whether they appear to be reported estimates. (Instead of lengths of rivers, you could use heights of mountains, heights of the tallest buildings, lengths of bridges, and so on.)

Technology Project

Use STATDISK, or Minitab, or Excel, or a TI-83 Plus calculator, or any other software package or calculator capable of generating equally likely random digits between 0 and 9 inclusive. Generate 500 digits and record the results in the accompanying table. Use a 0.05 significance level to test the claim that the sample digits come from a population with a

uniform distribution (so that all digits are equally likely). Does the random number generator appear to be working as it should?

Digit	0	1	2	3	4	5	6	7	8	9
Frequency										

from DATA *to* DECISION

Critical Thinking: Is the defendant guilty of fraud?

In the trial of *State of Arizona vs. Wayne James Nelson,* the defendant was accused of issuing checks to a vendor that did not really exist. The amounts of the checks are listed below in order by row.

Analyzing the Results

Do the leading digits conform to Benford's law described in the Chapter Problem? When testing for goodness-of-fit with the proportions expected with Benford's law, it is necessary to combine categories because not all expected values are at least 5. Use one category with leading digits of 1, a second category with leading digits of 2, 3, 4, 5,

and a third category with leading digits of 6, 7, 8, 9. Are the expected values for these three categories all at least 5? Is there sufficient evidence to conclude that the leading digits on the checks do not conform to Benford's law? Apart from the leading digits, are there any other patterns suggesting that the check amounts were created by the defendant instead of being the result of typical and real transactions? Based on the evidence, if you were a juror, would you conclude that the check amounts are the result of fraud? What would be one argument that you might present if you were the attorney for the defendant?

$1,927.48	$27,902.31	$86,241.90	$72,117.46	$81,321.75	$97,473.96
$93,249.11	$89,658.16	$87,776.89	$92,105.83	$79,949.16	$87,602.93
$96,879.27	$91,806.47	$84,991.67	$90,831.83	$93,766.67	$88,336.72
$94,639.49	$83,709.26	$96,412.21	$88,432.86	$71,552.16	

INTERNET PROJECT   Contingency Tables

An important characteristic of tests of independence with contingency tables is that the data collected need not be quantitative in nature. A contingency table summarizes observations by the categories or labels of the rows and columns. As a result, characteristics such as gender, race, and political party all become fair game for formal hypothesis testing procedures. The Internet Project for this chapter is found at the *Elementary Statistics* Web site:

http://www.aw.com/triola

You will find links to a variety of demographic data. With these data sets, you will conduct tests in areas as diverse as academics, politics, and the entertainment industry. In each test, you will draw conclusions related to the independence of interesting characteristics.

Statistics @ Work

"Even if you're not a numbers cruncher, [statistical] knowledge can be helpful in any situation that requires prediction, decision making, or evaluation."

Nabil Lebbos

Graphics illustrator, Published Image

As analyst for Standard & Poor's *Published Image,* Nabil's studies on investment performance are published in newspapers read by over one million investors.

Please describe your occupation.

I work for *Published Image* where I use statistics to generate the charts and data that we use in our financial publications—using loads of statistics and applications. We write newsletters for banks and mutual funds.

What concepts of statistics do you use?

I use standard deviation to measure risk, regression to measure an investment's relationship to its benchmark, and correlation to determine an investment's movement in relation to other investments.

How do you use statistics on the job?

I start with a given set of raw data. These are usually monthly, daily, or annual returns on an investment. I then use Excel to chart the data so I can get a picture of what I'm dealing with. From there I proceed to perform an analysis. Sometimes, the results do not back up a point that the accompanying article is trying to make strongly enough. In such situations, I look at other possibilities.

Please describe one specific example illustrating how the use of statistics was successful in improving a product or service.

One of our clients wanted to make the point that although their mutual fund did not outperform all others, it did succeed in consistently avoiding large negative returns. I ran some tests on skewness and downside risk and showed that, in fact, the fund's returns were positively skewed. We created histograms comparing this fund with an average of all funds, and that clearly made the point.

In terms of statistics, what would you recommend for prospective employees?

It's a logical tool that, when used informatively, can convince you and your audience of the point you're trying to make much more effectively than words. Even if you're not a numbers cruncher, [statistical] knowledge can be helpful in any situation that requires prediction, decision making, or evaluation.

Do you feel job applicants are viewed more favorably if they have studied some statistics?

Yes.

While a college student, did you expect to be using statistics on the job?

No. I studied architecture as an undergrad and business as a grad student.

11

Analysis of Variance

11-1 Overview

11-2 One-Way ANOVA

11-3 Two-Way ANOVA

Clancy, Rowling, and Tolstoy: Is there a difference in reading level?

Data Set 14 in Appendix B includes data obtained from 12 randomly selected pages in each of three different books: Tom Clancy's *The Bear and the Dragon,* J. K. Rowling's *Harry Potter and the Sorcerer's Stone,* and Leo Tolstoy's *War and Peace.* The Flesch Reading Ease score was obtained for each of those pages, and the results are listed in Data Set 14. The Flesch Reading Ease scoring system results in higher scores for text that is easier to read. Low scores result from works that are difficult to read.

In the spirit of exploring data by investigating center, variation, distribution, outliers, and changing patterns over time (CVDOT), we obtain the sample statistics included in Table 11-1. Also, histograms of the three data sets suggest that the samples come from populations with distributions that are approximately normal. When investigating outliers, the only candidates are the lowest Clancy score, which is 2.37 standard deviations below the mean, and the lowest Rowling score, which is 2.10 standard deviations below the mean. Judged in the context of the other scores, those two values do not appear to be dramatically far from the other values, so we will assume

that there are no outliers. If we want to be very cautious, we might analyze the data with and without those two values to see if the final conclusions are affected. (In fact, the results are not dramatically affected by those values.) Because the books will never change, the issue of a changing pattern over time is not relevant here.

When thinking of a comparison of readability of books by Tom Clancy, J. K. Rowling, and Leo Tolstoy, we might expect that Rowling's book should be easiest to read because it was written for children. We might also expect that Tolstoy's book should be the most difficult because it is a translation of a Russian classic. Now look at the mean readability scores and see that they appear to support our expectations, because Rowling has the highest readability score and Tolstoy has the lowest. But can we really conclude that the means are different? Here is the key issue that we will address in this chapter: Do the sample Flesch Ease of Reading scores in Data Set 14 provide sufficient evidence to support the claim that the books by Clancy, Rowling, and Tolstoy have different means?

Table 11-1	Readability Scores for Three Books		
	Flesch Reading Ease Score		
	Clancy	Rowling	Tolstoy
n	12	12	12
$\bar{x}$	70.73	80.75	66.15
s	11.33	4.68	7.86

Overview

Instead of "Analysis of Variance," a better title for this chapter might be "Testing for Equality of Three or More Population Means." Although it is not very catchy, the latter title does a better job of describing the objective of this chapter. We want to introduce a procedure for testing the hypothesis that three or more population means are equal, so a typical null hypothesis will be H_0: $\mu_1 = \mu_2 = \mu_3$, and the alternative hypothesis will be the statement that at least one mean is different from the others. In Section 8-3 we already presented procedures for testing the hypothesis that *two* population means are equal, but the methods of that section do not apply when three or more means are involved. Instead of referring to the main objective of testing for equal means, the term *analysis of variance* refers to the *method* we use, which is based on an analysis of sample variances.

Definition

Analysis of variance (ANOVA) is a method of testing the equality of three or more population means by analyzing sample variances.

ANOVA is used in applications such as the following:

- If we treat one group with two aspirin tablets each day and a second group with one aspirin tablet each day, while a third group is given a placebo each day, we can test to determine if there is sufficient evidence to support the claim that the three groups have different mean blood pressure levels.

- The claim has been made that supermarkets place high-sugar cereals on shelves that are at eye-level for children, so we can test the claim that the cereals on the shelves have the same mean sugar content.

Why Can't We Just Test Two Samples at a Time? Why do we need a new procedure when we can test for equality of two means by using the methods presented in Chapter 8? For example, if we want to use the sample data from Table 11-1 to test the claim that the three populations have the same mean, why not simply pair them off and do two at a time by testing H_0: $\mu_1 = \mu_2$, then H_0: $\mu_2 = \mu_3$, then H_0: $\mu_1 = \mu_3$? This approach (doing two at a time) requires three different hypothesis tests, so the degree of confidence could be as low as 0.95^3 (or 0.857). In general, as we increase the number of individual tests of significance, we increase the likelihood of finding a difference by chance alone (instead of a real difference in the means). The risk of a type I error—finding a difference in one of the pairs when no such difference actually exists—is far too high. The method of analysis of variance helps us avoid that particular pitfall (rejecting a true null hypothesis) by using one test for equality of several means.

F Distribution

The ANOVA methods of this chapter require the F distribution, which was first introduced in Section 8-5. In Section 8-5 we noted that the F distribution has the following important properties (see Figure 11-1):

1. The F distribution is not symmetric; it is skewed to the right.
2. The values of F can be 0 or positive, but they cannot be negative.
3. There is a different F distribution for each pair of degrees of freedom for the numerator and denominator.

Critical values of F are given in Table A-5.

Analysis of variance (ANOVA) is based on a comparison of two different estimates of the variance common to the different populations. Those estimates (the *variance between samples* and the *variance within samples*) will be described in Section 11-2. The term *one-way* is used because the sample data are separated into groups according to one characteristic, or factor. For example, the readability scores summarized in Table 11-1 are separated into three different groups according to the one characteristic (or factor) of author (Clancy, Rowling, Tolstoy). In Section 11-3 we will introduce two-way analysis of variance, which allows us to compare populations separated into categories using two characteristics (or factors). For example, we might separate heights of people using the following two factors: (1) gender (male or female) and (2) right- or left-handedness.

Suggested Study Strategy: Because the procedures used in this chapter require complicated calculations, we will emphasize the use and interpretation of computer software, such as STATDISK, Minitab, and Excel, or a TI-83 Plus calculator. We suggest that you begin Section 11-2 by focusing on this key concept: We are using a procedure to test a claim that three or more means are equal. Although the details of the calculations are complicated, our procedure will be easy because it is based on a P-value. If the P-value is small, such as 0.05 or lower, reject equality of means. Otherwise, fail to reject equality of means. After under-

Note to Instructor

When introducing (or reviewing) the F distribution, consider comparing it to the normal, t, and χ^2 distributions relative to the properties of symmetry, positive/negative values, and one standard curve versus a family of curves that correspond to values of *df*.

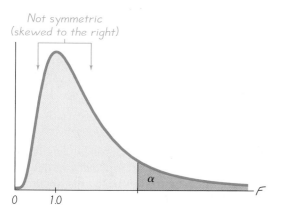

Not symmetric
(skewed to the right)

0   1.0

Nonnegative
values only

FIGURE 11-1

F Distribution

There is a different F distribution for each different pair of degrees of freedom for numerator and denominator.

standing that basic and simple procedure, proceed to understand the underlying rationale.

11-2 One-Way ANOVA

In this section we consider tests of hypotheses that three or more population means are all equal, as in H_0: $\mu_1 = \mu_2 = \mu_3$. The calculations are very complicated, so we recommend the following approach:

1. Understand that a small P-value (such as 0.05 or less) leads to rejection of the null hypothesis of equal means. With a large P-value (such as greater than 0.05), fail to reject the null hypothesis of equal means.

2. Develop an understanding of the underlying rationale by studying the example in this section.

3. Become acquainted with the nature of the SS (sum of square) and MS (mean square) values and their role in determining the F test statistic, but use statistical software packages or a calculator for finding those values.

The method we use is called **one-way analysis of variance** (or **single-factor analysis of variance**) because we use a single property, or characteristic, for categorizing the populations. This characteristic is sometimes referred to as a *treatment,* or *factor.*

> ### Definition
> A **treatment** (or **factor**) is a property, or characteristic, that allows us to distinguish the different populations from one another.

 For example, the readability scores summarized in Table 11-1 are distinguished according to the treatment (or factor) of author (Clancy, Rowling, Tolstoy). The term *treatment* is used because early applications of analysis of variance involved agricultural experiments in which different plots of farmland were treated with different fertilizers, seed types, insecticides, and so on. The accompanying box includes the required assumptions and the procedure we will use.

> ### Assumptions
> 1. The populations have distributions that are approximately normal. (This is a loose requirement, because the method works well unless a population has a distribution that is very far from normal. If a population does have a distribution that is far from normal, use the Kruskal-Wallis test described in Section 12-5.)
> 2. The populations have the same variance σ^2 (or standard deviation σ). (This is a loose requirement, because the method works well unless the population variances differ by large amounts. University of Wisconsin statistician George E. P.

Box showed that as long as the sample sizes are equal (or nearly equal), the variances can differ by amounts that make the largest up to nine times the smallest and the results of ANOVA will continue to be essentially reliable.)

3. The samples are simple random samples. (That is, samples of the same size have the same probability of being selected.)

4. The samples are independent of each other. (The samples are not matched or paired in any way.)

5. The different samples are from populations that are categorized in only one way. (This is the basis for the name of the method: *one-way* analysis of variance.)

Procedure for Testing H_0: $\mu_1 = \mu_2 = \mu_3 = \ldots$

1. **Use STATDISK, Minitab, Excel, or a TI-83 Plus calculator to obtain results.**

2. **Identify the *P*-value from the display.**

3. **Form a conclusion based on these criteria:**

   • **If the *P*-value $\leq \alpha$, reject the null hypothesis of equal means and conclude that at least one of the population means is different from the others.**

   • **If the *P*-value $> \alpha$, fail to reject the null hypothesis of equal means.**

Caution when interpreting results: When we conclude that there is sufficient evidence to reject the claim of equal population means, we cannot conclude from ANOVA that any particular mean is different from the others. (There are several other tests that can be used to identify the specific means that are different, and those procedures are called **multiple comparison procedures.** Comparison of confidence intervals, the Scheffé test, the extended Tukey test, and the Bonferroni test are common multiple comparison procedures.)

 EXAMPLE **Readability of Clancy, Rowling, Tolstoy** Given the readability scores summarized in Table 11-1 and a significance level of $\alpha = 0.05$, use STATDISK, Minitab, Excel, or a TI-83 PLUS calculator to test the claim that the three samples come from populations with means that are not all the same.

SOLUTION The null hypothesis is H_0: $\mu_1 = \mu_2 = \mu_3$ and the alternative hypothesis is the claim that at least one of the means is different from the others.

Step 1:   At the end of this section we will describe specific procedures for obtaining computer or calculator displays, but we will now consider the accompanying displayed results on the next page.

Step 2:   The displays all show that the *P*-value is 0.000562, or 0.001 when rounded.

Step 3:   Because the *P*-value is less than the significance level of $\alpha = 0.05$, we reject the null hypothesis of equal means.

continued

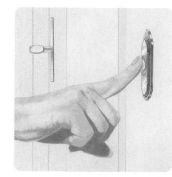

Poll Resistance

Surveys based on relatively small samples can be quite accurate, provided the sample is random or representative of the population. However, increasing survey refusal rates are now making it more difficult to obtain random samples. The Council of American Survey Research Organizations reported that in a recent year, 38% of consumers refused to respond to surveys. The head of one market research company said, "Everyone is fearful of self-selection and worried that generalizations you make are based on cooperators only." Results from the multi-billion-dollar market research industry affect the products we buy, the television shows we watch, and many other facets of our lives.

STATDISK

Source:	DF:	SS:	MS:	Test Stat, F:	Critical F:	P-Value:
Treatment:	2	1338.002	669.0011	9.469487	3.264914	0.0005621
Error:	33	2331.387	70.64808			
Total:	35	3669.389	104.8397			

Reject the Null Hypothesis
Reject equality of means

TI-83 Plus

```
One-way ANOVA
 F=9.469487401
 p=5.6213335E-4
 Factor
  df=2
  SS=1338.00222
↓ MS=669.001111
■
```

```
One-way ANOVA
↑ MS=669.001111
 Error
  df=33
  SS=2331.38667
  MS=70.6480808
 SxP=8.40524127
■
```

Excel

Anova: Single Factor

SUMMARY

Groups	Count	Sum	Average	Variance
Column 1	12	848.8	70.73333333	128.2806061
Column 2	12	969	80.75	21.91545455
Column 3	12	793.8	66.15	61.74818182

ANOVA

Source of Variation	SS	df	MS	F	P-value	F crit
Between Groups	1338.002222	2	669.0011111	9.469487401	0.000562133	3.284924333
Within Groups	2331.386667	33	70.64808081			
Total	3669.388889	35				

Minitab

```
Analysis of Variance
Source     DF        SS         MS        F         P
Factor      2     1338.0      669.0      9.47     0.001
Error      33     2331.4       70.6
Total      35     3669.4
```

INTERPRETATION There is sufficient evidence to support the claim that the three population means are not all the same. Based on randomly selected pages from Clancy's *The Bear and the Dragon,* Rowling's *Harry Potter and the Sorcerer's Stone,* and Tolstoy's *War and Peace,* we conclude that those books have readability levels that are not all the same. On the basis of this ANOVA test, we cannot conclude that any particular mean is different from the others.

Rationale

The method of analysis of variance is based on this fundamental concept: With the assumption that the populations all have the same variance σ^2, we estimate the common value of σ^2 using two different approaches. The F test statistic is the ratio of those estimates, so that a significantly *large* F test statistic (located far to the right in the F distribution graph) is evidence against equal population means. Figure 11-2 shows the relationship between the F test statistic and the P-value.

The two approaches for estimating the common value of σ^2 are as follows:

1. The **variance between samples** (also called **variation due to treatment**) is an estimate of the common population variance σ^2 that is based on the variation among the sample *means*.

2. The **variance within samples** (also called **variation due to error**) is an estimate of the common population variance σ^2 based on the sample *variances*.

Test Statistic for One-Way ANOVA

$$F = \frac{\text{variance between samples}}{\text{variance within samples}}$$

The numerator of the test statistic F measures variation between sample means. The estimate of variance in the denominator depends only on the sample variances and is not affected by differences among the sample means. Consequently, sample means that are close in value result in a small F test statistic and we

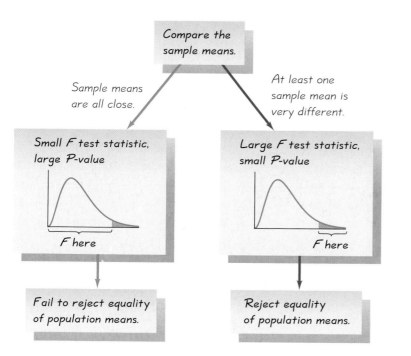

FIGURE 11-2 Relationship Between the F Test Statistic and P-Value

conclude that there are no significant differences among the sample means. But if the value of F is excessively *large*, then we reject the claim of equal means. (The vague terms "small" and "excessively large" are made objective by the corresponding P-value, which tells us whether the F test statistic is or is not in the critical region.) Because excessively large values of F reflect unequal means, the test is right-tailed.

Calculations with Equal Sample Sizes n

Refer to Data Set A in Table 11-2. If the data sets all have the same sample size (as in $n = 4$ for Data Set A in Table 11-2), the required calculations aren't overwhelmingly difficult. First, find the variance between samples by evaluating $ns_{\bar{x}}^2$, where $s_{\bar{x}}^2$ is the variance of the sample means and n is the size of each of the samples. That is, consider the sample means to be an ordinary set of values and calculate the variance. (From the central limit theorem, $\sigma_{\bar{x}} = \sigma/\sqrt{n}$ can be solved for σ to get $\sigma = \sqrt{n} \cdot \sigma_{\bar{x}}$, so that we can estimate σ^2 with $ns_{\bar{x}}^2$.) For example, the

Table 11-2	Effect of a Mean on the F Test Statistic					
	A	add 10		**B**		
	Sample 1	Sample 2	Sample 3	Sample 1	Sample 2	Sample 3
	7	6	4	17	6	4
	3	5	7	13	5	7
	6	5	6	16	5	6
	6	8	7	16	8	7
	$n_1 = 4$	$n_2 = 4$	$n_3 = 4$	$n_1 = 4$	$n_2 = 4$	$n_3 = 4$
	$\bar{x}_1 = 5.5$	$\bar{x}_2 = 6.0$	$\bar{x}_3 = 6.0$	$\bar{x}_1 = 15.5$	$\bar{x}_2 = 6.0$	$\bar{x}_3 = 6.0$
	$s_1^2 = 3.0$	$s_2^2 = 2.0$	$s_3^2 = 2.0$	$s_1^2 = 3.0$	$s_2^2 = 2.0$	$s_3^2 = 2.0$
Variance between samples	$ns_{\bar{x}}^2 = 4\,(0.0833) = 0.3332$			$ns_{\bar{x}}^2 = 4\,(30.0833) = 120.3332$		
Variance within samples	$s_p^2 = \dfrac{3.0 + 2.0 + 2.0}{3} = 2.3333$			$s_p^2 = \dfrac{3.0 + 2.0 + 2.0}{3} = 2.3333$		
F test statistic	$F = \dfrac{ns_{\bar{x}}^2}{s_p^2} = \dfrac{0.3332}{2.3333} = 0.1428$			$F = \dfrac{ns_{\bar{x}}^2}{s_p^2} = \dfrac{120.3332}{2.3333} = 51.5721$		
P-value (found from Excel)	P-value = 0.8688			P-value = 0.0000118		

sample means for Data Set A in Table 11-2 are 5.5, 6.0, and 6.0. Those three values have a variance of $s_{\bar{x}}^2 = 0.0833$, so that

$$\text{variance between samples} = ns_{\bar{x}}^2 = 4(0.0833) = 0.3332$$

Next, estimate the variance within samples by calculating s_p^2, which is the pooled variance obtained by finding the mean of the sample variances. The sample variances in Table 11-2 are 3.0, 2.0, and 2.0, so that

$$\text{variance within samples} = s_p^2$$
$$= \frac{3.0 + 2.0 + 2.0}{3} = 2.3333$$

Finally, evaluate the F test statistic as follows:

$$F = \frac{\text{variance between samples}}{\text{variance within samples}} = \frac{ns_{\bar{x}}^2}{s_p^2} = \frac{0.3332}{2.3333} = 0.1428$$

The critical value of F is found by assuming a right-tailed test, because large values of F correspond to significant differences among means. With k samples each having n values, the numbers of degrees of freedom are computed as follows.

Degrees of Freedom:
(k = number of samples and n = sample size)

$$\text{numerator degrees of freedom} = k - 1$$
$$\text{denominator degrees of freedom} = k(n - 1)$$

For Data Set A in Table 11-2, $k = 3$ and $n = 4$, so the degrees of freedom are 2 for the numerator and $3(4 - 1) = 9$ for the denominator. With $\alpha = 0.05$, 2 degrees of freedom for the numerator, and 9 degrees of freedom for the denominator, the critical F value from Table A-5 is 4.2565. If we were to use the traditional method of hypothesis testing with Data Set A in Table 11-2, we would see that this right-tailed test has a test statistic of $F = 0.1428$ and a critical value of $F = 4.2565$, so the test statistic is not in the critical region and we therefore fail to reject the null hypothesis of equal means.

To really see how the F test statistic works, consider both collections of sample data in Table 11-2. Note that the three samples in part A are identical to the three samples in part B, except that in part B we have added 10 to each value of Sample 1 from part A. The three sample means in part A are very close, but there are substantial differences in part B. The three sample variances in part A are identical to those in part B.

Adding 10 to each data value in the first sample of Table 11-2 has a dramatic effect on the test statistic, with F changing from 0.1428 to 51.5721. Adding 10 to each data value in the first sample also has a dramatic effect on the P-value, which changes from 0.8688 (not significant) to 0.0000118 (significant). Note that the variance between samples in part A is 0.3332, but for part B it is 120.3332 (indicating that the sample means in part B are farther apart). Note also that the variance within samples is 2.3333 in both parts, because the variance within a sample isn't affected when we add a constant to every sample value. *The change in the F test statistic and the P-value is attributable only to the change in $\bar{x}_1$.* This illustrates that the *F* test statistic is very sensitive to sample *means,* even though it is obtained through two different estimates of the common population *variance.*

Here is the key point of Table 11-2: Data Sets A and B are identical except that in Data Set B, 10 is added to each value of the first sample. Adding 10 to each value of the first sample causes the three sample means to grow farther apart, with the result that the *F* test statistic increases and the *P*-value decreases.

Calculations with Unequal Sample Sizes

While the calculations required for cases with equal sample sizes are reasonable, they become really complicated when the sample sizes are not all the same. The same basic reasoning applies because we calculate an *F* test statistic that is the ratio of two different estimates of the common population variance σ^2, but those estimates involve *weighted* measures that take the sample sizes into account, as shown below.

$$F = \frac{\text{variance between samples}}{\text{variance within samples}} = \frac{\left[\dfrac{\sum n_i(\bar{x}_i - \bar{\bar{x}})^2}{k - 1}\right]}{\left[\dfrac{\sum(n_i - 1)s_i^2}{\sum(n_i - 1)}\right]}$$

where
$\bar{\bar{x}}$ = mean of all sample values combined
k = number of population means being compared
n_i = number of values in the *i*th sample
$\bar{x}_i$ = mean of values in the *i*th sample
s_i^2 = variance of values in the *i*th sample

The factor of n_i is included so that larger samples carry more weight. The denominator of the test statistic is simply the mean of the sample variances, but it is a weighted mean with the weights based on the sample sizes.

Because calculating this test statistic can lead to large rounding errors, the various software packages typically use a different (but equivalent) expression that involves SS (for sum of squares) and MS (for mean square) notation. Although the following notation and components are complicated and involved, the basic idea is the same: The test statistic *F* is a ratio with a numerator reflecting variation *between* the means of the samples and a denominator reflecting variation *within* the samples. If the populations have equal means, the *F* ratio tends to be small, but if the population means are not equal, the *F* ratio tends to be significantly large. Key components in our ANOVA method are described as follows.

SS(total), or total sum of squares, is a measure of the total variation (around $\bar{\bar{x}}$) in all of the sample data combined.

Formula 11-1 $SS(total) = \Sigma(x - \bar{\bar{x}})^2$

SS(total) can be broken down into the components of SS(treatment) and SS(error), described as follows.

SS(treatment), also referred to as SS(factor) or SS(between groups) or (SS between samples), is a measure of the variation *between* the sample means.

Formula 11-2

$$SS(treatment) = n_1(\bar{x}_1 - \bar{\bar{x}})^2 + n_2(\bar{x}_2 - \bar{\bar{x}})^2 + \cdots + n_k(\bar{x}_k - \bar{\bar{x}})^2$$
$$= \Sigma n_i(\bar{x}_i - \bar{\bar{x}})^2$$

If the population means ($\mu_1, \mu_2, \ldots, \mu_k$) are equal, then the sample means $\bar{x}_1, \bar{x}_2 \ldots, \bar{x}_k$ will all tend to be close together and also close to $\bar{\bar{x}}$. The result will be a relatively small value of SS(treatment). If the population means are not all equal, however, then at least one of $\bar{x}_1, \bar{x}_2, \ldots, \bar{x}_k$ will tend to be far apart from the others and also far apart from $\bar{\bar{x}}$. The result will be a relatively large value of SS(treatment).

SS(error), also referred to as SS(within groups) or SS(within samples), is a sum of squares representing the variation that is assumed to be common to all the populations being considered.

Formula 11-3

$$SS(error) = (n_1 - 1)s_1^2 + (n_2 - 1)s_2^2 + \cdots + (n_k - 1)s_k^2$$

$$= \Sigma(n_i - 1)s_i^2$$

Given the preceding expressions for SS(total), SS(treatment), and SS(error), the following relationship will always hold.

Formula 11-4   SS(total) = SS(treatment) + SS(error)

SS(treatment) and SS(error) are both sums of squares, and if we divide each by its corresponding number of degrees of freedom, we get *mean* squares. Some of the following expressions for mean squares include the notation N:

N = total number of values in all samples combined

MS(treatment) is a mean square for treatment, obtained as follows:

Formula 11-5 $$MS(treatment) = \frac{SS(treatment)}{k-1}$$

MS(error) is a mean square for error, obtained as follows:

Formula 11-6 $$MS(error) = \frac{SS(error)}{N-k}$$

MS(total) is a mean square for the total variation, obtained as follows:

Formula 11-7 $$MS(total) = \frac{SS(total)}{N-1}$$

Test Statistic for ANOVA with Unequal Sample Sizes

In testing the null hypothesis $H_0: \mu_1 = \mu_2 = \ldots = \mu_k$ against the alternative hypothesis that these means are not all equal, the test statistic

Formula 11-8 $$F = \frac{MS(treatment)}{MS(error)}$$

has an F distribution (when the null hypothesis H_0 is true) with degrees of freedom given by

numerator degrees of freedom = $k - 1$

denominator degrees of freedom = $N - k$

This test statistic is essentially the same as the one given earlier, and its interpretation is also the same as described earlier. The denominator depends only on the sample variances that measure variation within the treatments and is not affected by the differences among the sample means. In contrast, the numerator is affected by differences among the sample means. If the differences among the sample means are extreme, they will cause the numerator to be excessively large, so F will also be excessively large. Consequently, very large values of F suggest unequal means, and the ANOVA test is therefore right-tailed.

Tables are a convenient format for summarizing key results in ANOVA calculations, and Table 11-3 has a format often used in computer displays. (See the preceding Minitab and Excel displays.) The entries in Table 11-3 result from the readability data in Table 11-1.

Table 11-3	ANOVA Table for Readability Data			
Source of Variation	Sum of Squares (SS)	Degrees of Freedom	Mean Square (MS)	F Test Statistic
Treatments	1338.00	2	669.000	9.4695
Error	2331.39	33	70.648	
Total	3669.39	35		

Designing the Experiment: When we use one-way (or single-factor) analysis of variance and conclude that the differences among the means are significant, we can't be absolutely sure that the given factor is responsible for the differences. It is possible that the variation of some other unknown factor is responsible. One way to reduce the effect of the extraneous factors is to design the experiment so that it has a **completely randomized design,** in which each element is given the same chance of belonging to the different categories, or treatments. For example, you might assign subjects to a treatment group, placebo group, and control group through a process of random selection equivalent to picking slips from a bowl. Another way to reduce the effect of extraneous factors is to use a **rigorously controlled design,** in which elements are carefully chosen so that all other factors have no variability. For example, you might treat a healthy 7-year-old girl from Texas, while another healthy 7-year-old girl from Texas is given a placebo, while a third healthy 7-year-old girl from Texas is put in a control group that is given nothing. But in addition to health, age, gender, and state of residence, you might have to identify other relevant factors that should be considered. In general, good results require that the experiment be carefully designed and executed.

Using Technology

STATDISK Enter the data in columns of the Data Window. Select **Analysis** from the main menu bar, then select **One-Way Analysis of Variance,** and proceed to select the columns of sample data. Click **Evaluate** when done.

Minitab First enter the sample data in columns C1, C2, C3, ... Next, select **Stat, ANOVA, ONEWAY (UNSTACKED),** and enter C1 C2 C3 . . . in the box identified as "Responses" (in separate columns).

Excel First enter the data in columns A, B, C, ...

Next select **Tools** from the main menu bar, then select **Data Analysis,** followed by **Anova: Single Factor.** In the dialog box, enter the range containing the sample data. (For example, enter A1:C12 if the first value is in row 1 of column A and the last entry is in row 12 of column C.)

TI-83 Plus First enter the data as lists in L1, L2, L3, . . . , then press **STAT,** select **TESTS,** and choose the option **ANOVA.** Enter the column labels. For example, if the data are in columns L1, L2, and L3, enter those columns to get **ANOVA (L1, L2, L3),** and press the **ENTER** key.

11-2 Basic Skills and Concepts

 1. Readability of Authors The Chapter Problem uses the Flesch Reading Ease scores for randomly selected pages from books by Tom Clancy, J. K. Rowling, and Leo Tolstoy. If the Flesch-Kincaid Grade Level scores are used instead (see Data Set 14 in Appendix B), the analysis of variance results from Minitab are as shown in the accompanying table. Assume that we want to use a 0.05 significance level in testing the null hypothesis that the three authors have Flesch-Kincaid Grade Level scores with the same mean.
   a. What is the null hypothesis?
   b. What is the alternative hypothesis?
   c. Identify the value of the test statistic.
   d. Find the critical value for a 0.05 significance level.
   e. Identify the P-value.
   f. Based on the preceding results, what do you conclude about equality of the population means?

Minitab

Analysis of Variance

Source	DF	SS	MS	F	P
Factor	2	68.19	34.09	8.98	0.001
Error	33	125.31	3.80		
Total	35	193.50			

2. Fabric Flammability Tests in Different Laboratories Flammability tests were conducted on children's sleepwear. The Vertical Semirestrained Test was used, in which pieces of fabric were burned under controlled conditions. After the burning stopped, the length of the charred portion was measured and recorded. The same fabric samples were tested at five different laboratories. The analysis of variance results from Excel are shown below.
   a. What is the null hypothesis?
   b. What is the alternative hypothesis?
   c. Identify the value of the test statistic.
   d. Find the critical value for a 0.05 significance level.
   e. Identify the P-value.
   f. Is there sufficient evidence to support the claim that the means for the different laboratories are not all the same?

Excel

Source of Variation	SS	df	MS	F	P-value	F crit
Between Groups	2.087194264	4	0.521798566	2.949333035	0.030665893	2.588834036
Within Groups	7.607597403	43	0.17692087			
Total	9.694791667	47				

3. Marathon Times A random sample of males who finished the New York marathon is partitioned into three categories with ages of 21–29, 30–39, and 40 or over. The times (in seconds) are obtained from Data Set 8 in Appendix B. The analysis of variance results obtained from Excel are shown below.

a. What is the null hypothesis?
b. What is the alternative hypothesis?
c. Identify the value of the test statistic.
d. Find the critical value for a 0.05 significance level.
e. Identify the *P*-value.
f. Is there sufficient evidence to support the claim that men in the different age categories have different mean times?

Excel

Source of Variation	SS	df	MS	F	P-value	F crit
Between Groups	3532063.284	2	1766031.642	0.188679406	0.828324293	3.080387501
Within Groups	1010875649	108	9359959.71			
Total	1014407712	110				

4. **Systolic Blood Pressure in Different Age Groups** A random sample of 40 women is partitioned into three categories with ages of below 20, 20 through 40, and over 40. The systolic blood pressure levels are obtained from Data Set 1 in Appendix B. The analysis of variance results obtained from Minitab are shown below.
a. What is the null hypothesis?
b. What is the alternative hypothesis?
c. Identify the value of the test statistic.
d. Identify the *P*-value.
e. Is there sufficient evidence to support the claim that women in the different age categories have different mean blood pressure levels?

4. a. $\mu_1 = \mu_2 = \mu_3$
b. At least one of the three means is different from the others.
c. $F = 1.65$
d. 0.205
e. No

Minitab

Source	DF	SS	MS	F	P
Factor	2	938	469	1.65	0.205
Error	37	10484	283		
Total	39	11422			

In Exercises 5 and 6, use the listed sample data from car crash experiments conducted by the National Transportation Safety Administration. New cars were purchased and crashed into a fixed barrier at 35 mi/h, and the listed measurements were recorded for the dummy in the driver's seat. The subcompact cars are the Ford Escort, Honda Civic, Hyundai Accent, Nissan Sentra, and Saturn SL4. The compact cars are Chevrolet Cavalier, Dodge Neon, Mazda 626 DX, Pontiac Sunfire, and Subaru Legacy. The midsize cars are Chevrolet Camaro, Dodge Intrepid, Ford Mustang, Honda Accord, and Volvo S70. The full-size cars are Audi A8, Cadillac Deville, Ford Crown Victoria, Oldsmobile Aurora, and Pontiac Bonneville.

T 5. Head Injury in a Car Crash The head injury data (in hic) are given below. Use a 0.05 significance level to test the null hypothesis that the different weight categories have the same mean. Do the data suggest that larger cars are safer?

Subcompact:	681	428	917	898	420
Compact:	643	655	442	514	525
Midsize:	469	727	525	454	259
Full-size:	384	656	602	687	360

5. $F = 0.9922$. CV: $F = 3.2389$. *P*-value: 0.4216. There is not sufficient evidence to support the claim that larger cars are safer.

T 6. Chest Deceleration in a Car Crash The chest deceleration data (g) are given below. Use a 0.05 significance level to test the null hypothesis that the different weight categories have the same mean. Do the data suggest that larger cars are safer?

Subcompact:	55	47	59	49	42
Compact:	57	57	46	54	51
Midsize:	45	53	49	51	46
Full-size:	44	45	39	58	44

T 7. Archaeology: Skull Breadths from Different Epochs The values in the table are measured maximum breadths of male Egyptian skulls from different epochs (based on data from *Ancient Races of the Thebaid,* by Thomson and Randall-Maciver). Changes in head shape over time suggest that interbreeding occurred with immigrant populations. Use a 0.05 significance level to test the claim that the different epochs do not all have the same mean.

4000 B.C.	1850 B.C.	150 A.D.
131	129	128
138	134	138
125	136	136
129	137	139
132	137	141
135	129	142
132	136	137
134	138	145
138	134	137

T 8. Solar Energy in Different Weather A student of the author lives in a home with a solar electric system. At the same time each day, she collected voltage readings from a meter connected to the system and the results are listed in the accompanying table. Use a 0.05 significance level to test the claim that the mean voltage reading is the same for the three different types of day. Is there sufficient evidence to support a claim of different population means? We might expect that a solar system would provide more electrical energy on sunny days than on cloudy or rainy days. Can we conclude that sunny days result in greater amounts of electrical energy?

Sunny Days	Cloudy Days	Rainy Days
13.5	12.7	12.1
13.0	12.5	12.2
13.2	12.6	12.3
13.9	12.7	11.9
13.8	13.0	11.6
14.0	13.0	12.2

T 9. Mean Weights of M&Ms Refer to Data Set 19 in Appendix B. At the 0.05 significance level, test the claim that the mean weight of M&Ms is the same for each of the six different color populations. If it is the intent of Mars, Inc. to make the candies so that the different color populations have the same mean weight, do these results suggest that the company has a problem requiring corrective action?

T 10. Home Run Distances Refer to Data Set 30 in Appendix B. Use a 0.05 significance level to test the claim that the home runs hit by Barry Bonds, Mark McGwire, and Sammy Sosa have mean distances that are not all the same. Do the home run distances explain the fact that as of this writing, Barry Bonds has the most home runs in one season, while Mark McGwire has the second highest number of runs?

T **11.** *Sugar in Cereal* Refer to Data Set 16 in Appendix B and combine the sugar amounts for shelves 3 and 4, which are the two highest shelves. Use a 0.05 significance level to test the null hypothesis that the mean sugar amounts on the different shelves are all the same. What do the results suggest about the common belief that supermarkets place high-sugar cereals on shelves that are at eye-level for children?

T **12.** *Secondhand Smoke in Different Groups* Refer to Data Set 6 in Appendix B. Use a 0.05 significance level to test the claim that the mean cotinine level is different for these three groups: nonsmokers who are not exposed to environmental tobacco smoke, nonsmokers who are exposed to tobacco smoke, and people who smoke. What do the results suggest about secondhand smoke?

11-2 Beyond the Basics

13. *Using t Test* Five independent samples of 50 values each are randomly drawn from populations that are normally distributed with equal variances. We wish to test the claim that $\mu_1 = \mu_2 = \mu_3 = \mu_4 = \mu_5$.
  a. If we used only the methods given in Section 8-3, we would test the individual claims $\mu_1 = \mu_2$, $\mu_1 = \mu_3$, and so on. How many ways can we pair off the five means?
  b. Assume that for each test of equality between two means, there is a 0.95 probability of not making a type I error. If all possible pairs of means are tested for equality, what is the probability of making no type I errors? (Although the tests are not actually independent, assume that they are.)
  c. If we use analysis of variance to test the claim that $\mu_1 = \mu_2 = \mu_3 = \mu_4 = \mu_5$ at the 0.05 significance level, what is the probability of not making a type I error?
  d. Compare the results of parts (b) and (c). Which approach is better in the sense of giving us a greater chance of not making a type I error?

14. *Equivalent Tests* In this exercise you will verify that when you have two sets of sample data, the *t* test for independent samples and the ANOVA method of this section are equivalent. Refer to the readability measurements in Table 11-1, but use only the data for Clancy and Rowling. The original data are listed in Data Set 14 in Appendix B.
  a. Use a 0.05 significance level and the method of Section 8-3 to test the claim that the two samples come from populations with the same mean. (Assume that both populations have the same variance.)
  b. Use a 0.05 significance level and the ANOVA method of this section to test the claim made in part (a).
  c. Verify that the squares of the *t* test statistic and critical value from part (a) are equal to the *F* test statistic and critical value from part (b).

11. $F = 9.0646$. CV: $F = 3.8056$. *P*-value: 0.0034. Reject the claim that the mean sugar amounts on the different shelves are all the same. Shelf 2 appears to have a considerably higher mean, which would support the claim that high-sugar cereals are placed on shelves that are at eye-level for children.

12. $F = 20.8562$. CV: $F = 3.0718$ approx. *P*-value: 0.0000. Support the claim that the three populations have different means. The sample data appear to support the claim that secondhand smoke is harmful, although this conclusion is not justified by the ANOVA test results.

13. a. 10
  b. 0.599
  c. 0.95
  d. Analysis of variance

14. a. $t = -2.831291425$. CV: $t = \pm 2.074$. Reject H_0: $\mu_1 = \mu_2$.
  b. $F = 8.0162$. CV: $F = 4.3009$. Reject H_0: $\mu_1 = \mu_2$.
  c. $t^2 = F = 8.0162$. CV: $t^2 = F = 4.301$.

11-3 Two-Way ANOVA

In Section 11-2 we used analysis of variance to decide whether three or more populations have the same mean. That section used procedures referred to as *one*-way analysis of variance (or single-factor analysis of variance) because the data are categorized into groups according to a *single* factor (or treatment). Recall that a factor, or treatment, is a property that is the basis for categorizing the different groups of data. See Table 11-4, which lists times (in seconds) of runners who finished a recent New York City marathon. The listed times were randomly selected from Data Set 8 in Appendix B and they are partitioned into six categories according to two variables: (1) the row variable of gender and (2) the column variable of age category. **Two-way analysis of variance** involves *two* factors, such as gender and age in Table 11-4. The six subcategories in Table 11-4 are often called *cells*, so Table 11-4 has six cells containing five values each.

Table 11-4	Times (in seconds) for New York Marathon Runners		
		Age	
	21–29	30–39	40 and over
Male	13,615	14,677	14,528
	18,784	16,090	17,034
	14,256	14,086	14,935
	10,905	16,461	14,996
	12,077	20,808	22,146
Female	16,401	15,357	17,260
	14,216	16,771	25,399
	15,402	15,036	18,647
	15,326	16,297	15,077
	12,047	17,636	25,898

In analyzing the sample data in Table 11-4, we have already discussed the one-way analysis of variance for a single factor, so it might seem reasonable to simply proceed with one-way ANOVA for the factor of gender and another one-way ANOVA for the factor of age. Unfortunately, conducting two separate one-way ANOVA tests wastes information and totally ignores a very important feature: the effect of an interaction between the two factors.

Definition

There is an **interaction** between two factors if the effect of one of the factors changes for different categories of the other factor.

As an example of an *interaction* between two factors, consider the pairings of food and wine at a quality restaurant. It is known that certain foods and wines interact well to produce an enjoyable taste, while others interact poorly to produce an unpleasant taste. There is a good interaction between Chablis wine and oysters; the limestone in the soil where Chablis is made leaves a residue in the wine that interacts well with oysters. Peanut butter and jelly also interact well. In contrast, chocolate syrup and hot dogs interact in a way that results in a bad taste. In using two-way ANOVA for the data of Table 11-4, we consider three possible effects on the marathon times: (1) the effects of an interaction between gender and age; (2) the effects of gender; (3) the effects of age. The calculations are quite involved, so *we will assume that a software package or TI-83 Plus calculator is being used.* (Procedures for using technology are described at the end of this section.) The Minitab display for the data in Table 11-4 is shown here.

Minitab

```
Analysis of Variance for TIME
Source        DF        SS          MS        F        P
GENDER         1  15225413    15225413     1.69    0.206
AGE            2  92086979    46043490     5.10    0.014
Interaction    2  21042069    10521034     1.17    0.329
Error         24 216683456     9028477
Total         29 345037917
```

Note to Instructor
When using Minitab for ANOVA, it's wise to *name* variables so that you can keep track of which results go with which variables, thereby reducing confusion. In this display, the column of sample data is labeled "TIME," the row values are labeled "GENDER," and the column values are labeled "AGE."

The Minitab display includes SS (sum of squares) components similar to those described in Section 11-2. Because the circumstances of Section 11-2 involved only a single factor, we used SS(treatment) as a measure of the variation due to the different treatment categories, and we used SS(error) as a measure of the variation due to sampling error. Here we use SS(gender) as a measure of variation among the gender means. We use SS(age) as a measure of variation among the age means. We continue to use SS(error) as a measure of variation due to sampling error. Similarly, we use MS(gender) and MS(age) for the two different mean squares and continue to use MS(error) as before. Also, we use df(gender) and df(age) for the two different degrees of freedom.

Here are the required assumptions and basic procedure for two-way analysis of variance (ANOVA). The procedure is also summarized in Figure 11-3.

Assumptions

1. For each cell, the sample values come from a population with a distribution that is approximately normal.

2. The populations have the same variance σ^2 (or standard deviation σ).

3. The samples are simple random samples. (That is, samples of the same size have the same probability of being selected.)

4. The samples are independent of each other. (The samples are not matched or paired in any way.)

5. The sample values are categorized two ways. (This is the basis for the name of the method: *two-way* analysis of variance.)

6. All of the cells have the same number of sample values. (This is called a *balanced* design.)

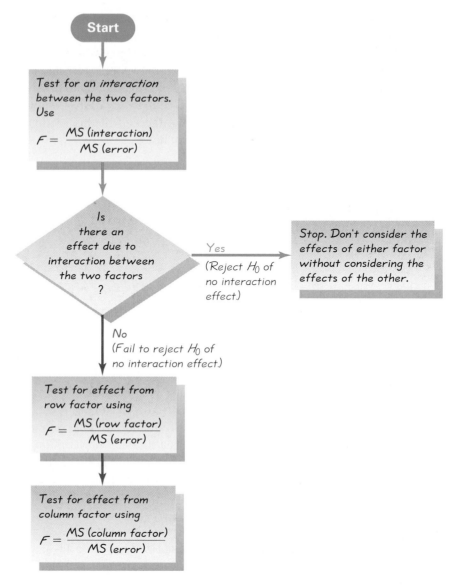

FIGURE 11-3 **Procedure for Two-Way ANOVA**

Procedure for Two-Way ANOVA (See Figure 11-3.)

Step 1: *Interaction Effect:* In two-way analysis of variance, begin by testing the null hypothesis that there is no interaction between the two factors. Using Minitab for the data in Table 11-4, we get the following test statistic:

$$F = \frac{\text{MS(interaction)}}{\text{MS(error)}} = \frac{10,521,034}{9,028,477} = 1.17$$

Interpretation: The corresponding *P*-value is shown in the Minitab display as 0.329, so we fail to reject the null hypothesis of no interaction between the two factors. It does not appear that the marathon times are affected by an interaction between gender and age category.

Step 2:   *Row/Column Effects:* If we do reject the null hypothesis of no interaction between factors, then we should stop now; we should not proceed with the two additional tests. (If there is an interaction between factors, we shouldn't consider the effects of either factor without considering those of the other.)

If we fail to reject the null hypothesis of no interaction between factors, then we should proceed to test the following two hypotheses:

H_0:   There are no effects from the row factor (that is, the row means are equal).

H_0:   There are no effects from the column factor (that is, the column means are equal).

In Step 1, we failed to reject the null hypothesis of no interaction between factors, so we proceed with the next two hypothesis tests identified in Step 2.

For the row factor of gender we get

$$F = \frac{MS(\text{gender})}{MS(\text{error})} = \frac{15{,}225{,}413}{9{,}028{,}477} = 1.69$$

Interpretation: This value is not significant because the corresponding *P*-value is shown in the Minitab display as 0.206. We fail to reject the null hypothesis of no effects from gender. That is, the gender of the runner does not appear to have an effect on time. Because the winner of such marathons is almost always male, we might have expected to find an effect from gender, but we did not. Perhaps there aren't enough sample values for the effect to be considered significant.

For the column factor of age we get

$$F = \frac{MS(\text{age})}{MS(\text{error})} = \frac{46{,}043{,}490}{9{,}028{,}477} = 5.10$$

Interpretation: This value is significant because the corresponding *P*-value is shown as 0.014. (With a *P*-value of 0.014, we have significance at the 0.05 level, but not at the 0.01 level.) We therefore reject the null hypothesis of no effects from age. The age of the runner does appear to have an effect on the time. Based on the sample data in Table 11-4, we conclude that times do appear to have unequal means for the different age categories, but the times appear to have equal means for both genders.

Special Case: One Observation per Cell and No Interaction
Table 11-4 contains 5 observations per cell. If our sample data consist of only one observation per cell, we lose MS(interaction), SS(interaction), and df(interaction) because those values are based on sample variances computed for each individual cell. If

there is only one observation per cell, there is no variation within individual cells and those sample variances cannot be calculated. Here's how we proceed when there is one observation per cell: *If it seems reasonable to assume (based on knowledge about the circumstances) that there is no interaction between the two factors, make that assumption and then proceed as before to test the following two hypotheses separately:*

H_0:   There are no effects from the row factor.

H_0:   There are no effects from the column factor.

As an example, suppose that we have only the first value in each cell of Table 11-4. Using only those first values, the two row means are 14,273.3 and 16,339.3. Is that difference significant, suggesting that there is an effect due to gender? Again using only the first value in each cell, the three column means are 15,008.0, 15,017.0, and 15,894.0. Are those differences significant, suggesting that there is an effect due to age? It is reasonable to believe that marathon running times are not affected by some interaction between gender and age. (If we believe there is an interaction, the method described here does not apply.) Following is the Minitab display for the data in Table 11-4, with only the first value from each cell.

```
Minitab

Analysis of Variance for TIME
Source      DF        SS          MS         F          P
GENDER       1     6402534     6402534      8.88      0.097
AGE          2     1036137      518069      0.72      0.582
Error        2     1441476      720738
Total        5     8880147
```

We first use the results from the Minitab display to test the null hypothesis of no effects from the row factor of gender.

$$F = \frac{MS(\text{gender})}{MS(\text{error})} = \frac{6{,}402{,}534}{720{,}738} = 8.88$$

This test statistic is not significant, because the corresponding P-value in the Minitab display is 0.0972. We fail to reject the null hypothesis; it appears that the marathon times are not affected by the gender of the runner.

We now use the Minitab display to test the null hypothesis of no effect from the column factor of age category. The test statistic is

$$F = \frac{MS(\text{age})}{MS(\text{error})} = \frac{518{,}069}{720{,}738} = 0.72$$

This test statistic is not significant because the corresponding P-value is given in the Minitab display as 0.582. We fail to reject the null hypothesis, so it appears that the marathon running time is not affected by the age category of the runner. Using only the first value from each cell, we conclude that the marathon running times do not appear to be affected by either gender or age category, but when we used 5 values from each cell, we concluded that the times appear to be affected by age category. Such is the power of larger samples.

In this section we have briefly discussed an important branch of statistics. We have emphasized the interpretation of computer displays while omitting the manual calculations and formulas, which are quite formidable.

STATDISK Click on **Analysis** and select **Two-Way Analysis of Variance.** Make the required entries in the window, then click on **Continue.** Proceed to enter the data in the "Values" column, then click on **Evaluate.**

Minitab First enter all of the sample values in column C1. Enter the corresponding row numbers in column C2. Enter the corresponding column numbers in column C3. From the main menu bar, select **Stat,** then select **ANOVA,** then **Two-Way.** In the dialog box, enter C1 for Response, enter C2 for Row factor, and enter C3 for Column factor. Click **OK.** *Hint:* Avoid confusion by labeling the columns C1, C2, and C3 with meaningful names.

Excel For two-way tables with more than one entry per cell: Entries from the same cell must be listed down a column, not across a row. Enter the labels corresponding to the data set in column A and row 1, as in this example, which corresponds to Table 11-4:

	A	B	C	D
1		21–29	30–39	40 and over
2	Male	13615	14677	14528
3	Male	18784	16090	17034
⋮	⋮	⋮	⋮	⋮

After entering the sample data and labels, select **Tools** from the main menu bar, then **Data Analysis,** then **Anova: Two-Factor With Replication.** In the dialog box, enter the input range. For the data in Table 11-4, enter A1:D11. For "rows per sample," en-

ter the number of values in each cell; enter 5 for the data in Table 11-4. Click **OK.**

For two-way tables with exactly one entry per cell: The labels are not required. Enter the sample data as they appear in the table. Select **Tools,** then **Data Analysis,** then **Anova: Two-factor Without Replication.** In the dialog box, enter the input range of the sample values only; do not include labels in the input range. Click **OK.**

TI-83 Plus The TI-83 Plus program A1ANOVA can be downloaded from the CD-ROM included with this book. Select the *software* folder. The program must be downloaded to your calculator, then the sample data must first be entered as matrix D with three columns. Press **2nd,** and the x^{-1} key, scroll to the right for **EDIT,** scroll down for **[D],** then press **ENTER** and proceed to enter the total number of data values followed by 3 (for 3 columns). The first column of D lists all of the sample data, the second column lists the corresponding row number, and the third column lists the corresponding column number. After entering all of the data and row numbers and column numbers in matrix D, press **PRGM,** select **A1ANOVA** and press **ENTER** twice, then select **RAN BLOCK DESI** (for random block design) and press **ENTER** twice. Select **CONTINUE** and press **ENTER.** After a while, the results will be displayed. F(A) is the *F* test statistic for the row factor, and it will be followed by the corresponding *P*-value. F(B) is the *F* test statistic for the column factor, and it is followed by the corresponding *P*-value. (It is necessary to press **ENTER** to see the remaining part of the display.) F(AB) is the *F* test statistic for the interaction effect, and it is followed by the corresponding *P*-value.

11-3 Basic Skills and Concepts

Interpreting a Computer Display. Some of Exercises 1–7 require the given Minitab display, which results from the amounts of the pesticide DDT measured in falcons in three different age categories (young, middle-aged, old) at three different locations (United States, Canada, Arctic region). The data set is included with the Minitab software as the file FALCON.MTW.

```
Minitab
Analysis of Variance for DDT
Source          DF          SS          MS          F          P
Site             2    17785.41     8892.70     2581.75      0.000
Age              2     1721.19      860.59      249.85      0.000
Interaction      4       17.70        4.43        1.28      0.313
Error           18       62.00        3.44
Total           26    19586.30
```

1. **Meaning of Two-Way ANOVA** The method of this section is referred to as two-way analysis of variance, or two-way ANOVA. Why is the term *two-way* used? Why is the term *analysis of variance* used?

2. **Why Not One-Way ANOVA?** The given Minitab display results from measured amounts of DDT in falcons partitioned into nine cells according to one factor of location and another factor of age of the falcon. Each cell includes three DDT measurements. Why can't we conduct a thorough analysis of the data by simply executing two separate tests using one-way ANOVA (described in Section 11-2), where one test addresses the differences in sites and the other test addresses the differences in age? That is, why is two-way ANOVA required instead of two separate applications of one-way ANOVA?

3. **Interaction Effect** Assume that two-way analysis of variance reveals that there is a significant effect from an interaction between two factors. Why should we *not* proceed to test the effect from the row factor?

4. **Why Not Use Two-Way ANOVA?** Why can't we use the method of two-way analysis of variance with two-way tables described in Section 10-3?

5. **Interaction Effect** Refer to the Minitab display and test the null hypothesis that the amounts of DDT are not affected by an interaction between site and age. What do you conclude?

6. **Effect of Site** Refer to the Minitab display and assume that the amounts of DDT in the falcons are not affected by an interaction between site and age. Is there sufficient evidence to support the claim that site has an effect on the amount of DDT?

7. **Effect of Age** Refer to the Minitab display and assume that the amounts of DDT in the falcons are not affected by an interaction between site and age. Is there sufficient evidence to support the claim that age has an effect on the amount of DDT?

Interpreting a Computer Display. In Exercises 8–10, use the Minitab display, which results from the scores listed in the accompanying table. The sample data are SAT scores on the verbal and math portions of SAT-I and are based on reported statistics from the College Board.

Verbal

Female	646	539	348	623	478	429	298	782	626	533
Male	562	525	512	576	570	480	571	555	519	596

Math

Female	484	489	436	396	545	504	574	352	365	350
Male	547	678	464	651	645	673	624	624	328	548

Minitab

```
Analysis of Variance for SAT
Source        DF        SS        MS         F        P
Gender         1     52635     52635      5.03    0.031
Ver/Math       1      6027      6027      0.58    0.453
Interaction    1     31528     31528      3.01    0.091
Error         36    376748     10465
Total         39    466938
```

8. **Interaction Effect** Test the null hypothesis that SAT scores are not affected by an interaction between gender and test (verbal/math). What do you conclude?

9. **Effect of Gender** Assume that SAT scores are not affected by an interaction between gender and the type of test (verbal/math). Is there sufficient evidence to support the claim that gender has an effect on SAT scores?

10. **Effect of Type of SAT Test** Assume that SAT scores are not affected by an interaction between gender and the type of test (verbal/math). Is there sufficient evidence to support the claim that the type of test (verbal/math) has an effect on SAT scores?

Interpreting a Computer Display. In Exercises 11 and 12, refer to the given Minitab display. This display results from a study in which 24 subjects were given hearing tests using four different lists of words. The 24 subjects had normal hearing and the tests were conducted with no background noise. The main objective was to determine whether the four lists are equally difficult to understand. In the original table of hearing test scores, each cell has one entry. The original data are from A Study of the Interlist Equivalency of the CID W-22 Word List Presented in Quiet and in Noise, *by Faith Loven, University of Iowa. The original data are available on the Internet through DASL (Data and Story Library).*

Minitab

```
Analysis of Variance for Hearing
Source        DF        SS        MS         F        P
Subject       23    3231.6     140.5      3.87    0.000
List           3     920.5     306.8      8.45    0.000
Error         69    2506.5      36.3
Total         95    6658.6
```

11. **Hearing Tests: Effect of Subject** Assuming that there is no effect on hearing test scores from an interaction between subject and list, is there sufficient evidence to support the claim that the choice of subject has an effect on the hearing test score? Interpret the result by explaining why it makes practical sense.

12. **Hearing Tests: Effect of Word List** Assuming that there is no effect on hearing test scores from an interaction between subject and list, is there sufficient evidence to support the claim that the choice of word list has an effect on the hearing test score?

(T) 13. **Pulse Rates** The following table lists pulse rates from Data Set 1 in Appendix B. Are pulse rates affected by an interaction between gender and age? Are pulse rates affected by gender? Are pulse rates affected by age?

continued

8. $F = 3.01$. P-value: 0.091. No significant effect from the interaction between gender and test.

9. $F = 5.03$. P-value: 0.031. Support the claim that gender has an effect on SAT scores.

10. $F = 0.58$. P-value: 0.453. There is not sufficient evidence to support the claim that the type of test has an effect on SAT scores.

11. $F = 3.87$. P-value: 0.000. Support the claim that the choice of subject has an effect on the hearing test score.

12. $F = 8.45$. P-value: 0.000. Support the claim that the choice of word list has an effect on the hearing test scores.

13. Interaction: $F = 0.36$ and P-value $= 0.701$, so there is no significant interaction effect. Gender: $F = 0.09$ and P-value $= 0.762$, so there is no significant effect from gender. Age: $F = 0.36$ and P-value $= 0.701$, so there is no significant effect from age.

	Age		
	Under 20	20–40	Over 40
Male	96 64 68 60	64 88 72 64	68 72 60 88
Female	76 64 76 68	72 88 72 68	60 68 72 64

14. Car Fuel Consumption The following table lists highway fuel consumption amounts (in mi/gal) from Data Set 22 in Appendix B. Assume that fuel consumption amounts are not affected by an interaction between the type of transmission (manual or automatic) and the number of cylinders. Are fuel consumption amounts affected by the type of transmission? Are fuel consumption amounts affected by the number of cylinders?

	Cylinders		
	4	6	8
Manual	33	30	28
Automatic	31	27	24

11-3 Beyond the Basics

15. Transformations of Data Assume that two-way ANOVA is used to analyze sample data consisting of more than one entry per cell. How are the ANOVA results affected in each of the following cases?
   a. The same constant is added to each sample value.
   b. Each sample value is multiplied by the same nonzero constant.
   c. The format of the table is transposed, so that the row and column factors are interchanged.
   d. The first sample value in the first cell is changed so that it becomes an outlier.

Review

In Section 8-3 we presented a procedure for testing equality between *two* population means, but in Section 11-2 we used analysis of variance (or ANOVA) to test for equality of three or more population means. This method requires (1) normally distributed populations, (2) populations with the same standard deviation (or variance), and (3) simple random samples that are independent of each other. The methods of one-way analysis of variance are used when we have three or more samples taken from populations that are characterized according to a single factor. The following are key features of one-way analysis of variance:

- The F test statistic is based on the ratio of two different estimates of the common population variance σ^2, as shown below.

$$F = \frac{\text{variance between samples}}{\text{variance within samples}} = \frac{MS(treatment)}{MS(error)}$$

- Critical values of F can be found in Table A-5, but we focused on the interpretation of P-values that are included as part of a computer display.

In Section 11-3 we considered two-way analysis of variance, with data categorized according to two different factors. One factor is used to arrange the sample data in different rows, while the other factor is used for different columns. The procedure for two-way analysis of variance is summarized in Figure 11-3, and it requires that we first test for an interaction between the two factors. If there is no significant interaction, we then proceed to conduct individual tests for effects from each of the two factors. We also considered the use of two-way analysis of variance for the special case in which there is only one observation per cell.

Because of the nature of the calculations required throughout this chapter, we emphasized the interpretation of computer displays.

Review Exercises

1. Drinking and Driving The Associated Insurance Institute sponsors studies of the effects of drinking on driving. In one such study, three groups of adult men were randomly selected for an experiment designed to measure their blood alcohol levels after consuming five drinks. Members of group A were tested after one hour, members of group B were tested after two hours, and members of group C were tested after four hours. The results are given in the accompanying table; the Minitab display for these data is also shown. At the 0.05 significance level, test the claim that the three groups have the same mean level.

A	B	C
0.11	0.08	0.04
0.10	0.09	0.04
0.09	0.07	0.05
0.09	0.07	0.05
0.10	0.06	0.06
		0.04
		0.05

Minitab

```
Analysis of Variance
Source     DF          SS          MS        F       P
Factor      2    0.0076571   0.0038286    46.90   0.000
Error      14    0.0011429   0.0000816
Total      16    0.0088000
```

2. Location, Location, Location The accompanying list shows selling prices (in thousands of dollars) for homes located on Long Beach Island in New Jersey. Different mean selling prices are expected for the different locations. Do these sample data support the claim of different mean selling prices? Use a 0.05 significance level.

Oceanside:	235	395	547	469	369	279
Oceanfront:	538	446	435	639	499	399
Bayside:	199	219	239	309	399	190
Bayfront:	695	389	489	489	599	549

Interpreting a Computer Display. In Exercises 3–5, use the Minitab display, which results from the values listed in the accompanying table. The sample data are student estimates (in feet) of the length of their classroom. The actual length of the classroom is 24 ft 7.5 in.

1. $F = 46.90$. P-value: 0.000. Reject the claim of equal population means.

2. $F = 9.4827$. CV: $F = 3.0984$. Support the claim of different mean selling prices.

	Math	Business	Liberal Arts
		Major	
Female	28 25 30	35 25 20	40 21 30
Male	25 30 20	30 24 25	25 20 32

Minitab

```
Analysis of Variance for LENGTH
Source         DF          SS          MS         F          P
GENDER          1        29.4        29.4      0.78      0.395
MAJOR           2        10.1         5.1      0.13      0.876
Interaction     2        14.1         7.1      0.19      0.832
Error          12       453.3        37.8
Total          17       506.9
```

3. $F = 0.19$. P-value: 0.832. No significant effect from the interaction between gender and major.
4. $F = 0.78$. P-value: 0.395. There is not sufficient evidence to support the claim that estimated length is affected by gender.
5. $F = 0.13$. P-value: 0.876. There is not sufficient evidence to support the claim that estimated length is affected by major.
6. a. $F = 1.00$. P-value: 0.423. There is not sufficient evidence to support the claim that amounts of emitted greenhouse gases are affected by the type of transmission.
   b. $F = 7.00$. P-value: 0.125. There is not sufficient evidence to support the claim that amounts of greenhouse gases are affected by the number of cylinders.
   c. Perhaps greenhouse gases *are* affected by the type of transmission and/or the number of cylinders; however, the given sample data do not provide sufficient evidence to support such claims.

3. Interaction Effect Test the null hypothesis that the estimated lengths are not affected by an interaction between gender and major.

4. Effect of Gender Assume that estimated lengths are not affected by an interaction between gender and major. Is there sufficient evidence to support the claim that estimated length is affected by gender?

5. Effect of Major Assume that estimated lengths are not affected by an interaction between gender and major. Is there sufficient evidence to support the claim that estimated length is affected by major?

6. Auto Pollution The accompanying table lists the amounts of greenhouse gases emitted by different cars in one year. (See Data Set 22 in Appendix B.) The Minitab display results from this table.

   a. Assuming that there is no interaction effect, is there sufficient evidence to support the claim that amounts of emitted greenhouse gases are affected by the type of transmission (automatic/manual)?

   b. Assuming that there is no interaction effect, is there sufficient evidence to support the claim that amounts of emitted greenhouse gases are affected by the number of cylinders?

   c. Based on the results from parts (a) and (b), can we conclude that greenhouse gas emissions are not affected by the type of transmission or the number of cylinders? Why or why not?

Emission of Greenhouse Gases (tons/year)

	4 Cylinders	6 Cylinders	8 Cylinders
Automatic	10	12	14
Manual	10	12	12

Minitab

```
Analysis of Variance for GASES
Source         DF          SS          MS         F          P
TRANS           1       0.667       0.667      1.00      0.423
CYL             2       9.333       4.667      7.00      0.125
Error           2       1.333       0.667
Total           5      11.333
```

Cumulative Review Exercises

1. Boston Rainfall Statistics Refer to the Boston rainfall amounts for Monday, as listed in Data Set 11 in Appendix B.
  a. Find the mean.
  b. Find the standard deviation.
  c. Find the 5-number summary.
  d. Identify any outliers.
  e. Construct a histogram.
  f. Assume that you want to test the null hypothesis that the mean amount of rainfall is the same for the seven days of the week. Can you use one-way ANOVA? Why or why not?
  g. Based on the sample data, estimate the probability that precipitation will fall on a randomly selected Monday in Boston.

2. M&M Treatment The table below lists 60 SAT scores separated into categories according to the color of the M&M candy used as a treatment. The SAT scores are based on data from the College Board, and the M&M color element is based on author whimsy.
  a. Find the mean of the 20 SAT scores in each of the three categories. Do the three means appear to be approximately equal?
  b. Find the median of the 20 SAT scores in each of the three categories. Do the three medians appear to be approximately equal?
  c. Find the standard deviation of the 20 SAT scores in each of the three categories. Do the three standard deviations appear to be approximately equal?
  d. Test the null hypothesis that there is no difference between the mean SAT score of subjects treated with red M&Ms and the mean SAT score of subjects treated with green M&Ms.
  e. Construct a 95% confidence interval estimate of the mean SAT score for the population of subjects receiving the red M&M treatment.
  f. Test the null hypothesis that the three populations (red, green, and blue M&M treatments) have the same mean SAT score.

Red	1130	621	813	996	1030	1257	898	743	921	1179
	1092	855	896	858	1095	1133	896	1190	908	699
Green	996	630	583	828	1121	993	1025	907	1111	1147
	780	916	793	1188	499	1180	1229	1450	1071	1153
Blue	706	1068	1013	892	1370	1611	939	1004	821	915
	866	848	1408	793	1097	1244	996	1131	1039	1159

3. Weights of Babies: Finding Probabilities In the United States, weights of newborn babies are normally distributed with a mean of 7.54 lb and a standard deviation of 1.09 lb (based on data from "Birth Weight and Prenatal Mortality," by Wilcox, Skjaerven, Buekens, and Kiely, *Journal of the American Medical Association*, Vol. 273, No. 9).
  a. If a newborn baby is randomly selected, what is the probability that he or she weighs more than 8.00 lb?
  b. If 16 newborn babies are randomly selected, what is the probability that their mean weight is more than 8.00 lb?
  c. What is the probability that each of the next three babies will have a birth weight greater than 7.54 lb?

1. a. 0.100 in.
   b. 0.263 in.
   c. 0.00, 0.00, 0.00, 0.010, 1.41
   d. 0.92 in., 1.41 in.
   e. Answer varies, depending on the number of classes used, but the histogram should depict a distribution that is skewed to the right.
   f. No, because the data do not appear to come from a normally distributed population.
   g. 19/52 or 0.365

2. a. 960.5, 980.0, 1046.0; no
   b. 914.5, 1010.5, 1008.5; no
   c. 174.6, 239.6, 226.8; no
   d. $t = -0.294$. CV: $t = \pm 2.093$ (assuming a 0.05 significance level). Fail to reject H_0: $\mu_1 = \mu_2$.
   e. $878.8 < \mu < 1042.2$
   f. $F = 0.8647$. P-value: 0.4266. There is not sufficient evidence to reject the claim that the three populations have the same mean SAT score.

3. a. 0.3372
   b. 0.0455
   c. 1/8

Cooperative Group Activities

1. Out-of-class activity The *World Almanac and Book of Facts* includes a section called "Noted Personalities," with subsections comprised of architects, artists, business leaders, cartoonists, social scientists, military leaders, philosophers, political leaders, scientists, writers, composers, entertainers, and others. Design and conduct an observational study that begins with the selection of samples from select groups, to be followed by a comparison of life spans of people from the different categories. Do any particular groups appear to have life spans that are different from the other groups? Can you explain such differences?

2. In-class activity Begin by asking each student in the class to estimate the length of the classroom. Specify that the length is the distance between the chalkboard and the opposite wall. (See Review Exercises 3–5.) On the same sheet of paper, each student should also write his or her gender (male/female) and major. Then divide into groups of three or four, and use the data from the entire class to address these questions:
   - Is there a significant difference between the mean estimate for males and the mean estimate for females?
   - Is there sufficient evidence to reject equality of the mean estimates for different majors? Describe how the majors were categorized.
   - Does an interaction between gender and major have an effect on the estimated length?
   - Does gender appear to have an effect on estimated length?
   - Does major appear to have an effect on estimated length?

3. Out-of-class activity Divide into groups of three or four students. Each group should survey other students at the same college by asking them to identify their major and gender. You might include other factors, such as employment (none, part-time, full-time) and age (under 21, 21–30, over 30). For each surveyed subject, determine the accuracy of the time on his or her wristwatch. First set your own watch to the correct time using an accurate and reliable source ("At the tone, the time is . . . "). For watches that are ahead of the correct time, record positive times. For watches that are behind the correct time, record negative times. Use the sample data to address questions such as these:
   - Does gender appear to have an effect on the accuracy of the wristwatch?
   - Does major have an effect on wristwatch accuracy?
   - Does an interaction between gender and major have an effect on wristwatch accuracy?

Technology Project

For U.S. presidents and the popes and British monarchs since 1690, the accompanying table lists the numbers of years that they lived after their inauguration, election, or coronation. Use boxplots and analysis of variance to determine whether the survival times for the different groups differ. Conduct the analysis of variance by using STATDISK, Minitab, Excel, a TI-83 Plus calculator, or some other statistical software package. Obtain printed copies of the computer displays and write your observations and conclusions.

Presidents		Popes		Kings and Queens	
Washington	10	Alex VIII	2	James II	17
J. Adams	29	Innoc XII	9	Mary II	6
Jefferson	26	Clem XI	21	William III	13
Madison	28	Innoc XIII	3	Anne	12
Monroe	15	Ben XIII	6	George I	13
J. Q. Adams	23	Clem XII	10	George II	33
Jackson	17	Ben XIV	18	George III	59
Van Buren	25	Clem XIII	11	George IV	10
Harrison	0	Clem XIV	6	William IV	7
Tyler	20	Pius VI	25	Victoria	63
Polk	4	Pius VII	23	Edward VII	9
Taylor	1	Leo XII	6	George V	25
Fillmore	24	Pius VIII	2	Edward VIII	36
Pierce	16	Greg XVI	15	George VI	15
Buchanan	12	Pius IX	32		
Lincoln	4	Leo XIII	25		
A. Johnson	10	Pius X	11		
Grant	17	Ben XV	8		
Hayes	16	Pius XI	17		
Garfield	0	Pius XII	19		
Arthur	7	John XXIII	5		
Cleveland	24	Paul VI	15		
Harrison	12	John Paul I	0		
McKinley	4				
T. Roosevelt	18				
Taft	21				
Wilson	11				
Harding	2				
Coolidge	9				
Hoover	36				
F. Roosevelt	12				
Truman	28				
Kennedy	3				
Eisenhower	16				
L. Johnson	9				
Nixon	25				

Based on data from *Computer-Interactive Data Analysis,* by Lunn and McNeil, John Wiley & Sons.

from DATA *to* DECISION

Critical Thinking: Should you approve this drug?

Drugs must undergo thorough testing before being approved for general use. In addition to testing for adverse reactions, they must also be tested for their effectiveness, and the analysis of such test results typically involves methods of statistics. Consider the development of Xynamine—a new drug designed to lower pulse rates. In order to obtain more consistent results that do not have a confounding variable of gender, the drug is tested using males only. Given below are pulse rates for a placebo group, a group of men treated with Xynamine in 10-mg doses, and a group of men treated with 20-mg doses of Xynamine. The project manager for the drug conducts research and finds that for adult males, pulse rates are normally distributed with a mean around 70 beats per minute and a standard deviation of approximately 11 beats per minute. His summary report states that the drug is effective, based on this evidence: The placebo group has a mean pulse rate of 68.9, which is close to the value of 70 beats per minute for adult males in general, but the group treated with 10-mg doses of Xynamine has a lower mean of 66.2, and the group treated with 20-mg doses of Xynamine has the lowest mean of 65.2.

Analyzing the Results

Analyze the data using the methods of this chapter. Based on the results, does it appear that there

Placebo Group	10-mg Treatment Group	20-mg Treatment Group
77	67	72
61	48	94
66	79	57
63	67	63
81	57	69
75	71	59
66	66	64
79	85	82
66	75	34
75	77	76
48	57	59
70	45	53

is sufficient evidence to support the claim that the drug lowers pulse rates? Are there any serious problems with the design of the experiment? Given that only males were involved in the experiment, do the results also apply to females? The project manager compared the posttreatment pulse rates to the mean pulse rate for adult males. Is there a better way to measure the drug's effectiveness in lowering pulse rates? How would you characterize the overall validity of the experiment? Based on the available results, should the drug be approved? Write a brief report summarizing your findings.

INTERNET PROJECT   Analysis of Variance

Go the Web site for this textbook at

http://www.aw.com/triola

Follow the link to the Internet Project for this chapter. The project provides the background for experiments in areas as varied as athletic performance,

consumer product labeling, and biology of the human body. In each case, the associated data will lend itself naturally to groupings ideal for application of this chapter's techniques. You will formulate the appropriate hypotheses, then conduct and summarize ANOVA tests.

Statistics @ Work

"Basic knowledge of statistics is critical in finance."

Joseph Marvin

Portfolio Manager

Principal and the Unit Head of

Portfolio Management and

Trading for the bond group at

State Street Global Advisors

(SSGA)

As a portfolio manager, Joseph specializes in the trading of fixed income derivatives. In his work, he uses statistical methods in assessing relative values among various financial instruments. SSGA is one of the country's largest institutional money managers with over $580 billion in assets under management.

What is your job?

I am Portfolio Manager and Unit Head of Portfolio Management for State Street Global Advisor (SSGA). SSGA is one of the nation's largest investment management companies and is a subsidiary of State Street Corporation: With over $580 billion in assets under management, SSGA's primary focus is managing the assets of public and private pension and retirement accounts.

Do you recommend statistics for today's college students? Why?

I absolutely recommend it to ALL college students. Statistics provides students with an excellent foundation for making better decisions. In positions dealing with economics and finance, individuals with statistics backgrounds are looked at more favorably.

What concepts of statistics do you use?

We actively use probability analysis and hypothesis testing. We also use both linear and non-linear regression analysis. For portfolio construction we use mean variance optimization. We use these statistics in assessing a bond's value. Without a good understanding of the basics and foundations of statistics, I would not be able to effectively perform my responsibilities. Basic knowledge of statistics is critical in finance.

Please describe one specific example illustrating how the use of statistics was successful in improving a product or service.

In determining which bond to buy, we have used simple hypothesis testing. The yield difference between a corporate bond's yield to maturity and the risk free rate, as represented by a similar maturity U.S. Treasury bond, represents a bond's risk premium, or "spread." A bond's spread is what we as managers use to compare the value of one bond to another. Bond spreads tend to be "mean reverting" over time, which results in a somewhat normal distribution. Assuming mean reversion and a normal distribution, we can use simple hypothesis testing to look for statistical significance. In other words, if a bond's spread is statistically significantly wide versus another bond's spread, with all else being equal, we might consider the bond to be "cheap." A wider spread means a higher yield. When spreads tighten, the price or value of a bond goes up.

Managers in the bond market increase their returns by buying bonds that have spreads that are expected to tighten. We want to buy bonds that have spreads that are 1.0 to 2.0 standard deviations wider than the average and sell bonds that have spreads that are 1.0 to 2.0 standard deviations away (tighter) from the average. Simple hypothesis testing!

12

Nonparametric Statistics

12-1 Overview

12-2 Sign Test

12-3 Wilcoxon Signed-Ranks Test for Matched Pairs

12-4 Wilcoxon Rank-Sum Test for Two Independent Samples

12-5 Kruskal-Wallis Test

12-6 Rank Correlation

12-7 Runs Test for Randomness

It rains more on weekends!?

In an article for the Knight Ridder News Service, Usha Lee McFarling writes that "your worst weather fears are true. It does rain more on weekends. Scientists who pored through years of rainfall data have discovered a clear and dismaying pattern. Fridays, Saturdays, and Sundays are the rainiest days of the week all along America's eastern coast, from Maine to Florida." The article refers to a study conducted by Arizona State University scientists Randall S. Cerveny and Robert C. Balling. But are their conclusions correct? Have they been reported and interpreted correctly?

Data Set 11 in Appendix B includes rainfall amounts for a recent year in Boston. In selecting a city for verifying the weekend rain phenomenon, Boston should be a good choice because it is located on the eastern coast. If we use STATDISK to enter the rainfall amount for each day of the week, we get the accompanying boxplots. Going from top to bottom, the boxplots represent Monday, Tuesday, . . . , Sunday. The Monday boxplot at the top is unusual because it appears to be missing the box, but the Monday rainfall amounts have so many 0s that the

minimum, first quartile, and median are all 0s, with the result that the box gets crunched to the left. Greater amounts of rainfall on Friday, Saturday, and Sunday would be visible with the distributions positioned farther to the right. Is that really the case? Are the differences really significant?

We might consider using methods presented in earlier chapters to investigate this issue. Analysis of variance (Section 11-2) would be a good candidate, but that method requires that the samples come from populations having a normal distribution. The STATDISK-generated histogram for the Monday rainfall amounts shows clearly that these values do not come from a normally distributed population. The other days have similar distributions that are clearly not normal. A major advantage of the methods discussed in this chapter is that they do not require a normal distribution or any other particular distribution.

Can we use the Boston data to support the claim of more rainfall on weekends? Are the rainfall amounts for the different days significantly different? We will address these questions later in the chapter.

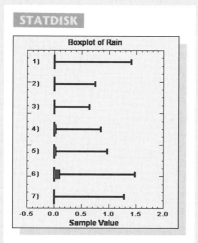

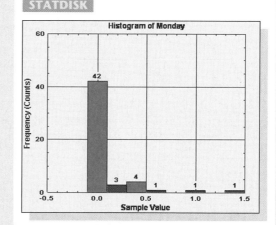

Overview

Note to Instructor

This chapter can be covered as one unit devoted to nonparametric methods, or some of its individual sections can be covered along with the related parametric methods found in the preceding chapters. Here is a guide to the sections relating the nonparametric methods of this chapter to the corresponding parametric methods.

Nonparametric	↔	Parametric
• 12-2, 12-3	↔	8-4
• 12-4	↔	8-3
• 12-5	↔	11-2
• 12-6	↔	9-2

The methods of inferential statistics presented in Chapters 6, 7, 8, 9, and 11 are called *parametric methods* because they are based on sampling from a population with specific parameters, such as the mean μ, standard deviation σ, or proportion p. Those parametric methods usually must conform to some fairly strict conditions, such as a requirement that the sample data come from a normally distributed population. This chapter introduces nonparametric methods, which do not have such strict requirements.

Definitions

Parametric tests require assumptions about the nature or shape of the populations involved; **nonparametric tests** do not require assumptions about the population distributions. Consequently, nonparametric tests of hypotheses are often called **distribution-free tests.**

Although the term *nonparametric* suggests that the test is not based on a parameter, there are some nonparametric tests that do depend on a parameter such as the median. The nonparametric tests do not, however, require a particular distribution, so they are sometimes referred to as *distribution-free* tests. Although *distribution-free* is a more accurate description, the term *nonparametric* is more commonly used. The following are major advantages and disadvantages of nonparametric methods.

Advantages of Nonparametric Methods

1. Nonparametric methods can be applied to a wide variety of situations because they do not have the more rigid requirements of the corresponding parametric methods. In particular, nonparametric methods do not require normally distributed populations.

2. Unlike parametric methods, nonparametric methods can often be applied to categorical data, such as the genders of survey respondents.

3. Nonparametric methods usually involve simpler computations than the corresponding parametric methods and are therefore easier to understand and apply.

Disadvantages of Nonparametric Methods

1. Nonparametric methods tend to waste information because exact numerical data are often reduced to a qualitative form. For example, in the nonparametric sign test (described in Section 12-2), weight losses by dieters are recorded simply as negative signs; the actual magnitudes of the weight losses are ignored.

2. Nonparametric tests are not as efficient as parametric tests, so with a nonparametric test we generally need stronger evidence (such as a larger sample or greater differences) in order to reject a null hypothesis.

When the requirements of population distributions are satisfied, nonparametric tests are generally less efficient than their parametric counterparts, but the reduced efficiency can be compensated for by an increased sample size. For example, Section 12-6 will present a concept called *rank correlation,* which has an efficiency rating of 0.91 when compared to the linear correlation presented in Chapter 9. This means that with all other things being equal, nonparametric rank correlation requires 100 sample observations to achieve the same results as 91 sample observations analyzed through parametric linear correlation, assuming the stricter requirements for using the parametric method are met. Table 12-1 lists the nonparametric methods covered in this chapter, along with the corresponding parametric approach and **efficiency** rating. Table 12-1 shows that several nonparametric tests have efficiency ratings above 0.90, so the lower efficiency might not be a critical factor in choosing between parametric and nonparametric methods. However, because parametric tests do have higher efficiency ratings than their nonparametric counterparts, it's generally better to use the parametric tests when their required assumptions are satisfied.

Ranks

Sections 12-3 through 12-6 use methods based on ranks, which we now describe.

Definition

Data are *sorted* when they are arranged according to some criterion, such as smallest to largest or best to worst. A **rank** is a number assigned to an individual sample item according to its order in the sorted list. The first item is assigned a rank of 1, the second item is assigned a rank of 2, and so on.

Note to Instructor

It sometimes becomes necessary to analyze data that are in the form of ranks. Parametric tests typically require that populations have normal distributions, but that requirement cannot be met when original data consist of ranks. Such data sets cannot be analyzed with parametric tests, and they require nonparametric methods, such as those discussed in this chapter.

Table 12-1	Efficiency: Comparison of Parametric and Nonparametric Tests		
Application	Parametric Test	Nonparametric Test	Efficiency Rating of Nonparametric Test with Normal Population
Matched pairs of sample data	*t* test or *z* test	Sign test	0.63
		Wilcoxon signed-ranks test	0.95
Two independent samples	*t* test or *z* test	Wilcoxon rank-sum test	0.95
Several independent samples	Analysis of variance (*F* test)	Kruskal-Wallis test	0.95
Correlation	Linear correlation	Rank correlation test	0.91
Randomness	No parametric test	Runs test	No basis for comparison

EXAMPLE The numbers 5, 3, 40, 10, and 12 can be sorted (arranged from lowest to highest) as 3, 5, 10, 12, and 40, and these numbers have ranks of 1, 2, 3, 4, and 5, respectively:

5	3	40	10	12	Original values
3	5	10	12	40	Values sorted (arranged in order)
↑	↑	↑	↑	↑	
1	2	3	4	5	Ranks

Handling ties in ranks: If a tie in ranks occurs, the usual procedure is to find the mean of the ranks involved and then assign this mean rank to each of the tied items, as in the following example.

Note to Instructor

This is one common method for dealing with ties in ranks. For other approaches, see Exercise 15 in Section 12-2.

EXAMPLE The numbers 3, 5, 5, 10, and 12 are given ranks of 1, 2.5, 2.5, 4, and 5, respectively. In this case, ranks 2 and 3 were tied, so we found the mean of 2 and 3 (which is 2.5) and assigned it to the values that created the tie:

3	5	5	10	12	Original values
↑	↑	↑	↑	↑	
1	2.5	2.5	4	5	Ranks

2 and 3 are tied

12-2 Sign Test

The main objective of this section is to understand the *sign test* procedure, which is among the easiest for nonparametric tests.

Note to Instructor

The sign test procedure is summarized in the flowchart of Figure 12-1. The sign test deals with sample data that are matched pairs. Matched pairs are also considered in Section 8-4, but the procedure used in that section requires that if the number of pairs of sample data is small ($n \leq 30$), then the population of differences in the paired values must be approximately *normally distributed*. This section has no such requirement of a normal or any other particular distribution.

> **Definition**
>
> The **sign test** is a nonparametric (distribution-free) test that uses plus and minus signs to test different claims, including:
>
> 1. Claims involving matched pairs of sample data
> 2. Claims involving nominal data
> 3. Claims about the median of a single population

Basic Concept of the Sign Test The basic idea underlying the sign test is to analyze the frequencies of the plus and minus signs to determine whether they are significantly different. For example, suppose that we test a treatment designed to lower blood pressure. If 100 subjects are treated and 51 of them experience lower blood pressure while the other 49 have increased blood pressure, common sense suggests that there is not sufficient evidence to say that the drug is effective, be-

FIGURE 12-1 **Sign Test Procedure**

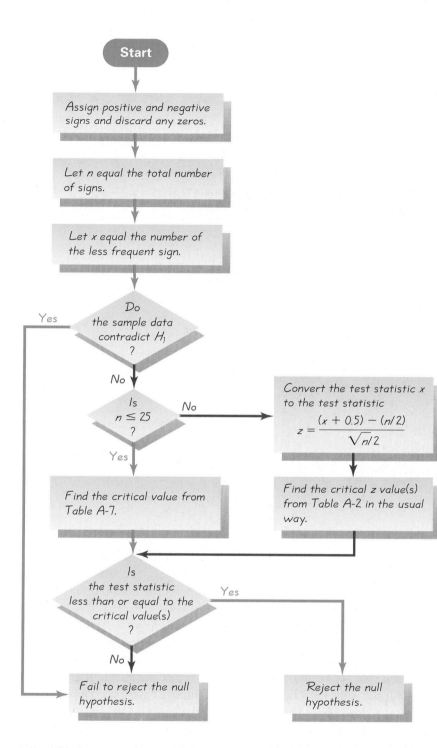

cause 51 decreases out of 100 is not significant. But what about 52 decreases and 48 increases? Or 90 decreases and 10 increases? The sign test allows us to determine when such results are significant.

For consistency and simplicity, we will use a test statistic based on the number of times that the less frequent sign occurs. The relevant assumptions, notation, test statistic, and critical values are summarized in the accompanying box. Figure 12-1 summarizes the sign test procedure, which will be illustrated with examples that follow.

Sign Test

Assumptions

1. The sample data have been randomly selected.
2. There is *no* requirement that the sample data come from a population with a particular distribution, such as a normal distribution.

Notation

x = the number of times the *less frequent* sign occurs

n = the total number of positive and negative signs combined

Test Statistic

For $n \leq 25$: x (the number of times the less frequent sign occurs)

For $n > 25$: $z = \dfrac{(x + 0.5) - \left(\dfrac{n}{2}\right)}{\dfrac{\sqrt{n}}{2}}$

Critical values

1. For $n \leq 25$, critical x values are found in Table A-7.
2. For $n > 25$, critical z values are found in Table A-2.

Note to Instructor

This is a very important point, and it should be made clearly. We should *never* apply procedures blindly without *thinking* about what we are doing. This text explains that in applying the sign test, it is important to determine whether the sense of the sample data conflicts with the alternative hypothesis. Whenever sample data oppose the alternative hypothesis, we can terminate the test by failing to reject the null hypothesis. For example, you need not formally test a claim that a coin favors heads if 100 tosses result in 48 heads. There is no way that we could ever support the claim of heads being favored if we get heads in fewer than 50% of the trials.

Caution: When applying the sign test in a one-tailed test, we need to be very careful to avoid making the wrong conclusion when one sign occurs significantly more often than the other, but the sample data contradict the alternative hypothesis. For example, suppose we are testing the claim that a gender selection technique favors boys, but we get a sample of 10 boys and 90 girls. With a sample proportion of boys equal to 0.10, the data contradict the alternative hypothesis H_1: $p > 0.5$. There is no way we can support a claim of $p > 0.5$ with any sample proportion less than 0.5, so we immediately fail to reject the null hypothesis and don't proceed with the sign test. Figure 12-1 summarizes the procedure for the sign test and includes this check: Do the sample data contradict H_1? If the sample data are in the opposite direction of H_1, fail to reject the null hypothesis. *It is always important to think about the data and to avoid relying on blind calculations or computer results.*

Claims Involving Matched Pairs

When using the sign test with data that are matched by pairs, we convert the raw data to plus and minus signs as follows:

1. We subtract each value of the second variable from the corresponding value of the first variable.

2. We record only the *sign* of the difference found in Step 1. We *exclude ties:* that is, we exclude any matched pairs in which both values are equal.

The key concept underlying this use of the sign test is this:

If the two sets of data have equal medians, the number of positive signs should be approximately equal to the number of negative signs.

EXAMPLE **Measuring Intelligence in Children** Mental measurements of young children are made by giving them blocks and telling them to build a tower as tall as possible. One experiment of block building was repeated a month later, with the times (in seconds) listed in Table 12-2 (based on data from "Tower Building," by Johnson and Courtney, *Child Development,* Vol. 3). Use a 0.05 significance level to test the claim that there is no difference between the times of the first and second trials.

SOLUTION   Here's the basic idea: If there is no difference between the times of the first trial and the times of the second trial, the numbers of positive and negative signs should be approximately equal. In Table 12-2 we have 12 positive signs and 2 negative signs. Are the numbers of positive and negative signs approximately equal, or are they significantly different? We follow the same basic steps for testing hypotheses as outlined in Figure 7-8, and we apply the sign test procedure summarized in Figure 12-1.

Steps 1, 2, 3:   The null hypothesis is the claim of no difference between the times of the first trial and the second trial, and the alternative hypothesis is the claim that there is a difference.

H_0:   There is no difference. (The median of the differences is equal to 0.)

H_1:   There is a difference. (The median of the differences is not equal to 0.)

Step 4:   The significance level is $\alpha = 0.05$.

Step 5:   We are using the nonparametric sign test.

continued

Table 12-2	Times For Building Towers of Blocks														
Child	A	B	C	D	E	F	G	H	I	J	K	L	M	N	O
First trial	30	19	19	23	29	178	42	20	12	39	14	81	17	31	52
Second trial	30	6	14	8	14	52	14	22	17	8	11	30	14	17	15
Sign of difference	0	+	+	+	+	+	+	−	−	+	+	+	+	+	+

Class Attendance and Grades

In a study of 424 undergraduates at the University of Michigan, it was found that students with the worst attendance records tended to get the lowest grades. (Is anybody surprised?) Those who were absent less than 10% of the time tended to receive grades of B or above. The study also showed that students who sit in the front of the class tend to get significantly better grades.

Step 6:   The test statistic x is the number of times the less frequent sign occurs. Table 12-2 includes differences with 12 positive signs and 2 negative signs, and we discard the one case with a difference of zero. We let x equal the smaller of 12 and 2, so $x = 2$. Also, $n = 14$ (the total number of positive and negative signs combined). Our test is two-tailed with $\alpha = 0.05$. We refer to Table A-7 where the critical value of 2 is found for $n = 14$ and $\alpha = 0.05$ in two tails. (See Figure 12-1.)

Step 7:   With a test statistic of $x = 2$ and a critical value of 2, we reject the null hypothesis of no difference. [See Note 2 included with Table A-7: "The null hypothesis is rejected if the number of the less frequent sign (x) is less than or equal to the value in the table." Because $x = 2$ is less than or equal to the critical value of 2, we reject the null hypothesis.]

Step 8:   There is sufficient evidence to warrant rejection of the claim that the median of the differences is equal to 0; that is, there is sufficient evidence to warrant rejection of the claim that there is no difference between the times of the first trial and the times of the second trial. This is the same conclusion that would be reached using the parametric t test with matched pairs in Section 8-4, but sign test results do not always agree with parametric test results.

Claims Involving Nominal Data

Recall that nominal data consist of names, labels, or categories only. Although such a nominal data set limits the calculations that are possible, we can identify the *proportion* of the sample data that belong to a particular category, and we can test claims about the corresponding population proportion p. The following example uses nominal data consisting of genders (male/female). The sign test is used by representing men with positive ($+$) signs and women with negative ($-$) signs. (Those signs are chosen arbitrarily, honest.) Also note the procedure for handling cases in which $n > 25$.

EXAMPLE **Gender Discrimination** The Hatters Restaurant Chain has been charged with discrimination based on gender because only 30 men were hired along with 70 women. A company official concedes that qualified applicants are about half men and half women, but she claims that "Hatters does not discriminate and the fact that 30 of the last 100 new employees are men is just a fluke." Use the sign test and a 0.05 significance level to test the null hypothesis that men and women are hired equally by this company.

SOLUTION   Let p denote the population proportion of hired men. The claim of no discrimination implies that the proportions of hired men and women are both equal to 0.5, so that $p = 0.5$. The null and alternative hypotheses can therefore be stated as follows:

H_0: $p = 0.5$ (the proportion of hired men is equal to 0.5)
H_1: $p \neq 0.5$

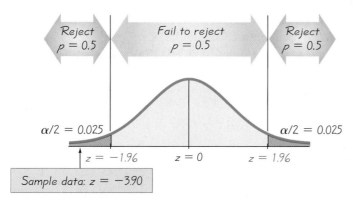

FIGURE 12-2 **Testing the Claim of Fair Hiring Practices**

Denoting hired men by + and hired women by −, we have 30 positive signs and 70 negative signs. Refer now to the sign test procedure summarized in Figure 12-1. The test statistic x is the smaller of 30 and 70, so $x = 30$. This test involves two tails because a disproportionately low number of either gender will cause us to reject the claim of equality. The sample data do not contradict the alternative hypothesis because 30 and 70 are not precisely equal. (That is, the sample data are consistent with the alternative hypothesis of a difference.) Continuing with the procedure in Figure 12-1, we note that the value of $n = 100$ is above 25, so the test statistic x is converted (using a correction for continuity) to the test statistic z as follows:

$$z = \frac{(x + 0.5) - \left(\dfrac{n}{2}\right)}{\dfrac{\sqrt{n}}{2}}$$

$$= \frac{(30 + 0.5) - \left(\dfrac{100}{2}\right)}{\dfrac{\sqrt{100}}{2}} = -3.90$$

With $\alpha = 0.05$ in a two-tailed test, the critical values are $z = \pm1.96$. The test statistic $z = -3.90$ is less than -1.96 (see Figure 12-2), so we reject the null hypothesis that the proportion of hired men is equal to 0.5. There is sufficient sample evidence to warrant rejection of the claim that the hiring practices are fair, with the proportions of hired men and women both equal to 0.5. This company appears to discriminate by not hiring equal proportions of men and women.

Claims About the Median of a Single Population

The next example illustrates the procedure for using the sign test in testing a claim about the median of a single population. See how the negative and positive signs are based on the claimed value of the median.

EXAMPLE Body Temperatures Data Set 4 in Appendix B includes measured body temperatures of adults. Use the 106 temperatures listed for 12 AM on day 2 with the sign test to test the claim that the median is less than 98.6°F. The data set has 106 subjects—68 subjects with temperatures below 98.6°F, 23 subjects with temperatures above 98.6°F, and 15 subjects with temperatures equal to 98.6°F.

SOLUTION The claim that the median is less than 98.6°F is the alternative hypothesis, while the null hypothesis is the claim that the median is equal to 98.6°F.

H_0: Median is equal to 98.6°F.   (median = 98.6°F)

H_1: Median is less than 98.6°F.   (median < 98.6°F)

Following the procedure outlined in Figure 12-1, we discard the 15 zeros, we use the negative sign ($-$) to denote each temperature that is below 98.6°F, and we use the positive sign ($+$) to denote each temperature that is above 98.6°F. We therefore have 68 negative signs and 23 positive signs, so $n = 91$ and $x = 23$ (the number of the less frequent sign). The sample data do not contradict the alternative hypothesis, because most of the 91 temperatures are below 98.6°F. (If the sample data did conflict with the alternative hypothesis, we could immediately terminate the test by concluding that we fail to reject the null hypothesis.) The value of n exceeds 25, so we convert the test statistic x to the test statistic z:

$$z = \frac{(x + 0.5) - \left(\frac{n}{2}\right)}{\frac{\sqrt{n}}{2}}$$

$$= \frac{(23 + 0.5) - \left(\frac{91}{2}\right)}{\frac{\sqrt{91}}{2}} = -4.61$$

In this one-tailed test with $\alpha = 0.05$, we use Table A-2 to get the critical z value of -1.645. From Figure 12-3 we can see that the test statistic of $z = -4.61$ does fall within the critical region. We therefore reject the null hypothesis. On the basis of the available sample evidence, we support the claim that the median body temperature of healthy adults is less than 98.6°F.

In this sign test of the claim that the median is below 98.6°F, we get a test statistic of $z = -4.61$ with a P-value of 0.00000202, but a parametric test of the claim that $\mu < 98.6$°F results in a test statistic of $t = -6.611$ with a P-value of 0.000000000813. Because the P-value from the sign test is not as low as the

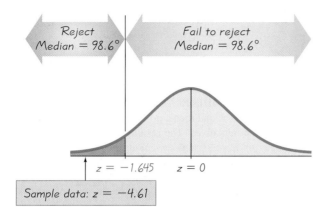

FIGURE 12-3 **Testing the Claim That the Median Is Less Than 98.6°F**

P-value from the parametric test, we see that the sign test isn't as sensitive as the parametric test. Both tests lead to rejection of the null hypothesis, but the sign test doesn't consider the sample data to be as extreme, partly because the sign test uses only information about the *direction* of the data, ignoring the *magnitudes* of the data values. The next section introduces the Wilcoxon signed-ranks test, which largely overcomes that disadvantage.

Rationale for the test statistic used when *n* > 25: When finding critical values for the sign test, we use Table A-7 only for *n* up to 25. When *n* > 25, the test statistic *z* is based on a normal approximation to the binomial probability distribution with $p = q = 1/2$. Recall that in Section 5-6 we saw that the normal approximation to the binomial distribution is acceptable when both $np \geq 5$ and $nq \geq 5$. Recall also that in Section 4-4 we saw that $\mu = np$ and $\sigma = \sqrt{npq}$ for binomial probability distributions. Because this sign test assumes that $p = q = 1/2$, we meet the $np \geq 5$ and $nq \geq 5$ prerequisites whenever $n \geq 10$. Also, with the assumption that $p = q = 1/2$, we get $\mu = np = n/2$ and $\sqrt{npq} = \sqrt{n/4} = \sqrt{n}/2$, so

$$z = \frac{x - \mu}{\sigma}$$

becomes

$$z = \frac{x - \left(\dfrac{n}{2}\right)}{\dfrac{\sqrt{n}}{2}}$$

Finally, we replace *x* by *x* + 0.5 as a correction for continuity. That is, the values of *x* are discrete, but because we are using a continuous probability distribution, a discrete value such as 10 is actually represented by the interval from 9.5 to 10.5. Because *x* represents the less frequent sign, we act conservatively by concerning ourselves only with *x* + 0.5; we thus get the test statistic *z*, as given in the equation and in Figure 12-1.

STATDISK Select **Analysis** from the main menu bar, then select **Sign Test.** Select the option **Given Number of Signs** if you know the number of plus and minus signs, or select **Given Pairs of Values** if the data are in the Data Window. After making the required entries in the dialog box, the displayed results will include the test statistic, critical value, and conclusion.

Minitab You must first create a column of values representing the differences between matched pairs of data or the number of plus and minus signs. (See the *Minitab Student Laboratory Manual and Workbook* for details.) Select **Stat,** then **Nonparametrics,** then **1-Sample Sign.** Click on the button for **Test Median.** Enter the median value and select the type of test, then click **OK.** Minitab will provide the *P*-value, so reject the null hypothesis if the *P*-value is less than or equal to the significance level. Otherwise, fail to reject the null hypothesis.

Excel Excel does not have a built-in function dedicated to the sign test, but you can use Excel's BINOMDIST function to find the *P*-value for a sign test. Click *fx* on the main menu bar, then select the function category **Statistical** and then

BINOMDIST. In the dialog box, first enter *x,* then the number of trials *n,* and then a probability of 0.5. Enter **TRUE** in the box for "cumulative." The resulting value is the probability of getting *x* or fewer successes among *n* trials. *Double this value for two-tailed tests.* The final result is the *P*-value, so reject the null hypothesis if the *P*-value is less than or equal to the significance level. Otherwise, fail to reject the null hypothesis.

TI-83 Plus The TI-83 Plus calculator does not have a built-in function dedicated to the sign test, but you can use the binomcdf function to find the *P*-value for a sign test. Press **2nd, VARS** (to get the **DISTR** menu); then scroll down to select **binomcdf.** Complete the entry of **binomcdf(n, p, x)** with *n* for the total number of plus and minus signs, 0.5 for *p,* and the number of the less frequent sign for *x.* Now press **ENTER,** and the result will be the probability of getting *x* or fewer successes among *n* trials. *Double this value for two-tailed tests.* The final result is the *P*-value, so reject the null hypothesis if the *P*-value is less than or equal to the significance level. Otherwise, fail to reject the null hypothesis.

Recommended Assignment:
Exercises 1–6, 9, 10, 12, 13.

12-2 Basic Skills and Concepts

In Exercises 1–4, assume that matched pairs of data result in the given number of signs when the value of the second variable is subtracted from the corresponding value of the first variable. Use the sign test with a 0.05 significance level to test the null hypothesis of no difference.

1. Positive signs: 10; negative signs: 5; ties: 3

2. Positive signs: 6; negative signs: 16; ties: 2

3. Positive signs: 50; negative signs: 40; ties: 5

4. Positive signs: 10; negative signs: 30; ties: 3

In Exercises 5–16, use the sign test.

5. Testing for a Difference Between Reported and Measured Male Heights As part of the National Health and Nutrition Examination Survey conducted by the Department of Health and Human Services, self-reported heights and measured heights were obtained for males aged 12–16. Listed below are sample results. Is there sufficient evidence to support the claim that there is a difference between self-reported heights and measured heights of males aged 12–16? Use a 0.05 significance level.

Reported height	68	71	63	70	71	60	65	64	54	63	66	72
Measured height	67.9	69.9	64.9	68.3	70.3	60.6	64.5	67.0	55.6	74.2	65.0	70.8

6. Testing for a Difference Between Reported and Measured Male Heights The table below lists matched pairs of measured heights of 12 male statistics students. Use a 0.05 significance level to test the claim that there is no difference between reported height and measured height.

Reported height	68	74	82.25	66.5	69	68	71	70	70	67	68	70
Measured height	66.8	73.9	74.3	66.1	67.2	67.9	69.4	69.9	68.6	67.9	67.6	68.8

7. Testing for a Median Body Temperature of 98.6°F A premed student in a statistics class is required to do a class project. Intrigued by the body temperatures in Data Set 4 Appendix B, she plans to collect her own sample data to test the claim that the mean body temperature is less than 98.6°F. Because of time constraints, she finds that she has time to collect data from only 12 people. After carefully planning a procedure for obtaining a simple random sample of 12 healthy adults, she measures their body temperatures and obtains the results listed below. Use a 0.05 significance level to test the claim that these body temperatures come from a population with a median that is less than 98.6°F.

97.6   97.5   98.6   98.2   98.0   99.0   98.5   98.1   98.4   97.9   97.9   97.7

8. Testing for Median Underweight The Prince County Bottling Company supplies bottles of lemonade labeled 12 oz. When the Prince County Department of Weights and Measures tests a random sample of bottles, the amounts listed below are obtained. Using a 0.05 significance level, is there sufficient evidence to file a charge that the bottling company is cheating consumers by giving amounts with a median less than 12 oz?

11.4   11.8   11.7   11.0   11.9   11.9   11.5   12.0   12.1   11.9   10.9   11.3   11.5   11.5   11.6

9. Nominal Data: Survey of Voters In a survey of 1002 people, 701 said that they voted in the recent presidential election (based on data from ICR Research Group). Is there sufficient evidence to support the claim that the majority of people say that they voted in the election?

10. Nominal Data: Smoking and Nicotine Patches In one study of 71 smokers who tried to quit smoking with nicotine patch therapy, 41 were smoking one year after the treatment (based on data from "High-Dose Nicotine Patch Therapy," by Dale et al., *Journal of the American Medical Association,* Vol. 274, No. 17). Use a 0.05 significance level to test the claim that among smokers who try to quit with nicotine patch therapy, the majority are smoking a year after the treatment.

11. Testing for Median Volume of Coke Cans Refer to Data Set 17 in Appendix B and use the volumes of regular Coke. Test the claim that cans of regular Coke have volumes with a median greater than 12 oz. Does it appear that the Coke cans are being filled correctly?

12. Testing for Median Amount of Domino Sugar in Packets Refer to Data Set 28 in Appendix B and use the sample data to test the claim that the median amount of sugar in the packets is equal to 3.5 oz. Does it appear that the sugar packets are being filled correctly?

6. $x = 1$ is less than or equal to the critical value of 2. Reject the claim of no difference.

7. $x = 1$ is less than or equal to the critical value of 2. Support the claim that the population has a median less than 98.6°F.

8. $x = 1$ is less than or equal to the critical value of 3. Support a claim that the company is cheating consumers.

9. Convert $x = 301$ to $z = -12.60$. CV: $z = -1.645$ (assuming a 0.05 significance level). Support the claim that the majority of people say that they voted in the election.

10. Convert $x = 30$ to $z = -1.19$. CV: $z = -1.645$ (assuming a 0.05 significance level). There is not sufficient evidence to support the claim that among smokers who try to quit with nicotine patch therapy, the majority are smoking a year after the treatment.

11. Convert $x = 1$ to $z = -5.32$. CV: $z = -1.645$ (assuming a 0.05 significance level). Support the claim that the Coke cans have volumes with a median greater than 12 oz.

12. Convert $x = 10$ to $z = -5.86$. CV: $z = \pm 1.96$ (assuming a 0.05 significance level). Reject the claim that the median amount equals 3.5 oz.

13. Testing for Median Time Interval for Old Faithful Geyser Refer to Data Set 13 in Appendix B. Test the claim that the intervals between eruptions of the Old Faithful geyser have a median greater than 77 min, which was the median about 30 years ago.

14. Testing for Difference Between Forecast and Actual Temperatures Refer to Data Set 10 in Appendix B and use the actual high temperatures and the three-day forecast high temperatures. Does there appear to be a difference?

12-2 Beyond the Basics

15. Procedures for Handling Ties In the sign test procedure described in this section, we excluded ties (represented by 0 instead of a sign of + or −). A second approach is to treat half of the 0s as positive signs and half as negative signs. (If the number of 0s is odd, exclude one so that they can be divided equally.) With a third approach, in two-tailed tests make half of the 0s positive and half negative; in one-tailed tests make all 0s either positive or negative, whichever supports the null hypothesis. Assume that in using the sign test on a claim that the median value is less than 100, we get 60 values below 100, 40 values above 100, and 21 values equal to 100. Identify the test statistic and conclusion for the three different ways of handling ties (with differences of 0). Assume a 0.05 significance level in all three cases.

16. Finding Critical Values Table A-7 lists critical values for limited choices of α. Use Table A-1 to add a new column in Table A-7 (down to $n = 15$) that represents a significance level of 0.03 in one tail or 0.06 in two tails. For any particular n, use $p = 0.5$, because the sign test requires the assumption that $P(\text{positive sign}) = P(\text{negative sign}) = 0.5$. The probability of x or fewer like signs is the sum of the probabilities for values up to and including x.

17. Normal Approximation Error The Compulife.com company has hired 18 women among the last 54 new employees. Job applicants are about half men and half women, all of whom are qualified. Using a 0.01 significance level with the sign test, is there sufficient evidence to charge bias? Does the conclusion change if the binomial distribution is used instead of the approximating normal distribution?

12-3 Wilcoxon Signed-Ranks Test for Matched Pairs

In Section 12-2 we used the sign test to analyze three different types of data, including sample data consisting of matched pairs. The sign test used only the signs of the differences and did not use their actual magnitudes (how large the numbers are). This section introduces the *Wilcoxon signed-ranks test,* which is also used with sample paired data. By using ranks, this test takes the magnitudes of the differences into account. (See Section 12-1 for a description of ranks.) Because the Wilcoxon signed-ranks test incorporates and uses more information than the sign test, it tends to yield conclusions that better reflect the true nature of the data.

Definition

The **Wilcoxon signed-ranks test** is a nonparametric test that uses ranks of sample data consisting of matched pairs. It is used to test for differences in the population distributions, so the null and alternative hypotheses are as follows:

H_0:   The two samples come from populations with the same distribution.

H_1:   The two samples come from populations with different distributions.

(The Wilcoxon signed-ranks test can also be used to test the claim that a sample comes from a population with a specified median. See Exercise 9 for this application.)

Matched pairs are also considered in Section 8-4, but the procedure used in that section requires that if the sample size is small, the population of differences in the paired values must be approximately *normally distributed*. This section does not require any particular distribution.

Wilcoxon Signed-Ranks Procedure

Step 1:   For each pair of data, find the difference d by subtracting the second value from the first value. Keep the signs, but discard any pairs for which $d = 0$.

Step 2:   *Ignore the signs of the differences,* then sort the differences from lowest to highest and replace the differences by the corresponding rank value (as described in Section 12-1). When differences have the same numerical value, assign to them the mean of the ranks involved in the tie.

Step 3:   Attach to each rank the sign of the difference from which it came. That is, insert those signs that were ignored in Step 2.

Step 4:   Find the sum of the absolute values of the negative ranks. Also find the sum of the positive ranks.

Step 5:   Let T be the *smaller* of the two sums found in Step 4. Either sum could be used, but for a simplified procedure we arbitrarily select the smaller of the two sums. (See the notation for T in the accompanying box.)

Step 6:   Let n be the number of pairs of data for which the difference d is not 0.

Step 7:   Determine the test statistic and critical values based on the sample size, as shown in the accompanying box.

Step 8:   When forming the conclusion, reject the null hypothesis if the sample data lead to a test statistic that is in the critical region—that is, the test statistic is less than or equal to the critical value(s). Otherwise, fail to reject the null hypothesis.

Note to Instructor

This procedure requires that you sort data, then assign ranks. When working with larger data sets, sorting and ranking becomes tedious, but software or a TI-83 Plus calculator can be used to automate that process. Stem-and-leaf plots can also be very helpful in sorting data.

Wilcoxon Signed-Ranks Test

Assumptions

1. The data consist of matched pairs that have been randomly selected.

2. The population of differences (found from the pairs of data) has a distribution that is approximately *symmetric,* meaning that the left half of its histogram is roughly a mirror image of its right half. (There is *no* requirement that the data have a normal distribution.)

continued

Notation

See the accompanying procedure steps for finding the rank sum T.

T = the smaller of the following two sums:

1. The sum of the absolute values of the negative ranks of the nonzero differences d
2. The sum of the positive ranks of the nonzero differences d

Test Statistic

If $n \le 30$, the test statistic is T.

If $n > 30$, the test statistic is $z = \dfrac{T - \dfrac{n(n+1)}{4}}{\sqrt{\dfrac{n(n+1)(2n+1)}{24}}}$

Critical values

1. If $n \le 30$, the critical T value is found in Table A-8.
2. If $n > 30$, the critical z values are found in Table A-2.

EXAMPLE Measuring Intelligence in Children The data in Table 12-3 are matched pairs of times (in seconds) obtained from a random sample of children who were given blocks and instructed to build a tower as tall as possible (based on data from "Tower Building," by Johnson and Courtney, *Child Development*, Vol. 3). This procedure is used to measure intelligence in children. Use the Wilcoxon signed-ranks test and a 0.05 significance level to test the claim that there is no difference between the times of the first and second trials.

SOLUTION   The null and alternative hypotheses are as follows:

H_0:   There is no difference between the times of the first and second trials.

H_1:   There is a difference between the times of the first and second trials.

Table 12-3   Times For Building Towers of Blocks

Child	A	B	C	D	E	F	G	H	I	J	K	L	M	N	O
First trial	30	19	19	23	29	178	42	20	12	39	14	81	17	31	52
Second trial	30	6	14	8	14	52	14	22	17	8	11	30	14	17	15
Differences d	0	13	5	15	15	126	28	−2	−5	31	3	51	3	14	37
Ranks of |differences|		6	4.5	8.5	8.5	14	10	1	4.5	11	2.5	13	2.5	7	12
Signed ranks		6	4.5	8.5	8.5	14	10	−1	−4.5	11	2.5	13	2.5	7	12

The significance level is $\alpha = 0.05$. We are using the Wilcoxon signed-ranks test procedure, so the test statistic is calculated by using the eight-step procedure presented earlier in this section.

Step 1:   In Table 12-3, the row of differences is obtained by computing this difference for each pair of data:

$$d = \text{time of first trial} - \text{time of second trial}$$

Step 2:   Ignoring their signs, we rank the absolute differences from lowest to highest. Note that ties in ranks are handled by assigning the mean of the involved ranks to each of the tied values, and the difference of 0 is discarded.

Step 3:   The bottom row of Table 12-3 is created by attaching to each rank the sign of the corresponding difference. If there really is no difference between the times of the first trial and the times of the second trial (as in the null hypothesis), we expect the number of positive ranks to be approximately equal to the number of negative ranks.

Step 4:   We now find the sum of the absolute values of the negative ranks, and we also find the sum of the positive ranks.

Sum of absolute values of negative ranks: 5.5

Sum of positive ranks: 99.5

Step 5:   Letting T be the smaller of the two sums found in Step 4, we find that $T = 5.5$.

Step 6:   Letting n be the number of pairs of data for which the difference d is not 0, we have $n = 14$.

Step 7:   Because $n = 14$, we have $n \leq 30$, so we use a test statistic of $T = 5.5$ (and we do not calculate a z test statistic). Also, because $n \leq 30$, we use Table A-8 to find the critical value of 21.

Step 8:   The test statistic $T = 5.5$ is less than or equal to the critical value of 21, so we reject the null hypothesis. It appears that there is a difference between times for the first trial and times for the second trial.

If we use the sign test with the preceding example, we will arrive at the same conclusion. Although the sign test and the Wilcoxon signed-ranks test agree in this particular case, there are other cases in which they do not agree.

Rationale: In this example the unsigned ranks of 1 through 14 have a total of 105, so if there are no significant differences, each of the two signed-rank totals should be around $105 \div 2$, or 52.5. That is, the negative ranks and positive ranks should split up as 52.5–52.5 or something close, such as 51–54. The table of critical values shows that at the 0.05 significance level with 14 pairs of data, a 21–84 split represents a significant departure from the null hypothesis, and any split that is farther apart (such as 20–85 or 5.5–99.5) will also represent a significant departure from the null hypothesis. Conversely, splits like 22–83 do not represent significant departures from a 52.5–52.5 split, and they would not justify rejecting the null hypothesis. The Wilcoxon signed-ranks test is based on the lower rank total, so instead of analyzing both numbers constituting the split, we consider only the lower number.

The sum $1 + 2 + 3 + \cdots + n$ of all the ranks is equal to $n(n + 1)/2$, and if this is a rank sum to be divided equally between two categories (positive and negative), each of the two totals should be near $n(n + 1)/4$, which is half of $n(n + 1)/2$. Recognition of this principle helps us understand the test statistic used when $n > 30$. The denominator in that expression represents a standard deviation of T and is based on the principle that

$$1^2 + 2^2 + 3^3 + \cdots + n^2 = \frac{n(n + 1)(2n + 1)}{6}$$

The Wilcoxon signed-ranks test can be used only for matched pairs of data. The next section will describe a rank-sum test that can be applied to two sets of independent data that are not matched in pairs.

Using Technology

STATDISK   Enter the paired data in columns of the Data Window. Select **Analysis** from the main menu bar, then select **Wilcoxon Tests.** Now select **Signed-Ranks Test,** and proceed to select the columns of data. Click on **Evaluate.**

Minitab   Enter the paired data in columns C1 and C2. Click on **Editor,** then **Enable Command Editor,** and enter the command **LET C3 = C1 − C2.** Press the **Enter** key. Select the options **Stat, Nonparametrics,** and **1-Sample Wilcoxon.** Enter C3 for the variable and click on the button for **Test Median.** The

Minitab display will include the *P*-value. Reject the null hypothesis of equal distributions if the *P*-value is less than or equal to the significance level. Fail to reject the null hypothesis if the *P*-value is greater than the significance level.

Excel   Excel is not programmed for the Wilcoxon signed-ranks test.

TI-83 Plus   The TI-83 Plus calculator is not programmed for the Wilcoxon signed-ranks test.

12-3 Basic Skills and Concepts

Recommended Assignment:
Exercises 1–6.

1. $T = 1$. CV: $T = 2$. Reject the null hypothesis that both samples come from the same population distribution.

2. $T = 6$. CV: $T = 4$. Fail to reject the null hypothesis that both samples come from the same population distribution.

Using the Wilcoxon Signed-Ranks Test. In Exercises 1 and 2, refer to the given paired sample data and use the Wilcoxon signed-ranks test to test the claim that both samples come from populations having the same distribution. Use a 0.05 significance level.

1.

x	12	14	17	19	20	27	29	30
y	12	15	15	14	12	18	19	20

2.

x	8	6	9	12	22	31	34	35	37
y	8	8	12	17	29	39	24	47	49

Using the Wilcoxon Signed-Ranks Test. In Exercises 3 and 4, refer to the sample data for the given exercises in Section 12-2. Instead of the sign test, use the Wilcoxon signed-ranks test to test the claim that both samples come from populations having the same distribution.

3. Exercise 5   **4.** Exercise 6

In Exercises 5–8, use the Wilcoxon signed-ranks test.

5. Testing for Difference Between Sitting Measurements and Lying Measurements In a study of techniques used to measure lung volumes, physiological data were collected for 10 subjects. The values given in the table are in liters and represent the measured functional residual capacities of the 10 subjects both in a sitting position and in a supine (lying) position. At the 0.05 significance level, test the claim that there is no significant difference between the measurements taken in the two positions.

Sitting	2.96	4.65	3.27	2.50	2.59	5.97	1.74	3.51	4.37	4.02
Supine	1.97	3.05	2.29	1.68	1.58	4.43	1.53	2.81	2.70	2.70

Based on "Validation of Esophageal Balloon Technique at Different Lung Volumes and Postures," by Baydur, Cha, and Sassoon, *Journal of Applied Physiology,* Vol. 62, No. 1.

6. Testing for Drug Effectiveness Captopril is a drug designed to lower systolic blood pressure. When subjects were tested with this drug, their systolic blood pressure readings (in mm of mercury) were measured before and after the drug was taken, with the results given in the accompanying table. Is there sufficient evidence to support the claim that the drug has an effect? Does captopril appear to lower systolic blood pressure?

Subject	A	B	C	D	E	F	G	H	I	J	K	L
Before	200	174	198	170	179	182	193	209	185	155	169	210
After	191	170	177	167	159	151	176	183	159	145	146	177

Based on data from "Essential Hypertension: Effect of an Oral Inhibitor of Angiotensin-Converting Enzyme," by MacGregor et al., *British Medical Journal,* Vol. 2.

7. Testing for Difference Between Forecast and Actual Temperatures Refer to Data Set 10 in Appendix B and use the actual high temperatures and the three-day forecast high temperatures. Does there appear to be a difference?

8. Testing for Difference Between Times Depicting Alcohol Use and Tobacco Use Refer to Data Set 7 in Appendix B. Use only those movies that showed some use of tobacco or alcohol. (That is, ignore those movies with times of zero for both tobacco use and alcohol use.) Does there appear to be a difference?

12-3 Beyond the Basics

9. Using the Wilcoxon Signed-Ranks Test for Claims About a Median The Wilcoxon signed-ranks test can be used to test the claim that a sample comes from a population with a specified median. The procedure used is the same as the one described in this section, except that the differences (Step 1) are obtained by subtracting the value of the hypothesized median from each value. Use the sample data consisting of the 106

3. $T = 34$. CV: $T = 14$. Fail to reject the null hypothesis that both samples come from the same population distribution.

4. $T = 6$. CV: $T = 14$. Reject the null hypothesis that both samples come from the same population distribution.

5. $T = 0$. CV: $T = 8$. Reject the claim that there is no difference.

6. $T = 0$. CV: $T = 14$ (assuming a 0.05 significance level). Support the claim that the drug has an effect.

7. $T = 178$. CV: $T = 117$ (assuming a 0.05 significance level). There does not appear to be a difference.

8. Convert $T = 207$ to $z = -1.55$. CV: $z = \pm 1.96$ (assuming a 0.05 significance level). There does not appear to be a difference.

9. Convert $T = 661$ to $z = -5.67$. CV: $z = \pm 1.96$. Reject the claim that healthy adults have a mean body temperature that is equal to 98.6°F.

body temperatures listed for 12 AM on day 2 in Data Set 4 in Appendix B. At the 0.05 significance level, test the claim that healthy adults have a median body temperature that is equal to 98.6°F.

12-4 Wilcoxon Rank-Sum Test for Two Independent Samples

Note to Instructor
Because this section deals with two independent samples, it could be covered along with Section 8-3. The test described in this section is equivalent to the Mann-Whitney U test described in some other textbooks.

This section introduces the *Wilcoxon rank-sum test,* which is a nonparametric test that two independent sets of sample data come from populations with the same distribution. Two samples are independent if the sample values selected from one population are not related or somehow matched or paired with the sample values from the other population. (To avoid confusion between the Wilcoxon rank-sum for independent samples and the Wilcoxon signed-ranks test for matched pairs, consider using the Internal Revenue Service for the mnemonic of IRS to remind us of "independent: rank sum".)

Definition

The **Wilcoxon rank-sum test** is a nonparametric test that uses ranks of sample data from two independent populations. It is used to test the null hypothesis that the two independent samples come from populations with the same distribution. (That is, the two populations are identical.) The alternative hypothesis is the claim that the two population distributions are different in some way.

H_0: The two samples come from populations with the same distribution. (That is, the two populations are identical.)

H_1: The two samples come from populations with different distributions. (That is, the two populations are different in some way.)

Basic Concept: The Wilcoxon rank-sum test is equivalent to the **Mann-Whitney U test** (see Exercise 11), which is included in some other textbooks and software packages (such as Minitab). The key idea underlying the Wilcoxon rank-sum test is this: If two samples are drawn from identical populations and the individual values are all ranked as one combined collection of values, then the high and low ranks should fall evenly between the two samples. If the low ranks are found predominantly in one sample and the high ranks are found predominantly in the other sample, we suspect that the two populations are not identical. This key idea is reflected in the following procedure for finding the value of the test statistic.

Procedure for Finding the Value of the Test Statistic

1. Temporarily combine the two samples into one big sample, then replace each sample value with its rank. (The lowest value gets a rank of 1, the next lowest value gets a rank of 2, and so on. If values are tied, assign to them the mean of

the ranks involved in the tie. See Section 12-1 for a description of ranks and the procedure for handling ties.)

2. Find the sum of the ranks for either one of the two samples.

3. Calculate the value of the z test statistic as shown in the following box, where either sample can be used as "Sample 1." (If testing the null hypothesis of identical populations and if both sample sizes are greater than 10, then the sampling distribution of R is approximately normal with mean μ_R and standard deviation σ_R, and the test statistic is as shown in the following box.)

Wilcoxon Rank-Sum Test

Assumptions

1. There are two independent samples of randomly selected data.

2. Each of the two samples has more than 10 values. (For samples with 10 or fewer values, special tables are available in reference books, such as *CRC Standard Probability and Statistics Tables and Formulae,* published by CRC Press.)

3. There is *no* requirement that the two populations have a normal distribution or any other particular distribution.

Notation

n_1 = size of Sample 1

n_2 = size of Sample 2

R_1 = sum of ranks for Sample 1

R_2 = sum of ranks for Sample 2

R = same as R_1 (sum of ranks for Sample 1)

μ_R = mean of the sample R values that is expected when the two populations are identical

σ_R = standard deviation of the sample R values that is expected when the two populations are identical

Test Statistic

$$z = \frac{R - \mu_R}{\sigma_R}$$

where

$$\mu_R = \frac{n_1(n_1 + n_2 + 1)}{2}$$

$$\sigma_R = \sqrt{\frac{n_1 n_2 (n_1 + n_2 + 1)}{12}}$$

n_1 = size of the sample from which the rank sum R is found

n_2 = size of the other sample

R = sum of ranks of the sample with size n_1

Critical values: Critical values can be found in Table A-2 (because the test statistic is based on the normal distribution).

Note to Instructor

As in the preceding sections of this chapter, there is no requirement of a normal distribution or any other particular distribution. This is why nonparametric methods are often called "distribution-free" methods.

Note that unlike the corresponding hypothesis tests in Section 8-3, the Wilcoxon rank-sum test does *not* require normally distributed populations. Also, the Wilcoxon rank-sum test can be used with data at the ordinal level of measurement, such as data consisting of ranks. In contrast, the parametric methods of Section 8-3 cannot be used with data at the ordinal level of measurement. In Table 12-1 we noted that the Wilcoxon rank-sum test has a 0.95 efficiency rating when compared with the parametric *t* test or *z* test. Because this test has such a high efficiency rating and involves easier calculations, it is often preferred over the parametric tests presented in Section 8-3, even when the requirement of normality is satisfied.

The expression for μ_R is based on the following result of mathematical induction: The sum of the first n positive integers is given by $1 + 2 + 3 + \cdots + n = n(n + 1)/2$. The expression for σ_R is based on a result stating that the integers $1, 2, 3, \ldots, n$ have standard deviation $\sqrt{(n^2 - 1)/12}$.

EXAMPLE Rowling and Tolstoy Data Set 14 in Appendix B includes the Flesch Reading Ease scores for randomly selected pages from each of two books: *Harry Potter and the Sorcerer's Stone* by J. K. Rowling and *War and Peace* by Leo Tolstoy. Table 12-4 includes values from Data Set 14 along with one additional value designed to better illustrate the Wilcoxon rank-sum procedure. (The value of 71.4 was added at the end of the Rowling list so that a tie would be created and the data sets would have different numbers of values.) Use the two sets of independent sample data in Table 12-4 with a 0.05 significance level to test the claim that reading scores for pages from the two books have the same distribution.

SOLUTION   The null and alternative hypotheses are as follows:

H_0: The Rowling and Tolstoy books have Flesch Reading Ease scores with the same distribution.

H_1: The two populations have distributions of Flesch Reading Ease scores that are different in some way.

Rank all 25 reading scores combined, beginning with a rank of 1 (assigned to the lowest value of 51.9). Ties in ranks are handled as described in Section 12-1: Find the mean of the ranks involved and assign this mean rank to each of the tied values. The 9th and 10th values are both 71.4, so assign the rank of 9.5 to each of those values. The ranks corresponding to the individual sample values are shown in parentheses in Table 12-4. *R* denotes the sum of the ranks for the sample we choose as Sample 1. If we choose the Rowling scores, we get

$$R = 24 + 22 + 18 + \cdots + 9.5 = 236.5$$

Because there are 13 Rowling values, we have $n_1 = 13$. Also, $n_2 = 12$ because there are 12 values for Tolstoy. We can now determine the values of μ_R, σ_R, and the test statistic z.

Table 12-4
Reading Scores

Rowling	Tolstoy
85.3 (24)	69.4 (7)
84.3 (22)	64.2 (4)
79.5 (18)	71.4 (9.5)
82.5 (20)	71.6 (11)
80.2 (19)	68.5 (6)
84.6 (23)	51.9 (1)
79.2 (17)	72.2 (12)
70.9 (8)	74.4 (15)
78.6 (16)	52.8 (2)
86.2 (25)	58.4 (3)
74.0 (14)	65.4 (5)
83.7 (21)	73.6 (13)
71.4 (9.5)	
$n_1 = 13$	$n_2 = 12$
$R_1 = 236.5$	$R_2 = 88.5$

$$\mu_R = \frac{n_1(n_1 + n_2 + 1)}{2} = \frac{13(13 + 12 + 1)}{2} = 169$$

$$\sigma_R = \sqrt{\frac{n_1 n_2 (n_1 + n_2 + 1)}{12}} = \sqrt{\frac{(13)(12)(13 + 12 + 1)}{12}} = 18.385$$

$$z = \frac{R - \mu_R}{\sigma_R} = \frac{236.5 - 169}{18.385} = 3.67$$

The test is two-tailed because a large positive value of z would indicate that the higher ranks are found disproportionately in the first sample, and a large negative value of z would indicate that the first sample had a disproportionate share of lower ranks. In either case, we would have strong evidence against the claim that the two samples come from populations with the same distribution.

The significance of the test statistic z can be treated in the same manner as in previous chapters. We are now testing (with $\alpha = 0.05$) the hypothesis that the two populations have the same distribution, so we have a two-tailed test with critical z values of 1.96 and -1.96. The test statistic of $z = 3.67$ does fall within the critical region, so we reject the null hypothesis that the Rowling and Tolstoy books have the same reading scores. It appears that the Rowling and Tolstoy pages come from populations with different distributions. Because the lower ranks appear to occur mostly in the Tolstoy values, it appears that Tolstoy has significantly lower reading ease scores, suggesting that Tolstoy's *War and Peace* is generally more difficult to read than J. K. Rowling's *Harry Potter and the Sorcerer's Stone*.

We can verify that if we interchange the two sets of sample values and consider the Tolstoy sample to be first, $R = 88.5$, $\mu_R = 156$, $\sigma_R = 18.385$, and $z = -3.67$, so the conclusion is exactly the same.

EXAMPLE Wednesday and Saturday Rainfall The Chapter Problem referred to the Boston rainfall amounts listed in Data Set 11 in Appendix B. The Chapter Problem included boxplots of the rainfall amounts for the seven days of the week, starting with Monday at the top. Comparisons of those boxplots show that Wednesday and Saturday appear to be the two days that differ most. But are those differences significant? Use the Wilcoxon rank-sum test to test the claim that the rainfall amounts for Wednesdays and Saturdays come from the same distribution.

SOLUTION   The null and alternative hypotheses are as follows:

H_0:   The Wednesday and Saturday rainfall amounts come from populations with the same distribution.

H_1:   The two distributions are different in some way.

Instead of manually calculating the rank sums, we refer to the Minitab display shown here. In that Minitab display, "ETA1" and "ETA2" denote the median

continued

Gender Gap in Drug Testing

A study of the relationship between heart attacks and doses of aspirin involved 22,000 male physicians. This study, like many others, excluded women. The General Accounting Office recently criticized the National Institutes of Health for not including both sexes in many studies because results of medical tests on males do not necessarily apply to females. For example, women's hearts are different from men's in many important ways. When forming conclusions based on sample results, we should be wary of an inference that extends to a population larger than the one from which the sample was drawn.

of the first sample and the median of the second sample, respectively. The display suggests that we are testing the null hypothesis of equal medians, but the Wilcoxon rank-sum test is based on the entire distributions, not just the medians. Here are the key components of the Minitab display: The rank sum for Wednesday is W = 2639.0, the P-value is 0.2773 (or 0.1992 after an adjustment for ties), and the conclusion is that we cannot reject (the null hypothesis) with a significance level of 0.05. Bottom line: The differences between Wednesday and Saturday are not significant. This seems to contradict the media reports that it rains more on weekends, but we will consider this issue more in the following section.

```
Minitab

WED       N = 53      Median =      0.0000
SAT       N = 52      Median =      0.0000
Point estimate for ETA1-ETA2 is      0.0000
95.1 Percent CI for ETA1-ETA2 is (0.0000,0.0000)
W = 2639.0
Test of ETA1 = ETA2  vs  ETA1 not = ETA2 is significant at 0.2773
The test is significant at 0.1992 (adjusted for ties)

Cannot reject at alpha = 0.05
```

Using Technology

STATDISK   Enter the sample data in columns of the Statdisk Data Window. Select **Analysis** from the main menu bar, then select **Wilcoxon Tests,** followed by the option **Rank-Sum Test.** Select the columns of data, then click on **Evaluate** to get a display that includes the rank sums, sample size, test statistic, critical value, and conclusion.

Minitab   First enter the two sets of sample data in columns C1 and C2. Then select the options **Stat, Nonparametrics,** and **Mann-Whitney,** and proceed to enter C1 for the first sample and C2 for the second sample. The confidence level of 95.0 corresponds to a significance level of $\alpha = 0.05$, and the "alternate: not equal" box refers to the alternative hypothesis, where "not equal" corresponds to a two-tailed hypothesis test. Minitab provides the P-value and conclusion. See the sample Minitab display included with the preceding example.

Excel   Excel is not programmed for the Wilcoxon rank-sum test.

TI-83 Plus   The TI-83 Plus calculator is not programmed for the Wilcoxon rank-sum test.

12-4 Basic Skills and Concepts

Exercises 1–4, 8.

1. $R_1 = 120.5$, $R_2 = 155.5$, $\mu_R = 132$, $\sigma_R = 16.248$, $z = -0.71$. CV: $z = \pm 1.96$. Fail to reject the null hypothesis that the populations have the same distribution.

Identifying Rank Sums. In Exercises 1 and 2, use a 0.05 significance level with the methods of this section to identify the rank sums R_1 and R_2, μ_R, σ_R, the test statistic z, the critical z values, and then state the conclusion.

1.

Sample 1 values:	1	3	4	6	8	12	15	16	17	22	26
Sample 2 values:	2	5	7	9	11	13	14	18	19	20	25	26

2. Sample 1 values: 1   3   4   6   8   12   15   16   17   22   26

   Sample 2 values: 22   25   28   33   34   35   37   39   41   43   45

Using the Wilcoxon Rank-Sum Test. In Exercises 3–10, use the Wilcoxon rank-sum test.

3. Are Severe Psychiatric Disorders Related to Biological Factors? One study used x-ray computed tomography (CT) to collect data on brain volumes for a group of patients with obsessive-compulsive disorders and a control group of healthy persons. The accompanying list shows sample results (in milliliters) for volumes of the right cordate (based on data from "Neuroanatomical Abnormalities in Obsessive-Compulsive Disorder Detected with Quantitative X-Ray Computed Tomography," by Luxenberg et al., *American Journal of Psychiatry,* Vol. 145, No. 9). Use a 0.01 significance level to test the claim that obsessive-compulsive patients and healthy persons have the same brain volumes. Based on this result, can we conclude that obsessive-compulsive disorders have a biological basis?

Obsessive-compulsive patients				Control group			
0.308	0.210	0.304	0.344	0.519	0.476	0.413	0.429
0.407	0.455	0.287	0.288	0.501	0.402	0.349	0.594
0.463	0.334	0.340	0.305	0.334	0.483	0.460	0.445

4. Testing the Anchoring Effect Randomly selected statistics students were given five seconds to estimate the value of a product of numbers with the results given in the accompanying table. (See the Cooperative Group Activities at the end of Chapter 2.) Is there sufficient evidence to support the claim that the two samples come from populations with different distributions?

Estimates from Students Given $1 \times 2 \times 3 \times 4 \times 5 \times 6 \times 7 \times 8$

1560	169	5635	25	842	40,320	5000	500	1110	10,000
200	1252	4000	2040	175	856	42,200	49,654	560	800

Estimates from Students Given $8 \times 7 \times 6 \times 5 \times 4 \times 3 \times 2 \times 1$

100,000	2000	42,000	1500	52,836	2050	428	372	300	225	64,582
23,410	500	1200	400	49,000	4000	1876	3600	354	750	640

5. Does the Arrangement of Test Items Affect the Score? The arrangement of test items was studied for its effect on anxiety. Sample results are listed below. Using a 0.05 significance level, test the claim that the two samples come from populations with the same scores. (The data are based on "Item Arrangement, Cognitive Entry Characteristics, Sex and Test Anxiety as Predictors of Achievement Examination Performance," by Klimko, *Journal of Experimental Education,* Vol. 52, No. 4.)

Easy to difficult				Difficult to easy			
24.64	39.29	16.32	32.83	33.62	34.02	26.63	30.26
28.02	33.31	20.60	21.13	35.91	26.68	29.49	35.32
26.69	28.90	26.43	24.23	27.24	32.34	29.34	33.53
7.10	32.86	21.06	28.89	27.62	42.91	30.20	32.54
28.71	31.73	30.02	21.96				
25.49	38.81	27.85	30.29				
30.72							

2. $R_1 = 68.5$, $R_2 = 184.5$, $\mu_R = 126.5$, $\sigma_R = 15.229$, $z = -3.81$. CV: $z = \pm 1.96$. Support the claim that the two samples come from populations with different distributions.

3. $\mu_R = 150$, $\sigma_R = 17.321$, $R = 96.5$, $z = -3.09$. CV: $z = \pm 2.575$. Reject the claim that the two samples come from populations with identical populations.

4. $\mu_R = 430.0$, $\sigma_R = 39.707$, $R = 411$, $z = -0.48$. CV: $z = \pm 1.96$ (assuming a 0.05 significance level). There is not sufficient evidence to support the claim that the two samples come from populations with different distributions.

5. $\mu_R = 525$, $\sigma_R = 37.417$, $R = 437$, $z = -2.35$. CV: $z = \pm 1.96$. There is sufficient evidence to warrant rejection of the claim that the two samples come from identical populations.

6. $\mu_R = 577.5$, $\sigma_R = 56.358$,
$R = 569$, $z = -0.15$.
CV: $z = \pm 1.96$. They appear to
be the same.

6. Testing Red and Brown M&Ms for Identical Populations Listed below are weights
(in grams) of M&M candies taken from Data Set 19 in Appendix B. Use a 0.05 significance level to test the claim that red and brown M&M plain candies have weights
with the same distribution. That is, test the claim that the populations of red and
brown M&M plain candies are identical.

Red: 0.870 0.933 0.952 0.908 0.911 0.908 0.913 0.983 0.920
        0.936 0.891 0.924 0.874 0.908 0.924 0.897 0.912 0.888
        0.872 0.898 0.882

Brown: 0.932 0.860 0.919 0.914 0.914 0.904 0.930 0.871 1.033
        0.955 0.876 0.856 0.866 0.858 0.988 0.936 0.930 0.923
        0.867 0.965 0.902 0.928 0.900 0.889 0.875 0.909 0.976
        0.921 0.898 0.897 0.902 0.920 0.909

7. $\mu_R = 150$, $\sigma_R = 17.321$,
$R = 86.5$, $z = -3.67$.
CV: $z = \pm 1.96$. Reject the null
hypothesis that the Rowling and
Tolstoy samples are from populations with the same distribution.

T 7. Readability of Rowling and Tolstoy An example from this section used the Flesch
Reading Ease scores for pages randomly selected from J. K. Rowling's *Harry Potter
and the Sorcerer's Stone* and Leo Tolstoy's *War and Peace*. (That example included
an additional sample value not listed in Appendix B.) Refer to Data Set 14 in Appendix B and use the Flesch-Kincaid Grade Level scores for the Rowling and Tolstoy
pages. Use a 0.05 significance level to test the claim that the two samples are from
populations with the same distribution.

8. $\mu_R = 5256$, $\sigma_R = 247.63$,
$R = 4827.5$, $z = -1.73$.
CV: $z = \pm 1.96$. Fail to reject
the null hypothesis that the
Bonds and McGwire samples
come from populations with the
same distribution.

T 8. Home Run Record Breakers Refer to the distances of the home runs hit by Barry
Bonds and Mark McGwire in Data Set 30 in Appendix B. Considering those distances
to be sample data, use a 0.05 significance level to test the claim that the Bonds and
McGwire samples come from populations with the same distribution.

9. $\mu_R = 3696$, $\sigma_R = 214.94$,
$R = 3861$, $z = 0.77$.
CV: $z = \pm 1.96$. Fail to reject
the null hypothesis that the two
populations of ages have the
same distribution.

T 9. Queen Mary Stowaways Refer to Data Set 15 in Appendix B and use a 0.05 significance level to test the claim that the ages of westbound stowaways and the ages of
eastbound stowaways come from populations with the same distribution.

10. $\mu_R = 1620$, $\sigma_R = 103.92$,
$R = 1727.5$, $z = 1.03$.
CV: $z = \pm 1.96$. Fail to reject
the null hypothesis that men
and women have BMI values
with the same distribution.

T 10. Body Mass Index Refer to Data Set 1 in Appendix B for the body mass index values
for men and women. Using a 0.05 significance level, test the claim that the two samples of BMI values come from populations with the same distribution.

11. $z = -3.67$; the test statistic is
the same number with opposite
sign.

12-4 Beyond the Basics

11. Using the Mann-Whitney U Test The Mann-Whitney U test is equivalent to the
Wilcoxon rank-sum test for independent samples in the sense that they both apply to
the same situations and always lead to the same conclusions. In the Mann-Whitney U
test we calculate

$$z = \frac{U - \frac{n_1 n_2}{2}}{\sqrt{\frac{n_1 n_2 (n_1 + n_2 + 1)}{12}}}$$

where

$$U = n_1 n_2 + \frac{n_1 (n_1 + 1)}{2} - R$$

Using the Rowling and Tolstoy readability measures listed in Table 12-4 in this section, find the z test statistic for the Mann-Whitney U test and compare it to the z test statistic of 3.67 that was found using the Wilcoxon rank-sum test.

12. Finding Critical Values Assume that we have two treatments (A and B) that produce quantitative results, and we have only two observations for treatment A and two observations for treatment B. We cannot use the test statistic given in this section because both sample sizes do not exceed 10.

	Rank			Rank sum for
1	2	3	4	treatment A
A	A	B	B	3

a. Complete the accompanying table by listing the five rows corresponding to the other five cases, and enter the corresponding rank sums for treatment A.

b. List the possible values of R, along with their corresponding probabilities. [Assume that the rows of the table from part (a) are equally likely.]

c. Is it possible, at the 0.10 significance level, to reject the null hypothesis that there is no difference between treatments A and B? Explain.

12. a. ABAB | 4
       ABBA | 5
       BBAA | 7
       BAAB | 5
       BABA | 6
   b. The R values of 3, 4, 5, 6, 7 have probabilities of 1/6, 1/6, 2/6, 1/6, 1/6.
   c. No.

12-5 Kruskal-Wallis Test

This section introduces the *Kruskal-Wallis test,* which is used to test the null hypothesis that three or more independent samples come from identical populations. In Section 11-2 we used one-way analysis of variance (ANOVA) to test the null hypothesis that three or more populations have the same mean, but ANOVA requires that all of the involved populations have normal distributions. The Kruskal-Wallis test does not require normal distributions.

Note to Instructor

This section deals with comparisons of three or more sets of sample data, so it could be covered along with Section 11-2 (ANOVA). The advantage of the Kruskal-Wallis test is that unlike the ANOVA methods of Section 11-2, there is no requirement that the populations have a normal distribution or any other particular distribution. If the class lacks sufficient time or energy to study analysis of variance, this section might be used as a quicker and easier alternative.

Definition

The **Kruskal-Wallis Test** (also called the *H* **test**) is a nonparametric test that uses ranks of sample data from three or more independent populations. It is used to test the null hypothesis that the independent samples come from populations with the same distribution; the alternative hypothesis is the claim that the population distributions are different in some way.

H_0:   The samples come from populations with the same distribution.

H_1:   The samples come from populations with different distributions.

In applying the Kruskal-Wallis test, we compute the *test statistic H, which has a distribution that can be approximated by the chi-square distribution as long as each sample has at least five observations.* When we use the chi-square distribution in this context, the number of degrees of freedom is $k - 1$, where k is the number of samples. (For a quick review of the key features of the chi-square distribution, see Section 6-5.)

Procedure for Finding the Value of the Test Statistic *H*

1. Temporarily combine all samples into one big sample and assign a rank to each sample value. (Sort the values from lowest to highest, and in cases of ties, assign to each observation the mean of the ranks involved.)

2. For each sample, find the sum of the ranks and find the sample size.

3. Calculate *H* by using the results of Step 2 and the notation and test statistic given in the following box.

Kruskal-Wallis Test

Assumptions

1. We have at least three independent samples, all of which are randomly selected.

2. Each sample has at least five observations. (If samples have fewer than five observations, refer to special tables of critical values, such as *CRC Standard Probability and Statistics Tables and Formulae*, published by CRC Press.)

3. There is *no* requirement that the populations have a normal distribution or any other particular distribution.

Notation

N = total number of observations in all samples combined

k = number of samples

R_1 = sum of ranks for Sample 1

n_1 = number of observations in Sample 1

For Sample 2, the sum of ranks is R_2 and the number of observations is n_2, and similar notation is used for the other samples.

Test Statistic

$$H = \frac{12}{N(N+1)} \left(\frac{R_1^2}{n_1} + \frac{R_2^2}{n_2} + \cdots + \frac{R_k^2}{n_k} \right) - 3(N+1)$$

Critical values

1. The test is *right-tailed.*

2. df = $k - 1$. (Because the test statistic *H* can be approximated by a chi-square distribution, use Table A-4 with $k - 1$ degrees of freedom, where *k* is the number of different samples.)

The test statistic *H* is basically a measure of the variance of the rank sums R_1, R_2, . . . , R_k. If the ranks are distributed evenly among the sample groups, then *H* should be a relatively small number. If the samples are very different, then the ranks will be excessively low in some groups and high in others, with the net effect that *H* will be large. Consequently, only large values of *H* lead to rejection of the null hypothesis that the samples come from identical populations. *The Kruskal-Wallis test is therefore a right-tailed test.*

EXAMPLE **Clancy, Rowling, Tolstoy** Data Set 14 in Appendix B includes data obtained from 12 randomly selected pages in each of three different books: Tom Clancy's *The Bear and the Dragon*, J. K. Rowling's *Harry Potter and the Sorcerer's Stone*, and Leo Tolstoy's *War and Peace*. The Flesch Reading Ease score was obtained for each of those pages, and the results are listed in Table 12-5. The Flesch Reading Ease scoring system results in higher scores for text that is easier to read. Low scores result from works that are difficult to read. In Section 11-2 we used analysis of variance to test the null hypothesis that the three samples of reading scores come from populations with the same mean. We will now use the Kruskal-Wallis test of the null hypothesis that the three samples come from populations with the same distribution.

Table 12-5	Readability Scores	
Clancy	Rowling	Tolstoy
58.2 (4)	85.3 (34)	69.4 (10.5)
73.4 (19)	84.3 (32)	64.2 (6)
73.1 (18)	79.5 (28)	71.4 (13)
64.4 (7)	82.5 (30)	71.6 (14)
72.7 (16)	80.2 (29)	68.5 (9)
89.2 (36)	84.6 (33)	51.9 (2)
43.9 (1)	79.2 (27)	72.2 (15)
76.3 (23)	70.9 (12)	74.4 (22)
76.4 (24)	78.6 (25)	52.8 (3)
78.9 (26)	86.2 (35)	58.4 (5)
69.4 (10.5)	74.0 (21)	65.4 (8)
72.9 (17)	83.7 (31)	73.6 (20)
$n_1 = 12$	$n_2 = 12$	$n_3 = 12$
$R_1 = 201.5$	$R_2 = 337$	$R_3 = 127.5$

SOLUTION   The null and alternative hypotheses are as follows:

H_0:   The populations of readability scores for pages from the three books are identical.

H_1:   The three populations are not identical.

In determining the value of the test statistic H, we must first rank all of the data. We begin with the lowest value of 43.9, which is assigned a rank of 1. Ranks are shown in parentheses with the original readability scores in Table 12-5. Next we find the sample size, n, and sum of ranks, R, for each sample,

continued

and those values are listed at the bottom of Table 12-5. Because the total number of observations is 36, we have $N = 36$. We can now evaluate the test statistic as follows:

$$H = \frac{12}{N(N + 1)}\left(\frac{R_1^2}{n_1} + \frac{R_2^2}{n_2} + \cdots + \frac{R_k^2}{n_k}\right) - 3(N + 1)$$

$$= \frac{12}{36(36 + 1)}\left(\frac{201.5^2}{12} + \frac{337^2}{12} + \frac{127.5^2}{12}\right) - 3(36 + 1)$$

$$= 16.949$$

Because each sample has at least five observations, the distribution of H is approximately a chi-square distribution with $k - 1$ degrees of freedom. The number of samples is $k = 3$, so we have $3 - 1 = 2$ degrees of freedom. Refer to Table A-4 to find the critical value of 5.991, which corresponds to 2 degrees of freedom and a 0.05 significance level (with an area of 0.05 in the right tail).

The test statistic $H = 16.949$ is in the critical region bounded by 5.991, so we reject the null hypothesis of identical populations. (In Section 11-2, we rejected the null hypothesis of equal means.)

INTERPRETATION There is sufficient evidence to support a conclusion that the populations of readability scores for pages from the three books are not identical. The books appear to have different readability scores. Examining the rank sums, we see that Tolstoy has the lowest rank sum, suggesting that his book is the most difficult to read. Rowling has the highest rank sum, suggesting that her book is the easiest of the three to read.

EXAMPLE **Rains More on Weekends?** In the Chapter Problem we noted that the media reported that it rains more on weekends all along America's eastern coast, from Maine to Florida. Data Set 11 in Appendix B includes rainfall amounts for a recent year in Boston. Using that data set, test the claim that the seven weekdays have distributions that are not all the same.

SOLUTION   Data Set 11 might seem like it could be analyzed using analysis of variance methods introduced in Section 11-2, but those methods require that the sample values come from populations having distributions that are approximately normal. The Chapter Problem includes a histogram for the Monday rainfall amounts, and it is obvious that this is not a normal distribution. The histograms for the other days of the week have the same basic shape as Monday. Because the data do not come from normal distributions, analysis of variance cannot be used and the Kruskal-Wallis test is an ideal alternative. Data Set 11 includes data for each of 365 days, so we are dealing with large data sets and manual calculations would be too cumbersome. We will use software instead. Shown below are the bottom two lines of the Minitab display. (See Exercise 11 for the correction to be used when there are many tied sample values.)

Minitab

```
H = 2.78  DF = 6  P = 0.836
H = 3.85  DF = 6  P = 0.697 (adjusted for ties)
```

We fail to reject the null hypothesis of identical distributions because the Minitab P-value is greater than a reasonable significance level of 0.05. STAT-DISK yields a test statistic of $H = 2.7806$, a critical value of 12.592, and includes the conclusion of "fail to reject the null hypothesis." There isn't enough evidence to support a claim that rainfall amounts on the seven weekdays have distributions that are not all the same. The rainfall amounts appear to be the same on the different days of the week.

INTERPRETATION Based on the Boston rainfall amounts, there does not appear to be evidence to support the claim that it rains more on weekends. So how did the newspaper, magazine, and television reports lead us to believe that it does rain more on weekends? The original study was conducted by Arizona State University scientists Randall S. Cerveny and Robert C. Balling. Can they be blamed for providing false information? No. The author contacted Randall Cerveny who stated that the original paper concerned rainfall *off the coast* of the Atlantic seaboard—not the rainfall associated with any particular city. Cerveny and Balling used satellite precipitation estimates and found that *areas in the ocean* and near the coast did get more rainfall on weekends, and they explain this phenomenon by its relationship with pollution coming from coastal regions. Their findings are interesting and significant. The media misinterpreted the Cerveny/Balling conclusions with reports implying that it rains more on weekends for those of us who live on land along the Atlantic coast. It's an interesting case of the media misinterpreting the study results.

Rationale: The test statistic H, as presented earlier, is the rank version of the test statistic F used in the analysis of variance discussed in Chapter 11. When we deal with ranks R instead of original values x, many components are predetermined. For example, the sum of all ranks can be expressed as $N(N + 1)/2$, where N is the total number of values in all samples combined. The expression

$$H = \frac{12}{N(N + 1)} \Sigma n_i (\overline{R}_i - \overline{\overline{R}})^2$$

where

$$\overline{R}_i = \frac{R_i}{n_i} \quad \overline{\overline{R}} = \frac{\Sigma R_i}{\Sigma n_i}$$

combines weighted variances of ranks to produce the test statistic H given here. This expression for H is algebraically equivalent to the expression for H given earlier as the test statistic. The earlier form of H (not the one given here) is easier to work with. In comparing the procedures of the parametric F test for analysis of variance and the nonparametric Kruskal-Wallis test, we see that in the absence of computer software, the Kruskal-Wallis test is much simpler to apply. We need not compute the sample variances and sample means. We do not require normal population distributions. Life becomes so much easier. However, the Kruskal-Wallis test is not as efficient as the F test, so it might require more dramatic differences for the null hypothesis to be rejected.

Using Technology

STATDISK Enter the data in columns of the Data Window. Select **Analysis** from the main menu bar, then select **Kruskal-Wallis Test** and proceed to select the columns of data. STATDISK will display the sum of the ranks for each sample, the H test statistic, the critical value, and the conclusion.

Minitab Refer to the *Minitab Student Laboratory Manual and Workbook* for the procedure required to use the options **Stat, Nonparametrics,** and **Kruskal-Wallis.** The basic idea is to list all of the sample data in one big column, with another column identifying the sample for the corresponding values. For the readability data of Table 12-5, enter the 36 scores in Minitab's column

C1; enter the 12 Clancy values, followed by the 12 Rowling values values, followed by the 12 Tolstoy values. In column C2, enter twelve 1s followed by twelve 2s followed by twelve 3s. Now select **Stat, Nonparametrics,** and **Kruskal-Wallis.** In the dialog box, enter C1 for response, C2 for factor, then click **OK.** The Minitab display includes the H test statistic and the P-value.

Excel Excel is not programmed for the Kruskal-Wallis test.

TI-83 Plus The TI-83 Plus calculator is not programmed for the Kruskal-Wallis test.

12-5 Basic Skills and Concepts

Recommended Assignment:
Exercises 1, 2, 3, 6, 8.

1. Yes. The *P*-value of 0.747 indicates that we fail to reject the null hypothesis that the three age categories have identical populations.

2. The *P*-value of 0.003 indicates that we should reject the null hypothesis that the different years have time intervals with identical populations. The behavior of Old Faithful appears to be changing over time.

3. $H = 1.1914$. CV: $\chi^2 = 7.815$. The given data do not provide sufficient evidence to conclude that heavier cars are safer in a crash.

4. $H = 14.749$. CV: $\chi^2 = 5.991$. There is sufficient evidence to warrant rejection of the claim that the voltage readings are the same for the three different types of day.

Interpreting Kruskal-Wallis Test Results. In Exercises 1 and 2, interpret the given Kruskal-Wallis test results and address the given question.

1. Marathon Times Running times for the males in the New York City marathon are listed in Data Set 8 in Appendix B. When those running times are partitioned into categories with ages of 21–29, 30–39, and 40 or older, the Minitab Kruskal-Wallis test results are as shown below. Do the running times for the different age groups appear to come from identical populations?

```
H = 0.58   DF = 2   P = 0.747
```

2. Is Old Faithful Changing Over Time? Twelve different time intervals (in minutes) between eruptions of the Old Faithful geyser were recorded for each of the years 1951, 1985, and 1996. (The data are from geologist Rick Hutchinson and the National Park Service.) When Minitab is used with the Kruskal-Wallis test, the results are as shown below. Do the different years have time intervals with identical populations? Does it appear that the eruption behavior of Old Faithful is changing over time?

Minitab
```
H = 11.93   DF = 2   P = 0.003
```

Using the Kruskal-Wallis Test. In Exercises 3–8, use the Kruskal-Wallis test.

3. Does the Weight of a Car Affect Head Injuries in a Crash? Data were obtained from car crash experiments conducted by the National Transportation Safety Administration. New cars were purchased and crashed into a fixed barrier at 35 mi/h, and measurements were recorded for the dummy in the driver's seat. Use the sample data listed below to test for differences in head injury measurements (in hic) among the four weight categories. Is there sufficient evidence to conclude that head injury measurements for the four car weight categories are not all the same? Do the data suggest that heavier cars are safer in a crash?

Subcompact:	681	428	917	898	420
Compact:	643	655	442	514	525
Midsize:	469	727	525	454	259
Full-size:	384	656	602	687	360

4. Is Solar Energy the Same Every Day? A student of the author lives in a home with a solar electric system. At the same time each day, she collected voltage readings from a meter connected to the system and the results are listed in the accompanying table. Use a 0.05 significance level to test the claim that voltage readings are the same for the three different types of day. Is there sufficient evidence to support a claim of different population distributions? We might expect that a solar system would provide more electrical energy on sunny days than on cloudy or rainy days. Can we conclude that sunny days result in greater amounts of electrical energy?

5. Testing for Skull-Breadth Differences in Different Times The accompanying values are measured maximum breadths of male Egyptian skulls from different epochs (based on data from *Ancient Races of the Thebaid*, by Thomson and Randall-Maciver). Changes in head shape over time suggest that interbreeding occurred with immigrant populations. Use a 0.05 significance level to test the claim that the three samples come from identical populations. Is interbreeding of cultures suggested by the data?

6. Laboratory Testing of Flammability of Children's Sleepwear Flammability tests were conducted on children's sleepwear. The Vertical Semirestrained Test was used, in which pieces of fabric were burned under controlled conditions. After the burning stopped, the length of the charred portion was measured and recorded. Results are given in the margin for the same fabric tested at different laboratories. Because the same fabric was used, the different laboratories should have obtained the same results. Did they?

T 7. Do All Colors of M&Ms Weigh the Same? Refer to Data Set 19 in Appendix B. At the 0.05 significance level, test the claim that the weights of M&Ms are the same for each of the six different color populations. If it is the intent of Mars, Inc. to make the candies so that the different color populations are the same, do your results suggest that the company has a problem that requires corrective action?

T 8. Home Run Distances Refer to Data Set 30 in Appendix B. Consider the home run distances to be samples randomly selected from populations. Use a 0.05 significance level to test the claim that the populations of distances of home runs hit by Barry Bonds, Mark McGwire, and Sammy Sosa are identical.

5. $H = 6.631$. CV: $\chi^2 = 5.991$. There is sufficient evidence to warrant rejection of the claim that the three samples come from identical populations.

6. $H = 9.129$. CV: $\chi^2 = 9.488$ (assuming a 0.05 significance level). There is not sufficient evidence to warrant rejection of the claim that the different laboratories obtain the same results.

Data for Exercise 4

Sunny	Cloudy	Rainy
13.5	12.7	12.1
13.0	12.5	12.2
13.2	12.6	12.3
13.9	12.7	11.9
13.8	13.0	11.6
14.0	13.0	12.2

Data for Exercise 5

4000 B.C.	1850 B.C.	150 A.D.
131	129	128
138	134	138
125	136	136
129	137	139
132	137	141
135	129	142
132	136	137
134	138	145
138	134	137

Data for Exercise 6

Laboratory

1	2	3	4	5
2.9	2.7	3.3	3.3	4.1
3.1	3.4	3.3	3.2	4.1
3.1	3.6	3.5	3.4	3.7
3.7	3.2	3.5	2.7	4.2
3.1	4.0	2.8	2.7	3.1
4.2	4.1	2.8	3.3	3.5
3.7	3.8	3.2	2.9	2.8
3.9	3.8	2.8	3.2	
3.1	4.3	3.8	2.9	
3.0	3.4	3.5		
2.9	3.3			

12-5 Beyond the Basics

9. Testing the Effect of Transforming the Sample Data
   a. In general, how is the value of the test statistic H affected if a constant is added to (or subtracted from) each sample value?
   b. In general, how is the value of the test statistic H affected if each sample value is multiplied (or divided) by a positive constant?
   c. In general, how is the value of the test statistic H affected if a single sample value is changed to become an outlier?

10. Finding Values of the Test Statistic For three samples, each of size 5, find the largest and smallest possible values of the test statistic H.

11. Correcting the H Test Statistic for Ties In using the Kruskal-Wallis test, there is a correction factor that should be applied whenever there are many ties: Divide H by

$$1 - \frac{\Sigma T}{N^3 - N}$$

For each group of tied observations in the combined set of all sample data, calculate $T = t^3 - t$, where t is the number of observations that are tied within the individual group. Find t for each group of tied values, then compute the value of T for each group, then add the T values to get ΣT. The total number of observations in all samples combined is N. Use this procedure to find the corrected value of H for Exercise 4. Does the corrected value of H differ substantially from the value found in Exercise 4?

12. Equivalent Tests Show that for the case of two samples, the Kruskal-Wallis test is equivalent to the Wilcoxon rank-sum test. This can be done by showing that for the case of two samples, the test statistic H equals the square of the test statistic z used in the Wilcoxon rank-sum test. Also note that with 1 degree of freedom, the critical values of χ^2 correspond to the square of the critical z score.

12-6 Rank Correlation

In this section we describe how the nonparametric method of rank correlation is used with paired data to test for an association between two variables. In Chapter 9 we used paired sample data to compute values for the linear correlation coefficient r, but in this section we use *ranks* as the basis for measuring the strength of the correlation between two variables.

Definition

The **rank correlation test** (or **Spearman's rank correlation test**) is a nonparametric test that uses ranks of sample data consisting of matched pairs. It is used to test for an association between two variables, so the null and alternative hypotheses are as follows (where ρ_s denotes the rank correlation coefficient for the entire population):

H_0: $\rho_s = 0$ (There is *no* correlation between the two variables.)
H_1: $\rho_s \neq 0$ (There is a correlation between the two variables.)

Advantages: Rank correlation has some distinct advantages over the parametric methods discussed in Chapter 9:

1. The nonparametric method of rank correlation can be used in a wider variety of circumstances than the parametric method of linear correlation. With rank correlation, we can analyze paired data that are ranks or can be converted to ranks. For example, if two judges rank 30 different gymnasts, we can use rank correlation, but not linear correlation. Unlike the parametric method of Chapter 9, the method of rank correlation does *not* require a normal distribution for any population.
2. Rank correlation can be used to detect some (not all) relationships that are not linear. (An example will be given later in this section.)

Disadvantage: A disadvantage of rank correlation is its efficiency rating of 0.91, as described in Section 12-1. This efficiency rating shows that with all other circumstances being equal, the nonparametric approach of rank correlation requires 100 pairs of sample data to achieve the same results as only 91 pairs of sample observations analyzed through the parametric approach, assuming that the stricter requirements of the parametric approach are met.

The assumptions, notation, test statistic, and critical values are summarized in the following box. We use the notation r_s for the rank correlation coefficient so that we don't confuse it with the linear correlation coefficient r. The subscript s has nothing to do with standard deviation; it is used in honor of Charles Spearman (1863–1945), who originated the rank correlation approach. In fact, r_s is often called **Spearman's rank correlation coefficient.** The rank correlation procedure is summarized in Figure 12-4.

Rank Correlation

Assumptions
1. The sample paired data have been randomly selected.
2. Unlike the parametric methods of Section 9-2, there is *no* requirement that the sample pairs of data have a bivariate normal distribution (as described in Section 9-2). There is *no* requirement of a normal distribution for any population.

Notation

r_s = rank correlation coefficient for sample paired data (r_s is a sample statistic)

ρ_s = rank correlation coefficient for all the population data (ρ_s is a population parameter)

n = number of pairs of sample data

d = difference between ranks for the two values within a pair

continued

Direct Link Between Smoking and Cancer

When we find a statistical correlation between two variables, we must be extremely careful to avoid the mistake of concluding that there is a cause-effect link. The tobacco industry has consistently emphasized that correlation does not imply causality. However, Dr. David Sidransky of Johns Hopkins University now says that "we have such strong molecular proof that we can take an individual cancer and potentially, based on the patterns of genetic change, determine whether cigarette smoking was the cause of that cancer." Based on his findings, he also said that "the smoker had a much higher incidence of the mutation, but the second thing that nailed it was the very distinct pattern of mutations . . . so we had the smoking gun." Although statistical methods cannot prove that smoking *causes* cancer, such proof can be established with physical evidence of the type described by Dr. Sidransky.

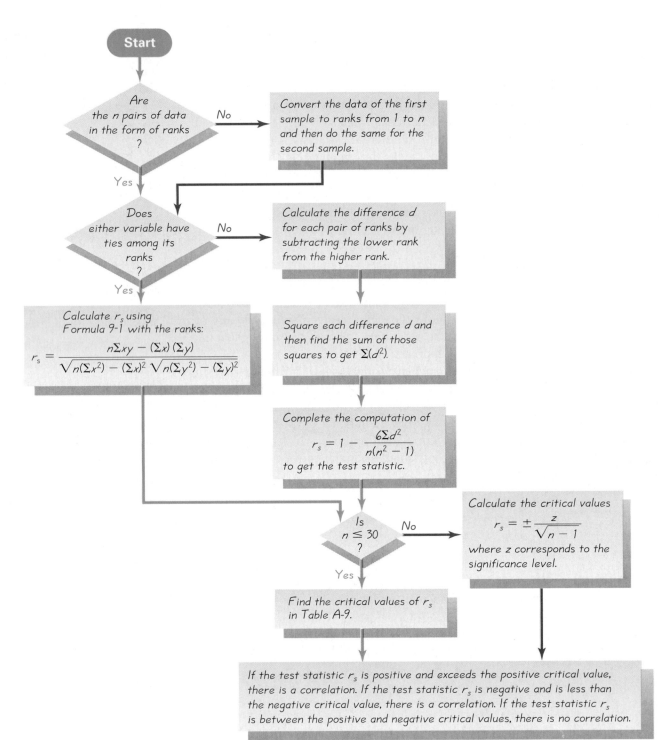

FIGURE 12-4 **Rank Correlation Procedure for Testing H_0: $\rho_s = 0$**

Test Statistic

No ties: After converting the data in each sample to ranks, if there are no ties among ranks for the first variable and there are no ties among ranks for the second variable, the exact value of the test statistic can be calculated using this formula:

$$r_s = 1 - \frac{6\Sigma d^2}{n(n^2 - 1)}$$

Ties: After converting the data in each sample to ranks, if either variable has ties among its ranks, the exact value of the test statistic r_s can be found by using Formula 9-1 with the ranks:

$$r_s = \frac{n\Sigma xy - (\Sigma x)(\Sigma y)}{\sqrt{n(\Sigma x^2) - (\Sigma x)^2}\ \sqrt{n(\Sigma y^2) - (\Sigma y)^2}}$$

Critical values

1. If $n \leq 30$, critical values are found in Table A-9.
2. If $n > 30$, critical values of r_s are found by using Formula 12-1.

Formula 12-1 $r_s = \dfrac{\pm z}{\sqrt{n - 1}}$ (critical values when $n > 30$)

where the value of z corresponds to the significance level.

Note to Instructor
Caution the class that Tables A-6 and A-9 are somewhat similar and care should be taken to avoid confusion between them. Table A-6 is used with the methods of Section 9-2, but Table A-9 is used for the methods of this section.

EXAMPLE **Perceptions of Beauty** *marie claire* magazine asked men and women to rank the beauty of ten different women, all of whom are quite attractive. ("Are you often surprised by what men and women find attractive? We asked 100 men and 100 women to rank these great-looking faces and explain exactly what they find appealing.") Table 12-6 lists the resulting ranks. Is there a correlation between the rankings of the men and women? The magazine asked "Do men and women agree?" Do they? Use a significance level of $\alpha = 0.05$.

Table 12-6	Rankings of the Beauty of Ten Women									
Men	4	2	5	1	3	6	7	8	9	10
Women	2	6	7	3	1	10	4	8	5	9
d	2	4	2	2	2	4	3	0	4	1
d^2	4	16	4	4	4	16	9	0	16	1 → Total = 74

SOLUTION   The linear correlation coefficient r (Section 9-2) should not be used because it requires normal distributions, but the data consist of ranks, which are not normally distributed. Instead, we use the rank correlation coefficient to test for a relationship between the ranks of men and women.

The null and alternative hypotheses are as follows:

$$H_0: \quad \rho_s = 0$$
$$H_1: \quad \rho_s \neq 0$$

continued

Following the procedure of Figure 12-4, the data are in the form of ranks and neither of the two variables (men and women) has ties among ranks, so the exact value of the test statistic can be calculated as shown below. We use $n = 10$ (for 10 pairs of data) and $\Sigma d^2 = 74$ (as shown in Table 12-4) to get

$$r_s = 1 - \frac{6\Sigma d^2}{n(n^2 - 1)} = 1 - \frac{6(74)}{10(10^2 - 1)}$$

$$= 1 - \frac{444}{990} = 0.552$$

Now we refer to Table A-9 to determine that the critical values are ± 0.648 (based on $\alpha = 0.05$ and $n = 10$). Because the test statistic $r_s = 0.552$ does not exceed the critical value of 0.648, we fail to reject the null hypothesis. There is not sufficient evidence to support a claim of a correlation between the rankings of the men and women. It appears that when it comes to beauty, men and women do not agree. (If they did agree, there would be a significant correlation, but there is not.)

EXAMPLE **Large Sample Case** Assume that the preceding example is expanded by including a total of 40 women and that the test statistic r_s is found to be 0.291. If the significance level is $\alpha = 0.05$, what do you conclude about the correlation?

SOLUTION   Because there are 40 pairs of data, we have $n = 40$. Because n exceeds 30, we find the critical values from Formula 12-1 instead of Table A-9. With $\alpha = 0.05$ in two tails, we let $z = 1.96$ to get

$$r_s = \frac{\pm 1.96}{\sqrt{40 - 1}} = \pm 0.314$$

The test statistic of $r_s = 0.291$ does not exceed the critical value of 0.314, so we fail to reject the null hypothesis. There is not sufficient evidence to support the claim of a correlation between men and women.

The next example is intended to illustrate the principle that rank correlation can sometimes be used to detect relationships that are not linear.

EXAMPLE **Detecting a Nonlinear Pattern** A *Raiders of the Lost Ark* pinball machine (model L-7) is used to measure learning that results from repeating manual functions. Subjects were selected so that they are similar in important characteristics of age, gender, intelligence, education, and so on. Table 12-7 lists the numbers of games played and the last scores (in millions) for subjects randomly selected from the group with similar characteristics. We expect that there should be an association between the number of games played and the pinball score. Is there sufficient evidence to support the claim that there is such an association?

SOLUTION   We will test the null hypothesis of no rank correlation ($\rho_s = 0$).

$$H_0: \quad \rho_s = 0 \text{ (no correlation)}$$
$$H_1: \quad \rho_s \neq 0 \text{ (correlation)}$$

Table 12-7	Pinball Scores (Ranks in parentheses)							
Number of Games Played	9 (2)	13 (4)	21 (5)	6 (1)	52 (7)	78 (8)	33 (6)	11 (3)	120 (9)
Score	22 (2)	62 (4)	70 (6)	10 (1)	68 (5)	73 (8)	72 (7)	58 (3)	75 (9)
d	0	0	1	0	2	0	1	0	0
d^2	0	0	1	0	4	0	1	0	0

Refer to Figure 12-4, which we follow in this solution. The original scores are not ranks, so we converted them to ranks and entered the results in parentheses in Table 12-7. (Section 12-1 describes the procedure for converting scores into ranks.)

After expressing all data as ranks, we calculate the differences, d, and then square them. The sum of the d^2 values is 6. We now calculate

$$r_s = 1 - \frac{6\Sigma d^2}{n(n^2 - 1)} = 1 - \frac{6(6)}{9(9^2 - 1)}$$

$$= 1 - \frac{36}{720} = 0.950$$

Proceeding with Figure 12-4, we have $n = 9$, so we answer yes when asked if $n \leq 30$. We use Table A-9 to get the critical values of ± 0.683. Finally, the sample statistic of 0.950 exceeds 0.683, so we conclude that there is significant correlation. Higher numbers of games played appear to be associated with higher scores. Subjects appeared to better learn the game by playing more.

In the preceding example, if we compute the linear correlation coefficient r (using Formula 9-1) for the original data, we get $r = 0.586$, which leads to the conclusion that there is not enough evidence to support the claim of a significant linear correlation at the 0.05 significance level. If we examine the Excel scatter diagram, we can see that the pattern of points is not a straight-line pattern. This last example illustrates an advantage of the nonparametric approach over the parametric approach: With rank correlation, we can sometimes detect relationships that are not linear.

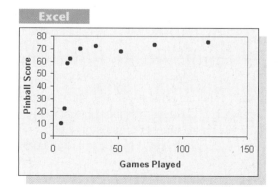

Using Technology

STATDISK   Enter the data in columns of the Statdisk Data Window. Select **Analysis** from the main menu bar, then select **Rank Correlation.** Select the columns of data, then click **Evaluate.** The STATDISK results include the exact value of the test statistic r_s, the critical value, and the conclusion.

Minitab   Enter the paired data in columns C1 and C2. If the data are not already ranks, use Minitab's **Data** (or **Manip**) and **Rank** options to convert the data to ranks, then select **Stat,** followed by **Basic Statistics,** followed by **Correlation.** Minitab will display the exact value of the test statistic r_s. Although Minitab identifies it as the Pearson correlation coefficient described in Section 9-2, it is actually the Spearman correlation coefficient described in this section (because it is based on ranks).

Excel   Excel does not have a function that calculates the rank correlation coefficient from original sample values, but the exact value of the test statistic r_s can be found as follows.

First replace each of the original sample values by its corresponding rank. Enter those ranks in columns A and B. Click on the *fx* function key located on the main menu bar. Select the function category **Statistical** and the function name **CORREL,** then click **OK.** In the dialog box, enter the cell range of values for *x*, such as A1:A10. Also enter the cell range of values for *y*, such as B1:B10. Excel will display the exact value of the rank correlation coefficient r_s.

TI-83 Plus   If using a TI-83 Plus calculator or any other calculator with 2-variable statistics, you can find the exact value of r_s as follows: (1) Replace each sample value by its corresponding rank, then (2) calculate the value of the linear correlation coefficient *r* with the same procedures used in Section 9-2. Enter the paired ranks in lists L1 and L2, then press **STAT** and select **TESTS.** Using the option **LinRegTTest** will result in several displayed values, including the exact value of the rank correlation coefficient r_s.

12-6 Basic Skills and Concepts

1. Finding the Test Statistic and Critical Value   For each of the following samples of paired ranks, sketch a scatter diagram, estimate the value of r_s, calculate the value of r_s, and state whether there appears to be a correlation between *x* and *y*.

a.

x	1	3	5	4	2
y	1	3	5	4	2

b.

x	1	2	3	4	5
y	5	4	3	2	1

c.

x	1	2	3	4	5
y	2	5	3	1	4

2. Finding Critical Values   Find the critical value(s) for r_s by using either Table A-9 or Formula 12-1, as appropriate. Assume two-tailed cases, where α represents the significance level and *n* represents the number of pairs of data.
   a. $n = 20, \alpha = 0.05$ **b.** $n = 50, \alpha = 0.05$
   c. $n = 40, \alpha = 0.02$ **d.** $n = 15, \alpha = 0.01$
   e. $n = 82, \alpha = 0.04$

Testing for Rank Correlation. In Exercises 3–12, use the rank correlation coefficient to test for a correlation between the two variables. Use a significance level of α = 0.05.

3. Correlation Between Salary and Stress The accompanying table lists salary rankings and stress rankings for randomly selected jobs (based on data from *The Jobs Rated Almanac*). Does it appear that salary increases as stress increases?

Job	Salary Rank	Stress Rank
Stockbroker	2	2
Zoologist	6	7
Electrical engineer	3	6
School principal	5	4
Hotel manager	7	5
Bank officer	10	8
Occupational safety inspector	9	9
Home economist	8	10
Psychologist	4	3
Airline pilot	1	1

3. $r_s = 0.855$. CV: $r_s = \pm0.648$. There appears to be a correlation between salary and stress.

4. Correlation Between Salary and Physical Demand Exercise 3 includes paired salary and stress level ranks for 10 randomly selected jobs. The physical demands of the jobs were also ranked; the salary and physical demand ranks are given below (based on data from *The Jobs Rated Almanac*). Does there appear to be a relationship between the salary of a job and its physical demands?

Salary	2	6	3	5	7	10	9	8	4	1
Physical demand	5	2	3	8	10	9	1	7	6	4

4. $r_s = 0.261$. CV: $r_s = \pm0.648$. No correlation between salary and physical demand.

5. Business School Rankings *Business Week* magazine ranked business schools two different ways. Corporate rankings were based on surveys of corporate recruiters, and graduate rankings were based on surveys of MBA graduates. The table below is based on the results for 10 schools. Is there a correlation between the corporate rankings and the graduate rankings? Use a significance level of α = 0.05.

School	PA	NW	Chi	Sfd	Hvd	MI	IN	Clb	UCLA	MIT
Corporate ranking	1	2	4	5	3	6	8	7	10	9
Graduate ranking	3	5	4	1	10	7	6	8	2	9

5. $r_s = 0.103$. CV: $r_s = \pm0.648$. No correlation between corporate and graduate rankings of business schools.

6. Correlation Between Restaurant Bills and Tips Students of the author collected sample data consisting of amounts of restaurant bills and the corresponding tip amounts. The data are listed below. Use rank correlation to determine whether there is a correlation between the amount of the bill and the amount of the tip.

Bill (dollars)	33.46	50.68	87.92	98.84	63.60	107.34
Tip (dollars)	5.50	5.00	8.08	17.00	12.00	16.00

6. $r_s = 0.829$. CV: $r_s = \pm0.886$. No correlation between the bill and the amount of tip.

7. Correlation Between Heights and Weights of Supermodels Listed below are heights (in inches) and weights (in pounds) for supermodels Niki Taylor, Nadia Avermann, Claudia Schiffer, Elle MacPherson, Christy Turlington, Bridget Hall, Kate Moss, Valerie Mazza, and Kristy Hume.

Height	71	70.5	71	72	70	70	66.5	70	71
Weight	125	119	128	128	119	127	105	123	115

7. $r_s = 0.557$. CV: $r_s = \pm0.683$. No correlation between height and weight.

8. $r_s = -0.228$. CV: $r_s = \pm 0.738$. No correlation between salary and the number of viewers.

8. Buying a TV Audience The *New York Post* published the annual salaries (in millions) and the number of viewers (in millions), with results given below for Oprah Winfrey, David Letterman, Jay Leno, Kelsey Grammer, Barbara Walters, Dan Rather, James Gandolfini, and Susan Lucci, repsectively. Is there a correlation between salary and number of viewers?

Salary	100	14	14	35.2	12	7	5	1
Viewers	7	4.4	5.9	1.6	10.4	9.6	8.9	4.2

9. $r_s = 0.506$. CV: $r_s = \pm 0.507$. No correlation between the amounts of fat and the calorie counts.

T **9.** Cereal Killers Refer to Data Set 16 in Appendix B and use the amounts of fat and the measured calorie counts. Is there a correlation?

10. $r_s = 0.520$. CV: $r_s = \pm 0.314$. Correlation between body mass index and cholesterol level.

T **10.** Cholesterol and Body Mass Index Refer to Data Set 1 in Appendix B and use the cholesterol levels and body mass index values of the 40 women. Is there a correlation between cholesterol level and body mass index?

11. a. $r_s = 0.918$. CV: $r_s = \pm 0.370$. Correlation between tar and nicotine.
   b. $r_s = 0.739$. CV: $r_s = \pm 0.370$. Correlation between carbon monoxide and nicotine.
   c. Tar, because it has a higher correlation with nicotine.

T **11.** Bad Stuff in Cigarettes Refer to Data Set 5 in Appendix B.
   a. Use the paired data consisting of tar and nicotine. Based on the result, does there appear to be a significant correlation between cigarette tar and nicotine? If so, can researchers reduce their laboratory expenses by measuring only one of these two variables?
   b. Use the paired data consisting of carbon monoxide and nicotine. Based on the result, does there appear to be a significant correlation between cigarette carbon monoxide and nicotine? If so, can researchers reduce their laboratory expenses by measuring only one of these two variables?
   c. Assume that researchers want to develop a method for predicting the amount of nicotine, and they want to measure only one other item. In choosing between tar and carbon monoxide, which is the better choice? Why?

12. a. $r_s = 0.584$. CV: $r_s = \pm 0.358$. Significant correlation.
   b. $r_s = 0.648$. CV: $r_s = \pm 0.358$. Significant correlation.
   c. One-day forecast temperatures should have a higher correlation with the actual temperatures. The results are as expected. A high correlation does not necessarily mean that the forecast temperatures are accurate.

T **12.** Forecasting Weather Refer to Data Set 10 in Appendix B.
   a. Use the five-day forecast high temperatures and the actual high temperatures. Is there a correlation? Does a significant correlation imply that the five-day forecast temperatures are accurate?
   b. Use the one-day forecast high temperatures and the actual high temperatures. Is there a correlation? Does a significant correlation imply that the one-day forecast temperatures are accurate?
   c. Which would you expect to have a higher correlation with the actual high temperatures: the five-day forecast high temperatures or the one-day forecast high temperatures? Are the results from parts (a) and (b) what you would expect? If there is a very high correlation between forecast temperatures and actual temperatures, does it follow that the forecast temperatures are accurate?

12-6 Beyond the Basics

13. a. ± 0.707
   b. ± 0.514
   c. ± 0.361
   d. ± 0.463
   e. ± 0.834

13. Finding Critical Values One alternative to using Table A-9 to find critical values is to compute them using this approximation:

$$r_s = \pm \sqrt{\frac{t^2}{t^2 + n - 2}}$$

Here t is the t score from Table A-3 corresponding to the significance level and $n - 2$ degrees of freedom. Apply this approximation to find critical values of r_s for the following cases.

a. $n = 8, \alpha = 0.05$ **b.** $n = 15, \alpha = 0.05$

c. $n = 30, \alpha = 0.05$ **d.** $n = 30, \alpha = 0.01$

e. $n = 8, \alpha = 0.01$

14. *Effect of Ties on* r_s Refer to Data Set 7 in Appendix B for the times (in seconds) of tobacco use and alcohol use depicted in animated children's movies. Calculate the value of the test statistic r_s by using each of the two formulas given in this section. Is there a substantial difference between the two results? Which result is better? Is the conclusion affected by the formula used?

14. First formula: 0.486. Second formula (Formula 9-1): 0.426. The difference is substantial. The value of 0.426 is better because it is exact (except for rounding), whereas 0.486 is not exact. In this case, the conclusion that there is a correlation is not affected by the formula used, but it could be in other cases.

12-7 Runs Test for Randomness

The main objective of this section is to introduce the runs test for randomness, which can be used to determine whether the sample data in a sequence are in a random order. The importance of randomness has been stressed throughout this book, and we now address one method for determining whether that characteristic is present.

Definitions

A **run** is a sequence of data having the same characteristic; the sequence is preceded and followed by data with a different characteristic or by no data at all.

The **runs test** uses the number of runs in a sequence of sample data to test for randomness in the order of the data.

Fundamental Principle of the Runs Test

The fundamental principle of the runs test can be briefly stated as follows:

Reject randomness if the number of runs is very low or very high.

- Example: The sequence of genders FFFFFMMMMM is not random because it has only 2 runs, so the number of runs is very *low*.

- Example: The sequence of genders FMFMFMFMFM is not random because there are 10 runs, which is very *high*.

The exact criteria for determining whether a number of runs is very high or low are found in the accompanying box, which summarizes the key elements of the runs test for randomness. The procedure for the runs test for randomness is also summarized in Figure 12-5.

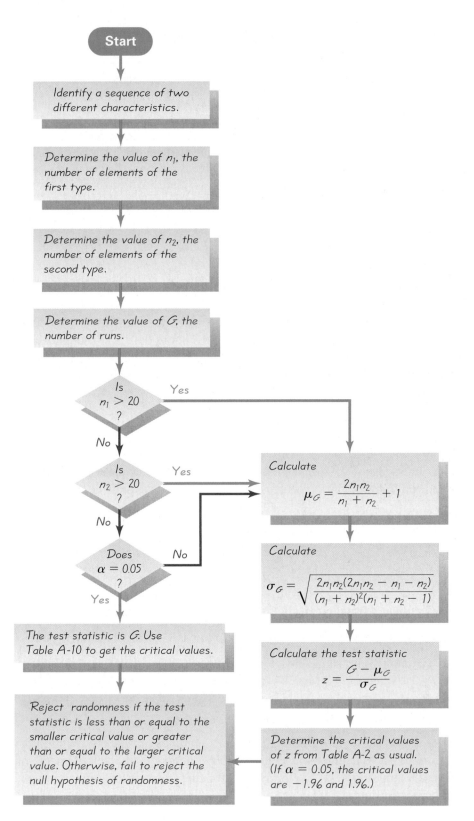

FIGURE 12-5 Runs Test for Randomness

Runs Test for Randomness

Assumptions

1. The sample data are arranged according to some ordering scheme, such as the order in which the sample values were obtained.

2. Each data value can be categorized into one of *two* separate categories.

3. The runs test for randomness is based on the *order* in which the data occur; it is *not* based on the *frequency* of the data. (For example, a sequence of 3 men and 20 women might appear to be random, but the issue of whether 3 men and 20 women constitute a *biased* sample is not addressed by the runs test.)

Notation

n_1 = number of elements in the sequence that have one particular characteristic. (The characteristic chosen for n_1 is arbitrary.)

n_2 = number of elements in the sequence that have the other characteristic

G = number of runs

Test Statistic

For Small Samples and $\alpha = 0.05$: If $n_1 \le 20$ and $n_2 \le 20$ and the significance level is $\alpha = 0.05$, the test statistic is the number of runs G. Critical values are found in Table A-10. Here is the decision criterion:

Reject randomness if the number of runs G is

- less than or equal to the smaller critical value found in Table A-10.

- or greater than or equal to the larger critical value found in Table A-10.

For Large Samples or $\alpha \ne 0.05$: If $n_1 > 20$ or $n_2 > 20$ or $\alpha \ne 0.05$, use the following test statistic and critical values.

Test statistic: $z = \dfrac{G - \mu_G}{\sigma_G}$

where $\mu_G = \dfrac{2n_1 n_2}{n_1 + n_2} + 1$

and $\sigma_G = \sqrt{\dfrac{(2n_1 n_2)(2n_1 n_2 - n_1 - n_2)}{(n_1 + n_2)^2 (n_1 + n_2 - 1)}}$

Critical values of z: Use Table A-2.

Sports Hot Streaks

It is a common belief that athletes often have "hot streaks"—that is, brief periods of extraordinary success. Stanford University psychologist Amos Tversky and other researchers used statistics to analyze the thousands of shots taken by the Philadelphia 76ers for one full season and half of another. They found that the number of "hot streaks" was no different than you would expect from random trials with the outcome of each trial independent of any preceding results. That is, the probability of a hit doesn't depend on the preceding hit or miss.

EXAMPLE Small Samples: Basketball Foul Shots In the course of a game, WNBA player Cynthia Cooper shoots 12 free throws. Denoting shots made by H (for "hit") and denoting missed shots by M, her results are as follows: H, H, H, M, H, H, H, H, M, M, M, H. Use a 0.05 significance level to test for randomness in the sequence of hits and misses.

continued

SOLUTION   Refer to the procedure summarized in Figure 12-5. The sequence of two characteristics (hit and miss) has been identified. We must now find the values of n_1, n_2, and the number of runs G. The sequence is shown below with spacing used to better identify the separate runs.

$$\underbrace{\text{HHH}}_{\text{1st run}} \quad \underbrace{\text{M}}_{\text{2nd run}} \quad \underbrace{\text{HHHH}}_{\text{3rd run}} \quad \underbrace{\text{MMM}}_{\text{4th run}} \quad \underbrace{\text{H}}_{\text{5th run}}$$

Because there are 8 hits and 4 misses and 5 runs, we have

$$n_1 = \text{number of shots made (H)} = 8$$
$$n_2 = \text{number of shots missed (M)} = 4$$
$$G = \text{number of runs} = 5$$

Because $n_1 \leq 20$ and $n_2 \leq 20$ and $\alpha = 0.05$, the test statistic is $G = 5$, and we refer to Table A-10 to find the critical values of 3 and 10. Because $G = 5$ is not less than or equal to 3, nor is it greater than or equal to 10, *we do not reject randomness*. There is not sufficient evidence to warrant rejection of the claim that the hits and misses occur randomly. It appears that the sequence of hits and misses is random.

EXAMPLE Large Samples: Boston Rainfall On Mondays Refer to the rainfall amounts for Boston as listed in Data Set 11 in Appendix B. Is there sufficient evidence to support the claim that rain on Mondays is not random? Use a 0.05 significance level.

SOLUTION   Let D (for dry) represent Mondays with no rain (indicated by values of 0.00), and let R represent Mondays with some rain (any value greater than 0.00). The 52 consecutive Mondays are represented by this sequence:

D D D D R D R D D R D D R D D D R D D R R R D D D D

R D R D R R R D R D D D R D D D R D R D D R D D D R

The null and alternative hypotheses are as follows:

H_0:   The sequence is random.

H_1:   The sequence is not random.

The test statistic is obtained by first finding the number of Ds, the number of Rs, and the number of runs. It's easy to examine the sequence to find that

$$n_1 = \text{number of Ds} = 33$$
$$n_2 = \text{number of Rs} = 19$$
$$G = \text{number of runs} = 30$$

As we follow Figure 12-5, we answer yes to the question "Is $n_1 > 20$?" We therefore need to evaluate the test statistic z given in the box summarizing the key elements of the runs test for randomness. We must first evaluate μ_G and σ_G. We get

$$\mu_G = \frac{2n_1 n_2}{n_1 + n_2} + 1 = \frac{2(33)(19)}{33 + 19} + 1 = 25.115$$

$$\sigma_G = \sqrt{\frac{(2n_1 n_2)(2n_1 n_2 - n_1 - n_2)}{(n_1 + n_2)^2(n_1 + n_2 - 1)}}$$

$$= \sqrt{\frac{(2)(33)(19)[2(33)(19) - 33 - 19]}{(33 + 19)^2(33 + 19 - 1)}} = 3.306$$

We can now find the test statistic:

$$z = \frac{G - \mu_G}{\sigma_G} = \frac{30 - 25.115}{3.306} = 1.48$$

Because the significance level is $\alpha = 0.05$ and we have a two-tailed test, the critical values are $z = -1.96$ and $z = 1.96$. The test statistic of $z = 1.48$ does not fall within the critical region, so we fail to reject the null hypothesis of randomness. The given sequence does appear to be random.

Numerical Data: Randomness Above and Below the Mean or Median

In each of the preceding examples, the data clearly fit into two categories, but we can also test for randomness in the way numerical data fluctuate above or below a mean or median. To test for randomness above and below the median, for example, use the sample data to find the value of the median, then replace each individual value with the letter A if it is *above* the median, and replace it with B if it is *below* the median. Delete any values that are equal to the median. It is helpful to write the As and Bs directly above the numbers they represent because this makes checking easier and also reduces the chance of having the wrong number of letters. After finding the sequence of A and B letters, we can proceed to apply the runs test as described. Economists use the runs test for randomness above and below the median in an attempt to identify trends or cycles. An upward economic trend would contain a predominance of Bs at the beginning and As at the end, so the number of runs would be small. A downward trend would have As dominating at the beginning and Bs at the end, with a low number of runs. A cyclical pattern would yield a sequence that systematically changes, so the number of runs would tend to be large. (See Exercise 11.)

Using Technology

STATDISK   STATDISK is programmed for the runs test but, because of the nature of the data, you must first determine the values of n_1, n_2, and the number of runs G. Select **Analysis** from the main menu bar, then select **Runs Test** and proceed to enter the required data in the dialog box. The STATDISK display will include the test statistic (G or z as appropriate), critical values, and conclusion.

Minitab   Minitab will do a runs test with a sequence of numerical data only, but see the *Minitab Student Laboratory Manual and Workbook* for ways to circumvent that constraint.

Enter numerical data in column C1, then select **Stat, Nonparametrics**, and **Runs Test.** In the dialog box, enter C1 for the variable, then either choose to test for randomness above and below the mean, or enter a value to be used. Click **OK.** The Minitab results include the number of runs and the *P*-value ("test is significant at . . .").

Excel   Excel is not programmed for the runs test for randomness.

TI-83 Plus   The TI-83 Plus calculator is not programmed for the runs test for randomness.

12-7 Basic Skills and Concepts

Recommended Assignment:
Exercises 1–8, 14.

1. $n_1 = 11$, $n_2 = 7$, $G = 3$, CV: 5, 14; reject randomness.

2. $n_1 = 10$, $n_2 = 6$, $G = 7$, CV: 4, 13; fail to reject randomness.

3. $n_1 = 10$, $n_2 = 10$, $G = 10$, CV: 6, 16; fail to reject randomness.

4. $n_1 = 10$, $n_2 = 10$, $G = 4$, CV: 6, 16; reject randomness.

5. $n_1 = 12$, $n_2 = 8$, $G = 10$, CV: 6, 16; fail to reject randomness.

6. $n_1 = 10$, $n_2 = 11$, $G = 10$, CV: 6, 17; fail to reject randomness.

Identifying Runs and Finding Critical Values. In Exercises 1–4, use the given sequence to determine the values of n_1, n_2, the number of runs G, and the critical values from Table A-10, and use those results to determine whether the sequence appears to be random.

1. H H H H M M M M M M M H H H H H H H

2. M M M M F F F M M M F F M M F M

3. A A B B A A B B A A B B A A B B A A B B

4. T T T T T F F F F F T T T T T F F F F F

Using the Runs Test for Randomness. In Exercises 5–12, use the runs test of this section to determine whether the given sequence is random. Use a significance level of $\alpha = 0.05$. (All data are listed in order by row.)

5. Randomness of Roulette Wheel Outcomes In conducting research for this book, the author recorded the outcomes of a roulette wheel in the Stardust Casino. (Yes, it was hard work, but somebody had to do it.) Test for randomness of odd (O) and even (E) numbers for the results given in the following sequence. What would a lack of randomness mean to the author? To the casino?

O O E E E E O O E O E O O O O O O E O E

6. Testing for Randomness of Survey Respondents When selecting subjects to be surveyed about Infograme's *Roller Coaster Tycoon* game, the subjects were selected in a sequence with the genders listed below. Does it appear that the subjects were randomly selected according to gender?

M M F F F M F M M M M M F F M M F F F F M F

7. Testing for Randomness in Dating Prospects Fred has had difficulty getting dates with women, so he is abandoning his strategy of careful selection and replacing it with a desperate strategy of random selection. In pursuing dates with randomly selected women, Fred finds that some of them are unavailable because they are married. Fred, who has an abundance of time for such activities, records and analyzes his observations. Given the results listed below (where M denotes married and S denotes single), what should Fred conclude about the randomness of the women he selects?

M M M M S S S S S S M M M M M S
S S M M M M M M M M M M S S S S

8. Testing for Randomness of Baseball World Series Victories Test the claim that the sequence of World Series wins by American League and National League teams is random. Given below are recent results with American and National league teams represented by A and N, respectively. What does the result suggest about the abilities of the two leagues?

A N A N A A A N N A A N N N N A A
A N A N A N A A A N A N A A A N

9. Testing for Randomness of Presidential Election Winners For a recent sequence of presidential elections, the political party of the winner is indicated by D for Democrat and R for Republican. Does it appear that we elect Democrat and Republican candidates in a sequence that is random?

R R D R D R R R R D D R R R D D
D D D R R D D R R D R R R D D R

10. Testing for Randomness of Selected Military Draft Dates Men were once drafted into the U.S. Army by using a process that was supposed to randomly select birthdays. Suppose the first few selections are as listed below. Test the sequence for randomness before and after the middle of the year.

| Nov. 27 | July 7 | Aug. 3 | Oct. 19 | Dec. 19 | Sept. 21 | May 3 |
| Mar. 5 | June 10 | May 15 | June 27 | Jan. 5 | | |

11. Stock Market: Testing for Randomness Above and Below the Median Trends in business and economics applications are often analyzed with the runs test. Data Set 25 in Appendix B lists the annual high points of the Dow-Jones Industrial Average for a recent sequence of years. First find the median of the values, then replace each value by A if it is above the median and B if it is below the median. Then apply the runs test to the resulting sequence of As and Bs. What does the result suggest about the stock market as an investment consideration? (Acts of terrorism and adverse economic conditions caused a dramatic drop in the DJIA in 2001.)

12. Testing for Randomness of Motor Vehicle Deaths Refer to Data Set 25 in Appendix B for the numbers of motor vehicle deaths in the United States for the preceding two decades. Test for randomness above and below the mean. Do the numbers of motor vehicle deaths appear to be random? If not, is there a trend? Can the trend be explained?

13. Large Sample: Testing for Randomness of Odd and Even Digits in Pi A *New York Times* article about the calculation of decimal places of π noted that "mathematicians are pretty sure that the digits of π are indistinguishable from any random sequence."

7. $n_1 = 19$, $n_2 = 13$, $G = 6$, CV: 10, 23; reject randomness.

8. $n_1 = 19$, $n_2 = 14$, $G = 20$, CV: 11, 23; fail to reject randomness.

9. $n_1 = 18$, $n_2 = 14$, $G = 15$, CV: 10, 23; fail to reject randomness.

10. The median corresponds to July 2; $n_1 = 6$, $n_2 = 6$, $G = 2$, CV: 3, 11; reject randomness.

11. $n_1 = 10$, $n_2 = 10$, $G = 2$, CV: 6, 16; reject randomness. Because the trend is upward, the stock market appears to be a good medium for investments.

12. $\bar{x} = 45{,}418$, $n_1 = 10$, $n_2 = 11$, $G = 4$, CV: 6, 17; reject randomness. There appears to be a downward trend, which is good. Possible reasons: Tougher drunk driving laws and better safety equipment on cars.

13. $n_1 = 49$, $n_2 = 51$, $G = 43$, $\mu_G = 50.98$, $\sigma_G = 4.9727$. TS: $z = -1.60$. CV: $z = \pm 1.96$. Fail to reject randomness.

14. $n_1 = 57$, $n_2 = 40$, $G = 54$, $\mu_G = 48.010$, $\sigma_G = 4.7467$. TS: $z = 1.26$. CV: $z = \pm 1.96$. Fail to reject randomness.

15. $n_1 = 111$, $n_2 = 39$, $G = 54$, $\mu_G = 58.720$, $\sigma_G = 4.6875$. TS: $z = -1.01$. CV: $z = \pm 1.96$. There is not sufficient evidence to support the claim that male runners tend to finish before the female runners.

16. $n_1 = 80$, $n_2 = 70$, $G = 68$, $\mu_G = 75.667$, $\sigma_G = 6.0758$. TS: $z = -1.26$. CV: $z = \pm 1.96$. There is not sufficient evidence to support the claim that younger runners tend to finish before the older runners.

17. Minimum is 2, maximum is 4. Critical values of 1 and 6 can never be realized so that the null hypothesis of randomness can never be rejected.

18. b. The 84 sequences yield two runs of 2, seven runs of 3, twenty runs of 4, twenty-five runs of 5, twenty runs of 6, and ten runs of 7.

c. With $P(2 \text{ runs}) = 2/84$, $P(3 \text{ runs}) = 7/84$, $P(4 \text{ runs}) = 20/84$, $P(5 \text{ runs}) = 25/84$, $P(6 \text{ runs}) = 20/84$, and $P(7 \text{ runs}) = 10/84$, the G values of 3, 4, 5, 6, 7 can easily occur by chance, whereas $G = 2$ is unlikely because $P(2 \text{ runs})$ is less than 0.025. The lower critical value of G is therefore 2, and there is no upper critical value that can be equaled or exceeded.

d. Critical value of $G = 2$ agrees with Table A-10. The table lists 8 as the upper critical value, but it is impossible to get 8 runs using the given elements.

Given below are the first 100 decimal places of π. Test for randomness of odd (O) and even (E) digits.

1 4 1 5 9 2 6 5 3 5 8 9 7 9 3 2 3 8 4 6 2 6 4 3 3 8 3 2 7 9 5 0 2 8 8 4 1 9 7 1
6 9 3 9 9 3 7 5 1 0 5 8 2 0 9 7 4 9 4 4 5 9 2 3 0 7 8 1 6 4 0 6 2 8 6 2 0 8 9 9
8 6 2 8 0 3 4 8 2 5 3 4 2 1 1 7 0 6 7 9

14. **Large Sample: Testing for Randomness of Baseball World Series Victories** Test the claim that the sequence of World Series wins by American League and National League teams is random. Given below are recent results, with American and National League teams represented by A and N, respectively.

A N A N N N A A A A N A A A A N A N N A A N N A A A A N A N
N A A A A A N A N A N A N A A A A A A A N N A N A N N A A N
N N A N A N A N A A A N N A A N N N N A A A N A N A N A A A
N A N A A A N

T 15. Large Sample: Testing for Randomness of Marathon Runners Refer to Data Set 8 in Appendix B for the random sample of runners who finished the New York City marathon. The runners are listed in the order in which they finished. Test the sequence of genders for randomness. Is there sufficient evidence to support the claim of a reporter who writes that the male runners tend to finish before the female runners?

T 16. Large Sample: Testing for Randomness of Marathon Runners Refer to Data Set 8 in Appendix B for the random sample of runners who finished the New York City marathon. The runners are listed in the order in which they finished. Test for randomness of ages above and below the mean age. Is there sufficient evidence to support the claim of a reporter who writes that the younger runners tend to finish before the older runners?

12-7 Beyond the Basics

17. **Finding Critical Numbers of Runs** Using the elements A, A, B, B, what is the minimum number of possible runs that can be arranged? What is the maximum number of runs? Now refer to Table A-10 to find the critical G values for $n_1 = n_2 = 2$. What do you conclude about this case?

18. **Finding Critical Values**
a. Using all of the elements A, A, A, B, B, B, B, B, B, list the 84 different possible sequences.
b. Find the number of runs for each of the 84 sequences.
c. Use the results from parts (a) and (b) to find your own critical values for G.
d. Compare your results to those given in Table A-10.

Review

In this chapter we examined six different nonparametric tests for analyzing sample data. Nonparametric tests are also called distribution-free tests because they do not require that the populations have a particular distribution, such as a normal distribution. However, nonparametric tests are not as efficient as parametric tests, so we generally need stronger evidence before we reject a null hypothesis.

Table 12-8 lists the nonparametric tests presented in this chapter, along with their functions. The table also lists the corresponding parametric tests.

Table 12-8	Summary of Nonparametric Tests	

Nonparametric Test	Function	Parametric Test
Sign test (Section 12-2)	Test for claimed value of average with one sample	z test or t test (Sections 7-4, 7-5)
	Test for differences between matched pairs	t test (Section 8-4)
	Test for claimed value of a proportion	z test (Section 7-3)
Wilcoxon signed-ranks test (Section 12-3)	Test for differences between matched pairs	t test (Section 8-4)
Wilcoxon rank-sum test (Section 12-4)	Test for difference between two independent samples	t test or z test (Section 8-3)
Kruskal-Wallis test (Section 12-5)	Test that more than two independent samples come from identical populations	Analysis of variance (Section 11-2)
Rank correlation (Section 12-6)	Test for relationship between two variables	Linear correlation (Section 9-2)
Runs test (Section 12-7)	Test for randomness of sample data	(No parametric test)

Review Exercises

Using Nonparametric Tests. In Exercises 1–8, use a 0.05 significance level with the indicated test. If no particular test is specified, use the appropriate nonparametric test from this chapter.

1. Testing Effectiveness of SAT Prep Courses Does it pay to take preparatory courses for standardized tests such as the SAT? Using a 0.05 significance level, test the claim that the Allan Preparation Course has no effect on SAT scores. Use the sign test with the sample data in the accompanying table (based on data from the College Board and "An Analysis of the Impact of Commercial Test Preparation Courses on SAT Scores," by Sesnowitz, Bernhardt, and Knain, *American Educational Research Journal,* Vol. 19, No. 3).

Subject	A	B	C	D	E	F	G	H	I	J
SAT score before course	700	840	830	860	840	690	830	1180	930	1070
SAT score after course	720	840	820	900	870	700	800	1200	950	1080

2. Testing Effectiveness of SAT Prep Courses Do Exercise 1 using the Wilcoxon signed-ranks test.

3. Testing for Gender Discrimination The Tektronics Internet Company claims that hiring is done without any gender bias. Among the last 66 new employees hired, 1/3 are women. Job applicants are about half men and half women, who are all qualified. Is there sufficient evidence to charge bias in favor of men? Use a 0.01 significance level, because we don't want to make such a serious charge unless there is very strong evidence.

1. $x = 2$ is not less than or equal to the critical value of 1. There is not sufficient evidence to reject the claim that the course has no effect.

2. $T = 9.5$. CV: $T = 6$. There is not sufficient evidence to reject the claim that the course has no effect.

3. Sign test: Convert $x = 22$ to $z = -2.58$. CV: $z = -2.33$. Support the claim of bias in favor of men.

4. Wilcoxon rank-sum test:
$\mu_R = 162$, $\sigma_R = 19.442$,
$R = 89.5$, $z = -3.73$.
CV: $z = \pm1.96$. Reject the
claim that beer drinkers and
liquor drinkers have the same
BAC levels.

5. Rank correlation: $r_s = -0.796$.
CV: $r_s = \pm0.648$. Correlation
between weight and highway
fuel consumption.

6. Runs test: $n_1 = 22$, $n_2 = 18$,
$G = 18$, $\mu_G = 20.8$,
$\sigma_G = 3.0894$. TS: $z = -0.91$.
CV: $z = \pm1.96$. Odd and even
digits appear to occur randomly.

7. Kruskal-Wallis test: $H = 4.234$.
CV: $\chi^2 = 7.815$. There is not
sufficient evidence to support
the claim that the injury measurements are not the same for
the four categories. There is not
sufficient evidence to support
the claim that heavier cars are
safer.

8. Rank correlation: $r_s = 0.190$.
CV: $r_s = \pm0.738$. There is not
sufficient evidence to support
the claim of a correlation between performance and price.
Buy the least expensive tapes.

Cumulative Review Exercises

1. a. $n_1 = 20$, $n_2 = 7$, $G = 15$,
CV: 6, 16; fail to reject
randomness.

b. $z = -2.50$. CV: $z = \pm1.96$.
Reject the null hypothesis
that the proportion of
women equals 0.5.

c. Convert $x = 7$ to $z = -2.31$.
CV: $z = \pm1.96$. Support the
claim that the proportion of
women is different from 0.5.

d. $0.0940 < p < 0.425$

e. The sequence appears to
be in a random order, but
the subjects appear to be
biased against women.
Further research should be

4. Do Beer Drinkers and Liquor Drinkers Have Different BAC Levels? The sample data in the following list show BAC (blood alcohol concentration) levels at arrest of randomly selected jail inmates who were convicted of DWI or DUI offenses. The data are categorized by the type of drink consumed (based on data from the U.S. Department of Justice). Test the claim that beer drinkers and liquor drinkers have the same BAC levels. Based on these results, do both groups seem equally dangerous, or is one group more dangerous than the other?

Beer				Liquor			
0.129	0.146	0.148	0.152	0.220	0.225	0.185	0.182
0.154	0.155	0.187	0.212	0.253	0.241	0.227	0.205
0.203	0.190	0.164	0.165	0.247	0.224	0.226	0.234
				0.190	0.257		

5. Correlation Between Car Weight and Fuel Consumption The accompanying table lists weights (in hundreds of pounds) and highway fuel consumption amounts (in mi/gal) for a sample of domestic new cars (based on data from the EPA). Based on the result, can you expect to pay more for gas if you buy a heavier car? How do the results change if the weights are entered as 2900, 3500, . . . , 2400?

x Weight	29	35	28	44	25	34	30	33	28	24
y Fuel	31	27	29	25	31	29	28	28	28	33

6. Is the Lottery Random? Listed below are the first digits selected in 40 consecutive drawings of the New York State Win 4 lottery game. (See Data Set 26 in Appendix B.) Do odd and even digits appear to be drawn in a sequence that is random?

9	7	0	7	5	5	1	9	0	0	8	7	6	0	1	6	7	2	4	7
5	5	5	2	0	4	4	9	9	0	5	3	3	1	9	2	5	6	8	2

7. Does the Weight of a Car Affect Leg Injuries in a Crash? Data were obtained from car crash experiments conducted by the National Transportation Safety Administration. New cars were purchased and crashed into a fixed barrier at 35 mi/h, and measurements were recorded for the dummy in the driver's seat. Use the sample data listed below to test for differences in left femur load measurements (in lb) among the four weight categories. Is there sufficient evidence to conclude that leg injury measurements for the four car weight categories are not all the same? Do the data suggest that heavier cars are safer in a crash?

Subcompact:	595	1063	885	519	422
Compact:	1051	1193	946	984	584
Midsize:	629	1686	880	181	645
Full-size:	1085	971	996	804	1376

8. Testing for Correlation Between Performance and Price *Consumer Reports* tested VHS tapes used in VCRs. Following are performance scores and prices (in dollars) of randomly selected tapes. Is there a correlation between performance and price? What does the conclusion suggest about buying VHS tapes?

Performance	91	92	82	85	87	80	94	97
Price	4.56	6.48	5.99	7.92	5.36	3.32	7.32	5.27

Cumulative Review Exercises

1. **Analyzing Poll Results** A market researcher for American Airlines was instructed to randomly select passengers waiting to board. (The author was one of the selected subjects.) The passengers were asked several questions about the airline service. Responses were recorded along with their genders, which are listed in the order in which they were selected.
   a. Use a 0.05 significance level to test the claim that the sequence is random.
   b. Use a 0.05 significance level to test the claim that the proportion of women is different from 0.5. Use the parametric test described in Section 7-3.
   c. Use a 0.05 significance level to test the claim that the proportion of women is different from 0.5. Use the sign test described in Section 12-2.
   d. Use the sample data to construct a 95% confidence interval for the proportion of women.
   e. What do the preceding results suggest? Is the sample biased against either gender? Was the sample obtained in a random sequence? If you are the manager, do you have any problems with these results?

   M M M M M F M M M F M M F M F M M F M M F M M F M M M

2. **Heights of Presidential Winners and Losers** The accompanying table shows the heights of presidents matched with the heights of the candidates they beat. All heights are in inches, and only the second-place candidates are included. Use a 0.05 significance level for the following.
   a. Use the linear correlation coefficient r to test for a significant linear correlation between the heights of the winners and the heights of the candidates they beat. (See Section 9-2.) Does there appear to be a correlation?
   b. Use the rank correlation coefficient r_s to test for a significant linear correlation between the heights of the winners and the heights of the candidates they beat. (See Section 12-6.) Does there appear to be a correlation?
   c. Use the sign test to test the claim that there is a difference between the heights of the winning candidates and the heights of the losing candidates.
   d. Use the Wilcoxon signed-ranks test to test the claim that there is a difference between the heights of the winning candidates and the heights of the corresponding losing candidates.
   e. Use the parametric t test (see Section 8-4) to test the claim that there is a difference between the heights of the winning candidates and the heights of the corresponding losing candidates.
   f. What do the preceding results suggest about the heights of winning presidential candidates and the heights of the corresponding losing presidential candidates?

Winner	76	66	70	70	74	71.5	73	74
Runner-up	64	71	72	72	68	71	69.5	74

conducted to determine whether the population has a proportion of women that is less than 0.5.

2. a. $r = -0.515$. CV: $r = \pm 0.707$. No significant linear correlation.
   b. $r_s = -0.463$. CV: $r_s = \pm 0.738$. No significant correlation.
   c. $x = 3$ is not less than or equal to the critical value of 0. There is not sufficient evidence to support the claim that there is a difference between the heights of the winning and losing candidates.
   d. $T = 10$. CV: $T = 2$. There is not sufficient evidence to support the claim that there is a difference between the heights of the winning and losing candidates.
   e. $t = 0.851$. CV: $t = \pm 2.365$. There is not sufficient evidence to support the claim that there is a difference between the heights of the winning and losing candidates.
   f. There is not sufficient evidence to conclude that the heights of the winning candidates and the heights of the losing candidates are related, and there is not sufficient evidence to conclude that there is a difference between the heights of the winning candidates and the heights of the losing candidates.

Cooperative Group Activities

1. **In-class activity** Use the existing seating arrangement in your class and apply the runs test to determine whether the students are arranged randomly according to gender. After recording the seating arrangement, analysis can be done in subgroups of three or four students.

2. In-class activity Divide into groups of 8 to 12 people. For each group member, *measure* his or her height and *measure* his or her arm span. For the arm span, the subject should stand with arms extended, like the wings on an airplane. It's easy to mark the height and arm span on a chalkboard, then measure the distances there. Divide the following tasks among subgroups of three or four people.
   a. Use rank correlation with the paired sample data to determine whether there is a correlation between height and arm span.
   b. Use the sign test to test for a difference between the two variables.
   c. Use the Wilcoxon signed-ranks test to test for a difference between the two variables.

3. In-class activity Do Activity 2 using pulse rate instead of arm span. Measure pulse rates by counting the number of heartbeats in 1 minute.

4. Out-of-class activity Divide into groups of three or four students. Investigate the relationship between two variables by collecting your own paired sample data and using the methods of Section 12-6 to determine whether there is a significant correlation. Suggested topics:
   - Is there a relationship between taste and cost of different brands of chocolate chip cookies (or colas)? (Taste can be measured on some number scale, such as 1 to 10.)
   - Is there a relationship between salaries of professional baseball (or basketball, or football) players and their season achievements?
   - Rates versus weights: Is there a relationship between car fuel consumption rates and car weights?
   - Is there a relationship between the lengths of men's (or women's) feet and their heights?
   - Is there a relationship between student grade-point averages and the amount of television watched?
   - Is there a relationship between heights of fathers (or mothers) and heights of their first sons (or daughters)?

5. Out-of-class activity See the "From Data to Decision" project, which involves analysis of the 1970 lottery used for drafting men into the U.S. Army. Because the 1970 results raised concerns about the randomness of selecting draft priority numbers, design a new procedure for generating the 366 priority numbers. Use your procedure to generate the 366 numbers and test your results by using the techniques suggested in parts (a), (b), and (c) of the "From Data to Decision" project. How do your results compare to those obtained in 1970? Does your random selection process appear to be better than the one used in 1970? Write a report that clearly describes the process you designed. Also include your analyses and conclusions.

6. Out-of-class activity Divide into groups of three or four. Survey students by asking them to identify their major and gender. For each surveyed subject, determine the accuracy of the time on his or her wristwatch. First set your own watch to the correct time using an accurate and reliable source ("At the tone, the time is . . ."). For watches that are ahead of the correct time, record positive times. For watches that are behind the correct time, record negative times. Use the sample data to address these questions:
   - Do the errors appear to be the same for both genders?
   - Do the errors appear to be the same for the different majors?

7. In-class activity Divide into groups of 8 to 12 people. For each group member, measure the person's height and also measure his or her navel height, which is the height from the floor to the navel. Use the rank correlation coefficient to determine whether there is a correlation between height and navel height.

8. In-class activity Divide into groups of three or four people. Appendix B includes many data sets not yet addressed by the methods of this chapter. For example, using Data Set 25, we can investigate the correlation between high values of the Dow-Jones Industrial Average and the numbers of U.S. car sales. Search Appendix B for variables of interest, then investigate using appropriate methods of nonparametric statistics. State your conclusions and try to identify practical applications.

Technology Project

Past attempts to identify extraterrestrial intelligent life have involved efforts to send radio messages carrying information about us earthlings. Dr. Frank Drake of Cornell University developed such a radio message that could be transmitted as a series of 1271 pulses and gaps. The pulses and gaps can be thought of as 1s and 0s. If we factor 1271 into the prime numbers of 41 and 31 and then make a 41 × 31 grid and put a dot at those positions corresponding to a pulse or 1, we get the pattern shown in the accompanying figure. This pattern contains information including the location of Earth in the solar system; the symbols for hydrogen, carbon, and oxygen; and drawings of a man, woman, child, fish, and water. Try to identify the man, woman, fish, and water in the pattern.

Assume that the sequence of 1271 1s and 0s is sent as a radio message that is intercepted by extraterrestrial life with enough intelligence to have studied this book. If the radio message is tested using the methods of this chapter, will the sequence appear to be "random noise" or will it be identified as a pattern that is not random? Use STATDISK or Minitab for the analysis.

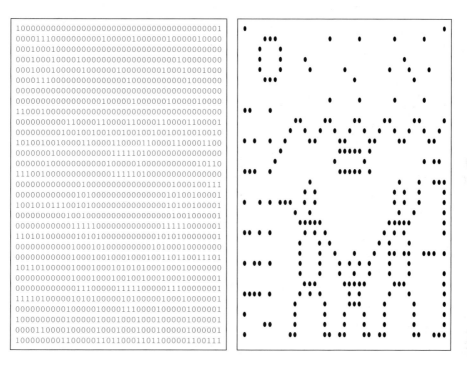

from DATA to DECISION

Critical Thinking: Was the draft lottery random?

In 1970, a lottery was used to determine who would be drafted into the U.S. Army. The 366 dates in the year were placed in individual capsules. First, the 31 January capsules were placed in a box; then the 29 February capsules were added and the two months were mixed. Then the 31 March capsules were added and the three months were mixed. This process continued until all months were included. The first capsule selected was September 14, so men born on that date were drafted first. The accompanying list shows the 366 dates in the order of selection.

Analyzing the Results

a. Use the runs test to test the sequence for randomness above and below the median of 183.5.

b. Use the Kruskal-Wallis test to test the claim that the 12 months had priority numbers drawn from the same population.

c. Calculate the 12 monthly means. Then plot those 12 means on a graph. (The horizontal scale lists the 12 months, and the vertical scale ranges from 100 to 260.) Note any pat-
continued

tern suggesting that the original priority numbers were not randomly selected.

d. Based on the results from parts (a), (b), and (c), decide whether this particular draft lottery was fair. Write a statement either supporting your position that it was fair or explaining why you believe that it was not fair. If you decided that this lottery was unfair, describe a process for selecting lottery numbers that would have been fair.

```
Jan:  305  159  251  215  101  224  306  199  194  325  329  221  318  238  017  121
      235  140  058  280  186  337  118  059  052  092  355  077  349  164  211

Feb:  086  144  297  210  214  347  091  181  338  216  150  068  152  004  089  212
      189  292  025  302  363  290  057  236  179  365  205  299  285

Mar:  108  029  267  275  293  139  122  213  317  323  136  300  259  354  169  166
      033  332  200  239  334  265  256  258  343  170  268  223  362  217  030

Apr:  032  271  083  081  269  253  147  312  219  218  014  346  124  231  273  148
      260  090  336  345  062  316  252  002  351  340  074  262  191  208

May:  330  298  040  276  364  155  035  321  197  065  037  133  295  178  130  055
      112  278  075  183  250  326  319  031  361  357  296  308  226  103  313

Jun:  249  228  301  020  028  110  085  366  335  206  134  272  069  356  180  274
      073  341  104  360  060  247  109  358  137  022  064  222  353  209

Jul:  093  350  115  279  188  327  050  013  277  284  248  015  042  331  322  120
      098  190  227  187  027  153  172  023  067  303  289  088  270  287  193

Aug:  111  045  261  145  054  114  168  048  106  021  324  142  307  198  102  044
      154  141  311  344  291  339  116  036  286  245  352  167  061  333  011

Sep:  225  161  049  232  082  006  008  184  263  071  158  242  175  001  113  207
      255  246  177  063  204  160  119  195  149  018  233  257  151  315

Oct:  359  125  244  202  024  087  234  283  342  220  237  072  138  294  171  254
      288  005  241  192  243  117  201  196  176  007  264  094  229  038  079

Nov:  019  034  348  266  310  076  051  097  080  282  046  066  126  127  131  107
      143  146  203  185  156  009  182  230  132  309  047  281  099  174

Dec:  129  328  157  165  056  010  012  105  043  041  039  314  163  026  320  096
      304  128  240  135  070  053  162  095  084  173  078  123  016  003  100
```

INTERNET PROJECT   Nonparametric Tests

This chapter introduced hypothesis-testing methods of the nonparametric or distribution-free variety. Nonparametric methods allow you to test hypotheses without making assumptions regarding the underlying distribution of the population being sampled. To continue your work with this important class of statistical testing methods, go to the *Elementary Statistics* Web site:

http://www.aw.com/triola

The Internet Project for this chapter will ask you to revisit some of the data sets from earlier projects, specifically data sets used in parametric tests. This time, however, you will apply the appropriate nonparametric test and compare the outcomes. In addition, you will conduct an investigation of randomness by applying the runs test.

Statistics @ Work

> *"If I didn't have any background in statistics, I would not be able to fully understand the data my company produces . . . help protect our workers and customer."*

Jeffrey Foy

Jeffrey Foy is a toxicologist working for the Cabot Corporation, a chemical company.

Jeffrey Foy is also responsible for the hazard evaluation of the chemicals Cabot Corporation produces. It is his job to understand how the company's products may affect humans or the environment and help decide on the best ways to protect both.

What do you do in your job?

My responsibilities include arranging and evaluating toxicological studies, writing material safety data sheets, and helping our research and development groups produce materials that are safe for both people and the environment or to understand what potential hazards the materials might have.

What concepts of statistics do you use?

The primary concept I use is hypothesis testing (probability testing).

How do you use statistics on the job?

I use statistics daily. Statistical methods have been and are used in two ways in my work. First, statistics is used to help determine how I design my experiments. Second, statistics is used to determine if the data generated are significant or sometimes if they're even good enough to use.

Studies that I am involved in can cost from as little as $1000 to as much as $500,000 or more, and if you don't properly determine how you are going to evaluate the data, you could cost your company a great deal of time and money. If the experiment is done properly, then we move on to analyze the data. The data from the studies we perform are used to assess any potential health effects our products may have on our workers, customers, or the environment. The results are used to determine how chemicals can be sold or handled. When performing experiments at a testing laboratory or drug company you want to determine if your materials have an effect, whether desired (a drug curing a disease) or undesired (that same drug being toxic). Statistics plays an enormous role in our evaluation of the significance of the effects.

Please describe one specific example illustrating how the use of statistics was successful in improving a product or service.

A toxicology study costing about $300,000 was recently conducted. The data from the study were to be used to help determine if a particular chemical caused any effects in the subjects studied. After the study was performed, flaws were found in both the data and statistics used. It took an additional two years to properly review the data and finish the health evaluation. If the proper methods and endpoints had been chosen, then the additional time and money may not have been necessary. It was the understanding of the data and correct statistical evaluation that helped prevent the failure and potential repeat of the study.

13

Statistical Process Control

....

13-1 Overview

13-2 Control Charts for Variation and Mean

13-3 Control Charts for Attributes

Is the production of aircraft altimeters dangerous for those who fly?

The Altigauge Manufacturing Company produces aircraft altimeters, which provide pilots with readings of their heights above sea level. The accuracy of altimeters is important because pilots rely on them to maintain altitudes with safe vertical clearance above mountains, towers, and tall buildings, as well as vertical separation from other aircraft. The accuracy of altimeters is especially important when pilots fly approaches to landing while not being able to see the ground. In the past, pilots and passengers have been killed in crashes caused by wrong altimeter readings that led pilots to believe that they were safely above the ground when they were actually flying dangerously low.

Because aircraft altimeters are so critically important to aviation safety, their accuracy is carefully controlled by government regulations. According to Federal Aviation Administration Regulation Part 43, Appendix E, an altimeter must give a reading with an error of no more than 20 ft when tested for an altitude of 1000 ft.

At the Altigauge Manufacturing Company, four altimeters are randomly selected from production on each of 20 consecutive business days, and Table 13-1 lists the errors (in feet) when they are tested in a pressure chamber that simulates an altitude of 1000 ft. On day 1, for example, the actual altitude readings for the four selected altimeters are 1002 ft, 992 ft, 1005 ft, and 1011 ft, so the corresponding errors (in feet) are 2, −8, 5, and 11.

In this chapter we will evaluate this altimeter manufacturing process by analyzing the behavior of the errors over time. We will see how methods of statistics can be used to monitor a manufacturing process with the goal of identifying and correcting any serious problems. In addition to helping companies stay in business, methods of statistics can positively affect our safety in very significant ways.

Table 13-1	Aircraft Altimeter Errors (in feet)						

Day	Error				Mean	Median	Range	Standard Deviation
1	2	−8	5	11	2.50	3.5	19	7.94
2	−5	2	6	8	2.75	4.0	13	5.74
3	6	7	−1	−8	1.00	2.5	15	6.98
4	−5	5	−5	6	0.25	0.0	11	6.08
5	9	3	−2	−2	2.00	0.5	11	5.23
6	16	−10	−1	−8	−0.75	−4.5	26	11.81
7	13	−8	−7	2	0.00	−2.5	21	9.76
8	−5	−4	2	8	0.25	−1.0	13	6.02
9	7	13	−2	−13	1.25	2.5	26	11.32
10	15	7	19	1	10.50	11.0	18	8.06
11	12	12	10	9	10.75	11.0	3	1.50
12	11	9	11	20	12.75	11.0	11	4.92
13	18	15	23	28	21.00	20.5	13	5.72
14	6	32	4	10	13.00	8.0	28	12.91
15	16	−13	−9	19	3.25	3.5	32	16.58
16	8	17	0	13	9.50	10.5	17	7.33
17	13	3	6	13	8.75	9.5	10	5.06
18	38	−5	−5	5	8.25	0.0	43	20.39
19	18	12	25	−6	12.25	15.0	31	13.28
20	−27	23	7	36	9.75	15.0	63	27.22

Overview

In Chapter 2 we noted that when describing, exploring, or comparing data sets, the following characteristics are usually extremely important. (We suggested that the sentence "Computer Viruses Destroy Or Terminate" could be used as a mnemonic device for remembering CVDOT, which summarizes these five characteristics.)

1. *Center:* Measure of center, which is a representative or average value that gives us an indication of where the middle of the data set is located.
2. *Variation:* A measure of the amount that the values vary among themselves.
3. *Distribution:* The nature or shape of the distribution of the data, such as bell-shaped, uniform, or skewed.
4. *Outliers:* Sample values that lie very far away from the vast majority of the other sample values.
5. *Time:* Changing characteristics of the data over time.

The main objective of this chapter is to address the fifth item: changing characteristics of data over *time.* When investigating characteristics such as center and variation, it is important to know whether we are dealing with a stable population or one that is changing with the passage of time.

There is currently a strong trend toward trying to improve the quality of American goods and services, and the methods presented in this chapter are being used by growing numbers of businesses. Evidence of the increasing importance of quality is found in its greater role in advertising and the growing number of books and articles that focus on the issue of quality. In many cases, job applicants (you?) have a definite advantage when they can tell employers that they have studied statistics and methods of quality control. This chapter will present some of the basic tools commonly used to monitor quality.

Minitab, Excel, and other software packages include programs for automatically generating charts of the type discussed in this chapter, and we will include several examples of such displays. Control charts, like histograms, boxplots, and scatterplots, are among those wonderful graphic devices that allow us to *see* and *understand* some property of data that would be more difficult or impossible to understand without graphs. The world needs more people who can construct and interpret important graphs, such as the control charts described in this chapter.

 Control Charts for Variation and Mean

The main objective of this section is to monitor important features of data over time. Such data are often referred to as *process data.*

Definition

Process data are data arranged according to some time sequence. They are measurements of a characteristic of goods or services that result from some combination of equipment, people, materials, methods, and conditions.

For example, Table 13-1 includes process data consisting of the measured error (in feet) in altimeter readings over 20 consecutive days of production. Each day, four altimeters were randomly selected and tested. Because the data in Table 13-1 are arranged according to the time at which they were selected, they are process data. It is very important to recognize this point:

Important characteristics of process data can change over time.

In making altimeters, a manufacturer might use competent and well-trained personnel along with good machines that are correctly calibrated, but if the personnel are replaced or the machines wear with use, the altimeters might begin to become defective. Companies have gone bankrupt because they unknowingly allowed manufacturing processes to deteriorate without constant monitoring.

Run Charts

There are various methods that can be used to monitor a process to ensure that the important desired characteristics don't change—analysis of a *run chart* is one such method.

Definition

A **run chart** is a sequential plot of *individual* data values over time. One axis (usually the vertical axis) is used for the data values, and the other axis (usually the horizontal axis) is used for the time sequence.

EXAMPLE Manufacturing Aircraft Altimeters Treating the 80 altimeter errors in Table 13-1 as a string of consecutive measurements, construct a run chart by using a vertical axis for the errors and a horizontal axis to identify the order of the sample data.

SOLUTION Figure 13-1 is the Minitab-generated run chart for the data in Table 13-1. The vertical scale is designed to be suitable for altimeter errors ranging from −27 ft to 38 ft, which are the minimum and maximum values in Table 13-1. The horizontal scale is designed to include the 80 values arranged in sequence. The first point represents the first value of 2 ft, the second point represents the second value of −8 ft, and so on.

continued

Note to Instructor
Minitab has excellent programs for generating run charts and control charts, and the DDXL add-in for Excel can be used as well. If neither Minitab nor Excel are available, consider using STATDISK or a TI-83 Plus calculator to generate graphs as *scatterplots*. To generate the run chart for the data of Table 13-1, use the paired data (1, 2), (2, −8), (3, 5), . . . , (80, 36) where the first coordinates are consecutive positive integers and the second coordinates are the sample data. Manual construction of graphs in this chapter is much more time consuming, but students do profit from more direct involvement.

FIGURE 13-1 **Run Chart of Individual Altimeter Errors in Table 13-1**

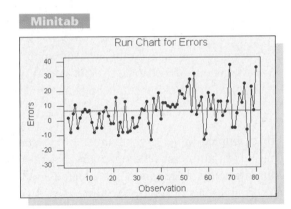

In Figure 13-1, the horizontal scale identifies the sample number, so the number 20 indicates the 20th sample item. The vertical scale represents the altimeter error (in feet). Now examine Figure 13-1 and try to identify any *patterns* that jump out begging for attention. Figure 13-1 does reveal this problem: As time progresses from left to right, the heights of the points appear to show a pattern of increasing variation. See how the points at the left fluctuate considerably less than the points farther to the right. The Federal Aviation Administration regulations require errors less than 20 ft (or between 20 ft and −20 ft), so the altimeters represented by the points at the left are OK, whereas several of the points farther to the right correspond to altimeters not meeting the required specifications. It appears that the manufacturing process started out well, but deteriorated as time passed. If left alone, this manufacturing process will cause the company to go out of business.

Interpreting Run Charts   Only when a process is *statistically stable* can its data be treated as if they came from a population with a constant mean, standard deviation, distribution, and other characteristics.

> **Definition**
>
> A process is **statistically stable** (or **within statistical control**) if it has only natural variation, with no patterns, cycles, or unusual points.

Figure 13-2 illustrates typical patterns showing ways in which the process of filling 16-oz soup cans may not be statistically stable.

- **Figure 13-2(a):** There is an obvious *upward trend* that corresponds to values that are increasing over time. If the filling process were to follow this type of pattern, the cans would be filled with more and more soup until they began to overflow, eventually leaving the employees swimming in soup.
- **Figure 13-2(b):** There is an obvious *downward trend* that corresponds to steadily decreasing values. The cans would be filled with less and less soup

Note to Instructor

Ask students to identify possible causes for each of the patterns in Figure 13-2. For example, the upward trend in Figure 13-2(a) might result from a nozzle with a hole that is growing larger through wear. As the soup makes the nozzle opening larger, more soup is poured in the same amount of time, resulting in an upward trend. This reinforces the importance of using statistics to help in making practical changes so that the process continues to have a high level of quality.

until they were extremely underfilled. Such a process would require a complete reworking of the cans in order to get them full enough for distribution to consumers.

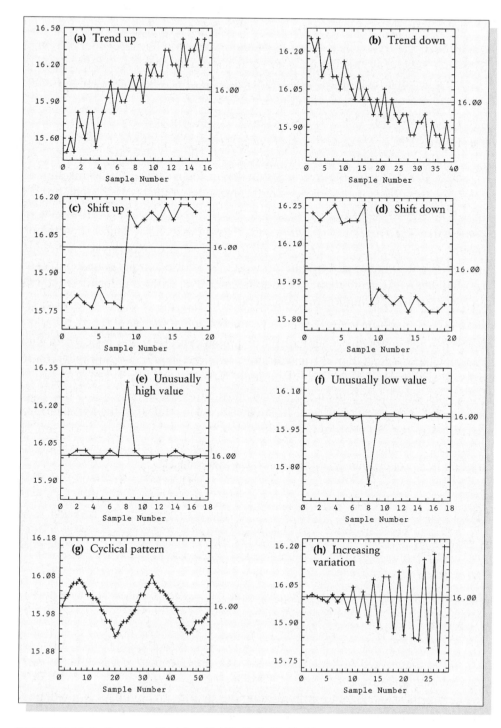

FIGURE 13-2 **Processes That Are Not Statistically Stable**

The Flynn Effect: Upward Trend in IQ Scores

A run chart or control chart of IQ scores would reveal that they exhibit an upward trend, because IQ scores have been steadily increasing since they began to be used about 70 years ago. The trend is worldwide, and it is the same for different types of IQ tests, even those that rely heavily on abstract and nonverbal reasoning with minimal cultural influence. This upward trend has been named the *Flynn effect,* because political scientist James R. Flynn discovered the trend in his studies of U.S. military recruits. The amount of the increase is quite substantial: Based on a current mean IQ score of 100, it is estimated that the mean IQ in 1920 would be about 77. The typical student of today is therefore brilliant when compared to his or her great-grandparents. So far, there is no generally accepted explanation for the Flynn effect.

- **Figure 13-2(c):** There is an *upward shift.* A run chart such as this one might result from an adjustment to the filling process, making all subsequent values higher.

- **Figure 13-2(d):** There is a *downward shift*—the first few values are relatively stable, and then something happened so that the last several values are relatively stable, but at a much lower level.

- **Figure 13-2(e):** The process is stable except for one *exceptionally high value.* The cause of that unusual value should be investigated. Perhaps the cans became temporarily stuck and one particular can was filled twice instead of once.

- **Figure 13-2(f):** There is an *exceptionally low value.*

- **Figure 13-2(g):** There is a *cyclical pattern* (or repeating cycle). This pattern is clearly nonrandom and therefore reveals a statistically unstable process. Perhaps periodic overadjustments are being made to the machinery, with the effect that some desired value is continually being chased but never quite captured.

- **Figure 13-2(h):** The *variation is increasing over time.* This is a common problem in quality control. The net effect is that products vary more and more until almost all of them are worthless. For example, some soup cans will be overflowing with wasted soup, and some will be underfilled and unsuitable for distribution to consumers.

A common goal of many different methods of quality control is this: *reduce variation* in the product or service. For example, Ford became concerned with variation when it found that its transmissions required significantly more warranty repairs than the same type of transmissions made by Mazda in Japan. A study showed that the Mazda transmissions had substantially less variation in the gearboxes; that is, crucial gearbox measurements varied much less in the Mazda transmissions. Although the Ford transmissions were built within the allowable limits, the Mazda transmissions were more reliable because of their lower variation. Variation in a process can result from two types of causes.

Definitions

Random variation is due to chance; it is the type of variation inherent in any process that is not capable of producing every good or service exactly the same way every time.

Assignable variation results from causes that can be identified (such factors as defective machinery, untrained employees, and so on).

Later in the chapter we will consider ways to distinguish between assignable variation and random variation.

The run chart is one tool for monitoring the stability of a process. We will now consider *control charts,* which are also extremely useful for that same purpose.

Control Chart for Monitoring Variation: The *R* Chart

In the article "The State of Statistical Process Control as We Proceed into the 21st Century" (by Stoumbos, Reynolds, Ryan, and Woodall, *Journal of the American Statistical Association,* Vol. 95, No. 451), the authors state that "control charts are among the most important and widely used tools in statistics. Their applications have now moved far beyond manufacturing into engineering, environmental science, biology, genetics, epidemiology, medicine, finance, and even law enforcement and athletics." We begin with the definition of a control chart.

> **Definition**
>
> A **control chart** of a process characteristic (such as mean or variation) consists of values plotted sequentially over time, and it includes a **centerline** as well as a **lower control limit** (LCL) and an **upper control limit** (UCL). The centerline represents a central value of the characteristic measurements, whereas the control limits are boundaries used to separate and identify any points considered to be *unusual*.

We will assume that the population standard deviation σ is not known as we consider only two of several different types of *control charts:* (1) *R* charts (or range charts) used to monitor variation and (2) $\bar{x}$ charts used to monitor means. When using control charts to monitor a process, it is common to consider *R* charts and $\bar{x}$ charts together, because a statistically unstable process may be the result of increasing variation or changing means or both.

An ***R* chart** (or **range chart**) is a plot of the sample ranges instead of individual sample values, and it is used to monitor the *variation* in a process. (It might make more sense to use standard deviations, but range charts are used more often in practice. This is a carryover from times when calculators and computers were not available. See Exercise 13 for a control chart based on standard deviations.) In addition to plotting the range values, we include a centerline located at $\bar{R}$, which denotes the mean of all sample ranges, as well as another line for the lower control limit and a third line for the upper control limit. Following is a summary of notation for the components of the *R* chart.

> **Notation**
>
> Given: Process data consisting of a sequence of samples all of the same size, and the distribution of the process data is essentially normal.
>
> n = size of each sample, or *subgroup*
>
> $\bar{R}$ = mean of the sample ranges (that is, the sum of the sample ranges divided by the number of samples)

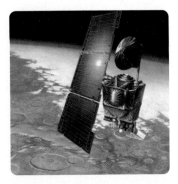

Costly Assignable Variation

The Mars Climate Orbiter was launched by NASA and sent to Mars, but it was destroyed when it flew too close to its destination planet. The loss was estimated at $125 million. The cause of the crash was found to be confusion between the use of units used for calculations. Acceleration data were provided in the English units of pounds of force, but the Jet Propulsion Laboratory assumed that those units were in metric "newtons" instead of pounds. The thrusters of the spacecraft subsequently provided wrong amounts of force in adjusting the position of the spacecraft. The errors caused by the discrepancy were fairly small at first, but the cumulative error over months of the spacecraft's journey proved to be fatal to its success.

In 1962, the rocket carrying the Mariner 1 satellite was destroyed by ground controllers when it went off course due to a missing minus sign in a computer program.

Monitoring Process Variation: Control Chart for R

Points plotted: Sample ranges

Centerline: $\overline{R}$

Upper control limit (UCL): $D_4\overline{R}$ (where D_4 is found in Table 13-2)

Lower control limit (LCL): $D_3\overline{R}$ (where D_3 is found in Table 13-2)

Table 13-2   Control Chart Constants

n: Number of Observations in Subgroup	$\overline{x}$		s		R	
	A_2	A_3	B_3	B_4	D_3	D_4
2	1.880	2.659	0.000	3.267	0.000	3.267
3	1.023	1.954	0.000	2.568	0.000	2.574
4	0.729	1.628	0.000	2.266	0.000	2.282
5	0.577	1.427	0.000	2.089	0.000	2.114
6	0.483	1.287	0.030	1.970	0.000	2.004
7	0.419	1.182	0.118	1.882	0.076	1.924
8	0.373	1.099	0.185	1.815	0.136	1.864
9	0.337	1.032	0.239	1.761	0.184	1.816
10	0.308	0.975	0.284	1.716	0.223	1.777
11	0.285	0.927	0.321	1.679	0.256	1.744
12	0.266	0.886	0.354	1.646	0.283	1.717
13	0.249	0.850	0.382	1.618	0.307	1.693
14	0.235	0.817	0.406	1.594	0.328	1.672
15	0.223	0.789	0.428	1.572	0.347	1.653
16	0.212	0.763	0.448	1.552	0.363	1.637
17	0.203	0.739	0.466	1.534	0.378	1.622
18	0.194	0.718	0.482	1.518	0.391	1.608
19	0.187	0.698	0.497	1.503	0.403	1.597
20	0.180	0.680	0.510	1.490	0.415	1.585
21	0.173	0.663	0.523	1.477	0.425	1.575
22	0.167	0.647	0.534	1.466	0.434	1.566
23	0.162	0.633	0.545	1.455	0.443	1.557
24	0.157	0.619	0.555	1.445	0.451	1.548
25	0.153	0.606	0.565	1.435	0.459	1.541

Source: Adapted from *ASTM Manual on the Presentation of Data and Control Chart Analysis,* © 1976 ASTM, pp. 134–136. Reprinted with permission of American Society for Testing and Materials.

The values of D_4 and D_3 were computed by quality-control experts, and they are intended to simplify calculations. The upper and lower control limits of $D_4\overline{R}$ and $D_3\overline{R}$ are values that are roughly equivalent to 99.7% confidence interval limits. It is therefore highly unlikely that values from a statistically stable process would fall beyond those limits. If a value does fall beyond the control limits, it's very likely that the process is not statistically stable.

EXAMPLE **Manufacturing Aircraft Altimeters** Refer to the altimeter errors in Table 13-1. Using the samples of size $n = 4$ collected each day of manufacturing, construct a control chart for R.

SOLUTION   We begin by finding the value of $\overline{R}$, the mean of the sample ranges.

$$\overline{R} = \frac{19 + 13 + \cdots + 63}{20} = 21.2$$

The centerline for our R chart is therefore located at $\overline{R} = 21.2$. To find the upper and lower control limits, we must first find the values of D_3 and D_4. Referring to Table 13-2 for $n = 4$, we get $D_3 = 0.000$ and $D_4 = 2.282$, so the control limits are as follows:

Upper control limit: $D_4\overline{R} = (2.282)(21.2) = 48.4$
Lower control limit: $D_3\overline{R} = (0.000)(21.2) = 0.0$

Using a centerline value of $\overline{R} = 21.2$ and control limits of 48.4 and 0.0, we now proceed to plot the sample ranges. The result is shown in the Minitab display.

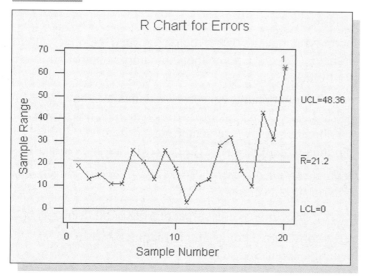

Don't Tamper!

Nashua Corp. had trouble with its paper-coating machine and considered spending a million dollars to replace it. The machine was working well with a stable process, but samples were taken every so often and, based on the results, adjustments were made. These overadjustments, called *tampering*, caused shifts away from the distribution that had been good. The effect was an increase in defects. When statistician and quality expert W. Edwards Deming studied the process, he recommended that no adjustments be made unless warranted by a signal that the process had shifted or had become unstable. The company was better off with no adjustments than with the tampering that took place.

Interpreting Control Charts

When interpreting control charts, the following point is extremely important:

> **Upper and lower control limits of a control chart are based on the *actual* behavior of the process, not the *desired* behavior. Upper and lower control limits are totally unrelated to any process *specifications* that may have been decreed by the manufacturer.**

When investigating the quality of some process, there are typically two key questions that need to be addressed:

1. Based on the current behavior of the process, can we conclude that the process is within statistical control?
2. Do the process goods or services meet design specifications?

The methods of this chapter are intended to address the first question, but not the second. That is, we are focusing on the behavior of the process with the objective of determining whether the process is within statistical control. Whether the process results in goods or services that meet some stated specifications is another issue not addressed by the methods of this chapter. For example, the Minitab *R* chart shown here includes upper and lower control limits of 48.36 and 0, which result from the sample values listed in Table 13-1. Government regulations require that altimeters have errors between −20 ft and 20 ft, but those desired (or required) specifications are not included in the control chart for *R*.

Also, we should clearly understand the specific criteria for determining whether a process is in statistical control (that is, whether it is statistically stable). So far, we have noted that a process is not statistically stable if its pattern resembles any of the patterns shown in Figure 13-2. This criterion is included with some others in the following list.

Criteria for Determining When a Process Is Not Statistically Stable (Out of Statistical Control)

1. There is a pattern, trend, or cycle that is obviously not random (such as those depicted in Figure 13-2).
2. There is a point lying outside of the region between the upper and lower control limits. (That is, there is a point above the upper control limit or below the lower control limit.)
3. *Run of 8 Rule:* There are eight consecutive points all above or all below the centerline. (With a statistically stable process, there is a 0.5 probability that a point will be above or below the centerline, so it is very unlikely that eight consecutive points will all be above the centerline or all below it.)

We will use only the three out-of-control criteria listed above, but some businesses use additional criteria such as these:

- There are six consecutive points all increasing or all decreasing.
- There are 14 consecutive points all alternating between up and down (such as up, down, up, down, and so on).

- Two out of three consecutive points are beyond control limits that are 2 standard deviations away from the centerline.
- Four out of five consecutive points are beyond control limits that are 1 standard deviation away from the centerline.

EXAMPLE **Statistical Process Control** Examine the R chart shown in the Minitab display for the preceding example and determine whether the process variation is within statistical control.

SOLUTION   We can interpret control charts for R by applying the three out-of-control criteria just listed. Applying the three criteria to the Minitab display of the R chart, we conclude that variation in this process is out of statistical control. There are not eight consecutive points all above or all below the centerline, so the third condition is not violated, but the first two conditions are violated.

1. There is a pattern, trend, or cycle that is obviously not random: Going from left to right, there is a pattern of upward trend, as in Figure 13-2(a).
2. There is a point (the rightmost point) that lies above the upper control limit.

INTERPRETATION   We conclude that the variation (not necessarily the mean) of the process is out of statistical control. Because the variation appears to be increasing with time, immediate corrective action must be taken to fix the *variation* among the altimeter errors.

Control Chart for Monitoring Means: The $\bar{x}$ Chart

An $\bar{x}$ **chart** is a plot of the sample means, and it is used to monitor the *center* in a process. In addition to plotting the sample means, we include a centerline located at $\bar{\bar{x}}$, which denotes the mean of all sample means (equal to the mean of all sample values combined), as well as another line for the lower control limit and a third line for the upper control limit. Using the approach common in business and industry, the centerline and control limits are based on ranges instead of standard deviations. See Exercise 14 for an $\bar{x}$ chart based on standard deviations.

Monitoring Process Mean: Control Chart for $\bar{x}$

Points plotted: Sample means
Centerline: $\bar{\bar{x}}$ = mean of all sample means
Upper control limit (UCL): $\bar{\bar{x}} + A_2\bar{R}$ (where A_2 is found in Table 13-2)
Lower control limit (LCL): $\bar{\bar{x}} - A_2\bar{R}$ (where A_2 is found in Table 13-2)

Bribery Detected with Control Charts

Control charts were used to help convict a person who bribed Florida jai alai players to lose. (See "Using Control Charts to Corroborate Bribery in Jai Alai," by Charnes and Gitlow, *The American Statistician,* Vol. 49, No. 4.) An auditor for one jai alai facility noticed that abnormally large sums of money were wagered for certain types of bets, and some contestants didn't win as much as expected when those bets were made. R charts and $\bar{x}$ charts were used in court as evidence of highly unusual patterns of betting. Examination of the control charts clearly shows points well beyond the upper control limit, indicating that the process of betting was out of statistical control. The statistician was able to identify a date at which assignable variation appeared to stop, and prosecutors knew that it was the date of the suspect's arrest.

Section 13-2 Answers

1. a. Process data are data arranged according to some
   time sequence.

   b. A process is out of statistical control if it has variation
   other than natural variation,
   and it has patterns, cycles,
   or unusual points.

   c. There is a pattern, trend, or
   cycle that is obviously not
   random, or there is a point
   lying beyond the upper or
   lower control limits, or there
   are eight consecutive points
   all above or all below the
   centerline.

   d. Random variation is due to
   chance, but assignable variation results from causes
   that can be identified.

   e. An R chart shows the pattern of sample ranges and is
   used to determine whether
   variation is within statistical
   control, whereas an $\bar{x}$ chart
   shows the pattern of sample
   means and is used to determine whether the mean in a
   process is within statistical
   control.

2. There is a cyclic pattern consisting of a rough "W" shape for
   each year. The pattern can be
   explained by the temperature
   changes throughout the year.

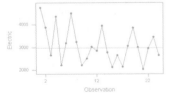

3. The process variation appears
   to be within statistical control.

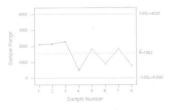

EXAMPLE **Manufacturing Aircraft Altimeters** Refer to the altimeter errors in Table 13-1. Using samples of size $n = 4$ collected each working day, construct a control chart for $\bar{x}$. Based on the control chart for $\bar{x}$ only, determine whether the process mean is within statistical control.

SOLUTION   Before plotting the 20 points corresponding to the 20 values of $\bar{x}$, we must first find the value for the centerline and the values for the control limits. We get

$$\bar{\bar{x}} = \frac{2.50 + 2.75 + \cdots + 9.75}{20} = 6.45$$

$$\bar{R} = \frac{19 + 13 + \cdots + 63}{20} = 21.2$$

Referring to Table 13-2, we find that for $n = 4$, $A_2 = 0.729$. Knowing the values of $\bar{\bar{x}}$, A_2, and $\bar{R}$, we can now evaluate the control limits.

Upper control limit: $\bar{\bar{x}} + A_2\bar{R} = 6.45 + (0.729)(21.2) = 21.9$

Lower control limit: $\bar{\bar{x}} - A_2\bar{R} = 6.45 - (0.729)(21.2) = -9.0$

INTERPRETATION The resulting control chart for $\bar{x}$ will be as shown in the accompanying Excel display. Examination of the control chart shows that the process mean is out of statistical control because at least one of the three out-of-control criteria is not satisfied. Specifically, the third criterion is not satisfied because there are eight (or more) consecutive points all below the centerline. Also, there does appear to be a pattern of an upward trend. Again, immediate corrective action is required to fix the production process.

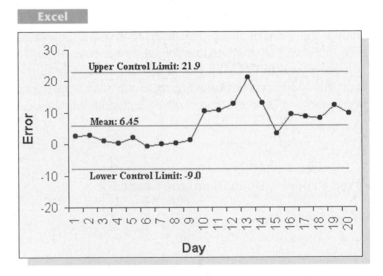

Using Technology

STATDISK   See the *STATDISK Student Laboratory Manual and Workbook* that is a supplement to this book.

Minitab   **Run Chart:** Begin by entering all of the sample data in column C1. Select the option **Stat,** then **Quality Tools,** then **Run Chart.** In the indicated boxes, enter C1 for the single column variable, enter 1 for the subgroup size, and then click on **OK.**

R Chart: First enter the individual sample values sequentially in column C1. Next, select the options **Stat, Control Charts, Variables Chart for Subgroups,** and **R.** Enter C1, enter the sample size in the box for the subgroup size, click on **R Options,** and click on **Estimate.** Select **Rbar.** (Selection of the *R* bar estimate causes the variation of the population distribution to be estimated with the sample ranges instead of the sample standard deviations, which is the default.) Click **OK** twice.

$\bar{x}$ Chart: First enter the individual sample values sequentially in column C1. Next, select the options **Stat, Control Charts, Variables Chart for Subgroups,** and **Xbar.** Enter C1, enter the size of each of the samples in the "subgroup size box," click **Xbar Options,** and click on **Estimate;** then select **Rbar.** Click **OK** twice.

Excel   To use the Data Desk XL add-in, click on **DDXL** and select **Process Control.** Proceed to select the type of chart you want. (You must first enter the data in column A with sample identifying codes entered in column B. For the data of Table 13-1, for example, enter a 1 in column B adjacent to each value from day 1, enter a 2 for each value from day 2, and so on.)

To use Excel's built-in graphics features instead of Data Desk XL, see the following:

Run chart: Enter all of the sample data in column A. On the main menu bar, click on the **Chart Wizard** icon, which looks like a bar graph. For the chart type, select **Line.** For the chart subtype, select the first graph in the second row, then click **Next.** Continue to click **Next,** then **Finish.** The graph can be edited to include labels, delete grid lines, and so on.

R Chart *Step 1:* Enter the sample data in rows and columns corresponding to the data set. For example, enter the data in Table 13-1 in four columns (A, B, C, D) and 20 rows as shown in the table.

Step 2: Next, create a column of the range values using the following procedure. Position the cursor in the first empty cell to the right of the block of sample data, then enter this expression in the formula box: = MAX(A1:D1)−MIN(A1:D1), where the range A1:D1 should be modified to describe the first row of your data set. After pressing the **Enter** key, the range for the first row should appear. Use the mouse to click and drag the lower right corner of this cell, so that the whole column fills up with the ranges for the different rows.

Step 3: Next, produce a graph by following the same procedure described for the run charts, but be sure to refer to the column of *ranges* when entering the input range. You can insert the required centerline and upper and lower control limits by editing the graph. Click on the line on the bottom of the screen, then click and drag to position the line correctly.

$\bar{x}$ Chart: *Step 1:* Enter the sample data in rows and columns corresponding to the data set. For example, enter the data in Table 13-1 in four columns (A, B, C, D) and 20 rows as shown in the table.

Step 2: Next, create a column of the sample means using the following procedure. Position the cursor in the first empty cell to the right of the block of sample data, then enter this expression in the formula box: = AVERAGE(A1:D1), where the range A1:D1 should be modified to describe the first row of your data set. After pressing the **Enter** key, the mean for the first row should appear. Use the mouse to click and drag the lower right corner of this cell, so that the whole column fills up with the means for the different rows.

Step 3: Next, produce a graph by following the same procedure described for the run chart, but be sure to refer to the column of *means* when entering the input range. You can insert the required centerline and upper and lower control limits by editing the graph. Click on the line on the bottom of the screen, then click and drag to position the line correctly. It's not easy.

4. The process mean appears to be within statistical control.

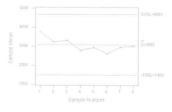

5. The process variation appears to be within statistical control.

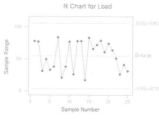

6. The process mean appears to be within statistical control.

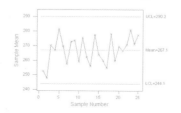

7. There is a pattern of increasing variation, so the process is out of statistical control. The increasing variation will result in more and more defects.

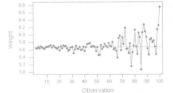

8. There is a pattern of increasing ranges, there are points beyond the upper control limit, and there are eight consecutive points lying below the centerline, so the process variation is out of statistical control.

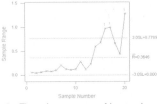

9. There is a pattern of increasing variation, there are points lying beyond the upper control limit, and there are eight consecutive points all below the centerline, so the process mean is out of statistical control. This process does need corrective action.

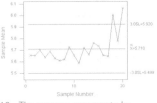

10. The process appears to be within statistical control.

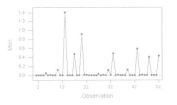

13-2 Basic Skills and Concepts

1. **a.** What are *process data?*
   b. What does it mean for a process to be out of statistical control?
   c. What are the three criteria for determining whether a process is out of statistical control?
   d. What is the difference between random variation and assignable variation?
   e. What is the difference between an R chart and an $\bar{x}$ chart?

Monitoring Home Energy Consumption. In Exercises 2–4, use the following information: The author recorded his electrical energy consumption (in kilowatt-hours) for his home in upstate New York for two-month intervals over four years, and the results are listed in the table.

	Jan.–Feb.	Mar.–Apr.	May–June	July–Aug.	Sept.–Oct.	Nov.–Dec.
Year 1	4762	3875	2657	4358	2201	3187
Year 2	4504	3237	2198	2511	3020	2857
Year 3	3952	2785	2118	2658	2139	3071
Year 4	3863	3013	2023	2953	3456	2647

2. Energy Consumption: Constructing a Run Chart Construct a run chart for the 24 values. Does there appear to be a pattern suggesting that the process is not within statistical control? Is there any pattern or variation that can be explained?

3. Energy Consumption: Constructing an R Chart Use samples of size 3 by combining the first three values for each year and combining the last three values for each year. Using the eight samples of size 3, construct an R chart and determine whether the process variation is within statistical control. If it is not, identify which of the three out-of-control criteria lead to rejection of statistically stable variation.

4. Energy Consumption: Constructing an $\bar{x}$ Chart Use samples of size 3 by combining the first three values for each year and combining the last three values for each year. Using the eight samples of size 3, construct an $\bar{x}$ chart and determine whether the process mean is within statistical control. If it is not, identify which of the three out-of-control criteria lead to rejection of a statistically stable mean. What is a practical effect of not having this process in statistical control? Give an example of a cause that would make the process go out of statistical control.

Constructing Control Charts for Aluminum Cans. Exercises 5 and 6 are based on the axial loads (in pounds) of aluminum cans that are 0.0109 in. thick, as listed in Data Set 20 in Appendix B. An axial load of a can is the maximum weight supported by its side, and it is important to have an axial load high enough so that the can isn't crushed when the top lid is pressed into place. The data are from a real manufacturing process, and they were provided by a student who used an earlier edition of this book.

5. On each day of production, seven aluminum cans with thickness 0.0109 in. were randomly selected and the axial loads were measured. The ranges for the different days are listed below, but they can also be found from the values given in Data Set 20 in Appendix B. Construct an R chart and determine whether the process variation is within statistical control. If it is not, identify which of the three out-of-control criteria lead to rejection of statistically stable variation.

78	77	31	50	33	38	84	21	38	77	26	78	78
17	83	66	72	79	61	74	64	51	26	41	31	

6. On each day of production, seven aluminum cans with thickness 0.0109 in. were randomly selected and the axial loads were measured. The means for the different days are listed below, but they can also be found from the values given in Data Set 20 in Appendix B. Construct an $\bar{x}$ chart and determine whether the process mean is within statistical control. If it is not, identify which of the three out-of-control criteria lead to rejection of statistically stable variation.

252.7 247.9 270.3 267.0 281.6 269.9 257.7 272.9 273.7 259.1 275.6 262.4 256.0
277.6 264.3 260.1 254.7 278.1 259.7 269.4 266.6 270.9 281.0 271.4 277.3

Monitoring the Minting of Quarters. *In Exercises 7–9, use the following information: The U.S. Mint has a goal of making quarters with a weight of 5.670 g, but any weight between 5.443 g and 5.897 g is considered acceptable. A new minting machine is placed into service and the weights are recorded for a quarter randomly selected every 12 min for 20 consecutive hours. The results are listed in the accompanying table.*

7. Minting Quarters: Constructing a Run Chart   Construct a run chart for the 100 values. Does there appear to be a pattern suggesting that the process is not within statistical control? What are the practical implications of the run chart?

8. Minting Quarters: Constructing an *R* Chart   Construct an *R* chart and determine whether the process variation is within statistical control. If it is not, identify which of the three out-of-control criteria lead to rejection of statistically stable variation.

9. Minting Quarters: Constructing an $\bar{x}$ Chart   Construct an $\bar{x}$ chart and determine whether the process mean is within statistical control. If it is not, identify which of the three out-of-control criteria lead to rejection of a statistically stable mean. Does this process need corrective action?

Weights (in grams) of Minted Quarters

Hour	Weight (g)					$\bar{x}$	s	Range
1	5.639	5.636	5.679	5.637	5.691	5.6564	0.0265	0.055
2	5.655	5.641	5.626	5.668	5.679	5.6538	0.0211	0.053
3	5.682	5.704	5.725	5.661	5.721	5.6986	0.0270	0.064
4	5.675	5.648	5.622	5.669	5.585	5.6398	0.0370	0.090
5	5.690	5.636	5.715	5.694	5.709	5.6888	0.0313	0.079
6	5.641	5.571	5.600	5.665	5.676	5.6306	0.0443	0.105
7	5.503	5.601	5.706	5.624	5.620	5.6108	0.0725	0.203
8	5.669	5.589	5.606	5.685	5.556	5.6210	0.0545	0.129
9	5.668	5.749	5.762	5.778	5.672	5.7258	0.0520	0.110
10	5.693	5.690	5.666	5.563	5.668	5.6560	0.0534	0.130
11	5.449	5.464	5.732	5.619	5.673	5.5874	0.1261	0.283
12	5.763	5.704	5.656	5.778	5.703	5.7208	0.0496	0.122
13	5.679	5.810	5.608	5.635	5.577	5.6618	0.0909	0.233
14	5.389	5.916	5.985	5.580	5.935	5.7610	0.2625	0.596
15	5.747	6.188	5.615	5.622	5.510	5.7364	0.2661	0.678
16	5.768	5.153	5.528	5.700	6.131	5.6560	0.3569	0.978
17	5.688	5.481	6.058	5.940	5.059	5.6452	0.3968	0.999
18	6.065	6.282	6.097	5.948	5.624	6.0032	0.2435	0.658
19	5.463	5.876	5.905	5.801	5.847	5.7784	0.1804	0.442
20	5.682	5.475	6.144	6.260	6.760	6.0642	0.5055	1.285

11. The process variation appears to be out of statistical control. There are points that lie beyond the control limits.

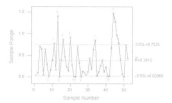

12. The process mean appears to be out of statistical control. There is a point lying beyond the upper control limit.

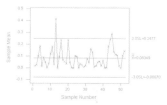

13. The process variation appears to be out of statistical control. There is a point beyond the upper control limit and there appears to be an upward trend.

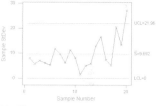

14. The process mean appears to be out of statistical control. There appears to be an upward trend, and there are eight consecutive points lying below the centerline.

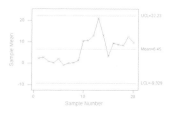

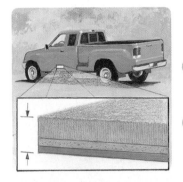

Quality Control at Perstorp

Perstorp Components, Inc. uses a computer that automatically generates control charts to monitor the thicknesses of the floor insulation the company makes for Ford Rangers and Jeep Grand Cherokees. The $20,000 cost of the computer was offset by a first-year savings of $40,000 in labor, which had been used to manually generate control charts to ensure that insulation thicknesses were between the specifications of 2.912 mm and 2.988 mm. Through the use of control charts and other quality-control methods, Perstorp reduced its waste by more than two-thirds.

Note to Instructor

The examples and exercises in this section assume that all samples have the same size *n*. There are ways to deal with unequal sample sizes, but we do not consider them in this book. (See the *Minitab Reference Manual*.)

Constructing Control Charts for Boston Rainfall. *In Exercises 10–12, refer to the daily amounts of rainfall in Boston for one year, as listed in Data Set 11 in Appendix B. Omit the last entry for Wednesday so that each day of the week has exactly 52 values.*

T **10.** Boston Rainfall: Constructing a Run Chart Using only the 52 rainfall amounts for Monday, construct a run chart. Does the process appear to be within statistical control?

T **11.** Boston Rainfall: Constructing an *R* Chart Using the 52 samples of seven values each, construct an *R* chart and determine whether the process variation is within statistical control. If it is not, identify which of the three out-of-control criteria lead to rejection of statistically stable variation.

T **12.** Boston Rainfall: Constructing an $\bar{x}$ Chart Using the 52 samples of seven values each, construct an $\bar{x}$ chart and determine whether the process mean is within statistical control. If it is not, identify which of the three out-of-control criteria lead to rejection of a statistically stable mean. If not, what can be done to bring the process within statistical control?

13-2 Beyond the Basics

13. Constructing an *s* Chart In this section we described control charts for *R* and $\bar{x}$ based on ranges. Control charts for monitoring variation and center (mean) can also be based on standard deviations. An *s chart* for monitoring variation is made by plotting sample standard deviations with a centerline at $\bar{s}$ (the mean of the sample standard deviations) and control limits at $B_4\bar{s}$ and $B_3\bar{s}$, where B_4 and B_3 are found in Table 13-2. Construct an *s* chart for the data of Table 13-1. Compare the result to the *R* chart given in this section.

14. Constructing an $\bar{x}$ Chart Based on Standard Deviations An $\bar{x}$ chart based on standard deviations (instead of ranges) is made by plotting sample means with a centerline at $\bar{\bar{x}}$ and control limits at $\bar{\bar{x}} + A_3\bar{s}$ and $\bar{\bar{x}} - A_3\bar{s}$, where A_3 is found in Table 13-2 and $\bar{s}$ is the mean of the sample standard deviations. Use the data in Table 13-1 to construct an $\bar{x}$ chart based on standard deviations. Compare the result to the $\bar{x}$ chart based on sample ranges (shown in this section).

13-3 Control Charts for Attributes

The main objective of this section is to develop the ability to monitor an *attribute* by constructing and interpreting an appropriate control chart. In Section 13-2 we monitored *quantitative* data, but we now consider *qualitative* data, investigating questions such as whether an item is defective, whether an item weighs less than a prescribed amount, or whether an item is nonconforming. (A good or a service is nonconforming if it doesn't meet specifications or requirements; nonconforming goods are sometimes discarded, repaired, or called "seconds" and sold at reduced prices.) As in Section 13-2, we select samples of size *n* at regular time intervals and plot points in a sequential graph with a centerline and control limits. (There are ways to deal with samples of different sizes, but we don't consider them here.)

The **control chart for p** (or **p chart**) is a control chart used to monitor the proportion p for some attribute. The notation and control chart values are as follows (where the attribute of "defective" can be replaced by any other relevant attribute).

Notation

$\bar{p}$ = pooled estimate of the proportion of defective items in the process

$\quad = \dfrac{\text{total number of defects found among all items sampled}}{\text{total number of items sampled}}$

$\bar{q}$ = pooled estimate of the proportion of process items that are *not* defective

$\quad = 1 - \bar{p}$

n = size of each sample (not the number of samples)

Control Chart for p

Centerline: $\bar{p}$

Upper control limit: $\bar{p} + 3\sqrt{\dfrac{\bar{p}\,\bar{q}}{n}}$

Lower control limit: $\bar{p} - 3\sqrt{\dfrac{\bar{p}\,\bar{q}}{n}}$

(If the calculation for the lower control limit results in a negative value, use 0 instead. If the calculation for the upper control limit exceeds 1, use 1 instead.)

We use $\bar{p}$ for the centerline because it is the best estimate of the proportion of defects from the process. The expressions for the control limits correspond to 99.7% confidence interval limits as described in Section 6-2.

EXAMPLE **Deaths from Infectious Diseases** Physicians report that infectious diseases should be carefully monitored over time because they are much more likely to have sudden changes in trends than are other diseases, such as cancer. In each of 13 consecutive and recent years, 100,000 subjects were randomly selected and the number who died from respiratory tract infections is recorded, with the results given here (based on data from "Trends in Infectious Diseases Mortality in the United States," by Pinner et al., *Journal of the American Medical Association*, Vol. 275, No. 3). Construct a control chart for p and determine whether the process is within statistical control. If not, identify which of the three out-of-control criteria apply.

Number of deaths: 25   24   22   25   27   30   31   30   33   32   33   32   31

continued

Six Sigma in Industry

Six Sigma is the term used in industry to describe a process that results in a rate of no more than 3.4 defects out of a million. The reference to Six Sigma suggests six standard deviations away from the center of a normal distribution, but the assumption of a perfectly stable process is replaced with the assumption of a process that drifts slightly, so the defect rate is no more than 3 or 4 defects per million.

Started around 1985 at Motorola, Six Sigma programs now attempt to improve quality and increase profits by reducing variation in processes. Motorola saved more than $940 million in three years. Allied Signal reported a savings of $1.5 billion. GE, Polaroid, Ford, Honeywell, Sony, and Texas Instruments are other major companies that have adopted the Six Sigma goal.

High Cost of Low Quality

The Federal Drug Administration recently reached an agreement whereby a pharmaceutical company, the Schering-Plough Corporation, would pay a record $500 million for failure to correct problems in manufacturing drugs. According to a *New York Times* article by Melody Petersen, "Some of the problems relate to the lack of controls that would identify faulty medicines, while others stem from outdated equipment. They involve some 200 medicines, including Claritin, the allergy medicine that is Schering's top-selling product."

SOLUTION   The centerline for our control chart is located by the value of $\bar{p}$:

$$\bar{p} = \frac{\text{total number of deaths from all samples combined}}{\text{total number of subjects sampled}}$$

$$= \frac{25 + 24 + 22 + \cdots + 31}{13 \cdot 100{,}000} = \frac{375}{1{,}300{,}000} = 0.000288$$

Because $\bar{p} = 0.000288$, it follows that $\bar{q} = 1 - \bar{p} = 0.999712$. Using $\bar{p} = 0.000288$, $\bar{q} = 0.999712$, and $n = 100{,}000$, we find the control limits as follows:

Upper control limit:

$$\bar{p} + 3\sqrt{\frac{\bar{p}\,\bar{q}}{n}} = 0.000288 + 3\sqrt{\frac{(0.000288)(0.999712)}{100{,}000}} = 0.000449$$

Lower control limit:

$$\bar{p} - 3\sqrt{\frac{\bar{p}\,\bar{q}}{n}} = 0.000288 - 3\sqrt{\frac{(0.000288)(0.999712)}{100{,}000}} = 0.000127$$

Having found the values for the centerline and control limits, we can proceed to plot the yearly proportion of deaths from respiratory tract infections. The Excel control chart for p is shown in the accompanying display.

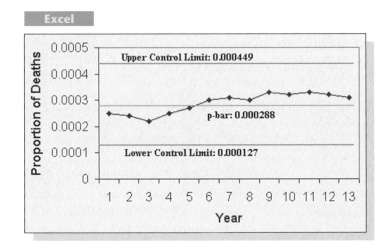

INTERPRETATION We can interpret the control chart for p by considering the three out-of-control criteria listed in Section 13-2. Using those criteria, we conclude that this process is out of statistical control for these reasons: There appears to be an upward trend, and there are eight consecutive points all lying above the centerline (Run of 8 Rule). Based on these data, public health policies affecting respiratory tract infections should be modified to cause a decrease in the death rate.

Using Technology

Minitab   Enter the numbers of defects (or items with any particular attribute) in column C1. Select the option **Stat,** then **Control Charts, Attributes Charts,** then **P.** Enter C1 in the box identified as variable, and enter the size of the samples in the box identified as subgroup size, then click **OK.**

Excel   **Using DDXL:** To use the DDXL add-in, begin by entering the numbers of defects or successes in column A, and enter the sample sizes in column B. For the example of this section, the first three items would be entered in the Excel spreadsheet as shown below.

	A	B
1	25	100000
2	24	100000
3	22	100000

Click on **DDXL,** select **Process Control,** then select **Summ Prop Control Chart** (for summary proportions control chart). A dialog box should appear. Click on the pencil icon for "Success Variable" and enter the range of values for column A, such as A1:A13. Click on the pencil icon for "Totals Variable" and enter the range of values for column B, such as B1:B13. Click **OK.** Next click on the **Open Control Chart** bar and the control chart will be displayed.

Using Excel's Chart Wizard: Enter the sample proportions in column A. (You could enter the actual numbers of defects in column A, then use Excel to create a column B consisting of the proportions. In the formula box, enter $=A1/n,$ where n is replaced by the size of each sample. After pressing Enter, cell B1 should contain the first sample proportion. Click and drag the lower right corner of cell B1 so that the entire B column has sample proportions corresponding to the actual numbers of defects in column A.) Having the data entered, proceed to generate the graph by first clicking on the **Chart Wizard** icon, which looks like a bar graph. For the chart type, select **Line.** For the chart subtype, select the first graph in the second row, then click **Next.** Continue to click **Next,** then **Finish.** The graph can be edited to include labels, delete grid lines, and so on. You can insert the required centerline and upper and lower control limits by editing the graph. Click on the line on the bottom of the screen, then click and drag to position the line correctly.

13-3 Basic Skills and Concepts

Determining Whether a Process Is in Control. In Exercises 1–4, examine the given control chart for p and determine whether the process is within statistical control. If it is not, identify which of the three out-of-control criteria apply.

1. Process appears to be within statistical control.

2. Process appears to be within statistical control.

3. Process appears to be out of statistical control because there is a pattern of an upward trend and there is a point that lies beyond the upper control limit.

4. Process appears to be out of statistical control because there are points lying beyond the control limits.

5. The process is out of statistical control because there is a downward trend and there are eight consecutive points all lying below the centerline. This downward trend is good and its causes should be identified so that it can continue.

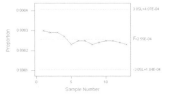

6. The process is within statistical control, but this describes the actual behavior of the process, not the desired behavior. People generally feel that there is too much violent crime, and the stable levels are too high.

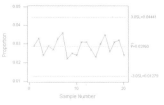

7. The process appears to be statistically stable.

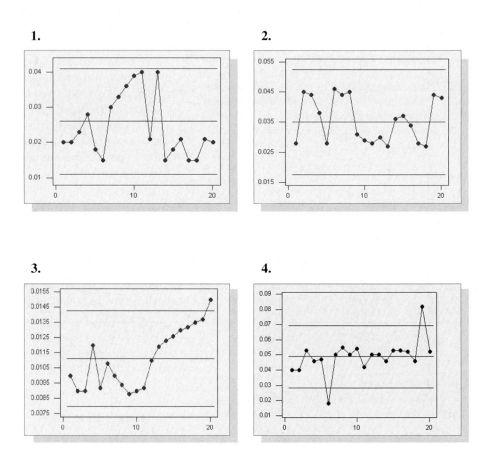

Constructing Control Charts for p. In Exercises 5–8, use the given process data to construct a control chart for p. In each case, use the three out-of-control criteria listed in Section 13-2 and determine whether the process is within statistical control. If it is not, identify which of the three out-of-control criteria apply.

5. *p* Chart for Deaths from Infectious Diseases In each of 13 consecutive and recent years, 100,000 children aged 0–4 years were randomly selected and the number who died from infectious diseases is recorded, with the results given below (based on data from "Trends in Infectious Diseases Mortality in the United States," by Pinner et al., *Journal of the American Medical Association*, Vol. 275, No. 3). Do the results suggest a problem that should be corrected?

   Number who died: 30   29   29   27   23   25   25   23   24   25   25   24   23

6. *p* Chart for Victims of Crime In each of 20 consecutive and recent years, 1000 adults were randomly selected and surveyed. Each value below is the number who were vic-

tims of violent crime (based on data from the U.S. Department of Justice, Bureau of Justice Statistics). Do the data suggest a problem that should be corrected?

29	33	24	29	27	33	36	22	25	24
31	31	27	23	30	35	26	31	32	24

7. *p* Chart for Boston Rainfall Refer to the Boston rainfall amounts in Data Set 11 in Appendix B. For each of the 52 weeks, let the sample proportion be the proportion of days that it rained. (Delete the 53rd value for Wednesday). In the first week, for example, the sample proportion is $3/7 = 0.429$. Do the data represent a statistically stable process?

8. *p* Chart for Marriage Rates Use *p* charts to compare the statistical stability of the marriage rates of Japan and the United States. In each year, 10,000 people in each country were randomly selected, and the numbers of marriages are given for eight consecutive and recent years (based on United Nations data).

Japan:	58	60	61	64	63	63	64	63
United States:	98	94	92	90	91	89	88	87

13-3 Beyond the Basics

9. Constructing an *np* Chart A variation of the control chart for *p* is the ***np* chart** in which the *actual numbers* of defects are plotted instead of the *proportions* of defects. The *np* chart will have a centerline value of $n\bar{p}$, and the control limits will have values of $n\bar{p} + 3\sqrt{n\bar{p}\,\bar{q}}$ and $n\bar{p} - 3\sqrt{n\bar{p}\,\bar{q}}$. The *p* chart and the *np* chart differ only in the scale of values used for the vertical axis. Construct the *np* chart for the example given in this section. Compare the result with the control chart for *p* given in this section.

10. Identifying Effect of Sample Size on *p* Chart
   a. Identify the locations of the centerline and control limits for a *p* chart representing a process that has been having a 5% rate of nonconforming items, based on samples of size 100.
   b. Repeat part (a) after changing the sample size to 300.
   c. Compare the two sets of results. Name an advantage and a disadvantage of using the larger sample size. Which chart would be better in detecting a shift from 5% to 10%?

Review

Whereas earlier chapters of this book focused on the important data characteristics of center, variation, distribution, and outliers, this chapter focused on pattern over time. Process data were defined to be data arranged according to some time sequence, and such data can be analyzed with run charts and control charts. Control charts have a centerline, an upper control limit, and a lower control limit. A process is statistically stable (or within statistical control) if it has only natural variation with no patterns, cycles, or unusual points. Decisions about statistical stability are based on how a process is actually behaving, not how we might like it to behave because of such factors as manufacturer specifications. The following graphs were described:

- *Run chart:* a sequential plot of *individual* data values over time
- *R chart:* a control chart that uses ranges in an attempt to monitor the *variation* in a process

8. Both processes appear to be statistically stable, but there appears to be a very slight upward trend in Japan, and a very slight downward trend in the United States.

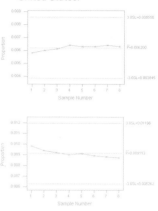

9. Except for the scale used, the charts are identical.

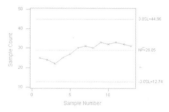

10. a. LCL = 0; centerline is at 0.05; UCL = 0.1154.
   b. LCL = 0.01225; centerline is at 0.05; UCL = 0.08775.
   c. As the sample size *n* increases, the control lines become closer together. The advantage is that smaller shifts in the process are detected. A disadvantage is that more items must be sampled. The chart in part (b) would be better for detecting a change from 5% to 10%.

1. The process appears to be within statistical control.

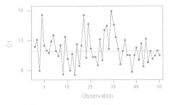

2. The process variation appears to be within statistical control.

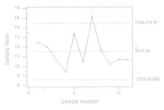

3. Because there is a point that lies beyond the upper control limit, the process mean is not within statistical control.

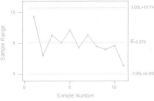

4. The process is out of control because there is a shift up and there are points beyond the control limits.

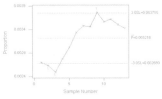

5. The process is out of control because there are points beyond the control limits. Also, there is a cyclical pattern.

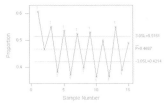

- $\bar{x}$ *chart:* a control chart used to determine whether the process *mean* is within statistical control
- *p chart:* a control chart used to monitor the proportion of some process *attribute,* such as whether items are defective

Review Exercises

Constructing Control Charts for Acid Rain. *In Exercises 1–3, use the following information. As part of a study monitoring acid rain, measurements of sulfate deposits (kg/hectare) are recorded for different locations on the East Coast (based on data from the U.S. Department of Agriculture). The results are listed in the following table for 11 recent and consecutive years.*

Acid Rain: Sulfate Deposits (kg/hectare)

Year	Location 1	Location 2	Location 3	Location 4	Location 5
1	11.94	13.09	7.96	17.29	12.12
2	11.28	10.88	12.84	13.87	11.21
3	10.38	12.19	7.38	13.64	9.95
4	8.00	10.75	7.26	12.37	8.77
5	12.12	17.21	10.12	15.73	11.68
6	10.27	10.26	8.89	13.21	9.71
7	14.80	15.49	11.60	17.94	15.59
8	13.52	11.61	9.02	11.22	13.05
9	10.55	10.53	7.78	10.57	11.77
10	9.81	12.50	8.70	13.29	9.37
11	11.27	9.94	10.50	11.28	10.54

1. Sulfate Deposits: Constructing a Run Chart Construct a run chart for the 55 values. Does there appear to be a pattern suggesting that the process is not within statistical control?

2. Sulfate Deposits: Constructing an *R* Chart Construct an *R* chart and determine whether the process variation is within statistical control. If it is not, identify which of the three out-of-control criteria lead to rejection of statistically stable variation.

3. Sulfate Deposits: Constructing an $\bar{x}$ Chart Construct an $\bar{x}$ chart and determine whether the process mean is within statistical control. Does the process appear to be statistically stable? How should this process behave if we implement effective programs to reduce the amount of acid rain?

4. Constructing a Control Chart for Infectious Diseases In each of 13 consecutive and recent years, 100,000 adults 65 years of age or older were randomly selected and the number who died from infectious diseases is recorded, with the results given below (based on data from "Trends in Infectious Diseases Mortality in the United States," by Pinner et al., *Journal of the American Medical Association,* Vol. 275, No. 3). Construct an appropriate control chart and determine whether the process is within statistical control. If not, identify which criteria lead to rejection of statistical stability.

Number who died: 270 264 250 278 302 334 348 347 377 357 362 351 343

5. Constructing a Control Chart for Voter Turnout In a continuing study of voter turnout, 1000 people of voting age are randomly selected in each year when there is a national election, and the numbers who actually voted are listed below (based on data

from the *Time Almanac*). Construct an appropriate control chart and determine whether the process is within statistical control. If not, identify which criteria lead to rejection of statistical stability.

Number who voted:	608	466	552	382	536	372	526	398
	531	364	501	365	551	388	491	

Cumulative Review Exercises

1. Analyzing Fuse Production Process The Telektronic Company produces 20-amp fuses used to protect car radios from too much electrical power. Each day 400 fuses are randomly selected and tested; the results (numbers of defects per 400 fuses tested) for 20 consecutive days are as follows:

10   8   7   6   6   9   12   5   4   7   9   6   11   4   6   5   10   5   9   11

  a. Use a control chart for p to verify that the process is within statistical control, so the data can be treated as coming from a population with fixed variation and mean.
  b. Using all of the data combined, construct a 95% confidence interval for the proportion of defects.
  c. Using a 0.05 significance level, test the claim that the rate of defects is more than 1%.

2. Using Probability in Control Charts When interpreting control charts, one of the three out-of-control criteria is that there are eight consecutive points all above or all below the centerline. For a statistically stable process, there is a 0.5 probability that a point will be above the centerline and there is a 0.5 probability that a point will be below the centerline. In each of the following, assume that sample values are independent and the process is statistically stable.
  a. Find the probability that when eight consecutive points are randomly selected, they are all above the centerline.
  b. Find the probability that when eight consecutive points are randomly selected, they are all below the centerline.
  c. Find the probability that when eight consecutive points are randomly selected, they are all above or all below the centerline.

3. Using Control Charts for Temperatures In Exercises 2–4 in Section 13-2, the amounts of electrical energy consumption were listed for the author's home during a period of four recent years. The accompanying table lists the average temperature (in degrees Fahrenheit) for the same time periods. Use appropriate control or run charts to determine whether the data appear to be part of a statistically stable process.

	Jan.–Feb.	Mar.–Apr.	May–June	July–Aug.	Sept.–Oct.	Nov.–Dec.
Year 1	32	35	59	76	66	42
Year 2	22	33	56	70	63	42
Year 3	30	38	55	71	61	38
Year 4	32	40	57	72	65	45

4. Relationship Between Energy Consumption and Temperature Refer to the data in Exercise 3 and the data used for Exercises 2–4 in Section 13-2. Match the data in pairs according to the corresponding time periods.
  a. Is there a significant linear correlation between the amounts of electrical energy consumption and the temperatures? Explain.

1. a. The process appears to be within statistical control.

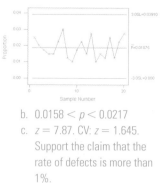

  b. $0.0158 < p < 0.0217$
  c. $z = 7.87$. CV: $z = 1.645$. Support the claim that the rate of defects is more than 1%.
2. a. 1/256
  b. 1/256
  c. 1/128
3. The run chart reveals very clear cycles, so that the process is not statistically stable.

4. a. $r = -0.484$. CV: $r = \pm 0.396$ (approximately, assuming a 0.05 significance level). Support the claim of a significant linear correlation between temperature and energy consumption.
  b. $\hat{y} = 4278 - 23.9x$
  c. 2844 kilowatt-hours

b. Identify the linear regression equation that relates electrical energy consumption (y) and the temperature (x).

c. What is the best predicted amount of electrical energy consumption for a two-month period with an average temperature of 60°F?

Cooperative Group Activities

1. Out-of-class activity Collect your own set of process data and analyze them using the methods of this section. It would be ideal to collect data from a real manufacturing process, but that may be difficult to accomplish. If so, consider using a simulation or referring to published data, such as those found in an almanac. Here are some suggestions:
   - Shoot five basketball foul shots (or shoot five crumpled sheets of paper into a wastebasket) and record the number of shots made; then repeat this procedure 20 times, and use a p chart to test for statistical stability in the proportion of shots made.
   - Your pulse rate can be measured by counting the number of times your heart beats in 1 minute. Measure your pulse rate four times each hour for several hours, then construct appropriate control charts. What factors contribute to random variation? Assignable variation?
   - Go through newspapers for the past 12 weeks and record the closing of the Dow-Jones Industrial Average for each business day. Use run and control charts to explore the statistical stability of the Dow-Jones Industrial Average. Identify at least one practical consequence of having this process statistically stable, and identify at least one practical consequence of having this process out of statistical control.

   - Find the divorce rate in terms of divorces per 1000 population for several years. (See the *Information Please Almanac* or the *Statistical Abstract of the United States.*) Assume that in each year 1000 people were randomly selected and surveyed to determine whether they became divorced. Use a p chart to test for statistical stability of the divorce rate. (Other possible rates: marriage, birth, death, accident fatality.)

   Obtain a printed copy of computer results, and write a report summarizing your conclusions.

2. In-class activity If the instructor can distribute the numbers of absences for each class meeting, groups of three or four students can analyze them for statistical stability and make recommendations based on the conclusions.

3. Out-of-class activity Conduct research to identify *Deming's funnel experiment,* then use a funnel and marbles to collect data for the different rules for adjusting the funnel location. Construct appropriate control charts for the different rules of funnel adjustment. What does the funnel experiment illustrate? What do you conclude?

Technology Project

a. Simulate the following process for 20 days: Each day, 200 heart pacemakers are manufactured with a 1% rate of defective units, and the proportion of defects is recorded for each of the 20 days. The pacemakers for one day are simulated by randomly generating 200 numbers, where each number is between 1 and 100. Consider an outcome of 1 to be a defect, with 2 through 100 being acceptable. This corresponds to a 1% rate of defects. (Parts, b, c, and d follow the technology instructions.)

STATDISK   Select **Data** from the main menu bar, then select **Uniform Generator** and proceed to generate 200 values with a minimum of 1 and a maximum of 100. Repeat

this procedure until results for 20 days have been simulated.

Minitab   From the main menu bar, select **Calc,** then **Random Data,** then **Integer.** Enter 200 in the box for the number of rows of data, enter C1 as the column to be used for storing the data, enter 1 for the minimum value, and enter 100 for the maximum value. Repeat this procedure until results for 20 days have been simulated.

Excel   Click on the *fx* icon on the main menu bar, then select the function category **Math & Trig,** followed by **RANDBETWEEN.** In the dialog box, enter 1 for bottom and 100 for top. A random value should appear in the first row of column A. Use the mouse to click and drag the lower right corner of that cell, then pull down the cell to cover the first 200 rows of column A. When you release the mouse button, column A should contain 200 random numbers. You can also click/drag the lower right corner of the bottom cell by moving the mouse to the right so that you get 20 columns of 200 numbers each. The different columns represent the different days of manufacturing.

TI-83 Plus   Press the **MATH** key. Select **PRB,** then select the 5th menu item, **randInt(,** and proceed to enter 1, 100, 200; then press the **ENTER** key. Press **STO** and **L1** to store the data in list L1. After recording the number of defects, repeat this procedure until results for 20 days have been simulated.

b. Construct a p chart for the proportion of defective pacemakers, and determine whether the process is within statistical control. Since we know the process is actually stable with $p = 0.01$, the conclusion that it is not stable would be a type I error; that is, we would have a false positive signal, causing us to believe that the process needed to be adjusted when in fact it should be left alone.

c. The result from part (a) is a simulation of 20 days. Now simulate another 10 days of manufacturing pacemakers, but modify these last 10 days so that the nonconforming rate is 3% instead of 1%.

d. Combine the data generated from parts (a) and (c) to represent a total of 30 days of sample results. Construct a p chart for this combined data set. Is the process out of control? If we concluded that the process was not out of control, we would be making a type II error; that is, we would believe that the process was okay when in fact it should be repaired or adjusted to correct the shift to the 3% nonconforming rate.

from DATA *to* DECISION

Critical Thinking: Are the axial loads within statistical control?

Is the process of manufacturing cans proceeding as it should?

Exercises 5 and 6 in Section 13-2 used process data from a New York company that manufactures 0.0109-in. thick aluminum cans for a major beverage supplier. Refer to Data Set 20 in Appendix B and conduct an analysis of the process data for the cans that are 0.0111 in. thick. The values in the data set are the measured axial loads of cans, and the top lids are pressed into place with pressures that vary between 158 lb and 165 lb.

Analyzing the Results

Should you take any corrective action? Write a report summarizing your conclusions. Address not only the issue of statistical stability, but also the ability of the cans to withstand the pressures applied when the top lids are pressed into place. Also compare the behavior of the 0.0111-in. cans to the behavior of the 0.0109-in. cans and recommend which thickness should be used.

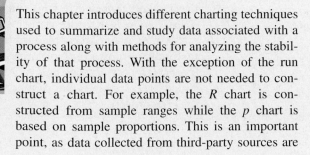

INTERNET PROJECT   Control Charts

This chapter introduces different charting techniques used to summarize and study data associated with a process along with methods for analyzing the stability of that process. With the exception of the run chart, individual data points are not needed to construct a chart. For example, the R chart is constructed from sample ranges while the p chart is based on sample proportions. This is an important point, as data collected from third-party sources are often given in terms of summarizing statistics. Go to the *Elementary Statistics* Web site:

http://www.aw.com/triola

Locate the Internet Project dealing with control charts. There you will be directed to data sets and sources of data for use in constructing control charts. From the resulting charts you will be asked to interpret and discuss trends in the underlying processes.

Statistics @ Work

"There is a certain amount of respect that is given to someone who knows statistics and can explain it to someone who doesn't know it."

Dan O'Toole

Account Executive: A. C. Nielsen

In his work in the Advanced Analytics Group at A. C. Nielsen, Dan develops statistical solutions to help clients like Polaroid, Ocean Spray, and Gillette understand which of their marketing vehicles drives sales most profitably. Dan has a Masters Degree in Business Economics from Bentley College.

What concepts of statistics do you use?

I have worked with analyses as simple as correlation and general significance tests, all the way to multiple regression, factor analysis, correspondence analysis, and cluster analysis.

How do you use statistics on the job?

My job is to discover or uncover client issues, and then find out if we can apply one of our statistical techniques to their specific issue. If a technique won't help a client, then you need to know that. An example of how I use stats: A client may say, I sell product "X," whether it is juice, bread, or a camera. Right now, they may control 20% of the market. They may come to us to see if they can increase market share by lowering their price. My job would be to design a study to analyze this question. To do this I have to design a study that will take into account everything that affects the sales of a product. Using techniques like regression, if I am able to create a model with good significance, I will be able to isolate specific influences on the sales and offer recommendations. Things like seasonality distribution as well as any marketing efforts that may have taken place, must be included. In addition, I will have to take into account complementary products' price (butter is complementary to bread; while film is for a camera) and also competitive products. For instance bread may compete with English muffins (I know it does for me).

Could you describe one specific example illustrating how the use of statistics was successful in improving a product or service?

We were modeling a juice product. The client felt that store brands, or private label brands, affected their sales, and the client felt they needed to drop price to save their market share. When we finished modeling, it seemed that the two products did not compete with each other based on price. So, if store brands lowered their price, it wouldn't affect their sales. This seemed not to make sense. How could this be true? What we discovered was that, when Private Label entered a market, they would steal all of the value customers—the people who will buy the lowest product no matter what. However, everyone else would remain. So, although the client lost some of their customers, if they were to lower the price, it wouldn't gain them any more sales or buyers.

Do you feel job applicants are viewed more favorably if they have studied some statistics?

By far. There is a certain amount of respect that is given to someone who knows statistics and can explain it to someone who doesn't know it (because it means you really know it and aren't reciting from a textbook). Almost every job uses statistics (particularly correlations and regressions). People will say things like, "Oh, check if they're correlated."

Is your use of probability and statistics increasing, decreasing, or remaining stable?

It definitely is increasing. In this business (consulting), You are constantly challenged to learn a new technique or look at an old technique in order to improve it. In addition, since we are constantly coming out with new products, our understanding of statistics has to increase to use these techniques effectively.

How beneficial do you find your knowledge of statistics for performing your responsibilities?

It is not a question of beneficial, but rather it is a necessity. In fact, we find that we have to know it so well, so that we can explain it in "layman's" terms to our clients.

In terms of statistics, what would you recommend for prospective employees?

Many people have taken statistics, but the level of retention of most concepts is low. If you had to focus on some core concepts, I would say you would need to understand correlation and linear regression—understanding these concepts will help you interpret other concepts you bump into, like multiple regression. I would also go with factor analysis—just a general understanding would even put you ahead of the curve in most respects.

14 Projects, Procedures, Perspectives

14-1 Projects

The main objective of this section is to provide some suggestions for a study that can be used as a capstone project for the introductory statistics course. One fantastic advantage of this course is that it deals with skills and concepts that can be applied immediately to the real world. After only one fun semester, students are able to conduct their own studies. Some of the suggested topics can be addressed by actually conducting experiments, whereas others might be observational studies that require research of results already available. For example, testing the effectiveness of air bags by actually crashing cars is strongly discouraged, but destructive taste tests of chocolate chip cookies can be an easy and somewhat enjoyable experiment. Here is a suggested format, followed by a list of suggested topics.

Group/Individual   Topics can be assigned to individuals, but group projects are particularly effective because they help develop the interpersonal skills that are so necessary in today's working environment. One study showed that the "inability to get along with others" is the main reason for firing employees, so a group project can be very helpful in preparing students for their future work environments.

Oral Report   A 10- to 15-minute-long class presentation should involve all group members in a coordinated effort to clearly describe the important components of the study. Students typically have some reluctance to speak in public, so a brief oral report can be very helpful in building the confidence that they so well deserve. Again, the oral report is an activity that helps students to be better prepared for future professional activities.

Written Report   The main objective of the project is not to produce a written document equivalent to a term paper, but a written report should be submitted, and it should include the following components:

1. List of data collected
2. Description of the method of analysis
3. Relevant graphs and/or statistics, including STATDISK, Minitab, Excel, or TI-83 Plus displays
4. Statement of conclusions
5. Reasons why the results might not be correct, along with a description of ways in which the study could be improved, given sufficient time and money

Suggested Topics   In addition to the topics suggested in the following list, also see the *Cooperative Group Activities* listed near the end of each chapter.

1. Graph from a newspaper or magazine redrawn to better describe the data
2. Newspaper article about a survey rewritten to better inform the reader
3. Using coin toss to get better survey results from sensitive question
4. Ages of student cars compared to faculty/staff cars
5. Proportion of foreign cars driven by students compared to the proportion of foreign cars driven by faculty
6. Car ages in the parking lot of a discount store compared to car ages in the parking lot of an upscale department store
7. Are husbands older than their wives?
8. Are husband/wife age differences the same for young married couples as for older married couples?
9. Analysis of the ages of books in the college library
10. How do the ages of books in the college library compare with those in the library of a nearby college?
11. Comparison of the ages of science books and English books in the college library
12. Estimate the hours that students study each week
13. Is there a relationship between hours studied and grades earned?
14. Is there a relationship between hours worked and grades earned?
15. A study of *reported* heights compared to *measured* heights
16. A study of the accuracy of wristwatches
17. Is there a relationship between taste and cost of different brands of chocolate chip cookies?
18. Is there a relationship between taste and cost of different brands of peanut butter?
19. Is there a relationship between taste and cost of different brands of cola?
20. Is there a relationship between salaries of professional baseball (or basketball, or football) players and their season achievements?

21. Rates versus weights: Is there a relationship between car fuel-consumption rates and car weights? If so, what is it?

22. Is there a relationship between the lengths of men's (or women's) feet and their heights?

23. Are there differences in taste between ordinary tap water and different brands of bottled water?

24. Were auto fatality rates affected by laws requiring the use of seat belts?

25. Were auto fatality rates affected when the national speed limit of 55 mi/h was eliminated?

26. Were auto fatality rates affected by the presence of air bags?

27. Is there a difference in taste between Coke and Pepsi?

28. Is there a relationship between student grade-point averages and the amount of television watched? If so, what is it?

29. Is there a relationship between the selling price of a home and its living area (in square feet), lot size (in acres), number of rooms, number of baths, and the annual tax bill?

30. Is there a relationship between the height of a person and the height of his or her navel?

31. Is there support for the theory that the ratio of a person's height to his or her navel height is the Golden Ratio of about 1.6:1?

32. A comparison of the numbers of keys carried by males and females

33. A comparison of the numbers of credit cards carried by males and females

34. Are murderers now younger than they were in the past?

35. Do people who exercise vigorously tend to have lower pulse rates than those who do not?

36. Do people who exercise vigorously tend to have reaction times that are different from those of people who do not?

37. Do people who smoke tend to have higher pulse rates than those who do not?

38. For people who don't exercise, how is pulse rate affected by climbing a flight of stairs?

39. Do statistics students tend to have pulse rates that are different from those of people not studying statistics?

40. A comparison of GPAs of statistics students with those of students not taking statistics

41. Do left-handed people tend to be involved in more car crashes?

42. Do men have more car crashes than women?

43. Do young drivers have more car crashes than older drivers?

44. Are drivers who get tickets more likely to be involved in crashes?

45. Do smokers tend to be involved in more car crashes?

46. Do people with higher pulse rates tend to be involved in more/fewer car crashes?

47. A comparison of reaction times measured with right and left hands

48. Are the proportions of male and female smokers equal?

49. Do statistics students tend to smoke more (or less) than the general population?

50. Are people more likely to smoke if their parents smoked?

51. Evidence to support/refute the belief that smoking tends to stunt growth

52. Does a sports team have an advantage by playing at home instead of away?

53. Analysis of service times (in seconds) for a car drive-up window at a bank

54. A comparison of service times for car drive-up windows at two different banks

55. Analysis of times that McDonald's' patrons are seated at a table

56. Analysis of times that McDonald's' patrons wait in line

57. Analysis of times cars require for refueling

58. Is the state lottery a wise investment?

59. Comparison of casino games: craps versus roulette

60. Starting with $1, is it easier to win a million dollars by playing casino craps or by playing a state lottery?

61. Bold versus cautious strategies of gambling: When gambling with $100, does it make any difference if you bet $1 at a time or if you bet the whole $100 at once?

62. Designing and analyzing results from a test for extrasensory perception

63. Analyzing paired data consisting of heights of fathers (or mothers) and heights of their first sons (or daughters)

64. Gender differences in preferences of dinner partners among the options of Brad Pitt, Tiger Woods, the President, Nicole Kidman, Cameron Diaz, Julia Roberts, and the Pope

65. Gender differences in preferences of activities among the options of dinner, movie, watching television, reading a book, golf, tennis, swimming, attending a baseball game, attending a football game

66. Is there support for the theory that cereals with high sugar content are placed on shelves at eye level with children?

67. Is there support for the claim that the mean body temperature is less than 98.6°F?

68. Is there a relationship between smoking and drinking coffee?

69. Is there a relationship between course grades and time spent playing video games?

70. Is there support for the theory that a Friday is unlucky if it falls on the 13th day of a month?

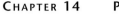 Procedure

Data Collection   You can collect your own data through experiments or observational studies. It is absolutely essential to critique the method used to collect the data, because data carelessly collected may be so completely useless that no amount of statistical torturing can salvage them. Look carefully for bias in the way data are collected, as well as bias on the part of the person or group collecting the data. Many of the procedures in this book are based on the assumption that we are working with a simple random sample, meaning that every possible sample of the same size has the same chance of being selected. If a sample is self-selected (voluntary response), it is worthless for making inferences about a population.

Exploring, Comparing, Describing   After collecting data, first consider exploring, describing, and comparing data sets using the basic tools included in Chapter 2. Be sure to address the following:

1. *Center:* Find the mean and median, which are measures of center that are representative or average values giving us an indication of where the middle of the data set is located.

2. *Variation:* Find the range and standard deviation, which are measures of the amount that the sample values vary among themselves.

3. *Distribution:* Construct a histogram to see the nature or shape of the distribution of the data, and determine if the distribution is bell-shaped, uniform, or skewed.

4. *Outliers:* Identify any sample values that lie very far away from the vast majority of the other sample values.

5. *Time:* Determine if the population is stable or if its characteristics are changing over time.

Inferences: Estimating Parameters and Hypothesis Testing   When trying to use sample data for making inferences about a population, it is often difficult to choose the particular procedure that should be applied. This text includes a wide variety of procedures that apply to many different circumstances. Here are some key questions that should be answered:

- What is the level of measurement (nominal, ordinal, interval, ratio) of the data?
- Does the study involve one, two, or more populations?
- Is there a claim to be tested or a parameter to be estimated?
- What is the relevant parameter (mean, standard deviation, proportion)?
- Is the population standard deviation known? (The answer is almost always "no.")
- Is there reason to believe that the population is normally distributed?
- What is the basic question or issue that you want to address?

In Figure 14-1 we list the major methods included in this book, along with a scheme for determining which of those methods should be used. To use Figure 14-1, start at the extreme left side of the figure and begin by identifying the level of measurement of the data. Proceed to follow the path suggested by the level of measurement, the number of populations, and the claim or parameter being considered.

Note: This figure applies to a fixed population. If the data are from a process that may change over time, construct a control chart (see Chapter 13) to determine whether the process is statistically stable. This figure applies to process data only if the process is statistically stable.

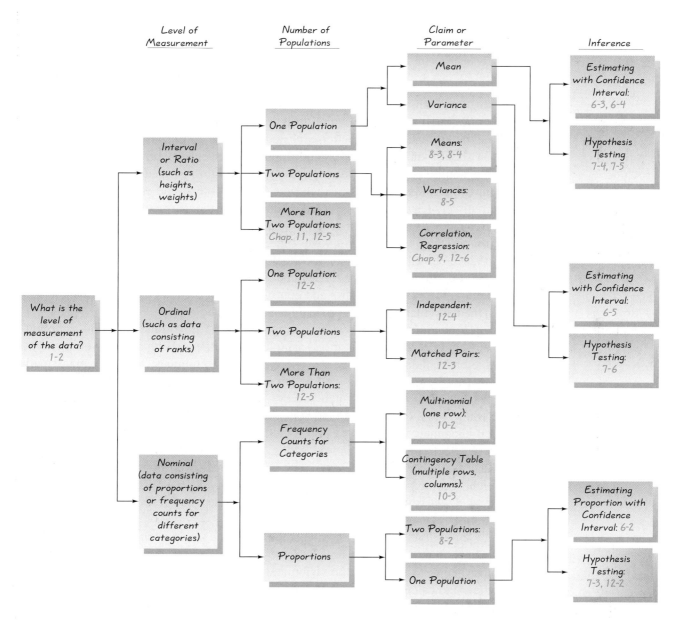

FIGURE 14-1   **Selecting the Appropriate Procedure**

Figure 14-1 can be used for statistical methods presented in this book, but there may be other methods that might be more suitable for a particular statistical analysis. Consult your friendly professional statistician for help with other methods.

14-3 Perspective

No one expects a single introductory statistics course to transform anyone into an expert statistician. After studying several of the chapters in this book, it is natural for students to feel that they have not mastered the material to the extent necessary for confidently using statistics in real applications. Many important topics (such as factor analysis and discriminant analysis) are not included in this text because they are too advanced for this introductory level. Some easier topics (such as time series) have been excluded for other reasons. It is important to know that professional help is available from expert statisticians, and this introductory statistics course will help you in discussions with one of these experts.

Although this course is not designed to make you an expert statistician, it is designed to make you a better educated person with improved job marketability. You should know and understand the basic concepts of probability and chance. You should know that in attempting to gain insight into a set of data, it is important to investigate measures of center (such as mean and median), measures of variation (such as range and standard deviation), the nature of the distribution (via a frequency distribution or graph), the presence of outliers, and whether the population is stable or is changing over time. You should know and understand the importance of estimating population parameters (such as a mean, standard deviation, and proportion), as well as testing claims made about population parameters. You should realize that the nature and configuration of the data have a dramatic effect on the particular statistical procedures that are used.

Throughout this text we have emphasized the importance of good sampling. You should recognize that a bad sample may be beyond repair by even the most expert statisticians using the most sophisticated techniques. There are many mail, magazine, and telephone call-in surveys that allow respondents to be "self-selected." The results of such surveys are generally worthless when judged according to the criteria of sound statistical methodology. Keep this in mind when you encounter voluntary response (self-selected) surveys, so that you don't let them affect your beliefs and decisions. You should also recognize, however, that many surveys and polls obtain very good results, even though the sample sizes might seem to be relatively small. Although many people refuse to believe it, a nationwide survey of only 1700 voters can provide good results if the sampling is carefully planned and executed.

Throughout this text we have emphasized the *interpretation* of results. A final conclusion to "reject the null hypothesis" is basically worthless to all of those other people who lacked the vision and wisdom to take a statistics course. Computers and calculators are quite good at yielding results, but such results typically require the careful interpretation that breathes life into an otherwise meaningless result. We should recognize that a result is not automatically valid and good simply because it was computer-generated. Computers don't think, and they are quite capable of providing results that are quite ridiculous when considered in the context of the real world. We should always apply the most important and indispensable tool in all of statistics: *common sense!*

There was once a time that a person was considered to be educated if he or she could simply read. But we are now in an era that demands so much more. Today, an educated person must be able to read, write, use computer software, speak a foreign language, and know basic algebra. A truly educated person is capable of combining disciplines with common goals, including the quest for truth. The study of statistics helps us see the truth that is sometimes distorted by others or concealed by data that are disorganized or perhaps not yet collected. Understanding principles of statistics is now essential for every educated person. H. G. Wells once said that "statistical thinking will one day be as necessary for efficient citizenship as the ability to read and write." That day is now.

Appendix A: Tables

Table A-1   Binomial Probabilities

Table A-2   Standard Normal Distribution

Table A-3 t Distribution

Table A-4   Chi-Square (χ^2) Distribution

Table A-5 F Distribution

Table A-6   Critical Values of the Pearson Correlation Coefficient r

Table A-7   Critical Values for the Sign Test

Table A-8   Critical Values of T for the Wilcoxon Signed-Ranks Test

Table A-9   Critical Values of Spearman's Rank Correlation Coefficient r_s

Table A-10   Critical Values for Number of Runs G

TABLE A-1	Binomial Probabilities

| | | | | | | | | p | | | | | | | |
n	x	.01	.05	.10	.20	.30	.40	.50	.60	.70	.80	.90	.95	.99	x
2	0	.980	.902	.810	.640	.490	.360	.250	.160	.090	.040	.010	.002	0+	0
	1	.020	.095	.180	.320	.420	.480	.500	.480	.420	.320	.180	.095	.020	1
	2	0+	.002	.010	.040	.090	.160	.250	.360	.490	.640	.810	.902	.980	2
3	0	.970	.857	.729	.512	.343	.216	.125	.064	.027	.008	.001	0+	0+	0
	1	.029	.135	.243	.384	.441	.432	.375	.288	.189	.096	.027	.007	0+	1
	2	0+	.007	.027	.096	.189	.288	.375	.432	.441	.384	.243	.135	.029	2
	3	0+	0+	.001	.008	.027	.064	.125	.216	.343	.512	.729	.857	.970	3
4	0	.961	.815	.656	.410	.240	.130	.062	.026	.008	.002	0+	0+	0+	0
	1	.039	.171	.292	.410	.412	.346	.250	.154	.076	.026	.004	0+	0+	1
	2	.001	.014	.049	.154	.265	.346	.375	.346	.265	.154	.049	.014	.001	2
	3	0+	0+	.004	.026	.076	.154	.250	.346	.412	.410	.292	.171	.039	3
	4	0+	0+	0+	.002	.008	.026	.062	.130	.240	.410	.656	.815	.961	4
5	0	.951	.774	.590	.328	.168	.078	.031	.010	.002	0+	0+	0+	0+	0
	1	.048	.204	.328	.410	.360	.259	.156	.077	.028	.006	0+	0+	0+	1
	2	.001	.021	.073	.205	.309	.346	.312	.230	.132	.051	.008	.001	0+	2
	3	0+	.001	.008	.051	.132	.230	.312	.346	.309	.205	.073	.021	.001	3
	4	0+	0+	0+	.006	.028	.077	.156	.259	.360	.410	.328	.204	.048	4
	5	0+	0+	0+	0+	.002	.010	.031	.078	.168	.328	.590	.774	.951	5
6	0	.941	.735	.531	.262	.118	.047	.016	.004	.001	0+	0+	0+	0+	0
	1	.057	.232	.354	.393	.303	.187	.094	.037	.010	.002	0+	0+	0+	1
	2	.001	.031	.098	.246	.324	.311	.234	.138	.060	.015	.001	0+	0+	2
	3	0+	.002	.015	.082	.185	.276	.312	.276	.185	.082	.015	.002	0+	3
	4	0+	0+	.001	.015	.060	.138	.234	.311	.324	.246	.098	.031	.001	4
	5	0+	0+	0+	.002	.010	.037	.094	.187	.303	.393	.354	.232	.057	5
	6	0+	0+	0+	0+	.001	.004	.016	.047	.118	.262	.531	.735	.941	6
7	0	.932	.698	.478	.210	.082	.028	.008	.002	0+	0+	0+	0+	0+	0
	1	.066	.257	.372	.367	.247	.131	.055	.017	.004	0+	0+	0+	0+	1
	2	.002	.041	.124	.275	.318	.261	.164	.077	.025	.004	0+	0+	0+	2
	3	0+	.004	.023	.115	.227	.290	.273	.194	.097	.029	.003	0+	0+	3
	4	0+	0+	.003	.029	.097	.194	.273	.290	.227	.115	.023	.004	0+	4
	5	0+	0+	0+	.004	.025	.077	.164	.261	.318	.275	.124	.041	.002	5
	6	0+	0+	0+	0+	.004	.017	.055	.131	.247	.367	.372	.257	.066	6
	7	0+	0+	0+	0+	0+	.002	.008	.028	.082	.210	.478	.698	.932	7
8	0	.923	.663	.430	.168	.058	.017	.004	.001	0+	0+	0+	0+	0+	0
	1	.075	.279	.383	.336	.198	.090	.031	.008	.001	0+	0+	0+	0+	1
	2	.003	.051	.149	.294	.296	.209	.109	.041	.010	.001	0+	0+	0+	2
	3	0+	.005	.033	.147	.254	.279	.219	.124	.047	.009	0+	0+	0+	3
	4	0+	0+	.005	.046	.136	.232	.273	.232	.136	.046	.005	0+	0+	4
	5	0+	0+	0+	.009	.047	.124	.219	.279	.254	.147	.033	.005	0+	5
	6	0+	0+	0+	.001	.010	.041	.109	.209	.296	.294	.149	.051	.003	6
	7	0+	0+	0+	0+	.001	.008	.031	.090	.198	.336	.383	.279	.075	7
	8	0+	0+	0+	0+	0+	.001	.004	.017	.058	.168	.430	.663	.923	8

NOTE: 0+ represents a positive probability less than 0.0005.

(continued)

n	x	.01	.05	.10	.20	.30	.40	.50	.60	.70	.80	.90	.95	.99	x
9	0	.914	.630	.387	.134	.040	.010	.002	0+	0+	0+	0+	0+	0+	0
	1	.083	.299	.387	.302	.156	.060	.018	.004	0+	0+	0+	0+	0+	1
	2	.003	.063	.172	.302	.267	.161	.070	.021	.004	0+	0+	0+	0+	2
	3	0+	.008	.045	.176	.267	.251	.164	.074	.021	.003	0+	0+	0+	3
	4	0+	.001	.007	.066	.172	.251	.246	.167	.074	.017	.001	0+	0+	4
	5	0+	0+	.001	.017	.074	.167	.246	.251	.172	.066	.007	.001	0+	5
	6	0+	0+	0+	.003	.021	.074	.164	.251	.267	.176	.045	.008	0+	6
	7	0+	0+	0+	0+	.004	.021	.070	.161	.267	.302	.172	.063	.003	7
	8	0+	0+	0+	0+	0+	.004	.018	.060	.156	.302	.387	.299	.083	8
	9	0+	0+	0+	0+	0+	0+	.002	.010	.040	.134	.387	.630	.914	9
10	0	.904	.599	.349	.107	.028	.006	.001	0+	0+	0+	0+	0+	0+	0
	1	.091	.315	.387	.268	.121	.040	.010	.002	0+	0+	0+	0+	0+	1
	2	.004	.075	.194	.302	.233	.121	.044	.011	.001	0+	0+	0+	0+	2
	3	0+	.010	.057	.201	.267	.215	.117	.042	.009	.001	0+	0+	0+	3
	4	0+	.001	.011	.088	.200	.251	.205	.111	.037	.006	0+	0+	0+	4
	5	0+	0+	.001	.026	.103	.201	.246	.201	.103	.026	.001	0+	0+	5
	6	0+	0+	0+	.006	.037	.111	.205	.251	.200	.088	.011	.001	0+	6
	7	0+	0+	0+	.001	.009	.042	.117	.215	.267	.201	.057	.010	0+	7
	8	0+	0+	0+	0+	.001	.011	.044	.121	.233	.302	.194	.075	.004	8
	9	0+	0+	0+	0+	0+	.002	.010	.040	.121	.268	.387	.315	.091	9
	10	0+	0+	0+	0+	0+	0+	.001	.006	.028	.107	.349	.599	.904	10
11	0	.895	.569	.314	.086	.020	.004	0+	0+	0+	0+	0+	0+	0+	0
	1	.099	.329	.384	.236	.093	.027	.005	.001	0+	0+	0+	0+	0+	1
	2	.005	.087	.213	.295	.200	.089	.027	.005	.001	0+	0+	0+	0+	2
	3	0+	.014	.071	.221	.257	.177	.081	.023	.004	0+	0+	0+	0+	3
	4	0+	.001	.016	.111	.220	.236	.161	.070	.017	.002	0+	0+	0+	4
	5	0+	0+	.002	.039	.132	.221	.226	.147	.057	.010	0+	0+	0+	5
	6	0+	0+	0+	.010	.057	.147	.226	.221	.132	.039	.002	0+	0+	6
	7	0+	0+	0+	.002	.017	.070	.161	.236	.220	.111	.016	.001	0+	7
	8	0+	0+	0+	0+	.004	.023	.081	.177	.257	.221	.071	.014	0+	8
	9	0+	0+	0+	0+	.001	.005	.027	.089	.200	.295	.213	.087	.005	9
	10	0+	0+	0+	0+	0+	.001	.005	.027	.093	.236	.384	.329	.099	10
	11	0+	0+	0+	0+	0+	0+	0+	.004	.020	.086	.314	.569	.895	11
12	0	.886	.540	.282	.069	.014	.002	0+	0+	0+	0+	0+	0+	0+	0
	1	.107	.341	.377	.206	.071	.017	.003	0+	0+	0+	0+	0+	0+	1
	2	.006	.099	.230	.283	.168	.064	.016	.002	0+	0+	0+	0+	0+	2
	3	0+	.017	.085	.236	.240	.142	.054	.012	.001	0+	0+	0+	0+	3
	4	0+	.002	.021	.133	.231	.213	.121	.042	.008	.001	0+	0+	0+	4
	5	0+	0+	.004	.053	.158	.227	.193	.101	.029	.003	0+	0+	0+	5
	6	0+	0+	0+	.016	.079	.177	.226	.177	.079	.016	0+	0+	0+	6
	7	0+	0+	0+	.003	.029	.101	.193	.227	.158	.053	.004	0+	0+	7
	8	0+	0+	0+	.001	.008	.042	.121	.213	.231	.133	.021	.002	0+	8
	9	0+	0+	0+	0+	.001	.012	.054	.142	.240	.236	.085	.017	0+	9
	10	0+	0+	0+	0+	0+	.002	.016	.064	.168	.283	.230	.099	.006	10
	11	0+	0+	0+	0+	0+	0+	.003	.017	.071	.206	.377	.341	.107	11
	12	0+	0+	0+	0+	0+	0+	0+	.002	.014	.069	.282	.540	.886	12

NOTE: 0+ represents a positive probability less than 0.0005.

(*continued*)

								p								
n	x	.01	.05	.10	.20	.30	.40	.50	.60	.70	.80	.90	.95	.99	x
13	0	.878	.513	.254	.055	.010	.001	0+	0+	0+	0+	0+	0+	0+	0
	1	.115	.351	.367	.179	.054	.011	.002	0+	0+	0+	0+	0+	0+	1
	2	.007	.111	.245	.268	.139	.045	.010	.001	0+	0+	0+	0+	0+	2
	3	0+	.021	.100	.246	.218	.111	.035	.006	.001	0+	0+	0+	0+	3
	4	0+	.003	.028	.154	.234	.184	.087	.024	.003	0+	0+	0+	0+	4
	5	0+	0+	.006	.069	.180	.221	.157	.066	.014	.001	0+	0+	0+	5
	6	0+	0+	.001	.023	.103	.197	.209	.131	.044	.006	0+	0+	0+	6
	7	0+	0+	0+	.006	.044	.131	.209	.197	.103	.023	.001	0+	0+	7
	8	0+	0+	0+	.001	.014	.066	.157	.221	.180	.069	.006	0+	0+	8
	9	0+	0+	0+	0+	.003	.024	.087	.184	.234	.154	.028	.003	0+	9
	10	0+	0+	0+	0+	.001	.006	.035	.111	.218	.246	.100	.021	0+	10
	11	0+	0+	0+	0+	0+	.001	.010	.045	.139	.268	.245	.111	.007	11
	12	0+	0+	0+	0+	0+	0+	.002	.011	.054	.179	.367	.351	.115	12
	13	0+	0+	0+	0+	0+	0+	0+	.001	.010	.055	.254	.513	.878	13
14	0	.869	.488	.229	.044	.007	.001	0+	0+	0+	0+	0+	0+	0+	0
	1	.123	.359	.356	.154	.041	.007	.001	0+	0+	0+	0+	0+	0+	1
	2	.008	.123	.257	.250	.113	.032	.006	.001	0+	0+	0+	0+	0+	2
	3	0+	.026	.114	.250	.194	.085	.022	.003	0+	0+	0+	0+	0+	3
	4	0+	.004	.035	.172	.229	.155	.061	.014	.001	0+	0+	0+	0+	4
	5	0+	0+	.008	.086	.196	.207	.122	.041	.007	0+	0+	0+	0+	5
	6	0+	0+	.001	.032	.126	.207	.183	.092	.023	.002	0+	0+	0+	6
	7	0+	0+	0+	.009	.062	.157	.209	.157	.062	.009	0+	0+	0+	7
	8	0+	0+	0+	.002	.023	.092	.183	.207	.126	.032	.001	0+	0+	8
	9	0+	0+	0+	0+	.007	.041	.122	.207	.196	.086	.008	0+	0+	9
	10	0+	0+	0+	0+	.001	.014	.061	.155	.229	.172	.035	.004	0+	10
	11	0+	0+	0+	0+	0+	.003	.022	.085	.194	.250	.114	.026	0+	11
	12	0+	0+	0+	0+	0+	.001	.006	.032	.113	.250	.257	.123	.008	12
	13	0+	0+	0+	0+	0+	0+	.001	.007	.041	.154	.356	.359	.123	13
	14	0+	0+	0+	0+	0+	0+	0+	.001	.007	.044	.229	.488	.869	14
15	0	.860	.463	.206	.035	.005	0+	0+	0+	0+	0+	0+	0+	0+	0
	1	.130	.366	.343	.132	.031	.005	0+	0+	0+	0+	0+	0+	0+	1
	2	.009	.135	.267	.231	.092	.022	.003	0+	0+	0+	0+	0+	0+	2
	3	0+	.031	.129	.250	.170	.063	.014	.002	0+	0+	0+	0+	0+	3
	4	0+	.005	.043	.188	.219	.127	.042	.007	.001	0+	0+	0+	0+	4
	5	0+	.001	.010	.103	.206	.186	.092	.024	.003	0+	0+	0+	0+	5
	6	0+	0+	.002	.043	.147	.207	.153	.061	.012	.001	0+	0+	0+	6
	7	0+	0+	0+	.014	.081	.177	.196	.118	.035	.003	0+	0+	0+	7
	8	0+	0+	0+	.003	.035	.118	.196	.177	.081	.014	0+	0+	0+	8
	9	0+	0+	0+	.001	.012	.061	.153	.207	.147	.043	.002	0+	0+	9
	10	0+	0+	0+	0+	.003	.024	.092	.186	.206	.103	.010	.001	0+	10
	11	0+	0+	0+	0+	.001	.007	.042	.127	.219	.188	.043	.005	0+	11
	12	0+	0+	0+	0+	0+	.002	.014	.063	.170	.250	.129	.031	0+	12
	13	0+	0+	0+	0+	0+	0+	.003	.022	.092	.231	.267	.135	.009	13
	14	0+	0+	0+	0+	0+	0+	0+	.005	.031	.132	.343	.366	.130	14
	15	0+	0+	0+	0+	0+	0+	0+	0+	.005	.035	.206	.463	.860	15

NOTE: 0+ represents a positive probability less than 0.0005.

From Frederick C. Mosteller, Robert E. K. Rourke, and George B. Thomas, Jr., *Probability with Statistical Applications*, 2nd ed., © 1970 Addison-Wesley Publishing Co., Reading, MA. Reprinted with permission.

NEGATIVE z Scores

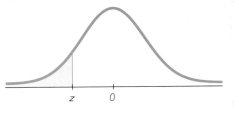

TABLE A-2	Standard Normal (z) Distribution: Cumulative Area from the LEFT								

z	.00	.01	.02	.03	.04	.05	.06	.07	.08	.09
−3.50 and lower	.0001									
−3.4	.0003	.0003	.0003	.0003	.0003	.0003	.0003	.0003	.0003	.0002
−3.3	.0005	.0005	.0005	.0004	.0004	.0004	.0004	.0004	.0004	.0003
−3.2	.0007	.0007	.0006	.0006	.0006	.0006	.0006	.0005	.0005	.0005
−3.1	.0010	.0009	.0009	.0009	.0008	.0008	.0008	.0008	.0007	.0007
−3.0	.0013	.0013	.0013	.0012	.0012	.0011	.0011	.0011	.0010	.0010
−2.9	.0019	.0018	.0018	.0017	.0016	.0016	.0015	.0015	.0014	.0014
−2.8	.0026	.0025	.0024	.0023	.0023	.0022	.0021	.0021	.0020	.0019
−2.7	.0035	.0034	.0033	.0032	.0031	.0030	.0029	.0028	.0027	.0026
−2.6	.0047	.0045	.0044	.0043	.0041	.0040	.0039	.0038	.0037	.0036
−2.5	.0062	.0060	.0059	.0057	.0055	.0054	.0052	.0051 *	.0049	.0048
−2.4	.0082	.0080	.0078	.0075	.0073	.0071	.0069	.0068	.0066	.0064
−2.3	.0107	.0104	.0102	.0099	.0096	.0094	.0091	.0089	.0087	.0084
−2.2	.0139	.0136	.0132	.0129	.0125	.0122	.0119	.0116	.0113	.0110
−2.1	.0179	.0174	.0170	.0166	.0162	.0158	.0154	.0150	.0146	.0143
−2.0	.0228	.0222	.0217	.0212	.0207	.0202	.0197	.0192	.0188	.0183
−1.9	.0287	.0281	.0274	.0268	.0262	.0256	.0250	.0244	.0239	.0233
−1.8	.0359	.0351	.0344	.0336	.0329	.0322	.0314	.0307	.0301	.0294
−1.7	.0446	.0436	.0427	.0418	.0409	.0401	.0392	.0384	.0375	.0367
−1.6	.0548	.0537	.0526	.0516	.0505 *	.0495	.0485	.0475	.0465	.0455
−1.5	.0668	.0655	.0643	.0630	.0618	.0606	.0594	.0582	.0571	.0559
−1.4	.0808	.0793	.0778	.0764	.0749	.0735	.0721	.0708	.0694	.0681
−1.3	.0968	.0951	.0934	.0918	.0901	.0885	.0869	.0853	.0838	.0823
−1.2	.1151	.1131	.1112	.1093	.1075	.1056	.1038	.1020	.1003	.0985
−1.1	.1357	.1335	.1314	.1292	.1271	.1251	.1230	.1210	.1190	.1170
−1.0	.1587	.1562	.1539	.1515	.1492	.1469	.1446	.1423	.1401	.1379
−0.9	.1841	.1814	.1788	.1762	.1736	.1711	.1685	.1660	.1635	.1611
−0.8	.2119	.2090	.2061	.2033	.2005	.1977	.1949	.1922	.1894	.1867
−0.7	.2420	.2389	.2358	.2327	.2296	.2266	.2236	.2206	.2177	.2148
−0.6	.2743	.2709	.2676	.2643	.2611	.2578	.2546	.2514	.2483	.2451
−0.5	.3085	.3050	.3015	.2981	.2946	.2912	.2877	.2843	.2810	.2776
−0.4	.3446	.3409	.3372	.3336	.3300	.3264	.3228	.3192	.3156	.3121
−0.3	.3821	.3783	.3745	.3707	.3669	.3632	.3594	.3557	.3520	.3483
−0.2	.4207	.4168	.4129	.4090	.4052	.4013	.3974	.3936	.3897	.3859
−0.1	.4602	.4562	.4522	.4483	.4443	.4404	.4364	.4325	.4286	.4247
−0.0	.5000	.4960	.4920	.4880	.4840	.4801	.4761	.4721	.4681	.4641

NOTE: For values of z below −3.49, use 0.0001 for the area.
*Use these common values that result from interpolation:

z score	Area
−1.645	0.0500
−2.575	0.0050

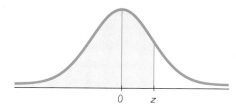

POSITIVE z Scores

TABLE A-2		(*continued*) Cumulative Area from the LEFT								
z	.00	.01	.02	.03	.04	.05	.06	.07	.08	.09
0.0	.5000	.5040	.5080	.5120	.5160	.5199	.5239	.5279	.5319	.5359
0.1	.5398	.5438	.5478	.5517	.5557	.5596	.5636	.5675	.5714	.5753
0.2	.5793	.5832	.5871	.5910	.5948	.5987	.6026	.6064	.6103	.6141
0.3	.6179	.6217	.6255	.6293	.6331	.6368	.6406	.6443	.6480	.6517
0.4	.6554	.6591	.6628	.6664	.6700	.6736	.6772	.6808	.6844	.6879
0.5	.6915	.6950	.6985	.7019	.7054	.7088	.7123	.7157	.7190	.7224
0.6	.7257	.7291	.7324	.7357	.7389	.7422	.7454	.7486	.7517	.7549
0.7	.7580	.7611	.7642	.7673	.7704	.7734	.7764	.7794	.7823	.7852
0.8	.7881	.7910	.7939	.7967	.7995	.8023	.8051	.8078	.8106	.8133
0.9	.8159	.8186	.8212	.8238	.8264	.8289	.8315	.8340	.8365	.8389
1.0	.8413	.8438	.8461	.8485	.8508	.8531	.8554	.8577	.8599	.8621
1.1	.8643	.8665	.8686	.8708	.8729	.8749	.8770	.8790	.8810	.8830
1.2	.8849	.8869	.8888	.8907	.8925	.8944	.8962	.8980	.8997	.9015
1.3	.9032	.9049	.9066	.9082	.9099	.9115	.9131	.9147	.9162	.9177
1.4	.9192	.9207	.9222	.9236	.9251	.9265	.9279	.9292	.9306	.9319
1.5	.9332	.9345	.9357	.9370	.9382	.9394	.9406	.9418	.9429	.9441
1.6	.9452	.9463	.9474	.9484	.9495 *	.9505	.9515	.9525	.9535	.9545
1.7	.9554	.9564	.9573	.9582	.9591	.9599	.9608	.9616	.9625	.9633
1.8	.9641	.9649	.9656	.9664	.9671	.9678	.9686	.9693	.9699	.9706
1.9	.9713	.9719	.9726	.9732	.9738	.9744	.9750	.9756	.9761	.9767
2.0	.9772	.9778	.9783	.9788	.9793	.9798	.9803	.9808	.9812	.9817
2.1	.9821	.9826	.9830	.9834	.9838	.9842	.9846	.9850	.9854	.9857
2.2	.9861	.9864	.9868	.9871	.9875	.9878	.9881	.9884	.9887	.9890
2.3	.9893	.9896	.9898	.9901	.9904	.9906	.9909	.9911	.9913	.9916
2.4	.9918	.9920	.9922	.9925	.9927	.9929	.9931	.9932	.9934	.9936
2.5	.9938	.9940	.9941	.9943	.9945	.9946	.9948	.9949 *	.9951	.9952
2.6	.9953	.9955	.9956	.9957	.9959	.9960	.9961	.9962	.9963	.9964
2.7	.9965	.9966	.9967	.9968	.9969	.9970	.9971	.9972	.9973	.9974
2.8	.9974	.9975	.9976	.9977	.9977	.9978	.9979	.9979	.9980	.9981
2.9	.9981	.9982	.9982	.9983	.9984	.9984	.9985	.9985	.9986	.9986
3.0	.9987	.9987	.9987	.9988	.9988	.9989	.9989	.9989	.9990	.9990
3.1	.9990	.9991	.9991	.9991	.9992	.9992	.9992	.9992	.9993	.9993
3.2	.9993	.9993	.9994	.9994	.9994	.9994	.9994	.9995	.9995	.9995
3.3	.9995	.9995	.9995	.9996	.9996	.9996	.9996	.9996	.9996	.9997
3.4	.9997	.9997	.9997	.9997	.9997	.9997	.9997	.9997	.9997	.9998
3.50 and up	.9999									

NOTE: For values of z above 3.49, use 0.9999 for the area.
*Use these common values that result from interpolation:

z score	Area
1.645	0.9500
2.575	0.9950

Common Critical Values

Confidence Level	Critical Value
0.90	1.645
0.95	1.96
0.99	2.575

TABLE A-3	t Distribution: Critical t Values				

			Area in One Tail		
	0.005	0.01	0.025	0.05	0.10
Degrees of Freedom			Area in Two Tails		
	0.01	0.02	0.05	0.10	0.20
1	63.657	31.821	12.706	6.314	3.078
2	9.925	6.965	4.303	2.920	1.886
3	5.841	4.541	3.182	2.353	1.638
4	4.604	3.747	2.776	2.132	1.533
5	4.032	3.365	2.571	2.015	1.476
6	3.707	3.143	2.447	1.943	1.440
7	3.499	2.998	2.365	1.895	1.415
8	3.355	2.896	2.306	1.860	1.397
9	3.250	2.821	2.262	1.833	1.383
10	3.169	2.764	2.228	1.812	1.372
11	3.106	2.718	2.201	1.796	1.363
12	3.055	2.681	2.179	1.782	1.356
13	3.012	2.650	2.160	1.771	1.350
14	2.977	2.624	2.145	1.761	1.345
15	2.947	2.602	2.131	1.753	1.341
16	2.921	2.583	2.120	1.746	1.337
17	2.898	2.567	2.110	1.740	1.333
18	2.878	2.552	2.101	1.734	1.330
19	2.861	2.539	2.093	1.729	1.328
20	2.845	2.528	2.086	1.725	1.325
21	2.831	2.518	2.080	1.721	1.323
22	2.819	2.508	2.074	1.717	1.321
23	2.807	2.500	2.069	1.714	1.319
24	2.797	2.492	2.064	1.711	1.318
25	2.787	2.485	2.060	1.708	1.316
26	2.779	2.479	2.056	1.706	1.315
27	2.771	2.473	2.052	1.703	1.314
28	2.763	2.467	2.048	1.701	1.313
29	2.756	2.462	2.045	1.699	1.311
30	2.750	2.457	2.042	1.697	1.310
31	2.744	2.453	2.040	1.696	1.309
32	2.738	2.449	2.037	1.694	1.309
34	2.728	2.441	2.032	1.691	1.307
36	2.719	2.434	2.028	1.688	1.306
38	2.712	2.429	2.024	1.686	1.304
40	2.704	2.423	2.021	1.684	1.303
45	2.690	2.412	2.014	1.679	1.301
50	2.678	2.403	2.009	1.676	1.299
55	2.668	2.396	2.004	1.673	1.297
60	2.660	2.390	2.000	1.671	1.296
65	2.654	2.385	1.997	1.669	1.295
70	2.648	2.381	1.994	1.667	1.294
75	2.643	2.377	1.992	1.665	1.293
80	2.639	2.374	1.990	1.664	1.292
90	2.632	2.368	1.987	1.662	1.291
100	2.626	2.364	1.984	1.660	1.290
200	2.601	2.345	1.972	1.653	1.286
300	2.592	2.339	1.968	1.650	1.284
400	2.588	2.336	1.966	1.649	1.284
500	2.586	2.334	1.965	1.648	1.283
750	2.582	2.331	1.963	1.647	1.283
1000	2.581	2.330	1.962	1.646	1.282
2000	2.578	2.328	1.961	1.646	1.282
Large	2.576	2.326	1.960	1.645	1.282

Left tail

α

Critical t value (negative)

Right tail

α

Critical t value (positive)

Two tails

$\alpha/2$ $\alpha/2$

Critical t value (negative)   Critical t value (positive)

TABLE A-4	Chi-Square (χ^2) Distribution

Area to the Right of the Critical Value

Degrees of Freedom	0.995	0.99	0.975	0.95	0.90	0.10	0.05	0.025	0.01	0.005
1	—	—	0.001	0.004	0.016	2.706	3.841	5.024	6.635	7.879
2	0.010	0.020	0.051	0.103	0.211	4.605	5.991	7.378	9.210	10.597
3	0.072	0.115	0.216	0.352	0.584	6.251	7.815	9.348	11.345	12.838
4	0.207	0.297	0.484	0.711	1.064	7.779	9.488	11.143	13.277	14.860
5	0.412	0.554	0.831	1.145	1.610	9.236	11.071	12.833	15.086	16.750
6	0.676	0.872	1.237	1.635	2.204	10.645	12.592	14.449	16.812	18.548
7	0.989	1.239	1.690	2.167	2.833	12.017	14.067	16.013	18.475	20.278
8	1.344	1.646	2.180	2.733	3.490	13.362	15.507	17.535	20.090	21.955
9	1.735	2.088	2.700	3.325	4.168	14.684	16.919	19.023	21.666	23.589
10	2.156	2.558	3.247	3.940	4.865	15.987	18.307	20.483	23.209	25.188
11	2.603	3.053	3.816	4.575	5.578	17.275	19.675	21.920	24.725	26.757
12	3.074	3.571	4.404	5.226	6.304	18.549	21.026	23.337	26.217	28.299
13	3.565	4.107	5.009	5.892	7.042	19.812	22.362	24.736	27.688	29.819
14	4.075	4.660	5.629	6.571	7.790	21.064	23.685	26.119	29.141	31.319
15	4.601	5.229	6.262	7.261	8.547	22.307	24.996	27.488	30.578	32.801
16	5.142	5.812	6.908	7.962	9.312	23.542	26.296	28.845	32.000	34.267
17	5.697	6.408	7.564	8.672	10.085	24.769	27.587	30.191	33.409	35.718
18	6.265	7.015	8.231	9.390	10.865	25.989	28.869	31.526	34.805	37.156
19	6.844	7.633	8.907	10.117	11.651	27.204	30.144	32.852	36.191	38.582
20	7.434	8.260	9.591	10.851	12.443	28.412	31.410	34.170	37.566	39.997
21	8.034	8.897	10.283	11.591	13.240	29.615	32.671	35.479	38.932	41.401
22	8.643	9.542	10.982	12.338	14.042	30.813	33.924	36.781	40.289	42.796
23	9.260	10.196	11.689	13.091	14.848	32.007	35.172	38.076	41.638	44.181
24	9.886	10.856	12.401	13.848	15.659	33.196	36.415	39.364	42.980	45.559
25	10.520	11.524	13.120	14.611	16.473	34.382	37.652	40.646	44.314	46.928
26	11.160	12.198	13.844	15.379	17.292	35.563	38.885	41.923	45.642	48.290
27	11.808	12.879	14.573	16.151	18.114	36.741	40.113	43.194	46.963	49.645
28	12.461	13.565	15.308	16.928	18.939	37.916	41.337	44.461	48.278	50.993
29	13.121	14.257	16.047	17.708	19.768	39.087	42.557	45.722	49.588	52.336
30	13.787	14.954	16.791	18.493	20.599	40.256	43.773	46.979	50.892	53.672
40	20.707	22.164	24.433	26.509	29.051	51.805	55.758	59.342	63.691	66.766
50	27.991	29.707	32.357	34.764	37.689	63.167	67.505	71.420	76.154	79.490
60	35.534	37.485	40.482	43.188	46.459	74.397	79.082	83.298	88.379	91.952
70	43.275	45.442	48.758	51.739	55.329	85.527	90.531	95.023	100.425	104.215
80	51.172	53.540	57.153	60.391	64.278	96.578	101.879	106.629	112.329	116.321
90	59.196	61.754	65.647	69.126	73.291	107.565	113.145	118.136	124.116	128.299
100	67.328	70.065	74.222	77.929	82.358	118.498	124.342	129.561	135.807	140.169

From Donald B. Owen, *Handbook of Statistical Tables,* ©1962 Addison-Wesley Publishing Co., Reading, MA. Reprinted with permission of the publisher.

Degrees of Freedom

$n - 1$	for confidence intervals or hypothesis tests with a standard deviation or variance
$k - 1$	for multinomial experiments or goodness-of-fit with k categories
$(r - 1)(c - 1)$	for contingency tables with r rows and c columns
$k - 1$	for Kruskal-Wallis test with k samples

TABLE A-5	*F* Distribution ($\alpha = 0.025$ in the right tail)

Numerator degrees of freedom (df$_1$)

	1	2	3	4	5	6	7	8	9
1	647.79	799.50	864.16	899.58	921.85	937.11	948.22	956.66	963.28
2	38.506	39.000	39.165	39.248	39.298	39.331	39.335	39.373	39.387
3	17.443	16.044	15.439	15.101	14.885	14.735	14.624	14.540	14.473
4	12.218	10.649	9.9792	9.6045	9.3645	9.1973	9.0741	8.9796	8.9047
5	10.007	8.4336	7.7636	7.3879	7.1464	6.9777	6.8531	6.7572	6.6811
6	8.8131	7.2599	6.5988	6.2272	5.9876	5.8198	5.6955	5.5996	5.5234
7	8.0727	6.5415	5.8898	5.5226	5.2852	5.1186	4.9949	4.8993	4.8232
8	7.5709	6.0595	5.4160	5.0526	4.8173	4.6517	4.5286	4.4333	4.3572
9	7.2093	5.7147	5.0781	4.7181	4.4844	4.3197	4.1970	4.1020	4.0260
10	6.9367	5.4564	4.8256	4.4683	4.2361	4.0721	3.9498	3.8549	3.7790
11	6.7241	5.2559	4.6300	4.2751	4.0440	3.8807	3.7586	3.6638	3.5879
12	6.5538	5.0959	4.4742	4.1212	3.8911	3.7283	3.6065	3.5118	3.4358
13	6.4143	4.9653	4.3472	3.9959	3.7667	3.6043	3.4827	3.3880	3.3120
14	6.2979	4.8567	4.2417	3.8919	3.6634	3.5014	3.3799	3.2853	3.2093
15	6.1995	4.7650	4.1528	3.8043	3.5764	3.4147	3.2934	3.1987	3.1227
16	6.1151	4.6867	4.0768	3.7294	3.5021	3.3406	3.2194	3.1248	3.0488
17	6.0420	4.6189	4.0112	3.6648	3.4379	3.2767	3.1556	3.0610	2.9849
18	5.9781	4.5597	3.9539	3.6083	3.3820	3.2209	3.0999	3.0053	2.9291
19	5.9216	4.5075	3.9034	3.5587	3.3327	3.1718	3.0509	2.9563	2.8801
20	5.8715	4.4613	3.8587	3.5147	3.2891	3.1283	3.0074	2.9128	2.8365
21	5.8266	4.4199	3.8188	3.4754	3.2501	3.0895	2.9686	2.8740	2.7977
22	5.7863	4.3828	3.7829	3.4401	3.2151	3.0546	2.9338	2.8392	2.7628
23	5.7498	4.3492	3.7505	3.4083	3.1835	3.0232	2.9023	2.8077	2.7313
24	5.7166	4.3187	3.7211	3.3794	3.1548	2.9946	2.8738	2.7791	2.7027
25	5.6864	4.2909	3.6943	3.3530	3.1287	2.9685	2.8478	2.7531	2.6766
26	5.6586	4.2655	3.6697	3.3289	3.1048	2.9447	2.8240	2.7293	2.6528
27	5.6331	4.2421	3.6472	3.3067	3.0828	2.9228	2.8021	2.7074	2.6309
28	5.6096	4.2205	3.6264	3.2863	3.0626	2.9027	2.7820	2.6872	2.6106
29	5.5878	4.2006	3.6072	3.2674	3.0438	2.8840	2.7633	2.6686	2.5919
30	5.5675	4.1821	3.5894	3.2499	3.0265	2.8667	2.7460	2.6513	2.5746
40	5.4239	4.0510	3.4633	3.1261	2.9037	2.7444	2.6238	2.5289	2.4519
60	5.2856	3.9253	3.3425	3.0077	2.7863	2.6274	2.5068	2.4117	2.3344
120	5.1523	3.8046	3.2269	2.8943	2.6740	2.5154	2.3948	2.2994	2.2217
∞	5.0239	3.6889	3.1161	2.7858	2.5665	2.4082	2.2875	2.1918	2.1136

Denominator degrees of freedom (df$_2$)

Numerator degrees of freedom (df_1)

	10	12	15	20	24	30	40	60	120	∞
1	968.63	976.71	984.87	993.10	997.25	1001.4	1005.6	1009.8	1014.0	1018.3
2	39.398	39.415	39.431	39.448	39.456	39.465	39.473	39.481	39.490	39.498
3	14.419	14.337	14.253	14.167	14.124	14.081	14.037	13.992	13.947	13.902
4	8.8439	8.7512	8.6565	8.5599	8.5109	8.4613	8.4111	8.3604	8.3092	8.2573
5	6.6192	6.5245	6.4277	6.3286	6.2780	6.2269	6.1750	6.1225	6.0693	6.0153
6	5.4613	5.3662	5.2687	5.1684	5.1172	5.0652	5.0125	4.9589	4.9044	4.8491
7	4.7611	4.6658	4.5678	4.4667	4.4150	4.3624	4.3089	4.2544	4.1989	4.1423
8	4.2951	4.1997	4.1012	3.9995	3.9472	3.8940	3.8398	3.7844	3.7279	3.6702
9	3.9639	3.8682	3.7694	3.6669	3.6142	3.5604	3.5055	3.4493	3.3918	3.3329
10	3.7168	3.6209	3.5217	3.4185	3.3654	3.3110	3.2554	3.1984	3.1399	3.0798
11	3.5257	3.4296	3.3299	3.2261	3.1725	3.1176	3.0613	3.0035	2.9441	2.8828
12	3.3736	3.2773	3.1772	3.0728	3.0187	2.9633	2.9063	2.8478	2.7874	2.7249
13	3.2497	3.1532	3.0527	2.9477	2.8932	2.8372	2.7797	2.7204	2.6590	2.5955
14	3.1469	3.0502	2.9493	2.8437	2.7888	2.7324	2.6742	2.6142	2.5519	2.4872
15	3.0602	2.9633	2.8621	2.7559	2.7006	2.6437	2.5850	2.5242	2.4611	2.3953
16	2.9862	2.8890	2.7875	2.6808	2.6252	2.5678	2.5085	2.4471	2.3831	2.3163
17	2.9222	2.8249	2.7230	2.6158	2.5598	2.5020	2.4422	2.3801	2.3153	2.2474
18	2.8664	2.7689	2.6667	2.5590	2.5027	2.4445	2.3842	2.3214	2.2558	2.1869
19	2.8172	2.7196	2.6171	2.5089	2.4523	2.3937	2.3329	2.2696	2.2032	2.1333
20	2.7737	2.6758	2.5731	2.4645	2.4076	2.3486	2.2873	2.2234	2.1562	2.0853
21	2.7348	2.6368	2.5338	2.4247	2.3675	2.3082	2.2465	2.1819	2.1141	2.0422
22	2.6998	2.6017	2.4984	2.3890	2.3315	2.2718	2.2097	2.1446	2.0760	2.0032
23	2.6682	2.5699	2.4665	2.3567	2.2989	2.2389	2.1763	2.1107	2.0415	1.9677
24	2.6396	2.5411	2.4374	2.3273	2.2693	2.2090	2.1460	2.0799	2.0099	1.9353
25	2.6135	2.5149	2.4110	2.3005	2.2422	2.1816	2.1183	2.0516	1.9811	1.9055
26	2.5896	2.4908	2.3867	2.2759	2.2174	2.1565	2.0928	2.0257	1.9545	1.8781
27	2.5676	2.4688	2.3644	2.2533	2.1946	2.1334	2.0693	2.0018	1.9299	1.8527
28	2.5473	2.4484	2.3438	2.2324	2.1735	2.1121	2.0477	1.9797	1.9072	1.8291
29	2.5286	2.4295	2.3248	2.2131	2.1540	2.0923	2.0276	1.9591	1.8861	1.8072
30	2.5112	2.4120	2.3072	2.1952	2.1359	2.0739	2.0089	1.9400	1.8664	1.7867
40	2.3882	2.2882	2.1819	2.0677	2.0069	1.9429	1.8752	1.8028	1.7242	1.6371
60	2.2702	2.1692	2.0613	1.9445	1.8817	1.8152	1.7440	1.6668	1.5810	1.4821
120	2.1570	2.0548	1.9450	1.8249	1.7597	1.6899	1.6141	1.5299	1.4327	1.3104
∞	2.0483	1.9447	1.8326	1.7085	1.6402	1.5660	1.4835	1.3883	1.2684	1.0000

From Maxine Merrington and Catherine M. Thompson, "Tables of Percentage Points of the Inverted Beta (*F*) Distribution," *Biometrika 33* (1943): 80–84. Reproduced with permission of the Biometrika Trustees.

(continued)

Denominator degrees of freedom (df_2)

| TABLE A-5 | | F Distribution ($\alpha = 0.05$ in the right tail) |

Numerator degrees of freedom (df_1)

	1	2	3	4	5	6	7	8	9
1	161.45	199.50	215.71	224.58	230.16	233.99	236.77	238.88	240.54
2	18.513	19.000	19.164	19.247	19.296	19.330	19.353	19.371	19.385
3	10.128	9.5521	9.2766	9.1172	9.0135	8.9406	8.8867	8.8452	8.8123
4	7.7086	6.9443	6.5914	6.3882	6.2561	6.1631	6.0942	6.0410	6.9988
5	6.6079	5.7861	5.4095	5.1922	5.0503	4.9503	4.8759	4.8183	4.7725
6	5.9874	5.1433	4.7571	4.5337	4.3874	4.2839	4.2067	4.1468	4.0990
7	5.5914	4.7374	4.3468	4.1203	3.9715	3.8660	3.7870	3.7257	3.6767
8	5.3177	4.4590	4.0662	3.8379	3.6875	3.5806	3.5005	3.4381	3.3881
9	5.1174	4.2565	3.8625	3.6331	3.4817	3.3738	3.2927	3.2296	3.1789
10	4.9646	4.1028	3.7083	3.4780	3.3258	3.2172	3.1355	3.0717	3.0204
11	4.8443	3.9823	3.5874	3.3567	3.2039	3.0946	3.0123	2.9480	2.8962
12	4.7472	3.8853	3.4903	3.2592	3.1059	2.9961	2.9134	2.8486	2.7964
13	4.6672	3.8056	3.4105	3.1791	3.0254	2.9153	2.8321	2.7669	2.7144
14	4.6001	3.7389	3.3439	3.1122	2.9582	2.8477	2.7642	2.6987	2.6458
15	4.5431	3.6823	3.2874	3.0556	2.9013	2.7905	2.7066	2.6408	2.5876
16	4.4940	3.6337	3.2389	3.0069	2.8524	2.7413	2.6572	2.5911	2.5377
17	4.4513	3.5915	3.1968	2.9647	2.8100	2.6987	2.6143	2.5480	2.4943
18	4.4139	3.5546	3.1599	2.9277	2.7729	2.6613	2.5767	2.5102	2.4563
19	4.3807	3.5219	3.1274	2.8951	2.7401	2.6283	2.5435	2.4768	2.4227
20	4.3512	3.4928	3.0984	2.8661	2.7109	2.5990	2.5140	2.4471	2.3928
21	4.3248	3.4668	3.0725	2.8401	2.6848	2.5727	2.4876	2.4205	2.3660
22	4.3009	3.4434	3.0491	2.8167	2.6613	2.5491	2.4638	2.3965	2.3419
23	4.2793	3.4221	3.0280	2.7955	2.6400	2.5277	2.4422	2.3748	2.3201
24	4.2597	3.4028	3.0088	2.7763	2.6207	2.5082	2.4226	2.3551	2.3002
25	4.2417	3.3852	2.9912	2.7587	2.6030	2.4904	2.4047	2.3371	2.2821
26	4.2252	3.3690	2.9752	2.7426	2.5868	2.4741	2.3883	2.3205	2.2655
27	4.2100	3.3541	2.9604	2.7278	2.5719	2.4591	2.3732	2.3053	2.2501
28	4.1960	3.3404	2.9467	2.7141	2.5581	2.4453	2.3593	2.2913	2.2360
29	4.1830	3.3277	2.9340	2.7014	2.5454	2.4324	2.3463	2.2783	2.2229
30	4.1709	3.3158	2.9223	2.6896	2.5336	2.4205	2.3343	2.2662	2.2107
40	4.0847	3.2317	2.8387	2.6060	2.4495	2.3359	2.2490	2.1802	2.1240
60	4.0012	3.1504	2.7581	2.5252	2.3683	2.2541	2.1665	2.0970	2.0401
120	3.9201	3.0718	2.6802	2.4472	2.2899	2.1750	2.0868	2.0164	1.9588
∞	3.8415	2.9957	2.6049	2.3719	2.2141	2.0986	2.0096	1.9384	1.8799

Denominator degrees of freedom (df_2)

(continued)

Numerator degrees of freedom (df$_1$)

df$_2$	10	12	15	20	24	30	40	60	120	∞
1	241.88	243.91	245.95	248.01	249.05	250.10	251.14	252.20	253.25	254.31
2	19.396	19.413	19.429	19.446	19.454	19.462	19.471	19.479	19.487	19.496
3	8.7855	8.7446	8.7029	8.6602	8.6385	8.6166	8.5944	8.5720	8.5494	8.5264
4	5.9644	5.9117	5.8578	5.8025	5.7744	5.7459	5.7170	5.6877	5.6581	5.6281
5	4.7351	4.6777	4.6188	4.5581	4.5272	4.4957	4.4638	4.4314	4.3985	4.3650
6	4.0600	3.9999	3.9381	3.8742	3.8415	3.8082	3.7743	3.7398	3.7047	3.6689
7	3.6365	3.5747	3.5107	3.4445	3.4105	3.3758	3.3404	3.3043	3.2674	3.2298
8	3.3472	3.2839	3.2184	3.1503	3.1152	3.0794	3.0428	3.0053	2.9669	2.9276
9	3.1373	3.0729	3.0061	2.9365	2.9005	2.8637	2.8259	2.7872	2.7475	2.7067
10	2.9782	2.9130	2.8450	2.7740	2.7372	2.6996	2.6609	2.6211	2.5801	2.5379
11	2.8536	2.7876	2.7186	2.6464	2.6090	2.5705	2.5309	2.4901	2.4480	2.4045
12	2.7534	2.6866	2.6169	2.5436	2.5055	2.4663	2.4259	2.3842	2.3410	2.2962
13	2.6710	2.6037	2.5331	2.4589	2.4202	2.3803	2.3392	2.2966	2.2524	2.2064
14	2.6022	2.5342	2.4630	2.3879	2.3487	2.3082	2.2664	2.2229	2.1778	2.1307
15	2.5437	2.4753	2.4034	2.3275	2.2878	2.2468	2.2043	2.1601	2.1141	2.0658
16	2.4935	2.4247	2.3522	2.2756	2.2354	2.1938	2.1507	2.1058	2.0589	2.0096
17	2.4499	2.3807	2.3077	2.2304	2.1898	2.1477	2.1040	2.0584	2.0107	1.9604
18	2.4117	2.3421	2.2686	2.1906	2.1497	2.1071	2.0629	2.0166	1.9681	1.9168
19	2.3779	2.3080	2.2341	2.1555	2.1141	2.0712	2.0264	1.9795	1.9302	1.8780
20	2.3479	2.2776	2.2033	2.1242	2.0825	2.0391	1.9938	1.9464	1.8963	1.8432
21	2.3210	2.2504	2.1757	2.0960	2.0540	2.0102	1.9645	1.9165	1.8657	1.8117
22	2.2967	2.2258	2.1508	2.0707	2.0283	1.9842	1.9380	1.8894	1.8380	1.7831
23	2.2747	2.2036	2.1282	2.0476	2.0050	1.9605	1.9139	1.8648	1.8128	1.7570
24	2.2547	2.1834	2.1077	2.0267	1.9838	1.9390	1.8920	1.8424	1.7896	1.7330
25	2.2365	2.1649	2.0889	2.0075	1.9643	1.9192	1.8718	1.8217	1.7684	1.7110
26	2.2197	2.1479	2.0716	1.9898	1.9464	1.9010	1.8533	1.8027	1.7488	1.6906
27	2.2043	2.1323	2.0558	1.9736	1.9299	1.8842	1.8361	1.7851	1.7306	1.6717
28	2.1900	2.1179	2.0411	1.9586	1.9147	1.8687	1.8203	1.7689	1.7138	1.6541
29	2.1768	2.1045	2.0275	1.9446	1.9005	1.8543	1.8055	1.7537	1.6981	1.6376
30	2.1646	2.0921	2.0148	1.9317	1.8874	1.8409	1.7918	1.7396	1.6835	1.6223
40	2.0772	2.0035	1.9245	1.8389	1.7929	1.7444	1.6928	1.6373	1.5766	1.5089
60	1.9926	1.9174	1.8364	1.7480	1.7001	1.6491	1.5943	1.5343	1.4673	1.3893
120	1.9105	1.8337	1.7505	1.6587	1.6084	1.5543	1.4952	1.4290	1.3519	1.2539
∞	1.8307	1.7522	1.6664	1.5705	1.5173	1.4591	1.3940	1.3180	1.2214	1.0000

Denominator degrees of freedom (df$_2$)

Use this to find critical r for Correlation Coefficient

TABLE A-6	Critical Values of the Pearson Correlation Coefficient r	
n	$\alpha = .05$	$\alpha = .01$
4	.950	.999
5	.878	.959
6	.811	.917
7	.754	.875
8	.707	.834
9	.666	.798
10	.632	.765
11	.602	.735
12	.576	.708
13	.553	.684
14	.532	.661
15	.514	.641
16	.497	.623
17	.482	.606
18	.468	.590
19	.456	.575
20	.444	.561
25	.396	.505
30	.361	.463
35	.335	.430
40	.312	.402
45	.294	.378
50	.279	.361
60	.254	.330
70	.236	.305
80	.220	.286
90	.207	.269
100	.196	.256

NOTE: To test $H_0: \rho = 0$ against $H_1: \rho \neq 0$, reject H_0 if the absolute value of r is greater than the critical value in the table.

| TABLE A-7 | Critical Values for the Sign Test | | | |

	α			
n	.005 (one tail) .01 (two tails)	.01 (one tail) .02 (two tails)	.025 (one tail) .05 (two tails)	.05 (one tail) .10 (two tails)
1	*	*	*	*
2	*	*	*	*
3	*	*	*	*
4	*	*	*	*
5	*	*	*	0
6	*	*	0	0
7	*	0	0	0
8	0	0	0	1
9	0	0	1	1
10	0	0	1	1
11	0	1	1	2
12	1	1	2	2
13	1	1	2	3
14	1	2	2	3
15	2	2	3	3
16	2	2	3	4
17	2	3	4	4
18	3	3	4	5
19	3	4	4	5
20	3	4	5	5
21	4	4	5	6
22	4	5	5	6
23	4	5	6	7
24	5	5	6	7
25	5	6	7	7

NOTES:

1. * indicates that it is not possible to get a value in the critical region.

2. Reject the null hypothesis if the number of the less frequent sign (x) is less than or equal to the value in the table.

3. For values of n greater than 25, a normal approximation is used with

$$z = \frac{(x + 0.5) - \left(\frac{n}{2}\right)}{\frac{\sqrt{n}}{2}}$$

TABLE A-8	Critical Values of *T* for the Wilcoxon Signed-Ranks Test

	α			
n	.005 (one tail) .01 (two tails)	.01 (one tail) .02 (two tails)	.025 (one tail) .05 (two tails)	.05 (one tail) .10 (two tails)
5	*	*	*	1
6	*	*	1	2
7	*	0	2	4
8	0	2	4	6
9	2	3	6	8
10	3	5	8	11
11	5	7	11	14
12	7	10	14	17
13	10	13	17	21
14	13	16	21	26
15	16	20	25	30
16	19	24	30	36
17	23	28	35	41
18	28	33	40	47
19	32	38	46	54
20	37	43	52	60
21	43	49	59	68
22	49	56	66	75
23	55	62	73	83
24	61	69	81	92
25	68	77	90	101
26	76	85	98	110
27	84	93	107	120
28	92	102	117	130
29	100	111	127	141
30	109	120	137	152

NOTES:

1. * indicates that it is not possible to get a value in the critical region.

2. Reject the null hypothesis if the test statistic *T* is less than or equal to the critical value found in this table. Fail to reject the null hypothesis if the test statistic *T* is greater than the critical value found in the table.

From *Some Rapid Approximate Statistical Procedures*, Copyright ©1949, 1964 Lederle Laboratories Division of American Cyanamid Company. Reprinted with the permission of the American Cyanamid Company.

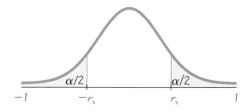

TABLE A-9	Critical Values of Spearman's Rank Correlation Coefficient r_s			
n	$\alpha = 0.10$	$\alpha = 0.05$	$\alpha = 0.02$	$\alpha = 0.01$
5	.900	—	—	—
6	.829	.886	.943	—
7	.714	.786	.893	—
8	.643	.738	.833	.881
9	.600	.683	.783	.833
10	.564	.648	.745	.794
11	.523	.623	.736	.818
12	.497	.591	.703	.780
13	.475	.566	.673	.745
14	.457	.545	.646	.716
15	.441	.525	.623	.689
16	.425	.507	.601	.666
17	.412	.490	.582	.645
18	.399	.476	.564	.625
19	.388	.462	.549	.608
20	.377	.450	.534	.591
21	.368	.438	.521	.576
22	.359	.428	.508	.562
23	.351	.418	.496	.549
24	.343	.409	.485	.537
25	.336	.400	.475	.526
26	.329	.392	.465	.515
27	.323	.385	.456	.505
28	.317	.377	.448	.496
29	.311	.370	.440	.487
30	.305	.364	.432	.478

NOTE: For $n > 30$, use $r_s = \pm z/\sqrt{n-1}$, where z corresponds to the level of significance. For example, if $\alpha = 0.05$, then $z = 1.96$.

To test $H_0: \rho_s = 0$

against $H_1: \rho_s \neq 0$

From "Distribution of sums of squares of rank differences to small numbers of individuals," *The Annals of Mathematical Statistics*, Vol. 9, No. 2. Reprinted with permission of the Institute of Mathematical Statistics.

TABLE A-10 Critical Values for Number of Runs G

Value of n_2

Value of n_1		2	3	4	5	6	7	8	9	10	11	12	13	14	15	16	17	18	19	20
2		1	1	1	1	1	1	1	1	1	1	2	2	2	2	2	2	2	2	2
		6	6	6	6	6	6	6	6	6	6	6	6	6	6	6	6	6	6	6
3		1	1	1	1	2	2	2	2	2	2	2	2	2	3	3	3	3	3	3
		6	8	8	8	8	8	8	8	8	8	8	8	8	8	8	8	8	8	8
4		1	1	1	2	2	2	3	3	3	3	3	3	3	3	4	4	4	4	4
		6	8	9	9	9	10	10	10	10	10	10	10	10	10	10	10	10	10	10
5		1	1	2	2	3	3	3	3	3	4	4	4	4	4	4	4	5	5	5
		6	8	9	10	10	11	11	12	12	12	12	12	12	12	12	12	12	12	12
6		1	2	2	3	3	3	3	4	4	4	4	5	5	5	5	5	5	6	6
		6	8	9	10	11	12	12	13	13	13	13	14	14	14	14	14	14	14	14
7		1	2	2	3	3	3	4	4	5	5	5	5	5	6	6	6	6	6	6
		6	8	10	11	12	13	13	14	14	14	14	15	15	15	16	16	16	16	16
8		1	2	3	3	3	4	4	5	5	5	6	6	6	6	6	7	7	7	7
		6	8	10	11	12	13	14	14	15	15	16	16	16	16	17	17	17	17	17
9		1	2	3	3	4	4	5	5	5	6	6	6	7	7	7	7	8	8	8
		6	8	10	12	13	14	14	15	16	16	16	17	17	18	18	18	18	18	18
10		1	2	3	3	4	5	5	5	6	6	7	7	7	7	8	8	8	8	9
		6	8	10	12	13	14	15	16	16	17	17	18	18	18	19	19	19	20	20
11		1	2	3	4	4	5	5	6	6	7	7	7	8	8	8	9	9	9	9
		6	8	10	12	13	14	15	16	17	17	18	19	19	19	20	20	20	21	21
12		2	2	3	4	4	5	6	6	7	7	7	8	8	8	9	9	9	10	10
		6	8	10	12	13	14	16	16	17	18	19	19	20	20	21	21	21	22	22
13		2	2	3	4	5	5	6	6	7	7	8	8	9	9	9	10	10	10	10
		6	8	10	12	14	15	16	17	18	19	19	20	20	21	21	22	22	23	23
14		2	2	3	4	5	5	6	7	7	8	8	9	9	9	10	10	10	11	11
		6	8	10	12	14	15	16	17	18	19	20	20	21	22	22	23	23	23	24
15		2	3	3	4	5	6	6	7	7	8	8	9	9	10	10	11	11	11	12
		6	8	10	12	14	15	16	18	18	19	20	21	22	22	23	23	24	24	25
16		2	3	4	4	5	6	6	7	8	8	9	9	10	10	11	11	11	12	12
		6	8	10	12	14	16	17	18	19	20	21	21	22	23	23	24	25	25	25
17		2	3	4	4	5	6	7	7	8	9	9	10	10	11	11	11	12	12	13
		6	8	10	12	14	16	17	18	19	20	21	22	23	23	24	25	25	26	26
18		2	3	4	5	5	6	7	8	8	9	9	10	10	11	11	12	12	13	13
		6	8	10	12	14	16	17	18	19	20	21	22	23	24	25	25	26	26	27
19		2	3	4	5	6	6	7	8	8	9	10	10	11	11	12	12	13	13	13
		6	8	10	12	14	16	17	18	20	21	22	23	23	24	25	26	26	27	27
20		2	3	4	5	6	6	7	8	9	9	10	10	11	12	12	13	13	13	14
		6	8	10	12	14	16	17	18	20	21	22	23	24	25	25	26	27	27	28

NOTE:

1. The entries in this table are the critical G values, assuming a two-tailed test with a significance level of $\alpha = 0.05$.

2. The null hypothesis of randomness is rejected if the total number of runs G is less than or equal to the smaller entry or greater than or equal to the larger entry.

From "Tables for testing randomness of groupings in a sequence of alternatives," *The Annals of Mathematical Statistics*, Vol. 14, No. 1. Reprinted with permission of the Institute of Mathematical Statistics.

Appendix B: Data Sets

Data Set 1:   Health Exam Results
Data Set 2:   Parent/Child Heights
Data Set 3:   Head Circumferences
Data Set 4:   Body Temperatures of Healthy Adults
Data Set 5:   Cigarette Tar, Nicotine, and Carbon Monoxide
Data Set 6:   Passive and Active Smoke
Data Set 7:   Alcohol and Tobacco Use in Animated Children's Movies
Data Set 8:   New York City Marathon Finishers
Data Set 9:   Bears (wild bears anesthetized)
Data Set 10:   Forecast and Actual Temperatures
Data Set 11:   Rainfall in Boston for One Year
Data Set 12:   Everglades Temperature, Rain, Conductivity
Data Set 13:   Old Faithful Geyser
Data Set 14:   Clancy, Rowling, and Tolstoy Books
Data Set 15:   Ages of Queen Mary Stowaways
Data Set 16:   Cereal
Data Set 17:   Weights and Volumes of Cola
Data Set 18:   Diamonds
Data Set 19:   Weights of a Sample of M&M Plain Candies
Data Set 20:   Axial Loads of Aluminum Cans
Data Set 21:   Movies
Data Set 22:   Cars
Data Set 23:   Weights of Discarded Garbage for One Week
Data Set 24:   Homes Sold in Dutchess County
Data Set 25:   Miscellaneous: DJIA, Car Sales, Motor Vehicle Deaths, Murders, Sunspots, Super Bowl
Data Set 26:   New York State Lottery
Data Set 27:   Solitaire Results
Data Set 28:   Weights of Domino Sugar Packets
Data Set 29:   Weights of Quarters
Data Set 30:   Homerun Distances

Data Set 1: Health Exam Results

AGE is in years, HT is height (inches), WT is weight (pounds), WAIST is circumference (cm), PULSE is pulse rate (beats per minute), SYS is systolic blood pressure (mmHg), DIAS is diastolic blood pressure (mmHg), CHOL is cholesterol (mg), BMI is body mass index, LEG is upper leg length (cm), ELBOW is elbow breadth (cm), WRIST is wrist breadth (cm), and ARM is arm circumference (cm). Data are from the U.S. Department of Health and Human Services, National Center for Health Statistics, Third National Health and Nutrition Examination Survey.

STATDISK: Data set name for males is Mhealth.
Minitab: Worksheet name for males is MHEALTH.MTW.
Excel: Workbook name for males is MHEALTH.XLS.
TI-83 Plus: App name for male data is MHEALTH and the file names are the same as for text files.
Text file names for males: MAGE, MHT, MWT, MWAST, MPULS, MSYS, MDIAS, MCHOL, MBMI, MLEG, MELBW, MWRST, MARM.

Male Age	HT	WT	Waist	Pulse	SYS	DIAS	CHOL	BMI	Leg	Elbow	Wrist	Arm
58	70.8	169.1	90.6	68	125	78	522	23.8	42.5	7.7	6.4	31.9
22	66.2	144.2	78.1	64	107	54	127	23.2	40.2	7.6	6.2	31.0
32	71.7	179.3	96.5	88	126	81	740	24.6	44.4	7.3	5.8	32.7
31	68.7	175.8	87.7	72	110	68	49	26.2	42.8	7.5	5.9	33.4
28	67.6	152.6	87.1	64	110	66	230	23.5	40.0	7.1	6.0	30.1
46	69.2	166.8	92.4	72	107	83	316	24.5	47.3	7.1	5.8	30.5
41	66.5	135.0	78.8	60	113	71	590	21.5	43.4	6.5	5.2	27.6
56	67.2	201.5	103.3	88	126	72	466	31.4	40.1	7.5	5.6	38.0
20	68.3	175.2	89.1	76	137	85	121	26.4	42.1	7.5	5.5	32.0
54	65.6	139.0	82.5	60	110	71	578	22.7	36.0	6.9	5.5	29.3
17	63.0	156.3	86.7	96	109	65	78	27.8	44.2	7.1	5.3	31.7
73	68.3	186.6	103.3	72	153	87	265	28.1	36.7	8.1	6.7	30.7
52	73.1	191.1	91.8	56	112	77	250	25.2	48.4	8.0	5.2	34.7
25	67.6	151.3	75.6	64	119	81	265	23.3	41.0	7.0	5.7	30.6
29	68.0	209.4	105.5	60	113	82	273	31.9	39.8	6.9	6.0	34.2
17	71.0	237.1	108.7	64	125	76	272	33.1	45.2	8.3	6.6	41.1
41	61.3	176.7	104.0	84	131	80	972	33.2	40.2	6.7	5.7	33.1
52	76.2	220.6	103.0	76	121	75	75	26.7	46.2	7.9	6.0	32.2
32	66.3	166.1	91.3	84	132	81	138	26.6	39.0	7.5	5.7	31.2
20	69.7	137.4	75.2	88	112	44	139	19.9	44.8	6.9	5.6	25.9
20	65.4	164.2	87.7	72	121	65	638	27.1	40.9	7.0	5.6	33.7
29	70.0	162.4	77.0	56	116	64	613	23.4	43.1	7.5	5.2	30.3
18	62.9	151.8	85.0	68	95	58	762	27.0	38.0	7.4	5.8	32.8
26	68.5	144.1	79.6	64	110	70	303	21.6	41.0	6.8	5.7	31.0
33	68.3	204.6	103.8	60	110	66	690	30.9	46.0	7.4	6.1	36.2
55	69.4	193.8	103.0	68	125	82	31	28.3	41.4	7.2	6.0	33.6
53	69.2	172.9	97.1	60	124	79	189	25.5	42.7	6.6	5.9	31.9
28	68.0	161.9	86.9	60	131	69	957	24.6	40.5	7.3	5.7	32.9
28	71.9	174.8	88.0	56	109	64	339	23.8	44.2	7.8	6.0	30.9
37	66.1	169.8	91.5	84	112	79	416	27.4	41.8	7.0	6.1	34.0
40	72.4	213.3	102.9	72	127	72	120	28.7	47.2	7.5	5.9	34.8
33	73.0	198.0	93.1	84	132	74	702	26.2	48.2	7.8	6.0	33.6
26	68.0	173.3	98.9	88	116	81	1252	26.4	42.9	6.7	5.8	31.3
53	68.7	214.5	107.5	56	125	84	288	32.1	42.8	8.2	5.9	37.6
36	70.3	137.1	81.6	64	112	77	176	19.6	40.8	7.1	5.3	27.9
34	63.7	119.5	75.7	56	125	77	277	20.7	42.6	6.6	5.3	26.9
42	71.1	189.1	95.0	56	120	83	649	26.3	44.9	7.4	6.0	36.9
18	65.6	164.7	91.1	60	118	68	113	26.9	41.1	7.0	6.1	34.5
44	68.3	170.1	94.9	64	115	75	656	25.6	44.5	7.3	5.8	32.1
20	66.3	151.0	79.9	72	115	65	172	24.2	44.0	7.1	5.4	30.7

(continued)

Data Set 1: Health Exam Results (*continued*)

STATDISK: Data set name for females is Fhealth.
Minitab: Worksheet name for females is FHEALTH.MTW.
Excel: Workbook name for females is FHEALTH.XLS.
TI-83 Plus: App name for female data is FHEALTH and the file names are the same as for text files.
Text file names for females: FAGE, FHT, FWT, FWAST, FPULS, FSYS, FDIAS, FCHOL, FBMI, FLEG, FELBW, FWRST, FARM.

Female	Age	HT	WT	Waist	Pulse	SYS	DIAS	CHOL	BMI	Leg	Elbow	Wrist	Arm
	17	64.3	114.8	67.2	76	104	61	264	19.6	41.6	6.0	4.6	23.6
	32	66.4	149.3	82.5	72	99	64	181	23.8	42.8	6.7	5.5	26.3
	25	62.3	107.8	66.7	88	102	65	267	19.6	39.0	5.7	4.6	26.3
	55	62.3	160.1	93.0	60	114	76	384	29.1	40.2	6.2	5.0	32.6
	27	59.6	127.1	82.6	72	94	58	98	25.2	36.2	5.5	4.8	29.2
	29	63.6	123.1	75.4	68	101	66	62	21.4	43.2	6.0	4.9	26.4
	25	59.8	111.7	73.6	80	108	61	126	22.0	38.7	5.7	5.1	27.9
	12	63.3	156.3	81.4	64	104	41	89	27.5	41.0	6.8	5.5	33.0
	41	67.9	218.8	99.4	68	123	72	531	33.5	43.8	7.8	5.8	38.6
	32	61.4	110.2	67.7	68	93	61	130	20.6	37.3	6.3	5.0	26.5
	31	66.7	188.3	100.7	80	89	56	175	29.9	42.3	6.6	5.2	34.4
	19	64.8	105.4	72.9	76	112	62	44	17.7	39.1	5.7	4.8	23.7
	19	63.1	136.1	85.0	68	107	48	8	24.0	40.3	6.6	5.1	28.4
	23	66.7	182.4	85.7	72	116	62	112	28.9	48.6	7.2	5.6	34.0
	40	66.8	238.4	126.0	96	181	102	462	37.7	33.2	7.0	5.4	35.2
	23	64.7	108.8	74.5	72	98	61	62	18.3	43.4	6.2	5.2	24.7
	27	65.1	119.0	74.5	68	100	53	98	19.8	41.5	6.3	5.3	27.0
	45	61.9	161.9	94.0	72	127	74	447	29.8	40.0	6.8	5.0	35.0
	41	64.3	174.1	92.8	64	107	67	125	29.7	38.2	6.8	4.7	33.1
	56	63.4	181.2	105.5	80	116	71	318	31.7	38.2	6.9	5.4	39.6
	22	60.7	124.3	75.5	64	97	64	325	23.8	38.2	5.9	5.0	27.0
	57	63.4	255.9	126.5	80	155	85	600	44.9	41.0	8.0	5.6	43.8
	24	62.6	106.7	70.0	76	106	59	237	19.2	38.1	6.1	5.0	23.6
	37	60.6	149.9	98.0	76	110	70	173	28.7	38.0	7.0	5.1	34.3
	59	63.5	163.1	104.7	76	105	69	309	28.5	36.0	6.7	5.1	34.4
	40	58.6	94.3	67.8	80	118	82	94	19.3	32.1	5.4	4.2	23.3
	45	60.2	159.7	99.3	104	133	83	280	31.0	31.1	6.4	5.2	35.6
	52	67.6	162.8	91.1	88	113	75	254	25.1	39.4	7.1	5.3	31.8
	31	63.4	130.0	74.5	60	113	66	123	22.8	40.2	5.9	5.1	27.0
	32	64.1	179.9	95.5	76	107	67	596	30.9	39.2	6.2	5.0	32.8
	23	62.7	147.8	79.5	72	95	59	301	26.5	39.0	6.3	4.9	31.0
	23	61.3	112.9	69.1	72	108	72	223	21.2	36.6	5.9	4.7	27.0
	47	58.2	195.6	105.5	88	114	79	293	40.6	27.0	7.5	5.5	41.2
	36	63.2	124.2	78.8	80	104	73	146	21.9	38.5	5.6	4.7	25.5
	34	60.5	135.0	85.7	60	125	73	149	26.0	39.9	6.4	5.2	30.9
	37	65.0	141.4	92.8	72	124	85	149	23.5	37.5	6.1	4.8	27.9
	18	61.8	123.9	72.7	88	92	46	920	22.8	39.7	5.8	5.0	26.5
	29	68.0	135.5	75.9	88	119	81	271	20.7	39.0	6.3	4.9	27.8
	48	67.0	130.4	68.6	124	93	64	207	20.5	41.6	6.0	5.3	23.0
	16	57.0	100.7	68.7	64	106	64	2	21.9	33.8	5.6	4.6	26.4

Data Set 2: Parent/Child Heights (inches)

Data are from the U.S. Department of Health and Human Services, National Center for Health Statistics, Third National Health and Nutrition Examination Survey.

STATDISK: Data set name is Parent Ht.
Minitab: Worksheet name is PARENTHT.MTW.
Excel: Workbook name is PARENTHT.XLS.
TI-83 Plus: App name is PARENTHT and the file names are the same as for STATDISK and text files.
Text file names: CHDHT, MOMHT, DADHT.

Gender	Height	Mother's Height	Father's Height
M	62.5	66	70
M	64.6	58	69
M	69.1	66	64
M	73.9	68	71
M	67.1	64	68
M	64.4	62	66
M	71.1	66	74
M	71.0	63	73
M	67.4	64	62
M	69.3	65	69
M	64.9	64	67
M	68.1	64	68
M	66.5	62	72
M	67.5	69	66
M	66.5	62	72
M	70.3	67	68
M	67.5	63	71
M	68.5	66	67
M	71.9	65	71
M	67.8	71	75
F	58.6	63	64
F	64.7	67	65
F	65.3	64	67
F	61.0	60	72
F	65.4	65	72
F	67.4	67	72
F	60.9	59	67
F	63.1	60	71
F	60.0	58	66
F	71.1	72	75
F	62.2	63	69
F	67.2	67	70
F	63.4	62	69
F	68.4	69	62
F	62.2	63	66
F	64.7	64	76
F	59.6	63	69
F	61.0	64	68
F	64.0	60	66
F	65.4	65	68

Data Set 3: Head Circumferences (cm) of Two-Month-Old Babies

Data are from the U.S. Department of Health and Human Services, National Center for Health Statistics, Third National Health and Nutrition Examination Survey.

STATDISK: Data set name is Headcirc.
Minitab: Worksheet name is HEADCIRC.MTW.
Excel: Workbook name is HEADCIRC.XLS.
TI-83 Plus: App name is HEADCIRC and the file names are the same as for text files.
Text file names: MHED, FHED.

Male

40.1	39.8	42.3	41.0	42.5	40.9	35.5	35.7	41.1	41.4
42.2	42.3	43.2	42.2	42.4	43.2	39.9	40.9	40.7	41.7
41.7	41.0	40.4	42.0	41.2	39.7	41.9	41.3	40.2	41.0
41.1	40.4	39.2	42.8	41.9	42.8	41.0	40.9	42.0	42.6
41.0	39.6	40.2	40.9	40.2	41.8	41.7	41.7	40.9	42.8

Female

39.3	40.2	41.3	38.1	39.6	40.6	38.6	40.5	40.5	40.3
39.5	40.7	40.2	38.2	40.3	42.6	39.9	40.0	40.7	38.6
41.0	43.7	40.0	40.1	41.0	40.8	41.0	40.3	40.2	39.2
34.4	41.0	39.6	40.9	36.9	43.6	40.2	40.8	37.8	41.2
42.0	38.3	39.6	38.9	36.3	39.9	40.3	40.1	42.0	41.6

Data Set 4: Body Temperatures (in degrees Fahrenheit) of Healthy Adults

Data provided by Dr. Steven Wasserman, Dr. Philip Mackowiak, and Dr. Myron Levine of the University of Maryland.

Subject	Age	Sex	Smoke	Temperature Day 1		Temperature Day 2	
				8 AM	12 AM	8 AM	12 AM
1	22	M	Y	98.0	98.0	98.0	98.6
2	23	M	Y	97.0	97.6	97.4	—
3	22	M	Y	98.6	98.8	97.8	98.6
4	19	M	N	97.4	98.0	97.0	98.0
5	18	M	N	98.2	98.8	97.0	98.0
6	20	M	Y	98.2	98.8	96.6	99.0
7	27	M	Y	98.2	97.6	97.0	98.4
8	19	M	Y	96.6	98.6	96.8	98.4
9	19	M	Y	97.4	98.6	96.6	98.4
10	24	M	N	97.4	98.8	96.6	98.4
11	35	M	Y	98.2	98.0	96.2	98.6
12	25	M	Y	97.4	98.2	97.6	98.6
13	25	M	N	97.8	98.0	98.6	98.8
14	35	M	Y	98.4	98.0	97.0	98.6
15	21	M	N	97.6	97.0	97.4	97.0
16	33	M	N	96.2	97.2	98.0	97.0
17	19	M	Y	98.0	98.2	97.6	98.8
18	24	M	Y	—	—	97.2	97.6
19	18	F	N	—	—	97.0	97.7
20	22	F	Y	—	—	98.0	98.8
21	20	M	Y	—	—	97.0	98.0
22	30	F	Y	—	—	96.4	98.0
23	29	M	N	—	—	96.1	98.3
24	18	M	Y	—	—	98.0	98.5
25	31	M	Y	—	98.1	96.8	97.3
26	28	F	Y	—	98.2	98.2	98.7
27	27	M	Y	—	98.5	97.8	97.4
28	21	M	Y	—	98.5	98.2	98.9
29	30	M	Y	—	99.0	97.8	98.6
30	27	M	N	—	98.0	99.0	99.5
31	32	M	Y	—	97.0	97.4	97.5
32	33	M	Y	—	97.3	97.4	97.3
33	23	M	Y	—	97.3	97.5	97.6
34	29	M	Y	—	98.1	97.8	98.2
35	25	M	Y	—	—	97.9	99.6
36	31	M	N	—	97.8	97.8	98.7
37	25	M	Y	—	99.0	98.3	99.4
38	28	M	N	—	97.6	98.0	98.2
39	30	M	Y	—	97.4	—	98.0
40	33	M	Y	—	98.0	—	98.6
41	28	M	Y	98.0	97.4	—	98.6
42	22	M	Y	98.8	98.0	—	97.2
43	21	F	Y	99.0	—	—	98.4
44	30	M	N	—	98.6	—	98.6

(continued)

Data Set 4: Body Temperatures (*continued*)

Subject	Age	Sex	Smoke	Temperature Day 1		Temperature Day 2	
				8 AM	12 AM	8 AM	12 AM
45	22	M	Y	—	98.6	—	98.2
46	22	F	N	98.0	98.4	—	98.0
47	20	M	Y	—	97.0	—	97.8
48	19	M	Y	—	—	—	98.0
49	33	M	N	—	98.4	—	98.4
50	31	M	Y	99.0	99.0	—	98.6
51	26	M	N	—	98.0	—	98.6
52	18	M	N	—	—	—	97.8
53	23	M	N	—	99.4	—	99.0
54	28	M	Y	—	—	—	96.5
55	19	M	Y	—	97.8	—	97.6
56	21	M	N	—	—	—	98.0
57	27	M	Y	—	98.2	—	96.9
58	29	M	Y	—	99.2	—	97.6
59	38	M	N	—	99.0	—	97.1
60	29	F	Y	—	97.7	—	97.9
61	22	M	Y	—	98.2	—	98.4
62	22	M	Y	—	98.2	—	97.3
63	26	M	Y	—	98.8	—	98.0
64	32	M	N	—	98.1	—	97.5
65	25	M	Y	—	98.5	—	97.6
66	21	F	N	—	97.2	—	98.2
67	25	M	Y	—	98.5	—	98.5
68	24	M	Y	—	99.2	97.0	98.8
69	25	M	Y	—	98.3	97.6	98.7
70	35	M	Y	—	98.7	97.5	97.8
71	23	F	Y	—	98.8	98.8	98.0
72	31	M	Y	—	98.6	98.4	97.1
73	28	M	Y	—	98.0	98.2	97.4
74	29	M	Y	—	99.1	97.7	99.4
75	26	M	Y	—	97.2	97.3	98.4
76	32	M	N	—	97.6	97.5	98.6
77	32	M	Y	—	97.9	97.1	98.4
78	21	F	Y	—	98.8	98.6	98.5
79	20	M	Y	—	98.6	98.6	98.6
80	24	F	Y	—	98.6	97.8	98.3
81	21	F	Y	—	99.3	98.7	98.7
82	28	M	Y	—	97.8	97.9	98.8
83	27	F	N	98.8	98.7	97.8	99.1
84	28	M	N	99.4	99.3	97.8	98.6
85	29	M	Y	98.8	97.8	97.6	97.9
86	19	M	N	97.7	98.4	96.8	98.8
87	24	M	Y	99.0	97.7	96.0	98.0

(continued)

Data Set 4: Body Temperatures (*continued*)

Subject	Age	Sex	Smoke	Temperature Day 1		Temperature Day 2	
				8 AM	12 AM	8 AM	12 AM
88	29	M	N	98.1	98.3	98.0	98.7
89	25	M	Y	98.7	97.7	97.0	98.5
90	27	M	N	97.5	97.1	97.4	98.9
91	25	M	Y	98.9	98.4	97.6	98.4
92	21	M	Y	98.4	98.6	97.6	98.6
93	19	M	Y	97.2	97.4	96.2	97.1
94	27	M	Y	—	—	96.2	97.9
95	32	M	N	98.8	96.7	98.1	98.8
96	24	M	Y	97.3	96.9	97.1	98.7
97	32	M	Y	98.7	98.4	98.2	97.6
98	19	F	Y	98.9	98.2	96.4	98.2
99	18	F	Y	99.2	98.6	96.9	99.2
100	27	M	N	—	97.0	—	97.8
101	34	M	Y	—	97.4	—	98.0
102	25	M	N	—	98.4	—	98.4
103	18	M	N	—	97.4	—	97.8
104	32	M	Y	—	96.8	—	98.4
105	31	M	Y	—	98.2	—	97.4
106	26	M	N	—	97.4	—	98.0
107	23	M	N	—	98.0	—	97.0

Data Set 5: Cigarette Tar, Nicotine, and Carbon Monoxide

All measurements are in milligrams per cigarette, and all cigarettes are 100 mm long, filtered, and not menthol or light types. Data are from the Federal Trade Commission.

STATDISK: Data set name is Cigaret.
Minitab: Worksheet name is CIGARET.MTW.
Excel: Workbook name is CIGARET.XLS.
TI-83 Plus: App name is CIGARET and the file names are the same as for text files.
Text file names: TAR, NICOT, CO.

Brand	Tar	Nicotine	CO
American Filter	16	1.2	15
Benson & Hedges	16	1.2	15
Camel	16	1.0	17
Capri	9	0.8	6
Carlton	1	0.1	1
Cartier Vendome	8	0.8	8
Chelsea	10	0.8	10
GPC Approved	16	1.0	17
Hi-Lite	14	1.0	13
Kent	13	1.0	13
Lucky Strike	13	1.1	13
Malibu	15	1.2	15
Marlboro	16	1.2	15
Merit	9	0.7	11
Newport Stripe	11	0.9	15
Now	2	0.2	3
Old Gold	18	1.4	18
Pall Mall	15	1.2	15
Players	13	1.1	12
Raleigh	15	1.0	16
Richland	17	1.3	16
Rite	9	0.8	10
Silva Thins	12	1.0	10
Tareyton	14	1.0	17
Triumph	5	0.5	7
True	6	0.6	7
Vantage	8	0.7	11
Viceroy	18	1.4	15
Winston	16	1.1	18

Data Set 6: Passive and Active Smoke

All values are measured levels of serum cotinine (in ng/ml), a metabolite of nicotine. (When nicotine is absorbed by the body, cotinine is produced.) Data are from the U.S. Department of Health and Human Services, National Center for Health Statistics, Third National Health and Nutrition Examination Survey.

STATDISK: Data set name is Cotinine.
Minitab: Worksheet name is COTININE.MTW.
Excel: Workbook name is COTININE.XLS.
TI-83 Plus: App name is COTININE and the file names are the same as for text files.
Text file names: NOETS, ETS, SMKR.

Smokers (subjects reported tobacco use)

1	0	131	173	265	210	44	277	32	3
35	112	477	289	227	103	222	149	313	491
130	234	164	198	17	253	87	121	266	290
123	167	250	245	48	86	284	1	208	173

ETS (nonsmokers exposed to environmental tobacco smoke at home or work)

384	0	69	19	1	0	178	2	13	1
4	0	543	17	1	0	51	0	197	3
0	3	1	45	13	3	1	1	1	0
0	551	2	1	1	1	0	74	1	241

NOETS (nonsmokers with no exposure to environmental tobacco smoke at home or work)

0	0	0	0	0	0	0	0	0	0
0	9	0	0	0	0	0	0	244	0
1	0	0	0	90	1	0	309	0	0
0	0	0	0	0	0	0	0	0	0

Data Set 7: Alcohol and Tobacco Use in Animated Children's Movies

Movie lengths are in minutes, tobacco use times are in seconds, and alcohol use times are in seconds. The data are based on "Tobacco and Alcohol Use in G-Rated Children's Animated Films," by Goldstein, Sobel, and Newman, *Journal of the American Medical Association,* Vol. 281, No. 12.

STATDISK: Data set name is Chmovie.
Minitab: Worksheet name is CHMOVIE.MTW.
Excel: Workbook name is CHMOVIE.XLS.
TI-83 Plus: App name is CHMOVIE and the file names are the same as for text files.
Text file names: CHLEN, CHTOB, CHALC.

Movie	Company	Length (min)	Tobacco Use (sec)	Alcohol Use (sec)
Snow White	Disney	83	0	0
Pinocchio	Disney	88	223	80
Fantasia	Disney	120	0	0
Dumbo	Disney	64	176	88
Bambi	Disney	69	0	0
Three Caballeros	Disney	71	548	8
Fun and Fancy Free	Disney	76	0	4
Cinderella	Disney	74	37	0
Alice in Wonderland	Disney	75	158	0
Peter Pan	Disney	76	51	33
Lady and the Tramp	Disney	75	0	0
Sleeping Beauty	Disney	75	0	113
101 Dalmatians	Disney	79	299	51
Sword and the Stone	Disney	80	37	20
Jungle Book	Disney	78	0	0
Aristocats	Disney	78	11	142
Robin Hood	Disney	83	0	39
Rescuers	Disney	77	0	0
Winnie the Pooh	Disney	71	0	0
Fox and the Hound	Disney	83	0	0
Black Cauldron	Disney	80	0	34
Great Mouse Detective	Disney	73	165	414
Oliver and Company	Disney	72	74	0
Little Mermaid	Disney	82	9	0
Rescuers Down Under	Disney	74	0	76
Beauty and the Beast	Disney	84	0	123
Aladdin	Disney	90	2	3
Lion King	Disney	89	0	0
Pocahontas	Disney	81	6	7
Toy Story	Disney	81	0	0
Hunchback of Notre Dame	Disney	90	23	46
James and the Giant Peach	Disney	79	206	38
Hercules	Disney	92	9	13
Secret of NIMH	MGM	82	0	0
All Dogs Go to Heaven	MGM	89	205	73
All Dogs Go to Heaven 2	MGM	82	162	72
Babes in Toyland	MGM	74	0	0
Thumbelina	Warner Bros	86	6	5
Troll in Central Park	Warner Bros	76	1	0
Space Jam	Warner Bros	81	117	0
Pippi Longstocking	Warner Bros	75	5	0
Cats Don't Dance	Warner Bros	75	91	0
An American Tail	Universal	77	155	74
Land Before Time	Universal	70	0	0
Fievel Goes West	Universal	75	24	28
We're Back: Dinosaur Story	Universal	64	55	0
Land Before Time 2	Universal	73	0	0
Balto	Universal	74	0	0
Once Upon a Forest	20th Century Fox	71	0	0
Anastasia	20th Century Fox	94	17	39

Data Set 8: New York City Marathon Finishers

Sample is 150 runners randomly selected from the population of 29,373 runners who finished the New York City Marathon in a recent year.

STATDISK: Data set name is Marathon.
Minitab: Worksheet name is MARATHON.MTW.
Excel: Workbook name is MARATHON.XLS
TI-83 Plus: App name is MARATHON and the file names are the same as for text files.
Text file names: MRORD, MRAGE, MRTIM.

Order	Age	Gender	Time (sec)	Order	Age	Gender	Time (sec)
130	32	M	9631	7082	38	M	13851
265	39	M	10209	7093	32	F	13854
314	39	M	10351	7933	50	M	14057
490	36	M	10641	7966	43	M	14066
547	34	M	10723	8011	25	M	14078
708	28	M	10905	8027	39	M	14082
834	42	M	11061	8042	31	M	14086
944	46	M	11188	8186	37	M	14121
1084	32	M	11337	8225	46	M	14128
1086	34	M	11338	8609	23	F	14216
1132	41	M	11382	8707	30	F	14235
1593	36	M	11738	8823	24	M	14256
1625	50	M	11761	9451	29	M	14375
1735	36	M	11830	9630	30	M	14402
1792	40	M	11874	10130	36	M	14512
1826	33	M	11897	10191	40	M	14528
2052	29	F	12047	10556	51	M	14617
2108	28	M	12077	10585	51	M	14623
2167	40	M	12115	10643	51	M	14632
2505	30	F	12289	10821	30	M	14677
2550	28	M	12312	10910	38	M	14698
3344	44	M	12639	10979	59	M	14720
3376	45	M	12652	10982	28	F	14721
4115	45	M	12940	11091	49	M	14752
4252	54	M	12986	11413	55	M	14836
4459	33	M	13063	11699	53	M	14919
4945	49	M	13217	11769	53	M	14935
5269	45	M	13315	11792	40	M	14942
5286	40	M	13322	11869	38	M	14964
5559	26	M	13408	11896	35	M	14971
6169	23	F	13593	11997	54	M	14996
6235	21	M	13615	12019	21	M	15002
6552	50	F	13704	12160	33	F	15036
6618	33	M	13722	12306	58	F	15077
6904	38	M	13802	12683	43	M	15167
6996	40	M	13829	12845	33	M	15210

(continued)

Data Set 8: New York City Marathon Finishers (*continued*)

Order	Age	Gender	Time (sec)	Order	Age	Gender	Time (sec)
12942	35	M	15232	21013	38	M	17396
13226	31	M	15309	21017	47	M	17397
13262	38	M	15318	21524	34	M	17563
13297	28	F	15326	21787	37	F	17636
13434	30	F	15357	22009	37	M	17711
13597	23	F	15402	22042	31	F	17726
14391	40	M	15608	22258	29	F	17799
14633	43	M	15671	22285	49	M	17807
14909	43	M	15741	22638	31	M	17918
15282	29	M	15825	22993	52	M	18041
16030	34	F	16013	23092	38	M	18080
16324	30	M	16090	24018	30	F	18469
16723	65	M	16194	24283	31	F	18580
16840	50	M	16229	24290	40	M	18583
17104	37	F	16297	24417	50	F	18647
17298	30	F	16352	24466	29	M	18677
17436	32	M	16389	24649	21	M	18784
17483	19	F	16401	24845	53	M	18906
17487	42	M	16402	25262	41	M	19164
17694	33	M	16461	25287	50	F	19177
18132	42	M	16582	25956	45	M	19669
18765	51	M	16752	26471	27	F	20084
18783	54	F	16758	26545	32	M	20164
18825	32	F	16771	26637	53	M	20269
18897	34	F	16792	27035	42	F	20675
19002	31	M	16812	27046	45	M	20698
19210	50	F	16871	27133	39	M	20808
19264	60	M	16886	27152	31	M	20841
19278	49	M	16889	27196	68	F	20891
19649	51	F	16991	27277	51	M	20970
19789	45	M	17034	27800	51	M	21649
20425	40	F	17211	27955	31	F	21911
20558	30	M	17245	27995	25	F	21983
20562	25	M	17246	28062	25	M	22087
20580	32	M	17252	28085	61	M	22146
20592	34	M	17257	28578	31	M	23545
20605	42	F	17260	28779	32	M	24384
20700	34	F	17286	28986	47	F	25399
20826	52	M	17327	29045	61	F	25898

Data Set 9: Bears (wild bears anesthetized)

AGE is in months, MONTH is the month of measurement (1 = January), SEX is coded with 1 = male and 2 = female, HEADLEN is head length (inches), HEADWTH is width of head (inches), NECK is distance around neck (in inches), LENGTH is length of body (inches), CHEST is distance around chest (inches), and WEIGHT is measured in pounds. Data are from Gary Alt and Minitab, Inc.

STATDISK: Data set name is Bears
Minitab: Worksheet name is BEARS.MTW.
Excel: Workbook name is BEARS.XLS.
TI-83 Plus: App name is BEARS and the file names are the same as for text files.
Text file names: BAGE, BMNTH, BSEX, BHDLN, BHDWD, BNECK, BLEN, BCHST, BWGHT.

Age	Month	Sex	Headlen	Headwth	Neck	Length	Chest	Weight
19	7	1	11.0	5.5	16.0	53.0	26.0	80
55	7	1	16.5	9.0	28.0	67.5	45.0	344
81	9	1	15.5	8.0	31.0	72.0	54.0	416
115	7	1	17.0	10.0	31.5	72.0	49.0	348
104	8	2	15.5	6.5	22.0	62.0	35.0	166
100	4	2	13.0	7.0	21.0	70.0	41.0	220
56	7	1	15.0	7.5	26.5	73.5	41.0	262
51	4	1	13.5	8.0	27.0	68.5	49.0	360
57	9	2	13.5	7.0	20.0	64.0	38.0	204
53	5	2	12.5	6.0	18.0	58.0	31.0	144
68	8	1	16.0	9.0	29.0	73.0	44.0	332
8	8	1	9.0	4.5	13.0	37.0	19.0	34
44	8	2	12.5	4.5	10.5	63.0	32.0	140
32	8	1	14.0	5.0	21.5	67.0	37.0	180
20	8	2	11.5	5.0	17.5	52.0	29.0	105
32	8	1	13.0	8.0	21.5	59.0	33.0	166
45	9	1	13.5	7.0	24.0	64.0	39.0	204
9	9	2	9.0	4.5	12.0	36.0	19.0	26
21	9	1	13.0	6.0	19.0	59.0	30.0	120
177	9	1	16.0	9.5	30.0	72.0	48.0	436
57	9	2	12.5	5.0	19.0	57.5	32.0	125
81	9	2	13.0	5.0	20.0	61.0	33.0	132
21	9	1	13.0	5.0	17.0	54.0	28.0	90
9	9	1	10.0	4.0	13.0	40.0	23.0	40
45	9	1	16.0	6.0	24.0	63.0	42.0	220
9	9	1	10.0	4.0	13.5	43.0	23.0	46
33	9	1	13.5	6.0	22.0	66.5	34.0	154
57	9	2	13.0	5.5	17.5	60.5	31.0	116
45	9	2	13.0	6.5	21.0	60.0	34.5	182
21	9	1	14.5	5.5	20.0	61.0	34.0	150
10	10	1	9.5	4.5	16.0	40.0	26.0	65
82	10	2	13.5	6.5	28.0	64.0	48.0	356
70	10	2	14.5	6.5	26.0	65.0	48.0	316
10	10	1	11.0	5.0	17.0	49.0	29.0	94
10	10	1	11.5	5.0	17.0	47.0	29.5	86
34	10	1	13.0	7.0	21.0	59.0	35.0	150
34	10	1	16.5	6.5	27.0	72.0	44.5	270
34	10	1	14.0	5.5	24.0	65.0	39.0	202
58	10	2	13.5	6.5	21.5	63.0	40.0	202
58	10	1	15.5	7.0	28.0	70.5	50.0	365
11	11	1	11.5	6.0	16.5	48.0	31.0	79
23	11	1	12.0	6.5	19.0	50.0	38.0	148
70	10	1	15.5	7.0	28.0	76.5	55.0	446
11	11	2	9.0	5.0	15.0	46.0	27.0	62
83	11	2	14.5	7.0	23.0	61.5	44.0	236
35	11	1	13.5	8.5	23.0	63.5	44.0	212
16	4	1	10.0	4.0	15.5	48.0	26.0	60
16	4	1	10.0	5.0	15.0	41.0	26.0	64
17	5	1	11.5	5.0	17.0	53.0	30.5	114
17	5	2	11.5	5.0	15.0	52.5	28.0	76
17	5	2	11.0	4.5	13.0	46.0	23.0	48
8	8	2	10.0	4.5	10.0	43.5	24.0	29
83	11	1	15.5	8.0	30.5	75.0	54.0	514
18	6	1	12.5	8.5	18.0	57.3	32.8	140

Data Set 10: Forecast and Actual Temperatures

Temperatures are in degrees Fahrenheit and the precipitation amounts are in inches. All measurements were recorded near the author's home.

STATDISK: Data set name is Weather.
Minitab: Worksheet name is WEATHER.MTW.
Excel: Workbook name is WEATHER.XLS.
TI-83 Plus: App name is WEATHER and the file names are the same as for text files.
Text file names: ACTHI, ACTLO, PHI1, PLO1, PHI3, PLO3, PHI5, PLO5, and PREC.

Date	Actual High	Actual Low	1 Day Predicted High	1 Day Predicted Low	3 Day Predicted High	3 Day Predicted Low	5 Day Predicted High	5 Day Predicted Low	Precip. (in.)
Jan. 1	30	1	28	13	30	18	28	16	0
Jan. 2	25	−5	29	13	26	17	27	16	0
Jan. 3	31	−5	32	14	30	13	28	20	0
Jan. 4	33	23	29	13	32	19	30	22	0
Jan. 5	29	9	30	19	35	26	26	15	0
Jan. 6	36	14	35	23	36	24	35	24	0.26
Jan. 7	36	12	35	21	38	25	34	23	0
Jan. 8	37	18	32	18	35	22	34	22	0.01
Jan. 9	32	26	27	17	33	18	33	21	0.21
Jan. 10	28	13	25	16	34	21	35	24	0.02
Jan. 11	43	7	41	22	37	26	38	26	0
Jan. 12	37	10	30	7	37	20	37	28	0
Jan. 13	36	6	33	20	31	14	36	16	0
Jan. 14	37	10	40	27	44	35	36	22	0
Jan. 15	34	29	34	24	38	29	45	26	0.02
Jan. 16	41	33	38	24	39	25	36	22	0.05
Jan. 17	40	36	33	15	37	21	33	21	0
Jan. 18	33	18	35	28	37	20	34	25	0
Jan. 19	35	32	40	25	39	32	36	21	0.01
Jan. 20	33	24	27	15	28	20	33	22	0.02
Jan. 21	31	19	27	10	30	16	31	18	0.21
Jan. 22	33	1	30	15	31	18	30	15	0.08
Jan. 23	35	0	37	19	40	23	38	20	0
Jan. 24	38	6	38	18	40	24	39	22	0
Jan. 25	37	26	29	14	31	18	33	17	0.01
Jan. 26	31	5	36	23	32	24	36	23	0
Jan. 27	38	20	34	16	36	24	37	21	0.01
Jan. 28	35	24	30	14	36	18	35	24	0
Jan. 29	33	9	36	28	39	25	40	26	0
Jan. 30	39	26	41	31	42	35	36	29	0
Jan. 31	46	32	42	26	42	30	40	18	0

Data Set 11: Rainfall (in inches) in Boston for One Year

STATDISK: Data set name is Bostrain.

Minitab: Worksheet name is BOSTRAIN.MTW.

Excel: Workbook name is BOSTRAIN.XLS.

TI-83 Plus: App name is BOSTRAIN and the file names are the same as for text files.

Text file names: RNMON, RNTUE, RNWED, RNTHU, RNFRI, RNSAT, RNSUN.

Mon	Tues	Wed	Thurs	Fri	Sat	Sun
0	0	0	0.04	0.04	0	0.05
0	0	0	0.06	0.03	0.1	0
0	0	0	0.71	0	0	0
0	0.44	0.14	0.04	0.04	0.64	0
0.05	0	0	0	0.01	0.05	0
0	0	0.64	0	0	0	0
0.01	0	0	0	0.3	0.05	0
0	0	0.01	0	0	0	0
0	0.01	0.01	0.16	0	0	0.09
0.12	0.06	0.18	0.39	0	0.1	0
0	0	0	0	0.78	0.49	0
0	0.02	0	0	0.01	0.17	0
1.41	0.65	0.31	0	0	0.54	0
0	0	0	0	0	0	0
0	0	0	0	0	0.4	0.28
0	0	0	0.3	0.87	0.49	0
0.47	0	0	0	0	0	0
0	0.09	0	0.24	0	0.05	0
0	0.14	0	0	0.04	0.07	0
0.92	0.36	0.02	0.09	0.27	0	0
0.01	0	0.06	0	0	0	0.27
0.01	0	0	0	0	0	0.01
0	0	0	0	0	0	0
0	0	0	0	0.71	0	0
0	0	0.27	0.08	0	0	0.33
0	0	0	0	0	0	0
0.03	0	0.08	0.14	0	0	0
0	0.11	0.06	0.02	0	0	0
0.01	0.05	0	0.01	0	0	0
0	0	0	0	0.12	0	0
0.11	0.03	0	0	0	0	0.44
0.01	0.01	0	0	0.11	0.18	0
0.49	0	0.64	0.01	0	0	0.01
0	0	0.08	0.85	0.01	0	0
0.01	0.02	0	0	0.03	0	0
0	0	0.12	0	0	0	0
0	0	0.01	0.04	0.26	0.04	0
0	0	0	0	0	0.4	0
0.12	0	0	0	0	0	0
0	0	0	0	0.24	0	0.23
0	0	0	0.02	0	0	0
0	0	0	0.02	0	0	0
0.59	0	0	0	0	0.68	0
0	0.01	0	0	0	1.48	0.21
0.01	0	0	0	0.05	0.69	1.28
0	0	0	0	0.96	0	0.01
0	0	0	0	0	0.79	0.02
0.41	0	0.06	0.01	0	0	0.28
0	0	0	0.08	0.04	0	0
0	0	0	0	0	0	0
0	0.74	0	0	0	0	0
0.43	0.3	0	0.26	0	0.02	0.01
		0				

Data Set 12: Everglades Temperature, Rain, Conductivity

Temperatures are in degrees Celsius and are measured at the bottom.
Conductivity is specific conductance and has a very high correlation with
salinity. All measurements are from the Garfield Bight hydrology outpost in
the Florida Everglades. Data are from Kevin Kotun and the National Park
Service.

STATDISK: Data set name is Everglades.
Minitab: Worksheet name is EVERGLADE.MTW.
Excel: Workbook name is EVERGLADE.XLS.
TI-83 Plus: App name is EVERGLADE and the file names are the
             same as for text files.
Text file names: EVTMP, EVCON, EVRN.

Temp	Conductivity	Rainfall (in.)	Temp	Conductivity	Rainfall (in.)
27.6	57.8	0.10	29.2	30.2	0.07
29.1	57.8	0.17	28.2	33.5	0.03
29.4	57.1	0.65	29.1	40.5	0.00
28.5	57.0	0.00	29.9	42.4	0.01
28.6	57.3	0.00	29.9	46.7	0.00
28.0	58.4	0.00	30.6	46.7	0.00
27.9	59.2	0.65	30.6	46.5	0.00
29.0	57.7	0.00	30.9	45.6	0.00
30.6	56.8	0.67	30.0	47.1	0.00
31.2	56.8	0.03	30.7	48.1	0.00
30.7	55.2	1.72	31.9	50.5	0.00
28.0	53.6	0.00	31.5	51.2	0.02
28.3	52.0	0.84	31.2	50.4	0.00
30.1	51.9	0.00	30.9	49.9	0.94
31.3	49.8	0.00	30.6	49.0	0.00
31.0	49.8	0.06	30.1	48.5	0.38
30.8	51.7	0.50	31.1	51.3	0.05
28.5	48.6	1.50	31.5	52.1	0.34
25.9	44.3	1.40	31.8	52.4	0.02
28.5	43.2	0.00	32.0	51.0	0.00
31.9	41.5	0.00	32.6	52.2	0.34
31.3	40.6	0.18	32.9	50.3	0.02
29.4	35.9	2.77	32.7	48.5	0.00
30.0	33.8	0.04	33.5	49.7	0.00
30.1	32.8	0.00	33.8	49.9	0.09
28.8	30.5	1.11	33.7	48.5	0.00
29.5	32.7	0.00	33.6	48.3	0.00
30.5	32.1	0.04	32.3	49.0	0.00
29.2	30.3	1.72	31.6	49.9	0.00
28.8	28.1	0.00	32.0	51.0	0.00
30.1	29.3	0.05			

Data Set 13: Old Faithful Geyser

Durations are in seconds, time intervals are in minutes to the next eruption, and heights of eruptions are in feet. Data are courtesy of the National Park Service and research geologist Rick Hutchinson.

STATDISK: Data set name is OldFaith.
Minitab: Worksheet name is OLDFAITH.MTW.
Excel: Workbook name is OLDFAITH.XLS.
TI-83 Plus: App name is OLDFAITH and the file names are the same as for text files.
Text file names: OFDTN, OFINT, OFHT.

Duration	Interval	Height
240	86	140
237	86	154
122	62	140
267	104	140
113	62	160
258	95	140
232	79	150
105	62	150
276	94	160
248	79	155
243	86	125
241	85	136
214	86	140
114	58	155
272	89	130
227	79	125
237	83	125
238	82	139
203	84	125
270	82	140
218	78	140
226	91	135
250	89	141
245	79	140
120	57	139
267	100	110
103	62	140
270	87	135
241	70	140
239	88	135
233	82	140
238	83	139
102	56	100
271	81	105
127	74	130
275	102	135
140	61	131
264	83	135
134	73	153
268	97	155
124	67	140
270	90	150
249	84	153
237	82	120
235	81	138
228	78	135
265	89	145
120	69	130
275	98	136
241	79	150

Data Set 14: Clancy, Rowling, and Tolstoy Books

Each row of data represents a randomly selected page.

STATDISK: Data set names are Clancy, Rowling, Tolstoy.
Minitab: Worksheet names are CLANCY.MTW, ROWLING.MTW,
           TOLSTOY.MTW.
Excel: Workbook names are CLANCY.XLS, ROWLING.XLS, TOLSTOY.XLS.
TI-83 Plus: App names are CLANCY, ROWLING, TOLSTOY, and the file names
           are the same as for text files.
Text file names: CLWDS, CLCHR, CLFRE, CLFKG, RWWDS, RWCHR, RWFRE,
           RWFKG, TLWDS, TLCHR, TLFRE, TLFKG.

Tom Clancy: *The Bear and the Dragon*

Words/sentence	Characters/word	Flesch Reading Ease	Flesch-Kincaid Grade Level
15.0	4.8	58.2	8.8
9.8	4.5	73.4	5.4
8.1	4.6	73.1	5.0
13.5	4.5	64.4	7.6
24.0	4.0	72.7	9.0
9.8	4.0	89.2	3.2
33.0	4.6	43.9	12.0
9.4	4.5	76.3	4.9
8.3	4.4	76.4	4.6
11.3	4.4	78.9	5.0
11.4	4.3	69.4	6.4
12.4	4.3	72.9	6.1

J. K. Rowling: *Harry Potter and the Sorcerer's Stone*

Words/sentence	Characters/word	Flesch Reading Ease	Flesch-Kincaid Grade Level
15.7	4.1	85.3	5.2
9.0	4.2	84.3	3.7
16.3	4.2	79.5	6.1
14.5	4.4	82.5	4.9
9.7	4.3	80.2	4.4
7.4	4.2	84.6	3.2
14.0	4.5	79.2	5.6
16.1	4.5	70.9	6.9
13.9	4.3	78.6	5.7
12.5	4.0	86.2	4.1
17.2	4.4	74.0	6.7
11.5	4.3	83.7	4.4

Leo Tolstoy: *War and Peace*

Words/sentence	Characters/word	Flesch Reading Ease	Flesch-Kincaid Grade Level
20.6	4.3	69.4	8.6
28.0	4.5	64.2	9.8
12.0	4.5	71.4	6.1
11.5	4.5	71.6	5.9
17.4	4.5	68.5	7.7
19.7	4.8	51.9	10.9
20.3	4.3	72.2	8.2
17.8	4.2	74.4	7.2
22.1	4.7	52.8	11.0
31.4	4.3	58.4	11.5
18.3	4.4	65.4	8.4
11.7	4.5	73.6	5.9

Data Set 15: Ages of Stowaways on the Queen Mary

Data are from the Cunard Steamship Co., Ltd.

STATDISK: Data set name is Stowaway.
Minitab: Worksheet name is STOWAWAY.MTW.
Excel: Workbook name is STOWAWAY.XLS.
TI-83 Plus: App name is STOWAWAY, and the file names are the same as for text files.
Text file names: WEST, EAST.

Westbound

41	24	32	26	39	45	24	21	22	21
40	18	33	33	19	31	16	16	23	19
16	20	18	22	26	22	38	42	25	21
29	24	18	17	24	18	19	30	18	24
31	30	48	29	34	25	23	41	16	17
15	19	18	66	27	43				

Eastbound

24	24	34	15	19	22	18	20	20	17
17	20	18	23	37	15	25	28	21	15
48	18	12	15	23	25	22	21	30	19
20	20	35	19	38	26	19	20	19	41
31	20	19	18	42	25	19	47	19	22
20	23	24	37	23	30	32	28	32	48
27	31	22	34	26	20	22	15	19	20
18	26	36	31	35					

Data Set 16: Cereal

STATDISK: Data set name is Cereal.
Minitab: Worksheet name is CEREAL.MTW.
Excel: Workbook name is CEREAL.XLS.
TI-83 Plus: App name is CEREAL, and the file names are the same
            as for text files.
Text file names: CRCST, CRCAL, CRFAT, CRSGR, CRCHO, CRSOD,
            CRPRO, CRSHL.

Cereal	Cost ($) per 100 grams of cereal	Calories per gram of cereal	Grams of fat per gram of cereal	Grams of sugar per gram of cereal	Cholesterol per gram of cereal	Sodium (mg) per gram of cereal	Protein (g) per gram of cereal	Shelf Location
Cheerios	0.67	3.7	0.07	0.03	0	9.3	0.10	1
Harmony	0.82	3.6	0.02	0.24	0	6.4	0.09	3
Smart Start	0.78	3.6	0.01	0.30	0	6.6	0.06	4
Cocoa Puffs	1.03	4.0	0.03	0.47	0	5.7	0.03	2
Lucky Charms	0.83	4.0	0.03	0.43	0	7.0	0.07	2
Corn Flakes	0.55	3.6	0.00	0.07	0	7.1	0.07	1
Fruit Loops	0.68	3.8	0.03	0.47	0	4.7	0.03	2
Wheaties	0.78	3.7	0.03	0.13	0	7.3	0.10	1
Cap'n Crunch	0.73	4.1	0.06	0.44	0	7.4	0.04	1
Frosted Flakes	0.65	3.9	0.00	0.39	0	4.8	0.03	1
Apple Jacks	0.81	3.9	0.02	0.48	0	4.5	0.03	2
Bran Flakes	0.70	3.3	0.02	0.17	0	7.0	0.10	4
Special K	0.78	3.5	0.00	0.13	0	7.1	0.23	1
Rice Krispies	0.95	3.6	0.00	0.09	0	9.7	0.06	4
Corn Pops	0.84	3.9	0.00	0.45	0	3.9	0.03	2
Trix	0.94	4.0	0.03	0.43	0	6.3	0.03	2

Data Set 17: Weights and Volumes of Cola

Weights are in pounds and volumes are in ounces.

STATDISK: Data set name is Cola.
Minitab: Worksheet name is COLA.MTW.
Excel: Workbook name is COLAL.XLS.
TI-83 Plus: App name is COLA, and the file names are the same as
           for text files.
Text file names: CRGWT, CRGVL, CDTWT, CDTVL, PRGWT, PRGVL,
           PDTWT, PDTVL.

Weight Regular Coke	Volume Regular Coke	Weight Diet Coke	Volume Diet Coke	Weight Regular Pepsi	Volume Regular Pepsi	Weight Diet Pepsi	Volume Diet Pepsi
0.8192	12.3	0.7773	12.1	0.8258	12.4	0.7925	12.3
0.8150	12.1	0.7758	12.1	0.8156	12.2	0.7868	12.2
0.8163	12.2	0.7896	12.3	0.8211	12.2	0.7846	12.2
0.8211	12.3	0.7868	12.3	0.8170	12.2	0.7938	12.3
0.8181	12.2	0.7844	12.2	0.8216	12.2	0.7861	12.2
0.8247	12.3	0.7861	12.3	0.8302	12.4	0.7844	12.2
0.8062	12.0	0.7806	12.2	0.8192	12.2	0.7795	12.2
0.8128	12.1	0.7830	12.2	0.8192	12.2	0.7883	12.3
0.8172	12.2	0.7852	12.2	0.8271	12.3	0.7879	12.2
0.8110	12.1	0.7879	12.3	0.8251	12.3	0.7850	12.3
0.8251	12.3	0.7881	12.3	0.8227	12.2	0.7899	12.3
0.8264	12.3	0.7826	12.3	0.8256	12.3	0.7877	12.2
0.7901	11.8	0.7923	12.3	0.8139	12.2	0.7852	12.2
0.8244	12.3	0.7852	12.3	0.8260	12.3	0.7756	12.1
0.8073	12.1	0.7872	12.3	0.8227	12.2	0.7837	12.2
0.8079	12.1	0.7813	12.2	0.8388	12.5	0.7879	12.2
0.8044	12.0	0.7885	12.3	0.8260	12.3	0.7839	12.2
0.8170	12.2	0.7760	12.1	0.8317	12.4	0.7817	12.2
0.8161	12.2	0.7822	12.2	0.8247	12.3	0.7822	12.2
0.8194	12.2	0.7874	12.3	0.8200	12.2	0.7742	12.1
0.8189	12.2	0.7822	12.2	0.8172	12.2	0.7833	12.2
0.8194	12.2	0.7839	12.2	0.8227	12.3	0.7835	12.2
0.8176	12.2	0.7802	12.1	0.8244	12.3	0.7855	12.2
0.8284	12.4	0.7892	12.3	0.8244	12.2	0.7859	12.2
0.8165	12.2	0.7874	12.2	0.8319	12.4	0.7775	12.1
0.8143	12.2	0.7907	12.3	0.8247	12.3	0.7833	12.2
0.8229	12.3	0.7771	12.1	0.8214	12.2	0.7835	12.2
0.8150	12.2	0.7870	12.2	0.8291	12.4	0.7826	12.2
0.8152	12.2	0.7833	12.3	0.8227	12.3	0.7815	12.2
0.8244	12.3	0.7822	12.2	0.8211	12.3	0.7791	12.1
0.8207	12.2	0.7837	12.3	0.8401	12.5	0.7866	12.3
0.8152	12.2	0.7910	12.4	0.8233	12.3	0.7855	12.2
0.8126	12.1	0.7879	12.3	0.8291	12.4	0.7848	12.2
0.8295	12.4	0.7923	12.4	0.8172	12.2	0.7806	12.2
0.8161	12.2	0.7859	12.3	0.8233	12.4	0.7773	12.1
0.8192	12.2	0.7811	12.2	0.8211	12.3	0.7775	12.1

Data Set 18: Diamonds

Price is in dollars. Depth is 100 times the ratio of height to diameter. Table is size of the upper flat surface. (Depth and table determine "cut.") Color indices are on a standard scale, with 1 = colorless and increasing numbers indicating more yellow. On the clarity scale, 1 = flawless and 6 indicates inclusions that can be seen by eye.

STATDISK: Data set name is Diamonds.
Minitab: Worksheet name is DIAMONDS.MTW.
Excel: Workbook name is DIAMONDS.XLS.
TI-83 Plus: App name is DIAMONDS, and the file names are the
                    same as for text files.
Text file names: PRICE, CARAT, DEPTH, TABLE, COLOR, CLRTY.

Price	Carat	Depth	Table	Color	Clarity
6958	1.00	60.5	65	3	4
5885	1.00	59.2	65	5	4
6333	1.01	62.3	55	4	4
4299	1.01	64.4	62	5	5
9589	1.02	63.9	58	2	3
6921	1.04	60.0	61	4	4
4426	1.04	62.0	62	5	5
6885	1.07	63.6	61	4	3
5826	1.07	61.6	62	5	5
3670	1.11	60.4	60	9	4
7176	1.12	60.2	65	2	3
7497	1.16	59.5	60	5	3
5170	1.20	62.6	61	6	4
5547	1.23	59.2	65	7	4
18596	1.25	61.2	61	1	2
7521	1.29	59.6	59	6	2
7260	1.50	61.1	65	6	4
8139	1.51	63.0	60	6	4
12196	1.67	58.7	64	3	5
14998	1.72	58.5	61	4	3
9736	1.76	57.9	62	8	2
9859	1.80	59.6	63	5	5
12398	1.88	62.9	62	6	2
25322	2.03	60.1	62	2	3
11008	2.03	62.0	63	8	3
38794	2.06	58.2	63	2	2
66780	3.00	63.3	62	1	3
46769	4.01	57.1	51	3	4
28800	4.01	63.0	63	6	5
28868	4.05	59.3	60	7	4

Data Set 19: Weights of a Sample of M&M Plain Candies

Weights are in grams.

STATDISK: Data set name is M and M's.
Minitab: Worksheet name is M&M.MTW.
Excel: Workbook name is M&M.XLS.
TI-83 Plus: App name is MM, and the file names are the same as
              for text files.
Text file names: RED, ORNG, YLLW, BROWN, BLUE, GREEN.

Red	Orange	Yellow	Brown	Blue	Green
0.870	0.903	0.906	0.932	0.838	0.911
0.933	0.920	0.978	0.860	0.875	1.002
0.952	0.861	0.926	0.919	0.870	0.902
0.908	1.009	0.868	0.914	0.956	0.930
0.911	0.971	0.876	0.914	0.968	0.949
0.908	0.898	0.968	0.904		0.890
0.913	0.942	0.921	0.930		0.902
0.983	0.897	0.893	0.871		
0.920		0.939	1.033		
0.936		0.886	0.955		
0.891		0.924	0.876		
0.924		0.910	0.856		
0.874		0.877	0.866		
0.908		0.879	0.858		
0.924		0.941	0.988		
0.897		0.879	0.936		
0.912		0.940	0.930		
0.888		0.960	0.923		
0.872		0.989	0.867		
0.898		0.900	0.965		
0.882		0.917	0.902		
		0.911	0.928		
		0.892	0.900		
		0.886	0.889		
		0.949	0.875		
		0.934	0.909		
			0.976		
			0.921		
			0.898		
			0.897		
			0.902		
			0.920		
			0.909		

Data Set 20: Axial Loads of Aluminum Cans

Axial loads are measured in pounds.

STATDISK: Data set name is Cans.
Minitab: Worksheet name is CANS.MTW.
Excel: Workbook name is CANS.XLS.
TI-83 Plus: App name is CANS, and the file names are the same as
            for text files.
Text file names: CN109, CN111.

| | Aluminum cans 0.0109 in. | | | | | | | | Aluminum cans 0.0111 in. | | | | | | |
Sample	Load (pounds)							Sample	Load (pounds)						
1	270	273	258	204	254	228	282	1	287	216	260	291	210	272	260
2	278	201	264	265	223	274	230	2	294	253	292	280	262	295	230
3	250	275	281	271	263	277	275	3	283	255	295	271	268	225	246
4	278	260	262	273	274	286	236	4	297	302	282	310	305	306	262
5	290	286	278	283	262	277	295	5	222	276	270	280	288	296	281
6	274	272	265	275	263	251	289	6	300	290	284	304	291	277	317
7	242	284	241	276	200	278	283	7	292	215	287	280	311	283	293
8	269	282	267	282	272	277	261	8	285	276	301	285	277	270	275
9	257	278	295	270	268	286	262	9	290	288	287	282	275	279	300
10	272	268	283	256	206	277	252	10	293	290	313	299	300	265	285
11	265	263	281	268	280	289	283	11	294	262	297	272	284	291	306
12	263	273	209	259	287	269	277	12	263	304	288	256	290	284	307
13	234	282	276	272	257	267	204	13	273	283	250	244	231	266	504
14	270	285	273	269	284	276	286	14	284	227	269	282	292	286	281
15	273	289	263	270	279	206	270	15	296	287	285	281	298	289	283
16	270	268	218	251	252	284	278	16	247	279	276	288	284	301	309
17	277	208	271	208	280	269	270	17	284	284	286	303	308	288	303
18	294	292	289	290	215	284	283	18	306	285	289	292	295	283	315
19	279	275	223	220	281	268	272	19	290	247	268	283	305	279	287
20	268	279	217	259	291	291	281	20	285	298	279	274	205	302	296
21	230	276	225	282	276	289	288	21	282	300	284	281	279	255	210
22	268	242	283	277	285	293	248	22	279	286	293	285	288	289	281
23	278	285	292	282	287	277	266	23	297	314	295	257	298	211	275
24	268	273	270	256	297	280	256	24	247	279	303	286	287	287	275
25	262	268	262	293	290	274	292	25	243	274	299	291	281	303	269

Data Set 21: Movies

STATDISK: Data set name is Movies.
Minitab: Worksheet name is MOVIES.MTW.
Excel: Workbook name is MOVIES.XLS.
TI-83 Plus: App name is MOVIES, and the file names are the same
            as for text files.
Text file names: MVBUD, MVGRS, MVLEN, MVRAT.

Title	Year	Rating	Budget ($) in Millions	Gross ($) in Millions	Length in Minutes	Viewer Rating
Aliens	1986	R	18.5	81.843	137	8.2
Armageddon	1998	PG-13	140	194.125	144	6.7
As Good As It Gets	1997	PG-13	50	147.54	138	8.1
Braveheart	1995	R	72	75.6	177	8.3
Chasing Amy	1997	R	0.25	12.006	105	7.9
Contact	1997	PG	90	100.853	153	8.3
Dante's Peak	1997	PG-13	104	67.155	112	6.7
Deep Impact	1998	PG-13	75	140.424	120	6.4
Executive Decision	1996	R	55	68.75	129	7.3
Forrest Gump	1994	PG-13	55	329.691	142	7.7
Ghost	1990	PG-13	22	217.631	128	7.1
Gone with the Wind	1939	G	3.9	198.571	222	8.0
Good Will Hunting	1997	R	10	138.339	126	8.5
Grease	1978	PG	6	181.28	110	7.3
Halloween	1978	R	0.325	47	93	7.7
Hard Rain	1998	R	70	19.819	95	5.2
I Know What You Did Last Summer	1997	R	17	72.219	100	6.5
Independence Day	1996	PG-13	75	306.124	142	6.6
Indiana Jones and the Last Crusade	1989	PG-13	39	197.171	127	7.8
Jaws	1975	PG	12	260	124	7.8
Men in Black	1997	PG-13	90	250.147	98	7.4
Multiplicity	1996	PG-13	45	20.1	117	6.8
Pulp Fiction	1994	R	8	107.93	154	8.3
Raiders of the Lost Ark	1981	PG	20	242.374	115	8.3
Saving Private Ryan	1998	R	70	178.091	170	9.1
Schindler's List	1993	R	25	96.067	197	8.6
Scream	1996	R	15	103.001	111	7.7
Speed 2: Cruise Control	1997	PG-13	110	48.068	121	4.3
Terminator	1984	R	6.4	36.9	108	7.7
The American President	1995	PG-13	62	65	114	7.6
The Fifth Element	1997	PG-13	90	63.54	126	7.8
The Game	1997	R	50	48.265	128	7.6
The Man in the Iron Mask	1998	PG-13	35	56.876	132	6.5
Titanic	1997	PG-13	200	600.743	195	8.4
True Lies	1994	R	100	146.261	144	7.2
Volcano	1997	PG-13	90	47.474	102	5.8

Data Set 22: Cars

CITY is the city fuel consumption in mi/gal, HWY is the highway fuel consumption in mi/gal, WEIGHT is the weight of the car in pounds, CYLINDER is the number of cylinders, DISPLACEMENT is the engine displacement in liters, MAN/AUT indicates manual or automatic transmission, GHG is the amount of emitted greenhouse gases (in tons/yr), and NOX is the amount of tailpipe emissions of NO_x (in lb/yr).

STATDISK: Data set name is Cars.
Minitab: Worksheet name is CARS.MTW.
Excel: Workbook name is CARS.XLS.
TI-83 Plus: App name is CARS, and the file names are the same as for text files.
Text file names: CRCTY, CRHWY, CRWT, CRCYL, CRDSP, CRGHG, CRNOX.

Car	City	HWY	Weight	Cylinder	Displacement	MAN/AUT	GHG	NOX
Chev. Camaro	19	30	3545	6	3.8	M	12	34.4
Chev. Cavalier	23	31	2795	4	2.2	A	10	25.1
Dodge Neon	23	32	2600	4	2	A	10	25.1
Ford Taurus	19	27	3515	6	3	A	12	25.1
Honda Accord	23	30	3245	4	2.3	A	11	25.1
Lincoln Cont.	17	24	3930	8	4.6	A	14	25.1
Mercury Mystique	20	29	3115	6	2.5	A	12	34.4
Mitsubishi Eclipse	22	33	3235	4	2	M	10	25.1
Olds. Aurora	17	26	3995	8	4	A	13	34.4
Pontiac Grand Am	22	30	3115	4	2.4	A	11	25.1
Toyota Camry	23	32	3240	4	2.2	M	10	25.1
Cadillac DeVille	17	26	4020	8	4.6	A	13	34.4
Chev. Corvette	18	28	3220	8	5.7	M	12	34.4
Chrysler Sebring	19	27	3175	6	2.5	A	12	25.1
Ford Mustang	20	29	3450	6	3.8	M	12	34.4
BMW 3-Series	19	27	3225	6	2.8	A	12	34.4
Ford Crown Victoria	17	24	3985	8	4.6	A	14	25.1
Honda Civic	32	37	2440	4	1.6	M	8	25.1
Mazda Protege	29	34	2500	4	1.6	A	9	25.1
Hyundai Accent	28	37	2290	4	1.5	A	9	34.4

Data Set 23: Weights of Discarded Garbage for One Week

Weights are in pounds. HHSIZE is the household size. Data provided by
Masakuza Tani, the Garbage Project, University of Arizona.

STATDISK: Data set name is Garbage.
Minitab: Worksheet name is GARBAGE.MTW.
Excel: Workbook name is GARBAGE.XLS.
TI-83 Plus: App name is GARBAGE, and the file names are the
same as for text files.
Text file names: HHSIZ, METAL, PAPER, PLAS, GLASS, FOOD, YARD,
TEXT, OTHER, TOTAL.

Household	HHSize	Metal	Paper	Plas	Glass	Food	Yard	Text	Other	Total
1	2	1.09	2.41	0.27	0.86	1.04	0.38	0.05	4.66	10.76
2	3	1.04	7.57	1.41	3.46	3.68	0.00	0.46	2.34	19.96
3	3	2.57	9.55	2.19	4.52	4.43	0.24	0.50	3.60	27.60
4	6	3.02	8.82	2.83	4.92	2.98	0.63	2.26	12.65	38.11
5	4	1.50	8.72	2.19	6.31	6.30	0.15	0.55	2.18	27.90
6	2	2.10	6.96	1.81	2.49	1.46	4.58	0.36	2.14	21.90
7	1	1.93	6.83	0.85	0.51	8.82	0.07	0.60	2.22	21.83
8	5	3.57	11.42	3.05	5.81	9.62	4.76	0.21	10.83	49.27
9	6	2.32	16.08	3.42	1.96	4.41	0.13	0.81	4.14	33.27
10	4	1.89	6.38	2.10	17.67	2.73	3.86	0.66	0.25	35.54
11	4	3.26	13.05	2.93	3.21	9.31	0.70	0.37	11.61	44.44
12	7	3.99	11.36	2.44	4.94	3.59	13.45	4.25	1.15	45.17
13	3	2.04	15.09	2.17	3.10	5.36	0.74	0.42	4.15	33.07
14	5	0.99	2.80	1.41	1.39	1.47	0.82	0.44	1.03	10.35
15	6	2.96	6.44	2.00	5.21	7.06	6.14	0.20	14.43	44.44
16	2	1.50	5.86	0.93	2.03	2.52	1.37	0.27	9.65	24.13
17	4	2.43	11.08	2.97	1.74	1.75	14.70	0.39	2.54	37.60
18	4	2.97	12.43	2.04	3.99	5.64	0.22	2.47	9.20	38.96
19	3	1.42	6.05	0.65	6.26	1.93	0.00	0.86	0.00	17.17
20	3	3.60	13.61	2.13	3.52	6.46	0.00	0.96	1.32	31.60
21	2	4.48	6.98	0.63	2.01	6.72	2.00	0.11	0.18	23.11
22	2	1.36	14.33	1.53	2.21	5.76	0.58	0.17	1.62	27.56
23	4	2.11	13.31	4.69	0.25	9.72	0.02	0.46	0.40	30.96
24	1	0.41	3.27	0.15	0.09	0.16	0.00	0.00	0.00	4.08
25	4	2.02	6.67	1.45	6.85	5.52	0.00	0.68	0.03	23.22
26	6	3.27	17.65	2.68	2.33	11.92	0.83	0.28	4.03	42.99
27	11	4.95	12.73	3.53	5.45	4.68	0.00	0.67	19.89	51.90
28	3	1.00	9.83	1.49	2.04	4.76	0.42	0.54	0.12	20.20
29	4	1.55	16.39	2.31	4.98	7.85	2.04	0.20	1.48	36.80
30	3	1.41	6.33	0.92	3.54	2.90	3.85	0.03	0.04	19.02
31	2	1.05	9.19	0.89	1.06	2.87	0.33	0.01	0.03	15.43
32	2	1.31	9.41	0.80	2.70	5.09	0.64	0.05	0.71	20.71
33	2	2.50	9.45	0.72	1.14	3.17	0.00	0.02	0.01	17.01
34	4	2.35	12.32	2.66	12.24	2.40	7.87	4.73	0.78	45.35
35	6	3.69	20.12	4.37	5.67	13.20	0.00	1.15	1.17	49.37

(continued)

Data Set 23: Weights of Discarded Garbage for One Week
(*continued*)

Household	HHSize	Metal	Paper	Plas	Glass	Food	Yard	Text	Other	Total
36	2	3.61	7.72	0.92	2.43	2.07	0.68	0.63	0.00	18.06
37	2	1.49	6.16	1.40	4.02	4.00	0.30	0.04	0.00	17.41
38	2	1.36	7.98	1.45	6.45	4.27	0.02	0.12	2.02	23.67
39	2	1.73	9.64	1.68	1.89	1.87	0.01	1.73	0.58	19.13
40	2	0.94	8.08	1.53	1.78	8.13	0.36	0.12	0.05	20.99
41	3	1.33	10.99	1.44	2.93	3.51	0.00	0.39	0.59	21.18
42	3	2.62	13.11	1.44	1.82	4.21	4.73	0.64	0.49	29.06
43	2	1.25	3.26	1.36	2.89	3.34	2.69	0.00	0.16	14.95
44	2	0.26	1.65	0.38	0.99	0.77	0.34	0.04	0.00	4.43
45	3	4.41	10.00	1.74	1.93	1.14	0.92	0.08	4.60	24.82
46	6	3.22	8.96	2.35	3.61	1.45	0.00	0.09	1.12	20.80
47	4	1.86	9.46	2.30	2.53	6.54	0.00	0.65	2.45	25.79
48	4	1.76	5.88	1.14	3.76	0.92	1.12	0.00	0.04	14.62
49	3	2.83	8.26	2.88	1.32	5.14	5.60	0.35	2.03	28.41
50	3	2.74	12.45	2.13	2.64	4.59	1.07	0.41	1.14	27.17
51	10	4.63	10.58	5.28	12.33	2.94	0.12	2.94	15.65	54.47
52	3	1.70	5.87	1.48	1.79	1.42	0.00	0.27	0.59	13.12
53	6	3.29	8.78	3.36	3.99	10.44	0.90	1.71	13.30	45.77
54	5	1.22	11.03	2.83	4.44	3.00	4.30	1.95	6.02	34.79
55	4	3.20	12.29	2.87	9.25	5.91	1.32	1.87	0.55	37.26
56	7	3.09	20.58	2.96	4.02	16.81	0.47	1.52	2.13	51.58
57	5	2.58	12.56	1.61	1.38	5.01	0.00	0.21	1.46	24.81
58	4	1.67	9.92	1.58	1.59	9.96	0.13	0.20	1.13	26.18
59	2	0.85	3.45	1.15	0.85	3.89	0.00	0.02	1.04	11.25
60	4	1.52	9.09	1.28	8.87	4.83	0.00	0.95	1.61	28.15
61	2	1.37	3.69	0.58	3.64	1.78	0.08	0.00	0.00	11.14
62	2	1.32	2.61	0.74	3.03	3.37	0.17	0.00	0.46	11.70

Data Set 24: Homes Sold in Dutchess County

STATDISK: Data set name is Homes.
Minitab: Worksheet name is HOMES.MTW.
Excel: Workbook name is HOMES.XLS.
TI-83 Plus: App name is HOMES, and the file names are the same
            as for text files.
Text file names: HMSP, HMLST, HMLA, HMRMS, HMBRS, HMBTH,
            HMAGE, HMACR, HMTAX.

Selling Price (thousands)	List Price (thousands)	Living Area (hundreds of sq. ft.)	Rooms	Bedrooms	Bathrooms	Age (years)	Acres	Taxes (dollars)
142.0	160	28	10	5	3	60	0.28	3167
175.0	180	18	8	4	1	12	0.43	4033
129.0	132	13	6	3	1	41	0.33	1471
138.0	140	17	7	3	1	22	0.46	3204
232.0	240	25	8	4	3	5	2.05	3613
135.0	140	18	7	4	3	9	0.57	3028
150.0	160	20	8	4	3	18	4.00	3131
207.0	225	22	8	4	2	16	2.22	5158
271.0	285	30	10	5	2	30	0.53	5702
89.0	90	10	5	3	1	43	0.30	2054
153.0	157	22	8	3	3	18	0.38	4127
86.5	90	16	7	3	1	50	0.65	1445
234.0	238	25	8	4	2	2	1.61	2087
105.5	116	20	8	4	1	13	0.22	2818
175.0	180	22	8	4	2	15	2.06	3917
165.0	170	17	8	4	2	33	0.46	2220
166.0	170	23	9	4	2	37	0.27	3498
136.0	140	19	7	3	1	22	0.63	3607
148.0	160	17	7	3	2	13	0.36	3648
151.0	153	19	8	4	2	24	0.34	3561
180.0	190	24	9	4	2	10	1.55	4681
293.0	305	26	8	4	3	6	0.46	7088
167.0	170	20	9	4	2	46	0.46	3482
190.0	193	22	9	5	2	37	0.48	3920
184.0	190	21	9	5	2	27	1.30	4162
157.0	165	20	8	4	2	7	0.30	3785
110.0	115	16	8	4	1	26	0.29	3103
135.0	145	18	7	4	1	35	0.43	3363
567.0	625	64	11	4	4	4	0.85	12192
180.0	185	20	8	4	2	11	1.00	3831
183.0	188	17	7	3	2	16	3.00	3564
185.0	193	20	9	3	2	56	6.49	3765
152.0	155	17	8	4	1	33	0.70	3361
148.0	153	13	6	3	2	22	0.39	3950

(continued)

Data Set 24: Homes Sold in Dutchess County (*continued*)

Selling Price (thousands)	List Price (thousands)	Living Area (hundreds of sq. ft.)	Rooms	Bedrooms	Bathrooms	Age (years)	Acres	Taxes (dollars)
152.0	159	15	7	3	1	25	0.59	3055
146.0	150	16	7	3	1	31	0.36	2950
170.0	190	24	10	3	2	33	0.57	3346
127.0	130	20	8	4	1	65	0.40	3334
265.0	270	36	10	6	3	33	1.20	5853
157.0	163	18	8	4	2	12	1.13	3982
128.0	135	17	9	4	1	25	0.52	3374
110.0	120	15	8	4	2	11	0.59	3119
123.0	130	18	8	4	2	43	0.39	3268
212.0	230	39	12	5	3	202	4.29	3648
145.0	145	18	8	4	2	44	0.22	2783
129.0	135	10	6	3	1	15	1.00	2438
143.0	145	21	7	4	2	10	1.20	3529
247.0	252	29	9	4	2	4	1.25	4626
111.0	120	15	8	3	1	97	1.11	3205
133.0	145	26	7	3	1	42	0.36	3059

Data Set 25: Miscellaneous: DJIA, Car Sales, Motor Vehicle Deaths, Murders, Sunspots, Super Bowl

STATDISK: Data set name is Misc.
Minitab: Worksheet name is MISC.MTW.
Excel: Workbook name is MISC.XLS.
TI-83 Plus: App name is MISC, and the file names are the same as for text files.
Text file names: DJIA, CRSLS, MVDTH, MURDR, SNSPT, SUPER.

Year	DJIA High	US Car Sales (thousands)	US Motor Vehicle Deaths	US Murders and Non-Negligent Homicides	Sunspot Number	Super Bowl Points
1980	1000	8979	53172	23040	154.6	50
1981	1024	8536	51385	22520	140.5	37
1982	1071	7982	45779	21010	115.9	57
1983	1287	9182	44452	19310	66.6	44
1984	1287	10390	46263	18690	45.9	47
1985	1553	11042	45901	18980	17.9	54
1986	1956	11460	47865	20610	13.4	56
1987	2722	10277	48290	20100	29.2	59
1988	2184	10530	49078	20680	100.2	36
1989	2791	9773	47575	21500	157.6	65
1990	3000	9300	46814	23440	142.6	39
1991	3169	8175	43536	24700	145.7	61
1992	3413	8213	40982	23760	94.3	69
1993	3794	8518	41893	24530	54.6	43
1994	3978	8991	42524	23330	29.9	75
1995	5216	8635	43363	21610	17.5	44
1996	6561	8527	43649	19650	8.6	56
1997	8259	8272	43458	18210	21.5	55
1998	9374	8142	43501	16970	64.3	53
1999	11568	8698	41300	15522	93.3	39
2000	11401	8847	43000	15517	119.6	41

Data Set 26: New York State Lottery

STATDISK: Data set name is Lottery.
Minitab: Worksheet name is LOTTO.MTW.
Excel: Workbook name is LOTTO.XLS.
TI-83 Plus: App name is LOTTO, and the file names are the same as for text files.
Text file names: LOT1, LOT2, LOT3, LOT4, LOT5, LOT6, WIN1, WIN2, WIN3, WIN4.

New York State Lotto						NYS Win 4			
19	22	26	38	44	48	9	2	5	4
4	6	16	24	37	49	7	7	5	4
22	31	35	38	41	48	0	1	7	5
4	11	22	31	34	35	7	3	7	6
9	15	19	23	24	51	5	5	7	1
3	8	21	28	30	45	5	2	6	4
1	2	15	32	33	48	1	5	4	3
18	29	30	32	38	43	9	3	5	0
1	8	13	35	44	46	0	6	2	7
11	21	25	32	37	49	0	7	2	7
6	8	9	33	34	40	8	9	1	9
11	14	20	25	31	33	7	0	0	9
6	11	25	30	42	49	6	6	6	2
19	32	33	41	50	51	0	0	1	5
8	13	24	42	43	47	1	6	6	0
12	13	16	25	27	31	6	1	9	3
3	23	26	36	40	45	7	1	5	6
11	18	20	24	25	41	2	7	5	9
2	10	17	19	42	43	4	4	1	0
5	18	20	23	46	49	7	2	8	6
12	13	17	31	32	35	5	7	4	5
5	23	26	32	45	46	5	9	3	3
8	12	27	39	40	50	5	7	7	6
6	21	41	43	50	51	2	4	0	4
18	19	21	23	38	49	0	8	7	2
13	14	32	39	44	51	4	3	5	7
17	19	21	22	31	35	4	0	4	7
6	12	19	41	47	49	9	6	1	5
5	15	38	41	42	50	9	2	9	5
3	4	6	14	24	46	0	6	4	7
6	28	29	46	47	51	5	4	6	9
8	9	29	30	33	50	3	0	6	0
3	6	22	26	41	45	3	7	4	7
2	15	33	36	38	46	1	9	1	6
10	16	36	37	46	51	9	0	9	8
8	10	13	23	33	45	2	6	7	6
20	23	26	39	48	50	5	2	2	9
12	22	31	33	43	50	6	8	6	8
22	30	31	40	45	49	8	7	4	7
9	23	25	27	37	38	2	4	0	7

Data Set 27: Solitaire Results

Results from the Microsoft Solitaire game (Vegas rules of "draw 3" with $52 bet and return of $5 per card. Amounts are in dollars and represent net gain or loss by player.)

STATDISK: Data set name is Solitaire.
Minitab: Worksheet name is SOLITAIRE.MTW.
Excel: Workbook name is SOLITAIRE.XLS.
TI-83 Plus: App name is SOLITAIRE, and the file name is SOLTR.
Text file name: SOLTR.

−42	−17	−17	−37	−12	−22	−32	−12	−27	−27
−27	−17	−47	−47	−12	−2	−22	−27	−37	−27
−12	−37	−2	23	−17	−42	23	−42	−32	53
−27	−17	18	−2	−37	−32	−2	−7	23	−7
−7	−42	−42	−27	38	−27	−42	−47	−32	−32
−27	28	3	−7	−47	13	−42	3	−17	−27
−37	−27	23	−47	−47	−27	−27	−42	−17	−37
−32	−22	−12	208	−27	3	−17	−42	−32	−22
−52	−12	−47	−47	−22	−42	−12	−12	−32	−37
13	208	−42	−7	−47	−47	−42	−7	−32	−47
−42	−27	23	−52	−37	−42	−32	−42	−52	−17
−37	−47	−42	−47	−47	−37	−17	−17	−52	33
−17	18	−52	208	−47	−47	−37	−42	108	−2
−47	−42	23	18	−37	−7	−27	43	−27	−12
8	−27	8	−27	−2	−47	−12	18	−47	−17
−32	−52	−32	28	−2	−12	−27	208	−7	−37
−22	−12	−42	−7	−12	−27	−22	−27	−42	−42
−7	−12	−27	−27	−22	−12	8	−27	−32	3
8	−47	−37	−17	8	−42	−7	−47	−47	13
−7	−17	−22	−12	208	−2	−37	−32	−32	−27
−47	−12	53	−42	−32	−47	3	−42	−42	−37
−12	−27	−37	−37	−42	208	−22	−17	−32	−37
18	3	−42	−22	−22	−42	−42	−27	−32	−17
−17	−32	−2	−2	−2	−27	−27	−22	−22	−52
−32	−17	3	58	−22	−37	−42	−27	−52	−7
−32	−37	−12	−42	−52	−47	−37	208	−47	−27
−22	−12	−42	−32	−27	−37	−12	−37	−32	−7
−47	−42	−47	−37	−37	−37	−52	73	−32	53
−12	−27	−2	−17	−37	−37	−52	−22	−17	8
−32	−42	−2	−22	−32	−37	−47	68	−32	−17
−2	−12	−7	−37	−42	−42	−22	−37	−47	−32
−22	−47	−42	−47	−27	−47	208	−17	−17	−7
−37	−37	−32	−17	−32	−37	−37	−32	−37	−12
−47	−22	−27	−32	−42	−32	−37	−7	−22	−52
−22	−27	3	−47	−27	−42	208	−22	−7	−32
−27	−42	−22	−37	−52	−27	−42	−17	−37	18
−2	3	−32	−37	−42	18	−12	−17	43	−17
−42	−12	−22	−37	−17	18	−32	8	−12	208
−37	−32	−32	−2	−52	−42	−37	−27	−37	−27
8	−52	−32	208	13	−22	48	−27	−37	−42
−32	−37	−47	−7	−22	−17	−52	−2	−37	−37
−17	−42	−22	98	−37	−37	−22	13	−22	3
−47	33	−37	−47	23	−12	−17	−22	23	8
3	−32	−32	−17	3	−37	−27	−22	−7	18
−47	−37	−32	−12	−12	−42	−2	−27	−32	−37
−22	−32	−7	−12	3	−52	−22	−42	8	−7
−47	−2	−27	−7	−32	−17	−37	−17	−32	−42
−7	−42	−32	−52	−42	−27	208	−32	−32	−42
−42	−52	208	−37	−7	−27	−2	−37	23	−37
−12	−22	−7	−2	−32	13	8	−47	−37	−32

Data Set 28: Weights of Domino Sugar Packets

Weights are in grams.

STATDISK: Data set name is Sugar.
Minitab: Worksheet name is SUGAR.MTW.
Excel: Workbook name is SUGAR.XLS.
TI-83 Plus: App name is SUGARWT, and the file name is SUGAR.
Text file name: SUGAR.

3.647	3.638	3.635	3.645	3.521	3.617	3.666
3.588	3.545	3.590	3.621	3.532	3.511	3.516
3.531	3.678	3.643	3.583	3.723	3.673	3.588
3.600	3.611	3.580	3.667	3.506	3.632	3.450
3.660	3.569	3.573	3.526	3.494	3.601	3.604
3.407	3.522	3.598	3.585	3.577	3.522	3.464
3.604	3.508	3.718	3.635	3.643	3.507	3.687
3.582	3.622	3.654	3.482	3.494	3.475	3.492
3.542	3.625	3.688	3.468	3.639	3.582	3.491
3.535	3.548	3.671	3.665	3.726	3.576	3.725

Data Set 29: Weights of Quarters

Weights are in grams.

STATDISK: Data set name is Quarters.
Minitab: Worksheet name is QUARTERS.MTW.
Excel: Workbook name is QUARTERS.XLS.
TI-83 Plus: App name is QUARTERS, and the file name is QRTRS.
Text file name: QRTRS.

5.60	5.63	5.58	5.56	5.66	5.58	5.57	5.59	5.67	5.61
5.84	5.73	5.53	5.58	5.52	5.65	5.57	5.71	5.59	5.53
5.63	5.68	5.62	5.60	5.53	5.58	5.60	5.58	5.59	5.66
5.73	5.59	5.63	5.66	5.67	5.60	5.74	5.57	5.62	5.73
5.60	5.60	5.57	5.71	5.62	5.72	5.57	5.70	5.60	5.49

Data Set 30: Homerun Distances

Homerun distances are in feet for Mark McGwire (1998),
Sammy Sosa (1998), and Barry Bonds (2001).

STATDISK: Data set name is Homeruns.
Minitab: Worksheet name is HOMERUNS.MTW.
Excel: Workbook name is HOMERUNS.XLS.
TI-83 Plus: App name is HOMERUNS, and the file names are the
           same as for text files.
Text file names: MCGWR, SOSA, BONDS.

McGwire

360	370	370	430	420	340	460	410	440	410
380	360	350	527	380	550	478	420	390	420
425	370	480	390	430	388	423	410	360	410
450	350	450	430	461	430	470	440	400	390
510	430	450	452	420	380	470	398	409	385
369	460	390	510	500	450	470	430	458	380
430	341	385	410	420	380	400	440	377	370

Sosa

371	350	430	420	430	434	370	420	440	410
420	460	400	430	410	370	370	410	380	340
350	420	410	415	430	380	380	366	500	380
390	400	364	430	450	440	365	420	350	420
400	380	380	400	370	420	360	368	430	433
388	440	414	482	364	370	400	405	433	390
480	480	434	344	410	420				

Bonds

420	417	440	410	390	417	420	410	380	430
370	420	400	360	410	420	391	416	440	410
415	436	430	410	400	390	420	410	420	410
410	450	320	430	380	375	375	347	380	429
320	360	375	370	440	400	405	430	350	396
410	380	430	415	380	375	400	435	420	420
488	361	394	410	411	365	360	440	435	454
442	404	385							

Appendix C: TI-83 Plus

CLEAR

To **CLEAR** data in list L_1:

$$\boxed{\text{STAT}} \quad [\text{4:ClrList}] \quad \boxed{\text{2nd}} \quad \overset{L_1}{\boxed{1}} \quad \boxed{\text{ENTER}}$$

ENTER

To **ENTER** data in list L_1:

$$\boxed{\text{STAT}} \quad [\text{1:Edit}] \quad \text{ENTER value, press } \boxed{\text{ENTER}},\ldots$$

When all data have been ENTERed, press $\boxed{\text{2nd}} \quad \overset{\text{Quit}}{\boxed{\text{Mode}}}$

STATS

To get **STATISTICS** for data in list L_1:

$$\boxed{\text{STAT}} \quad [\text{CALC}] \quad [\text{1:1-Var Stats}] \quad \boxed{\text{2nd}} \quad \overset{L_1}{\boxed{1}} \quad \boxed{\text{ENTER}}$$

Notes: Sx is the sample standard deviation s.
Q_1 and Q_3 may be different from textbook.

GRAPH

To get **HISTOGRAM** or **BOXPLOT** for data in L_1:

1. $\boxed{\text{2nd}} \quad \overset{\text{STAT PLOT}}{\boxed{\text{Y=}}} \quad \boxed{\text{ENTER}} \quad \boxed{\text{ENTER}}$
2. Select "Type" (for boxplot, middle of second row).
3. $\boxed{\text{ZOOM}} \quad [\text{9: ZoomStat}]$

FREQ. DIST.

To get **STATISTICS FROM A FREQUENCY DISTRIBUTION:**

1. Clear L_1 and L_2: $\boxed{\text{STAT}}$ [4:ClrList] $\boxed{\text{2nd}}$ $\overset{L_1}{\boxed{1}}$ $\boxed{,}$ $\boxed{\text{2nd}}$ $\overset{L_2}{\boxed{2}}$ $\boxed{\text{ENTER}}$
2. ENTER the data in L_1 and L_2: ENTER CLASS MIDPOINTS IN L_1.
   ENTER FREQUENCIES IN L_2.
3. To get the statistics:

$$\boxed{\text{STAT}} \quad [\text{CALC}] \quad [\text{1:1-VarStats}] \quad \boxed{\text{2nd}} \quad \overset{L_1}{\boxed{1}} \quad \boxed{,} \quad \boxed{\text{2nd}} \quad \overset{L_2}{\boxed{2}} \quad \boxed{\text{ENTER}}$$

BINOM.

To find **BINOMIAL PROBABILITIES:**

1. Clear L_1 and L_2: $\boxed{\text{STAT}}$ [4:ClrList] $\boxed{\text{2nd}}$ $\overset{L_1}{\boxed{1}}$ $\boxed{,}$ $\boxed{\text{2nd}}$ $\overset{L_2}{\boxed{2}}$ $\boxed{\text{ENTER}}$
2. $\boxed{\text{2nd}}$ $\overset{\text{DISTR}}{\boxed{\text{Vars}}}$ [0:binompdf(] $\overset{\text{number of trials}}{\searrow}$ n $\boxed{,}$ $\overset{\text{prob.}}{\downarrow}$ p $\boxed{\text{ENTER}}$

   $\boxed{\text{STO}\rightarrow}$ $\boxed{\text{2nd}}$ $\overset{L_2}{\boxed{2}}$ $\boxed{\text{ENTER}}$
3. Now use $\boxed{\text{STAT}}$ $\boxed{\text{Edit}}$ to view the probabilities in list L_2 and to ENTER the x-values (such as 0, 1, 2, ...) in L_1.
4. You can get the mean μ and the standard deviation σ with

$$\boxed{\text{STAT}} \quad [\text{CALC}] \quad [\text{1:1-Var Stats}] \quad \boxed{\text{2nd}} \quad \overset{L_1}{\boxed{1}} \quad \boxed{,} \quad \boxed{\text{2nd}} \quad \overset{L_2}{\boxed{2}} \quad \boxed{\text{ENTER}}$$

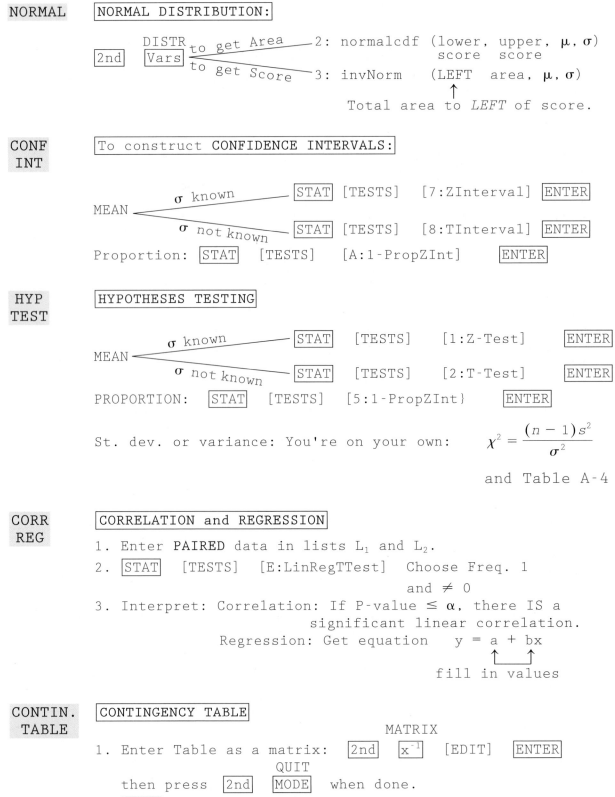

NORMAL

NORMAL DISTRIBUTION:

2nd Vars DISTR to get Area   2: normalcdf (lower, upper, μ, σ)
                                         score score
                  to get Score 3: invNorm (LEFT area, μ, σ)
                                             ↑
                              Total area to *LEFT* of score.

**CONF
INT**

To construct CONFIDENCE INTERVALS:

MEAN   σ known   STAT [TESTS] [7:ZInterval] ENTER

       σ not known STAT [TESTS] [8:TInterval] ENTER

Proportion: STAT [TESTS] [A:1-PropZInt] ENTER

**HYP
TEST**

HYPOTHESES TESTING

MEAN   σ known   STAT [TESTS] [1:Z-Test] ENTER

       σ not known STAT [TESTS] [2:T-Test] ENTER

PROPORTION: STAT [TESTS] [5:1-PropZInt} ENTER

St. dev. or variance: You're on your own: $\chi^2 = \dfrac{(n-1)s^2}{\sigma^2}$

and Table A-4

**CORR
REG**

CORRELATION and REGRESSION

1. Enter PAIRED data in lists L_1 and L_2.
2. STAT [TESTS] [E:LinRegTTest] Choose Freq. 1
                                and ≠ 0
3. Interpret: Correlation: If P-value ≤ α, there IS a
                           significant linear correlation.
              Regression: Get equation   y = a + bx
                                              ↑   ↑
                                          fill in values

**CONTIN.
TABLE**

CONTINGENCY TABLE

                                    MATRIX
1. Enter Table as a matrix: 2nd x^{-1} [EDIT] ENTER
                  QUIT
   then press 2nd MODE when done.
2. STAT [TESTS] [C:χ^2-Test]

Appendix D: Glossary

Absolute deviation   The measure of variation equal to the sum of the deviations of each value from the mean, divided by the number of values

Acceptance sampling   Sampling items without replacement and rejecting the whole batch based on the number of defects obtained

Actual odds against   The ratio $P(\overline{A})/P(A)$, usually expressed in the form of $a{:}b$ (or "a to b")

Actual odds in favor   The reciprocal of the actual odds against an event

Addition rule   Rule for determining the probability that, on a single trial, either event A occurs, or event B occurs, or they both occur

Adjusted coefficient of determination   Multiple coefficient of determination R^2 modified to account for the number of variables and sample size

Alpha (α)   Symbol used to represent the probability of a type I error. *See also* significance level.

Alternative hypothesis   Statement that is equivalent to the negation of the null hypothesis; denoted by H_1

Analysis of variance   Method of analyzing population variances in order to test hypotheses about means of populations

ANOVA   *See* analysis of variance.

Arithmetic mean   Sum of a set of values divided by the number of values; usually referred to as the mean

Assignable variation   Type of variation in a process that results from causes that can be identified

Attribute data   Data that can be separated into different categories distinguished by some nonnumeric characteristic

Average   Any one of several measures designed to reveal the center of a collection of data

Beta (β)   Symbol used to represent the probability of a type II error

Bimodal   Having two modes

Binomial experiment   Experiment with a fixed number of independent trials, where each outcome falls into exactly one of two categories

Binomial probability formula   Expression used to calculate probabilities in a binomial experiment (see Formula 4-5 in Section 4-3)

Bivariate data   Data arranged in pairs

Bivariate normal distribution   Distribution of paired data in which, for any fixed value of one variable, the values of the other variable are normally distributed

Blinding   Procedure used in experiments whereby the subject doesn't know whether he or she is receiving a treatment or a placebo

Block   A group of subjects that are similar in the ways that might affect the outcome of an experiment.

Box-and-whisker diagram   *See* boxplot.

Boxplot   Graphical representation of the spread of a set of data

Case-control study   Study in which data are collected from the past by going back in time (through examination of records, interviews, and so on).

Categorical data   Data that can be separated into different categories that are distinguished by some nonnumeric characteristic

Cell   Category used to separate qualitative (or attribute) data

Census   Collection of data from every element in a population

Center line   Line used in a control chart to represent a central value of the characteristic measurements

Central limit theorem   Theorem stating that sample means tend to be normally distributed with mean μ and standard deviation $\sigma/\sqrt{n}$

Centroid   The point $(\overline{x}, \overline{y})$ determined from a collection of bivariate data

Chebyshev's theorem   Theorem that uses the standard deviation to provide information about the distribution of data

Chi-square distribution   A continuous probability distribution (first introduced in Section 6-5)

Class boundaries   Values obtained from a frequency distribution by increasing the upper class limits and decreasing the lower class limits by the same amount so that there are no gaps between consecutive classes

Classical approach to probability   Approach in which the probability of an event is determined by dividing the number of ways the event can occur by the total number of possible outcomes

Classical method of testing hypotheses   Method of testing hypotheses based on a comparison of the test statistic and critical values

Class midpoint   In a class of a frequency distribution, the value midway between the lower class limit and the upper class limit

Class width   The difference between two consecutive lower class limits in a frequency distribution

Cluster sampling   Dividing the population area into sections (or clusters), then randomly selecting a few of those sections, and then choosing *all* the members from those selected sections

Coefficient of determination   Amount of the variation in y that is explained by the regression line

Coefficient of variation (or CV)   The ratio of the standard deviation to the mean, expressed as a percent.

Cohort study   Study of subjects in identified groups sharing common factors (called *cohorts*), with data collected in the future.

Combinations rule   Rule for determining the number of different combinations of selected items

Complement of an event   All outcomes in which the original event does not occur

Completely randomized design   Procedure in an experiment whereby each element is given the same chance of belonging to the different categories or treatments

Compound event   Combination of simple events

Conditional probability   The probability of an event, given that some other event has already occurred

Confidence coefficient   Probability that a population parameter is contained within a particular confidence interval; also called confidence level or degree of confidence

Confidence interval   Range of values used to estimate some population parameter with a specific confidence level; also called an interval estimate

Confidence interval limits   Two numbers that are used as the high and low boundaries of a confidence interval

Confidence level   Probability that a population parameter is contained within a particular confidence interval.

Confounding   A situation that occurs when the effects from two or more variables cannot be distinguished from each other

Contingency table   Table of observed frequencies where the rows correspond to one variable of classification and the columns correspond to another variable of classification; also called a two-way table

Continuity correction   Adjustment made when a discrete random variable is being approximated by a continuous random variable (Section 5-6)

Continuous data   Data resulting from infinitely many possible values that correspond to some continuous scale that covers a range of values without gaps, interruptions, or jumps.

Continuous random variable   A random variable with infinite values that can be associated with points on a continuous line interval

Control chart   Any one of several types of charts (Chapter 13) depicting some characteristic of a process in order to determine whether there is statistical stability

Control group   A group of subjects in an experiment who are not given a particular treatment

Control limit   Boundary used in a control chart for identifying unusual points

Convenience sampling   Sampling in which data are selected because they are readily available

Correlation   Statistical association between two variables

Correlation coefficient   Measurement of the strength of the relationship between two variables

Critical region   The set of all values of the test statistic that would cause rejection of the null hypothesis

Critical value   Value separating the critical region from the values of the test statistic that would not lead to rejection of the null hypothesis

Cross-sectional study   Study in which data are observed, measured, and collected at one point in time.

Cumulative frequency   Sum of the frequencies for a class and all preceding classes

Cumulative frequency distribution   Frequency distribution in which each class and frequency represents cumulative data up to and including that class

Data   Numbers or information describing some characteristic

Degree of confidence   Probability that a population parameter is contained within a particular confidence interval; also called level of confidence

Degrees of freedom   Number of values that are free to vary after certain restrictions have been imposed on all values

Denominator degrees of freedom   Degrees of freedom corresponding to the denominator of the F test statistic

Density curve   Graph of a continuous probability distribution

Dependent events   Events for which the occurrence of any one event affects the probabilities of the occurrences of the other events

Dependent sample   Sample whose values are related to the values in another sample

Dependent variable y variable in a regression or multiple regression equation

Descriptive statistics   Methods used to summarize the key characteristics of known data

Deviation   Amount of difference between a value and the mean; expressed as $x - \bar{x}$

Discrete data   Data with the property that the number of possible values is either a finite number or a "countable" number, which results in 0 possibilities, or 1 possibility, or 2 possibilities, and so on

Discrete random variable   Random variable with either a finite number of values or a countable number of values

Disjoint events   Events that cannot occur simultaneously

Distribution-free tests   Tests not requiring a particular distribution, such as the normal distribution. *See also* nonparametric tests.

Dotplot   Graph in which each data value is plotted as a point (or dot) along a scale of values

Double-blind Procedure used in an experiment whereby the subject doesn't know whether he or she is receiving a treatment or placebo, and the person administering the treatment also does not know

Efficiency Measure of the sensitivity of a nonparametric test in comparison to a corresponding parametric test

Empirical rule Rule that uses standard deviation to provide information about data with a bell-shaped distribution (Section 2-5)

Estimate Specific value or range of values used to approximate some population parameter

Estimator Sample statistic (such as the sample mean $\bar{x}$) used to approximate a population parameter

Event Result or outcome of an experiment

Expected frequency Theoretical frequency for a cell of a contingency table or multinomial table

Expected value For a discrete random variable, the mean value of the outcomes

Experiment Application of some treatment followed by observation of its effects on the subjects

Experimental units Subjects in an experiment

Explained deviation For one pair of values in a collection of bivariate data, the difference between the predicted y value and the mean of the y values

Explained variation Sum of the squares of the explained deviations for all pairs of bivariate data in a sample

Exploratory data analysis (EDA) Branch of statistics emphasizing the investigation of data

Factor In analysis of variance, a property or characteristic that allows us to distinguish the different populations from one another

Factorial rule Rule stating that n different items can be arranged $n!$ different ways

F distribution Continuous probability distribution first introduced in Section 8-5

Finite population correction factor Factor for correcting the standard error of the mean when a sample size exceeds 5% of the size of a finite population

Five-number summary Minimum value, maximum value, median, and the first and third quartiles of a set of data

Fractiles Numbers that partition data into parts that are approximately equal in size

Frequency distribution Listing of data values (either individually or by groups of intervals), along with their corresponding frequencies (or counts).

Frequency polygon Graphical representation of the distribution of data using connected straight-line segments

Frequency table List of categories of values along with their corresponding frequencies

Fundamental counting rule Rule stating that, for a sequence of two events in which the first event can occur m ways and the second can occur n ways, the events together can occur a total of $m \cdot n$ ways

Goodness-of-fit test Test for how well some observed frequency distribution fits some theoretical distribution

Histogram Graph of vertical bars representing the frequency distribution of a set of data

H test The nonparametric Kruskal-Wallis test

Hypothesis Statement or claim about some property of a population

Hypothesis test Method for testing claims made about populations; also called test of significance

Independent events Events for which the occurrence of any one of the events does not affect the probabilities of the occurrences of the other events

Independent sample Sample whose values are not related to the values in another sample

Independent variable The x variable in a regression equation, or one of the x variables in a multiple regression equation

Inferential statistics Methods involving the use of sample data to make generalizations or inferences about a population

Influential point Point that strongly affects the graph of a regression line

Interaction In two-way analysis of variance, the effect when one of the factors changes for different categories of the other factor

Interquartile range The difference between the first and third quartiles

Interval Level of measurement of data; characterizes data that can be arranged in order and for which differences between data values are meaningful

Interval estimate Range of values used to estimate some population parameter with a specific level of confidence; also called a confidence interval

Kruskal-Wallis test Nonparametric hypothesis test used to compare three or more independent samples; also called an H test

Least-squares property Property stating that, for a regression line, the sum of the squares of the vertical deviations of the sample points from the regression line is the smallest sum possible

Left-tailed test Hypothesis test in which the critical region is located in the extreme left area of the probability distribution

Level of confidence Probability that a population parameter is contained within a particular confidence interval; also called degree of confidence

Linear correlation coefficient   Measure of the strength of the relationship between two variables

Longitudinal study   Study of subjects in identified groups sharing common factors (called *cohorts*), with data collected in the future.

Lower class limits   Smallest numbers that can actually belong to the different classes in a frequency distribution

Lower control limit   Boundary used in a control chart to separate points that are unusually low

Lurking variable   Variable that affects the variables being studied, but is not itself included in the study

Mann-Whitney *U* test   Hypothesis test equivalent to the Wilcoxon rank-sum test for two independent samples

Marginal change   For variables related by a regression equation, the amount of change in the dependent variable when one of the independent variables changes by one unit and the other independent variables remain constant

Margin of error   Maximum likely (with probability $1 - \alpha$) difference between the observed sample statistic and the true value of the population parameter

Matched pairs   With two samples, there is some relationship so that each value in one sample is paired with a corresponding value in the other sample.

Mathematical model   Mathematical function that "fits" or describes real-world data

Maximum error of estimate   *See* margin of error.

Mean   The sum of a set of values divided by the number of values

Mean absolute deviation   Measure of variation equal to the sum of the deviations of each value from the mean, divided by the number of values

Measure of center   Value intended to indicate the center of the values in a collection of data

Measure of variation   Any of several measures designed to reflect the amount of variation or spread for a set of values

Median   Middle value of a set of values arranged in order of magnitude

Midquartile   One-half of the sum of the first and third quartiles

Midrange   One-half the sum of the highest and lowest values

Mode   Value that occurs most frequently

MS(error)   Mean square for error; used in analysis of variance

MS(total)   Mean square for total variation; used in analysis of variance

MS(treatment)   Mean square for treatments; used in analysis of variance

Multimodal   Having more than two modes

Multinomial experiment   Experiment with a fixed number of independent trials, where each outcome falls into exactly one of several categories

Multiple coefficient of determination   Measure of how well a multiple regression equation fits the sample data

Multiple comparison procedures   Procedures for identifying which particular means are different, after concluding that three or more means are not all equal

Multiple regression   Study of linear relationships among three or more variables

Multiple regression equation   Equation that expresses a linear relationship between a dependent variable y and two or more independent variables $(x_1, x_2, \ldots, x_k)$

Multiplication rule   Rule for determining the probability that event A will occur on one trial and event B will occur on a second trial

Mutually exclusive events   Events that cannot occur simultaneously

Negatively skewed   Skewed to the left

Nominal   Level of measurement of data; characterizes data that consist of names, labels, or categories only

Nonparametric tests   Statistical procedures for testing hypotheses or estimating parameters, where there are no required assumptions about the nature or shape of population distributions; also called distribution-free tests

Nonsampling errors   Errors from external factors not related to sampling

Normal distribution   Bell-shaped probability distribution described algebraically by Formula 5-1 in Section 5-1

Normal quantile plot   Graph of points (x, y), where each x value is from the original set of sample data, and each y value is a z score corresponding to a quantile value of the standard normal distribution

np chart   Control chart in which numbers of defects are plotted so that a process can be monitored

Null hypothesis   Claim made about some population characteristic, usually involving the case of no difference; denoted by H_0

Numerator degrees of freedom   Degrees of freedom corresponding to the numerator of the F test statistic

Numerical data   Data consisting of numbers representing counts or measurements

Observational study   Study in which we observe and measure specific characteristics, but don't attempt to manipulate or modify the subjects being studied

Observed frequency   Actual frequency count recorded in one cell of a contingency table or multinomial table

Odds against   Ratio of the probability of an event not occurring to the event occurring, usually expressed in the form of $a{:}b$ where a and b are integers having no common factors

Odds in favor   Ratio of the probability of an event occurring to the event not occurring, usually expressed as the ratio of two integers with no common factors

Ogive   Graphical representation of a cumulative frequency distribution

One-way analysis of variance   Analysis of variance involving data classified into groups according to a single criterion only

Ordinal   Level of measurement of data; characterizes data that may be arranged in order, but differences between data values either cannot be determined or are meaningless

Outliers   Values that are very unusual in the sense that they are very far away from most of the data

Paired samples   Two samples that are dependent in the sense that the data values are matched by pairs

Parameter   Measured characteristic of a population

Parametric tests   Statistical procedures, based on population parameters, for testing hypotheses or estimating parameters

Pareto chart   Bar graph for qualitative data, with the bars arranged in order according to frequencies

Payoff odds   Ratio of net profit (if you win) to the amount bet

p chart   Control chart used to monitor the proportion p for some attribute in a process

Pearson's product moment correlation coefficient   *See* linear correlation coefficient.

Percentile   The 99 values that divide ranked data into 100 groups with approximately 1% of the values in each group

Permutations rule   Rule for determining the number of different arrangements of selected items

Pie chart   Graphical representation of data in the form of a circle containing wedges

Placebo effect   Effect that occurs when an untreated subject incorrectly believes that he or she is receiving a real treatment and reports an improvement in symptoms

Point estimate   Single value that serves as an estimate of a population parameter

Poisson distribution   Discrete probability distribution that applies to occurrences of some event over a specified interval of time, distance, area, volume, or some similar unit

Pooled estimate of p_1 and p_2   Probability obtained by combining the data from two sample proportions and dividing the total number of successes by the total number of observations

Pooled estimate of σ^2   Estimate of the variance σ^2 that is common to two populations, found by computing a weighted average of the two sample variances

Population   Complete and entire collection of elements to be studied

Positively skewed   Skewed to the right

Power of a test   Probability $(1 - \beta)$ of rejecting a false null hypothesis

Predicted values   Values of a dependent variable found by using values of independent variables in a regression equation

Prediction interval   Confidence interval estimate of a predicted value of y

Predictor variables   Independent variables in a regression equation

Probability   Measure of the likelihood that a given event will occur; expressed as a number between 0 and 1

Probability distribution   Collection of values of a random variable along with their corresponding probabilities

Probability histogram   Histogram with outcomes listed along the horizontal axis and probabilities listed along the vertical axis

Probability value   *See P*-value.

Process data   Data, arranged according to some time sequence, that measure a characteristic of goods or services resulting from some combination of equipment, people, materials, methods, and conditions

Prospective study   Study of subjects in identified groups sharing common factors (called *cohorts*), with data collected in the future.

P-value   Probability that a test statistic in a hypothesis test is at least as extreme as the one actually obtained

Qualitative data   Data that can be separated into different categories distinguished by some nonnumeric characteristic

Quantitative data   Data consisting of numbers representing counts or measurements

Quartiles   The three values that divide ranked data into four groups with approximately 25% of the values in each group

Randomized block design   Design in which a measurement is obtained for each treatment on each of several individuals matched according to similar characteristics

Random sample   Sample selected in a way that allows every member of the population to have the same chance of being chosen

Random selection   Selection of sample elements in such a way that all elements available for selection have the same chance of being selected

Random variable Variable (typically represented by x) that has a single numerical value (determined by chance) for each outcome of an experiment

Random variation Type of variation in a process that is due to chance; the type of variation inherent in any process not capable of producing every good or service exactly the same way every time

Range The measure of variation that is the difference between the highest and lowest values

Range chart Control chart based on sample ranges; used to monitor variation in a process

Range rule of thumb Rule based on the principle that for typical data sets, the difference between the lowest typical value and the highest typical value is approximately 4 standard deviations ($4s$)

Rank Numerical position of an item in a sample set arranged in order

Rank correlation coefficient Measure of the strength of the relationship between two variables; based on the ranks of the values

Rare event rule If, under a given assumption, the probability of a particular observed result is extremely small, we conclude that the assumption is probably not correct.

Ratio Level of measurement of data; characterizes data that can be arranged in order, for which differences between data values are meaningful, and there is an inherent zero starting point

R chart Control chart based on sample ranges; used to monitor variation in a process

Regression equation Algebraic equation describing the relationship among variables

Regression line Straight line that best fits a collection of points representing paired sample data

Relative frequency Frequency for a class, divided by the total of all frequencies

Relative frequency approximation of probability Estimated value of probability based on actual observations

Relative frequency distribution Variation of the basic frequency distribution in which the frequency for each class is divided by the total of all frequencies

Relative frequency histogram Variation of the basic histogram in which frequencies are replaced by relative frequencies

Replication Repetition of an experiment

Residual Difference between an observed sample y value and the value of y that is predicted from a regression equation

Response variable y variable in a regression or multiple regression equation

Retrospective study Study in which data are collected from the past by going back in time (through examination of records, interviews, and so on).

Right-tailed test Hypothesis test in which the critical region is located in the extreme right area of the probability distribution

Rigorously controlled design Design of experiment in which all factors are forced to be constant so that effects of extraneous factors are eliminated

Run Sequence of data exhibiting the same characteristic; used in runs test for randomness

Run chart Sequential plot of individual data values over time, where one axis (usually the vertical axis) is used for the data values and the other axis (usually the horizontal axis) is used for the time sequence

Runs test Nonparametric method used to test for randomness

Sample Subset of a population

Sample size Number of items in a sample

Sample space Set of all possible outcomes or events in an experiment that cannot be further broken down

Sampling distribution of proportion The probability distribution of sample proportions, with all samples having the same sample size n.

Sampling distribution of sample means Distribution of the sample means that is obtained when we repeatedly draw samples of the same size from the same population

Sampling error Difference between a sample result and the true population result; results from chance sample fluctuations

Sampling variability Variation of a statistic in different samples.

Scatter diagram Graphical display of paired (x, y) data

Scatterplot Graphical display of paired (x, y) data

s chart Control chart, based on sample standard deviations, that is used to monitor variation in a process

Self-selected sample Sample in which the respondents themselves decide whether to be included; also called voluntary response sample

Semi-interquartile range One-half of the difference between the first and third quartiles

Significance level Probability of making a type I error when conducting a hypothesis test

Sign test Nonparametric hypothesis test used to compare samples from two populations

Simple event Experimental outcome that cannot be further broken down

Simple random sample Sample of a particular size selected so that every possible sample of the same size has the same chance of being chosen

Simulation   Process that behaves in a way that is similar to some experiment so that similar results are produced

Single factor analysis of variance   *See* one-way analysis of variance.

Skewed   Not symmetric and extending more to one side than the other

Slope   Measure of steepness of a straight line

Sorted data   Data arranged in order

Spearman's rank correlation coefficient   *See* rank correlation coefficient.

SS(error)   Sum of squares representing the variability that is assumed to be common to all the populations being considered; used in analysis of variance

SS(total)   Measure of the total variation (around $\bar{\bar{x}}$) in all of the sample data combined; used in analysis of variance

SS(treatment)   Measure of the variation between the sample means; used in analysis of variance

Standard deviation   Measure of variation equal to the square root of the variance

Standard error of estimate   Measure of spread of sample points about the regression line

Standard error of the mean   Standard deviation of all possible sample means $\bar{x}$

Standard normal distribution   Normal distribution with a mean of 0 and a standard deviation equal to 1

Standard score   Number of standard deviations that a given value is above or below the mean; also called z score

Statistic   Measured characteristic of a sample

Statistically stable process   Process with only natural variation and no patterns, cycles, or unusual points

Statistical process control (SPC)   Use of statistical techniques such as control charts to analyze a process or its outputs so as to take appropriate actions to achieve and maintain a state of statistical control and to improve the process capability

Statistics   Collection of methods for planning experiments, obtaining data, organizing, summarizing, presenting, analyzing, interpreting, and drawing conclusions based on data

Stem-and-leaf plot   Method of sorting and arranging data to reveal the distribution

Stepwise regression   Process of using different combinations of variables until the best model is obtained; used in multiple regression

Stratified sampling   Sampling in which samples are drawn from each stratum (class)

Student t distribution   *See* t distribution.

Subjective probability   Guess or estimate of a probability based on knowledge of relevant circumstances

Symmetric   Property of data for which the distribution can be divided into two halves that are approximately mirror images by drawing a vertical line through the middle

Systematic sampling   Sampling in which every kth element is selected

t distribution   Bell-shaped distribution usually associated with sample data from a population with an unknown standard deviation.

10–90 percentile range   Difference between the 10th and 90th percentiles

Test of homogeneity   Test of the claim that different populations have the same proportion of some characteristic

Test of independence   Test of the null hypothesis that for a contingency table, the row variable and column variable are not related

Test of significance   *See* hypothesis test.

Test statistic   Sample statistic based on the sample data; used in making the decision about rejection of the null hypothesis

Time-series data   Data that have been collected at different points in time.

Total deviation   Sum of the explained deviation and unexplained deviation for a given pair of values in a collection of bivariate data

Total variation   Sum of the squares of the total deviation for all pairs of bivariate data in a sample

Traditional method of testing hypotheses   Method of testing hypotheses based on a comparison of the test statistic and critical values

Treatment   Property or characteristic that allows us to distinguish the different populations from one another; used in analysis of variance

Treatment group   Group of subjects given some treatment in an experiment

Tree diagram   Graphical depiction of the different possible outcomes in a compound event

Two-tailed test   Hypothesis test in which the critical region is divided between the left and right extreme areas of the probability distribution

Two-way analysis of variance   Analysis of variance involving data classified according to two different factors

Two-way table   *See* contingency table.

Type I error   Mistake of rejecting the null hypothesis when it is true

Type II error   Mistake of failing to reject the null hypothesis when it is false

Unbiased estimator   Sample statistic that tends to target the population parameter that it is used to estimate

Unexplained deviation   For one pair of values in a collection of bivariate data, the difference between the y coordinate and the predicted value

Unexplained variation   Sum of the squares of the unexplained deviations for all pairs of bivariate data in a sample

Uniform distribution   Probability distribution in which every value of the random variable is equally likely

Upper class limits   Largest numbers that can belong to the different classes in a frequency distribution

Upper control limit   Boundary used in a control chart to separate points that are unusually high

Variance   Measure of variation equal to the square of the standard deviation

Variance between samples   In analysis of variance, the variation among the different samples

Variation due to error   *See* variation within samples.

Variation due to treatment   *See* variance between samples.

Variation within samples   In analysis of variance, the variation that is due to chance

Voluntary response sample   Sample in which the respondents themselves decide whether to be included.

Weighted mean   Mean of a collection of values that have been assigned different degrees of importance

Wilcoxon rank-sum test   Nonparametric hypothesis test used to compare two independent samples

Wilcoxon signed-ranks test   Nonparametric hypothesis test used to compare two dependent samples

Within statistical control   *See* statistically stable process.

$\bar{x}$ **chart**   Control chart used to monitor the mean of a process

y**-intercept**   Point at which a straight line crosses the y-axis

z **score**   Number of standard deviations that a given value is above or below the mean

Appendix E: Bibliography

*An asterisk denotes a book recommended for reading. Other books are recommended as reference texts.

Andrews D., and A. Herzberg. 1985. *Data: A Collection of Problems from Many Fields for the Student and Research Worker.* New York: Springer.

Bell, M. 2003. *The TI-83 Plus Companion to Accompany Elementary Statistics.* 9th ed. Reading, Mass.: Addison-Wesley.

Bennett, D. 1998. *Randomness.* Cambridge: Harvard University Press.

*Best, J. 2001. *Damned Lies and Statistics.* Berkeley: University of California Press.

Beyer, W. 1991. *CRC Standard Probability and Statistics Tables and Formulae.* Boca Raton, Fla.: CRC Press.

*Campbell, S. 1974. *Flaws and Fallacies in Statistical Thinking.* Englewood Cliffs, N.J.: Prentice-Hall.

*Crossen, C. 1994. *Tainted Truth: The Manipulation of Fact in America.* New York: Simon & Schuster.

Devore, J., and R. Peck. 1997. *Statistics: The Exploration and Analysis of Data.* 3rd ed. St. Paul, Minn.: West Publishing.

*Fairley, W., and F. Mosteller. 1977. *Statistics and Public Policy.* Reading, Mass.: Addison-Wesley.

Fisher, R. 1966. *The Design of Experiments.* 8th ed. New York: Hafner.

*Freedman, D., R. Pisani, R. Purves, and A. Adhikari. 1991. *Statistics.* 2nd ed. New York: Norton.

*Gonick, L., and W. Smith. 1993. *The Cartoon Guide to Statistics.* New York: HarperCollins.

Halsey, J., and E. Reda. 2003. *Excel Student Laboratory Manual and Workbook.* Reading, Mass.: Addison-Wesley.

*Heyde, C., and E. Seneta, editors. 2001. *Statisticians of the Centuries.* New York: Springer-Verlag.

Hoaglin, D., F. Mosteller, and J. Tukey, eds. 1983. *Understanding Robust and Exploratory Data Analysis.* New York: Wiley.

*Hollander, M., and F. Proschan. 1984. *The Statistical Exorcist: Dispelling Statistics Anxiety.* New York: Marcel Dekker.

*Holmes, C. 1990. *The Honest Truth About Lying with Statistics.* Springfield, Ill.: Charles C Thomas.

*Hooke, R. 1983. *How to Tell the Liars from the Statisticians.* New York: Marcel Dekker.

*Huff, D. 1993. *How to Lie with Statistics.* New York: Norton.

*Jaffe, A., and H. Spirer. 1987. *Misused Statistics.* New York: Marcel Dekker.

*Kimble, G. 1978. *How to Use (and Misuse) Statistics.* Englewood Cliffs, N.J.: Prentice-Hall.

Kotz, S., and D. Stroup. 1983. *Educated Guessing—How to Cope in an Uncertain World.* New York: Marcel Dekker.

*Loyer, M. 2003. *Student Solutions Manual to Accompany Elementary Statistics.* 9th ed. Reading, Mass.: Addison-Wesley.

*Moore, D. 1997. *Statistics: Concepts and Controversies.* 4th ed. San Francisco: Freeman.

Mosteller, F., R. Rourke, and G. Thomas, Jr. 1970. *Probability with Statistical Applications.* 2nd ed. Reading, Mass.: Addison-Wesley.

Ott, L., and W. Mendenhall. 1994. *Understanding Statistics.* 6th ed. Boston: Duxbury Press.

Owen, D. 1962. *Handbook of Statistical Tables.* Reading, Mass.: Addison-Wesley.

*Paulos, J. 1988. *Innumeracy: Mathematical Illiteracy and Its Consequences.* New York: Hill and Wang.

Peck, R. 2003. *SPSS Student Laboratory Manual and Workbook.* Reading, Mass.: Addison-Wesley.

*Reichard, R. 1974. *The Figure Finaglers.* New York: McGraw-Hill.

*Reichmann, W. 1962. *Use and Abuse of Statistics.* New York: Oxford University Press.

*Rossman, A. 1996. *Workshop Statistics: Discovery with Data.* New York: Springer.

Ryan, T., B. Joiner, and B. Ryan. 1995. *MINITAB Handbook.* 3rd ed. Boston: Duxbury.

*Salsburg, D. 2000. *The Lady Tasting Tea: How Statistics Revolutionized the Twentieth Century.* New York: W. H. Freeman.

Schaeffer, R., M. Gnanadesikan, A. Watkins, and J. Witmer. 1996. *Activity-Based Statistics: Student Guide.* New York: Springer.

Schmid, C. 1983. *Statistical Graphics.* New York: Wiley.

Sheskin, D. 1997. *Handbook of Parametric and Nonparametric Statistical Procedures.* Boca Raton, Fla.: CRC-Press.

Simon, J. 1992. *Resampling: The New Statistics.* Belmont, Calif.: Duxbury Press.

Smith, G. 1995. *Statistical Process Control and Quality Improvement.* 2nd ed. Columbus: Merrill.

*Stigler, S. 1986. *The History of Statistics.* Cambridge, Mass.: Harvard University Press.

*Tanur, J., ed. 1989. *Statistics: A Guide to the Unknown.* 3rd ed. Belmont, Calif.: Wadsworth.

Triola, M. 2003. *Minitab Student Laboratory Manual and Workbook.* 9th ed. Reading, Mass.: Addison-Wesley.

Triola, M. 2003. *STATDISK 9.0 Student Laboratory Manual and Workbook.* 9th ed. Reading, Mass.: Addison-Wesley.

Triola, M., and L. Franklin. 1994. *Business Statistics.* Reading, Mass.: Addison-Wesley.

*Tufte, E. 1983. *The Visual Display of Quantitative Information.* Cheshire, Conn.: Graphics Press.

Tukey, J. 1977. *Exploratory Data Analysis.* Reading, Mass.: Addison-Wesley.

Utts, J. 1996. *Seeing Through Statistics.* Belmont, Calif.: Wadsworth.

Appendix F: Answers to Odd-Numbered Exercises (and ALL Review Exercises and Cumulative Review Exercises)

Section 1-2

1. Parameter
3. Statistic
5. Discrete
7. Discrete
9. Ratio
11. Interval
13. Ordinal
15. Ratio
17. Sample: the 10 selected adults; population: all adults; not representative
19. Sample: the 1059 selected adults; population: all adults; representative
21. With no natural starting point, the temperatures are at the interval level of measurement; ratios such as "twice" are meaningless.
23. Either ordinal or interval are reasonable answers, but ordinal makes more sense because differences between values are not likely to be meaningful. For example, the difference between a food rated 1 and a food rated 2 is not likely to be the same as the difference between a food rated 9 and a food rated 10.

Section 1-3

1. Truck drivers are often forced to dine in fast-food establishments, which provide diets that are higher in fat content. It is probably the fast-food diet that causes higher weights, not the trucks themselves. Avoid causality altogether by saying that driving trucks is associated with higher weights.
3. One possible alternative: Racial profiling is used so that Orange County police tend to stop and ticket more minorities than whites.
5. Because the study was funded by a candy company and the Chocolate Manufacturers Association, there is a real possibility that researchers were somehow encouraged to obtain results favorable to the consumption of chocolate.
7. No, she used a voluntary response sample.
9. People with unlisted numbers and without telephones are excluded.
11. Motorcyclists who were killed
13. No. Each of the 29 cigarettes is given an equal weight, but some cigarettes are consumed in much greater numbers than others. Also, there are other cigarettes not included in Data Set 5.
15. The results would not be good because you would be sampling only those people who died at a relatively young age. Many people born after 1945 are still alive.
17. a. 68%
    b. 0.352
    c. 855
    d. 48.6%

19. a. 540
    b. 5%
21. 62% of 8% of 1875 is only 93
23. All percentages of success should be multiples of 5. The given percentages cannot be correct.
25. Answer varies.

Section 1-4

1. Experiment
3. Observational study
5. Retrospective
7. Cross-sectional
9. Convenience
11. Random
13. Cluster
15. Systematic
17. Stratified
19. Cluster
21. Yes; yes
23. No; no
25. Yes; no
27. Answers vary.
29. No, not every voter has the same chance of being selected. Voters in less populated states have a better chance of being selected.
31. Asking drivers to use cell phones might possibly place them in danger. The population of drivers not having cell phones might be fundamentally different from the population of drivers that have cell phones. The magnitude of cell phone use might vary considerably, so the effects of using a cell phone might not be clear. The cell phone users know that they are in the treatment group and might behave differently, and they might tend to blame driving problems or crashes on the presence of the cell phone.

Chapter 1 Review Exercises

1. No, because it is a voluntary response sample, it might not be representative of the population.
2. Answer varies.
3. a. Ratio
   b. Ordinal
   c. Nominal
   d. Interval
4. a. Discrete
   b. Ratio
   c. Stratified
   d. Statistic
   e. The largest values because they represent stockholders that could potentially gain control of the company.
   f. The voluntary response sample is likely to be biased.

5. a. Systematic; representative
   b. Convenience; not representative
   c. Cluster; not representative
   d. Random; representative
   e. Stratified; not representative
6. a. Design the experiment so that the subjects don't know whether they are using Sleepeze or a placebo, and also design it so that those who observe and evaluate the subjects do not know which subjects are using Sleepeze and which are using a placebo.
   b. Blinding will help to distinguish between the effectiveness of Sleepeze and the placebo effect, whereby subjects and evaluators tend to believe that improvements are occurring just because some treatment is given.
   c. Subjects are put into different groups through a process of *random selection.*
   d. Subjects are very *carefully chosen* for the different groups so that those groups are made to be similar in the ways that are important.
   e. Replication is used when the experiment is repeated. It is important to have a sample of subjects that is large enough so that we can see the true nature of any effects. It is important so that we are not misled by erratic behavior of samples that are too small.

Chapter 1 Cumulative Review Exercises

1. 163.85
2. −0.64516129
3. −6.6423420
4. 216.09
5. 4.3588989
6. 18.647867
7. 0.47667832
8. 0.89735239
9. 0.0000000000072744916
10. 4,398,046,500,000
11. 282,429,540,000
12. 0.000000000058207661

Chapter 2 Answers

Section 2-2

1. Class width: 10. Class midpoints: 94.5, 104.5, 114.5, 124.5, 134.5, 144.5, 154.5.
   Class boundaries: 89.5, 99.5, 109.5, 119.5, 129.5, 139.5, 149.5, 159.5.
3. Class width: 200. Class midpoints: 99.5, 299.5, 499.5, 699.5, 899.5, 1099.5, 1299.5.
   Class boundaries: −0.5, 199.5, 399.5, 599.5, 799.5, 999.5, 1199.5, 1399.5.

5.

Systolic Blood Pressure of Men	Relative Frequency
90–99	2.5%
100–109	10.0%
110–119	42.5%
120–129	30.0%
130–139	12.5%
140–149	0.0%
150–159	2.5%

7.

Cholesterol of Men	Relative Frequency
0–199	32.5%
200–399	27.5%
400–599	12.5%
600–799	20.0%
800–999	5.0%
1000–1199	0.0%
1200–1399	2.5%

9.

Systolic Blood Pressure of Men	Cumulative Frequency
Less than 100	1
Less than 110	5
Less than 120	22
Less than 130	34
Less than 140	39
Less than 150	39
Less than 160	40

11.

Cholesterol of Men	Cumulative Frequency
Less than 200	13
Less than 400	24
Less than 600	29
Less than 800	37
Less than 1000	39
Less than 1200	39
Less than 1400	40

13. Change the heading of "Frequency" to "Relative Frequency," and enter these relative frequencies: 13.5%, 15.5%, 21.0%, 20.0%, 14.0%, and 16.0%. The relative frequencies appear to vary somewhat. (Using methods described later in the book, the differences are not significant.)

15.

Weight (lb)	Frequency
0–49	6
50–99	10
100–149	10
150–199	7
200–249	8
250–299	2
300–349	4
350–399	3
400–449	3
450–499	0
500–549	1

17. The circumferences for females appear to be slightly lower, but the difference does not appear to be significant.

Circumference (cm)	Males	Females
34.0–35.9	2	1
36.0–37.9	0	3
38.0–39.9	5	14
40.0–41.9	29	27
42.0–43.9	14	5

19. The female runners appear to be a few years younger.

Age	Male	Female
19–28	9.9%	20.5%
29–38	38.7%	46.2%
39–48	27.9%	10.3%
49–58	19.8%	17.9%
59–68	3.6%	5.1%

21. An outlier can dramatically affect the frequency table.

Weight (lb)	With Outlier	Without Outlier
200–219	6	6
220–239	5	5
240–259	12	12
260–279	36	36
280–299	87	87
300–319	28	28
320–339	0	
340–359	0	
360–379	0	
380–399	0	
400–419	0	
420–439	0	
440–459	0	
460–479	0	
480–499	0	
500–519	1	

Section 2-3

1. 26 years
3. 71%
5. 40%; 200
7. The distribution of faculty/staff cars is weighted slightly more to the left, so their cars are slightly newer.

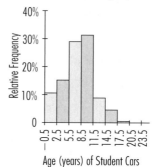

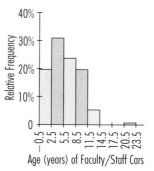

9. 183 pounds

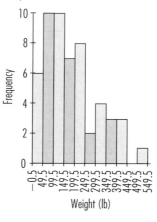

11. There does not appear to be a significant difference.

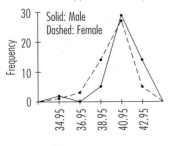

Solid: Male
Dashed: Female

Circumference (cm)

13. The ages of men appear to have a distribution weighted a little more to the right, so they tend to be slightly older.

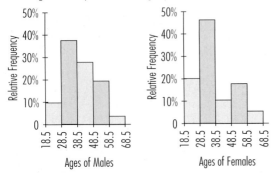

Ages of Males

Ages of Females

15. 200, 200, 200, 205, 216, 219, 219, 219, 219, 222, 222, 223, 223, 223, 223, 223, 241, 241, 247, 247

17.

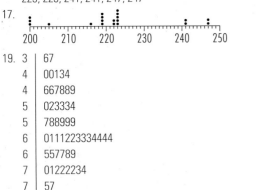

19.
```
3 | 67
4 | 00134
4 | 667889
5 | 023334
5 | 788999
6 | 0111223334444
6 | 557789
7 | 01222234
7 | 57
```

21. Networking appears to be the most effective approach in getting a job.

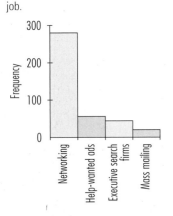

23.

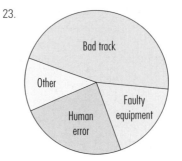

25. As the amount of tar in cigarettes increases, the amount of carbon monoxide also increases.

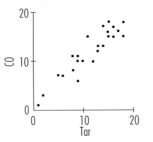

27. There appears to be an upward trend, suggesting that the stock market makes a good investment.

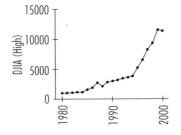

29. 10,000/422,000: 2.4%

31. 13,000 (from 37,000 to 24,000)

33. a.

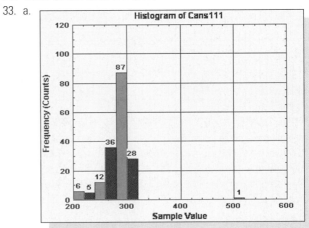

b.

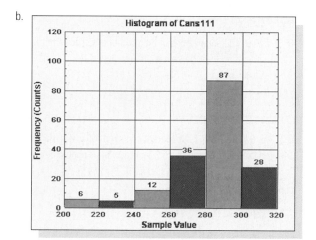

Histogram of Cans111

c. The outlier can have a dramatic effect on the histogram. Using a class width larger than the one used in parts (a) and (b) could hide the true nature of the distribution.

Section 2-4

1. $\bar{x} = 157.8$ sec; median $= 88.0$ sec; mode $= 0$ sec; midrange $= 274.0$ sec.
   Yes, young children should not be influenced by exposure to tobacco use.
3. $\bar{x} = 0.295$ g; median $= 0.345$ g; mode: 0.13 g, 0.43 g, 0.47 g; midrange $= 0.255$ g.
   Not necessarily. There are other cereals not listed, and Americans might consume much more of some brands than others.
5. $\bar{x} = 0.187$ g; median $= 0.170$; mode: 0.16, 0.17; midrange $= 0.205$.
   Yes.
7. $\bar{x} = 18.3$; median $= 18.0$; mode $= 17$; midrange $= 18.0$.
   The results are very consistent, so the mean should be a good estimate.
9. Jefferson Valley: $\bar{x} = 7.15$ min; median $= 7.20$ min; mode $= 7.7$ min; midrange $= 7.10$ min.
   Providence: same results as Jefferson Valley.
   Although the measures of center are the same, the Providence times are much more varied than the Jefferson Valley times.
11. McDonald's: $\bar{x} = 186.3$ sec; median $= 184.0$ sec; mode $=$ none; midrange $= 189.5$ sec.
    Jack in the Box: $\bar{x} = 262.5$ sec; median $= 262.5$ sec; mode $= 109$ sec; midrange $= 277.5$ sec.
    McDonald's appears to be significantly faster.
13. Males: $\bar{x} = 41.10$ cm; median $= 41.10$ cm
    Females: $\bar{x} = 40.05$ cm; median $= 40.20$ cm
    There does appear to be a small difference.
15. Thursday: $\bar{x} = 0.069$ in.; median $= 0.000$ in.
    Sunday: $\bar{x} = 0.068$ in.; median $= 0.000$ in.
    There does not appear to be a substantial difference.
17. 74.4 min

19. 46.8 mi/h; the mean is significantly higher than the posted limit of 30 mi/h.
21. a. 182.9 lb
    b. 171.0 lb
    c. 159.2 lb
    The results differ by substantial amounts, suggesting that the mean of the original set of weights is strongly affected by extreme values.
23. a. 52
    b. $n - 1$
25. 84.5
27. 48.0 mi/h
29. 62.9 volts

Section 2-5

1. range $= 548.0$ sec; $s^2 = 46308.2$ sec^2; $s = 215.2$ sec; they vary widely.
3. range $= 0.450$ g; $s^2 = 0.028$ g^2; $s = 0.168$ g
5. range $= 0.170$; $s^2 = 0.003$; $s = 0.051$
   No, the intent is to lower all of the individual values, which would result in a lower mean.
7. range $= 6.0$; $s^2 = 2.5$; $s = 1.6$
   The measures of variation are low values.
9. Jefferson Valley: range $= 1.20$ min; $s^2 = 0.23$ min^2; $s = 0.48$ min
   Providence: range $= 5.80$ min; $s^2 = 3.32$ min^2; $s = 1.82$ min
11. McDonald's: range $= 195.0$ sec; $s^2 = 4081.7$ sec^2; $s = 63.9$ sec
    Jack in the Box: range $= 407.0$ sec; $s^2 = 16644.3$ sec^2; $s = 129.0$ sec
13. Males: 1.50 cm; females: 1.64 cm; difference does not appear to be very substantial.
15. Thursday: 0.167 in.; Sunday: 0.200 in.
17. 14.7 min
19. 4.1 mi/h
21. Approximately 12 years (based on minimum of 23 years and maximum of 70 years)
23. Minimum: 31.30 cm; maximum: 46.42 cm; yes
25. a. 68%
    b. 99.7%
27. Percentage is at least 75%.
29. Calories: 5.9%; sugar 56.9%. The sugar content has much greater variation when compared to calories.
31. All of the values are the same.
33. Everlast batteries are better in the sense that they are more consistent and predictable.
35. Section 1: range $= 19.0$; $s = 5.7$
    Section 2: range $= 17.0$; $s = 6.7$
    The ranges suggest that Section 2 has less variation, but the standard deviations suggest that Section 1 has less variation.
37. 1.44
39. 15.8

41. a. 6
    b. 6
    c. 3.0
    d. $n-1$
    e. No The mean of the sample variances (6) equals the population variance (6), but the mean of the sample standard deviations (1.9) does not equal the mean of the population standard deviation (2.4).

Section 2-6

1. a. 60
   b. 3.75
   c. 3.75
   d. Unusual
3. a. −3.21
   b. 5.71
   c. 0.26
5. 2.56; unusual
7. 4.52; yes; patient is ill.
9. Psychology test, because $z = -0.50$ is greater than $z = -2.00$.
11. −3.56; yes
13. 43
15. 15
17. 46
19. 251.5
21. 121
23. 0
25. 25
27. 65
29. 415.5
31. 117.5
33. 98
35. 254
37. The z score remains the same.
39. a. Uniform
    b. Bell-shaped
    c. The shape of the distribution remains the same.
41. a. 165
    b. 169
    c. 279.5
    d. Yes; yes
    e. No; no
43. a. P_{10}, P_{50}, P_{80}
    b. 10, 46, 107.5, 130.5, 170, 209, 239.5, 265.5, 289.5
    c. 46, 130.5, 209, 265.5

Section 2-7

1. 0, 2, 5, 7, 9. The separations in the boxplot are approximately the same, indicating that the values are equally likely.

3. 3.3, 3.6, 3.75, 3.95, 4.1. No, cereal consumption is not uniformly distributed among the brands, so weighted values should be used.

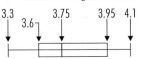

5. 0.870, 0.891, 0.908, 0.924, 0.983; yes

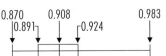

7. 0, 0, 1.5, 39, 414. Skewed.

9. Actors: 31, 37, 43, 51, 76.
   Actresses: 21, 30, 34, 41, 80.
   Actresses appear to be younger.

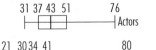

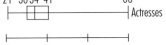

11. Smokers: 0, 86.5, 170, 251.5, 491. ETS: 0, 1, 1.5, 32, 551. NOETS: 0, 0, 0, 0, 309. Differences are significant and show that cotinine increases with exposure or use of tobacco.

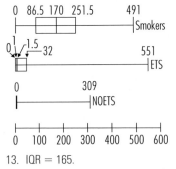

13. IQR = 165.
    Mild outliers: values x such as that $-408.5 \le x < -161$ or $499 < x \le 746.5$.
    Extreme outliers: values x such that $x < -408.5$ or $x > 746.5$.
    There are no mild outliers or extreme outliers.

Chapter 2 Review Exercises

1. a. 54.8 years
   b. 55.0 years
   c. 51 years, 54 years
   d. 55.5 years
   e. 27.0 years
   f. 6.2 years
   g. 38.7 years^2
   h. 51 years
   i. 58 years
   j. 47 years
2. a. -1.90
   b. No, because the z score is within two standard deviations of the mean.
   c. 42, 68, 69
   d. Yes; yes
3.

Age	Frequency
40–44	2
45–49	6
50–54	13
55–59	12
60–64	7
65–69	3

4. Bell-shaped

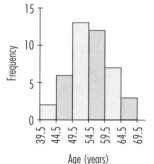

5. 42, 51, 55, 58, 69

6. a. The percentage is 68%.
   b. The percentage is 95%.
7. The score of 19 is better, because $z = -0.20$ is greater than $z = -0.67$.
8. a. Answer varies, but 7 or 8 years is reasonable.
   b. 5 years (based on minimum of 0 years and maximum of 20 years)
9. a. 140 min
   b. 15 min
   c. 225 square minutes

10.

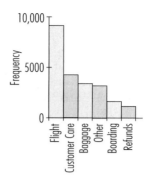

Chapter 2 Cumulative Review Exercises

1. a. $\bar{x} = 20.5$ sec; median $= 27.0$ sec; mode $= 20$ sec; midrange $= 42.0$ sec
   b. $s = 142.2$ sec; $s^2 = 20216.4$ sec^2; range $= 566.0$ sec
   c. The original exact times are continuous, but the given data appear to have been rounded to discrete values.
   d. Ratio
2. a. Mode, because the other measures of center require calculations that cannot (or should not) be done with data at the nominal level of measurement.
   b. Convenience
   c. Cluster
   d. Standard deviation; lowered
3. No, the 50 values should be weighted, with the state populations used as weights.

Chapter 3 Answers

Section 3-2

1. a. 0.5
   b. 0.20
   c. 0
3. $-1, 2, 5/3, \sqrt{2}$
5. a. 3/8
   b. 3/8
   c. 1/8
7. 0.153; yes
9. a. 1/17 or 0.0588
   b. No
11. a. 0.0501
    b. No
13. a. 0.0154 (not 0.0156)
    b. Yes
15. a. 1/365
    b. Yes
    c. He already knew.
    d. 0

17. a. 1/365
    b. 31/365
    c. 1
19. 0.130
21. a. 77/500 or 0.154
    b. 13/500 or 0.026
23. a. bb, bg, gb, gg
    b. 1/4
    c. 1/2
25. a. $21
    b. 21:2
    c. 14:1
    d. $30
27. Because the probability of showing improvement with an ineffective
    drug is so small (0.04), it appears that the drug is effective.
29. 5/8
31. a. 4/1461
    b. 400/146,097
33. 1/4

Section 3-3

1. a. No
   b. No
   c. Yes
3. a. 0.95
   b. 0.782
5. 5/7 or 0.714
7. 364/365 or 0.997
9. 0.239
11. 0.341
13. 0.600
15. 0.490
17. 0.140
19. 0.870
21. 0.5
23. 0.290
25. a. They are disjoint.
    b. They are not disjoint.
27. $P(A \text{ or } B) = P(A) + P(B) - 2P(A \text{ and } B)$

Section 3-4

1. a. independent
   b. independent
   c. dependent
3. 1/12
5. a. 9/49
   b. 1/7

7. a. 0.288
   b. 0.288
   c. Although the results are slightly different, they are the same
      when rounded to three decimal places.
   d. Sample without replacement so that duplication is avoided.
9. a. 1/1024
   b. No, because there are other ways to pass.
11. a. 1/133225 or 0.00000751
    b. 1/365
13. 0.694
15. 1/1024; yes, because the probability of getting 10 girls by chance is
    so small.
17. 1/64
19. 0.739 (or 0.738 assuming dependence); no
21. 0.702
23. 0.736
25. a. 0.992
    b. 0.973
    c. 0.431
27. 0.0192

Section 3-5

1. None of the students has Group A blood.
3. At least one of the returns is found to be correct.
5. 0.97; no
7. 31/32; yes
9. 0.410
11. 0.5; no
13. 11/14; get another test.
15. 0.999999; yes because the likelihood of being awakened increases
    from 0.99 to 0.999999.
17. 0.271
19. 0.897
21. 0.0793
23. 1/12; 35
25. a.

	Positive	Negative
HIV Infected	285	15
Not HIV Infected	4985	94,715

    b. 0.0541
27. 1/3

Section 3-6

1. F, F, T, T, F
3. good, good, defective, good, good
5. With odd = girl: 17/20 or 0.85. The result is reasonably close to
   0.813.
7. Among the 20 rows, there is at least one 0 in 7 rows, so the estimated probability is 7/20 or 0.35, which is reasonably close to the
   correct result of 0.410.

9. Approximately 0.813
11. Approximately 0.410
13. Switch: $P(\text{win}) = 2/3$; stick: $P(\text{win}) = 1/3$.
15. No; no

Section 3-7

1. 720
3. 600
5. 300
7. 2,598,960
9. 1/13,983,816
11. 1/45,057,474
13. 1/35,960; it appears that the oldest employees were selected.
15. 1/3,776,965,920
17. 1/5005; yes
19. 4; 40,320
21. 10
23. 720; satire; 1/720
25. 1/125,000
27. a. 256
    b. 70
    c. 70/256 = 0.273
29. 144
31. 1/41,416,353
33. 2,095,681,645,538 (about 2 trillion)
35. a. Calculator: 3.0414093×10^{64}; approximation: 3.0363452×10^{64}
    b. 615

Chapter 3 Review Exercises

1. 0.2
2. 0.32
3. 0.35
4. 0.83
5. 0.638
6. 0.100
7. 15/32 or 0.469
8. 15/80 = 3/16 or 0.188
9. a. 0.248
    b. 0.0615
    c. 0.575
10. 0.0777
11. 1/4096; yes
12. a. 1/120
    b. 720
13. a. 9/19
    b. 10:9
    c. $5
14. 0.000000531; no
15. 0.979

16. a. 1/20,358,520
    b. 1/142,506
    c. 1/76,275,360

Chapter 3 Cumulative Review Exercises

1. a. 4.0
    b. 4.0
    c. 2.2
    d. 4.7
    e. Yes
    f. 6/7
    g. 0.729
    h. 1/262,144; yes
2. a. 63.6 in.
    b. 1/4
    c. 3/4
    d. 1/16
    e. 5/16

Chapter 4 Answers

Section 4-2

1. a. continuous
    b. discrete
    c. continuous
    d. discrete
    e. discrete
3. $\mu = 1.5$, $\sigma = 0.9$
5. Not a probability distribution because $\Sigma P(x) = 0.94 \neq 1$.
7. $\mu = 0.7$, $\sigma = 0.9$
9. $\mu = 5.8$, $\sigma = 1.1$; no
11. −7.07¢; 1.4¢
13. a. Lives: −$250 (a loss); dies: $99,750 (a gain)
    b. −$100
    c. $150
    d. The negative expected value is a relatively small price to pay for insuring for the financial security of his heirs.
15. a. 10,000
    b. 0.0001
    c. $2787.50
    d. −22.12¢
    e. Pick 4, because −22.12¢ is greater than −22.5¢.
17. a. 0.122
    b. 0.212
    c. Part (b). The occurrence of 9 girls among 14 would be unusually high if the probability of 9 or more girls is very small (such as less than 0.05).
    d. No, because the probability of 9 or more girls is not very small (0.212). The result of 9 or more girls could easily occur by chance.

19. a. 0.029
    b. Yes, because the probability of 11 or more girls is very small (0.029). The result of 11 or more girls could not easily occur by chance.
21. Because the probability of correctly guessing 8 or more answers is 0.395, that outcome could easily occur, so there is no evidence that Bob has special powers.
23. The A bonds are better because the expected value is $49.40, which is higher than the expected value of $26 for the B bonds. She should select the A bond because the expected value is positive, which indicates a likely gain.
25. $\mu = 0.6, \sigma = 0.6$
27. a. 3
    b. $\sqrt{2}$
    c. $\mu = 10.5, \sigma = 5.8$

Section 4-3

1. Not binomial; more than two outcomes; not a fixed number of trials.
3. Not binomial; more than two outcomes
5. Binomial
7. Not binomial; more than two outcomes
9. a. 0.128
    b. WWC, WCW, CWW; 0.128 for each
    c. 0.384
11. 0.980
13. 0.171
15. 0+
17. 0.278
19. 0.208
21. 0.4711; no
23. 0.9925 (or 0.9924); yes
25. 0.0833
27. a. 0+ (or 0.00000980)
    b. 0+ (or 0.00000985)
    c. They are probably being targeted.
29. 0.0874; no
31. a. 0.107
    b. 0.893
    c. 0.375 (or 0.376)
    d. No, because with a 20% rate, the probability of at most one is high (it's greater than 0.05).
33. 0.000201; yes
35. P(9 or more girls) = 0.073, so 9 girls could easily occur by chance. There is not enough evidence to conclude that the gender technique is effective.
37. 0.0524
39. 0.000535

Section 4-4

1. $\mu = 80.0, \sigma = 8.0$, minimum = 64.0, maximum = 96.0
3. $\mu = 1488.0, \sigma = 19.3$, minimum = 1449.4, maximum = 1526.6
5. a. $\mu = 5.0, \sigma = 1.6$
    b. No, because 7 is within two standard deviations of the mean.
7. a. $\mu = 2.6, \sigma = 1.6$
    b. No, because 0 wins is within two standard deviations of the mean.
9. a. The probabilities for 0, 1, 2, 3, . . . , 15 are 0+, 0+, 0.003, 0.014, . . . , 0+ (from Table A-1).
    b. $\mu = 7.5, \sigma = 1.9$
    c. No, because 10 is within two standard deviations of the mean. Also, P(10 or more girls) = 0.151, showing that it is easy to get 10 or more girls by chance.
11. a. $\mu = 27.2, \sigma = 5.1$
    b. Yes, it appears that the training program had an effect.
13. a. $\mu = 142.8, \sigma = 11.9$
    b. No, 135 is not unusual because it is within two standard deviations of the mean.
    c. Based on the given results, cell phones do not pose a health hazard that increases the likelihood of cancer of the brain or nervous system.
15. a. 901
    b. $\mu = 506.0, \sigma = 15.9$
    c. Yes, because 901 is more than two standard deviations above the mean.
17. a. Yes (based on the probability histogram)
    b. Probability is 0.95.
    c. Probability is 0.997.
    d. At least 75% of such groups of 100 will have between 40 and 60 girls.

Section 4-5

1. 0.180
3. 0.0399
5. a. 62.2
    b. 0.0155 (0.0156 using rounded mean)
7. a. 0.497
    b. 0.348
    c. 0.122
    d. 0.0284
    e. 0.00497
    The expected frequencies of 139, 97, 34, 8, and 1.4 compare reasonably well to the actual frequencies, so the Poisson distribution does provide good results.
9. a. 0.00518 (using binomial: 0.00483)
    b. 0.995
    c. 0.570
    d. 0.430
11. 4.82×10^{-64} is so small that, for all practical purposes, we can consider it to be zero.

Chapter 4 Review Exercises

1. a. A random variable is a variable that has a single numerical value (determined by chance) for each outcome of some procedure.
   b. A probability distribution gives the probability for each value of the random variable.
   c. Yes, because each probability value is between 0 and 1 and the sum of the probabilities is 1.
   d. 4.2 days
   e. 2.1 days
   f. No, because the probability of 0.08 shows that it easy to get 0 days by chance.
2. a. 3.0
   b. 3.0
   c. 1.6
   d. 0.103
   e. Yes, because the probability of 0 sets is 0.0388, which shows that it is very unlikely that no sets would be tuned to *West Wing*.
3. a. 0.026
   b. 0.992 (or 0.994)
   c. $\mu = 8.0$, $\sigma = 1.3$
   d. No, because 6 is within two standard deviations of the mean.
4. a. 0.00361
   b. This company appears to be very different because the event of at least four firings is so unlikely, with a probability of only 0.00361.
5. a. 7/365
   b. 0.981
   c. 0.0188
   d. 0.0002
   e. No, because the event is so rare.

Chapter 4 Cumulative Review Exercises

1. a. $\bar{x} = 1.8$, $s = 2.6$
   b.

x	Relative Frequency
0	64.4%
1	4.1%
2	1.4%
3	0.0%
4	4.1%
5	15.1%
6	4.1%
7	4.1%
8	1.4%
9	1.4%

   c. $\mu = 4.5$, $\sigma = 2.9$
   d. The excessively large number of zeros suggests that the distances were estimated, not measured. The digits do not appear to be randomly selected.

2. a. 0.2
   b. $\mu = 0.2$, $\sigma = 0.4$
   c. 0.182
   d. Yes, because 3 is more than two standard deviations above the mean.
   e. 0.0001

Chapter 5 Answers

Section 5-2

1. 0.15
3. 0.15
5. 1/3
7. 1/2
9. 0.4013
11. 0.5987
13. 0.0099
15. 0.9901
17. 0.2417
19. 0.1359
21. 0.8959
23. 0.6984
25. 0.0001
27. 0.5
29. 68.26%
31. 99.74%
33. 0.9500
35. 0.9950
37. 1.28
39. −1.645
41. a. 68.26%
    b. 95%
    c. 99.74%
    d. 81.85%
    e. 4.56%
43. a. 1.23
    b. 1.50
    c. 1.52
    d. −2.42
    e. −0.13
45. a.
    b.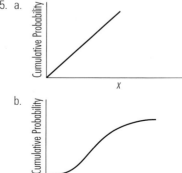

Section 5-3

1. 0.8413
3. 0.4972
5. 87.4
7. 115.6
9. a. 0.01%; yes
   b. 99.2°
11. a. 69.15%
    b. 1049; if the top 40% of scores from the applicant pool are se-
       lected, nobody would know about acceptance or rejection until all
       applicant scores have been obtained.
13. a. 0.0018
    b. 5.6 years
15. a. 25% agrees
    b. 0.8895; close
17. 0.52%
19. 0.1222; 12.22%; yes, they are all well above the mean.
21. a. The z scores are real numbers that have no units of measure-
       ment.
    b. $\mu = 0$; $\sigma = 1$; distribution is normal.
    c. $\mu = 64.9$ kg, $\sigma = 13.2$ kg, distribution is normal.
23. a. 75; 5
    b. No, the conversion should also account for variation.
    c. 31.4, 27.6, 22.4, 18.6
    d. Part (c), because variation is included in the conversion.
25. a. 1087; 22.9
    b. 26.0

Section 5-4

1. No, because of sampling variability, sample proportions will naturally
   vary from the true population proportion, even if the sampling is
   done with a perfectly valid procedure.
3. No, the histogram represents the distribution shape of one sample,
   but a sampling distribution includes all possible samples of the same
   size, such as all of the means computed from all possible samples of
   106 people.
5. a. 10-10; 10-6; 10-5; 6-10; 6-6; 6-5; 5-10; 5-6; 5-5; the means are
      listed in part (b).
   b.

Mean	10.0	8.0	7.5	8.0	6.0	5.5	7.5	5.5	5.0
Probability	1/9	1/9	1/9	1/9	1/9	1/9	1/9	1/9	1/9

   c. 7.0
   d. Yes; yes
7. a. Means: 85.0, 82.0, 83.5, 79.0, 81.5, 82.0, 79.0, 80.5, 76.0, 78.5,
      83.5, 80.5, 82.0, 77.5, 80.0, 79.0, 76.0, 77.5, 73.0, 75.5, 81.5, 78.5,
      80.0, 75.5, 78.0
   b. The probability of each mean is 1/25. The sampling distribution
      consists of the 25 sample means paired with the probability of
      1/25.
   c. 79.4
   d. Yes; yes

9. a. 0, 0.5, 0.5, 0.5, 0.5, 1, 1, 1, 0.5, 1, 1, 1, 0.5, 1, 1, 1
   b. The sampling distribution consists of the 16 proportions paired
      with the probability of 1/16.
   c. 0.75
   d. Yes; yes
11. a. Answer varies.
    b. Answer varies, but it must be one of these: 0, 0.2, 0.4, 0.6, 0.8, 1.
    c. Statistic
    d. No; no
    e. It must be 10/13 or 0.769.
13. a. 59.4; 4.6
    b. 59.4; 3.1
    c. 59.4; 1.9
    d. Yes. Each sampling distribution has a mean of 59.4, which is the
       mean of the population.
    e. As the sample size increases, the variation of the sampling distri-
       bution of sample means decreases.
15. Medians: 2.5; means: 2.7. The sample means again target the popu-
    lation mean, but the medians do not. The median is not a good
    statistic for estimating the population mean.

Section 5-5

1. a. 0.4325
   b. 0.1515
3. a. 0.0677
   b. 0.5055
5. a. 0.9808
   b. If the original population has a normal distribution, the central
      limit theorem provides good results for any sample size.
7. a. 0.5302
   b. 0.7323
   c. Part (a), because the seats will be occupied by individual women,
      not groups of women.
9. a. 0.0119
   b. No; yes
11. a. 0.0001
    b. No, but consumers are not being cheated because the cans are
       being overfilled, not underfilled.
13. a. 0.0051
    b. Yes
15. a. 0.1170
    b. No, because the probability of 0.1170 shows that it is easy to get
       a mean such as 0.882 g, assuming that the nicotine amounts
       have not been changed.
17. 0.0069; level is acceptable.
19. 2979 lb
21. a. 0.9750
    b. 1329 lb

23. 0.0240. We could conclude that the random number generator is defective if we obtained a sample mean that differs from 0.500 in such a way that there is a very small probablity of getting a sample mean "at least as exreme" as the value of the sample mean that was obtained. With a sample size of 100, there is no sample mean between 0.499 and 0.501 that would meet that criterion, so we should not conclude that the random number generator is somehow defective.

Section 5-6

1. The area to the right of 15.5
3. The area to the left of 99.5
5. The area to the left of 4.5
7. The area between 7.5 and 10.5
9. Table: 0.122; normal approximation: 0.1218
11. Table: 0.549; normal approximation is not suitable.
13. 0.1357; no
15. 0.0287; no
17. 0.2676; no
19. 0.7389; no, not very confident.
21. 0.0708; yes
23. 0.0080; yes
25. 0.6368; the pool is likely to be sufficient, but the probability should be much higher. It would be better to increase the pool of volunteers.
27. 0.0026; yes
29. 6; 0.4602
31. a. 0.821
    b. 0.9993
    c. 0.0000165
    d. 0.552

Section 5-7

1. Not normal
3. Not normal
5. Not normal
7. Normal
9. Not normal
11. Normal
13. Heights appear to be normal, but cholesterol levels do not appear to be normal. Cholesterol levels are strongly affected by diet, and diets might vary in dramatically different ways that do not yield normally distributed results.
15. -1.28, -0.52, 0, 0.52, 1.28; normal

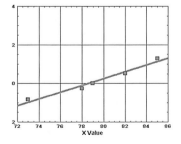

17. No, the transformation to z scores involves subtracting a constant and dividing by a constant, so the plot of the (x, z) points will always be a straight line, regardless of the nature of the distribution.

Chapter 5 Review Exercises

1. a. 0.0222
   b. 0.2847
   c. 0.6720
   d. 254.6
2. a. 0.69% of 900 = 6.21 babies
   b. 2405 g
   c. 0.0119
   d. 0.9553
3. 0.1020; no, assuming that the correct rate is 25%, there is a high probability (0.1020) that 19 or fewer offspring will have blue eyes. Because the observed event could easily occur by chance, there is not strong evidence against the 25% rate.
4. a. 0.9626
   b. 63.3 in., 74.7 in.
   c. 0.9979
5. a. 0.5
   b. 1
   c. 0
   d. 0.25
6. a. Normal distribution
   b. 51.2 lb
   c. Normal distribution
7. Normal approximation: 0.0436; exact value: 0.0355. Because the probability of getting only 2 women by chance is so small, it appears that the company is discriminating based on gender.
8. Yes. The histogram is very roughly bell-shaped, and the normal quantile plot has points that reasonably approximate a straight-line pattern. Also, there are no outliers.

Chapter 5 Cumulative Review Exercises

1. a. 63.0 mm
   b. 64.5 mm
   c. 66 mm
   d. 4.2 mm
   e. -0.95
   f. 75%
   g. 82.89%
   h. ratio
   i. continuous
2. a. 0.001
   b. 0.271
   c. The requirement that $np \geq 5$ is not satisfied, indicating that the normal approximation would result in errors that are too large.
   d. 5.0
   e. 2.1
   f. No, 8 is within two standard deviations of the mean and is within the range of values that could easily occur by chance.

Chapter 6 Answers

Section 6-2

1. 2.575
3. 2.33
5. $p = 0.250 \pm 0.030$
7. $p = 0.654 \pm 0.050$
9. $\hat{p} = 0.464$; $E = 0.020$
11. $\hat{p} = 0.655$; $E = 0.023$
13. 0.0300
15. 0.0405
17. $0.708 < p < 0.792$
19. $0.0887 < p < 0.124$
21. 461
23. 232
25. a. There is 95% confidence that the limits of 0.0489 and 0.0531 contain the population proportion.
    b. Yes, about 5% of males aged 18–20 drive while impaired.
    c. 5.31%
27. a. 29%
    b. $25.4\% < p < 32.6\%$
    c. 32.6%
29. a. $22.6\% < p < 29.8\%$
    b. No, the confidence interval limits include 25%.
31. a. $0.134\% < p < 6.20\%$ using $x = 7$, $n = 221$
    b. Ziac does not appear to cause dizziness as an adverse reaction.
33. 4145
35. a. 473
    b. 982
    c. Because they are based on a voluntary response sample, the results would not be valid.
37. a. $0.0355 < p < 0.139$
    b. 373
    c. Yes
39. a. $0.0267\% < p < 0.0376\%$
    b. No, because 0.0340% is in the confidence interval.
41. a. $1.07\% < p < 8.68\%$
    b. $70.1\% < p < 75.3\%$
    c. Yes, if wearing orange had no effect, we would expect that the percentage of injured hunters wearing orange should be between 70.1% and 75.3%, but it is much lower.
43. $13.0\% < p < 29.0\%$; yes
45. $x = 419$ results in the confidence interval (0.471, 0.539) and $x = 426$ results in the confidence interval (0.480, 0.548). They do not differ by substantial amounts.
47. $p > 0.818$; 81.8%
49. $0.894 < p < 1.006$; the upper confidence interval limits exceeds 1; use an upper limit of 1.
51. 602

Section 6-3

1. 2.33
3. 2.05
5. Yes
7. Yes
9. $2419.62; $92,580 < \mu < $97,420$
11. 0.823 sec; 4.42 sec $< \mu <$ 6.06 sec
13. 62
15. 250
17. 318.1
19. $\mu = 318.10 \pm 56.01$
21. $30.0^\circ C < \mu < 30.8^\circ C$; it is unrealistic to know σ.
23. $141.4 < \mu < 203.6$; it is unrealistic to know σ.
25. 217
27. 601
29. 80,770; no, increase the margin of error.
31. The range is 40, so σ is estimated to be $40/4 = 10$ by the range rule of thumb, and the sample size is 97. The sample standard deviation is $s = 11.3$, which results in a sample size of 123. The sample size of 123 is likely to be better because s is a better estimate of σ than range/4.
33. $105 < \mu < 115$

Section 6-4

1. $t_{\alpha/2} = 2.776$
3. Neither the normal nor the t distibution applies.
5. $t_{\alpha/2} = 1.662$
7. $z_{\alpha/2} = 2.33$
9. 60; $436 < \mu < 556$
11. $112.84 < \mu < 121.56$; there is 95% confidence that the interval from 112.84 to 121.56 contains the true value of the population mean μ.
13. $16,142 < \mu < $36,312$; there is 95% confidence that the interval from $16,142 to $36,312 contains the true value of the population mean μ.
15. a. $-2.248^\circ < \mu < 1.410^\circ$
    b. The CI does include 0°. The claim does not appear to be valid because a mean of 0° represents no difference between the actual high temperatures and the three-day forecast high temperatures, and the CI does include 0° indicating that 0° is a very possible value of the difference.
17. $0.075 < \mu < 0.168$; no, it is possible that the requirement is being met, but it is also very possible that the mean is not less than 0.165 grams/mile.

19. a. $164 < \mu < 186$
    b. $111 < \mu < 137$
    c. 186
    d. Because definitive conclusions about equality of means should not be based on overlapping of confidence intervals, this is a tentative conclusion: The confidence intervals do not overlap at all, suggesting that the two population means are likely to be significantly different, and the mean heart rate for those who manually shovel snow appears to be higher than the mean heart rate for those who use the electric snow thrower.

21. 95% CI for 4000 B.C.: $125.7 < \mu < 131.6$
    95% CI for 150 A.D.: $130.1 < \mu < 136.5$
    Because definitive conclusions about equality of means should not be based on overlapping of confidence intervals, this is a tentative conclusion: The two CIs overlap, so it is possible that the two population means are the same, and we cannot conclude that head sizes appear to have changed.

23. a. $0.82217 \text{ lb} < \mu < 0.82603 \text{ lb}$
    b. $0.78238 \text{ lb} < \mu < 0.78533 \text{ lb}$
    c. Because definitive conclusions about equality of means should not be based on overlapping of confidence intervals, this is a tentative conclusion: The cans of diet Pepsi appear to have a mean weight that is significantly less than the mean weight of cans of regular Pepsi, probably due to the sugar content.

25. $-12.244 < \mu < 29.613$; the CI is dramatically different with the outlier. The CI limits are very sensitive to outliers. Outliers should be carefully examined and discarded if they are found to be errors.

27. a. E is multiplied by $5/9$.
    b. $\frac{5}{9}(a - 32), \frac{5}{9}(b - 32)$
    c. Yes

Section 6-5

1. 6.262, 27.488
3. 51.172, 116.321
5. $\$9388 < \sigma < \$18{,}030$
7. $2.06 \text{ sec} < \sigma < 3.20 \text{ sec}$
9. 191
11. 133,448; no
13. $\$11{,}244 < \sigma < \$26{,}950$; there is 95% confidence that the limits of $\$11{,}244$ and $\$26{,}950$ contain the true value of the population standard deviation σ.
15. $1.195 < \sigma < 4.695$; yes, the confidence interval is likely to be a poor estimate because the value of 5.40 appears to be an outlier, suggesting that the assumption of a normally distributed population is not correct.
17. a. $10 < \sigma < 27$
    b. $12 < \sigma < 33$
    c. Because definitive conclusions about equality of standard deviations should not be based on overlapping of confidence intervals, this is a tentative conclusion: The variation does not appear to be significantly different.

19. a. $0.33 \text{ min} < \sigma < 0.87 \text{ min}$
    b. $1.25 \text{ min} < \sigma < 3.33 \text{ min}$
    c. Because definitive conclusions about equality of standard deviations should not be based on overlapping of confidence intervals, this is a tentative conclusion: The variation appears to be significantly lower with a single line. The single line appears to be better.

21. a. 98%
    b. 27.0

Chapter 6 Review Exercises

1. a. 9.00%
   b. $7.40\% < p < 10.6\%$
   c. 2653
2. a. $5.47 \text{ years} < \mu < 8.55 \text{ years}$
   b. $2.92 \text{ years} < \sigma < 5.20 \text{ years}$
   c. 1484
   d. No, the sample would not be representative of the population of all car owners.
3. a. 50.4%
   b. $45.7\% < p < 55.1\%$
   c. No, perhaps respondents are trying to impress the pollsters, or perhaps their memories have a tendency to indicate that they voted for the winner.
4. a. $4.94 < \mu < 8.06$
   b. $4.33 < \mu < 5.82$
   c. $7.16 < \mu < 9.71$
   d. Because definitive conclusions about equality of means should not be based on overlapping of confidence intervals, this is a tentative conclusion: Tolstoy has a significantly higher mean, so that his work is more difficult to read than Clancy or Rowling.
5. 65
6. $0.83 < \sigma < 1.99$
7. 2944
8. 221

Chapter 6 Cumulative Review Exercises

1. a. 121.0 lb
   b. 123.0 lb
   c. 119 lb, 128 lb
   d. 116.5 lb
   e. 23.0 lb
   f. 56.8 lb^2
   g. 7.5 lb
   h. 119.0 lb
   i. 123.0 lb
   j. 127.0 lb
   k. ratio

l.

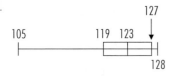

m. 112.6 lb $< \mu <$ 129.4 lb
n. 4.5 lb $< \sigma <$ 18.4 lb
o. 95
p. The individual supermodel weights do not appear to be considerably different from weights of randomly selected women, because they are all within 1.31 standard deviations of the mean of 143 lb. However, when considered as a group, their mean is significantly less than the mean of 143 lb (see part [m]).

2. a. 0.0089
    b. 0.260 $< p <$ 0.390
    c. Because the confidence interval limits do not contain 0.25, it is unlikely that the expert is correct.

3. a. 39.0%
    b. 36.1% $< p <$ 41.9%
    c. Yes, because the entire confidence interval is below 50%.
    d. The required sample size depends on the confidence level and the sample proportion, not the population size.

Chapter 7 Answers

Section 7-2

1. There is not sufficient evidence to support the claim that the gender selection method is effective.

3. There does appear to be sufficient evidence to support the claim that the majority of adult Americans like pizza.

5. $H_0: \mu =$ \$50,000. $H_1: \mu >$ \$50,000.

7. $H_0: p = 0.5$. $H_1: p > 0.5$.

9. $H_0: \sigma = 2.8$. $H_1: \sigma < 2.8$.

11. $H_0: \mu = 12$. $H_1: \mu < 12$.

13. $z = \pm 1.96$

15. $z = 2.33$

17. $z = \pm 1.645$

19. $z = -2.05$

21. -13.45

23. 1.85

25. 0.2912

27. 0.0512

29. 0.0244

31. 0.4412

33. There is sufficient evidence to support the claim that the proportion of married women is greater than 0.5.

35. There is not sufficient evidence to support the claim that the proportion of fatal commercial aviation crashes is different from 0.038.

37. Type I: Conclude that there is sufficient evidence to support $p > 0.5$ when in reality $p = 0.5$. Type II: Fail to reject $p = 0.5$ (and therefore fail to support $p > 0.5$) when in reality $p > 0.5$.

39. Type I: Conclude that there is sufficient evidence to support $p \neq 0.038$ when in reality $p = 0.038$. Type II: Fail to reject $p = 0.038$ (and therefore fail to support $p \neq 0.038$) when in reality $p \neq 0.038$.

41. P-value $= 0.9999$. With an alternative hypothesis of $p > 0.5$, it is impossible for a sample statistic of 0.27 to fall in the critical region. No sample proportion less than 0.5 can ever support a claim that $p > 0.5$.

43. 0.01, because this lowest P-value would correspond to sample data that are most supportive of the claim that the defect rate is lower.

45. There are no finite critical values corresponding to $\alpha = 0$, and it is impossible to have a P-value ≤ 0. With $\alpha = 0$, the null hypothesis will never be rejected.

Section 7-3

1. a. $z = -0.12$
    b. $z = \pm 1.96$
    c. 0.9044
    d. There is not sufficient evidence to warrant rejection of the claim that green-flowered peas occur at a rate of 25%.
    e. No, a hypothesis test cannot be used to prove that a proportion is equal to some claimed value.

3. $H_0: p = 0.62$. $H_1: p < 0.62$. Test statistic: $z = -2.06$. Critical value: $z = -2.33$. P-value: 0.0197. Fail to reject H_0. There is not sufficient evidence to support the claim that less than 62% of brides spend less than \$750. If the responses were volunteered by readers, the sample is a voluntary response sample and the hypothesis test results are not valid.

5. $H_0: p = 0.15$. $H_1: p > 0.15$. Test statistic: $z = 1.60$. Critical value: $z = 1.645$. P-value: 0.0548. Fail to reject H_0. There is not sufficient evidence to support the claim that more than 15% of U.S. households use e-mail. The conclusion is not valid today because the population characteristics (use of e-mail) are changing quickly over time.

7. $H_0: p = 0.5$. $H_1: p > 0.5$. Test statistic: $z = 0.58$. Critical value: $z = 1.28$. P-value: 0.2810. Fail to reject H_0. There is not sufficient evidence to support the claim that the proportion is greater than 0.5.

9. $H_0: p = 0.01$. $H_1: p \neq 0.01$. Test statistic: $z = 2.19$. Critical values: $z = \pm 1.96$. P-value: 0.0286. Reject $H_0: p = 0.01$. There is sufficient evidence to warrant rejection of the claim that 1% of sales are overcharges. Because 1.62% of the sampled items are overcharges, it appears that the error rate is worse with scanners, not better.

11. $H_0: p = 0.61$. $H_1: p > 0.61$. Test statistic: $z = 1.60$. Critical value: $z = 1.645$. P-value: 0.0548. Fail to reject H_0. There is not sufficient evidence to support the claim that Morrison's strike rate is greater than 61%.

13. $H_0: p = 0.000340$. $H_1: p \neq 0.000340$. Test statistic: $z = -0.66$. Critical values: $z = \pm 2.81$. P-value: 0.5092. Fail to reject H_0. There is not sufficient evidence to support the claim that the rate is different from 0.0340%. Cell phone users should not be concerned about cancer of the brain or nervous system.

15. H_0: $p = 0.27$. H_1: $p < 0.27$. Test statistic: $z = -5.46$. Critical value: $z = -2.33$. P-value: 0.0001. Reject H_0. There is sufficient evidence to support the claim that the rate of smoking among those with four years of college is less than 27%.

17. H_0: $p = 0.75$. H_1: $p > 0.75$. Test statistic: $z = 8.26$. Critical value: $z = 2.33$. P-value: 0.0000 when rounded to four decimal places. Reject H_0. There is sufficient evidence to support the claim that more than three-fourths of aviation accidents result in fatalities.

19. H_0: $p = 0.10$. H_1: $p \neq 0.10$. Test statistic: $z = -1.67$. Critical values assuming a 0.05 significance level: $z = \pm 1.96$. P-value: 0.0950. Fail to reject H_0. There is not sufficient evidence to warrant rejection of the claim that 10% of the plain M&M candies are blue.

21. a. H_0: $p = 0.10$. H_1: $p \neq 0.10$. Test statistic: $z = 2.00$. Critical values: $z = \pm 1.96$. Reject H_0. There is sufficient evidence to warrant rejection of the claim that the proportion of zeros is 0.1.

    b. H_0: $p = 0.10$. H_1: $p \neq 0.10$. Test statistic: $z = 2.00$. P-value: 0.0456. There is sufficient evidence to warrant rejection of the claim that the proportion of zeros is 0.1.

    c. $0.0989 < p < 0.139$; because 0.1 is contained within the confidence interval, fail to reject H_0: $p = 0.10$. There is not sufficient evidence to warrant rejection of the claim that the proportion of zeros is 0.1.

    d. The traditional and P-value methods both lead to rejection of the claim, but the confidence interval method does not lead to rejection.

23. Original claim: $p \leq c$; H_0: $p = c$; H_1: $p > c$. We can either reject or fail to reject the original claim. The sample data cannot "support" the claim that children who live near high voltage power lines are no more likely to get leukemia than other children.

25. H_0: $p = 0.10$. H_1: $p \neq 0.10$. Test statistic: $z = -2.36$. Critical values: $z = \pm 2.575$. P-value: 0.0182. Fail to reject H_0. Even though no blue candies are obtained, there is not sufficient evidence to warrant rejection of the claim that 10% of the candies are blue.

27. 47% is not a possible result because, with 20 mice, the only possible success rates are 0%, 5%, 10%, . . . , 100%.

Section 7-4

1. Yes

3. No

5. $z = 1.18$, P-value: 0.1190; critical value: $z = 1.645$. There is not sufficient evidence to support the claim that the mean is greater than 118.

7. $z = 0.89$, P-value: 0.3734; critical values: $z = \pm 2.575$. There is not sufficient evidence to warrant rejection of the claim that the mean equals 5.00 sec.

9. H_0: $\mu = 30.0$. H_1: $\mu > 30.0$. Test statistic: $z = 1.84$. P-value: 0.0329. (Critical value: $z = 1.645$.) Reject H_0. There is sufficient evidence to support the claim that the mean is greater than 30.0°C.

11. H_0: $\mu = 200.0$. H_1: $\mu \neq 200.0$. Test statistic: $z = -1.46$. P-value: 0.1442. (Critical values: $z = \pm 2.575$.) Fail to reject H_0. There is not sufficient evidence to warrant rejection of the claim that the mean equals 200.0.

13. H_0: $\mu = 0.9085$. H_1: $\mu \neq 0.9085$. Test statistic: $z = 1.68$. P-value: 0.093. Fail to reject H_0. There is not sufficient evidence to support the claim that the mean is different from 0.9085 g.

15. H_0: $\mu = 0$. H_1: $\mu \neq 0$. Test statistic: $z = -0.63$. P-value: 0.5288. Fail to reject H_0. There is not sufficient evidence to support a claim that the mean is different from 0. These results suggest that the three-day forecast high temperatures are reasonably accurate, because they do not appear to differ from the actual high temperatures by a significant amount.

17. a. It is not likely that σ is known.

    b. 2.10

    c. No, so the assumption that $\sigma = 0.62$ is a safe assumption in the sense that if σ is actually some value other than 0.62, it is very unlikely that the result of the hypothesis test would be affected.

19. a. 0.6178

    b. 0.0868

Section 7-5

1. Student t

3. Normal

5. Between 0.005 and 0.01

7. Less than 0.01

9. $t = 0.745$. P-value is greater than 0.10. Critical value: $t = 1.729$. There is not sufficient evidence to support the claim that the mean is greater than 118.

11. $t = 0.900$; P-value is greater than 0.20. Critical values: $t = \pm 2.639$. There is not sufficient evidence to warrant rejection of the claim that the mean equals 5.00 sec.

13. H_0: $\mu = 4$. H_1: $\mu > 4$. Test statistic: $t = 3.188$. P-value is less than 0.005. Critical value: $t = 1.796$. Reject H_0. There is sufficient evidence to support the claim that the mean is greater than 4.

15. H_0: $\mu = 0$. H_1: $\mu \neq 0$. Test statistic: $t = -0.63$. P-value is greater than 0.20. Critical values: $t = \pm 2.042$. Fail to reject H_0. There is not sufficient evidence to support the claim that the mean is different from 0°. Based on these results, it appears that the three-day forecast high temperatures are reasonably accurate.

17. H_0: $\mu = 0$. H_1: $\mu \neq 0$. Test statistic: $t = 4.010$. P-value is less than 0.01. Critical values: $t = \pm 2.704$. Reject H_0. There is sufficient evidence to warrant rejection of the claim that the mean is equal to 0 sec. The watches do not appear to be reasonably accurate.

19. H_0: $\mu = 69.5$. H_1: $\mu > 69.5$. Test statistic: $t = 2.652$. P-value is between 0.005 and 0.01. Critical value: $t = 1.691$. Reject H_0. There is sufficient evidence to support the claim that the mean is greater than 69.5 years. However, men do not become conductors until they are at least 25 years old, and the life expectancy for such men is naturally greater than the life expectancy for males at birth.

21. H_0: $\mu = \$1000$. H_1: $\mu < \$1000$. Test statistic: $t = -1.83$. P-value: 0.071. Using a 0.05 significance level, fail to reject H_0 and conclude that there is not sufficient evidence to support the claim that the mean is less than $1000.

23. H_0: $\mu = 3.39$. H_1: $\mu \neq 3.39$. Test statistic: $t = 1.734$. P-value: 0.1034. Fail to reject H_0. There is not sufficient evidence to warrant rejection of the claim that the mean equals 3.39 kg (assuming a significance level of 0.05). There is not sufficient evidence to conclude that the vitamin supplement has an effect on birth weight.

25. H_0: $\mu = 1.5$. H_1: $\mu > 1.5$. Test statistic: $t = 0.049$. P-value is greater than 0.10. Critical value: $t = 2.015$. Fail to reject H_0. There is not sufficient evidence to support a claim that the mean is greater than 1.5 $\mu g/m^3$. The assumption of a normal distribution is questionable because 5.40 appears to be an outlier.

27. H_0: $\mu = 11$. H_1: $\mu < 11$. Test statistic: $t = -0.095$. The critical values depend on the significance level, but the test statistic will not fall in the critical region for any reasonable choices. P-value is greater than 0.10. Fail to reject H_0. There is not sufficient evidence to support the claim that the mean is less than 11 sec. Because the data are taken in consecutive Olympic games, the population mean is changing as athletes become faster. We cannot conclude that future times should be around 10.5 sec.

29. H_0: $\mu = 6$. H_1: $\mu > 6$. Test statistic: $t = 0.707$. P-value is greater than 0.10. Critical value: $t = 1.796$ (assuming $\alpha = 0.05$). Fail to reject H_0. There is not sufficient evidence to support the claim that the mean is greater than 6.

31. H_0: $\mu = 12$. H_1: $\mu > 12$. Test statistic: $t = 10.166$. P-value is less than 0.005. Critical value: $t = 2.441$ (approximately). Reject H_0. There is sufficient evidence to support the claim that the mean is greater than 12 oz. The production method could be adjusted to be less wasteful.

33. The P-value becomes 0.070. The test statistic does not change. The claim that the mean is different from 420 h is not rejected at the 0.05 significance level.

35. The test statistic changes to $t = 0.992$ and the P-value changes to 0.182. An outlier can change the test statistic and P-value substantially. Although the conclusion does not change here, it could change in other cases.

37. 0.10

Section 7-6

1. Test statistic: $\chi^2 = 8.444$. Critical values: $\chi^2 = 8.907, 32.852$. P-value: Between 0.02 and 0.05. Reject H_0. There is sufficient evidence to support the claim that $\sigma \neq 15$.

3. Test statistic: $\chi^2 = 10.440$. Critical value: $\chi^2 = 14.257$. P-value: Less than 0.005. Reject H_0. There is sufficient evidence to support the claim that $\sigma < 50$.

5. H_0: $\sigma = 0.04$. H_1: $\sigma > 0.04$. Test statistic: $\chi^2 = 2342.438$. Critical value: $\chi^2 = 63.691$ (approximately). (P-value: Less than 0.005.) Reject H_0. There is sufficient evidence to support the claim that the weights of peanut M&Ms vary more than the weights of plain M&Ms.

7. H_0: $\sigma = 43.7$. H_1: $\sigma \neq 43.7$. Test statistic: $\chi^2 = 114.586$. Critical values: $\chi^2 = 57.153, 106.629$. (P-value: Between 0.01 and 0.02.) Reject H_0. There is sufficient evidence to support the claim that the standard deviation is different from 43.7 ft. Because the sample standard deviation is larger than in the past, it appears that the new production method is worse than in the past.

9. H_0: $\sigma = 6.2$. H_1: $\sigma < 6.2$. Test statistic: $\chi^2 = 9.016$. Critical value: $\chi^2 = 13.848$. (P-value: Less than 0.005.) Reject H_0. There is sufficient evidence to support the claim that the single line corresponds to lower variation. These results do not necessarily imply that the single line results in a shorter wait.

11. H_0: $\sigma = 29$. H_1: $\sigma < 29$. Test statistic: $\chi^2 = 0.540$. Critical value: $\chi^2 = 1.646$. (P-value: Less than 0.005.) Reject H_0. There is sufficient evidence to support the claim that the weights of supermodels vary less than the weights of women in general.

13. H_0: $\sigma = 0.10$. H_1: $\sigma < 0.10$. Test statistic: $\chi^2 = 28.750$. Critical value of χ^2 is between 18.493 and 26.509. Fail to reject H_0. There is not sufficient evidence to support the claim that the volumes have a standard deviation less than 0.10 oz.

15. H_0: $\sigma = 28.7$. H_1: $\sigma \neq 28.7$. Test statistic: $\chi^2 = 32.818$. Critical values: $\chi^2 = 24.433, 59.342$ (approximately). (P-value: Greater than 0.20.) Fail to reject H_0. There is not sufficient evidence to warrant rejection of the claim that the standard deviation is 28.7 lb.

17. Using interpolation, estimate the critical value as $\chi^2 = 22.501$ to get $s = 0.08$ oz.

19. a. Estimated values: 74.216, 129.565; Table A-5 values: 74.222, 129.561.
    b. 117.093, 184.690

21. a. The standard deviation will be lower.
    b. The requirement of a normally distributed population is not satisfied.

Chapter 7 Review Exercises

1. a. No, the sample is a voluntary response sample, so the results do not necessarily apply to the population of adult Americans.
   b. No, even though there does appear to be statistically significant weight loss, the average amount of lost weight is so small that the drug is not practical.
   c. 0.001, because this P-value corresponds to results that provide the most support for the effectiveness of the cure.
   d. There is not sufficient evidence to support the claim that the mean is greater than 12 oz.
   e. rejecting a true null hypothesis.

2. a. H_1: $\mu < \$10,000$; Student t distribution.
   b. H_1: $\sigma > 1.8$ sec; chi-square distribution.
   c. H_1: $p > 0.5$; normal distribution.
   d. H_1: $\mu \neq 100$; normal distribution.

3. a. H_0: $\mu = 100$. H_1: $\mu \neq 100$. Test statistic: $z = -0.75$. P-value: 0.4532. (Critical values: $z = \pm 1.645$.) Fail to reject H_0. There is not sufficient evidence to warrant rejection of the claim that the mean is equal to 100.

b. $H_0: \mu = 100$. $H_1: \mu \neq 100$. Test statistic: $t = -0.694$. P-value: Greater than 0.20. (Critical values: $t = \pm 1.676$ approximately.) Fail to reject H_0. There is not sufficient evidence to warrant rejection of the claim that the mean is equal to 100.

c. $H_0: \sigma = 15$. $H_1: \sigma \neq 15$. Test statistic: $\chi^2 = 57.861$. Critical values: $\chi^2 = 34.764, 67.505$. (P-value: Greater than 0.20.) Fail to reject H_0. There is not sufficient evidence to warrant rejection of the claim that the standard deviation is equal to 15.

d. Yes

4. $H_0: p = 0.5$. $H_1: p < 0.5$. Test statistic: $z = -1.47$. Critical value: $z = -1.645$. P-value: 0.0708. Fail to reject H_0. There is not sufficient evidence to support the claim that less than half of all executives identify "little or no knowledge of the company" as the most common interview mistake.

5. $H_0: \mu = 5.670$ g. $H_1: \mu \neq 5.670$ g. Test statistic: $t = -4.991$. P-value is less than 0.01. Critical values: $t = \pm 2.678$ approximately. Reject H_0. There is sufficient evidence to warrant rejection of the claim that the mean weight is 5.670 g. A possible explanation is that quarters lose weight as they wear from handling in circulation.

6. $H_0: \mu = 0.9085$ g. $H_1: \mu < 0.9085$ g. Test statistic: $t = -0.277$. P-value is greater than 0.10. Critical value: $t = -2.132$. Fail to reject H_0. There is not sufficient evidence to support a claim that the mean is less than 0.9085 g. The claimed weight appears to be correct as printed on the label.

7. $H_0: p = 0.10$. $H_1: p < 0.10$. Test statistic: $z = -1.17$. Critical value: $z = -1.645$. P-value: 0.1210. Fail to reject H_0. There is not sufficient evidence to support the claim that less than 10% of trips include a theme park visit.

8. $H_0: p = 0.43$. $H_1: p \neq 0.43$. Test statistic: $z = 3.70$. Critical values: $z = \pm 2.05$. P-value: 0.0002. Reject H_0. There is sufficient evidence to warrant rejection of the claim that the percentage of voters who say that they voted for the winning candidate is equal to 43%.

9. $H_0: \mu = 12$ oz. $H_1: \mu < 12$ oz. Test statistic: $t = -4.741$. P-value is less than 0.005. Critical value: $t = -1.714$ (assuming $\alpha = 0.05$). Reject H_0. There is sufficient evidence to support the claim that the mean is less than 12 oz. Windsor's argument is not valid.

10. $H_0: p = 0.10$. $H_1: p < 0.10$. Test statistic: $z = -2.36$. Critical value: $z = -2.33$. P-value: 0.0091. Reject H_0. There is sufficient evidence to support the claim that the true percentage is less than 10%. The phrase "almost 1 out of 10" is not justified.

11. $H_0: \sigma = 0.15$. $H_1: \sigma < 0.15$. Test statistic: $\chi^2 = 44.800$. Critical value: $\chi^2 = 51.739$. (P-value: Between 0.005 and 0.01.) Reject H_0. There is sufficient evidence to support the claim that variation is lower with the new machine. The new machine should be purchased.

12. $H_0: \mu = 3.5$ g. $H_1: \mu \neq 3.5$ g. Test statistic: $t = 9.720$. P-value is less than 0.01. Critical values: $t = \pm 1.994$ (approximately, assuming $\alpha = 0.05$). Reject H_0. There is sufficient evidence to warrant rejection of the claim that the mean is equal to 3.5 g. It appears that the packets have more sugar than is indicated on the label.

Chapter 7 Cumulative Review Exercises

1. a. 0.0793 ng/m^3
   b. 0.044 ng/m^3
   c. 0.0694 ng/m^3
   d. 0.0048
   e. 0.158 ng/m^3
   f. $0.0259 < \mu < 0.1326$
   g. $H_0: \mu = 0.16$. $H_1: \mu < 0.16$. Test statistic: $t = -3.491$. P-value is less than 0.005. Critical value: $t = -1.860$. Reject H_0. There is sufficient evidence to support the claim that the mean is less than 0.16 ng/m^3.
   h. Yes, the data are listed in order and there appears to be a trend of decreasing values. The population is changing over time.

2. a. 0.4840
   b. 0.0266 (from 0.4840^5)
   c. 0.4681
   d. 634

3. a. 6.3
   b. 2.2
   c. Binom: 0.0034; normal: 0.0019
   d. Based on the low probability value in part (c), reject $H_0: p = 0.25$. There is sufficient evidence to reject the claim that the subject made random guesses.
   e. 423

Chapter 8 Answers

Section 8-2

1. 30

3. 85

5. a. 0.417
   b. 2.17
   c. ± 1.96
   d. 0.0300

7. $H_0: p_1 = p_2$. $H_1: p_1 > p_2$. Test statistic: $z = 2.17$. P-value: 0.0150. Critical value: $z = 1.645$. Reject H_0. There is sufficient evidence to support the claim that for those saying that monitoring e-mail is seriously unethical, the proportion of employees is greater than the proportion of bosses.

9. $0.00216 < p_1 - p_2 < 0.00623$; it appears that physical activity corresponds to a lower rate of coronary heart disease.

11. $H_0: p_1 = p_2$. $H_1: p_1 \neq p_2$. Test statistic: $z = -0.21$. P-value: 0.8336. Critical values: $z = \pm 1.96$. Fail to reject H_0. There is not sufficient evidence to warrant rejection of the claim that the reversal rate is the same in both years.

13. $H_0: p_1 = p_2$. $H_1: p_1 < p_2$. Test statistic: $z = -12.39$. P-value: 0.0001. Critical value for $\alpha = 0.05$: $z = -1.645$. Reject H_0. There is sufficient sample evidence to support the stated claim.

15. $H_0: p_1 = p_2$. $H_1: p_1 > p_2$. Test statistic: $z = 1.07$. P-value: 0.1423. Critical value: $z = 1.645$. Fail to reject H_0. No. Based on the available evidence, delay taking any action.

17. With a test statistic of $z = 4.94$ and a P-value of 0.000, reject H_0. There is sufficient evidence to support the claim that the conviction rate for wives is less than the conviction rate for husbands.

19. $H_0: p_1 = p_2$. $H_1: p_1 \neq p_2$. Test statistic: $z = -2.01$. P-value: 0.0444. Critical values: $z = \pm 1.96$. Reject H_0. There appears to be a significant difference. Because the Autozone failure rate is lower, it appears to be the better choice.

21. $-0.135 < p_1 - p_2 < 0.0742$ (using $x_1 = 49$ and $x_2 = 70$); there does not appear to be a gender gap.

23. $-0.0144 < p_1 - p_2 < 0.0086$; yes

25. a. $H_0: p_1 = p_2$. $H_1: p_1 \neq p_2$. Test statistic: $z = -3.06$. Critical values: $z = \pm 2.575$ (assuming a 0.01 significance level). P-value: 0.0022. Reject $H_0: p_1 = p_2$. There is sufficient evidence to support the claim that the two population percentages are different.
    b. $-0.0823 < p_1 - p_2 < -0.00713$; because the confidence interval limits do not contain 0, there appears to be a significant difference (although it would be better to use a hypothesis test of the null hypothesis $p_1 = p_2$).

27. $H_0: p_1 = p_2$. $H_1: p_1 \neq p_2$. Test statistic: $z = 4.46$. P-value: 0.0002. Critical values: $z = \pm 2.575$. Reject H_0. There is sufficient evidence to warrant rejection of the claim that the central-city refusal rate is the same as the refusal rate in other areas.

29. $H_0: p_1 = p_2$. $H_1: p_1 < p_2$. Test statistic: $z = -0.60$. P-value: 0.2743. There is not sufficient evidence to support the claim that the proportion of children's movies showing alcohol use is less than the proportion showing tobacco use. The results don't apply to Data Set 7 because the samples are not independent.

31. a. $0.0227 < p_1 - p_2 < 0.217$; because the confidence interval limits do not contain 0, it appears that $p_1 = p_2$ can be rejected.
    b. $0.491 < p_1 < 0.629$; $0.371 < p_2 < 0.509$; because the confidence intervals do overlap, it appears that $p_1 = p_2$ cannot be rejected.
    c. $H_0: p_1 = p_2$. $H_1: p_1 \neq p_2$. Test statistic: $z = 2.40$. P-value: 0.0164. Critical values: $z = \pm 1.96$. Reject H_0. There is sufficient evidence to reject $p_1 = p_2$.
    d. Reject $p_1 = p_2$. Least effective: Using the overlap between the individual confidence intervals.

33. The test statistic changes to $z = 2.03$ and the 90% confidence interval limits change to 0.00231 and 0.0277, so there is now sufficient evidence to support the given claim.

35. a. Test statistic: $z = 1.48$. P-value: 0.1388. Critical values: $z = \pm 1.96$. Fail to reject $H_0: p_1 = p_2$.
    b. Test statistic: $z = 1.63$. P-value: 0.1032. Critical values: $z = \pm 1.96$. Fail to reject $H_0: p_2 = p_3$.
    c. Test statistic: $z = 3.09$. P-value: 0.0020. Critical values: $z = \pm 1.96$. Reject $H_0: p_1 = p_3$.
    d. No

37. a. No, because the conditions $np \geq 5$ and $nq \geq 5$ are not satisfied for both samples.
    b. With 144 people in the placebo group, 1.8% is not a possible result.

Section 8-3

1. Independent samples

3. Matched pairs

5. $H_0: \mu_1 = \mu_2$. $H_1: \mu_1 > \mu_2$. Test statistic: $t = 2.790$. Critical value: $t = 2.660$. P-value < 0.01. (Using TI-83: df = 122, P-value = 0.003.) Reject H_0. There is sufficient evidence to support the claim that the population of heavy marijuana users has a lower mean than the light users. Heavy users of marijuana should be concerned about deteriorating cognitive abilities.

7. $-0.65 < \mu_1 - \mu_2 < 3.03$ (TI-83: df = 69 and $-0.61 < \mu_1 - \mu_2 < 2.99$.) Because the confidence interval does contain zero, we should not conclude that the two population means are different. The treatment does not appear to be effective, so paroxetine should not be prescribed.

9. $H_0: \mu_1 = \mu_2$. $H_1: \mu_1 > \mu_2$. Test statistic: $t = 0.132$. Critical value: $t = 1.729$. P-value: 0.461 approximately. (Using TI-83: df = 34, P-value = 0.448.) Fail to reject H_0. There is not sufficient evidence to support the claim that the magnets are effective in reducing pain. It is valid to argue that the magnets might appear to be effective if the sample sizes are larger.

11. a. $H_0: \mu_1 = \mu_2$. $H_1: \mu_1 \neq \mu_2$. Test statistic: $t = 22.098$. Critical values: $t = \pm 2.728$. P-value < 0.01. (Using TI-83: df = 56, P-value = 0.000.) Reject H_0. There is sufficient evidence to warrant rejection of the claim that regular Coke and diet Coke have the same mean weight. The difference is probably due to the sugar that is in regular Coke but not in diet Coke.
    b. $0.02808 < \mu_1 - \mu_2 < 0.03598$ (TI-83: df = 56 and $0.02817 < \mu_1 - \mu_2 < 0.03589$.)

13. $-0.01 < \mu_1 - \mu_2 < 0.23$; because this CI contains zero, there does not appear to be a significant difference between the two population means, so it does not appear that obsessive-compulsive disorders have a biological basis. (With a TI-83 calculator, df = 18 and $0.01 < \mu_1 - \mu_2 < 0.21$, which does not contain zero, suggesting that there is a significant difference so that obsessive-compulsive disorders appear to have a biological basis. This is a rare case where the simple and conservative estimate of df leads to a different conclusion than the more accurate Formula 8-1.)

15. $1.46 < \mu_1 - \mu_2 < 3.52$ (TI-83: df = 25 and $1.47 < \mu_1 - \mu_2 < 3.51$.) Because the confidence interval does not contain zero, there appears to be a significant difference between the two population means. It does appear that there are significantly more errors made by those treated with alcohol.

17. $H_0: \mu_1 = \mu_2$. $H_1: \mu_1 > \mu_2$. Test statistic: $t = 2.879$. Critical value: $t = 2.429$. P-value = 0.006 approximately. (Using TI-83: df = 77, P-value = 0.003.) Reject H_0. There is sufficient evidence to support the claim that the "stress" population has a lower mean than the "nonstress" population. However, we cannot conclude that stress *decreases* the amount recalled.

19. H_0: $\mu_1 = \mu_2$. H_1: $\mu_1 \neq \mu_2$. Test statistic: $t = 1.130$. Critical values: $t = \pm 1.983$. P-value = 0.261. Fail to reject H_0. There is not sufficient evidence to support the claim that there is a significant difference between the two population means.

21. Filtered: $n_1 = 21$, $\bar{x}_1 = 13.3$, $s_1 = 3.7$. Nonfiltered: $n_2 = 8$, $\bar{x}_2 = 24.0$, $s_2 = 1.7$. H_0: $\mu_1 = \mu_2$. H_1: $\mu_1 < \mu_2$. Test statistic: $t = -10.585$. Critical value: $t = -1.895$. P-value = 0.000. (Using TI-83: df = 26, P-value = 0.0000.) Reject H_0. There is sufficient evidence to support the claim that the mean amount of tar in filtered king-size cigarettes is less than the mean amount of tar in nonfiltered king-size cigarettes.

23. Men: $n_1 = 40$, $\bar{x}_1 = 25.9975$, $s_1 = 3.4307$. Women: $n_2 = 40$, $\bar{x}_2 = 25.7400$, $s_2 = 6.1656$. H_0: $\mu_1 = \mu_2$. H_1: $\mu_1 \neq \mu_2$. Test statistic: $t = 0.231$. Critical values: $t = \pm 2.024$ (assuming a 0.05 significance level). P-value = 0.842 approximately. (Using TI-83: df = 61, P-value = 0.818.) Fail to reject H_0. There is not sufficient evidence to warrant rejection of the claim that the mean BMI of men is equal to the mean BMI of women.

25. $-0.62 < \mu_1 - \mu_2 < 3.00$ (TI-83: $-0.60 < \mu_1 - \mu_2 < 2.98$.) The results did not change much.

27. These new results are very close to those found in Exercise 9: Critical value: $t = 1.686$. P-value = 0.460 approximately. (Using TI-83: P-value = 0.448.) The other results are the same.

29. a. The test statistic changes substantially from $t = 1.130$ to $t = 1.508$, but it is not enough of a change to cause a change in the conclusion.
    b. The numerator of the test statistic does increase substantially because the sample means do have a much greater difference, but the denominator also increases dramatically because of the increase in the variance of the first sample.

31. a. 50/3
    b. 2/3
    c. $50/3 + 2/3 = 52/3$
    d. The range of the x-y values equals the range of the x values plus the range of the y values.

33. df = 18 (instead of 9), the critical values become $t = \pm 2.878$ (instead of ± 3.250), and the confidence interval limits become 0.007 and 0.213, and the P-value is less than 0.01 (instead of between 0.01 and 0.02). Using Formula 8-1, the confidence interval is a little narrower, the critical value is a little smaller, and the P-value is a little smaller. With df = 9 it does not appear that obsessive-compulsive disorders have a biological basis; with df = 18 from Formula 8-1, it does appear that obsessive-compulsive disorders have a biological basis. Using the smaller of $n_1 - 1$ and $n_2 - 1$ for df is more conservative (than the use of Formula 8-1) in the sense that the sample data need to be more extreme to be considered significant, as can be seen by the different conclusions.

Section 8-4

1. a. -0.2
   b. 2.8
   c. $t = -0.161$
   d. ± 2.776

3. $-3.6 < \mu_d < 3.2$

5. a. H_0: $\mu_d = 0$. H_1: $\mu_d \neq 0$. Test statistic: $t = -0.831$. Critical values: $t = \pm 2.201$. P-value: 0.440 approximately. Fail to reject H_0. There is not sufficient evidence to support the claim that there is a difference between self-reported heights and measured heights.
   b. $-1.7 < \mu_d < 0.8$; because the confidence interval limits contain 0, there is not sufficient evidence to support the claim that there is a difference between self-reported heights and measured heights.

7. a. H_0: $\mu_d = 0$. H_1: $\mu_d < 0$. Test statistic: $t = -1.718$. Critical value: $t = -1.833$. P-value: 0.062. Fail to reject H_0. There is not sufficient evidence to conclude that the preparatory course is effective in raising scores.
   b. $-25.5 < \mu_d < 3.5$; we have 95% confidence that the interval from -25.5 to 3.5 actually contains the true population mean difference.

9. a. $0.69 < \mu_d < 5.56$
   b. H_0: $\mu_d = 0$. H_1: $\mu_d > 0$. Test statistic: $t = 3.036$. Critical value: $t = 1.895$. P-value: 0.007. Reject H_0. There is sufficient evidence to support the claim that the sensory measurements are lower after hypnosis.
   c. Yes

11. a. H_0: $\mu_d = 0$. H_1: $\mu_d \neq 0$. Test statistic: $t = -1.690$. Critical values: $t = \pm 2.228$. P-value: 0.120. Fail to reject H_0. There is not sufficient evidence to warrant rejection of the claim that there is no difference between the yields from the two types of seed.
    b. $-78.2 < \mu_d < 10.7$
    c. No

13. a. H_0: $\mu_d = 0$. H_1: $\mu_d \neq 0$. Test statistic: $t = -0.41$. P-value: 0.691. Fail to reject H_0. There is not sufficient evidence to support the claim that astemizole has an effect. Don't take astemizole for motion sickness.
    b. 0.3455; there is not sufficient evidence to support the claim that astemizole prevents motion sickness.

15. H_0: $\mu_d = 0$. H_1: $\mu_d \neq 0$. Test statistic: $t = -0.501$. Critical values: $t = \pm 2.201$. P-value: 0.626. Fail to reject H_0. There is not sufficient evidence to support the claim that there is a difference between self-reported weights and measured weights of males aged 12–16.

17. a. $-1.40 < \mu_d < -0.17$
    b. H_0: $\mu_d = 0$. H_1: $\mu_d \neq 0$. Test statistic: $t = -2.840$. Critical values: $t = \pm 2.228$. P-value < 0.02. Reject H_0. There is sufficient evidence to warrant rejection of the claim that the mean difference is 0. Morning and night body temperatures do not appear to be about the same.

19. a. H_0: $\mu_d = 0$. H_1: $\mu_d \neq 0$.Test statistic: $t = -2.966$. Critical values: $t = \pm 2.042$. P-value: 0.006. Reject H_0. There is sufficient evidence to support the claim that there is a difference between the actual low temperatures and the low temperatures that were forecast five days earlier.

 b. $-10.1 < \mu_d < -1.9$

 c. With the larger data set consisting of 31 matched pairs, there is sufficient evidence to conclude that there is a significant difference between the actual low temperatures and the low temperatures that were forecast five days earlier.

21. a. Yes

 b. The hypothesis test is not affected. The confidence interval limits will change from the Fahrenheit scale to equivalent values on the Celsius scale.

23. a. Test statistic: $t = 1.861$. Critical value: $t = 1.833$. P-value: 0.045. Reject H_0. There is sufficient evidence to support $\mu_d > 0$.

 b. Test statistic: $t = 1.627$. Critical value: $t = 1.833$. P-value: 0.072. Fail to reject H_0. There is not sufficient evidence to support $\mu_1 > \mu_2$.

 c. Yes, the conclusion is affected by the test that is used.

Section 8-5

1. H_0: $\sigma_1^2 = \sigma_2^2$. H_1: $\sigma_1^2 \neq \sigma_2^2$. Test statistic: $F = 2.2500$. Upper critical value: $F = 2.1540$. Reject H_0. There is sufficient evidence to support the claim that the treatment and placebo populations have different variances.

3. H_0: $\sigma_1^2 = \sigma_2^2$. H_1: $\sigma_1^2 > \sigma_2^2$. Test statistic: $F = 2.1267$. The critical F value is between 2.1555 and 2.2341. Fail to reject H_0. There is not sufficient evidence to support the claim that the pain reductions for the sham treatment group vary more than the pain reductions for the magnet treatment group.

5. H_0: $\sigma_1 = \sigma_2$. H_1: $\sigma_1 \neq \sigma_2$. Test statistic: $F = 2.9228$. The upper critical value of F is between 1.8752 and 2.0739. Reject H_0. There is sufficient evidence to support the claim that the populations have different standard deviations.

7. H_0: $\sigma_1^2 = \sigma_2^2$. H_1: $\sigma_1^2 > \sigma_2^2$. Test tatistic: $F = 3.7539$. Critical value: $F = 3.4445$. Reject H_0. There is sufficient evidence to support the claim that king-size cigarettes with filters have amounts of nicotine that vary more than the amounts of nicotine in nonfiltered king-size cigarettes.

9. H_0: $\sigma_1^2 = \sigma_2^2$. H_1: $\sigma_1^2 > \sigma_2^2$. Test statistic: $F = 1.0110$. The critical value of F is less than 1.3519 (assuming a 0.05 significance level). (Although the conclusion is not clear from the test statistic and critical value, the values of the standard deviations [3.67 and 3.65] suggest that the difference is not significant. Using a TI-83 Plus calculator results in a P-value of 0.4745.) Fail to reject H_0. There is not sufficient evidence to support the claim that the ages of faculty cars vary less than the ages of student cars.

11. a. Test statistic: $F = 2.1722$. Upper critical value of F is between 1.6668 and 1.8752. Reject H_0. There is sufficient evidence to warrant rejection of the claim that Wednesday and Sunday rainfall amounts have the same standard deviation.

 b. Because they have so many zeros as the lowest values, neither the Wednesday rainfall amounts nor the Sunday rainfall amounts are normally distributed.

 c. Because the populations do not appear to be normally distributed, the conclusion given in part (a) is not necessarily valid. The methods of Section 8-5 do not apply.

13. H_0: $\sigma_1 = \sigma_2$. H_1: $\sigma_1 \neq \sigma_2$. Test statistic: $F = 1.2478$. P-value: 0.6852. Fail to reject H_0. There is not sufficient evidence to warrant rejection of the claim that the two sample groups come from populations with equal standard deviations. Yes.

15. H_0: $\sigma_1^2 = \sigma_2^2$. H_1: $\sigma_1^2 \neq \sigma_2^2$. Test statistic: $F = 2.8176$. Upper critical value of F is between 3.5257 and 3.4296. Fail to reject H_0. There is not sufficient evidence to warrant rejection of the claim that the two samples come from populations with the same variation.

17. The test statistic changes from $F = 1.5824$ to $F = 1.0000$. Fail to reject H_0. There is not sufficient evidence to warrant rejection of the claim that the populations have the same standard deviation. The conclusion changes. The outlier does have a dramatic effect on the results.

19. a. $F_L = 0.2484$, $F_R = 4.0260$

 b. $F_L = 0.2315$, $F_R = 5.5234$

 c. $F_L = 0.1810$, $F_R = 4.3197$

Chapter 8 Review Exercises

1. a. H_0: $p_1 = p_2$. H_1: $p_1 < p_2$. Test statistic: $z = -2.82$. Critical value: $z = -1.645$. P-value: 0.0024. Reject H_0. There is sufficient evidence to support the stated claim. It appears that surgical patients should be routinely warmed.

 b. 90%

 c. $-0.205 < p_1 - p_2 < -0.0543$

 d. No, the conclusions may be different.

2. a. H_0: $\mu_d = 0$. H_1: $\mu_d \neq 0$. Test statistic: $t = -1.532$. Critical values: $t = \pm 2.228$. P-value: 0.164 approximately. Fail to reject H_0. There is not sufficient evidence to warrant rejection of the claim that there is no difference.

 b. $-2.7 < \mu_d < 0.5$

 c. No, there is not a significant difference.

3. a. $-27.80 < \mu_1 - \mu_2 < 271.04$; TI-83 uses df $= 17.7$ to get limits of -17.32 and 260.56.

 b. H_0: $\mu_1 = \mu_2$. H_1: $\mu_1 \neq \mu_2$. Test statistic: $t = 1.841$. Critical values: $t = \pm 2.262$. P-value: 0.106 approximately. Fail to reject H_0. There is not sufficient evidence to warrant rejection of the claim of no difference.

 c. No

4. H_0: $\sigma_1 = \sigma_2$. H_1: $\sigma_1 \neq \sigma_2$. Test statistic: $F = 1.2922$. Upper critical value: $F = 4.0260$. Fail to reject H_0. There is not sufficient evidence to support the claim that the two populations have different amounts of variation.

5. H_0: $\mu_1 = \mu_2$. H_1: $\mu_1 \neq \mu_2$. Test statistic: $t = -3.500$. Critical values: $t = \pm 2.365$. P-value: 0.010. Reject H_0. There is sufficient evidence to warrant rejection of the claim that the means are equal. Filters appear to be effective in reducing carbon monoxide.

6. $H_0: \mu_1 = \mu_2$. $H_1: \mu_1 > \mu_2$. Test statistic: $t = 2.169$. Critical values: $t = \pm 1.968$ approximately. P-value is between 0.01 and 0.025. Reject H_0. There is sufficient evidence to support the claim that zinc supplementation is associated with increased birth weights.

7. $H_0: p_1 = p_2$. $H_1: p_1 > p_2$. Test statistic: $z = 2.41$. Critical value: $z = 1.645$. P-value: 0.0080. Reject H_0. There is sufficient evidence to support the stated claim.

8. a. $H_0: \mu_d = 0$. $H_1: \mu_d \neq 0$. Test statistic: $t = 2.301$. Critical values: $t = \pm 2.262$ (assuming a 0.05 significance level). P-value: 0.046. Reject H_0 (assuming a 0.05 significance level). There is sufficient evidence to conclude that there is a difference between pretraining and posttraining weights.

   b. $0.0 < \mu_d < 4.0$

Chapter 8 Cumulative Review Exercises

1. a. 0.0707
   b. 0.369
   c. 0.104
   d. 0.0540
   e. $H_0: p_1 = p_2$. $H_1: p_1 < p_2$. Test statistic: $z = -2.52$. Critical value: $z = -1.645$. P-value: 0.0059. Reject H_0. There is sufficient evidence to support the claim that the percentage of women ticketed for speeding is less than the percentage for men.

2. There must be an error, because the rates of 13.7% and 10.6% are not possible with sample sizes of 100.

3. a. $0.0254 < p < 0.0536$ (using $x = 29$)
   b. $0.0103 < p < 0.0311$ (using $x = 15$)
   c. $0.00133 < p_1 - p_2 < 0.0363$
   d. method (iii)

4. a. $H_0: p = 0.5$. $H_1: p < 0.5$. Test statistic: $z = -5.88$. Critical value: $z = -1.645$ (assuming a 0.05 significance level). P-value: 0.0001. Reject H_0. There is sufficient evidence to support the claim that the proportion of females is less than 0.5.
   b. $\bar{x} = 17{,}198.3$ sec; median $= 16{,}792$ sec; $s = 3107.2$; distribution is approximately normal; no outliers
   c. $H_0: \mu = 18{,}000$ sec. $H_1: \mu < 18{,}000$ sec. Test statistic: $t = -1.611$. Critical value: $t = -1.686$. P-value: 0.059. Fail to reject H_0. There is not sufficient evidence to support the claim that females have a mean running time less than 5 hours.
   d. $H_0: \mu_1 = \mu_2$. $H_1: \mu_1 \neq \mu_2$. Test statistic: $t = 3.101$. Critical values: $t = \pm 2.024$. P-value: 0.004. Reject H_0. There is sufficient evidence to support the claim that the mean time of males is different from the mean time of females.
   e. Using the sample proportions of 39/150 and 111/150 results in the pooled value of $\bar{p} = 150/300$, which assumes that the total sample size is 300 instead of 150. The sample of 150 values is from one population, not two populations.

Chapter 9 Answers

Section 9-2

1. a. Yes, because the absolute value of the test statistic exceeds the critical values $r = \pm 0.707$.
   b. 0.986

3. a. No, because the absolute value of the test statistic does not exceed the critical values $r = \pm 0.444$ (approximately).
   b. 0.0177

5. The scatterplot suggests that there is a correlation, but it is not linear. With $r = 0$ and critical values of $r = \pm 0.878$ (for a 0.05 significance level), there is not a significant linear correlation.

7. a. There appears to be a linear correlation.
   b. $r = 0.906$. Critical values: $r = \pm 0.632$ (for a 0.05 significance level). There is a significant linear correlation.
   c. $r = 0$. Critical values: $r = \pm 0.666$ (for a 0.05 significance level). There does not appear to be a significant linear correlation.
   d. The effect from a single pair of values can be very substantial, and it can change the conclusion.

9. $r = -0.118$. Critical values: $r = \pm 0.707$. There is not a significant linear correlation. Lowest: Susan Lucci. Highest: Kelsey Grammer.

11. $r = 0.658$. Critical values: $r = \pm 0.532$. There is a significant linear correlation. Another issue is the accuracy of the measurements, which appear to vary widely. A study might be conducted to determine whether the subject's blood pressure really does vary considerably, or whether the measurements are in error because of other factors.

13. $r = 0.262$. Critical values: $r = \pm 0.576$. There is a not significant linear correlation. Because the cigarette counts were reported, subjects may have given wrong values. Subjects may have been exposed to varying levels of secondhand smoke.

15. $r = 0.359$. Critical values: $r = \pm 0.497$. There is not a significant linear correlation.

17. $r = 0.482$. Critical values: $r = \pm 0.312$. There is a significant linear correlation.

19. a. $r = 0.997$. Critical values: $r = \pm 0.279$. There is a significant linear correlation.
    b. $r = 0.899$. Critical values: $r = \pm 0.279$. There is a significant linear correlation. There is a correlation between the tax bill and the value of the house.

21. a. $r = 0.574$. Critical values: $r = \pm 0.361$ approximately. There is a significant linear correlation. No, there can be a high correlation even though the forecast temperatures are very inaccurate.
    b. $r = 0.685$. Critical values: $r = \pm 0.361$ approximately. There is a significant linear correlation. No, there can be a high correlation even though the forecast temperatures are very inaccurate.
    c. The one-day forecast temperatures are better because they have a higher correlation with the actual temperatures. However, a high correlation does not imply that the forecast temperatures are accurate.

23. a. $r = 0.870$. Critical values: $r = \pm 0.279$. There is a significant linear correlation.
   b. $r = -0.010$. Critical values: $r = \pm 0.279$. There is not a significant linear correlation.
   c. Duration, because it has a significant linear correlation with interval.
25. With a linear correlation coefficient very close to 0, there does not appear to be a correlation, but the conclusion suggests that there is a correlation.
27. Although there is no *linear* correlation, the variables may be related in some other *nonlinear* way.
29. $r = 0.819$ (approximately). Critical values: $r = \pm 0.553$. There is a significant linear correlation.
31. a. ± 0.279
   b. ± 0.191
   c. -0.378
   d. 0.549
   e. 0.658
33. $0.386 < \rho < 0.753$

Section 9-3

1. a. 18.00
   b. 5.00
3. 401 lb
5. $\hat{y} = 2 + 0x$ (or $\hat{y} = 2$)
7. a. $\hat{y} = 0.264 + 0.906x$
   b. $\hat{y} = 2 + 0x$ (or $\hat{y} = 2$)
   c. The results are very different, indicating that one point can dramatically affect the regression equation.
9. $\hat{y} = 6.76 - 0.0111x$; 6.5 million. The predicted value of 6.5 million is very far from the actual value of 24 million.
11. $\hat{y} = -14.4 + 0.769x$; 79
13. $\hat{y} = 139 + 2.48x$; 175.2
15. $\hat{y} = 3.68 + 3.78x$; 3.76
17. $\hat{y} = 21.9 + 0.0160x$; 29.9
19. a. $\hat{y} = 7.26 + 0.914x$; $190,060
   b. $\hat{y} = 380 + 19.5x$; $8180
21. a. $\hat{y} = 13.8 + 0.611x$; 31°
   b. $\hat{y} = 13.8 + 0.634x$; 32°
   c. Part (b), because the correlation coefficient is higher.
23. a. $\hat{y} = 41.9 + 0.179x$; 79 min
   b. $\hat{y} = 81.9 - 0.009x$; 81 min
   c. Part (a), because there is a significant linear correlation between durations and intervals, but not between heights and intervals.
25. Yes; no. The point is far away from the others, but it doesn't have a dramatic effect on the regression line.
27. $\hat{y} = -182 + 0.000351x$; $\hat{y} = -182 + 0.351x$. The slope is multiplied by 1000 and the y-intercept doesn't change. If each y entry is divided by 1000, the slope and the y-intercept are both divided by 1000.

29. The equation $\hat{y} = -49.9 + 27.2x$ is better because it has $r = 0.997$, which is higher than $r = 0.963$ for $\hat{y} = -103.2 + 134.9 \ln x$.
31. No.

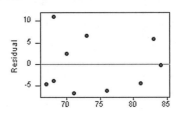

Section 9-4

1. 0.64; 64%
3. 0.253; 25.3%
5. 0.961; yes
7. 1.3
9. a. 287.37026
   b. 166.62974
   c. 454
   d. 0.63297415
   e. 4.8789597
11. a. 3696.9263
   b. 1690.5830
   c. 5387.5093
   d. 0.68620324
   e. 11.869369
13. a. 116 lb
   b. 103.7 lb $< y <$ 128.8 lb
15. a. 44 ft
   b. 6.2 ft $< y <$ 81.4 ft
17. 54.2 $< y <$ 107.0
19. 71.7 $< y <$ 112.2
21. $-170.8 < \beta_0 < -54.6$; $1.5 < \beta_1 < 3.1$
23. a. $(n - 2)s_e^2$
   b. $\dfrac{r^2 \cdot (\text{unexplained variation})}{1 - r^2}$
   c. $r = -0.949$

Section 9-5

1. $\hat{y} = -272 - 0.870x_1 + 0.554x_2 + 12.2\,x_3$
3. Yes, because the P-value is 0.000 and the adjusted R^2 value is 0.924.
5. Highway fuel consumption, because it has the highest adjusted R^2.
7. The regression equation with highway fuel consumption and weight has the highest adjusted R^2 of 0.861, but a good argument can be made for using the single independent variable of highway fuel consumption, because its adjusted R^2 is 0.853, which is only slightly less.

9. a. $\hat{y} = 21.6 + 0.690x$
   b. $\hat{y} = 45.7 + 0.293x$
   c. $\hat{y} = 9.80 + 0.658x_1 + 0.200x_2$ (where x_1 = mother's height)
   d. Part (c), because the adjusted R^2 is highest
   e. No, because the highest value of adjusted R^2 is only 0.366, which isn't very high.
11. a. $\hat{y} = 3.68 + 3.78x$
   b. $\hat{y} = 3.46 + 1.01x$
   c. $\hat{y} = 3.40 + 3.20x_1 + 0.982x_2$ (where x_1 = fat)
   d. Part (c), because the adjusted R^2 is highest
   e. Yes, but the predicted values will not necessarily be very accurate.
13. $\hat{y} = 0.154 + 0.0651x$, where x is the amount of tar. Even though the multiple regression equation with tar and nicotine as independent variables has the highest adjusted R^2 of 0.928, the adjusted R^2 using the single independent variable of tar has an adjusted R^2 of 0.921, which is very close. With values of R^2 that close, it is better to select the equation with one independent variable instead of two.
15. $\hat{y} = 7.26 + 0.914x_1$ (where x_1 represents list price); by using more variables, adjusted R^2 can be raised from 0.995 to 0.996, but the small increase in adjusted R^2 does not justify the inclusion of additional variables.

Section 9-6

1. Quadratic: $y = 2x^2 - 12x + 18$
3. Exponential: $y = 3^x$
5. Quadratic: $y = 0.0516657x^2 + 1.50881x + 18.6857$, where x is coded as 1 for 1980, 2 for 1981, and so on. Predicted value: 77.
7. Quadratic: $y = 1.21445x^2 + 42.4084x + 371.958$, where x is coded as 1 for 1990, 2 for 1991, and so on. Predicted value: 1361.
9. a. Exponential: $y = 2^{\frac{2}{3}(x-1)}$ [or $y = (0.629961)(1.587401)^x$ for an initial value of 1 that doubles every 1.5 years]
   b. Exponential: $y = (2.32040)(1.36587)^x$
   c. Moore's law does appear to be working reasonably well.
11. a. 189.1
    b. 0.9979
    c. The quadratic R^2 value of 0.9992 is higher than $R^2 = 0.9979$ for the logistic model, and the sum of squares of the residuals is lower for the quadratic model (73.2) than for the logistic model (189.1).

Chapter 9 Review Exercises

1. $r = -0.069$. Critical values: $r = \pm 0.707$ (assuming a 0.05 significance level). There is no significant linear correlation. The BAC level does not appear to be related to the age of the person tested.
2. a. $r = 0.828$. Critical values: $r = \pm 0.811$ (assuming a 0.05 significance level). There is a significant linear correlation.
   b. The regression equation of $\hat{y} = -0.347 + 0.149x$ (with x representing the bill) shows that the predicted amount of tip is 35 cents less than 15% of the bill, which is approximately 15% of the bill.

3. a. $r = 0.338$. Critical values: $r = \pm 0.632$; no significant linear correlation.
   b. 11%
   c. $\hat{y} = -0.488 + 0.611x$
   d. 0.347 pints per capita per week
4. a. $r = 0.116$. Critical values: $r = \pm 0.632$; no significant linear correlation.
   b. 1.3%
   c. $\hat{y} = 0.0657 + 0.000792x$
   d. 0.347 pints per capita per week
5. a. $r = 0.777$. Critical values: $r = \pm 0.632$. There is a significant linear correlation.
   b. 60%
   c. $\hat{y} = 0.193 + 0.00293x$
   d. 0.286 pints per capita per week
6. $\hat{y} = -0.0526 + 0.747x_1 - 0.00220x_2 + 0.00303x_3$; $R^2 = 0.726$; adjusted $R^2 = 0.589$; P-value = 0.040. Because the overall P-value of 0.040 is less than 0.05, the equation can be used to predict ice cream consumption. Using the consumption/temperature data, the adjusted R^2 is 0.554. Although the adjusted R^2 is slightly higher using all three variables, the slight increase in adjusted R^2 does not justify the inclusion of additional variables, so the best regression equation appears to result from using temperature as the only independent variable.

Chapter 9 Cumulative Review Exercises

1. a. $r = -0.884$. Critical values: $r = \pm 0.576$ (assuming a 0.05 significance level). There is a significant linear correlation.
   b. $\hat{y} = 95.3 - 3.46x$
   c. It is possible to conduct the test of two equal population means, but the test would make no sense because the two variables measure readability using different criteria and different scales.
   d. $61.16 < \mu < 71.14$
2. a. $\bar{x} = 99.1$, $s = 8.5$
   b. $\bar{x} = 102.8$, $s = 8.7$
   c. No, but a better comparison would involve treating the data as matched pairs instead of two independent samples.
   d. $H_0: \mu = 100$. $H_1: \mu \neq 100$. Test statistic: $t = 0.546$. Critical values: $t = \pm 2.069$. P-value is greater than 0.20. Fail to reject H_0. There is not sufficient evidence to support the claim that mean IQ score of twins reared apart is different from the mean IQ of 100.
   e. Yes. $r = 0.702$ and the critical values are $r = \pm 0.576$ (assuming a 0.05 significance level). There is a significant linear correlation.

Chapter 10 Answers

Section 10-2

1. a. $H_0: p_1 = p_2 = p_3 = p_4$
   b. 8, 8, 8, 8
   c. $\chi^2 = 4.750$
   d. $\chi^2 = 7.815$
   e. There is not sufficient evidence to warrant rejection of the claim that the four categories are equally likely.

3. a. df $= 37$, so $\chi^2 = 51.805$ (approximately).

   b. $0.10 <$ P-value < 0.90

   c. There is not sufficient evidence to warrant rejection of the claim that the roulette slots are equally likely.

5. Test statistic: $\chi^2 = 5.860$. Critical value: $\chi^2 = 11.071$. There is not sufficient evidence to support the claim that the outcomes are not equally likely. The loaded die does not appear to behave differently from a fair die.

7. Test statistic: $\chi^2 = 9.233$. Critical value: $\chi^2 = 12.592$. There is not sufficient evidence to warrant rejection of the claim that accidents occur with equal frequency on the different days.

9. Test statistic: $\chi^2 = 10.653$. Critical value: $\chi^2 = 9.488$ (assuming a 0.05 significance level). There is sufficient evidence to warrant rejection of the claim that accidents occur with equal proportions on the five workdays.

11. Test statistic: $\chi^2 = 16.333$. Critical value: $\chi^2 = 14.067$ (assuming a 0.05 significance level). There is sufficient evidence to support the claim that the probabilities of winning in the different post positions are not all the same.

13. Test statistic: $\chi^2 = 18.500$. Critical value: $\chi^2 = 16.919$. There is sufficient evidence to warrant rejection of the claim that the digits occur with the same frequency. The conclusion changes with a 0.01 significance level. The selection process would have to be changed immediately if there is strong evidence suggesting that the digits are not equally likely.

15. Test statistic: $\chi^2 = 5.950$. Critical value: $\chi^2 = 11.071$. There is not sufficient evidence to warrant rejection of the distribution claimed by Mars, Inc.

17. Test statistic: $\chi^2 = 4.200$. Critical value: $\chi^2 = 16.919$. There is not sufficient evidence to warrant rejection of the claim that the digits are uniformly distributed.

19. Test statistic: $\chi^2 = 14.421$. Critical value: $\chi^2 = 15.507$. There is not sufficient evidence to warrant rejection of the claim that the digits come from a population of leading digits that conform to Benford's law.

21. The test statistic changes from 4.600 to 76.638, so the outlier has a dramatic effect.

23. a. Critical value is $\chi^2 = 3.841$ and test statistic is

$$\chi^2 = \frac{\left(f_1 - \dfrac{f_1 + f_2}{2}\right)^2}{\dfrac{f_1 + f_2}{2}} + \frac{\left(f_2 - \dfrac{f_1 + f_2}{2}\right)^2}{\dfrac{f_1 + f_2}{2}}$$

$$= \frac{(f_1 - f_2)^2}{f_1 + f_2}$$

   b. Critical values: The χ^2 critical value is 3.841 and it is approximately equal to the square of $z = 1.96$.

25. a. 0.0853, 0.2968, 0.3759, 0.1567, 0.0853

   b. 17.06, 59.36, 75.18, 31.34, 17.06

   c. Test statistic: $\chi^2 = 60.154$. Critical value: $\chi^2 = 13.277$. Reject H_0: The IQ scores come from a normally distributed population with the given mean and standard deviation. There is sufficient evidence to warrant rejection of the claim that the IQ scores were randomly selected from a normally distributed population with mean 100 and standard deviation 15.

Section 10-3

1. Test statistic: $\chi^2 = 0.413$. P-value: 0.521. There is not sufficient evidence to warrant rejection of the claim that race and ethnicity are independent of whether someone is stopped by police. There is not sufficient evidence to support a claim of racial profiling.

3. Test statistic: $\chi^2 = 4.698$. Critical value: $\chi^2 = 3.841$. There is sufficient evidence to warrant rejection of independence between response and whether the subject is a worker or senior-level boss. The conclusion changes if the significance level of 0.01 is used.

5. Test statistic: $\chi^2 = 51.458$. Critical value: $\chi^2 = 6.635$. There is sufficient evidence to warrant rejection of the claim that the proportions of agree/disagree responses are the same for the subjects interviewed by men and the subjects interviewed by women.

7. Test statistic: $\chi^2 = 63.908$. Critical value: $\chi^2 = 3.841$. There is sufficient evidence to warrant rejection of the claim that gender is independent of the fear of flying.

9. Test statistic: $\chi^2 = 3.062$. Critical value: $\chi^2 = 5.991$. There is not sufficient evidence to warrant rejection of the claim that success is independent of the method used. The evidence does not suggest that any method is significantly better than the others.

11. Test statistic: $\chi^2 = 65.524$. Critical value: $\chi^2 = 7.815$ (assuming a 0.05 significance level). There is sufficient evidence to warrant rejection of the claim that occupation is independent of whether the cause of death was homicide. Cashiers appear to be most vulnerable to homicide.

13. Test statistic: $\chi^2 = 20.271$. Critical value: $\chi^2 = 15.086$. There is sufficient evidence to warrant rejection of the claim that cooperation of the subject is independent of the age category.

15. Test statistic: $\chi^2 = 119.330$. Critical value: $\chi^2 = 5.991$. There is sufficient evidence to warrant rejection of the claim that the type of crime is independent of whether the criminal is a stranger.

17. Test statistic: $\chi^2 = 42.557$. Critical value: $\chi^2 = 3.841$. There is sufficient evidence to warrant rejection of the claim that the sentence is independent of the plea. The results encourage pleas for guilty defendants.

19. Test statistic: $\chi^2 = 1.199$. Critical value: $\chi^2 = 7.815$. There is not sufficient evidence to warrant rejection of the claim that getting a headache is independent of the amount of atorvastatin used as a treatment.

21. Without Yates' correction: $\chi^2 = 0.413$. With Yates' correction: $\chi^2 = 0.270$. Yates' correction decreases the test statistic so that sample data must be more extreme to be considered significant.

Chapter 10 Review Exercises

1. $\chi^2 = 16.747$. Critical value: $\chi^2 = 9.488$. There is sufficient evidence to warrant rejection of the claim that calls are uniformly distributed over the days of the business week.
2. $\chi^2 = 6.780$. Critical value: $\chi^2 = 12.592$. There is not sufficient evidence to support the theory that more gunfire deaths occur on weekends.
3. $\chi^2 = 49.731$. Critical value: $\chi^2 = 11.071$ (assuming a 0.05 significance level). There is sufficient evidence to support the claim that the type of crime is related to whether the criminal drinks or abstains.
4. $\chi^2 = 5.297$. Critical value: $\chi^2 = 3.841$. There is sufficient evidence to warrant rejection of the claim that whether a newborn is discharged early or late is independent of whether the newborn was re-hospitalized within a week of discharge. The conclusion changes if the significance level is changed to 0.01.

Chapter 10 Cumulative Review Exercises

1. $\bar{x} = 80.9$; median: 81.0; range: 28.0; $s^2 = 75.4$; $s = 8.6$; 5-number summary: 66, 76.0, 81.0, 86.5, 94.
2. a. 0.272
   b. 0.468
   c. 0.614
   d. 0.282
3. Contingency table; see Section 10-3. Test statistic: $\chi^2 = 0.055$. Critical value: $\chi^2 = 7.815$ (assuming a 0.05 significance level). There is not sufficient evidence to warrant rejection of the claim that men and women choose the different answers in the same proportions.
4. Use correlation; see Section 9-2. Test statistic: $r = 0.978$. Critical values: $r = \pm 0.950$ (assuming a 0.05 significance level). There is sufficient evidence to support the claim that there is relationship between the memory and reasoning scores.
5. Use the test for matched pairs; see Section 8-4. $\bar{d} = -10.25$; $s_d = 1.5$. Test statistic: $t = -13.667$. Critical value: $t = -2.353$ (assuming a 0.05 significance level). Reject H_0. There is sufficient evidence to support the claim that the training session is effective in raising scores.
6. Test for the difference between two independent samples; see Section 8-3. Test statistic: $t = -2.014$. Critical values: $t = \pm 3.182$ (assuming a 0.05 significance level). Fail to reject H_0: $\mu_1 = \mu_2$. There is not sufficient evidence to warrant rejection of the claim that men and women have the same mean score.

Chapter 11 Answers

Section 11-2

1. a. $\mu_1 = \mu_2 = \mu_3$
   b. At least one of the three means is different from the others.
   c. $F = 8.98$
   d. $F = 3.3158$ approximately
   e. 0.001

   f. There is sufficient evidence to warrant rejection of the claim that the three authors have the same mean Flesch-Kincaid Grade Level score.
3. a. $\mu_1 = \mu_2 = \mu_3$
   b. At least one of the three means is different from the others.
   c. $F = 0.1887$
   d. 3.0804
   e. 0.8283
   f. No
5. Test statistic: $F = 0.9922$. Critical value: $F = 3.2389$. P-value: 0.4216. Fail to reject H_0: $\mu_1 = \mu_2 = \mu_3 = \mu_4$. There is not sufficient evidence to support the claim that larger cars are safer.
7. Test statistic: $F = 4.0497$. Critical value: $F = 3.4028$. P-value: 0.0305. Reject H_0: $\mu_1 = \mu_2 = \mu_3$. There is sufficient evidence to support the claim that the mean breadth is not the same for the different epochs.
9. Test statistic: $F = 0.5083$. Critical value: $F = 2.2899$ (approximately). P-value: 0.7694. Fail to reject H_0: $\mu_1 = \mu_2 = \mu_3 = \mu_4 = \mu_5 = \mu_6$. There is not sufficient evidence to warrant rejection of the claim that the populations of different colors of M&Ms have the same mean.
11. Test statistic: $F = 9.0646$. Critical value: $F = 3.8056$. P-value: 0.0034. Reject H_0: $\mu_1 = \mu_2 = \mu_3$. There is sufficient evidence to warrant rejection of the claim that the mean sugar amounts on the different shelves are all the same. Shelf 2 appears to have a considerably higher mean, which would support the claim that high-sugar cereals are placed on shelves that are at eye-level for children.
13. a. 10
    b. 0.599
    c. 0.95
    d. Analysis of variance

Section 11-3

1. "Two-way" refers to the inclusion of two different factors, which are properties or characteristics used to distinguish different populations from one another. "Analysis of variance" refers to the method used, which is based on two different estimates of the assumed common population variance.
3. If there is an interaction between factors, we should not consider the effects of either factor without considering those of the other.
5. Test statistic: $F = 1.28$. P-value: 0.313. Fail to reject the null hypothesis of no interaction. There does not appear to be a significant effect from the interaction between site and age.
7. Test statistic: $F = 249.85$. P-value: 0.000. Reject the null hypothesis that age has no effect on the amount of DDT. There is sufficient evidence to support the claim that age has an effect on the amount of DDT.
9. Test statistic: $F = 5.03$. P-value: 0.031. Reject the null hypothesis that gender has no effect on SAT scores. There is sufficient evidence to support the claim that gender has an effect on SAT scores.

11. Test statistic: $F = 3.87$. P-value: 0.000. Reject the null hypothesis that the choice of subject has no effect on the hearing test score. There is sufficient evidence to support the claim that the choice of subject has an effect on the hearing test score.

13. For interaction, the test statistic is $F = 0.36$ and the P-value is 0.701, so there is no significant interaction effect. For gender, the test statistic is $F = 0.09$ and the P-value is 0.762, so there is no significant effect from gender. For age, the test statistic is $F = 0.36$ and the P-value is 0.701, so there is no significant effect from age.

15. a. The test statistics, critical values, P-values, and conclusions do not change.
    b. The test statistics, critical values, P-values, and conclusions do not change.
    c. The test statistics, critical values, P-values, and conclusions do not change.
    d. An outlier can dramatically affect and change all of the results and conclusions.

Chapter 11 Review Exercises

1. Test statistic: $F = 46.90$. P-value: 0.000. Reject H_0: $\mu_1 = \mu_2 = \mu_3$. There is sufficient evidence to warrant rejection of the claim of equal population means.

2. Test statistic: $F = 9.4827$. Critical value: $F = 3.0984$. Reject H_0: $\mu_1 = \mu_2 = \mu_3 = \mu_4$. There is sufficient evidence to support the claim of different mean selling prices.

3. Test statistic: $F = 0.19$. P-value: 0.832. Fail to reject the null hypothesis of no interaction. There does not appear to be a significant effect from the interaction between gender and major.

4. Test statistic: $F = 0.78$. P-value: 0.395. Fail to reject the null hypothesis that gender has no effect on SAT scores. There is not sufficient evidence to support the claim that estimated length is affected by gender.

5. Test statistic: $F = 0.13$. P-value: 0.876. Fail to reject the null hypothesis that major has no effect on SAT scores. There is not sufficient evidence to support the claim that estimated length is affected by major.

6. a. Test statistic: $F = 1.00$. P-value: 0.423. There is not sufficient evidence to support the claim that amounts of emitted greenhouse gases are affected by the type of transmission.
   b. Test statistic: $F = 7.00$. P-value: 0.125. There is not sufficient evidence to support the claim that amounts of greenhouse gases are affected by the number of cylinders.
   c. Perhaps greenhouse gases *are* affected by the type of transmission and/or the number of cylinders; however, the given sample data do not provide sufficient evidence to support such claims.

Chapter 11 Cumulative Review Exercises

1. a. 0.100 in.
   b. 0.263 in.
   c. 0.00, 0.00, 0.00, 0.010, 1.41
   d. 0.92 in., 1.41 in.

e. Answer varies, depending on the number of classes used, but the histogram should depict a distribution that is skewed to the right.
   f. No, because the data do not appear to come from a normally distributed population.
   g. 19/52 or 0.365

2. a. 960.5, 980.0, 1046.0; no
   b. 914.5, 1010.5, 1008.5; no
   c. 174.6, 239.6, 226.8; no
   d. Test statistic: $t = -0.294$. Critical values: $t = \pm 2.093$ (assuming a 0.05 significance level). Fail to reject H_0: $\mu_1 = \mu_2$.
   e. $878.8 < \mu < 1042.2$
   f. Test statistic: $F = 0.8647$. P-value: 0.4266. Fail to reject H_0: $\mu_1 = \mu_2 = \mu_3$. There is not sufficient evidence to warrant rejection of the claim that the three populations have the same mean SAT score.

3. a. 0.3372
   b. 0.0455
   c. 1/8

Chapter 12 Answers

Section 12-2

1. The test statistic of $x = 5$ is not less than or equal to the critical value of 3. There is not sufficient evidence to warrant rejection of the claim of no difference.

3. The test statistic of $z = -0.95$ is not less than or equal to the critical value of -1.96. There is not sufficient evidence to warrant rejection of the claim of no difference.

5. The test statistic of $x = 5$ is not less than or equal to the critical value of 2. There is not sufficient evidence to support the claim that there is a difference between self-reported heights and measured heights.

7. The test statistic of $x = 1$ is less than or equal to the critical value of 2. There is sufficient evidence to support the claim that the population has a median less than 98.6°F.

9. Convert $x = 301$ to the test statistic $z = -12.60$. Critical value: $z = -1.645$ (assuming a 0.05 significance level). There is sufficient evidence to support the claim that the majority of people say that they voted in the election.

11. Convert $x = 1$ to the test statistic $z = -5.32$. Critical value: $z = -1.645$ (assuming a 0.05 significance level). There is sufficient evidence to support the claim that the Coke cans have volumes with a median greater than 12 oz.

13. (Instead of a right-tailed test to determine whether $x = 37$ is large enough to be significant, use a left-tailed test to determine whether $x = 13$ is small enough to be significant.) Convert $x = 13$ to the test statistic $z = -3.25$. Critical value: $z = -1.645$ (assuming a 0.05 significance level). There is sufficient evidence to support the claim that the median is greater than 77 min.

15. First approach: $z = -1.90$; reject H_0.
    Second approach: $z = -1.73$; reject H_0.
    Third approach: $z = 0$; fail to reject H_0.

17. Convert $x = 18$ to the test statistic $z = -2.31$. Critical value: $z = -2.33$. There is not sufficient evidence to support a charge of gender bias. If the binomial distribution is used instead of the normal approximation, the P-value is 0.0099, which is less than 0.01, so there is sufficient evidence to support a charge of gender bias. Using the normal approximation, the test statistic is just barely outside of the critical region; using the binomial distribution, the test statistic is just barely inside the critical region.

Section 12-3

1. Test statistic: $T = 1$. Critical value: $T = 2$. Reject the null hypothesis that both samples come from the same population distribution.
3. Test statistic: $T = 34$. Critical value: $T = 14$. Fail to reject the null hypothesis that both samples come from the same population distribution.
5. Test statistic: $T = 0$. Critical value: $T = 8$. Reject the null hypothesis that both samples come from the same population distribution. There is sufficient evidence to warrant rejection of the claim that there is no difference.
7. Test statistic: $T = 178$. Critical value: $T = 117$ (assuming a 0.05 significance level). Fail to reject the null hypothesis that both samples come from the same population distribution. There does not appear to be a difference.
9. Convert $T = 661$ to test statistic $z = -5.67$. Critical values: $z = \pm 1.96$. There is sufficient evidence to warrant rejection of the claim that healthy adults have a mean body temperature that is equal to 98.6°F.

Section 12-4

1. $R_1 = 120.5$, $R_2 = 155.5$, $\mu_R = 132$, $\sigma_R = 16.248$. Test statistic: $z = -0.71$. Critical values: $z = \pm 1.96$. Fail to reject the null hypothesis that the populations have the same distribution.
3. $\mu_R = 150$, $\sigma_R = 17.321$, $R = 96.5$, $z = -3.09$. Test statistic: $z = -3.09$. Critical values: $z = \pm 2.575$. There is sufficient evidence to warrant rejection of the claim that the two samples come from populations with identical populations.
5. $\mu_R = 525$, $\sigma_R = 37.417$, $R = 437$, $z = -2.35$. Test statistic: $z = -2.35$. Critical values: $z = \pm 1.96$. There is sufficient evidence to warrant rejection of the claim that the two samples come from identical populations.
7. $\mu_R = 150$, $\sigma_R = 17.321$, $R = 86.5$, $z = -3.67$. Test statistic: $z = -3.67$. Critical values: $z = \pm 1.96$. Reject the null hypothesis that the Rowling and Tolstoy samples are from populations with the same distribution.
9. $\mu_R = 3696$, $\sigma_R = 214.94$, $R = 3861$, $z = 0.77$. Test statistic: $z = 0.77$. Critical values: $z = \pm 1.96$. Fail to reject the null hypothesis that the two populations of ages have the same distribution.
11. $z = -3.67$; the test statistic is the same number with opposite sign.

Section 12-5

1. Yes. The P-value of 0.747 indicates that we fail to reject the null hypothesis that the three age categories have identical populations.
3. Test statistic: $H = 1.1914$. Critical value: $\chi^2 = 7.815$. There is not sufficient evidence to support the claim that the head injuries for the four weight categories are not all the same. The given data do not provide sufficient evidence to conclude that heavier cars are safer in a crash.
5. Test statistic: $H = 6.631$. Critical value: $\chi^2 = 5.991$. There is sufficient evidence to warrant rejection of the claim that the three samples come from identical populations.
7. Test statistic: $H = 2.075$. Critical value: $\chi^2 = 11.071$. There is not sufficient evidence to warrant rejection of the claim that the weights are the same for each of the six different color populations.
9. a. The test statistic H does not change.
   b. The test statistic H does not change.
   c. The value of the test statistic does not change much (because the rank is used instead of the magnitude of the outlier).
11. 14.840 (using $T = 6, 6, 24$); no

Section 12-6

1. a. $r_s = 1$ and there appears to be a correlation between x and y.
   b. $r_s = -1$ and there appears to be a correlation between x and y.
   c. $r_s = 0$ and there does not appear to be a correlation between x and y.
3. $r_s = 0.855$. Critical values: $r_s = \pm 0.648$. Significant correlation. There appears to be a correlation between salary and stress.
5. $r_s = 0.103$. Critical values: $r_s = \pm 0.648$. No significant correlation. There does not appear to be a correlation between corporate and graduate rankings of business schools.
7. $r_s = 0.557$. Critical values: $r_s = \pm 0.683$. No significant correlation. There does not appear to be a correlation between height and weight.
9. $r_s = 0.506$. Critical values: $r_s = \pm 0.507$. No significant correlation. There does not appear to be a correlation between the amounts of fat and the calorie counts.
11. a. $r_s = 0.918$. Critical values: $r_s = \pm 0.370$. Significant correlation. There appears to be a correlation between tar and nicotine.
    b. $r_s = 0.739$. Critical values: $r_s = \pm 0.370$. Significant correlation. There appears to be a correlation between carbon monoxide and nicotine.
    c. Tar, because it has a higher correlation with nicotine.
13. a. ± 0.707
    b. ± 0.514
    c. ± 0.361
    d. ± 0.463
    e. ± 0.834

Section 12-7

1. $n_1 = 11$, $n_2 = 7$, $G = 3$, critical values: 5, 14; reject randomness.
3. $n_1 = 10$, $n_2 = 10$, $G = 10$, critical values: 6, 16; fail to reject randomness.
5. $n_1 = 12$, $n_2 = 8$, $G = 10$, critical values: 6, 16; fail to reject randomness.
7. $n_1 = 19$, $n_2 = 13$, $G = 6$, critical values: 10, 23; reject randomness.
9. $n_1 = 18$, $n_2 = 14$, $G = 15$, critical values: 10, 23; fail to reject randomness.
11. $n_1 = 10$, $n_2 = 10$, $G = 2$, critical values: 6, 16; reject randomness. Because the trend is upward, the stock market appears to be a good medium for investments.
13. $n_1 = 49$, $n_2 = 51$, $G = 43$, $\mu_G = 50.98$, $\sigma_G = 4.9727$. Test statistic: $z = -1.60$. Critical values: $z = \pm 1.96$. Fail to reject randomness.
15. $n_1 = 111$, $n_2 = 39$, $G = 54$, $\mu_G = 58.720$, $\sigma_G = 4.6875$. Test statistic: $z = -1.01$. Critical values: $z = \pm 1.96$. Fail to reject randomness. There is not sufficient evidence to support the claim that male runners tend to finish before the female runners.
17. Minimum is 2, maximum is 4. Critical values of 1 and 6 can never be realized so that the null hypothesis of randomness can never be rejected.

Chapter 12 Review Exercises

1. The test statistic $x = 2$ is not less than or equal to the critical value of 1. There is not sufficient evidence to warrant rejection of the claim that the course has no effect.
2. Test statistic: $T = 9.5$. Critical value: $T = 6$. There is not sufficient evidence to warrant rejection of the claim that the course has no effect.
3. Sign test: Convert $x = 22$ to the test statistic $z = -2.58$. Critical value: $z = -2.33$. There is sufficient evidence to support the claim of bias in favor of men.
4. Wilcoxon rank-sum test: $\mu_R = 162$, $\sigma_R = 19.442$, $R = 89.5$, $z = -3.73$. Test statistic: $z = -3.73$. Critical values: $z = \pm 1.96$. Reject the null hypothesis that the two samples come from identical populations. There is sufficient evidence to warrant rejection of the claim that beer drinkers and liquor drinkers have the same BAC levels.
5. Rank correlation: $r_s = -0.796$. Critical values: $r_s = \pm 0.648$. Significant correlation. There appears to be a correlation between weight and highway fuel consumption.
6. Runs test: $n_1 = 22$, $n_2 = 18$, $G = 18$, $\mu_G = 20.8$, $\sigma_G = 3.0894$. Test statistic: $z = -0.91$. Critical values: $z = \pm 1.96$. Fail to reject randomness. Odd and even digits appear to occur randomly.
7. Kruskal-Wallis test: Test statistic: $H = 4.234$. Critical value: $\chi^2 = 7.815$. There is not sufficient evidence to support the claim that the injury measurements are not the same for the four categories. There is not sufficient evidence to support the claim that heavier cars are safer.
8. Rank correlation: $r_s = 0.190$. Critical values: $r_s = \pm 0.738$. There is not sufficient evidence to support the claim of a correlation between performance and price. Buy the least expensive tapes.

Chapter 12 Cumulative Review Exercises

1. a. $n_1 = 20$, $n_2 = 7$, $G = 15$, critical values: 6, 16; fail to reject randomness.
   b. Test statistic: $z = -2.50$. Critical values: $z = \pm 1.96$. Reject the null hypothesis that the proportion of women equals 0.5.
   c. Convert $x = 7$ to the test statistic $z = -2.31$. Critical values: $z = \pm 1.96$. There is sufficient evidence to support the claim that the proportion of women is different from 0.5.
   d. $0.0940 < p < 0.425$
   e. The sequence appears to be in a random order, but the subjects appear to be biased against women. Further research should be conducted to determine whether the population has a proportion of women that is less than 0.5.
2. a. Test statistic: $r = -0.515$. Critical values: $r = \pm 0.707$. No significant linear correlation.
   b. Test statistic: $r_s = -0.463$. Critical values: $r_s = \pm 0.738$. No significant correlation.
   c. The test statistic $x = 3$ is not less than or equal to the critical value of 0. There is not sufficient evidence to support the claim that there is a difference between the heights of the winning and losing candidates.
   d. Test statistic: $T = 10$. Critical value: $T = 2$. There is not sufficient evidence to support the claim that there is a difference between the heights of the winning and losing candidates.
   e. Test statistic: $t = 0.851$. Critical values: $t = \pm 2.365$. Fail to reject H_0: $\mu_1 = \mu_2$. There is not sufficient evidence to support the claim that there is a difference between the heights of the winning and losing candidates.
   f. There is not sufficient evidence to conclude that the heights of the winning candidates and the heights of the losing candidates are related, and there is not sufficient evidence to conclude that there is a difference between the heights of the winning candidates and the heights of the losing candidates.

Chapter 13

Section 13-2

1. a. Process data are data arranged according to some time sequence.
   b. A process is out of statistical control if it has variation other than natural variation, and it has patterns, cycles, or unusual points.
   c. There is a pattern, trend, or cycle that is obviously not random, or there is a point lying beyond the upper or lower control limits, or there are eight consecutive points all above or all below the centerline.
   d. Random variation is due to chance, but assignable variation results from causes that can be identified.
   e. An R chart shows the pattern of sample ranges and is used to determine whether variation is within statistical control, whereas an $\bar{x}$ chart shows the pattern of sample means and is used to determine whether the mean in a process is within statistical control.

3. The process variation appears to be within statistical control.

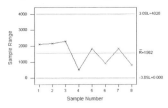

5. The process variation appears to be within statistical control.

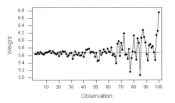

7. There is a pattern of increasing variation, so the process is out of statistical control. The increasing variation will result in more and more defects.

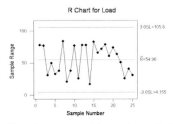

9. There is a pattern of increasing variation, there are points lying beyond the upper control limit, and there are eight consecutive points all below the centerline, so the process mean is out of statistical control. This process does need corrective action.

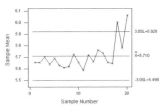

11. The process variation appears to be out of statistical control. There are points that lie beyond the control limits.

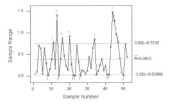

13. The process variation appears to be out of statistical control. There is a point beyond the upper control limit and there appears to be an upward trend.

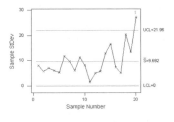

Section 13-3

1. Process appears to be within statistical control.
3. Process appears to be out of statistical control because there is a pattern of an upward trend and there is a point that lies beyond the upper control limit.
5. The process is out of statistical control because there is a downward trend and there are eight (or more) consecutive points all lying below the centerline. This downward trend is good and its causes should be identified so that it can continue.

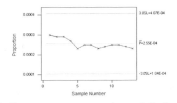

7. The process appears to be statistically stable.

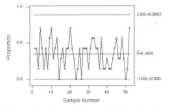

9. Except for the scale used, the charts are identical.

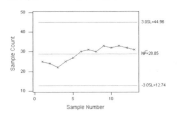

Chapter 13 Review Exercises

1. The process appears to be within statistical control.

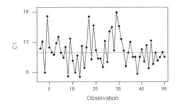

2. The process variation appears to be within statistical control.

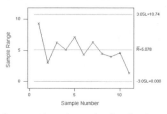

3. Because there is a point that lies beyond the upper control limit, the process mean is not within statistical control.

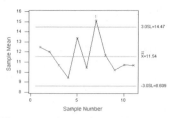

4. The process is out of control because there is a shift up and there are points beyond the control limits.

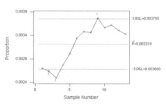

5. The process is out of control because there are points beyond the control limits. Also, there is a cyclical pattern.

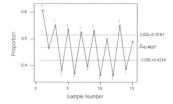

Chapter 13 Cumulative Review Exercises

1. a. The process appears to be within statistical control.

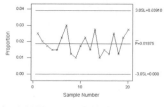

   b. $0.0158 < p < 0.0217$
   c. Test statistic: $z = 7.87$. Critical value: $z = 1.645$. There is suffi-cient evidence to support the claim that the rate of defects is more than 1%.

2. a. 1/256
   b. 1/256
   c. 1/128

3. The run chart reveals very clear cycles, so that the process is not sta-tistically stable.

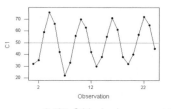

4. a. $r = -0.484$. Critical values: $r = \pm 0.396$ (approximately, assum-ing a 0.05 significance level). There is sufficient evidence to support the claim of a significant linear correlation between tem-perature and energy consumption.
   b. $\hat{y} = 4278 - 23.9x$
   c. 2844 kilowatt-hours

Credits

Photographs

Chapter Openers

Chapter 1: Students Reading © Getty

Chapter 2: Person Smoking in Elevator © Getty

Chapter 3: Pregnancy Test © PictureQuest

Chapter 4: Couple on Beach with Child © Corbis

Chapter 5: Gondola on Snowy Mountain © Corbis

Chapter 6: Red Light Enforcement Camera © AP/Wide World Photos

Chapter 7: Busy Intersection at Night © Macduff Everton/Corbis

Chapter 8: People at Airport © AFP/Corbis

Chapter 9: Manatees © Douglas Faulkner/Corbis

Chapter 10: Volumes of Tax Revenue Code © PhotoEdit, Inc.

Chapter 11: Harry Potter Books © AP/Wide World Photos

Chapter 12: Crowd of People with Umbrellas © PhotoDisc

Chapter 13: Airplane Instrument Panel © Chris Sorensen/Corbis

The following margin photos are all copyright PhotoDisc, Inc.

Chapter 1

p. 5, The State of Statistics

p. 6, Interesting Statistics

p. 13, Literary Digest Poll

p. 17, Sampling Rejected for the Census

Chapter 2

p. 61, Class Size Paradox

p. 66, Not at Home

p. 75, More Stocks, Less Risk

Chapter 3

p. 126, How Probable?

p. 127, You Bet

p. 150, Composite Sampling

p. 153, Bayes' Theorem

p. 159, The Random Secretary

p. 163, How Many Shuffles?

p. 165, Making Cents of the Lottery

p. 166, The Number Crunch

Chapter 4

p. 188, Is Parachuting Safe?

p. 190, Picking Lottery Numbers

p. 198, Prophets for Profits

p. 218, Probability of an Event That Has Never Occurred

Chapter 5

p. 229, Reliability and Validity

p. 273, Boys and Girls Are Not Equally Likely

Chapter 6

p. 300, Small Sample

p. 320, Estimating Wildlife Population Sizes

Chapter 7

p. 371, Lie Detectors

p. 374, Large Sample Size Isn't Good Enough

p. 393, Polls and Psychologists

Chapter 8

p. 443, Does Aspirin Help Prevent Heart Attacks?

p. 445, The Lead Margin of Error

p. 457, Commercials

p. 458, Using Statistics to Identify Thieves

p. 468, Research in Twins

p. 480, Lower Variation; Higher Quality

Chapter 9

p. 498, Student Ratings of Teachers

p. 520, Cell Phones and Crashes

p. 544, Making Music with Multiple Regression

p. 546, Predictors for Success

Chapter 10

p. 586, Home Field Advantage

Chapter 12

p. 659, Gender Gap in Drug Testing

Misc. photos

Chapter 1

p. 8, Gambling for Science; Copyright Henk Binnendijk

p. 12, Should You Believe a Statistical Study?; Copyright Jeff Greenberg/Photo Edit

p. 12, Statistics and Land Mines; Copyright Corbis/Nevada Weir

Chapter 2

p. 40, Growth Charts Updated; Copyright Kelly/Mooney Photography/Corbis

p. 77, Where Are the 0.400 Hitters; Copyright AP/Wide World Photos

p. 103, An Outlier Tip; Copyright Reuters NewMedia, Inc./Corbis

Chapter 3

p. 122, Most Common Birthday; October 5; Copyright Dora Kingsley

p. 141, Perfect SAT Scores; Taxi/Getty Images

Chapter 4

p. 208, Do Boys or Girls Run in the Family; Copyright Anthony Redpath/Corbis

Chapter 5

p. 227, Changing Populations; Copyright Taxi/Getty Images

p. 241, Clinical Trial Cut Short; Copyright Seits/Photo Researchers, Inc.

Chapter 7

p. 389, Ethics in Reporting; Copyright Corbis/Robert Maas

p. 394, Test of Touch Therapy; Copyright Jon Feingersh/Corbis

Chapter 8

p. 444, Author as a Witness; Copyright Bill Wisser/Getty Images

p. 455 Expensive Diet Pill; Copyright Lance Manning/Corbis

p. 542, NBA Salaries and Performance; Copyright Marc Asnin/Corbis Saba

Chapter 12

p. 681, Sports Hot Streaks; Copyright 2002 NBAE Photo by Noren Trotman/NBAE/Getty Images

Chapter 13

p. 702, Costly Assignable Variation; Copyright NASA

p. 703, Don't Tamper! Copyright Davis Lees/Corbis

p. 705, Bribery Detected with Control Charts; Copyright Owen Franken/Corbis

Illustrations

The following margin illustrations were rendered by Bob Giuliani.

Chapter 1
p. 16, Detecting Phony Data
p. 22, Hawthorne and Experimenter Effects

Chapter 2
p. 42, Authors Identified
p. 52, Florence Nightingale
p. 60, Six Degrees of Separation
p. 95, Cost of Laughing Index

Chapter 3
p. 128, Probabilities That Challenge Intuition
p. 136, Shakespeare's Vocabulary
p. 157, Monkey Typist

Chapter 4
p. 201, Voltaire Beats the Lottery
p. 202, Sensitive Surveys

Chapter 6
p. 319, Captured Tank Serial Numbers Reveal Population Size
p. 331, Estimating Sugar in Oranges
p. 332, Excerpts from a Department of Transportation Circular

Chapter 7
p. 409, Death Penalty as a Deterrent
p. 422, Ethics in Experiments

Chapter 8
p. 469, Crest and Dependent Samples

Chapter 9
p. 522, Pizza Correlates with Crisis

Chapter 10
p. 583, Delaying Death

Chapter 11
p. 607, Poll Resistance

Chapter 13
p. 699, The Flynn Effect
p. 711, Six Sigma in Industry

The following margin illustrations were rendered by James Bryant.

Chapter 3
p. 140, Convicted by Probability
p. 144, Independent Jet Engines
p. 151, Coincidences?
p. 164, Choosing Personal Security Codes

Chapter 5
p. 243, Survey Medium Affects Results
p. 260, The Fuzzy Central Limit Theorem

Chapter 6
p. 302, Push Polling

Chapter 9
p. 501, Palm Reading

Chapter 12
p. 644, Class Attendance and Grades

Chapter 13
p. 710, Quality Control at Perstop

The following illustrations were rendered by Darwen Hennings.

Chapter 5
p. 242, Queues

Chapter 7
p. 401, Statistics: Jobs and Employers
p. 408, Better Results with Small Class Size

Chapter 8
p. 456, The Placebo Effect

USA Today Snapshot
p. 491, Just plane scared. Copyright 1997, USA TODAY. Reprinted with permission.

Tufte Illustration
p. 53, Losses of Soldiers in Napoleon's Army During the Russian Campaign (1812-1813): Edward R. Tufte, *The Visual Display of Quantitative Information* (Cheshire, CT: Graphics Press, 1983). Reprinted with permission.

Index

A

Accuracy, 16
Addition rule, 132–139, 145
Adjusted coefficient of determination, 544
Air bag study, 470
Aircraft altimeter production problem, 695, 697–698, 703, 705, 706
Airline weight problem, 273–275
Airplane seat problem, 243–244, 571
Alternative hypothesis, 372–374
Analysis of variance (ANOVA), 602–629. *See also* Variance
  definition, 604
  degrees of freedom, 611–612
  experiment design, 615
  F distribution, 605–606
  interaction, 620–621
  MS(treatment), 614
  one-way, 606–619
  sample size, 610–615
  SS(total), 613–614
  treatment (factor), 606
  two samples, 604
  two-way, 619–628
Arithmetic mean. *See* Mean
Aspirin, 443
Assignable variation, 700
"at least one," 150–151
Attributes, control charts, 710–715
Average, 64

B

Bank customer line problem, 73–74, 76–79
Baseball problems, 77, 455–460, 567–568, 570–572
Basketball foul shot problem, 681–682
Bayes' theorem, 153
Bear study, 542–543
Beauty perception problem, 673–674
Bell-shaped distribution, 83–84
Bimodal data, 63
Binomial distribution, 196–207
  binomial probability formula, 199–200, 201–202
  definition, 196–197
  mean, variance, and standard deviation, 207–209
  normal distribution approximation, 271–281
  notation, 197
  Poisson approximation, 214–215
  technology, 200–201
Binomial probability formula, 199–200, 201–202
Birthday problem, 157
Bivariate data, 496
Blinding, 22
Blocks, experimental, 22
Body temperature problem, 265–266, 320, 322–323, 352–353, 409–410, 646
Bolt manufacturing problem, 73, 74, 76–79
Bombing problem, 213–214
Boundary, class, 39–40
Box-and-whisker diagram, 104–107, 108
Boxplots, 104–107, 108
  modified, 110
Burger, Joanna, 365

C

Calculator. *See* Software/calculator results
Cancer and smoking study, 671
Car design problems, 241–242, 244–245
Carvalho, Barbara, 223
Causality, 16
  vs. correlation, 503–504
Census, 4, 228
Center, 38, 59–68
  best measure, 66–67
  mean, 60–61, 62, 65–66, 67
  median, 61–62, 67
  midrange, 62, 63–65, 67
  mode, 62, 63, 67
Centerline, 701
Central limit theorem, 259–271
  application, 262–266
  continuity corrections, 275–278
  finite population correction, 266–267
Centroid, 508, 509
Chebyshev's theorem, 85
Chi-square distribution, 348–350, 566–567
  goodness-of-fit test, 567–581
  hypothesis testing, 420–423
Cigarette smoking problem, 37, 41, 43–44, 46–47, 81–83, 96–98, 102–105, 163

Class
    boundary, 39–40
    midpoint, 40
    width, 40
Class attendance, 644
Class length problems, 227–229
Cluster sampling, 24, 25
    vs. stratified sampling, 26
Coefficient of determination, 533–534, 544
Coefficient of variation, 79–80
Coke versus Pepsi problem, 480–481
Combinations rule, 167–169
Compact disk manufacture, 544
Complementary events, 125–126, 136–137
Complements, 150–151
Completely randomized experimental design, 22
Compound event, 132
Conditional probability, 142, 151–153
    intuitive approach, 151
Confidence interval (interval estimate), 298–311
    alternative method, 323–324
    bootstrap method, 363
    data comparison, 340–343
    definition, 301
    interpretation, 302–303, 322–323
    limits, 321
    margin of error, 305–308, 320, 333, 339–340
    matched pairs, 452–453
    mean, 320–324, 330–347, 403, 454, 457–458, 459
    mean (standard deviation not known), 413
    overlapping, 340–343, 352
    point estimate, 300, 310–311, 331
    prediction intervals, 534–537
    proportion, 301–303, 305–308, 392–395, 443–445
    rationale, 323, 353–354
    round-off rule, 306, 321
    standard deviation, 423
    variance, 351–354
Confidence level (confidence coefficient, degree of
        confidence), 301
Confounding, 21
Contingency tables, 582–596
    definition, 582
    expected frequency, 584–586
    P-values, 588
    test of homogeneity, 588–590
    test of independence, 583
Continuity correction, 275–278
Continuous (numerical) data, 6, 7
Control charts, 694–721
    attributes, 710–715

centerline, 701
    interpretation, 704–705
    mean, 705–706
    p chart, 710–715
    R chart, 701–703
    run charts, 697–700
    upper/lower control limits, 701, 704–705
    variation, 701–703
    $\bar{x}$ chart, 705–706
Convenience sampling, 24, 25
Correlation, 16, 496–517
    centroid, 508, 509
    definition, 496
    errors, 503–504
    explained variation, 502–503
    formal hypothesis test, 504–510
    linear correlation coefficient, 499–503
    rank, 670–679
    scatterplots, 497–499
Cost of Laughing Index, 95
Counting, 162–172
    combinations rule, 167–169
    factorial rule, 164
    fundamental counting rule, 162
    permutations rule, 165–167
Crimean War, 52, 54
Critical region (rejection region), 375
Critical thinking, 11–19, 38–39
    frequency distribution, 42–44
    sample, 108
Critical value, 303–305, 375–376
    chi-square distribution, 349
    definition, 304
    goodness-of-fit tests, 569–570
    Kruskal-Wallis test, 664
    linear correlation, 504–507
    matched pairs, 467–468, 469
    mean, 401
    notation, 304
    standard deviation/variance, 478
    test of independence, 583
    two variances, 478–479
    variance, 349–350
    Wilcoxon tests, 652
Cross-sectional study, 20, 21
Cumulative frequency distribution, 42

D
Damaged goods problem, 144–145
Data
    bell-shaped, 83–84

bimodal, 63
center, 38, 59–68. *See also* Center
characteristics, 38
continuous (numerical), 6, 7
definition, 4
discrete, 6, 7
distribution, 38
interval, 8–9, 10
multimodal, 63
nominal, 7, 10
ordinal, 8, 10
outliers, 38
paired/bivariate, 496
process, 696–697
qualitative, 6
quantitative, 6
range, 74–75
ratio, 9, 10
runs test, 679–687
skewness, 67–68
symmetric, 67
time, 38
types, 5–11
variation, 38
visualization, 46–55
Death
    infectious disease study, 711–712
    postponement study, 583
Death penalty problem, 124, 409
Decimal, percentage conversion from, 15
Degrees of freedom, 332–333, 453–454
    chi-square distribution, 348–350, 420–421
    definition, 332
Density curve (probability density function), 228–229
Dependent events, 142–143
Deviation. *See* Standard deviation; Variation
Dice simulation, 159
Discrete data, 6, 7
Disjoint event, 135
Disobedience experiment, 7
Dispersion. *See* Variation
Distortion, 16–17
Dotplots, 48
Double-blind study, 22
Drug testing problem, 453

E
Ejection seat problem, 225, 242, 243–244
Elected office problem, 167–168
E-mail survey problem, 309
Empirical rule, 83–84

Ergonomics problem, 225, 242, 243–244, 262–264
Errors
    controlling, 382–383
    notation, 382
    sampling, 26–27
    type I, 381–383
    type II, 381–383
Estimating, 296–361
    chi-square distribution, 348–350
    confidence intervals, 301. *See also* Confidence interval (interval estimate)
    critical value, 303–305. *See also* Critical value
    degrees of freedom, 332, 348
    margin of error, 305–308. *See also* Margin of error
    mean, 318–347
    point estimate, 300, 310–311, 339–340, 350–351
    probability, 121–122
    proportion, 298–318
    sample proportion, 300
    sample size, 308–311, 354–355
    sample variance, 350–351
    sampling distributions, 255–256
    standard deviation, 82–83
    Student *t* distribution, 332, 335–339
    variance, 347–358
Estimator, 255–256, 319
Event(s)
    complementary, 125–126, 136–137
    compound, 132
    definition, 120
    dependent, 142
    disjoint, 135
    independent, 142, 153
Excel. *See* Software/calculator results
Expected frequency, 568–576, 584–586
Expected value, 190–191
Experiment(s), 20–31
    confounding, 21
    definition, 20
    replication, 22–23
    sample size, 23
    sampling errors, 26–27
    sampling strategies, 23–26
    types, 20–21
    variables, 21–22
Explained deviation, 532–533
Exploratory data analysis (EDA), 102–108

F
F distribution, 477–479, 605–606, 609–615
Factorial rule, 164

Factorial symbol (!), 163
False negative, 119
False positive, 119
Five-number summary, 104
Flowchart, 135
Flynn effect, 699
Forest temperature problem, 468–470
Formal addition rule, 134
Formal multiplication rule, 143
Foy, Jeffrey, 693
Fraction, percentage conversion from, 14
Fraud detection problem, 565, 573–575
Frequency, expected, 568–576, 584–586
Frequency distribution, 39–44
    classes, 39–40
    construction, 40–41
    cumulative, 42
    definition, 39
    mean, 65–66
    relative, 41–42
    standard deviation, 80–81
Frequency polygon, 47, 48, 54
Fundamental counting rule, 162

G
Gender problems
    birth probabilities, 124, 125–126, 150–151, 156
    discrimination, 644–645
    selection methods, 181, 183, 188–190, 208–209, 369, 370–371
    surveys, 589–590
    wage gap, 532
Genetics problems, 133, 141–142, 393–394
Geometric mean, 72
Gillespie, Angela, 563
Goodness-of-fit test, 567–581
Graph(s), 13, 14, 46–55, 184–186
    boxplot (box-and-whisker), 104–107, 108
    dotplot, 48
    frequency polygon, 47, 48
    histogram, 46–47
    ogive, 47–48
    Pareto chart, 50, 51
    probability histogram, 184–186
    purposes, 54
    relative frequency histogram, 47
    scatter diagram, 51
    stem-and-leaf plot, 48–50
    time-series, 52
Graunt, John, 5
Growth charts, 40

H
H test, 663–670
Harmonic mean, 72
Hat size, 522
Hawthorne effect, 22
Head circumference problem, 83
Height and weight problems, 80, 83, 241–242
Histogram, 46–47, 54, 184–186
Home alarm system problem, 162–163
Home run distance problem, 455–460, 567–568, 570–572
Hydroxyurea study, 241
Hypothesis. See also Hypothesis testing
    alternative, 372–374
    definition, 367
    null, 372–374, 377–385
Hypothesis testing, 366–435. See also Test statistic
    chi-square distribution, 420–423, 427, 567–581
    components, 384, 385
    conclusions, 379–381
    confidence interval, 384, 385, 403, 443–445, 457–458
    critical region, 375
    critical value, 375, 376
    errors, 381–383
    left-tailed, 377, 378
    linear correlation, 504–510
    matched pairs, 467–470
    mean, 400–407, 427, 453–454, 455–460
    mean (standard deviation not known), 407–419, 427
    multiple-negative conclusions, 380
    power, 383, 385
    proportion, 388–400, 427, 440–443
    P-value method, 377–380, 384, 391–392, 401–402, 411–414, 423
    right-tailed, 377, 378
    significance level, 376, 378
    standard deviation/variance, 419–427, 476–481
    summary, 427
    test statistic, 374–375. See also Test statistic
    traditional method, 384, 390–391, 402–403
    two-tailed, 376, 377, 378

I
Independent events, 142–143, 153
Inferences from two samples, 436–493
    confidence interval, 443–445, 454, 457–458, 459
    F distribution, 477–481
    hypothesis testing, 438–443, 453–460, 477–481
    matched pairs, 466–476
    means, 452–466
    proportions, 438–452
    standard deviation/variance, 476–486

Interaction, 620–621
Interval data, 8–9, 10
Intuitive addition rule, 134
Intuitive multiplication rule, 143
IQ score problems, 84–85, 229, 326–327, 421–423, 522, 643–644, 652–653, 699

J
Jet ejection seat problem, 225, 242, 243–244

K
Kruskal-Wallis test, 663–670

L
Landmines, 15
Law of large numbers, 122–123
Lead pollution problem, 60–64
Least-squares property, 525–526
Lebbos, Nabil, 601
Left-tailed test, 377, 378, 508
Lie detectors, 371
Linear correlation coefficient, 499–503
    definition, 499
    interpretation, 501–503
    properties, 502
    rounding, 500–501
Linearity, vs. correlation, 504
Loaded questions, 15
Lottery problems, 168–169, 191, 198, 201, 209, 215, 274
Lower class limits, 39
Lurking variable, 504
Lycett, Mark T., 493

M
Madison, James, 42
Manatee killing problem, 495, 497, 502, 503, 506–507, 520, 521–522, 535–537
Mann-Whitney U test, 656
Margin of error
    mean, 320, 333, 339–340
    poll results, 445
    proportion, 305–308
Mars Climate Orbiter, 702
Marvin, Joseph, 635
Matched pairs
    confidence intervals, 452–453
    hypothesis testing, 467–470
    sign test, 642–644
    Wilcoxon signed-ranks test, 650–656

Mathematical modeling, 551–555
Mean, 60–61, 186, 318–330. *See also* Estimating; Standard deviation
    alternative method, 341–342
    binomial distribution, 207–209
    confidence intervals, 320–324, 330–347, 413
    control chart, 705–706
    definition, 62, 67
    expected value, 190–191
    frequency distribution, 65–66
    geometric, 72
    harmonic, 72
    inferences about two means, 452–466
    margin of error, 320, 333, 339–340
    point estimate, 339–340
    quadratic, 72–73
    sample mean, 319–320, 331
    sample size, 319, 324–327
    sampling distribution, 251–252
    standard deviation not known, 330–347, 407–419
    standard error, 261
    testing a claim, 400–419, 427
    trimmed, 71–72
    weighted, 66
Mean absolute deviation, 86
Mean square (MS), 614
Measure of center, 59
Measurement, 7–9
Median, 61–62, 67
    sign test, 645–647
Mesnick, Sarah, 27
Meta-analysis, 353
Meteorite problem, 123
Midpoint, class, 40
Midrange, 62, 63–65, 67
Milgram, Stanley, 7
Minitab. *See* Software/calculator results
Miringoff, Lee, 223
Mode, 62, 63, 67
Modeling, 551–555
Morrison, Kathleen, 493
MS(treatment), 614
Multimodal data, 63
Multinomial experiments
    definition, 567
    expected frequencies, 568–576
    goodness-of-fit test, 567–581
    P-value, 576
Multiple coefficient of determination, 544
Multiple comparison procedures, 607
Multiple regression, 541–547

Multiple-choice question problem, 198, 199–200, 201–202
Multiplication rule, 139–149

N
Napoleon's Army, 52, 53
NBA salaries, 542
Nicotine patch problem, 310–311
Nielsen survey, 310
Nightingale, Florence, 52, 54
Nominal data, 7, 10
    sign test, 644–645
Nonparametric methods, 636–687
    advantages, 638
    definition, 638
    disadvantages, 638–639
    efficiency rating, 639
    Kruskal-Wallis test, 663–670
    matched pairs, 642–644, 650–656
    median of single population, 645–647
    nominal data, 644–645
    rank correlation, 670–679
    ranks, 639–640
    runs test for randomness, 679–687
    sign test, 640–650
    summary, 687
    ties in ranks, 640
    Wilcoxon rank-sum test, 656–663
    Wilcoxon signed-ranks test, 650–656
Nonsampling error, 26
Normal distribution, 224–287. *See also* Standard normal
        distribution
    applications, 240–249
    binomial distribution approximation, 271–281
    bivariate, 499
    central limit theorem, 259–271
    definition, 224–225
    determination, 282–287
    finding values, 243–246
    nonstandard, 240–242
    normal quantile plot, 282–286
    sampling distribution, 249–259
Normal quantile plot, 282–286
Null hypothesis, 372–374, 377–385
    accept/fail to reject, 379–380
Numerical data, 6, 7

O
Observational study, 20–21
Odds, 126–127
Ogive, 47–48

Old Faithful, 43
One-tailed test, 507–508
Ordinal data, 8, 10
O'Toole, Dan, 720
Outliers, 102–103, 523–524
    extreme, 110
    mild, 110

P
p charts, 710–715
Paired data, 496
Palmistry, 501
Parameter, 5
Pareto charts, 50, 51
Payoff odds, 127
Pearson product moment correlation coefficient,
        499–500
Pearson's index of skewness, 91
Percentages, 14–15
    proportion conversion, 300
Percentiles, 95–99
Permutations rule, 165–167
Photo-cop problem, 297, 300, 302–303, 306–307, 367
Pictographs, 13–14
Pie charts, 50, 51
Pinball problem, 674–675
Pizza sales, 522
Placebo effect, 22, 456
Point estimate, 300, 310–311, 339–340, 350–351
Poisson distribution, 212–218
Polio experiment, 441
Polling. *See* Surveys
Pooled estimate, 458
Population
    definition, 4
    size, 310
    standard deviation, 78
Population mean. *See* Mean
Population model, 552–553
Population proportion. *See* Proportion
Population variance. *See* Variance
Power, test, 383, 385
Precision, 16
Prediction, regression equation, 521–523
Prediction intervals, 534–537
Predictor variable, 542
Pregnancy test problem, 119, 135–136, 152–153
Presidents' ages problem, 283–284
Probability, 118–173
    addition rule, 132–139, 145

"at least one," 150–151
    classical approach, 121, 122, 124–125
    combinations rule, 167–169
    complementary events, 125–126, 136–137
    complements, 150–151
    conditional, 142, 151–153
    counting, 162–172
    factorial rule, 164–165
    law of large numbers, 122–123
    mean, variance, standard deviation, 186–187
    multiplication rule, 139–149
    notation, 121
    odds, 126–127
    permutations rule, 165–167
    rare event rule, 120
    relative frequency approximation, 121, 122
    rounding off, 126
    rules, 121–122
    simulations, 156–162
    subjective, 121–122
    unusual results, 189–190
    z score, 230–237
Probability distribution, 180–217. *See also* Normal
        distribution; Standard normal distribution
    binomial, 196–207
    definition, 183
    expected value, 190–191
    Poisson distribution, 212–218
    probability histogram, 184–186
    random variables, 183–196
    requirements, 185–186
    unusual results, 187–190
Probability histogram, 184–186
Process data, 696–697
Project topics, 722–725
Proportion, 298–318
    confidence interval, 301–303, 305–308, 392–395
    critical values, 303–304
    hypothesis testing, 388–400, 427, 440–443
    inferences about two proportions, 438–452
    margin of error, 305–308
    notation, 299
    point estimate, 300, 310–311
    sample size, 308–311
    sampling distribution of proportions, 252–254
    test statistic, 389, 394–395
    testing a claim, 388–400, 427
Prospective (longitudinal or cohort) study, 20, 21
Pulse rate problem, 42–43
Push polling, 302

P-value method
    contingency tables, 588
    hypothesis testing, 377–380, 384, 391–392, 401–402,
        411–414, 423
    multinomial experiments, 576
    multiple regression, 545–547

Q
Quadratic mean, 72–73
Qualitative data, 6
Quality control, 694–721
    p chart, 710–715
    problem, 145
    R chart, 701–703
    run charts, 697–700
    $\bar{x}$ chart, 705–706
Quantitative data, 6
Quartiles, 94–99
Questions
    loaded, 15
    order of, 15

R
R chart, 701–703
Racial profiling problem, 437, 441–443, 444–445
Rainfall problem, 106–107, 284–285, 637, 659–660,
        666–667, 682–683
Random numbers, 157–159, 259–260, 261–262
Random sampling, 23–24, 25, 27
Random selection, 5, 22
Random variables, 183–196
    binomial probability distribution, 196–207
    continuous, 183
    definition, 183
    discrete, 183
    expected value, 190–191
    probability histogram, 184–186
    uniform distribution, 227–229
Random variation, 700
Randomness, runs test, 679–687
Range, 74–75
Range chart, 701–703
Range rule of thumb, 82, 187–188, 325
Rank, 639–640
Rank correlation test, 670–679
Rare event rule, 120, 265, 365–366
Ratio data, 9, 10
Ratio test, 9
Reading level problem, 338–339, 603, 607–608, 615,
        658–659, 665–666

Red-light running problem, 367, 373–374, 375, 389–390
Refusals, 15
Regression, 517–526
    coefficient of determination, 533–534, 544
    guidelines, 523
    influential points, 523–524
    interpretation, 523
    least-squares property, 525–526
    marginal change, 523
    multiple, 541–547
    outliers, 523–524
    predictions, 521–523
    regression equation/line, 518–521
    residuals, 524–525
    standard error of estimate, 535–536
    stepwise, 546–547
Relative frequency distribution, 41–42
Relative frequency histogram, 46, 47
Relative standing, 92–99
    percentiles, 95–99
    quartiles, 94–99
    z score, 92–94. See also z score
Reliability, 229
Residual, 524–525, 532
Residual plot, 531
Retrospective (case-control) study, 20, 21
Right-tailed test, 377, 378, 508
Rigorously controlled experimental design, 22
Roulette problem, 123, 127
Round-off rule, 65, 79
    confidence interval, 306, 321
    linear correlation coefficient, 500–501
    mean, 321
    mean, variance, standard deviation, 187
    probabilities, 126
    proportion, 306
    regression, 519–521
    sample size, 308, 325
Routing problem, 164–165
Run charts, 697–700
Runs test for randomness, 679–687

S
Saccucci, Michael, 434–435
Sample(s)
    definition, 4
    dependent (matched pairs), 452–453
    independent, 452–453
    mean, 319–320, 331
    multistage design, 26
    random, 23–24, 25

    small, 13
    space, 120
    standard deviation, 75. See also Standard deviation
    variance, 350–351
    variance between, 609–610
    variance within, 609–610
    voluntary response (self-selected), 12–13
Sample proportion, 300
Sample size (n), 23, 60
    equal, 610–612
    mean, 319–320, 324–327
    proportion, 308–311
    round-off rule, 308, 325
    standard deviation, 325–327
    unequal, 612–615
    variance, 354–355
Sampling
    cluster, 24, 25, 26
    convenience, 24, 25
    distribution, 249–259
    errors, 26–27
    random, 23–24, 25, 27
    with replacement, 250–251
    stratified, 24, 25, 26
    systematic, 24, 25
    variability, 252
    without replacement, 266
Sampling distribution, 249–259
    central limit theorem, 259–271
    estimators, 255–256
    mean, 251–252
    proportion, 252–254
Scatterplot (scatter diagram), 51, 497–499, 504
    outlier/influential point, 524
    quadrants, 509
Secondhand smoke problem, 37, 41, 43–44, 46–47, 81–83, 96–98, 102–105, 163
Self-interest study, 16
Sickle cell anemia study, 241
Sign test, 640–650
Signal-to-noise ratio, 91
Significance level, 376, 378
Simple event, 120, 121
Simple random sample, 23–24, 25, 26
Simulations, 125, 156–162
Six Sigma, 711
Skewness, 67–68
Ski gondola problem, 262–264
Slope, regression equation, 518–521
Software/calculator results, 33
    binomial probabilities, 200–201, 203

boxplots, 108
confidence interval, 311, 327, 342–343, 355
contingency tables, 590
correlation, 510
descriptive statistics, 68
graphs, 55
hypothesis testing, 395, 404, 414, 423
Kruskal-Wallis test, 668
matched pairs, 471
mathematical modeling, 554
multinomial experiments, 576
multiple regression, 47, 543
normal distribution, 231, 245–246
normal quantile plot, 285–286
one-way ANOVA, 615
p chart, 713
Poisson distribution, 215
random number generation, 157–159
rank correlation, 676
regression, 526
regression equation, 520–521
run chart, 707
runs test for randomness, 684
sample size, 311
sign test, 648
two means, 460–461
two population proportions, 446
two variances, 482
two-way ANOVA, 625
variation, 537–538
Wilcoxon tests, 654, 660
z score, 237
Spearman's rank correlation coefficient, 671
Spearman's rank correlation test, 670–679
Spread. *See* Variation
SS(total), 613–614
Standard deviation, 75, 186
binomial distribution, 207–209
calculation, 76–77, 85–87
Chebyshev's theorem, 85
comparing two samples, 476–486
confidence intervals, 423
definition, 75
empirical rule, 83–84
estimation, 82–83
frequency distribution, 80–81
interpretation, 81–85
mean absolute deviation, 86
population, 78
sample size, 325–327
testing a claim, 419–427, 429

Standard error of estimate, 534–535
Standard error of the mean, 261
Standard normal distribution, 227–240
definition, 230
density curve, 229
finding probabilities, 230–235
finding z scores, 235–237
Standard score. *See* z score
STATDISK. *See* Software/calculator results
Statistic, 5
Statistics, 4
Stem-and-leaf plots, 48–50
Stepwise regression, 546–547
Stock investing problem, 75, 166–167
Stowaway problem, 340
Stratified sampling, 24, 25
vs. cluster sampling, 26
Student t distribution, 331, 332, 335–336, 408–409
P-value, 411–414
vs. z distribution, 336–339
Student teacher rating, 498
Subjective probability, 121–122
Successes, 439
Surveys, 3, 4–5, 302
ethics, 389
gender effects, 589–590
margin of error, 445
resistance, 607
Symmetric data, 67
Systematic sampling, 24, 25

T
t distribution. *See* Student t distribution
Tampering, 703
Television programming problem, 165
Television ratings problem, 276–278
Test of homogeneity, 588–590
Test of independence, 583
Test statistic, 374–375
ANOVA, 609–615
critical region, 375
critical value, 376
goodness-of-fit tests, 569–572, 574
Kruskal-Wallis test, 664
linear correlation, 506–509
matched pairs, 467–470
mean, 374, 401, 408, 453–454, 458–459
proportion, 374, 389, 394–395
P-value, 377
rank correlation, 670–679
runs test for randomness, 681

Test statistic (*Continued*)
  sign test, 641, 647
  significance level, 376
  standard deviation/variance, 374, 420
  test of homogeneity, 588–590
  test of independence, 583
  two means, 453–454, 457–459
  two proportions, 440, 441–443, 445
  two variances, 477–479, 480–481
  Wilcoxon tests, 652, 656–657
Thanksgiving Day problem, 125
Thermometer testing problem, 232–234, 235–236
TI-83 Plus. *See* Software/calculator results
Time-series graph, 52
Titanic survivor problem, 586–587
Total deviation, 532–533
Touch therapy, 394
Traveling salesman problem, 164–165
Treatment (factor), 606
Tree diagram, 141
Trimmed mean, 71–72
Twin research, 468, 469
Two-tailed test, 376, 377, 378, 504–507
Type I errors, 381–383
Type II errors, 381–383

U
Unbiased estimator, 78
Unexplained deviation, 532–533
Uniform distribution, 227–229
Unusual results, 187–190
Unusual values, 93–94
Upper class limits, 39

V
Validity, 229
Variable, 21–22
  lurking, 504
  prediction, 521–523
  random. *See* Random variables
Variance, 78–79, 186, 347–358. *See also* Analysis of variance (ANOVA)
  between/within samples, 609–610
  binomial distribution, 207–209
  chi-square distribution, 348–350
  comparing two samples, 476–486
  confidence interval, 351–354

  definition, 78
  sample, 350–351
  sample size, 354–355
  sample variance, 350–351
  testing a claim, 419–427, 429
Variation, 531–541. *See also* Standard deviation; Variance
  assignable, 700
  coefficient, 79–80
  coefficient of determination, 533–534
  control chart, 701–703
  explained/unexplained deviation, 531–537
  mean absolute deviation, 86
  measures, 73–87
  prediction intervals, 534–537
  random, 700
  range, 74–75
  standard deviation, 75
  standard error of estimate, 534–535
Venn diagram, 134
Voluntary response sample (self-selected sample), 12, 108
Voting behavior, 444

W
Wage gender gap, 532
Weighted mean, 66
Width, class, 40
Wilcoxon rank-sum test, 656–663
Wilcoxon signed-ranks test, 650–656
Wildlife population size, 320

X
$\bar{x}$, sampling distribution, 260–261
$\bar{x}$ chart, 705–706

Y
y-intercept, 518–521

Z
z score, 92–94
  vs. area, 243
  critical value, 303–305
  definition, 93
  finding, 235–237
  notation, 304
  standard normal distribution, 230–235
  vs. student t distribution, 336–339

Symbol Table

$\overline{A}$	complement of event A		H	Kruskal-Wallis test statistic
H_0	null hypothesis		R	sum of the ranks for a sample; used in the Wilcoxon rank-sum test
H_1	alternative hypothesis			
α	alpha; probability of a type I error or the area of the critical region		μ_R	expected mean rank; used in the Wilcoxon rank-sum test
β	beta; probability of a type II error		σ_R	expected standard deviation of ranks; used in the Wilcoxon rank-sum test
r	sample linear correlation coefficient		G	number of runs in runs test for randomness
ρ	rho; population linear correlation coefficient		μ_G	expected mean number of runs; used in runs test for randomness
r^2	coefficient of determination			
R^2	multiple coefficient of determination		σ_G	expected standard deviation for the number of runs; used in runs test for randomness
r_s	Spearman's rank correlation coefficient			
b_1	point estimate of the slope of the regression line		$\mu_{\overline{x}}$	mean of the population of all possible sample means $\overline{x}$
b_0	point estimate of the y-intercept of the regression line		$\sigma_{\overline{x}}$	standard deviation of the population of all possible sample means $\overline{x}$
$\hat{y}$	predicted value of y		E	margin of error of the estimate of a population parameter, or expected value
d	difference between two matched values		Q_1, Q_2, Q_3	quartiles
$\overline{d}$	mean of the differences d found from matched sample data		$D_1, D_2, \ldots, D_9$	deciles
s_d	standard deviation of the differences d found from matched sample data		$P_1, P_2, \ldots, P_{99}$	percentiles
s_e	standard error of estimate		x	data value
T	rank sum; used in the Wilcoxon signed-ranks test			

Symbol Table

f	frequency with which a value occurs		$z_{\alpha/2}$	critical value of z
Σ	capital sigma; summation		t	t distribution
Σx	sum of the values		$t_{\alpha/2}$	critical value of t
Σx^2	sum of the squares of the values		df	number of degrees of freedom
$(\Sigma x)^2$	square of the sum of all values		F	F distribution
Σxy	sum of the products of each x value multiplied by the corresponding y value		χ^2	chi-square distribution
n	number of values in a sample		χ_R^2	right-tailed critical value of chi-square
$n!$	n factorial		χ_L^2	left-tailed critical value of chi-square
N	number of values in a finite population; also used as the size of all samples combined		p	probability of an event or the population proportion
k	number of samples or populations or categories		q	probability or proportion equal to $1 - p$
$\overline{x}$	mean of the values in a sample		$\hat{p}$	sample proportion
$\overline{R}$	mean of the sample ranges		$\hat{q}$	sample proportion equal to $1 - \hat{p}$
μ	mu; mean of all values in a population		$\overline{p}$	proportion obtained by pooling two samples
s	standard deviation of a set of sample values		$\overline{q}$	proportion or probability equal to $1 - \overline{p}$
σ	lowercase sigma; standard deviation of all values in a population		$P(A)$	probability of event A
s^2	variance of a set of sample values		$P(A\mid B)$	probability of event A, assuming event B has occurred
σ^2	variance of all values in a population		$_nP_r$	number of permutations of n items selected r at a time
z	standard score		$_nC_r$	number of combinations of n items selected r at a time